Global Groundwater
Source, Scarcity, Sustainability, Security, and Solutions

Global Groundwater
Source, Scarcity, Sustainability, Security, and Solutions

Edited by

Abhijit Mukherjee
Department of Geology and Geophysics, Indian Institute of Technology Kharagpur, Kharagpur, India
Applied Policy Advisory for Hydrosciences (APAH) group,
School of Environmental Science and Engineering, Indian Institute of Technology Kharagpur,
Kharagpur, India

Bridget R. Scanlon
Bureau of Economic Geology, Jackson School of Geosciences, University of Texas, Austin, TX,
United States

Alice Aureli
Groundwater Systems and Settlements Section, International Hydrological Programme,
United Nations Educational, Scientific and Cultural Organization (UNESCO), Paris, France

Simon Langan
International Water Management Institute (IWMI), Colombo, Sri Lanka

Huaming Guo
School of Water Resources and Environment, China University of Geosciences, Beijing, P.R. China

Andrew A. McKenzie
British Geological Survey, Oxfordshire, United Kingdom

ELSEVIER

Elsevier
Radarweg 29, PO Box 211, 1000 AE Amsterdam, Netherlands
The Boulevard, Langford Lane, Kidlington, Oxford OX5 1GB, United Kingdom
50 Hampshire Street, 5th Floor, Cambridge, MA 02139, United States

Copyright © 2021 Elsevier Inc. All rights reserved.

No part of this publication may be reproduced or transmitted in any form or by any means, electronic or mechanical, including photocopying, recording, or any information storage and retrieval system, without permission in writing from the publisher. Details on how to seek permission, further information about the Publisher's permissions policies and our arrangements with organizations such as the Copyright Clearance Center and the Copyright Licensing Agency, can be found at our website: www.elsevier.com/permissions.

This book and the individual contributions contained in it are protected under copyright by the Publisher (other than as may be noted herein).

Notices
Knowledge and best practice in this field are constantly changing. As new research and experience broaden our understanding, changes in research methods, professional practices, or medical treatment may become necessary.

Practitioners and researchers must always rely on their own experience and knowledge in evaluating and using any information, methods, compounds, or experiments described herein. In using such information or methods they should be mindful of their own safety and the safety of others, including parties for whom they have a professional responsibility.

To the fullest extent of the law, neither the Publisher nor the authors, contributors, or editors, assume any liability for any injury and/or damage to persons or property as a matter of products liability, negligence or otherwise, or from any use or operation of any methods, products, instructions, or ideas contained in the material herein.

British Library Cataloguing-in-Publication Data
A catalogue record for this book is available from the British Library

Library of Congress Cataloging-in-Publication Data
A catalog record for this book is available from the Library of Congress

ISBN: 978-0-12-818172-0

For Information on all Elsevier publications
visit our website at https://www.elsevier.com/books-and-journals

Publisher: Candice Janco
Acquisitions Editor: Louisa Munro
Editorial Project Manager: Andrea Dulberger
Production Project Manager: Joy Christel Neumarin Honest Thangiah
Cover Designer: Victoria Pearson

Typeset by MPS Limited, Chennai, India

Contents

List of Contributors — xix
About the Editors — xxiii
Forewords — xxv
Claudia Sadoff, John W. Hess, Franklin W. Schwartz, József Toth, António Chambel
Preface — xxxi
Acknowledgment — xxxiii
Disclaimer — xxxv
Introduction: Why Study Global Groundwater? — xxxvii
Tom Gleeson, Mark Cuthbert, Grant Ferguson and Debra Perrone

Theme 1
Global groundwater

1. Global groundwater: from scarcity to security through sustainability and solutions — 3
Abhijit Mukherjee, Bridget R. Scanlon, Alice Aureli, Simon Langan, Huaming Guo and Andrew McKenzie

1.1 Introduction — 3
1.2 Groundwater source and availability — 4
1.3 Groundwater scarcity — 6
   1.3.1 Quantity — 6
   1.3.2 Groundwater quality — 9
1.4 Groundwater sustainability and security — 11
   1.4.1 Groundwater–food–energy nexus — 12
   1.4.2 Urbanization — 12
   1.4.3 Groundwater trade and hydro-economics — 13
1.5 Solutions — 14
   1.5.1 Enhancing irrigation and urban groundwater efficiency — 15
   1.5.2 Groundwater rejuvenation — 15
   1.5.3 Desalination — 15
1.6 Conclusion — 15
References — 16

Theme 2
Groundwater sources

2. Groundwater of carbonate aquifers — 23
Alan E. Fryar

2.1 Introduction — 23
2.2 Carbonate geochemistry and hydrochemical evolution — 23
2.3 Porosity and permeability — 25
2.4 Recharge and flow — 26
2.5 Water supply and environmental issues — 30
2.6 Challenges in monitoring and modeling — 31
2.7 Conclusion — 32
References — 32

3. Groundwater resources in Australia—their occurrence, management, and future challenges — 35
Steve Barnett, Craig T. Simmons and Rebecca Nelson

3.1 Introduction — 35
3.2 Groundwater resources in Australia — 35
3.3 Historical development of groundwater — 36
3.4 Evolution of groundwater management — 38
3.5 Current groundwater usage — 38

3.6 Groundwater management issues　40
　3.6.1 Overuse and overallocation of groundwater　40
　3.6.2 Groundwater-dependent ecosystems　40
　3.6.3 Impacts of groundwater extraction on surface-water systems　40
　3.6.4 Effect of climate change on groundwater resources　41
　3.6.5 Impacts of mining on groundwater resources　41
　3.6.6 Land and groundwater salinization　42
　3.6.7 Seawater intrusion　43
3.7 Future challenges　43
　3.7.1 Managed aquifer recharge　44
　3.7.2 Declining resources for understanding and managing groundwater　45
3.8 Conclusion　45
References　45
Further reading　46

4. Groundwater storage dynamics in the Himalayan river basins and impacts of global change in the Anthropocene　47

M. Shamsudduha

4.1 Introduction　47
4.2 Hydrology and climate of Himalayan river basins　49
　4.2.1 The Indus river basin　49
　4.2.2 The Ganges—Brahmaputra—Meghna river basin　50
　4.2.3 The Irrawaddy river basin　52
4.3 Groundwater for drinking and agricultural use　53
4.4 Groundwater storage dynamics in Himalayan river basins　53
　4.4.1 Gravity Recovery and Climate Experiment: Earth observation satellite monitoring　53
　4.4.2 Dynamics in Gravity Recovery and Climate Experiment terrestrial water storage　55
　4.4.3 Mapping groundwater storage using Gravity Recovery and Climate Experiment　56
　4.4.4 Reported changes of groundwater storage and impacts of global change　57
4.5 Concluding discussion　59
Acknowledgments　61
References　61

5. Groundwater variations in the North China Plain: monitoring and modeling under climate change and human activities toward better groundwater sustainability　65

Wenting Yang, Long Di and Zhangli Sun

5.1 Introduction　65
5.2 Impacts of human activities on groundwater in the North China Plain　65
5.3 Climate change impact on groundwater in the North China Plain　66
5.4 China's South-to-North Water Diversion　66
5.5 Review on groundwater storage assessment in the North China Plain　68
Acknowledgment　70
References　70

6. Emerging groundwater and surface water trends in Alberta, Canada　73

Soumendra Nath Bhanja and Junye Wang

6.1 Introduction　73
6.2 Data and methods　74
　6.2.1 Study region　74
　6.2.2 Groundwater level observation　75
　6.2.3 Observations of surface water　75
　6.2.4 Rainfall and snowmelt water　75
6.3 Results and discussions　76
　6.3.1 Rainfall and snowmelt water　76
　6.3.2 Surface water level changes　77
　6.3.3 Groundwater level changes　78
6.4 Summary　78
Acknowledgments　78
References　78

7. Groundwater irrigation and implication in the Nile river basin　81

Anjuli Jain Figueroa and Mikhail Smilovic

7.1 Introduction　81
7.2 Surface water in the Nile basin　81
7.3 Land use and irrigation in the Nile basin　84
7.4 Groundwater in the Nile basin　86
7.5 Aquifers in Nile riparian countries　88
　7.5.1 Groundwater in Egypt　88
　7.5.2 Groundwater in Sudan and South Sudan　90
　7.5.3 Groundwater in Ethiopia　90

7.5.4 Groundwater in the Extended Lake Victoria basin	91
7.6 Discussion and conclusion	91
References	93

8. Groundwater availability and security in the Kingston Basin, Jamaica — 97

Arpita Mandal, Debbie-Ann D.S. Gordon-Smith and Peta-Gay Harris

8.1 Introduction	97
8.2 The Kingston Hydrologic Basin	99
8.2.1 Population and water supply	99
8.2.2 Hydrogeology of the KHB	102
8.2.3 Climate of the KHB	103
8.3 Methodology and analytical procedures	103
8.3.1 Field work	103
8.3.2 Water quality analysis	106
8.4 Results and discussion	110
8.5 Conclusion	111
Acknowledgments	112
References	112

9. Transboundary aquifers: a shared subsurface asset, in urgent need of sound governance — 113

Shaminder Puri

9.1 Introduction	113
9.2 Definition of transboundary aquifer: international and intranational	113
9.3 Governance—collaboration, potential dispute resolution	114
9.4 Water availability as a driver for governance	114
9.5 Current global inventory and classification of transboundary aquifers	114
9.6 Review of recent developments—the Red Queen effect	116
9.7 The place of transboundary aquifers in national priorities	117
9.8 SDGs as a driver toward sound governance of transboundary aquifers	119
9.9 The climate change megatrend and relevance to transboundary aquifers	120
9.10 Transboundary aquifers under high developmental stress	120
9.11 Estimating the urgency of sound governance as a function of water abundance/water scarcity	122
9.12 Case history: the Stampriet aquifer—Botswana, Namibia, and South Africa	124
9.13 Hurdles to progress in intercountry dialogue—the "invisibility cape"?	126
9.14 The hiatus in the progress to adoption of the Draft Articles	126
9.15 Conclusion: light at the end of the tunnel	127
Conflict of interest	127
Acknowledgment	127
References	127

10. Transboundary groundwater of the Ganges–Brahmaputra–Meghna River delta system — 129

Madhumita Chakraborty, Abhijit Mukherjee and Kazi Matin Ahmed

10.1 Introduction	129
10.2 Geologic and geomorphologic setting	130
10.3 Aquifer framework	131
10.4 Groundwater flow system	131
10.5 Hydrogeochemistry	133
10.6 Groundwater arsenic contamination	134
10.7 Policy interventions and management options for arsenic mitigation	135
References	138
Further reading	141

Theme 3
Groundwater scarcity: quantity and quality

11. Groundwater drought: environmental controls and monitoring — 145

Bailing Li and Matthew Rodell

11.1 Introduction	145
11.2 Environmental controls on groundwater	146
11.2.1 Precipitation	146
11.2.2 Subsurface hydrogeological conditions	148
11.2.3 Large-scale climate phenomena	148
11.3 Groundwater drought monitoring	151
11.3.1 Gravity Recovery and Climate Experiment data assimilation for groundwater drought monitoring	151
11.3.2 Other groundwater drought indicators	153
11.4 Characteristics of groundwater drought at the global domain	153
11.5 Discussions and future research	156
References	158

12. Groundwater scarcity in the Middle East — 163

Ahmed A. Al-Taani, Yousef Nazzal and Fares M. Howari

- 12.1 Introduction — 163
- 12.2 Water resources: current use and future trends — 163
- 12.3 Impacts of water scarcity — 164
  - 12.3.1 Water resources and climate change — 164
  - 12.3.2 Water quality — 164
- 12.4 Water resources management — 165
  - 12.4.1 Mitigation to water scarcity — 165
- 12.5 Case studies — 166
  - 12.5.1 Jordan River — 166
  - 12.5.2 Tigris–Euphrates River — 168
  - 12.5.3 Nile River — 170
- References — 173

13. Groundwater scarcity and management in the arid areas in East Africa — 177

Seifu Kebede and Meron Teferi Taye

- 13.1 Introduction — 177
- 13.2 Typical characteristics of the dryland areas — 179
- 13.3 Typologies of hydrogeology difficulties in arid areas in the East Africa — 179
  - 13.3.1 Arid volcanic mountains (old rugged volcanics) — 180
  - 13.3.2 Rift volcanics and pyroclastics — 180
  - 13.3.3 Nazareth series ignimbrites — 180
  - 13.3.4 Extensive limestone and sandstone plateaus, rocky hills, and plains in arid environments — 180
  - 13.3.5 Extensive loose inland alluvio-lacustrine, inland deltaic, and coastal plain aquifers — 181
- 13.4 Current and past drinking water delivery practices — 181
- 13.5 Securing water in difficult hydrogeological environments — 182
  - 13.5.1 Identifying and protecting viable aquifers — 182
  - 13.5.2 Adaptation of customary water schemes — 183
  - 13.5.3 Enhancing water availability by water harvesting — 183
  - 13.5.4 Water quality management — 183
  - 13.5.5 Long distance and interbasin water transfer — 184
  - 13.5.6 Investing in sustainability of existing systems — 184
- 13.6 Policy and practice implication — 184
- Acknowledgment — 185
- References — 185
- Further reading — 186

14. Global geogenic groundwater pollution — 187

Poulomee Coomar and Abhijit Mukherjee

- 14.1 Introduction — 187
- 14.2 Global distribution of geogenic groundwater pollutants — 187
  - 14.2.1 Arsenic — 187
  - 14.2.2 Fluoride — 190
  - 14.2.3 Selenium — 192
  - 14.2.4 Uranium — 194
  - 14.2.5 Salinity — 196
- 14.3 Conclusion — 198
- References — 198

15. Out of sight, but not out of mind: Per- and polyfluoroalkyl substances in groundwater — 215

Ruth Marfil-Vega, Brian C. Crone, Marc A. Mills and Susan T. Glassmeyer

- 15.1 Introduction — 215
- 15.2 Analytical methods for monitoring per- and polyfluoroalkyl substances — 216
- 15.3 Sources of per- and polyfluoroalkyl substances to the environment — 218
  - 15.3.1 Aqueous film-forming foam — 218
  - 15.3.2 Landfill leachate — 219
  - 15.3.3 Industrial sources — 219
  - 15.3.4 Other sources — 220
- 15.4 Occurrence studies — 220
- 15.5 Removal of per- and polyfluoroalkyl substances from groundwater — 221
  - 15.5.1 Granular activated carbon — 222
  - 15.5.2 Ion-exchange resins — 222
  - 15.5.3 Nanofiltration and reverse osmosis — 223
- 15.6 Conclusion — 224
- References — 224

16. Geogenic-contaminated groundwater in China — 229

Yongfeng Jia

- 16.1 Introduction — 229
- 16.2 The distribution and formation of geogenic-contaminated groundwater — 230
  - 16.2.1 High-salinity groundwater — 230
  - 16.2.2 High-Fe and -Mn groundwater — 230
  - 16.2.3 High-As groundwater — 231
  - 16.2.4 High-fluoride groundwater — 231
  - 16.2.5 High-/low-iodine groundwater — 234
  - 16.2.6 High-nitrogen groundwater — 234
  - 16.2.7 Other trace elements — 234
- 16.3 Cooccurrence of different geogenic-contaminated groundwater components — 235
  - 16.3.1 High salinity and fluoride — 235
  - 16.3.2 Arsenic and fluoride — 235
  - 16.3.3 Iron, manganese, and ammonia — 235
- 16.4 Geogenic-contaminated groundwater affected by anthropogenic activities — 235
  - 16.4.1 Further salinization of groundwater — 235
  - 16.4.2 Elevated groundwater hardness — 237
  - 16.4.3 Cross contamination of aquifers — 237
  - 16.4.4 Trace element release/sequester due to redox change — 237
- 16.5 Conclusion — 237
- References — 238

17. Screening of emerging organic pollutants in the typical hygrogeological units of China — 243

Xiaopeng Qin, Tian Zhou, Shengzhang Zou and Fei Liu

- 17.1 Introduction — 243
- 17.2 Materials and methods — 243
  - 17.2.1 Study area and sample collection — 243
  - 17.2.2 Chemicals — 244
  - 17.2.3 Analytical method — 245
  - 17.2.4 Risk characterization — 245
- 17.3 Results and discussion — 245
  - 17.3.1 Presence of antibiotics in groundwater — 245
  - 17.3.2 Statistical analysis — 246
  - 17.3.3 Environmental risk assessment — 247
  - 17.3.4 Screening of antibiotics in groundwater — 247
- 17.4 Conclusion and further research — 248
- Acknowledgments — 248
- References — 249

18. Groundwater pollution of Pearl River Delta — 251

Guanxing Huang, Lingxia Liu, Chunyan Liu, Wenzhong Wang and Dongya Han

- 18.1 Introduction — 251
- 18.2 Study area — 251
  - 18.2.1 Hydrogeological and geological conditions — 251
  - 18.2.2 Characteristics of urbanization and industrialization in the Pearl River Delta — 252
- 18.3 Materials and methods — 253
- 18.4 Results and discussion — 253
  - 18.4.1 Groundwater chemistry — 253
  - 18.4.2 Groundwater quality and main impact chemicals — 255
  - 18.4.3 Groundwater contamination — 255
- 18.5 Conclusion — 259
- Acknowledgments — 259
- References — 259

19. Hydrochemical characteristics and quality assessment of water from different sources in Northern Morocco — 261

Lahcen Benaabidate, Ahmed Zian and Othman Sadki

- 19.1 Introduction — 261
- 19.2 Material and methods — 262
- 19.3 Hydrochemistry — 262
  - 19.3.1 Source water chemical facies — 262
  - 19.3.2 Quality of source waters for irrigation — 265
- 19.4 Control of chemical element concentrations — 266
  - 19.4.1 Binary ion correlations — 266
  - 19.4.2 Cl–SO$_4$–HCO$_3$ diagram — 268
  - 19.4.3 Index of base exchange — 268
  - 19.4.4 Water standards and potability — 269
  - 19.4.5 Sodium and potassium — 269
  - 19.4.6 Calcium and magnesium — 269
  - 19.4.7 Chlorides — 269
  - 19.4.8 Sulfates and bicarbonates — 269
- 19.5 Principal component analysis — 270
  - 19.5.1 Variable space — 270
  - 19.5.2 Individual space — 270
- 19.6 Water minerals equilibrium — 272
  - 19.6.1 Carbonates equilibrium — 272
  - 19.6.2 Silica equilibrium — 272
  - 19.6.3 N$_2$–Ar–CH$_4$ gases diagram — 273
- 19.7 Conclusion — 273
- References — 274

20. Arsenic in groundwater in the United States: research highlights since 2000, current concerns and next steps 275

Madeline E. Schreiber

20.1 Introduction 275
20.2 Research on arsenic in groundwater: 2000–20 276
    20.2.1 Sources of Arsenic in groundwater 276
    20.2.2 Key biogeochemical processes that influence As cycling 276
    20.2.3 Tools for studying arsenic 277
    20.2.4 Mechanisms of arsenic release to groundwater 279
20.3 Hydrogeochemical settings for arsenic in groundwater in the United States 279
    20.3.1 Sand and gravel aquifers 279
    20.3.2 Basaltic rock aquifers 282
    20.3.3 Glacial aquifers 282
    20.3.4 Sedimentary rock aquifers 282
    20.3.5 Crystalline and meta-sedimentary rock aquifers 282
    20.3.6 Coastal plain (semiconsolidated) aquifers 282
    20.3.7 Geothermal areas (western United States) 282
20.4 Research highlights from 2000 to 2020 283
    20.4.1 Nationwide datasets show statistical and spatial patterns of groundwater As 283
    20.4.2 Statistical models yield can predict drivers of arsenic release to groundwater 284
    20.4.3 Statistical models can produce probability maps of arsenic risk 284
    20.4.4 Arsenic concentrations may (but do not always) change over time 284
    20.4.5 Human activities can promote arsenic release to groundwater 285
    20.4.6 Research leads to improved technology for arsenic detection and treatment 286
20.5 Current concerns about arsenic in groundwater in the United States 287
    20.5.1 Most, but not all, public water supplies are meeting the drinking water standard 287
    20.5.2 Homeowners are responsible for testing of private well water 287
20.6 Next steps 288
    20.6.1 Required testing would improve identification of wells with elevated As 288
    20.6.2 More support is needed for homeowners, especially in areas of high risk 288
    20.6.3 More data are needed for prediction of spatial and temporal patterns 289
    20.6.4 Education and effective communication can improve awareness and action 289
References 290

21. Hydrogeochemical characterization of groundwater quality in the states of Texas and Florida, United States 301

Shama E. Haque

21.1 Groundwater quality in Texas 301
    21.1.1 Edwards–Trinity plateau aquifer 301
    21.1.2 Ogallala aquifer 302
    21.1.3 Seymour aquifer 302
    21.1.4 Pecos Valley Aquifer 303
    21.1.5 Carrizo aquifer 303
    21.1.6 Barnett Shale aquifer 303
21.2 Aquifers in Florida 304
    21.2.1 Floridan aquifer system 304
    21.2.2 Sand-and-gravel aquifer 305
    21.2.3 Biscayne aquifer 305
Acknowledgments 306
References 306

22. Groundwater pollution in Pakistan 309

Noshin Masood, Shehla Batool and Abida Farooqi

22.1 Introduction 309
22.2 Groundwater quality 310
    22.2.1 Biological contamination of groundwater 310
22.3 Chemical contamination 312
    22.3.1 Organic pollution of groundwater 312
22.4 Inorganic pollution of groundwater 313
    22.4.1 Trace and heavy metals 313
    22.4.2 Major anions 319
References 320

23. Groundwater of Afghanistan (potential capacity, scarcity, security issues, and solutions) 323

Abdul Qayeum Karim and Sayed Hashmat Sadat

23.1 Introduction 323
23.2 Topography and hydrogeology of Afghanistan 323
23.3 Scarcity of groundwater quality and quantity 324
    23.3.1 Quality challenges of groundwater in Afghanistan 324
    23.3.2 Quantity challenges of groundwater in Afghanistan 325
23.4 Afghanistan groundwater sustainability 326
23.5 Afghanistan groundwater security 327
23.6 Solutions 327
References 328

Theme 4
Groundwater sustainability and security

24. Groundwater resources sustainability 331

Jac van der Gun

24.1 Sustainability and sustainable development 331
24.2 Sustainability of groundwater services 332
    24.2.1 Groundwater services 332
    24.2.2 Potential threats to groundwater services 334
24.3 Approaches to pursuing, restoring, or enhancing groundwater resources sustainability 335
    24.3.1 The umbrella: groundwater governance and management 335
    24.3.2 Hydrogeological approaches to defining sustainability limits of abstraction 335
    24.3.3 Enhancing groundwater recharge 336
    24.3.4 Water demand management 337
    24.3.5 Groundwater quality management 337
    24.3.6 Adaptation to climate change and sea-level rise 337
    24.3.7 Environmental management 338
24.4 Geographic variation of groundwater resources sustainability 338
    24.4.1 General comments 338
    24.4.2 Groundwater resources sustainability endangered or disrupted by progressive storage depletion 339
    24.4.3 Groundwater resources sustainability endangered or disrupted by water quality degradation 341
    24.4.4 Groundwater resources sustainability constrained by environmental considerations 341
24.5 Conclusion 343
References 344

25. Sustainability of groundwater used in agricultural production and trade worldwide 347

Carole Dalin

25.1 Introduction 347
    25.1.1 Water use for global food production and virtual water flows via international food trade 348
    25.1.2 Sustainability of groundwater use overall and in particular for global food production 350
    25.1.3 Quantification of groundwater depletion for food trade 352
25.2 Conclusion 355
Financial support 356
References 356

26. Groundwater and society: enmeshed issues, interdisciplinary approaches 359

Flore Lafaye de Micheaux and Mukherjee Jenia

26.1 Introduction 359
26.2 Socio-hydrology and socio-geohydrology: modeling of the groundwater–society interactions improved with stakeholders' perspectives 360
    26.2.1 Introduction to socio-hydrology 360
    26.2.2 Socio-hydrology and groundwater 360
    26.2.3 Incorporating stakeholders' perspectives: a "public" turn for socio-hydrology 361
26.3 Political ecology and the hydrosocial cycle: paying attention to power relations and discourses embedded in water circulation 361

26.3.1 Political ecology of water	361
26.3.2 The hydrosocial cycle: a critical rethinking of "water"	362
26.4 Mobilizing hydrosocial analyses to capture ground (water) realities	362
26.4.1 Dispossession of irrigating farmers through institutions and infrastructures	363
26.4.2 State and "scientific" versus local knowledge of water	363
26.4.3 Groundwater and politics of scale	363
26.4.4 Trajectories from "safe and good" groundwater to "bad" citizens	364
26.5 Discussion: what interdisciplinarity for enmeshed issues?	364
26.6 Conclusion	367
References	367

27. Groundwater sustainability in cold and arid regions 371

Rui Ma and Yanxin Wang

27.1 Importance of groundwater in hydrological systems	371
27.1.1 Cold regions	371
27.1.2 Arid and semi-arid regions	373
27.2 The characteristics of the hydrological cycle	373
27.2.1 The effect of permafrost distribution, snow and /or ice on groundwater systems in cold regions	373
27.2.2 Hydrological processes and its effect on groundwater quality in arid and semi-arid regions	375
27.3 Groundwater modeling and challenges	376
27.3.1 Model development in the cold regions	376
27.3.2 Model application and challenges in the arid and semi-arid regions	377
27.4 The effect of climate change	377
27.4.1 Cold regions	378
27.4.2 Arid and semi-arid regions	378
27.5 Integrated water management for groundwater sustainability	379
Acknowledgements	379
References	379

28. Groundwater in Australia—understanding the challenges of its sustainable use 383

Basant Maheshwari

28.1 Introduction	383
28.2 Aquifers in Australia	383
28.3 The Great Artesian Basin	384
28.4 The Murray—Darling Basin	385
28.5 The Perth Basin	385
28.6 The Canning Basin	386
28.7 The Daly Basin	386
28.8 The Otway Basin	386
28.9 Groundwater uses	387
28.10 Groundwater entitlements and extractions	387
28.11 Groundwater salinity	388
28.12 Australian ecosystems and groundwater	389
28.13 Concluding remarks	391
References	392
Further reading	392

29. Groundwater recharge and sustainability in Brazil 393

Paulo Tarso S. Oliveira, Murilo Cesar Lucas, Raquel de Faria Godoi and Edson Wendland

29.1 Insights from groundwater availability in Brazil	393
29.2 Overview of global groundwater recharge dynamics	396
29.3 Studies on recharge in Brazil	397
29.3.1 Recharge methods used in Brazilian studies	400
29.4 Challenges and future directions toward a groundwater sustainability in Brazil	402
Acknowledgments	403
References	404

30. Groundwater management in Brazil: current status and challenges for sustainable utilization 409

Prafulla Kumar Sahoo, Paulo Rógenes Monteiro Pontes, Gabriel Negreiros Salomão, Mike A Powell, Sunil Mittal, Pedro Walfir Martins e Souza Filho and José Tasso Felix Guimarães

30.1 Introduction	409
30.2 Groundwater resources of Brazil	410
30.2.1 Physical and climatic characteristics	410
30.2.2 Hydrogeological features of aquifers	411
30.3 Groundwater resource management in Brazil	414
30.3.1 Background of water resource management	414

30.3.2 National laws/legislation	415
30.3.3 Integrated management of surface water and groundwater	415
30.3.4 Management of transboundary groundwater	415
30.3.5 Management of mineral water resources	416
30.3.6 Groundwater monitoring and assessment	416
30.4 Alternatives for groundwater management and water sourcing	**417**
30.4.1 Adopting rainwater harvesting	417
30.4.2 Artificial groundwater recharge and reuse of wastewater	418
30.4.3 Desalination	418
30.5 The hydroschizophrenia of groundwater management	**418**
30.6 Final considerations and current challenges	**419**
References	**420**

31. Challenges of sustainable groundwater development and management in Bangladesh: vision 2050 — 425

K.M. Ahmed

31.1 Introduction	**425**
31.2 Groundwater occurrences in Bangladesh	**425**
31.3 Groundwater quality and concerns	**426**
31.3.1 Occurrences and distribution of arsenic	428
31.3.2 Occurrences and distribution of salinity	428
31.4 Groundwater uses and impacts of abstractions	**428**
31.4.1 Domestic uses in rural and urban areas	430
31.4.2 Irrigation uses	431
31.4.3 Industrial uses	431
31.5 Major challenges	**432**
31.5.1 Meeting increased demands in 2050	432
31.5.2 Impacts of climate change	432
31.5.3 Arsenic and other contamination issues	433
31.5.4 Transboundary issues	433
31.6 Sustainable groundwater management: vision 2050	**433**
31.6.1 Surface water harnessing	433

31.6.2 Better irrigation water management	434
31.6.3 Groundwater monitoring, abstraction controls, and licensing	434
31.6.4 Pollution abatement and control	434
31.6.5 Applications of managed aquifer recharge	434
31.6.6 Wastewater reuse	435
31.6.7 Awareness building	435
31.6.8 Judicial use of deep groundwater	435
31.6.9 Groundwater governance	435
31.6.10 Research and development activities	435
31.7 Groundwater: resource out of sight but not to be out of mind	**435**
Acknowledgments	**436**
References	**436**

32. Integrating groundwater for water security in Cape Town, South Africa — 439

G. Thomas LaVanchy, James K. Adamson and Michael W. Kerwin

32.1 Introduction	**439**
32.2 Situating Cape Town	**440**
32.2.1 The Day Zero drought	441
32.2.2 Water provision and security	442
32.3 Groundwater opportunities	**442**
32.3.1 Table Mountain Group aquifers	443
32.3.2 Sandveld Group aquifers	444
32.4 Groundwater management challenges	**445**
32.4.1 Physical dimensions	445
32.4.2 Human dimensions	446
32.5 Conclusion	**447**
References	**448**

33. Drivers for progress in groundwater management in Lao People's Democratic Republic — 451

Cécile A. Coulon, Paul Pavelic and Evan Christen

33.1 Introduction	**451**
33.2 Groundwater resources in Lao People's Democratic Republic	**452**
33.2.1 Groundwater systems	452
33.2.2 Groundwater use	453

33.3 Major groundwater challenges 454
    33.3.1 Quantity and quality-related issues 454
    33.3.2 State of groundwater knowledge and information systems 454
    33.3.3 Other barriers to groundwater management 455
33.4 Recent efforts to strengthen groundwater governance 455
    33.4.1 Overview of policy, institutional, and legal changes 455
    33.4.2 Enhancing groundwater knowledge and data management 457
    33.4.3 Mechanisms of stakeholder coordination and involvement 460
    33.4.4 Development of human resources and groundwater-management capacity 461
33.5 Outlook: pathways forward for Lao People's Democratic Republic 463
    33.5.1 Effective policy making and implementation 463
    33.5.2 Strengthening institutional and human resource capacity 464
    33.5.3 Continuing efforts in applied research 464
    33.5.4 Participation of stakeholders 465
Acknowledgments 465
Acronyms 465
References 466

34. Groundwater sustainability and security in South Asia 469

Soumendra Nath Bhanja and Abhijit Mukherjee

34.1 Introduction 469
34.2 Data 469
    34.2.1 Study region 469
    34.2.2 WaterGAP3 model 470
34.3 Results and discussions 470
    34.3.1 Evapotranspiration and groundwater recharge 470
    34.3.2 Contamination issues 471
    34.3.3 Population 472
34.4 Summary and way forward 472
Acknowledgments 475
References 475

35. Role of measuring the aquifers for sustainably managing groundwater resource in India 477

Dipankar Saha, Sanjay Marwaha and S.N. Dwivedi

35.1 Introduction 477
35.2 Regional aquifer framework 477
35.3 Spatiotemporal behavior of hydraulic heads and replenishable resources 478
35.4 How much groundwater we are extracting 479
35.5 Expanding groundwater contamination 480
35.6 Measuring and understanding the aquifers 482
35.7 The sustainable management plan—an example 483
35.8 Way forward 485
References 485
Further reading 486

36. Balancing livelihoods and environment: political economy of groundwater irrigation in India 487

Tushaar Shah, Abhishek Rajan and Gyan P Rai

36.1 Evolution of Indian irrigation 487
36.2 Changing organization of the irrigation economy 488
36.3 Energy-irrigation nexus 489
36.4 Socioeconomic significance of the groundwater boom 490
36.5 The sustainability challenge 491
36.6 Sustainable groundwater governance 493
    36.6.1 Direct regulation through legal framework and administrative action 493
    36.6.2 Community-based groundwater management 494
    36.6.3 Indirect instruments—energy pricing and rationing 495
    36.6.4 The advent of solar irrigation 495
36.7 Conclusion: from resource development to management mode 496
References 497

Theme 5
Future of groundwater and solutions

37. The future of groundwater science and research 503

David K. Kreamer, David M. Ball, Viviana Re, Craig T. Simmons, Thomas Bothwell, Hanneke J.M. Verweij, Abhijit Mukherjee and Magali F. Moreau

37.1 Introduction 503
37.2 How are fundamental groundwater perspectives changing?—"Darcy is dead" 504

37.3 Fossil fuel energy, geothermal energy, and mineral resources—the groundwater connection and the future 505
37.4 Groundwater can be a deep subject 506
37.5 The subterranean biological world and groundwater-dependent ecosystems 507
37.6 Coast to coast 508
37.7 Under the ocean 508
37.8 Extraterrestrial hydrology—the sky's not the limit 508
37.9 Groundwater quality and emerging contaminants 509
37.10 The new tools 510
37.11 Laws, regulation, guidance, and governance of groundwater 511
37.12 Socio-hydrogeology in the future of groundwater science 511
37.13 Education and outreach 512
37.14 The unexpected challenges 512
Acknowledgments 513
References 513
Further reading 517

38. Technologies to enhance sustainable groundwater use 519

Roger Sathre

38.1 Technology levers to enhance groundwater security 519
38.2 Groundwater mapping and management 519
38.3 Managing aquifer recharge 520
38.4 Managing saline groundwater intrusion 521
38.5 Improving groundwater-use efficiency 522
    38.5.1 Improving irrigation and agricultural efficiency 522
    38.5.2 Improving household water distribution and use efficiency 523
    38.5.3 Improving industrial water-use efficiency 523
38.6 Purifying contaminated groundwater 524
    38.6.1 Removing salt from brackish groundwater 524
    38.6.2 Removing arsenic from groundwater 526
    38.6.3 Removing fluoride from groundwater 526
    38.6.4 Killing biological pathogens in groundwater 527
38.7 Improving groundwater access 527
    38.7.1 Well digging and drilling 527
    38.7.2 Groundwater pumping 528
38.8 Conclusion 528
References 528

39. Applications of Gravity Recovery and Climate Experiment (GRACE) in global groundwater study 531

Jianli Chen and Matt Rodell

39.1 Introduction 531
39.2 GRACE and GFO missions and data products 532
39.3 Quantification of groundwater change using Gravity Recovery and Climate Experiment 533
39.4 Gravity recovery and climate experiment applications in groundwater storage change 534
39.5 Major error sources of Gravity Recovery and Climate Experiment—estimated groundwater change 537
39.6 Gravity Recovery and Climate Experiment data assimilation 539
39.7 Summary 539
References 540

40. Use of machine learning and deep learning methods in groundwater 545

Pragnaditya Malakar, Soumyajit Sarkar, Abhijit Mukherjee, Soumendra Bhanja and Alexander Y. Sun

40.1 Introduction 545
    40.1.1 Importance of advanced data-driven methods in groundwater resources 545
40.2 Global literature review 546
    40.2.1 Groundwater quantity 546
    40.2.2 Groundwater quality 546
40.3 Application of some of the widely used artificial intelligence methods in India 550
    40.3.1 Methods description 551
    40.3.2 Case studies from India 551
References 554

41. Desalination of brackish groundwater to improve water quality and water supply 559

Yvana D. Ahdab and John H. Lienhard

41.1 Introduction 559
    41.1.1 Brackish groundwater composition 559
    41.1.2 Desalination 560
41.2 Desalination process 560
    41.2.1 Membrane fouling and pretreatment 561

41.2.2 Reverse osmosis	561
41.2.3 Electrodialysis	563
41.2.4 Energy consumption using conventional energy sources	564
41.2.5 Economics of desalination	565
41.2.6 Brine management	567
41.2.7 Brine disposal	567
41.2.8 Brine treatment	567
41.2.9 Desalination using renewable energy sources	568
41.2.10 Emerging desalination technologies	568
41.2.11 Nanofiltration	569
41.2.12 Semibatch reverse osmosis	569
41.3 Global and national trends in desalination	569
41.3.1 Global trends	569
41.3.2 National trends	572
Acknowledgments	573
References	573

42. Desalination of deep groundwater for freshwater supplies 577

Veera Gnaneswar Gude and Anand Maganti

42.1 Introduction	577
42.2 Groundwater desalination—influencing factors	577
42.2.1 Motivation for groundwater desalination	577
42.2.2 Considerations for groundwater desalination	578
42.2.3 Environmental impacts of groundwater desalination	579
42.3 Desalination technology assessment	579
42.4 Groundwater desalination in the United States	580
42.5 Groundwater desalination in developing countries	581
42.6 Decision-making for municipal desalination plants	581
42.7 Conclusion	582
References	582

43. Quantifying future water environment using numerical simulations: a scenario-based approach for sustainable groundwater management plan in Medan, Indonesia 585

Pankaj Kumar, Binaya Kumar Mishra, Ram Avtar and Shamik Chakraborty

43.1 Introduction	585
43.2 Study area	586
43.3 Methodology	586
43.3.1 Different drivers	587
43.3.2 Urban flood	590
43.3.3 Water quality	591
43.4 Results and discussion	592
43.4.1 Precipitation change	592
43.4.2 Land use change	593
43.4.3 Urban flood	594
43.4.4 Water quality	594
43.5 Conclusion and recommendation	595
References	595

44. Managed aquifer recharge with various water sources for irrigation and domestic use: a perspective of the Israeli experience 597

Daniel Kurtzman and Joseph Guttman

44.1 Introduction	597
44.1.1 Why Israel has a significant managed aquifer recharge experience?	597
44.1.2 The Israeli Coastal Aquifer	598
44.2 Managed aquifer recharge of ephemeral stream floods in the coastal aquifer through infiltration basins, increasing freshwater supply (1959–present)	598
44.3 Managed aquifer recharge of groundwater and especially lake water through wells for freshwater supply (1965–90 and reexamination 2012–20)	600
44.3.1 Technical considerations concerning managed aquifer recharge through wells	601
44.3.2 Some history and experience from the managed aquifer recharge through well period 1965–90	602
44.3.3 New thoughts and experiments on managed aquifer recharge through wells due to availability of water of better quality today (2012–20)	602
44.4 Managed aquifer recharge of secondary effluents in infiltration basins—the Shafdan water reclamation system for irrigation (1987–present)	604
44.5 Managed aquifer recharge of surplus desalinated seawater through infiltration basins (2014–present)	605
References	606

45. MAR model: a blessing adaptation for hard-to-reach livelihood in thirsty Barind Tract, Bangladesh   609

Chowdhury Sarwar Jahan, Md. Ferozur Rahaman, Quamrul Hasan Mazumder and Md. Iquebal Hossain

45.1	Introduction	609
45.2	Challenges of groundwater resource management plan	612
45.3	Groundwater resource potentiality	613
45.4	Potential zones for groundwater recharge and selection of sites for artificial recharge of groundwater	616
45.5	Implementation of managed aquifer recharge model	616
45.5.1	Piloting of managed aquifer recharge model at household level—pioneer attempt during 2013–16	616
45.5.2	Managed aquifer recharge model as integrated water resource management strategy in Barind Tract since 2015	618
45.5.3	Impact assessment of managed aquifer recharge model as integrated water resource management strategy	621
45.6	Conclusion	623
	Acknowledgments	624
	References	624

Index   627

List of Contributors

James K. Adamson Northwater International, Chapel Hill, NC, United States

Yvana D. Ahdab Rohsenow Kendall Heat Transfer Laboratory, Massachusetts Institute of Technology, Cambridge, MA, United States

K.M. Ahmed Department of Geology, Faculty of Earth and Environmental Sciences, University of Dhaka, Dhaka, Bangladesh

Kazi Matin Ahmed Department of Geology, University of Dhaka, Curzon Hall Campus, Dhaka, Bangladesh

Ahmed A. Al-Taani College of Natural and Health Sciences, Zayed University, Abu Dhabi, United Arab Emirates; Department of Earth and Environmental Sciences, Faculty of Science, Yarmouk University, Irbid, Jordan

Alice Aureli Groundwater Systems and Settlements Section, International Hydrological Programme, United Nations Educational, Scientific and Cultural Organization (UNESCO), Paris, France

Ram Avtar Faculty of Environmental Earth Science, Hokkaido University, Sapporo, Japan

David M. Ball Independent Hydrogeological Consultant, Dublin, Ireland

Steve Barnett Department of Environment, Water and Natural Resources, Adelaide, SA, Australia

Shehla Batool Department of Environmental Sciences, Faculty of Biological Sciences, Quaid-i-Azam University, Islamabad, Pakistan

Lahcen Benaabidate Laboratory of Functional Ecology and Environment Engineering, University of Sidi Mohammed Ben Abdellah, Fez, Morocco

Soumendra Bhanja Interdisciplinary Centre for Water Research, Indian Institute of Science, Bangalore, India

Soumendra Nath Bhanja Interdisciplinery Centre for Water Research, Indian Institute of Science, Bangalore, India

Thomas Bothwell Rosetta Stone Consulting, Perth, WA, Australia

Madhumita Chakraborty Department of Geology and Geophysics, Indian Institute of Technology (IIT) Kharagpur, Kharagpur, India

Shamik Chakraborty Faculty of Sustainability Studies, Hosei University, Tokyo, Japan

Jianli Chen Center for Space Research, University of Texas at Austin, Austin, TX, United States

Evan Christen Penevy Services Pty Ltd, Huskisson, NSW, Australia

Poulomee Coomar Department of Geology and Geophysics, Indian Institute of Technology (IIT) Kharagpur, Kharagpur, India

Cécile A. Coulon International Water Management Institute, Vientiane, Lao PDR; Department of Geology and Geological Engineering, Université Laval, Québec, Québec, Canada

Brian C. Crone United States Environmental Protection Agency, Office of Research and Development, Center of Environmental Solutions and Emergency Response, Cincinnati, OH, United States

Mark Cuthbert School of Earth and Ocean Sciences & Water Research Institute, Cardiff University, Cardiff, United Kingdom; Connected Waters Initiative Research Centre, University of New South Wales, Sydney, NSW, Australia

Carole Dalin University College London, London, United Kingdom

Raquel de Faria Godoi Faculty of Engineering, Architecture and Urbanism and Geography, Federal University of Mato Grosso do Sul, Campo Grande, Brazil

Long Di State Key Laboratory of Hydroscience and Engineering, Department of Hydraulic Engineering, Tsinghua University, Beijing, China

S.N. Dwivedi Central Ground Water Board, Faridabad, India

Abida Farooqi Department of Environmental Sciences, Faculty of Biological Sciences, Quaid-i-Azam University, Islamabad, Pakistan

Grant Ferguson Department of Civil, Geological and Environmental Engineering, University of Saskatchewan, Saskatoon, SK, Canada

Anjuli Jain Figueroa Postdoctoral Scholar, School of Earth, Energy and Environmental Science, Stanford University, Stanford, CA, United States

Alan E. Fryar Department of Earth and Environmental Sciences, University of Kentucky, Lexington, KY, United States

Susan T. Glassmeyer United States Environmental Protection Agency, Office of Research and Development, Center of Environmental Solutions and Emergency Response, Cincinnati, OH, United States

Tom Gleeson Department of Civil Engineering and School of Earth and Ocean Sciences, University of Victoria, Victoria, BC, Canada

Debbie-Ann D.S. Gordon-Smith Department of Chemistry, The University of the West Indies, Mona, Jamaica, West Indies

Veera Gnaneswar Gude Department of Civil and Environmental Engineering, Mississippi State University, Mississippi State, MS, United States

José Tasso Felix Guimarães Vale Institute of Technology (ITV), Belém, Brazil

Huaming Guo School of Water Resources and Environment, China University of Geosciences, Beijing, P.R. China

Joseph Guttman Mekorot, Israel National Water Company Ltd., Tel Aviv, Israel

Dongya Han Institute of Hydrogeology and Environmental Geology, Chinese Academy of Geological Sciences, Shijiazhuang, P.R. China; Hebei GEO University, Shijiazhuang, P.R. China

Shama E. Haque North South University, Dhaka, Bangladesh

Peta-Gay Harris Department of Geography and Geology, The University of the West Indies, Mona, Jamaica, West Indies

Md. Iquebal Hossain Barind Multi-Purpose Development Authority, Rajshahi, Bangladesh

Fares M. Howari College of Natural and Health Sciences, Zayed University, Abu Dhabi, United Arab Emirates

Guanxing Huang Institute of Hydrogeology and Environmental Geology, Chinese Academy of Geological Sciences, Shijiazhuang, P.R. China

Chowdhury Sarwar Jahan Department of Geology & Mining, University of Rajshahi, Rajshahi, Bangladesh

Mukherjee Jenia Indian Institute of Technology, Kharagpur, India

Yongfeng Jia State Key Laboratory of Environmental Criteria and Risk Assessment, Chinese Research Academy of Environmental Sciences, Beijing, P.R. China; State Environmental Protection Key Laboratory of Simulation and Control of Groundwater Pollution, Chinese Research Academy of Environmental Sciences, Beijing, P.R. China

Abdul Qayeum Karim Department of Civil Engineering, Faculty of Engineering, Kabul University, Kabul, Afghanistan

Seifu Kebede Seifu Kebede Gurmessa, School of Agricultural Earth and Environmental Sciences, Center for Water Resources Research, University of KwaZulu Natal, Pietermaritzburg, South Africa

Michael W. Kerwin Department of Geography & the Environment, University of Denver, Denver, CO, United States

David K. Kreamer Department of Geosciences, University of Nevada, Las Vegas, NV, United States

Pankaj Kumar Natural Resources and Ecosystem Services, Institute for Global Environmental Strategies, Hayama, Japan

Daniel Kurtzman Institute of Soil, Water and Environmental Sciences, Agricultural Research Organization, The Volcani Center, Rishon LeZion, Israel

Flore Lafaye de Micheaux University of Lausanne, Lausanne, Switzerland; International Union for Conservation of Nature, Gland, Switzerland; French Institute of Pondicherry, Puducherry, India

Simon Langan International Water Management Institute (IWMI), Colombo, Sri Lanka

G. Thomas LaVanchy Department of Geography, Oklahoma State University, Stillwater, OK, United States

Bailing Li ESSIC, University of Maryland, College Park, MD, United States; Hydrological Sciences Laboratory, NASA Goddard Space Flight Center, Greenbelt, MD, United States

John H. Lienhard Rohsenow Kendall Heat Transfer Laboratory, Massachusetts Institute of Technology, Cambridge, MA, United States

Chunyan Liu Institute of Hydrogeology and Environmental Geology, Chinese Academy of Geological Sciences, Shijiazhuang, P.R. China

Fei Liu MOE Key Laboratory of Groundwater Circulation and Environmental Evolution, Beijing Key Laboratory of Water Resources and Environmental Engineering, School of Water Resources and Environment, China University of Geosciences (Beijing), Beijing, P.R. China

Lingxia Liu Institute of Hydrogeology and Environmental Geology, Chinese Academy of Geological Sciences, Shijiazhuang, P.R. China

Murilo Cesar Lucas Department of Civil Engineering, Federal University of Technology-Paraná, Pato Branco, Brazil

Rui Ma School of Environmental Studies & State Key Laboratory of Biogeology and Environmental Geology, China University of Geosciences, Wuhan, Hubei 430074, China

Anand Maganti Department of Civil and Environmental Engineering, Mississippi State University, Mississippi State, MS, United States

Basant Maheshwari Western Sydney University, Hawkesbury Campus, Penrith, NSW, Australia

Pragnaditya Malakar Department of Geology and Geophysics, Indian Institute of Technology (IIT) Kharagpur, Kharagpur, India

Arpita Mandal Department of Geography and Geology, The University of the West Indies, Mona, Jamaica, West Indies

Ruth Marfil-Vega Shimadzu Scientific Instruments, Columbia, MD, United States

Pedro Walfir Martins e Souza Filho Vale Institute of Technology (ITV), Belém, Brazil

Sanjay Marwaha Central Ground Water Board, Faridabad, India

Noshin Masood Department of Environmental Sciences, Faculty of Biological Sciences, Quaid-i-Azam University, Islamabad, Pakistan

Quamrul Hasan Mazumder Department of Geology & Mining, University of Rajshahi, Rajshahi, Bangladesh

Andrew McKenzie British Geological Survey, Oxfordshire, United Kingdom

Marc A. Mills United States Environmental Protection Agency, Office of Research and Development, Center of Environmental Solutions and Emergency Response, Cincinnati, OH, United States

Binaya Kumar Mishra School of Engineering, Pokhara University, Lekhnath, Nepal

Sunil Mittal Department of Environmental Science and Technology, Central University of Punjab, Bathinda, India

Paulo Rógenes Monteiro Pontes Vale Institute of Technology (ITV), Belém, Brazil

Magali F. Moreau GNS Science, Wairakei Research Center, Taupo, New Zealand

Abhijit Mukherjee Department of Geology and Geophysics, Indian Institute of Technology (IIT) Kharagpur, Kharagpur, India; Applied Policy Advisory for Hydrosciences (APHA) group, School of Environmental Science and Engineering, Indian Institute of Technology (IIT) Kharagpur, Kharagpur, India

Yousef Nazzal College of Natural and Health Sciences, Zayed University, Abu Dhabi, United Arab Emirates

Rebecca Nelson Melbourne Law School, University of Melbourne, Melbourne, VIC, Australia

Paulo Tarso S. Oliveira Faculty of Engineering, Architecture and Urbanism and Geography, Federal University of Mato Grosso do Sul, Campo Grande, Brazil

Paul Pavelic International Water Management Institute, Vientiane, Lao PDR

Debra Perrone Environmental Studies Program, University of California at Santa Barbara, Santa Barbara, CA, United States

Mike A Powell Department of Renewable Resources, Faculty of Agriculture, Life and Environmental Sciences (ALES), University of Alberta, Edmonton, AB, Canada

Shaminder Puri Sustainable Solutions in Practical Hydrogeology, Oxford, United Kingdom; IAH Commission on Transboundary Aquifers, Oxford, United Kingdom; International Association of Hydrogeologists, Reading, United Kingdom

Xiaopeng Qin Department of Technology Assessment, Technical Centre for Soil, Agricultural and Rural Ecology and Environment, Ministry of Ecology and Environment, Beijing, P.R. China

Md. Ferozur Rahaman Institute of Environmental Science, University of Rajshahi, Rajshahi, Bangladesh

Gyan P Rai International Water Management Institute (IWMI)-Tata Water Policy Program, Anand, India

Abhishek Rajan International Water Management Institute (IWMI)-Tata Water Policy Program, Anand, India

Viviana Re Department of Earth Sciences, University of Pisa, Pisa, Italy

Matt Rodell NASA Goddard Space Flight Center, Greenbelt, MD, United States

Matthew Rodell Hydrological Sciences Laboratory, NASA Goddard Space Flight Center, Greenbelt, MD, United States

Sayed Hashmat Sadat Department of Civil Engineering, Faculty of Engineering, Kabul University, Kabul, Afghanistan

Othman Sadki Department of Geochemistry, National Office of Hydrocarbons and Mines, Rabat, Morocco

Dipankar Saha Formerly at the Central Ground Water Board, Government of India, Faridabad, India

Prafulla Kumar Sahoo Department of Environmental Science and Technology, Central University of Punjab, Bathinda, India; Vale Institute of Technology (ITV), Belém, Brazil

Gabriel Negreiros Salomão Vale Institute of Technology (ITV), Belém, Brazil; Geology and Geochemistry Graduate Program (PPGG), Geosciences Institute (IG), Federal University of Pará (UFPA), Belém, Brazil

Soumyajit Sarkar Applied Policy Advisory for Hydrosciences (APHA) group, School of Environmental Science and Engineering, Indian Institute of Technology (IIT) Kharagpur, Kharagpur, India

Roger Sathre Institute for Transformative Technologies, Berkeley, CA, United States; Linnaeus University, Växjö, Sweden

Bridget R. Scanlon Bureau of Economic Geology, Jackson School of Geosciences, University of Texas, Austin, TX, United States

Madeline E. Schreiber Department of Geosciences, Virginia Tech, Blacksburg, VA, United States

Tushaar Shah Institute of Rural Management Anand, Anand, India

M. Shamsudduha Department of Geography, University of Sussex, Brighton, United Kingdom; Institute for Risk and Disaster Reduction, University College London, London, United Kingdom

Craig T. Simmons National Centre for Groundwater Research and Training, College of Science and Engineering, Flinders University, Adelaide, SA, Australia

Mikhail Smilovic Research Scholar, Water program, IIASA – Institute of Applied Systems Analysis, Laxenburg, Austria

Alexander Y. Sun Bureau of Economic Geology, The University of Texas at Austin, Austin, TX, United States

Zhangli Sun State Key Laboratory of Hydroscience and Engineering, Department of Hydraulic Engineering, Tsinghua University, Beijing, China

Meron Teferi Taye IWMI, East Africa and Nile basin Office, Addis Ababa, Ethiopia

Jac van der Gun Van der Gun Hydro-Consulting, Schalkhaar, The Netherlands

Hanneke J.M. Verweij Independent Expert Pressure and Fluid Flow Systems, Delft, The Netherlands

Junye Wang Athabasca River Basin Research Institute (ARBRI), Athabasca University, Athabasca, AB, Canada

Wenzhong Wang Institute of Hydrogeology and Environmental Geology, Chinese Academy of Geological Sciences, Shijiazhuang, P.R. China

Yanxin Wang School of Environmental Studies & State Key Laboratory of Biogeology and Environmental Geology, China University of Geosciences, Wuhan, Hubei 430074, China

Edson Wendland Department of Hydraulics and Sanitary Engineering, University of São Paulo, São Carlos, Brazil

Wenting Yang State Key Laboratory of Hydroscience and Engineering, Department of Hydraulic Engineering, Tsinghua University, Beijing, China

Tian Zhou MOE Key Laboratory of Groundwater Circulation and Environmental Evolution, Beijing Key Laboratory of Water Resources and Environmental Engineering, School of Water Resources and Environment, China University of Geosciences (Beijing), Beijing, P.R. China

Ahmed Zian National School of Applied Sciences of Al Hoceima, University Abdelmalek Essaadi, Tétouan, Morocco

Shengzhang Zou Institute of Karst Geology, CAGS, Karst Dynamics Laboratory, MLR & GZAR, Guilin, P.R. China

About the Editors

Abhijit Mukherjee has a PhD in Hydrogeology from the University of Kentucky, USA and has been a Postdoctoral Fellow at the Jackson School of Geoscience, the University of Texas at Austin, United States. He has also served as the Physical Hydrogeologist at the Alberta Geological Survey in Canada. He is presently an Associate Professor at the Department of Geology and Geophysics and the School of Environmental Science and Engineering at the Indian Institute of Technology (IIT) Kharagpur in India. He has over 20 years of teaching and research experience. He is a globally renowned expert in groundwater contamination and one of the pioneer in application of data science and AI in groundwater studies. He is author of over a hundred journal articles. Among many awards and recognitions, in 2016, he was conferred the National Geoscience Award by the President of India. He has also received the prestigious Shanti Swarup Bhatnagar Prize, the highest science award in India, for the year 2020. He has been in Editorial role in several journals, including the Journal of Hydrology, Applied Geochemistry, ES&T Engineering, Scientific Report, Groundwater for Sustainable Development, Frontiers in Environmental Science, and Journal of Earth System Science.

Bridget Scanlon has a PhD in Hydrogeology from the University of Kentucky, United States, and is presently the Fisher Endowed Chair in Geological Sciences and a Senior Research Scientist at the Bureau of Economic Geology, Jackson School of Geosciences, the University of Texas at Austin, United States. As a world-leading authority on water research, her career has been characterized by a commitment to data as well as innovative approaches that cut across disciplines. During her ~40 years academic career, she has published articles in numerous peer-reviewed journals, and has been involved with US Department of Energy scientific endeavors, and has been a member of the NASA GRACE satellite Science team. In 2016 she was elected as a member of the National Academy of Engineering, one of the highest US scientific professional honors. A Fellow of both Geological Society of America (GSA) and American Geophysical Union (AGU), Bridget has received many awards including the GSA O.E. Meinzer Award and the National Ground Water Association's M. King Hubbert Award. She is widely considered as one of the foremost authorities on global groundwater resources and besides being an Associate Editor of several subject journals, she is the former Managing Editor of Hydrogeology Journal.

Alice Aureli has a PhD in Hydrogeology from the University of Rome, Italy and has worked in the UNESCO Water Sciences Division since 1989. She is the Chief of the Groundwater Resources and Aquifer Systems Section of UNESCO's International Hydrological Programme (IHP). She is coordinator for the International Shared Aquifers Resources Management (ISARM) programme of the UNESCO. This role has led her to supervise the work of the interdisciplinary group that advised the UN International Law Commission to prepare the Draft Articles on the Law of Transboundary Aquifers. An important aspect of her work has been on scientific and policy-related issues surrounding groundwater governance. She is the author of a large number of publications and has also served as editor of various international journals.

Simon Langan received his PhD from University of St. Andrews, United Kingdom, followed by a postdoctoral fellowship at Imperial College, London, United Kingdom. He was the Director of IIASA's Water Program and the Water Futures and Solutions Initiative. He is presently serving as the Director, Digital Innovation and Country Manager, Sri Lanka of the International Water Management Institute (IWMI). Throughout his career, he has won grants and secured funding from regional and international donor projects, including from the private sector, the EU 7th Framework, Natural Environment Research Council, National Power, Scottish Environment Protection Agency, USAID, and Canadian Government. He has an extensive number of publications in peer-reviewed journals, as well as experience in policy-related analyses, including numerous technical reports, books/chapters, and conference proceedings.

Huaming Guo has a PhD from the China University of Geosciences, Wuhan, Hubei, China, followed by a Postdoctoral Fellowship at Tsinghua University, Beijing, China. He has also been an Alexander van Humboldt Research Fellow at the Karlsruhe Institute of Technology, Germany. He is currently a Professor at the School of Water Resources and Environment, China University of Geosciences, Beijing, China. He has been also a Senior Visiting Professor to Columbia University, United States. He has over 20 years of teaching and research experience. He has been Associate Editor of several journals and presently serves as an Editor-in-Chief of Journal of Hydrology.

Andrew McKenzie has a BA (Hons.) from Oxford University and MSc from University College London in Hydrogeology. He worked as an exploration geologist and hydrogeologist in Africa, the Middle East and the United Kingdom before joining the British Geological Survey (BGS) in 1988, working on groundwater issues, in Central America. As a hydrogeologist in BGS's Groundwater Directorate he has been responsible for managing the survey's databases on groundwater, focusing on field data collection, data processing, and developing systems to disseminate data to stakeholders. This includes contributing to the NERC systems for monitoring groundwater status, investigating drought and floods, and, more recently, developing forecasts of groundwater resources at a national level. He has extensive international experience principally in Africa and South Asia, where he was Senior Hydrogeologist for the World Bank India Hydrology Project, and coinvestigator on research projects in the Ganga and Cauvery basins. He is currently Platform Lead for the BGS ODA Project "Sustainable Asian Cities" which is building networks for urban geoscience across several Asian countries. He has over 35 years of research experience. He is a Fellow of Geological Society of London.

Forewords

1 Foreword on groundwater as a resource

Groundwater is the most abundant freshwater resource available on earth. More than 95% of all liquid freshwater is groundwater. More than 2.5 billion people rely on groundwater for their basic drinking water. More than 40% of the water we use for irrigated agriculture is groundwater. Groundwater feeds the baseflows of our lakes and rivers and sustains biodiversity. Groundwater is often the water of last resort, in remote communities, in conflict-affected contexts, and during droughts.

In many places, however, these precious groundwater resources are managed blindly and are being very dangerously overdrawn. As an "unseen" resource, its invisibility puts it at great risk of mismanagement and makes it immensely complex to govern. We know that many groundwater sources, from the most arid to the most humid regions, are facing serious risks of overabstraction and contamination.

Population and economic growth relentlessly drive global water usage and pollution, creating mounting pressures on groundwater resources. Moreover the climate crisis further compounds the groundwater challenge. As rainfall and surface water flows become more unpredictable, people turn to groundwater abstraction. Groundwater quality is also threatened by climate change as sea levels rise and threatens coastal freshwater aquifers with saltwater intrusion which, like other forms of groundwater pollution, is extremely difficult to remediate.

This book provides insights and evidence from eminent groundwater researchers on how we can manage these resources more wisely. The Editors have curated a range of groundwater scholarship from scarcities to solutions, at global- to country-specific scales, across all of the major groundwater-using nations. I hope that the book will help water managers and policymakers better understand and balance the competing and interconnected needs of groundwater usage—including biodiversity preservation and ecosystem function, food production, and poverty reduction and livelihoods development. Global initiatives, including the Sendai Framework and Agenda 2030, lay the groundwork for collective action. We need to build on these initiatives, inform ourselves with the best available science, and redouble our efforts to manage this crucial resource. This book is an important contribution toward that end.

Claudia Sadoff
Director General
International Water Management Institute (IWMI)
Colombo, Sri Lanka

II Foreword on groundwater for society

Global groundwater is under growing stress from climate change, overdevelopment, and pollution leading to decrease clean groundwater supplies for domestic, agricultural, and industrial uses. This in turn increases concerns with global security and sustainability of clean water supplies. Overstressed groundwater systems can lead to land subsidence and water quality degradation from saltwater intrusion and mobilization of natural and man-made pollutants.

Global groundwater withdrawal rates are estimated to be on the order of 982 km^3/year Currently, groundwater supplies approximately 50% of global drinking water supply. About 70% of groundwater withdrawal is meant for agriculture uses. Withdrawals of groundwater are expected to continue to increase as the world's population continues to increase.

Complicating situation is the large differences across the globe in the legal and managerial frameworks that govern water supply development and distribution. Most groundwater basins cross one or more political boundaries.

This book stresses the availability of safe and sustainable groundwater across the world by providing insights into the issues of stressed groundwater systems and offers unique insights and knowledge for groundwater scientists, mangers, and policy makers. Chapters written by global experts and researchers include groundwater studies on quantity, exploration, quality and pollution, economics, management and policies, groundwater and society, and sustainable sources and efficient solutions.

John W. Hess
President
Geological Society of America (GSA) Foundation
Boulder, CO, United States

III Foreword on groundwater for sustainability

I applaud the initiative of this leading group of groundwater scholars in organizing this effort on sustainability, tackling what is the most serious collection of problems in the groundwater arena. One look at the numbers in the global water balance makes it clear how important groundwater is as the ultimate source of freshwater on the planet. Yet evidence from groundwater investigations and space-based measurements makes the extent of continuing overutilization of the resource clear, especially in populous countries in arid-zone settings. In the absence of significant natural recharge, even modest withdrawals of groundwater are unsustainable. Moreover, a large and growing numbers of wells in Asian countries also have the possibility of even overwhelming aquifer systems that receive abundant seasonal recharge.

The need for action to address the problems of groundwater sustainability is long overdue. Only in a few places such as California, United States there has been significant progress to manage groundwater resources sustainably. Elsewhere, progress has been limited. For groundwater users in India and Pakistan, the window for sustainability appears to be closing with continuing impacts from overpumping and salinization having emerged as major threats. In China, groundwater utilization continues to increase with evident impacts reflected in terms of subsidence, seawater intrusion, and water-level declines.

A book of this kind is important because it contributes the essential knowledge needed to support country-specific solutions and provides examples of progress. However, one cannot discount the immense problems and roadblocks to substantive progress in areas of sustainability. Chief among them are politics associated with water/agricultural issues, data limitations, and difficulties in managing millions of existing extraction wells. Most countries have an appropriate legal framework in terms of policies and regulatory implements but lack the capacity in rural areas to manage the resource effectively. Other problems include government policies aligned to promote food production, for example, with subsidies in electric power or water-hungry crops, which work against efforts to reduce pumping. There are also concerns that the verifiable sustainability practices already demonstrated in Singapore and Orange County, California may be neither affordable nor scalable to places with larger populations. Success in these places depends on expensive infrastructure providing for the advanced purification of urban wastewaters, and in the case of Orange County managed aquifer recharge.

What has proven feasible for some countries are traditional approaches to groundwater management like tanks and recharge ponds, which are locally managed. In China, their "sponge city" initiatives to reduce urban flooding have the potential to contribute to sustainability. Yet, there is a need for research to establish their efficiency with time and their ability to meet sustainability needs.

On behalf of hydrogeologists all around the world, I would like to express our gratitude to all those who contributed to this wonderful volume.

Franklin W. Schwartz
Professor, Ohio Eminent Scholar
Ohio State University
Columbus, OH, United States

IV Foreword on groundwater for future

The book on *Global Groundwater: Source, Scarcity, Sustainability, Security, and Solutions* presents a unique collection of empirical studies of groundwater-dependent scientific, technical, and societal problems in a large area of the earth's surface of greatly variable geologic, topographic, and climatic conditions. Apart from the real-life importance of such studies in their specific locations, the collection provides a rich assemblage of case histories for practicing as well as budding hydrogeologists.

József Toth
Professor Emeritus of Hydrogeology
University of Alberta
Edmonton, AB, Canada

V Foreword on groundwater research

Groundwater is vital for the survival and health of the population all over the world. But it is especially important in areas where groundwater resources are the main or even the only possibility to get freshwater, like in many arid or semiarid areas of the globe.

But the world georesources, not only groundwater, have been exploited at a level that causes enormous environmental stress, including the disappearance of species at unprecedented rates. Our need for more and more resources, as for the fuel, led to new techniques for old resources and has been appointed as a problem for groundwater resources, like fracking for oil and gas prospecting and abstraction. Other mining resources are also a source of contamination and depletion of groundwater. Agriculture uses more and more water resources, and fertilization, pesticides, and herbicides have been widely used to increment production and reduce the losses caused by harmful fauna and flora.

With all these needs and uses, aquifers are being depleted all around the world. Groundwater levels have been decreasing in many of our biggest aquifers, and groundwater has been mined in areas where recharge is not possible (deserts for example), compromising the water supply for future generations. The unsustainable water abstraction from aquifers will need to be controlled, or we will see in the future much more migrations of population due to the shortage of water resources in their own regions. We are facing a time where more powerful countries deplete groundwater resources of other countries under the borders, calling for the need for transboundary agreements that can help countries manage their waters near its borders.

Contamination is also affecting our water resources. With this, a part of the world population is struggling against water-related diseases that lead to the death of thousands of people all the years. In some parts of the world, groundwater is no longer even suitable for agriculture. Salinization processes occur not only near shore, with saline intrusion being a problem in overexploited areas, but also heavy metal contamination from mines and industries is a strong problem in other areas. Moreover water is essential for food security in the world; without water for irrigation, we could not support the actual number of inhabitants on earth.

With all these problems, ecosystems have disappeared from the face of earth and will continue to disappear; the respect for the environment has been lost. And we need the ecosystems to have a better life. When ecosystems suffer due to water problems, it is the first indicator that something is not right with our water resources and humans will be the next to suffer.

The solution for these issues is a tremendous effort of the international community of groundwater experts to maintain the vital resource for future generations. Hydrogeologists are many times involved in decisions that can affect other people, other countries, or groundwater-dependent ecosystems, for example, but are sometimes tided to politic or economic decisions taken by others.

But we are the persons that can help change the world. Hydrogeologists must be wise in their decisions. A book like this about *Global Groundwater: Source, Scarcity, Sustainability, Security, and Solutions* is a great contribution for the understanding of this kind of problems and to indicate solutions that can be used or can be appointed for the future. It is an essential book to help address the issues of groundwater scarcity and contamination and the way we can guarantee its future sustainability. A book that builds the bridge between local case studies and global-scale studies is essential for the understanding of groundwater issues at regional and global scale and possible future solutions and commitments.

António Chambel
President
International Association of Hydrogeologists (IAH)
Évora, Portugal

Preface

Groundwater is the largest, usable, freshwater resource available to humans. It plays a crucial role in our livelihoods by making itself available for drinking and by providing food security through groundwater-fed irrigation. At the same time, groundwater supports significant aspects of ecosystem functioning and the wider environment. Dependence on groundwater across our societies is rapidly increasing worldwide. In addition to human consumption for domestic needs, a large volume of groundwater is required for agricultural and industrial purposes. In 2002 the UN Committee on Economic, Social and Cultural Rights recognized the human right to water as "... indispensable for leading a life in human dignity." And, today this is further reinforced by the UN 2015 sustainable development goals, which requires all countries, by 2030, to achieve universal and equitable access to safe and affordable drinking water, improve water quality by reducing pollution, and deliver integrated water resources management. In meeting these goals, an essential component is groundwater that forms the largest source of liquid freshwater in this planet. There is a significant need to improve the way we utilize and manage our groundwaters in a more sustainable manner to benefit our societies and the ecosystems on which we all depend. The dynamics of groundwater resources are sensitive to recharge, influenced by spatiotemporal precipitation patterns and the intensification of extreme climate events, along with hydrogeological properties of subsurface soils and geology. Earlier, groundwater scarcity was prevalent only in arid and semiarid areas (e.g., Middle East); however, in present times, from Asia to the Americas, numerous countries are suffering from significant groundwater shortage. For that, improved management of our groundwater may improve societal resilience against shocks such as droughts and floods as well as make them more sustainable for society and nature. Even the traditionally water-affluent countries are facing acute shortages of usable water as those are witnessing a combination of rapidly rising population, urbanization, and change in societal water use, agricultural cropping patterns along with lifestyle changes. Hence, the interactions of the society and human strategies with nature and groundwater resources, in a changing world with new sociopolitical alignments, are becoming the new normal for the present and impending future times.

In this context, this book provides a unique opportunity that integrates the current knowledge on groundwater, ranging from availability to pollution, nation-level groundwater management to transboundary aquifer governance, and global-scale reviews to local-scale case studies. Area-specific studies range from the extremely groundwater-stressed regions of North Africa, Middle East, South Asia, and Australia to the relatively groundwater-sufficient regions of Brazil, Canada, China, East Asia, and the United States. In these studied areas, several countries comprise some of the densest populations across the world and the highest global users of groundwater, for example, India, the United States, China, and Pakistan. Many of them are drained by some of the largest and most important transboundary river systems of the world, such as the Amazon, Nile Indus, Ganges, Brahmaputra, Kabul, Irrawaddy, Mississippi, Pearl, Yellow River, reflecting large reserves of subsurface water. However, groundwater availability in these regions is extremely heterogeneous, with aquifers ranging from high-yielding unconsolidated sedimentary formations to low-yielding crystalline bedrocks. Further, the seasonal precipitation–based aquifer recharges are spatially and temporally variable, thereby influencing the formation of climate zones that range from extremely arid to some of the wettest places on earth. Moreover, the available groundwater is often excessively abstracted, thereby characterizing much of the areas as under very high water stressed. In addition, geogenic and anthropogenic pollution of groundwater poses larger uncertainties and constraints even on the available groundwater. Hence, the global- to local-scale challenges highlight the need of creating solid evidence and knowledge-bases for integrated scientific and technological advances, as well as building policy and management capacities in order to adapt to and evolve for the present-day groundwater needs and potential groundwater demand for future generations in a sustainable manner.

In this book, we have attempted to integrate those evidence and knowledge that exist in various studies on groundwater across the globe, extending from the extensively and intricately studied aquifers of the United States and South Asia to the less-studied regions of Africa, Middle East, and Afghanistan, along with an emphasis on the need of understanding transboundary aquifers. Authored by leading experts across the world, the studies compiled in this book range from high-resolution, field-scale studies to global-scale gross estimates, thereby attempting to bridge the gap of scale-of-observation.

In today's world, strategies and solutions for groundwater management and policy need to be scale and condition dependent in order to achieve present-day sustainability and security for future times. Therefore we have arranged the chapters following logical, thematic areas, such that the readers can easily find out their subject of interest. Thematic topics included are groundwater studies on quantity, exploration, quality and pollution, economics, management and policies, groundwater and society, and sustainable sources and efficient solutions, presented in subsections based on (1) sources, (2) scarcity, (3) sustainability and security, and (4) solutions.

We hope, the book provides the initial step for the integration of ideas and knowledge of this invaluable resource for our planet, with knowledge from immensely populous counties, diplomatically important areas, and some of the fastest growing global economies, so that we would be able to effectively manage and preserve the water security sustainably for our future generations and for the humankind.

May 2020

Abhijit Mukherjee[1], Bridget R. Scanlon[2], Alice Aureli[3], Simon Langan[4], Huaming Guo[5] and Andrew A. McKenzie[6]

[1]*Indian Institute of Technology (IIT) Kharagpur, Kharagpur, India*, [2]*University of Texas at Austin, Austin, TX, United States*, [3]*UNESCO, Paris, France*, [4]*IWMI, Colombo, Sri Lanka*, [5]*China University of Geosciences, Beijing, P.R. China*, [6]*British Geological Survey, Oxfordshire, United Kingdom*

Acknowledgment

This work would not have been possible without constant inspiration from our students, lessons from our teachers, enthusiasm from our colleagues and collaborators, and support of our families.

We are indebted to Ms. Andrea Dulberger (Elsevier) for her diligent editorial management and Ms. Louisa Munro (Elsevier) for her support.

The book is dedicated to the people living in groundwater-stressed counties across the world.

Disclaimer

The authors of individual chapters are solely responsible for ideas, views, data, figures, and geographical boundaries presented in the respective chapters of this book, and these have not been endorsed, in any form, by the publisher, editors, or the authors of forewords, introduction and preface. The boundaries between the different countries depicted in the figures of various chapters are presented for illustration purpose only and no other inference should be drawn from them.

Introduction: Why Study Global Groundwater?

(Excerpted from Gleeson et al., 2020 with permission)

Groundwater is the largest store of liquid freshwater on the Earth (Alley et al., 2002; Gleeson et al., 2016). Despite this large volume of groundwater, the fluxes between groundwater and other compartments of the hydrologic cycle are relatively small compared to those on and above the Earth's surface. While these fluxes are often small, they can be of critical local importance in nutrient (Hayashi and Rosenberry, 2002) and elemental cycling (Ferguson and McIntosh, 2019; Stahl, 2019), regulation of temperature (Power et al., 1999), and maintaining streamflow during low flow periods (Winter, 2007). The various inputs, outputs, boundaries, and stores described previously interact with each other in a dynamic and nonlinear way, creating globally diverse, complex, and dynamic groundwater systems, with multiple feedbacks with other parts of the Earth system.

Just a decade ago, groundwater was generally ignored in global hydrology models before global groundwater recharge was estimated for the first time (Döll and Fiedler, 2008). Seminal groundwater sustainability reviews highlighted that groundwater was critical to agriculture and people and locally overused or contaminated but that groundwater recharge, use, and quality was largely unknown for vast parts of the world, or at least not synthesized into a cohesive and consistent global perspective (Foster and Chilton, 2003; Giordano, 2009; Moench, 2004; Zektser and Everett, 2000). Continental- to global-scale studies of groundwater systems, resources, and sustainability have proliferated in the last decade.

Considering groundwater at continental- to global-scales (Gleeson et al., 2019) allows us to (1) understand and quantify the two-way interactions between groundwater and the rest of the hydrologic cycle, as well as the broader Earth system; (2) inform water governance and management for large, and often transboundary, groundwater systems (Wada and Heinrich, 2013) in an increasingly globalized world with virtual water trade (Dalin et al., 2017); (3) consistently and systematically analyze problems and solutions globally regardless of local context, which could enable prioritization of regions or knowledge transfer between regions; and (4) create visualizations and interactive opportunities that are consistent across the globe to improve understanding and appreciation of groundwater resources.

It is important to simultaneously view groundwater globally and regionally because groundwater does not operate solely on global scales or regional scales, but at both scales simultaneously. Groundwater depletion is considered a global problem owing to its widespread distribution and its potential consequences for water and food security and for sea-level rise (Aeschbach-Hertig and Gleeson, 2012; Konikow and Kendy, 2005; Famiglietti, 2014). Even more broadly, groundwater is a global issue, connected to other global issues such as environmental degradation, climate change, and food security. Yet unlike integrated, well-mixed physical systems (e.g., climate), groundwater storage, flow, and pumping are focused locally in aquifers that occur in specific locations. Groundwater flow and pumping in one location is likely to have a negligible effect on an aquifer across the world since the system is poorly mixed. Therefore herein "global-scale" implies aggregated, characteristic or representative processes rather than suggesting that groundwater acts as an integrated, well-mixed physical system. The impact of groundwater pumping is most acute and obvious at local scales, and groundwater resources also have strong local characteristics related to specific hydrology, politics, laws, culture, etc. (Foster et al., 2013).

Unfortunately, groundwater resources are threatened globally in a number of different regions where both quantity and quality issues are common (Aeschbach-Hertig and Gleeson, 2012; Bierkens and Wada, 2019; Foster and Chilton, 2003). The direct impacts of groundwater use can be land subsidence, enhancement of hydrological drought, sea-level rise, groundwater salinization, and impact on groundwater-dependent ecosystems [see references in Bierkens and Wada (2019)]. These direct impacts can have broader sustainability impacts on water, food and energy security, infrastructure, social well-being, and local economies. In addition, there can be broader impacts on Earth systems such as oceans (e.g., coastal eutrophication), climate (e.g., groundwater–climate interactions), or lithosphere (e.g., critical zone or petroleum resources); these broader impacts, generally, have not been as well recognized or described as the direct impacts.

Three approaches have been pioneered in the last 10 years to characterize the continental- to global-scale interactions between groundwater and the other components of the Earth systems and quantify various aspects of groundwater sustainability:

1. Several global hydrological models, land surface models, and Earth system models now incorporate groundwater processes to varying degrees of complexity as has been recently reviewed by Gleeson et al. (2019) who catalog model characteristics. These numerical models are built for diverse purposes but share the ability to carry out water balance calculations at the land surface in order to estimate groundwater recharge. Several regional–global models also explicitly include the two-dimensional (2D), transient, redistribution of groundwater flow.
2. Currently such global numerical models are computationally "expensive" and do not include all physical processes that are relevant to some Earth system or sustainability problems, such as density-dependent groundwater flow. The computational expense makes it challenging to rigorously quantify the uncertainties due to the coarse-scale global datasets on which they are based, and the choice of model structure simulated. Hence, another approach that has recently been adopted is the use of mathematical "analytical" models (Cuthbert et al., 2019); although they are inherently more simple in their assumptions and processes that can be included, these allow for a much more extensive uncertainty analysis to be carried out due to being much more computationally efficient. Another recent approach is synthesizing a large number local numerical models that include all crucial physical processes (Luijendijk et al., 2020). These numerical models were geometrically simple (2D cross sections rather than 3D groundwater flow) but included all the relevant physical processes and allow for sensitivity analysis.
3. Remote sensing, in particular measurements of changes in the Earth's gravity field by NASA's GRACE satellite, has provided insights into changes in groundwater storage due to pumping (Rodell et al., 2009; Famiglietti, 2014) and changes in climate (Thomas and Famiglietti, 2019). Previous estimates of large-scale shifts in groundwater storage typically required integration of large numbers of point measurements from individual wells or estimation through numerical modeling (Konikow, 2013). Working between the low-resolution ($\sim 200,000$ km^2) data from GRACE and the higher resolution required for many hydrogeological studies remains a challenge but progress is currently being made on downscaling techniques (Miro and Famiglietti, 2018).

Considering groundwater at continental- to global-scales allows us to inform water management and governance for large, and often transboundary, groundwater systems in an increasingly globalized world with virtual water trade. A global perspective provides for a consistent and systematic framework that could enable prioritization of regions or knowledge transfer between regions. Finally, a global perspective allows for the creation of visualizations and interactive opportunities to improve understanding and appreciation of groundwater resources relevant to the population at large.

Tom Gleeson[1], Mark Cuthbert[2,3], Grant Ferguson[4] and Debra Perrone[5]

[1]Department of Civil Engineering and School of Earth and Ocean Sciences, University of Victoria, Victoria, BC, Canada, [2]School of Earth and Ocean Sciences & Water Research Institute, Cardiff University, Cardiff, United Kingdom, [3]Connected Waters Initiative Research Centre, University of New South Wales, Sydney, NSW, Australia, [4]Department of Civil, Geological and Environmental Engineering, University of Saskatchewan, Saskatoon, SK, Canada, [5]Environmental Studies Program, University of California at Santa Barbara, Santa Barbara, CA, United States

References

Aeschbach-Hertig, W., Gleeson, T., 2012. Regional strategies for the accelerating global problem of groundwater depletion. Nat. Geosci. 5 (12), 853–861. Available from: https://doi.org/10.1038/ngeo1617.

Alley, W.M., Healy, R.W., LaBaugh, J.W., Reilly, T.E., 2002. Flow and storage in groundwater systems. Science 296 (5575), 1985–1990. Available from: https://doi.org/10.1126/science.1067123.

Bierkens, M.F., Wada, Y., 2019. Non-renewable groundwater use and groundwater depletion: a review. Environ. Res. Lett. 14 (6), 063002.

Cuthbert, M.O., Gleeson, T., Moosdorf, N., Befus, K.M., Schneider, A., Hartmann, J., et al., 2019. Global patterns and dynamics of climate–groundwater interactions. Nature Climate Change 1. Available from: https://doi.org/10.1038/s41558-018-0386-4.

Dalin, C., Wada, Y., Kastner, T., Puma, M.J., 2017. Groundwater depletion embedded in international food trade. Nature 543 (7647), 700–704. Available from: https://doi.org/10.1038/nature21403.

Döll, P., Fiedler, K., 2008. Global-scale modeling of groundwater recharge. Hydrol. Earth Syst. Sci. 12, 863–885. Available from: https://doi.org/10.5194/hess-12-863-2008.

Famiglietti, J.S., 2014. The global groundwater crisis. Nat. Clim. Change 4 (11), 945.

Ferguson, G., McIntosh, J.C., 2019. Comment on "Groundwater pumping is a significant unrecognized contributor to global anthropogenic element cycles.". Groundwater 57 (1), 82. Available from: https://doi.org/10.1111/gwat.12840.

Foster, S.S.D., Chilton, P.J., 2003. Groundwater: the processes and global significance of aquifer degradation. Philos. Trans. R. Soc. B 358, 1957–1972. Retrieved from: <http://www.pubmedcentral.nih.gov/articlerender.fcgi?artid = 1693287>.

Foster, S., Chilton, J., Nijsten, G.-J., Richts, A., 2013. Groundwater—a global focus on the 'local resource'. Curr. Opin. Environ. Sustain. 5 (6), 685–695. Available from: https://doi.org/10.1016/j.cosust.2013.10.010.

Giordano, M., 2009. Global groundwater? Issues and solutions. Annu. Rev. Environ. Resour. 34 (1), 153–178. Available from: https://doi.org/10.1146/annurev.environ.030308.100251.

Gleeson, T., Befus, K.M., Jasechko, S., Luijendijk, E., Cardenas, M.B., 2016. The global volume and distribution of modern groundwater. Nat. Geosci. 9 (2), 161–167. Available from: https://doi.org/10.1038/ngeo2590.

Gleeson, T., Wagener, T., Döll, P., Bierkens, M., Wada, Y., Lo, M.-H., et al., 2019. Groundwater representation in continental to global hydrologic models: a call for open and holistic evaluation, conceptualization and classification. EarthArXiv. https://doi.org/10.31223/osf.io/zxyku.

Gleeson, T., Cuthbert, M., Ferguson, G., Perrone, D., 2020. Global groundwater sustainability, resources and systems in the Anthropocene. Annu. Rev. Earth Planet. Sci. . Available from: https://doi.org/10.1146/annurev-earth-071719-055251.

Hayashi, M., Rosenberry, D.O., 2002. Effects of ground water exchange on the hydrology and ecology of surface water. Ground Water 40 (3), 309–316. Available from: https://doi.org/10.1111/j.1745-6584.2002.tb02659.x.

Konikow, L., 2013. Groundwater Depletion in the United States (1900–2008) (USGS SIR 2013–5079). United States Geological Survey, p. 75.

Konikow, L.F., Kendy, E., 2005. Groundwater depletion: a global problem. Hydrogeol. J. 13 (1), 317–320. Available from: https://doi.org/10.1007/s10040-004-0411-8.

Luijendijk, E., Gleeson, T., Moosdorf, N., 2020. Fresh groundwater discharge insignificant for the world's oceans but important for coastal ecosystems. Nat. Commun. 11 (1), 1–12.

Miro, M., Famiglietti, J., 2018. Downscaling GRACE remote sensing datasets to high-resolution groundwater storage change maps of California's Central Valley. Remote Sens. 10 (1), 143.

Moench, M., 2004. Groundwater: the challenge of monitoring and management, The World's Water, vol. 2005. Island Press, Washington, DC, pp. 79–100.

Power, G., Brown, R.S., Imhof, J.G., 1999. Groundwater and fishhter and fish fish99. G North America. Hydrol. Processes 13 (3), 401ces.

Rodell, M., Velicogna, I., Famiglietti, J.S., 2009. Satellite-based estimates of groundwater depletion in India. Nature 460 (7258), 999–1002. Available from: https://doi.org/10.1038/nature08238.

Stahl, M.O., 2019. Groundwater pumping is a significant unrecognized contributor to global anthropogenic element cycles. Groundwater 57 (3), 455–464.

Thomas, B.F., Famiglietti, J.S., 2019. Identifying climate-induced groundwater depletion in GRACE observations. Sci. Rep. 9 (1), 4124.

Wada, Y., Heinrich, L., 2013. Assessment of transboundary aquifers of the world—vulnerability arising from human water use. Environ. Res. Lett. 8 (2), 024003.

Winter, T.C., 2007. The role of ground water in generating streamflow in headwater areas and in maintaining base blow. J. Am. Water Resour. Assoc. 43 (1), 15–25. Available from: https://doi.org/10.1111/j.1752-1688.2007.00003.x.

Zektser, I.S., Everett, L.G., 2000. Groundwater and the Environment: Applications for the Global Community. CRC Press.

Theme 1

Global groundwater

Chapter 1

Global groundwater: from scarcity to security through sustainability and solutions

Abhijit Mukherjee[1,2], Bridget R. Scanlon[3], Alice Aureli[4], Simon Langan[5], Huaming Guo[6] and Andrew McKenzie[7]

[1]*Department of Geology and Geophysics, Indian Institute of Technology (IIT) Kharagpur, Kharagpur, India,* [2]*Applied Policy Advisory for Hydrosciences (APHA) group, School of Environmental Science and Engineering, Indian Institute of Technology (IIT) Kharagpur, Kharagpur, India,* [3]*Bureau of Economic Geology, Jackson School of Geosciences, University of Texas, Austin, TX, United States,* [4]*Groundwater Systems and Settlements Section, International Hydrological Programme, United Nations Educational, Scientific and Cultural Organization (UNESCO), Paris, France,* [5]*International Water Management Institute (IWMI), Colombo, Sri Lanka,* [6]*School of Water Resources and Environment, China University of Geosciences, Beijing, P.R. China,* [7]*British Geological Survey, Oxfordshire, United Kingdom*

1.1 Introduction

Globally, groundwater represents the largest volume of freshwater readily available to society and ecosystems (Alley and Leake, 2004). Groundwater plays a critical role in providing drinking water and domestic water along with irrigation water for food security. Large-scale use of groundwater, particularly for irrigated agriculture, and also for industry and drinking water, has resulted in water scarcity and water quality issues (lack of availability of potable and safe drinking water for human intake) (Margat and Van der Gun, 2013). More than half of drinking water is sourced from groundwater. About 40% of global cultivated land is equipped with irrigation based on Shiklomanov (1998), Alley et al. (2002), and Siebert et al. (2010). About 70% of groundwater abstraction globally is used for irrigation, and irrigation wells are abstracting ~1000 km^3 of groundwater annually to support water needs for agricultural production and other societal needs (Famiglietti, 2014; Gleeson et al, 2011, 2016).

Groundwater availability is highly heterogeneous and varies substantially among different geological and geomorphological terrains. This skewed distribution of groundwater resources, subjected to substantial stress for sustaining a large groundwater-dependent population of more than 5 billion people, has resulted in increasing water crises for both society and ecosystems supported by groundwater, as well as public health and politics (Margat and Van der Gun, 2013). In addition, studies suggest that recent changes in climatic patterns may intensify the problem in some regions. The unaccounted loss of groundwater along with processes such as outflow to oceans, evapotranspiration, and seepage to deeper levels or pipeline leakages, add up to ~50% of the total volume of groundwater that leaves basins annually. Out of the residual groundwater resources that are accounted for, >60% is stored in porous, alluvial deposits of large fluvial systems, for example, the Amazon, Nile, Indus–Ganges–Brahmaputra, Yellow, Tigris–Euphrates, Mekong, Murray-Darling, which aerially constitutes only ~40% of the continental landmass. Thus it implies that the remaining 60% of the areal landmass shares a much smaller volume of the total annual water budget, resulting in an extremely inequitable distribution of groundwater resources within the landmass (Aeschbach-Hertig and Gleeson, 2012).

Depending on the rates of groundwater level decline and aquifer productivity, groundwater overextractions can sustain society at current rates of overuse only for a few more decades (Margat and Van der Gun, 2013). With advanced technologies, groundwater exploration and abstraction from greater depths are a plausible option. The need for additional energy for groundwater pumping as well as questions related to deteriorating groundwater quality within the context of drinking water standards are emerging as critical issues. Hence, the future availability of freshwater for the development of this planet would pivot between imprudent abstractions and governance options to reserve groundwater for present and impending needs, and achieving sustainable development goals (SDGs), and eventually leading to water security for drinking water, food production, and societal growth while maintaining ecosystems for future generations through access to clean water (Mukherjee et al., 2020).

1.2 Groundwater source and availability

Groundwater, as a natural resource, is sourced in underground geological units, also known as aquifers. The upper 2 km of Earth's crust contains about 23×10^6 km^3 of groundwater, of which only $\sim 0.1-5.0 \times 10^6$ km^3 is modern groundwater that is recharged within the last 50 years (Gleeson et al., 2016). The global-scale distribution of the depth of the groundwater table is highly heterogeneous, depending on the dynamic equilibrium between geology/geomorphology of the aquifers, hydrological conditions leading to recharge, regional to local-scale groundwater flow systems, and human impacts in terms of groundwater abstraction. In a natural setting the groundwater level can elucidate patterns of vegetation distribution and wetland extents. Shallow groundwater is believed to influence $\sim 30\%$ of the global continental area, including contributing $\sim 15\%$ to the surface water bodies, for example, perennial or seasonal rivers, and much of the remainder contributing to the plant root zone, through the capillary fringe (Fan et al., 2013).

Understanding groundwater resources from regional to global scales, along with aquifer parameters, recharge, and discharge volumes are still evolving (Margat and Van der Gun, 2013). While, many of the regions, such as a large part of the United States, had detailed knowledge of groundwater extent from the mid-last century, several other parts of the world have little or no data related to groundwater. Hence, developing a global-scale picture of the resource and its regional-scale sustainability and security is still in its nascent state. Notwithstanding that the first efforts of global-scale assessments were initiated by the United Nations in the 1960s (Giordano, 2009). However, it was not until relatively recently that more realistic and pervasive estimates have become available through estimates from the UN Food and Agricultural Organization (FAO)'s AQUASTAT database (Shiklomanov, 1998), the United Nations Educational, Scientific and Cultural Organization (UNESCO) hosted World-wide Hydrological Mapping and Assessment Program's (WHYMAP) data, and maps of major recharge areas (http://www.whymap.org/) (Döll and Fiedler, 2008). In more recent times the advent of advanced observational and numerical techniques, for example, GRACE mission and global modeling, has resulted in broader estimates of global groundwater resources (Döll and Fiedler, 2008; Rodell et al., 2009, 2018; Wada et al., 2010; Bhanja et al., 2016; Chen and Rodell, 2020).

Renewable groundwater resources are naturally recharged, partly, over time by precipitation, for example, rainwater infiltrating into the ground (Bierkens and Wada, 2019). Groundwater recharge is heavily dependent on spatiotemporal variability in precipitation and other land-based and hydrogeologic factors (Fig. 1.1). Groundwater discharges by natural

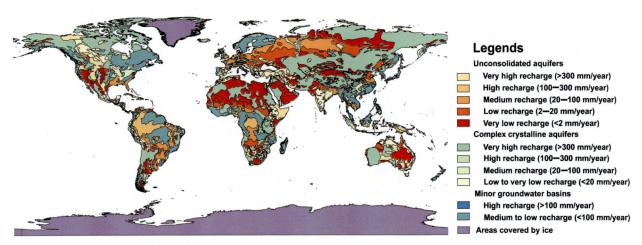

FIGURE 1.1 Global aquifer map, also classified on basis of groundwater recharge rates. *Adapted and modified from BGR/WHYMAP, 2015. Bundesanstalt für Geowissenschaften und Rohstoffe (BGR) and UNESCO World-wide Hydrogeological Mapping and Assessment Programme (WHYMAP) Maps/Data. <https://www.whymap.org/whymap/EN/Maps_Data/maps_data_node_en.html> (accessed 20.05.20.).*

drainage to springs and streams are also abstracted by human (Wada et al., 2010). When the rate of discharge exceeds the recharge rate, groundwater storage decreases, as evidenced by declining groundwater tables (Scanlon et al., 2010). Because of the geological evolution of the continental landmass, there are two predominant types of aquifers that form major groundwater systems that typically exist globally prolific, sedimentary rock-sourced, porous fluvial aquifers, and fractured, crystalline/hard rock aquifers (Fig. 1.1). A third type, carbonate aquifers, which also plays a major role in some parts of the world, can be either porous or fractured systems (Fryar, 2020).

Porous media aquifers, dominated by alluvial aquifers, are typically characterized by hundreds of meters of unconsolidated sediments that are weathered and rapidly deposited in large sedimentary basins, thus forming some of the most prolific groundwater aquifers in the world [major basin aquifers (MBA)]. They can also be present as local, shallow, surficial aquifers [local and shallow aquifers (LSA)] (Fig. 1.2). These aquifers are often regarded as the "breadbasket" for adjoining regions. These thick, unconsolidated alluvial aquifers are conducive to recharge (Bhanja and Mukherjee, 2020; Shamsudduha, 2020). Coastal and eolian-deposited aquifers also sometimes host a few prolific and replenishable aquifers (Barnett et al., 2020). Overabstraction in these regions leads to a relatively slower decline of groundwater levels, albeit eventually necessitating exploitation from greater depths and increased abstraction rates.

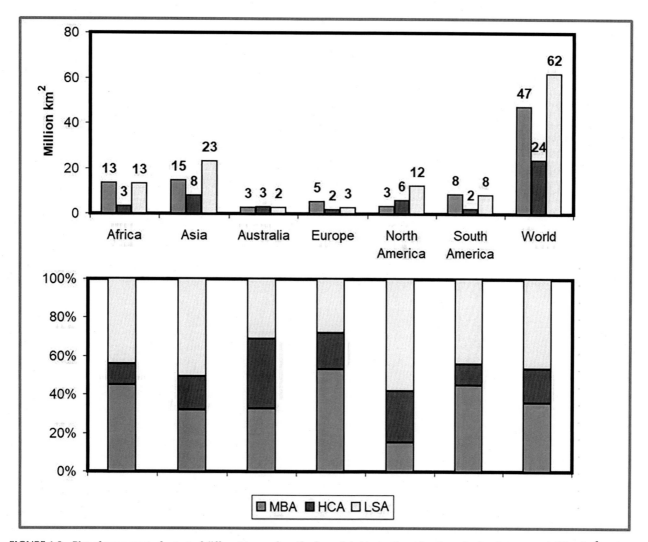

FIGURE 1.2 Plot of assessment of extent of different types of aquifer from global to continental scale, as land surface extent (million km^2) and percentage. Estimates are provided for the World (excluding Antarctica) and the six continents. *HCA*, Hydrogeologically complex aquifers; *LSA*, local and shallow aquifers; *MBA*, major basin aquifer. *Data source: Groundwater Resources of the World Statistics of BGR/WHYMAP, 2015. Bundesanstalt für Geowissenschaften und Rohstoffe (BGR) and UNESCO World-wide Hydrogeological Mapping and Assessment Programme (WHYMAP) Maps/Data. <https://www.whymap.org/whymap/EN/Maps_Data/maps_data_node_en.html> (accessed 20.05.20.).*

However, as a result of the large groundwater storage in many of these aquifers, they are also among the most heavily depleted in many regions (Yang et al., 2020).

Hard rock aquifers are prevalent mostly in cratonic parts of the globe. These aquifers typically have limited groundwater storage in weathered soil and subsoil horizons of few to several meter thickness and are underlain by less-porous, unweathered bedrock aquifers. Thus they provide a complex hydrogeological setting and sometimes are called hydrogeologically complex aquifers (HCA). The bedrock is typically devoid of primary porosity but is characterized by variable quantities of secondary porosity formed by fractures, cracks, and fissures, where groundwater flows through preferential pathways. Groundwater in these hard rock aquifers is characterized by lower productivity of abstracting wells, less predictable spatiotemporal storage and productivity, and generally with poor quality. Thus the groundwater resource availability in hard rock regions is limited and discontinuous, resulting in unreliable prediction and estimation of groundwater availability for human usage. Unsustainable groundwater abstraction from these hard rock aquifers leads to rapidly declining groundwater levels, imposing severe limitations on groundwater extraction rates. Consequently, exponentially increasing groundwater demands in several parts of these hard rock regions result in rapid groundwater storage declines from severe overdraft, with rates of abstraction substantially exceeding recharge rates. As a result, over time there is a substantial increase in wells becoming dry.

The WHYMAP and associated initiatives (BGR/WHYMAP, 2015) have developed a global to continental-scale assessment of the extent of the areal distribution of the major aquifer types (Fig. 1.2). The estimate suggests that on a global scale, almost of half of the land surface ($\sim$62 million km^2 or $\sim$47%) of the continents (excluding the Antarctic) is underlain by the local scale, shallow aquifer or LSA, followed by about one-third of the Earth's surface represented by relatively homogeneous aquifers of the large sedimentary basins or MBA ($\sim$47 million km^2 or $\sim$35%), and about one-fifth of the land subsurface representing geologically complex aquifers or HCA ($\sim$24 million km^2 or $\sim$18%), which are sometimes highly productive in local scale in heterogeneous folded or faulted regions.

The efficiency of managing and utilizing groundwater as a resource is not limited to natural systems and groundwater users but extends from physical to social scientists, to stakeholders at various levels of society, to law and policy makers, and is understandably multifaceted. Management becomes more complex when groundwater-rich aquifers are shared across international borders, and even sometimes across state borders, termed transboundary aquifers (TBA) (Aureli and Eckstein, 2011; Giordano and Shah, 2014; Puri, 2020, Puri et al, 2001). At present, there are 592 identified TBAs, which have been identified by the International Shared Aquifer Resource Management initiative of the UNESCO-IHP. These include 72 TBAs in Africa, 73 in the Americas, 129 in Asia and Oceania, and 318 in Europe. Although there are hundreds of treaties for water sharing among the member states, management and exploitation of the groundwater resources across these TBAs can result in international conflicts (Puri and Arnold, 2002; Wada and Heinrich, 2013; Puri, 2020). Some of the best examples of such conflicts include the Nubian Sandstone (North Africa), Guarani (South America), and the Ganges–Brahmaputra-Meghna basins (South Asia) (Chakraborty et al., 2020).

1.3 Groundwater scarcity

1.3.1 Quantity

Water table in several major global aquifers, specifically those in (semi)arid zones are declining rapidly (Rodell et al., 2018). Groundwater abstraction is taking place at a much higher level than that they (Wada et al., 2010) are naturally replenished by recharge processes, losing groundwater at a nonrenewable level. These include High Plains and Central Valley aquifers in the United States, Northern Sahara and Nubian aquifers in Africa, North China Plain, most of the Middle East Aquifers in the Persian Peninsula, and the aquifers in the northwestern and eastern parts of India and Bangladesh (Chakraborty et al., 2020; Figueroa and Smilovic, 2020; Yang et al., 2020). However, currently, even some traditionally water-affluent regions from Asia to the Americas are facing acute shortage of usable water, because of rapid population growth, urbanization, changes in societal water use, and cropping patterns. Many other regions are also constrained by their limited extent of groundwater reservoir or accessibility, for example, the cold and boreal regions (Bhanja and Wang, 2020; Ma and Wang, 2020), and island aquifers (Mandal et al., 2020). Groundwater conditions in many parts of the world that are subjected to substantial groundwater stresses have become global paradigms for water scarcity, in terms of both groundwater quantity and quality. While many of these regions host some of the richest fluvial aquifers of the world, they are also highly productive agricultural regions with groundwater contributing to their high productivity. Unsustainable "abstraction of the largest groundwater volume in human history" from some of the largest aquifers has led to "groundwater droughts," due to overdraft across major parts of the globe (Mukherjee, 2018; Li and Rodell, 2020).

Human civilization has been flourishing in many groundwater basins (e.g., Nile, Yellow, Indus−Ganges, Tigris−Euphrates river valleys) from Paleolithic times and, at present, is some of the most densely populated parts of the globe. The fluvial plains and deltas of these basins are extremely fertile, and hence conducive for cultivation for food production. The rapidly increasing demands for irrigation sourced by groundwater are directly proportional to the exponential increase in food demand for the proliferating population (Mukherjee et al., 2007a,b). Consequently, ever-increasing population, cultivation of water-intensive crops (e.g., high yield rice paddy cultivation), cropping pattern changes (e.g., food crop replacement by cash crop) are some of the most acute causes of severe groundwater water depletion (Fig. 1.3). In situ and satellite observations have demonstrated emerging trends of widespread groundwater depletion (Fig. 1.4). Further, recent studies have demonstrated that rapid groundwater exploitation and overdraft have also significantly impacted environmental flows by reducing groundwater discharge to adjoining rivers (Mukherjee et al., 2018; de Graaf et al., 2019), resulting in even seasonal drying of the some largest riverine systems (e.g., the Ganges river; Mukherjee et al., 2018). Hence, interactions of society with groundwater resources, in a changing world with new sociopolitical alignments, are negatively impacting groundwater resources and highlight concerns for future times.

Groundwater abstraction in only a few countries (e.g., India, China, United States, Pakistan, and Iran) accounts for ∼65% of total global groundwater abstraction (Fig. 1.5). India is currently the largest consumer of groundwater, with abstraction volume exceeding the sum of groundwater abstraction in the United States (second highest groundwater user) and China (third) (Mukherjee et al., 2015; Mukherjee, 2018), together accounting for >40% of the global population [Food and Agriculture Organization of the United Nations (FAO), 2013]. These countries also cover ∼70% of the global groundwater irrigated land and have rapidly rising population, urbanization, and changes in anthropogenic water use, cropping pattern, and lifestyle leading to unsustainable abstraction of available groundwater. The resultant stress calculated from the ratio of groundwater withdrawal to availability is high in many parts of these countries, and these regions are now identified as one of the largest exporter of nonrenewable groundwater through international food trade (Dalin et al., 2017, 2020). Future reductions in groundwater availability in water-stressed regions of the world could trigger civil uprising along with violent law-and-order situations that may be inter-or transboundary in nature, eventually culminating in regional to international conflicts (Famiglietti,2014). Thus improved understanding and management of declining groundwater resources would not be limited to scientific interests, but could greatly impact the socioeconomic fabric of vast parts of the world.

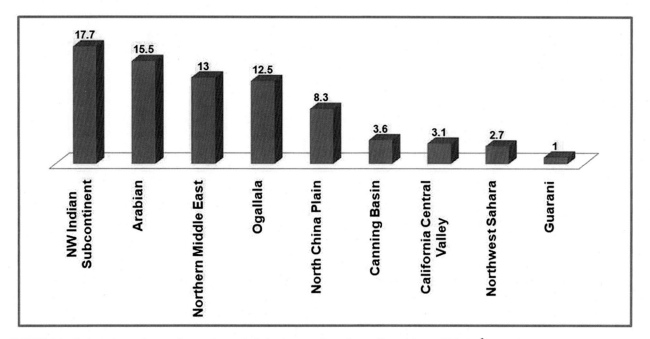

FIGURE 1.3 Estimated annual rates of groundwater depletion in some the major aquifers of the world (in km^3/year). *Data source: Famiglietti, J.S., 2014. The global groundwater crisis. Nat. Clim. Change 4 (11), 945−948.*

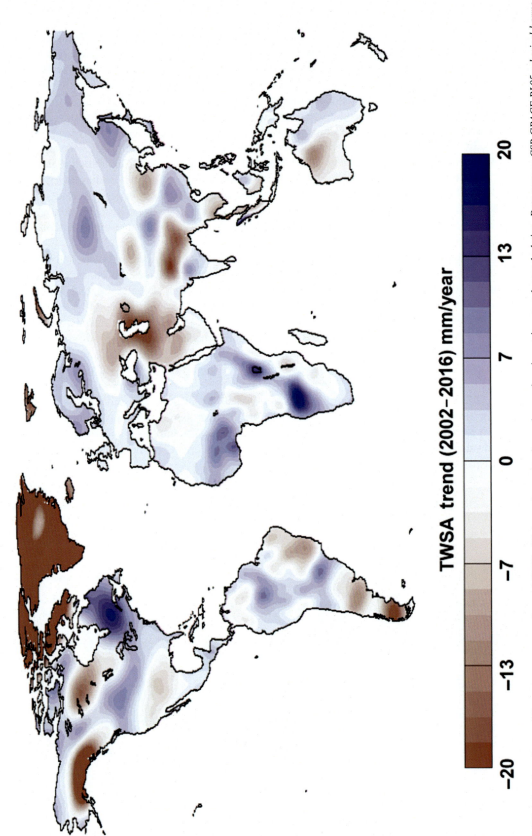

FIGURE 1.4 Map of terrestrial water storage anomaly trend (2002–16) in mm/year, showing regions undergoing major groundwater depletion. *Data source: CSR GRACE RL05 spherical harmonics solution with 300 km Gaussian filter.* <http://geoid.colorado.edu/grace/dataportal.html> *and* <http://geoid.colorado.edu/grace/documentation.html>.

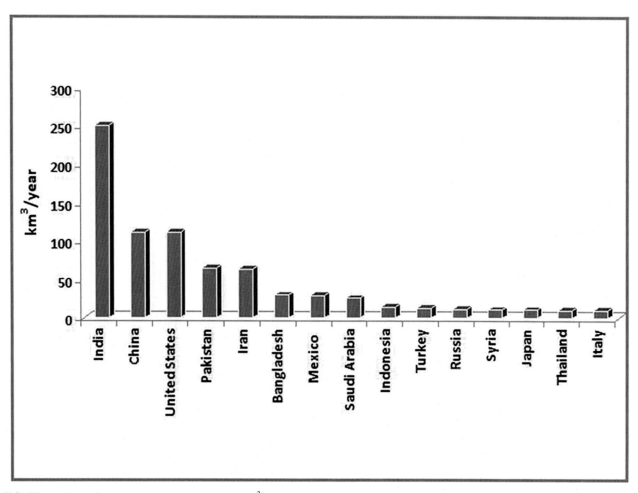

FIGURE 1.5 Plot of annual groundwater abstraction (km^3/year) of the major groundwater using countries of the world. c.2010. Data from www.ngwa.org and Margat, J., Van der Gun, J., 2013. Groundwater Around the World: A Geographic Synopsis. CRC Press.

1.3.2 Groundwater quality

Globally, the pervasive exploitation of surface water bodies, such as rivers and lakes for disposal of industrial waste and sewage, has rendered them nonpotable. Use of surface water has resulted in recurrent epidemics from sewage and water-borne diseases up to the early 20th century. Hence, policy implementing authorities in numerous countries in many urban and rural areas are gradually switching over to groundwater sources to meet drinking and domestic water needs for their inhabitants. Presently, >80% of domestic water supplies for many populous countries are met by groundwater (e.g., South Asian countries). However, identification of the presence of natural contaminants, along with delineation of human-sourced contaminants, has resulted in a growing concern about the availability of safe groundwater in many regions globally (Benaabidate et al., 2020), with water quality emerging as another facet of groundwater scarcity. While there is a general consensus that the extent and processes related to natural pollutions are more pervasive and nonpoint source, the extent and effect of other emerging and unidentified groundwater contaminants, mostly sourced from agriculture, human sewage, industrial, and medical waste (e.g., nitrate, pesticides, radiogens, antibiotics) are yet to be accounted for. Intensive agriculture is associated with high inputs of chemical and synthetic pesticides that leach into groundwater systems. More acute, but barely reported and discussed, groundwater pollution can be linked to improper sanitation, which has resulted in widespread concerns about public health.

The combination of declining groundwater levels and limited availability of safe, potable quality groundwater amplifies water stress relative to that from water quantity issues alone. For example, recent estimates from the Indus−Ganges−Brahmaputra river basin, regarded as one of the largest and most prolific groundwater systems globally, indicates that ∼60% of the 300 km^3 of its groundwater as unusable and unsafe (MacDonald et al., 2016).

High concentrations of naturally occurring, geogenic contaminants, such as arsenic (As) and fluoride (F) in groundwater are common across all continents (Fig. 1.6) (Coomar and Mukherjee, 2020). Together, they have exposed more than ~500 million people globally to a potential severe human health hazards. Groundwater arsenic contamination has exposed >100 million inhabitants just in eastern parts of South Asia, United States, and China (Haque, 2020; Jia, 2020; Karim and Sadat, 2020; Masood et al., 2020; Schreiber, 2020), resulting in "the largest mass poisoning in human history" (Smith et al., 2000). Prolonged exposure to water with elevated arsenic levels can result in serious health conditions, including skin lesions, hyperkeratosis, melanosis, and cancer in different organs, which in some cases can be fatal (Mukherjee et al., 2009; Chakraborty et al., 2015). These contaminants can be sourced from aquifer geologic units, where As and F exist as trace elements in rocks. While previous studies suggest that As is linked to the sediments in the host aquifers that are derived from the adjoining young mountain chains through global plate tectonics (e.g., the Himalayas in South Asian countries such as India, Bangladesh, Pakistan, Myanmar, and the South American Andes in Argentina, Chile, Bolivia; and the North American Rockies) (Mukherjee et al., 2014, 2019b; Raychaowdhury et al., 2014; Coomar et al., 2019) and deposited in alluvial plains and deltas of fluvial system (Huang et al., 2020), F is believed to be linked with aquifers that are mostly hosted in crystalline or weathered bedrock systems. They are primarily leached from the host, mostly granitites type rocks due to surface weathering. Highly saline groundwater is also prevalent in inland aquifers in several countries (e.g., Argentina, Middle East, Afghanistan, Pakistan, India). Such inland salinization may be linked with paleo-sea water captured as connate water (Mukherjee and Fryar, 2008), mineral dissolution through water−rock interactions, and agricultural pollution/sewage infiltration.

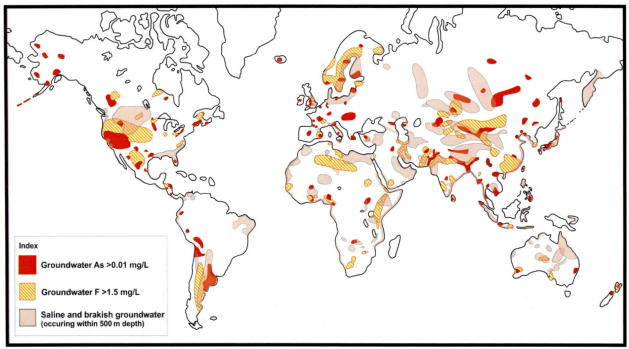

FIGURE 1.6 Map of global distribution of geogenic groundwater pollution by salinitation, arsenic, and fluoride. *Data from Smedley, P.L., Kinniburgh, D.G., 2002. A review of the source, behavior and distribution of arsenic in natural waters. Appl. Geochem. 17 (5), 517−568 (Smedley and Kinniburgh, 2002); Edmunds, M., Smedley, P., 2013. Fluoride in natural waters. In: Selinus, O., Alloway, B., Centeno, J.A., Finkelman, R.B., Fuge, R., Lindh, U., Smedley, P. L. (Eds.), Essentials of Medical Geology. Springer, pp. 311−336 (Edmunds and Smedley, 2013); Mukherjee, A., Bhattacharya, P., Savage, K., Foster, A., Bundschuh, J., 2008. Distribution of geogenic arsenic in hydrologic systems: controls and challenges. J. Contam. Hydrol. 99, 1−7; Mukherjee, A., Fryar, A.E., O'Shea, B.M., 2009. Major occurrences of elevated arsenic in groundwater and other natural waters. In: Henke, K.R. (Ed.), Arsenic—Environmental Chemistry, Health Threats and Waste Treatment. John Wiley & Sons, Chichester, pp. 303−350; Mukherjee, A., Verma, S., Gupta, S., Henke, K.R., Bhattacharya, P., 2014. Influence of tectonics, sedimentation and aqueous flow cycles on the origin of global groundwater arsenic: paradigms from three continents. J. Hydrol. 518, 284−299; Mukherjee, A., Gupta, S., Coomar, P., Fryar, A.E., Guillot, S., Verma, S., et al., 2019b. Plate tectonics influence on geogenic arsenic cycling: from primary sources to global groundwater enrichment. Sci. Total Environ. 683, 793−807 (Mukherjee et al., 2019b); Ravenscroft, P., Brammer, H., Richards, K.S., 2009. Arsenic Pollution: A Global Synthesis. Wiley-Blackwell Publication, 588 p. (Ravenscroft et al., 2009); van Weert, F., van der Gun, J., Reckman, J., 2009. Global Overview of Saline Groundwater Occurrence and Genesis. IGRAC Report Nr. GP 2009-1. International Groundwater Resources Assessment Center, Utrecht (van Weert et al., 2009); Coomar, P., Mukherjee, A., 2020. Global geogenic groundwater pollution. In: Mukherjee, A., Scanlon, B.R., Aureli, A., Langan, S., Guo, H., McKenzie, A. (Eds.), Global Groundwater: Source, Scarcity, Sustainability, Security, Solutions. Elsevier. ISBN:9780128181720.*

Rapid industrial and agricultural expansion in the last century, across major parts of the globe, has resulted in the exposure to a diverse range of chemical contaminants that are increasingly polluting surface water and groundwater [Institute for Transformative Technologies (ITT), 2018]. Agricultural runoff is characterized as diffuse, nonpoint source pollution, originating from large areas. Agricultural practices across the world, over the years, in quest of higher productivity, have used large amounts of agrochemicals to improve crop production. Common agriculturally sourced contaminants leaching into underlying aquifers include fertilizers (e.g., nitrate, phosphate-based or synthetic) and pesticides. Many agricultural pesticides, for example, the persistent organic pollutants (PoPs) and their by-products, persist in groundwater for a prolonged period. They tend to bioaccumulate as well as become biomagnified in organisms drinking the polluted groundwater.

In developed countries, such as United States and Europe, industrial pollution has been known for decades (Marfil-Vega et al., 2020). However, in the developing world and emerging economies, for example, China and India, industrial pollution is increasing with rising industrial production in numerous industrial facilities (Qin et al., 2020). Industrial and medical waste effluent is typically concentrated through "point source" pollution from individual facilities. This effluent contains a wide range of chemical and biological contaminants (e.g., PoPs, polycyclic aromatic hydrocarbons, antibiotics) that are extremely toxic, ignitable, corrosive, and reactive and can cause severe human health effects (Mukherjee et al., 2019a, Duttagupta et al., 2020a,b; Marfil-Vega et al., 2020; Qin et al., 2020). Further, two particularly concerning types of toxins discharged through industrial effluent to groundwater include heavy metals and endocrine disruptors (Mondal et al., 2012). Heavy metals, such as lead, mercury, cadmium, and chromium, are very common contaminants in industrial effluent. These contaminants cannot be degraded or destroyed and also bioaccumulate in the body over time.

Less commonly acknowledged groundwater quality deterioration is related to inadequate channelization of human excreta to nearby surface water and groundwater systems due to improper sanitation infrastructure. Sanitation issues are most common in poverty-stricken populations, leading to severe groundwater pollution and public health hazards. Thus a large portion of global human diseases comprises groundwater-based drinking water sourced from improper sanitation-borne diseases, such as diarrhea, cholera, as more than 1 billion of the global poor still have to resort to open defecation due to lack of suitable basic sanitation structures (Mukherjee et al, 2019a). The major sanitation-sourced polluted groundwater affected areas are present across major parts of Africa and South Asia.

Groundwater scarcity thus reflects depletion of the groundwater resource integrated with suitable water quality for human usage. There is a need to integrate such evidence of groundwater scarcity and knowledge that exists from groundwater scarcity across the globe. Strategies and solutions for groundwater management and policy need to be scaled to achieve present-day groundwater sustainability and security in the future.

1.4 Groundwater sustainability and security

Groundwater sustainability and security are complex issues in a world with increasing connectivity promoted by international trade, cultural exchanges, and information sharing. To understand the routes to develop sustainability pathways and aspire to security, there is a need to integrate social perspectives into groundwater availability, with management and governance (Van Der Gun, 2020). To do these, it is also imperative to integrate physical and chemical perspectives of groundwater science with socioeconomic and engineering solutions. It is thus required to delineate how human perspective at regional scales may be integrated into a global framework, to develop a global sustainability framework to influence the UN SDGs by the year 2030 (UN, 2017). Groundwater is an integral component of SDG 6, which proposes to ensure access to clean water and sanitation for all. However, groundwater also contributes to several other goals, for example, SDG 1 (poverty eradication), SDG 2 (food security), SDG 5 (gender equality), SDG 11 (sustainability of cities and human settlement), SDG 13 (combating climate change), and SDG 15 (protecting terrestrial ecosystems) (Gleeson et al., 2020).

Regional specific solutions are required to attain sustainability and security to accommodate diverse needs related to natural conditions (e.g., geology, geomorphology, and hydrology), human intervention, as well as political, cultural, and socioeconomic realities (Van Der Gun, 2020). To achieve these, it is important to understand local societies and related groundwater management and governance (Lafaye de Michaeux and Mukherjee, 2020). While groundwater governance relates to processes of decision-making through multiple institutions and legal actors, management is more related to implementation of methods and rules that are required to attain sustainability goals. Groundwater governance can be summarized in four primary pathways (Gleeson et al., 2020): (1) initiate institutions that integrate stakeholders; (2) develop policies to attain local, regional, and global resource goals; (3) develop legal systems with capacity to institute and implement laws; and (4) integrate local knowledge with scientific understanding of groundwater systems with respect to regional, social and cultural contexts (FAO, 2016; Kreamer et al., 2020). Thus the concept of sustainability and security are intricately related to human survival through food security and energy consumption, urbanization, and intra-and international economic exchanges, which are highlighted in the following subsections.

1.4.1 Groundwater–food–energy nexus

Food production across major parts of the world largely depends on groundwater-based irrigation (Fig. 1.7). The cumulative stress on global groundwater reserves due to exponentially increasing population of ~7 billion, coupled with seasonal variation in rainfall, has consequently resulted in a critically vulnerable state (Fig. 1.8) of water and food security in the most populous parts of the world (Scanlon et al., 2010, 2017). There has been a steady increase in total irrigated area with groundwater-fed irrigation, from about 30% in the 1960s to more than 60% currently, which is estimated to rise further in coming years as surface water reserves decline due to rapid depletion of many glaciers [Scanlon et al., 2010; Food and Agriculture Organization of the United Nations (FAO), 2013; Mukherjee et al., 2018]. Effects of climate change can result in overall reduction in net groundwater recharge (Taylor et al., 2013). Groundwater abstraction technology through tube-well pumping is largely dependent on power source type and availability and, hence, is crucial for sustainable water abstraction (Ahmed, 2020). The basic requirement for groundwater pumps, to abstract water from a few meters to hundreds of meters deep, is a source of power, typically electrical current in submersible electric irrigation pumps. Shallow pumps are common for limited agricultural and domestic purposes (KPMG, 2014). Thus restricting the power source for these pumps can result in limiting rampant groundwater exploitation. However, the advent of solar-powered pumps has sufficient potential to provide a long-term renewable energy source but with lesser control and regulation. Thus, in spite of their efficiency, such uninterrupted power supply without the constraints of operational energy costs could result in a catastrophic condition because of uninhibited, enhanced groundwater abstraction [Institute for Transformative Technologies (ITT), 2018].

1.4.2 Urbanization

Exponential growth of urban populations across large parts of the developing world, currently and projected within the next few decades could exacerbate unsustainable groundwater abstraction in many areas (Foster et al., 1994;

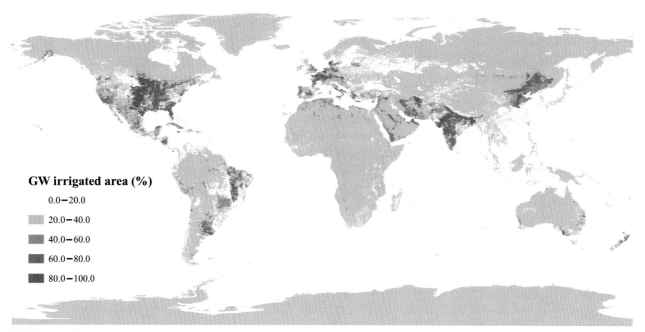

FIGURE 1.7 Global map of area equipped with groundwater irrigation (percentage of total area equipped for irrigation). *Data source: www.fao.org and Siebert, S., Henrich, V., Frenken, K., Burke, J., 2013a. Global Map of Irrigation Areas Version 5. Rheinische Friedrich-Wilhelms-University, Bonn; Food and Agriculture Organization of the United Nations, Rome; Siebert, S., Henrich, V., Frenken, K., Burke, J., 2013b. Update of the Digital Global Map of Irrigation Areas to Version 5. Rheinische Friedrich-Wilhelms-Universität, Bonn; Food and Agriculture Organization of the United Nations, Rome (Siebert et al., 2013a).*

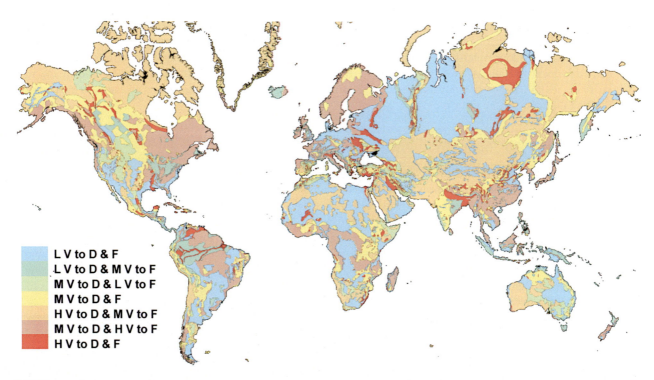

FIGURE 1.8 Groundwater vulnerability map to drought and flood. *D*, Drought; *F*, flood; *H*, high; *L*, low; *M*, medium; *V*, vulnerabilities. *Adapted and modified from BGR/WHYMAP, 2015. Bundesanstalt für Geowissenschaften und Rohstoffe (BGR) and UNESCO World-wide Hydrogeological Mapping and Assessment Programme (WHYMAP) Maps/Data. <https://www.whymap.org/whymap/EN/Maps_Data/maps_data_node_en.html> (accessed 20.05.20.) through https://resourcewatch.org/data/explore/wat022-Groundwater-Vulnerability-to-Floods-and-Droughts_replacement.*

Sharp et al., 2003). Enhanced groundwater demand for urban amenities is becoming an emerging concern. Estimated per capita urban water demand (~150 L, in most developing nations) is almost three times that of the per capita rural water requirement. Most of this excess water has to be sourced from groundwater. Further, the upgraded socioeconomic conditions within the urban population lead to improved living standards with rapid transformation of food habits (Moench, 2004). Thus more water-intensive food (including meat and dairy products) is replacing traditional, local hydroclimatically favored agriculture-based food products. Consequently, the need for enhanced groundwater supply is increasing (Bhanja and Mukherjee, 2019). Moreover, collaterally, newer industries and manufacturing hubs are being developed in these neo-developed periurban hinterlands, which are largely thriving on mining groundwater resources and are primarily responsible for point-sourced industrial pollution, resulting in groundwater scarcity from multiple facets. Thus rapidly rising urban groundwater demand from developing cities eventually results in groundwater overdraft, level decline, and ultimately scarcity. Multiple examples of such urban groundwater scarcity are available from many parts of Australia, Brazil, India, South Africa, etc. (Maheswari, 2020; Oliveira et al., 2020; Sahoo et al., 2020) and could transform to become one of the most acute urban crises in near future. A very relevant example from Cape Town has received many attention (La Vanchy et al., 2020).

1.4.3 Groundwater trade and hydro-economics

Groundwater trade happens through embedded "virtual water" that is used to produce crops or industrial products that are transported elsewhere. Import and export of virtual water between countries act as a potential switch between enhancing future global food security and jeopardizing groundwater sustainability, mostly driven by strong socioeconomic factors. At present, India, United States, and Pakistan are regarded as the most prominent water exporting countries globally (Dalin et al., 2017, 2020), suggesting food or industrial products from these countries are being exported internationally, along with the embedded water footprint, which is typically sourced in groundwater [Institute for Transformative Technologies (ITT), 2018]. Even within nations, virtual water can flow from largely unsustainable groundwater regions with high water-scarcity to water-rich regions with abundant groundwater resources, when water-intensive agricultural commodities are traded (Hoekstra, 2017). Currently, water usage mostly encourages increased groundwater abstraction through motorized pumping, which is promoted through imprudent incentives to the agricultural sector to overexploit groundwater. Developing suitable groundwater pricing structures and subsidies can act as efficient incentives to adopt water-saving behavior, processes, and technologies. Such institutional reforms have to be structured from local to national governing agencies. Suitable valuation of water resources may be one of the most effective strategies to achieve groundwater security. Realistic pricing should stimulate efforts to increase groundwater efficiency and improved water security.

1.5 Solutions

As groundwater demand and related stress increase with agriculture, industry, and domestic needs (Fig. 1.9), sustainability through governance and management is not sufficient to attain security. Heterogeneous groundwater distribution and population demand further complicate the already constrained local groundwater reserves. Even in some of the most water-stressed regions, some areas are groundwater abundant but lack accessibility or quality (e.g., eastern South Asia) (Bhanja and Mukherjee, 2020), while others are being depleted due to limited reserves (western South Asia, Middle East, western and southern United States), and because of lack of proper methods and plans of utilization, groundwater scarcity is amplified (Al-Taani et al., 2020; Kebede and Taye, 2020). In more recent times, emergence of advanced observational for example, GRACE mission (Rodell et al., 2009, 2018; Richey et al., 2015; Chen and Rodell, 2020) and numerical techniques, for example, application of AI (Sun et al., 2019; Kumar et al., 2020; Malakar et al., 2020) has resulted in better and broader estimates and predictions of the global groundwater extent, and thus provide identities of the problem target areas. In all these cases, levers and pathways of solution interventions are required to secure and sometimes, rejuvenate groundwater reservoirs. Deployment of suitable technologies can achieve the desired groundwater sustainability at a local scale, which eventually forms part of the network of global sustainability (Coulon et al., 2020). Proper utilization of the scientific knowledge-base with suitable intervention-based engineering solutions can be effectively applied to augment and replenish groundwater recharge through large-scale rejuvenation exercises or

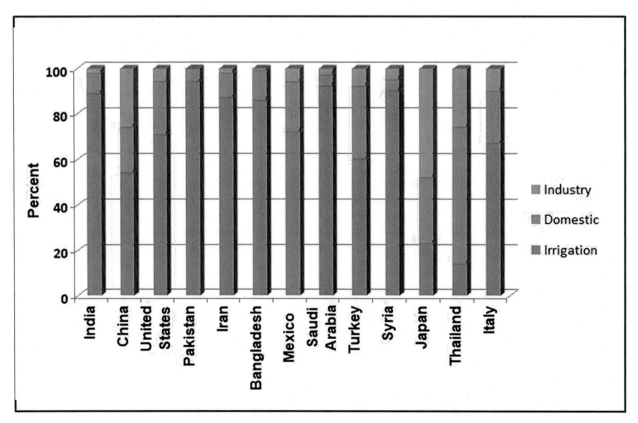

FIGURE 1.9 Plot of percentage of annual groundwater usage (irrigation/industry/domestic) for few of major groundwater using countries of the world. c. 2010. Data from www.ngwa.org and Margat, J., Van der Gun, J., 2013. Groundwater Around the World: A Geographic Synopsis. CRC Press.

local-scale aquifer recharge through identification of aquifer architecture (Bhanja et al., 2017; Saha et al., 2020), improvement of efficiency of groundwater use in agriculture by crop planning and using modern irrigation techniques (Shah et al., 2020), applying innovative solutions to reduce groundwater exploitation required for domestic and industrial demands, producing freshwater by desalinization of brackish and saline water, and applying contaminant specific treatment to reduce pollution load for supply of potable groundwater. Some of these primary levers are outlined in the following subsections [Institute for Transformative Technologies (ITT), 2018; Sathre, 2020].

1.5.1 Enhancing irrigation and urban groundwater efficiency

In several parts of the world that generate most of the global food required, farmers are still using traditional irrigation methods, which have barely evolved within the last few centuries to millennia (e.g., South and East Asia, Africa, parts of South America). Ideally, in order to preserve groundwater security in these areas, enhanced irrigation efficiency is needed (Ahmed, 2020; Sathre, 2020). Currently, there is an array of existing methods and emerging technological solutions that increase crop water use efficiency in irrigation. Traditionally, many of these aforesaid regions use flood or surface irrigation methods, but, interestingly, many of the farmed crops can be irrigated with sprinkler and drip irrigation, which only needs a fraction of the groundwater than that used for flood irrigation. Further, modern techniques of slope leveling on moisture meters, sometimes linked with weather satellites can hugely benefit water conservation. Further, less water-hungry crops can also alleviate the growing irrigation water need in the future (Kreamer et al., 2020).

Similarly, in urban regimes, reduction of groundwater demand can be achieved at the household level with appropriate water management. Modern and improved household devices can significantly reduce domestic water use. In several countries, including major parts of the developed world (e.g., Australia, southern United States), it is now mandated that water-efficient appliances be used, for example, low flow toilets and faucets, and garden sprinklers. Technological interventions in such urban settings can balance out some of the negative effects of ongoing rapid urbanization globally.

1.5.2 Groundwater rejuvenation

Groundwater replenishment at local to regional scales can be achieved by scientific application of a suite of technologies that can increase the groundwater recharge and storage according to local hydrogeological conditions. These technologies are together referred to as managed aquifer recharge (MAR). MAR can be achieved through enhanced infiltration of surface runoff from rainwater or storm flow, and also by channelizing excess surface water from local water bodies. There are large numbers of engineering interventions to develop a successful MAR, including rainwater harvesting structures, check dams, percolation tanks, recharge canals, and dams. Recent observations suggest that regional application of MAR may be even effective in reversing groundwater declines in extremely water-stressed areas (Kurtzman and Guttman, 2020; Chowdhury et al., 2020), as well as for flushing contaminated groundwater or intruding seawater in coastal areas.

1.5.3 Desalination

A significant volume of global groundwater, as well as ocean water, are nonpotable due to higher concentrations of total dissolved solids (TDSs). In terrestrial settings, such brackish or saline groundwater exists in large volumes in deep basins and in confined aquifer systems (Gude and Maganti, 2020). These high TDS groundwater have so far remained mostly underutilized and actually, sometimes can result in pollution of adjoining freshwater sources. Emergence of technologies is showing promise to sustainably desalinate this brackish groundwater (Ahdab and Leinhard, 2020). However, the impediments for wide-scale application of desalination technologies are considerations around environmental as well as socioeconomic sustainability. For example, the energy requirements to extract and utilize such waters are largely prohibitive from a cost angle. However, innovative technologies, mostly using unconventional power sources can possibly efficiently breach the energy-cost barrier, and increase the possibility of providing sufficient groundwater globally.

1.6 Conclusion

As various nations and regions of the world grow more and more groundwater dependent, securing sustainable groundwater sources for present and future use is essential to the sustainability and growth of these countries. However, groundwater resources across the world are undergoing acute water scarcity, both in terms of available water volume and unpolluted, safe water quality. The exponentially increasing population, along with heterogeneous distribution of groundwater resources, seasonal fluctuations in precipitation, unpredictable climate-dependent recharge, transboundary water sources, archaic water-intensive irrigational techniques, groundwater pollution by nonpoint geogenic contaminants and/or industrial effluents and sewage water, unplanned changes in land use, random urbanization and the recent effects of climate change have all cumulatively contributed to the growing groundwater stress. Thus proper, scientifically prudent governance plan and solution interventions are required to restore, sustain, and replenish the abstracted groundwater. Emerging methods and technologies together with an improved understanding of groundwater dynamics do provide optimistic opportunities to preserve the natural resource, but it is largely dependent on socio-politico-economic judgment and decisions from local to national level.

References

Aeschbach-Hertig, W., Gleeson, T., 2012. Regional strategies for the accelerating global problem of groundwater depletion. Nat. Geosci. 5 (12), 853–861.

Ahdab, Y.D., Leinhard, J.H., 2020. Desalination of brackish groundwater to improve water quality and water supply. In: Mukherjee, A., Scanlon, B. R., Aureli, A., Langan, S., Guo, H., McKenzie, A. (Eds.), Global Groundwater: Source, Scarcity, Sustainability, Security, Solutions. Elsevier, ISBN: 9780128181720.

Ahmed, K.M., 2020. Challenges of sustainable groundwater management in Bangladesh. In: Mukherjee, A., Scanlon, B.R., Aureli, A., Langan, S., Guo, H., McKenzie, A. (Eds.), Global Groundwater: Source, Scarcity, Sustainability, Security, Solutions. Elsevier, ISBN: 9780128181720.

Alley, W.M., Leake, S.A., 2004. The journey from safe yield to sustainability. Ground Water 42 (1), 12–16.

Alley, W.M., Healy, R.W., LaBaugh, J.W., Reilly, T.E., 2002. Flow and storage in groundwater systems. Science 296 (5575), 1985–1990. Available from: https://doi.org/10.1126/science.1067123.

Al-Taani, A.A., Nazzal, Y., Howari, F.M., 2020. Groundwater scarcity in the Middle East. In: Mukherjee, A., Scanlon, B.R., Aureli, A., Langan, S., Guo, H., McKenzie, A. (Eds.), Global Groundwater: Source, Scarcity, Sustainability, Security, Solutions. Elsevier, ISBN: 9780128181720.

Aureli, A., Eckstein, G., 2011. Strengthening cooperation on transboundary groundwater resources. Water Int. 36 (5), 549–556.

Barnett, S., Simmons, C.T., Nelson, R., 2020. Groundwater resources in Australia – their occurrence, management and future challenges. In: Mukherjee, A., Scanlon, B.R., Aureli, A., Langan, S., Guo, H., McKenzie, A. (Eds.), Global Groundwater: Source, Scarcity, Sustainability, Security, Solutions. Elsevier, ISBN: 9780128181720.

Benaabidate, L., Zian, A., Sadki, O., 2020. Hydrochemical characteristics and quality assessment of water from different sources in Northern Morocco. In: Mukherjee, A., Scanlon, B.R., Aureli, A., Langan, S., Guo, H., McKenzie, A. (Eds.), Global Groundwater: Source, Scarcity, Sustainability, Security, Solutions. Elsevier, ISBN: 9780128181720.

Bhanja, S.N., Mukherjee, A., 2019. In situ and satellite-based estimates of usable groundwater storage across India: implications for drinking water supply and food security. Adv. Water Resour. 126, 15–23.

Bhanja, S.N., Mukherjee, A., 2020. Groundwater sustainability and security in South Asia. In: Mukherjee, A., Scanlon, B.R., Aureli, A., Langan, S., Guo, H., McKenzie, A. (Eds.), Global Groundwater: Source, Scarcity, Sustainability, Security, Solutions. Elsevier, ISBN: 9780128181720.

Bhanja, S.N., Wang, J., 2020. Emerging groundwater and surface water trends in Alberta, Canada. In: Mukherjee, A., Scanlon, B.R., Aureli, A., Langan, S., Guo, H., McKenzie, A. (Eds.), Global Groundwater: Source, Scarcity, Sustainability, Security, Solutions. Elsevier, ISBN: 9780128181720.

Bhanja, S.N., Mukherjee, A., Saha, D., Velicogna, I., Famiglietti, J.S., 2016. Validation of GRACE based groundwater storage anamaly using in-situ groundwater level measurement in India. J. Hydrol. 543, 729–738.

Bhanja, S.N., Mukherjee, A., Wada, Y., Chattopadhyay, S., Velicogna, I., Pangaluru, K., et al., 2017. Groundwater rejuvenation in parts of India influenced by water-policy change implementation. Sci. Rep. 7.

Bierkens, M.F., Wada, Y., 2019. Non-renewable groundwater use and groundwater depletion: a review. Environ. Res. Lett. 14 (6), 063002.

BGR/WHYMAP, 2015. Bundesanstalt für Geowissenschaften und Rohstoffe (BGR) and UNESCO World-wide Hydrogeological Mapping and Assessment Programme (WHYMAP) Maps/Data. <https://www.whymap.org/whymap/EN/Maps_Data/maps_data_node_en.html> (accessed 20.05.20.).

Chakraborty, M., Mukherjee, A., Ahmed, K.M., 2015. A review of groundwater arsenic in the Bengal Basin, Bangladesh and India: from source to sink. Curr. Pollut. Rep. 1 (4), 220–247.

Chakraborty, M., Mukherjee, A., Ahmed, K.M., 2020. Transboundary Groundwater of the Ganges-Bramphaputra-Meghna River Delta System, India and Bangladesh. In: Mukherjee, A., Scanlon, B.R., Aureli, A., Langan, S., Guo, H., McKenzie, A. (Eds.), Global Groundwater: Source, Scarcity, Sustainability, Security, Solutions. Elsevier, ISBN: 9780128181720.

Chen, J., Rodell, M., 2020. Applications of GRACE in Global groundwater study. In: Mukherjee, A., Scanlon, B.R., Aureli, A., Langan, S., Guo, H., McKenzie, A. (Eds.), Global Groundwater: Source, Scarcity, Sustainability, Security, Solutions. Elsevier, ISBN: 9780128181720.

Chowdhury, S.J., Rahaman, M.F., Mazumder, Q.H., Hossain, M.I., 2020. MAR model: a blessing adaptation for hard-to-reach livelihood in thirsty barind tract, Bangladesh. In: Mukherjee, A., Scanlon, B.R., Aureli, A., Langan, S., Guo, H., McKenzie, A. (Eds.), Global Groundwater: Source, Scarcity, Sustainability, Security, Solutions. Elsevier, ISBN: 9780128181720.

Coomar, P., Mukherjee, A., 2020. Global geogenic groundwater pollution. In: Mukherjee, A., Scanlon, B.R., Aureli, A., Langan, S., Guo, H., McKenzie, A. (Eds.), Global Groundwater: Source, Scarcity, Sustainability, Security, Solutions. Elsevier, ISBN: 9780128181720.

Coomar, P., Mukherjee, A., Bhattacharya, P., Bundschuh, J., Verma, S., Fryar, A.E., et al., 2019. Contrasting controls on hydrogeochemistry of arsenic-enriched groundwater in the homologous tectonic settings of Andean and Himalayan basin aquifers, Latin America and South Asia. Sci. Total Environ. 689, 1370–1387.

Coulon, C.A., Pavelic, P., Christen, E., 2020. Drivers for progress in groundwater management in Lao PDR. In: Mukherjee, A., Scanlon, B.R., Aureli, A., Langan, S., Guo, H., McKenzie, A. (Eds.), Global Groundwater: Source, Scarcity, Sustainability, Security, Solutions. Elsevier, ISBN: 9780128181720.

Dalin, C., 2020. Sustainability of groundwater used in agricultural production and trade. In: Mukherjee, A., Scanlon, B.R., Aureli, A., Langan, S., Guo, H., McKenzie, A. (Eds.), Global Groundwater: Source, Scarcity, Sustainability, Security, Solutions. Elsevier, ISBN: 9780128181720.

Dalin, C., Wada, Y., Kastner, T., Puma, M.J., 2017. Groundwater depletion embedded in international food trade. Nature 543 (7647), 700–704.

de Graaf, I.E.M., Gleeson, T., van Beek, L. P. H. (Rens), Sutanudjaja, E.H., Bierkens, M.F.P., 2019. Environmental flow limits to global groundwater pumping. Nature 574 (7776), 90–94.

Döll, P., Fiedler, K., 2008. Global-scale modeling of groundwater recharge. Hydrol. Earth Syst. Sci. 12, 863–885.

Duttagupta, S., Mukherjee, A., Bhattacharya, A., Bhattacharya, J., 2020a. Wide exposure of persistent organic pollutants (PoPs) in natural waters and sediments of the densely populated Western Bengal basin, India. Sci. Total Environ. 717, 137187.

Duttagupta, S., Mukherjee, A., Routh, J., Devi, L.G., Bhattacharya, A., Bhattacharya, J., 2020b. Role of aquifer media in determining the fate of polycyclic aromatic hydrocarbons in the natural water and sediments along the lower Ganges river basin. J. Environ. Sci. Health, A 55 (4), 354–373.

Edmunds, W.M., Smedley, P.L., 2013. Fluoride in natural waters. In: Selinus, O., Alloway, B., Centeno, J.A., Finkelman, R.B., Fuge, R., Lindh, U., Smedley, P.L. (Eds.), Essentials of Medical Geology. Springer, pp. 311–336.

Famiglietti, J.S., 2014. The global groundwater crisis. Nat. Clim. Change 4 (11), 945–948.

Fan, Y., Li, H., Miguez-Macho, G., 2013. Global patterns of groundwater table depth. Science 339 (6122), 940–943.

Figueroa, A.J., Smilovic, M., 2020. Groundwater in the Nile River Basin. In: Mukherjee, A., Scanlon, B.R., Aureli, A., Langan, S., Guo, H., McKenzie, A. (Eds.), Global Groundwater: Source, Scarcity, Sustainability, Security, Solutions. Elsevier, ISBN: 9780128181720.

Food and Agriculture Organization of the United Nations (FAO), 2013. FAO Statistical Yearbook 2013: World Food and Agriculture, 289 pp.

Food and Agriculture Organization (FAO), 2016. The state of food and agriculture. Climate change, agriculture and food security. Rome, Italy. ISBN 978-92-5-109374-0.

Foster, S.S.D., Morris, B.L., Lawrence, A.R., 1994. Effects of urbanization on groundwater recharge. Groundwater Problems in Urban Areas: Proceedings of the International Conference Organized by the Institution of Civil Engineers and Held in London, June 2–3, 1993. Thomas Telford Publishing, pp. 43–63.

Fryar, A.E., 2020. Groundwater of carbonate aquifers. In: Mukherjee, A., Scanlon, B.R., Aureli, A., Langan, S., Guo, H., McKenzie, A. (Eds.), Global Groundwater: Source, Scarcity, Sustainability, Security, Solutions. Elsevier, ISBN: 9780128181720.

Giordano, M., 2009. Global groundwater? Issues and solutions. Annu. Rev. Environ. Resour. 34, 153–178.

Giordano, M., Shah, T., 2014. From IWRM back to integrated water resources management. Int. J. Water Resour. Dev. 30 (3), 364–376.

Gleeson, T., Marklund, L., Smith, L., Manning, A.H., 2011. Classifying the water table at regional to continental scales. Geophys. Res. Lett. 38 (5), L05401.

Gleeson, T., Befus, K.M., Jasechko, S., Luijendijk, E., Cardenas, M.B., 2016. The global volume and distribution of modern groundwater. Nat. Geosci. 9, 161–167.

Gleeson, T., Cuthbert, M., Ferguson, G., Perrone, D., 2020. Global groundwater sustainability, resources and systems in the Anthropocene. Annu. Rev. Earth Planet. Sciences. <https://doi.org/10.1146/annurev-earth-071719-055251>.

Gude, V.G., Maganti, A., 2020. Desalination of deep groundwater for freshwater supplies. In: Mukherjee, A., Scanlon, B.R., Aureli, A., Langan, S., Guo, H., McKenzie, A. (Eds.), Global Groundwater: Source, Scarcity, Sustainability, Security, Solutions. Elsevier, ISBN: 9780128181720.

Haque, S.E., 2020. Hydrogeochemical characterization of groundwater quality in the states of Texas and Florida. In: Mukherjee, A., Scanlon, B.R., Aureli, A., Langan, S., Guo, H., McKenzie, A. (Eds.), Global Groundwater: Source, Scarcity, Sustainability, Security, Solutions. Elsevier, ISBN: 9780128181720.

Hoekstra, A.Y., 2017. Global food and trade dimensions of groundwater governance. Advances in Groundwater Governance. CRC Press, pp. 353–366.

Huang, G., Liu, L., Liu, C., Wang, W., Han, D., 2020. Groundwater pollution of Pearl River Delta. In: Mukherjee, A., Scanlon, B.R., Aureli, A., Langan, S., Guo, H., McKenzie, A. (Eds.), Global Groundwater: Source, Scarcity, Sustainability, Security, Solutions. Elsevier, ISBN: 9780128181720.

Institute for Transformative Technologies (ITT), 2018. Technology Breakthroughs for Global Water Security: A Deep Dive Into South Asia, 185 pp.

Jia, Y., 2020. Geogenic contaminated groundwater in China. In: Mukherjee, A., Scanlon, B.R., Aureli, A., Langan, S., Guo, H., McKenzie, A. (Eds.), Global Groundwater: Source, Scarcity, Sustainability, Security, Solutions. Elsevier, ISBN: 9780128181720.

Karim, A.Q., Sadat, S.H., 2020. Groundwater pollution of Afghanistan. In:. In: Mukherjee, A., Scanlon, B.R., Aureli, A., Langan, S., Guo, H., McKenzie, A. (Eds.), Global Groundwater: Source, Scarcity, Sustainability, Security, Solutions. Elsevier, ISBN: 9780128181720.

Kebede, S., Taye, M.T., 2020. Groundwater scarcity and management in the arid areas in East Africa. In: Mukherjee, A., Scanlon, B.R., Aureli, A., Langan, S., Guo, H., McKenzie, A. (Eds.), Global Groundwater: Source, Scarcity, Sustainability, Security, Solutions. Elsevier, ISBN: 9780128181720.

KPMG, 2014. Feasibility Analysis for Solar Agricultural Water Pumps in India.

Kreamer, D.K., Ball., D.M., Re, V., Simmons, C.T., Bothwell., T., Verweij, H.J.M., et al., 2020. Future of groundwater science and research. In: Mukherjee, A., Scanlon, B.R., Aureli, A., Langan, S., Guo, H., McKenzie, A. (Eds.), Global Groundwater: Source, Scarcity, Sustainability, Security, Solutions. Elsevier, ISBN: 9780128181720.

Kumar, P., Mishra, B.K., Avtar, R., Chakraborty, S., 2020. Quantifying future water environment using numerical simulations: a scenario based approach for sustainable groundwater management plan in Medan, Indonesia. In: Mukherjee, A., Scanlon, B.R., Aureli, A., Langan, S., Guo, H., McKenzie, A. (Eds.), Global Groundwater: Source, Scarcity, Sustainability, Security, Solutions. Elsevier, ISBN: 9780128181720.

Kurtzman, D., Guttman, J., 2020. Managed aquifer recharge with various water sources for irrigation and domestic use:a perspective of the Israeli experience. In: Mukherjee, A., Scanlon, B.R., Aureli, A., Langan, S., Guo, H., McKenzie, A. (Eds.), Global Groundwater: Source, Scarcity, Sustainability, Security, Solutions. Elsevier, ISBN: 9780128181720.

Lafaye de Michaeux, F., Mukherjee, J., 2020. Groundwater and society: enmeshed issues, interdisciplinary approaches. In: Mukherjee, A., Scanlon, B.R., Aureli, A., Langan, S., Guo, H., McKenzie, A. (Eds.), Global Groundwater: Source, Scarcity, Sustainability, Security, Solutions. Elsevier, ISBN: 9780128181720.

Li, B., Rodell, M., 2020. Groundwater Drought: Environmental Controls and Monitoring. In: Mukherjee, A., Scanlon, B.R., Aureli, A., Langan, S., Guo, H., McKenzie, A. (Eds.), Global Groundwater: Source, Scarcity, Sustainability, Security, Solutions. Elsevier, ISBN: 9780128181720.

Ma, R., Wang, Y., 2020. Groundwater sustainability in cold and arid regions. In: Mukherjee, A., Scanlon, B.R., Aureli, A., Langan, S., Guo, H., McKenzie, A. (Eds.), Global Groundwater: Source, Scarcity, Sustainability, Security, Solutions. Elsevier, ISBN: 9780128181720.

MacDonald, A.M., et al., 2016. Groundwater quality and depletion in the Indo-Gangetic Basin mapped from in situ observations. Nat. Geosci. 9, 762–766.

Malakar, P., Sarkar, S., Mukherjee, A., Bhanja, S.N., Sun, A.Y., 2020. Use of machine learning and deep learning methods in groundwater. In: Mukherjee, A., Scanlon, B.R., Aureli, A., Langan, S., Guo, H., McKenzie, A. (Eds.), Global Groundwater: Source, Scarcity, Sustainability, Security, Solutions. Elsevier, ISBN: 9780128181720.

Maheswari, B., 2020. Groundwater sustainability and solutions of Australia. In: Mukherjee, A., Scanlon, B.R., Aureli, A., Langan, S., Guo, H., McKenzie, A. (Eds.), Global Groundwater: Source, Scarcity, Sustainability, Security, Solutions. Elsevier, ISBN: 9780128181720.

Mandal, A., Gordon-Smith, D.D.S., Harris, P., 2020. Groundwater availability and security in the Kingston Basin, Jamaica. In: Mukherjee, A., Scanlon, B.R., Aureli, A., Langan, S., Guo, H., McKenzie, A. (Eds.), Global Groundwater: Source, Scarcity, Sustainability, Security, Solutions. Elsevier, ISBN: 9780128181720.

Marfil-Vega, R., Crone, C.B., Glassmeyer, S.T., 2020. Out of sight, but not out of mind: per- and polyfluoroalkyl substances in groundwater. In: Mukherjee, A., Scanlon, B.R., Aureli, A., Langan, S., Guo, H., McKenzie, A. (Eds.), Global Groundwater: Source, Scarcity, Sustainability, Security, Solutions. Elsevier, ISBN: 9780128181720.

Margat, J., Van der Gun, J., 2013. Groundwater Around the World: A Geographic Synopsis. CRC Press.

Masood, N., Batool., S., Farooqi., A., 2020. Groundwater pollution in Pakistan. In: Mukherjee, A., Scanlon, B.R., Aureli, A., Langan, S., Guo, H., McKenzie, A. (Eds.), Global Groundwater: Source, Scarcity, Sustainability, Security, Solutions. Elsevier, ISBN: 9780128181720.

Moench, M., 2004. Groundwater: the challenge of monitoring and management, The World's Water, vol. 2005. Island Press, Washington, DC, pp. 79–100.

Mondal, D., Lopez-Espinosa, M.J., Armstrong, B., Stein, C.R., Fletcher, T., 2012. Relationships of perfluorooctanoate and perfluorooctane sulfonate serum concentrations between mother-child pairs in a population with perfluorooctanoate exposure from drinking water. Environ. Health Perspect. 120 (5), 752–757.

Mukherjee, A., 2018.Groundwater of South Asia. Springer Nature, ISBN:978-981-10-3888-4, Singapore, 799 p.

Mukherjee, A., Fryar, A.E., Howell, P.D., 2007a. Regional hydrostratigraphy and groundwater flow modeling in the arsenic-affected areas of the western Bengal basin, West Bengal, India, Hydrogeol. J 15 (7), 1397.

Mukherjee, A., Fryar, A.E., Rowe, H.D., 2007b. Regional-scale stable'isotopic signatures of recharge and deep groundwater in the arsenic affected areas of West Bengal, India. J. Hydrol 334 (1-2), 151–161.

Mukherjee, A., Bhattacharya, P., Savage, K., Foster, A., Bundschuh, J., 2008. Distribution of geogenic arsenic in hydrologic systems: controls and challenges. J. Contam. Hydrol. 99, 1–7.

Mukherjee, A., Fryar, A.E., 2008. Deeper groundwater chemistry and geochemical modeling of the arsenic affected western Bengal basin, West Bengal, India. Applied Geochemistry 23 (4), 863–894.

Mukherjee, A., Fryar, A.E., O'Shea, B.M., 2009. Major occurrences of elevated arsenic in groundwater and other natural waters. In: Henke, K.R. (Ed.), Arsenic—Environmental Chemistry, Health Threats and Waste Treatment. John Wiley & Sons, Chichester, pp. 303–350.

Mukherjee, A., Verma, S., Gupta, S., Henke, K.R., Bhattacharya, P., 2014. Influence of tectonics, sedimentation and aqueous flow cycles on the origin of global groundwater arsenic: paradigms from three continents. J. Hydrol. 518, 284–299.

Mukherjee, A., Saha, D., Harvey, C.F., Taylor, R.G., Ahmed, K.M., 2015. Groundwater systems of the Indian Sub-continent. J. Hydrol.-Reg. Stud. 4A, 1–14.

Mukherjee, A., Bhanja, S.N., Wada, Y., 2018. Groundwater depletion causing reduction of baseflow triggering Ganges river summer drying. Sci. Rep. 8 (1), 12049.

Mukherjee, A., Duttagupta, S., Chattopadhyay, S., Bhanja, S.N., Bhattacharya, A., Chakraborty, S., et al., 2019a. Impact of sanitation and socioeconomy on groundwater fecal pollution and human health towards achieving sustainable development goals across India from groundobservations and satellite-derived nightlight. Sci. Rep. 9 (1), 1–11.

Mukherjee, A., Gupta, S., Coomar, P., Fryar, A.E., Guillot, S., Verma, S., et al., 2019b. Plate tectonics influence on geogenic arsenic cycling: from primary sources to global groundwater enrichment. Sci. Total Environ. 683, 793–807.

Mukherjee, A., Scanlon, B.R., Aureli, A., Langan, S., Guo, H., McKenzie, A. (Eds.), 2020. Global Groundwater: Source, Scarcity, Sustainability, Security, Solutions. Elsevier, ISBN: 9780128181720.

Oliveira, P.T.S., Lucas, M.C., Godoi, R., Wendland, E., 2020. Groundwater recharge and sustainability in Brazil. In: Mukherjee, A., Scanlon, B.R., Aureli, A., Langan, S., Guo, H., McKenzie, A. (Eds.), Global Groundwater: Source, Scarcity, Sustainability, Security, Solutions. Elsevier, ISBN: 9780128181720.

Puri, S. and Arnold, G. Challenges to management of transboundary aquifers: The ISARM Programme. In Second International Conference on Sustainable Management of Transboundary Waters in Europe (Vol.'21). 2002, April.

Puri, S., 2020. Transboundary aquifers: a shared sub surface asset, in urgent need of sound governance. In: Mukherjee, A., Scanlon, B.R., Aureli, A., Langan, S., Guo, H., McKenzie, A. (Eds.), Global Groundwater: Source, Scarcity, Sustainability, Security, Solutions. Elsevier, ISBN: 9780128181720.

Puri, S., Appelgren, B., Arnold, G., Aureli, A., Burchi, S., et al., Internationally shared (transboundary) aquifer resources management, their significance and sustainable management: a framework document. vIHP-VI, IHP Non Serial Publ. Hydrol. SC-2001/WS/40, UN Educ. Sci. Cult. Organ, 2001, UNESCO, Paris.

Qin, X., Zhou, T., Liu, F., 2020. Screening of emerging organic pollutants in the typical hygrogeological units of China. In: Mukherjee, A., Scanlon, B.R., Aureli, A., Langan, S., Guo, H., McKenzie, A. (Eds.), Global Groundwater: Source, Scarcity, Sustainability, Security, Solutions. Elsevier, ISBN: 9780128181720.

Ravenscroft, P., Brammer, H., Richards, K.S., 2009. Arsenic Pollution: A Global Synthesis. Wiley-Blackwell Publication, 588 p.

Raychowdhury, N., Mukherjee, A., Bhattacharya, P., Johannesson, K., Bundschuh, J., Sifuentes, G.B., et al., 2014. Provenance and fate of arsenic and other solutes in the Chaco-Pampean Plain of the Andean foreland, Argentina: from perspectives of hydrogeochemical modeling and regional tectonic setting, J. Hydrol., 518. pp. 300–316.

Richey, A.S., Thomas, B.F., Lo, M.-H., Reager, J.T., Famiglietti, J.S., Voss, K., et al., 2015. Quantifying renewable groundwater stress with GRACE. Water Resour. Res. 51 (7), 5217–5238.

Rodell, M., Velicogna, I., Famiglietti, J.S., 2009. Satellite-based estimates of groundwater depletion in India. Nature 460, 999–1002.

Rodell, M., Famiglietti, J.S., Wiese, D.N., Reager, J.T., Beaudoing, H.K., Landerer, F.W., et al., 2018. Emerging trends in global freshwater availability. Nature 557 (7707), 651–659.

Siebert, S., Henrich, V., Frenken, K., Burke, J., 2013a. Global Map of Irrigation Areas Version 5. Rheinische Friedrich-Wilhelms-University, Bonn; Food and Agriculture Organization of the United Nations, Rome.

Saha, D., Marawaha, S., Dwivedi, S.N., 2020. Role of measuring the aquifers for sustainably managing groundwater resource in India. In: Mukherjee, A., Scanlon, B.R., Aureli, A., Langan, S., Guo, H., McKenzie, A. (Eds.), Global Groundwater: Source, Scarcity, Sustainability, Security, Solutions. Elsevier, ISBN: 9780128181720.

Sahoo, P.K., Pontes, P.R., Salomão, G.N., Powell, M.A., Mittal, S., Souza Filho, P.W.M., et al., 2020. Groundwater resource conditions and management in Brazil: current status and challenges for its sustainable utilization. In: Mukherjee, A., Scanlon, B.R., Aureli, A., Langan, S., Guo, H., McKenzie, A. (Eds.), Global Groundwater: Source, Scarcity, Sustainability, Security, Solutions. Elsevier. ISBN: 9780128181720.

Sathre, R., 2020. Technologies to enhance sustainable groundwater use. In: Mukherjee, A., Scanlon, B.R., Aureli, A., Langan, S., Guo, H., McKenzie, A. (Eds.), Global Groundwater: Source, Scarcity, Sustainability, Security, Solutions. Elsevier, ISBN: 9780128181720.

Scanlon, B.R., Mukherjee, A., Gates, J.B., Reedy, R.C., Sinha, A.N., 2010. Groundwater recharge in natural dune systems and agricultural ecosystems in the Thar Desert region, Rajasthan, India. Hydrogeol. J. 18 (4), 959–972.

Scanlon, B.R., Ruddell, Ben, L., Reed, Patrick, M., Hook, Ruth, I., Zheng, Chunmiao, Tidwell, Vince, C., et al., 2017. The food-energy-water nexus: transforming science for society. Water Resour. Res. 53 (5), 3550–3556.

Schreiber, M.E., 2020. Arsenic in groundwater in the United States: research highlights since 2000, current concerns and next steps. In: Mukherjee, A., Scanlon, B.R., Aureli, A., Langan, S., Guo, H., McKenzie, A. (Eds.), Global Groundwater: Source, Scarcity, Sustainability, Security, Solutions. Elsevier, ISBN: 9780128181720.

Shah, T., Rajan, A., Rai, G.P., 2020. Balancing livelihoods and environment: political economy of groundwater irrigation in India. In: Mukherjee, A., Scanlon, B.R., Aureli, A., Langan, S., Guo, H., McKenzie, A. (Eds.), Global Groundwater: Source, Scarcity, Sustainability, Security, Solutions. Elsevier, ISBN: 9780128181720.

Shamsudduha, M., 2020. Groundwater storage dynamics in the Himalayan River Basins and impacts of global change in the Anthropocene. In: Mukherjee, A., Scanlon, B.R., Aureli, A., Langan, S., Guo, H., McKenzie, A. (Eds.), Global Groundwater: Source, Scarcity, Sustainability, Security, Solutions. Elsevier, ISBN: 9780128181720.

Sharp Jr, J.M., Krothe, J.N., Mather, J.D., Garcia-Fresca, B., Stewart, C.A., 2003. Effects of urbanization on groundwater systems. Earth Sci. City 257–278.

Shiklomanov, I.A., 1998. World water resources: an appraisal for the 21st century. IHP Rep. UN Educ.Sci. Cult. Organ. UNESCO, Paris, <http://www.fao.org/nr/water/aquastat/main/index.stm>.

Siebert, S., Burke, J., Faures, J.M., Frenken, K., Hoogeveen, J., Doll, P., et al., 2010. Groundwater use for irrigation—a global inventory. Hydrol. Earth Syst. Sci. 7, 3977–4021. Available from: https://doi.org/10.5194/hessd-3977-3977-2010.

Smedley, P.L., Kinniburgh, D.G., 2002. A review of the source, behavior and distribution of arsenic in natural waters. Appl. Geochem. 17 (5), 517–568.

Smith, A.H., Lingas, E.O., Rahman, M., 2000. Contamination of drinking-water by arsenic in Bangladesh: a public health emergency. Bull. World Health Organ. 78 (9), 1093–1103.

Sun, A.Y., Scanlon, B.R., Zhang, Z., Walling, D., Bhanja, S.N., Mukherjee, A., et al., 2019. Combining physically based modeling and deep learning for fusing GRACE satellite data: Can we learn from mismatch? Water Resources Res 55 (2), 1179–1195.

Taylor, R.G., Scanlon, B., Doll, P., Rodell, M., van Beek, R., Wada, Y., et al., 2013. Ground water and climate change. Nat. Clim. Change 3 (4), 322–329.

UN, 2017. Report of the Secretary-General, "Progress Towards the Sustainable Development Goals" E/2017/66(2017).

Van Der Gun, J., 2020. Groundwater resources sustainability. In: Mukherjee, A., Scanlon, B.R., Aureli, A., Langan, S., Guo, H., McKenzie, A. (Eds.), Global Groundwater: Source, Scarcity, Sustainability, Security, Solutions. Elsevier, ISBN: 9780128181720.

van Weert, F., van der Gun, J., Reckman, J., 2009. Global Overview of Saline Groundwater Occurrence and Genesis. IGRAC Report Nr. GP 2009-1. International Groundwater Resources Assessment Center, Utrecht.

Wada, Y., Heinrich, L., 2013. Assessment of transboundary aquifers of the world—vulnerability arising from human water use. Environ. Res. Lett. 8 (2), 024003.

Wada, Y., Van Beek, L.P., Van Kempen, C.M., Reckman, J.W., Vasak, S., Bierkens, M.F., 2010. Global depletion of groundwater resources. Geophys. Res. Lett. 37 (20).

Yang, W., Long, D., Sun, Z., 2020. Global Groundwater: from scarcity to solutions. In: Mukherjee, A., Scanlon, B.R., Aureli, A., Langan, S., Guo, H., McKenzie, A. (Eds.), Global Groundwater: Source, Scarcity, Sustainability, Security, Solutions. Elsevier, ISBN: 9780128181720.

Siebert, S., Henrich, V., Frenken, K., Burke, J., 2013b. Update of the Digital Global Map of Irrigation Areas to Version 5. Rheinische Friedrich-Wilhelms-Universität, Bonn; Food and Agriculture Organization of the United Nations, Rome.

La Vanchy, G.T., Adamson, J.K., Kerwin, M.W., 2020. Integrating groundwater for water security in Cape Town, South Africa. In: Mukherjee, A., Scanlon, B.R., Aureli, A., Langan, S., Guo, H., McKenzie, A. (Eds.), Global Groundwater: Source, Scarcity, Sustainability, Security, Solutions. Elsevier, ISBN: 9780128181720.

Theme 2

Groundwater sources

Chapter 2

Groundwater of carbonate aquifers

Alan E. Fryar
Department of Earth and Environmental Sciences, University of Kentucky, Lexington, KY, United States

2.1 Introduction

Carbonate rocks (especially limestone, but also dolomite, chalk, marl, and marble) are exposed over ~13% of the Earth's surface (Amiotte Suchet et al., 2003) (Fig. 2.1). They underlie 22% of Europe (Chen et al., 2017) and ~40% of the Eastern United States (White et al., 1995). Between ~10% and 25% of the Earth's population depends on carbonate aquifers for water supply (Ford and Williams, 2007; Stevanović, 2018), particularly in North America, Europe, the Middle East, and Southeast Asia. Because calcite dissolves relatively rapidly in acidic water, carbonate aquifers frequently are distinguished by solution-enhanced flowpaths, with apertures orders of magnitude greater than in porous or fractured silicate media (i.e., centimeters or larger vs millimeters or smaller). These flowpaths are marked by karst features, such as sinkholes (dolines), stream sinks (swallets or ponors), conduits or caves, and springs, which integrate surface and subsurface drainage. Groundwater velocities can be much faster than in noncarbonate media—up to hundreds of meters per hour, comparable to surface streams—and carbonate aquifers can be highly transmissive. For example, the highest capacity drilled well in the world (2.5 m^3/s) occurs in the Edwards aquifer of Texas (United States; Kresic, 2013), and the Floridan aquifer hosts 33 first-magnitude springs [with discharge $Q \geq 2.83$ m^3/s; U.S. Geological Survey (USGS), n.d.]. The heterogeneity of carbonate rocks can result in discontinuous zones of saturation, distortion of the water table owing to preferential flow, and mass transfer under turbulent rather than laminar conditions. Consequently, delineation of groundwater flowpaths can be more complicated than in conventional porous media and carbonate aquifers can be more susceptible to contamination. Carbonate aquifers can also host distinctive aquatic ecosystems. This chapter highlights reactions in carbonate aquifers; development of porosity and permeability; mechanisms of recharge and flow; water supply and environmental issues; and challenges in monitoring and modeling, with emphasis on examples from the United States.

2.2 Carbonate geochemistry and hydrochemical evolution

As exemplified by calcite, carbonate minerals tend to dissolve via reaction with protons (H$^+$):

$$CaCO_3 + H^+ \rightleftarrows Ca^{2+} + HCO_3^- \quad (2.i)$$

although calcite can also dissolve via direct reaction with hydrated CO$_2$ and with H$_2$O (Plummer et al., 1978). In general, the primary source of acidity in soil and groundwater is the weak acid H$_2$CO$_3$ (carbonic acid), which is generated by oxidation of organic matter (represented by the simplified composition CH$_2$O) and by dissolution of CO$_2$ gas as follows:

$$CH_2O + O_2 \rightarrow H_2CO_3 \quad (2.ii)$$

$$CO_2\ (g) \rightleftarrows CO_2\ (aq) \quad (2.iii)$$

$$CO_2\ (aq) + H_2O \rightleftarrows H_2CO_3 \quad (2.iv)$$

The equilibrium constant for the combination of reactions (2.iii) and (2.iv) is:

$$K_{CO_2} = \frac{a_{H_2CO_3}}{P_{CO_2}} \quad (2.1)$$

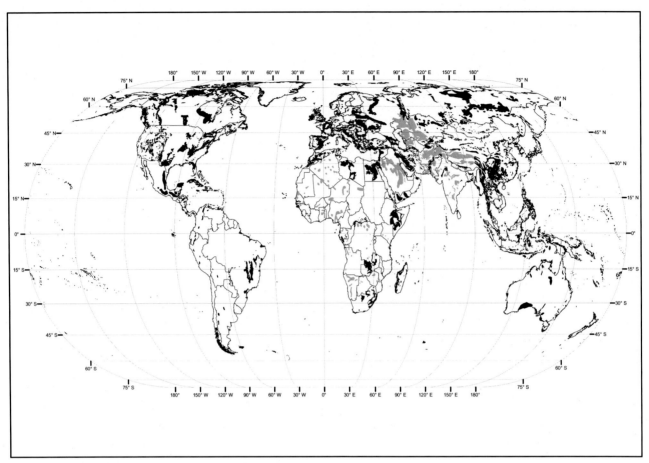

FIGURE 2.1 World map of carbonate rock outcrops. Dark shading represents relatively pure, continuous carbonates; light shading represents discontinuous or impure carbonates. *From Williams, P., Fong, Y.T. 2010. World map of carbonate rock outcrops v. 3.0. Available at <https://www.fos.auckland.ac.nz/our_research/karst/index.html> (accessed 28.03.20.).*

where a is solute activity and P is partial pressure of the gas phase (Hess and White, 1989). Carbonic acid dissociates as:

$$H_2CO_3 \rightleftarrows H^+ + HCO_3^- \tag{2.v}$$

$$HCO_3^- \rightleftarrows H^+ + CO_3^{2-} \tag{2.vi}$$

The net reactions for dissolution of calcite and dolomite can be written as:

$$CaCO_3 + H_2CO_3 \rightleftarrows Ca^{2+} + 2HCO_3^- \tag{2.vii}$$

$$CaMg(CO_3)_2 + 2H_2CO_3 \rightleftarrows Ca^{2+} + Mg^{2+} + 4HCO_3^- \tag{2.viii}$$

For the pH range (from 5 to 9) encountered in most carbonate groundwaters, the dominant form of dissolved inorganic carbon is H_2CO_3 or HCO_3^- (Hess and White, 1989). In some geologic settings, carbonate dissolution (exemplified by calcite) is driven by sulfuric acid, which is formed by sulfide oxidation, resulting in precipitation of gypsum (Martin, 2017):

$$H_2S + 2O_2 \rightarrow H_2SO_4 \rightleftarrows 2H^+ + SO_4^{2-} \tag{2.ix}$$

$$H_2SO_4 + CaCO_3 \rightarrow CaSO_4 \cdot 2H_2O + CO_2 \text{ (aq)} \tag{2.x}$$

For specified P_{CO_2}, calcite dissolution is generally assumed to be greater under open-system conditions (e.g., in the vadose zone and at the water table, where CO_2 can exchange between aqueous and gas phases) than under closed-system conditions (Ford and Williams, 2007). However, this assumption may not always be accurate (Covington and

Vaughn, 2019). The solubility of calcite, like that of CO_2, increases with decreasing temperature (i.e., is retrograde). The rate of calcite dissolution was conceptualized by Plummer et al. (1978) as:

$$R = k_1 a_{H^+} + k_2 a_{H_2CO_3} + k_3 a_{H_2O} - k_4 a_{Ca^{2+}} a_{HCO_3^-} \quad (2.2)$$

where values of k are rate constants, with k_1 being mass-transport controlled and dominant under acidic conditions, whereas k_2 is surface-reaction controlled (Hess and White, 1989). Far from equilibrium, the calcite dissolution rate increases abruptly at the transition from laminar to turbulent flow, because more efficient mixing enhances the rate of CO_2 hydration (Buhmann and Dreybrodt, 1985a,b; White, 1988). Although calcite and dolomite solubilities are similar (Brahana et al., 1988), dolomite dissolution is slower (Hess and White, 1989).

Overall rates of carbonate rock dissolution depend on mineralogy, reactive surface area, and mass-transport rates. Modern marine sediments are composed of aragonite, Mg calcite, and calcite, whereas chalk consists almost entirely of low-Mg calcite admixed with clay (Brahana et al., 1988). Dolomite is typically formed by postdepositional replacement of limestones in shallow-water carbonate platforms, particularly by reflux of brines and open-system thermal (Kohout-type) convection (Machel, 2004). Slower rates of limestone dissolution in the field relative to laboratory conditions may result from insoluble clay and silica coatings (White, 1988). Notwithstanding retrograde solubility, hot and humid environments (i.e., wet tropical climates) provide optimal conditions for carbonate weathering: high precipitation enhances water fluxes and high biological productivity generates CO_2 and organic acids (Palmer and Palmer, 2009).

As noted by Hanshaw and Back (1979), the chemical evolution of groundwater in carbonate aquifers can follow different compositional trends. Where meteoric water displaces seawater (e.g., during early diagenesis of carbonate sediments), freshening occurs and the major-ion composition becomes Ca-HCO_3 dominated. In coastal aquifers, mixing of fresh and saline groundwaters can promote undersaturation with respect to calcite and, thereby, dissolution of limestone (Brahana et al., 1988). Downgradient of the recharge zone in regional carbonate aquifers, total dissolved solids (TDS) increase with water–rock interaction along the flowpath, and hydrochemical facies can evolve from Ca-HCO_3 to Ca-Mg–HCO_3–SO_4 (Hanshaw and Back, 1979). This evolution can result from dedolomitization in confined aquifers: calcite dissolves prior to dolomite and gypsum, then, as Ca^{2+} concentrations increase, calcite reprecipitates, causing undersaturation with respect to dolomite (Plummer et al., 1990).

2.3 Porosity and permeability

A distinctive feature of carbonate aquifers is the evolution of porosity and permeability (and thus hydraulic conductivity) over much shorter timescales than in silicate aquifers (Hess and White, 1989). Depositional environment (e.g., the extent of terrigenous clastic sedimentation on carbonate platforms, which inhibits dissolution) and tectonic setting (because of the tendency of carbonate rocks to fracture with deformation) have a profound influence on the hydraulic properties of carbonate aquifers (Brahana et al., 1988; Kresic, 2013). Porosity and permeability are initially reduced by compaction and cementation before becoming enhanced by dissolution. Karstic rocks are marked by three types of porosity: matrix or granular, fracture, and conduit (White, 1989; Ford and Williams, 2007). Porosity can vary from 40%–70% in modern carbonate sediment to 0.1% in carbonate rocks (Fig. 2.2; Brahana et al., 1988), although chalk can retain relatively high porosity (30%; Worthington, 1999; Ford and Williams, 2007). Values of hydraulic conductivity for various carbonate aquifers range from 10^{-4} to 10^7 m/day (Fig. 2.2; Brahana et al., 1988), and permeability values have been reported to range over nine orders of magnitude within a single karst aquifer (the Edwards aquifer of Texas), depending on scale and direction of measurement (Halihan et al., 1999; Ford and Williams, 2007).

Development of conduits is commonly initiated by focused flow along fractures such as joints, bedding planes, and faults (Brahana et al., 1988), although carbonate rocks with high matrix porosity, such as in the Biscayne aquifer of Florida (United States), can also have well-developed karst features (Kresic, 2013). Where carbonate rock is subaerially exposed or mantled by soil, infiltration of meteoric water forms epikarst (a network of solution-enhanced fractures and pores; Palmer and Palmer, 2009). In contrast to this epigenetic porosity, hypogenetic porosity can be created by dissolution unrelated to near-surface activity; a prominent example is sulfuric acid speleogenesis in sedimentary basins (Palmer, 1991). Karstification begins with slow enlargement of continuous openings with apertures on the order of 10–100 μm (Ford and Williams, 2007). Palmer (1981) calculated the maximum dissolution rate for fracture enlargement to be 0.14 cm/year. When apertures exceed ~1 cm, conduit development in limestone is promoted by three interrelated phenomena: dissolution kinetics shift from fourth to first order, flow shifts from laminar to turbulent, and velocities become sufficient for entrainment and transport of insoluble clastic sediment (White, 2002). As conduits are enlarged (i.e., become caves), the rate of wall retreat approaches a maximum of ~0.01–0.1 cm/year, which is determined primarily by chemical kinetics rather than by further increases in discharge (Palmer, 1991). In dolomite, pores tend to be fewer but larger and

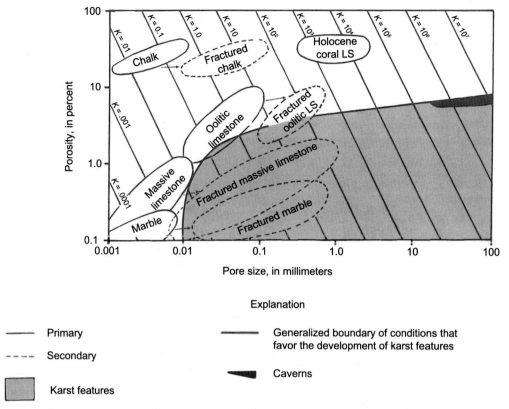

FIGURE 2.2 Primary and secondary porosity, pore size, and theoretical hydraulic conductivity K (m/day) of selected carbonate rocks, karst features, and caverns. Values of K assume the rock mass behaves as a bundle of straight, parallel capillary tubes. *Modified from Brahana, J.V., Thrailkill, J., Freeman, T., Ward, W.C., 1988. Carbonate rocks. In: Back, W., Rosenshein, J.S., Seaber, P.R. (Eds.), Hydrogeology: The Geology of North America, vol. O-2. Geological Society of America, Boulder, CO, pp. 333–352.*

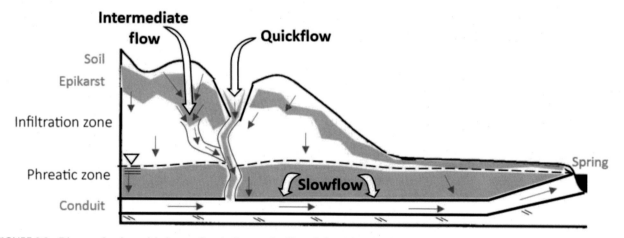

FIGURE 2.3 Diagram showing quick, intermediate (epikarst and soil), and slow (phreatic) flow pathways. *Reprinted by permission from Husic, A., Fox, J., Adams, E., Backus, J., Pollock, E., Ford, W., Agouridis, C., 2019. Inland impacts of atmospheric river and tropical cyclone extremes on nitrate transport and stable isotope measurements. Environ. Earth Sci. 78, 36. ©2019, Springer.*

more homogeneously distributed (Brahana et al., 1988). Because the rates of pore broadening and the advance of the reaction front are much slower, channelization is less pronounced in dolomite than in limestone (Hoefner and Fogler, 1988).

2.4 Recharge and flow

Recharge to carbonate aquifers can be either focused or diffuse. Infiltration can be focused through depressions and solution-enhanced openings in epikarst (Figs. 2.3 and 2.4) and by quickflow through sinkholes and swallets (Fig. 2.3).

FIGURE 2.4 Soil-filled, solution-enhanced joint, Puding County, Guizhou province (China). *Photo by A.E. Fryar.*

Recharge in karst terrains is defined as autogenic if precipitation directly infiltrates into carbonate rocks and allogenic if runoff originates elsewhere (e.g., from sinking streams on clastic caprock) (Ford and Williams, 2007). Temporally and spatially discontinuous zones of saturation (perched aquifers) can form within epikarst because of the lower permeability of underlying, less weathered rock, which constitutes a leaky capillary barrier (Ford and Williams, 2007). In the phreatic zone, water and solutes are exchanged between fast-flow pathways through conduits and slow-flow storage in the matrix (Fig. 2.3). Within conduits, Darcy's law becomes invalid for describing groundwater flow, although laminar, non-Darcian flow can occur in pipes up to ~ 0.5 m diameter if velocity (v) is not >1 mm/s (Ford and Williams, 2007). Beyond this threshold, flow is described by the Darcy–Weisbach equation:

$$Q = \left(\frac{2dgA^2}{f}\right)^{1/2} \left(\frac{dh}{dl}\right)^{1/2} \qquad (2.3)$$

where d is the conduit diameter, g is acceleration due to the Earth's gravity (9.81 m/s^2), A is the cross-sectional area, dh/dl is the hydraulic gradient, and the friction factor $f = 8\tau/(\rho_f v^2)$, where τ is shear stress and ρ_f is freshwater density (Ford and Williams, 2007). Responses to recharge in conduits can be much more rapid than in porous media. In open channels, kinematic waves move $\sim 30\%$ faster than the water itself (on the order of $10-10^3$ m/h; Ford and Williams, 2007), whereas under pipe-full conditions, pressure pulses move at the speed of sound (~ 1500 m/s; Greenspan and Tschiegg, 1957). Consequently, water levels can rise tens of meters within hours following rainfall or flooding (White, 1988).

Discharge from karst aquifers commonly occurs via springs, including by gravity flow where the water table intersects the land surface (Fig. 2.5) and by artesian flow where water levels are elevated under pressure (Fig. 2.6). The outlet of a surficial groundwater basin (springshed) is typically a large permanent spring, which occurs

FIGURE 2.5 Gravity-flow spring, Preston's Cave, Lexington, Kentucky (United States). *Photo by A.E. Fryar.*

FIGURE 2.6 Artesian spring (The Boils), McConnell Springs, Lexington, Kentucky (United States). *Photo by A.E. Fryar.*

at regional base level or at a stratigraphic contact or fault juxtaposing less soluble rocks (Quinlan and Ewers, 1989; Kresic, 2013). However, perched karst aquifers can also be drained by ephemeral springs (Brahana et al., 1988). Where surface and subsurface flowpaths are well-integrated and resistance to flow is low (White, 1989), conduits function as drains within the water table (Fig. 2.7), in contrast to semiconfined and confined aquifers, where discharge is more diffuse. In response to precipitation, springs in surficial karst aquifers can show not only relatively rapid increases in discharge, but also fluctuations in temperature (depending upon flowpath connectivity and the contrast between air and aquifer temperatures; Luhmann et al., 2011) and specific conductance, which is indicative of TDS (Ryan and Meiman, 1996). Because these fluctuations depend upon heat and mass transfer, rather than pressure-pulse propagation, they normally lag the discharge response (Fig. 2.8). TDS tends to decrease as a result of dilution by runoff then slowly rebound during hydrograph recession as the temporary hydraulic-gradient reversal (analogous to bank storage in riparian aquifers) ends and more mineralized groundwater seeps from the matrix into conduits.

At timescales of 10^3-10^6 years (Hess and White, 1989; Sullivan et al., 2019), karst drainage networks evolve both at the surface and in the subsurface. Sinkholes can develop where vertical leakage along preferential flowpaths beneath epikarst creates a depression in the perched water table, analogous to a cone of depression around a pumped well (Ford and Williams, 2007). Where limestone is interbedded between clastic rocks, sinkholes can also form via depression-focused recharge. In both cases, biogenic CO_2 production in soils, accumulation of drainage, and piping of soil and rock promote dissolution in a positive-feedback loop (Ford and Williams, 2007). As conduits lengthen (at rates as rapid as ~0.2 m/year) and connect recharge points to springs, sinkhole development is further promoted (Ford and Williams, 2007). Because of relatively rapid fluxes, conduit-flow springs are normally undersaturated with respect to calcite, thereby promoting dissolution (White, 1988). Focused discharge results in enlargement of spring orifices and flow capture from deeper flowpaths (Brahana et al., 1988). Where orifices collapse or are blocked, flow piracy occurs, or rivers

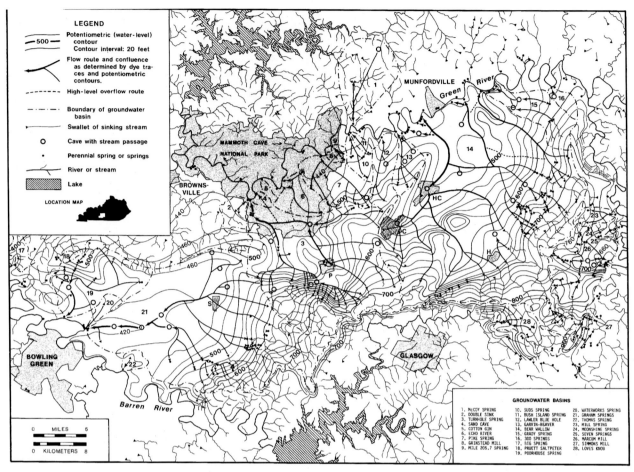

FIGURE 2.7 Map of groundwater basins, inferred flowpaths, water table, and surface drainage in the Mammoth Cave Region, Kentucky (United States). Towns are denoted by abbreviations as follows: Smiths Grove (S), Park City (P), Cave City (C), Horse Cave (HC), and Hiseville (H). Note 1.00 ft. = 0.305 m. *From Quinlan, J.F., Ewers, R.O., Ray, J.A., Powell, R.L., Krothe, N.C., 1983. Ground-water hydrology and geohydrology of the Mammoth Cave region, Kentucky, and of the Mitchell Plain, Indiana (Field trip 7). In: Shaver, R.H., Sunderman, J.A. (Eds.), Field Trips in Midwestern Geology, vol. 2. Geological Society of America, Boulder, CO, and Indiana Geological Survey, Bloomington, IN, pp. 1–85. Figure © Indiana Geological and Water Survey, Indiana University, Bloomington, Indiana.*

backflood, distributary spring networks can form (Quinlan and Ewers, 1989). Ultimately, as karstification progresses, phreatic conduits are dewatered and the regional water table is lowered (Ford and Williams, 2007). This process can be accelerated by increases in hydraulic gradient resulting from base-level lowering or uplift (Brahana et al., 1988; White, 1988) but can also be reversed by glacio-eustatic sea-level rise, as in the Floridan aquifer (Ford and Williams, 2007). Depending upon their geologic history, carbonate rocks can exhibit karstification at depths up to several kilometers (Kresic, 2013).

2.5 Water supply and environmental issues

Karst aquifers provide at least half of water supplies in parts of Central Europe and Southwest China (Chen et al., 2017) and fractured limestone is the largest source of groundwater in Great Britain (Sorensen et al., 2018). Carbonate aquifers provide significant water supplies in 15 US states (White, 1989) and are the primary water sources for several cities with populations >1 million [e.g., San Antonio and Miami (United States), Vienna (Austria), Rome (Italy), and Damascus (Syria)] (Chen et al., 2017; Prinos et al., 2014). Caves and springs in karst aquifers also host endemic ecosystems and vulnerable species (e.g., USGS, 2012; U.S. Fish and Wildlife Service, 2018; Krejca and Reddell, 2019). Consequently, concerns about exploitation of karst groundwater include the need to maintain spring flows for aquatic habitat conservation (Knight and Vick, 2018; Payne et al., 2019). In addition, pumping of karst aquifers can reduce

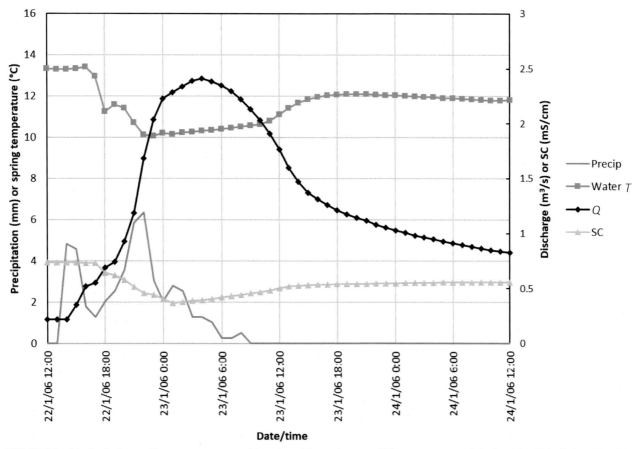

FIGURE 2.8 Hourly discharge (*Q*), water temperature (*T*), and specific conductance (SC) responses to precipitation, Blue Hole Spring, Versailles, Kentucky (United States) (J.W. Ward, T.M. Reed, and A.E. Fryar, unpublished data).

buoyant support of the matrix, thereby promoting the development of cover-subsidence and cover-collapse sinkholes, as in West Central Florida (Tihansky, 1999; Ford and Williams, 2007).

Because enhanced porosity and permeability facilitate rapid infiltration and transport, carbonate aquifers are susceptible to contamination (Vesper et al., 2000). Where conduit flow dominates, point-source pollutants are transported with minimal dispersion (White, 1989), whereas more conventional contaminant plumes develop in fractured dolomite (Ford and Williams, 2007). Carbonate aquifers are particularly vulnerable to pathogens and nutrients from agricultural activities (Panno et al., 2001; Katz et al., 2001), sewage and septage (Katz et al., 2009; Zirlewagen et al., 2016; Worthington and Smart, 2017; Sorensen et al., 2018), and urban runoff (Sharp, 2019). Such nonpoint-source pollution can have impacts on ecosystem health (e.g., as a consequence of eutrophication; Knight and Vick, 2018) as well as human health. Coastal carbonate aquifers are also vulnerable to saltwater intrusion resulting from groundwater pumping (Schoonmaker, 2007), inflow along drainage canals (Prinos et al., 2014), and sea-level rise (Langevin and Zygnerski, 2013). In addition, sea-level rise may cause water-table rises in coastal aquifers that would impair the functioning of septic systems, thereby increasing the potential for groundwater contamination (Miami-Dade County, 2018).

2.6 Challenges in monitoring and modeling

Because of pronounced heterogeneity and anisotropy, different strategies are necessary to delineate flowpaths in karst aquifers compared to nonkarst aquifers. The assumptions that the aquifer functions as an equivalent porous medium (EPM) and that groundwater basin divides coincide with topographic divides (Tóth, 1963) are problematic in karst aquifers, where as much as 99.7% of flow may occur through channels (Worthington, 1999; Ford and Williams, 2007). Conduit development can result in subsurface flow piracy and interbasin flow, and groundwater basin divides can migrate laterally depending upon flow conditions, with high-level overflow springs being activated during wet periods (Quinlan and Ewers, 1989; Hartmann et al., 2014). Consequently, the conventional strategy of installing monitoring wells to map the water table and infer flowpaths can give misleading results in karst aquifers (Quinlan and Ewers, 1989). Instead, flowpaths are typically delineated by injection of artificial tracers at inferred recharge points (e.g., sinkholes) and monitoring at discharge points (usually springs) (Fig. 2.7). Ideal tracers of groundwater flow have low background concentrations and low detection limits and are highly soluble, chemically conservative, and nontoxic (Goldscheider et al., 2008). Fluorescent dyes, which meet these criteria to a large extent, are commonly used, as are microscopic particulate tracers (e.g., latex microspheres and bacteriophages) to mimic transport of sediment and microbes (Goldscheider et al., 2008).

The accuracy and applicability of groundwater-flow models depend upon the validity of the underlying mechanistic assumptions. At the aquifer or basin scale, distributed-parameter models (with spatially explicit discretization of hydraulic parameters and system states) and lumped-parameter models (with values spatially averaged over the entire domain) are used (Hartmann et al., 2014). Distributed-parameter models include both single-continuum (EPM) and dual-continuum approaches. The EPM approach may be useful for regional water-resource assessment and management in carbonate aquifers, but it has difficulty simulating rates and directions of local groundwater flow (Scanlon et al., 2003; Ford and Williams, 2007; Hartmann et al., 2014; Green et al., 2019). The dual-continuum approach treats the conduit network and fractured matrix (or fracture network and porous matrix) as separate, overlapping domains, each with characteristic parameters and flow equations (Ford and Williams, 2007). Conduits can also be embedded as discrete elements within a matrix continuum (Shoemaker et al., 2008; Hartmann et al., 2014). Lumped-parameter models, which focus on dynamic input−output relationships, require continuously monitored discharge data (i.e., spring hydrographs) for karst aquifers (Ford and Williams, 2007; Hartmann et al., 2014). Distributed-parameter models are often limited by insufficient data, whereas lumped-parameter models are often too simplistic for predictive purposes (Hartmann et al., 2014). In modeling solute transport, breakthrough curves from tracer injections have been simulated by assuming one-dimensional advection with dispersion along conduits and solute exchange with matrix storage (Dewaide et al., 2016; Ender et al., 2018).

2.7 Conclusion

Because of the dynamic heterogeneity of carbonate rocks (particularly limestone), surface-water flow, vadose-zone hydrology, groundwater flow, water−rock interactions, and landscape evolution are more closely coupled than in any other lithologic setting. Consequently, the controls on groundwater flow and chemistry in carbonate aquifers are complex. Given their utilization for water supply in various parts of the world, their role in hosting distinctive and sensitive ecosystems, and their vulnerability to contamination, as well as the cultural significance of springs (e.g., for recreation; Knight and Vick, 2018), understanding how carbonate aquifers respond to environmental stresses is important. This is especially true where climate change is projected to cause increased temperatures and decreased precipitation (e.g., the Western Mediterranean Basin; Howell et al., 2019). Expanded monitoring of carbonate aquifers, integration of existing data into publicly accessible databases (Olarinoye et al., 2020), and large-scale application of mathematical models using multiple types of data (Hartmann et al., 2014) are all recommended for resource management.

References

Amiotte Suchet, P., Probst, J.-L., Ludwig, W., 2003. Worldwide distribution of continental rock lithology: implications for the atmospheric/soil CO_2 uptake by continental weathering and alkalinity river transport to the oceans. Global Biogeochem. Cycles 17 (2), 1038. Available from: https://doi.org/10.1029/2002GB001891.

Brahana, J.V., Thrailkill, J., Freeman, T., Ward, W.C., 1988. Carbonate rocks. In: Back, W., Rosenshein, J.S., Seaber, P.R. (Eds.), Hydrogeology: The Geology of North America, vol. O-2. Geological Society of America, Boulder, CO, pp. 333−352.

Buhmann, D., Dreybrodt, W., 1985a. The kinetics of calcite dissolution and precipitation in geologically relevant situations of karst areas, 1. Open system. Chem. Geol. 48, 189−211.

Buhmann, D., Dreybrodt, W., 1985b. The kinetics of calcite dissolution and precipitation in geologically relevant situations of karst areas, 2. Closed system. Chem. Geol. 53, 109−124.

Chen, Z., Auler, A.S., Bakalowicz, M., Drew, D., Griger, F., Hartmann, J., et al., 2017. The World Karst Aquifer Mapping project: concept, mapping procedure and map of Europe. Hydrogeol. J. 25, 771−785.

Covington, M.D., Vaughn, K.A., 2019. Carbon dioxide and dissolution rate dynamics within a karst underflow-overflow system, Savoy Experimental Watershed, Arkansas, USA. Chem. Geol. 527, 118689.

Dewaide, L., Bonniver, I., Rochez, G., Hallet, V., 2016. Solute transport in heterogeneous karst systems: dimensioning and estimation of the transport parameters via multi-sampling tracer-tests modelling using the OTIS (One-dimensional Transport with Inflow and Storage) program. J. Hydrol. 534, 567−578.

Ender, A., Goeppert, N., Goldscheider, N., 2018. Spatial resolution of transport parameters in a subtropical karst conduit system during dry and wet seasons. Hydrogeol. J. 26, 2241–2255.
Ford, D., Williams, P., 2007. Karst Hydrogeology and Geomorphology. John Wiley & Sons, Chichester, p. 562.
Goldscheider, N., Meiman, J., Pronk, M., Smart, C., 2008. Tracer tests in karst hydrogeology and speleology. Int. J. Speleol. 37 (1), 27–40.
Green, R.T., Winterle, J., Fratesi, B., 2019. Numerical groundwater models for Edwards Aquifer systems. In: Sharp Jr., J.M., Green, R.T., Schindel, G.M. (Eds.), The Edwards Aquifer: The Past, Present, and Future of a Vital Water Resource. Memoir 215. Geological Society of America, Boulder, CO, pp. 19–28.
Greenspan, M., Tschiegg, C.E., 1957. Speed of sound in water by a direct method. J. Res. Natl. Bur. Stand. 59 (4), 249–254.
Halihan, T., Sharp, J.M., Mace, R.E., 1999. Interpreting flow using permeability at multiple scales. In: Palmer, A.N., Palmer, M.V., Sasowsky, I.D. (Eds.), Karst Modeling. Karst Waters Institute, Charles Town, WV, pp. 82–96. Special Publication 5.
Hanshaw, B.B., Back, W., 1979. Major geochemical processes in the evolution of carbonate aquifer systems. J. Hydrol. 43, 287–312.
Hartmann, N., Goldscheider, N., Wagener, T., Lange, J., Weiler, M., 2014. Karst water resources in a changing world: review of hydrological modeling approaches. Rev. Geophys. 52, 218–242.
Hess, J.W., White, W.B., 1989. Chemical hydrology. In: White, W.B., White, E.L. (Eds.), Karst Hydrology: Concepts from the Mammoth Cave Area. Van Nostrand Reinhold, New York, pp. 145–174.
Hoefner, M.L., Fogler, H.S., 1988. Pore evolution and channel formation during flow and reaction in porous media. AIChE J. 34 (1), 45–54.
Howell, B.A., Fryar, A.E., Benaabidate, L., Bouchaou, L., Farhaoui, M., 2019. Variable responses of karst springs to recharge in the Middle Atlas region of Morocco. Hydrogeol. J. 27, 1693–1710.
Knight, R., Vick, H., 2018. Florida Springs Conservation Plan. Howard T. Odom Florida Springs Institute, High Springs, FL. Available from: <https://floridaspringsinstitute.org/wp-content/uploads/2018/11/Springs-Conservation-Plan-final-draft-FINAL.pdf> (accessed 28.03.20.).
Katz, B.G., Böhlke, J.K., Hornsby, H.D., 2001. Timescales for nitrate contamination of spring waters, northern Florida, USA. Chem. Geol. 179, 167–186.
Katz, B.G., Griffin, D.W., Davis, J.H., 2009. Groundwater quality impacts from the land application of treated municipal wastewater in a large karstic spring basin: chemical and microbiological indicators. Sci. Total Environ. 407, 2872–2886.
Krejca, J., Reddell, J., 2019. Biology and ecology of the Edwards Aquifer. In: Sharp Jr., J.M., Green, R.T., Schindel, G.M. (Eds.), The Edwards Aquifer: The Past, Present, and Future of a Vital Water Resource. Memoir 215. Geological Society of America, Boulder, CO, pp. 159–169.
Kresic, N., 2013. Water in Karst: Management, Vulnerability, and Restoration. McGraw-Hill, New York, 708 p.
Langevin, C.D., Zygnerski, M., 2013. Effect of sea-level rise on saltwater intrusion near a coastal well field in southeastern Florida. Groundwater 51, 781–803.
Luhmann, A.J., Covington, M.D., Peters, A.J., Alexander, S.C., Anger, C.T., Green, J.A., et al., 2011. Classification of thermal patterns at karst springs and cave streams. Groundwater 49, 324–335.
Machel, H.G., 2004. Concepts and models of dolomitization: a critical reappraisal. In: Braithwaite, C.J.R., Rizzi, G., Darke, G. (Eds.), Geology and Petrogenesis of Dolomite Hydrocarbon Reservoirs. Geological Society, London, pp. 7–63. Special Publication 235.
Martin, J.B., 2017. Carbonate minerals in the global carbon cycle. Chem. Geol. 449, 58–72.
Miami-Dade County, 2018. Septic Systems Vulnerable to Sea-Level Rise. Final Report in Support of Resolution No. R-911-16. Miami-Dade County Department of Regulatory and Economic Services, Miami-Dade County Water and Sewer Department, and Florida Department of Health. Available from: <https://www.miamidade.gov/green/library/vulnerability-septic-systems-sea-level-rise.pdf> (accessed 29.03.20.).
Olarinoye, T., et al., 2020. Global karst springs hydrograph dataset for research and management of the world's fastest-flowing groundwater. Sci. Data 7, 59. Available from: https://doi.org/10.1038/s41597-019-0346-5.
Palmer, A.N., 1981. Hydrochemical factors in the origin of limestone caves. In: Proceedings 8th International Congress of Speleology, Bowling Green, KY, pp. 120–122.
Palmer, A.N., 1991. Origin and morphology of limestone caves. Geol. Soc. Am. Bull. 103, 1–21.
Palmer, A.N., Palmer, M.V., 2009. Geologic overview. In: Palmer, A.N., Palmer, M.V. (Eds.), Caves and Karst of the USA. National Speleological Society, Huntsville, AL, pp. 1–16.
Panno, S.V., Hackley, K.C., Hwang, H.H., Kelly, W.R., 2001. Determination of the sources of nitrate contamination in karst springs using isotopic and chemical indicators. Chem. Geol. 179, 113–128.
Payne, S., Pence, N., Furl, C., 2019. The Edwards Aquifer Habitat Conservation Plan: its planning and implementation. In: Sharp Jr., J.M., Green, R.T., Schindel, G.M. (Eds.), The Edwards Aquifer: The Past, Present, and Future of a Vital Water Resource. Memoir 215. Geological Society of America, Boulder, CO, pp. 199–206.
Plummer, L.N., Wigley, T.M.L., Parkhurst, D.L., 1978. The kinetics of calcite dissolution in CO_2-water systems at 5° to 60°C and 0.0 to 1.0 atm CO_2. Am. J. Sci. 278, 179–216.
Plummer, L.N., Busby, J.F., Lee, R.W., Hanshaw, B.B., 1990. Geochemical modeling of the Madison aquifer in parts of Montana, Wyoming, and South Dakota. Water Resour. Res. 26, 1981–2014.
Prinos, S.T., Wacker, M.A., Cunningham, K.J., Fitterman, D.V., 2014. Origins and delineation of saltwater intrusion in the Biscayne aquifer and changes in the distribution of saltwater in Miami-Dade County, Florida. In: U.S. Geological Survey Scientific Investigations Report 2014–5025.
Quinlan, J.F., Ewers, R.O., 1989. Subsurface drainage in the Mammoth Cave area. In: White, W.B., White, E.L. (Eds.), Karst Hydrology: Concepts From the Mammoth Cave Area. Van Nostrand Reinhold, New York, pp. 65–103.
Ryan, M., Meiman, J., 1996. An examination of short-term variations in water quality at a karst spring in Kentucky. Groundwater 34, 23–30.

Scanlon, B.R., Mace, R.E., Barrett, M.E., Smith, B., 2003. Can we simulate regional groundwater flow in a karst system using equivalent porous media models? Case study, Barton Springs Edwards Aquifer, USA. J. Hydrol. 276, 137–158.

Schoonmaker, D., 2007. The pump don't work. Am. Sci. 95, 315–316.

Sharp Jr., J.M., 2019. Effects of urbanization on the Edwards Aquifer. In: Sharp Jr., J.M., Green, R.T., Schindel, G.M. (Eds.), The Edwards Aquifer: The Past, Present, and Future of a Vital Water Resource. Memoir 215. Geological Society of America, Boulder, CO, pp. 213–222.

Shoemaker, W.B., Kuniansky, E.L., Birk, S., Bauer, S., Swain, E.D., 2008. Chapter A24: Documentation of a conduit-flow process (CFP) for MODFLOW-2005. In: U.S. Geological Survey Techniques and Methods, Book 6.

Sorensen, J.P.R., Vivanco, A., Ascott, M.J., Gooddy, D.C., Lapworth, D.J., Read, D.S., et al., 2018. Online fluorescence spectroscopy for the real-time evaluation of the microbial quality of drinking water. Water Res. 137, 301–309.

Stevanović, Z., 2018. Global distribution and use of water from karst aquifers. In: Parise, M., Gabrovsek, F., Kaufmann, G., Ravbar, N. (Eds.), Advances in Karst Research: Theory, Fieldwork and Applications. Geological Society, London, pp. 217–236. Special Publication 466.

Sullivan, P.L., Macpherson, G.L., Martin, J.B., Price, R.M., 2019. Evolution of carbonate and karst critical zones. Chem. Geol. 527, 119223.

Tihansky, A.B., 1999. Sinkholes, west-central Florida: a link between surface water and ground water. In: Galloway, D., Jones, D.R., Ingebritsen, S.E. (Eds.), Land Subsidence in the United States. U.S. Geological Survey Circular 1182, pp. 121–140.

Tóth, J., 1963. A theoretical analysis of groundwater flow in small drainage basins. J. Geophys. Res. 68, 4795–4812.

U.S. Fish and Wildlife Service, 2018. Devils Hole Pupfish. Available from: <https://www.fws.gov/nevada/protected_species/fish/species/dhp/dhp.html> (accessed 24.03.20.).

U.S. Geological Survey, 2012. Devils Hole, Nevada—a primer. In: Fact Sheet 2012-3021. 6 p. Available from: <https://pubs.usgs.gov/fs/2012/3021/pdf/fs2012-3021.pdf> (accessed 24.03.20.).

U.S. Geological Survey, n.d. Floridan aquifer system groundwater availability study: introduction. Available from: <https://fl.water.usgs.gov/floridan/intro.html> (accessed 19.03.20.).

Vesper, D.J., Loop, C.M., White, W.B., 2000. Contaminant transport in karst aquifers. Theor. Appl. Karstol. 13, 63–73.

White, W.B., 1988. Geomorphology and Hydrology of Karst Terrains. Oxford University Press, New York, p. 464.

White, W.B., 1989. Introduction to the karst hydrology of the Mammoth Cave area. In: White, W.B., White, E.L. (Eds.), Karst Hydrology: Concepts From the Mammoth Cave Area. Van Nostrand Reinhold, New York, pp. 145–174.

White, W.B., 2002. Karst hydrology: recent developments and open questions. Eng. Geol. 65, 85–105.

White, W.B., Culver, D.C., Herman, J.S., Kane, T.C., Mylroie, J.E., 1995. Karst lands. Am. Sci. 450–459.

Worthington, S.R.H., 1999. A comprehensive strategy for understanding flow in carbonate aquifers. In: Palmer, A.N., Palmer, M.V., Sasowsky, I.D. (Eds.), Karst Modeling. Karst Waters Institute, Charles Town, WV, pp. 30–37. Special Publication 5.

Worthington, S.R.H., Smart, C.C., 2017. Transient bacterial contamination of the dual-porosity aquifer at Walkerton, Ontario, Canada. Hydrogeol. J. 25, 1003–1016.

Zirlewagen, J., Licha, T., Schiperski, F., Nödler, K., Scheytt, T., 2016. Use of two artificial sweeteners, cyclamate and acesulfame, to identify and quantify wastewater contributions in a karst spring. Sci. Total Environ. 547, 356–365.

Chapter 3

Groundwater resources in Australia—their occurrence, management, and future challenges

Steve Barnett[1], Craig T. Simmons[2] and Rebecca Nelson[3]

[1]Department of Environment, Water and Natural Resources, Adelaide, SA, Australia, [2]National Centre for Groundwater Research and Training, College of Science and Engineering, Flinders University, Adelaide, SA, Australia, [3]Melbourne Law School, University of Melbourne, Melbourne, VIC, Australia

3.1 Introduction

Although Australia lies across several climatic zones ranging from tropical in the north to temperate in the south, an arid climate prevails over much of the interior of the country, with some areas recording below 100 mm annual rainfall and annual evaporation exceeding 4000 mm. Australia has therefore the lowest rainfall of any inhabited continent on Earth. Most of the highly urbanized population of 23.4 million resides in areas of higher rainfall, which mainly occurs due to topographic effects along the east coast, and in the southwestern and southeastern regions of the country (Fig. 3.1).

Because of very limited surface-water resources, groundwater is one of Australia's most significant natural resources and over 60% of the continent's total area is the only reliable and cost-effective source of water available, supplying urban areas, agriculture, industry, and mining developments. More than 30% of Australia's total water consumption is provided by groundwater, which generates economic activity worth in excess of $A34 billion a year (Marsden Jacob Associates, 2012).

This chapter provides an overview of the groundwater resources in Australia and traces the history of their development. Although many countries have severely depleted groundwater resources (Mukherjee et al., 2020), the evolution and timely implementation of the current management approaches have prevented this from happening in Australia. Future challenges are also discussed in this chapter.

3.2 Groundwater resources in Australia

Groundwater occurs in both sedimentary and fractured rock aquifers, although most extraction occurs from the higher yielding sedimentary aquifers, which cover about 65% of the continent. In the higher rainfall highland areas along the eastern coastline, shallow unconsolidated alluvial sediments contain good quality groundwater; however, further inland in drier areas, these sediments contain more saline groundwater due to higher evaporation and lower rainfall.

Aquifers within large sedimentary basins contain significant volumes of groundwater in consolidated sandstones and limestones. Aeolian sediments and karstic limestone formations also form important aquifers.

Fig. 3.2 presents the major groundwater resources in Australia, which include:

- the Great Artesian Basin (GAB), which covers 20% of the continent;
- the major alluvial aquifers of the Murray Basin, which produce much of Australia's food;
- the Otway Basin aquifers of southeast South Australia and Southwest Victoria;
- the Perth Basin, which supplies most of Perth's water demands;
- the Canning Basin in northern Western Australia; and
- the Daly River Basin of the Northern Territory.

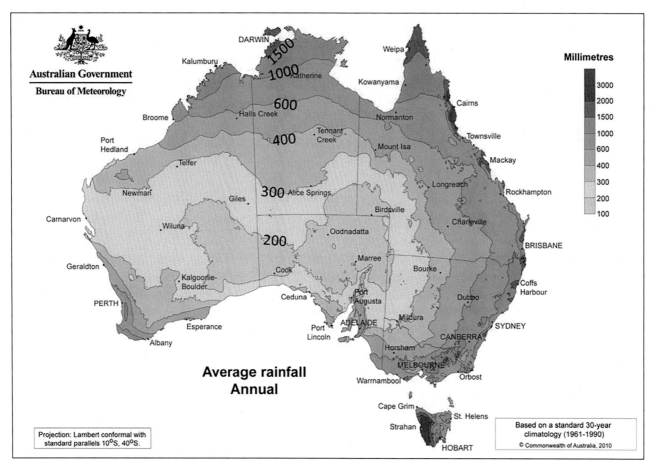

FIGURE 3.1 Average annual rainfall of Australia. *Courtesy Bureau of Meteorology.*

Fractured rock aquifers occur throughout much of the Great Dividing Range of Eastern Australia, Tasmania, the Mt Lofty and Flinders Ranges in South Australia, and the ancient hills and ranges of Western Australia and the Northern Territory (Fig. 3.2). Due to the variability of water-bearing joints and fractures in these aquifers, well yields and salinities can vary greatly over short distances. Generally, lower salinity groundwater is found in regions with higher rainfall where recharge can occur more readily. Sandstone, quartzite, siltstone, and basalt aquifers tend to have higher yields because the joints and fractures are open and permeable, unlike metamorphic and intrusive igneous rock aquifers where they are often poorly developed.

3.3 Historical development of groundwater

It is certain that near-surface groundwater has been accessed in Australia by aborigines for thousands of years according to the artifact evidence found adjacent to shallow wells and springs in several locations (Australian Water Resources Council, 1975). However in a global historical context, the colonization and development of Australia is a relatively recent process. The first European settlements in the late 18th century were located adjacent to surface-water resources, but shallow hand-dug wells were also used when surface water was unavailable during drought, or of unsuitable quality due to contamination. These original colonies were separated by vast distances and their isolation led to the development of strongly independent "nation" states who later initiated management of their own water resources. As settlement spread into the drier interior from the well-watered coast, groundwater became increasingly important as a reliable source of water. In 1857 the first known investigation into groundwater resources was carried out in the state of Victoria, and in 1871 an artesian well was drilled to a depth of 52 m near Perth in Western Australia (AWRC, 1975).

The first well drilled in the iconic GAB was in 1879 in north-western New South Wales. This basin is the largest in the world, covering 1.7 million km^2 and underlying 20% of the Australian land mass. Additional wells were drilled in

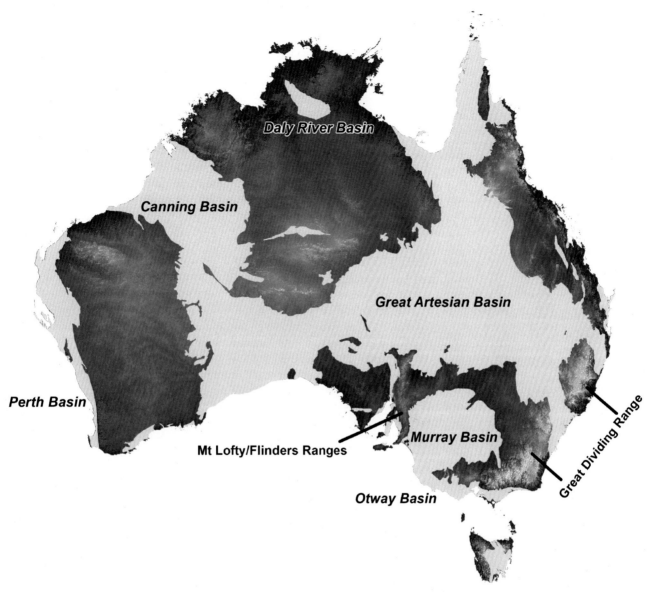

FIGURE 3.2 Major groundwater resources of Australia.

adjacent states over the next few years, with over 1500 artesian wells being drilled throughout the GAB by 1910. Concerns about marked reductions in flows in many wells and that extraction from the Basin may be unsustainable led to a series of Interstate Conferences on Artesian Water, which occurred between 1912 and 1928. These conferences were very beneficial in clarifying many of the issues and instigated the systematic collection and interpretation of data from other artesian basins around the country. They also recognized the need for the controlled development of artesian groundwater resources into the future. Subsequently, over 30,000 wells have been drilled in the GAB, which have been essential for the expansion of the pastoral and mining industries into the dry interior of Australia. In response to gradually declining pressure levels, an extensive program has been undertaken to cap uncontrolled flowing wells.

After the World War II, more thorough investigations of groundwater resources occurred through the increased use of the rapid rotary drilling technique and downhole geophysical logging tools that could detect permeable layers. Demand for groundwater increased in the 1950s and 1960s to support agricultural development in the following regions:

- Burdekin Delta—alluvial sediments deposited by the Burdekin River in northern Queensland;
- Lachlan Valley—alluvial sediments in tributary valleys in the New South Wales part of the Murray Basin;

- Western Port Basin—Tertiary sands, limestones, and volcanics bordering the coastline in Victoria; and
- Northern Adelaide Plains—Tertiary limestones of the St Vincent Basin in South Australia.

In order to initiate the huge task of assessing Australia's water resources, the federal and state governments formed the Australian Water Resources Council in 1962. Although now no longer in existence, the council was responsible for a significant increase in investment into the investigation of major groundwater resources, which continued through the 1980s and 1990s. More recently, the focus has changed since 2000 to the monitoring and management of the developed groundwater resources where groundwater extractions for irrigation purposes have generally stabilized.

3.4 Evolution of groundwater management

Originally, the state colonies of Australia inherited the British doctrine of "rule of capture," which generally regulated early groundwater development, whereby landowners have the right to take all the water they can capture from under their property. While this approach was acceptable during the days of limited means to extract groundwater, it was later found to be inappropriate in areas of high groundwater development.

During the rapid groundwater development in the 1960s and 1970s, extractions reached unsustainable levels in some regions, which prompted the enactment of the first legislation which enabled management of the groundwater resources in specific areas. During this period, inadequate resources for the management of several groundwater systems, and occasionally some political interference, led to too many licenses being issued and agricultural users being granted allocations in excess of the sustainable extraction limits (overallocation) and in some cases, the volumes of groundwater extracted were also above the sustainable extraction limits (overuse) (National Water Commission, 2012).

The World Commission on Environment and Development (1987) released the Brundtland report, which outlined the principle of sustainable development. The working plan for action (Agenda 21) was ratified in Rio de Janeiro in 1992 and was agreed by Australia. Central to this plan are the principles of ecologically sustainable development, which underpin the current management approaches for groundwater resources in Australia.

The foundation of the current management approach is groundbreaking legislation that replaced the long-established English doctrine of "riparian rights" for surface water and also rejected the Western United States' doctrine of "prior appropriation" (first in time, first in right). This legislation (the *Irrigation Act*) was passed in the state of Victoria in 1886 and vested the right to the use, flow, and the control of water in any watercourse exclusively to the state, so that the cardinal rights of the state took precedence over the riparian rights of the individual. The act also instituted a system whereby the state would administer the granting of water rights to water users.

The development of formal national-level policy about water began with the 1994 National Water Reform Framework Agreement [Council of Australian Governments (CoAG), 1994], motivated chiefly by principles of efficient competition and sustainable use. The framework agreement sought to reform water pricing, separate land title from the right to use water, ensure that the environment received allocations of water, deal with overallocation of water, establish systems for water trading and increase public consultation. This agreement marked the first notable federal government involvement in water resource management issues, for which the states had primary responsibility. The framework agreement also noted the need to consider management arrangements for groundwater. A subsequent national-level report provided "advice" to the states and territories on groundwater management, including the concept of "sustainable yield," in 1996 [Agriculture and Resource Management Council of Australia and New Zealand and Standing Committee on Agriculture and Resource Management (ARMCANZ and SCARM), 1996].

The National Water Initiative (NWI), which all the federal and state governments signed between 2004 and 2006, aimed to build on the framework agreement. A central pillar of the NWI was the concept of legally binding water management plans, which would guide the granting and management of water rights. The plans constituted a move away from the traditional case-by-case assessment of the impacts of granting an individual application to take groundwater, toward a regional view of resource sustainability. A key function of the plans was to control the aggregate impacts of groundwater extraction through volumetric limits on licensed withdrawals (Gardner et al., 2018).

3.5 Current groundwater usage

Australia has undergone a rapid development of the use of groundwater resources over the past 40–50 years. This was caused by a rise in population, and also the development of groundwater resources by irrigators wanting to replace surface-water sources that had become depleted and more tightly managed (Harrington and Cook, 2014). Fig. 3.3 shows that Western Australia, the Northern Territory, and South Australia have the most reliance on groundwater. The highest

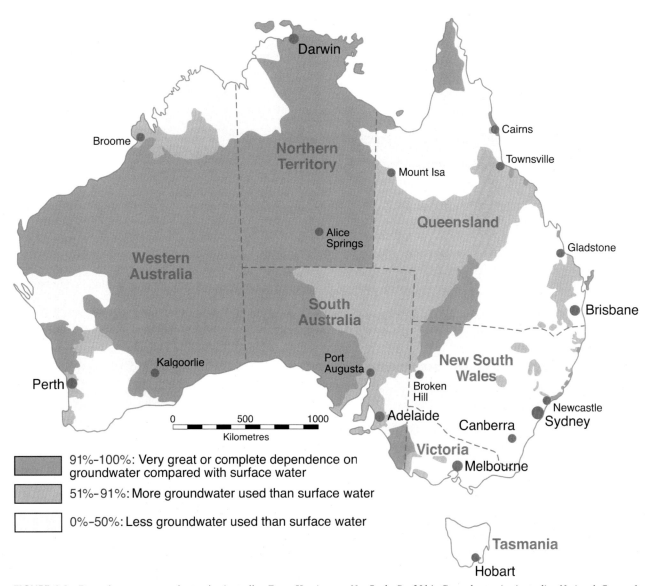

FIGURE 3.3 Dependence on groundwater in Australia. *From Harrington, N., Cook, P., 2014. Groundwater in Australia. National Centre for Groundwater Research and Training.*

occurrence of groundwater use occurs in the Murray Basin, primarily to support irrigated agriculture. Accurate estimates of groundwater abstraction in Australia have been difficult to obtain in the past due to limited monitoring infrastructure and different water accounting methods. However in recent years, more consistent national reporting procedures required by the National Water Initiative have resulted in more accurate estimates.

For the 2017/18 water use year an estimated 5250 GL was extracted, which represents about 60% of the total volume of entitlements (Bureau of Meteorology, 2018). This volume does not include stock and domestic use but amounts to approximately 30% of Australia's total water consumption. Because of the generally dry climate experienced over most of Australia, 70% of groundwater extraction is for irrigation. The mining industry in Australia is estimated to extract an additional 1500 GL/year.

A major factor controlling groundwater use and development in Australia is the salinity of the groundwater resources, which varies widely in response to the differing climate zones. In unconfined aquifers, low-salinity groundwater that can be used for human consumption is generally found in higher rainfall areas around the northern and eastern coasts, as well as the southeastern and southwestern portions of the continent (Fig. 3.1). Older low-salinity groundwater may be found in deeper confined aquifers beneath arid areas after traveling along very long flow paths, for example, in the Great Artesian and Murray–Darling Basin (Harrington and Cook, 2014). These "ancient" resources, sometimes over 1 million years old in the GAB, provide essential supplies for towns, livestock and irrigation, were recharged thousands of years ago during climates that were much wetter than are being experienced today. Over large areas of the arid interior where evaporation considerably exceeds rainfall, high salinities occur that are too high for drinking or agricultural purposes. However, groundwater over 200,000 mg/L sourced from Tertiary palaeochannels is used by the gold mining industry in Western Australia for mineral processing.

3.6 Groundwater management issues

3.6.1 Overuse and overallocation of groundwater

The impacts of groundwater overuse are widely recognized (unacceptable declines in groundwater levels, increased pumping costs, seawater intrusion; and increased salinity of aquifers through inter-aquifer leakage).

The National Water Initiative committed the states to address these issues. In 2018 there were 279 groundwater management areas (GMAs) throughout Australia. Of these, 127 GMAs had volumetric limits for extraction, with 27% of these classified as overallocated, but more importantly, only 6% considered overused. While some of the GMAs that do not have volumetric limits could be considered "at risk" or "overallocated," the management plans generally have rules that keep extractions at sustainable levels (Bureau of Meteorology, 2018). Despite the uncertainties in determining extraction volumes in some GMAs, it can be concluded that overextraction is not a widespread problem in Australia, although there may be localized issues in some areas.

3.6.2 Groundwater-dependent ecosystems

Throughout Australia, groundwater supports a number of ecosystems. Clifton et al. (2007) describes three main types of these groundwater-dependent ecosystems (GDEs):

- ecosystems reliant on groundwater discharging at the surface such as springs and wetlands,
- ecosystems such as vegetation that access groundwater through root systems, and
- ecosystems such as stygofauna that occur within the aquifer pore spaces and solution features.

Several systems have been developed to assist in the location of GDEs and the assessment of their water dependency (Clifton et al., 2007; Richardson et al., 2011a,b). As most groundwater management plans in Australia must take into account the water requirements of GDEs, a key advance has been the development of a web-based GDE Atlas for Australia (http://www.bom.gov.au/water/groundwater/gde/map.shtml). This atlas has collated ecological and hydrogeological information on GDEs from a number of sources across Australia, including published research and remote sensing data.

3.6.3 Impacts of groundwater extraction on surface-water systems

Although the interconnection between surface-water and groundwater resources has long been recognized by hydrologists and hydrogeologists, a study found that in some of Australia's catchments that have high groundwater extraction, reductions in groundwater discharge to rivers and streams have caused reduced in surface-water flows or complete drying out of streams (Evans, 2007).

There has been limited recognition of this impact of groundwater abstraction in policy and management frameworks. Previously, there has been separate management of groundwater and surface-water resources; however, when usage from many surface-water resources was restricted due to low rainfall and increases in demand, the extraction was transferred to nearby groundwater resources.

An important case study of this issue is the Murray−Darling Basin, where there is often a strong interaction between surface-water and the alluvial aquifers. In 2008, 16% of total water use in the basin came from groundwater, but this is expected to increase to over 25% by 2030 under the existing management arrangements that restrict surface-water use. It was estimated that 25% of the groundwater extraction would ultimately be derived from induced streamflow leakage (CSIRO, 2008). An important and sometimes controversial instance of government involvement in achieving effective conjunctive management of groundwater resources and surface water is the development of the Murray−Darling Basin Plan.

3.6.4 Effect of climate change on groundwater resources

Analysis of 80-year climate records has shown warming over most of Australia: an increase in rainfall over Central, Northern, and North Western Australia and a decrease in rainfall in Eastern, South Eastern, and South Western Australia (Barron et al., 2011). Large areas of Southern Australia, particularly the southern Murray−Darling Basin, endured the prolonged millennium drought from 1997 to 2009, which was the most severe drought in recorded rainfall history extending over 110 years.

Most global climate models predict that Southern Australia is likely to experience drier conditions in the future, which has important implications for the future sustainability of groundwater resources. This is because the magnitude of changes in groundwater recharge can be two to four times greater than the magnitude of changes in rainfall, with the impact being very significant in areas of low recharge (Barron et al., 2011). Sixteen global climate models were used to make predictions of future changes in recharge and the results have been scaled according to three global warming scenarios (low, medium, and high) for both 2030 and 2050 (Barron et al., 2011). These predictions show a reduction in diffuse recharge across most of the west, center, and southeast of Australia, and an increase across Northern Australia.

As well as a reduction in the availability of water, a warmer and drier climate may increase the demand for water resources from irrigated agriculture, urban populations, wetlands, and other water-dependent ecosystems. From a groundwater perspective, given the uncertainties in how and when climate change will manifest itself, an adaptive and flexible management approach is essential to deal with these changes, together with comprehensive monitoring of the condition of the groundwater resources.

3.6.5 Impacts of mining on groundwater resources

Mining is a large industrial user of water that has a high gross value per gigaliter of water consumed compared with irrigation. Improvements in water use efficiency in the industry have resulted in water use being relatively stable up until recently, despite exponential increases in production. However, possible underreporting of water use may have also contributed (Prosser et al., 2011). Most mining developments occur in arid or semiarid regions where groundwater is often saline and there are few competing users. However because mining is occurring more often in agricultural areas, the competition for water resources and the potential impacts on existing water users is becoming more controversial.

The mining industry generally builds its own infrastructure, and so water requirements tend to be considered as part of the mining approval process and in remote arid areas, these are not always included in a licensing regime under a water management plan that would normally include other agricultural users. Exceptions are the states of New South Wales and Victoria, which have state-wide licensing requirements, and in GMAs in other states where all extraction for mining is included under the water planning process and requires a water access entitlement (Harrington and Cook, 2014).

Queensland and New South Wales have experienced a major expansion in coal seam gas (CSG) developments, which presents major challenges in understanding and managing impacts of mining on other water users and the environment (Comino et al., 2014). Extraordinary CSG production from areas previously considered to be economically nonviable has been triggered by new technology to extract methane from deep coal beds. Over 1000 gas production wells are in operation, requiring a peak extraction of about 95 GL to lower groundwater pressures, which will allow the gas to be released from the coal deposits (Department of Natural Resources and Mines, 2016).

The current boom in the production of CSG has highlighted some key water management challenges, namely, (1) the effect of depressurisation on surrounding aquifers, (2) the likelihood and impacts of inter-aquifer leakage caused by aquifer depressurisation and hydraulic fracturing, and (3) chemical processes affecting the quality and safe disposal of the released water (Prosser et al., 2011). In Queensland, the possible interaction of CSG developments with usable aquifers (e.g., the GAB) that can occur above or below the coal seams is a matter of concern. The locations of the potential development areas are shown in Fig. 3.4.

A fundamental issue is the uncertainty regarding the cumulative regional impacts of multiple developments on groundwater levels and pressures, and inter-aquifer leakage. Groundwater flow rates are slow in many of the aquifers of interest, and any unforeseen impacts of the mining process could take decades or centuries to appear. Groundwater flow models are essential for this kind of analysis and require a good characterization of the basin geology and groundwater flow systems (Prosser et al., 2011).

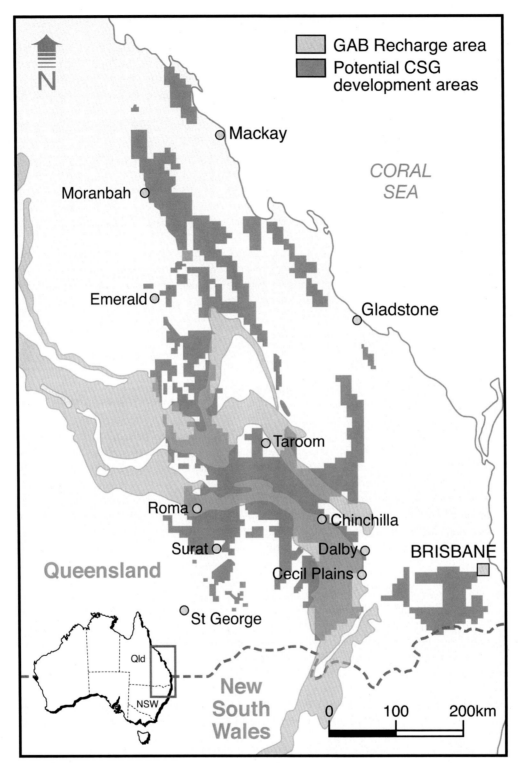

FIGURE 3.4 Potential coal seam gas developments in Queensland. *From Harrington, N., Cook, P., 2014. Groundwater in Australia. National Centre for Groundwater Research and Training.*

3.6.6 Land and groundwater salinization

The salinization of land and groundwater resources is the major Australian natural resource management issue. Salt occurs naturally in the Australian landscape and originates mainly from oceanic salt deposited by rain and wind. Over thousands of years, this salt has been concentrated in soil water by high evaporation rates and transpiration by plants. Following European settlement, wide-ranging changes in land use have changed the water balance, which has resulted in mobilization of the naturally occurring salt, causing salinization of land and water resources. These land use changes involve the clearing of native vegetation for dryland farming systems (dryland salinity) and the establishment of large irrigation areas (irrigation salinity).

Dryland salinity occurs when deep-rooted native vegetation is replaced with shallow-rooted annual crops and pastures, resulting in a significant increase in groundwater recharge rates, which can cause a rise in the watertable, which transports salts up into the root zones of plants and often to the surface where evaporation concentration occurs, impacting upon vegetation and rural infrastructure, including buildings, roads, and pipes. Watertable rises can also cause an increase in hydraulic gradient toward surface-water bodies, which can increase the discharge of saline groundwater into rivers and streams, thereby increasing river salinity. The National Land and Water Resources Audit investigated the distribution and impacts of dryland salinity across Australia, with an estimation of 5.7 million hectares being affected or having a high potential for the future development of dryland salinity (National Land and Water Resources Audit, 2001). The National Dryland Salinity Program, which operated between 1993 and 2004, invested $A40 million on research into the causes, costs, consequences, possible solutions, and management of dryland salinity in Australia.

Since this audit, the previously discussed millennium drought from 1997 to 2009 affected large areas of Southern Australia, reducing recharge and causing widespread declines in watertable levels, which has reduced the risk of dryland salinity in many areas. It remains to be seen whether the predicted drying impacts of climate change will also reduce the risk of expansion of the current areas of dryland salinity.

Irrigation salinity (or recycling) occurs when groundwater is extracted from a shallow aquifer and is applied to a crop by irrigation, and as water is absorbed through the root system, much of the dissolved salts are not taken up by the plant and consequently, accumulate in the root zone. This salt is then flushed back down into the shallow aquifer during subsequent irrigation applications or by rainfall recharge. This process results in rising groundwater salinity over time due to the recycling of the irrigation drainage water.

3.6.7 Seawater intrusion

Seawater intrusion can be defined as the landward movement of seawater into fresh coastal aquifers in response to changes in the groundwater regime, such as extraction, reductions in recharge and sea-level rise. The threat of seawater intrusion has increased in Australia due to increased development of coastal aquifers, caused by increasing demand from coastal populations, and drier climates (Werner, 2010).

An assessment of coastal aquifers on a national-scale (Ivkovic et al., 2012) identified those groundwater resources that are most vulnerable to seawater intrusion, taking into account the future consequences of overextraction, sea-level rise, and climate change. The risk has been identified to be highest in Queensland, although smaller areas of Victoria, South Australia, and Western Australia have also been identified as being at risk (Ivkovic et al., 2012). To date, targeted investigations into seawater intrusion have been mostly limited to the highly developed groundwater resources for agricultural production in Queensland, and important aquifers used for public water supplies.

3.7 Future challenges

Australia's future water requirements will increasingly depend on groundwater. It is expected that climate change and a potential doubling of the population in the next 50 years will be important future drivers of demand, which will place further demand on already fully developed groundwater resources (Simmons, 2016). Although Australia has made solid gains with water reform over the past decade or so, with advances in groundwater science, education, management, and policy reform (National Water Commission, 2014), the attention given by policy makers to groundwater management is declining at an alarming rate, especially at a time when there are many nationally important groundwater issues described previously in this chapter that must be dealt with. The universal "hydro-illogical cycle" (Wilhite, 2012), whereby political attention and funding rains down during droughts but evaporates when water in plentiful, is being implemented in Australia.

There must be a clear understanding that droughts are a fundamental component of the Australian climate. An ongoing and proactive approach to water reform in Australia is urgently needed, which rejects short-term political and drought-driven interventions. The benefits for current and future generations will be enormous. For current purposes, this section will now focus on two emerging responses, managed aquifer recharge (MAR) and general resourcing and education.

3.7.1 Managed aquifer recharge

In order to meet future demand, Australia needs to diversify its water sources. MAR is a growing alternative that may be available to water resource managers in some circumstances. MAR is the deliberate recharge of water to aquifers using injection wells, infiltration basins or galleries, using water sourced from stormwater, rainwater, recycled effluent, mains water, or water from other aquifers. With suitable levels of treatment before recharge and after extraction, the recovered water may then be used for potable water supplies, irrigation or industrial water. Fig. 3.5 shows a simple MAR operation using a confined aquifer.

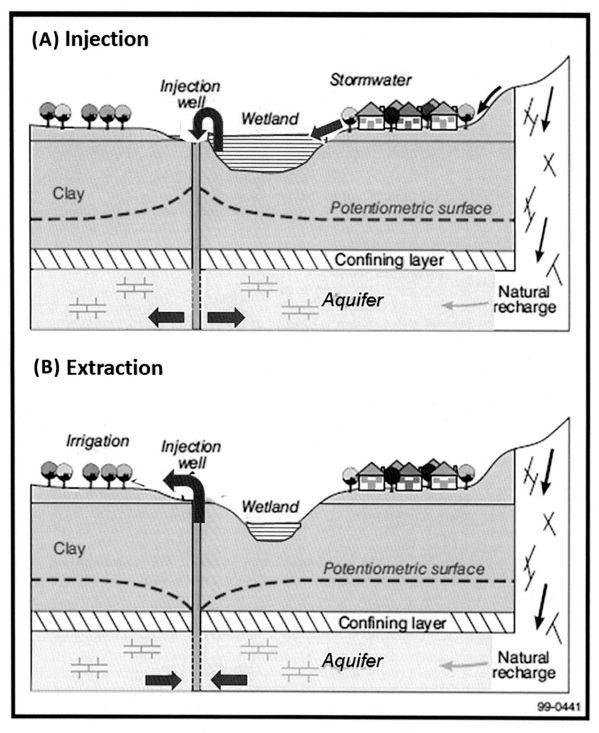

FIGURE 3.5 A simple managed aquifer recharge scheme (A) the injection phase (B) the extraction phase.

MAR has several advantages: it is far cheaper than dam construction and covers far less land area, minimizes evaporation losses, and can make use of the natural attenuation properties of aquifer materials.

Dillon et al. (2009) provided a comprehensive overview of MAR. The number of MAR projects in Australia has increased in recent years, especially since the millennium drought. In 2016 five states had operational MAR projects that contributed 410 GL/year to water supplies across Queensland, South Australia, Western Australia, and the Northern Territory. The Western Australian Water Corporation has completed a trial of injecting more than 2.5 GL of highly treated recycled water into a deep confined aquifer over three years and it has been estimated that by 2023, groundwater recharge of 28 GL/year could be possible during winter for subsequent extraction during the dry summer months for potable supplies.

There are some issues that need to be recognized and managed in the development of MAR schemes. These include water entitlement issues in water management plans (who owns the water), ensuring appropriate treatment of the source water (normally stormwater or recycled water) to prevent any health risk or environmental degradation, including making an aquifer unusable for certain purposes (e.g., drinking or irrigation), and also to prevent water quality changes or a reduction in aquifer permeability. The National Water Quality Management Strategy provides guidance for ensuring that MAR projects protect human and environmental health. Specific guidelines for MAR were developed in 2009 and form an integral part of the Australian guidelines for water recycling along with other relevant guidelines for end uses of recycled water (Natural Resource Management Ministerial Council, Environmental Protection and Heritage Council and National Health and Medical Research Council, 2009).

3.7.2 Declining resources for understanding and managing groundwater

Despite being required as part of the National Water Initiative, the full cost recovery of management costs from water users has not been determinedly pursued by most states, with the result that the funding available for the management and monitoring of groundwater resources is declining. This trend will increase the risk of inadequate monitoring being carried out, leading to a general decline in capacity and capability in management agencies, and the non-replacement of aging monitoring infrastructure (Simmons, 2016).

In addition to addressing this decline in resources, there are a range of additional educational options that would enhance better understanding and management of groundwater in order to break the "hydro-illogical cycle" and initiate further reform. These include:

- continuing to increase the profile and awareness of groundwater issues in Australia to politicians, policy makers, and water users to combat ignorance and misunderstanding;
- establishing a national groundwater policy and planning forum where policy makers, water resource managers, and industry and research scientists work together to define and solve important issues relevant to groundwater policy, management and technology;
- integrating all of the essential interdisciplinary subjects that relate to hydrogeology such as climate science, ecology, socioeconomics, public policy, and law; and
- quantifying and reducing uncertainty in analyses and explaining confidence levels to policy makers.

3.8 Conclusion

The impacts of climate change and an increasing population will result in Australia's future water requirements being more dependent on groundwater resources. Despite solid progress being made with groundwater science, education, management, and policy reform, there are many challenges ahead. Awareness of the importance of groundwater needs to be raised among politicians and policy makers so that current complacency and the resultant decline in budgets and resources within groundwater management agencies does not compromise their capability to address these issues in the future.

References

Agriculture and Resource Management Council of Australia and New Zealand and Standing Committee on Agriculture and Resource Management (ARMCANZ and SCARM), 1996. Allocation and Use of Groundwater: A National Framework for Improved Groundwater Management in Australia – Policy Position Paper for Advice to States and Territories.

Australian Water Resources Council, 1975. Groundwater Resources of Australia. Commonwealth of Australia, Department of Environment and Conservation.

Barron, O.V., Crosbie, R.S., Charles, S.P., Dawes, W.R., Ali, R., Evans, W.R., et al., 2011. Climate change impact on groundwater resources in Australia. Waterlines Report Series No 67. National Water Commission, Canberra, ACT.

Bureau of Meteorology, 2018. Australian groundwater insight. Available from: <http://www.bom.gov.au/water/groundwater/insight>.

Comino, M., Tan, P.L., George, D., 2014. Between the cracks: water governance in Queensland, Australia and potential cumulative impacts from mining coal seam gas. J. Water Law 23 (6), 219–228.

Council of Australian Governments (CoAG), 1994. Communiqué of February Meeting. Attachment A. National Competition Council. Available from: <http://ncp.ncc.gov.au/docs/PIAg-001.pdf>.

Clifton, C., Cossens, B., McAuley, C., 2007. A framework for assessing the environmental water requirements of groundwater dependent ecosystems. In: Report Prepared for Land & Water Australia, Canberra.

CSIRO, 2008. Water Availability in the Murray–Darling Basin: A Report to the Australian Government From the CSIRO Murray–Darling Basin Sustainable Yields Project. CSIRO, Australia.

Department of Natural Resources and Mines, 2016. Underground Water Impact Report for the Surat Cumulative Management Area – 2016. Government of Queensland.

Dillon, P., Pavelic, P., Page, D., Beringen, H., Ward, J., 2009. Managed aquifer recharge: an introduction. Waterlines Report Series No. 13. National Water Commission, Canberra, ACT.

Evans, R., 2007. The impact of groundwater use on Australia's rivers. Land & Water Australia Technical Report. Land & Water Australia, Canberra, ACT.

Gardner, A., Bartlett, R., Gray, J., Nelson, R., 2018. Water Resources Law, second ed. LexisNexis Butterworths.

Harrington, N., Cook, P., 2014. Groundwater in Australia. National Centre for Groundwater Research and Training.

Ivkovic, K.M., Marshall, S.M., Morgan, L.K., Werner, A.D., Carey, H., Cook, S., et al., 2012. National-scale vulnerability assessment of seawater intrusion: summary report. Waterlines Report No. 85. National Water Commission, Canberra, ACT.

Marsden Jacob Associates, 2012. Assessing the value of groundwater. Waterlines Report. National Water Commission, Canberra, ACT.

Mukherjee, A., Scanlon, B., Aureli, A., Langan, S., Guo, H., McKenzie, A., 2020. Global Groundwater: Source, Scarcity, Sustainability, Security and Solutions, first ed. Elsevier, ISBN: 9780128181720.

National Land and Water Resources Audit, 2001. Australian Water Resources Assessment 2000. Commonwealth of Australia, Canberra, ACT, <http://nrmonline.nrm.gov.au/catalog/mql:1674> (accessed 26.01.14.).

Natural Resource Management Ministerial Council, Environmental Protection and Heritage Council and National Health and Medical Research Council, 2009. Australian Guidelines for Water Recycling: Managing Health and Environmental Risks (Phase 2) – Managed Aquifer Recharge. National Water Quality Management Strategy, NRMMC, EPHC and NHRMC, Canberra, ACT.

National Water Commission, 2012. Groundwater Essentials. Commonwealth of Australia, Canberra, ACT.

National Water Commission, 2014. Australia's Water Blueprint: National Reform Assessment 2014. Commonwealth of Australia, Canberra, ACT.

Prosser, I., Wolf, L., Littleboy, A., 2011. Water in mining and industry. In: Prosser, I. (Ed.), Water: Science and Solutions for Australia. CSIRO, Australia.

Richardson, S., Irvine, E., Froend, R., Boon, P., Barber, S., Bonneville, B., 2011a. Australian groundwater-dependent ecosystems toolbox part 1: assessment framework. Waterlines Report. National Water Commission, Canberra, ACT.

Richardson, S., Irvine, E., Froend, R., Boon, P., Barber, S., Bonneville, B., 2011b. Australian groundwater-dependent ecosystems toolbox part 2: assessment tools. Waterlines Report. National Water Commission, Canberra, ACT.

Simmons, C.T., 2016. Groundwater down under. Ground Water 54 (4), 459–460.

Werner, A.D., 2010. A review of seawater intrusion and its management in Australia. Hydrogeol. J. 18, 281–285.

Wilhite, D.A., 2012. Breaking the hydro-illogical cycle: changing the paradigm for drought management. EARTH Mag. 57 (7), 71–72.

World Commission on Environment and Development, 1987. Our Common Future. Oxford University Press.

Further reading

Commonwealth of Australia, et al., 2004. Intergovernmental Agreement on a National Water Initiative (as amended 2006).

Chapter 4

Groundwater storage dynamics in the Himalayan river basins and impacts of global change in the Anthropocene

M. Shamsudduha[1,2]
[1]*Department of Geography, University of Sussex, Brighton, United Kingdom,* [2]*Institute for Risk and Disaster Reduction, University College London, London, United Kingdom*

4.1 Introduction

Water is a precondition to life on Earth and is critical for socioeconomic development, food security, and functioning of healthy ecosystems (UN Water, 2019). Achieving the United Nation's 2030 Agenda of Sustainable Development Goals (SDGs), especially SDG 1 (to end poverty) and SDG 2 (to ensure food security and nutrition) and SDG 6 (to ensure clean water and sanitation for all), is critically dependent on the availability of clean freshwater. Ensuring food security and providing access to safe drinking water for all remain a key challenge for Asia's SDG, particularly in South Asia due to high population density, natural variability of water resources, and lack of infrastructure. The challenge is especially great in the South Asian countries, namely Afghanistan, Bangladesh, Bhutan, India, Maldives, Nepal, Pakistan, and Sri Lanka where more than 40% of the world's poor live (Rasul, 2014).

Groundwater is the largest store of freshwater representing nearly 99% of all liquid freshwater on earth, and is also the source for almost half of world's irrigation water (Villholth and Signs, 2019). Much of the world's groundwater is modern (i.e., <50 years old) and an active component of the global hydrological system (Gleeson et al., 2016). Globally, groundwater has been playing a pivotal role in achieving food security and development of agricultural economy—transforming vast swathes of drylands into flourishing croplands (Turner et al., 2019; Mukherjee et al., 2020). In many developing nations (e.g., India and Bangladesh), groundwater use has accelerated economic development (e.g., agricultural and industrial growth) over the past 20 years and led to major social and economic transformations (Foster and Chilton, 2003). Globally, ~75% groundwater is used in irrigated agriculture (Wada et al., 2014). Nearly one-fifth of the Earth's total freshwater is estimated to be stored in the Himalayan region of southern Asia (Fig. 4.1) that is the home to some 1.7 billion people (Shamsudduha, 2018). Being situated in the subtropical climate region, South Asian countries (Bangladesh, Bhutan, India, Myanmar, Nepal, Pakistan, and Sri Lanka) are generally characterized by abundant water resources (e.g., seasonally available surface water). These South Asian countries with ~24% of the global population contain only 4.6% of the global annual renewable water resources (FAO, 2016a) that are unevenly distributed in the floodplains of large river basins located in these transboundary nations (Hirji et al., 2017).

The Himalayan region is characterized by highly seasonal surface water storage (SWS) (i.e., water in river channels, wetlands, and floodplains), which is abundant during the monsoon season (generally from June to September); however, during the dry season (generally from March to May), surface water is generally scarce when there is little or no rainfall. In contrast a vast amount of freshwater is stored as the "hidden resource" of groundwater beneath the alluvial floodplains of the Ganges, Brahmaputra, Meghna, and Indus river systems and groundwater is thought to be much more resilient to climate change (MacDonald et al., 2016). However, this abundant groundwater storage (GWS) in the Himalayan river basins is highly dynamic in nature. GWS and flow dynamics have been altered by human abstractions and changes in global climate—together termed as the global change, and these changes to groundwater systems will continue to increase (Taylor et al., 2013) in the Anthropocene.

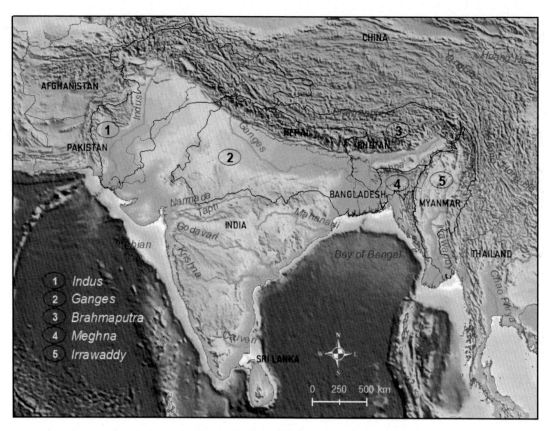

FIGURE 4.1 Location of the major Himalayan river basins (1—Indus River basin, 2—Ganges River basin, 3—Brahmaputra River basin, 4—Meghna River basin, and 5—Irrawaddy River basin) and countries in South Asia. The background image is a shaded relief map showing low-elevated area in green and elevated areas in dark brown colors; bathymetry of seafloor is shown as shaded blue with darker shades indicating the deeper part of seafloors.

Much of world's surface water is highly polluted (Damania et al., 2019), especially in the Global South. Groundwater is a safer alternative to often-polluted surface water, and is generally available all year round. For these qualities, groundwater is increasingly becoming the main source of domestic, industrial, and irrigation water supplies in the Himalayas and worldwide too (Shamsuddha and Panda, 2019). Groundwater-fed irrigation has become the mainstay of irrigated agriculture over much of India, Bangladesh, Pakistan (Punjab and Sindh provinces), Nepal (Terai plains), and Myanmar (Central Dry Zone) (Shah et al., 2004). Groundwater abstraction from the transboundary Indo-Gangetic basin (IGB) that stretches from the west (Pakistan) to the east (Bangladesh) and traverses the entire length of northern India comprises 25% of global groundwater withdrawals, sustaining irrigated agriculture in Pakistan, India, Nepal, and Bangladesh (MacDonald et al., 2016). Groundwater has been playing a pivotal role in food production of these transboundary river basins of the Himalayan region (Fig. 4.2); for example, Bangladesh is now nearly self-sufficient in food grains (Shamsuddha, 2018). However, intensive and unsustainable use of groundwater in South Asia, particularly in northern India, Pakistan, and central and northwestern Bangladesh, has led to rapid depletion of alluvial aquifers in recent years.

Monitoring of groundwater resources is critical to sustainable development and effective management of the "hidden resource." However, ground-based monitoring infrastructure of groundwater is limited in the Himalayan region, especially in remote as well as rural areas. National-scale monitoring of groundwater levels in India and Bangladesh has revealed localized areas of depletion and rising trends in some areas (MacDonald et al., 2016). To overcome the limitation of ground-based monitoring, NASA's Earth Observation system, GRACE (Gravity Recovery and Climate Experiment) satellite mission, offers an opportunity to map changes in terrestrial water storage (TWS), including groundwater (Tapley et al., 2019). GRACE observations show that northern India has lost ~ 109 km^3 of groundwater between 2002 and 2008 (Rodell et al., 2009; Jin, 2013). Over the same period, India's neighbor Bangladesh, which has an equivalent of 4.5% of India's landmass, has depleted nearly 3 km^3 (between 2003 and 2007) of its groundwater due

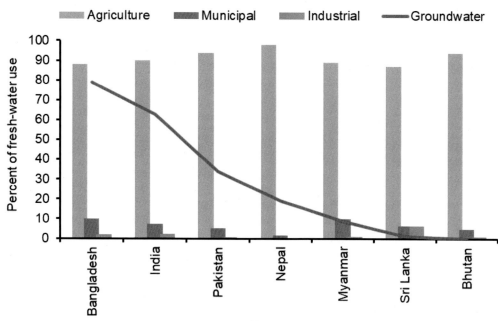

FIGURE 4.2 Use of total freshwater use in various sectors, including agricultural, industry, and municipal sectors and proportion of area equipped for irrigation by groundwater in seven South Asian countries using statistics from various years ranging from 2000 to 2012. *FAO, 2016a. AQUASTAT website. In: FAO (Ed.), Land and Water Division, Food and Agriculture Organization of the United Nations, Rome, Italy; http://www.fao.org/nr/water/aquastat/water_use/index.stm.*

to overabstraction for dry-season rice cultivation (Shamsudduha et al., 2012). It is reported that the long-term groundwater depletion has contributed substantially to global sea-level rise (Taylor et al., 2013); its depletion in Asia is estimated to have contributed to a global sea-level rise of 2.2 mm over the period 2001–08. Recent sea-level rise in the Bay of Bengal can be attributed, at least in part, to overabstraction of groundwater to supply irrigation and municipal water over the last few decades (Shamsudduha et al., 2009).

This chapter reports decadal (2003–14) changes in total terrestrial water and GWS in the large Himalayan river basins—critical information for assessing the sustainability of irrigated agriculture and food security in South and Southeast Asia. GRACE satellite observations and global models are used here to map the spatiotemporal changes in GWS in the river basins. Potential application of this earth observation tool and its limitations are also discussed in relation to GWS mapping in the Himalayan region—vital source of water in achieving the SDGs.

4.2 Hydrology and climate of Himalayan river basins

The Himalayan region is drained by a number of large river basins (Fig. 4.1) located in transboundary settings that are collectively home to over 905 million people (Table 4.1). These are the Ganges, Brahmaputra, Meghna, Indus, and Irrawaddy river basins, all of which are sourced in the Himalayan or the Tibetan Plateau, and their hydrology is largely influenced by the Asian monsoonal climate and meltwater from Himalayan glaciers. Hydrology of these basins and TWS variations are very dynamic due to the strong influence of the Asian monsoon climate and the shifting of the Intertropical Convergence Zone seasonality (Shamsudduha and Panda, 2019). The climate is characterized by intensive rainfall during the monsoon season but little or no rainfall during the dry season. The spatial distribution of rainfall in the major river basins of the Himalayan region is seasonal and highly variable due to variations in surface elevation (i.e., orographic effect), distance from the sea, and the location of the mountain ranges. Descriptions of precipitation and changes in TWS—seasonal and long-term are provided below, which is mostly drawn from a recent review by Shamsudduha and Panda (2019) starting with a brief outline of hydrology and climate conditions.

4.2.1 The Indus river basin

The Indus river is a transboundary river system that flows through several countries, including India, China, Afghanistan, and Pakistan. The Indus river basin (Fig. 4.1) covers a land area of ~1 million km^2 and encompasses

TABLE 4.1 The major river basins in the Himalayan region and the respective countries where they are located, population and cropped area within these river basins.

Rivers	Basin area (km^2)	Population	Country (% land area of each country)	Land-area cropped (km^2)
Indus	1,039,560	177,877,830	Afghanistan (11), China (1), India (14), Pakistan (65)	317,189
Ganges	1,097,000	561,726,420	India (26), Bangladesh (33), China (<1), Nepal (100)	781,452
Brahmaputra	539,410	102,744,980	India (6), Bangladesh (27), Bhutan (100), China (3)	106,612
Meghna	75,060	33,532,360	Bangladesh (24), India (1)	31,642
Irrawaddy	387,700	29,512,090	China (<1), India (<1), Myanmar (53)	99,666

Pakistan (65%), India (14%), China (1%), and Afghanistan (11%) (Table 4.1) (Shamsudduha and Panda, 2019). The Indus river basin stretches in the north from the Himalayan Mountains to in the south into the dry alluvial plains of Sindh Province in Pakistan before finally flowing out into the Arabian Sea (FAO, 2016a). The length of the Indus river is ∼3180 km (1976 mi) and the total basin area is about 1,039,560 km^2. The main tributaries of the Indus river are the Jhelum, Chenab, Ravi, and Sutlej rivers, from the Indian states of Jammu, Kashmir and Himachal Pradesh, and the Kabul, Swat, and Chitral Rivers from the Hindu Kush Mountains (Shamsudduha and Panda, 2019). Annual flow through the Indus river from China to India is ∼180 km^3; it is estimated that the flow generated within India is 50 km^3, resulting in a total flow from India to Pakistan of 230 km^3. The total inflow from Afghanistan to Pakistan in the Indus basin is estimated at 22 km^3. Annual discharge in the Indus river varies from place to place with an average discharge of ∼7610 m^3/s (Gaurav et al., 2011). The Indus river basin is characterized by an arid to semiarid climatic condition (FAO, 2016a), from the east of the region, with varying rainfall (annual rainfall around 500 mm, see Fig. 4.3A). The mean annual rainfall over the Indus River basin varies spatially from 130 to 1380 mm/year. The climate varies from semiarid on the plains to temperate subhumid in the mountainous parts of the basin in the north. Overall, the upper part of the Indus basin (northern part) receives slightly more rainfall than the lower part (southern). This spatial variability in rainfall plays an important role in surface water availability over the entire basin throughout the year (Shamsudduha and Panda, 2019).

4.2.2 The Ganges–Brahmaputra–Meghna river basin

The Ganges–Brahmaputra–Meghna (GBM) river basin is one of the world's mega river basins and the largest river basin in the Himalayan region (Fig. 4.1). The GBM river system is the third largest freshwater outlet to the world's oceans after the Amazon and the Congo river systems (Chowdhury and Ward, 2004). Three individual river basins, namely, the Ganges, Brahmaputra, and Meghna rivers jointly form the GBM basin. The basin covers a significant land proportion of India, China, Bangladesh, Nepal, and Myanmar (Shamsudduha and Panda, 2019). The GBM basin covers an area of nearly 1.7 million km^2 and has a population of roughly 700 million. The GBM river system is a transboundary system even though the three rivers of this system have distinct characteristics and flow through different regions within the Himalayas (FAO, 2012). The headwaters of both the Ganges and Brahmaputra rivers originate in the Himalayan Mountains. The Ganges river flows southwest into India after being originated from the high western Himalayas (Gangotri and Satopanth Glaciers); it then turns southeast and flows through the entire length of the Indo-Gangetic Plain and enters Bangladesh from the west (Shamsudduha and Panda, 2019). The total length of the Ganges river is ∼2600 km. The Ganges river is known as the Padma in Bangladesh that forms a large confluence with Brahmaputra (Jamuna) river in central Bangladesh. The annual flow of the Ganges river from China to Nepal is 12 km^3. All river channels in Nepal ultimately drain into the Ganges river to the south with an annual flow of 210 km^3 to India. The combined annual flow of the Ganges river from India into Bangladesh is 525 km^3 (Shamsudduha and Panda, 2019). The mean annual rainfall over the Ganges river basin ranges spatially from 330 to 3200 mm/year (Fig. 4.3A) with a substantial variability from east to west of the basin (Fig. 4.3B).

In the Himalayan region the Brahmaputra river is the largest transboundary river with a total length of nearly 2900 km. The Brahmaputra river originates in the Angsi glacier, which is located on the northern side of the Himalayas in Tibet. The river then flows through southern Tibet (China) and breaks through the Himalayas through many great

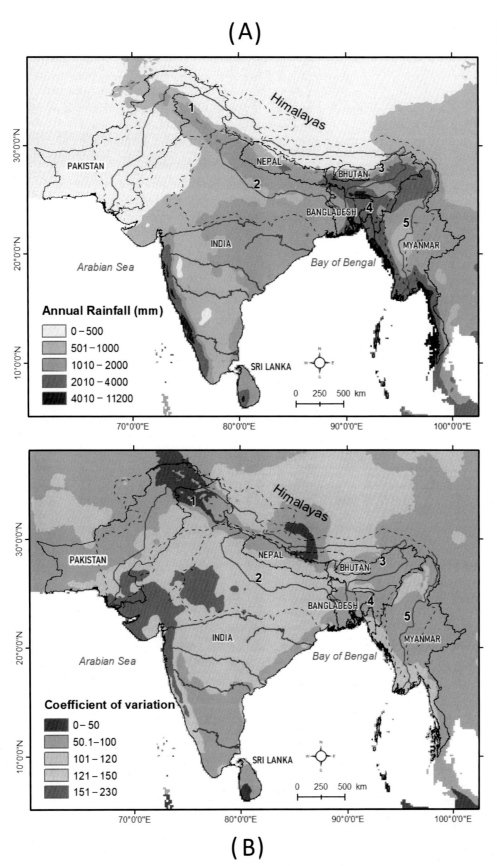

FIGURE 4.3 Long-term (1970–2000) mean (mm) (A) and coefficient of variation (%) (B) in precipitation provided by the WorldClim (version 2.1; https://worldclim.org/).

gorges. The river eventually flows southwest through the Assam Valley and turns toward the south to enter Bangladesh. In Bangladesh, the Brahmaputra river is known as Jamuna before merging with the Ganges (Padma) and Meghna rivers in south central part of Bangladesh. This course of the Brahmaputra (Jamuna) river is roughly 720 km within Bangladesh (Shamsudduha and Panda, 2019). The annual flow of the Brahmaputra river from China to India is $\sim$165 km^3 and from Bhutan to India is 78 km^3. The combined annual flow of the Brahmaputra river from India into Bangladesh is 537 km^3 (FAO, 2016a). The annual rainfall ranges spatially from 310 to 4800 mm/year (Fig. 4.3A) over the Brahmaputra river basin.

The smallest of the three rivers in GBM system, the Meghna originates in the hills of Shillong and Meghalaya of India. The main course is known as the Barak river, which has an extensive catchment in the ridge and valley terrains in eastern Assam State of India. Upon reaching the border with Bangladesh near Sylhet district, the Barak river bifurcates into two separate courses—the Surma and the Kushiyara rivers. The tributaries are locally deep and incised and highly flashy in nature, and these rivers often generates flash floods during the monsoon season. The Ganges (Padma) forms a confluence with the Meghna, and their combined course ultimately flows into the Bay of Bengal as the lower Meghna. The total length of the river is nearly 930 km and, for the most part, the river is mainly meandering in nature (FAO, 2016a). The discharge through the Meghna river is much lower than that of the Ganges and Brahmaputra rivers. The annual flow of the Meghna river from India to Bangladesh is 48 km^3. The mean annual rainfall ranges spatially from 1600 to 4800 mm/year (Fig. 4.3A) over the Meghna river basin. Although the smallest in size, the Meghna river basin is the wettest basin in the Himalayan region (Shamsudduha and Panda, 2019).

The combined annual flow through the GBM river basin from India into Bangladesh is nearly 1110 km^3. The integrated annual discharge of the Ganges, Brahmaputra, and Meghna rivers is one the highest in the world. Mean peak discharges during the monsoon seasons through these Himalayan rivers are 100,000 m^3/s in the Brahmaputra, 75,000 m^3/s in the Ganges, 20,000 m^3/s in the upper Meghna, and 160,000 m^3/s in the lower Meghna (FAO, 2012; Shamsudduha and Panda, 2019). Mean annual rainfall (Fig. 4.3A) is the highest in the east and lowest in the west. The hydroclimatology of the GBM river basin is characterized by a humid, subtropical climate with a seasonal, unimodal distribution of rainfall that is strongly influenced by the monsoon. Between June and September, $\sim$80% of the annual rainfall is observed over these river basin (Webster et al., 1998; Panda and Wahr, 2015)—making this one of the highly seasonal hydrological systems in the world.

4.2.3 The Irrawaddy river basin

The Irrawaddy river basin (Fig. 4.1) is located mainly in Myanmar (formerly known as Burma), which is drained by the Irrawaddy river (locally known as the Ayeyarwady river). There are many tributaries in the north and distributaries of the Irrawaddy river in the south (FAO, 2016a). The Irrawaddy river is formed by the combined flow of the Nmai and Mali rivers that originate from the Himalayan glaciers of northern mountains in Myanmar. Irrawaddy river is the main waterway of Myanmar and is about 2170 km (1350 mi) long from its origin in the eastern Himalayas to the south into the Andaman Sea. It flows through the vast plains of the Irrawaddy Delta located in the south. The important tributaries of the Irrawaddy river are the Chindwin and Shweli rivers. The Irrawaddy river is the main course of communication between important locations in the interior and the southern port cities, Yangon (formerly known as Rangoon). The drainage area of the Irrawaddy river basin is $\sim$387,700 km^2 (Fig. 4.3) with a population of nearly 30 million (Table 4.1) (Shamsudduha and Panda, 2019). The discharge rates through the Irrawaddy river vary from as small as 2000 m^3/s in the headwaters of the eastern Himalayas and Tibetan Plateau to an average of 13,000 m^3/s to the Andaman Sea. Discharge is highest (average 35,000 m^3/s) during the monsoon season (August and September) and lowest (average 4000 m^3/s) in the dry season (January–April) (Furuichi et al., 2009). The Irrawaddy river basin is characterized primarily by a subtropical climate with the lower part being a humid subtropical climate. The upper part has a warm, humid subtropical climate, and the middle part is characterized by a tropical dry zone. Both climatic zones are dominated by the Asian summer monsoon that brings monsoonal rainfalls from May to October (Shamsudduha and Panda, 2019). According to the Department of Meteorology and Hydrology of Myanmar, between 1988 and 1997, nearly 92% of the annual rainfall occurs during the southwest monsoon between May and October. In the Irrawaddy river basin the mean annual rainfall varies from 900 to 3700 mm/year. The middle part of the basin, which lies in the rain shadow of the Rakhine Mountains, is called the Central Dry Zone with an average annual rainfall ranging from 600 to 800 mm (Fig. 4.3A). Monsoonal rainfall progressively increases both upstream and downstream away from the dry zone ranging from about 4000 mm in the north to 2500 mm in the south (Furuichi et al., 2009).

4.3 Groundwater for drinking and agricultural use

Traditionally, surface water had been used for drinking and water supplies in irrigation in all South Asian countries (Shamsuddha and Panda, 2019). Over the course of time, most surface water sources such as ponds, rivers, and lakes got polluted due to waste dumping from household and industrial sources, poor sanitation, open defecation, and storm surge inundation of saltwater in coastal region (Shamsuddha, 2013a; Hasan et al., 2019; Mukherjee et al., 2019). During the 1980s and 1990s, groundwater has largely replaced surface water−fed water supplies (van Geen et al., 2002), especially for drinking water because of its natural "good" quality; however, the recent discovery of elevated arsenic concentrations in shallow groundwater in the Indus and GBM basins has been recognized as one of the largest mass poisoning in human history (Smith et al., 2000). Compared to their close neighbors—Bangladesh and India, groundwater-fed irrigation in Bhutan and Sri Lanka is negligible. In Bangladesh, currently 97% of drinking and nearly 80% of irrigation water supplies come from groundwater (Fig. 4.2). Currently, the groundwater use for dry-season irrigation in India, Pakistan, Nepal, and Myanmar is ∼64%, 23%, 20%, and 5% of the total withdrawal of freshwater (FAO, 2016a). By volume, India is the largest groundwater user in the world with an estimated annual withdrawal of ∼250 km^3 compared to the global total of 982 km^3/year (Margat and van der Gun, 2013). About 62% of the region's 555 km^3 of annual renewable groundwater is pumped—making South Asia the world's largest abstractor of groundwater (Hirji et al., 2017). India alone withdraws ∼25% of the world's total groundwater withdrawal that is more than 80% of the country's available GWS; as a result of overabstraction, groundwater levels in some parts of the country are being depleted fast (Bhanja et al., 2017). A significant share of this groundwater is used to irrigate rice crop, which is the staple food for many south Asian nations. Recently, Bangladesh has become nearly self-sufficient in food grains and has sustained through groundwater-irrigated agriculture (Shamsuddha, 2013b).

Land-use patterns in the Himalayan river basins are highly diverse and seasonal. Rice is most widely grown in the arable lands as this is the staple food in the South Asian countries. Rice and wheat account for 84% of total cereal production (rice 48% and wheat 36%) in South Asia (ADB, 2015). India is the largest producer of rice and wheat in the region. The next largest producer is Bangladesh for rice and Pakistan for wheat. Rice requires substantial water to grow, and thus irrigation is required that often comes from both surface water and groundwater as well as rainwater during the monsoon season. However, how much of water is used for irrigation in these river basins is largely unknown due to limitations of abstraction information throughout the basin. This is one of the biggest challenges when it comes to managing irrigation water resources. Groundwater use if often estimated from area of irrigation or the number of irrigation pumps used during each season. Maps of cropped areas across the Indian Subcontinent (Fig. 4.4) show not only areas of irrigated and nonirrigated areas but also areas where single, double, or triple crops are grown annually. Among the five transboundary river basins in the Himalayan region, the proportion of irrigated lands is the highest (72%) in the Ganges river basin, followed by the Meghna (40%), Indus (29%), Irrawaddy (26%), and the Brahmaputra (20%) river basins (Table 4.2). These maps (Fig. 4.4A and 4.4B) reveal that double cropping pattern is the dominant land-use type in these river basins. There are some areas in northern Bangladesh, and some parts of West Bengal and Bihar states of India where irrigated crops are grown three times a year; one or two of these crops are irrigated rice crops. These maps also reveal that the proportion of irrigated lands in the Irrawaddy river basin in Myanmar is the smallest where ∼13% arable land is rainfed.

4.4 Groundwater storage dynamics in Himalayan river basins

4.4.1 Gravity Recovery and Climate Experiment: Earth observation satellite monitoring

Monitoring of the changes in total TWS, which critically includes GWS in time and space, is critical for better understanding of global hydrological cycles as well as regional-scale water budget. The temporal and spatial variabilities in TWS over the Himalayan river basins are substantial that impact the annual hydrological budgets, and, thus, livelihoods of millions of people who are critically dependent on water supplies for irrigation. Ground-based monitoring of individual terrestrial water stores (e.g., surface water, soil moisture, and groundwater) in the Himalayan river basins varies spatially and temporally. Earth observation satellite monitoring of earth's gravitational field by a pair GRACE satellites makes it possible to map TWS variations over the entire globe at monthly timescale since early 2002. The first GRACE mission has been successfully operated jointly by NASA and DLR (German Aerospace Centre) under the NASA's Earth System Science Pathfinder Program (Tapley et al., 2004). The first GRACE mission consists of two identical satellites that flew some 220 km apart in a polar orbit and 500 km above the earth surface. The GRACE satellites mapped earth's gravity field with great accuracy by making accurate measurements of the orbiting distance between the two satellites, using Global Positioning System and a microwave ranging system (Tapley et al., 2004). The first

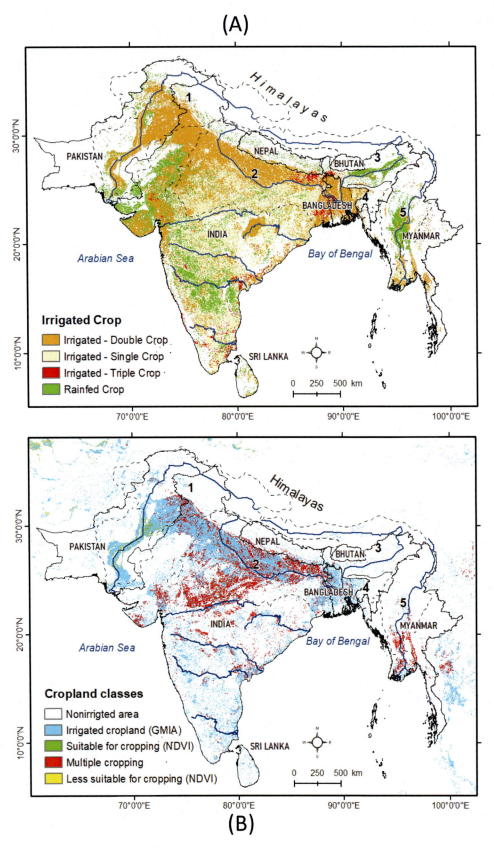

FIGURE 4.4 Map (A, top) showing irrigated crops (i.e., type of irrigation) in South Asian countries based on Irrigated Area Map of Asia (2000–10) provided by IWMI (www.iwmi.org); and map (B, bottom) showing cropland classes based on a downscaled GMIA dataset (Siebert et al., 2013; Meier et al., 2018). *GMIA*, Global map of irrigation areas.

TABLE 4.2 The major river basins in the Himalayan region and proportion (in %) of noncropped and cropped area including single, double, triple, and rainfed cropping types.

Area (%)	Indus	Ganges	Brahmaputra	Meghna	Irrawaddy
Noncropped area	70.72	28.26	80.43	60.36	73.54
Single cropped area	4.30	29.19	4.45	17.82	6.56
Double cropped area	20.05	34.82	7.06	19.89	6.00
Triple cropped area	0.14	2.47	1.11	0.97	0.44
Rainfed cropped area	4.78	5.26	6.94	0.97	13.46
Total area cropped	29.28	71.74	19.57	39.64	26.46

GRACE satellite mission ended in late 2017 and a second GRACE Follow-On (GRACE-FO) mission has begun in May 2018 that will continue to measure the earth's gravity field and, therefore, the movement of terrestrial water masses (e.g., soil moisture, groundwater, surface water in large lakes and rivers, and ice sheets and glaciers) at the global scale (Rodell et al., 2018). The measurements of the geospatial "big data" from GRACE satellite mission have provided information about the spatiotemporal distribution of TWS within earth's surface and also the subsurface environments (e.g., GWS). TWS variations within the Himalayan river basins are particularly important as water resources are vital for growing food to feed nearly a billion people in the area (Shamsudduha and Panda, 2019). A number of studies on GRACE-derived observations of TWS anomalies over the Himalayan river basins has been conducted over the last 10 years, of which geospatial mapping of spatiotemporal changes in GWS has been the key focus (Rodell et al., 2009; Tiwari et al., 2009; Cazenave and Chen, 2010; Shamsudduha et al., 2012; Chen et al., 2014; Dasgupta et al., 2014; Panda and Wahr, 2015; Papa et al., 2015; Bhanja et al., 2016; Khandu et al., 2016; Long et al., 2016).

4.4.2 Dynamics in Gravity Recovery and Climate Experiment terrestrial water storage

Earth's 97.5% water is saline and found in salty lakes, seas, and oceans. The remaining 2.5% freshwater is stored in various individual components of the hydrosphere: 30% is stored as groundwater and the rest of the freshwater is stored in glaciers and ice caps (68.6%), snow and ice (0.95%), surface-water bodies (0.3%) (i.e., rivers, lakes and wetlands), soil ($\sim$0.05%), and atmosphere ($\sim$0.03%) (Shiklomanov, 1993). All saline water and freshwater collectively represent the total TWS on the Earth's surface. TWS is an important part of the global hydrological cycle (Tapley et al., 2004), and its changes through time and space, which demonstrate the dynamic nature of its individual hydrological components, are often influenced by anthropogenic activities (e.g., groundwater pumping for irrigation) and climatic variations.

GRACE satellites have provided spatiotemporal (i.e., generally at monthly time steps at the typical spatial footprint of 100,000 km^2) scenes of the earth's terrestrial water mass/TWS movement from month to month—season to season at the global scale from 2002 to 2016. The newly launched GRACE-FO satellites are now continuing the mapping of earth's TWS, and data (https://gracefo.jpl.nasa.gov/data/grace-fo-data/) have started to come in for scientific analysis. Here, spatiotemporal variations in TWS over the large five river basins in the Himalayan region have been analyzed and mapped using the time-series data from the first GRACE mission.

GRACE observations over the Himalayan river basins for the period of 2003–14 have revealed the dynamic seasonality in TWS variations due to climate and anthropogenic changes (i.e., pumping for irrigation). TWS varies substantially from one month to the next, with April being, hydrologically, the driest and September being the wettest month in these river basins (Shamsudduha and Panda, 2019). The greatest fluctuations in monthly TWS anomalies are observed over the GBM basin where the range in average annual amplitudes can be as high as 400–500 mm. Seasonal variations in TWS anomalies over the Irrawaddy river basin are smaller compared to the GBM basin and fluctuations between dry and wet seasons. Monthly variations in TWS are smallest in the Indus river basin where annual rainfall is also the lowest (130—1380 mm/year) among the five basins.

Temporal trends in GRACE TWS at the basin scale can be linear or nonlinear in nature as the seasonality or episodic events (e.g., extreme rainfall and droughts) can dominate over the variability in time-series components (Bonsor et al., 2018; Shamsudduha and Taylor, 2018). Calculation of trends requires careful considerations of these extreme

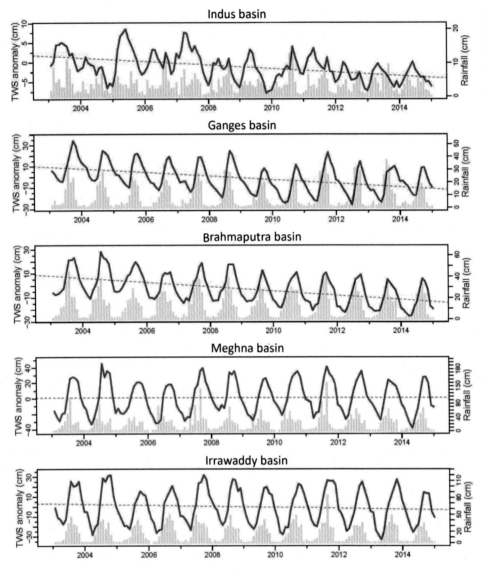

FIGURE 4.5 Monthly time-series (2003–14) records of the JPL-Mascons GRACE-derived ΔTWS anomalies for the five river basins in the southern Asia. Monthly ΔTWS data are averaged over the respective river basin to produce a time series (shown in *thick blue line*); monthly rainfall records from CRU dataset (University of East Anglia, United Kingdom) are extracted and processed for each river basin (shown as *yellow bars*). Linear trends on seasonally adjusted time-series are shown as dashed red lines. *CRU*, Climatic Research Unit; *GRACE*, Gravity Recovery and Climate Experiment; *TWS*, terrestrial water storage.

events as well as the presence of strong seasonality in the time-series data. One of the robust ways to derive trends in highly seasonal time-series is the use of multiple linear regression with seasonal covariates that explicitly represent seasonality (Chandler and Scott, 2011). Here, trends in monthly GRACE-TWS anomalies (2003–14) are calculated using multiple linear regression method considering sine and cosine functions of time as additional factors. This way of calculating trends deems to be much for robust statically. Trends in the Himalayan river basins are shown in Fig. 4.5. Trends (2003–14) in TWS of the Indus basin is estimated at −0.45 (±0.12) cm/year (rate of annual trend ± standard error × 2); Ganges basin: −1.70 (±0.18); Brahmaputra basin: −1.80 (±0.16); Meghna basin: 0.00 (±0.38); and Irrawaddy basin: −0.52 (±0.30) for the entire period of 2003–14 and −1.61 (±1.06) for the recent period of 2010–14. All the reported trends are statistically significant (P value <.001) and the overall fit of the multiple linear models for all five river basins is robust (R^2 ranges from 0.51 to 0.92). These secular trends suggest that TWS is generally declining in the Indus, Ganges, and Brahmaputra basins but TWS has remained stable in Meghna and Irrawaddy river basins for the GRACE period.

4.4.3 Mapping groundwater storage using Gravity Recovery and Climate Experiment

The aquifers in the Indus, Ganges, and Brahmaputra basins (i.e., the floodplains of the Indo-Gangetic Plains) have immense natural storage capacity estimated to be 30,000 km^3—which is over a hundred times the total storage

(~280 km³) in all South Asian dams, reservoirs, and tanks (Hirji et al., 2017). Due to the lack of ground-based monitoring of groundwater in the remote parts of these river basins, Earth observation satellites (GRACE) are currently being used more widely. Furthermore, information needs to be made available to all levels including water-management authorities in transboundary countries so that better regional cooperation can be devised for effective groundwater monitoring and promoting sustainable development and achieving food security in the region (Hirji et al., 2017).

To map groundwater storage changes (ΔGWS), postprocessed, gridded (1° × 1°) monthly GRACE TWS data JPL-Mascon (Watkins et al., 2015; Wiese et al., 2016) solutions from NASA (http://grace.jpl.nasa.gov/data) are used in this analysis. These time-series of GRACE and GLDAS data were processed as part of a global-scale analysis of groundwater-storage dynamics in the world's large aquifer systems (Shamsudduha and Taylor, 2020). A brief description of the datasets and data-processing is summarized here from the work by Shamsudduha and Taylor (2020). JPL-Mascons (version RL05M_1. MSCNv01) data processing involves correction for glacial isostatic adjustment where no spatial filtering is applied as the JPL-RL05M data are directly related to the intersatellite ranging to mass concentration blocks (mascons) to estimate monthly gravity fields in terms of equal area 3° × 3° mass concentration functions in order to minimize measurement errors. Gridded mascon fields are provided at a spatial sampling of 0.5° in both latitude and longitude (~56 km at the equator) along with dimensionless scaling factors at the same spatial grids (Shamsudduha et al., 2017) to apply to the JPL-Mascons product that derived from the Community Land Model (CLM4.0) (Wiese et al., 2016). The dimensionless scaling factors are multiplicative coefficients that minimize the difference between the unfiltered and smoothed (filtered) monthly ΔTWS variations from the CLM4.0 model (Wiese et al., 2016).

Over the Himalayan river basins, monthly measurements GRACE ΔTWS and simulated records of soil moisture storage (ΔSMS), surface runoff or SWS (ΔSWS) and snow water equivalent (ΔSNS) from NASA's Global Land Data Assimilation System (GLDAS version 1.0) at 1° × 1° grids for the period of January 2003—December 2014 are applied to estimate using the following formula:

$$\Delta GWS = \Delta TWS - (\Delta SWS + \Delta SMS + \Delta SNS) \tag{4.1}$$

This approach is consistent with previous global (Thomas et al., 2017; Shamsudduha and Taylor, 2019) and basin-scale (Bhanja et al., 2016; Asoka et al., 2017; Feng et al., 2018) analyses of ΔGWS from GRACE satellite measurements. This study applies GRACE ΔTWS from JPL-Mascons product and monthly anomalies of ΔSMS (soil moisture storage) and ΔSWS (surface water storage) from four land surface models [LSMs: CLM, Noah, variable infiltration capacity (VIC), Mosaic], and ΔSNS (snow water storage) time-series data from the Noah model (GLDAS version 2.1). GRACE and LSMs datasets are processed and analyzed in R programming language (R Core Team, 2017). NASA's GLDAS system (https://ldas.gsfc.nasa.gov/gldas/) drives multiple offline LSMs globally (Rodell et al., 2004), at variable grid resolutions (from 2.5° to 1 km), enabled by the land information system (Kumar et al., 2006). NASA's GLDAS (version 1) drives four LSMs: Mosaic, Noah, the CLM, and the VIC. This study applies monthly ΔSMS (sum of all soil layers from models) and ΔSWS data at a spatial resolution of 1° × 1° from 4 GLDAS LSMs: the CLM, version 2.0 (Dai et al., 2003), Noah (version 2.7.1) (Ek et al., 2003), the VIC model (version 1.0) (Liang et al., 2003), and Mosaic (version 1.0) (Koster and Suarez, 1992). The depths of simulated soil layers are 3.4, 2.0, 1.9, and 3.5 m, respectively, in CLM (10 vertical layers), Noah (4 vertical layers), VIC (3 vertical layers), and Mosaic (3 vertical layers) (Rodell et al., 2004). Simulated time-series data on snow water equivalent (ΔSNS) derive from Noah (v.2.1) model (GLDAS version 2.1).

Similar to the trend analysis of ΔTWS, trends in GRACE-derived ΔGWS are calculated using multiple linear regression method. Results show that from 2003 to 2014, GWS has changed in the Indus river basin at -0.69 (± 0.08) cm/year, Ganges basin at -1.53 (± 0.18) cm/year, Brahmaputra basin at -2.11 (± 0.16) cm/year, Meghna basin at -1.22 (± 0.4), and the Irrawaddy basin at -2.87 (± 0.44) cm/year for the period of 2003–14 and -0.11 (± 0.88) for the period of 2010–14. Of the special interest, it is to note that the abrupt change in ΔGWS in Irrawaddy river basin after 2009 is caused by a shift in the time-series data of simulated soil moisture in all four GLDAS LSMs (all time-series data are provided as supplementary information for the reproduction of the analyses undertaken in this study).

4.4.4 Reported changes of groundwater storage and impacts of global change

Using the JPL-Mascons GRACE product and five GLDAS LSMs, this study evaluates spatiotemporal changes in ΔGWS in fine large river basins in the Himalayan region applying multiple linear regression models (Fig. 4.6). Calculated trends for the period of 2003–14 in ΔGWS are as follows: Indus river basin (-7.2 km³/year), Ganges basin (-16.8 km³/year), Brahmaputra basin (-11.4 km³/year), Meghna basin (-0.92 km³/year), and Irrawaddy basin (-0.43 km³/year from 2010 to 2014). How do these trends compare to the estimates reported in recent studies using GRACE satellite data and in situ

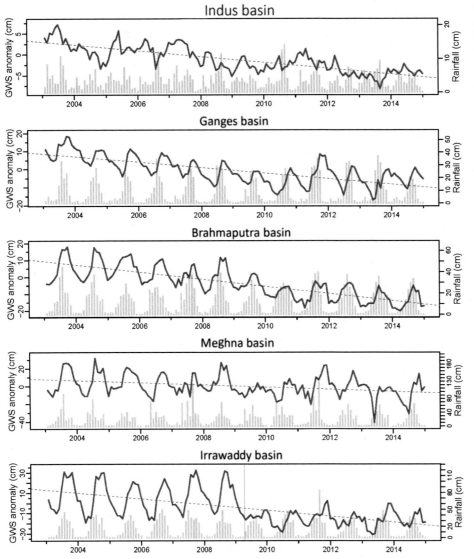

FIGURE 4.6 Monthly time-series (2003–14) records of the JPL-Mascons GRACE-derived ΔTWS anomalies for the five river basins in the southern Asia. Resolved monthly ΔGWS data are averaged over the respective river basin to produce a time series (shown in *thick blue line*); monthly rainfall records from CRU dataset (University of East Anglia, United Kingdom) are extracted and processed for each river basin (shown as *yellow bars*). Linear trends on seasonally adjusted time-series are shown as dashed red lines. The vertical gray line in the last panel (Irrawaddy basin) demarcates a sharp change in the time-series plot. *CRU*, Climatic Research Unit; *GRACE*, Gravity Recovery and Climate Experiment; *GWS*, groundwater storage; *TWS*, terrestrial water storage.

estimates of groundwater storage change in the region? A series of studies (Rodell et al., 2009; Tiwari et al., 2009; Chen et al., 2014; Panda and Wahr, 2015; Bhanja et al., 2016; Asoka et al., 2017) mapped the spatiotemporal changes in ΔGWS in the Himalayan region, particularly, northwest of India using GRACE ΔTWS and auxiliary information water storage from simulated LSMs and global hydrological models. These studies reported that GWS has declined substantially in the northern part of India. For example, Rodell et al. (2009) reported groundwater storage loss of 4.0 ± 1.0 cm/year (equivalent volumetric loss of 17.7 ± 4.5 km^3/year) over the Indian states of Rajasthan, Punjab, and Haryana (including Delhi) for the period of August 2002–October 2008. Another GRACE analysis (Tiwari et al., 2009) encompassing the entire Indo-Gangetic Plain and part of the Himalayan Mountains reported a total water-mass loss of 54 ± 9 km^3/year and they attributed the entire rate to GWS loss. These estimates suggested that groundwater-loss rate over the Ganga–Brahmaputra basin was 34 km^3/year compared to $\sim$10 km^3/year over the Indus River basin and $\sim$10 km^3/year over western Pakistan and mountainous areas in Afghanistan. A recent GRACE-based study in India (Chen et al., 2014) reports a total GWS loss at a rate of 20 ± 7.1 km^3/year over a 10-year period (January 2003–December 2012). Using both GRACE and in situ groundwater-level time-series records, Panda and Wahr (2015) report that GRACE-derived ΔGWS loss over the Ganges river basin is at a rate of 1.25 cm/year ($\sim$14 km^3/year). Yet, another study (Long et al., 2016) evaluates ΔGWS loss over northern of India using GRACE and global hydrological models and reports that ΔGWS has declined at a rate of 3.1 ± 0.1 cm/year ($\sim$14 $\pm$ 0.4 km^3/year) for the period of January 2005–December 2010. This is consistent with the ΔGWS loss of 2.8 cm/year ($\sim$12.3 km^3/year) based on estimates from observation data collated through a groundwater-level monitoring network. Using in situ groundwater-level time-series data over the IGB aquifer system, a recent study (MacDonald et al., 2016) reports a net mean annual groundwater depletion at a rate of 8.0 km^3/year (range: 4.7−11.0 km^3/year) over the IGB with significant variation observed across the basin over the period of 2000−12. According to MacDonald et al. (2016), the largest depletion of GWS occurred in areas of high abstraction and consumptive use in northern India (Punjab 2.6 ± 0.9 km^3/year, Haryana 1.4 ± 0.5 km^3/year, and Uttar Pradesh 1.2 ± 0.5 km^3/year) and Pakistan (Punjab Province 2.1 ± 0.8 km^3/year). In the Lower Indus, within the Sindh Province, groundwater is accumulating at a rate of 0.3 ± 0.15 km^3/year. This rise in GWS has led to increased waterlogging of land and substantial reduction in the outflow from the River Indus. Across the rest of the IGB, changes in GWS are generally small to moderate (within ±1 cm/year). Time-series records of seasonally monitored boreholes from the Ganges river basin reveal that the long-term dry-season levels have declined but substantial variation is observed during the wet season due to variability in monsoon precipitation (Panda and Wahr, 2015).

4.5 Concluding discussion

The transboundary Himalayan river basins are characterized by very high seasonality in both surface water and groundwater storage. Groundwater is plentiful in alluvial aquifers found beneath the floodplains of the Himalayan river basins at shallow depths (tens of meters below ground). These properties of groundwater have enabled the region nearly self-sufficient in food grains (ADB, 2015). Groundwater, largely considered as a common pool resource, provides greater drought and climate resilience than surface water (Cuthbert et al., 2019), but remains highly undervalued and featured in national policies (Hirji et al., 2017). On the contrary, global warming is accelerating the loss of alpine glaciers in the Himalayas through melting at faster rates and adversely affecting the pattern and behavior of monsoonal rains, riverflow regimes, and water demand patterns. Consequently, groundwater resources are under unprecedented pressure (Hirji et al., 2017). Furthermore, quality of surface water and groundwater is far more critical than storage (MacDonald et al., 2016) when it comes to drinking and industrial water supplies—posing greater challenges to achieving the UN Sustainable Development Goals (e.g., SDG 6 Clean water for all).

Agriculture is the mainstay for South Asian economies and serves as the most important livelihood option for the poor (ADB, 2015). Agriculture provides employment to about 50% of the population in South Asia and the sector contributes ∼20% to the region's gross domestic product (ADB, 2015). Dependence of groundwater is increasing among various sectors, for example, industrial usage of groundwater in Bangladesh (e.g., garments and textile industries) is increasing rapidly as the country is becoming a service-dominant economy than agricultural-based economy. Similarly, over the recent years, dependence on groundwater has increased substantially in India, Pakistan, and Nepal while the potential in slowly being realized in Myanmar as the country has started to increase groundwater-fed irrigation for food production (Pavelic et al., 2015). Thus food security in the region means (ground) water security in the time of Anthropocene.

GRACE-based assessment of groundwater storage change has the potential to provide near real-time situation analysis; however, the coarse spatial resolution (∼100,000 km^2) of the satellite footprint is the biggest limitation. There also remain large uncertainties in the estimation of ΔGWS using GRACE satellites and over-reliance on uncalibrated LSMs (e.g., GLDAS models). For example, the declining trends in the Irrawaddy river basins is largely erroneous due to large uncertainty in the ancillary water storage (e.g., soil moisture storage) derived from GLDAS models. In addition, the choice of statistical methods for estimating trends in GRACE-derived GWS needs to be robust as variability in TWS across the highly seasonal Himalayan river basins is great.

Nevertheless, the reported trends (2003−14) in ΔGWS in the transboundary river basins clearly indicate impacts of anthropogenic influences through recent climate change on land-water budgets (e.g., irrigation abstraction, trends in precipitation, and land-use change) in the Himalayan river basins. Declining trends in the Indus and Brahmaputra river basins are clearly indicating a case of overabstraction of groundwater used increasingly for irrigation water supplies. To note, the years with greater rainfall help stop or halt a secular declining trend and stabilize the trend, for example, declining trend in the Ganges river basins was interrupted in 2011 (Fig. 4.6) when annual rainfall was 132 cm—much higher than the average of 104 cm/year.

One of the key challenges in the estimation of groundwater storage to evaluate its full potential is the lack of continuous monitoring of groundwater levels and quality across the entire region. Another constraint in devising an effective management of groundwater is the lack of abstraction data over the region. Spatially modeled abstraction data from limited observations show (Fig. 4.7) the extent of groundwater-fed irrigation in South Asia. It is also clear from national statistics that some countries (Bangladesh, India, and Pakistan) are now greatly dependent on groundwater than surface water sources for irrigation (Fig. 4.8).

60 Theme | 2 Groundwater sources

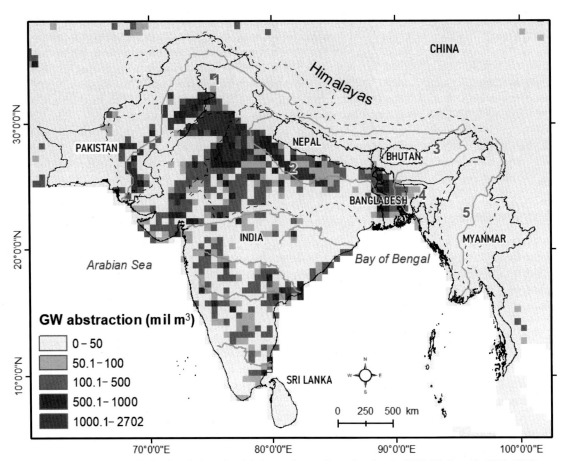

FIGURE 4.7 Groundwater abstraction in South Asia from the global groundwater abstraction data for 2000 (Wada et al., 2010, 2012).

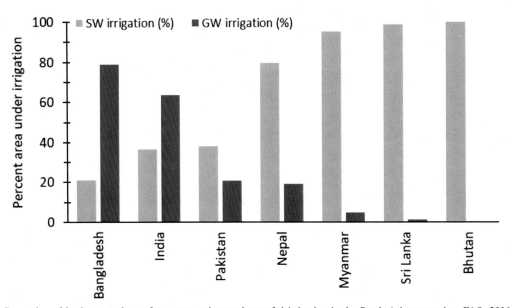

FIGURE 4.8 Proportion of land area under surface water and groundwater-fed irrigation in the South Asian countries. *FAO, 2016a. AQUASTAT website. In: FAO (Ed.). Land and Water Division, Food and Agriculture Organization of the United Nations, Rome, Italy.*

In summary, demand for food commodities in South Asia is projected to rise at a much higher rate than population growth as there is a large consumption deficit (ADB, 2015). The region is highly dependent on its groundwater resources and the dependence is likely to increase due to global change (i.e., directly from increased anthropogenic pumping of groundwater and indirectly from climate variability) as the surface water becomes highly unpredictable and dry-season flows become less due to decrease in meltwater from alpine glaciers under climate change (Taylor et al., 2013). Climate-resilient water-resource management (e.g., conjunctive use of surface and groundwater as opposed to overreliance solely on groundwater for irrigation) is a potentially powerful mechanism to achieve local, regional, and possibly global, food security (UN Water, 2019). Sharing of monitoring data and regional cooperations are the key to promoting the sustainable development of transboundary waters and achieving water and food security in the Himalayan region of South Asia (FAO, 2016b).

Acknowledgments

NASA's MEaSUREs Program for providing GRACE land solutions and the Goddard Earth Sciences Data is acknowledged. Support from Bangladesh Water Development Board (BWDB) and the Indian Central Ground Water Board (CGWB) for providing groundwater-level monitoring records is also acknowledged. Support from the UK Government's NERC-ESRC-DFID UPGro *GroFutures* (Ref. NE/M008932/1; www.grofutures.org), and DFID funded project, *Indo-Gangetic Basin Groundwater Resilience* (Ref. 202125−108) is duly acknowledged for processing the GRACE and GLDAS model data and earlier work upon which this paper is partly based.

References

ADB, 2015. Food Security in South Asia: Developing Regional Supply Chains for the Food Processing Industry. Asian Development Bank, Delhi, India.

Asoka, A., Gleeson, T., Wada, Y., Mishra, V., 2017. Relative contribution of monsoon precipitation and pumping to changes in groundwater storage in India. Nat. Geosci. 10, 109−117.

Bhanja, S.N., Mukherjee, A., Saha, D., Velicogna, I., Famiglietti, J.S., 2016. Validation of GRACE based groundwater storage anomaly using in-situ groundwater level measurements in India. J. Hydrol. 543, 729−738.

Bhanja, S.N., Mukherjee, A., Rodell, M., Wada, Y., Chattopadhyay, S., Velicogna, I., et al., 2017. Groundwater rejuvenation in parts of India influenced by water-policy change implementation. Sci. Rep. 7, 7453.

Bonsor, H.C., Shamsudduha, M., Marchant, B.P., MacDonald, A.M., Taylor, R.G., 2018. Seasonal and decadal groundwater changes in African sedimentary aquifers estimated using GRACE products and LSMs. Remote Sens. 10.

Cazenave, A., Chen, J., 2010. Time-variable gravity from space and present-day mass redistribution in the Earth System. Earth Planet. Sci. Let. 298, 263−274.

Chandler, R.E., Scott, M., 2011. Statistical Methods for Trend Detection and Analysis in the Environmental Sciences. John Wiley & Sons, Inc.

Chen, J., Li, J., Zhang, Z., Ni, S., 2014. Long-term groundwater variations in Northwest India from satellite gravity measurements. Glob. Planet. Change 116, 130−138.

Chowdhury, M.R., Ward, N., 2004. Hydro-meteorological variability in the greater Ganges-Brahmaputra-Meghna basins. Int. J. Climatol. 24, 1495−1508.

Cuthbert, M.O., Taylor, R.G., Favreau, G., Todd, M.C., Shamsudduha, M., Villholth, K.G., et al., 2019. Observed controls on resilience of groundwater to climate variability in sub-Saharan Africa. Nature 572, 230−234.

Dai, Y., Zeng, X., Dickinson, R.E., Baker, I., Bonan, G.B., Bosilovich, M.G., et al., 2003. The common land model (CLM). Bull. Am. Meteorol. Soc. 84, 1013−1023.

Damania, R., Desbureaux, S., Rodella, A.-S., Russ, J., Zaveri, E., 2019. Quality Unknown: The Invisible Water Crisis. World Bank, Washington, DC.

Dasgupta, S., Das, I.C., Subramanian, S.K., Dadhwal, V.K., 2014. Space-based gravity data analysis for groundwater storage estimation in the Gangetic plain, India. Curr. Sci. 107, 832−844.

Ek, M.B., Mitchell, K.E., Lin, Y., Rogers, E., Grunmann, P., Koren, V., et al., 2003. Implementation of Noah land surface model advances in the National Centers for Environmental Prediction operational mesoscale Eta model. J. Geophys. Res. 108 (D22), 8851.

FAO, 2012. In: Frenken, K. (Ed.), Irrigation in Southern and Eastern Asia in Figures, AQUASTAT Survey − 2011. Food and Agriculture Organization of the United Nations, Rome, Italy.

FAO, 2016a. In: FAO (Ed.), AQUASTAT website. Land and Water Division, Food and Agriculture Organization of the United Nations, Rome, Italy.

FAO, 2016b. Thematic Papers on Groundwater. Food and Agriculture Organization of the United Nations (FAO), Rome, Italy, p. 776.

Feng, W., Shum, C.K., Zhong, M., Pan, Y., 2018. Groundwater storage changes in China from satellite gravity: an overview. Remote Sens. 10, 674.

Foster, S.S.D., Chilton, P.J., 2003. Groundwater: the processes and global significance of aquifer degradation. Philos. Trans. R. Soc. London, Ser. B 358, 1957−1972.

Furuichi, T., Win, Z., Wasson, R.J., 2009. Discharge and suspended sediment transport in the Ayeyarwady River, Myanmar: centennial and decadal changes. Hydrol. Processes 23, 1631−1641.

Gaurav, K., Sinha, R., Panda, P.K., 2011. The Indus flood of 2010 in Pakistan: a perspective analysis using remote sensing data. Nat. Hazard. 59, 1815–1826.

Gleeson, T., Befus, K.M., Jasechko, S., Luijendijk, E., Cardenas, M.B., 2016. The global volume and distribution of modern groundwater. Nat. Geosci. 9, 161–167.

Hasan, M.K., Shahriar, A., Jim, K.U., 2019. Water pollution in Bangladesh and its impact on public health. Heliyon 5, e02145.

Hirji, R., Nicol, A., Davis, R., 2017. South Asia Climate Change Risks in Water Management. World Bank and International Water Management Institute (IWMI), Washington, DC and Colombo.

Jin, S., 2013. Satellite gravimetry: mass transport and redistribution in the Earth system. In: Shuanggen, J. (Ed.), Geodetic Sciences – Observations, Modeling and Applications. InTech, p. 344.

Khandu, Forootan, E., Schumacher, M., Awange, J.L., Schmied, H.M., 2016. Exploring the influence of precipitation extremes and human water use on total water storage (TWS) changes in the Ganges-Brahmaputra-Meghna River basin. Water Resour. Res. 52, 2240–2258.

Koster, R.D., Suarez, M.J., 1992. Modeling the land surface boundary in climate models as a composite of independent vegetation stands. J. Geophys. Res. 97, 2697–2715.

Kumar, S.V., Peters-Lidard, C.D., Tian, Y., Houser, P.R., Geiger, J., Olden, S., et al., 2006. Land information system: an interoperable framework for high resolution land surface modeling. Environ. Model. Softw. 21, 1402–1415.

Liang, X., Xie, Z., Huang, M., 2003. A new parameterization for surface and groundwater interactions and its impact on water budgets with the variable infiltration capacity (VIC) land surface model. J. Geophys. Res. 108 (D16), 8613.

Long, D., Chen, X., Scanlon, B.R., Wada, Y., Hong, Y., Singh, V.P., et al., 2016. Have GRACE satellites overestimated groundwater depletion in the Northwest India Aquifer? Nat. Sci. Rep. 6, 24398.

MacDonald, A.M., Bonsor, H.C., Ahmed, K.M., Burgess, W.G., Basharat, M., Calow, R.C., et al., 2016. Groundwater quality and depletion in the Indo-Gangetic basin mapped from in situ observations. Nat. Geosci. 9, 762–766.

Margat, J., van der Gun, J., 2013. Groundwater Around the World: A Geographic Synopsis. CRC Press/Balkema, Boca Raton, FL.

Meier, J., Zabel, F., Mauser, W., 2018. A global approach to estimate irrigated areas – a comparison between different data and statistics. Hydrol. Earth Syst. Sci. 22.

Mukherjee, A., Duttagupta, S., Chattopadhyay, S., Bhanja, S.N., Bhattacharya, A., Chakraborty, S., et al., 2019. Impact of sanitation and socioeconomy on groundwater fecal pollution and human health towards achieving sustainable development goals across India from ground-observations and satellite-derived nightlight. Sci. Rep. 9, 15193.

Mukherjee, A., Scanlon, B., Aureli, A., Langan, S., Guo, H., McKenzie, A., 2020. Global Groundwater: Source, Scarcity, Sustainability, Security and Solutions, first ed. Elsevier, ISBN. 9780128181720.

Panda, D.K., Wahr, J., 2015. Spatiotemporal evolution of water storage changes in India from the updated GRACE-derived gravity records. Water Resour. Res. 51, 135–149.

Papa, F., Frappart, F., Malbeteau, Y., Shamsuddha, M., Venugopal, V., Sekhar, M., et al., 2015. Satellite-derived surface and sub-surface water storage in the Ganges-Brahmaputra river basin. J. Hydrol.: Reg. Stud. 4, 15–35.

Pavelic, P., Sellamuttu, S.S., Johnston, R., McCartney, M., Sotoukee, T., Balasubramanya, S., et al., 2015. Integrated Assessment of Groundwater Use for Improving Livelihoods in the Dry Zone of Myanmar. International Water Management Institute (IWMI), Colombo, Sri Lanka, p. 47.

Rasul, G., 2014. Food, water, and energy security in South Asia: a nexus perspective from the Hindu Kush Himalayan region. Environ. Sci. Policy 39, 35–48.

R Core Team, 2017. R: A Language and Environment for Statistical Computing (R Version 3.4.3). R Foundation for Statistical Computing, Vienna, Austria, URL <http://www.R-project.org/>.

Rodell, M., Houser, P.R., Jambor, U., Gottschalck, J., Mitchell, K., Meng, C.-J., et al., 2004. The global land data assimilation system. Bull. Am. Meteorol. Soc. 85, 381–394.

Rodell, M., Velicogna, I., Famiglietti, J.S., 2009. Satellite-based estimates of groundwater depletion in India. Nature 460, 999–1003.

Rodell, M., Famiglietti, J.S., Wiese, D.N., Reager, J.T., Beaudoing, H.K., Landerer, F.W., et al., 2018. Emerging trends in global freshwater availability. Nature 557, 651–659.

Shah, T., Scott, C., Kishore, A., Sharma, A., 2004. Energy-Irrigation Nexus in South Asia: Improving Groundwater Conservation and Power Sector Viability, Improving Groundwater Conservation and Power Sector Viability.

Shamsuddha, M., 2013a. Groundwater-fed irrigation and drinking water supply in Bangladesh: challenges and opportunities. In: Zahid, A., Hassan, M.Q., Islam, R., Samad, Q.A. (Eds.), Adaptation to the Impact of Climate Change on Socio-economic Conditions of Bangladesh. Alumni Association of German Universities in Bangladesh, German Academic Exchange Service (DAAD), Dhaka, pp. 150–169.

Shamsuddha, M., 2013b. Groundwater resilience to human development and climate change in South Asia, In: GWF Discussion Paper 1332. Global Water Forum, Canberra, Australia.

Shamsuddha, M., 2018. Impacts of human development and climate change on groundwater resources in Bangladesh. In: Mukherjee, A. (Ed.), Groundwater of South Asia. Springer Hydrogeology, Singapore, pp. 523–544.

Shamsuddha, M., Chandler, R.E., Taylor, R.G., Ahmed, K.M., 2009. Recent trends in groundwater levels in a highly seasonal hydrological system: the Ganges-Brahmaputra-Meghna Delta. Hydrol. Earth Syst. Sci. 13, 2373–2385.

Shamsuddha, M., Panda, D.K., 2019. Spatio-temporal changes in terrestrial water storage in the Himalayan river basins and risks to water security in the region: a review. Int. J. Disaster Risk Reduct. 35, 101068.

Shamsuddha, M., Taylor, R.G., 2018. Sustainability of the World's large aquifer systems from GRACE and the role of extreme precipitation. AGU Fall Meeting 2018. American Geophysical Union, Washington DC, H11T-1726.

Shamsudduha, M., Taylor, R.G., 2020. Groundwater storage dynamics in the world's large aquifer systems from GRACE: uncertainty and role of extreme precipitation. Earth Syst. Dyn. https://doi.org/10.5194/esd-2019-43.

Shamsudduha, M., Taylor, R.G., Longuevergne, L., 2012. Monitoring groundwater storage changes in the highly seasonal humid tropics: validation of GRACE measurements in the Bengal basin. Water Resour. Res. 48, W02508.

Shamsudduha, M., Taylor, R.G., Jones, D., Longuevergne, L., Owor, M., Tindimugaya, C., 2017. Recent changes in terrestrial water storage in the Upper Nile basin: an evaluation of commonly used gridded GRACE products. Hydrol. Earth Syst. Sci. 21, 4533–4549.

Shiklomanov, I.A., 1993. World fresh water resources. In: Gleick, P.H. (Ed.), Water in Crisis: A Guide to the World's Fresh Water Resources. Oxford University Press, New York, pp. 13–25.

Siebert, S., Henrich, V., Frenken, K., Burke, J., 2013. Update of the Digital Global Map of Irrigation Areas to Version 5. Rheinische Friedrich-Wilhelms-University, Bonn, Germany/Food and Agriculture Organization (FAO) of the United Nations, Rome.

Smith, A.H., Lingas, E.O., Rahman, M., 2000. Contamination of drinking water by arsenic in Bangladesh: a public health emergency. Bull. World Health Organ. 78, 1093–1103.

Tapley, B.D., Bettadpur, S., Ries, J.C., Thompson, P.F., Watkins, M.M., 2004. GRACE measurements of mass variability in the Earth system. Science 305, 503–505.

Tapley, B.D., Watkins, M.M., Flechtner, F., Reigber, C., Bettadpur, S., Rodell, M., et al., 2019. Contributions of GRACE to understanding climate change. Nat. Clim. Change 9, 358–369.

Taylor, R.G., Scanlon, B., Doll, P., Rodell, M., van Beek, R., Wada, Y., et al., 2013. Ground water and climate change. Nat. Clim. Change 3, 322–329.

Thomas, B.F., Caineta, J., Nanteza, J., 2017. Global assessment of groundwater sustainability based on storage anomalies. Geophys. Res. Lett. 44, 11445–11455.

Tiwari, V.M., Wahr, J., Swenson, S., 2009. Dwindling groundwater resources in northern India, from satellite gravity observations. Geophys. Res. Lett. 36, L18401.

Turner, S.W.D., Hejazi, M., Yonkofski, C., Kim, S.H., Kyle, P., 2019. Influence of groundwater extraction costs and resource depletion limits on simulated global nonrenewable water withdrawals over the twenty-first century. Earth's Future 7, 123–135.

UN Water, 2019. Climate Change and Water UN-Water Policy Brief. UN-Water Technical Advisory Unit, Genève, p. 28.

van Geen, A., Ahsan, H., Horneman, A.H., Dhar, R.K., Zheng, Y., Hussain, I., et al., 2002. Promotion of well-switching to mitigate the current arsenic crisis in Bangladesh. Bull. World Health Organ. 80, 732–737.

Villholth, K.G., Signs, M., 2019. Groundwater Solutions Initiative for Policy and Practice (GRIPP). Groundwater: Critical for Sustainable Development. GRIPP Infographic. International Water Management Institute (IWMI) and CGIAR Research Program on Water, Land and Ecosystems (WLE), p. 16.

Wada, Y., van Beek, L.P., van Kempen, C.M., Reckman, J.W., Vasak, S., Bierkens, M.F., 2010. Global depletion of groundwater resources. Geophys. Res. Lett. 37.

Wada, Y., van Beek, L.P.H., Sperna Weiland, F.C., Chao, B.F., Wu, Y.-H., Bierkens, M.F.P., 2012. Past and future contribution of global groundwater depletion to sea-level rise. Geophys. Res. Lett. 39, L09402.

Wada, Y., Wisser, D., Bierkens, M.F.P., 2014. Global modeling of withdrawal, allocation and consumptive use of surface water and groundwater resources. Earth Syst. Dyn. 5, 15–40.

Watkins, M.M., Wiese, D.N., Yuan, D.-N., Boening, C., Landerer, F.W., 2015. Improved methods for observing Earth's time variable mass distribution with GRACE using spherical cap mascons. J. Geophys. Res. Solid Earth 120, 2648–2671.

Webster, P.J., Magaña, V., Palmer, T.N., Shukla, J., Tomas, R.A., Yanai, M., et al., 1998. Monsoons: Processes, predictability and the prospects for prediction. J. Geophys. Res. 103, 14451–14510.

Wiese, D.N., Landerer, F.W., Watkins, M.M., 2016. Quantifying and reducing leakage errors in the JPL RL05M GRACE mascon solution. Water Resour. Res. 52, 7490–7502.

Chapter 5

Groundwater variations in the North China Plain: monitoring and modeling under climate change and human activities toward better groundwater sustainability

Wenting Yang, Long Di and Zhangli Sun
State Key Laboratory of Hydroscience and Engineering, Department of Hydraulic Engineering, Tsinghua University, Beijing, China

5.1 Introduction

The North China Plain (NCP) is the political, economic, and cultural center in China and among the most densely populated areas in the world (Zheng et al., 2010). As the most important crop production base in China, the NCP produces ~20% of the total crop production each year. To ensure food security, ~70% of freshwater resources in the NCP are used for agriculture (Qiu et al., 2018; Zheng et al., 2010). Meanwhile, the NCP is also among regions that are subject to most severe water shortage in the world. The water resources in the NCP are unevenly distributed throughout a year, with more than 70% of the annual precipitation falling in summer (Zheng et al., 2010). In North China, only 4% of the water resources sustain 25% of the population and 27% of the GDP throughout the country (Zhu, 2018). To meet the need for agricultural irrigation, population growth, and socioeconomic development, groundwater has been largely overexploited over the past decades. Moreover, most irrigation methods are inefficient, which intensifies the exploitation of groundwater in the NCP (Qiu et al., 2018; Sun et al., 2006).

5.2 Impacts of human activities on groundwater in the North China Plain

The NCP is characterized by long-term intensive human activities. Groundwater storage (GWS) is largely influenced by human activity, that is, groundwater withdrawal and water diversion. The groundwater overdraft has led to significant decreases in groundwater level. By the late 1990s groundwater levels fell at a rate of ~1 m/year, making the NCP among the most depleted aquifers globally (Liu et al., 2008; Zheng et al., 2010). Hebei Province with its plain areas belonging to a primary portion of the NCP suffered the most severe groundwater overexploitation, with 150 km^3 of groundwater extracted over the past 30 years (Chen and Li, 2004). Groundwater depth in the piedmont areas of the NCP has dropped from ~10 m in the late 1970s to ~40 m in 2010, forming the largest groundwater funnel area in China (Shen et al., 2002; Zhang et al., 2011). The excessive extraction of groundwater further leads to a series of eco-environmental and geologic issues, such as drying up of rivers, shrinkage of lakes and wetlands, land subsidence, seawater intrusion, and deterioration of groundwater quality. In 2010 the annual runoff had fallen to a third of it in the 1950s in the Haihe River basin, the plain area of which is highly overlapped with the NCP defined in this study, and more than 4,000 km of the river dried up. The wetland area has shrunk from 10,000 km^2 in the 1970s to less than 2,000 km^2 in 2010. In 2012 areas affected by land subsidence were over 120,000 km^2 in the NCP (~86% of the total area), and an area covers about 17,500 km^2 with subsidence over 2 cm. Based on observed land subsidence in Beijing, the average land subsidence rate was 15–25 mm/year from 1999 to 2009, and some areas reached a peak of ~140 mm/year in 2009 (Gong et al., 2018; Pool, 2005; Ye et al., 2016).

To ensure sustainable development of economy, society, and the environment, it is imperative to address a series of issues caused by groundwater overexploitation in the NCP. China's South-to-North Water Diversion (SNWD) Project could partially alleviate the water crisis and reduce the reliance on groundwater in the NCP. This project transports water from the Yangtze River in the humid south to the arid regions in the north. More details on the SNWD will be provided in Section 5.4.

5.3 Climate change impact on groundwater in the North China Plain

Climate variability/change will directly affect groundwater recharge and indirectly affect the withdrawal of groundwater (Asoka et al., 2017; Havril et al., 2018; Taylor et al., 2013b; Woldeamlak et al., 2007). The direct influence of climate on groundwater is reflected in two aspects. First, in semiarid areas, groundwater recharge is often controlled by heavy precipitation. The concentrated recharge occurs below surface waterbodies (Döll and Fiedler, 2008; Favreau et al., 2009; Pool, 2005; Small, 2005; Taylor et al., 2013a). In humid areas, more precipitation does not always indicate more groundwater recharge. An increase in precipitation may promote vegetation growth and thus lead to an increase in evapotranspiration (Taylor et al., 2013b). Second, climate variability will affect the instantaneous groundwater flow system and also the mechanism of interactions between groundwater and surface water (Havril et al., 2018).

In addition, as irrigation plays a main role in groundwater use globally (i.e., ~70% of groundwater extraction is used for crop irrigation), climate variability/change will largely affect GWS by indirectly affecting irrigation water demand (Döll, 2002). Meteorological drought caused by decreases in rainfall and/or high temperatures will increase water demand for irrigation, and the water source for irrigation during drought often shifts from surface water to groundwater. The indirect impact on groundwater, caused by the change of irrigation demand and irrigation water sources through climate variability, is more significant than its direct impact (Taylor et al., 2013b). The Intergovernmental Panel on Climate Change report published in 2012 indicates that the frequency and intensity of extreme weather or climate events, such as floods, droughts, and heat waves, are likely to increase in the future (Taylor et al., 2013b). Therefore it is necessary to explore the influence of climate variability/change on groundwater in the NCP.

5.4 China's South-to-North Water Diversion

To alleviate the contradiction between the water shortage and rapid development of society and economy in North China, the SNWD Project has been built to transport water from the water-rich Yangtze River basin in the humid south to relatively dry regions in the north. This project transports water from the Yangtze River (including the upper, middle, and lower reaches) to the north of China through three routes, the eastern, central, and western. The first phase of the eastern and central routes covers major provinces and cities in North China, including Beijing, Tianjin, Hebei, Shandong, Henan, and Jiangsu (Fig. 5.1) (Dou et al., 2010). The eastern route transports water from Yangzhou City, located in the lower reaches of the Yangtze River, flowing through the Beijing–Hangzhou Grand Canal and its parallel channels, Lake Hongze, Lake Luoma, Lake Nansi, and Lake Dongping, then divided into two routes. One transports water to Tianjin in the north and the other to Yantai and Weihai in Shandong Province in the east. The eastern route has been running since November 15, 2013. The western route is still being planned, which is quite complicated due to difficulties in engineering and infrastructure, and potential influences on ecosystems of the upper Yangtze River.

The central route transports water from the Danjiangkou Reservoir, located in the middle reaches of the Han River, flowing through Henan and Hebei Provinces, and finally reaches Beijing. The central route is initially planned to provide 9.5 km^3 of water annually to North China and has started to transfer water on December 12, 2014. By the end of 2019 totaling 30 km^3 of water has been transported through the eastern and central routes. This has benefited more than ~120 million people and greatly changed the water supply structure in the NCP. Furthermore, the SNWD Project has opened a new chapter for the history of groundwater that may have been stabilized and recovered to some degree, providing an historical opportunity for groundwater sustainability in the NCP that is exemplified for heavily stressed aquifers globally.

Meanwhile, water-receiving areas of the SNWD also take active measures to restrict groundwater pumping. As a pilot area for groundwater recovery, a series of measures have been taken in Hebei Province, including (1) using water from outside areas as substitution for groundwater abstraction, and constructing water storage projects and (2) adjusting the agricultural planting structure. For example, single cropping instead of traditional double cropping has been

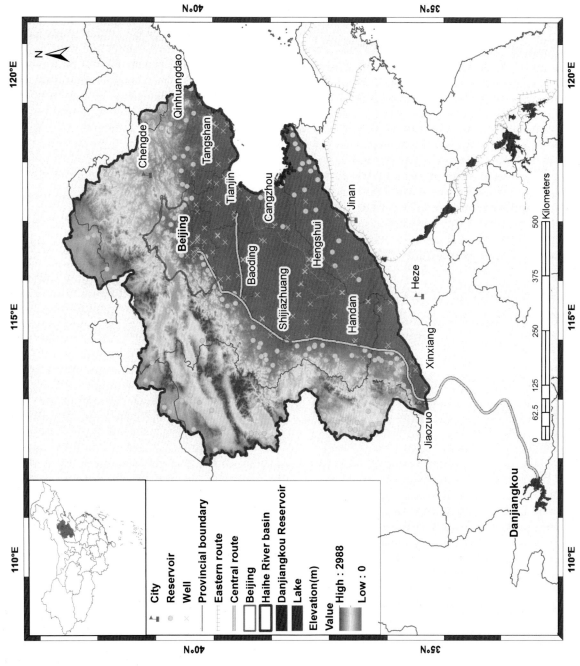

FIGURE 5.1 Schematic of the eastern and central routes of the South-to-North Water Diversion Project, including main cities, reservoirs, locations of key groundwater monitoring wells, Danjiangkou Reservoir, and lakes.

implemented, and efficient irrigation technology has been developed, such as sprinkler irrigation and microirrigation. Third, restrictions of groundwater use have been implemented. There were 7,063 machines used to pump water from wells that were shut down in 2014 and an area of 2,496 km^2 was prohibited to pump groundwater in 2017. In 2016 the groundwater level in Hebei Province has risen by 0.58 m in shallow groundwater and by 0.70 m in deep groundwater. By the end of 2018 a total suppression of ~1.5 km^3 of groundwater abstraction has been achieved in water-receiving areas of both the eastern and central routes.

The groundwater level has remained stable or even risen in some areas of these affected provinces or cities, for example, Beijing, Tianjin, Hebei, Henan, and Jiangsu. In the Beijing Plain the groundwater level rose by 3 m (~0.6 m/year) during the 5-year period from December 2014 to December 2019. On June 30, 2018, water replenishment for the eco-environment was conducted for 30 rivers in Hebei, Henan, Tianjin, and other provinces supported by the central route, with a total of ~0.9 km^3 water replenished. Streamflow increased and water quality improved after being replenished. Areas of Lake Baiyangdian that was ever the largest lake in the NCP and other rivers, lakes, and wetlands were largely expanded. The shallow groundwater level rose by ~0.5 m in areas lying within 5 km along the river channel in Hebei Province.

By the end of December 2019 the central SNWD route had transported 5.2 km^3 water in Beijing (~1 km^3/year), accounting for 80% of total water supply in the urban areas of Beijing, and thus has become the main source of water supply for Beijing. In 2019 the Ministry of Water Resources of the People's Republic of China together with relevant departments worked out the plan for comprehensive treatment of overextraction of groundwater in North China. This plan is focused on the Beijing–Tianjin–Hebei region (mainly the Haihe River basin) and involves comprehensive measures of "one decrease and one increase." "One decrease" means restricting groundwater extraction by water saving and adjusting agricultural structure. "One increase" means elevating regional water resources by increasing water supply and groundwater recharge. The ultimate goal of this plan is to gradually achieve a dynamic balance of groundwater between exploitation and replenishment in the NCP.

5.5 Review on groundwater storage assessment in the North China Plain

Many studies in terms of GWS change have been carried out in the NCP. GWS change can be inferred by mass balance, that is, subtracting water components (e.g., surface water, soil water, snow, and ice) from total water storage (TWS) change that can be observed by the GRACE satellites (Ebead et al., 2017; Feng et al., 2013; Huang et al., 2015). However, the time span of GWS change derived from GRACE data is relatively short (after 2002 when GRACE data were available). Based on groundwater level observations, GRACE data, and interpolation method, long-term changes in GWS could be obtained. For example, Gong et al. (2018) concluded that the groundwater depletion rate in the NCP was estimated to be −17.8 ± 0.1 mm/year during 1971−2015 based on in situ groundwater monitoring and GRACE data.

Regarding the influence of climate change and human activities on GWS in the NCP, Leng et al. (2015) modeled the GWS using Community Land Model version 4.0 (CLM4.0) by setting up two scenarios considering precipitation and temperature change and evaluated the impacts of climate and irrigation on groundwater and surface water (e.g., runoff, soil moisture storage, and groundwater level). Results show that irrigation exerts more influences on the groundwater than on the surface water. Irrigation and climate change have greater impacts on surface soil moisture and deep soil moisture, respectively. However, the climate scenarios were too simple to reflect the impacts of climate change in terms of spatial and temporal variability. Moreover, this study calculated the irrigation water demand based on the deficit of soil water content. This is not quite applicable in the NCP, that is, the time and amount of irrigation are determined largely by farmers (Lei et al., 2018).

Tang et al. (2013) analyzed GWS change in North China using GRACE data and concluded that human activities, for example, reservoir operation, interbasin water diversion, and large-scale commodity transportation (such as coal), would influence the mass variations of the entire region and thus affect the change in GWS detected by GRACE. From 2003 to 2011 the loss of GWS (112−136 mm) in the NCP has been balanced by the increase of the mass in this area caused by the increased reservoir storage (26 mm), diverted water (46 mm), and coal transportation (16.6 mm). Therefore the trend in TWS change monitored by GRACE was not that obvious. Xu et al. (2018) evaluated the GWS change by calculating electricity consumption per unit of groundwater pumping after implementing groundwater pumping restrictions. Results showed that reductions in groundwater withdrawal were 10.54 million m^3 in 2014 and 5.65 million m^3 in 2015, and the GWS in this area recovered to some degree. Ebead et al. (2017) calculated the GWS change during 2003 and 2012 using GRACE and CLM4.5 outputs (including river water storage, soil moisture, canopy interception, and snow water-equivalent) in North China and evaluated the influence of human activities on the GWS by comparing GWS derived from the CLM4.5 model (climate driving) and GRACE observations (climate and human activities driving).

A feasible solution to quantify GWS change is to look at model simulation output or to set up a hydrologic model or groundwater model for groundwater simulations (Ebead et al., 2017; Mukherjee et al., 2020). GRACE data are also often used to evaluate GWS change jointly with hydrologic models. Feng et al. (2018) estimated GWS changes of three main aquifers in China, including the NCP based on GRACE data and two global hydrologic models [i.e., WaterGAP Global Hydrological Model (WGHM) and PCRaster Global Water Balance (PCR-GLOBWB)] between 2002 and 2014 in China. Results show that the groundwater depletion rate was 7.2 ± 1.1 km^3/year in the NCP, which is relatively consistent with the estimate derived from in situ groundwater measurements (7.8 km^3/year). However, the rates of groundwater depletion from WGHM (12.8 ± 0.2 km^3/year) and PCR-GLOBWB (9.7 ± 0.2 km^3/year) were largely overestimated, and the peak of the simulated GWS by both models appeared to be ahead of time compared with the in situ measurements. Cao et al. (2013) established the MODFLOW-2000 model to simulate GWS changes and calibrated the model using the contour map of groundwater levels and in situ observations in the NCP. Contributions to groundwater recovery were also evaluated by increasing groundwater recharge, using brackish water to replace groundwater, improving water use efficiency, and the operation of the SNWD Project. However, this evaluation was realized by numerically adjusting the recharge and groundwater withdrawals, as opposed to taking water supply and water use as a whole in the analysis.

Other simulation methods, such as the Finite Element subsurface FLOW (FEFLOW) system and the method of coupling surface and groundwater models, have also been used to analyze the GWS change under different climate and water diversion scenarios in the NCP (Li et al., 2017; Xia et al., 2018; Zhang et al., 2018). However, effects of policy (e.g., groundwater pumping restrictions) on GWS are often overlooked in these methods. In addition, groundwater recharge (e.g., water percolation from the deep soil layer and preferential flow) is often not well described in these models. Since precipitation is the main source of groundwater recharge, the absence of these critical physical processes might lead to unrealistic analyses of the influence of climate variability/change on GWS.

In addition to physical models, machine learning and data-driven methods have also been used as alternatives to examining GWS changes, especially in regions where observed data are difficult to obtain (e.g., topographic, soil property, and geological data). Sun et al. (2020) used three machine learning methods (i.e., deep neural network, multiple linear regression, and seasonal autoregressive integrated moving average with exogenous variables models) to simulate TWS changes and compared with in situ GWS changes in the NCP. Results show that positive correlation coefficients accounted for most of the observation wells (~70%). However, the correlation was generally weaker in regions around megacities (e.g., Beijing and Tianjin) and some piedmont areas. This area has been subject to severe groundwater depletion due to intensive agriculture irrigation and urban water use (Feng et al., 2013; Shen et al., 2015). Possible reasons for the weaker correlation include (1) boring wells are often unable to reach the deep confined aquifers where a portion of groundwater is extracted and (2) large-scale coal mining activities in Shanxi Province (on the west of the Haihe River basin) would destroy the aquifers, which can be detected by the GRACE satellites but cannot be monitored by in situ measurements (Sun et al., 2020).

Machine learning approaches lack the description and simulation of physical mechanisms of GWS changes, and thus both uncertainties and risks exist (e.g., overfitting) (Sahoo et al., 2017). However, these approaches can be effective in improving our understanding of relationships among water storage change, climate variability/change, and human activities. Also, the learning-based models can remedy the mismatch between TWS anomalies derived from GRACE and land surface models by combining physically based modeling and deep learning (Sun et al., 2019). In addition, these approaches can simulate water storage changes accurately and efficiently and even perform better than global hydrologic models in some cases (Humphrey and Gudmundsson, 2019; Sahoo et al., 2017). Encouragingly, big data from remote sensing, the hydrologic field, and other state-of-the-art technology grow rapidly in recent years, which promotes machine learning as a valuable tool to explore complicated relationships within the fields of remote sensing and hydrology (Shen, 2018). In fact, numerous studies have successfully resolved hydrologic questions in the real world (Assem et al., 2017; Marçais and De Dreuzy, 2017; Reichle et al., 2001; Shi et al., 2015). Furthermore, with the rapid growth of remote sensing and big data, hydrologists should explore data analytics and artificial intelligence to continue our rapid progress into the future (AGU, 2020).

However, the scope and extent to which groundwater recovery in the NCP remains largely unknown, which relies on various monitoring approaches to be further assessed. In particular, there are no consensuses as to how various factors such as climate change/variability and human activities, including water diversion, managed aquifer recharge, and groundwater pumping restrictions contribute to the stability and recovery of groundwater, and how groundwater will change under the joint impacts of climate change/variability and intensive human activities, which is important to the development of the later phase of the SNWD Project and food security in the future. All of these urgent issues need to be properly assessed and fully addressed to provide important reference for water managers, practitioners, and policymakers toward a more sustainable water resources management and agricultural production.

Acknowledgment

This work was supported by the National Nature Science Foundation of China (51722903 and 51579128).

References

AGU, 6 April 2020. Konar, Long, and Madani receive 2019 hydrologic sciences early career award. Eos, 101. https://doi.org/10.1029/2020EO140823.

Asoka, A., Gleeson, T., Wada, Y., Mishra, V., 2017. Relative contribution of monsoon precipitation and pumping to changes in groundwater storage in India. Nat. Geosci. 10 (2), 109–117.

Assem, H., Ghariba, S., Makrai, G., Johnston, P., Gill, L., Pilla, F., 2017. Urban water flow and water level prediction based on deep learning. In: Paper Presented at Joint European Conference on Machine Learning and Knowledge Discovery in Databases. Springer.

Cao, G., Zheng, C., Scanlon, B.R., Liu, J., Li, W., 2013. Use of flow modeling to assess sustainability of groundwater resources in the North China Plain. Water Resour. Res. 49 (1), 159–175.

Chen, J., Li, X., 2004. Simulation of hydrological response to land-cover changes. Ying Yong Sheng Tai Xue Bao 15 (5), 833–836.

Döll, P., 2002. Impact of climate change and variability on irrigation requirements: a global perspective. Clim. Change 54 (3), 269–293.

Döll, P., Fiedler, K., 2008. Global-scale modeling of groundwater recharge. Hydro. Earth Sys. Sci. 12 (3), 863–885.

Dou, M., Zhao, H., Guan, F., Yao, B., Geng, Z., 2010. Design of supervision and management system of groundwater exploitation control in the South-to-North Water Transferred area. China Water Resour. 19, 43–45.

Ebead, B.M., Ahmed, M.E., Niu, Z., Huang, N., 2017. Quantifying the anthropogenic impact on groundwater resources of North China using Gravity Recovery and Climate Experiment data and land surface models. J. Appl. Remote. Sens. 11 (2), 026029.

Favreau, G., Cappelaere, B., Massuel, S., Leblanc, M., Boucher, M., Boulain, N., et al., 2009. Land clearing, climate variability, and water resources increase in semiarid southwest Niger: a review. Water Resour. Res. 45 (7).

Feng, W., Zhong, M., Lemoine, J.-M., Biancale, R., Hsu, H.-T., Xia, J., 2013. Evaluation of groundwater depletion in North China using the Gravity Recovery and Climate Experiment (GRACE) data and ground-based measurements. Water Resour. Res. 49 (4), 2110–2118. Available from: https://doi.org/10.1002/wrcr.20192.

Feng, W., Shum, C., Zhong, M., Pan, Y., 2018. Groundwater storage changes in China from satellite gravity: an overview. Remote. Sens. 10 (5), 674.

Gong, H., Pan, Y., Zheng, L., Li, X., Zhu, L., Zhang, C., et al., 2018. Long-term groundwater storage changes and land subsidence development in the North China Plain (1971–2015). Hydrogeol. J. 26 (5), 1417–1427.

Havril, T., Tóth, Á., Molson, J.W., Galsa, A., Mádl-Szőnyi, J., 2018. Impacts of predicted climate change on groundwater flow systems: can wetlands disappear due to recharge reduction? J. Hydrol. 563, 1169–1180.

Huang, Z., Pan, Y., Gong, H., Yeh, P.J.F., Li, X., Zhou, D., et al., 2015. Subregional-scale groundwater depletion detected by GRACE for both shallow and deep aquifers in North China Plain. Geophys. Res. Lett. 42 (6), 1791–1799.

Humphrey, V., Gudmundsson, L., 2019. GRACE-REC: a reconstruction of climate-driven water storage changes over the last century. Earth Syst. Sci. Data Discuss. 2019, 1–41. Available from: https://doi.org/10.5194/essd-2019-25.

Lei, H., Gong, T., Zhang, Y., Yang, D., 2018. Biological factors dominate the interannual variability of evapotranspiration in an irrigated cropland in the North China Plain. Agric. For. Meteorol. 250, 262–276.

Leng, G., Tang, Q., Huang, M., Leung, L.-y R., 2015. A comparative analysis of the impacts of climate change and irrigation on land surface and subsurface hydrology in the North China Plain. Reg. Environ. Change 15 (2), 251–263.

Li, X., Ye, S.-Y., Wei, A.-H., Zhou, P.-P., Wang, L.-H., 2017. Modelling the response of shallow groundwater levels to combined climate and water-diversion scenarios in Beijing-Tianjin-Hebei Plain, China. Hydrogeol. J. 25 (6), 1733–1744.

Liu, J., Zheng, C., Zheng, L., Lei, Y., 2008. Ground water sustainability: methodology and application to the North China Plain. Groundwater 46 (6), 897–909. Available from: https://doi.org/10.1111/j.1745-6584.2008.00486.x.

Marçais, J., De Dreuzy, J.-R., 2017. Prospective interest of deep learning for hydrological inference. Groundwater 55 (5), 688–692. Available from: https://doi.org/10.1111/gwat.12557.

Mukherjee, A., Scanlon, B., Aureli, A., Langan, S., Guo, H., McKenzie, A., 2020. Global Groundwater: Source, Scarcity, Sustainability, Security and Solutions, first ed. Elsevier, ISBN: 9780128181720.

Pool, D., 2005. Variations in climate and ephemeral channel recharge in southeastern Arizona, United States. Water Resour. Res. 41, 11.

Qiu, G.Y., Zhang, X., Yu, X., Zou, Z., 2018. The increasing effects in energy and GHG emission caused by groundwater level declines in North China's main food production plain. Agric. Water Manage. 203, 138–150. Available from: https://doi.org/10.1016/j.agwat.2018.03.003.

Reichle, R.H., Entekhabi, D., McLaughlin, D.B., 2001. Downscaling of radio brightness measurements for soil moisture estimation: a four-dimensional variational data assimilation approach. Water Resour. Res. 37 (9), 2353–2364.

Sahoo, S., Russo, T.A., Elliott, J., Foster, I., 2017. Machine learning algorithms for modeling groundwater level changes in agricultural regions of the U.S. Water Resour. Res. 53 (5), 3878–3895. Available from: https://doi.org/10.1002/2016wr019933.

Shen, C., 2018. Deep learning: a next-generation big-data approach for hydrology. Eos 99. Available from: https://doi.org/10.1029/2018EO095649.

Shen, Y., Kondoh, A., Tang, C., Zhang, Y., Chen, J., Li, W., et al., 2002. Measurement and analysis of evapotranspiration and surface conductance of a wheat canopy. Hydrol. Processes 16 (11), 2173−2187.

Shen, H., Leblanc, M., Tweed, S., Liu, W., 2015. Groundwater depletion in the Hai River Basin, China, from in situ and GRACE observations. Hydrol. Sci. J. 60 (4), 671−687. Available from: https://doi.org/10.1080/02626667.2014.916406.

Shi, X., Chen, Z., Wang, H., Yeung, D., Wong, W., Woo, W., 2015. Convolutional LSTM network: a machine learning approach for precipitation now casting. In: Paper Presented at Advances in Neural Information Processing Systems.

Small, E.E., 2005. Climatic controls on diffuse groundwater recharge in semiarid environments of the southwestern United States. Water Resour. Res. 41, 4.

Sun, H.-Y., Liu, C.-M., Zhang, X.-Y., Shen, Y.-J., Zhang, Y.-Q., 2006. Effects of irrigation on water balance, yield and WUE of winter wheat in the North China Plain. Agric. Water Manage. 85 (1), 211−218. Available from: https://doi.org/10.1016/j.agwat.2006.04.008.

Sun, A.Y., Scanlon, B.R., Zhang, Z., Walling, D., Bhanja, S.N., Mukherjee, A., et al., 2019. Combining physically based modeling and deep learning for fusing GRACE satellite data: can we learn from mismatch? Water Resour. Res. 55, 1179−1195. Available from: https://doi.org/10.1029/2018WR023333.

Sun, Z., Long, D., Yang, W., Li, X., Pan, Y., 2020. Reconstruction of GRACE data on changes in total water storage over the global land surface and 60 basins. Water Resour. Res. 56 (4), e2019WR026250. Available from: https://doi.org/10.1029/2019wr026250.

Tang, Q., Zhang, X., Tang, Y., 2013. Anthropogenic impacts on mass change in North China. Geophys. Res. Lett. 40 (15), 3924−3928.

Taylor, R.G., Todd, M.C., Kongola, L., Maurice, L., Nahozya, E., Sanga, H., et al., 2013a. Evidence of the dependence of groundwater resources on extreme rainfall in East Africa. Nat. Clim. Change 3 (4), 374−378.

Taylor, R.G., Scanlon, B., Döll, P., Rodell, M., Van Beek, R., Wada, Y., et al., 2013b. Ground water and climate change. Nat. Clim. Change 3 (4), 322−329.

Woldeamlak, S., Batelaan, O., De Smedt, F., 2007. Effects of climate change on the groundwater system in the Grote-Nete catchment, Belgium. Hydrogeol. J. 15 (5), 891−901.

Xia, J., Wang, Q., Zhang, X., Wang, R., She, D., 2018. Assessing the influence of climate change and inter-basin water diversion on Haihe River basin, eastern China: a coupled model approach. Hydrogeol. J. 26 (5), 1455−1473.

Xu, T., Yan, D., Weng, B., Bi, W., Do, P., Liu, F., et al., 2018. The Effect evaluation of comprehensive treatment for groundwater overdraft in Quzhou County, China. Water 10 (7), 874.

Ye, S., Xue, Y., Wu, J., Yan, X., Yu, J., 2016. Progression and mitigation of land subsidence in China. Hydrogeol. J. 24 (3), 685−693.

Zhang, Y., Shen, Y., Sun, H., Gates, J.B., 2011. Evapotranspiration and its partitioning in an irrigated winter wheat field: a combined isotopic and micrometeorologic approach. J. Hydrol. 408 (3−4), 203−211.

Zhang, M., Hu, L., Yao, L., Yin, W., 2018. Numerical studies on the influences of the South-to-North Water Transfer Project on groundwater level changes in the Beijing Plain, China. Hydrol. Processes 32 (12), 1858−1873.

Zheng, C., Liu, J., Cao, G., Kendy, E., Wang, H., Jia, Y., 2010. Can China cope with its water crisis?—perspectives from the North China Plain. Groundwater 48 (3), 350−354. Available from: https://doi.org/10.1111/j.1745-6584.2010.00695_3.x.

Zhu, Z., 2018. The process of water replenishment for the eco-environment in China's South-to-North Water Diversion. Available from: <https://mp.weixin.qq.com/s/Wm82PhS2fKwHyNxMa4OQfQ>.

Chapter 6

Emerging groundwater and surface water trends in Alberta, Canada

Soumendra Nath Bhanja[1] and Junye Wang[2]

[1]*Interdisciplinery Centre for Water Research, Indian Institute of Science, Bangalore, India*, [2]*Athabasca River Basin Research Institute (ARBRI), Athabasca University, Athabasca, AB, Canada*

6.1 Introduction

Freshwater (groundwater and surface water) is a crucial natural resource to support the basic water requirements in the world. While surface water is often used for water supply to cities, groundwater is common to support household water supply in rural areas, where surface water supply is not available (Alberta Environment, 2010). Groundwater wells are particularly an essential resource during winter in cold regions because the top surface of the surface water bodies is covered by ice and snowpack.

Understanding groundwater resources under changing climate is not easy due to the uncertainty associated with the responses of different hydrological components (Taylor et al., 2013). Infiltration and groundwater discharge in cold regions are complex hydrological processes due to permafrost, freeze−thaw cycles, and snow melt. Snowmelt can affect the infiltration and runoff rates because the snowmelt water movement in either vertical (infiltration) or horizontal (runoff) direction depends on several factors such as permafrost thickness, soil physical properties, and soil vertical structure. Snowmelt could therefore facilitate groundwater recharge at a warming scenario as one of the major sources of groundwater recharge and surface water in cold regions (Bales et al., 2006; Shrestha et al., 2017). However, this could give rise to rapid loss of stored soil organic carbon and have thus increasing atmospheric carbon dioxide concentration (Bhanja et al., 2019a; Bhanja and Wang, 2020).

Soil freeze−thaw cycle can essentially control the infiltration and runoff rates because infiltration can be inhibited in frozen soil and it reaches maximum during thaw conditions in spring and summer (Hayashi, 2013). Diminishing the extent of permafrost and frozen soil ground could also make way for other greenhouse gases such as nitrous oxide production (Bhanja et al., 2019b). The enhanced rates of soil greenhouse gas production at a changing climate might alter our understanding about the interlinked, hydro-biogeochemical processes in the future. Furthermore, a significant change in groundwater recharge and flow processes has been projected under a warming scenario in the near future (Evans et al., 2015). Remote sensing−based observations show that the temporal variation of seasonally frozen grounds is reduced in recent years, more specifically ~34 days during 1988 and 2007 in parts of Tibetan plateau (Li et al., 2012). The areal extent of permafrost regime might be diminishing substantially in 2100 (Slater and Lawrence, 2013). The change of state and regime of permafrost will be degrading, when they are subjected to the projected rise in air temperature (Lawrence and Slater, 2005).

Apart from the heterogeneous subsurface structure, the variations in thickness and duration of the seasonally frozen grounds due to ongoing climatic conditions make the groundwater studies very challenging in these areas. Scarcity of in situ data availability leads the scenario to more complex. Other proxies such as vegetation data should also be taken into consideration for groundwater level prediction at areas with seasonally frozen grounds (Bhanja et al., 2019c). Our understanding in this aspect can be improved with advancement in various available remote sensing techniques. Both remotely sensed and in situ hydrologic data assimilation are proving to be the best available solution at present (Girotto et al., 2017; Li et al., 2019; Sun et al., 2019). Quantifying uncertainty in multiple remote sensing and reanalysis products should be cautiously estimated in these cold regions (Yoon et al., 2019; Bhanja et al., 2018). Otherwise, groundwater security issues could exacerbate with increasing population and climate change−related concerns in near future

(Mukherjee et al., 2020). Here we show the groundwater and surface water level patterns from in situ estimates in Alberta, Canada. We also discuss their influencing factors such as rainfall and snowmelt patterns.

6.2 Data and methods

6.2.1 Study region

We have selected the entire province of Alberta, Canada. The study region is a part of cold climate zone (Peel et al., 2007) and good quality data are available here. Based on the annual soil temperature, less than 0°C occurs during some time of the year. Therefore most parts of Alberta can be characterized as seasonally frozen grounds with parts of shallow surface being subjected to freeze and thaw cycle (Willmott and Matsuura, 2012). Precipitation patterns show strong spatial heterogeneity in parts of Alberta with precipitation rates varying from <150 to >800 mm/year (Fig. 6.1). Most parts of the study area are characterized as semiarid climatic zone. Forest is the dominant land cover type with >70% of the area coverage (Bhanja et al., 2018).

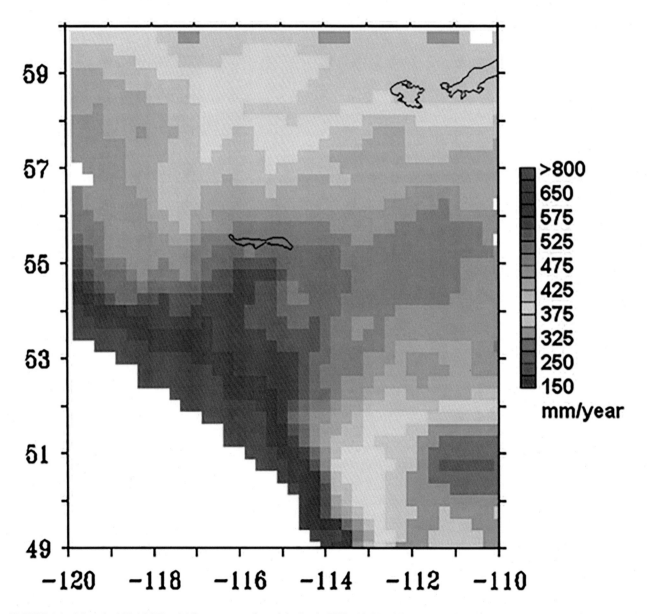

FIGURE 6.1 Map of gridded 0.25° × 0.25°, mean annual precipitation in 2003−14. Monthly mean precipitation data are obtained from the archives of Climatic Research Unit (CRU), University of East Anglia. We used the TS4.0 total precipitation product (Harris et al., 2014).

6.2.2 Groundwater level observation

Continuously monitored groundwater level data are retrieved from the Groundwater Observation Well Network, Alberta Environment and Parks, Government of Alberta. Initially, we obtained observed data of 470 wells. To maintain quality and continuity, we select the locations ($n = 157$) with at least 80% of the temporal coverage in 2003–14. The distribution of wells in different types of aquifers is categorized as follows: unconfined of 24, semiconfined of 17, confined of 100, and unclassified of 16.

6.2.3 Observations of surface water

We retrieve the daily time series of surface water level from Government of Canada's Water Office. Out of 393 locations, 65 locations are selected based on continuous data availability (at least 80% of the temporal coverage in 2003–14). Monthly mean data are computed from the daily data.

6.2.4 Rainfall and snowmelt water

Rainfall and snowmelt water data are obtained from the Global Land Data Assimilation System simulation (Rodell et al., 2004). We used three-model ensemble outputs of community land model, noah and variable infiltration capacity models. The ensemble outputs show a better performance than the individual model's estimates (Bhanja et al., 2016).

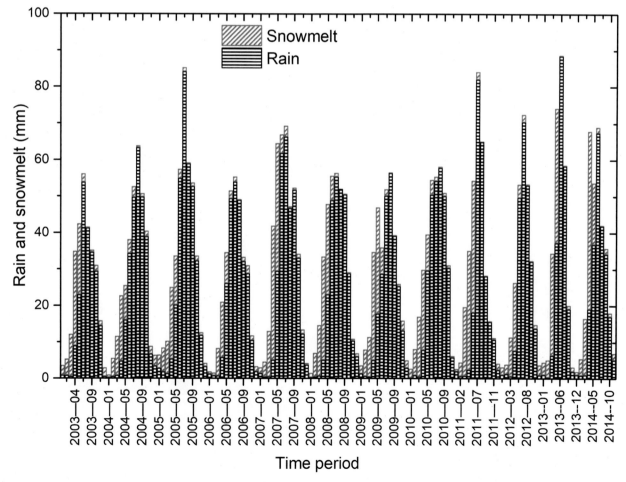

FIGURE 6.2 Time series of rainfall and snowmelt water in Alberta during 2003–14.

6.3 Results and discussions

6.3.1 Rainfall and snowmelt water

The two most natural factors of groundwater storage in cold regions are rainfall and snowmelt water. Infiltration rates are dominantly controlled by these two parameters. Snowmelt occurred throughout the years with highest magnitude in March to May. Being a semiarid region, snowfall rates are not too high in Alberta and most of the accumulated water

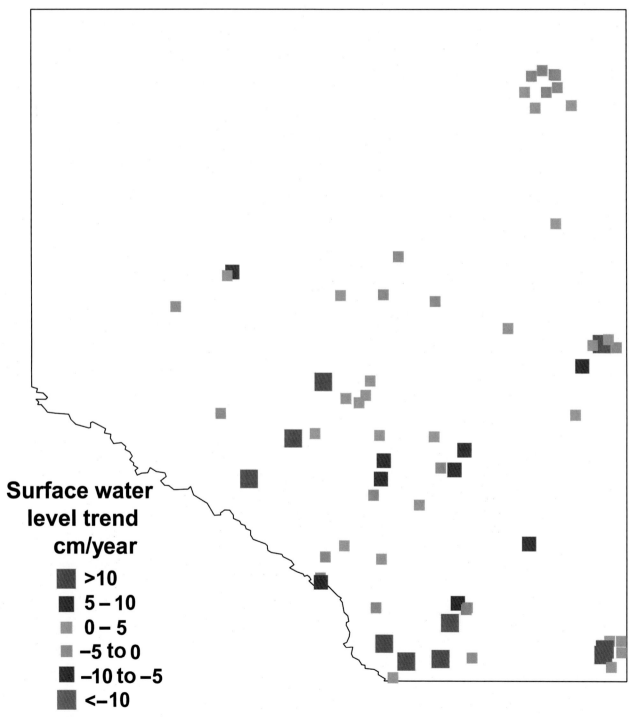

FIGURE 6.3 Surface water level trend (cm/year) in 2003–14.

of snowmelt run away within May. The observation shows spatial averaged estimates in Fig. 6.2, the time period would little differ across the regions from north to south direction. Rainfall rates show the highest values in June–July (Fig. 6.2).

6.3.2 Surface water level changes

In situ observation shows the highest surface water level during May–July. Rainfall and snowmelt runoff started peaking during this time (Fig. 6.2). Surface water levels are not kept to their highest values during August–September due

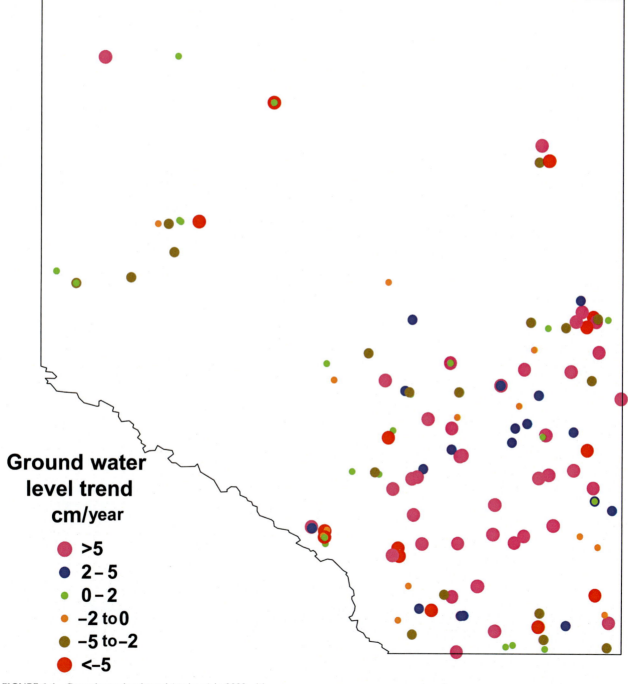

FIGURE 6.4 Groundwater level trend (cm/year) in 2003–14.

to increasing evapotranspiration levels linked to comparatively higher air temperature. Most of the locations show increasing surface water level over the time period 2003−14 (Fig. 6.3). As precipitation data have not shown any significant increasing patterns (Bhanja et al., 2018), the increase of surface water level may point toward higher depletion rates of glaciers or thawing soils. Another probable reason could be the potential lowering of evapotranspiration linked to a decrease in forest cover due to anthropogenic activities.

6.3.3 Groundwater level changes

Groundwater level trends show high spatial variability (Fig. 6.4). Most of the locations show a rise in groundwater table during 2003−14 at rates of >5 cm/year. One of the major reasons for this observed groundwater table rise might be linked to the increase in both rainfall and snowmelt water across Alberta. Area-averaged rainfall and snowmelt rates are increased at a value of 0.53 mm/year. Another crucial reason might be associated with the rising atmospheric temperature. Thawing soil during spring−summer leads to enhance permeability and thus induces groundwater recharge (Ge et al., 2011; Frampton et al., 2013). The value looks smaller in magnitude; however, considering the large area average, it can make way for substantial rise in available water for groundwater recharge as well as surface water stock rise. In situ observations show that very few negative trends might be linked to rapid groundwater usage in those regions (Fig. 6.4).

6.4 Summary

We analyze the variations in groundwater and surface water with rainfall and snowmelt. Our results show that the groundwater as well as surface water level data shows an increasing trend during 2003−14 across Alberta, Canada. Major reason for this observation might be linked to the combined rise in rainfall and snowmelt water during the same period. Other controlling factors include rising temperature-linked melting of permafrost and seasonally frozen grounds and their enhanced contribution toward groundwater and surface water storage. Rise in both groundwater and surface water levels warrants more research in this area.

Acknowledgments

S.N.B. acknowledges support from the Indian Institute of Science in the form of CV Raman Postdoctoral Fellowship for partly carrying out the study. We obtained the groundwater level data from the Groundwater Observation Well Network (GOWN), Alberta Environment and Parks (http://environment.alberta.ca/apps/GOWN). Surface water data were available from the Water Office, Government of Canada (https://wateroffice.ec.gc.ca/). The precipitation data were obtained from the Climatic Research Unit (CRU) archives (http://www.cru.uea.ac.uk/data) from the University of East Anglia, United Kingdom.

References

Alberta Environment, 2010. Facts About Water in Alberta. Government of Alberta, Edmonton, AB, p. 66.

Bales, R.C., Molotch, N.P., Painter, T.H., Dettinger, M.D., Rice, R., Dozier, J., 2006. Mountain hydrology of the western United States. Water Resour. Res. 42 (8). Available from: https://doi.org/10.1029/2005WR004387.

Bhanja, S.N., Wang, J., 2020. Estimating influences of environmental drivers on terrestrial heterotrophic respiration. Environ. Pollut. 257, 113630.

Bhanja, S.N., Mukherjee, A., Saha, D., Velicogna, I., Famiglietti, J., 2016. Validation of GRACE based groundwater storage anomaly using in-situ groundwater level measurements in India. J. Hydrol. 543 (B), 729−738.

Bhanja, S.N., Zhang, X., Wang, J., 2018. Estimating long-term groundwater storage and its controlling factors in Alberta, Canada. Hydrol. Earth Syst. Sci. 22, 6241−6255.

Bhanja, S.N., Wang, J., Shrestha, N., Zheng, X., 2019a. Microbial kinetics and thermodynamic (MKT) processes for soil organic matter decomposition and dynamic oxidation-reduction potential: model descriptions and applications to soil N_2O emissions. Environ. Pollut. 247, 812−823.

Bhanja, S.N., Wang, J., Shrestha, N., Zheng, X., 2019b. Modelling microbial kinetics and thermodynamic processes for quantifying soil CO_2 emission. Atmos. Environ. 209, 125−135.

Bhanja, S.N., Malakar, P., Mukherjee, A., Rodell, M., Mitra, P., Sarkar, S., 2019c. Using satellite-based vegetation cover as indicator of groundwater storage. Geophys. Res. Lett. 46 (14), 8082−8092.

Evans, S.G., Ge, S., Liang, S., 2015. Analysis of groundwater flow in mountainous, headwater catchments with permafrost. Water Resour. Res. 51, 9127−9140. Available from: https://doi.org/10.1002/2014WR016259.

Frampton, A., Painter, S.L., Destouni, G., 2013. Permafrost degradation and subsurface-flow changes caused by surface warming trends. Hydrogeol. J. 21 (1), 271−280. Available from: https://doi.org/10.1007/s10040-012-0938-z.

Ge, S., McKenzie, J., Voss, C., Wu, Q., 2011. Exchange of groundwater and surface-water mediated by permafrost response to seasonal and long term air temperature variation. Geophys. Res. Lett. 38 (14). Available from: https://doi.org/10.1029/2011GL047911.

Girotto, M., De Lannoy, G.J., Reichle, R.H., Rodell, M., Draper, C., Bhanja, S.N., et al., 2017. Benefits and pitfalls of GRACE data assimilation: a case study of terrestrial water storage depletion in India. Geophys. Res. Lett. 44 (9), 4107–4115.

Harris, I.P.D.J., Jones, P.D., Osborn, T.J., Lister, D.H., 2014. Updated high-resolution grids of monthly climatic observations – the CRU TS3. 10 Dataset. Int. J. Climatol. 34, 623–642.

Hayashi, M., 2013. The cold vadose zone: hydrological and ecological significance of frozen-soil processes. Vadose Zone J. 12, 2136. Available from: https://doi.org/10.2136/vzj2013.03.0064.

Lawrence, D.M., Slater, A.G., 2005. A projection of severe near-surface permafrost degradation during the 21st century. Geophys. Res. Lett. 32 (24), 1–5. Available from: https://doi.org/10.1029/2005GL025080.

Li, X., Jin, R., Pan, X., Zhang, T., Guo, J., 2012. Changes in the near-surface soil freeze-thaw cycle on the Qinghai-Tibetan Plateau. Int. J. Appl. Earth Obs. Geoinf. 17 (1), 33–42. Available from: https://doi.org/10.1016/j.jag.2011.12.002.

Li, B., Rodell, M., Kumar, S., Beaudoing, H., Getirana, A., Zaitchik, B., et al., 2019. Global GRACE data assimilation for groundwater and drought monitoring: advances and challenges. Water Resour. Res. 55 (9), 7564–7586.

Mukherjee, A., Scanlon, B., Aureli, A., Langan, S., Guo, H., McKenzie, A., 2020. Global Groundwater: Source, Scarcity, Sustainability, Security and Solutions, first ed. Elsevier, ISBN: 9780128181720.

Peel, M.C., Finlayson, B.L., McMahon, T.A., 2007. Updated world map of the Köppen-Geiger climate classification. Hydrol. Earth Syst. Sci. 11, 1633–1644. Available from: https://doi.org/10.5194/hess-11-1633-2007.

Rodell, M., Houser, P.R., Jambor, U.E.A., Gottschalck, J., Mitchell, K., Meng, C.J., et al., 2004. The Global Land Data Assimilation System, B. Am. Meteorol. Soc. 85, 381–394. Available from: https://doi.org/10.1175/BAMS-85-3-381.

Shrestha, N.K., Du, X., Wang, J., 2017. Assessing climate change impacts on fresh water resources of the Athabasca River Basin, Canada. Sci. Total Environ. 601, 425–440.

Slater, A.G., Lawrence, D.M., 2013. Diagnosing present and future permafrost from climate models. J. Clim. 26 (15), 5608–5623. Available from: https://doi.org/10.1175/JCLI-D-12-00341.1.

Sun, A.Y., Walling, D., Scanlon, B.R., Zhang, Z., Bhanja, S.N., Mukherjee, A., et al., 2019. Combining physically-based modeling and deep learning for fusing GRACE satellite data: can we learn from mismatch? Water Resour. Res. 55 (2), 1179–1195.

Taylor, R.G., Scanlon, B., Döll, P., Rodell, M., Van Beek, R., Wada, Y., et al., 2013. Ground water and climate change. Nat. Clim. Change 3 (4), 322–329.

Willmott, C.J., Matsuura, K., 2012. Terrestrial Air Temperature and Precipitation: Monthly and Annual Time Series (1900–2010).

Yoon, Y., Kumar, S.V., Forman, B.A., Zaitchik, B., Kwon, Y., Qian, Y., et al., 2019. Evaluating the uncertainty of terrestrial water budget components over High Mountain Asia. Front. Earth Sci. 7, 120.

Chapter 7

Groundwater irrigation and implication in the Nile river basin

Anjuli Jain Figueroa[1] and Mikhail Smilovic[2]

[1]Postdoctoral Scholar, School of Earth, Energy and Environmental Science, Stanford University, Stanford, CA, United States, [2]Research Scholar, Water program, IIASA — Institute of Applied Systems Analysis, Laxenburg, Austria

7.1 Introduction

The Nile is a well-known case of a transboundary river basin. Such basins require international cooperation to alleviate water scarcity and create opportunity (United Nations Development Program, 2006). Water scarcity is a serious problem across many nations (Mukherjee et al., 2020). Disputes among Nile riparian countries often revolve around how to divide the river among competing uses, including irrigation. Awulachew et al. (2012) estimated that satisfying the planned future irrigation projects in the basin would require 1.5 Nile rivers. In 2011 Ethiopia launched the Grand Ethiopian Renaissance Dam (GERD). Although primarily intended for hydropower, this project has sparked controversy as downstream countries fear a change in river flow. This has created an occasion for cooperation. The Nile Basin Initiative (NBI), an organization for joint management of the Nile, aims at equitable water use to prevent overallocation of the water, where the uncoordinated, individual, country-based, water-consumptive projects would leave the river dry. So far, the focus of international management organizations in the Nile Basin has been on surface water, but recently, emphasis is being placed on groundwater (IAEA and GEF, 2016). The importance of groundwater and surface water interactions has gained attention (Gleeson et al., 2019; Lewandowski et al., 2020) but has not yet been systematized in collaboration frameworks, in part because of limited information. Understanding groundwater and its interactions with surface water in this region is crucial to appreciating if developing groundwater will alleviate or intensify competition and conflicts in the basin.

Questions related to transboundary surface water management and rights of use extend to groundwater: Who should develop their groundwater resources? Where is groundwater and surface water linked? Could groundwater and surface water be treated as one resource to gain regional-level efficiencies? These questions are important because the groundwater recharge areas and the water use areas are not necessarily coincident and may fall in different countries.

In this chapter, we present a review of the transboundary Nile basin, exploring the current water use and land use along with the added dimension of groundwater availability. Sections 7.2 and 7.3 assess surface water and land use to understand riparian countries' dependency on rainfall, river water, and irrigation. Sections 7.4–7.6 describe the variable groundwater context and its connections to the surface water. The purpose of this chapter is twofold: first, to raise the possibility that a coordinated management of the regional hydrologic system (both surface and groundwater) could increase water for agricultural irrigation and second to identify aquifers with groundwater irrigation potential and indicate the implications (i.e., downstream effects) of development.

7.2 Surface water in the Nile basin

The Nile basin covers a surface area of around 3 million km^2, representing over 10% of the African continent. Fig. 7.1 shows the path of the Nile river flowing northward as it traverses about 6900 km and passes through 11 countries: Burundi, Democratic Republic of Congo, Egypt, Eritrea, Ethiopia, Kenya, Rwanda, South Sudan, Sudan, Tanzania, and Uganda. The Nile Basin Water Resources Atlas [Nile Basin Initiative (NBI), 2016] provides a good overview of surface water in the Nile basin. Several researchers such as Hurst (1952), Shahin (1985), Sutcliffe and Parks (1999) have

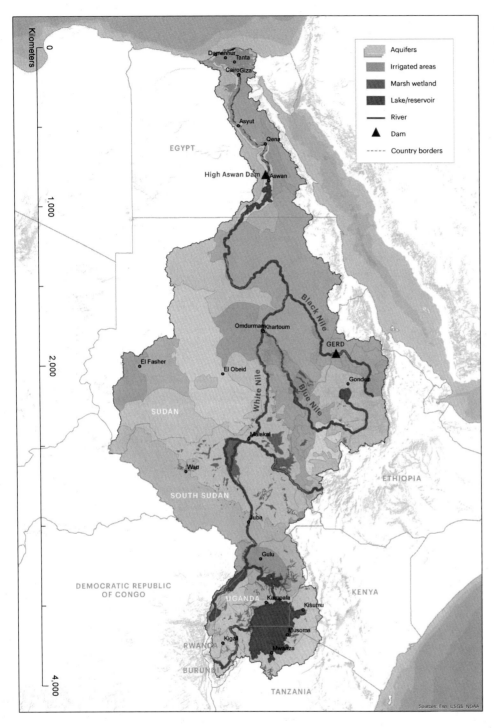

FIGURE 7.1 Water in the Nile basin: the Nile river and its main tributaries, including the White Nile, Blue Nile, and Atbara (Black Nile) rivers, are placed in the context of its different aquifers (Richts et al., 2011), wetlands (Lehner and Döll, 2004), irrigated areas (Meier et al., 2018), riparian nations, the High Aswan Dam and GERD. *GERD*, Grand Ethiopian Renaissance Dam.

provided a historical view of hydrology and climatology in the region, albeit neglecting the role of groundwater (Kebede et al., 2017). Fig. 7.1 shows the main Nile tributaries and their paths: The White Nile originates in the generally rainfall abundant Equatorial Lakes Region around Lake Victoria (Kenya, Tanzania, Uganda), where the wetlands and shallow aquifer systems moderate the flow. The Sobat river joins the White Nile toward the end of the Sudd wetlands and continues until its confluence with the Blue Nile near Khartoum, Sudan. The Blue Nile along with the Black Nile originate in the Ethiopian highlands. When they join, the Nile continues a dry path crossing Sudan and Egypt, to end in one of the largest river deltas in the world. The High Aswan Dam highly regulates the outflow to the delta with little Nile water reaching the Mediterranean Sea.

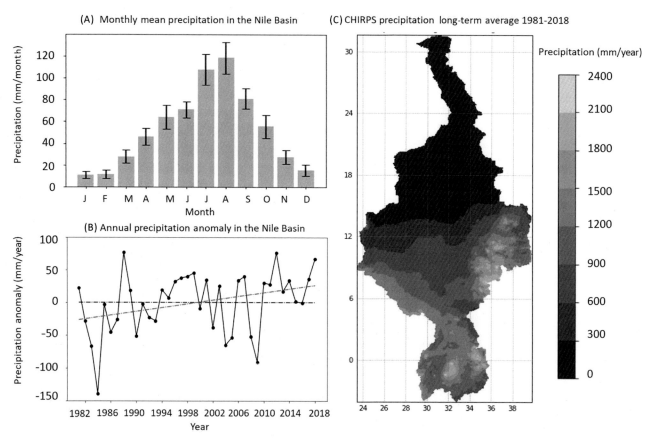

FIGURE 7.2 Temporal and spatial variability of rainfall in the Nile basin. Using data from the Climate Hazards Group Infrared Precipitation with Stations (CHIRPS) (Funk et al., 2015), (A) shows the monthly means with the effect of the monsoon in June to September, (B) shows the increasing trend in annual rainfall anomaly, and (C) shows the spatial distribution of the long-term average from 1981 to 2018. The abundant rainfall areas reflect the Nile river source with rain over the Ethiopian Highlands and Equatorial Lakes.

The water accounting of the basin has been estimated with both remote sensing (Karimi et al., 2012) and hydrological models (Kirby et al., 2010); for a review see Johnston (2012) and Johnston and Smakhtin (2014). On average, around 2000 km^3 of rain falls over the entire Nile basin. Fig. 7.2 shows the spatial and temporal rainfall variability throughout the basin. The monthly distribution of rainfall reflects the impact of the African Monsoon (Conway 2000) with rainfall peaking between June and September. Analysis of precipitation anomalies suggest an increase in rainfall between 1981 and 2018. The wetter regions (upstream), although variable, generally offer enough rainfall to support rainfed agriculture. The drier regions have adapted to living without rain and often irrigate and recycle water. Irrigation occurs almost exclusively downstream on around one-seventh of the agricultural land (Figs. 7.1 and 7.4). Nearly 80% of irrigation in the basin occurs in Egypt and the other 20% in Sudan (Johnston, 2012). These drier regions, particularly Egypt, have become entirely dependent on the conversion of upstream rainfall into the Nile's river flow.

A striking feature of the Nile, given its great length and importance, is the low annual stream flow. The long-term average flow is typically taken around 85 km^3 as measured at the High Aswan Dam in Egypt. Senay et al. (2014) identifies the flow of water, sources, and sinks, along the river. Most of the runoff generated in the Equatorial Lakes Region is evapotranspired in a series of swamps in the Sudd, where the river experiences a 50% loss in flows of the White Nile (Senay et al., 2014). The Blue Nile is more seasonal, experiencing high flows in June to September and running dry in between rains. The Sobat river, the White Nile, and the Atbara river each contribute about one-seventh of the flow, with the Blue Nile contributing four-sevenths (Blackmore and Whittington, 2008). The Ethiopian Highlands comprise 15%–20% of the land area and contribute 60%–80% of the annual flow of the lower Nile (Drooger and Immerzeel, 2010) The Nile river is a highly regulated system with several dams and subject to major abstractions. Fig. 7.3 shows the historical data from Global Runoff Data Center (GRDC) for 1960–80s at three dams in Sudan: Malakal, Roseires, and Dongola.

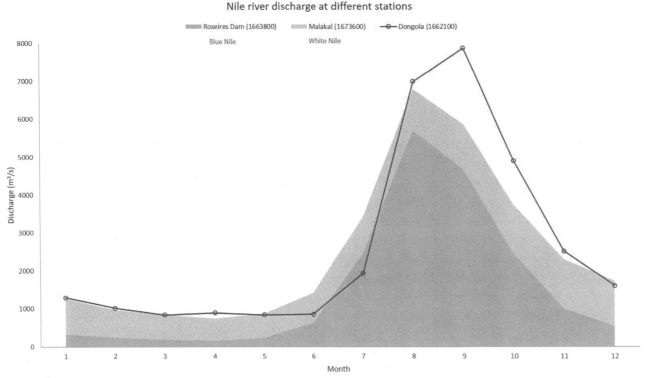

FIGURE 7.3 Average runoff based on Global Runoff Data Center data for three stations along the Nile basin representative of the White Nile (Malakal), the Blue Nile (Roseires), and the confluence of these at Dongola. Note the Blue Nile contributes 57% of flow the White Nile loses water in month June. North of Dongola rainfall is minimal. The High Aswan Dam regulates the outflow to the sea. *Credit: GRDC Data can be downloaded from: <http://www.compositerunoff.sr.unh.edu/>.*

7.3 Land use and irrigation in the Nile basin

Agriculture is an important economic sector in the Nile basin, making up about 20% of GDP in the region [Nile Basin Initiative (NBI), 2016]. Rainfed agriculture dominates in the Nile basin (Johnston, 2012) and covers around 35 million ha. Land cover change analysis between 2005 and 2009 suggest a decrease in forest areas and nearly a 6% increase in agricultural and cultivated lands across the basin [Nile Basin Initiative (NBI), 2016]. Productivity from rainfed agriculture varies significantly but is generally relatively low to moderate (Karimi et al., 2012), often averaging less than 1 t/ha [Nile Basin Initiative (NBI), 2016]. The Equatorial Lakes Region, covering 464,000 km^2 (Tramberend et al., 2019) experiences a relatively steady climate throughout the year and two rainy seasons, with long rains around April and shorter rains around November. Cropland covers one-third of the Equatorial Lakes Region or around 15 million ha and is almost exclusively rainfed for a variety of crops, including maize, root crops, and pulses. In Ethiopia, rainfed agriculture also dominates and the main agricultural cash crop for export is coffee. Irrigation is prominent in Egypt and Sudan. On average, 82 km^3 of water are withdrawn for irrigation, and officially about 8.8 km^3 (and potentially 2−4 km^3 more) is reused agricultural drainage water in Egypt (Barnes, 2014). Sudan has large-scale surface water irrigation covering about 1.5 million ha (Senay et al., 2014). The Gezira irrigation scheme fed by the Rosaries and Sennar dams is the largest and accounts for nearly 85% of irrigated areas in Sudan. Fig. 7.4 shows landcover from GlobCover data (ESA and U.C. Louvain, 2010) for the year 2009 over the Nile basin and helps explain why 70% of rainfall is evaporated from unmanaged vegetation including grasslands, shrublands, and forests, while 10% is evaporated from rainfed agriculture, and only 3% from irrigated agriculture. The remaining rainfall is partitioned as follows: 8% evaporates from water bodies and swamps, 7% goes to local runoff, and only 2% flows out into the sea (Johnston, 2012).

The riparian countries rely to different degrees on the Nile river (Oestigaard, 2012) as shown in Table 7.1. Over 200 million people rely on the Nile's waters (Johnston, 2012), and the population of the basin is projected to continue growing significantly. The Nile basin has large hydropower and agricultural potentials that have not yet been tapped; these great potentials are particularly true for groundwater, which has been used for irrigation in limited quantities.

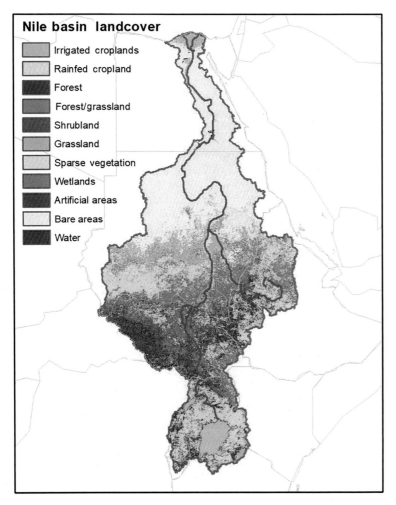

FIGURE 7.4 Irrigation in the Nile basin. Irrigation in the Nile basin which occurs primarily in Sudan and Egypt. Rainfed croplands are spread across riparian nations, but the dominant landcover is still natural noncrop vegetation, including forests, shrublands, and grasslands which evaporate nearly 70% of the rainfall. *Credit: Data from GlobCover 2009 by European Space Agency (ESA) and the Université Catholique de Louvain Public dataset under CC-BY-4.0.*

TABLE 7.1 Surface water use overview.

Nile region	Surface water consumption level and primary use	Downstream effect of surface water use and impact
Egypt	High Irrigation	Medium Little water reaches the delta
Ethiopia	Low Rainfed agriculture	Medium Increased use reduces flow
Sudan and South Sudan	Medium Irrigation	Medium Increased use reduces flow
Equatorial Lakes	Low Rainfed agriculture	Low-medium Increase use could reduce flow to the Sudd swamp

The development of groundwater irrigation in the generally upstream and predominately rainfed regions may be an opportunity to buffer rainfall variability, stabilize yields, and prevent crop failures (Abiye and Mmayi, 2014), while not necessarily challenging downstream water demands for irrigation (Johnston, 2012). It is still an open question whether using groundwater can ameliorate or exacerbate transboundary surface water issues. On the one hand, in systems with linked surface and groundwater, increases in groundwater pumping can drain the river exacerbating the surface water conflict. On the other hand, tapping a fossil aquifer disconnected from the surface water, could supplement water resources and help reduce dependency on rainfall and surface water. Table 7.1 summarizes the surface water use and its effect downstream.

7.4 Groundwater in the Nile basin

Groundwater in the basin has some salient properties. The groundwater aquifers follow neither the river basin bounds nor the administrative bounds (Fig. 7.1) and the aquifers may be shared with additional countries outside of the basin (MacAlister et al., 2012). In fact, there are several transboundary aquifers within the Nile basin, with the largest being the Nubian Sandston Aquifer (Nijsten et al., 2018). Groundwater plays a buffering role in times of no rain, as well as recharging aquifers in wet years, and still discharging to rivers and lakes in dry years. This helps reduce the shocks of erratic rainfall.

Groundwater in the Nile occurs primarily in four hydrogeological formations: (1) the crystalline hard rock basements of the Equatorial Lakes and South Sudan, (2) the volcanic rocks of the Ethiopian highlands, (3) the consolidated sedimentary rocks, mostly sandstone and limestone, in Egypt and Sudan, and (4) the unconsolidated sedimentary rock aquifers which follow the course of the Nile river. Fig. 7.5 shows where these formations are in the Nile Basin. Furthermore, the figure shows estimated groundwater productivity from the Africa Groundwater Atlas (Upton et al., 2018; Gadelmula et al., 2018; Owor et al., 2018; Kebede et al., 2018). The maps reveal the spatial distribution of groundwater productivity, with high productivity levels under Egypt and quite low in the Equatorial Lakes; this is

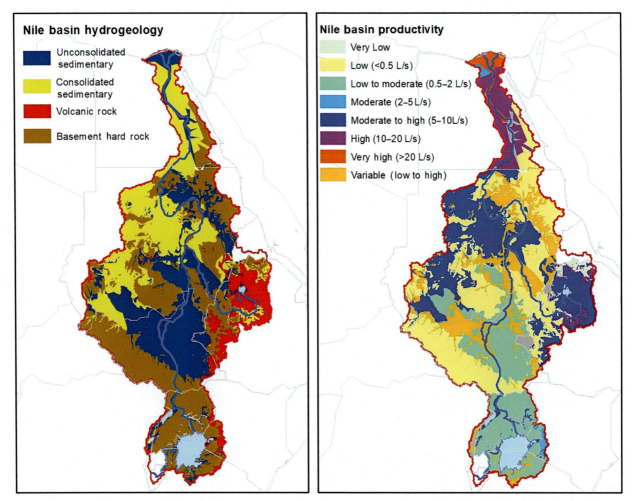

FIGURE 7.5 Hydrogeology and estimates for groundwater productivity for aquifers in the Nile basin. *Credit: Data from The Africa Groundwater Atlas under a Creative Commons 'Attribution/Share Alike' (CC BY SA) license.*

almost the reverse of the spatial distribution of the rainfall availability in Fig. 7.2 with a weak but statistically significant negative correlation between the spatial rainfall map and spatial groundwater productivity map (spearman $\rho = -0.33$ and $P < .05$).

Generally, the aquifers in Egypt's portion of the Nile are consolidated sedimentary basins with low recharge (<2 mm/year) and large depths to the groundwater [100−250 m below ground level (mbgl)]. Sudan's portion of the Nile has major, local, and shallow aquifers with medium to low recharge (<100 mm/year) and varying depths to groundwater generally decreasing depths as you move south. South Sudan has major groundwater basins with medium recharge (20−100 mm/year) and depths to groundwater between 7 and 25 mbgl. The Ethiopian highlands have a volcanic complex hydrogeological structure with high recharge (100−300 mm/year) and water depths less than 25 mbgl. WHYMAP data of the Equatorial Lakes Region show shallow and local aquifers with water depths less than 25 mbgl and with medium recharge (<100 mm) (Richts et al., 2011). Maps of groundwater data can be obtained from the British Geological Survey Africa Groundwater Atlas described in MacDonald et al. (2012).

Data from the Gravity Recovery and Climate Experiment (GRACE) provide an estimate of groundwater trends (Wiese, 2015; Watkins et al., 2015; Wiese et al., 2016, Swenson 2012, Landerer and Swenson, 2012, Swenson and Wahr, 2006) and are shown in Fig. 7.6 for the Nile basin. The data suggest a gain in groundwater storage of 3.5 mm/year at the basin level. This increasing trend could be related to the increased precipitation trend (Fig. 7.2). The longer term trend relative to precipitation is minimal, representing less than 1% of the average basin precipitation (Senay et al., 2014). It is worth noting that the basin-wide view of groundwater trends hides some important regional variability and local depletions.

Surface water can be spatially characterized by limited upstream use, where rainfed agriculture dominates, and extensive downstream use, where irrigated agriculture dominates. The aquifer systems in the Nile basin can be similarly distinguished. Downstream sedimentary basins have considerable storage, high productivity, and are generally more tapped while upstream crystalline basement and volcanic rock basins have varying groundwater productivity and limited

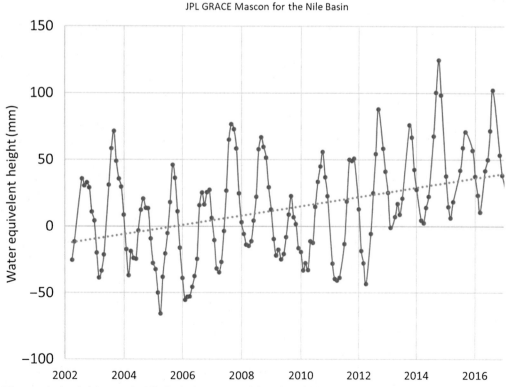

FIGURE 7.6 Water equivalent height over the Nile Basin from the Land GRACE-FO JPL mass concentrations. The data suggest an increasing trend in groundwater in the Nile Basin. The large scale necessary for Grace data to be meaningful can hide many local depletion problems. *GRACE*, Gravity Recovery and Climate Experiment. *Credit: GRACE land data are available at http://grace.jpl.nasa.gov, supported by the NASA MEaSUREs Program. GRACE data can be accessed and visualized for river basins with NASA's GRACE data analysis visualization tool (DAT) (https://grace.jpl.nasa.gov/data/data-analysis-tool/) or the Colorado University visualization tool (https://ccar.colorado.edu/grace/jpl.html).*

use (Abiye and Mmayi, 2014). Groundwater is currently used for some irrigation and drinking water needs. Expansion of groundwater development could be an important additional source of water in the region that faces water stress. Alternative methods to increase water supply include increasing irrigation efficiency, which is unlikely to be enough to meet demands (Multsch et al., 2017) and converting water that is lost to evaporation into productive agricultural water, which can have ecological implications (Mohamed et al., 2005). Identifying aquifers for increased groundwater irrigation with limited downstream impacts could enhance food and water security and could lead to regional economic growth.

7.5 Aquifers in Nile riparian countries

Fig. 7.7 shows seven selected aquifers that are described in order to identify the opportunities and limitations of increased groundwater use. The aquifers considered include the Nile delta, the Nile valley, the Nubian Sandstone, the Umm Ruwaba (Sudd), the alluvial Rift Valley, Lake Tana, and the weathered basement aquifer of the Equatorial Lakes region. Groundwater pumping is projected to continue to increase, and the effects that additional pumping may have on the basin and river are still unclear. Kebede et al. (2017) describe a historical state change that occurred after the desertification of the Sahara in which reduced recharge to groundwater shifted the Nile from a groundwater gaining to a groundwater losing river in Sudan and Egypt. Could unmonitored pumping lead to another state change? Proper planning should seek to better understand the surface and groundwater interactions that could occur from increased water use.

7.5.1 Groundwater in Egypt

El Tahlawi et al. (2008) describe in detail eight aquifers that underlie Egypt. The three main aquifers are the Nile delta, the Nile Valley, and the Nubian Sandston Aquifer (Abiye and Mmayi, 2014).

The Nile delta is one of the world's most stressed aquifers (Gleeson et al., 2012). The aquifer is bounded by the Mediterranean Sea on the north, the Nile river in the south, and canals in the east and west. While the Nile delta represents 5% of Egypt's area, it is densely populated holding 25 large cities. Recharge into the aquifer is primarily from

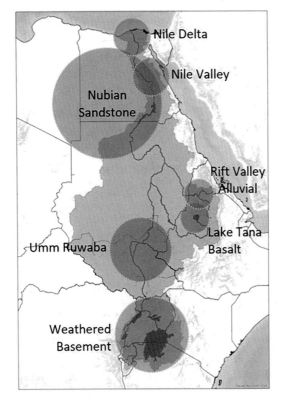

FIGURE 7.7 The general locations of the aquifers discussed in the chapter.

deep percolation; excess irrigation water; and seepage from the river branches, canals and drainage system. Estimates of the recharge rate vary ranging from 0.25−0.8 mm/day (Negm, 2019) to 3.5−7 km^3/year (El-Agha et al., 2015). The aquifer has an estimated depth of 50 m. Most groundwater abstractions (about 85%) occur in this aquifer with an estimated volume of 6.1 km^3/year (Negm, 2019). MacAlister et al. (2012) suggests that over 31,000 productive deep wells and 1700 observations wells are distributed in Nile delta. These serve a fraction of the 2 million farmers in the region, but according to El-Agha et al. (2017) a massive tube-well drilling industry is silently booming and conjunctive use of surface water and groundwater is becoming the norm in the Nile delta. Fig. 7.8 shows that the trend of irrigation expansion is also occurring outside of the Nile delta, with many wells largely unmonitored. The increased use of groundwater for irrigation in the Nile delta, which is recharged by waters from the Nile river, can exacerbate surface water tensions. Continued groundwater expansion within the Nile delta is facing constraints because of saline groundwater (van Engelen et al., 2018), ocean intrusion, and land subsidence (Richts et al., 2011). Consequently, this aquifer does not present the best opportunity for irrigation expansion. A more detailed description of the Nile delta is provided by El-Agha et al. (2015).

The Nile Valley quaternary aquifer underlies the banks of the Nile river. It includes the floodplain with sand, gravel, and loamy soil. Recharge occurs primarily from irrigation seepage. The aquifer is directly connected to the Nile river, canals, and drains. Groundwater is abstracted for both irrigation and domestic purposes, with an estimate of 1.932 km^3/year for irrigation and 0.56 km^3/year for domestic use (Dawoud and Ismail, 2013). The aquifer thickness ranges from 50 m at the desert fringes to 300 m in the valley center. Dawoud and Ismail (2013) provide an estimated water balance

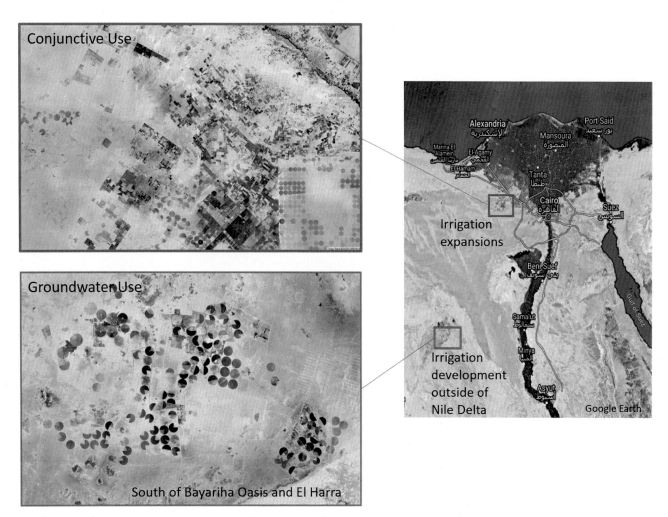

FIGURE 7.8 Agricultural irrigation expansion in Egypt into dry regions far from the Nile river and canal system where irrigation is possible by using groundwater. *Credit: Google Earth Satellite Images.*

of the Nile Valley aquifer. Since the Nile Valley is hydraulically connected to the Nile river, increased pumping in this aquifer could eventually lead to using water from the river and leaving it dry. Thus, this aquifer provides limited opportunity for irrigation expansion.

The Nubian Sandstone Aquifer System (NSAS) is the largest fossil aquifer in the Nile basin, at about 259 km^3. Shared among several countries, it covers an area of 2.2 million km^2 with 38% lying under Egypt. The quantity of water in the NSAS is uncertain with estimates ranging from 15,000 to over 500,000 km^3 (Negm, 2019). Sediment thickness varies from few hundred meters in the south to four km in the west. Heads vary from 78 to 570 m. Recharge is limited and comes primarily from the Nile river by influx from the Blue Nile. Annual recharge is estimated at 1.54 km^3 and total abstraction estimates are less than 3 km^3/year (Zekri, 2020), both minimal compared to its significant storage (Abiye and Mmayi, 2014). Expansion of irrigation in the NSAS could offer some opportunities due to its large size. Use has already begun in the New Valley. Although recharge is connected to the Nile river, if treated as a fossil aquifer (negligible recharge), groundwater could be mined. However, additional information regarding the aquifer properties would be important to prevent wells from drying out and having to dig deeper. The main constraint in NSAS may be the large depths making it costly to tap and the expansion of unsustainable use. Proper management and coordination with neighbors would be needed to prevent subsidence.

7.5.2 Groundwater in Sudan and South Sudan

The vast lowland of Sudan and South Sudan provides a suitable environment for groundwater storage. According to Omer (2002), 80% of the population of Sudan (including South Sudan) uses groundwater for day-to-day activities. FAO AQUASTAT (2016) estimated the area equipped for groundwater irrigation at 74,000 ha. Omer (2002) provides a detailed description of the geology and groundwater in Sudan and South Sudan. He estimates groundwater abstractions of 1.43 km^3/year and describes four main water-bearing formations in Sudan with the most important being the Nubian sandstone formations and the Umm Ruwaba formation. The Umm Ruwaba basin has two major basins, the Sudd and the Keaster Kordofan and an estimated storage of 4150 km^3. Omer (2002) identifies the southern margin of the Sahara area as a promising area for groundwater development.

South Sudan has limited infrastructure for irrigated agriculture with FAO AQUASTAT (2016) estimating an agricultural area equipped for groundwater irrigation of only 1525 ha. A large part of the country's area is covered by the protected Sudd wetland and the southwestern part of the country is low-productivity hard rock basement with unconfined aquifer of 4–60 m depth. Some shallow water aquifers of unconsolidated alluvium have depths less than 15 mbgl. Part of the Nubian Sandstone aquifer underlies South Sudan but hydrogeological information is limited. The aquifer most likely to currently offer opportunities is the Umm Ruwaba formation, an unconsolidated aquifer shared with Sudan that covers a large area and has low to moderate productivity (well yields of 1.4–4.5 L/s). Recharge is primarily from rainfall. Salinity can be an issue in this aquifer [Nile Basin Initiative (NBI), 2016]. In South Sudan the main limitation to develop groundwater potential is civil conflict and limited data. Further information is needed regarding the aquifers in Sudan and South Sudan to better understand the extent to which they are connected to the river system.

7.5.3 Groundwater in Ethiopia

Although in Ethiopia groundwater is not used for agriculture, drinking water supply depends on it. Nearly 70% of the rural population and some medium and large cities, including the capital, Addis Ababa (40%), are supplied by groundwater (MacAlister et al., 2012). For Ethiopia, groundwater for agricultural irrigation could assuage the dependence on rainfall that has often led to crop failure and famines. Supplementary groundwater irrigation at the end of the wet season could reduce losses and failures in crop yield by 50%–80% (Araya and Stroosnijder, 2011; Johnston, 2012). Water-centered development is viewed as a way for growth, reduction in poverty, and improvement in livelihood (Kebede, 2013).

The state of groundwater in Ethiopia is still not well understood. Estimates for groundwater potential in the region have a large range, between 3 and 185 km^3 (Moges, 2012). Kebede (2013) estimates the total groundwater storage, accounting for the total surface area of Ethiopia, to be about 1000 km^3 with 55% freshwater and 45% brackish and salty. Areas with low groundwater potential have deep water tables, high salinity, or low recharge. For a detailed description of groundwater hydrogeology in Ethiopia, see Kebede (2013). The hydrogeology in Ethiopia reflects all four formations found in the Nile basin (Fig. 7.5): (1) volcanic rock, (2) hard basement rock, (3) unconsolidated alluvial sediments, and (4) consolidated sedimentary rock. Volcanic rock underlies 60% of Ethiopia and is the predominant formation in the Ethiopian part of the Nile basin. The hard basement rock underlies 18% of Ethiopia's surface area. These areas have little rainfall leading to limited weathering. In the highlands area, most groundwater movement occurs in the hard rock open fractures (Abiye and Kebede, 2011). Alluvial sediments cover less than 25% of the land surface, particularly in the Rift valley, with mostly shallow water tables. Wells in these formation range in depth from 20 to 100 m (Moges, 2012). These can be prolific aquifers with wells yielding 1–5 L/s and can support large groundwater storage, thus making it a possible region for groundwater development. The majority of Ethiopia's recharge is localized and indirect. Rather than recharge from excess soil moisture, recharge is mainly from heavy rainfall in the highlands and from rapid infiltration through fractures in areas covered by fractured volcanic rock. Kebede et al. (2017) estimate that groundwater in the Ethiopian highlands can contribute up to 50% of the surface water flows emphasizing the importance of understanding where surface water and groundwater are linked.

Aquifers made of sedimentary rocks have highly variable aquifer properties and occur in the Blue Nile basin. Many of these aquifers require deep wells, averaging 250 m and as high as 400 m (Moges, 2012), and, thus, it is unlikely that these would be suitable for large-scale groundwater irrigation development. The extraction of shallow groundwater could be economical for domestic use and small-scale farming, such as in the basalt aquifer south of Lake Tana showing high groundwater potential (Kebede, 2013). Exploitation of groundwater resources in Ethiopia has been limited by human capacity. Furthermore, well failure is a major problem, since, in some areas, up to 70% of developed water wells fail to deliver water after construction (Kebede, 2013).

7.5.4 Groundwater in the Extended Lake Victoria basin

Groundwater in the extended Lake Victoria basin (defined as the subbasin upstream from the border between Uganda and South Sudan) is generally held in shallow aquifer systems found in the weathered mantle and fractured sections of underlying crystalline basement rock with low to moderate flows (Abiye and Mmayi, 2014; Johnston, 2012). Groundwater abstraction from shallow wells is limited and intended for domestic use and livestock, while abstraction for irrigation is not widely practiced. However, the 230,000 ha of rainfed agriculture practised in and around wetlands and valleys depends on the interaction with underlying groundwater systems, which moderate flows and act as sinks for surface runoff.

Projections based on different development scenarios for the extended Lake Victoria basin suggest increasing population and rapid urbanization will increase water demand across all sectors (Tramberend et al., 2019). Expanding irrigation in the region could provide an opportunity for the extended Lake Victoria basin to stabilize and increase yields, as well as support economic development and food security. Different studies suggest that increasing the irrigated areas from 50,000 to 400,000 ha is unlikely to have any significant impact on downstream flows (Tramberend et al., 2019). Given the relatively low flows leaving the extended Lake Victoria basin and high evapotranspiration from the Sudd wetlands, an increase to 700,000 ha of irrigated land could result in an estimated reduction of 1.5 km^3 to High Aswan Dam, representing less than 2% of Egypt's annual irrigation use (Johnston, 2012). Whether these increases come from surface or groundwater irrigation will depend on the different levels of support provided for irrigation infrastructure, as well as each country's complex land-tenure systems (Wanyama et al., 2017). Although the connection between the surface water and shallow groundwater systems is not fully understood, the shallow system and series of lakes and wetlands suggest that abstractions from either could affect the availability of the other. The development of either surface water or groundwater, however, could affect the flows to the Sudd swamps and potentially interrupt ecosystems (Mohamed et al., 2005; Angarita., et al, 2017). Groundwater offers the benefit of potentially buffering periods of drought and offering stability amongst changing rainfall onset and duration, potentially reducing the risk of rainfed agriculture (Johnston 2012; Wanyama et al., 2017). Although shallow, the storage capacity is "enormous" (Abiye and Mmayi, 2014) and seasonally recharged from local runoff (Johnston, 2012). Abiye and Mmayi (2014) suggest that groundwater abstraction could provide an opportunity of storing excess seasonal rainfall. Scenario development for the region up to 2050 using both hydrological modeling and economic optimization suggests that surface water will continue to satisfy most water demands for the region, with groundwater estimated to provide 7%–9% (Tramberend et al., 2019). Although groundwater may offer a significant opportunity, the necessary costs in the context of the availability and current predominant use of surface water suggest that investment in groundwater in the short term may be limited.

7.6 Discussion and conclusion

The groundwater opportunity in many regions around the Nile basin is high, but it requires integrated planning to avoid exacerbating conflicts regarding surface water (Mukherjee et al., 2020). In the basin, Egypt uses the most groundwater, while South of Sudan, large-scale groundwater development has yet to take off. Groundwater productivity has the

TABLE 7.2 Overview of select aquifers in the Nile river basin.

Region	Aquifer and hydrogeology	Groundwater irrigation opportunity	Groundwater abstraction and use	Effect of groundwater development downstream and on surface water
Egypt	Nile Delta Unconsolidated sedimentary	Limited opportunity because development could exacerbate sea-water intrusion and subsidence.	6.1 km^3/year (Negm, 2019) Irrigation, primarily rice	Development could decrease downstream flow into the delta.
	Nile Valley Unconsolidated sedimentary	Limited opportunity due to groundwater–surface water interactions.	~2.5 km^3/year (Dawoud and Ismail, 2013) Irrigation and domestic	Development could potentially decrease downstream flows into the delta.
	Nubian Sandstone Consolidated sedimentary	Opportunity to tap a large store and reduce dependency on the Nile river by mining water.	<3 km^3/year (Zekri, 2020) Irrigation	Development does not affect Nile surface water downstream. It has limited recharge from the river.
Ethiopia	Lake Tana Volcanic (Basalt)	Opportunity to supplement rainfed agriculture through irrigation.	Unknown quantity Primarily domestic	Development could affect surface water availability if water is taken from the lake that feeds the Nile river.
	Rift Valley Alluvial sediments	Opportunity to supplement rainfed agriculture by tapping a shallow, prolific, and large storage for irrigation.	0.005 km^3/year (Abiye and Mmayi, 2014) Domestic	Development could affect rivers that drain north and south of the uplift but may not affect the Nile. A better understanding of the fragile Rift ecosystem is necessary.
Sudan and South Sudan	Umm Ruwaba	Unknown. Information is limited including interaction with protected Sudd wetland.	0.012 km^3/year (Abiye and Mmayi, 2014) Domestic and livestock	Extended development could affect surface water flows. There could be ecological implications for the wetlands.
Extended Lake Victoria	Weathered basement	Opportunity to supplement rainfed agriculture by tapping a shallow aquifer with significant storage and seasonal recharge but limited by transmissivity.	~2.7 km^3/year (Abiye and Mmayi, 2014) Domestic and livestock	Extended development could affect flow into the Sudd Swamps. Limited effect downstream into Egypt as most of the water is consumed in the swamps.

opposite spatial pattern as rainfall with downstream Egypt exhibiting high productivities and the Equatorial Lakes Region having lower productivities. Including groundwater in the Nile basin management would stimulate debate regarding the regional hydrologic system and may focus the problem of water scarcity, water sources and water uses in a different way. Downstream Egypt may not seem as water scarce and upstream Ethiopia may find alternatives that do not take from the surface water. Many of the potential development areas are hydrologically linked to surface water, for example, in Ethiopia aquifers under the Blue Nile are connected to Lake Tana. Similarly, in Egypt, the Nile delta and Nile Valley are recharged by irrigation seepage from the Nile river. Table 7.2 summarizes the groundwater development opportunity in the aquifers described in this chapter. The interaction between surface water and groundwater adds to the already contentious transboundary water issues in the region. For example, if an upstream country uses river water, it can affect the sustainability of groundwater use in a downstream country even if the downstream country does not increase abstraction. This is because the diminished recharge from the river allows for less water to be extracted sustainably and the same extraction level may now mean overexploitation. Furthermore, it is important to understand the impact that water infrastructure, both dams and wells, will have on surface water groundwater interaction. In management of the new GERD dam, these groundwater interactions could be substantial, albeit local. Kebede et al. (2017) describes the impact that construction of HAD dam had in recharging the nearby aquifers. The strong surface water––groundwater interactions suggest that the water should be managed as a single resource.

Each Nile riparian country has potential of developing groundwater use but for different purposes. Ethiopia can benefit from increased irrigation to reduce dependence on rainfall, though it is constrained by limited human capacity. Developing groundwater irrigation could be less expensive than surface water irrigation, and it requires less coordinated government effort. In selecting where to develop, Ethiopia should plan with its drier, river-dependent neighbors in mind to ensure that groundwater pumping does not take water from the Nile river. Egypt and Sudan can develop fossil aquifers that are not linked to the Nile, such as in the Nubian Sandstone aquifer. This is an opportunity to reduce dependence on the Nile river (for which they do not control the source) but may require transboundary aquifer management with other neighbors. The main risk of developing fossil aquifers is mining them and exploitation leading to subsidence. South Sudan and the countries of the extended Lake Victoria basin have shallow aquifers and few irrigation schemes. There is opportunity to benefit from both surface and groundwater development and the impact downstream in terms of quantity would be limited, but a proper understanding of the ecological effect would still need to be carried out.

This chapter has argued that groundwater should not be overlooked in the management of transboundary surface water basins such as the Nile, and the potential of aquifer development should be considered in the context of its surface water implications and downstream effects. Integrating groundwater into basin management can reveal which sources, when tapped, can intensify surface water competition and which can supplement and enhance water and food security. Groundwater is an important buffer for surface streams, lakes, and crops. In dry years, groundwater can bolster the water levels and help sustain trees and plots when precipitation is low. Information about aquifers is still scant, and there is no regional monitoring of groundwater development or pumping. Uncontrolled and unplanned pumping beyond sustainable limits could remove the safety net for lakes, streams, and crops, as aquifers can take a long time to recharge, and overexploitation can lead to subsidence and saltwater intrusion. Development of groundwater resources must occur in a coordinated way, especially in a region facing surface water pressures as is the Nile basin. A first step toward this is better information about the role of groundwater, the hydrogeology, and the surface water−groundwater links. Several questions remain to be addressed in the future including: To what extent is groundwater taking or giving water to the river? What shifts could the river experience with increased pumping in the basin?

References

Abiye, A., Kebede, S., 2011. The role of geodiversity on the groundwater resource potential in the upper Blue Nile River Basin, Ethiopia. Environ. Earth Sci. 64 (5), 1283−1291. Available from: https://doi.org/10.1007/s12665-011-0946-7.

Abiye, T.A., Mmayi, P., 2014. Groundwater as a viable resource under climate change in the Nile basin: A rapid hydrogeological assessment. South Afr. J. Geol. 117, 97−108.

Angarita, H., Wickel, A.J., Sieber, J., Chavarro, J., Maldonado-Ocampo, J.A., Herrera-R, G.A., et al., 2017. Large-scale impacts of hydropower development on the Mompós Depression wetlands, Colombia. Hydrol. Earth Syst. Sci. Discuss. 1−39.

Araya, A., Stroosnijder, L., 2011. Assessing drought risk and irrigation need in northern Ethiopia. Agric. For. Meteorol. 151 (4), 425−436.

Awulachew, S., Smakhtin, V., Molden, D., Peden, D., Cascão, A., 2012. The Nile River Basin: Water, Agriculture, Governance and Livelihoods.

Barnes, J., 2014. Mixing waters: The reuse of agricultural drainage water in Egypt. Geoforum 57, 181−191. Available from: https://doi.org/10.1016/j.geoforum.2012.11.019.

Blackmore, D., Whittington, D., 2008. Opportunities for cooperative water resources development on the eastern Nile: risks and rewards. In: Report to the Eastern Nile Council of Ministers, Nile Basin Initiative, Entebbe.

Conway, D., 2000. Some aspects of climate variability in the north east Ethiopian highlands - Wollo and Tigray. Sinet Ethiopian J. Sci. 23, 139−161.

Dawoud, A., Ismail, S., 2013. Saturated and unsaturated River Nile/groundwater aquifer interaction systems in the Nile Valley, Egypt. Arab. J. Geosci. 6 (6), 2119−2130. Available from: https://doi.org/10.1007/s12517-011-0483-4.

Drooger and Immerzeel, 2010. Preliminary Data Compilation for the Nile Basin Decision Support System: Analysis Phase Report. Future Water Report 92. Nile Basin Initiative.

El Tahlawi, M.R., Farrag, A.A., Ahmed, S.S., 2008. Groundwater of Egypt: "an environmental overview". Environ. Geol. 55 (3), 639−652. Available from: https://doi.org/10.1007/s00254-007-1014-1.

El-Agha, D.E., Closas, A., Molle, F., 2015. Survey of Groundwater use in the Central Part of the Nile Delta. International Water Management Institute, IWMI, Report No. 6.

El-Agha, D.E., Closas, A., Molle, F., 2017. Below the radar: the boom of groundwater use in the central part of the Nile Delta in Egypt. Hydrogeol. J. 25 (6), 1621−1631. Available from: https://doi.org/10.1007/s10040-017-1570-8.

ESA and U.C. Louvain (2010). GlobCover 2009 Land Cover Map Dataset. <https://datacatalog.worldbank.org/dataset/global-land-cover-2009> (accessed on 25.04.20.).

Funk, C., Peterson, P., Landsfeld, M., et al., 2015. The climate hazards infrared precipitation with stations—a new environmental record for monitoring extremes. Sci. Data 2, 150066Retrieved from. Available from: https://doi.org/10.1038/sdata.2015.66.

FAO, 2016. AQUASTAT Main Database, Food and Agriculture Organization of the United Nations. <http://www.fao.org/nr/water/aquastat/data/query/index.html?lang = en> (website accessed on 20.04.20.).

Gadelmula A.H., Upton K., Dochartaigh B.É.Ó., Bellwood-Howard, I., 2018. Africa Groundwater Atlas: Hydrogeology of Sudan. British Geological Survey. <http://earthwise.bgs.ac.uk/index.php/Hydrogeology_of_Sudan> (accessed 25.04.20.).

Gleeson, T., Wada, Y., Bierkens, M., van Beek, Ludovicus, P.H., 2012. Water balance of global aquifers revealed by groundwater footprint. Nature 488 (7410), 197–200. Available from: https://doi.org/10.1038/nature11295.

Gleeson, T., Villholth, K., Taylor, R., Perrone, D., Hyndman, D., 2019. Groundwater: a call to action. Nature 576 (7786), 213. Available from: https://doi.org/10.1038/d41586-019-03711-0.

Hurst, H.E., 1952. The Nile. Constable, London.

International Atomic Energy Agency (IAEA) and Global Environment Facility (GEF). (2016). Mainstreaming groundwater considerations into the integrated management of the Nile River Basin, RAF/8/042.

Johnston, R., 2012. Availability of Water for Agriculture in the Nile Basin. <https://www.iwmi.cgiar.org/Publications/Books/PDF/H045312.pdf> (accessed 17.11.19.).

Johnston, R., Smakhtin, V., 2014. Hydrological modeling of large river basins: how much is enough? Water Resour. Manage. 28 (10), 2695–2730. Available from: https://doi.org/10.1007/s11269-014-0637-8.

Karimi, P., Molden, D., Bastiaanssen, W., Xueliang, C., 2012. Water accounting to assess use and productivity of water: evolution of a concept and new frontiers. In: Godfrey, J.M., Chalmers, K. (Eds.), Water Accounting. International Approaches to Policy and Decision-Making. Edward Elgar, Cheltenham and Northampton, MA.

Kebede, S., 2013. Groundwater in Ethiopia. Springer Berlin Heidelberg, Berlin, Heidelberg.

Kebede, S., Abdalla, O., Sefelnasr, A., Tindimugaya, C., Mustafa, O., 2017. Interaction of surface water and groundwater in the Nile River basin: isotopic and piezometric evidence. Hydrogeol. J. 25 (3), 707–726. Available from: https://doi.org/10.1007/s10040-016-1503-y.

Kebede, S., Hailu, A., Crane, E., Ó Dochartaigh, B.É., Bellwood-Howard, I., 2018. Africa Groundwater Atlas: Hydrogeology of Ethiopia. British Geological Survey. <http://earthwise.bgs.ac.uk/index.php/Hydrogeology_of_Ethiopia> (accessed 25.04.20.).

Kirby, M., Eastham, J., Mainuddin, M., 2010. Water-use accounts in CPWF basins: simple water-use accounting of the Nile Basin. In: CPWF Working Paper: Basin Focal Project series, BFP03. The CGIAR Challenge Program on Water and Food, Colombo, Sri Lanka, 29 pp.

Landerer, F.W., Swenson, S.C., 2012. Accuracy of scaled GRACE terrestrial water storage estimates. Water Resour. Res. 48 (W04531), 11. Available from: https://doi.org/10.1029/2011WR011453.

Lehner, B., Döll, P., 2004. Development and validation of a global database of lakes, reservoirs and wetlands. J. Hydrol. 296/1–4, 1–22.

Lewandowski, J., Meinikmann, K., Krause, S., 2020. Groundwater–surface water interactions: recent advances and interdisciplinary challenges. Water 12 (1), 296. Available from: https://doi.org/10.3390/w12010296.

MacAlister, C., Tindimugaya, C., Ayenewd, T., Elhassan Ibrahim, M., Abdel Megui, M., 2012. Overview of groundwater in the Nile River Basin. In: Awulachew, S.B., Vladimir Smahktin, D.M., Peden, D. (Eds.), (Eds.), The Nile River Basin Water, Agriculture, Governance and Livelihoods.

MacDonald, A.M., Bonsor, H.C., Dochartaigh, B.É.Ó., Taylor, R.G., 2012. Quantitative maps of groundwater resources in Africa. Environ. Res. Lett. 7 (2), 024009. Available from: https://doi.org/10.1088/1748-9326/7/2/024009.

Meier, J., Zabel, F., Mauser, W., 2018. A global approach to estimate irrigated areas – a comparison between different data and statistics. Hydrol. Earth Syst. Sci. 22 (2), 1119–1133. Available from: https://doi.org/10.5194/hess-22-1119-2018.

Moges, S., 2012. Ag Water Solutions Project Case Study. Agricultural Use of Ground Water in Ethiopia: Assessment of Potential and Analysis of Economics, Policies, Constraints and Opportunities. International Water Management Institute. Retrieved from: <https://www.agriknowledge.org/file_downloads/wp988j905> (Website accessed 20.04.20.).

Mohamed, Y.A., Van den Hurk, B.J.J.M., Savenije, H.H.G., Bastiaanssen, W.G.M., 2005. Impact of the Sudd wetland on the Nile hydroclimatology. Water Resour. Res. 41 (8).

Mukherjee, A., Scanlon, B., Aureli, A., Langan, S., Guo, H., McKenzie, A., 2020. Global Groundwater: Source, Scarcity, Sustainability, Security and Solutions, 1st Ed. Elsevier, ISBN. 9780128181720.

Multsch, S., Elshamy, M.E., Batarseh, S., Seid, A.H., Frede, H.G., Breuer, L., 2017. Improving irrigation efficiency will be insufficient to meet future water demand in the Nile Basin. J. Hydrol.: Regional Stud. 12, 315–330.

Negm, A.M. (Ed.), 2019. Groundwater in the Nile Delta. Springer.

Nijsten, G., Christelis, G., Villholth, K., Braune, E., Gaye, C., 2018. Transboundary aquifers of Africa: review of the current state of knowledge and progress towards sustainable development and management. J. Hydrol.: Reg. Stud. 20, 21–34. Available from: https://doi.org/10.1016/j.ejrh.2018.03.004.

Nile Basin Initiative (NBI), 2016. Nile Basin Water Resources Atlas. <http://atlas.nilebasin.org/> (website accessed 20.04.20.).

Oestigaard, T., 2012. Water Scarcity and Food Security Along the Nile. Politics, Population Increase and Climate Change. Nordiska Afrikainstitutet, Uppsala (Current African issues, 49).

Omer, A., 2002. Focus on groundwater in Sudan. Environ. Geol. 41 (8), 972–976. Available from: https://doi.org/10.1007/s00254-001-0476-9.

Owor, M., Tindimugaya, C., Brown, L., Upton, K., Ó Dochartaigh, B.É., Bellwood-Howard, I., 2018. Africa Groundwater Atlas: Hydrogeology of Uganda. British Geological Survey. <http://earthwise.bgs.ac.uk/index.php/Hydrogeology_of_Uganda> (accessed 25.04.20.).

Richts, A., Struckmeir, W., Zaepke, M., 2011. WHYMAP and the Groundwater Resources of the World 1:25,000,000. In: Jones, J. (Ed.), Sustaining Groundwater Resources. International Year of Planet Earth. Springer. Available from: http://dx.doi.org/10.1007/978-90-481-3426-7_10.

Senay, G., Velpuri, N., Bohms, S., Demissie, Y., Gebremichael, M., 2014. Understanding the hydrologic sources and sinks in the Nile Basin using multisource climate and remote sensing data sets. Water Resour. Res. 50 (11), 8625–8650. Available from: https://doi.org/10.1002/2013WR015231.

Shahin, M.M.A., 1985. Hydrology of the Nile Basin. Developments in Water Science, 21. Elsevier, Amsterdam.

Sutcliffe, J.V., Parks, Y.P., 1999. The Hydrology of the Nile. International Association of Hydrological Sciences. Special Publication No. 5. Feb, 1999 ISBN 1-910502-75-9.

Swenson, S.C., 2012. GRACE monthly land water mass grids NETCDF RELEASE 5.0. Ver. 5.0. PO.DAAC, CA, USA. Dataset accessed from <https://doi.org/10.5067/TELND-NC005>.

Swenson, S.C., Wahr, J., 2006. Post-processing removal of correlated errors in GRACE data. Geophys. Res. Lett. 33, L08402. Available from: https://doi.org/10.1029/2005GL025285.

The Africa Groundwater Atlas <https://www.bgs.ac.uk/africagroundwateratlas/index.cfm> Data released under a Creative Commons 'Attribution/Share Alike' (CC BY SA) licence. <https://creativecommons.org/licenses/by-sa/4.0/legalcode>.

The Global Runoff Data Centre (GRDC), 2020. Retrieved from <http://www.compositerunoff.sr.unh.edu/>.

Tramberend, S., Burtscher, R., Burek, P., Kahil, T., Fischer, G., Mochizuki, J., et al., 2019. East Africa Water Scenarios to 2050. Retrieved from <http://pure.iiasa.ac.at/id/eprint/15904/>.

United Nations Development Program, 2006. Human Development Report. Retrieved from <https://www.undp.org/content/dam/undp/library/corporate/HDR/2006%20Global%20HDR/HDR-2006-Beyond%20scarcity-Power-poverty-and-the-global-water-crisis.pdf>.

Upton K., Ó Dochartaigh B.É., Bellwood-Howard, I., 2018. Africa Groundwater Atlas: Hydrogeology of Egypt, South Sudan. British Geological Survey. <http://earthwise.bgs.ac.uk/index.php/> (accessed 25.04.20.).

van Engelen, J., Essink, G.H.O., Kooi, H., Bierkens, M.F., 2018. On the origins of hypersaline groundwater in the Nile Delta aquifer. J. Hydrol. 560, 301–317.

Wanyama, J., Ssegane, H., Kisekka, I., Komakech, A.J., Banadda, N., Zziwa, A., et al., 2017. Irrigation development in Uganda: constraints, lessons learned, and future perspectives. J. Irrig. Drain. Eng. 143 (5), 4017003. Available from: https://doi.org/10.1061/(ASCE)IR.1943-4774.0001159.

Watkins, M.M., Wiese, D.N., Yuan, D.N., Boening, C., Landerer, F.W., 2015. Improved methods for observing Earth's time variable mass distribution with GRACE using spherical cap mascons. J. Geophys. Res.: Solid Earth 120 (4), 2648–2671.

Wiese, D.N., 2015. GRACE Monthly Global Water Mass Grids NETCDF RELEASE 5.0. Ver. 5.0. PO.DAAC, CA, USA.

Wiese, D.N., Landerer, F.W., Watkins, M.M., 2016. Quantifying and reducing leakage errors in the JPL RL05M GRACE mascon solution. Water Resour. Res. 52, 7490–7502. Available from: https://doi.org/10.1002/2016WR019344.

Zekri, S., 2020. Water Policies in MENA Countries. Springer, Cham, p. 56.

Chapter 8

Groundwater availability and security in the Kingston Basin, Jamaica

Arpita Mandal[1], Debbie-Ann D.S. Gordon-Smith[2] and Peta-Gay Harris[1]

[1]*Department of Geography and Geology, The University of the West Indies, Mona, Jamaica, West Indies,* [2]*Department of Chemistry, The University of the West Indies, Mona, Jamaica, West Indies*

8.1 Introduction

The global demand for water has been increasing at a rate of about 1% per year as a function of population growth, economic development, and changing consumption patterns among other factors, and it will continue to grow significantly over the next two decades. At the same time, the global water cycle is intensifying due to climate change, with wetter regions generally becoming wetter and drier regions becoming even drier. The trends in water availability and quality are accompanied by projected changes in flood and drought risks.

United Nations World Water Development Report (2018).

Global groundwater has been under immense stress in recent times and sustainable management of water resources is important to meet the sectoral demand for any country (Mukherjee et al., 2020). Variations in the rainfall projections, due to variability in the climate pattern, poor management of the existing resources, and increased urbanization coupled with improper disposal of sewage and solid waste, have affected the water resources in the Caribbean causing disproportionate adjustments between the demand and supply. This was the case for many of the islands in the Caribbean (e.g., Jamaica, Antigua and Barbuda, Barbados and Grenada) during the years 2009–10 and 2014–15, when the El Niño Southern Oscillation caused prolonged below normal rainfall leading to hydrometeorological drought. With climate projections showing an increase in the drying trend for the Caribbean, the water resources are expected to be directly impacted and, with an increase in urbanization, lead to more severe water shortages (Meehl et al., 2007). The cumulative effect of the above will pose challenges for the islands in meeting their Sustainable Development Goals (Goals 6, 12, and 15) (Fig. 8.1).

Jamaica is the third largest island in the Caribbean and is centered on latitude 18°15′N and longitude 77°20′W. It is approximately 230 km long and 80 km wide at its broadest point. The island's topography is characterized by several rugged mountain ranges along the northeastern section, central highlands, and a narrow coastal plain on which most of its major cities, towns, and critical infrastructure are located. It relies heavily on rainfall as the primary source of water to rivers and direct recharge to its limestone and alluvium aquifers (Mandal and Haiduk, 2011). The island is divided into 10 hydrologic basins (Fig. 8.2), which are drained by surface and/or ground water (Water Resources Authority Jamaica, 1990). The basin boundaries are both surface and ground water divides with the ground water divide used mainly in the karstic areas. There are 14 parishes in Jamaica with a total population estimated by the Statistical Institute of Jamaica (STATIN) in 2018 at just over 2.7 million.

Jamaica has a tropical maritime climate with a bimodal rainfall pattern with the peak rainfall seasons in the months of April–May (early wet season) and September–October (late wet season), the latter coinciding with the Atlantic hurricane season. Rainfall ranges from convective systems during the wet season due to tropical storms, tropical depressions, and hurricanes to orographic from the NE trade winds. The 30-year mean annual rainfall map (Fig. 8.2) shows that there is significant variability in the rainfall distribution pattern across the island with the northeast, that is, the Blue Mountain North basin, receiving greater than 6900 mm/year of rainfall, on average, as opposed to the dry south and southeast of the island, lying in the rain shadow belt of the Blue Mountain range. The hydrologic basins of Kingston and St Andrew and Blue Mountain South receive an average of only ~690 mm/year.

FIGURE 8.1 Sustainable Development Goals. From https://www.un.org/sustainabledevelopment/. The content of this publication has not been approved by the United Nations and does not reflect the views of the United Nations or its officials or Member States.

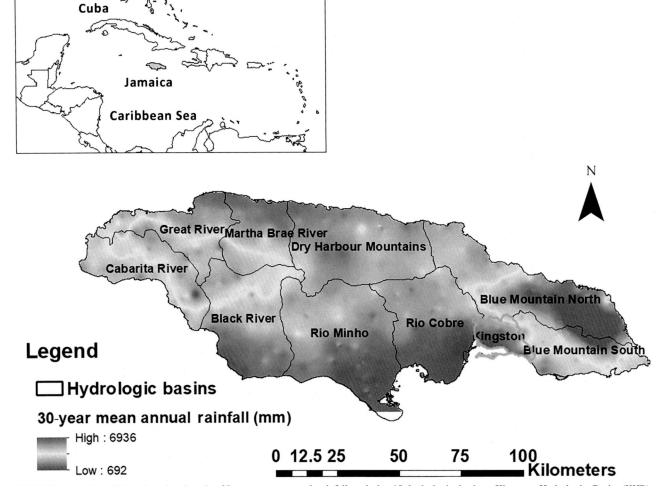

FIGURE 8.2 Map of Jamaica showing the 30-year mean annual rainfall and the 10 hydrologic basins—Kingston Hydrologic Basin (KHB) highlighted. Also shown inset is the location of Jamaica with respect to Cuba and the Caribbean Sea. *Meteorological Service of Jamaica.*

Both surface and ground water are used for water supply to all the sectors on the island (domestic, industrial, agriculture, and tourism). Surface water comes from more than 120 rivers flowing from the mountains to the coast and is primarily used for domestic water supply in the Kingston basin and for irrigation in the Rio Cobre basin (Fig. 8.2). Groundwater is the major source of water for the island, accounting for 90% of the water supply, and is stored mainly in the limestone aquifers (Mandal and Haiduk, 2011; Miller et al., 2001). About 60% of the island's bedrock is white limestone, 25% Cretaceous volcanics, 10% alluvial, and 5% yellow limestone (Taylor et al., 2014) (Fig. 8.3). The groundwater resources have shown marked deterioration in quality in several basins due to anthropogenic impacts, including increased demand by agricultural, industrial, and domestic users. Saline water intrusion and elevated levels of nitrate are the major types of contamination in the groundwater with the former in the coastal aquifers and the latter observed in urban areas (Fernandez, 1982).

8.2 The Kingston Hydrologic Basin

8.2.1 Population and water supply

The Kingston Hydrologic Basin (KHB) covers the parishes of Kingston and St. Andrew (Fig. 8.2) with the city of Kingston, the capital of Jamaica, being the most populous area on the island; the combined population of Kingston and St Andrew is ~670,000 (Statistical Institute of Jamaica, 2012). It is the smallest of the 10 basins, covering an area of 258 km^2, but accounts for the majority of the water demand on the island (~70% for domestic use) due to its high population density (see next). Water supply in the KHB is disproportionate to demand as the current water demand is

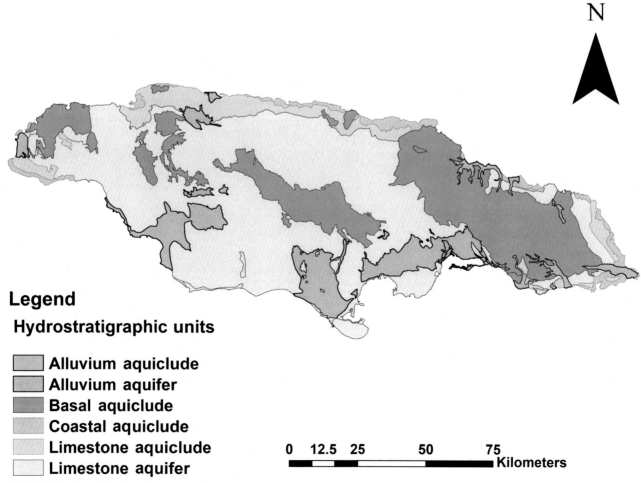

FIGURE 8.3 Map of Jamaica showing the different hydrostratigraphic units.

estimated to be 272,640 m³/day while supply is only 90,880 m³/day, based on data from the National Water Commission (NWC), the primary domestic (potable) water supply agency in Jamaica. The KHB receives potable water from a combination of surface and groundwater sources; surface water is derived from the Hope and Yallahs Rivers in the east which are diverted to the Hope Filter Plant and the Mona Reservoir, and the Wagwater River and other streams flowing into the Hermitage Dam in the northwestern section of the basin (Fig. 8.4) (Barnett, 2010). Additional water is abstracted from 17 operational wells (private and public) in the basin and used for both domestic and industrial purposes. Potable water is also obtained from the Rio Cobre Basin located to the west of the KHB via a pipeline supplied by surface and ground water (Fig. 8.4). Approximately 60% (104.3 million m³) of available water in the KHB is lost due to nitrate contamination as 50%–80% of wastewater in the basin is disposed of via septic tanks and absorption pits directly into the aquifers (Water Task Force, 2009).

The population density (per km²) for the area based on the 2001 and 2011 census years conducted by STATIN is shown in Fig. 8.5A and B, respectively. The population of Kingston and St Andrew increased from 645,534 in 2001 to 661,584 in 2011, indicating a 2.5% growth. The data show that, in 2001, the central and southern sections of the basin comprised the maximum population density, up to 14865 per km², in the constituencies of St Andrew South, South West, and West Central (Fig. 8.5A). By 2011, St. Andrew South (9587 in 2001 to 13479 in 2011) and St Andrew West Central (14865 in 2001 to 15743 in 2011) recorded the most significant growth (Fig. 8.5B). Being the capital city of Jamaica, Kingston experiences migration of people from neighboring parishes in search of livelihood, education, and health services. The population growth over the past decade, as seen from the post census estimation archived from the STATIN data set (https://statinja.gov.jm/Demo_SocialStats/PopulationStats.aspx), showed a moderate-to-steep increase in population from 2001 to 2014 followed by gentle growth until 2016 to a slight decrease in 2018 (Fig. 8.6).

The increase in population in the KHB has been accompanied by an increase in urbanization that, along with inadequate sewage treatment, has resulted in severe deterioration of the groundwater quality and subsequent closure of many groundwater wells. Many of the productive wells in the alluvium aquifer contain nitrate concentrations greater than Jamaica's ambient water standard of 7.5 mg NO_3-N/L, thus limiting their usage for potable water supply. Currently, only five limestone wells in the basin are abstracted for potable water. This has resulted in increased abstraction from surface sources that have been affected by consecutive dry seasons which, coupled with inadequate water storage infrastructure, inefficiency and unreliability of the existing water supply systems (losses from leakage), have exacerbated the

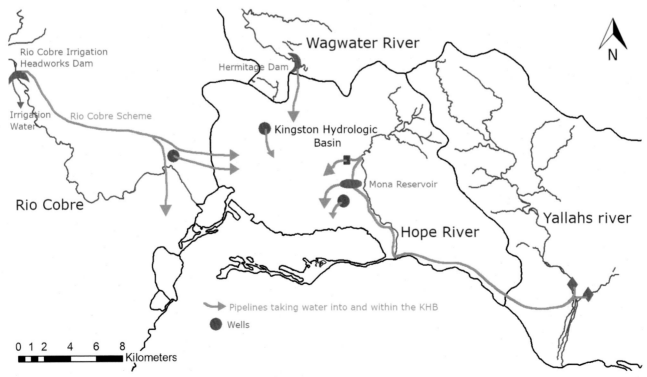

FIGURE 8.4 Schematics of the public water supply system in the Kingston Hydrologic Basin (KHB).

Groundwater availability and security in the Kingston Basin, Jamaica **Chapter | 8** 101

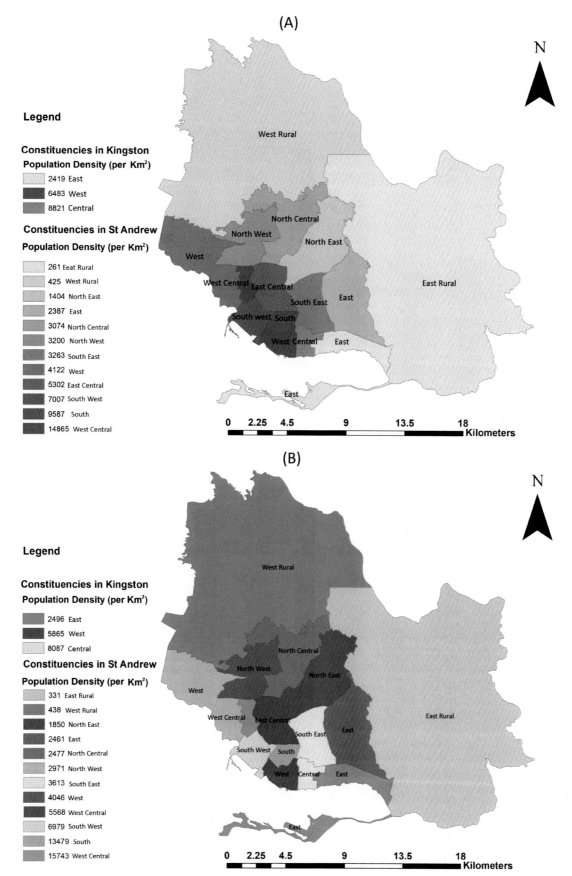

FIGURE 8.5 Population density per sq km for Kingston and St Andrew for the census years (A) 2001 and (B) 2011. *STATIN Jamaica.*

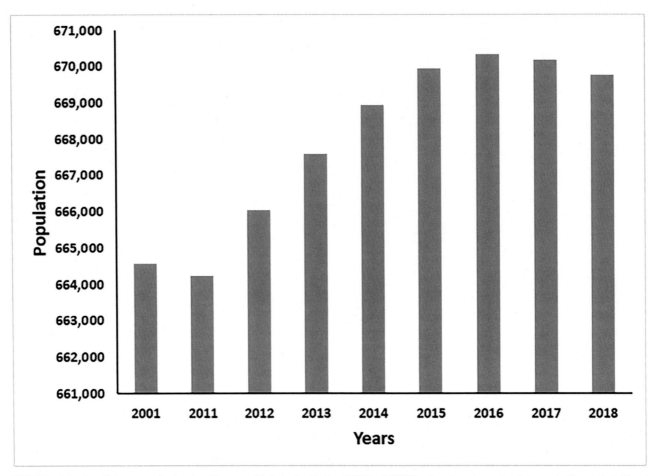

FIGURE 8.6 Population growth in the Kingston Hydrologic Basin (KHB). *STATIN, Jamaica. https://statinja.gov.jm/Demo_SocialStats/PopulationStats.aspx.*

water supply problems (Miller et al., 2001). This was particularly observed during the drought years of 2009–10 when the water levels in the two main storage systems, the Mona Reservoir and Hermitage Dam, fell to below normal levels, from 93% in January 2009 to 41% in February 2010. Further decline was observed during 2014–15 when the storage in the Mona Reservoir and the Hermitage Dam fell to <20% and 30% of their capacity, respectively. This resulted in partial to complete water lock-off of the domestic water supply in Kingston and St. Andrew during these periods.

Due to the KHB's proximity to the coast and overpumping of wells in the past, saline intrusion has been observed along the coast (Fernandez, 1982; Mandal and Haiduk, 2011). High salinity is also observed in groundwater near the western boundary of the basin (limestone aquifer, wells L1–L3) (Fig. 8.7A), which may or may not be related to seawater intrusion or overpumping of the wells. Mandal and Haiduk (2011), in an early study of the hydrochemistry of the basin, showed that the Cl^- concentrations in these three limestone wells (L1–L3) exceeded the Jamaican groundwater standard (5.0–20.0 mg/L) and suggested possible saline water intrusion along fault lines. Municipal solid waste is disposed of at an unlined site just west of the KHB and the fate of the leachate from this site is unknown. Therefore, potential contaminants in the alluvium aquifer include nutrients and pathogens, likely derived from improper disposal and/or treatment of wastewater, as well as, saline intrusion.

8.2.2 Hydrogeology of the KHB

The KHB ranges in elevation from 0 to 1926 m above sea level as seen from the available digital elevation model of 6 m horizontal and 1 m vertical resolution (Fig. 8.7A). The hydrogeology of the basin, as delineated from the Kingston geological map, shows that the basin comprises an alluvium aquifer, a limestone aquifer, and a basal aquiclude (Fig. 8.7B). The basal aquiclude, located along the northern and eastern sections of the basin, comprises the Wagwater Group of volcanics and volcano-sedimentary rocks and conglomerates (Mandal and Haiduk, 2011) with insignificant groundwater resources and thus are the sites of origin of major rivers (Hope and Wag water Rivers and their tributaries) (Fig. 8.7B). The Liguanea Plains cover an area of 111 km^2 with the greater part underlain by secondary alluvial material and a primary limestone aquifer. This Liguanea alluvial fan was developed from the sediments of the Hope River, the largest river in the Kingston Basin. The plains are bordered to the northeast by the Wagwater Trough, to the east by the limestones of the Dallas Mountain, in the north by limestone, and to the west by alluvium (Mandal and Haiduk, 2011). The limestone aquifers located in the southeast and northwest sections cover an area of 34 km^2 and comprise the white-cream-colored limestones of the White Limestone Group. The sand and gravel deposits of the Liguanea formation rise to 230 m above sea level, and the thickness of the sequence, estimated by Robinson (1969) at 300 m, is the main aquifer in the study area (Robinson, 1969). The coastal aquiclude comprises the August Town Formation of the Coastal Group and consists mainly of yellowish sandy marls and limestones (Chubb, 1958).

The alluvium aquifer represents the major hydrostratigraphic unit in which groundwater yields are very heterogeneous due to variable distributions of clay-rich zones. According to the draft Water Resources Authority (WRA) Master Plan (Water Resources Authority Jamaica, 1990), the KHB has a total surface water yield of 15.8 m^3 and a groundwater safe yield of 8.4 m^3, which is low when compared to the other basins in the island. The basin consists of 168 wells of which 59 are active (domestic and industrial). Of these only five wells in the limestone aquifer (three in the west and two in the east) are used for potable water supply. In the present study, only 12 of the operating pumping wells were made accessible for the hydrochemical study by the respective well owners (private and public). Fig. 8.7A and B show the locations of the sampled wells with respect to the elevation and hydrogeology of the basin. These include three limestone wells, used for potable water supply, and nine alluvium wells, used for irrigation and industrial purposes.

8.2.3 Climate of the KHB

As mentioned earlier, the KHB is located in the rain shadow of the Blue Mountain range and is thus challenged in its water resources during periods of below normal rainfall to drought, such as during the years 2009−10 and 2014−15. The average annual rainfall from 2000 to 2019 for the KHB was 1084 mm/year based on data obtained for all rainfall stations located in the basin from the Meteorological Service of Jamaica. It should be noted that the number of rainfall stations decreased from 20 to 12 after 2012 then to 10 in 2019 thereby limiting the geographical spread of the stations. The highest mean annual rainfall over the last two decades was recorded in 2005 at 1772 mm, and the lowest recorded in 2018 at 632 mm (Fig. 8.8A). The basin experienced extreme hydrometeorological drought conditions from late November 2009 to April 2010, where the mean monthly rainfall fell below the 30-year mean for the same months. Similar conditions were observed during the early wet season (April−May) to the dry period (June−July) in 2014 and 2015. Total monthly rainfall for the KHB for the active sampling years (2018−19; Fig. 8.8B) exhibited the typical bimodality in the annual rainfall pattern with the months of May and September/October recording the highest rainfall (1675 and 1076 mm in 2018; 899 and 1517 mm in 2019, respectively). Historical temperatures over the past century ranged from 13.4°C to 38.8°C.

8.3 Methodology and analytical procedures

The study included a combination of field work involving collection of water samples from the operational pumping wells of the KHB, laboratory analysis of the major cations and anions, and GIS-based spatial analysis to examine the overall groundwater quality and availability in the basin.

8.3.1 Field work

Field work involved sampling the 12 accessible private and public wells over ∼2 years from January 2018 to November 2019. A total of 143 untreated groundwater samples were collected within both aquifers, 3 wells in the limestone (L1−L3) and 9 wells in the alluvium (A1−A9) (Fig. 8.7). Samples were usually collected on a monthly basis with attempts made to collect samples post rain days, but the closure of some wells due to operational failure impacted the ability to collect samples at all sites on all occasions. The wells were first purged for 10 minutes followed by pH and conductivity measurements in the field using an Oakton Handheld pH 11 Series Meter and an Oakton Handheld CON 11 Series TDS/Conductivity Meter. Filtered (0.45 μm) well water samples were collected in acid-washed 60 mL polypropylene bottles and acidified (pH < 2) with concentrated HCl or HNO$_3$ (99.999% purity, trace metal basis) for trace metal analyses. Unfiltered water samples were collected in acid-washed 1 L polypropylene amber bottles for all remaining analyses and all water samples were immediately stored on ice until transported to the laboratory. At the

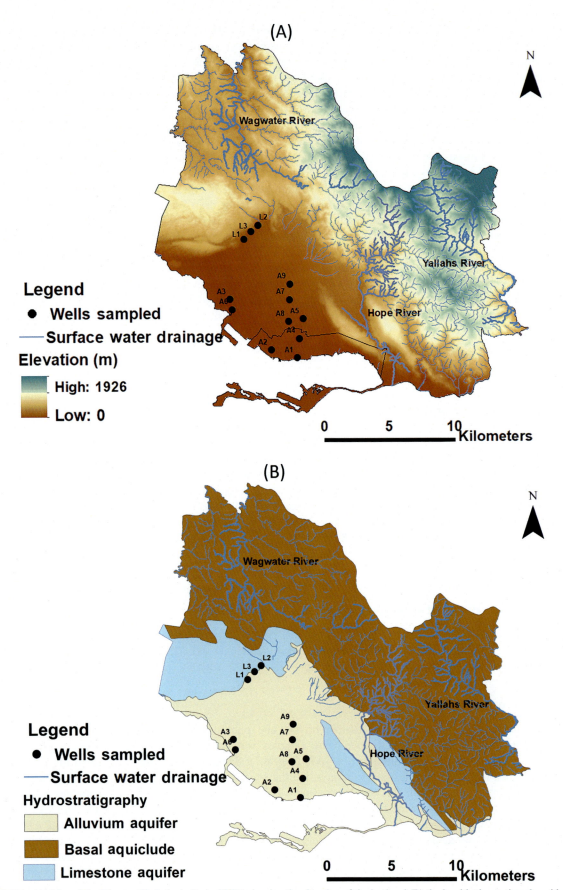

FIGURE 8.7 (A) Map of the Kingston Hydrologic Basin (KHB) showing the elevation of the land and (B) the local hydrostratigraphy with the location of the wells sampled.

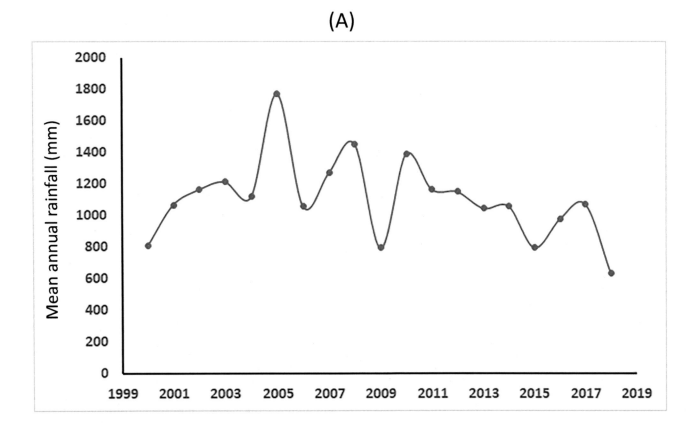

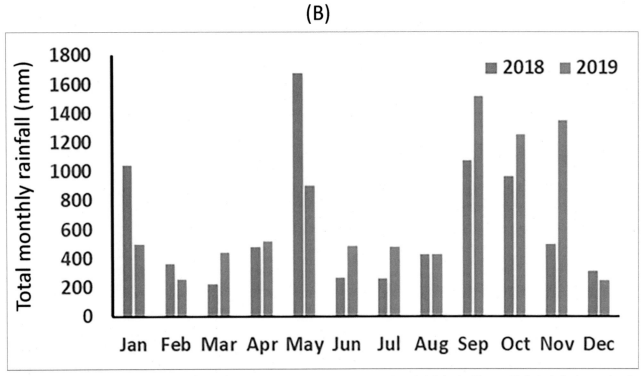

FIGURE 8.8 (A) Mean annual rainfall for the Kingston Hydrologic Basin. (B) Total monthly rainfall for the Kingston Hydrologic Basin covering the sampling years 2018−19.

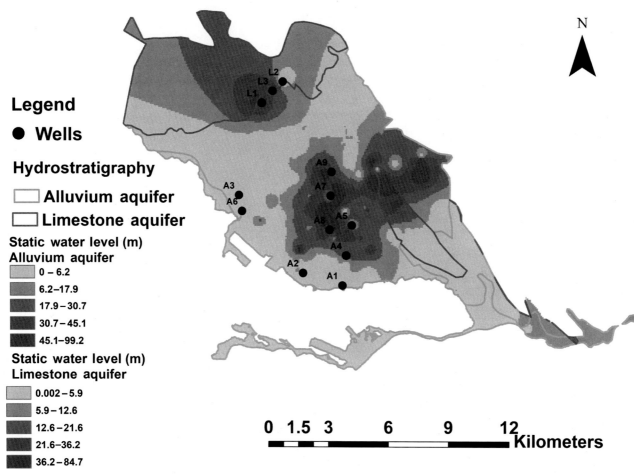

FIGURE 8.9 Static water level (m) contours of the limestone and alluvium aquifers.

laboratory, the 1-L water samples were filtered through 0.45 μm membrane filters and stored at <4°C until analysis. If analyses could not be completed within 48 hours of sample collection, the nutrient samples were kept frozen at −18°C until analysis.

In addition to the pumping wells, the KHB consists of a set of monitoring wells that are periodically monitored for their static water level (SWL) by the WRA, the primary agency responsible for water resource assessment, licensing, and allocation. SWLs were measured by the WRA in November 2018 in all 168 wells (both current and discontinued), as a one-time measurement plan. The data were used in the present study to show the overall slope of the water table with the locations of the sampled wells superimposed.

8.3.2 Water quality analysis

The well samples were analyzed for major cations and anions at the Environmental Research Laboratory located at the University of The West Indies (Mona). Analyses were carried out according to the standard methods outlined in Baird et al., 2017. Ammonium (NH_4^+), nitrate + nitrite ($NO_3^- + NO_2^-$, hereafter referred to as NO_3^-), and orthophosphate (PO_4^{3-}) were analyzed using colorimetric methods. Sulfate (SO_4^{2-}) was measured using the turbidimetric method while chloride (Cl^-) concentrations were determined by argentometric titration. Sodium (Na+) and potassium (K+) were measured using Flame Atomic Emission Spectrometry (FAES) while magnesium (Mg+) and calcium (Ca+) by Flame Atomic Absorption Spectrometry (FAAS). These methods were selected as suitable for these analyses due to their simplicity, sensitivity, precision, and ability to detect a wide range of concentrations.

The hydrochemical facies were determined using the (Piper, 1944) trilinear diagram and (Durov, 1948) plot. These diagrams allow for a better understanding of the water chemistry, with samples of similar qualities grouping together allowing for a comparative analysis (Todd, 2001). In the Piper diagram, the data are plotted on the subdivisions of

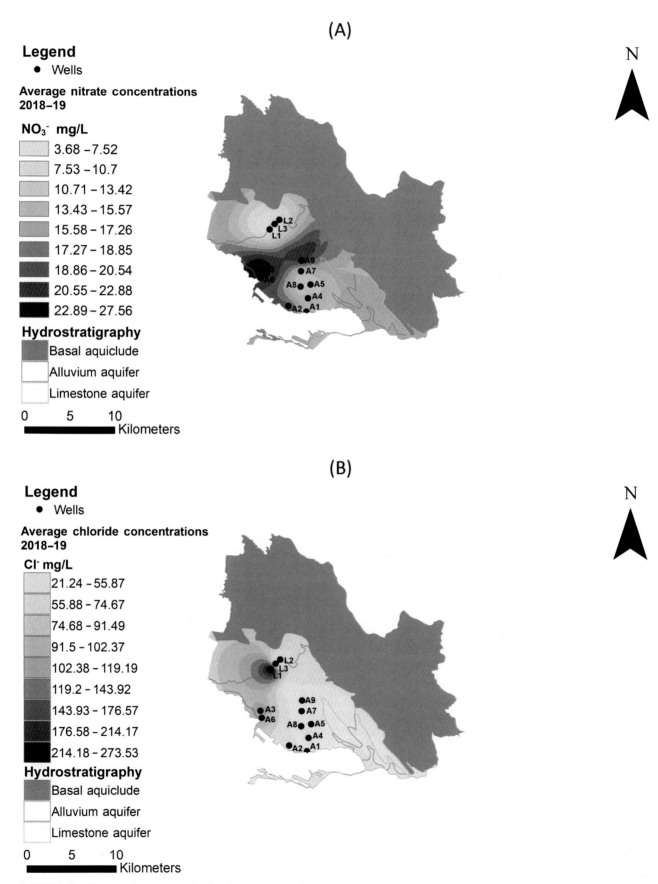

FIGURE 8.10 Concentration contours showing the average (A) nitrate and (B) chloride concentrations in the well waters of the KHB.

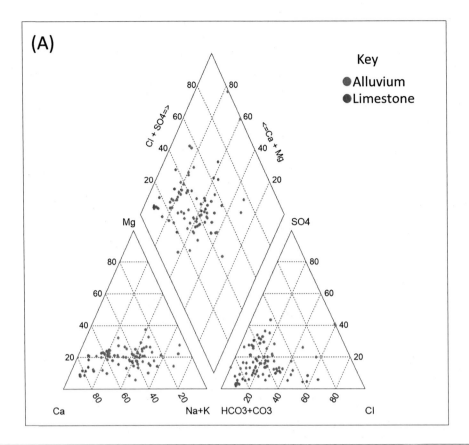
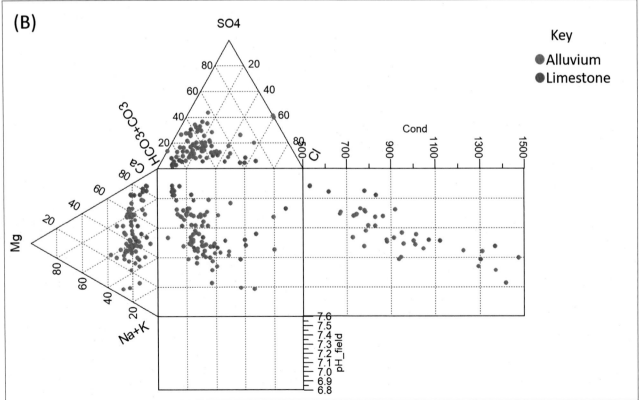

FIGURE 8.11 (A) Piper and (B) Durov diagrams showing the hydrochemical facies of the groundwaters of the Kingston Hydrologic Basin.

TABLE 8.1 Summary of results for the hydrochemical analyses carried out on well samples from the KHB.

	Temperature	Specific conductivity	pH	NO$_3$-N	PO$_4$-P	SO$_4$-S	Na$^+$	K$^+$	Ca^{2+}	Mg^{2+}	Cl$^-$
Limestone aquifer											
Minimum	25.0	419	5.98	2.76	n.d.	1.58	6.17	0.23	18.30	1.67	10.90
Maximum	28.7	1859	8.47	15.22	0.76	16.26	157.2	10.90	108.3	29.61	330.0
Alluvium aquifer											
Minimum	20.7	669	6.03	6.23	n.d.	4.76	13.65	0.53	3.31	6.60	8.60
Maximum	34.7	1752	7.75	69.5	0.32	64.7	229.5	12.23	208.8	45.95	187.8
Ambient water standard range[a]		150–600	7.00–8.40	0.1–7.5	0.01–0.8	3.0–10.0	4.5–12.0	0.74–5.0	40.0–101.0	3.6–27.0	5.0–20.0

Unit of conductivity = µS/cm; temperature in Celsius; other concentrations in mg/L; n.d. = below detection limit.
[a]Jamaica National Ambient Water Quality Standard (Freshwater) (NEPA, 2009).

diamond-shaped fields, which classify the water type/hydrochemical facies in the samples. In contrast, intersection of lines extended from the points in ternary diagrams and projected on the subdivisions of the binary plot of the Durov diagram defines the hydrochemical processes involved along with water type (Lloyd and Heathcote, 1985).

The average NO_3^- and Cl^- concentrations obtained at each well site were used to create concentration contour maps of the KHB using the statistical interpolation method, kriging in Arc Map version 10.5 of ArcGIS. Along with the SWL data used to create the water table map, the concentration contours are used to illustrate a possible trend in contaminant transport through the basin.

8.4 Results and discussion

The SWL map for the KHB (Fig. 8.9) shows that the water table slopes from the highlands of the limestone aquifers in the east [85 m bgl (below ground level)] toward the coast. The SWL for the limestone wells in the west (L1–L3) ranged from 83–55 m bgl with L2 at a shallower level than L1. The SWL in the alluvium wells ranged in depth from 121 m bgl for the inland wells to 22.8 m bgl at the coast with SWL sloping toward the south from 80 m to ~0.51–1 m bgl. Overall, the slope of the SWL follows the topography of the land with flow direction toward the coast.

The overall hydrochemistry of the KHB over the sampling period is presented in Table 8.1 and Fig. 8.10A and B; the ranges in cation and anion concentrations in the groundwater samples collected are given in Table 8.1 and concentration contour maps showing average NO_3^- and Cl^- concentrations (at each site) for the sampling period are shown in Fig. 8.10A and B. Concentration contours were created for all months sampled but no significant temporal variations were detected in the data; hence, average values are shown here.

NH_4^+ concentrations in the samples were typically below detection limit (<0.01 mg/L), which corresponds to an indication of oxic conditions and low levels of denitrification within the aquifers reported in a previous study (Harrison et al., 2019). PO_4^{3-} concentrations in both the limestone and alluvium aquifers were always low with maximum concentrations of 0.76 mg P/L in the limestone and 0.32 mg P/L in the alluvium. pH readings varied significantly within the groundwater: 5.98–8.47 in the limestone and 6.03–7.75 in the alluvium. Temperature was typically lower in the limestone wells (25.0°C–28.7°C) than in the alluvial wells (20.7°C–34.7°C). Conductivity levels varied from 419 to 1859 μS/cm in the limestone wells located in the west with site L1 exhibiting the highest levels. In the alluvium aquifer, conductivity levels ranged from 669 to 1752 μS/cm with the highest conductivities recorded along the coast at sites A3 and A6 (Table 8.1).

The concentration contours for NO_3^- and Cl^- show that both species vary spatially in the basin (Fig. 8.10A and B). The highest concentrations of NO_3^- were observed in the alluvial wells along the coastal and southwestern (SW) sections of the basin, with a maximum average concentration of 27.7 mg NO_3-N/L obtained at well A3, and the highest concentration obtained in a single well sample of 69.5 mg NO_3-N/L at well A9. These are privately owned wells used for industrial or irrigation purposes (Fig. 8.10A). The wells located inland also showed NO_3^- concentrations exceeding Jamaica's ambient water standard for freshwater (7.5 mg NO_3-N/L), possibly due to leakage from absorption pits. These results suggest that the alluvium aquifer is significantly contaminated with NO_3^-, especially in the southern regions, with the levels at some sites being unsuitable for human consumption (the drinking water standard for Jamaica is 10 mg NO_3-N/L). In comparison to the alluvium aquifer, the limestone aquifer typically contained lower concentrations of NO_3^-, which could be attributed to lower population densities in the sampled areas as well as the porosity of the limestone allowing for natural filtration and seepage into the alluvium aquifer. Unfortunately, it was not possible to assess the correlation between the number of absorption pits and NO_3^- concentration in adjacent wells as limited data were available on the locations and distribution of absorption pits in the basin.

Historical data for NO_3^- concentrations within the basin are limited and incomplete, which make it difficult to relate any changes in NO_3^- concentration to population change or the expansion of the central sewage system in parts of the KHB. Nonetheless, comparison with the population density maps (Fig. 8.5A and B) shows that elevated NO_3^- concentrations in the lower SW elevations of the Liguanea Plains correspond to the most populous areas where improper sewage disposal from informal residential communities could result in the leaching of inorganic nitrogen species (NH_4^+ and NO_3^-) into the alluvium aquifer. A similar spatial pattern was seen in a previous study (Mandal and Haiduk, 2011) in which higher concentrations of NO_3^- were also observed in the wells of the alluvium aquifer versus the limestone.

The Cl^- concentration map shows higher concentrations along the western side of the basin in the limestone aquifer (Fig. 8.10B). The highest concentrations were obtained at well L1 with concentrations often greater than 200 mg Cl^-/L, more than 10 times the ambient freshwater standard (Table 8.1). Anomalously high concentrations of chloride (>350 mg/L) obtained at this site during the latter half of 2019 were not included in the summary and further investigation of the water discharging from this well is required. These elevated Cl^- concentrations may be attributed to possible saline water intrusion along the Ferry Glade fault, which extends along the SW margin of the basin (Wright, 1990). Lower concentrations of Cl^- were observed in the inland alluvium wells with the coastal wells (A3 and A6) showing slightly higher concentrations, possibly due to seawater intrusion.

The SO_4^{2-} concentrations in the KHB ranged from 1.58 to 64.7 mg S/L with the highest concentrations obtained in samples collected near the coast (greater than six times the ambient freshwater standard; Table 8.1), corresponding to possible seawater intrusion in the southern section of the aquifer.

The Piper and Durov diagrams derived from the hydrochemical data are given in Fig. 8.11A and B. The results suggest that the water within the limestone aquifer can be classified as earth alkaline with prevailing bicarbonates and SO_4^{2-} resulting from ionic exchange and dissolution of the limestone rocks. Conversely, the alluvium aquifer illustrates water of two types ($Ca^{2+}-Mg^{2+}-Cl^--SO_4^{2-}$ and $Na^+-K^+-Cl^--SO_4^{2-}$). Similarly, this water is classified as earth alkaline water with increased portions of alkalis with prevailing bicarbonates. This type of water results from the dissolution of carbonates and weathering of silicate rocks.

The Durov diagram (Fig. 8.11B) also supports a dominant earth alkaline water type. Fourteen percent of the samples representing the limestone aquifer was dominant in Ca^{2+} indicating recharging waters deriving locally within the aquifer. About 33% of the samples representing the alluvium aquifer water was dominated by Ca^{2+} and SO_4^{2-} indicating water deriving from gypsiferous deposits, mixing, or dissolution. The hydrochemical facies and processes for the two aquifers were similar, suggesting the presence of an interactive boundary between the aquifers resulting in lateral flow from the limestone aquifer to the alluvium aquifer.

8.5 Conclusion

This chapter outlined the challenges associated with the water resources of the Kingston Hydrologic Basin, which has resulted from increased demand and limited supply from the surface and groundwater sources. Rapidly increasing urbanization, coupled with variability in the rainfall pattern and contamination of the groundwater, has resulted in water scarcity and thus impacted water security in this basin. Water demand in the KHB is met with the surface water resources from the reservoirs at the Mona Reservoir and Hermitage Dam and some private and public wells, but the basin has been experiencing prolonged periods of significant low supplies, due to drought conditions, which have affected the surface water sources. Studies have shown that contamination from inadequate sewage treatment has resulted in the closure of many of the wells in the alluvium aquifer, thus adding to the stresses on the surface water sources. Although water quality monitoring has been carried out by various agencies, there is a paucity of consistent data consisting of all relevant parameters and on a regular basis. Thus the present study attempted to conduct a 2-year continuous sampling of the wells (monthly), covering the wet and dry seasons, examining the major cation and anion concentrations and possible flow patterns by classifying the groundwater into different hydrochemical facies. The alluvium water in the basin is of two types, $Ca^{2+}-Mg^{2+}-Cl^--SO_4^{2-}$ and $Na^+-K^+-Cl^--SO_4^{2-}$, and indications are that water flow occurs from the limestone aquifer to the alluvium aquifer.

The chemical analysis of the groundwater of the Kingston Basin reveals the following salient findings:

The concentrations of NO_3^-, Cl^-, Na^+, and SO_4^{2-} consistently exceeded the Jamaican Ambient Water Standards in most of the wells studied, implying significant contamination from absorption pits in the alluvium aquifer, corresponding to high population density, and saline water intrusion in the western limestone aquifer and along the coast in the alluvium aquifer. NO_3^- and Cl^- contamination has affected the groundwater to such an extent (up to 10 times the ambient standard) that wells have been abandoned and should remain thus, unless more expensive water treatment technologies are employed.

It is clear that groundwater scarcity and security has become a serious issue for the Kingston Basin and is therefore in need of special attention. The Kingston Basin, by virtue of its location in the rain shadow of the Blue Mountain range, is relatively dry with high demand and limited supply. Climate change projections show a drying trend that would continue to directly impact the available water resources, especially since rainfall is the primary source of recharge to its rivers. Increasing population trends will result in increased pressure on the already scarce water resources and would imply an increase in the abstraction from other water sources already impacted by periods of below normal rainfall to drought. With only ~30% of the residences in Kingston and St. Andrew connected to the main sewer network, most of the growing population relies on septic tanks and absorption pits, which, through leakage, have led to contamination of the groundwater. To meet demand, there is need for improved sewage treatment to prevent further contamination of the aquifer, possible rehabilitation of abandoned wells, and investigation of water sources from adjacent basins, which have larger groundwater reserves and experience less impact from drought or contamination.

Acknowledgments

The authors greatly acknowledge the assistance of the Water Resources Authority and the National Water Commission in Jamaica for providing access to wells and spatial data for the Kingston Hydrologic Basin. Acknowledgment is also due to the Meteorological Service of Jamaica for providing access to rainfall data and to the Statistical Institute of Jamaica for the census data. This work was supported by a research grant provided by the Board of Graduate Studies and Research, University of the West Indies, Mona.

References

Baird, R.B., Eaton, A.D., Rice, E.W. (Eds.), 2017. Standard Methods for the Examination of Water and Wastewater, 23rd ed. American Public Health Association, American Water Works Association, Water Environment Federation.

Barnett, M., 2010. The Impact of the Recent Drought on the National Water Commission (NWC) Water Supply Services to Kingston & St. Andrew.

Chubb, L.J., 1958. Higher Miocene rocks of Southeast Jamaica. J. Geol. Soc. Jam. 102, 26–31.

Durov, S.A., 1948. Natural waters and graphic representation of their composition. *Dokl. Akad. Nauk SSSR* **59**, 87–90.

Fernandez, B., 1982. The pollution of Jamaica's groundwater resources – an islandwide overview by basins. Water Resources Authority. Kingston, Jamaica.

Harrison, W., Mandal, A., Gordon-Smith, D.-A., Selby, C., Street, M., Marshall, G., et al., 2019. Assessment of the Kingston Hydrologic Basin (JAM no.7003), National Water Commission Final Report prepared for the International Atomic Energy Agency. Kingston, Jamaica.

Lloyd, J.W., Heathcote, J.A.A., 1985. *Natural inorganic hydrochemistry in relation to groundwater: an introduction*. Clarendon Press, Oxford.

Mandal, A., Haiduk, A., 2011. Hydrochemical characteristics of groundwater in the Kingston Basin, Kingston, Jamaica. Environ. Earth Sci. 63, 415–424. Available from: https://doi.org/10.1007/s12665-010-0835-5.

Meehl, G.A., Stocker, T.F., Collins, W.D., Friedlingstein, P., Gaye, T., Gregory, J.M., et al., 2007. Global climate projections. In: Solomon, S., Qin, D., Manning, M., Chen, Z., Marquis, M., Averyt, K.B., et al., IPCC, 2007: Climate Change 2007: The Physical Science Basis. Contribution of Working Group I to the Fourth Assessment Report of the Intergovernmental Panel on Climate Change. Cambridge University Press, pp. 747–846.

Miller, N.K., Waite, L., Harlan, A.E., 2001. Water Resources Assessment of Jamaica. US Army Corps of Engineers Mobile District & Topographic Engineering Center.

Mukherjee, A., Scanlon, B., Aureli, A., Langan, S., Guo, H., McKenzie, A. (Eds.), 2020. Global Groundwater: Source, Scarcity, Sustainability, Security and Solutions. first ed. Elsevier, ISBN: 9780128181720.

NEPA, 2009. Ambient Water Quality Standard (Freshwater) (WWW Document). Available from: <https://www.nepa.gov.jm/new/legal_matters/policies_standards/docs/standards/water_quality_standard_freshwater.pdf>.

Piper A.M., 1944. A graphic procedure in the geochemical interpretation of water-analyses. *Trans. Am. Geophys. Union* **25**, 914–928. https://doi.org/10.1029/TR025i006p00914.

Robinson, E., 1969. Geological field guide to Neogene sections in Jamaica, West Indies. J. Geol. Soc. Jam. 10, 1–24.

Statistical Institute of Jamaica, 2012. National Population Census 2011.

Taylor, M., Mandal, A., Burgess, C., Stephenson, T., 2014. Flooding in Jamaica: causes and controls. In: Chadee, D., Sutherland, J., Agard, J. (Eds.), Flooding and Climate Change: Sectorial Impacts and Adaptation Strategies for the Caribbean Region. Nova Science Publishers, New York, p. 163.

Todd, D.K., 2001. Groundwater Hydrology. John Wiley & Sons, Ltd.

United Nations World Water Development Report, 2018. World Water Development Report 2018: Nature-Based Solutions for Water.

Water Resources Authority Jamaica, 1990. Water Resources Authority Master Plan.

Water Task Force, 2009. Vision 2030 Jamaica Sector Plan 2009–2030. Planning Institute of Jamaica, Jamaica.

Wright, E., 1990. Ferry Springs Utilization Project Report. Ministry of Public Utilities, Underground Water Authority, Jamaica.

Chapter 9

Transboundary aquifers: a shared subsurface asset, in urgent need of sound governance

Shaminder Puri[1,2,3]

[1]Sustainable Solutions in Practical Hydrogeology, Oxford, United Kingdom, [2]IAH Commission on Transboundary Aquifers, Oxford, United Kingdom, [3]International Association of Hydrogeologists, Reading, United Kingdom

9.1 Introduction

Transboundary aquifers are found in the subsurface space, in which water occurring in the rock pores is found in almost all types of rock formations. Through a function of the rock pore size and interpore connectivity, the volume and the rate of flow of water can be large (thus a very productive aquifer, a single well producing 100 L/s) or very small (less than 0.01 L/s), in which case they might be classed an aquiclude. An aquifer system may be a series of aquifers and aquicludes, in which the overall dynamics horizontal and vertical is determined by hydraulic heads, which results in the bulk flux of water, in many instances emerging as baseflow to river/lake systems. These are well-established hydrogeological principles (see Freeze and Chery, 1979) and provide essentially the unchallenged concepts underlying the discussion that follows. As in river basins, so also in aquifer systems, they can occur as hydrologically integrated, self-contained units, in which the principle of time-dependent mass balance always holds true. The 900-km^3 annual withdrawal of water used by 2 Bn people originates from this subsurface space—the urgency in its sound governance cannot be questioned, because it is being exhausted and contaminated at an alarming rate. Many examples and analyses supporting this statement can be found in the chapters of this volume (Mukherjee et al., 2020).

9.2 Definition of transboundary aquifer: international and intranational

A perceived lack of a formal definition of transboundary aquifers has been sometimes lamented (Sanchez et al., 2018, Rivera and Candela, 2018). As a response to this, the definition is clearly restated here: *An aquifer system is defined to be a transboundary aquifer, if the flow (as represented by the motion of particles of water) in the rock formation crosses from one sovereign country to another.* This is grounded in the Draft Articles on the law of transboundary aquifers (see Eckstein, 2020), which itself is based on a rigorous combination of science and law. In republican administrations, state boundaries may also create nationally "transboundary aquifers." Disputes in national, republican contexts can be resolved at the highest national courts of law (see, e.g., India's Inter-State River Water Disputes Act 1956, amended in 2019, and Panda and Agrawal, 2011; also interstate compacts in the United States, Katz and Moore, 2011; Larson, 2017). Not so between sovereign countries—dispute (unless resolved amicably, or through a mutually agreed arbitrator) can only be determined at an international court, where norms of international water law apply, but which remain deficient in addressing aquifers. Consequently the focus the IAH and UNESCO IHP-led 20-year initiative ISARM has been on the *international*, rather than the national aspects of aquifer governance (Puri and Aureli, 2005). The progress in the governance of national aquifers has been very thoroughly addressed in the Groundwater Governance project (Groundwater Governance, FAO, IAH, UNESCO, World Bank, 2015) and need not be repeated here.

9.3 Governance—collaboration, potential dispute resolution

Sound governance implies a bilateral or multilateral formal agreement or understanding (based on some common norms) between sovereign countries that share such a subsurface asset (connected or unconnected to surface water), in order to equitably and reasonably (and conjunctively with surface water) utilize their aquifer resources, while causing no quantifiable harm.

This sets the stage for the analysis of the "shared assets" that two or more sovereign countries may wish to benefit from, for their respective national needs. Transboundary aquifers provide a common pool resource, primarily for agriculture, and the classic tragedy of the commons applies here, both in the national (republican) and international contexts. It is also an undeniable fact that intercountry relationships are first a foreign policy matter and are led principally by diplomats, and only then by the science-based government officials—given the predominantly "silo style" governmental structures in many regions, transboundary waters so often take a distant back seat, well behind security, trade, transport, and finance. Where democracy may be poorly implemented, the often strongly hierarchical culture pushes this issue off the table, because water managers usually come low down in the bureaucratic arrangement of seniority.

Experience has shown that for sound governance of a transboundary aquifer, the predetermining factor is acceptance of a common set of complementary norms (ranging from diplomatic through to scientific) by each "aquifer system state"—in the absence of such acceptance, "sound governance" remains an unattainable chimera that achieves neither cooperation nor collaboration and does not help in eventual dispute resolution (see O'Brien and Gowen, 2012). An important observation to make in passing is that almost all countries have national aquifer management legislation, with at least a chapter devoted to transboundary waters (see Burchi, 2018)—though complementarity in the legal frameworks between neighboring countries is rare. Such complementarity is a binary precondition—contradictory or diverging norms cannot logically deliver "sound governance" in any sphere (e.g., trade, fisheries, and communications) let alone transboundary aquifers. There are however nuances. In a vast transboundary system (be it a river or an aquifer, say 2 Mkm^2 in area, such as the Nile river basin and the Nubian aquifer system), national activities thousands of kilometers away from the international border, along an aquifer flow line, will probably only have an impact in tens or hundreds of years. Countries may then argue that common international norms are irrelevant to actions taken far away from the border when the impact might occur in a hundred years, with some, but somewhat limited justification, as discussed later.

9.4 Water availability as a driver for governance

There are norms enshrined in the Draft Articles on transboundary aquifers, which some member states of the UN even now (in 2019) have some reservations to adopting explicitly. This unnecessary hesitancy is a hindrance to the further evolution of sound governance, (collaboration and potential dispute resolution). The information on the status of governance of the 570 or so self-declared transboundary aquifers is easily available (e.g., in IGRAC, UNESCO-IHP, 2016). Combining this with the findings of the WATER-GAP TWAP analysis (GEF-TWAP, 2015) for "water-abundant regions" (with high annual recharge >200 mm/an) and "water-scarce regions" (with annual recharge <50 mm/an) can help to appraise possible factors that create hurdles to joint sound governance. Loosening up of the current UN Member States' resistance to the Draft Articles would well lead to progress in sound joint governance for mutual benefits. Lastly, it is worth recognizing that attention to transboundary aquifers appears to be a low priority in many regions of the world. Some agencies (UNESCO and UNECE) wistfully anticipate that in the course of achieving SDG 6, countries may be stimulated to pay more attention to the national portions of their transboundary aquifer. This seems to be unlikely, unless that aquifer portion is significant for economic stability (through its water-based contribution to GDP). If this is clarified, countries may then undertake sound governance actions, at least within their own territories (see UNESCO-UNECE & UNEP surveys of countries on their mid-term status of SDG 6, 2017).

9.5 Current global inventory and classification of transboundary aquifers

The Global Groundwater Information System inventory of aquifers numbers 569, as identified by networks of national experts of the IHP–ISARM (IGRAC, UNESCO-IHP, 2016). However, this includes the 219 in EU's definition of "groundwater bodies" occurring *within* an EU member state or is shared with another member state, but the regulation of which would still benefit from the adoption of the provisions of the Draft Articles (Allana et al., 2011) in addition to conforming to the EU Framework Directive. There are also 46 "groundwater bodies" that are shared with nonmembers of the EU, where the EU directive is not binding, but the UNECE Convention may apply.

Across the continents their numerical distribution is shown in Table 9.1, together with the available international instruments under which countries might be able to collaborate or eventually resolve any dispute that might arise.

In areal distribution the world's transboundary aquifers range from about 100 km^2 to 2.8 Mkm^2, though the Amazonas aquifer is just over 4 Mkm^2 (Fig. 9.1). The vast majority of transboundary aquifers range between 1000 and 5000 km^2, but some remain approximately mapped due to the locally complex geology. Seven of the world's largest transboundary aquifers are shown on Table 9.2. It is notable that apart from Amazonas and Irtysh-Ob'sky all the rest are in arid or hyper arid regions, where availability of surface water resources is insignificant.

Classification of transboundary aquifers by superficial dimension or occurrence in continental groupings is not as helpful as classification by aquifer typology, volume of water held in storage or such critical factors as the balance of annual recharge and withdrawals for agriculture, industry and municipal supply, either of which would be among the major factors in determining how to ensure their sustainable use. A complementary classification could be according to the availability of some international or regional instrument under which some form of intercountry collaboration might be possible. As can be seen from Table 9.1 there are 129 transboundary aquifers in the Asian region, 49 in Africa, 22 in Central America and the Caribbean, and 30 in the South American continent that appear to have no obvious instrument under which collaboration or consideration of potential aquifer disputes may be addressed. The existing River Basin Organization agreements (e.g., Mekong, Chad, Senegal, and Trinational Lempa River Association of Municipalities) make barely any reference to the aquifers in their scope of regulation and principally focus on regulating surface water resources. Where reference is made, it has to date been ineffectual (see, e.g., the GWP transboundary groundwater fact sheet on Lake Chad aquifer system, October 2013).

Without at least some of the common basic norms (based on science or legislative frameworks) having been adopted, dialogue between potentially collaborating or disputing parties is unrewarding. The basic norms include at minimum an agreed *definition* of the transboundary aquifer (as indicated previously), and a degree of clarity on the aquifer flow directions and quality, as well as a mutual recognition of, at a minimum, one of the common global conventions on transboundary waters.

TABLE 9.1 Global distribution of transboundary aquifers and grouped according to applicable regulations.

Region	Number of transboundary aquifers identified	Remarks
EU member states—groundwater bodies	219[a]	Regulated under the EU Water Framework Directive and the Daughter Directive, including the exceptional Geneva Aquifer Agreement
EU member state groundwater bodies shared with nonmember states	46	Regulated under the UNECE Convention
The continent of Africa	72	24 Regulated under the SADC Protocol, 8 under the IGAD collaboration
Asian region (including Arabian Peninsula, archipelago of Indonesia/South China Sea)	129	55 of these are in the Former Soviet Union[b] and thus had historic, now outdated, sharing agreements, and includes Rum-Saq uniquely with a formal agreement
Central America and Caribbean	22	None
Balkan region—not interfaced with EU member states	24	Regulated under the UNECE Convention
North American continent	21	11 shared between the United States and Mexico—collaboration under the International Border Commission. 10 shared between the United States and Canada—collaboration under the International Joint Commission
South American continent	30	Includes the Guarani with a formal agreement—under the auspices of MERCOSUR

[a]With the United Kingdom leaving the EU in 2021, this number will change due to Northern Irelands aquifers shared with the Republic of Ireland will no longer be under the directive.
[b]The Aral Sea basin agreements make no reference to aquifers or their management.

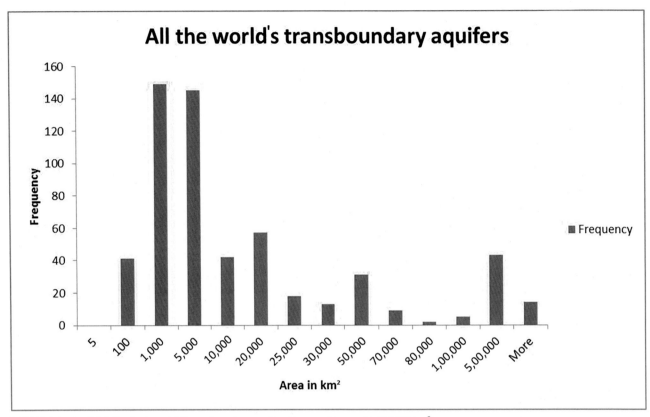

FIGURE 9.1 Surficial distribution of 571 transboundary aquifers (increased interval >100,000 km²).

TABLE 9.2 Seven of the largest transboundary aquifers and countries that share them.

Aquifer	Shared by	Area (km²)
Amazonas	Bolivia, Brazil, Colombia, Ecuador, Peru, Venezuela	4,056,605
Nubian Aquifer System	Chad, Egypt, Libya, Sudan	2,892,867
Lake Chad basin	Chad, Niger, Nigeria, Cameroon, Central African republic, Algeria	2,271,303
Sistema Aquifero Guarani	Argentina, Brazil, Paraguay, Uruguay	1,437,799
Irtysh-Ob'sky	Russia, Kazakhstan	1,368,453
Northwest Sahara Aquifer System	Algeria, Libya	1,279,963
Taoudeni basin	Algeria, Mali, Mauritania	1,260,940
Northern Great Plains	Canada, United States	1,077,798

9.6 Review of recent developments—the Red Queen effect

A comprehensive discussion of the recent status of transboundary aquifer governance was given in Puri and Villholth (2018), in the context of advances in groundwater governance, in the chapter devoted to transboundary aquifers (Chapter 19). Tracing the history of the development of the topic from the 1950s (when *regional*, rather than "transboundary," was the preferred terminology arising from diplomatic sensitivities of the Cold War), through till October 2016, when the UN General Assembly debated the issue, the authors conclude that at the UN a very cautious and restrained approach has been adopted by countries toward explicitly and progressively addressing the issue of their shared aquifer resources. This Lewis Carroll's "The Red Queen effect" approach was repeated at the 2019 UN General Assembly (held on October 22, 2019). Sindico and Nagle (2019) in summarizing the debate at the Sixth Committee note that this was the sixth time in 11 years that the member states of UN General Assembly addressed this issue and preferred to "wait and see" until 2022. Bearing in mind that there are about 350 transboundary aquifers declared by scientists in member states, found in practically every continental landmass (excluding single state Pacific Islands: New Zealand, Australia, Cuba, and Japan), where there is no suitable legal instrument, just nine members of the global community of nations made statements, giving their opinion to the Draft Articles. Sindico and Nagle consider that though only a handful of countries gave some opinions, there still does seem to be an emerging consensus on the normative value of the Draft Articles. As its conclusion, the UN General Assembly through its Resolution 71/150 "Commends to the attention of Governments the Draft Articles on the Law of Transboundary Aquifers annexed to its resolution 68/118 as guidance for bilateral or regional agreements and arrangements for the proper management of transboundary aquifers."

It would seem (from the commentary given by Sindico and Nagle) that the current apprehension that countries have is related to the "obligation not to cause significant harm." The use of the term "significant" as a threshold was thought to be too high, as well as the uncertainty relating to the meaning of "significant." There was a suggestion that a legal definition of "significant harm" may be required. Countries also felt that the relationship between the eventual adoption of the Draft Articles and the commitments made in relation to the SDGs as well as to combating the impact of climate change may create an additional financial and institutional burden. At its conclusion the UN Sixth Committee decided to defer further consideration of the issue until 2022—truly analogous to the Alice in Wonderland dilemma of the Red Queen, thus leaving the Draft Articles in limbo! With the onset of the COVID-19 across the globe in 2020, the state of limbo may well be further prolonged.

In their policy brief, Sindico and Nagle speculate, "How can the groundwater community remove the invisibility cape [cloak] from transboundary aquifers in the next 5 years in order to reveal the critical importance of aquifers and stimulate active progress on governance?." They further ask, "In addition, there is a second, possibly even more important, question. If the law of transboundary aquifers has not received adequate attention within the UN family, are there other international and national fora, public or private, where the law of transboundary aquifers and, more generally, transboundary aquifer management should be addressed?" Some responses to these are given in the later sections of this chapter.

9.7 The place of transboundary aquifers in national priorities

As has been noted in Puri and Villholth (2018), transboundary aquifers are in physical fact a continuation of national aquifers, demarcated by an international sovereign boundary, that nevertheless constitute a consistent geologically constrained aquifer system. It follows logically from this assertion that if the resources of the relevant aquifer in the national portion constitute a significant resource to the economy (in terms of generating the GDP), then the place of the transboundary aquifer could be high in national natural resource management priorities. This should be especially true in arid climates, where annual recharge is very limited, but the resources are deployed for agriculture as a key contributor to the national GDP (Fig. 9.2). Each of the countries in the figure shares a transboundary aquifer, with its resources being unquestionably vital for irrigation. Thus in the case of Chad, the Central African Republic, and Kenya (among the top countries where national GDP is created in agriculture)—attention to the sound governance of aquifers (national as well as transboundary) should be a high priority, because a collapse of access to water, through aquifer exhaustion, could impact on up to 50% of the GDP.

However, it is a sad, but true observation to make that aquifer management in national priorities appears quite low (Global Diagnostic on Groundwater Governance, 2015) especially in terms of its sound governance. In this context governance is taken to mean a set of national rules, regulations, practices, and institutions that jointly ensure sustainability of the aquifer (Global Diagnostic on Groundwater Governance, 2015). The saga of accelerating impact of unsatisfactory governance has been reported progressively over the decades, across the world. (see Shah et al., 2000, Aeschbach-Hertig et al., 2012, Gorelick et al., 2015) from the early 2000s till today, 2020—and the probability for serious, fully costed and financed national actions (included within the national expenditure budgets) to address this is a rarity (see Oates and Mwathunga, 2018). In this context the fate of transboundary aquifers as regards their sound governance would appear to be dire in the immediate future, when governments have been reluctant or unable to allocate suitable financial resources to the sector (e.g., the Guarani, Sindico et al., 2018). A worrying fact is that in 2020, even before the impact of COVID-19 became clear, the prospects of global economic growth are limited (World Bank, January 2020), and the probability for serious exacerbation due to the global outbreak, per capita growth will remain well below average, at a pace too slow to meet poverty eradication goals. The forecast for Latin America's decline in GDP for 2020 is up to 5% (CEPAL, 2020). The growth of income is expected to be slowest in sub-Saharan Africa,

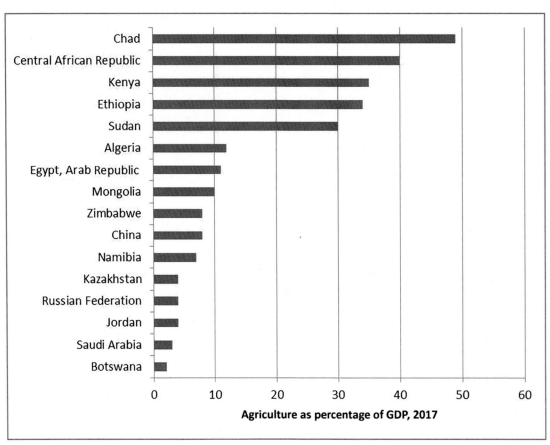

FIGURE 9.2 Agriculture as percentage of GDP in selected countries with annual recharge <50 mm. *Data from World Development Indicators (2018).*

where 56% of the world's poor live, but also in the developed world. It may be foreseen (in March 2020 with peaking coronavirus), there will be further constraints to expenditure on such natural resource governance in this decade.

As mentioned previously only exceptional transboundary aquifers are very high up in the national priorities and of these, Haute Savoie–Geneva aquifer and the Rum-Saq aquifer are two examples. The former is critical for the public supply of the city of Geneva (population less than 0.5 M, annual rainfall of c. 900 mm), the contribution of which to the national GDP in 2016 was €44.6 Bn, 7% of the national total (Swiss Federal Bureau of Statistics). Any break in the consistent supply of water would have immeasurable consequences for the economy and therefore the responsible agencies are prepared to finance the full costs of actions and fully coordinate across sovereign borders (see the text of the Convention, 2008), even though quite exceptionally, the international interrelationships are not at the national, but local level (see de Los Cobos, 2018). In the case of the latter, concerning Jordan and Saudi Arabia, up until the year 2000, the aquifer resource in both the sovereign countries was primarily devoted to purely national agricultural needs, though with the gradual recognition of the immensely high future value of these nonrenewable resources, each country independently started to review the use and benefit of the water (Puri et al., 2003). In Jordan, in a more or less complete *volte face*, the resources were redirected to the emergency critical, rocketing municipal demands of Amman (arising from the third wave of refugee crises in one decade), from the wellfields located about 350 km away (see Puri and Wong, 2018). In Saudi Arabia, some fine-tuning of the water use for agrouse has continued, in a sector that makes barely 2.2% contribution to the national GDP of $726 Bn (2018, World Bank stats).

Since Jordan's use of the water was a "life or death" solution that would ensure that the capital city and its environs remained viable (contributing up to 42% of the national GDP of €35.8 Bn (National Economic Policy Council, 2019), the significant capital investment of €1.2 Bn had to be made, subject to a managed cross-border risk, by means of an agreement with Saudi Arabia, in which both countries agreed to "live and let live"—but with a curious disregard of some of the prevailing science or the norms of international water law (Burchi, 2018, Eckstein, 2015), signed at the mutual ministerial levels. One overriding reason to have such an agreement in place was to manage the risk to private sector investment of the Disi−Amman water conveyance project that delivers 110 MCM/year water from the aquifer (Puri, 2017). The complementary withdrawals in Saudi Arabia, centered around Tabuk, have reached 800 MCM/year, providing irrigation for about 47,000 ha.

9.8 SDGs as a driver toward sound governance of transboundary aquifers

The interconnection between "water," "groundwater," and the SDGs has been much commented upon (Guppy et al., 2018). In all of the 17 SDGs, municipal water supply and sanitation are the relevant factor (see IAH Strategic Overview on SDGs 2017). Broadly the SDG indicators focus on the consumption side of the use of water—but not adequately on the source (or natural resource) side. The source side consists of attention to the natural reservoirs of water in the surface and in the aquifers—with the critical link to their annual replenishment (in recharging aquifers) and storage management in nonrecharging aquifers (i.e., those that receive 50 mm annual recharge or less), in which climate change is likely to have a strong impact, as discussed next. The one SDG that makes a passing reference to (groundwater) resource management in mildly explicit terms is SDG 6.6.1, as a subset of ecosystems (UN Environment, 2018). Even this requires a stretch of the imagination (set out in SDG 6.6.1) for countries to address resources management, let alone aquifer resource management in the transboundary context. Through some implicit interpretation, the SDGs 6.5.1 and 6.5.2 are intended to deliver "firm commitment to implementing IWRM" by countries and to undertake intercountry collaboration.

In order to monitor the progress made by countries in the achievement of SDG 6.5.2, the UNECE and UNESCO conducted a very detailed global survey in 2018 (UNESCO, 2018). The survey questionnaires were sent out to 153 countries, and although 107 responses were received, due to gaps and inconsistencies for the "aquifers' component," only 64 responses provided the required information, while 22 required further clarifications from responders, and 67 did not reply at all. The corresponding response on "lakes and river basins" was 87, 20, and 46, respectively. The number of country responses that contain the required amount of information on transboundary aquifers is 61. The prime objective of the questionnaire, and indeed the indicator to be monitored, is the "the proportion of transboundary basin area with an operational arrangement for cooperation." Further, the "arrangement" might include a bilateral or multilateral treaty, convention, agreement, or other formal arrangement among countries that provides a framework for their cooperation. The findings of the survey are summarized in Table 9.3, which classifies responding countries according to whether they have an operational agreement in place, or they have one, but it is ineffective. The "indicator" for this must be considered to be somewhat curious as it sums the national "portion" of a transboundary river, plus transboundary aquifer, for which an agreement is in place, as a ratio of the total sum of both, expressed as a percentage and not the full transboundary extent, extending into neighboring country. Given that boundaries of aquifers underlying river basins may not coincide ...., and that it is the subsurface extent of an aquifer (sometimes completely confined by overlying formations—such as in the Guarani, the Rum-Saq, and the Nubian) that such a ratio does not provide a sufficiently

TABLE 9.3 Summary of country responses to survey on SDG indicator 6.5.1—on transboundary cooperation.

Extent of cooperation among sharing countries	List of countries self-reporting
All transboundary aquifers covered by an operational agreement with sharing country	• Europe—18 (EU member states)
	• Botswana−Namibia
	• Tunisia
	• Ecuador
A low level of operational agreement with sharing country	• Angola, Gabon, Gambia, Kenya, Lesotho
	• Senegal, Somalia, Uganda and Zambia
	• Brazil, Chile, Dominican Republic
	• El Salvador, Honduras, Mexico, Paraguay, and Venezuela
	• Armenia, Georgia, Iraq, Jordan, Morocco, and Qatar
	• Canada, Montenegro, and the United Kingdom
	• Republic of Korea

meaningful measure of the importance that countries place on stewarding their shared natural resource, and could do with developing further.

To conclude from the above, the present structure and the philosophy of the monitoring of the progress of transboundary cooperation under the SDG 6.5.2 could be much improved and thus provide a stimulus for incentivizing flagging intercountry collaboration. The potentially available and lacking instruments by region are shown in Table 9.1 and could help to improve the SDG indicator monitoring.

An indicator couched in the decline of GDP through failure to cooperate and jointly sustain a transboundary aquifer is likely to have more promise and could be a strong national political driving force for improved attention to aquifers.

9.9 The climate change megatrend and relevance to transboundary aquifers

Climate change (among demography, health-related pandemics, economic shifts, resources stress, urbanization, and political disruptions) has been recognized as one of the megatrends that have started to affect the shape of the future. Extreme weather effects have started to cause global, including transboundary, infrastructure damages valued at $156 Bn in 2018 (Swiss Re, December 2018) and over the next decade this is expected to balloon by 50%, which, in turn, can be expressed at country level in measurable shocks to GDP. There is a clear climate-related link to increasing demand for water, transmitted through the agriculture sector. The pre-COVID-19 dynamics of the sector are expressed as loss of human capital (from 44% world employment in agriculture in 1990, this has declined to 28% in 2015—with corresponding increase in urbanization), and an increase in global food demand, with preference for calorie-rich diets in emerging economies. Water withdrawals from national and transboundary aquifers can be directly related to this dynamic now, and also in the future, despite the slowdown related to COVID-19. Increasing urbanization (from 10 megacities in 1990 to 23 in 2010 and projected to reach 43 in 2030—with an estimated population of 730 M) translates into less opportunity for good rural land management, and thus amplifying negative impacts on potential for managing and enhancing recharge in rural lands, transboundary groundwater management, and effective waste disposal.

For transboundary aquifers the current and future annual recharge determined by climate change should give countries cause for concern. As already mentioned, in water-scarce countries (annual recharge <50 mm/an) water is being drawn rapidly from long-term storage, and the impact of unreliable annual recharge will only be intensified (see Fig. 9.3). In such aquifers there is an urgency to address such a silent, ticking time bomb, at national, as well as transboundary collaboration. Arguably in water-rich countries the annual recharge (>200 mm/an) may not be a principal driver, as interannual high recharge could ameliorate the impact of excess abstraction. Of the 570 aquifers Fig. 9.3 shows the low-recharge (<50 mm/an) and high-recharge (>200 mm/an) transboundary aquifers, based on the TWAP (2015) evaluations. Of the former the stakeholders of the Stampriet, NW Sahara, the Nubian, and the Rum-Saq are in varying stages of collaboration over their shared aquifers (Burchi, 2018). Aside from the North American aquifers (especially United States—Canada where intercountry collaboration is feasible through the existing long-term bilateral agreements), there is clearly an urgent need for many countries to commence on their national assessments and start up an effective intercountry collaboration. In the latter group, recharge might not be a constraining factor, and thus there will other factors (e.g., trade and communications) that may drive transboundary collaboration. The case for the Amazonas aquifer is interesting in that Brazil may need a "subaquifer" specific approach, expressed through a series of somewhat independent bilateral agreements—a multilateral approach is likely to be ineffective and unnecessary—due to the hydrogeological contexts of the aquifers' flow systems (see Rivera and Candela, 2018).

Given the multidimensional nature of the issues that connect transboundary aquifers to SDGs and climate change, it is understandable, though hardly acceptable, that countries appear to put this issue off for later consideration. In aquifers with storage that far exceeds the current demand (e.g., the Guarani, with storage equivalent to 200 years of global drinking water demand, Sindico et al., 2018), this position might be somewhat justified, but certainly not where small aquifers with limited storage (e.g., the Stampriet) that receive less than 50 mm annual recharge, where investments for enhanced recharge through managed changes in land use are becoming critical (see Puri, 2017, reports for UNESCO—GGRETA).

9.10 Transboundary aquifers under high developmental stress

The question of "prioritization" of transboundary aquifers for interventions has attracted the interest of several authors (e.g., Davies et al., 2013) who have sought to find criteria (troublesomeness) under which countries would give priority to implementation of sound governance. Majority of the criteria have been semiqualitative, because suitable and relevant quantitative parameters are evasive—whether to consider dimension, production, possible competition for the same resource, etc. remains questionable. Some have resorted to linear distance from the border to the locations of

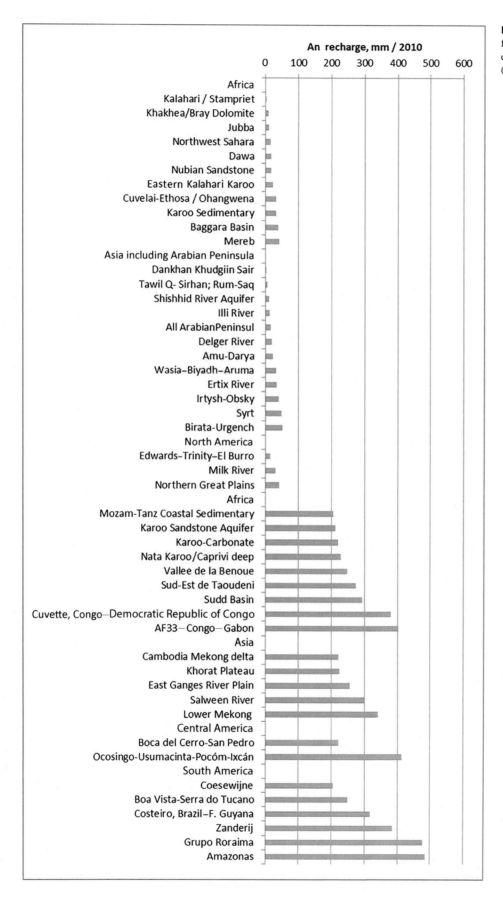

FIGURE 9.3 Transboundary aquifers with lowest (<50 mm) and highest (>200 mm) annual recharge (mm/year for 2010).

wellfields—but these disregard the fact that groundwater flow (and thus the magnitude of impact) is determined by the groundwater flow path—and not the land surface distances.

As of now, science-based quantitative hydrogeological impact analysis is rare in most of the world's transboundary aquifers—the exceptions include the Nubian (for which a flow model has been constructed and flow path calculations are possible), the Rum-Saq (similar modeling to the Nubian—but with more detail), the Stampriet (an updated groundwater model was created, Puri, 2017), the NWSAS (for which models have also been constructed), and the Guarani (for which modeling helped to steer the intercountry agreement). Despite these few models that could provide insights for a global overview (with various margins of error), the only consistent global approach is to analyze the mass balance of recharge, plus storage versus production. Such a "bulk approach" (though disregarding storage) has been used under the TWAP study (2015) to provide a global overview that can be objective to some extent—using globalized models of climate forecast—however, all estimates are infected with varying degrees of error.

In the GEF-TWAP (2015) study a series of indicators were used to make quantitative assessment of current conditions (2010–15) and make projections for 2030 and 2050. Data constraints and the remote sensing limitations meant that only aquifers greater than 20,000 km^2 were included—numbering 91, among which multilayer (sometimes hydraulically unconnected) aquifers were aggregated into a single "lump." From the outcome of the evaluations, the indicator of most interest to this paper is the "groundwater development stress," which is the ratio of withdrawals to recharge (Fig. 9.4). However, the age-old problem of "country limited data" (restricted to country units) from the various complementary national segments of the transboundary aquifers had to be made consistent, leading, at times, to somewhat unusual results. The finding of the TWAP study was that the highest groundwater depletions were not in transboundary aquifers—but rather occurred within territories of one of the sharing countries (referred to as "country units")—the Syrian Neogene aquifer and the Indus Plain aquifers are two examples. The surprising result of taking "country units" as the base (258 units were considered—which, in their transboundary context, reduce down to 91) was that 20 country units were suffering from modest to very high groundwater development stress.

The result for the Lake Chad basin aquifer provides the illustration of this peculiarity—the Nigeria and Niger country units (both of which receive about 100 mm recharge) show the groundwater stress factor of 1.7 and 1.2, while Algeria shows 19 and Libya 346. Clearly, as the latter two countries have annual recharge of about 17 mm, and all the abstracted water is taken from aquifer storage, the calculated country stress indicator is not helpful in assessing and jointly managing the contiguous transboundary Lake Chad aquifer system. The logical, but unhelpful conclusion at country unit level might well be that Nigeria and Niger have reduced role to play in mitigating future impacts—while in fact the lands in their territories are the ones where managed aquifer recharge is viable and indeed should be undertaken. It may also be noted that between 1966 and 1988 major irrigation schemes (67,000 ha) in the Nigerian portion of the basin have had a plethora of near successes and many failures—and the latest effort is another project, this time, dependent on the use of artesian discharge through a series of self-flowing wells (Abubakar, 2017), in which the lack of a plan for shared management of the transboundary resources is not sufficiently highlighted. If the new program would include (transboundary) managed aquifer recharge, as well as the production, logically, as beneficiaries of the recharge, Libya and Algeria would be encouraged to make suitable financial and in-kind contributions for the benefit they would receive, albeit over the long term. Such principles are set out in the philosophy of the Draft Articles on transboundary aquifers, which encourages the obligations of countries to collaborate and cooperate, and which could make a significant contribution to the sound design of such an investment.

9.11 Estimating the urgency of sound governance as a function of water abundance/water scarcity

Puri and Villholth (2018) propose a framework for promoting and supporting sound transboundary aquifer governance, which consists of three nested "boxes" (analogous to nested "Russian dolls"), in which the innermost box can only be accessed when the outer boxes have been opened in sequence. The three nested boxes proposed were an "outmost box" representing "firm and well-grounded knowledge base (hydrogeology, national legislation, and institutional structure)," into which is nested the second "box" representing "trust-building measures (capacity building and participatory analysis)," and a third, innermost "box" representing "hydro diplomacy (technical collaboration, harmonization of transboundary legal frameworks, and institutional approximation)." The logic being that the success of hydro diplomacy can only be assured once the knowledge base has been established and trust-building measures have been started in earnest. Efforts at engaging in hydro diplomacy, before knowledge and trust have reached a reasonable level of maturity, have been found to be abortive in several instances (e.g., the GGRETA efforts in Kazakhstan–Uzbekistan, where trust

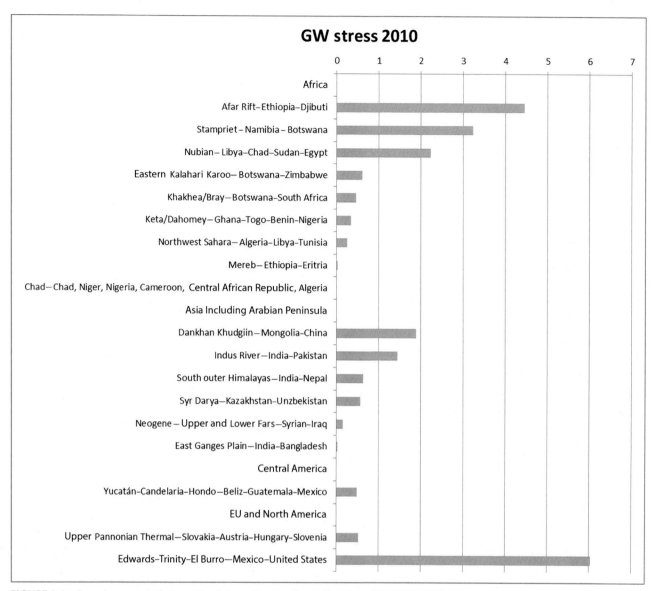

FIGURE 9.4 Groundwater stress factor: ratio of abstraction to recharge. *Data from GEF-TWAP 2015.*

remains low). The reason being, as mentioned previously, hydro diplomacy is the domain of experts in foreign affairs who are not well grounded in natural resources management, and thus they tend to resist any risk taking in intercountry relations, especially in connection with transboundary waters.

Using such a framework of embedded boxes, each box can be scored (from A to D) indicative of the degree of maturity and development of the contents represented by that box—thus A, indicative of high maturity and D, of low maturity. For example, the Stampriet aquifer system, through the GGRETA Project has a well-developed knowledge base (scoring A), trust-building measures are underway (scoring B, as still progressing), and therefore the first steps toward hydro diplomacy have been taken—still not yet completed (scoring B). In addition, it may be noted that the aquifer system is under stress (from significant abstraction) and has very low annual recharge. It may be concluded that due to water scarcity urgent measures by the sharing countries are needed to move toward full-scale hydro diplomacy, through which joint investments could be negotiated and implemented for gaining shared benefits.

Table 9.4 is a compilation of aquifers (listed previously, see Fig. 9.3) with the lowest annual recharge (thus the resources are stretched) with scores for each of the current (2020) status of the three "boxes." The scores have been given on the basis expert judgment. This expert judgment is derived from data submitted in national country responses to the in depth questionnaires circulated by UNESCO—UNECE in connection with SDG 6.5.2 (November 2017), in which the current status of transboundary cooperation has been estimated. Table 9.4 also includes aquifers receiving

TABLE 9.4 Scoring a framework for promoting and supporting sound transboundary aquifer governance.

Transboundary aquifer system	Resource status (natural recharge—2010 mm/year)	Knowledge base (hydrogeology, legislation, and institutional info)	Trust-building measures (capacity building and participatory analysis)	Hydro diplomacy—interactions (technical collaboration and legal harmonization)
Africa				
Stampriet—Namibia–Botswana	3	A	B	B
Khakhea/Bray—Botswana–South Africa	9	C	B	C
Northwest Sahara—Algeria–Libya–Tunisia	16	A	A	B
Nubian—Libya–Chad–Sudan–Egypt	17	A	B	C
Eastern Kalahari Karoo—Botswana–Zimbabwe	23	B	B	C
Mereb—Ethiopia–Eretria	42	C	D	D
Afar Rift—Ethiopia–Djibouti	81	B	C	D
Chad—Chad, Niger, Nigeria, Cameroon, Central African Republic, Algeria	109	B	C	D
Keta/Dahomey—Ghana–Togo–Benin–Nigeria	190	B	C	D
Asia including Arabian Peninsula				
Dankhan Khudgiin—Mongolia–China	3	C	C	C
Neogene—Upper and Lower Fars—Syria–Iraq	70	C	D	D
Indus river—India–Pakistan	123	B	D	D
South outer Himalayas—India–Nepal	124	B	B	D
Syr Darya—Kazakhstan–Uzbekistan	139	B	C	D
East Ganges Plain—India–Bangladesh	256	B	C	D
Central America				
Yucatán–Candelaria–Hondo—Belize–Guatemala–Mexico	108	B	C	C
EU and North America				
Upper Pannonian Thermal aquifer—Slovakia–Austria–Hungary–Slovenia	87	A	A	A
Edwards–Trinity–El Burro—Mexico–United States	14	A	B	C

Notes—the scoring A–D is indicated based on the following. *Hydrogeological appreciation*—A—well-developed understanding of the hydrogeology, such that a mathematical model has been built and some simulations have been conducted.... B, C, D—progressively less well mapped, understood hydrogeology. *Trust-building measures*—A—capacity-building measures undertaken, participatory analyses carried out.... B, C, D—progressively fewer measures and analysis, or none conducted. *Hydro diplomacy—interactions*—A—technical collaboration carried out, legal frameworks compared, and institutionalization efforts made ... B, C, D—progressively fewer or none carried out.

high annual recharge (also included in Fig. 9.3), where resources may not be at high stress at present, but are likely to need attention in the near future—as they occur in regions subject to severe climatic variations.

The scoring suggests that urgent actions are needed for the Stampriet, the Dankhan Khudgiin and the Edwards–Trinity aquifers.

9.12 Case history: the Stampriet aquifer—Botswana, Namibia, and South Africa

The case history of developing a consultation mechanism for the Stampriet aquifer system illustrates some of the hurdles and the successes of efforts in taking the issue of collaboration forward into the 2030s. The full background to the aquifer system, the geology, the hydrogeology, and socioeconomics has been given in UNESCO (2016). Consequently, this short summary will focus on the final definition of the aquifer system, its dynamics, and the identification of key issues that countries need to focus on to implement the sound governance of their subsurface space—which provides 20 Mm^3 of water to the three countries, though it is mainly used in Namibia. From in depth reviews of the extensive data that was collated—the aquifer typology formulated was "... a weakly recharging three layer aquifer system, characterized by low transmissivity, and low storage, primarily being utilized in Namibia for socio economic growth, where withdrawals from storage has caused local groundwater level declines; in the extension of the aquifers into South Africa, water quality constraints have restricted its utilization, while in Botswana the potential for available resources is likely, but insufficient data is available for making firm conclusions for investments." Mathematical model development was conducted in several phases—updating the original that modeled a single country (Namibia), extending it to cover the full flow system (including Botswana and South Africa) followed by simulations to characterize the aquifer. The conceptual interpretation of the hydrodynamic system is shown in Fig. 9.5, and this is a more effective illustration of "storage" and "hydraulic heads" to engage the nonspecialists.

The conceptual interpretation shows that storage is being depleted—and the solution is that managed aquifer recharge should be initiated in both the territories of Namibia and Botswana for mutual long-term benefit—else the GDP, generated from farming, will decline due to the onset of loss of access to water (see Fig. 9.2 for 2017 GDP gained from agriculture). Taking account of the hydrodynamics and the proposed scope of engagement between the countries, an institutional structure was formulated shown in Fig. 9.6, which was adopted by the countries in 2018. Further details of the responsibilities and obligations of such an institution have been given by Puri (2017) and Burchi (2018).

The institutional structure has been forwarded to the ORASECOM Secretariat and it has been accepted by the member states. The next step—formulation of the work program that should be implemented through the Stampriet Aquifer Consultation Mechanism is still awaited.

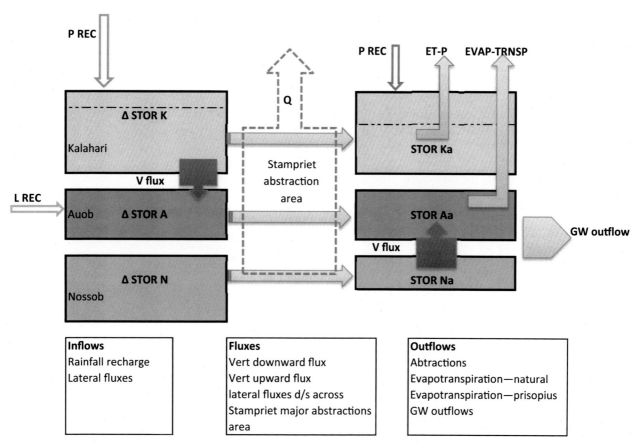

FIGURE 9.5 Conceptual representation of hydrodynamics—Stampriet aquifer system.

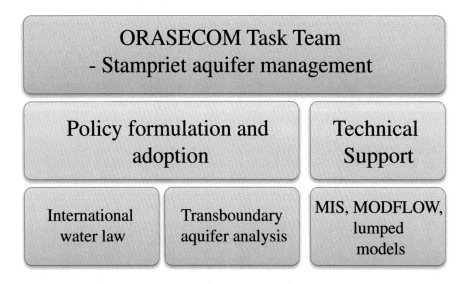

FIGURE 9.6 Consultation mechanism for stakeholders—Stampriet aquifer system.

9.13 Hurdles to progress in intercountry dialogue—the "invisibility cape"?

As noted in Puri and Villholth (2018), among the key lessons learned from the consideration of the issue from nearly two decades, it can be said that although the science community of hydrogeology practitioners may have raised the profile, they still have not managed to convince many national decision makers of the value of taking the science to policy, leading to actions on the ground. For scientists the "invisibility cloak" has been lifted (see the case history for Stampriet, as an example, as well as several others) but the foreign policy makers who are usually at the "chalk face" for leading intercountry diplomatic dialogues have yet to see the "face" that has been made bare from behind the cloak, despite significant efforts to do so (see Puri and Struckmieir, 2010).

Following on from the commentary by Sindico and Nagle (above), the "obligation not to cause significant harm" remains a stumbling block for the legal—diplomatic community as evidenced at the UN Meetings of the Sixth Committee (2019), with their suggestion that the term "significant" creates uncertainty, and proposing that it might require a legal definition. Reverting again to the case history of the Stampriet, it may be determined that inaction to replenish the aquifer through managed recharge (in Namibia and in Botswana) could amount to significant self-harm. This can be expressed as the decline of GDP if the farming output were to be adversely affected—the GDPs for 2017 are shown on Fig. 9.2. Economic analysis could provide a numerical value, if required, of the severity of the decline, which can be converted into the "degree" of significance. Similar efforts can be put in place in the case of most transboundary aquifers, to enable the diplomatic community to move forward toward progressing collaboration for mutual benefits.

9.14 The hiatus in the progress to adoption of the Draft Articles

The current hiatus in the progress with the adoption of the Draft Articles as commented on by Sindico et al. has to be seen through the prism of today's socio political reality—the world in the 2020s is more nationalistic and disruptions in social, political, and customary norms are common. Add to this the outbreak of global coronavirus (March 2020). This seems to lead to reduced momentum to collaborate and a latent tendency to create barriers and protectionist measures. In this light the analysis of the fate of 1997 UN Convention on transboundary water courses by Eckstein (2020) is only salutary. It would seem that the misconceptions, misgivings, and national self-interested stance of the later 1990s over transboundary water have not been overcome and even with the coming into force (in 2014) of the UN Convention, less than a fifth of the world's nations have become state party to it. Tracing some of the curiosities of the voting and abstention patterns at the UN General Assembly, summarized by Eckstein, may give some insights into the hiatus and the stagnancy in moving forward with the Draft Articles.

Today UN member states in discussing the Draft Articles have questioned the "obligation not to cause significant harm"—with the term "significant" being under the spotlight—exactly as it was in 1997 when the Pakistani delegate reacted to it (see Eckstein, 2020, p 21). The demand today that a legal definition for "significant harm" may be required is much the same as the 1997 comment that it "lacks precision and can become a bone of contention when one is considering the type of harm which should or should not be taken into account. One party's definition of 'significant' would be different from that of another …."

It may be fair to conclude that today's sociopolitical forces that have unleashed nationalism, self-interest, and introversion among so many of the world's countries (India, United States, United Kingdom, Poland, Hungary, Turkey, Portugal, and Brazil) will consign the issue of transboundary aquifers on the shelf for the next decade—despite the contrary perverse assertions made by politicians, the civil servants of the many UN agencies associated with water, and the plethora of NGOs that seek to further their own narrow cause.

9.15 Conclusion: light at the end of the tunnel

In conclusion of this critical perspective, it may be stated that despite all of the hurdles and constraints described previously, there is a prospect for moving ahead. The countries in which risks from lack of sound governance of their subsurface space are high can be objectively appraised and some indication of the onset of potential problems can be quantified in terms of cost and time. This quantification has to be in terms that are meaningful to the highest decision making authorities in countries and should be stated in such tangible quantities as the impact on the national GDP—even where there is some margin for error in long-term future forecasts. Time frames for their onset can also be developed, with reasonable margins of error. Other indicators, such as decline of water table and decrease in well production, while critical in the scientific analysis, have limited practical significance and implications for the nonspecialists, in order to enable them to make choices. From all of the earlier discussions, Draft Articles remain the primary source of direction with which such progress on intercountry dialogue can be made—if not through their formal adoption, then by their use as guidance for cooperation and collaboration.

Conflict of interest

None.

Acknowledgment

The author wishes to acknowledge the valuable and insightful discussions and consultations with his numerous colleagues who supported him during his tenure as Secretary General of the IAH, and as the founding Chair of the Commission on Transboundary Aquifers between 2001 and 2019.

References

Abubakar, B., 2017. Harnessing Lake Chad's ground water for improving livelihoods of lake chads indigenous communities threatened by climate change. Report accsible at Research Gate.

Allana, A., Louresb, F., Tigninoc, M., 2011. The role and relevance of the draft articles on the law of transboundary aquifers in the European context. J. Eur. Environ. Plan. Law v 8.3, 231−251.

Aeschbach-Hertig, W., Gleeson, T., 2012. Regional strategies for the accelerating global problem of groundwater depletion. Nat. Geosci. 5 (2012), 853−861. Available from: https://doi.org/10.1038/ngeo1617.

Burchi, S., 2018. Legal frameworks for the governance of international transboundary aquifers: Pre- and post-ISARM experience. J. Hydrol.: Regional Stud. v. 20, 15−20.

CEPAL, 2020. América Latina y el Caribe: Dimensionar los efectos del COVID-19 para pensar en la reactivación. Informe especial COVID-19, April 2020.

Convention, 2008. On the Protection, Utilisation, Recharge and Monitoring of the Franco-Swiss Genevois Aquifer. Accessed from: <https://www.internationalwaterlaw.org/documents/regionaldocs/2008Franko-Swiss-Aquifer-English.pdf>.

Davies, J., Robins, N.S., Farr, J., Sorensen, J., Beetlestone, P., Cobbing, J.E., 2013. Identifying transboundary aquifers in need of international resource management in the Southern African Development Community region. Hydrogeol. J. 21, 321−330.

De los Cobos, G., 2018. The Genevese transboundary aquifer (Switzerland-France): the secret of 40 years of successful management. J. Hydrol.: Regional Stud. v. 20, pp. 116−127.

Eckstein, G., 2020. The status of the UN watercourses convention: does it still hold water? Int. J. Water Resour. Dev. Available from: https://doi.org/10.1080/07900627.2019.1690979.

Eckstein, G., 2015. The Newest Transboundary Aquifer Agreement: Jordan and Saudi Arabia Cooperate Over the Al-Sag /Al-Disi Aquifer. Accessed https://www.internationalwaterlaw.org/blog/2015/08/31/the-newest-transboundary-aquifer-agreement-jordan-and-saudi-arabia-cooperate-over-the-al-sag-al-disi-aquifer/.

FAO, IAH, UNESCO, World Bank, 2015. Final Report of the Global Groundwater Governance Project. Accessed from: <http://www.groundwatergovernance.org>.

Freeze, R.A., Chery, J.A., 1979. Groundwater. Pub Prentice Hall Inc.

GEF-TWAP, 2015. Global scale modelling and quantification of indicators for assessing transboundary aquifers. (prepared by C Riedel & P Doll). Final Report, Goethe University, Frankfurt.

Global Diagnostic on Groundwater Governance, 2015. Developed Under the GEF Financing. Available at <http://www.groundwatergovernance.org/fileadmin/user_upload/gwg/documents/Global_Diagnostic_on_Groundwater_Governance_Draft.pdf>.

Gorelick, S.M, Zheng., C., 2015. Global change and groundwater management challenge. Water Resources Research 51, 3031–3051.

Guppy, L., Uyttendaele, P., Villholth, K.G., Smakhtin, V., 2018. Groundwater and Sustainable Development Goals: Analysis of Interlinkages. UNU-INWEH Report Series, Issue 04. United Nations University Institute for Water, Environment and Health, Hamilton, Canada.

IAH, 2017. The UN SDG's for 2030 – Essential Indicator for Groundwater. Strategic Overview Series.

IGRAC, UNESCO-IHP, 2016. TWAP groundwater data and information portal. In: Data From UNESCO-IHP, UNEP, 2016. Transboundary Aquifers and Groundwater Systems of Small Island Developing States: Status and Trends. Transboundary Waters Assessment Programme TWAP. Online. https://www.un-igrac.org/special-project/twap-groundwater.

Katz, D.I., Moore, M.R., 2011. Dividing the waters: an empirical analysis of interstate compact allocation of transboundary rivers. Water Resour. Res. 47 (6), First published: 15 June 2011. Available from: https://doi.org/10.1029/2010WR009736.

Larson, R.B., 2017. Inter-State Water Law in the United States of America in Brill Research Perspectives in International Water Law, ISSN: 23529369, Publication Date: 06 Sep 2017 In: Volume 2: Issue 3.

Mukherjee, A., Scanlon, B., Aureli, A., Langan, S., Guo, H., McKenzie, A., 2020. Global Groundwater: Source, Scarcity, Sustainability, Security and Solutions, first ed. Elsevier, ISBN. 9780128181720

National Economic Policy Council, 2019. Jordan Economic Growth Plans 2018–2022. Ministry of Planning. Government of Jordan.

Oates, N., Mwathunga, E., 2018. A political economy analysis of Malawi's rural water supply sector. In: Report Published by ODI, as Part of UpGro-Hidden Crises.

O'Brien, E., Gowan, R., 2012. What makes international agreements work: defining factors for success. Centre for International Cooperation, New York University, Accessed from: <http://www.cic.nyu.edu>.

Panda, A., Agrawal, V., 2011. Interstate River Water Disputes: Constitutional Mechanisms and Judicial Expositions (June 17, 2011). Available at SSRN: <https://ssrn.com/abstract = 1866563 or https://doi.org/10.2139/ssrn.1866563>.

Puri, S., Aureli, A., 2005. Transboundary Aquifers: a global programme to assess, evaluate and develop policy. Groundwater, 43, 661–668.

Puri, S., Wong, H., 2018. Wellfield optimisation and resource management – update to the model of the Disi-Rum aquifer, Jordan. In: Confidential Reports – Stage1, Stage 2, Stage 3 and Stage 4 Reports. Prepared for the Ministry of Water & Irrigation, Government of Jordan.

Puri, S., 2017. GGRETA: evaluation of the Regional Strategy for the management of the Stampriet Transboundary Aquifer National institutional efficiencies for the formulation and design of the Multi Country Cooperation Mechanism. In: Contract Reports, Deliverable 1, 2 & 3. Prepared for UNESCO-IHP, Paris

Puri, S., Villholth, K.G., 2018. Chapter 19: Governance and management of transboundary aquifers. Advances in Groundwater Governance. Pub Taylor & Francis, London, pp. 367–388.

Puri, S., Struckmieir, W., 2010. Chapter 6: Aquifer resources in a transboundary context: a hidden resource? – Enabling the practitioner to 'See It and Bank It' for good use. In: Earle, A., Jägerskog, A., Öjendal, J. (Eds.), Transboundary Water Management: Principles and Practice. Earthscan Publications for SIWI. ISBN 978-1-84971-137-1.

Puri, S., Elnaser, H., 2003. Intensive use of groundwater in transboundary aquifers. In: Llamas, R., Custodio, E. (Eds.), 2003. Intensive Use of Groundwater: Challenges and Opportunities. A.A. Balkema Publishers, the Netherlands, pp. 415–439, ISBN 9789058093905.

Rivera, A., Candela, L., 2018. Fifteen year experience of the internationally shared aquifer resources management initiative of UNESCO at the global scale. J. Hydrol. Regional Stud. v20, 5–14.

Sanchez, R., Rodriguez, L., Tortajada, C., 2018. Effective Transboundary aquifer areas: a potential approach for transboundary groundwater management. In: JAWRA Manuscript 1900523-P, Technical Paper.

Sindico, F., Hirata, R., Manganelli, A., 2018. The Guarani aquifer system: from a Beacon of hope to a question mark in the governance of transboundary aquifers. J. Hydrol.: Regional Stud. v20, 49–59.

Sindico, F., Nagle, R.M., 2019. Transboundary Aquifers at the 2019 UN General Assembly 6th Committee: in Uni of Strathclyde Policy Brief, CELG No. 13 twapviewer.un-igrac.org (accessed 15.05.17.).

Shah T., Molden D., Sakthivadivel R., Seckler D., 2000. The Global Groundwater Situation: Overview of Opportunities and Challenges. IWMI, Sri Lanka. ISBN 92-9090-402-X.

UNESCO, 2016. GGRETA – Ph I – Stampriet aquifer system assessment. In: Prepared Under the Support of the Swiss Development Agency. UNESCO, Paris

UNESCO, 2018. Progress on Transboundary Water Cooperation – Global Baseline for SDG Indicator 6.5.2.

UN Environment, 2018. Progress on Water Related Ecosystems – Piloting the Monitoring Methodology and Initial Findings for SDG Indicator 6.6.1.

World Development Indicators, 2018. World Bank, Washington DC, USA. Accessed from: <http://wdi.worldbank.org/tables>.

Chapter 10

Transboundary groundwater of the Ganges–Brahmaputra–Meghna River delta system

Madhumita Chakraborty[1], Abhijit Mukherjee[1,2] and Kazi Matin Ahmed[3]

[1]*Department of Geology and Geophysics, Indian Institute of Technology (IIT) Kharagpur, Kharagpur, India,* [2]*Applied Policy Advisory for Hydrosciences (APHA) group, School of Environmental Science and Engineering, Indian Institute of Technology (IIT) Kharagpur, Kharagpur, India,* [3]*Department of Geology, University of Dhaka, Curzon Hall Campus, Dhaka, Bangladesh*

10.1 Introduction

The Ganga–Brahmaputra–Meghna (GBM) River delta, synonymously known as the Bengal basin, covering most of Bangladesh and some states of eastern India (including most of West Bengal and some parts of Assam and Tripura), is regarded as the largest fluvio–deltaic system in the world. Three main river systems, that is, the Ganges and the Brahmaputra Rivers originating from the Himalayas and the Meghna River originating from the Tripura Hills converge to collectively form the GBM river delta, before meeting the Bay of Bengal (BoB) across the 380 km wide active delta mouth (Allison, 1998). The Brahmaputra and the Ganges Rivers are the world's fourth and fifth largest rivers by discharge, respectively (Mukherjee et al., 2009a). The GBM River system has the fourth highest riverine discharge (Milliman and Meade, 1983) and the largest sediment dispersal system in the world (Kuehl et al., 1989; Milliman et al., 1995; Goodbred et al., 2003). These rivers together carry more than 1 GT of sediment per year (Curray and Moore, 1971; Kuehl et al., 2005) from the Himalayan orogen, Indian shield, Indo-Burmese arc, and their surrounding areas (Alam et al., 2003) to the delta and then to the BoB, resulting in the formation of the world's largest submarine fan. However, much of the sediments are trapped within the deltaic landmass before reaching to the BoB that has led to the deposition of up to 22 km thick sediment sequence underlying the delta plains (Alam et al., 2003).

The unconsolidated sediments underlying the GBM River delta form a transboundary aquifer system that holds one of the largest groundwater reserves in the world and sustains a major portion of the freshwater demands for about 150 million people inhabiting the delta (Goodbred et al., 2014). The GBM River delta aquifer system is shared between India and Bangladesh and plays an integral part in the socioeconomic growth, public health, and political dialogue of both the countries. The 1970s encountered a major shift from surface water to groundwater usage in both these countries with the advent of installation of millions of tube wells within the delta (Mukherjee and Fryar, 2008), which have safeguarded majority of the population against the endemic occurrence of surface water–borne diseases, such as diarrhea. Much of the domestic and industrial water needs for the fast-growing population and the increasing number of urban centers within the delta are sustained by groundwater from these aquifers. The GBM River delta also hosts two of the world's largest megacities, Kolkata (India) and Dhaka (Bangladesh). Both India and Bangladesh are important agricultural exporters (mostly rice) in the world (FAO, http://www.fao.org/faostat/en/#data/QC/), much of which is derived from groundwater-fed irrigation within the fertile alluvial plains of the GBM River delta. Rice is widely cultivated within the delta plains using traditional water-intensive irrigational practices, exploiting huge volumes of groundwater annually from these aquifers.

Presently, millions of hand pump and mechanized tube wells are installed across the delta to yield groundwater from the aquifers all-round year. However, this extensive exploitation of groundwater of the GBM River delta aquifers has resulted in several water quality and quantity issues (Mukherjee et al., 2020). Although groundwater-fed irrigation has enabled all year round cultivation (in comparison to seasonally varying surface water availability) leading to the strengthening of the national agricultural economy and food security for both the countries, but unplanned and ill-managed exploitation of groundwater resource is leading to fast-declining water levels in some parts of the delta. Moreover, the shift from surface water to groundwater as the primary source of drinking and irrigation water has surely decreased the number of diarrhea cases but has also resulted in prolonged exposure of millions of people to several groundwater contaminants, especially arsenic (As) that has caused severe mass poisoning and long-term public health concerns for both the nations.

10.2 Geologic and geomorphologic setting

The delta is bounded by the Indian shield and the Rajmahal Traps to the west and northwest; the Garo, Khasi, and Jaintia Hills and Shillong Plateau to the northeast (successively from west to east); the Tripura Hills, Chittagong Hills, and Indo-Burma orogen to the east; and BoB to the south (Mukherjee et al., 2009a) (Fig. 10.1). The area has a very low topography, with very gentle slopes.

Tectonically, the GBM River delta constitutes a peripheral foreland basin (Raman et al., 1986) formed due to the subduction of the Indian plate to the south below the Eurasian and Burmese plates to the north and northeast, respectively. The formation of the basin was concurrently initiated with the breakup of the Gondwanaland during late

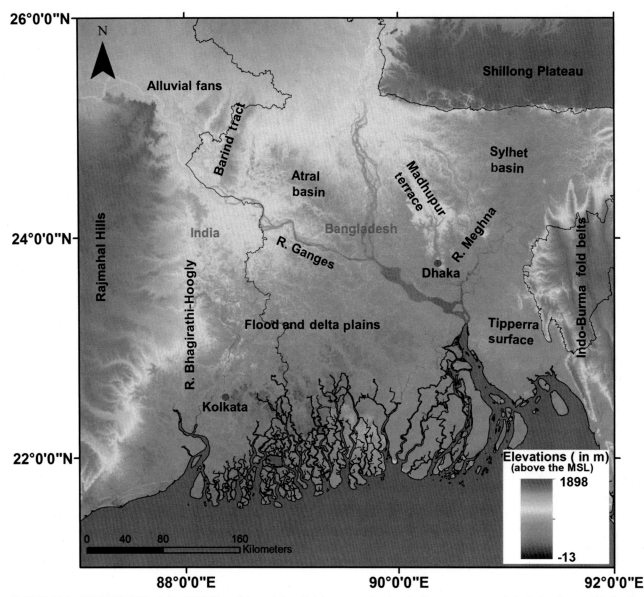

FIGURE 10.1 SRTM-90 DEM of the GBM River delta and the adjoining areas, showing the important geomorphological and geopolitical units. *DEM*, Digital Elevation Model; *GBM*, Ganga−Brahmaputra−Meghna.

Mesozoic, which was followed by the development of the proto-GBM River delta that gradually evolved to the modern-day fluvio−tidal GBM River delta during middle to late Miocene (Mukherjee et al., 2009a). Since then, the evolution of the GBM River delta has been controlled by fluvio−deltaic dynamics, glacio−eustatic changes, and active tectonics. The deltaic foreland basin can be divided into the stable shelf to the west and the Bengal foredeep to the east along the Calcutta−Mymensingh Hinge Zone, which is a deep-seated seismic zone of flexure that marks the eastern extension of the Indian plate below the foreland basin (Sengupta, 1966; Alam, 1989). The basin deepens toward the southeastern parts of the Bengal foredeep with up to 16−22 km of tertiary to recent alluvium (Alam et al., 2003), while the western stable shelf has 1−8 km of Permian to recent sediments (Imam and Shaw, 1985).

Geologically, the GBM River delta consists of Pleistocene uplands and Holocene lowlands (Morgan and McIntire, 1959; Umitsu, 1987, 1993). Of the four main Pleistocene units, the Barind and the Madhupur terraces lie in the northwest and north-central parts of the basin respectively, while the other two terraces border the basin, lying east of the Rajmahal Hills and west of the Tripura Hills (Morgan and McIntire, 1959). The Holocene sediments are found within the piedmont alluvial fans in the Himalayan foothills, the Tippera surface, the Sylhet basin, and the extensive stretches of the GBM flood and delta plains (which alone cover more than 80% of the areal extent) (Mukherjee et al., 2009a).

10.3 Aquifer framework

The basinal-scale aquifer−aquitard architecture of the transboundary GBM River delta aquifer system remains to be precisely delineated. However, numerous studies have characterized the aquifer units at a local scale in various parts of the delta (DPHE/DFID/JICA, 2006; Mukherjee et al., 2007a; Burgess et al., 2010). The intricate interplay of fluvial dynamics and eustatic changes involving multiple episodes of marine transgression and regression along with river channel migration has resulted in the formation of very complex and highly heterogeneous sediment architecture. Mukherjee et al. (2007a) delineated the aquifer geometry of the western part of the GBM River delta. Model results show the presence of a single continuous unconfined aquifer system to the northern parts of the delta, which is increasingly interspersed by discontinuous aquitard layers of varying thickness toward the active delta mouth (Fig. 10.2). Thus, to the south, the aquifer system breaks up into confined to semiconfined, poorly connected, and multilayered aquifer units. This aquifer system serves as the main source of fresh groundwater that can be sustainably withdrawn over the years. This aquifer unit is heavily exploited, mostly up to a depth of about 300 m for domestic, irrigational, and industrial purposes. The main aquifer is mostly connected to the atmosphere in the northern parts of the delta, while toward the south it is overlain by an overall increasingly thicker aquitard layer (with thickness up to several tens of meter at places). This aquifer deepens toward the active delta mouth, that is, it transitions from being up to 80 m thick in the north to being more than 450 m deep to the extreme south. The main aquifer is underlain by a thick basin-wide layer of marine clay that serves as a hydraulic basement to the main aquifer system of the GBM River delta. This basal aquitard hosts several deep, isolated aquifer units that mostly trap the saline and brackish groundwater (Mukherjee et al., 2007a).

The shallower aquifers within the Holocene flood and delta plains are mostly found to be gray in color and are believed to indicate reduced aquifer conditions (McArthur et al., 2004). In the eastern parts of the GBM delta, the Holocene sediments are reported to be underlain by the Pleistocene sediments that are characterized by their brown color and oxic nature [BGS/DPHE (British Geological Survey/Department of Public Health Engineering [Bangladesh]) et al., 2001; vanGeen et al., 2003a,b; 2008; Hossain et al., 2014]. The depth of the Pleistocene−Holocene boundary in sediments varies across the delta, ranging up to a maximum of 150 m below the surface (vanGeen et al., 2003a,b; 2008). However, Mukherjee et al. (2007a) reported that the presence of such brown sediments was not found in the western part of the delta.

10.4 Groundwater flow system

On a regional scale the GBM River delta shows the presence of a single continuous unconfined aquifer system to the north, which breaks down into multiple confined to semiconfined poorly connected aquifer units separated by several discontinuous aquitard lenses to the south (Mukherjee et al., 2007a). Although no delta-wide, laterally continuous confining layer has been reported from the delta aquifer system, the presence of the numerous local-scale laterally discontinuous aquitard units has resulted in a layered and heterogeneous nature of the aquifer fabric (Michael and Voss, 2009a) leading to the development of highly anisotropic hydraulic properties, where the horizontal hydraulic conductivity is several times greater than the vertical hydraulic conductivity. Modeling by Michael and Voss (2009a) suggests that the horizontal hydraulic conductivity is $\geq 10^4$ times greater than the vertical hydraulic conductivity. Thus under the low topography and natural hydraulic gradients of the GBM River delta, the multilayered aquifer−aquitard

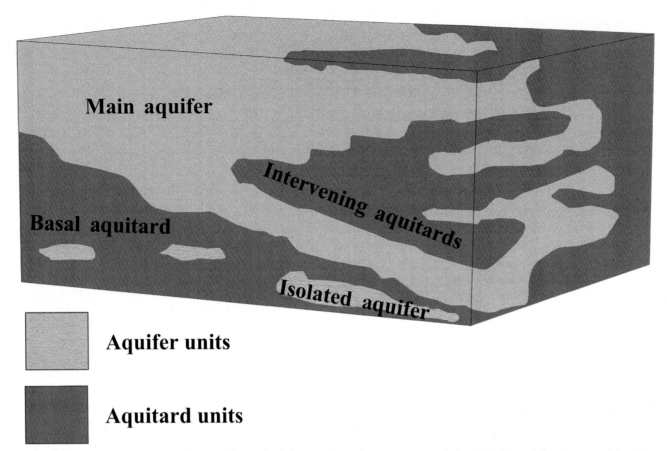

FIGURE 10.2 Conceptual model of the aquifer–aquitard framework on the western part of the GBM River delta (not to scale). *GBM*, Ganga–Brahmaputra–Meghna. *Modified from Mukherjee, A., Fryar, A.E., Howell, P., 2007a. Regional hydrostratigraphy and groundwater flow modeling of the arsenic contaminated aquifers of the western Bengal basin, West Bengal, India. Hydrogeol. J. 15, 1397–1418.*

framework generates a regional-scale hierarchical groundwater flow that is dominated by the inherent hydraulic anisotropy of the aquifer medium (Michael and Voss, 2009a; Hoque et al., 2017). The layered aquifer–aquitard framework within the GBM River delta has resulted in significantly low vertical hydraulic conductivity at a regional scale, such that despite the low hydraulic gradients, the delta aquifers are dominated by large-scale sub-basinal to basinal level flow systems, especially at depths of or greater than 100 m (Michael and Voss, 2009b). Under the natural hydraulic gradients the GBM River delta aquifer system experiences regional-scale groundwater flow from north to south across the delta (Mukherjee et al., 2007a). Factors controlling the lateral and vertical variation of hydraulic properties within the delta aquifer, that is, presence of surficial confining layer, lateral variation in hydraulic parameters, and depth trends of hydraulic conductivity, have insignificant control on the groundwater flow paths, although these factors seem to have a substantial impact on the groundwater residence time (age) within the delta aquifers. While deeper ($>$ 150 m) groundwater may have a residence time of over thousands of years, shallow ($\sim$ 100 m) groundwater can have a travel time for less than several hundreds of years (Hoque and Burgess, 2012).

The GBM River delta aquifers are primarily replenished by the meteoric recharge. Every year, the aquifers are recharged to full conditions in most parts of the delta after the annual flooding caused by the heavy monsoonal rainfall in this part of the world (Mukherjee et al., 2007a). However, in areas undergoing extensive pumping, monsoonal recharge fails to replenish back the aquifers, which is manifested as declining hydraulic heads over the years in these areas. The surface water bodies within the delta, which include a dense network of river channels, streams, lakes, ponds, and wetlands, closely interact with the groundwater system. In absence of extensive groundwater irrigation, the groundwater discharges to rivers under natural hydraulic gradients.

With the advent of large-scale pumping within the GBM River delta in the 1970s, more than 6–11 million tube wells were installed to abstract groundwater [BGS/DPHE (British Geological Survey/Department of Public Health Engineering [Bangladesh]) et al., 2001] for domestic, irrigational, and industrial needs. Among all the uses of groundwater, dry season irrigation consumes the largest share of abstracted groundwater per year for high-yielding rice cultivation (Shamsudduha et al., 2009). In most parts of the delta, irrigation is principally supported by groundwater from shallow tube wells, with the installation of about 25–75 wells per sq. km, along with surface water irrigation pumps (Shamsudduha et al., 2009). As per BADC (2003), of the total of 947,457 irrigational tube wells used for dry season irrigation in 2003 in Bangladesh, 2.5% wells tap the deep aquifer zone, while the rest were shallow (Harvey et al., 2006). This large-scale groundwater exploitation imposes a dominant control on the shallow groundwater flow system that overpowers the influence of any natural geological factors (Michael and Voss, 2009b). Pumping induces an increase in the vertical hydraulic gradients within the shallow groundwater flow system along with the development of local flows and recharge zones (Michael and Voss, 2009b). In addition, heavy pumping has led to the development of large cones of depression in many parts of the delta, ranging from about a kilometer to several kilometers in diameter (Mukherjee et al., 2007a). These cones of depression around the points of heavy groundwater abstraction induce a rapid inflow of groundwater from the surroundings, resulting in the development of local flow systems. Harvey et al. (2006) showed that pumping-induced increase in the flow rate can cause infiltration of groundwater up to a depth of 20–30 m below the surface in just about 20–30 years. Mukherjee et al. (2007a) argued that extensive deep pumping in the western part of the delta possibly has caused a significant change in the deeper groundwater chemistry, due to rapid mixing with the infiltrated shallow/surficial groundwater. Such mixing over time strictly limits the potential of the deeper groundwater as a long-term viable alternative for safe drinking water in areas where the shallow aquifers are As contaminated. Hence many researchers strongly advise restriction of irrigational pumping wells only within the shallow aquifer depths (Burgess et al., 2010; vanGeen et al., 2003a,b). The extensive groundwater pumping is reported to have resulted in steady decline in groundwater level in many parts of the delta (0.1–0.5 m/year), especially in and around the two largest metropolitan centers within the delta, that is, Kolkata (India) (Biswas and Saha, 1985, Sahuand Sikdar, 2011) and Dhaka (Bangladesh) (>1 m/year) (Shamsudduha et al., 2009). The increase in the vertical hydraulic gradient induced by large-scale shallow groundwater pumping is also accompanied by significant increase recharge in terms of increase in irrigation return flow, inflow/backflow from rivers, and intrusion of seawater. The coastal aquifers of the GBM River delta show an increase in the groundwater table height (0.5–2.5 cm/year) caused by enhanced seawater intrusion, possibly as a result of eustatic rise accompanied by coastal groundwater pumping (Shamsudduha et al., 2009).

10.5 Hydrogeochemistry

The groundwater chemistry of the GBM River delta aquifer system is controlled by the hydrostratigraphy and the groundwater flow paths and hence varies across the delta. The shallow aquifers are dominated by relatively young groundwater that bears groundwater isotopic ($\delta^{18}O$) signature of meteoric recharge coupled with some meteoric diagenesis and evaporation loss (Mukherjee, 2006; Mukherjee et al., 2007b). A significant fraction of the groundwater is also recharged as the irrigation return flow through the extensive stretches of agricultural fields that are cultivated using water-intensive irrigation practices. The groundwater in the main aquifer undergoes hydrogeochemical evolution along the groundwater flow paths by water–sediment interactions that are majorly dominated by carbonate dissolution and cation exchange along with a substantial influence of silicate weathering processes, all of which are guided by the aquifer redox states (Mukherjee and Fryar, 2008). The multi-depth hierarchical groundwater flow system within the delta has resulted in highly heterogeneous hydrogeochemical signatures that vary both spatially and vertically across the GBM River delta.

A total of seven distinct hydrogeochemical facies have been reported from the GBM river delta (namely, $Ca^{2+}-HCO_3$, $Ca^{2+}-Na^+-HCO_3$, $Na^+-Ca^{2+}-HCO_3$, $Ca^{2+}-Na^+-HCO_3-Cl$, $Na^+-Cl-HCO_3$, Na^+-Cl, and $Ca^{2+}-Cl$), which were found to vary both spatially and vertically (Mukherjee and Fryar, 2008). The unconfined continuous main aquifer system to the north of the delta is mostly dominated by $Ca-HCO_3^-$-rich groundwater, with the predominance of Ca^+ and Mg^+ and HCO_3^- ions as major solutes. The Ca^{2+} ions are suggested to be introduced into the groundwater majorly through carbonate dissolution, while the Mg^{2+} ions are derived from both carbonate dissolution and silicate weathering. The HCO_3^- ions are believed to be majorly derived from silicate weathering and oxidation of organic matter along with carbonate dissolution and root respiration (Mukherjee and Fryar, 2008). The multilayered aquifer system to the south shows a somewhat varied hydrogeochemical signature, broadly falling within the $Ca-Na-HCO_3^--Cl^-$ facies. The semiconfined and poorly connected aquifers show significantly higher proportions of brackish/saline water with a higher concentration of Na^+ and Cl^- ions indicating contribution from marine source (i.e., BoB). The isolated aquifer lenses majorly host connate waters dominated by Na^+ and Cl^- ions. These highly brackish/saline waters represent entrapped seawater from the proto-BoB, which have undergone limited diagenetic evolution through interaction with aquifer sediments and fresher groundwater leakage (Mukherjee and Fryar, 2008). On average, the delta groundwater is also characterized by the dominant presence of Fe and Mn as minor solutes along with a very low concentration of SO_4^{2-} and NO_3^- and undetectable levels of NO_2^- and H_2S (Mukherjee and Fryar, 2008; Mukherjee et al., 2008). The solute concentrations are found to vary with depth, for example, Fe, Mn, and SO_4^{2-} concentrations are found to decrease with depth, while CH_4 shows an increasing trend (Mukherjee and Fryar, 2008).

The aquifer redox states within the GBM river delta are found to be concurrently controlled by multiple geochemical processes that are hydrostratigraphy and depth dependent (Mukherjee and Fryar, 2008). Fe−S−C redox cycles strongly controls the aquifer redox states across the delta. The groundwater is mostly found to be postoxic in nature, and Fe(II)/Fe(III) is found to be the most predominant redox couple within the aquifers. The reducing nature of groundwater within the delta is also indicated by the prevalence of CH^4 and S^{2-} coupled with low dissolved oxygen content (Mukherjee and Fryar, 2008). However, the aquifers are found to be in redox disequilibria with multiple overlapping redox zones as indicated by the co-presence of multiple redox-sensitive solutes [e.g., Fe(II), As(III), SO_4^{2-}, CH_4, and NH_4^+]. The redox disequilibria also control the trace element cycling, that is, mobilization and sequestration of trace elements such as As. Microbially mediated metal (Fe/Mn/Al) (oxy)(hydro)oxides reduction coupled with oxidation of natural organic matter (Bhattacharya et al., 1997; Nickson et al., 2000; Mukherjee et al., 2008) is found to the widely prevalent within the reduced delta aquifers with anoxic groundwater. The partial redox equilibria within the aquifers have also led to the development of very local oxidized environments within aquifers at places (Mukherjee and Fryar, 2008).

10.6 Groundwater arsenic contamination

The widespread occurrence of groundwater As concentrations exceeding the World Health Organization (WHO) guideline for permissible limit in drinking water of 10 μg/L is found across the GBM River delta. Groundwater As concentration exceeding the WHO guideline value was first detected in some tube wells of North 24 Parganas district in West Bengal (India) in 1978 (GuhaMazumder et al., 1998). It was closely followed by the first detection of arsenicosis in a patient from South 24 Parganas district in West Bengal (India) in 1983 (Garai et al., 1984). Subsequently, of the millions of wells tested for As, 42%−48% wells were detected to have more than 10 μg/L of As, while 24%−25% wells were found to have more than 50 μg/L of As [BGS/DPHE (British Geological Survey/Department of Public Health Engineering [Bangladesh]) et al., 2001; Chakraborti et al., 2009]. A maximum of up to 4730 μg/L of As has been reported from the groundwater of the Bengal basin (Mukherjee et al., 2009b). The total population at risk of As poisoning is estimated to be more than 26 million people in India (Chakraborti et al., 2009) and ∼53 million in Bangladesh (BBS/UNICEF Bangladesh Bureau of Statistics/United Nations Children's Fund, 2011; Hossain et al., 2015). Prolonged consumption of such toxic levels of As in drinking water or through the food chain (mainly rice) (Chatterjee et al., 2010) by the mass population has resulted in prevalent cases of arsenicosis (arsenic poisoning) having severe health effects, including cardiovascular and pulmonary diseases, skin disorders [(leuco)melanosis, (hyper)keratosis)], and even skin and internal cancers. Of the people at risk, more than a million people in the GBM River delta is estimated to be already affected by arsenicosis (Yu et al., 2003). Based on the sheer magnitude of the arsenic crisis within the GBM River delta, Smith et al. (2000) declared it to be "the largest mass poisoning of a population in history," which was subsequently taken up by WHO (Havelaar and Melse, 2003). This has gathered a lot of global attention from the scientific and research communities and non-governmental organizations (NGOs) toward understanding and mitigation of the arsenic toxicity in the groundwater of the GBM River delta.

Over the past three decades, numerous researches have been done to understand the distribution and the causes of arsenic contamination within the aquifers of the GBM River delta (Chakraborty et al., 2015). The distribution of groundwater As is found to be highly heterogeneous across the basin at various scales (Mukherjee et al., 2018). Spatial patterns of groundwater As is found to vary with geology, with worst contamination found within the Holocene delta plains to the southeastern part of the basin, while those on the north and the central part are moderately contaminated and the Pleistocene units to the north of the basin are mostly unaffected with >1% of contaminated wells (BGS/DPHE (British Geological Survey/Department of Public Health Engineering [Bangladesh]) et al., 2001; Smedley and Kinniburgh, 2002). Even within a very local scale, groundwater As concentrations are found to show considerable well-to-well variation, much of which can be attributed to depth variation among the adjacent tube wells, considering similar geological setting among the wells. In Bangladesh, it is found that the shallow aquifer depths are significantly high in groundwater As with 46% of the shallow wells (<150 m) having more than 10 μg/L of As, while 27% have more than 50 μg/L of As [BGS/DPHE (British Geological Survey/Department of Public Health Engineering [Bangladesh]) et al., 2001; Kinniburgh et al., 2003]. The extent of As contamination substantially decreases toward the deeper aquifer zones, where only 5% of the deeper wells (>150 m) have more than 10 μg/L of As and 1% have more than 50 μg/L. However, to the western part of the delta, Mukherjee et al. (2009b) found that about 60% of deeper wells have more than 10 μg/L of As. Such heterogeneity in As concentrations among the aquifers in the different parts of the delta can be attributed to the difference in basin geology, sediment geochemistry, hydrostratigraphy, aquifer redox states, presence or absence of surficial clay layer, groundwater flow paths and flushing rates, and rate and depths of groundwater abstraction (Chakraborty et al., 2020).

Arsenic in the groundwater of the GBM River delta is geogenic in origin, being released from basin sediments (nonpoint source) that have been carried down by the GBM River system from the Himalayan and the Indo-Burma orogen, the Indian Shield, the Shillong Plateau, and their surrounding areas (Mukherjee et al., 2009a) (Fig. 10.3). Although the average solid-phase As contents within the delta sediments are not unusually high (1–20 mg/kg) (Hossain, 2006; Mukherjee et al., 2011), it is enough to contaminate the groundwater beyond the permissible limits for consumption under favorable conditions (vanGeen et al., 2003a,b). Among various types of lithologies, sands are found to contain low As concentrations, while silts, clays, and peat layers have significantly higher As content. Iron (oxy)(hydro)oxides and detrital pyrites (Polizzotto et al., 2006) act as the major hosts for solid-phase As within the delta sediments (Chakraborty et al., 2015), while silicates and carbonates also contain considerable amounts of As (Charlet et al., 2007; Pal and Mukherjee, 2009). The fate of As within the aquifers is controlled by multiple processes that may simultaneously operate under specific favorable conditions. The mobilization processes of solid-phase As within the aquifers depend on local biogeochemistry and the aquifer redox state. Dissimilatory reductive dissolution of iron (oxy)(hydro) oxides is widely accepted to be the most dominant process of arsenic mobilization within the reduced Holocene aquifers of the GBM River delta (Bhattacharya et al., 1997; Nickson et al., 2000; Mukherjee and Fryar, 2008). The anaerobic metal-reducing microbes within the reduced sediments catalyze and mediate the reduction of ferric [Fe(III)] (oxy) (hydro)oxides to ferrous [Fe(II)] ions for their metabolism, often coupled with oxidation of organic matter. The process results in the dissolution of the ferrous ions accompanied by the release of the bound As into the groundwater. The organic matter may occur as in situ peat layers or may be dispersed within the sediments or can also be leached from surficial sources through groundwater as dissolved organic carbon (DOC). On the other hand, under oxidizing aquifer conditions developed at local scales, pyrite oxidation can prove to be a potent process for mobilizing As to groundwater (Das et al., 1995; Mallick and Rajgopal, 1995; Mandal et al., 1996). Excessive and unsustainable groundwater abstraction often promotes oxidizing environment within the reduced Holocene aquifer sediments of the GBM river delta, either due to water table decline leading to the exposure of the reduced sediments to atmospheric oxygen and/or enhanced groundwater recharge resulting in an inflow of dissolved oxygen to the aquifer depths. Under such conditions, oxidation of pyrite (and arsenopyrite) may lead to mobilization of the bound As, along with the dissolution of the sulfate into the groundwater (Das et al., 1995; Mallick and Rajgopal, 1995). Several dissolved ions in groundwater also compete for the sorption sites and replace any existing weakly bound ionic species, including As(III), which may lead to the release of the sorbed As from any host mineral into the groundwater. Among the most potent ions for competitive ion exchange and mobilization of As in the groundwater of the GBM river delta, phosphate (released from fertilizers, latrines, and peat layers) is found to the strongest competitor, while chlorine (released from pit latrines and wastewater), silicates, carbonates, and bicarbonates are found to have a moderate-to-weak potential for As mobilization (Stollenwerk et al., 2007; McArthur et al., 2012; Chakraborty et al., 2015).

10.7 Policy interventions and management options for arsenic mitigation

Of all the groundwater contaminants found within the GBM River delta aquifers, As has received maximum focus in terms of mitigation efforts, both from the scientific community and the administrative organizations due to its extent and severity. The Agency of Toxic Substances and Disease Registry (ASTDR) ranks As to be of the highest priority in the ATSDR's Substance Priority List (2019) based on its toxicity, frequency of human exposure, and the potential effect on human health. The safe limits of As in drinking water have been revised over the years by the WHO, from being 2000 μg/L in 1958 to being 50 μg/L in 1963 (Yamamura, 2003). In the last edition of WHO guidelines for drinking water quality in 1993, the provisional guideline for As in drinking water was established to be 10 μg/L based on epidemiological responses and practically quantifiable limit. This identifies millions of tube wells across the globe that yields As more than 10 μg/L to be a serious threat to human health.

In both India and Bangladesh, there have been persistent efforts by the governments and several NGOs and research groups to mitigate the As crisis within the GBM River delta. In the second revision of drinking water specification by the Bureau of Indian Standards (BIS) in 2004 [Bureau of Indian Standards (BIS), 2005], the acceptable/desirable limit of As was revised down to 10 μg/L from being 50 μg/L in 1991 (first revision), although permissible limit, in the absence of any alternate source of drinking water, remains to be 50 μg/L [Bureau of Indian Standards (BIS), 2003]. Bangladesh still holds its national standard value of arsenic in drinking water to be 50 μg/L (Bangladesh, 2009). Tube well testing has been found to

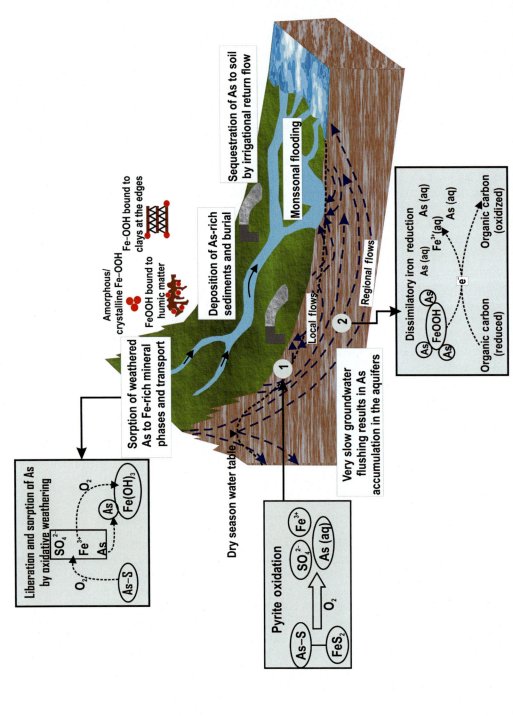

FIGURE 10.3 Schematic diagram illustrating the dominant processes from source to sink controlling the release and distribution of As in groundwater. Modified from BGS/DPHE (British Geological Survey/Department of Public Health Engineering [Bangladesh]), 2001. Arsenic contamination of groundwater in Bangladesh. In: Kinniburgh, D.G., Smedley, P.L. (Eds.). Report WC/00/19, Keyworth: British Geological Survey Technical; Chakraborty, M., Mukherjee, A., Ahmed, K.M., 2018. The Groundwater Flow, Chemistry and Pollutant Distribution in the Bengal Basin, Bangladesh and India. In: Mukherjee A., (Eds.), Groundwater of South Asia. Springer Hydrogeology (Chakraborty et al., 2018), Hoque, M.A., 2010. Models for Managing the Deep Aquifer in Bangladesh (Unpublished Ph.D. Thesis). University College London, London, 265 pp. (Hoque, 2010), Fendrof, S., Micheal, H.A., van Geen, A., 2010. Spatial and temporal variations of groundwater arsenic in South and Southeast Asia. Science 328, 1123–1127 (Fendrof et al., 2010).

be the most effective method of identification of As-affected wells within the delta. Moreover, seasonal changes in the hydrogeochemistry of groundwater (including As concentrations) make it essential for routine testing of the tube wells. The first nation-wide systematic and lab-tested hydrogeochemical survey that mapped the groundwater arsenic distribution across Bangladesh was conducted by the British Geological Survey (BGS) in collaboration with the Directorate of Public Health Engineering (DPHE), Bangladesh during 1998–99 [BGS/DPHE (British Geological Survey/Department of Public Health Engineering [Bangladesh]) et al., 2001]. This paved the way for large-scale blanket testing of more than 5 million tube wells in Bangladesh using field test kits in 270 chosen upazilas (administrative units) through the Bangladesh Arsenic Mitigation Water Supply Project (BAMWSP), Directorate of Public Health Engineering and United Nations Children's Fund (DPHE/UNICEF), Danish International Development Agency (DANIDA), and other national and international agencies, during 2000–2006 (Johnston and Sarker, 2007). In India, Public Health Engineering Department (PHED), West Bengal (India) and UNICEF conducted the "Joint Plan Of Action" (JPOA) that involved screening of tube wells in Assam and West Bengal and several other As-affected states during 2004–05 (PHED, 2006; Nickson et al., 2007). In West Bengal, 132,262 tube wells from 79 affected blocks in 8 districts were tested, while in Assam, 5792 tube wells were tested from 192 blocks in 22 districts. Apart from these surveys, tube wells are being intermittently tested for groundwater As at various scales within the delta by both governmental and NGOs.

Several mitigation measures have been adopted by both India and Bangladesh, which include installation of As removal filters (ARF), providing safe, filtered, and bacteria-free surface water as an alternate drinking water source, rainwater harvesting, and installation and identification of safe tube wells (Hossain et al., 2015). ARFs are a commonly implemented mitigation option for yielding As-free drinking water in the villages, wherein As-contaminated water is passed through different types of filtering media (e.g., activated alumina and alum) and techniques (such as coagulation–filtration, ion exchange, reverse osmosis, oxidation–filtration, adsorption, electrodialysis, and lime softening) for purification. However, these ARFs lose their efficiency and need to be replaced/regenerated after a given period. In most cases, gaps in monitoring and maintenance have rendered most of the ARF units to be non-operational. Dug wells are considered to be an alternative source of drinking water, based on the findings that groundwater in the dug wells is mostly found to be low in As [BGS/DPHE (British Geological Survey/Department of Public Health Engineering [Bangladesh]) et al., 2001]. However, the poor bacteriological purity of water from the dug wells severely impedes its utility as a viable drinking water source. Surface water, that is, river water and pond water can be an alternative source of drinking water in As-affected areas. However, the availability of potable surface water is extremely limited, and most of them have severe bacteriological contamination. The pond sand filter (PSF) is used in some rural parts of the delta where pond water is filtered by percolation through a media of sand and gravel packed within a concrete reservoir. The effectiveness of the PSFs depends on the availability of fresh ponds (high bacteriological purity, not exploited for pisciculture and domestic activities, such as washing) and periodic cleaning of the filtering media (Edmunds et al., 2015). Piped water supply (PWS) of treated surface water has been also found to be a viable long-term alternative. Numerous PWS schemes have been made operational that provide drinking water from large surface water treatment plants to the surrounding rural and urban areas. PWS schemes have a greater reach and are built to target a larger population. However, PWS schemes involve significant financial costs for its construction and maintenance. Rainwater harvesting (RWH) is considered to be an effective mitigation option in areas with high rainfall. The technology involves the collection of rainwater (mostly in the rooftop) and channelizing it to a storage container. Using the first rain in the monsoons to purge out the RWH unit is essential for avoiding any contamination. The unavailability of rainwater in dry seasons remains a major limitation for this technique. Identification and marking of As-affected tube wells are also a cheap and effective way of spreading As awareness and promoting well switching from As-affected to As-safe tube wells among communities (vanGeen et al., 2002). Both India and Bangladesh are working on color coding of As-affected tube wells by painting them with a specific color based on the reported As concentration found in the tube well groundwater. Construction of As-safe community tube wells in As-affected areas has been suggested to be a viable As-mitigation measure by vanGeen et al. (2003a,b). One community well is found to have the potential to cater to about 500 people. The abstraction of drinking water from deep tube wells has been reported to be the most successful mitigation option for As-safe drinking water [BGS/DPHE (British Geological Survey/Department of Public Health Engineering [Bangladesh]) et al., 2001; DPHE/APSU, 2005; Ravenscroft et al., 2009; Edmunds et al., 2015], although results from few other studies question the universal validity of such claims (Mukherjee et al., 2009b). PWS from deep (As safe) production wells has been found to be a dependable alternative for drinking water throughout the delta, where most of the shallow aquifers are As affected (Ravenscroft et al., 2009). However, the installation of deep tube wells is expensive. Moreover, overexploitation of deeper aquifers (specifically for irrigation) can eventually lead to infiltration of high-As water from the surficial/shallow aquifers into the deeper aquifers, rendering them toxic

(Burgess et al., 2010; vanGeen et al., 2003a,b). Installation of newer tube wells that draw groundwater from the As-safe aquifers can be an extremely viable mitigation strategy where other measures of mitigation are impractical. Hossain et al. (2014) developed a sediment color tool for the local drillers, aimed to target As-safe aquifers based on sediment color, even in the shallow depths. The authors suggested that black sands are highly reduced and have the highest risk of groundwater As contamination, while the white, off-white, and red sands have a decreasing risk of As.

A "National Arsenic Policy and Mitigation Action Plan" was formulated by the Bangladesh Government in 2004 that aims to build public awareness, provide As-free, safe drinking water to the population at risk, address the issue of ingestion of As through agricultural products, and offer medical facilities to arsenicosis patients (Edmunds et al., 2015). A total of more than 1 lakh new As-safe drinking water sources have been installed since 2000 in the As-contaminated areas in Bangladesh (Johnston and Sarker, 2007). On the other hand, the Government of West Bengal (India) established the "Arsenic Task Force" in 2005 that formulated the master plan for the strategic short, intermediate, and long-term measures for As mitigation based on the UNICEF/PHED As database. The reports stress on the fact that a holistic As-mitigation plan should focus on building public awareness, providing As-safe drinking water sources, developing health support infrastructure, and framing policy interventions (Bhavan, 2007). However, planning, implementation, and maintenance of the As-mitigation measures at a national scale can be technically complex, financially taxing, and challenging in terms of social acceptance. Despite rigorous efforts, millions of people remain out of reach of any mitigation measures. Significant involvement of all stakeholders, including the governmental organizations and NGOs, policymakers, facilitating, implementing, and monitoring agencies and end users, is important for the large-scale success of the As-mitigation measures across the delta.

References

Alam, M., 1989. Geology and depositional history of Cenozoic sediments of the Bengal Basin of Bangladesh. Palaeogeogr. Palaeoclimatol. Palaeoecol 69, 125–139.

Alam, M., Alam, M.M., Curray, J.R., Chowdhury, M.L.R., Gani, M.R., 2003. An overview of the sedimentary geology of the Bengal basin in relation to the regional tectonic framework and basin-fill history. Sediment. Geol 155, 179–208.

Allison, M.A., 1998. Geologic framework and environmental status of the Ganges–Brahmaputra delta. J. Coast. Res. 14 (3), 826–836.

BADC, 2003. Survey Report on Irrigation Equipment and Irrigated Area in Boro/2003 Season. Bangladesh Agricultural Development Corporation.

BBS/UNICEF (Bangladesh Bureau of Statistics/United Nations Children's Fund), 2011. Bangladesh National Drinking Water Quality Survey of 2009. UNICEF, Dhaka, p. 192.

BGS/DPHE (British Geological Survey/Department of Public Health Engineering [Bangladesh]), 2001. Arsenic contamination of groundwater in Bangladesh. In: Kinniburgh, D.G., Smedley, P.L. (Eds.), Report WC/00/19. British Geological Survey Technical, Keyworth.

Bhattacharya, P., Chatterjee, D., Jacks, G., 1997. Occurrence of arsenic contaminated groundwater in alluvial aquifers from delta plains, eastern India: options for safe drinking water supply. Int. J. Water Resour. Dev. 13, 79–92.

Bhavan, Y., 2007. Report of the Task Force on Formulating Action Plan for Removal of Arsenic Contamination in West Bengal. Government of India Planning Commission, New Delhi [http://www.arsenicnetwork.in/wp-content/uploads/2017/11/Report%20of%20Task%20Force%20on%20Arsenic%20Removal.pdf].

Biswas, A.B., Saha, A.K., 1985. Environmental hazards of the recession of piezometric surface of the groundwater under Calcutta. Proc. Indian Natl. Sci. Acad. U.S.A. 51 A (3), 610–621.

Bureau of Indian Standards (BIS), 2003. Indian Standard Drinking Water—Specification (First Revision) (Incorporating Amendment No. 1) UDC 628.1.033, IS 10500: 1991 Edition 2.1 (1993-01).

Bureau of Indian Standards (BIS), 2005. Indian Standard Drinking Water—Specification (First Revision) (Incorporating Amendment No. 1) UDC 628.1.033, IS 10500: 1991 Edition 2.1 (1993-01).

Burgess, W.G., Hoque, M.A., Michael, H.A., Voss, C.I., Breit, G.N., Ahmed, K.M., 2010. Vulnerability of deep groundwater in the Bengal aquifer system to contamination by arsenic. Nat. Geosci. 3, 83–87.

Chakraborti, D., Das, B., Rahman, M.M., Chowdhury, U.K., Biswas, B., Goswami, A.B., et al., 2009. Status of groundwater arsenic contamination in the state of West Bengal, India: a 20-year study report. Mol. Nutr. Food Res. 53 (5), 542–551.

Chakraborty, M., Mukherjee, A., Ahmed, K.M., 2015. A review of groundwater arsenic in the Bengal basin, Bangladesh and India: from source to sink. Curr. Pollut. Rep. 1 (4), 220–247.

Chakraborty, M., Mukherjee, A., Ahmed, K.M., 2018. The groundwater flow, chemistry and pollutant distribution in the Bengal Basin, Bangladesh and India. In: Mukherjee, A. (Ed.), Groundwater of South Asia. Springer Hydrogeology.

Chakraborty, M., Sarkar, S., Mukherjee, A., Shamsudduha, M., Ahmed, K.M., Bhattacharya, A., et al., 2020. Modeling regional-scale groundwater arsenic hazard in the transboundary Ganges River Delta, India and Bangladesh: infusing physically-based model with machine learning. Sci. Total Environ. doi.org/10.1016/j.scitotenv.2020.1411077.

Charlet, L., Chakraborty, S., Appelo, C.A.J., Roman-Ross, G., Nath, B., Ansari, A.A., et al., 2007. Chemodynamics of an arsenic "hotspot" in a West Bengal aquifer: a field and reactive transport modeling study. Appl. Geochem. 22, 1273–1292.

Chatterjee, D., Halder, D., Majumder, S., Biswas, A., Nath, B., Bhattacharya, P., et al., 2010. Assessment of arsenic exposure from groundwater and rice in Bengal Delta Region, West Bengal, India. Water Res. 44, 5803–5812.

Curray, J.R., Moore, D.G., 1971. Growth of the Bengal deep-sea fan and denudation in the Himalayas: Geological Society of America Bulletin, v. 82, p. 563–572, https://doi.org/10.1130/0016-7606(1971)82[563: GOTBDF]2.0.CO;2.

Das, D., Chatterjee, A., Mandal, B.K., Samanta, G., Chakraborti, D., Chanda, B., 1995. Arsenic in groundwater in six districts of West Bengal, India: the biggest arsenic calamity in the world. Part 2: Arsenic concentration in drinking water, hair, nails, urine, skinscale and liver tissue (biopsy) of the affected people. Analyst 120, 917–924.

DPHE/APSU, The response to arsenic contamination in Bangladesh. A Position Paper. Arsenic Policy Support Unit, Department of Public Health, Dhaka 1000, 2005.

DPHE/DFID/JICA. 2006. Development of deep aquifer database and preliminary deep aquifer map. In: Final Report of First Phase, Department of Public Health Engineering (DPHE), UK Department for International Development (DFID) and Japan International Cooperation Agency (JICA), pp: 165; http://dphe.gov.bd/aquifer/index.php.

Edmunds, W.M., Ahmed, K.M., Whitehead, P.G., 2015. A review of arsenic and its impact in groundwater of the Ganges-Brahmaputra Meghna delta, Bangladesh. Environ. Sci. Proc. Impact. Available from: https://doi.org/10.1039/c4em00673a.

Fendrof, S., Micheal, H.A., van Geen, A., 2010. Spatial and temporal variations of groundwater arsenic in South and Southeast Asia. Science 328, 1123–1127.

Garai, R., Chakraborty, A.K., Dey, S.B., Saha, K.C., 1984. Chronic arsenic poisoning from tube-well water. J. Indian Med. Assoc. 82 (1), 34–35.

Goodbred, S.L., Kuehl, S.A., Steckler, M.S., Sarkar, M.H., 2003. Controls on facies distribution and stratigraphic preservation in the Ganges–Brahmaputra delta sequence. Sed. Geol. 155, 301–316.

Goodbred, S.L., Paolo, P.M., Ullah, M.S., Pate, R.D., Khan, S.R., Kuehl, S.A., et al., 2014. Piecing together the Ganges–Brahmaputra–Meghna River delta: use of sediment provenance to reconstruct the history and interaction of multiple fluvial systems during Holocene delta evolution. Geol. Soc. Am. Bull. 126 (1495–1510), B30965. 1.

GuhaMazumder, D.N., Haque, R., Ghosh, N., De, B.K., Santra, A., Chakraborti, D., et al., 1998. Arsenic levels in drinking water and prevalence of skin lesions in West Bengal India. Int. J. Epidemiol. 27, 871–877.

Harvey, C.F., Ashfaque, K.N., Yu, W., et al., 2006. Groundwater dynamics and arsenic contamination in Bangladesh. Chem. Geol. 228 (1–3), 112–136.

Havelaar, A., Melse, J., 2003. Quantifying Public Health Risk in the WHO Guidelines for Drinking-Water Quality: A Burden of Disease Approach. RIVM Report 734301022/2003.

Hoque, M.A., 2010. Models for Managing the Deep Aquifer in Bangladesh (Unpublished PhD Thesis). University College London, London, 265 pp.

Hoque, M.A., Burgess, W.G., 2012. C dating of deep groundwater in the Bengal Aquifer System, Bangladesh: implications for aquifer anisotropy, recharge sources and sustainability, J. Hydrol., 44–445. pp. 209–220.

Hoque, M.A., Burgess, W.G., Ahmed, K.M., 2017. Integration of aquifer geology, groundwater flow and arsenic distribution in deltaic aquifers – a unifying concept. Hydrol. Process. Available from: https://doi.org/10.1002/hyp.11181.

Hossain, M.F., 2006. Arsenic contamination in Bangladesh—an overview. Agric. Ecosyst. Environ. 113, 1–16.

Hossain, M., Bhattacharya, P., Frape, S.K., Jacks, G., Islam, M.M., Rahman, M.M., et al., 2014. Sediment color tool for targeting arsenic-safe aquifers for the installation of shallow drinking water tubewells. Sci. Total. Environ 493, 615–625.

Hossain, M., Rahman, S.N., Bhattacharya, P., Jacks, G., Saha, R., Rahman, M., 2015. Sustainability of arsenic mitigation in-terventions—an evaluation of different alternative safe drinking water options provided in Matlab, an arsenic hot spot in Bangladesh. Front. Environ. Sci. 3 (30).

Imam, M.B., Shaw, H.F., 1985. The diagenesis of Neogene clastic sediments from Bengal basin, Bangladesh. J. Sed. Pet. 55, 665–671.

Johnston, R., Sarker, M.H., 2007. Arsenic mitigation in Bangladesh: national screening data and case studies in three upazilas. J. Environ. Sci. Health 42 (12), 1889–1896.

Kinniburgh, D.G., Smedley, P.L., Davies, J., Milne, C.J., Gaus, I., Trafford, J.M., et al., 2003. The scale and causes of the groundwater arsenic problem in Bangladesh. In: Welch, Alan H., Stollenwerk, K.G. (Eds.), Arsenic in Groundwater. Kluwer Academic Publishers, Boston, pp. 211–257.

Kuehl, S.A., Hairu, T.M., Moore, W.S., 1989. Shelf sedimentation off the Ganges–Brahmaputra river system: evidence of sediment bypassing to the Bengal fan. Geology 17, 1132–1135.

Kuehl, S.A., Allison, M.A., Goodbred, S.L., Kudrass, H.-R., 2005. The Ganges–Brahmaputra delta. In: Giosan, L., Bhattacharya, J., (Eds.), Deltas—Old and New: Society for Sedimentary Geology (SEPM) Special Publication 83, pp. 413–434.

Mallick, S., Rajgopal, N.R., 1995. Groundwater development in the arsenic affected alluvial belt of West Bengal—some questions. Curr. Sci. 70 (11), 956–958.

Mandal, B.K., Chowdhury, T.R., Samanta, G., et al., 1996. Arsenic in groundwater in seven districts of West Bengal, India the biggest arsenic calamity in the world. Curr. Sci. 70, 976–986.

McArthur, J.M., Banerjee, D.M., Hudson-Edwards, K.A., Mishra, R., Purohit, R., Ravenscroft, P., et al., 2004. Natural organic matter in sedimentary basins and its relation arsenic in anoxic ground water: the example of West Bengal and its worldwide implications. Appl. Geochem. 19, 1255–1293.

McArthur, J.M., Sikdar, P.K., Hoque, M.A., Ghosal, U., 2012. Waste-water impacts on groundwater: Cl/Br ratios and implications for arsenic pollution of groundwater in the Bengal Basin and Red River Basin, Vietnam. Sci. Total. Environ. 437, 390–402.

Michael, H.A., Voss, C.I., 2009a. Estimation of regional-scale groundwater flow properties in the Bengal Basin of India and Bangladesh. Hydrogeol. J. 17 (6), 1329–1346.

Michael, H.A., Voss, C.I., 2009b. Controls on groundwater flow in the Bengal Basin of India and Bangladesh: regional modeling analysis. Hydrogeol. J. 17 (7), 1561–1577.

Milliman, J.D., Meade, R.H., 1983. World-wide delivery of river sediments to the oceans. J. Geol. 91, 1–22.
Milliman J.D., Rutkowski C. and Meybeck M., River discharge to sea: a global river index (GLORI), *Land-Ocean Interactions in the Coastal Zone (LOICZ) Reports & Studies No. 2. Texel, The Netherlands, NIOZ* 1995, 125, p.
Morgan, J.P., McIntire, W.G., 1959. Quaternary geology of Bengal Basin, East Pakistan and India. Geol. Soc. Am. Bull. 70, 319–342.
Mukherjee, A., 2006. Deeper Groundwater Flow and Chemistry in the Arsenic Affected Western Bengal Basin, West Bengal, India (Ph.D. thesis). University of Kentucky, Lexington, USA, 248p.
Mukherjee, A., Fryar, A.E., 2008. Deeper groundwater chemistry and geochemical modeling of the arsenic affected western Bengal basin, West Bengal, India. Appl. Geochem. 23 (4), 863–894.
Mukherjee, A., Fryar, A.E., Howell, P., 2007a. Regional hydrostratigraphy and groundwater flow modeling of the arsenic contaminated aquifers of the western Bengal basin, West Bengal, India. Hydrogeol. J. 15, 1397–1418.
Mukherjee, A., Fryar, A.E., Rowe, H.D., 2007b. Regional-scale stable isotopic signatures of recharge and deep groundwater in the arsenic affected areas of West Bengal, India. J. Hydrol. 334 (1), 151–161.
Mukherjee, A., Fryar, A.E., Thomas, W.A., 2009a. Geologic, geomorphic and hydrologicframeworks and evolution of the Bengal basin, India and Bangladesh. J. Asian Earth Sci. 34, 227–244.
Mukherjee, A., Fryar, A.E., O'Shea, B.M., 2009b. Major occurrences of elevated arsenicin groundwater and other natural waters. In: Henke, K.R. (Ed.), Arsenic—Environmental Chemistry, Health Threats and Waste Treatment. John Wiley & Sons, Chichester, pp. 303–350.
Mukherjee, A., von Brömssen, M., Scanlon, B.R., Bhattacharya, P., Fryar, A.E., Hasan, M.A., et al., 2008. Hydrogeochemical comparison and effects of overlapping redox zones on groundwater arsenic near the western (Bhagirathi sub-basin, India) and eastern (Meghna sub-basin, Bangladesh) of the Bengal basin. J. Contam. Hydrol. 99 (1-4), 31–48.
Mukherjee, A., Fryar, A.E., Scanlon, B.R., Bhattacharya, P., Bhattacharya, A., 2011. Elevated arsenic in deeper groundwater of the western Bengal basin, India: extent and controls from regional to local scale. Appl. Geochem. 26 (4), 600–613.
Mukherjee, A., Fryar, A.E., Eastridge, E., Nally, R.S., Chakraborty, M., Scanlon, B.R., 2018. Controls on high and low groundwater arsenic on the opposite banks ofthe lower reaches of River Ganges, Bengal basin, India. Sci. Total Environ. 645, 1371–1387.
Mukherjee, A., Scanlon, B., Aureli, A., Langan, S., Guo, H., McKenzie, A., 2020. Global Groundwater: Source, Scarcity, Sustainability, Security and Solutions. Elsevier, first ed., ISBN. 9780128181720.
Nickson, R.T., McArthur, J.M., Ravenscroft, P., et al., 2000. Mechanism of arsenic release to groundwater, Bangladesh West Bengal. Appl. Geochem. 15 (4), 403–413.
Nickson, R., Sengupta, C., Mitra, P., Dave, S.N., Banerjee, A.K., Bhattacharya, A., et al., 2007. Current knowledge on the distribution of arsenic in groundwater in five states of India. J. Environ. Sci. Health 42, 1707–1718.
Pal, T., Mukherjee, P.K., 2009. Study of subsurface geology in locating arsenic-free groundwater in Bengal delta, West Bengal, India. Environ. Geol. 56, 1211–1225.
PHED, 2006. Results of Tubewells Tested for Arsenic Under UNICEF Supported JPOA for Arsenic Mitigation.
Polizzotto, M.L., Harvey, C.F., Li, G., Badruzzman, B., Ali, A., Newville, M., et al., 2006. Solid-phases and desorption processes of arsenic within Bangladesh sediments. Chem. Geol. 228, 97–111.
Raman, K.S., Kumar, S., Neogi, B.B., 1986. Exploration in Bengal Basin India—an overview. Offshore South East Asia Show. Society of Petroleum Engineers, Singapore, https://doi.org/10.2118/14598-MS.
Ravenscroft, P., Brammer, H., Richards, K.S., 2009. Arsenic Pollution: A Global Synthesis. Wiley-Blackwell, Chichester. Available from: https://doi.org/10.1002/9781444308785.
Sahu, P., Sikdar, P.K., 2011. Threat of land subsidence in and around Kolkata City and East Kolkata Wetlands, West Bengal, India. J. Earth Syst. Sci. 120 (3), 435–446.
Sengupta, S., 1966. Geological and geophysical studies in the western part of the Bengal Basin, India. Am. Assoc. Petrol. Geol. Bull. 50, 1001–1017.
Shamsuddha, M., Chandler, R.E., Taylor, R.G., Ahmed, K.M., 2009. Recent trends in groundwater levels in a highly seasonal hydrological system: the Ganges-Brahmaputra-Meghna Delta. Hydrol. Earth Syst. Sci. 13, 2373–2385.
Smedley, P.L., Kinniburgh, D.G., 2002. A review of the source, behavior and distribution of arsenic in natural waters. Appl. Geochem. 17 (5), 517–568.
Smith, A.H., Lingas, E.O., Rahman, M., 2000. Contamination of drinkingwater by arsenic in Bangladesh: a public health emergency. Bull. World Health Organ. 78 (9), 1093–1103.
Stollenwerk, K.G., Breit, N.G., Welch, A.H., Yount, J.C., Whitney, J.W., Foster, A.L., et al., 2007. Arsenic attenuation by oxidized aquifer sediments in Bangladesh. Sci. Total Environ. 379, 133–150.
Umitsu, M., 1987. Late Quaternary sedimentary environment and landform evolution in the Bengal Lowland. Geograph. Rev. Jap. Ser. B 60, 164–178.
vanGeen, A., Ahsan, H., Horneman, A.H., Dhar, R.K., Zheng, Y., Hussein, I., et al., 2002. Promotion of well-switching to mitigate the current arsenic crisis in Bangladesh. Bull. WHO 80, 732–737.
vanGeen, A., Ahmed, K.M., Seddique, A.A., Shamsuddha, M., 2003a. Community wells to mitigate the arsenic crisis in Bangladesh. Bull. WHO 81, 632–638.
vanGeen, A., Zheng, Y., Versteeg, R., Stute, M., Horneman, A., Dhar, R., et al., 2003b. Spatial variability of arsenic in 6000 tube wells in a 25 km^2 area of Bangladesh. Water Resour. Res. 39 (5), 1140. Available from: https://doi.org/10.1029/2002WR001617.
van Geen, A., Radloff, K., Aziz, Z., Cheng, Z., Huq, M.R., Ahmed, K.M., et al., 2008. Comparison of arsenic concentrations in simultaneously-collected groundwater and aquifer particles from Bangladesh, India, Vietnam, and Nepal. Appl. Geochem. 23, 3244–3325.

Yamamura, S., 2003. Drinking water guidelines and standards. In: Hashizume, H., Yamamura, S. (Eds.), Arsenic, Water and Health: The State of the Art. World Health Organization, Geneva, ch. 5.

Yu, W.H., Harvey, C.M., Harvey, C.F., 2003. Arsenic groundwater in Bangladesh: a geostatistical and epidemiological framework for evaluating health effects and potential remedies. Water Resour. Res. 39 (6), 1146.

Further reading

Johnston, R., 2011. Bangladesh National Drinking Water Quality Survey of 2009. https://www.unicef.org/bangladesh/knowledgecentre_6868.htm.

Mukherjee, A., 2018. Groundwater chemistry and arsenic enrichment of the Ganges river basin aquifer systems. In: Mukherjee, A. (Ed.), Groundwater of South Asia. Springer Hydrogeology.

Umitsu, M., 1985. Natural levees and landform evolution in the Bengal lowland. Geograph. Rev. Jap 58 (2), 149–164.

Theme 3

Groundwater scarcity: quantity and quality

Chapter 11

Groundwater drought: environmental controls and monitoring

Bailing Li[1,2] and Matthew Rodell[2]
[1]ESSIC, University of Maryland, College Park, MD, United States, [2]Hydrological Sciences Laboratory, NASA Goddard Space Flight Center, Greenbelt, MD, United States

11.1 Introduction

Drought, as initiated by below-normal precipitation over weeks or longer time periods, can propagate through the components of the hydrological cycle, including groundwater (Changnon, 1987). Groundwater drought, defined as conditions when water tables drop below their normal levels, is a class of drought distinct from hydrological and agricultural drought (Mishra and Singh, 2010). Integrating surface meteorological conditions over months to years, groundwater variations lag changes in precipitation longer than soil moisture or streamflow (Eltahir and Yeh, 1999; Li and Rodell, 2015). In addition, because overlying layers of soil and rock act as a low-pass filter on near-surface processes, groundwater storage changes reflect mainly the low-frequency hydrometeorological variations (Eltahir and Yeh, 1999; Wang, 2012), making it difficult to estimate groundwater drought conditions using indicators designed for other types of drought.

Groundwater has been under immense stress in many parts of the world in recent decades (Mukherjee et al., 2020). Globally, more than 2 billion people use groundwater exclusively for drinking water, and 43% of irrigation water comes from aquifers (UNESCO, https://en.unesco.org/themes/water-security/hydrology/groundwater). Groundwater depletion associated with groundwater withdrawals for irrigation has been reported in many agriculturally important regions (Rodell et al., 2009, 2018; Famiglietti et al., 2011; Feng et al., 2013; Voss et al., 2013). Changes in the groundwater systems also affect watershed health, as groundwater sustains streamflow in arid and semiarid regions and in wet regions during dry periods (Eltahir and Yeh, 1999; Hughes et al., 2012). Decreases in water table levels in recent decades have led to significant forest diebacks in the floodplains of Australia's managed rivers (Cunningham et al., 2011; Kath et al., 2014) and altered the landscape of riparian-dependent woodlands in the southwestern United States (Stromberg et al., 1992). Climate change may further stress groundwater systems through its impacts on precipitation quantities, types, and intensity, evapotranspiration (ET), and water demand and redistribution (Green et al., 2011; Taylor et al., 2012; Famiglietti, 2014; Van Loon et al., 2016; Condon et al., 2020).

Groundwater drought is generally not well monitored due to limited availability of groundwater observations around the world. The US Geological Survey (USGS) is one of a very few organizations that provide real-time groundwater conditions using well data (https://groundwaterwatch.usgs.gov/default.asp), but it is limited to the United States, and much of the country is undersampled in real-time groundwater observations. At the time of writing, there were 1793 wells in the real-time network. The USGS archives data from thousands more wells, but most of them are either infrequent or have short or discontinuous records. In addition, wells located in confined aquifers are unsuitable for groundwater storage monitoring because changes in hydraulic head are not necessarily caused by storage changes. A handful of other countries, including Australia and the Netherlands, also make groundwater well data available either directly (e.g., http://www.bom.gov.au/water/groundwater/explorer/; https://www.dinoloket.nl/) or through the International Groundwater Resources Assessment Centre (https://ggmn.un-igrac.org/) but they suffer from the same issues as the USGS data. Groundwater data are unavailable in most of the rest of the world due to lack of observation networks or measurements, records that are not digitized and centralized, or data access restrictions (e.g., Rodell et al., 2009).

Numerical modeling provides an alternative means for estimating groundwater variations and monitoring groundwater drought (Li and Rodell, 2015). When forced by temporally consistent meteorological data, hydrological models can provide spatially and temporally continuous groundwater fields at regional and global scales (Döll et al., 2014; Sutanudjaja et al., 2018; Li et al., 2019a). In addition, advanced hydrological models, constrained by water and energy balances, are capable of providing reasonable ET estimates (Mueller et al., 2011; Jiménez Cisneros et al., 2014), which, along with precipitation, have the largest impact on groundwater temporal variability (Eltahir and Yeh, 1999; Li and Rodell, 2015). In recent decades, satellite observations from the Gravity Recovery and Climate Experiment (GRACE, Tapley et al., 2004) and GRACE Follow On (GRACE-FO) missions have enabled new methods of monitoring groundwater storage and drought at regional to global scales (Rodell and Famiglietti, 2002; Rodell et al., 2007; Zaitchik et al., 2008; Houborg et al., 2012; Famiglietti and Rodell, 2013; Döll et al., 2014; Kumar et al., 2016; Girotto et al., 2017; Li et al., 2019b).

In this chapter, we first review groundwater temporal variability reflected in in situ groundwater observations and then describe an approach that infuses GRACE and GRACE-FO observations into a land surface model (LSM) to enable groundwater storage and drought monitoring. Finally, we discuss challenges facing groundwater drought monitoring and future research directions.

11.2 Environmental controls on groundwater

To understand groundwater drought and its environmental controls, we first examine groundwater temporal variability using in situ data from 181 wells located in four regions in the northeastern United States and four subbasins of the Mississippi River (Fig. 11.1). Three criteria were used for selecting well records from the larger set of available observations in the USGS archive (Rodell et al., 2007; Li and Rodell, 2015). First, water-level measurements must show seasonal variability that indicates communication with the atmosphere. This eliminates wells located in confined aquifers. Second, to exclude direct anthropogenic impacts such as withdrawals and injections, wells with large abrupt changes in their water-level time series were excluded. Third, we only selected wells with at least 10 years of temporally consistent observations (minimum four measurements per year) in order to allow examination of seasonality and interannual variability. For the purpose of examining temporal variability, water-level measurements were converted to anomalies by subtracting the long-term mean at each location. The resulting water-level anomalies were further converted to groundwater storage anomalies by multiplying them by the specific yield, which was estimated individually for each well (Rodell et al., 2007; Li et al., 2015) and provided in Li et al. (2019b). This enables comparison of groundwater storage changes across wells and regions.

11.2.1 Precipitation

Fig. 11.2 shows monthly groundwater storage anomalies from individual wells (*gray lines*) and their regional average in the eight regions (*dark black lines*). The bars at the top represent monthly precipitation from the North America

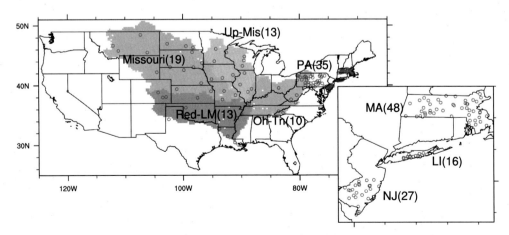

FIGURE 11.1 Locations of groundwater wells in LI, NJ, MA, PA, and the four subbasins of the Mississippi River: the Up-Mis, the Oh-Tn, the combined Red-LM, and the Missouri. Numbers in parentheses indicate the number of wells in each region. *LI*, Long Island; *MA*, Massachusetts; *NJ*, New Jersey; *Oh-Tn*, Ohio-Tennessee; *PA*, Pennsylvania; *Red-LM*, Red River and Lower Mississippi; *Up-Mis*, Upper Mississippi.

Groundwater drought: environmental controls and monitoring **Chapter | 11** 147

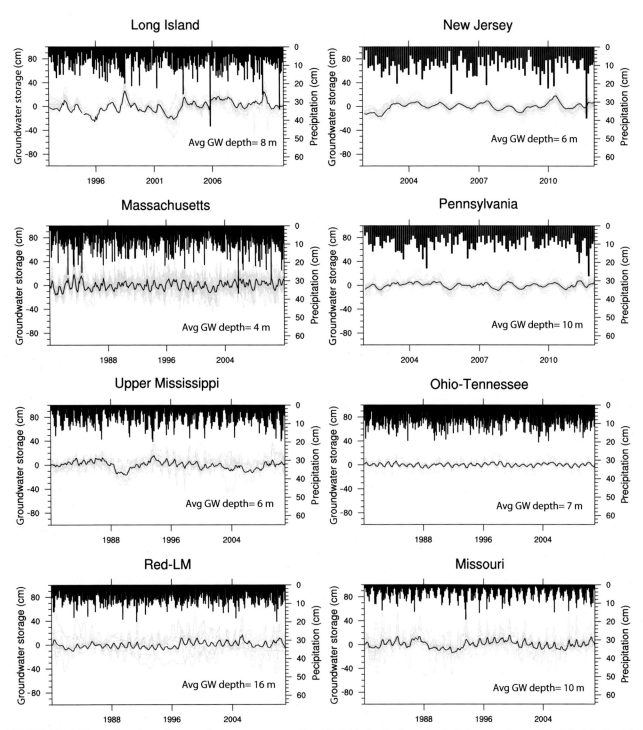

FIGURE 11.2 Time series of monthly groundwater storage anomalies at individual wells (*gray lines*), their regional mean (*black line*) in the eight regions. The top bars represent averaged monthly NLDAS-2 precipitation. The average groundwater depth in each region is also provided. *NLDAS-2*, North America Land Data Assimilation System-2.

Land Data Assimilation System-2 (NLDAS-2, Xia et al., 2012). NLDAS-2 precipitation is derived by temporally disaggregating daily gauge measurements using hourly radar images and is further spatially interpolated to a 0.125-degree grid (Cosgrove et al., 2003). Regional groundwater storage and conforming precipitation time series were computed by averaging in situ groundwater data and gridded precipitation data from all well locations within each region. The influence of precipitation on regional groundwater storage anomalies (relative to long-term mean) is apparent during prolonged wet and dry events (Fig. 11.2). For instance, the well-known drought that affected the Midwest and Northern Great Plains during 1987−89 produced large negative anomalies in the Upper Mississippi and Missouri basins, and a large positive anomaly is seen in the Upper Mississippi during the summer floods of 1993.

These groundwater measurements exhibit strong seasonal variation peaking in April to May and reaching annual minimum in fall and winter, despite weak seasonality in precipitation such as in the four northeast regions or that out of phase with that of groundwater such as in the Upper Mississippi basin (Fig. 11.3) (Li et al., 2015). The strong seasonality is shaped by combined influences of precipitation and ET. For instance, in the Upper Mississippi and Missouri where precipitation shows strong seasonality, maximum groundwater and precipitation occurs in the same month (Missouri) or 1 month apart (Upper Mississippi). In the northern and high altitudes of these regions, seasonal snowpack also stores a significant amount of annual precipitation and releases it through spring snowmelt (Perez-Valdivia et al., 2012), which balances the effect of increasing ET in the spring. On the other hand, in the Ohio-Tennessee and the combined Red River and Lower Mississippi (Red-LM) basins precipitation increases in late spring to early summer, but evaporative demand and plant root uptake also increase, reducing the water available for groundwater recharge and resulting in groundwater peaking before precipitation peaks.

Groundwater exhibits more noticeable lagged responses to precipitation when the seasonal cycles are removed from both time series (Fig. 11.4) than in Fig. 11.2. The maximum lagged correlation is greater than 0.5 in 7 of the 8 regions, with the lag of maximum correlation ranging from 2 to 7 months. In the Missouri basin, which is the driest among all regions/basins, groundwater storage reached the lowest level in 1992, much later than the precipitation minimum in 1988 (note precipitation in Fig. 11.4 was smoothed with a 6-month running average). The long lag likely reflects the delayed response and long recovery time needed in a dry climate where recharge is low and the water table is deep (Fig. 11.2). Such lagged groundwater responses to precipitation were postulated by Changnon (1987). Further, Eltahir and Yeh (1999), based on nearly 30 years of in situ data in Illinois, showed that groundwater drought persisted longer with higher intensity than pluvial events. This behavior was attributed to the nonlinear dependency of groundwater discharge on water levels, which allows faster dissipation of groundwater during floods.

11.2.2 Subsurface hydrogeological conditions

Hydrogeological conditions such as water table depths and aquifer properties may also impact groundwater temporal variability and groundwater drought through their impacts on recharge rates. Bloomfield et al. (2015) examined water-level measurements from three principle aquifers in a small region in England where climate conditions are relatively uniform. They found that the autocorrelation scale of groundwater time series increases with increasing water table depth and that longer autocorrelation scale is associated with longer duration and more intensive groundwater drought. The former can be attributed to the previously mentioned low-pass filtering effect of overlying soil and rock on hydrometeorological conditions (Wu et al., 2002). Also based on well data from England, Bloomfield and Marchant (2013) showed that groundwater autocorrelation scales are negatively correlated with soil diffusivity, more so in granular aquifers than in fractured geological formations where the unsaturated zone depth plays a stronger role on groundwater temporal variability.

The impacts of hydrogeological properties and conditions are not easily observable in the groundwater data from the eight US regions discussed earlier, where climate variability likely dominates groundwater temporal variability. However, some results may be attributed to hydrogeological conditions. For instance, the low correlation (0.17 at 7-month lag) between nonseasonal groundwater and precipitation in the Red-LM basin may be due to the thickness of the unsaturated zone above the water table (Fig. 11.4), whereas the correlation in Massachusetts, where the water table is shallow, was strong (0.85 at 3-month lag). In Pennsylvania, where the regional water table is also deep, nonseasonal groundwater reached a maximum correlation with precipitation, 0.74, at a 2-month lag. Most wells in Pennsylvania are located in fractured rock formations (Li and Rodell, 2015), which allow fast recharge during precipitation, hence strong correlation at a short lag. These results suggest complex and intertwined environmental controls on groundwater at the seasonal to interannual scales.

11.2.3 Large-scale climate phenomena

As with other types of drought, groundwater drought is often caused by large-scale climate signals and multiyear oscillations (Asong et al., 2018). These phenomena influence groundwater through their impacts on precipitation, mainly on interannual scales. Among them, the Pacific Decadal Oscillation (PDO) and the El Niño−Southern Oscillation (ENSO) are known to have the largest impacts on multiyear cyclic behaviors of groundwater in the western United States

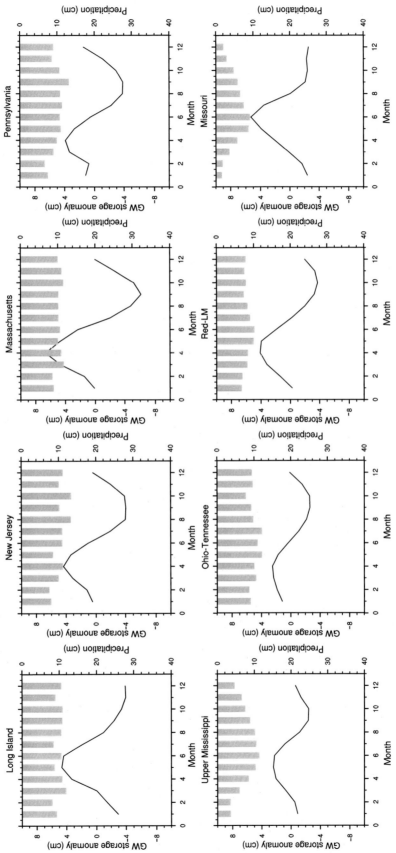

FIGURE 11.3 Monthly seasonal cycles of regional mean groundwater storage anomalies and precipitation (*gray bars*) for the eight study regions.

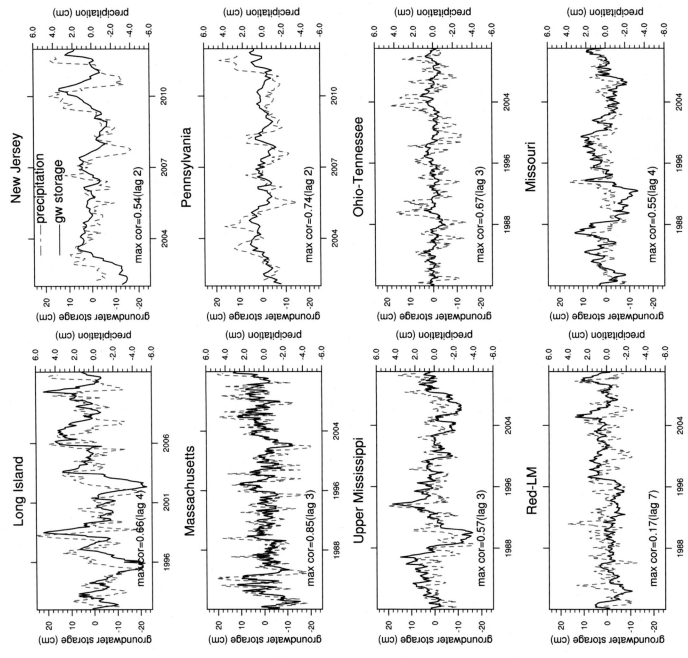

FIGURE 11.4 Time series of nonseasonal groundwater storage (*black lines*) and precipitation (*blue lines*, smoothed with 6-month running average) anomalies in the eight regions. Numbers represent maximum correlation and the lag of maximum correlation.

(Hanson et al., 2006; Barco et al., 2010; Velasco et al., 2015). Perez-Valdivia et al. (2012) observed effects of ENSO and PDO on shallow groundwater in the Canadian Prairies. Anderson and Emanuel (2008) reported that ENSO affected winter baseflow in North Carolina. Asoka et al. (2017) theorized that groundwater decline in northern Indian was partly attributable to Indian Ocean warming. Several studies have investigated the potential effects of climate change on groundwater storage and recharge (Green et al., 2011; Taylor et al., 2013; Kløve et al., 2014; Meixner et al., 2016; Niraula et al., 2017).

Subsurface properties and hydrogeological conditions have been shown to modulate groundwater responses to large-scale climate signals. For instance, Anderson and Emanuel (2008) found that groundwater in highly permeable aquifers shows stronger responses to climate signals than that in low permeability aquifers in the same climate regime. Hanson et al. (2006) showed that groundwater variations in wells close to mountain fronts or rivers where hydraulic gradients are large exhibit stronger influences of climate phenomena than those in other areas. These results reflect the more dominant control of climate on groundwater in aquifers where the overlying geology does not inhibit hydrological communication between the land surface and atmosphere.

11.3 Groundwater drought monitoring

Because of the scarcity of in situ groundwater observations in most of the world, global-scale groundwater drought monitoring has to rely on information from hydrological models and/or satellite observations. In particular, LSMs have been widely applied for drought monitoring at continental to global scales (e.g., Mo, 2008; Sheffield et al., 2012; Houborg et al, 2012; Li et al., 2019b). When forced by observation-based meteorological fields, LSMs are capable of simulating spatially and temporally continuous groundwater conditions suitable for drought monitoring (Houborg et al., 2012; Li et al., 2015). An important component of this approach is the skill of LSMs in simulating ET (Mueller et al., 2011; Jiménez Cisneros et al., 2014), which, along with precipitation, is the dominant driver of groundwater variability (Eltahir and Yeh, 1999; Li et al., 2015). Just as importantly, LSM simulations can provide the long time series of groundwater storage changes needed for identifying and quantifying groundwater drought in the context of historical variability (Houborg et al., 2012). However, simulated groundwater time series are subject to uncertainties stemming from simplified model physics (Koster et al., 2000; Yeh and Eltahir, 2005a,b; Niu et al., 2007; Koirala et al., 2014) and errors in the forcing data used to drive the model (Xia et al., 2017; Li et al., 2019a,b).

11.3.1 Gravity Recovery and Climate Experiment data assimilation for groundwater drought monitoring

The GRACE and GRACE-FO satellite missions were designed to map the Earth's gravity field on a monthly basis with sufficient precision to quantify variations in terrestrial water storage (TWS), among other mass changes (Tapley et al., 2004). TWS comprises soil moisture, groundwater, snow, surface waters, and ice (Rodell and Famiglietti, 2001). GRACE and GRACE-FO, which span from 2002 to present with an 11-month gap in 2017−18 (https://grace.jpl.nasa.gov/), provide the only satellite-based observations that can be used to infer changes in water stored below the surface soil layer, including groundwater. GRACE and GRACE-FO detect water storage changes throughout all seasons (Li et al., 2012), which is critical for continuous drought monitoring. Low levels of nonseasonal GRACE-derived TWS have been shown to correlate strongly with drought (Andersen et al., 2005; Yirdaw et al., 2008; Chen et al., 2009; Leblanc et al., 2009; Li et al., 2012; Long et al., 2013; Thomas et al., 2017; Zhao et al., 2017).

One of the challenges in applying GRACE/GRACE-FO data for groundwater drought monitoring lies in isolating groundwater from the vertically integrated TWS observations. In addition, GRACE/GRACE-FO data are provided monthly, often with 2−5 months of latency, with an effective spatial resolution around 150,000 km^2 at midlatitudes (Rowlands et al., 2005; Swenson et al., 2006). These characteristics complicate the use of GRACE and GRACE-FO data for operational applications such as drought monitoring, which requires near-real-time, fine-scale information.

To overcome these challenges, methods have been developed, tested, and refined for assimilating GRACE/GRACE-FO data into LSMs using ensemble Kalman smoothers and similar techniques (Zaitchik et al., 2008; Girotto et al., 2016; Kumar et al., 2016). The assimilated results have significantly higher spatial and temporal resolutions than the GRACE and GRACE-FO data with subweekly latency. The output states and fluxes, including soil moisture, groundwater, snow water equivalent, and streamflow, are in most cases superior to unassimilated (open loop) model output (Zaitchik et al., 2008; Su et al., 2010; Houborg et al., 2012; Li et al., 2012, 2019b; Girotto et al., 2017; Kumar et al., 2016). An ensemble smoother can be represented by the update equation (Zaitchik et al., 2008):

$$X^a = X^f + \underline{K}(Y_0 - Y_M) \tag{11.1}$$

where X^a and X^f represent the analysis and forecast of modeled states at the current time step, respectively; $\underline{K}$ is the Kalman gain matrix that is calculated from observation errors and model uncertainty represented in the ensemble; Y_o represents GRACE/GRACE-FO-derived TWS; Y_M represents model-simulated TWS. Because GRACE/GRACE-FO TWS are anomalies (relative to temporal mean), Y_o is created by adding the long-term temporal mean of modeled TWS to GRACE/GRACE-FO TWS. The states (X^f and X^a) are computed at the model's finer scale and daily time step, while Y_o and Y_M are evaluated monthly at the coarser scale of GRACE data; hence, the smoother enables spatial and temporal downscaling.

Next we discuss results obtained from assimilating GRACE/GRACE-FO data into the catchment LSM (CLSM, Koster et al., 2000) for groundwater storage and drought monitoring at the global scale (Li et al., 2019b). CLSM simulates subsurface water storage changes in three subsurface water zones, a surface layer (0–2 cm below the surface), the root zone (0–100 cm), and the full soil profile, the thickness of which is determined by CLSM's bedrock depth parameter. It also simulates seasonal snowpack changes in three snow layers and canopy interception. CLSM does not model water table variations; hence, groundwater storage is derived by subtracting the water stored in the root zone from that in the full soil profile. Based on this configuration, the state vector in Eq. (11.1), X, represents root zone soil moisture, groundwater, snow, and canopy interception and Y_M, represents the sum of the aforementioned variables.

Fig. 11.5 shows that monthly TWS from the open-loop simulation, that is, without assimilating GRACE data (top panel), GRACE data assimilation (second panel from the top), and GRACE alone (third panel from the top) for August 2015. It is clear that GRACE data assimilation brought simulated TWS closer to the GRACE observation while retaining the spatial details of the open-loop simulation, which in turn reflect spatial variability in the high-resolution model parameter and forcing data. The large negative values in the Canadian Archipelago and the Gulf of Alaska are associated with ice sheet and glacier mass losses (Gardner et al., 2011). CLSM does not simulate these processes and, therefore, did not simulate decreasing trends in those areas.

The bottom panel of Fig. 11.5 shows TWS time series for an area in Southern Germany, which experienced severe drought in 2003, 2015, and 2018, which has been studied by Van Loon et al. (2017) using in situ groundwater data. Open-loop CLSM indicates more severe drought in 2003 than in 2015, which is consistent with 2015 being a wetter year than 2003 (Van Loon et al., 2017). However, the region had experienced precipitation deficits for several years leading up to the 2015 drought, which caused a long-term deficit in groundwater storage (Van Loon et al., 2017). GRACE detected this long-term decline in groundwater storage and, hence, signified more severe drought in 2015. Open-loop CLSM TWS did not capture this multiyear water storage decline, likely due to deficiencies in model physics and/or errors in the forcing data. Regardless of the source of uncertainty, data assimilation adjusted simulated TWS toward GRACE observations, resulting in a more accurate depiction of drought in 2015. The same plot also demonstrates that the assimilation model fills data gaps including that between GRACE and GRACE-FO (August 2017 to June 2018) by relying on forcing data alone as input, while the soil moisture and groundwater "memory" ensures continuity from GRACE to GRACE-FO periods.

Fig. 11.6 presents an evaluation of the data assimilation results using the independent, in situ groundwater observations described in Section 11.2. GRACE data assimilation decreased the root mean square error (RMSE) in all regions and increased the correlation with in situ groundwater in six of the eight regions. Fig. 11.6 also shows that GRACE data assimilation does not guarantee agreement with the in situ observations. For instance, groundwater from both the open-loop and the GRACE data assimilation identified 2006 as a drought year in Red-LM, more than a year earlier than did the in situ observations. It is unknown whether this discrepancy was caused by errors in the forcing and GRACE data or perhaps by undersampling of the region by the 13 wells used to construct the average (Fig. 11.1). Despite the limitations of this type of comparison, Li et al. (2019b) reported generally positive impacts of GRACE data assimilation on simulated groundwater in regions that spanned five continents based on time series from nearly 4000 wells. In particular, RMSE decreased by 36% and 10% and correlation improved by 16% and 22% at the regional and point scales, respectively (Li et al., 2019b). These improvements demonstrate the value of GRACE/GRACE-FO data assimilation for groundwater storage change assessment and drought monitoring.

Fig. 11.7 shows global groundwater wetness percentile maps for Sept. 21, 2003 and Sept. 21, 2015 derived from GRACE/GRACE-FO data assimilation output. As shown in Fig. 11.5, the GRACE-based groundwater indictor shows more severe drought in 2015 than in 2003 in Southern Germany and the surrounding region. In addition to spatial downscaling, GRACE/GRACE-FO data assimilation enables temporal downscaling so that drought conditions can be provided daily (as opposed to monthly with GRACE/GRACE-FO observations). The percentiles were created by ranking groundwater estimates against a climatology derived from a 1948–2014 CLSM simulation forced by the Princeton meteorological forcing dataset (Sheffield et al., 2006). The model simulation is used for climatology because the

GRACE and GRACE-FO data period alone (2002–present) is too short to reveal the full range of variability in groundwater. Scale factors were derived, based on the overlapping period between GRACE/GRACE-FO data assimilation (which utilized a different meteorological forcing dataset, European Center for Medium-range Weather Forecasts, Dee et al., 2011) and the long-term simulation, and employed to ensure the consistency between the long-term model climatology and the GRACE/GRACE-FO data assimilation (Li and Rodell, 2015; Li et al., 2019b). The scale factors were derived for each pixel and each calendar month to avoid non-Gaussian behaviors in simulated subsurface states (Mo, 2008).

11.3.2 Other groundwater drought indicators

Groundwater drought indicators have also been derived by subtracting soil moisture and snow water equivalents (usually estimated by LSMs) from GRACE TWS, under the assumption that other TWS change components are negligible (e.g., Thomas et al., 2014; Van Loon et al., 2017). However, the resulting indicators have the same spatial and temporal (monthly) scales and latency as those of GRACE and GRACE-FO data, making them unsuitable for real-time operational drought monitoring.

Van Loon et al. (2017) developed a groundwater drought indicator based on the correlation scale between in situ groundwater and the Standard Precipitation Index (SPI). While allowing detection of groundwater drought using readily available precipitation data, the approach requires long records of groundwater data and, thus, is limited to regions where such records are available. In addition, nonlinear processes affecting groundwater such as groundwater discharge and recharge (Eltahir and Yeh,1999) are likely to be missed by the approach.

Standardized groundwater indices (SGIs) have been developed to identify groundwater drought (Bloomfield and Marchant, 2013) using the similar procedure as SPIs. However, as noted by Bloomfield and Marchant (2013), such indicators do not show advantages over the percentile approach because groundwater is a continuous variable in time and, therefore, there is no need to calculate SGIs at different accumulation periods as for SPIs.

Groundwater drought can also be identified using a threshold value approach (Peters et al., 2006). This results in binary drought or nondrought conditions being identified but does not provide information on drought severity (Bloomfield and Marchant, 2013). Further, variable thresholds for different months of the year may be necessary to account for groundwater seasonality.

11.4 Characteristics of groundwater drought at the global domain

Characteristics of groundwater drought are not well understood at the global domain due to limited spatial coverage of available in situ well records. Here we use CLSM-simulated groundwater to examine groundwater drought and its relationship with climate and other factors. Specifically, we use the SPI accumulation period at which maximum correlation between SPIs and groundwater is reached to examine temporal scales of groundwater and groundwater drought (Li and Rodell, 2015; Bloomfield et al., 2015; Li et al., 2019b). For this analysis, CLSM-simulated groundwater storage from the climatology run (see Section 11.3.1), which spans 1948–2014, was converted to standardized anomalies (with monthly mean removed). The Princeton precipitation data (Sheffield et al., 2006) used to force the climatology run were converted to SPIs at 3- to 36-month accumulation periods. Use of the longer record of simulated groundwater (as opposed to that from GRACE/GRACE-FO data assimilation) provides more robust characterization of the relationship between groundwater and large-scale climate variability.

CLSM-simulated groundwater correlates strongly with SPI3 in certain tropical and subtropical regions, including the northern South America, the East-Central Africa, Indonesia, and the southeastern Australia, where the climate is both wet and strongly affected by the ENSO (Dai and Wigley, 2000). Abundant precipitation in these regions leads to wet soils and shallow water tables and hence rapid groundwater recharge.

As the SPI accumulation period increases from 3 to 12 months, the correlation between SPI and CLSM groundwater storage increases in most regions where the influence of ENSO is weak. Simulated groundwater is best correlated with SPI12 over a large portion of the global land (Fig. 11.8D and E). Longer SPI integration periods are best correlated with groundwater storage in high-latitude regions where groundwater recharge is dominated by spring snowmelt (Perez-Valdivia et al., 2012), while shorter integration periods are best correlated in the tropics. Russo and Lall (2017) and Asoka et al. (2017) also identified 12 months as the optimal correlation scale based on in situ groundwater data in the United States and India, respectively. Bloomfield et al. (2015) reported that regional groundwater showed the strongest correlation with SPI12, even though the subregional groundwater correlated strongly with SPIs at 4- to 17-month

154 Theme | 3 Groundwater scarcity: quantity and quality

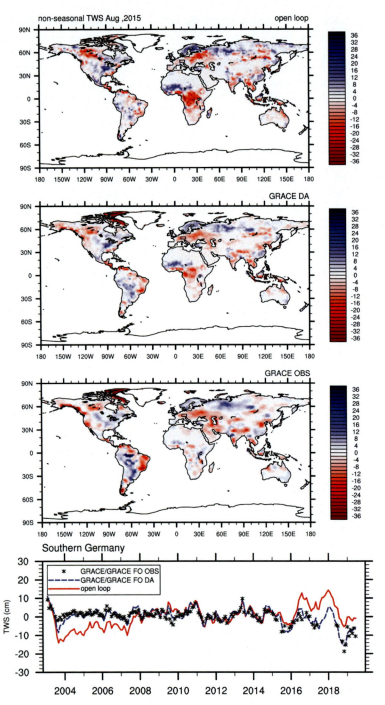

FIGURE 11.5 Monthly nonseasonal TWS from CLSM open-loop simulation, GRACE data assimilation, GRACE for August, 2015 (top three panels), and region-averaged monthly nonseasonal TWS time series in Southern Germany (identified by the black rectangle in the top three panels). GRACE/GRACE-FO data are based on the CSR RL06 product. *CLSM*, Catchment land surface model; *GRACE*, Gravity Recovery and Climate Experiment; *GRACE-FO*, GRACE Follow On; *TWS*, terrestrial water storage.

periods depending on vadose thickness. The impact of seasonal variation of ET on groundwater (Eltahir and Yeh, 1999) is likely to be a factor in most cases excepting cold lands. In the Sahara Desert, where groundwater recharge is nearly zero, a low correlation is observed at all timescales.

As discussed earlier, temporal variability of groundwater may be affected by hydrogeological conditions that are represented in a simplified way in CLSM by a bedrock depth parameter. In general, the deeper the bedrock, the deeper the water table and the slower the groundwater response. The impact of the bedrock depth can be observed in the Great Plains of the United States and in Northwestern Europe (Fig. 11.8E and F), where CLSM bedrock is particularly deep and groundwater correlates most strongly with SPI periods that often exceed 24 months. However, the effect of bedrock

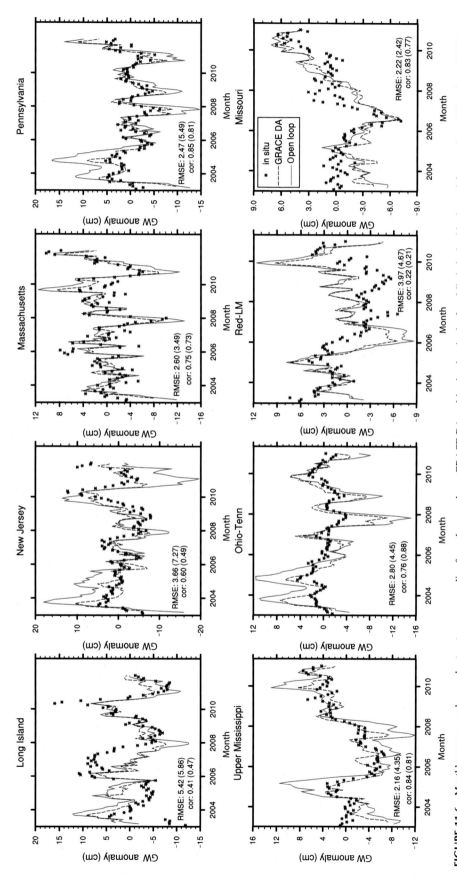

FIGURE 11.6 Monthly nonseasonal groundwater storage anomalies from the open-loop, GRACE DA and in situ data in the four Mississippi subbasins and four northeast US regions. RMSE and correlation between in situ and simulated DA and open loop (in parentheses) groundwater time series are provided. *DA*, Data assimilation; *GRACE*, Gravity Recovery and Climate Experiment; *RMSE*, root mean square error.

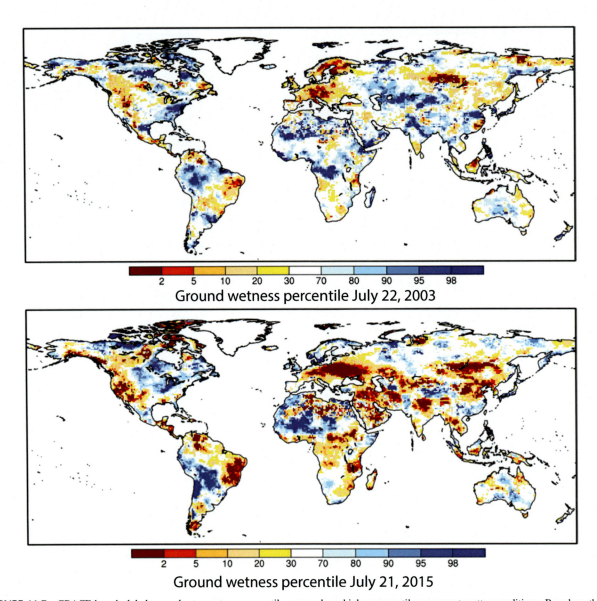

FIGURE 11.7 GRACE-based global groundwater wetness percentile maps where higher percentiles represent wetter conditions. Based on the US Drought Monitor, drought severity can be categorized using the following thresholds: exceptional drought: ≤2%, extreme drought: 2%–5%, severe drought: 5%–10%, moderate drought: 10%–20%, and abnormally dry: 20%–30%. *GRACE*, Gravity Recovery and Climate Experiment.

depth may be overridden by a wet climate, in which abundant precipitation keeps the water table shallow, thus enabling rapid recharge. This is the case in Central Africa, Southeast China, and Indonesia, where the bedrock is deep but CLSM groundwater correlates most strongly with short-period SPIs. In these regions, groundwater drought can develop during short-term precipitation deficits but can recover quickly once precipitation increases.

11.5 Discussions and future research

Groundwater drought is caused by an extended precipitation deficit and may be exacerbated by evaporative demand. Because it is driven by atmospheric conditions, it only occurs in aquifers that are in communication with the surface hydrology, that is, unconfined and semiconfined aquifers. Due to its lagged relationship with low-frequency hydrometeorological phenomena (Eltahir and Yeh, 1999), groundwater drought differs from agricultural and hydrological droughts and must be evaluated separately.

Analyses of in situ groundwater data reveal the strong influence of precipitation variations. In the northeastern United States and the Mississippi River basin, regional groundwater storage changes lag precipitation by 2–4 months

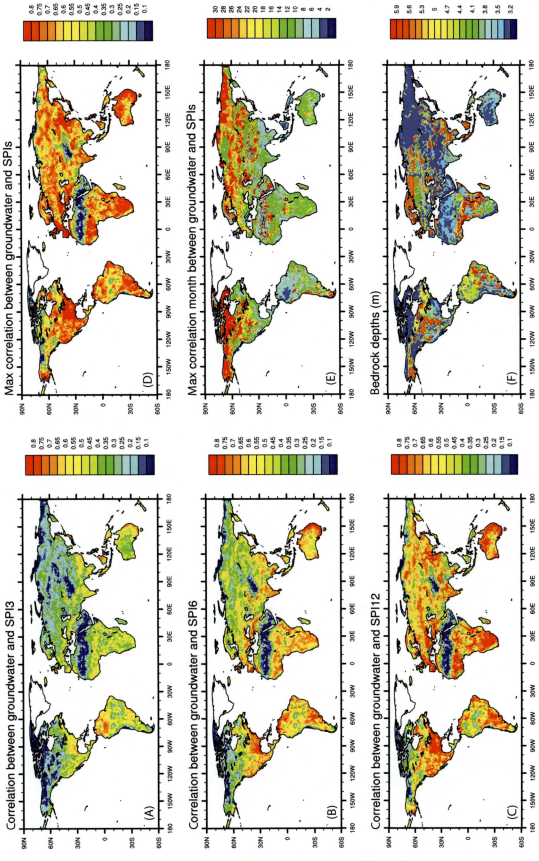

FIGURE 11.8 Correlation between the CLSM-simulated groundwater and SPI3, SPI6, and SPI12 (A–C), maximum correlation and CLSM bedrock depths (D–F). *CLSM*, Catchment land surface model; *SPI*, Standard Precipitation Index.

with maximum correlation generally exceeding 0.5, suggesting that groundwater drought occurs often even in the humid and semihumid climates. However, individual groundwater drought events may persist longer, well after precipitation has returned to normal, as observed in well time series from the Missouri subbasin and in Illinois (see Eltahir and Yeh, 1999). Subsurface hydrogeological conditions also affect groundwater drought onset and evolution through their partial control on recharge rates, but they can also be nullified in wet climates.

Considering the scarcity of groundwater observations around the globe, simulated or assimilated groundwater time series from hydrological models are a valuable alternative that has proven useful for global-scale groundwater drought monitoring. Forced by observation-based meteorological fields and constrained by water and energy balance equations, hydrological models are capable of simulating groundwater response to near-surface processes, including precipitation and evapotranspiration, with reasonable skill (Li et al., 2019a). In particular, LSMs, which are both physically based and computationally efficient, can run with high spatial and temporal resolution at the continental to global scales. Assimilation of GRACE and GRACE-FO data into LSMs provides much needed constraints on global groundwater simulation, especially in data-scarce regions where uncertainty in forcing and parameter data are likely high. Data assimilation also enables disaggregation of GRACE/GRACE-FO-derived TWS so that groundwater and the other components of TWS can be evaluated independently. Recent studies have demonstrated that GRACE/GRACE-FO data assimilation into CLSM reduces uncertainty in simulated groundwater time series at both regional and point scales (Li et al., 2019b), while also filling observational gaps and extending the information to near-real time. These improvements enable more reliable and comprehensive groundwater drought monitoring.

One of the significant limitations of LSMs such as CLSM is that they typically do not simulate relevant anthropogenic activities such as groundwater pumping and irrigation, though such capabilities are now under development (e.g., Nie et al., 2019). This has limited the ability of data assimilation to properly downscale GRACE observations in regions with intensive withdrawals such as the North China Plain (Li et al., 2019b). Consequently, a resulting GRACE/GRACE-FO-based groundwater drought indicator cannot separate drought conditions associated with climate variability from water depletion caused by groundwater extraction. While the need to account for anthropogenic activities in drought and water management has drawn considerable attention (e.g., Van Loon et al., 2017; Nie et al., 2018), doing so would present other challenges to drought identification. For instance, a percentile-based groundwater drought indicator would always yield exceptional drought in areas where intensive groundwater withdrawals have caused a long-term decline in the water table, such as California's Central Valley. This is due to the fact that most statistical methods for quantifying drought severity assume stationarity, while groundwater depletion is a nonstationary behavior. Under these circumstances a separate system for quantifying anthropogenic impacts on groundwater (Van Loon and Van Lanen, 2013) may be an alternative for understanding and communicating groundwater depletion. However, nonstationary statistical methods will continue to be needed to analyze groundwater drought, as climate change may also lead to nonstationary behaviors in groundwater.

Another challenge to using LSMs and other global-scale hydrological models for groundwater drought monitoring is that they do not simulate lateral flows of groundwater. Therefore they may misrepresent groundwater dynamics in aquifers such as those located at the base of mountains, which are mainly recharged by lateral flow from the adjacent mountain blocks (Wilson and Guan, 2004; Markovich et al., 2019). GRACE/GRACE-FO generally does not resolve water storage changes at these small scales. Dynamic flow models are needed to simulate three-dimensional flows and to account for complex geological and hydrogeological conditions and processes such as preferential flow and focused recharge (Wilson and Guan, 2004). While significant improvements have been made, including coupling sophisticated groundwater flow models with LSMs (Maxwell et al., 2014), it remains a challenge to calibrate such models for global-scale simulation due to the need for more hydrogeological information and in situ well data (de Graaf et al., 2015, 2017). Future studies should focus on combining successful modeling and data assimilation approaches and using all available data for parameterization, calibration, and evaluation, in order to achieve optimal global groundwater drought monitoring.

References

Anderson Jr., W.P., Emanuel, R.E., 2008. Effect of interannual and interdecadal climate oscillations on groundwater in North Carolina. Geophys. Res. Lett. 35, L23402. Available from: https://doi.org/10.1029/2008GL036054.

Andersen, O.B., Seneviratne, S.I., Hinderer, J., Viterbo, P., 2005. GRACE-derived terrestrial water storage depletion associated with the 2003 European heat wave. Geophy. Res. Lett. 32, L18405.

Asoka, A., Gleeson, T., Wada, Y., Mishra, V., 2017. Relative contribution of monsoon precipitation and pumping to changes in groundwater storage in India. Nat. Geosci. 10 (2), 109–117.

Asong, Z.E., Wheater, H.S., Bonsal, B., Razavi, S., Kurkute, S., 2018. Historical drought patterns over Canada and their teleconnections with large-scale climate signals. Hydrol. Earth Syst. Sci. 22, 3105–3124. Available from: https://doi.org/10.5194/hess-22-3105-2018.

Barco, J., Hogue, T.S., Girotto, M., Kendall, D.R., Putti, M., 2010. Climate signal propagation in southern California aquifers. Water Resour. Res. 46, W00F05. Available from: https://doi.org/10.1029/2009WR008376.

Bloomfield, J.P., Marchant, B.P., 2013. Analysis of ground-water drought building on the standardised precipitation index approach. Hydrol. Earth Syst. Sci. 17, 4769–4787. Available from: https://doi.org/10.5194/hess-17-4769-2013.

Bloomfield, J.P., Marchant, B.P., Bricker, S.H., Morgan, R.B., 2015. Regional analysis of groundwater droughts using hydro-graph classification. Hydrol. Earth Syst. Sci. 19, 4327–4344. Available from: https://doi.org/10.5194/hess-19-4327-2015.

Changnon Jr., S.A., 1987. Detecting drought conditions in Illinois. In: ISWS/CIR-169-87. Illinois State Water Survey, Champaign, IL, 36 pp.

Chen, J.L., Wilson, C.R., Tapley, B.D., Yang, Z.L., Niu, G.Y., 2009. 2005 Drought event in the Amazon River basin as measured by GRACE and estimated by climate models. J. Geophys. Res. Solid Earth 114 (B5).

Condon, L.E., Atchley, A.L., Maxwell, R.M., 2020. Evapotranspiration depletes groundwater under warming over the contiguous United States. Nat. Commun. 11 (1), 1–8.

Cosgrove, B.A., Lohmann, D., Mitchell, K.E., Houser, P., Wood, E.F., et al., 2003. Real-time and retrospective forcing in the North American Land Data Assimilation System (NLDAS) project. J. Geophys. Res. 108 (D22), 8842. Available from: https://doi.org/10.1029/2002JD003118.

Cunningham, S.C., Thomson, J.R., Mac Nally, R., Read, J., Baker, P.J., 2011. Groundwater change forecasts widespread forest dieback across an extensive floodplain system. Freshw. Biol. 56 (2011), 1494–1508.

Dai, A.D., Wigley, T.M.L., 2000. Global patterns of ENSO induced precipitation. Geophys. Res. Lett. 27, 1283–1286.

Dee, D.P., Uppala, S.M., Simmons, A.J., Berrisford, P., Poli, P., Kobayashi, S., et al., 2011. The ERA-Interim reanalysis: configuration and performance of the data assimilation system. Q. J. R. Meteorol. Soc. 137, 553–597. Available from: https://doi.org/10.1002/qj.828.

de Graaf, I.D., Sutanudjaja, E.H., Van Beek, L.P.H., Bierkens, M.F.P., 2015. A high-resolution global-scale groundwater model. Hydrol. Earth Syst. Sci. 19 (2), 823–837.

de Graaf, I.E.M., van Beek, R.L.P.H., Gleeson, T., Moosdorf, N., Schmitz, O., Sutanudjaja, E.H., et al., 2017. A global-scale two-layer transient groundwater model: development and application to groundwater depletion. Adv. Water Resour. 102, 53–67.

Döll, P., Schmied, H.M., Schuh, C., Portmann, F.T., Eicker, A., 2014. Global-scale assessment of groundwater depletion and related groundwater abstractions: combining hydrological modeling with information from well observations and GRACE satellites. Water Resour. Res. 50, 5698–5720. Available from: https://doi.org/10.1002/2014WR015595.

Eltahir, E.A.B., Yeh, P.J.-F., 1999. On the asymmetric response of aquifer water level to floods and droughts in Illinois. Water Resour. Res. 35 (4), 1199–1217.

Famiglietti, J.S., 2014. The global groundwater crisis. Nat. Clim. Change 4 (11), 945–948. Available from: https://doi.org/10.1038/nclimate2425.

Famiglietti, J.S., Rodell, M., 2013. Water in the balance. Science 340 (6138), 1300–1301. Available from: https://doi.org/10.1126/science.1236460.

Famiglietti, J.S., Lo, M., Ho, S.L., Bethune, J., Anderson, K.J., Syed, T.H., et al., 2011. Satellites measure recent rates of groundwater depletion in California's Central Valley. Geophys. Res. Lett. 38, L03403. Available from: https://doi.org/10.1029/2010GL046442.

Feng, W., Zhong, M., Lemoine, J.-M., Biancale, R., Hsu, H.-T., Xia, J., 2013. Evaluation of groundwater depletion in North China using the Gravity Recovery and Climate Experiment (GRACE) data and ground-based measurements. Water Resour. Res. 49, 2110–2118. Available from: https://doi.org/10.1002/wrcr.20192.

Gardner, A.S., Moholdt, G., Wouters, B., Wolken, G.J., Burgess, D.O., Sharp, M.J., et al., 2011. Sharply increased mass loss from glaciers and ice caps in the Canadian Arctic Archipelago. Nature 473, 357–360.

Girotto, M., de Lannoy, G.J.M., Reichle, R., Rodell, M., 2016. Assimilation of gridded terrestrial water storage observations from GRACE into a land surface model. Wat. Resour. Res. 52, 4164–4183. Available from: https://doi.org/10.1002/2015WR018417. 2016.

Girotto, M., De Lannoy, G.J.M., Reichle, R.H., Rodell, M., Draper, C., Bhanja, S.N., et al., 2017. Benefits and pitfalls of GRACE data assimilation: a case study of terrestrial water storage depletion in India. Geophys. Res. Lett. 44, 4107–4115. Available from: https://doi.org/10.1002/2017gl072994. 2017.

Green, T.R., Taniguchi, M., Kooi, H., Gurdak, J.J., Allen, D.M., Hiscock, K.M., et al., 2011. Beneath the surface of global change, impacts of climate change on groundwater. J. Hydrol. 405 (3), 532–560.

Hanson, R.T., Dettinger, M.D., Newhouse, M.W., 2006. Relations between climatic variability and hydrologic time series from four alluvial basins across the southwestern United States. Hydrogeol. J. 14, 1122–1146.

Houborg, R., Rodell, M., Li, B., Reichle, R., Zaitchik, B.F., 2012. Drought indicators based on model-assimilated Gravity Recovery and Climate Experiment (GRACE) terrestrial water storage observations. Water Resour. Res. 48, W07525. Available from: https://doi.org/10.1029/2011WR011291.

Hughes, J.D., Petrone, K.C., Silberstein, R.P., 2012. Drought, groundwater storage and stream flow decline in southwestern Australia. Geophys. Res. Lett. 39, L03408. Available from: https://doi.org/10.1029/2011GL050797.

Jiménez Cisneros, et al., 2014. Freshwater resources. In: Field, C.B., Barros, V.R., Dokken, D.J., Mach, K.J., Mastrandrea, M.D., Bilir, T.E., Chatterjee, M., Ebi, K.L., Estrada, Y.O., Genova, R.C., Girma, B., Kissel, E.S., Levy, A.N., MacCracken, S., Mastrandrea, P.R., White, L.L. (Eds.), Climate Change 2014: Impacts, Adaptation, and Vulnerability. Part A: Global and Sectoral Aspects. Contribution of Working Group II to the Fifth Assessment Report of the Intergovernmental Panel on Climate Change. Cambridge University Press, Cambridge and New York, pp. 229–269.

Kath, J., Reardon-Smith, K., Le Brocque, A.F., Dyer, F.J., Dafny, E., Fritz, L., et al., 2014. Groundwater decline and tree change in floodplain landscapes: Identifying non-linear threshold responses in canopy condition. Glob. Ecol. Conserv. 2, 148−160.

Kløve, B., Ala-Aho, P., Bertrand, G., Gurdak, J.J., Kupfersberger, H., Kværner, J., et al., 2014. Climate change impacts on groundwater and dependent ecosystems. J. Hydrol. 518, 250−266.

Koirala, S., Yeh, P.J.-F., Hirabayashi, Y., Kanae, S., Oki, T., 2014. Global-scale land surface hydrologic modeling with the representation of water table dynamics. J. Geophys. Res. Atmos. 118. Available from: https://doi.org/10.1002/2013JD020398.

Koster, R.D., Suarez, M.J., Ducharne, A., Stieglitz, M., Kumar, P., 2000. A catchment-based approach to modeling land surface processes in a general circulation model 1. Model structure. J. Geophys. Res. Atmos. 105, 24809−24822. Available from: https://doi.org/10.1029/2000JD900327.

Kumar, S.V., Zaitchik, B.F., Peters-Lidard, C., Rodell, M., Reichle, R., Li, B., et al., 2016. Assimilation of gridded GRACE terrestrial water storage estimates in the North American Land Data Assimilation System. J. Hydrometeor. 17, 1951−1972. Available from: https://doi.org/10.1175/JHM-D-15-0157.1.

Leblanc, M.J., Tregoning, P., Ramillien, G., Tweed, S.O., Fakes, A., 2009. Basin-scale, integrated observations of the early 21st century multiyear drought in southeast Australia. Water Resour. Res. 45, W04408. Available from: https://doi.org/10.1029/2008WR007333.

Li, B., Rodell, M., 2015. Evaluation of a model-based groundwater drought indicator in the conterminous U.S. J. Hydrol. 526, 78−88. Available from: https://doi.org/10.1016/j.jhydrol.2014.09.027].

Li, B., Rodell, M., Zaitchik, B.F., et al., 2012. Assimilation of GRACE terrestrial water storage into a land surface model: Evaluation and potential value for drought monitoring in western and central Europe. J. Hydrol. 446−447, 103−115. Available from: https://doi.org/10.1016/j.jhydrol.2012.04.035].

Li, B., Rodell, M., Famiglietti, J.S., 2015. Groundwater variability across temporal and spatial scales in the central and northeastern U.S. J. Hydrol. 525, 769−780. Available from: https://doi.org/10.1016/j.jhydrol.2015.04.033].

Li, B., Rodell, M., Sheffield, J., Wood, E., Sutanudjaja, E., 2019a. Long-term, non-anthropogenic groundwater storage changes simulated by three global-scale hydrological models. Sci. Rep. 9 (1), 10746. Available from: https://doi.org/10.1038/s41598-019-47219-z.

Li, B., Rodell, M., Kumar, S., et al., 2019b. Global GRACE data assimilation for groundwater and drought monitoring: advances and challenges. Water Resour. Res. WR024618. Available from: https://doi.org/10.1029/2018wr024618. 2018.

Long, D., Scanlon, B.R., Longuevergne, L., Sun, A.Y., Fernando, D.N., Save, H., 2013. GRACE satellite monitoring of large depletion in water storage in response to the 2011 drought in Texas. Geophys. Res. Lett. 40 (13), 3395−3401.

Markovich, K., Manning, A.H., Condon, L.E., McIntosh, J.C., 2019. Mountain-block recharge: a review of current understanding. Water Resour. Res. 55, 8278−8304. Available from: https://doi.org/10.1029/2019WR025676.

Maxwell, R.M., et al., 2014. Surface-subsurface model intercomparison: a first set of benchmark results to diagnose integrated hydrology and feedbacks. Water Resour. Res. 50, 1531−1549. Available from: https://doi.org/10.1002/2013WR013725.

Meixner, T., Manning, A.H., Stonestrom, D.A., Allen, D.M., Ajami, H., Blasch, K.W., et al., 2016. Implications of projected climate change for groundwater recharge in the western United States. J. Hydrol. 534, 124−138. Available from: https://doi.org/10.1016/j.jhydrol.2015.12.027.

Mishra, A.K., Singh, V.P., 2010. A review of drought concepts. J. Hydrol. 391 (2010), 202−216. Available from: https://doi.org/10.1016/j.jhydrol.2010.07.012.

Mo, K.C., 2008. Model-based drought indices over the United States. J. Hydrometeor. 9, 1212−1230. Available from: https://doi.org/10.1175/2008JHM1002.1.

Mueller, et al., 2011. Evaluation of global observations-based evapotranspiration datasets and IPCC AR4 simulations. Geophys. Res. Lett. 38, L06402. Available from: https://doi.org/10.1029/2010GL046230.

Mukherjee, A., Scanlon, B., Aureli, A., Langan, S., Guo, H., McKenzie, A., 2020. Global Groundwater: Source, Scarcity, Sustainability, Security and Solutions, first ed. Elsevier, ISBN: 9780128181720.

Niu, G.-Y., Yang, Z.-L., Dickinson, R.E., Gulden, L.E., Su, H., 2007. Development of a simple groundwater model for use in climate models and evaluation with Gravity Recovery and Climate Experiment data. J. Geophys. Res. 112, D07103. Available from: https://doi.org/10.1029/2006JD007522.

Nie, W., Zaitchik, B., Rodell, M., Kumar, S.V., Anderson, M.C., Hain, C., 2018. Groundwater withdrawals under drought: reconciling GRACE and land surface models in the United States High Plains aquifer. Wat. Resour. Res. 54. Available from: https://doi.org/10.1029/2017WR022178.

Nie, W., Zaitchik, B.F., Rodell, M., Kumar, S.V., Arsenault, K.R., Li, B., et al., 2019. Assimilating GRACE into a land surface model in the presence of an irrigation-induced groundwater trend. Wat. Resour. Res. 55 (12), 11274−11294. Available from: https://doi.org/10.1029/2019WR025363.

Niraula, R., Meixner, T., Dominguez, F., Rodell, M., Ajami, H., Gochis, D., et al., 2017. How might recharge change under projected climate change in western US? Geophys. Res. Lett. 44, 10,407−10,418. Available from: https://doi.org/10.1002/2017GL075421.

Peters, E., Bier, G., van Lanen, H.A.J., Torfs, P.J.J.F., 2006. Propagation and spatial distribution of drought in a groundwater catchment. J. Hydrol. 321, 257−275.

Perez-Valdivia, C., Sauchyn, D., Vanstone, J., 2012. Groundwater levels and teleconnection patterns in the Canadian Prairies. Water Resour. Res. 48, W07516. Available from: https://doi.org/10.1029/2011WR010930.

Rodell, M., Famiglietti, J.S., 2002. The potential for satellite-based monitoring of groundwater storage changes using GRACE: the high plains aquifer, central U.S. J. Hydrol. 263, 245−256.

Rodell, M., Famiglietti, J.S., 2001. An analysis of terrestrial water storage variations in Illinois with implications for the Gravity Recovery and Climate Experiment (GRACE). Water Resour. Res. 37, 1327−1340. Available from: https://doi.org/10.1029/2000WR900306.

Rodell, M., Chen, J., Kato, H., Famiglietti, J.S., Nigro, J., Wilson, C.R., 2007. Estimating groundwater storage changes in the Mississippi River basin (USA) using GRACE. Hydrologeol. J. 15, 159−166. Available from: https://doi.org/10.1007/s10040-006-0103-7.

Rodell, M., Velicogna, I., Famiglietti, J.S., 2009. Satellite-based estimates of groundwater depletion in India. Nature 460, 999−1002. Available from: https://doi.org/10.1038/nature08238.

Rodell, M., Famiglietti, J.S., Wiese, D.N., Reager, J.T., Beaudoing, H.K., Landerer, F.W., et al., 2018. Emerging trends in global freshwater availability. Nature 557, 651−659. Available from: https://doi.org/10.1038/s41586-018-0123-1.

Rowlands, D.D., Luthcke, S.B., Klosko, S.M., Lemoine, F.G.R., Chinn, D.S., McCarthy, et al., 2005. Resolving mass flux at high spatial and temporal resolution using GRACE intersatellite measurements. Geophys. Res. Lett. 32, L04310. Available from: https://doi.org/10.1029/2004GL021908.

Russo, T.A., Lall, U., 2017. Depletion and response of deep groundwater to climate-induced pumping variability. Nat. Geosci. 10 (2), 105. Available from: https://doi.org/10.1038/NGEO288.

Sheffield, J., Goteti, G., Wood, E.F., 2006. Development of a 50-yr high-resolution global dataset of meteorological forcings for land surface modeling. J. Clim. 19, 3088−3111.

Sheffield, J., Livneh, B., Wood, E., 2012. Representation of terrestrial hydrology and large scale drought of the continental US from the North American Regional Reanalysis. J. Hydrometeor. Available from: https://doi.org/10.1175/JHM-D-11-065.1.

Swenson, S., Yeh, P.J.F., Wahr, J., Famiglietti, J., 2006. A comparison of terrestrial water storage variations from GRACE with in situ measurements from Illinois. Geophys. Res. Lett. 33 (16). Available from: https://doi.org/10.1029/2006GL026962.

Stromberg, J.C., Tress, J.A., Wilkins, S.D., Clark, S.D., 1992. Response of velvet mesquite to groundwater decline. J. Arid. Environ. 23 (1992), 45−58.

Su, H., Yang, Z.-L., Dickinson, R.E., Wilson, C.R., Niu, G.-Y., 2010. Multisensor snow data assimilation at the continental scale: the value of Gravity Recovery and Climate Experiment terrestrial water storage information. J. Geophys. Res. 115, D10104. Available from: https://doi.org/10.1029/2009JD013035.

Sutanudjaja, et al., 2018. PCR-GLOBWB 2: a 5 arcmin global hydrological and water resources model. Geosci. Model. Dev. 11, 2429−2453. Available from: https://doi.org/10.5194/gmd-11-2429-2018.

Tapley, B.D., Bettadpur, S., Ries, J.C., Thompson, P.F., Watkins, M.M., 2004. GRACE measurements of mass variability in the Earth system. Science 305, 503−505.

Taylor, R.G., Scanlon, B., Döll, P., Rodell, M., van Beek, R., Wada, Y., et al., 2012. Ground water and climate change. Nat. Clim. Change 3 (4), 322−329. Available from: https://doi.org/10.1038/nclimate1744.

Thomas, A., Reager, J., Famiglietti, J., Rodell, M., 2014. A GRACE-based water storage deficit approach for hydrological drought characterization. Geophys. Res. Lett. 41 (5), 1537−1545. Available from: https://doi.org/10.1002/2014GL059323.

Thomas, B.F., Famiglietti, J.S., Landerer, F.W., Wiese, D.N., Molotch, N.P., Argus, D.F., 2017. GRACE groundwater drought index: evaluation of California Central Valley groundwater drought. Remote Sens. Environ. 198, 384−392. Available from: https://doi.org/10.1016/j.rse.2017.06.026.

Van Loon, A.F., et al., 2016. Drought in the Anthropocene. Nat. Geosci. 9, 89−91.

Van Loon, A., Van Lanen, H.A.J., 2013. Making the distinction between water scarcity and drought using observation-modeling framework. Water Resour. Res. 49, 1483−1502. Available from: https://doi.org/10.1002/wrcr.20147.

Van Loon, A., Kuman, R., Mishra, V., 2017. Testing the use of standardized indices and GRACE satellite data to estimate the European 2015 groundwater drought in near-real time. Hydrol. Earth Syst. Sci. Available from: https://doi.org/10.5194/hess-21-1947-2017. Available from: https://www.hydrol-earth-syst-sci.net/21/1947/2017/.

Velasco, E.M., Gurdak, J.J., Dickinson, J.E., Ferré, T.P.A., Corona, C.R., 2015. Interannual to multidecadal climate forcings on groundwater resources of the U.S. West Coast. J. Hydrol. Reg. Stud. 11, 250−265. Available from: https://doi.org/10.1016/j.ejrh.2015.11.018.

Voss, K.A., Famiglietti, J.S., Lo, M., de Linage, C., Rodell, M., Swenson, S.C., 2013. Groundwater depletion in the Middle East from GRACE with implications for transboundary water management in the Tigris-Euphrates-Western Iran region. Water Resour. Res. 49. Available from: http://doi.org/10.1002/wrcr.20078.

Wang, D., 2012. Evaluating interannual water storage changes at watersheds in Illinois based on long-term soil moisture and groundwater level data. Water Resour. Res. 48, W03502. Available from: https://doi.org/10.1029/2011WR010759.

Wilson, J., Guan, H., 2004. Mountain-block hydrology and mountain-front recharge, groundwater recharge in a desert environment: the southwestern United States. In: Phillips, F.M., Hogan, J., Scanlon, B. (Eds.), AGU, Washington, DC.

Wu, W., Geller, M.A., Dickinson, R.E., 2002. The response of soil moisture to long-term variability of precipitation. J. Hydrometeor. 3, 604−613.

Xia, Y., Mitchell, K., Ek, M., Sheffield, J., Cosgrove, B., Wood, E., et al., 2012. Continental-scale water and energy flux analysis and validation for the North American Land Data Assimilation System project phase 2 (NLDAS-2): 1. Intercomparison and application of model products. J. Geophys. Res. 117, D03109. Available from: https://doi.org/10.1029/2011JD016048.

Xia, Y., Mocko, D.M., Huang, M., Li, B., Rodell, M., Mitchell, K.E., et al., 2017. Comparison and assessment of three advanced land surface models in simulating terrestrial water storage components over the United States. J. Hydrometeor. 18 (3), 625−649. Available from: https://doi.org/10.1175/jhm-d-16-0112.1.

Yeh, P.J.-F., Eltahir, E.A.B., 2005a. Representation of water table dynamics in a land surface scheme: 1. Model development. J. Clim. 18 (12), 1861−1880.

Yeh, P.J.-F., Eltahir, E.A.B., 2005b. Representation of water table dynamics in a land surface scheme: 2. Subgrid heterogeneity. J. Clim. 18 (12), 1881−1901.

Yirdaw, S.Z., Snelgrove, K.R., Agboma, C.O., 2008. GRACE satellite observations of terrestrial moisture changes for drought characterization in the Canadian Prairie. J. Hydrol. 356 (1-2), 84–92.

Zaitchik, B.F., Rodell, M., Reichle, R.H., 2008. Assimilation of GRACE terrestrial water storage data into a land surface model: results for the Mississippi River basin. J. Hydrometeorol. 9, 535–548. Available from: https://doi.org/10.1175/2007JHM951.1.

Zhao, M., Velicogna, I., Kimball, J.S., 2017. A global gridded dataset of GRACE drought severity index for 2002–14: comparison with PDSI and SPEI and a case study of the Australia millennium drought. J. Hydrometeorol. 18 (8), 2117–2129.

Chapter 12

Groundwater scarcity in the Middle East

Ahmed A. Al-Taani[1,2], Yousef Nazzal[1] and Fares M. Howari[1]
[1]College of Natural and Health Sciences, Zayed University, Abu Dhabi, United Arab Emirates, [2]Department of Earth and Environmental Sciences, Faculty of Science, Yarmouk University, Irbid, Jordan

12.1 Introduction

The scarcity of natural water resources in the Middle East and North Africa (MENA) region is inherently a critical constraint to socioeconomic growth, developments, and stability. Two-thirds of the MENA's population face very high water-stressed conditions as compared to the global average of about one-third (Arab Countries Water Utilities Association, 2014; Wada et al., 2011). The region is home to 6% of the world's population with an average annual population growth rate of 1.7% (Kabbani and Kothari, 2005).

Water scarcity in this region is increasingly becoming a compelling problem from a national security perspective, particularly, because most of the surface water (9 river basins) and groundwater resources (20 aquifers) are transboundary and shared by multiple countries (United Nations, Economic and Social Commission for Western Asia, 2013; Mukherjee et al., 2020).

Traditionally, arid and semiarid climatic condition regions are the most water-scarce regions around the globe. The annual rainfall is approximately 300 mm that together constitutes around 1% of the world's freshwater resources (Wagdy and AbuZeid, 2006). Climate change has already exacerbated the water shortage and further disrupted the existing weak water supply–demand relations.

Combinations of declining rainfall (with frequent flash floods), prolonged drought cycles, high rates of water consumption, inefficient usage and mismanagement, poor water governance, and weak enforcement impacted water supplies and quality in the region (World Bank, 2017).

Inappropriate exchange in the water–energy–food nexus confers to overexploitation of water resources. In addition, geopolitical and the socioeconomic context (high population growth and economic development) are contributing factors to the severity of water scarcity in the region. Countries will be forced to adopt relatively expensive nonconventional alternatives to water supply problems, such as food and water imports, desalination, and treated wastewater, with significant economic, environmental, security, and social impacts. While some countries in the region have adopted a number of alternatives to address water scarcity and improve their resilience to cope with shocks, such as climate change or refugee influx, other options have remained relatively underused in the region.

12.2 Water resources: current use and future trends

Current trends in water resource use are unsustainable (Borgomeo et al., 2018), with total water production only half the world's average (World Bank, 2017). Water resources per capita in the region are one-sixth of the global average or about 720 m^3/capita/year (World Bank, 2007). Water availability is highly variable, spatially and temporally within the region. For example, the per capita water availability in Yemen and Jordan is less than 200 m^3/year.

The average renewable water resources per capita for 2005 are about 20,000, 11,000, and 1500 m^3/year for North America, Europe, and the MENA regions, respectively (WRI, 2005). More than 80% of the available freshwater reserves in the region is withdrawn for agriculture, industry, and urban needs, putting them at the highest level of water stress (Hofste et al., 2019). This will have a serious impact on long-term growth and stability in the region. Average annual renewable water resources were estimated at 360 km^3, more than two-thirds of which comes from surface resources (rainfall, rivers, springs, and lakes). Rainfall is highly variable in this region, though it is the only source of water available. Approximately, 65% of the area receives an average annual rainfall of less than 100 mm, while only 20% receives more than 300 mm of rainfall (Ahmed, 2007). Rainfall variation influences the recharge of groundwater and green water (soil moisture) with implications for rain-fed agriculture (World Bank, 2017).

Agriculture is the primary water resource user in the country, accounting for 80% of total water withdrawals. Groundwater is the main supply of irrigation water, where unsustainable abstraction becomes widespread (Food and Agriculture Organization of the United Nations, 2017). About 58% of the renewable water resources in the region will be used for food production by 2030 (FAO, 2006). The projected increases in the total water demand will rise from 270 km^3 in 2000 to 460 km^3 in 2050, with an increase in the demand–supply gap from 50 km^3 in 2000 to 150 km^3 in 2050 (Trieb and Müller-Steinhagen, 2008).

Droogers et al. (2009) demonstrated that the total renewable water resources and the recharge in the MENA region will significantly decline toward 2050, with a projected decrease of about 0.6 km^3/year compared to 250 km^3 (the average total MENA renewable water resources from 2000 to 2009).

In addition, it was projected that the total demand will increase by 132 km^3/year, while the total water shortage will grow by 157 km^3/year for the average climate change projection (Droogers et al., 2009). This increase in water shortage was attributed to the impact of increasing demand by 50% combined with the decreasing supply by 12%. Water demand in the MENA region is predicted to rise between 74 (Droogers et al., 2009) and 99 km^3 by 2030 (Droogers et al., 2012).

12.3 Impacts of water scarcity

12.3.1 Water resources and climate change

The MENA area is extremely vulnerable to climate change. The area is prone to extreme temperatures and water scarcity. It has been estimated that the MENA region will be affected by drastic climate change with a peak rise in temperatures and declining precipitation (Paasche, 2007). During the period from 1961 to 1990 a likable increase of approximately 0.2°C in each decade has been noticed in the MENA region (Waha et al., 2017).

The fluctuating precipitation patterns concerning their temporal scale and degrading hazardous frequency and quantity leading to flash floods and droughts would be likely to augment climate change more. Water resource scarcity would shoot up more as rainfall is the only source of freshwater in the region.

Besides the increase in population growth, climate change further complicates the regional water scarcity, with a shortage in water supply as precipitation becomes more erratic and rising temperatures increase the demand. The deficit is, therefore, the combined effect of a dramatically increasing demand for water (some forecasts expect a rise of 50 % by the middle of this century) and a decline in water supply of about 12%. Total water demand is apprehended to likely rise from 37% in 2020–30 to 51% by 2040–50 (World Bank, 2007).

The water shortage in the MENA area for the period 2000–09 was about 42 km^3/year, with annual fluctuations varying from 24 (2004) to 64 km^3 (2008) (Droogers et al., 2012). This substantial unmet demand indicates the current conditions in the MENA region, where water shortage is occurring in most of the countries. The climate-related water scarcity is expected to cause economic losses estimated at 6%–14% of gross domestic product (GDP) by 2050, which are the highest in the world. Regional instability has also impacted water resources. According to the reports of Ministry of Water and Irrigation in Jordan, per capita availability of water sums up to 140 m^3 that is way beneath the global water availability of 500 m^3, considered as water-scarce regions. Water demands grossly depend on the rise in population, urban developments, socioeconomic conditions, and variations of water usage for agricultural and industrial sections (Droogers et al., 2012; Howari and Banat, 2002).

12.3.2 Water quality

The continuous downgrade in the characteristic of water has increased in recent times (Howari and McDonnell, 2008; Howari et al., 2005). Improper applications of water and its overuse have led to decrease in groundwater levels (Howari and Banat, 2001; Jasechko et al., 2017). The amount of overuse vastly exceeds the available water level, leading to water scarcity (Al-Taani, 2014).

The quality of water in the region of North Africa and Middle East has been the cause of concern due to its use in agriculture, domestic use, and industries (Al-Taani et al., 2018; Banat and Howari, 2003; Batayneh and Al-Taani, 2016; Muhaidat et al., 2019; Al-Rawabdeh et al., 2013). The lack of proper treatment of wastewater, dumping of sewage that has been left untreated, disposal of waste without any regulation, and overuse of fertilizer are some of the reasons why the quality of water has suffered in this region (Al-Rawabdeh et al., 2014).

The ecosystem and the environment have suffered due to unregulated pumping of water and pollution. This has impacted the socioeconomic condition of countries such as Jordan and Morocco where 1% of GDP is responsible for such costs (Huh and Park, 2018).

Wastewater that has been treated partially or left untreated is used for industrial and irrigation applications (Al-Rawabdeh et al., 2013). This can lead to contamination and the solution (Al-Taani et al., 2018) lies in following an integrated management of water resource (Verner, 2012).

12.4 Water resources management

12.4.1 Mitigation to water scarcity

MENA contrived a wide array of choices to fulfill the water demands of the region such as desalination, water harvesting, and recycling and groundwater extraction. The region mostly depends on food imports that further lessen the water demand. Food imports meet about 50% of the domestic wheat and barley supply, 70% of rice consumption, and 60% of corn consumption (Hub, 2019). Despite these many initiatives, the region still does not satiate the water demand.

The following section explains a few initiatives considered to mitigate the huge difference between supply and demand for water with potentially high applicability in the region. All the initiatives are being followed across the globe depending on the countries requirement and potential resources, with many favorable outcomes.

12.4.1.1 Desalination

Currently holding about half of the world's desalination capacity, the region is about to become the largest desalination market as it adds more plants (World Bank, 2017). The Arabian Peninsula produces about 87.4% of the total desalinated water in the region (Amy et al., 2017). Although technology helps to meet the water requirements up to a large extent but bears high cost and energy that makes it difficult to put into practice by poor countries. On average, however, 71% of the region's GDP is produced in areas with high to very high surface water stress compared to a global average of around 22% (Waha et al., 2017). This implies that the desalination of seawater is generally a viable choice in this area.

MENA receives an ample amount of solar energy that can be exercised for installing solar energy–driven desalination plants, for cleaner, greener, and sustainable development of the environment (Jan Stuyfzand et al., 2019). Over the years the cost of desalination has descended because of advanced membrane technologies that serve as a favorable option for the region to practice and meet the water requirements. From 2005 the region is producing more than 20 km^3/year desalinated water (Ghaffour et al., 2013). The region made an impression of attaining the cost of water reuse and desalination comparable to conventional water treatment. While desalination gained as a prior choice for meeting water requirements, still the region has to consider revising the domestic water needs and its precise use for overall sustainable development (WHO/UNICEF Joint Water Supply and Sanitation Monitoring Programme, 2014).

12.4.1.2 Treated wastewater reuse

Treated wastewater became a valuable resource to meet the growing demand (Arab Countries Water Utilities Association, 2014; Lowder et al., 2014). It offers the potential for highly reliable nonconventional water supply independent of the effects of climate change. About 82% of wastewater in the region is not recycled (Vo et al., 2014). This water supply alternative is generally more cost-effective and less energy intensive than desalination.

To fulfill the growing demand for water consumption, wastewater is being treated. Wastewater treatment and reuse is a sustainable practice that provides exceedingly dependable water source for agricultural, industrial, and household use (Arab Countries Water Utilities Association, 2014; Lowder et al., 2014). In comparison to the desalination process, this technique is more cost-effective and less energy intensive. Approximately, more than three-fourths of the wastewater in the region is not treated (Jeuland, 2015). Most of the countries within the MENA region practice water treatment and reuse up to some extent. But the number of treatment plants is limited and hence concise use of treated water needs to be focused. Four-fifths of the total average of water delivered through pipe systems in MENA and around three-fifths for sewage water (Faour-Klingbeil and Todd, 2018). A large amount of the treated wastewater is generally dispensed into the sea. Moreover, the price of traditional water in this area (as subsidized by governments) is small, making reuse an unattractive choice

Around 22.3 km³/year amount of wastewater is produced in the MENA, of which nearly half of it is treated (Vo et al., 2014; Qadir et al., 2010). Approximately, half of the treated water is used for irrigation purposes. Jordan reuses more than three-fourths of the treated wastewater (El-Radaideh et al., 2017a,b). In the region, treated wastewater has been used for agricultural irrigation (cooked vegetables, with strict quality control), forestry, industrial crops, nonedible crops (alfalfa and barely), and landscaping. A mix of treated wastewater and freshwater has been also used for agricultural produce (e.g., in Jordan) (Al-Taani, 2013).

12.4.1.3 Rainwater harvesting and artificial aquifer recharge

Many countries in MENA region have invested heavily in infrastructure to capture and store water (Jiménez and Asano, 2008). It has one of the world's largest reservoir storage units for storing treated or freshwater of the region as a practice of sustainable water management. Approximately, more than three-fourths of the regions freshwater resources kept in reserve in comparison to one-tenth of the global average (Al Khateeb et al., 2019). As the region receives scanty precipitation, restricted reservoirs for water storage can be opted. The environmental and social impacts, dam safety issues, and increased sedimentation have received more attention in recent years (El-Radaideh et al., 2017b; Al-Taani, 2013; Tuinhof et al., 2002; Al-Maktoumi et al., 2016). However, water reservoir sediments can be used to improve soil properties and increase fertility (because of high nutrients and sand contents) with implications for restoring reservoir storage capacity (Al-Maktoumi et al., 2016).

Artificial aquifer recharge may be used to store excess flood/storm runoff or treated wastewater for later use. It has attracted attentions and is becoming a promising alternative in this region (Food and Agriculture Organization of the United Nations, 2014; Salameh et al., 2019). For example, about 80% of floodwaters were stored in Iran, implementing managed aquifer recharge (Salameh et al., 2019).

Artificial recharge can be utilized to restore the groundwater balance. It encourages controlled infiltration in suitable sites and is used during periods when excess water is available. It is relatively inexpensive option, reduces evaporative loss, provides protection from direct pollution, and improves groundwater conditions, compared to surface storage (Salameh et al., 2019). Groundwater recharge in coastal aquifers may also be used to mitigate or control saltwater intrusion. However, the controlled use of aquifers for subsurface storage requires treatment before recharge and after recovery from the aquifer, especially when treated wastewater is used.

In the United Arab Emirates, artificial groundwater recharge is being practiced by injecting the treated wastewater in the aquifers. This practice strengthens the adaptability of supply of water in arid areas as it experiences high evaporation rates (Escalante et al., 2014). For instance, artificial groundwater recharge aquifer has been set up in Liwa, Abu Dhabi bearing a capacity of 50 Mm³ (Dawoud and Sallam, 2012).

12.5 Case studies

12.5.1 Jordan River

The Jordan River is also addressed as Nahr Al-Urdun in Arabic and HaYarden in Hebrew and is located in a geographic depression of southwestern part of Asia in Middle East region (Fig. 12.1).

The Jordan River has its roots in Mount Hermon where it starts from the plains and then continues to the south via northern Israel to meet the Sea of Galilee. Its final destination after exiting the sea is the Dead Sea, the surface of which is 430 m below sea level and is Earth's lowest land point.

Although the direct distance is less than 200 km between the Dead Sea and its origin, the Jordan River traverses more than 360 km. The river used to be a border between Israel and Jordan in 1948. Later in 1967 when Israel's forces took over the West Bank, it served as the cease-fire line (Mimi and Sawalhi, 2003).

Since the quantity of water in the basin is very less, its inhabitants who live near the banks of the river face water scarcity. Increased use of water resources has led to lesser water flow in the river. The quality of the water in the region has suffered as is evident from the high salinity and increased pollution of the river in the southern part (Young, 2015).

Five countries (Fig.12.2) form the catchments of the Jordan River. The total catchment area of the Jordan River basin is about 18285 km², and the catchment distribution shared by each country has been depicted in Fig. 12.2.

Several climatic conditions prevail in the basin such as arid and semiarid that can be noted over a distance of 10–20 km. Due to higher altitude, the mountainous northern part of the basin witnesses Mediterranean climate whereas dry, temperate climate can be seen in the northeastern and eastern part of the bank. In summers the rate of evaporation is more than 50% that makes it difficult in maintaining the water quality and consumption rate. Rainfall pattern varies

Groundwater scarcity in the Middle East **Chapter | 12** 167

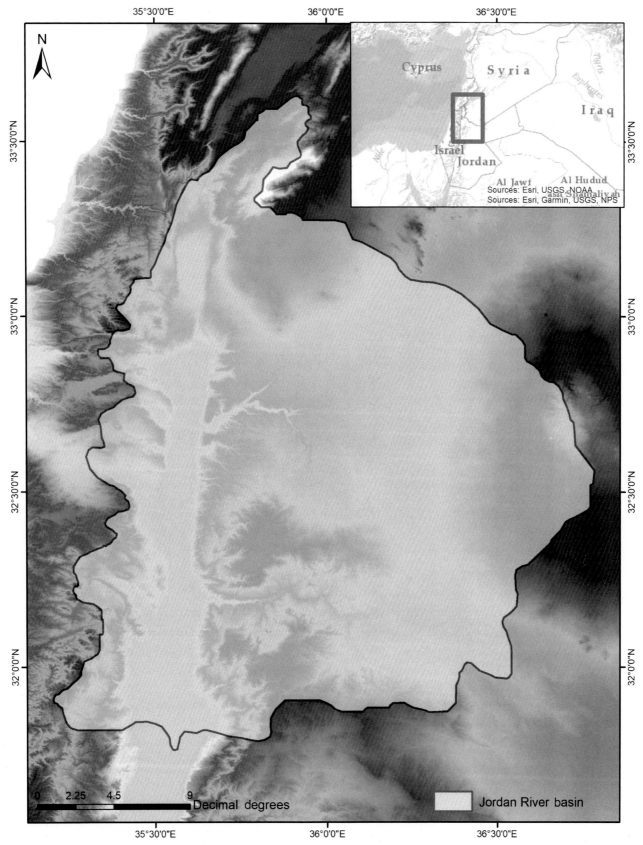

FIGURE 12.1 Map of the Jordan River drainage basin.

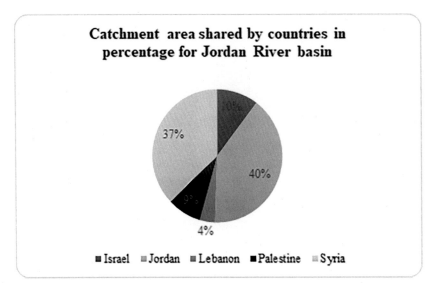

FIGURE 12.2 Distribution of the Jordan River basin area. *Modified after UN-ESCWA and BGR (United Nations Economic and Social Commission for Western Asia; Bundesanstalt für Geowissenschaften und Rohstoffe). 2013. Inventory of Shared Water Resources in Western Asia. Beirut.*

from 1000 mm/year in the northeastern to >200 mm/year in the west and >100 mm/year in the coast of Dead Sea of the basin. The major rainfall is mostly observed in winter months (Rosenthal and Sabel, 2009).

12.5.2 Tigris–Euphrates River

Tigris–Euphrates river system comprises the Tigris and Euphrates Rivers, in southwestern Asia with a length of 1900 and 2800 km, respectively (Fig. 12.3). Originating from eastern Turkey, the rivers travel via Syria and Iraq and join at the Persian Gulf. The basin is categorized as upper, middle, and lower courses with an altitude variation of 1800–3000 m in the upper basin to 370–50 m in the middle basin. The lower course lies on the alluvial plains where both the rivers meet and discharge into the Persian Gulf in the southeastern part of Iraq (El-Fadel et al., 2002).

Both the Tigris and Euphrates play a major role in improving the habitability and productivity of the region. The region has temperatures of more than 32°C in summer and less than 10°C in winter. Precipitation varies through the course of the river with high precipitation in source areas and very less rainfall in the lower areas. The annual total rainfall accounts for 200 mm/year. The major form of precipitation is snow. Rains generally occur during the months of winter mostly in the lower part of the basin.

Turkey and Syria contribute 90% and 10%, respectively, in the mean annual discharge of the Euphrates River of 32 billion m^3. The Tigris has a much larger annual discharge of 52 bcm with major contributors being Iraq (51%), Turkey (40%), and Iran (9%) (Wilson, 2012). Some of the major uses of the river water are irrigation, drinking, power generation, and industrial usage. The flow of Tigris has been affected in the recent years due to large-scale water development projects in Iraq and Turkey.

The catchment areas of the Tigris and Euphrates comprise Turkey, Iran, Saudi Arabia, Iraq, and Syria (Fig. 12.4). A number of 46 million people inhabit the catchment area of 917,103 km^2 of these rivers (United Nations, Economic and Social Commission for Western Asia, 2013). Turkey has become the hegemonic riparian party of the two basins since both the rivers originate from its southeastern part (Flint et al., 2011). The initial water projects of Syria and Turkey focused on regulating floods of the two rivers. Slowly, the projects grew to be used as hydroelectric power generation sources and drinking water sources.

Iraq, Syria, and Turkey have a scarcity of water resources. Studies showed that the per capita annual water shares for Iraq, Syria, and Turkey are 2.920, 1.440, and 2.950 m^3, respectively, as against the total amount of water resources as 75.42, 26.26, and 229.3 bcm/year, respectively (Williams, 2012). The water quality has suffered as a result of the return flow of water from the irrigation projects of Syria and Turkey, and this is set to only worsen as more areas of land are irrigated.

Groundwater scarcity in the Middle East **Chapter | 12** **169**

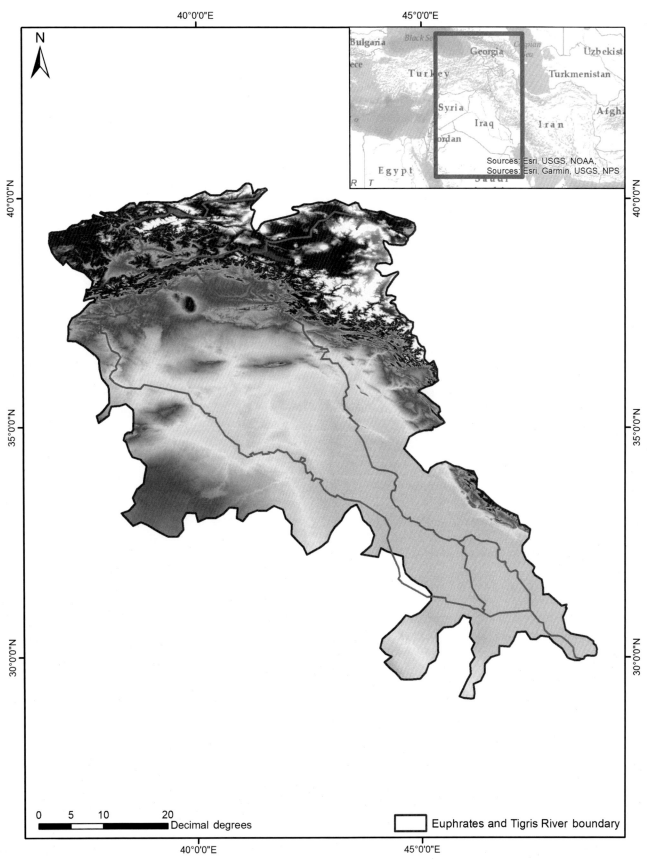

FIGURE 12.3 Map of the Tigris–Euphrates drainage basin.

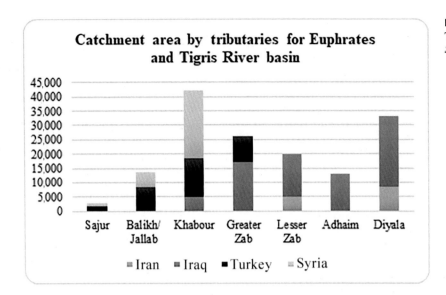

FIGURE 12.4 Distribution of the Euphrates and Tigris River basin. *Modified after Inventory of Shared Water Resources in Western Asia.*

12.5.3 Nile River

The Nile River is the longest in the world covering 6850 km^2 distance and comprising a catchment area of approximately 3 million km^2, and it covers 10 countries (Ethiopia, Sudan, Egypt, Rwanda, Tanzania, Uganda, Burundi, DRC, Eritrea, and Kenya) that contribute around 10% of Africa's landmass. The river has three tributaries: the White Nile, the Blue Nile, and the Atbara (Awulachew, 2012) originating from Burundi, Rwanda, Democratic Republic of the Congo, Ethiopia, Eritrea, Sudan, and Ethiopian highlands. The basin lies in Egypt and empties into the Mediterranean Sea (Fig. 12.5).

12.5.3.1 Victoria Nile or the White Nile

About four-fifths of the water in the While Nile are from rainfall, and the remaining one-fifth part of the water comprises from the rivers that channels from the neighboring basins. More than 85% of water is evaporated, and the other 15% of water empties to sea near Uganda. Kenya, Tanzania, and Uganda share the shoreline covering 6%, 51%, and 43%, respectively, whereas Burundi, DRC, Kenya, Rwanda, Tanzania, and Uganda share the basin. The region is one of the rapidly growing in East Africa with a population total of 30 million in 2011 (Hilhorst, 2011).

12.5.3.2 Blue Nile River basin

Higher differences in the terrain can be observed in the Blue Nile river basin aggravating uneven topography and precipitation patterns. The annual average rainfall ranges from 2049 to 794 mm in the southern part and the northeastern part of the basin, respectively (Fig. 12.6).

Declining level of water is agitating in the countries covering the southern part of the basin. Dams in Uganda have unbridled water with an average discharge rate of 1250 m^3/s that is double the rate of the permitted flow for maintenance of the water levels. Socioeconomic sectors such as fisheries and navigation are highly affected with declining water level. It also hampers the marine life.

The total population of the Nile basin is 238 million. Figs 12.7 and 12.8 present water use and population in the 10 countries for the year 2000. Due to increasing population, peak water demand in the urban sectors has tremendous consequences for water scarcity and water management. Urbanization can play a major role to create opportunities in water delivery and sanitation while also facing challenges of water delivery in far off rural places such as treated wastewater for sanitation and potable use (Merrill, 2008).

The average rainfall in the Nile basin varies greatly by country. Of the total average rainfall of 650 mm, Egypt contributes the lowest with an average rainfall of about 200 mm/year and Ethiopia the highest with rainfall of about 510−1525 mm. Nearly, one-third of Sudan shares an arid desert climate and is considered under one of the most drought-prone countries. The upstream countries contribute more to the precipitation (Fig. 12.9).

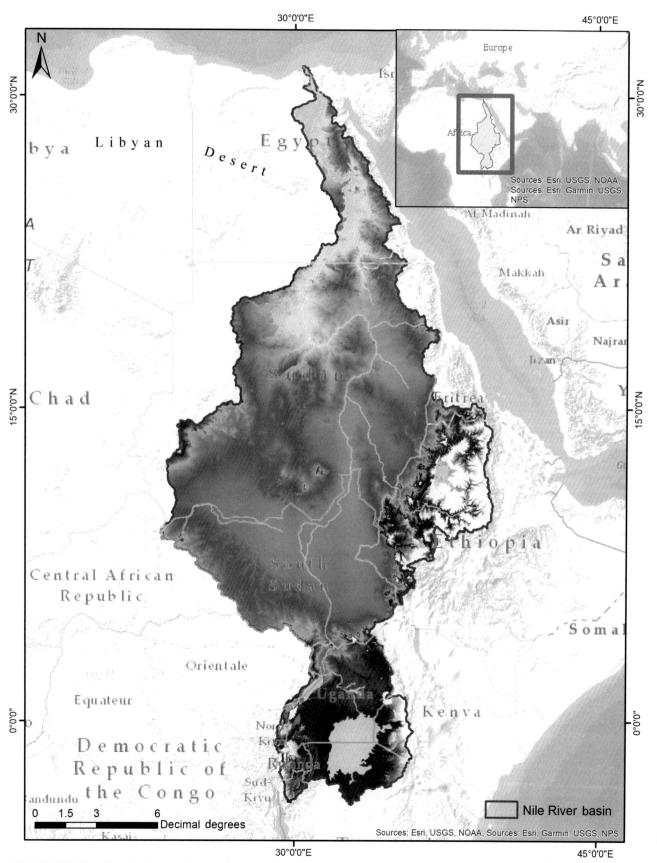
FIGURE 12.5 Map of the Nile River drainage basin.

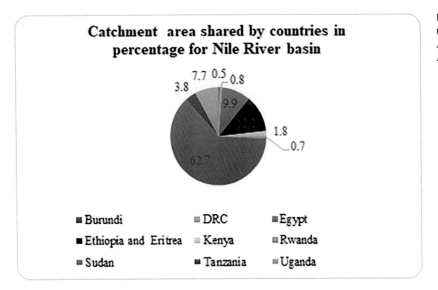

FIGURE 12.6 Catchment area shared by countries in percentage for the Nile River basin. *Modified after Inventory of Shared Water Resources in Western Asia.*

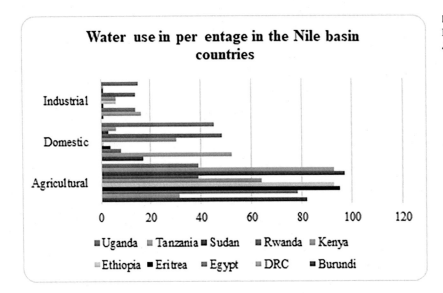

FIGURE 12.7 Distribution of water use in the Nile Basin countries. *Modified after Inventory of Shared Water Resources in Western Asia.*

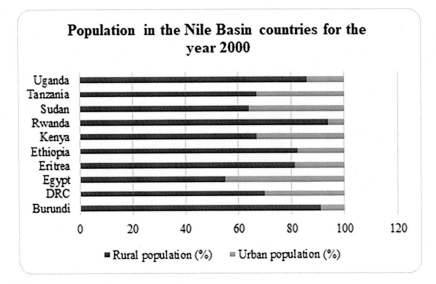

FIGURE 12.8 Population and water use in the Nile basin countries for the year 2000. *Modified after Inventory of Shared Water Resources in Western Asia.*

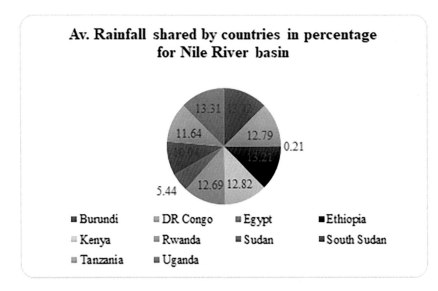

FIGURE 12.9 Average rainfall shared by countries for the Nile River basin. *Modified after Inventory of Shared Water Resources in Western Asia.*

The countries such as Egypt, Ethiopia, and Sudan are fully dependent on the Nile and its tributaries for water as they have no other water resources. As the total amount of rainfall is mainly concentrated in the upstream of the basin that leaves the lower part drier. This is the reason the countries sharing the basin should implement policies and decisions regarding the precise use of water consumption thereby taking a step forward in eliminating the harsh effects of water resource scarcity (Arjoon et al., 2014).

References

Ahmed, A., 2007. Technological transformation and sustainability in the MENA region. In: Science, Technology and Sustainability in the Middle East and North Africa. pp. 3–18.

Al Khateeb, W., Al-Taani, A., El-Radaideh, N., Tashtoush, Y., Al-Momani, R., 2019. Growth, yield and genetic integrity of spinach and chrysanthemum as affected by soil supplementation with dam sediments collected from King Talal and Al-Mujib Dams/Jordan. World Appl. Sci. J. 37 (1), 58–69.

Al-Maktoumi, A., El-Rawy, M., Zekri, S., 2016. Management options for a multipurpose coastal aquifer in Oman. Arab. J. Geosci. 9 (14), 636.

Al-Rawabdeh, A.M., Al-Ansari, N., Al-Taani, A.A., Knutsson, S., 2013. A GIS-based drastic model for assessing aquifer vulnerability in Amman-Zerqa groundwater basin, Jordan. Engineering 5, 490–504.

Al-Rawabdeh, A.M., Al-Ansari, N.A., Al-Taani, A.A., Al-Khateeb, F.L., Knutsson, S., 2014. Modeling the risk of groundwater contamination using modified DRASTIC and GIS in Amman-Zerqa Basin, Jordan. Cent. Eur. J. Eng. 4 (3), 264–280.

Al-Taani, A.A., 2013. Seasonal variations in water quality of Al-Wehda Dam North of Jordan and water suitability for irrigation in summer. Arab. J. Geosci. 6.

Al-Taani, A.A., 2014. Trend analysis in water quality of Al-Wehda Dam, north of Jordan. Environ. Monit. Assess. 186 (10), 6223–6239.

Al-Taani, A.A., El-Radaideh, N.M., Al Khateeb, W.M., 2018. Status of water quality in King Talal Reservoir Dam, Jordan. Water Resour. 45 (4), 603–614.

Amy, G., Ghaffour, N., Li, Z., Francis, L., Linares, R.V., Missimer, T., et al., 2017. Membrane-based seawater desalination: present and future prospects. Desalination 401, 16–21.

Arab Countries Water Utilities Association, 2014. Water Utilities Reform in the Arab Region. Lessons Learned and Guiding Principles. ACWUA, Amman, Jordan.

Arjoon, D., Mohamed, Y., Goor, Q., Tilmant, A., 2014. Hydro-economic risk assessment in the eastern Nile River basin. Water Resour. Econ. 8, 16–31.

Awulachew, S.B. (Ed.), 2012. The Nile River Basin: Water, Agriculture, Governance and Livelihoods. Routledge.

Banat, K.M., Howari, F.M., 2003. Pollution load of Pb, Zn, and Cd and mineralogy of the recent sediments of Jordan River/Jordan. Environ. Int. 28 (7), 581–586.

Batayneh, A.T., Al-Taani, A.A., 2016. Integrated resistivity and water chemistry for evaluation of groundwater quality of the Gulf of Aqaba coastal area in Saudi Arabia. Geosci. J. 20 (3), 403–413.

Borgomeo, E., Jagerskog, A., Talbi, A., Wijnen, M., Hejazi, M., Miralles-Wilhelm, F., 2018. The Water-Energy-Food Nexus in the Middle East and North Africa: Scenarios for a Sustainable Future. World Bank.

Dawoud, M.A., Sallam, O.M., 2012. Sustainable groundwater resources management in arid regions: Abu Dhabi case study. In: Environmental Agency Abu Dhabi Report.

Droogers, P., Immerzeel, W., Perry, C., 2009. Application of remote sensing in national water plans: demonstration cases for Egypt, Saudi-Arabia and Tunisia. In: Report Future Water: 80.

Droogers, P., Immerzeel, W.W., Terink, W., Hoogeveen, J., Bierkens, M.F.P., Van Beek, L.P.H., et al., 2012. Water resources trends in Middle East and North Africa towards 2050. Hydrol. Earth Syst. Sci. 16, 3101–3114.

El-Fadel, M., El Sayegh, Y., Abou Ibrahim, A., Jamali, D., El-Fadl, K., 2002. The Euphrates‒Tigris Basin: a case study in surface water conflict resolution. J. Nat. Resour. Life Sci. Educ. 31, 99–110.

El-Radaideh, N., Al-Taani, A.A., Al Khateeb, W.M., 2017a. Characteristics and quality of reservoir sediments, Mujib Dam, Central Jordan, as a case study. Environ. Monit. Assess. 189 (4), 143.

El-Radaideh, N., Al-Taani, A.A., Al Khateeb, W.M., 2017b. Status of sedimentation in King Talal Dam, case study from Jordan. Environ. Earth Sci. 76 (3), 132.

Escalante, E.F., Gil, R.C., San Miguel Fraile, M.Á., Serrano, F.S., 2014. Economic assessment of opportunities for managed aquifer recharge techniques in Spain using an Advanced Geographic Information System (GIS). Water 6 (7), 2021–2040.

Faour-Klingbeil, D., Todd, E.C., 2018. The impact of climate change on raw and untreated wastewater use for agriculture, especially in arid regions: a review. Foodborne Pathog. Dis. 15 (2), 61–72.

Flint, A., Flint, L., Curtis, J., Boesch, C., 2011. A preliminary water balance model for the Tigris and Euphrates river system. In: US Geological Survey, Water Budget Report.

Food and Agriculture Organization of the United Nations, 2014. FAO. The State of Food and Agriculture 2014. United Nations Publications.

Food and Agriculture Organization of the United Nations, 2017. Water for Sustainable Food and Agriculture, A Report Produced for the G20 Presidency of Germany. FAO, 2017.

Ghaffour, N., Missimer, T.M., Amy, G.L., 2013. Technical review and evaluation of the economics of water desalination: current and future challenges for better water supply sustainability. Desalination 309, 197–207.

Hilhorst, B., 2011. Information Products for Nile Basin Water Resources Management. Synthesis Report.

Hofste, R.W., Kuzma, S., Walker, S., Sutanudjaja, E.H., Bierkens, M.F., Kujiper, M.J.M., et al., 2019. Aqueduct 3.0: updated decision-relevant global water risk indicators. In: Technical Note. World Resources Institute.

Howari, F.M., Banat, K.M., 2001. Assessment of Fe, Zn, Cd, Hg, and Pb in the Jordan and Yarmouk river sediments in relation to their physicochemical properties and sequential extraction characterization. Water Air Soil Pollut. 132 (1–2), 43–59.

Howari, F.M., Banat, K.M., 2002. Hydrochemical characteristics of Jordan and Yarmouk river waters: effect of natural and human activities. J. Hydrol. Hydromech. 50 (1), 50–64.

Howari, F.M., McDonnell, R., 2008. The special conditions of water management in the Arab Gulf States: present and future challenges. *First Break*, 26(2).

Howari, F.M., Yousef, A.R., Rafie, S., 2005. Hydrochemical analyses and evaluation of groundwater resources of North Jordan. Water Resour. 32 (5), 555–564.

Hub, N.S.D.S., 2019. Preventing water wars—water diplomacy as a possible driver of stability.

Huh, H.S., Park, C.Y., 2018. Asia-Pacific regional integration index: construction, interpretation, comparison. J. Asian Econ. 54, 22–38.

Jan Stuyfzand, P., Grischek, T., Lluria, M., Jain, R.C., Wang, W., Sanchez-Fernandez, E., et al., 2019. Sixty years of global progress in managed aquifer recharge. Hydrogeol. J. 27 (1).

Jasechko, S., Perrone, D., Befus, K.M., Cardenas, M.B., Ferguson, G., Gleeson, T., et al., 2017. Global aquifers dominated by fossil groundwaters but wells vulnerable to modern contamination. Nat. Geosci. 10 (6), 425–429.

Jeuland, M., 2015. Challenges to wastewater reuse in the Middle East and North Africa. Middle East. Dev. J. 7 (1), 1–25.

Jiménez, B., Asano, T. (Eds.), 2008. Water Reuse: An International Survey of Current Practice, Issues and Needs. IWA, London.

Kabbani, N., Kothari, E., 2005. Youth employment in the MENA region: a situational assessment. In: World Bank, Social Protection Discussion Paper, 534.

Lowder, S.K., Skoet, J., Singh, S., 2014. What do we really know about the number and distribution of farms and family farms in the world? Backgr. Pap. State Food Agric. 2014.

Merrill, J.C., 2008. Water Management and Decision-Making in the Nile Basin: A Case Study of the Nile Basin Initiative.

Mimi, Z.A., Sawalhi, B.I., 2003. A decision tool for allocating the waters of the Jordan River Basin between all riparian parties. Water Resour. Manage. 17 (6), 447–461.

Muhaidat, R., Al-Qudah, K., Al-Taani, A.A., AlJammal, S., 2019. Assessment of nitrate and nitrite levels in treated wastewater, soil, and vegetable crops at the upper reach of Zarqa River in Jordan. Environ. Monit. Assess. 191 (3), 153.

Mukherjee, A., Scanlon, B., Aureli, A., Langan, S., Guo, H., McKenzie, A., 2020. Global Groundwater: Source, Scarcity, Sustainability, Security and Solutions, first ed. Elsevier. ISBN: 9780128181720.

Paasche, Ø., 2007. Climate Change 2007: The Physical Science Basis. Contribution of Working Group I to the Fourth Assessment Report of the IPCC. United Kingdom Cambridge University Press, Cambridge, IPCC [Intergovernmental Panel on Climate Change].

Qadir, M., Bahri, A., Sato, T., Al-Karadsheh, E., 2010. Wastewater production, treatment, and irrigation in Middle East and North Africa. Irrig. Drain. Syst. 24 (1-2), 37–51.

Rosenthal, E., Sabel, R., 2009. Water and diplomacy in the Jordan River basin. Isr. J. Foreign Aff. 3 (2), 95–115.

Salameh, E., Abdallat, G., van der Valk, M., 2019. Planning considerations of managed aquifer recharge (MAR) projects in Jordan. Water 11 (2), 182.

Trieb, F., Müller-Steinhagen, H., 2008. Concentrating solar power for seawater desalination in the Middle East and North Africa. Desalination 220 (1−3), 165−183.

Tuinhof, A., Heederik, J.P., Tuinhof, A., 2002. Management of Aquifer Recharge and Subsurface Storage: Making Better Use of Our Largest Reservoir: Seminar, Wageningen, 18-19 December 2002. Netherlands National Committee for the IAH.

United Nations, Economic and Social Commission for Western Asia, 2013. Inventory of Shared Water Resources in Western Asia. United Nations Publications.

Verner, D. (Ed.), 2012. Adaptation to a Changing Climate in the Arab Countries: A Case for Adaptation Governance and Leadership in Building Climate Resilience. The World Bank.

Vo, P.T., Ngo, H.H., Guo, W., Zhou, J.L., Nguyen, P.D., Listowski, A., et al., 2014. A mini-review on the impacts of climate change on wastewater reclamation and reuse. Sci. Total Environ. 494, 9−17.

Wada, Y., Van Beek, L.P.H., Bierkens, M.F., 2011. Modelling global water stress of the recent past: on the relative importance of trends in water demand and climate variability. Hydrol. Earth Syst. Sci. 15 (12), 3785−3805.

Wagdy, A., AbuZeid, K., 2006. Challenges of implementing IWRM in the Arab region. In: Fourth World Water Forum, Mexico, March 2006.

Waha, K., Krummenauer, L., Adams, S., Aich, V., Baarsch, F., Coumou, D., et al., 2017. Climate change impacts in the Middle East and Northern Africa (MENA) region and their implications for vulnerable population groups. Reg. Environ. Change 17 (6), 1623−1638.

WHO/UNICEF Joint Water Supply and Sanitation Monitoring Programme, 2014. Progress on Drinking Water and Sanitation: 2014 Update. World Health Organization.

Williams, P.A., 2012. Euphrates and Tigris Waters−Turkish-Syrian and Iraqi relations. In: Vajpeyi, D.K. (Ed.), Water Resource Conflicts and International Security: A Global Perspective, Lexington Books, Lanham, MD, p. 43.

Wilson, R., 2012. Water-shortage crisis escalating in the Tigris-Euphrates basin. Future Dir. Int. 28. Available from: http://www.futuredirections.org.au/publication/water-shortage-crisisescalating-in-the-tigris-euphrates-basin/.

World Bank, 2007. Making the Most of Scarcity: Accountability for Better Water Management Results in the Middle East and North Africa. World Bank, Washington, DC.

World Bank, 2017. Beyond Scarcity: Water Security in the Middle East and North Africa. World Bank Group.

WRI, 2005. World Resources Institute (WRI) in collaboration with United Nations Development Programme, United Nations Environment Programme, and World Bank, World Resources 2005: The Wealth of the Poor-Managing.

Young, M., 2015. Climate Change Implications on Transboundary Water Management in the Jordan River Basin: A Case Study of the Jordan River Basin and the Transboundary Agreements Between Riparians Israel, Palestine and Jordan.

Chapter 13

Groundwater scarcity and management in the arid areas in East Africa

Seifu Kebede[1] and Meron Teferi Taye[2]

[1]Seifu Kebede Gurmessa, School of Agricultural Earth and Environmental Sciences, Center for Water Resources Research, University of KwaZulu Natal, Pietermaritzburg, South Africa, [2]IWMI, East Africa and Nile basin Office, Addis Ababa, Ethiopia

13.1 Introduction

Global groundwater is undergoing immense stress in recent times (Mukherjee et al., 2020). Arid areas in the East African region are some of the most geographically disadvantaged regions in the world in terms of their aggregate surface and groundwater availability. Both interannual and intraannual variability in rainfall are remarkable. This hydrologic reality hampers economic growth and perpetuates poverty (Grey and Sadoff, 2007); makes incidence of communal violence more likely (Döring, 2020); hampers the attainment of Sustainable Development Goal (SDG) 6.1 on drinking water (Whitley et al., 2019; World Bank, 2017); triggers migration and internal displacement (Döring, 2020); governs livelihood outcomes such as labor allocation, school attendance and health (MacDonald et al., 2019; Tucker et al., 2014); and reduces the life spans of water schemes (Foster et al., 2018).

Groundwater resources in Africa are often considered as the remedy to bring about overall socioeconomic transformation (Foster et al., 2013; Kebede, 2013) to overcome the current hydrologic difficulty, and meet future demand. The most recent literature shows positive tones. For example, Cobbing and Hiller (2019) in their paper entitled *Waking a sleeping giant: realizing the potential of groundwater in Sub Saharan Africa* suggested that groundwater has the potential to be a foundational resource to support irrigated agriculture, urban and rural water security, and drought resilience across the region. Similarly extrapolating the findings in a district in north-western Ethiopia to the whole Ethiopia and to the Sub-Saharan Africa, Gowing et al. (2020) assert that shallow groundwater represents a neglected opportunity for promoting sustainable small-scale irrigation in Sub-Saharan Africa. MacDonald et al. (2012) show groundwater storage is 100 times greater than the annual renewable surface water flows. Groundwater is expected to be increasingly used as source of reliable water supplies throughout Africa (Giordano, 2009; Macdonald and Calow, 2009). With regard to future climate change, Cuthbert et al. (2019) concludes that future drying climate trends could affect surface water supplies but might not decrease groundwater supplies.

However, groundwater information can be prone to *the problem of survival bias* (Hora et al., 2019), with some difficult hydrogeology environments systematically excluded from the analysis because of lack of readily accessible information. The lack of information is in itself often caused by technical reasons such as the scarcity of groundwater, unreported unsuccessful wells, the high frequency of dry wells, the inaccessibility of the areas because of difficult terrains, the abandonment of water wells because of poor water quality, or because of socioeconomic reason such as violence/conflict, natural hazards, and lack of infrastructure, which deny access to information. Identification and characterization of such unique difficult hydrogeological environments is essential so as to design effective intervention strategies to address the United Nations (UN) SDG of leave no one behind.

This chapter focuses on the vast geographic area in Eastern Africa (Fig. 13.1) where the scarcity of both surface and groundwater is remarkable. The chapter describes the water features of the difficult hydrogeology environments and documents current water supply practices. In these challenging environments (Fig. 13.1), accessing groundwater requires the most *specialist approaches* in groundwater exploration, water supply systems design, and water resources management strategies. Past drillings for groundwater were mostly unsuccessful in this region, with drilling success rate in some cases as low as 10%–25% (Fig. 13.2). The study area covers approximately 1,000,000 km^2. It is home to an

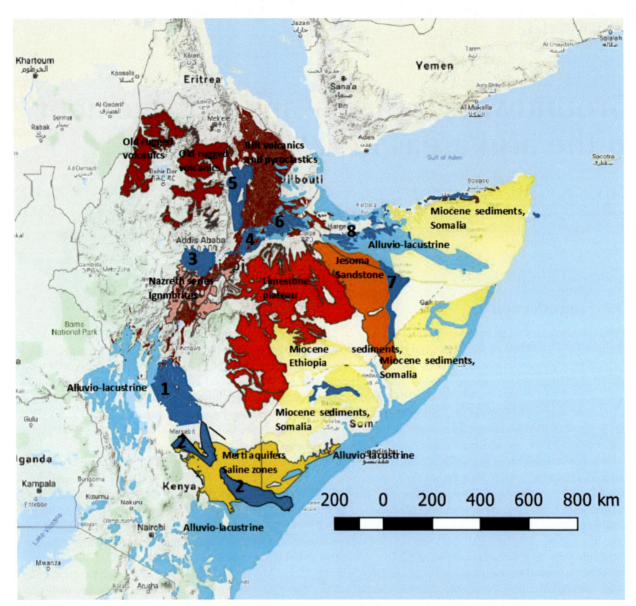

FIGURE 13.1 Distribution of some of the difficult hydrogeologic environments (names of aquifers written on in the figure) and locations of some strategic groundwater sources (shown in numbers 1–8) in the Horn of Africa region. Difficult hydrogeologic environments include (a) Jesoma sandstone, (b) Miocene sediments in Somali and Ethiopia, (c) limestone plateau in Eastern Ethiopia, (d) Nazareth series ignimbrites in central Ethiopia, (e) rift volcanics in Ethiopia, (f) old rugged volcanics in north Ethiopia. Relatively high-yielding, good quality groundwater bearing aquifers are also shown on the map. These aquifers include 1—Bulal transboundary aquifer, 2—Merti transboundary aquifer, 3—Upper Awash Volcanics, 4—Alidegie plain aquifers, 5—Afar stratoid basalts in Ethiopia, 6—Shinile marginal graben sediments, 7—Auradu limestone in Ethioipia, and 8—Auradu limestones in Somalia.

estimated population of 20–30 million people and stretches across six countries (Ethiopia, Somalia, Djibouti, Eritrea, part of North Eastern Sudan, and Kenya). The precarity of water security is highlighted by the prevalence of displaced people. There are 4–6 million refugees and internally displaced people sheltered in refugee camps across the region. Without recognition of the hydrogeological challenges, and adequate action to understand and manage these systems, the population risks to be "left behind" in the push to achieve the SDG 6.1 for water access.

The chapter starts with characterization of the groundwater status of arid lowland and highland regions in East Africa. We then reflect on the current water provision practices in these environments. We explore the potential water management approaches that can help ensure sustainable access to water in these areas. Finally, we propose policy and management approaches for a better water provision in difficult hydrogeology environments.

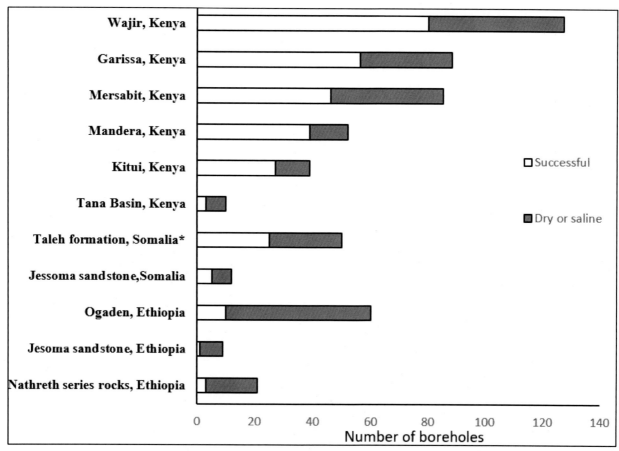

FIGURE 13.2 Reported drilling success (failure) rates in the difficult hydrogeology environments. *Compiled from Ethiopian Institute of Geological Survey, 1973. Geological map of Ethiopia at 1:2000000 scale. Ethiopian Geological Survey, Addis Ababa., Hadwen, P., Aytenfisu, M., Mengesha, G., 1973. Groundwater in the Ogaden. In: Geological Survey of Ethiopia, Report Number 880-551-14. pp. 59; FAO-SWALIM, 2012. Hydrogeological survey and assessment of selected areas in Somaliland and Puntland (Somalia). In: Project Report No. W-20. Nairobi, Kenya; and Krhoda, G., 1989. Groundwater assessment in sedimentary basins of eastern Kenya, Africa. Reg. Charact. Water Qual. 182, 111–124.*

13.2 Typical characteristics of the dryland areas

The dryland areas of Eastern Africa show some distinct characteristics when compared to other arid regions in the West, in Sahel Africa. Unlike the arid regions in Northern Africa, which are known to hold large aquifers with good quality groundwaters (Debie et al., 2019; MacDonald et al., 2012; Sultan et al., 2011; Nijsten et al., 2018), the arid areas in Eastern Africa benefit from few such aquifers. Unlike its equivalent in Northern Africa, the groundwaters in the multilayered sedimentary rocks of Eastern Africa are characterized by a very deep water table and saline water conditions (Hadwen et al., 1973). East Africa is underlain by volcanic rocks forming deep valleys, isolated mountain peaks, and narrow plateaus linked to the East African Rift System. These rugged mountains create a rain shadow effect that reduces moisture availability in the leeward side of the mountains (Maslin et al., 2014).

The climate and geology lead to distinct groundwater quality characteristics. The arid climate, volcanic, and geothermal activities have resulted in high-salinity groundwaters, and the abundance of multiple geogenic contaminants in groundwater (Kebede et al., 2010; Rango et al., 2010; Olaka et al., 2016). One notable contaminant in the rift valley is fluoride. According to Reimann et al. (2003), 86% of wells sampled from the Ethiopian Rift Valley yield waters that fail to pass EU and WHO quality standards set for drinking water when evaluated based by their trace element contents, salinity, and fluoride.

13.3 Typologies of hydrogeology difficulties in arid areas in the East Africa

The difficult hydrogeology regions are predominantly located in arid and semiarid lowland regions in the Horn of Africa region. Semiarid highlands are not exempt from demonstrating the attributes of difficult hydrogeology environment.

13.3.1 Arid volcanic mountains (old rugged volcanics)

Arid isolated volcanic mountains are common in Ethiopia, Kenya, and Eritrea. The high relief, combined with low rainfall, leads to short residence times for groundwater in the environment. The area in north-central Ethiopia (Fig. 13.1) is a known drought hotspot, with recurrent humanitarian crisis (e.g., the 1972, 1984, and 2025/2016 droughts). Groundwater scarcity is mainly attributed to the fact that the oldest Oligo-Miocene volcanic rocks (also known as Ashangie formation) cover the rugged terrain in this environment. This rock formation is known for its very low permeability and storage capacity because of the deep weathering of the rocks (Kebede, 2013). Unlike weathered basement aquifers, deeply weathered volcanic rocks are poor aquifers (Kebede, 2013), as they weather into clay materials. Deep boreholes drilled in this aquifer can turn dry upon drilling or lose their yield over a short period of time. For example, we have observed that out of 12 deep boreholes (up to 600 m) drilled over the last 10 years for water supply of Sekota town (located midway between Mekelle and Bahrdar towns in Fig. 13.1), only 2 were operational in 2019. In these areas, shallow hand pump—fitted wells in relatively humid higher elevation areas are the only viable option for groundwater development.

13.3.2 Rift volcanics and pyroclastics

In the rift valley environment, rather than groundwater availability, poor groundwater quality caused by geogenic pollution is the main reason for water insecurity. The rift valley aquifers in Kenya, Ethiopia, Eritrea, and Djibouti are known for their high natural fluoride levels exceeding the WHO guideline values (Olaka et al., 2016; Reimann et al., 2003; Rango et al., 2009; Kahsai et al., 2001; Kebede et al., 2010). A considerable number of drilled boreholes are abandoned because of high level of fluoride in Tanzania (Chacha et al., 2018), Ethiopia, and Kenya (Fig. 13.2); however, it is difficult to estimate the extent of unsuccessful and abandoned wells due to a lack of reporting. Reimann et al. (2003) show that 86% of water wells randomly sampled in the Ethiopian Rift Valley fail to meet the WHO criteria because of elevated fluoride as well as high levels of other undesirable trace elements and heavy metals.

13.3.3 Nazareth series ignimbrites

These rocks crop out in central Ethiopia both within the rift valley in the (Fig. 13.1) rift margin. Both the storage capacity and permeability of the rocks are very low. The thickness of the unit varies from 100 m to over 600 m. For productive high-capacity wells to be drilled, deep boreholes that penetrate through the Nazareth series rocks and tap the underlying basalt aquifer are required. In the drilling program in the early 1970s more than 70% of deep boreholes drilled to the depth of 300 m in Nazareth series ignimbrites turned to be dry (Fig. 13.2). Our own field experience reveals that well yields from recent successful deep boreholes drilled in the unit rarely exceeds 2 L/s.

13.3.4 Extensive limestone and sandstone plateaus, rocky hills, and plains in arid environments

These rocks are the most difficult of all the hydrogeological environments (Fig. 13.1). The combination of high hydraulic conductivity (e.g., because of karstification) in sedimentary rocks and limited recharge means that the water table is very deep. In the mountainous part, erosion is so deep that recharge water soon finds its way toward the valley floor, to become stream or wadi bed flow. Thick shale and gypsum beds isolate the limestones and sandstone where extensive plains exist. These result in both high-salinity groundwaters (Olago, 2019) and very deep water strike depth (Hadwen et al., 1973). The drilling success rate in these regions is extremely low. Extensive drilling programs in 1970s succeeded only in 20% of the cases (Hadwen et al., 1973); more recent data is lacking; however, low success rates are not uncommon in recent drilling programs. In these environments, drilling by itself is also a major challenge as drilling speeds are as slow as 2 m/day. In an attempt to drill eight boreholes in the Jesoma sandstone (Fig. 13.1) in Ethiopia, only one borehole was successfully drilled to the target depth (Hadwen et al., 1973), while all the others were abandoned because of drilling difficulty. More recently, of 12 boreholes drilled to a depth of 300—400 m in Somalia, 7 did not yield water while 4 have a very low yield and only 1 borehole was successful with a yield of 1 L/s (FAO-SWALIM, 2012). In some areas of Kenya, drilling success rates in this formation has been as low as 30% (Krhoda, 1989) and is reported to be equally low in Somalia (Somaliland) (FAO-SWALIM, 2012). High salinity encountered during drilling resulting in abandonment of boreholes is very common. For instance, in Somaliland out of 50 boreholes drilled in the Miocene sediments (Taleh formation), all have been abandoned because of high salinity. Perched groundwater may occur in these rocks, when local shale beds are encountered. Coping strategies in this environment include water trucking and use of customary water access technologies such as haffirs, bircados, wadi bed excavation, and ella wells.

13.3.5 Extensive loose inland alluvio-lacustrine, inland deltaic, and coastal plain aquifers

Unlike their equivalent elsewhere in the world, in the Horn of Africa, these loose sediments are characterized by complex hydrogeology. While some of these aquifers contain fresh water zones, high salinity and complex patterns in salinity limit the full exploitation of these aquifers for productive uses. The loose sediments are abundantly found in the coastal plains of Somalia, in the rift valley in all the countries of East Africa, in the Omo and Turkana valleys, and in the coastal plains of Kenya. In most of these aquifers, the extent of the high-salinity zones exceeds that of the fresh water zone. Fresh water zones are often restricted to near the axis of river beds, adjacent to wadi beds, near active alluvial fans, and at shallow levels. Groundwater development projects targeting the fresh water zones have been stalled in many places because of the risk of salt water intrusion (Oord et al., 2007; FAO-SWALIM, 2012).

13.3.5.1 Permissible hydrogeology environments

Exceptionally productive and fresh water aquifers exempt from water quality problems and low yields are known to occur in a few localities in this difficult hydrogeology environment. Their occurrence in an otherwise difficult environment makes them strategic aquifers. The widely known and well-described aquifers are shown in Fig. 13.1 and summarized in Table 13.1. (1) The Merti transboundary aquifer that contains fresh water zones, (2) the Upper Awash volcanics in Ethiopia, (3) the Alidegie volcanics, (4) the Afar stratoid basalt in Ethiopia, (5) Shinile aquifers, (6) Jurassic limestone aquifer in Somalia, fresh water zones in the coastal plains of Kenya, and (7) the Auradu–Taleh–Karkar multilayered aquifers in Ethiopia and Somalia. In a few occasions, these aquifers have been used as source of water for the large-scale rural water supply piped networks or have been considered as source of water for interbasin (intercountry) water transfer. One notable example is the Shinile multilayered (alluvial/basalt) aquifer that is currently tapped within Ethiopia to supply drinking water supply for Djibouti.

13.4 Current and past drinking water delivery practices

During the Millennium Development Goals period (2000–15), the emphasis was to quickly increase water supply coverage (Godfrey and Hailemichael, 2017). Water supply planners and policy makers selected low-cost "point source" water supply options such as hand-dug wells, springs, and shallow wells, with limited consideration of sustainability or water quality. While this facilitated strong progress in levels of drinking water access, as these approaches are tailored for simple hydrogeological environments, it has exacerbated inequalities between areas. In Ethiopia, for instance, regions covered by difficult hydrogeological environments continue to have low levels of improved water supply coverage (World Bank, 2017).

TABLE 13.1 Salient features of the high-potential aquifers in the difficult hydrogeology environments.

Aquifer name	Location	Reference in Fig. 13.1	Aquifer type	Current water use
Merti	Kenya	2	Multilayered loose sediments	U + IDP + Irr.
Upper Awash	Ethiopia	3	Fractured volcanics	U + Ind + Irr.
Alidegie	Ethiopia	4	Volcanics + loose sediments	Irr
Afar Stratoid basalts	Ethiopia	5	Fractured basalts	
Shinile aquifers	Ethiopia	6	Loose sediments basalts	Exp.
Bulal basalts	Ethiopia/Kenya	1	Basalts	Liv
Auradu Taleh Karaka formations	Somalia/Ethiopia	7	Limestones + shale	Liv + RW
Jurassic limestones	Ethiopia/Somalia	8	Limestones	Liv + RW

Current water use: *Exp*, Intercountry groundwater export; *IDP*, Water supply for Internally Displaced People Camps; *Ind*, Industrial Water Use; *Irr.*, Irrigation water use; *Liv*, Livestock Watering; *RW*, Rural Water Supply; *U*, Urban Water Supply.

Expanding water access in difficult hydrogeological environments requires a more specialized mapping and resource-intensive approaches. To date, this work has relied on the technical support by international nongovernmental organizations and donors for groundwater exploration, water infrastructure development, and water management. In high-fluoride groundwater-bearing zones of the rift valley, mitigation measures of fluoride problems remained largely the work of NGOs (Gebauer and Saul, 2014). In the most difficult hydrogeological environments, groundwater exploration to support drilling is only a recent phenomenon, and when present is largely supported by international development partners, including the USGS, UNESCO, and UNICEF.

There is no current practice of systematically addressing the water challenges of the difficult hydrogeological environments. Current projects and programs in the regions are restricted to taking a specific technical measure, for example, by drilling deep boreholes. Government programs are dominated by continuing the status quo of finding new sources by drilling more wells in the hope that the next well would be successful.

Without a strong groundwater science to inform drilling, the industry continues to rely on the limited expertise, or luck, of individual hydrogeologists leading sometimes to unintended severe consequences. Examples of the personal toll on hydrogeologists are rarely publicized but understood by those in the industry. For example, hydrogeologists have been imprisoned in Somali region in Ethiopia because of unsuccessful boreholes they sited. In some cases, hydrogeologists have been punched in the face because of unsuccessful boreholes. The maximum innovation around drilling may be its accompaniment with geophysical exploration or changing drilling depth from shallow to deep. Historical evidences show deep drilling is not the panacea in most of these regions. Other innovative measures such as conjunctive surface water groundwater use, increasing longevity of existing schemes, monitoring and water well asset management, and salinity and water quality management are rarely practiced but are being introduced.

13.5 Securing water in difficult hydrogeological environments

The conventional approach of securing water for communities by drilling shallow wells, developing springs, and digging hand-dug wells with little consideration to informed groundwater exploration and development, information, and institution is unlikely to succeed in these difficult hydrogeology regions. In the region finance needs for water source development were greater than budgetary allocations. Their exceptionally difficult hydrogeologic conditions require the most unusual management strategies for finding sustainable water sources, developing the meager available resources, and protecting available sources and resources. Choosing technologies merely based on cost, or resilience to future climate change, or safe sourcing against microbial contamination can be challenged in this environment. Measures to be taken in these environments should encompass integrated technical, financial, capacity building, and institutional measures.

There are signs of progress in addressing the challenges of water insecurity; however so far, they are too small scale to make any difference at national and regional levels. Some of the approaches are summarized in the following sections.

13.5.1 Identifying and protecting viable aquifers

Unlike the conventional approach of sitting wells based on visual field topographic evidences and geophysics (e.g., Vertical Electrical Sounding), finding sustainable groundwater sources in this environment requires the most advanced and integrated methodology, including remote sensing, integrated geophysics, field geological and hydrogeological mapping, water quality surveys, isotope hydrology, stratigraphic and geomorphologic analysis, and inventory of all successful and failed water schemes, so as to find sustainable source of groundwater even for rural water supply. There is evidence that ongoing advanced groundwater exploration efforts have yielded positive outcomes in finding sustainable sources (Godfrey et al., 2019; Olago, 2019) and in identifying suitable geologic conditions for groundwater occurrence (https://www.unicef.org/ethiopia/stories/combining-groundwater-mapping-capacity-building; https://en.acaciawater.com/nw-29143-7-3652306/nieuws/improving_drilling_success_rate_in_the_afar_and_tigray_regions.html?page = 1). The most successful proven mapping project is the groundwater exploration in arid areas in southern Ethiopia bordering Kenya and the subsequent identification of a strategic Bulal aquifer (Fig. 13.1), which is now under consideration for development for piped rural multicommunity scheme. The other strategic aquifers marked in Fig. 13.1 have been also identified through professional and integrated groundwater mapping programs. Although the scope for finding additional high-potential aquifers is meager, developing use and management strategies for the strategic aquifers is essential.

13.5.2 Adaptation of customary water schemes

Pastoralist communities in some difficult hydrology environments have developed, over hundred years, sophisticated water and pasture management strategies, allowing them to overcome severe droughts. These customary water management structures, such as very large diameter ella wells (aka singing wells in Borena region of Ethiopia), have not been given government policy support. Their important role in delivering reliable drinking water sources is not recognized in JMP and National WASH inventories. However, they remain the vital water system in the pastoralist areas. Since the 1970s these traditional water management strategies have been eroded by the introduction of water delivery approaches designed for sedentary communities. The strength of the customary water management practices is their capacity to enhance resilience of water sources by designs that match the low yields of the aquifers. For example, the ella wells are large diameter wells (around 2000 m^2 surface area; and depth of 20–30 m). The rim of the well is cut and a gentle slope is formed so that the cattle reach the vicinity of the water table for direct access for drinking. The large diameter of the well allows the pastoralists to deepen their wells far below the water table and to enhance yields of the normally low-yielding aquifers. Apart from their role as sources of water, the traditional water sources have other pivotal social function. Disregarding these schemes because of the risk of microbial contamination and human-induced pollution will put large pastoral community under risk of water insecurity. Methods to improve the schemes management for better water quality outcome need to be sought.

13.5.3 Enhancing water availability by water harvesting

Water harvesting programs that aim to enhance water availability have been in practice in the various geophysical setting (arid and humid areas alike) in the region. The implementation approaches, the scale of the programs, and technological designs of the programs vary greatly. Some designs are meant to take physical water conservation measures (e.g., terraces) on degraded land to conserve soil and water (Debie et al., 2019; Erkossa et al., 2018; Kosmowski, 2018; Yigzaw et al., 2019). Other designs construct weirs to reduce runoff and erosion during sporadic flash floods to retain eroded soil and spread incoming flood water over the valley floor to allow much water to infiltrate into the soils (Mehari et al., 2011; van Steenbergen et al., 2011). Others construct sand dams—reinforced wall on the bedrock of seasonal river beds so that the sand carried by the river is deposited behind the wall—so that water is stored in the sands (Quinn et al., 2019). Some others construct microdams and ponds to retain runoff (Teshome et al., 2010; Berhane et al., 2016).

Albeit the numerous concern, including lack of appropriate design, cost, scalability, poor water quality, low sustainability, leakage, and siltation, there are growing evidence that water harvesting practices increase water security (Edward and Tiruneh Claudia Ringler, 2019; Ertsen and Hut, 2009). Two main gaps in the current practices are (1) none of the current technologies, except the sand dams, are meant to increase drinking water security directly and (2) a particular intervention technology is suitable only under given environmental conditions, limiting feasibility for upscaling in other areas. The positive outcome of increased water security in relation to soil water harvesting practice, for example, in Ethiopia, is an indirect inadvertent outcome, as the current designs are not meant to intentionally enhance groundwater storage.

13.5.4 Water quality management

High salinity, fluoride, and other undesirable elements of geogenic origin cause abandonment of considerable amount of drilled wells yielding water. This problem is the most prevalent in the alluvio-lacustrine sediments covering vast area in the region. Current practice of water quality management includes shifting borehole sites in the hope of finding good quality groundwater, defluoridation, and desalination. Shifting borehole site and drilling new boreholes when a drilled borehole turns saline or of poor quality is the most widely practiced method of water quality managment. Selection of low-salinity zones are usually made based on the geophysical surveys, as the method allows to delineate high-salinity groundwater environments. However, in many instances, it is difficult to identify a saline zone from saturated fine-grained sediments. As the result, many boreholes drilled in the alluvio-lacustrine sediments have been abandoned because of high salinity. Water treatment technologies (e.g., desalination and defluoridation) are too expensive for communities to afford the construction and running costs. Currently such technologies are being introduced via international development partner organization and NGOs. Desalination technologies can be applicable, able to draw on large volumes of saline water, for example, in the rift valley, but at high cost, and potential impacts on surface water if wastewater is not carefully managed. Lack of awareness and capacity is currently limiting the potential application of other salinity management strategies such as safe sourcing. Safe sourcing is a hydrogeological approach of finding low-salinity groundwaters in otherwise high-salinity groundwater environments. This can be done after detailed mapping of salinity pattern and detecting low-salinity environment. Low-salinity groundwater can be found near streams and wadis, on apexes of alluvial fans, and in underlying bedrocks. Safe sourcing has shown some promising results and can be scaled up for salinity management as well as in fluoride management. This requires however detailed understanding of hydrogeology and patterns in groundwater water quality.

13.5.5 Long distance and interbasin water transfer

Water transfer, transferring groundwater from where it is available to where water is needed by long distance pipe network, is technologically attractive, but is politically challenging, and poses management challenges. The current rural water supply water management approaches, which largely rely on community-based management of water sources, are not suitable to manage complex schemes of this sort. There have been some past successful efforts of interbasin water transfers, mainly in Ethiopia, that have increased water security in water scarce areas in the Ethiopian Rift Valley. However, past attempts have in many places stalled because of complex technical, political, or capacity reasons. An attempt to transfer water from the Merti aquifers (Fig. 13.1) to supply the town of Wajir 100 km north of the aquifer has been envisaged for more than a decade now but has never materialized. A nearly 2000 km pipe network to tap the Bulal aquifer (Fig. 13.1) to supply the region east of the aquifer has never been functional for nearly 10 years. The Shinile aquifer (Fig. 13.1) is currently pumped to supply the rural areas in the vicinity and the water is being piped to supply Djibouti city but faces political challenges.

13.5.6 Investing in sustainability of existing systems

Sustaining existing system means sustaining the groundwater resources through taking management measures and sustaining the schemes (mainly boreholes) through affordable preventive and corrective maintenance.

In these difficult hydrogeological environments, viable aquifers (strategic aquifers listed in Fig. 13.1) need to be protected to ensure the sustainability of drinking water supplies. The high-yielding, good-quality aquifers exist, but their longevity is threatened if there is insufficient science and monitoring to define sustainable abstraction rates. As there is growing demand for food production, some of these aquifers are already targeted for irrigation and industrial water abstraction (Table 13.1). The increasing competition on these resources will pose greater drinking water security challenge.

Affordable operation and maintenance of the hard-won existing schemes (mainly boreholes) is critical to sustain water access and increase drought resilience. In addition to affordability, repairs of existing schemes are too often delayed because of a lack of information on service function, challenges with budgeting and revenue collection, limited technical capacity, and the priority given to installing new infrastructure at the expense of maintenance (USAID, 2018). While innovative, sensor (e.g., GSM mobile phone sensor) based approaches for monitoring of functionality of water schemes have been tested in the region through development partners interventions (as reported by USAID 2018 in https://www.usaid.gov/sites/default/files/documents/1860/Ethiopia_Lowland-WASH-Sensor-Brief_FINAL.pdf), affordable maintenance mechanism has been challenging. An ongoing research by the REACH research program of the Oxford University is testing an affordable maintenance model called FundiFix in Kenya and other countries. The FundiFix model has the potential to be out scaled in the difficult hydrogeology environments. Led by local entrepreneurs and powered by Africa's mobile network, the FundiFix model offers a performance-based approach working with government, communities and investors to keep water flowing. The details of the FundiFix model can be found at: https://reachwater.org.uk/wp-content/uploads/2016/11/Fundifix-booklet-WEB.pdf.

13.6 Policy and practice implication

Past and ongoing response to climate disasters in the region focuses on humanitarian response (Whitley et al., 2019) rather than on long-term development. The hydrology and hydrogeology difficulty of East African regions means that water−centered, long-term development requires special financial, institutional, and technical approaches.

The current discussion about financing the WASH sector focuses on how to balance capital expenditure on construction and operation and maintenance. While giving equal emphasis on financial resources to the operation and maintenance can lead to sustaining the hard-won successful water schemes, the difficult environments require capital-intensive engineering interventions (e.g., interbasin groundwater transfer through piped schemes, managed aquifer recharge, sand dams, desalination technologies, salinity management, and conjunctive surface water and groundwater use) if the SDG agenda of leave no one behind is to be achieved. Sufficient, reliable, and safe water can not be provided based on simplistic approach of selecting WASH technologies, which often targets a single problem (e.g., susceptibility of scheme to bacteria contamination, fragility in drought and flood events, cost, or institutional capacity to manage schemes). Technology choice and engineering measures to reach difficult environments should rather be outcome driven and should consider water access in the local context. To ensure universal access the most specialist (e.g., integrated exploration of the groundwater resources, protection of resources, specialized engineering, monitoring of information system, and affordable professionalized maintenance models) approaches that account for hydrological difficulty and the spatial heterogeneity of the problem need to be considered.

Countries within the Horn of Africa and IGAD region need develop strategies for the protection of the viable aquifer resources.

Acknowledgment

This document is supported by the REACH program funded by UK Aid from the UK Department for International Development (DFID) for the benefit of developing countries (Aries Code 201880). However, the views expressed and information contained in it are not necessarily those of or endorsed by DFID, which can accept no responsibility for such views or information or for any reliance placed on them. Katrina Charles from the REACH Oxford project provided valuable constructive comments on the draft manuscript.

References

Berhane, G., et al., 2016. Overview of micro-dam reservoirs (MDR) in Tigray (northern Ethiopia): challenges and benefits. J. Afr. Earth. Sci 123, 210−222. Available from: https://doi.org/10.1016/J.JAFREARSCI.2016.07.022.

Chacha, N., et al., 2018. Hydrogeochemical characteristics and spatial distribution of groundwater quality in Arusha well fields, Northern Tanzania. Appl. Water Sci. 8 (4), 1−23. Available from: https://doi.org/10.1007/s13201-018-0760-4.

Cobbing, J., Hiller, B., 2019. Waking a sleeping giant: Realizing the potential of groundwater in Sub-Saharan Africa. World Dev. 122, 597−613. Available from: https://doi.org/10.1016/j.worlddev.2019.06.024.

Cuthbert, M., et al., 2019. Observed controls on resilience of groundwater to climate variability in sub-Saharan Africa. Available from: https://doi.org/10.1038/s41586-019-1441-7.

Debie, E., Singh, K.N., Belay, M., 2019. Effect of conservation structures on curbing rill erosion in micro-watersheds, northwest Ethiopia. Int. Soil Water Conserv. Res. 7 (3), 239−247. Available from: https://doi.org/10.1016/j.iswcr.2019.06.001.

Döring, S., 2020. Come rain, or come wells: How access to groundwater affects communal violence. Pol. Geogr. 76, 102073. Available from: https://doi.org/10.1016/j.polgeo.2019.102073.

Edward, K., Tiruneh Claudia Ringler, S., 2019. Sustainable Land Management and its Effects on Water Security and Poverty Evidence from a Watershed Intervention Program in Ethiopia.

Erkossa, T., Williams, T.O., Laekemariam, F., 2018. Integrated soil, water and agronomic management effects on crop productivity and selected soil properties in Western Ethiopia. Int. Soil Water Conserv. Res. 6 (4), 305−316. Available from: https://doi.org/10.1016/j.iswcr.2018.06.001.

Ertsen, M., Hut, R., 2009. Two waterfalls do not hear each other. Sand-storage dams, science and sustainable development in Kenya. Phys. Chem. Earth, A/B/C 34 (1−2), 14−22. Available from: https://doi.org/10.1016/J.PCE.2008.03.009.

Ethiopian Institute of Geological Survey, 1973. Geological map of Ethiopia at 1:2000000 scale. Ethiopian Geological Survey, Addis Ababa.

FAO-SWALIM, 2012. Hydrogeological survey and assessment of selected areas in Somaliland and Puntland (Somalia). In: Project Report No. W-20. Nairobi, Kenya.

Foster, S., et al., 2013. Groundwater—a global focus on the "local resource". Curr. Opin. Environ. Sustain. 5 (6), 685−695. Available from: https://doi.org/10.1016/J.COSUST.2013.10.010.

Foster, T., et al., 2018. Risk factors associated with rural water supply failure: a 30-year retrospective study of handpumps on the south coast of Kenya. Sci. Total Environ. 626, 156−164. Available from: https://doi.org/10.1016/j.scitotenv.2017.12.302.

Gebauer, H., Saul, C.J., 2014. Business model innovation in the water sector in developing countries. Sci. Total Environ. 488−489, 512−520. Available from: https://doi.org/10.1016/J.SCITOTENV.2014.02.046.

Giordano, M., 2009. Global groundwater? Issues and solutions. Annu. Rev. Environ. Res. 34 (1), 153−178. Available from: https://doi.org/10.1146/annurev.environ.030308.100251.

Godfrey, S., Hailemichael, G., 2017. Life cycle cost analysis of water supply infrastructure affected by low rainfall in Ethiopia. J. Water Sanit. Hyg. Dev. 7 (4), 601−610. Available from: https://doi.org/10.2166/washdev.2017.026.

Godfrey, S., Hailemichael, G., Serele, C., 2019. Deep Groundwater as an Alternative Source of Water in the Ogaden Jesoma Sandstone Aquifers of Somali. *Water*, Available from: https://doi.org/10.3390/w11081735.

Gowing, J., et al., 2020. Can shallow groundwater sustain small-scale irrigated agriculture in sub-Saharan Africa? Evidence from N-W Ethiopia. Groundwater Sustain. Dev. 10, 100290. Available from: https://doi.org/10.1016/j.gsd.2019.100290.

Grey, D., Sadoff, C.W., 2007. Sink or Swim? Water security for growth and development. Water Policy 9 (6), 545−571. Available from: https://doi.org/10.2166/wp.2007.021.

Hadwen, P., Aytenfisu, M., Mengesha, G., 1973. Groundwater in the Ogaden. Geological Survey of Ethiopia, Report Number 880-551-14. pp. 59.

Hora, T., Srinivasan, V., Basu, N.B., 2019. The groundwater recovery paradox in South India. Geophys. Res. Lett. 46 (16), 9602–9611. Available from: https://doi.org/10.1029/2019gl083525.

Kahsai, F., Fisahatsion, A., Asmellash, M., 2001. Fluoride in Groundwater in Selected Villages in. pp. 169–177.

Kebede, S., 2013. Groundwater in Ethiopia: Features, Vital Numbers and Opportunities. Springer, Berlin, ISBN 978-3-642-30390-6, 297pp.

Kebede, S., Travi, Y., Stadler, S., 2010. Groundwaters of the Central Ethiopian Rift: diagnostic trends in trace elements, $d^{18}O$ and major elements. Environ. Earth Sci. 61, 1641–1655. doi:10.1007/s12665-010-0479-5.

Kosmowski, F., 2018. Soil water management practices (terraces) helped to mitigate the 2015 drought in Ethiopia. Agric. Water Manage. 204, 11–16. Available from: https://doi.org/10.1016/j.agwat.2018.02.025.

Krhoda, G., 1989. 'Groundwater assessment in sedimentary basins of eastern Kenya, Africa'. Reg. Charact. Water Qual. 182, 111–124.

MacDonald, A.M., et al., 2012. Quantitative maps of groundwater resources in Africa. Environ. Res. Lett. 7 (2). Available from: https://doi.org/10.1088/1748-9326/7/2/024009.

MacDonald, A.M., et al., 2019. Groundwater and resilience to drought in the Ethiopian highlands. Environ. Res. Lett. 14 (9), 095003. Available from: https://doi.org/10.1088/1748-9326/ab282f.

Macdonald, A.M., Calow, R.C., 2009. Developing groundwater for secure rural water supplies in Africa. Desalination 248, 546–556. Available from. Available from: http://nora.nerc.ac.uk/8460/1/AMM_desalination.pdf.

Maslin, M.A., et al., 2014. East African climate pulses and early human evolution. Quat. Sci. Rev. 101, 1–17. Available from: https://doi.org/10.1016/j.quascirev.2014.06.012.

Mehari, A., Van Steenbergen, F., Schultz, B., 2011. Modernization of spate irrigated agriculture: a new approach. Irrig. Drain. 60 (2), 163–173. Available from: https://doi.org/10.1002/ird.565.

Mukherjee, A., Scanlon, B., Aureli, A., Langan, S., Guo, H., McKenzie, A., 2020. Global Groundwater: Source, Scarcity, Sustainability, Security and Solutions, first ed. Elsevier, ISBN: 9780128181720.

Nijsten, G.J., et al., 2018. Transboundary aquifers of Africa: review of the current state of knowledge and progress towards sustainable development and management. J. Hydrol. Regional Stud. 20, 21–34. Available from: https://doi.org/10.1016/j.ejrh.2018.03.004.

Olago, D.O., 2019. Constraints and solutions for groundwater development, supply and governance in urban areas in Kenya. Hydrogeol. J. 27 (3), 1031–1050. Available from: https://doi.org/10.1007/s10040-018-1895-y.

Olaka, L.A., et al., 2016. Groundwater fluoride enrichment in an active rift setting: Central Kenya Rift case study. Sci. Total Environ. 545–546, 641–653. Available from: https://doi.org/10.1016/j.scitotenv.2015.11.161.

Oord, A., Collenteur, R., Tolk, L., 2007. Hydrogeological Assessment of the Merti Aquifer.

Quinn, R., Rushton, K., Parker, A., 2019. An examination of the hydrological system of a sand dam during the dry season leading to water balances. J. Hydrol. X 4, 100035. Available from: https://doi.org/10.1016/j.hydroa.2019.100035.

Rango, T., et al., 2009. Hydrogeochemical study in the Main Ethiopian Rift: new insights to the source and enrichment mechanism of fluoride. Environ. Geol. 58 (1), 109–118. Available from: https://doi.org/10.1007/s00254-008-1498-3.

Rango, T., et al., 2010. Geochemistry and water quality assessment of central Main Ethiopian Rift natural waters with emphasis on source and occurrence of fluoride and arsenic. J. Afr. Earth. Sci. 57 (5), 479–491. Available from: https://doi.org/10.1016/j.jafrearsci.2009.12.005.

Reimann, C., et al., 2003. Drinking water quality in the Ethiopian section of the East African Rift Valley I – Data and health aspects. Sci. Total Environ. 311 (1–3), 65–80. Available from: https://doi.org/10.1016/S0048-9697(03)00137-2.

van Steenbergen, F., et al., 2011. Status and Potential of Spate Irrigation in Ethiopia. Water Res. Manage. 25 (7), 1899–1913. Available from: https://doi.org/10.1007/s11269-011-9780-7.

Sultan, M., et al., 2011. Modern recharge to fossil aquifers: Geochemical, geophysical, and modeling constraints. J. Hydrol. 403 (1–2), 14–24. Available from: https://doi.org/10.1016/j.jhydrol.2011.03.036.

Teshome, A., Adgo, E., Mati, B., 2010. Impact of water harvesting ponds on household incomes and rural livelihoods in Minjar Shenkora district of Ethiopia. Ecohydrol. Hydrobiol. 10 (2–4), 315–322. Available from: https://doi.org/10.2478/V10104-011-0016-5.

Tucker, J., et al., 2014. Household water use, poverty and seasonality: Wealth effects, labour constraints, and minimal consumption in Ethiopia. Water Res. Rural Dev. 3, 27–47. Available from: https://doi.org/10.1016/J.WRR.2014.04.001.

Whitley, L., et al., 2019. A framework for targeting water, sanitation and hygiene interventions in pastoralist populations in the Afar region of Ethiopia. Int. J. Hyg. Environ. Health 222 (8), 1133–1144. Available from: https://doi.org/10.1016/j.ijheh.2019.08.001.

World Bank, 2017. Maintaining the Momentum While Addressing Service Quality and Equity. doi:10.1596/30562.

Yigzaw, N., et al., 2019. Data for the evaluation of irrigation development interventions in Northern Ethiopia. Data Brief 25, 104342. Available from: https://doi.org/10.1016/j.dib.2019.104342.

Further reading

Acworth, R.I., 1987. The development of crystalline basement aquifers in a tropical environment. Q. J. Eng. Geol. 20 (4), 265–272. Available from: https://doi.org/10.1144/gsl.qjeg.1987.020.04.02.

Deyassa, G., et al., 2014. Crystalline basement aquifers of Ethiopia: their genesis, classification and aquifer properties. J. Afr. Earth. Sci. 100, 191–202. Available from: https://doi.org/10.1016/j.jafrearsci.2014.06.002.

Chapter 14

Global geogenic groundwater pollution

Poulomee Coomar[1] and Abhijit Mukherjee[1,2]

[1]*Department of Geology and Geophysics, Indian Institute of Technology (IIT) Kharagpur, Kharagpur, India,* [2]*Applied Policy Advisory for Hydrosciences (APHA) group, School of Environmental Science and Engineering, Indian Institute of Technology (IIT) Kharagpur, Kharagpur, India*

14.1 Introduction

Natural subsurface water is never devoid of impurities and typically contains salts, organic and inorganic chemicals, sediments, and microorganisms either in solution or suspension. Pollution of water resources occurs when the concentration of one or more of these constituents increases to levels that it becomes hazardous for plants, animals, human life, and to the environment, thereby delimiting its use for any practical purpose (Morris et al., 2003). Groundwater has been sustaining civilizations from historical times since it is buffered against local climatic changes especially in arid regions where other sources of freshwater are scarce. However, in recent times, rampant use of surface-water bodies for the disposal of anthropogenically generated waste has rendered them unusable without proper treatment, and their use in regions without proper infrastructure has often resulted in the repeated occurrence of water-borne diseases in rural and urban communities of many underdeveloped and developing countries. Therefore, in the last few decades, inhabitants and administrative authorities of many countries worldwide have turned to groundwater to meet their domestic and industrial needs. However, its use, especially for drinking and agricultural purposes, has often been limited not only by the presence of various anthropogenic but many naturally occurring geogenic contaminants, creating an immense pressure on global groundwater (Mukherjee, 2018; Mukherjee et al., 2020). Such contaminants are often nonpoint sourced and can be traced back to geological formations. Though elevated concentrations above the permissible limit of these elements are well known from many aquifers worldwide, a plot of their global distribution suggests that their occurrence follows a distinctive pattern. Thus the aim of the present chapter is to elucidate how the conjunction of global-scale phenomenon (plate tectonics) which controls the distribution of their primary sources and local-scale forcings (geochemical conditions prevailing in the aquifer, residence time, soil, ambient climate) which controls their fate and transport gives rise to the observed geographical distribution of few of these selected geogenic groundwater pollutant.

14.2 Global distribution of geogenic groundwater pollutants

14.2.1 Arsenic

Arsenic (As) is a trace element and is found in the Earth's crust at a mean concentration of 1.8 mg/kg (Matschullat, 2000; Nordstrom, 2002). It is the 20th most abundant naturally occurring element and occurs pervasively in all spheres of the planet (Cullen and Reimer, 1989). It can enter the human body from various sources (air, food, drinking water), of which drinking water probably poses the greatest health threat. Though concentrations in freshwater can vary over four orders of magnitude (<0.5 to >5000 μg/L), the greatest concentrations are most likely to occur in groundwater due to extended water–rock interaction and the conducive environment that aquifers provide for its accumulation and mobilization (Smedley and Kinniburgh, 2002). In recent times the first cases of extensive arsenicosis had been reported about half a century ago from Argentina (Bado, 1939) and Taiwan (Tseng et al., 1968). However, it became a topic of extensive global research after its discovery in the Bengal Basin of eastern India and Bangladesh, where more than 40 million people are exposed to high groundwater As in the 1980s and 1990s. Gradually, over the past three decades elevated concentration of groundwater As above the WHO guideline value of 10 μg/L [World Health Organization (WHO), 2001] has been detected in more than forty geological provinces (Mukherjee et al., 2008a, 2009, 2014, 2019)

(Fig. 14.1), endangering more 200 million people worldwide (Mukherjee et al., 2009). In most natural systems, As is released to the groundwater by a combination of natural geologic process. Though anthropogenic activities such as burning of fossil fuels, smelting, wood preservation, glass, and paper making [United States Environmental Protection Agency (US EPA), 1998a], application of arsenical agricultural products [pesticides, herbicides, crop desiccant products (Li et al., 2016; Bencko and Yan Li Foong, 2017)], and mining (Komnitsas et al., 1995; Williams et al., 1996, 1997; Armienta et al., 2005) can cause localized increase in As levels, it is the natural mobilization by geochemical processes that tends to have a more widespread effect.

Arsenic has four oxidation states—arsenide (−II), elemental arsenic (0), arsenite (+III), and arsenate (+V), of which the protonated oxyanions of arsenite ($H_3AsO_3^0$, $H_2AsO_3^-$, $HAsO_3^{3-}$) and arsenate ($H_3AsO_4^0$, $H_2AsO_4^-$, $HAsO_4^{2-}$, AsO_4^{3-}) are most commonly found in freshwater (Smedley and Kinniburgh, 2002). The degree of protonation depends on the Eh and pH conditions that play an important role in controlling the behavior of As in aqueous systems. Oxic waters, which are characterized by a dominance of pentavalent species, $H_2AsO_4^-$, $HAsO_4^{2-}$ ions, dominate over the pH range of most groundwaters, thereby immobilizing As(V) by adsorption onto Fe, Mn, Al (oxy)hydroxides, clay minerals, and organic matter. However, in reducing conditions that favor trivalent forms, the uncharged $H_3AsO_3^0$ molecule is the main species at circum-neutral pH thus limiting the adsorption of As(III) (Clifford and Ghurye, 2002). Moreover, due to slow redox kinetics of As(III) oxidation, its species may often linger under oxidizing conditions (Boyle et al., 1998). The adsorbed As can be mobilized into the groundwater system in the pH range of 6.5−8.8 under both oxidizing and reducing conditions, thereby making it more deadly than other oxyanions that stay conveniently immobilized under reducing conditions. Thioarsenic and organic forms are rare and occur mostly in sulfidic and waters rich in microorganisms, respectively. Dissolved As(−III) and As(0) species are rarely encountered in natural waters (Mandal and Suzuki, 2002).

Arsenic is a chalcophile element and often tends to associate with sulfur (S) in zones of sulfide mineralization, organic-rich black shales, arc lithologies, and hydrothermal fluids (Mandal and Suzuki, 2002; Hattori and Guillot, 2003). However, it has often been observed that in most cases, As-enriched groundwaters are hosted by seemingly ordinary sedimentary aquifers, with such As-enriched lithologies to be found nowhere in their vicinity. Since, the average concentration of As in the upper crust if leached under favorable biogeochemical conditions is sufficient to produce toxic levels of groundwater As (Korte and Fernando, 1991; Stüben et al., 2003; Saunders et al., 2005; Guilliot and

FIGURE 14.1 Global distribution of groundwater provinces with naturally occurring as concentration greater than 10 μg/L (following Smedley and Kinniburgh, 2002; Mukherjee et al., 2008a, 2009, 2014, 2019; Ravenscroft et al., 2009).

Charlet, 2007) and not all aquifers are As enriched, it can be assumed that elevated levels of labile As in the provenance region of aquifers increase the possibility of As accumulation in the aquifer solids and associated groundwater (Mukherjee et al., 2019). A plot of global distribution of high groundwater As provinces shows that they are invariably located in large sedimentary basins in the vicinity of young orogens or active or ancient magmatic arcs. This observation has led Mukherjee et al. (2014, 2019) to propose that the source of groundwater As primarily lies in magmatic arcs located in zones of subduction and is eventually transported to downstream aquifers. Their proposal is based on the survey of As concentration from various rock types (Hattori and Guillot, 2003, and references therein) from around the world, which has shown that a close relationship exists between primary solid-phase As and arc magma. According to their model, high-temperature aqueous fluids released from the subducting oceanic plate (Tatsumi, 1989) metasomatize the depleted continental lithospheric wedge, thereby initiating the generation of hydrous arc magma. On melting the incompatible As in the mantle is preferentially partitioned into the partial melt making it rich in As (Noll et al., 1996). Sedimentary inputs from the trench can also contribute to its As content (Plank and Langmuir, 1993). Arsenic content is further enhanced as the rising partial melt assimilates surrounding rock material and scavenges As from the lower and middle continental crust and arc basement. Upon extrusion the hydrous arc magma deposits volcanic rocks and ash beds that tend to have the highest As content and act as immediate provenance of As rich detritus which are eventually deposited in adjacent foreland and retro-arc basins. Chemical weathering of metastable high-temperature mafic minerals facilitates the release of As. The liberated As is typically adsorbed and/or coprecipitated onto more labile secondary phases, which act as an immediate source of groundwater As in these foreland basin aquifers and are released into the liquid phase under both oxidizing and reducing conditions (Smedley and Kinniburgh, 2002; Mukherjee et al., 2014, 2018). With continued convergence over geological time, several generations of foreland basins form and merge with the main orogenic belt. As the agents of erosion cut through the orogenic belt, they erode these paleo-foreland deposits along with any further exposed arc rocks and deposits them in the most recent foreland basin which acts as aquifers for present-day groundwater exploitation.

The redox ubiquity of As has led Smedley and Kinniburgh (2002) to classify As polluted aquifers into two broad types on the basis of their hydrogeochemical condition and the redox reactions involved in the release and mobilization of As from aquifer solids. In both cases the As is initially adsorbed onto the surfaces of Fe, Mn (oxy)hydroxides. However, in one case the As is released on dissolution of the adsorbent phase, while in the other the As is simply desorbed leaving the solid carrier phase intact. Both these settings are usually characterized by relatively young sediments deposited in flat low lying areas, where groundwater recharge and flow is slow.

The first mechanism is characteristic of strongly reducing anoxic waters at near-neutral pH, where As is liberated by microbially assisted reductive dissolution of Fe, Mn (oxy)hydroxides in the presence of organic matter. Fe(II), Mn(II), As(III), and HCO_3^- are relatively more abundant and concentration of oxidized species (SO_4^{2-}, NO_3^-) is typically small in these waters. Moreover, large concentrations of HCO_3^-, PO_4^{3-}, silicate, and natural organic matter can promote As desorption due to competitive ion exchange. Such waters are most commonly reported from humid regions such as the riverine and deltaic aquifers of major Himalayan Rivers [the Ganges−Brahmaputra−Meghna in India, Nepal, and Bangladesh (Bhattacharya et al., 2002; Ahmed et al., 2004; Tandukar et al., 2005; Mukherjee et al., 2008b, 2012; Mukherjee and Fryar, 2008; Brikowski et al., 2014; Kumar et al., 2018; Verma et al., 2016, 2019; Coomar et al., 2019); Irrawaddy in Myanmar (Smedley, 2005; van Geen et al., 2014); Red in Vietnam (Berg et al., 2001, 2008; Postma et al., 2007, 2010, 2017); and Mekong in Cambodia (Berg et al., 2007; Buschmann et al., 2007), Vietnam (Buschmann et al., 2008), and Laos (Chanpiwat et al., 2011)]. Examples of other groundwater provinces where such processes are operating include the Choushui River alluvial fan of Taiwan (Liu et al., 2006); Yellow River Basin, China (Luo et al., 1997; Lin et al., 2002; Smedley et al., 2003); the Po River Basin of Italy (Rotiroti et al., 2014); and the Great Hungarian Plain of Hungary (Varsányi and Kovács, 2006) and Romania (Rowland et al., 2011).

The second mechanism is triggered by an increase in pH (typically >8.5) due to weathering and ion exchange reactions. Examples of groundwater provinces where this process is operating include the Chaco-Pampean plains of Argentina, the Bolivian Altiplano (Coomar et al., 2019, and references therein), the Indus River plains of Pakistan (Podgorski et al., 2017), the Basin and Range Province (Robertson, 1989), and the Southern High Plains aquifer, United States (Scanlon et al., 2009), the Madrid and Duero Basin, Spain (Hernández-García and Custodio, 2004; Gómez et al., 2006a) where As and other oxyanion forming elements (B, Mo, Se, V, and U) are desorbed from iron oxides under alkaline conditions. Moreover, in such setting an increase in Na content of groundwater further facilitates As release by counter-ion effect (Scanlon et al., 2009; Raychowdhury et al., 2014; Coomar et al., 2019). Though this mechanism is considered to be typical of inland or closed basins in arid and semiarid regions with high evaporation rates, As released via pH-dependent desorption has also been reported from regions experiencing a more temperate, humid climate such as the bedrock aquifers of New Hampshire, United States (Peters and Blum, 2003), British Columbia (Boyle et al.,

1998), and Nova Scotia and New Brunswick, Canada (Bottomley, 1984). Arsenic can also be released from labile secondary phases in aquifers experiencing variable redox conditions such as the Tulare Basin, San Joaquin Valley, United States (Fujii and Swain, 1995) and the Datong Basin of China (Guo and Wang, 2005), where high As concentrations occur in both reducing and oxidizing conditions.

Moreover, in such settings the exolution of late-stage volcanic fluids and heating of infiltrating recharge usually sets up a deep circulating hydrothermal system rich in As. Such fluids upon travel through preferential pathways can mix with freshwater resources and result in an end product which is high in As, as seen in Turkey (Aksoy et al., 2009; Baba et al., 2009), Greece (Tyrovola et al., 2006), the Niigata plains of central and Shinji lowland of western Japan (Kubota et al., 2003), Ethiopia (Reimann et al., 2003), southwestern Idaho (Welch et al., 2000), and the aquifers recharged from the Madison and Missouri Rivers, Montana, United States (Nimick, 1998). On cooling, these fluids precipitate As-bearing sulfides along their flow paths (Goldhaber et al., 2001) which upon oxidation by infiltrating water rich in oxidants [dissolved oxygen (DO), NO_3^-] or exposure to the atmosphere due to water table fluctuation can produce deleterious concentration of aqueous As(V) in acidic and SO_4^{2-} rich waters, as has been reported from the bedrock aquifers of central India (Acharyya et al., 2005), France (Bonnemaison, 2005), Finland (Roman and Peuraniemi, 1999), eastern Wisconsin (Schreiber et al., 2000; Schreiber et al., 2003), and the Goose River Basin, Maine, United States (Sidle et al., 2001).

Thus As can be released under a wide range of pH (6.8–8.5) and redox conditions by both biotic and abiotic processes under both humid and arid climate.

14.2.2 Fluoride

Fluorine (F), a member of the halogen group of elements, has been ranked 13th in terms of its abundance in the Earth's crust (625 mg/kg) (Koritnig, 1951). It is also the lightest and most electronegative of the halogens. Fluorine exists in natural water as negatively charged fluoride ion (F^-) (Hem, 1985). Fluoride is an essential nutrient required in the human body for tooth and skeletal development [World Health Organization (WHO), 1996]. Drinking water is the main source of F^- intake in the human body [World Health Organization (WHO), 2011]. Though it is preferably retained in solid phases as compared to chlorine, which is a highly fluid mobile element, trace amounts of it exit in waters of all type (Koritnig, 1951). Its concentration can span over four orders of magnitude but the typical range lies between 0.1 and 10 mg/L in most natural waters (Edmunds and Smedley, 2013). Though its deficiency has often been linked to the occurrence of dental caries, lack of enamel formation, and bone fragility [World Health Organization (WHO), 1996], any intake exceeding the WHO safe limit of 1.5 mg/L [World Health Organization (WHO), 2004] can cause dental (tooth discoloration, enamel pitting, early tooth loss) and in severe cases skeletal fluorosis (joint stiffening and deformation), and a variety of nonskeletal maladies depending on period of exposure and degree of exposure [Susheela et al., 1993; World Health Organization (WHO), 1996; Lu et al., 2000; Ozsvath, 2006]. It is estimated that over 200 million people globally consume water with excess F^- (Amini et al., 2008). Cases of fluorosis have been reported from various countries around the globe including India (Gupta et al., 2004; Ayoob and Gupta, 2006), Pakistan (Rafique et al., 2008), Sri Lanka (van der Hoek et al., 2003; Jayawardana et al., 2012), China (Fuhong and Shuqin, 1988; Wang and Huang, 1995; Luo et al., 1997, 2008; Smedley et al., 2003; Currell et al., 2011; Chen et al., 2012; Guo et al., 2012; Li et al., 2015; Su et al., 2015), Afghanistan (Houben et al., 2003; Broshears et al., 2005), Indonesia (Heikens et al., 2005), United States [Dean, 1933; Deering et al., 1983; Corbett and Manner, 1984; Robertson, 1985; United States Geological Survey (USGS), 1991–2003; United States Environmental Protection Agency (US EPA), 2012], Canada (Boyle and Chagnon, 1995; Hitchon, 1995; Boyle et al., 1996; Desbarats, 2009), Mexico (Razo et al., 1993; Diaz-Barriga et al., 1997; Ortiz et al., 1998; Planer-Friedrich, 2000), Honduras (Foletti and Paz, 2001), Argentina (Kruse and Ainchil, 2003; Gomez et al., 2009; Cid et al., 2011; Martinez et al., 2012), Russia (de Caritat et al., 1998), Norway (de Caritat et al., 1998), Sweden (Gierup, 2007), Spain (Schwartz and Friedrich, 1973), Estonia (Indermitte et al., 2009), New Zealand (Daughney and Reeves, 2005), Australia (Ivkovic et al., 1998; Larsen et al., 1998; Fitzgerald et al., 2000), and several African countries [Smedley, 1996; République du Sénégal, Ministère de l'Energie et de l'Hydraulique, Service de Gestion et de Planification des Ressources en Eau (SGPRE), 2001; Water Research Commission (WRC), 2001; Reimann et al., 2002; Smedley et al., 2002a; Srikanth et al., 2002; Näslund and Snell, 2005; Messaïtfa, 2008; Sajidu et al., 2008; Yidana et al., 2010; Thole, 2013] (Fig. 14.2).

While on a local scale, anthropogenic activities, such as improper use of phosphate fertilizers (Kundu and Mandal, 2009) and F^--enriched irrigation water (Pettenati et al., 2013), aluminum smelting (Fuge and Andrews, 1988), coal burning (Finkelman et al., 1999; Ando et al., 2001), glass (Pickering, 1985) and brick making (MacDonald, 1969; Datta et al., 1996) can introduce significant amounts of F^- to the natural system, its concentration in groundwater in

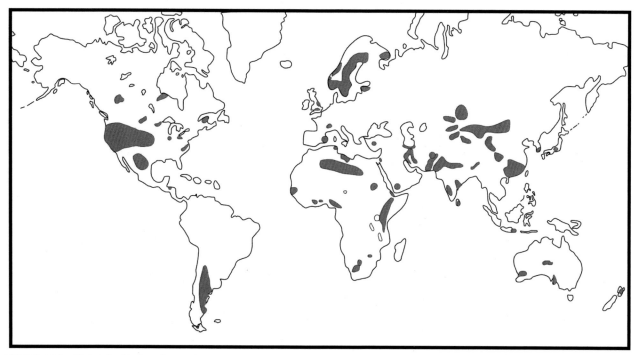

FIGURE 14.2 Global distribution of groundwater provinces with naturally occurring F⁻ concentration greater than 1.5 mg/L (following Amini et al., 2008 and Edmunds and Smedley, 2013).

controlled by geologic processes (Ayoob and Gupta, 2006; Edmunds and Smedley, 2013). A high concentration of F⁻ can accumulate in water by interaction with rocks and geothermal fluids (Edmunds and Smedley, 2013). Fluoride has an ionic radius (1.36 Å) similar to hydroxyl ion (OH⁻) (1.40 Å) and easily replaces it in micas, amphiboles, clay minerals, and apatites (Fleischer and Robinson, 1963). Moreover, owing to its incompatibility, high mobility under high temperature, lightness, and high volatility also accumulates in hydrothermal fluids (Hayes et al., 2017). Thus high-fluoride groundwater is often associated with exposures of acidic igneous rocks, hydrothermal fluids, vein deposits, and phosphate horizon bearing sedimentary formations (Edmunds and Smedley, 2013).

The release of F⁻ to groundwater bodies is controlled by water–rock interaction. The solubility of the mineral fluorite (CaF_2) predominantly controls its concentration (Handa, 1975; Saxena and Ahmed, 2001), making high F⁻ groundwater often calcium deficient and alkaline (Turner et al., 2005; Reddy et al., 2010). Such Ca-deficient, Na-HCO₃ type waters are often associated with rocks with low calcium content (alkaline igneous rocks) (Kilham and Hecky, 1973; Ashley and Burley, 1994) and aquifers where active cation exchange is taking (Handa, 1975; Edmunds and Walton, 1983). Moreover, solubility of fluorite increases with increasing salinity (Edmunds and Smedley, 2013). Though F dominantly occurs as free F⁻ ion, it can also form complexes with other elements (H⁺, Al, B, Be, Ca, Fe^{3+}, Mg, Na, Si, U, and V) depending on pH conditions and their availability (Hem, 1985; Edmunds and Smedley, 2013; Deng et al., 2011).

Based on the abovementioned discussions, it can be said that high-F⁻ groundwater generally occurs in three broad hydrogeological settings—granitoid basement aquifers, zones of active volcanism, and certain sedimentary basins (Edmunds and Smedley, 2013).

In igneous systems, fluorine is one of the many incompatible elements and hence is not easily incorporated in the early formed silicate phases (Hayes et al., 2017). As a result of this, it becomes concentrated in the residual magma during magmatic differentiation (Hildreth, 1981), giving rise to relatively high F concentration in more differentiated members of the series (granites, granodiorites, syenites, carbonatites), and their extrusive counterparts (Hayes et al., 2017), where it is usually incorporated in micas, amphiboles, and accessory phases (apatite, topaz, and fluorite) (Fleischer and Robinson, 1963). Thus F⁻-polluted waters are most commonly reported from granitoid basement or sedimentary aquifers derived from their erosion. For example, deleterious F⁻ concentrations have been reported from basement aquifers of India (Handa, 1975; Maithani et al., 1998; Suma Latha et al., 1999; Chakraborti et al., 2000; Rao, 2002; Kotoky et al., 2008; Reddy et al., 2010), Pakistan (Rafique et al., 2009; Naseem et al., 2010), Sri Lanka (Dissanayake and Weerasooriya, 1986; Dissanayake, 1991; Dharmagunawardhane and Dissanayake, 1993; Young et al., 2010), Yemen

(Fara et al., 1999), Japan (Abdelgawad et al., 2009), Iran (Keshavarzi et al., 2010), South Korea (Kim and Jeong, 2005; Chae et al., 2007; Kim et al., 2010), Norway (Bårdsen et al., 1996; Banks et al., 1998; Frengstad et al., 2000), United States (Ozsvath, 2006), Cameroon (Fantong et al., 2010), Nigeria (Dibal et al., 2012), Ethiopia (Ayenew, 2008), Malawi (Msonda et al., 2007), Senegal (Travi, 1993), Ghana (Apambire et al., 1997; Smedley et al., 1995; Dibal et al., 2012), Eritrea (Estifanos, 2005), and South Africa (McCaffrey, 1998).

As a result of its incompatibility, F also tends to accumulate in late-stage magmatic fluids. Moreover, F^- can also be dissolved in relatively high concentrations in meteoric waters circulating through F rich young volcanic rocks. Hydrolysis of such lithologies produces Na-HCO_3 type waters which are relatively deficient in Ca and Mg, thus facilitating F^- extraction from host rocks (Kilham and Hecky, 1973; Jones et al., 1977; Ayenew, 2008). Also, the fine components in such young volcanics that are known have the highest F concentration facilitate easy extraction of F^- owing to their inherent instability under surficial conditions (Hayes et al., 2017). Thus in regions of active volcanism, exolution of late-stage magmatic fluids and heating of percolating precipitation can set up geothermal systems rich in F^-. Such fluids upon migration via suitable pathways can enrich local groundwater resources in F^- as has been currently occurring in many countries in the East African Rift Valley [e.g., Ethiopia (Ashley and Burley, 1994; Tekle-Haimanot et al., 2006; Ayenew, 2008), Tanzania (Bugaisa, 1971), Kenya (Gaciri and Davies, 1993)], Mexico (Carrillo-Rivera et al., 1996), Yemen (Fara et al., 1999), and Turkey (Davraz et al., 2008). Moreover, rocks metasomatized by such fluids usually contain significant amount of F, and in some cases, fluorite may also be precipitated as hydrothermal vein deposit depending upon concentration.

Apart from these, certain sedimentary aquifers are also known to yield F^--enriched waters, such as the volcanic ash bearing La Pampean aquifers of Argentina (Nicolli et al., 1989; Cabrera et al., 2001; Smedley et al., 2002b; Paoloni et al., 2003; Buchhamer et al., 2012) and the High Plain aquifer, United States (Hudak, 1999; Hudak and Sanmanee, 2003; Scanlon et al., 2009), the phosphatic horizons (phosphorite deposits, microfossil bone beds) bearing Cretaceous to Quaternary aquifers of western Tunisia, Cretaceous to Tertiary aquifers of western Senegal (Travi, 1993), Cretaceous Black Creek formation of South Carolina, United States (Zack, 1980), and the Lincolnshire Limestone, Cretaceous Chalk deposits of England (Edmunds, 1973; Edmunds and Walton 1983; Edmunds et al., 1989), limestone aquifers of Turkey (Oruc, 2008), Iran (Looie and Moore, 2010) and Germany (Queste et al., 2001) with localized fluoroapatite accumulation. Fluoridation of aquifers with nominal solid-phase F content can also happen due to leakage of F^- rich water from underlying/adjoining marine formations [e.g., Upper Sirte Basin, Libya and the Cretaceous Nubian Sandstone aquifer, Sudan (Edmunds, 1994), High Plains Aquifer, United States (Nativ, 1988; Hopkins, 1993; Hudak, 2009)].

Apart from hydrochemistry, other factors such as residence time, climatic, and soil conditions can also influence F^- concentrations (Saxena and Ahmed, 2003; Valenzuela-Vasquez et al., 2006). High concentrations are often associated with deeper (older) waters, due to long residence time (Han et al., 2006; Vasquez et al., 2006; Farooqi et al., 2007; Nicolli et al., 2012). Exception to this observation is shallow groundwater in volcanic regions which receives inputs from geothermal sources (Fantong et al., 2010). Arid climate, where groundwater recharge and flow rates are low tends to promote high F^- concentrations by extending the water–rock reaction time (Smedley et al., 2002b) in contrast to humid tropics where high precipitation flushes and has a diluting effect on groundwater chemical composition (Zeng, 1997). Also, the properties of the soil horizon can influence the F^- concentration in the underlying aquifer (Lavado and Reinaudi, 1979), with acidic and fine-grained soil rich in clay, organic matter, and metal (oxy)hydroxides retaining more F^- than sandy soil types (Pickering, 1985; Fuge and Andrews, 1988; Wenzel and Blum, 1992; Wuyi et al., 2002).

14.2.3 Selenium

Selenium (Se) is a trace element with a crustal abundance of 50–90 µg/kg (Taylor and McLennan, 1985; Wang and Gao, 2001) and has been ranked as 78th amongst the 90 elements that make up the Earth's crust (Oldfield, 2002). Though it is a micronutrient essential for the survival of most flora and fauna, only a very narrow range exists between nutritional requirements (<40 µg/day) and toxic levels (>400 µg/day) [World Health Organization (WHO), 1996; Zingaro et al., 1997; Chen et al., 1999], giving it the title "a double-edged sword" (Conde and Alaejos, 1997; Fernández-Martínez and Charlet, 2009) or "essential toxin" (Stolz et al., 2002). Diseases caused both by Se deficiency (Keshan diseases, Kashin–Beck disease, and malfunctions in thyroid metabolism) and excess (lameness, loss of hair and finger nails, neurological damage, and liver cirrhosis) have been widely reported during the past decades (Chen et al., 1980; Yang et al., 1983; Arthur and Beckett, 1994; Rayman, 2000; Goldhaber, 2003). Though trace concentrations of Se occur in all natural water bodies (<0.1–100 µg/L), with values rarely exceeding 3 µg/L (Plant et al., 2004), values as high as 12 mg/L have been measured in agricultural drainage water from evaporation ponds in the Kesterson

Reservoir in San Joaquin Valley, California, United States (Presser and Barnes, 1984; Fujii et al., 1988). Aquatic systems such as large reservoirs, lakes, marshes, wetlands enriched with dissolved Se are considered to be the repository of its toxicity problems (Zingaro et al., 1997). Similarly, observed values in groundwater can range from negligible (0.07 µg/L) to very high values (1000 µg/L) (Jacobs, 1989; Alfthan et al., 1995; Fordyce et al., 2000; Bajaj et al., 2011). Portions of Finland, western United States, and some other countries (Fig. 14.3) are under the direct threat of Se concentration in groundwater higher than the present WHO guideline of 40 µg/L [World Health Organization (WHO), 2011]. Thus, even though Se ranks 52nd in the ATSDR list of hazardous substance [Agency of Toxic Substances and Dieses Registry (ATSDR), 2019], it is one of the most notorious metals in groundwater contamination, and its sources, processes of mobilization, and fate in aquatic environment have become a formidable environmental issue of international concern [Trelease and Beath, 1949; National Research Council (NRC), 1976, 1989; United States Environmental Protection Agency (US EPA), 1980, 1987, 1992, 1998b; Wilber, 1983; Hamilton, 1999; Hamilton et al., 2000].

In nature Se is often associated with S owing to similarity in valence structures (Shrift, 1954). Selenium occurs in the environment in four metastable oxidation states: (−II) (selenide [Se^{2-}]), (0) (elemental Se), (+IV) (selenite, [SeO_3^{2-}]), and (+VI) (selenate, [SeO_4^{2-}]) (Fernández-Martínez and Charlet, 2009). Owing to similarity of ionic radii, Se^{2-} (1.98 Å) often replaces for S^{2-} (1.84 Å) in sulfide minerals (Coleman et al., 1993). Selenium can also substitute for S in organic complexes forming selenomethionine (SeMet), and D-methyl-selenide (DMSe) (Coleman et al., 1993; Bailey, 2017). Oxidizing conditions favor the formation of water soluble oxyanionic species (SeO_4^{2-} and SeO_3^{2-}) that are leached from well-aerated aquifers to the groundwater system (Winkle et al., 2012). Selenate (the most toxic of all Se species) (Bailey, 2017) is the dominant species in oxygenated and near neutral to basic groundwater (Masscheleyn et al., 1989) owing to its weak adsorption characteristics (Ahlrichs and Hossner, 1987). Selenite on the other hand is strongly adsorbed by iron oxides (Balistrieri and Chao, 1987; Manceau and Charlet, 1994), clay minerals (BarYosef and Meek, 1987), and soil organic matter (Gustafsson and Johnsson, 1994). By contrast, insoluble forms (elemental Se and selenides) are often retained in water saturated, poorly drained soils, the reducing condition of which favors them (National Research Council (NRC), 1983; Zhang and Moore, 1997).

The cause of Se contamination of groundwater can be attributed to both geologic and anthropogenic [mining (Naftz and Rice, 1989; Mars and Crowley, 2003), irrigating lands underlain by shales (Fujii et al., 1988; Gates et al., 2009), irrigational runoff from Se rich soils (Cooke and Bruland, 1987), and irrigating with Se-contaminated water (Bajaj et al., 2011)] factors. In nature, Se occurs in fossil fuels, shales, and alkaline soils and in spite of a low crustal

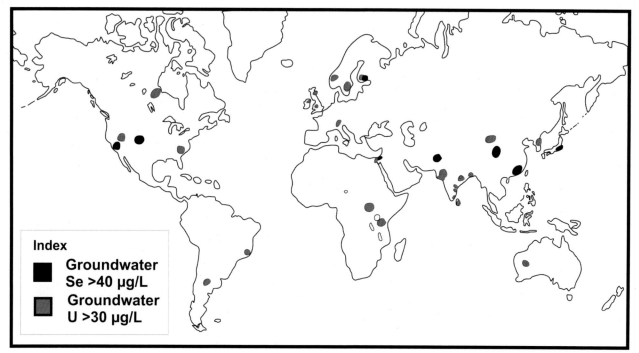

FIGURE 14.3 Global distribution of groundwater provinces with naturally occurring Se and U concentration greater than 40 µg/L and 30 µg/L, respectively.

abundance of about 0.0001% (Nazarenko and Ermakov, 1972), under suitable condition, it can accumulate in micromolar concentration (Stolz and Oremland, 1999).

Volcanoes introduce significant quantities of Se to the Earth's surface (180 t/year; Nriagu, 1989). It has been estimated that throughout the geological evolution of the planet, volcanoes have added ~ 0.1 g of Se per cm^2 of its surface (Fordyce, 2013). However, because of its incompatible and highly volatile nature, it escapes as high-temperature gases, as a result of which its concentration in igneous rocks does not usually exceed crustal values (10–50 μg/kg) (Fleming, 1980; Jacobs, 1989; Nriagu, 1989; Neal, 1995). Conversely, Se tends to concentrate in sedimentary environment, but its accumulation is limited to some specific rock types deposited in reducing conditions which favor the preservation organic matter, such as phospahatic rocks, coals, and shales, with concentrations in sandstones and limestones rarely exceeding 0.1 mg/kg (Neal, 1995). In phosphatic rocks a maximum concentration of 300 mg/kg have been reported by Fordyce (2005), and references therein, reflecting replacement of biogenic PO_4^{3-} by SeO_4^{2-} ion (Fleming, 1980; Jacobs, 1989; Nriagu, 1989; Neal, 1995). Concentrations in coals and shale typically range between 1 and 20 mg/kg, although values in excess of 6000 and 600 mg/kg have been reported, respectively (Fordyce, 2013, and references therein). In such lithotypes the reduced Se is sequestered microbially by both organic matter and pyrite (Herbel et al., 2003). In coals, it is primarily associated with organic matter (Coleman et al., 1993), whereas in shales, it is preferentially incorporated in pyrites (Matamoros-Veloza et al., 2011). Seleno-pyrite ($FeSe_xS_{2-x}$) bearing shales have been studied extensively in western United States (Williams and Byers, 1934; Presser and Barnes, 1984; Wright, 1999; Kulp and Pratt, 2004; Bailey et al., 2012), Finland (Alfthan et al., 1995), Japan (Mizutani et al., 2001), and Australia (Huston et al., 1995). Selenium is also associated with sulfide mineralization, where it substitutes for S in sulfide minerals (Boyle, 1979; Fleming, 1980; Neal, 1995; Tokunaga et al., 1996).

However, since these Se-concentrating rock types do not constitute bulk of the Earth's surface and have a very restricted geological occurrence, Se pollution of water resources is restricted to small pockets in their vicinity. Thus the groundwater Se content can have a high degree of spatial variability (Gates et al., 2009). Of the abovementioned lithologies, shale has the greatest geological abundance and thus seliniferous groundwater is often found in regions where marine shale occurs as confining units or outcrops. Release of Se from marine shale has been reported from United States (Deverel and Fujii, 1988; Fujii et al., 1988; Seiler, 1995; Seiler et al., 1999; Gates et al., 2009; Paschke et al., 2014; Mills et al., 2016), Pakistan (Afzal et al., 2000), Jordan (Al Kuisi and Abdel-Fattah, 2010), China (Fordyce et al., 1998, 2000), and Finland (Lahermo et al., 1998). Selenium is released to the groundwater system by oxidation of seleno-pyrite by oxygenated water (either containing DO or NO_3^-) in the presence of microbes as SeO_4^{2-} or SeO_3^{2-} (Wright, 1999; Stillings and Amacher, 2010). Moreover, it has been observed that areas with seleniferous groundwaters have high evaporation indices, suggesting that Se accumulation is more prone in arid and semiarid climates (Seiler, 1997; Hudak, 2009).

14.2.4 Uranium

Uranium (U) is a ubiquitous trace element of the Earth's crust with an average concentration of 2.6–2.8 mg/kg (Hu and Gao, 2008; Herring, 2013). It is the heaviest naturally occurring member of the periodic table. Though considered to be rare, it is more plentiful in nature than several other trace elements (Cd, Se, REEEs) (Kay and Laby, 1995). It has been classified as a lithophilic element and occurs naturally in varying degrees in all minerals, rocks, soils, and water (Waseem et al., 2015). It has three naturally occurring radioisotopes (^{238}U, ^{235}U, ^{234}U) with very low-specific activity, thereby making its radiological toxicity a less significant concern compared to its chemical toxicity [Health Canada, 1987; Agency of Toxic Substances and Dieses Registry (ATSDR), 2009; European Food Safety Authority (EFSA), 2009]. Nontoxic levels of U is found in almost all natural water bodies (Mkandawire, 2013), and exposure from drinking water is usually very mild (Smedley et al., 2006). However, at elevated levels, drinking water can become a chief source of nonprofessional U exposure (Smedley et al., 2006). Though concentrations in groundwater are usually less than 1 μg/L (Herring, 2013), it is susceptible to natural U accumulation due to extended water–rock interaction (Smedley et al., 2006) with elevated concentrations having dire health consequences on the exposed populace (Brugge et al., 2015; Konietzka, 2015). Toxic effects of U exposure have often been linked to nephritis [Health Canada, 1987; Zamora et al., 1998; Kurttio et al., 2002; Kurttio et al., 2006; Orloff et al., 2004; Giddings, 2005; Committee on Toxicity of Chemicals in Food, Consumer Products and the Environment (COT), 2006; European Food Safety Authority (EFSA), 2009; Seldén et al., 2009], changes in bone structure (Kurttio et al., 2005), and different types of cancers (Wagner et al., 2011; Radespiel-Tröger and Meyer, 2013). Though most studies over the past decades have focused on U mobilization from anthropogenic sources [mining and processing of U ores (Parkhurst et al., 1984; Goode and Wilder, 1987; Gómez et al., 2006b; Stone et al., 2007), conversion and enrichment of U for fuel and nuclear weapon production, improper handling of radioactive wastes, and application of phosphate fertilizers (Zielinski et al., 2000; Kawabata et al., 2006; Birke et al., 2010; Liesch et al., 2015)], many recent studies from around the globe have now recognized U concentration above 30 μg/L [World Health Organization (WHO), 2011] in their groundwater resources occurring from natural sources [Skeppström and Olofsson, 2007; European Food Safety Authority (EFSA), 2009] (Fig. 14.3). Since, groundwater is a major source of freshwater supply for domestic purposes, its contamination by U can cause a major human health crisis worldwide [Earth Sciences for Society Foundation (ESS), 2005; Ericson et al., 2008]. Therefore it is important to identify and develop an understanding of the sources and pathways of U release to the groundwater system (Liesch et al., 2015).

Uranium is introduced to the groundwater system naturally from aquifers solid or nearby formations by dissolution of U minerals and/or desorption of U complexes from mineral surfaces (Fanghanel and Neck, 2002; Liu et al., 2004, 2009; Chen et al., 2005; Qafoku et al., 2005; James and Sinha, 2006; Fox et al., 2012; Alam and Cheng, 2014). In nature U occurs in +2, +3, +4, +5, or +6 valence states (Waseem et al., 2015). Under reducing conditions U occurs in the +4 oxidation state in phases like uraninite and coffinite (Duff et al., 1999) and organic complexes (Bednar et al., 2007). Under such conditions the very low solubility of U(IV) minerals limits its dissolved concentration (Lovley and Phillips, 1992; Wiedemeier et al., 1995; Abdelouas et al., 1998; Fredrickson et al., 2000; Finneran et al., 2002). Oxidative dissolution of U(IV) bearing phases is the major mechanism of U release to the aqueous system (Finch and Ewing, 1992; Finch and Murakami, 1999). In water U usually forms complexes with HCO_3^-, PO_4^{3-}, F^- depending pH-pCO$_2$ conditions (Langmuir, 1978). Under oxidizing condition and low pH (<5), U exits in the +6 valence state and as uranyl oxycation (UO_2^{2+}) (Kelly et al., 2003). In spite of its mobilization, the relative mobility of U(VI) in liquid medium is controlled by adsorption/desorption of UO_2^{2+} onto/from clay minerals, natural organic matter, and iron (oxy)hydroxides (Bowman, 1997; Wang et al., 2005; Catalano et al., 2006). Adsorption is controlled by pH, Eh, and dissolved CO_3^{2-}, PO_4^{3-}, and organic matter content of water (Sanding and Bruno, 1992; Echevarria et al., 2001; Cheng et al., 2004; Kohler et al., 2004). Maximum sorption takes place under the near-neutral pH (Prikryl et al., 2001). Since many of the uranyl-carbonate complexes are anionic in nature, desorption usually takes place at pH >7–8, due to competition between mineral surface site and carbonate ligands for UO_2^{2+} (Villalobos et al., 2001). Thus high U concentration is expected in waters with DO and CO_3^{2-} rich and PO_4^{3-} free conditions (Kelly et al., 2006).

Though U content of rocks typically ranges from 1 to 4 mg/kg (Hess et al., 1985; Taylor and McLennan, 1985; Drever, 1997), relatively high U concentrations have been recognized in granitic and other late phase magmatics (20–54 mg/kg; Andrews et al., 1989), argillaceous and organic-rich deposits (clays, black shale, oil shale) (250 mg/kg; Lee et al., 2001), ferruginous (1.8–2.4 mg/kg; Zielinski et al., 1983) and phosphatic sediments (30–200 mg/kg; Merkel and Sperling, 1998), and metamorphic rocks derived from them (Langmuir, 1978). Therefore U-enriched groundwater is expected to occur in two broad hydrogeological settings—granitoid basement aquifers and certain sedimentary basins (Smedley et al., 2006).

Uranium is a highly incompatible element, and high concentrations of it have been measured from granites where late-stage accessory minerals such as monazite, zircon, sphene, uraninite, and hydroxyapatite tend to accumulate it (Porcelli and Swarzenski, 2003). Of these, uraninite is the most easily leachable and is considered to be an important source groundwater U as compared to other relatively insoluble phases (monazite, zircon, sphene) which are unlikely to add any significant amounts of U to the hydrogeological system of these aquifers (Smedley et al., 2006). High U groundwaters have been reported from many crystalline aquifers such as the Lac du Bonnet Granite in the Canadian Shield (Gascoyne, 2004); Leinster Granite, Ireland (Cullen, 2005); Stripa Granite, Sweden (Andrews et al., 1989); granitic basement in Uganda (BGS, unpublished data), Finland (Kurttio et al., 2002), and Norway (Frengstad et al., 2000; Reimann et al., 2005); granite–schist–greenstone assemblage in Tanzania (BGS, unpublished data); granite–charnockite–quartzite assemblage in Sri Lanka (BGS, unpublished data); Precambrian granites and metaigneous rocks of Brazil (Almeida et al., 2004); acidic igneous and metamorphic rocks in Korea (Kim et al., 2004; Cho and Choo, 2019); tertiary granites and alluvium in Nevada, United States (Cizdziel et al., 2005); biotite granite of the Inner Piedmont terrane, South Carolina, United States (Warner et al, 2011); granites and secondary mineralization in its vicinity in Australia (Appleyard, 1984); granite, metagranitoid and gniesses of Switzerland (Stalder et al., 2012); and Precambrian granitic basement of India (Coyte et al., 2018).

Upon hydrolysis the U is removed from its primary source and redistributed to sedimentary deposits where it can be adsorbed onto and/or coprecipitated with iron (oxy)hydroxides (Hsi and Langmuir, 1985; Bianconi and Kögler, 1992; Duff et al., 2002; Scott et al., 2005), phosphates (Cothern and Lappenbusch, 1983; Murray et al., 1983; Arey et al., 1999; Jerden et al., 2003), clay minerals (Bonotto and Andrews, 2000), and organic matter (Bednar et al., 2007) from which it can be released on changing geochemical environment. Therefore high U concentrations can be expected from sedimentary aquifers with an abundance of argillaceous sediments, iron oxides, organic matter, and phosphate minerals.

Examples include the phosphate horizon bearing limestone aquifer in Jordan (Smith et al., 1996), alluvial aquifers underlain by black shale in Colorado, United States (Zielinski et al., 1997) and Korea (Lee et al., 2001). Elevated concentrations have also been known to occur in the Holocene and Pleistocene alluvial aquifers of Bangladesh [British Geological Survey (BGS) and Department of Public Health Engineering (DPHE), 2001], Holocene lacustrine and riverine deposits in the Huhhot Basin, Inner Mongolia, China (Smedley et al., 2003), Pleistocene fluvio-lacustrine and aeolian sediments in the Carson Desert, United States (Welch and Lico, 1998), Quaternary losses of La Pampa, Argentina (Smedley et al., 2002b), and ferruginous sandstone units (Old Red Sandstone, Permo-Triassic Sandstone Aquifer, and Torridonian Sandstone) of England and Scotland (Smedley et al., 2006), where U release has been attributed to desorption from iron (oxy)hydroxides in an oxidizing environment in the presence of HCO_3^- ion.

14.2.5 Salinity

Salinity of water refers to its salt content and is often measured in terms of total dissolved solids (milligrams of dissolved solids per liter of water) or other proxies such as chloride (Cl^-) concentration (in mg/L) or electrical conductivity (in μS/cm). All natural water bodies contain some amount of dissolved solids and it can vary widely from place to place. Though dissolved solutes in water are essential for sustenance of life, any concentration over 1000 mg/L often restricts its use for any practical purposes without treatment (Mayer et al., 2005). Thus one of the most prominent processes leading to groundwater quality degradation is salinization of freshwater aquifers. It is a global environmental crisis affecting various aspects of our lives. Increase in salt content can either lead to abandonment of supply wells, thereby causing drinking water scarcity or severe ailments on consumption. Also, irrigation with salty water reduces soil fertility, interferes with uptake of nutrients by plants, and causes loss of biodiversity and taxonomic replacement of native species by halotolerant ones, all of which can ultimately result in the collapse of the agricultural sector (Shiklomanov, 1997; Postel, 1999; Jackson et al., 2001; Williams, 2001a,b; Williams et al., 2002; Dregne, 2002; Rengasamy, 2006). Furthermore, water of low salinity in the basic requirement for many industrial sectors. Moreover, salinization of freshwater resources not only increases the levels of chloride and sodium but is also often responsible for the increase in concentration of other inorganic contaminants [e.g., As (Scanlon et al., 2009; Nicolli et al., 2010; Rango et al., 2010b), B (Vengosh et al., 1999b, 2002), F (Shiklomanov, 1997; Misra and Mishra, 2007; Ayenew, 2008; D'Alessandro et al., 2008; Mor et al., 2009; Viero et al., 2009; Rango et al., 2009, 2010a,b,c), Ra (Kraemer and Reid, 1984; Hammond et al., 1988; Krishnaswami et al., 1991; Moise et al., 2000; Sturchio et al., 2001; Carvalho et al., 2005; Raanan et al., 2009; Moatar et al., 2010; Tomita et al., 2010; Otero et al., 2011; Vinson, 2011), Se (Miller et al., 1981; Deverel and Fujii, 1987; Tanji and Valoppi, 1989; Presser and Swain, 1990; Schroeder et al., 2002; Schoups et al., 2005; Deverel and Gallanthine, 2007)], and hence, it is often correctly referred to as the "tip of the iceberg" by many workers (Vengosh, 2014). Though groundwater salinization is a world-wide nuisance, its effects are most severe in arid and semiarid regions, where recharge and other sources of freshwater supply are scarce. In such regions it is the salinity of available water resources that restricts its usage and future prospect of water utilization is often complicated by increasing salinization (Vengosh and Rosenthal, 1994; Salameh, 1996). Groundwater salinization can be caused by both human activities on a shorter or by geological processes on longer time scale. Whatever be the rate, the net result is that many aquifers across the globe have become unsuitable for human consumption as a result of it (Fig. 14.4).

Salinization of freshwater resources can occur by interaction of potable water with waters of high salinity, with salt deposits or a combination of both.

Although saline-water—freshwater interaction can take place in a wide variety of settings, it is most commonly observed in coastal aquifers in hydraulic connection with the sea. Here, a lens of discharging fresh groundwater overlies a landward-thinning wedge of intruding seawater, owing to density differences. The two water bodies are separated by a zone of mixing or diffusion (Bear et al., 1999). Although the degree of intrusion is limited by the recharge and the volume of discharging water, encroachment is often enhanced by the presence of preferential pathways through more permeable layers, as seen in the Jeju volcanic island, South Korea, where saltwater has ingressed up to 2.5 km inland owing to heterogeneity and high conductance of basaltic aquifer (Kim et al., 2003; Park et al., 2005; Koh et al., 2009), or as seen in many krastic aquifers worldwide (Cotecchia et al., 1974; Cotecchia et al., 2005; Sanford and Konikow, 1989; Fidelibus and Tulipano, 1990, 1996; Price and Herman, 1991; Ghiglieri et al., 2012; Héry et al., 2014; Han et al., 2015). Moreover, intrusion is also aided by presence of surface water bodies, and long-term sea-level increase (transgression) (van Weert et al., 2009). Furthermore, instantaneous rise in sea-level during storm surges, tsunamis, and high tides is also capable of salinizing significant areal extent of the shallow aquifers. Although the rise in water level is temporary and occurs on a much shorter time scale as compared to marine transgression, salting of aquifers is due to

FIGURE 14.4 Global distribution of groundwater provinces naturally affected by salinization (following van Weert et al., 2009).

extended stagnation of seawater on the land surface. Also inundation of poorly maintained well heads during such events may lead to saltwater introduction in deeper parts of the aquifer through well screens (van Weert et al., 2009).

However, such interactions are not solely restricted to coastal sites and have also been documented in many inland settings worldwide. In such settings usually freshwater overlies a body of dense saline water and a delicate hydraulic equilibrium is maintained between the two Their interaction can be brought about by migration of fluids of high salinity via suitable pathways (Magaritz et al., 1984; Vengosh and Benzvi, 1994; Hsissou et al., 1999; Sánchez-Martos and Pulido-Bosch, 1999; Kloppmann et al., 2001; Sánchez-Martos et al., 2002; Vengosh et al., 2002; Bouchaou et al., 2009) or diffusion of solutes from adjacent or underlying formations (Herczeg and Edmunds, 2000). For instance, in the Upper Floridian aquifer in Georgia, United States, faults breaching the aquitards confining the deeper aquifer provide a suitable pathway for upward migration of pressurized saline groundwater (Maslia and Prowell, 1990). Similar fracturing events during and after deglaciation have led the migration of saline fluids from depths into the Devonian and Pleistocene shallow aquifers of the Michigan Basin in southwestern Ontario, Canada (Weaver et al., 1995). Other examples of such cross-aquifer saline flows include the Ogallala aquifer in the Southern High plains, Texas, United States (Mehta et al., 2000a, 2000b), aquifers along the Mediterranean coast of Israel (Vengosh et al., 1994, 1999a,b) and Gaza Strip (Vengosh et al., 2005), and the eastern Pennsylvanian shallow aquifers in the Appalachian Basin where salinization has been brought about by Middle Devonian Marcellus brine (Weaver et al., 1995; Warner et al., 2012). Salinity increment can also be attributed to diffusion of solutes from brines entrapped in low conductivity units that are in hydraulic connection with the active aquifer, as has been hypothesized by Hendry and Schwartz (1988) for gradual salinization of the Cretaceous sandstone aquifer hosted in the Milk River Formation in Canada.

Fluids with such high salt content in terrestrial settings can have numerous modes of origin. These include residual seawater left after evaporation of solutes, which may or may not be modified by further water–rock interactions (Starinsky, 1974; Marie and Vengosh, 2001; Farber et al., 2007), unflushed seawater trapped in sedimentary formations of marine origin (Nativ, 1988; Hopkins, 1993; Hudak, 1999; Mehta et al., 2000a,b), connate water trapped in sedimentary units during episodes of marine transgression (Mukherjee and Fryar, 2008), hydrothermal fluids (Ellis and Mahon, 1977), and brines left over after freezing of seawater during glaciation [deep saline groundwaters of Sweden and Finland (Nelson and Thompson, 1954; Richardson, 1976; Herut et al., 1990; Bein and Arad, 1992; Bottomley et al., 1999; Marion et al., 1999; Wang et al., 2000; Stotler et al., 2009)].

Groundwater can also become salty on dissolution of salts from evaporite deposits or salt accumulations in the soil or unsaturated zone. Examples include the Ogallala aquifers in the Southern High Plains, Texas, United States (Mehta et al., 2000a,b), Dammam aquifer, Kuwait (Al-Ruwaih, 1995), Nubian Sandstone aquifer located in the Sinai and

Negev Desert, Israel (Rosenthal et al., 1998; Vengosh et al., 2007), Great Artesian Basin, Australia, (Herczeg et al., 1991; Love et al., 2000), the coastal Hammamet-Nabeul aquifer in Tunisia (Moussa et al., 2011), Salinas Valley, California, United States (Vengosh et al., 2002), the High Atlas (Warner et al., 2010), and the Anti-Atlas mountains (Ettayfi et al., 2012), Morocco.

14.3 Conclusion

Groundwater has now proven to be the most reliable source of freshwater supply and is mostly favored over other resources for good microbial quality. However, it is seldom devoid of other dissolved and suspended phases and any natural occurrence of such solids above permissible limits often restricts its utility. Since groundwater is a major source of freshwater supply for domestic purposes, its contamination by geogenic pollutants can cause a major global human health issue. The geographic disposition of these primary solid-phase sources of groundwater contaminants and in turn contaminated groundwater provinces (which mostly occur in their vicinity) is controlled by large-scale tectonic processes, and their fate and transport in the hydrogeologic system are further modified by local climatic and geochemical factors. Therefore the occurrence of such pollutants in groundwater often exhibits a great degree of spatial variability, thereby making it important to identify and develop an understanding of the sources and pathways of their release to the groundwater system.

References

Abdelgawad, A.M., Watanabe, K., Takeuchi, S., Mizuno, T., 2009. The origin of fluoride rich groundwater in Mizunami area, Japan—mineralogy and geochemistry implications. Eng. Geol. 108, 76–85.

Abdelouas, A., Lu, Y., Lutze, W., Nuttall, H.E., 1998. Reduction of U(VI) to U(IV) by indigenous bacteria in contaminated ground water. J. Contam. Hydrol. 35 (1–3), 217–233.

Acharyya, S.K., Shah, B.A., Ashyiya, I.D., Pandey, Y., 2005. Arsenic contamination in groundwater from parts of Ambagarh-Chowki block, Chhattisgarh, India: source and release mechanism. Environ. Geol. 49 (1), 148–158.

Ahmed, K.M., Bhattacharya, P., Hasan, M.A., Akhter, S.H., Alam, S.M.M., Bhuyian, M.A.H., et al., 2004. Arsenic enrichment in groundwater of the alluvial aquifers in Bangladesh: an overview. Appl. Geochem. 19 (2), 181–200.

Afzal, S., Younas, M., Ali, K., 2000. Selenium speciation studies from Soan-Sakesar Valley, Salt Range, Pakistan. Water Int. 25 (3), 425–436.

Agency of Toxic Substances and Dieses Registry (ATSDR), 2009. Toxic Substances Portal: Uranium and Compounds.

Agency of Toxic Substances and Dieses Registry (ATSDR), 2019. Substance Priority List.

Ahlrichs, J.S., Hossner, L.R., 1987. Selenate and selenite mobility in overburden by saturated flow. J. Environ. Qual. 16 (2), 95–98.

Aksoy, N., Şimşek, C., Gunduz, O., 2009. Groundwater contamination mechanism in agrothermal field: a case study of Balcova, Turkey. J. Contam. Hydrol. 103, 13–28.

Amini, M., Mueller, K., Abbaspour, K. C., Rosenberg, T., Afyuni, M., Møller, K. N., Sarr, M., et al., 2008. Statistical modeling of global geogenic fluoride contamination in groundwaters. Environ. Sci. Tech. 42, 3662–3668.

Alam, M.S., Cheng, T., 2014. Uranium release from sediment to groundwater: Influence of water chemistry and insights into release mechanism. J. Contam. Hydrol. 164, 72–87. <https://doi.org/10.1016/j.jconhyd.2014.06.001>.

Alfthan, G., Wang, D., Aro, A., Soveri, J., 1995. The geochemistry of selenium in groundwaters in Finland. Sci. Total Environ. 162, 93–103.

Almeida, R.M.R., Lauria, D.C., Ferreira, A.C., Sracek, O., 2004. Groundwater radon, radium and uranium concentrations in Regiao dos Lagos, Rio de Janeiro state, Brazil. J. Environ. Radioact. 73, 323–334.

Al Kuisi, M., Abdel-Fattah, A., 2010. Groundwater vulnerability to selenium in semi-arid environments: Amman Zarqa Basin, Jordan. Environ. Geochem. Health 32, 107–128.

Al-Ruwaih, F.M., 1995. Chemistry of groundwater in the Dammam Aquifer, Kuwait. Hydrogeol. J. 3, 42–55.

Ando, M., Tadano, M., Yamamoto, S., Tamura, K., Asanuma, S., Watanabe, T., et al., 2001. Health effects of fluoride pollution caused by coal burning. Sci. Total Environ. 271 (1-3), 107–116.

Andrews, J.N., Ford, D.J., Hussain, N., Trivedi, D., Youngman, M.J., 1989. Natural radioelement solution by circulating groundwaters in the Stripa granite. Geochim. Cosmochim. Acta 53, 1791–1802.

Apambire, W.B., Boyle, D.R., Michel, F.A., 1997. Geochemistry, genesis, and health implications of fluoriferous groundwaters in the upper regions of Ghana. Environ. Geol. 33, 13–24.

Appleyard, S.J., 1984. Uranium in groundwater near Uaroo, Western Australia: distribution and relevance to mineralization, J. Geochem. Explor., 22. pp. 357–358.

Arey, J.S., Seaman, J.C., Bertsch, P.M., 1999. Immobilization of uranium in contaminated sediments by hydroxyapatite addition. Environ. Sci. Technol. 33, 337–342.

Armienta, M.A., Rodriguez, R., Cruz, O., Aguayo, A., Ceniceros, N., Villasenor, G., et al., 2005. Environmental behavior of arsenic in a mining zone: Zimapan, Mexico. In: Bundschuh, J., Bhattacharya, P., Chandrasekharam, D. (Eds.), Natural Arsenic in Groundwater: Occurrence, Remediation and Management: Proceedings of the Pre-Congress Workshop "Natural Arsenic in Groundwater", 32nd International Geological Congress, Florence, August 18–19, 2004. Taylor & Francis Group, London, pp. 125–130.

Arthur, R., Beckett, G.T., 1994. New metabolic roles for selenium. Proc. Nutr. Soc. 53, 615–624.

Ashley, P.P., Burley, M.J., 1994. Controls on the occurrence of fluoride in groundwater in the Rift Valley of Ethiopia. In: Nash, H., McCall, G.J.H. (Eds.), Groundwater Quality. Chapman & Hall, London, pp. 45–54.

Ayenew, T., 2008. The distribution and hydrogeological controls of fluoride in the groundwater of central Ethiopian rift and adjacent highlands. Environ. Geol. 54, 1313–1324.

Ayoob, S., Gupta, A.K., 2006. Fluoride in drinking waters: a review on the status and stress effects. Crit. Rev. Environ. Sci. Technol. 36, 433–487.

Baba, A., Yuce, G., Deniz, O., Ugurluoglu, D.Y., 2009. Hydrochemical and isotopic com-position of Tuzla geothermal field (Canakkale-Turkey) and its environmental impacts. Environ. Forensics 10, 144–161.

Bado, A.A., 1939. Composition of water and interpretation of analytical results. J. Am. Water Work. Assoc. 31, 1975–1977.

Bailey, R.T., Hunter, W.J., Gates, T.K., 2012. The influence of nitrate on selenium in irrigated agricultural groundwater systems. J. Environ. Qual. 41 (3), 783–792.

Bailey, R.T., 2017. Review: Selenium contamination, fate, and reactive transport in groundwater in relation to human health. Hydrogeol. J. 25 (Special Issue: Hydrogeology and Human Health), 1191–1217. Available from: https://doi.org/10.1007/s10040-016-1506-8.

Balistrieri, L.S., Chao, T.T., 1987. Selenium adsorption by goethite. Soil Sci. Soc. Am. J. 51 (5), 1145–1151.

Bajaj, M., Eiche, E., Neumann, T., Winter, J., Gallert, C., 2011. Hazardous concentrations of selenium in soil and groundwater in North-West India. J. Hazard. Mater. 189, 640–646.

Banks, D., Frengstad, B., Midtgard, A.K., Krog, J.R., Strand, T., 1998. The chemistry of Norwegian groundwaters: I. The distribution of radon, major and minor elements in 1604 crystalline bedrock groundwaters. Sci. Total Environ. 222, 71–91.

BarYosef, B., Meek, D., 1987. Selenium sorption by kaolinite and montmorillonite. Soil Sci. 144, 11–19.

Bårdsen, A., Bjorvatn, K., Selvig, K.A., 1996. Variability in fluoride content of subsurface water reservoirs. Acta Odontol. Scand. 54, 343–347.

Bear, J., Cheng, A.H.-D., Sorek, S., Ouazar, D., Herrera, I. (Eds.), 1999. Seawater Intrusion in Coastal Aquifers, Concepts, Methods and Practices. Kluwer Academic Publishers.

Bednar, A., Medina, V., Ulmer-Scholle, D., Frey, B., Johnson, B., Brostoff, W., et al., 2007. Effects of organic matter on the distribution of uranium in soil and plant matrices. Chemosphere 70 (2), 237–247.

Bein, A., Arad, A., 1992. Formation of saline groundwaters in the Baltic region through freezing of seawater during glacial periods. J. Hydrol. 140, 75–87.

Bencko, V., Yan Li Foong, F., 2017. The history of arsenical pesticides and health risks related to the use of Agent Blue. Ann. Agric. Environ. Med. 124 (2), 312–316. Available from: https://doi.org/10.26444/aaem/74715.

Berg, M., Tran, H.C., Nguyen, T.C., Pham, H.V., Schertenleib, R., Giger, W., 2001. Arsenic contamination of groundwater and drinking water in Vietnam: a human health threat. Environ. Sci. Technol. 35 (13), 2621–2626.

Berg, M., Stengel, C., Trang, P.T.K., Viet, P.H., Mikey, L.S., Moniphea, L., et al., 2007. Magnitude of arsenic pollution in the Mekong and Red River Deltas—Cambodia and Vietnam. Sci. Total Environ. 372, 413–425.

Berg, M., Trang, P.T.K., Stengel, C., Buschmann, J., Viet, P.H., Dan, N.V., et al., 2008. Hydrological and sedimentary controls leading to arsenic contamination of groundwater in the Hanoi area, Vietnam: the impact of iron-arsenic ratios, peat, river bank deposits, and excessive groundwater abstraction. Chem. Geol. 249, 91–112.

British Geological Survey (BGS), Department of Public Health Engineering (DPHE), 2001. Arsenic Contamination of Groundwater in Bangladesh, BGS Technical Report WC/00/19. British Geological Survey, Keyworth.

Bhattacharya, P., Jacks, G., Ahmed, K.M., Khan, A.A., Routh, J., 2002. Arsenic in groundwater of the Bengal delta plain aquifers in Bangladesh. Bull. Environ. Contam. Toxicol. 69, 538–545.

Bianconi, F., Kögler, K., 1992. Uranium exploration in tropical terrains. In: Butt, C.R.M., Zeegers, H. (Eds.), Regolith Exploration Geochemistry in Tropical and Subtropical Terrains, vol. 4. Elsevier, Amsterdam, pp. 439–460.

Birke, M., Rauch, U., Lorenz, H., Kringel, R., 2010. Distribution of uranium in German bottled and tap water. J. Geochem. Explor. 107, 272–282.

Bonotto, D.M., Andrews, J.N., 2000. The transfer of uranium isotopes U-234 and U-238 to the waters interacting with carbonates from Mendip Hills area (England). Appl. Radiat. Isotopes 52, 965–983.

Bonnemaison, M., 2005. L'eau, facteur de liberation de l'arsenic naturel. Geosciences 2, 54–59 (in French).

Bottomley, D.J., 1984. Origins of some arseniferous groundwaters in Nova Scotia and New Brunswick, Canada. J. Hydrol. 69, 223–257.

Bottomley, D.J., Katz, A., Chan, L.H., Starinsky, A., Douglas, M., Clark, I.D., et al., 1999. The origin and evolution of Canadian Shield brines: Evaporation or freezing of seawater? New lithium isotope and geochemical evidence from the Slave craton. Chem. Geol. 155, 295–320.

Bouchaou, L., Michelot, J.L., Qurtobi, M., Zine, N., Gaye, C.B., Aggarwal, P.K., et al., 2009. Origin and residence time of groundwater in the Tadla basin (Morocco) using multiple isotopic and geochemical tools. J. Hydrol. 379, 323–338.

Bowman, R.S., 1997. Aqueous environmental geochemistry. EOS Transcr. Am. Geophys. Union 78 (50), 586.

Boyle, R.W., 1979. The geochemistry of gold and its deposits: together with a chapter on geochemical prospecting for the element. Bulletin 280. Geological Survey of Canada, Ottawa.

Boyle, D.R., Chagnon, M., 1995. An incidence of skeletal fluorosis associated with groundwaters of the maritime Carboniferous basin, Gaspé region, Quebec, Canada. Environ. Geochem. Health 17, 5–12.

Boyle, D.R., Spirito, W.A., Adcock, S.W., 1996. Groundwater hydrogeochemical survey of central New Brunswick. In: Geological Survey of Canada, Open File 3306. <http://gdr.nrcan.gc.ca/geochem/metadata_pub_e.php?id=00241>

Boyle, D.R., Turner, R.J.W., Hall, G.E.M., 1998. Anomalous arsenic concentrations in groundwaters of an island community, Bowen Island, British Columbia. Environ. Geochem. Health 20 (4), 199–212.

Brikowski, T.H., Neku, A., Shrestha, S.D., Smith, L.S., 2014. Hydrologic control and temporal variability of groundwater arsenic in the Ganges floodplain of Nepal. J. Hydrol. 518 (Special Issue: Arsenic in hydrological processes: Speciation, bioavailability and management), 342–353. <https://doi.org/10.1016/j.jhydrol.2013.09.021>.

Broshears, R.E., Akbari, M.A., Chornack, M.P., Mueller, D.K., Ruddy, B.C., 2005. Inventory of ground-water resources in the Kabul Basin, Afghanistan. In: United States Geological Survey Scientific Investigations Report 2005–5090, 34 p. <http://pubs.usgs.gov/sir/2005/5090/>.

Brugge, D., de Lemos, J.L., Oldmixon, B., 2005. Exposure pathways and health effects associated with chemical and radiological toxicity of natural uranium: a review. Rev. Environ. Health 20, 177–193.

Buchhamer, E.E., Blanes, P.S., Osicka, R.M., Giménez, M.C., 2012. Environmental risk assessment of arsenic and fluoride in the Chaco province, Argentina: research advances. J. Toxicol. Environ. Health, A 75 (22–23), 1437–1450. Available from: https://doi.org/10.1080/15287394.2012.721178.

Bugaisa, S.L., 1971. Significance of fluorine in Tanzania drinking water. In: Tschannerl, G. (Ed.), Proceedings of a Conference on Rural Water Supply in East Africa. Bureau of Resource Assessment and Land Use Planning, Dar Es Salaam, pp. 107–113.

Buschmann, J., Berg, M., Stengel, C., Sampson, M.L., 2007. Arsenic and manganese contamination of drinking water resources in Cambodia: coincidence of risk areas with low relief topography. Environ. Sci. Technol. 41 (7), 2146–2152.

Buschmann, J., Berg, M., Stengel, C., Winkel, L., Sampson, M.L., Trang, P.T.K., et al., 2008. Contamination of drinking water resources in the Mekong delta floodplains: arsenic and other trace metals pose serious health risks to population. Environ. Int. 34, 756–764.

Cabrera, A., Blarasin, M., Villalba, G., 2001. Groundwater contaminated with arsenic and fluoride in the Argentine Pampean plain. J. Environ. Hydrol. 9 (6), 1–9. <http://www.hydroweb.com/jeh/jeh2001/blara.pdf>.

Carrillo-Rivera, J.J., Cardona, A., Moss, D., 1996. Importance of vertical component of groundwater flow: a hydrogeochemical approach in the valley of San Luis Potosi, Mexico. J. Hydrol. 185, 23–44.

Carvalho, I.G., Cidu, R., Fanfani, L., Pitsch, H., Beaucaire, C., Zuddas, P., 2005. Environmental impact of uranium mining and ore processing in the Lagoa Real District, Bahia, Brazil. Environ. Sci. Technol. 39, 8646–8652.

Catalano, J.G., McKinley, J.P., Zachara, J.M., Heald, S.M., Smith, S.C., Brown Jr., G.E., 2006. Changes in uranium speciation through a depth sequence of contaminated Hanford sediments. Environ. Sci. Technol. 40 (8), 2517–2524.

Chae, G.T., Yun, S.T., Mayer, B., Kim, K.H., Kim, S.Y., Kwon, J.S., et al., 2007. Fluorine geochemistry in bedrock groundwater of South Korea. Sci. Total Environ. 385, 272–283.

Chakraborti, D., Chanda, C.R., Samanta, G., Chowdhury, U.K., 2000. Fluorosis in Assam, India. Curr. Sci. 78 (12), 1421–1423.

Chanpiwat, P., Sthiannopkao, S., Cho, K.H., Kim, K.-W., San, V., Suvanthong, B., et al., 2011. Contamination by arsenic and other trace elements of tube-well water along the Mekong River in Lao PDR. Environ. Pollut. 159, 567–576.

Chen, X., Yang, G., Chen, J., Chen, X., Wen, Z., Ge, K., 1980. Studies on the relations of selenium and keshan disease. Biol. Trace Elem. Res. 2, 91–107.

Chen, F., Burns, P.C., Ewing, R.C., 1999. Se: geochemical and crystallo-chemical retardation mechanisms. J. Nucl. Mater. 275, 81–94.

Chen, B., Zhu, Y.-G., Zhang, X., Jakobsen, I., 2005. The influence of mycorrhiza on uranium and phosphorus uptake by barley plants from a field-contaminated soil. Environ. Sci. Pollut. Res. 12, 325–331.

Chen, H., Yan, M., Yang, X., Chen, Z., Wang, G., Schmidt-Vogt, D., et al., 2012. Spatial distribution and temporal variation of high fluoride contents in groundwater and prevalence of fluorosis in humans in Yuanmou County, Southwest China. J. Hazard. Mater. 235, 201–209. Available from: https://doi.org/10.1016/j.jhazmat.2012.07.042.

Cheng, T., Barnett, M.O., Roden, E.E., Zhuang, J., 2004. Effects of phosphate on uranium (VI) adsorption to goethite-coated sand. Environ. Sci. Technol. 38 (22), 6059–6065.

Cho, B.W., Choo, C.O., 2019. Geochemical behavior of uranium and radon in groundwater of Jurassic Granite Area, Icheon, Middle Korea. Water 11 (6), 1278. Available from: https://doi.org/10.3390/w11061278.

Cid, F.D., Antón, R.I., Pardo, R., Vega, M., Caviedes-Vidal, E., 2011. Modelling spatial and temporal variations in the water quality of an artificial water reservoir in the semiarid Midwest of Argentina. Anal. Chim. Acta 705 (1), 243–252. Available from: https://doi.org/10.1016/j.aca.2011.06.013.

Cizdziel, J., Farmer, D., Hodge, V., Lindley, K., Stetzenbach, K., 2005. U-234/U-238 isotope ratios in groundwater from Southern Nevada: a comparison of alpha counting and magnetic sector ICP-MS. Sci. Total Environ. 350, 248–260.

Clifford, D.A., Ghurye, G.L., 2002. Metal-oxide adsorption, ion exchange, and coagulation-microfiltration for arsenic removal from water. In: Frankenberger Jr., W.T. (Ed.), Environmental Chemistry of Arsenic. Marcel Dekker, New York, pp. 217–245.

Coleman, L., Bragg, L.J., Finkelman, R.B., 1993. Distribution and mode of occurrence of selenium in US coals. Environ. Geochem. Health 15 (4), 215–228.

Cooke, T.D., Bruland, K.W., 1987. Aquatic chemistry of selenium: evidence of biomethylation. Environ. Sci. Technol. 21, 1214–1219.

Committee on Toxicity of Chemicals in Food, Consumer Products and the Environment (COT), 2006. COT Statement on Uranium Levels in Water Used to Reconstitute Infant Formula. Committee on Toxicity of Chemicals in Food, Consumer Products and the Environment, London.

Conde, J.E., Alaejos, M.S., 1997. Selenium concentrations in natural and environmental waters. Chem. Rev. 97 (6), 1979–2004.

Coomar, P., Mukherjee, A., Bhattacharya, P., Bundschuh, J., Verma, S., Fryar, A.E., et al., 2019. Contrasting controls on hydrochemistry of arsenic-enriched groundwater in the homologous tectonic settings of Andean and Himalayan basin aquifers, Latin America and South Asia. Sci. Total Environ. 689, 1370–1387.

Corbett, R.G., Manner, B.M., 1984. Fluoride in ground water of northeastern Ohio. Ground Water 22, 13–17.

Cotecchia, V., Tazioli, G.S., Magri, G., 1974. Isotopic measurements in research on seawater ingression in the carbonate aquifer of the Salentine Peninsula, Southern Italy. In: Proceedings of a Symposium on Isotope Techniques in Groundwater Hydrology, vol. I. International Atomic Energy Agency, Vienna, pp. 459–477.

Cotecchia, V., Grassi, D., Polemio, M., 2005. Carbonate aquifers in Apulia and seawater intrusion. G. Geol. Appl. 1, 219–231. Available from: https://doi.org/10.1474/GGA.2005−01.0-22.0022.

Cothern, C.R., Lappenbusch, W.L., 1983. Occurrence of uranium in drinking water in the U.S. Health Phys. 45, 89–99.

Coyte, R.M., Jain, R.C., Srivastava, S.K., Sharma, K.C., Khalil, A., Ma, L., et al., 2018. Large-scale uranium contamination of groundwater resources in India. Environ. Sci. Technol. Lett. 5, 341–347.

Cullen, W.R., Reimer, K.J., 1989. Arsenic speciation in the environment. Chem. Rev. 89 (4), 713–764.

Cullen, K., 2005. Uranium in groundwater. Geol. Surv. Irel. Groundw. Newsl. 11–12.

Currell, M., Cartwright, I., Raveggi, M., Han, D., 2011. Controls on elevated fluoride and arsenic concentrations in groundwater from the Yuncheng Basin, China. Appl. Geochem. 26 (4), 540–552. Available from: https://doi.org/10.1016/j.apgeochem.2011.01.012.

D'Alessandro, W., Bellomo, S., Parello, F., Brusca, L., Longo, M., 2008. Survey on fluoride, bromide and chloride contents in public drinking water supplies in Sicily (Italy). Environ. Monit. Assess. 145, 303–313.

Datta, P.S., Deb, D.L., Tyagi, S.K., 1996. Stable isotope (^{18}O) investigations on the processes controlling fluoride contamination of groundwater. J. Contam. Hydrol. 24 (1), 85–96. Available from: https://doi.org/10.1016/0169-7722(96)00004-6.

Daughney, C.J., Reeves, R.R., 2005. Definition of hydrochemical facies in the New Zealand National Groundwater Monitoring Programme. J. Hydrol. N. Z. 44 (2), 105–253.

Davraz, A., Sener, E., Sener, S., 2008. Temporal variations of fluoride concentration in Isparta public water system and health impact assessment (SW-Turkey). Environ. Geol. 56, 159–170.

de Caritat, P., Danilova, S., Jager, Ø., Reimann, C., Storrø, G., 1998. Groundwater composition near the nickel-copper smelting industry on the Kola Peninsula, central Barents Region (NW Russia and NE Norway). J. Hydrol. 208 (1–2), 92–107.

Dean, H.T., 1933. Distribution of mottled enamel in the United States. Public. Health Rep. 48 (25), 703–734.

Deering, M.F., Mohr, E.T., Sypniewski, B.F., Carlson, E.H., 1983. Regional hydrogeochemical patterns in ground water of northwestern Ohio and their relation to Mississippi Valley-type mineral occurrences. J. Geochem. Explor. 19, 225–241.

Deng, Y., Nordstrom, D.K., McCleskey, R.B., 2011. Fluoride geochemistry of thermal waters in Yellowstone National Park: I. Aqueous fluoride speciation. Geochim. Cosmochim. Acta 75, 4476–4489.

Desbarats, A.J., 2009. On elevated fluoride and boron concentrations in groundwaters associated with the Lake Saint-Martin impact structure, Manitoba. Appl. Geochem. 24, 915–927.

Deverel, S.J., Fujii, R., 1987. Processes affecting the distribution of selenium in shallow groundwater of agricultural areas, western San Joaquin Valley, California. United States Geological Survey Open-File Report 87–220. United States Geological Survey, Menlo Park, CA.

Deverel, S.J., Fujii, R., 1988. Processes affecting the distribution of selenium in shallow groundwater in agricultural areas, western San Josquin Valley, California. Water Resour. Res. 24, 516–524.

Deverel, S.J., Gallanthine, S.K., 2007. Relation of salinity and selenium in shallow groundwater to hydrologic and geochemical processes, Western San Joaquin Valley, California. J. Hydrol. 335, 223–224.

Dharmagunawardhane, H.A., Dissanayake, C.B., 1993. Fluoride problems in Sri Lanka. Environ. Manage. Health 4 (2), 9–16. Available from: https://doi.org/10.1108/09566169310033422.

Diaz-Barriga, F., Leyva, R., Quistian, J., Loyola-Rodriguez, J.B., Pozos, A., Grimaldo, M., 1997. Endemic fluorosis in San Luis Potosi, Mexico. Fluoride 30, 219–222.

Dibal, H.U., Schoeneich, K., Garba, I., Lar, U.A., Bala, E.A., 2012. Overview of fluoride distribution in major aquifer units of northern Nigeria. Health 4, 1287–1294. Available from: https://doi.org/10.4236/health.2012.412189.

Dissanayake, C.B., Weerasooriya, S.V.R., 1986. Fluorine as an indicator of mineralization – Hydrogeochemistry of a Precambrian mineralized belt in Sri Lanka. Chem. Geol. 56, 257–270.

Dissanayake, C.B., 1991. The fluoride problem in the groundwater of Sri Lanka—environmental management and health. Int. J. Environ. Stud. 38, 137–156.

Dregne, H.E., 2002. Land degradation in the drylands. Arid. Land. Res. Manage. 16, 99–132.

Drever, J.I., 1997. The Geochemistry of Natural Waters, third ed. Prentice Hall, Upper Saddle River; NJ.

Duff, M., Hunter, D., Bertsch, P., Amrhein, C., 1999. Factors influencing uranium reduction and solubility in evaporation pond sediments. Biogeochemistry 45 (1), 95–114.

Duff, M.C., Coughlin, J.U., Hunter, D.B., 2002. Uranium co-precipitation with iron oxide minerals. Geochim. Cosmochim. Acta 66, 3533–3547.

Earth Sciences for Society Foundation (ESS), 2005. Groundwater—reservoir for a thirsty planet?. In: Earth Sciences for Society—A Prospectus for a Key Theme of the International Year of Planet Earth. International Union of Geological Sciences Secretariat, Geological Survey of Norway.

Echevarria, G., Sheppard, M.I., Morel, J., 2001. Effect of pH on the sorption of uranium in soils. J. Environ. Radioact. 53 (2), 257–264.

Edmunds, W.M., 1973. Trace element variations across an oxidation-reduction barrier in a limestone aquifer. In: Proceedings of the Symposium of Hydrogeochemistry, Clarke Co. Washington, Tokyo, 1970, pp. 500–526.

Edmunds, W.M., Walton, N., 1983. The Lincolnshire limestone—hydrogeochemical evolution over a ten-year period. J. Hydrol. 61, 201–211.

Edmunds, W.M., Cook, J.M., Kinniburgh, D.G., Miles, D.G., Trafford, J.M., 1989. Trace Element Occurrence in British Groundwaters. In: British Geological Survey Research Report SD/89/3. British Geological Survey, Keyworth, p. 424.

Edmunds, W.M., 1994. Characterization of groundwaters in semi-arid and arid zones using minor elements. In: Nash, H., McCall, G.J.H. (Eds.), Groundwater Quality. Chapman & Hall, London, pp. 19–30.

Edmunds, W.M., Smedley, P.L., 2013. Fluoride in natural waters. In: Selinus, O., Alloway, B., Centeno, J.A., Finkelman, R.B., Fuge, R., Lindh, U., Smedley, P.L. (Eds.), Essentials of Medical Geology, revised ed. Springer, pp. 311–336.

Ellis, A.J., Mahon, W.A.J., 1977. Chemistry and Geothermal Systems. Academic Press, New York.

Ericson, B., Hanrahan, D., Kong, V., 2008. The World's Worst Pollution Problems: The Top Ten of the Toxic Twenty. Blacksmith Institute and Green Cross.

Estifanos, H., 2005. Groundwater Chemistry and Recharge Rate in Crystalline Rocks: Case Study From the Eritrean Highland. Land and Water Resource Engineering, Stockholm. <http://urn.kb.se/resolve?urn = urn:nbn:se:kth:diva-4085> (30.05.07.).

Ettayfi, N., Bouchaou, L., Michelot, J.L., Tagma, T., Warner, N., Boutaleb, S., et al., 2012. Geochemical and isotopic (oxygen, hydrogen, carbon, strontium) constraints for the origin, salinity, and residence time of groundwater from a carbonate aquifer in the Western Anti-Atlas Mountains, Morocco. J. Hydrol. 438, 97–111.

European Food Safety Authority (EFSA), 2009. Scientific opinion of the panel on contaminants in the food chain on a request from German Federal Institute for Risk Assessment (BfR) on uranium in foodstuff, in particular mineral water. EFSA J. 1018, 1–59.

Fanghanel, T., Neck, V., 2002. Aquatic chemistry and solubility phenomena of actinide oxides/hydroxides. Pure Appl. Chem. 74 (10), 1895–1908.

Fantong, W.Y., Satake, H., Ayonghe, S.N., Suh, E.C., Adelana, S.M.A., Fantong, E.B.S., et al., 2010. Geochemical provenance and spatial distribution of fluoride in groundwater of Mayo Tsanaga River Basin, Far North Region, Cameroon: implications for incidence of fluorosis and optimal consumption dose. Environ. Geochem. Health 32, 147–163.

Fara, M., Chandrasekharam, D., Minissale, A., Minissale, I., 1999. Hydrogeochemistry of Damt thermal springs, Yemen Republic. Geothermics 28, 241–252.

Farber, E., Vengosh, A., Gavrieli, I., Marie, A., Bullen, T.D., Mayer, B., et al., 2007. The geochemistry of groundwater resources in the Jordan Valley: the impact of the Rift Valley brines. Appl. Geochem. 22, 494–514.

Farooqi, A., Masuda, H., Firdous, N., 2007. Toxic fluoride and arsenic contaminated groundwater in the Lahore and Kasur districts, Punjab, Pakistan and possible contaminant sources. Environ. Pollut. 145 (3), 839–849. Available from: https://doi.org/10.1016/j.envpol.2006.05.007.

Fernández-Martínez, A., Charlet, L., 2009. Selenium environmental cycling and bioavailability: a structural chemist point of view. Rev. Environ. Sci. Biotechnol. 8 (1), 81–110.

Fidelibus, M.D., Tulipano, L., 1990. Major and minor ions as natural tracers in mixing phenomena in coastal carbonate aquifers of Apulia. In: Proceeding of 11th Salt Water Intrusion Meeting, Gdansk, pp. 283–293.

Fidelibus, M.D., Tulipano, L., 1996. Regional flow of intruding sea water in the carbonate aquifers of Apulia (Southern Italy). In: Proceedings of the 14th Salt Water Intrusion Meeting, Malmo, pp. 230–240.

Finch, R.J., Ewing, R.C., 1992. The corrosion of uraninite under oxidizing conditions. J. Nucl. Mater. 190, 133–156.

Finch, R., Murakami, T., 1999. Systematics and paragenesis of uranium minerals. Rev. Mineral. Geochem. 38 (1), 91–179.

Finkelman, R.B., Belkin, H.E., Zheng, B., 1999. Health impacts of domestic coal use in China. Proc. Natl. Acad. Sci. U.S.A. 96, 3427–3431.

Finneran, K.T., Anderson, R.T., Nevin, K.P., Lovley, D.R., 2002. Potential for bioremediation of uranium-contaminated aquifers with microbial U (VI) reduction. Soil Sediment Contam. Int. J. 11 (3), 339–357.

Fitzgerald, J., Cunliffe, D., Rainow, S., Dodds, S., Hostetler, S., Jacobson, G., 2000. Groundwater Quality and Environmental Health Implications, Anangu Pitjantjatjara Lands, South Australia. Bureau of Rural Sciences, Canberra.

Fleming, G.A., 1980. Essential micronutrients II: iodine and selenium. In: Davis, B.E. (Ed.), Applied Soil Trace Elements. Wiley, New York, pp. 199–234.

Fleischer, M., Robinson, W.O., 1963. Some problems of the geochemistry of fluorine. In: Shaw, D.M. (Ed.), Studies in Analytical Geochemistry. Royal Society of Canada Special Publication No. 6, Canada. University of Toronto Press, Toronto, pp. 58–75.

Foletti, C., Paz, G., 2001. Diagnóstico de fluorosis dental en 39 comunidades del valle de Sula, Honduras. En Superación sanitaria y ambiental: et reto. AIDIS, Tegucigalpa.

Fordyce, F.M., Zhang, G., Green, K., Liu, X., 1998. Soil, Grain and water chemistry and human Selenium imbalances in Enshi District, Hubei Province, China. In: British Geological Survey Overseas Geology Series Technical Report WC/96/54.

Fordyce, F.M., Guangdi, Z., Green, K., Xinping, L., 2000. Soil, grain and water chemistry in relation to human selenium-responsive diseases in Enshi District, China. Appl. Geochem. 15 (1), 117–132.

Fordyce, F.M., 2013. Selenium deficiency and toxicity in the environment. In: Selinus, O., Alloway, B., Centeno, J.A., Finkelman, R.B., Fuge, R., Lindh, U., et al., Essentials of Medical Geology, revised ed. Springer, pp. 375–416. Available from: https://doi.org/10.1007/978-94-007-4375-5_16.

Fox, P.M., Davis, J.A., Hay, M.B., Conrad, M.E., Campbell, K.M., Williams, K.H., et al., 2012. Rate-limited U(VI) desorption during a small-scale tracer test in a heterogeneous uranium-contaminated aquifer. Water Resour. Res. 48 (5).

Fredrickson, J.K., Zachara, J.M., Kennedy, D.W., Duff, M.C., Gorby, Y.A., Li, S.-M.W., et al., 2000. Reduction of U(VI) in goethite (α-FeOOH) suspensions by a dissimilatory metal-reducing bacterium. Geochim. Cosmochim. Acta 64 (18), 3085–3098.

Frengstad, B., Skrede, A.K.M., Banks, D., Krog, J.R., Siewers, U., 2000. The chemistry of Norwegian groundwaters: III. The distribution of trace elements in 476 crystalline bedrock groundwaters, as analysed by ICP-MS techniques. Sci. Total Environ. 246 (1–3), 101–117.

Fuge, R., Andrews, M.J., 1988. Fluorine in the UK environment. Environ. Geochem. Health 10 (3/4), 96–104. Available from: https://doi.org/10.1007/BF01758677.

Fuhong, R., Shuqin, J., 1988. Distribution and formation of high fluorine groundwater in China. Environ. Geol. Water Sci. 12 (1), 3–10. Available from: https://doi.org/10.1007/BF02574820.

Fujii, R., Swain, W.C., 1995. Areal distribution of selected trace elements, salinity, and major ions in shallow ground water, Tulare Basin, Southern San Joaquin Valley, California. United States Geological Survey Water Resources Investigation Report 95-4048, USGS, Sacramento, California.

Fujii, R., Deverel, S.J., Hatfield, D.B., 1988. Distribution of selenium in soils of agricultural fields, Western San Joaquin Valley, California. Soil Sci. Soc. Am. J. 52 (5), 1274–1283.

Gaciri, S.J., Davies, T.C., 1993. The occurrence and geochemistry of fluoride in some natural waters of Kenya. J. Hydrol. 143, 395–412.

Gascoyne, M., 2004. Hydrogeochemistry, groundwater ages and sources of salts in a granitic batholith on the Canadian Shield, southeastern Manitoba. Appl. Geochem. 19, 519–560.

Gates, T.K., Cody, B.M., Donnelly, J.P., Herting, A.W., Bailey, R.T., Mueller Price, J., 2009. Assessing selenium contamination in the irrigated stream aquifer system of the Arkansas River, Colorado. J. Environ. Qual. 38 (6), 2344–2356.

Ghiglieri, G., Carletti, A., Pittalis, D., 2012. Analysis of salinization processes in the coastal carbonate aquifer of Porto Torres (NW Sardinia, Italy). J. Hydrol. 432–433 (11), 43–51. <https://doi.org/10.1016/j.jhydrol.2012.02.016>.

Giddings, M., 2005. Uranium in drinking-water: background document for development of WHO guidelines for drinking-water quality. WHO/SDE/WSH/03.04/118. World Health Organization.

Gierup, J., 2007. Personal Communication. Swedish Geological Survey.

Goldhaber, M.B., Irwin, E., Atkins, B., Lee, L., Black, D.D., Zappia, H., et al., 2001. Arsenic in Stream Sediments of Northern Alabama. United States Geological Survey Miscellaneous Field Studies Map, Reston, VA.

Goldhaber, S.B., 2003. Trace element risk assessment: essentiality vs. toxicity. Regul. Toxicol. Pharmacol. 38, 232–242.

Gómez, J.J., Lillo, J., Sahún, B., 2006a. Naturally occurring arsenic in groundwater and identification of the geochemical sources in the Duero Cenozoic Basin, Spain. Environ. Geol. 50 (8), 1151–1170.

Gómez, P., Garralón, A., Buil, B., Turrero, M.J., Sánchez, L., de la Cruz, B., 2006b. Modeling of geochemical processes related to uranium mobilization in the groundwater of a uranium mine. Sci. Total Environ. 366, 295–309.

Gomez, M.L., Blarasin, M.T., Martínez, D.E., 2009. Arsenic and fluoride in a loess aquifer in the central area of Argentina. Environ. Geol. 57 (1), 143–155. Available from: https://doi.org/10.1007/s00254-008-1290-4.

Goode, D.J., Wilder, R.J., 1987. Groundwater contamination near a uranium tailings disposal site in Colorado. Ground Water 25, 545–554.

Guilliot, S., Charlet, L., 2007. Bengal arsenic, an archive of Himalaya orogeny and paleohydrology. J. Environ. Sci. Health, A 42, 1785–1794.

Guo, H., Wang, Y., 2005. Geochemical characteristics of shallow groundwater in Datong basin, northwestern China. J. Geochem. Explor. 87 (3), 109–120.

Guo, H., Zhang, Y., Xing, L., Jia, Y., 2012. Spatial variation in arsenic and fluoride concentrations of shallow groundwater from the town of Shahai in the Hetao basin, Inner Mongolia. Appl. Geochem. 27 (11), 2187–2196. Available from: https://doi.org/10.1016/j.apgeochem.2012.01.016.

Gupta, S., Kumar, A., Ojha, C.K., Seth, G., 2004. Chemical analysis of ground water of Sanganer area, Jaipur in Rajasthan. J. Environ. Sci. Eng. 46 (1), 74–78.

Gustafsson, J.P., Johnsson, L., 1994. The association between selenium and humic substances in forested ecosystems – laboratory evidence. Appl. Organomet. Chem. 8, 141–147.

Hamilton, S.J., 1999. Hypothesis of historical effects from selenium on endangered fish in the Colorado River Basin. Hum. Ecol. Risk Assess. 5, 1153–1180.

Hamilton, S.J., Muth, R.T., Wadell, B., May, T.W., 2000. Hazard assessment of selenium and other trace elements in wild larval razorback sucker from the Green River, Utah. Ecotoxicol. Environ. Saf. 45, 132–147.

Hammond, D.E., Zukin, J.G., Ku, T.L., 1988. The kinetics of radioisotope exchange between brine and rock in geothermal system. J. Geophys. Res. 93, 13175–13186.

Han, Y., Yan, S.L., Ma, H.T., et al., 2006. Groundwater Resources and Environmental Issues Assessment in the Six Major Basins of Shanxi Province. Geological Publishing House, Beijing (in Chinese).

Han, D., Post, V.E.A., Song, X., 2015. Groundwater salinization processes and reversibility of seawater intrusion in coastal carbonate aquifers. J. Hydrol. 531 (3), 1067–1080. <https://doi.org/10.1016/j.jhydrol.2015.11.013>.

Handa, B.K., 1975. Geochemistry and genesis of fluoride-containing ground waters in India. Ground Water 13, 275–281.

Hattori, K.H., Guillot, S., 2003. Volcanic fronts form as a consequence of serpentinite dehydration in the forearc mantle wedge. Geology 31 (6), 525–528.

Hayes, T.S., Miller, M.M., Orris, G.J., Piatak, N.M., 2017. Fluorine, chap. G. In: Schulz, K.J., DeYoung Jr., J.H., Seal II, R.R., Bradley, D.C. (Eds.), Critical Mineral Resources of the United States—Economic and Environmental Geology and Prospects for Future Supply. United States Geological Survey Professional Paper 1802, pp. G1–G80. <https://doi.org/10.3133/pp1802G>.

Health Canada, 1987. Uranium. In: Guidelines for Canadian Drinking Water Quality: Supporting Documentation. pp. 1–10.

Heikens, A., Sumarti, S., Van Bergen, M., Widianarko, B., Fokkert, L., Van Leeuwen, K., et al., 2005. The impact of the hyperacidic Ijen Crater Lake: risks of excess fluoride to human health. Sci. Total Environ. 346 (1–3), 56–69.

Hem, J.D., 1985. Study and interpretation of the chemical characteristics of natural water. In: USGS Water Supply Paper 2254, p. 263.

Hendry, M.J., Schwartz, F.W., 1988. An alternative view on the origin of chemical and isotopic patterns in groundwater from the Milk River Aquifer, Canada. Water Resour. Res. 24, 1747–1763.

Herbel, M.J., Blum, J.S., Oremland, R.S., Borglin, S.E., 2003. Reduction of elemental selenium to selenide: experiments with anoxic sediments and bacteria that respire Se-oxyanions. Geomicrobiol. J. 20, 587–602.

Herczeg, A.L., Torgersen, T., Chivas, A.R., Habermehl, M.A., 1991. Geochemistry of ground waters from the Great Artesian Basin, Australia. J. Hydrol. 126, 225–245.

Herczeg, A.L., Edmunds, W.M., 2000. Inorganic ions as tracers. In: Cook, P., Herczeg, A.L. (Eds.), Environmental Tracers in Subsurface Hydrology. Kluwer Academic, Boston, pp. 31–77.

Hernández-García, M.E., Custodio, E., 2004. Natural baseline quality of Madrid Tertiary Detrital Aquifer groundwater (Spain): a basis for aquifer management. Environ. Geol. 46 (2), 173–188.

Herring, J.S., 2013. Uranium and thorium resources. In: Tsoulfanidis, N. (Ed.), Nuclear Energy. Springer, New York, pp. 463–490.

Herut, B., Starinsky, A., Katz, A., Bein, A., 1990. The role of seawater freezing in the formation of subsurface brines. Geochim. Cosmochim. Acta 54, 13–21.

Héry, M., Volant, A., Garing, C., Luquot, L., Poulichet, F.A., Gouze, P., 2014. Diversity and geochemical structuring of bacterial communities along a salinity gradient in a carbonate aquifer subject to seawater intrusion. FEMS Microbiol. Ecol. 90 (3), 922–934. <https://doi.org/10.1111/1574-6941.12445>.

Hess, C.T., Michel, J., Horton, T.R., Pritchard, H.M., Coniglio, W.A., 1985. The occurrence of radioactivity in public water supplies in the United States. Health Phys. 48, 553–586.

Hitchon, B., 1995. Fluorine in formation waters in Canada. Appl. Geochem. 10, 357–367.

Hildreth, W., 1981. Gradients in silicic magma chambers—Implications for lithospheric magmatism. J. Geophys. Res. 86 (B11), 10153–10192. <https://doi.org/10.1029/jb086ib11p10153>.

Hopkins, J., 1993. Water-Quality Evaluation of the Ogallala Aquifer, Texas. Texas Water Development Board, Austin, TX.

Houben, G., Tünnermeier, T., Himmelsbach, T., 2003. Hydrogeology of the Kabul Basin Part II: Groundwater geochemistry and microbiology. In: Federal Institute for Geoscience and Natural Resources (BGR), Record no. 10277/05. <http://www.bgr.de/app/projektspiegel/fachbeitraege/hydrogeology_kabul_basin_2.pdf>.

Hudak, P.F., 1999. Fluoride levels in Texas groundwater. J. Environ. Sci. Health, A 34 (8), 1659–1676. Available from: https://doi.org/10.1080/10934529909376919.

Hudak, P.F., Sanmanee, S., 2003. Spatial patterns of nitrate, chloride, sulfate, and fluoride concentrations in the woodbine aquifer of North-Central Texas. Environ. Monit. Assess. 82 (3), 311–320. Available from: https://doi.org/10.1023/A:1021946402095.

Hudak, P.F., 2009. Elevated fluoride and selenium in West Texas groundwater. Bull. Environ. Contam. Toxicol. 82, 39–42. Available from: https://doi.org/10.1007/s00128-008-9583-6.

Hsi, C.D., Langmuir, D., 1985. Adsorption of uranyl onto ferric oxyhydroxides: application of the surface complexation site-binding model. Geochim. Cosmochim. Acta 49, 1931–1941.

Hsissou, Y., Mudry, J., Mania, J., Bouchaou, L., Chauve, P., 1999. Utilisation du rapport Br/Cl pour déterminer l'origine de la salinité des eaux souterraines: exemple de la plaine du Souss (Maroc). C.R. Acad. Sci., Ser. IIA: Earth Planet. Sci. 328, 381–386.

Hu, Z., Gao, S., 2008. Upper crustal abundances of trace elements: a revision and update. Chem. Geol. 253, 205–221.

Huston, D.L., Sie, S.H., Suter, G.F., Cooke, D.R., Both, R.A., 1995. Trace elements in sulfide minerals from eastern Australian volcanic-hosted massive sulfide deposits: part I, proton microprobe analyses of pyrite, chalcopyrite, and sphalerite, and part II, selenium levels in pyrite—comparison with delta ^{34}S values and implications for the source of sulfur in volcanogenic hydrothermal systems. Econ. Geol. 90 (5), 1167–1196.

Indermitte, E., Saava, A., Karro, E., 2009. Exposure to high fluoride drinking water and risk of dental fluorosis in Estonia. Int. J. Environ. Res. Public Health 6, 710–721.

Ivkovic, K.M., Watkins, K.L., Cresswell, R.G., Bauld, J.A., 1998. Groundwater quality assessment of the fractured rock aquifer of the Picadilly Valley, South Australia. In: Australian Geological Survey Organisation, Record 1998/16.

Jackson, R.B., Carpenter, S.R., Dahm, C.N., McKinght, D.M., Naiman, R.J., Postel, S.L., et al., 2001. Water in a changing world. Ecol. Appl. 11, 1027–1045.

Selenium in agriculture and the environment. In: Jacobs, L.W. (Ed.), Soil Science Society of America Special Publication. Soil Science Society of America, Madison, WI.

James, L.J., Sinha, A.K., 2006. Geochemical coupling of uranium and phosphorous in soils overlying an unmined uranium deposit: Coles Hill, Virginia. J. Geochem. Explor. 91 (1–3), 56–70.

Jayawardana, D.T., Pitawala, H.M., Ishiga, H., 2012. Geochemical assessment of soils in districts of fluoride-rich and fluoride-poor groundwater, north-central Sri Lanka. J. Geochem. Explor. 114, 118–125. Available from: https://doi.org/10.1016/j.gexplo.2012.01.004.

Jerden, J.L., Sinha, A.K., Zelazny, L., 2003. Natural immobilization of uranium by phosphate mineralization in an oxidizing saprolite-soil profile: chemical weathering of the Coles Hill uranium deposit, Virginia. Chem. Geol. 199, 129–157.

Jones, B.F., Eugster, H.P., Reitig, S.L., 1977. Hydrochemistry of the Lake Magadi basin, Kenya. Geochim. Cosmochim. Acta 41, 53–72.

Kawabata, Y., Yamamoto, M., Aparin, V., Ko, S., Shiraishi, K., Nagai, M., et al., 2006. Uranium pollution of water in the western part of Uzbekistan. J. Radioanal. Nucl. Chem. 270, 137–141.

Kay, G.W.C., Laby, T.H., 1995. Tables of Physical and Chemical Constants and Some Mathematical Functions, 16th ed. National Physical Laboratory, Teddington.

Kelly, S.D., Newville, M.G., Cheng, L., Kemner, K.M., Sutton, S.R., Fenter, P., et al., 2003. Uranyl incorporation in natural calcite. Environ. Sci. Technol. 37 (7), 1284–1287.

Kelly, S.D., Rasbury, E.T., Chattopadhyay, S., Kropf, A.J., Kemner, K.M., 2006. Evidence of a stable uranyl site in ancient organic-rich calcite. Environ. Sci. Technol. 40 (7), 2262–2268.

Keshavarzi, B., Moore, F., Esmaeili, A., Rastmanesh, F., 2010. The source of fluoride toxicity in Muteh area, Isfahan, Iran. Environ. Earth Sci. 61, 777–786. Available from: https://doi.org/10.1007/s12665-009-0390-0.

Kilham, P., Hecky, R.E., 1973. Fluoride: geochemical and ecological significance in East African waters and sediments. Limnol. Oceanogr. 18, 932–945.

Kim, Y., Lee, K.-S., Koh, D.-C., Lee, D.-H., Lee, S.-G., Park, W.-B., et al., 2003. Hydrogeochemical and isotopic evidence of groundwater salinization in a coastal aquifer: a case study in Jeju volcanic island, Korea. J. Hydrol. 270, 282–294.

Kim, Y.S., Park, H.S., Kim, J.Y., Park, S.K., Cho, B.W., Sung, I.H., et al., 2004. Health risk assessment for uranium in Korean groundwater. J. Environ. Radioact. 77, 77–85.

Kim, K., Jeong, G.Y., 2005. Factors influencing natural occurrence of fluoride rich ground waters: a case study in the southeastern part of the Korean Peninsula. Chemosphere 58 (10), 1399–1408.

Kim, Y., Kim, J.Y., Kim, K., 2010. Geochemical characteristics of fluoride in groundwater of Gimcheon, Korea: lithogenic and agricultural origins. Environ. Earth Sci. 1139–1148. Available from: https://doi.org/10.1007/s12665-010-0789-7.

Kloppmann, W., Négrel, P., Casanova, J., Klinge, H., Schelkes, K., Guerrot, C., 2001. Halite dissolution derived brines in the vicinity of a Permian salt dome (N German Basin). Evidence from boron, strontium, oxygen, and hydrogen isotopes. Geochim. Cosmochim. Acta 65, 4087–4101.

Koh, D.C., Chae, G.T., Yoon, Y.Y., Kang, B.R., Koh, G.W., Park, K.H., 2009. Baseline geochemical characteristics of groundwater in the mountainous area of Jeju Island, South Korea: Implications for degree of mineralization and nitrate contamination. J. Hydrol. 376, 81–93.

Kohler, M., Curtis, G.P., Meece, D.E., Davis, J.A., 2004. Methods for estimating adsorbed uranium (VI) and distribution coefficients of contaminated sediments. Environ. Sci. Technol. 38 (1), 240–247.

Komnitsas, K., Xenidis, A., Adam, K., 1995. Oxidation of pyrite and arsenopyrite in sulphidic spoils in Lavrion. Miner. Eng. 12, 1443–1454.

Konietzka, R., 2015. Gastrointestinal absorption of uranium compounds—a review. Regul. Toxicol. Pharmacol. 71, 125–133.

Korte, N.E., Fernando, Q., 1991. A review of arsenic (III) in groundwater. Crit. Rev. Environ. Control 21, 1–39.

Koritnig, S., 1951. Ein Beitrag Zur Geochemie Des Fluor. Geochim. Cosmochim. Acta 1 (2), 89–116.

Kotoky, P., Barooah, P.K., Baruah, M.K., Goswami, A., Borah, G.C., Gogoi, H.M., et al., 2008. Fluoride and endemic fluorosis in the Karbianglong district, Assam, India. Fluoride 41, 72–75.

Kraemer, T.F., Reid, D.F., 1984. The occurrence and behavior of radium in saline formation water of the US Gulf coast region. Chem. Geol. 46, 153–174.

Krishnaswami, S., Bhushan, R., Baskaran, M., 1991. Radium isotopes and ^{222}Rn in shallow brines, Kharaghoda (India). Chem. Geol. 87, 125–136.

Kruse, E., Ainchil, J., 2003. Fluoride variations in groundwater of an area in Buenos Aires Province, Argentina. Environ. Geol. 44 (1), 86–89. Available from: https://doi.org/10.1007/s00254-002-0702-0.

Kubota, Y., Yokota, D., Ishiyama, Y., 2003. Arsenic concentration in hot spring waters from the Nigata Plain and Shinji Lowland, Japan. Part 2: Source supply of arsenic in arsenic contaminated ground water problem. Earth Sci. (Chikyukagaku) 55, 11–22.

Kulp, T.R., Pratt, L.M., 2004. Speciation and weathering of selenium in upper cretaceous chalk and shale from South Dakota and Wyoming, USA. Geochim. Cosmochim. Acta 68 (18), 3687–3701.

Kumar, M., Ramanathan, A.L., Mukherjee, A., Verma, S., Rahman, M.M., Naidu, R., 2018. Hydrogeomorphological influences for arsenic release and fate in the central Gangetic Basin, India. Environ. Technol. Innov. 12, 243–260.

Kundu, M.C., Mandal, B., 2009. Assessment of potential hazards of fluoride contamination in drinking groundwater of an intensively cultivated district in West Bengal, India. Environ. Monit. Assess. 152 (1–4), 97–103. Available from: https://doi.org/10.1007/s10661-008-0299-1.

Kurttio, P., Auvinen, A., Salonen, L., Saha, H., Pekkanen, J., Makelainen, I., et al., 2002. Renal effects of uranium in drinking water. Environ. Health Perspect. 110, 337–342.

Kurttio, P., Komulainen, H., Leino, A., Salonen, L., Auvinen, A., Saha, H., 2005. Bone as a possible target of chemical toxicity of natural uranium in drinking water. Environ. Health Perspect. 113, 68–72.

Kurttio, P., Harmoinen, A., Saha, H., Salonen, L., Karpas, Z., Komulainen, H., et al., 2006. Kidney toxicity of ingested uranium from drinking water. Am. J. Kidney Dis. 47, 972–982.

Lahermo, P., Alfthan, G., Wang, D., 1998. Selenium and arsenic in the environment in Finland. J. Environ. Pathol. Toxicol. Oncol. 17 (3–4), 205–216.

Langmuir, D., 1978. Uranium solution-mineral equilibria at low temperatures with applications to sedimentary ore deposits. Geochim. Cosmochim. Acta 42, 547–569.

Larsen, R.M., Watkins, K.W., Steel, N.A., Appleyard, S.J., Bauld, J.A., 1998. Groundwater quality assessment of the Jandakot Mound, Swan Coastal Plain, Western Australia. In: Australian Geological Survey Organisation, Record 1998/18.

Lavado, R.S., Reinaudi, N., 1979. Fluoride in salt affected soils of La Pampa (Republica Argentina). Fluoride 12, 28–32.

Lee, M.H., Choi, G.S., Cho, Y.H., Lee, C.W., Shin, H.S., 2001. Concentrations and activity ratios of uranium isotopes in the groundwater of the Okchun Belt in Korea. J. Environ. Radioact. 57, 105–116.

Li, C., Gao, X., Wang, Y., 2015. Hydrogeochemistry of high-fluoride groundwater at Yuncheng Basin, northern China. Sci. Total Environ. 508, 155–165. Available from: https://doi.org/10.1016/j.scitotenv.2014.11.045.

Li, Y., Ye, F., Wang, A., Wang, D., Yang, B., Zheng, Q., et al., 2016. Chronic arsenic poisoning probably caused by arsenic-based pesticides: findings from an investigation study of a household. Int. J. Environ. Res. Public Health 13 (1), 133. Available from: https://doi.org/10.3390/ijerph13010133.

Liesch, T., Hinrichsen, S., Goldscheider, N., 2015. Uranium in groundwater—fertilizers versus geogenic sources. Sci. Total Environ. 536, 981–995.

Lin, N.-F., Tang, J., Bian, J.-M., 2002. Characteristics of environmental geochemistry in the arseniasis area of the Inner Mongolia of China. Environ. Geochem. Health 24 (3), 249–259 (in Chinese).

Liu, C., Zachara, J.M., Qafoku, O., McKinley, J.P., Heald, S.M., Wang, Z., 2004. Dissolution of uranyl microprecipitates in subsurface sediments at Hanford Site, USA. Geochim. Cosmochim. Acta 68 (22), 4519–4537.

Liu, C.-W., Wang, S.-W., Jang, C.-S., Lin, K.-H., 2006. Occurrence of arsenic in ground water in the Choushui River alluvial fan, Taiwan. J. Environ. Qual. 35 (1), 68–75.

Liu, C., Shi, Z., Zachara, J.M., 2009. Kinetics of uranium(VI) desorption from contaminated sediments: effect of geochemical conditions and model evaluation. Environ. Sci. Technol. 43, 6560–6566.

Looie, S.B., Moore, F., 2010. A study of fluoride groundwater occurrence in Posht-e-Koohe Dashtestan, South of Iran. World Appl. Sci. J. 8 (11), 1317–1321.

Love, A.J., Herczeg, A.L., Sampson, L., Cresswell, R.G., Fifield, L.K., 2000. Sources of chloride and implications for 36Cl dating of old groundwater, Southwestern Great Artesian Basin, Australia. Water Resour. Res. 36, 1561–1574.

Lovley, D.R., Phillips, E.J., 1992. Bioremediation of uranium contamination with enzymatic uranium reduction. Environ. Sci. Technol. 26 (11), 2228–2234.

Lu, Y., Sun, Z.R., Wu, L.N., Wang, X., Lu, W., Liu, S.S., 2000. Effect of high fluoride water on intelligence in children. Fluoride 33 (2), 74–78.

Luo, Z.D., Zhang, Y.M., Ma, L., Zhang, G.Y., He, X., Wilson, R., et al., 1997. Chronic arsenicism and cancer in Inner Mongolia – consequences of well-water arsenic levels greater than 50 µg/l. In: Abernathy, C.O., Calderon, R.L., Chappell, W.R. (Eds.), Arsenic Exposure and Health Effects II. Elsevier, Oxford, pp. 54–72.

Luo, K., Feng, F., Li, H., Chou, C.L., Feng, Z., Yunshe, D., 2008. Studies on geological background and source of fluorine in drinking water in the North China Plate fluorosis areas. Toxicol. Environ. Chem. 90 (2), 237–246. Available from: https://doi.org/10.1080/02772240701456091.

Manceau, A., Charlet, L., 1994. The mechanism of selenate adsorption on goethite and hydrous ferric oxide. J. Colloid Interface Sci. 168, 87–93.

Mandal, B.K., Suzuki, K.T., 2002. Arsenic round the world: a review. Talanta 58 (1), 201–235.

MacDonald, H.E., 1969. Fluoride as air pollutant. Fluoride Q. Rep. 2 (1), 4–12. <http://www.fluorideresearch.org/21/files/214-12.pdf>.

Magaritz, M., Nadler, A., Kafri, U., Arad, A., 1984. Hydrogeochemistry of continental brackish waters in the southern Coastal Plain, Israel. Chem. Geol. 42, 159–176.

Maithani, P.B., Gurjar, R., Banerjee, R., Balaji, B.K., Ramachandran, S., Singh, R., 1998. Anomalous fluoride in groundwater from western part of Sirohi district, Rajasthan, and its crippling effects on human health. Curr. Sci. 74, 773–777.

Marie, A., Vengosh, A., 2001. Sources of salinity in ground water from Jericho Area, Jordan Valley. Ground Water 39, 240–248.

Marion, G.M., Farren, R.E., Komrowski, A.J., 1999. Alternative pathways for seawater freezing. Cold Reg. Sci. Technol. 29, 259–266.

Martinez, D.E., Quiroz Londono, O.M., Massone, H.E., Palacio Buitrago, P., Lima, L., 2012. Hydrogeochemistry of fluoride in the Quequen river basin: natural pollutants distribution in the argentine pampa. Environ. Earth Sci. 65, 411–420.

Mars, J.C., Crowley, J.K., 2003. Mapping mine wastes and analyzing areas affected by selenium-rich water runoff in southeast Idaho using AVIRIS imagery and digital elevation data. Remote Sens. Environ. 84, 422–436.

Maslia, M.L., Prowell, D.C., 1990. Effect of faults on fluid flow and chloride contamination in a carbonate aquifer system. J. Hydrol. 115, 1–49.

Masscheleyn, P.H., Delaune, R.D., Patrick, J.W.H., 1989. Transformations of selenium as affected by sediment oxidation-reduction potential and pH. Environ. Sci. Technol. 24 (1), 91–96.

Matamoros-Veloza, A., Newton, R.J., Benning, L.G., 2011. What controls selenium release during shale weathering? Appl. Geochem. 26 (Suppl.), S222–S226. <https://doi.org/10.1016/j.apgeochem.2011.03.109>.

Matschullat, J., 2000. Arsenic in the geosphere – a review. Sci. Total Environ. 249 (1–3), 297–312.

Mayer, X.M., Ruprecht, J.K., Bari, M.A., 2005. Stream salinity status and trends in south-west Western Australia. Salinity and land use impacts series. In: Department of Environment, Report No SLUI 38.

McCaffrey, L.P., 1998. Distribution and Causes of High Fluoride Groundwater in the Western Bushveld Area of South Africa (Ph.D. thesis). Faculty of Science, Department of Geological Sciences, University of Cape Town. <http://hdl.handle.net/11427/9597>

Mehta, S., Fryar, A.E., Banner, J.L., 2000a. Controls on the regional-scale salinization of the Ogallala aquifer, Southern High Plains, Texas, USA. Appl. Geochem. 15, 849–864.

Mehta, S., Fryar, A.E., Brady, R.M., Morin, R.H., 2000b. Modeling regional salinization of the Ogallala aquifer, Southern High Plains, Texas, USA. J. Hydrol. 238, 44–64.

Merkel, B.J., Sperling, B., 1998. Hydrogeochemische Stoffsysteme Teil II. Bonn: Kommissionsvertrieb Wirtschafts- und Verlagsgesellschaft Gas und Wasser mbH.

Messaïtfa, A., 2008. Fluoride contents in groundwaters and the main consumed foods (dates and tea) in Southern Algeria region. Environ. Geol. 55 (2), 377–383. Available from: https://doi.org/10.1007/s00254-007-0983-4.

Miller, M.R., Brown, P.L., Donovan, J.J., Bergatino, R.N., Sonderegger, J.L., Schmidt, F.A., 1981. Saline seep development and control in the North American Great Plains – Hydrogeological aspects. Agric. Water Manage. 4, 115–141.

Mills, T.J., Mast, M.A., Thomas, J., Keith, G., 2016. Controls on selenium distribution and mobilization in an irrigated shallow groundwater system underlain by Mancos Shale, Uncompahgre River Basin, Colorado, USA. Sci. Total Environ. 566-567 (1), 1621–1631.

Misra, A.K., Mishra, A., 2007. Study of quaternary aquifers in Ganga Plain, India: Focus on groundwater salinity, fluoride and fluorosis. J. Hazard. Mater. 144, 438–448.

Mizutani, T., Kanaya, K., Osaka, T., 2001. Map of seleniumcontent in soil in Japan. J. Health Sci. 47(4), 407-413.

Mkandawire, M., 2013. Biogeochemical behavior and bioremediation of uranium in waters of abandoned mines. Environ. Sci. Pollut. Res. 20, 7740–7767.

Moatar, F., Shadizadeh, S.R., Karbassi, A.R., Ardalani, E., Derakhshi, R.A., Asadi, M., 2010. Determination of naturally occurring radioactive materials (NORM) in formation water during oil exploration. J. Radioanal. Nucl. Chem. 283, 3–7.

Moise, T., Starinsky, A., Katz, A., Kolodny, Y., 2000. Ra isotopes and Rn in brines and ground waters of the Jordan-Dead Sea Rift Valley: Enrichment, retardation, and mixing. Geochim. Cosmochim. Acta 64, 2371–2388.

Mor, S., Singh, S., Yadav, P., Rani, V., Rani, P., Sheoran, M., et al., 2009. Appraisal of salinity and fluoride in a semi-arid region of India using statistical and multivariate techniques. Environ. Geochem. Health 31, 643–655.

Morris, B.L., Lawrence, A.R., Chilton, P., Adams, B., Calow, R.C., Klinck, B.A., 2003. Groundwater and its susceptibility to degradation: a global assessment of the problem and options for management. In: UNEP Early Warning and Assessment Vol Report Series RS.03–3. UNEP, Nairobi.

Moussa, A.B., Zouari, K., Marc, V., 2011. Hydrochemical and isotope evidence of groundwater salinization processes on the coastal plain of Hammamet-Nabeul, north-eastern Tunisia. Phys. Chem. Earth, A/B/C 36, 167–178.

Msonda, K.W.M., Masamba, W.R.L., Fabiano, E., 2007. A study of fluoride groundwater occurrence in Nathenje, Lilongwe, Malawi. Phys. Chem. Earth, A/B/C 32 (15), 1178–1184. Available from: https://doi.org/10.1016/j.pce.2007.07.050.

Mukherjee, A., Fryar, A.E., 2008. Deeper groundwater chemistry and geochemical modeling of the arsenic affected western Bengal basin, West Bengal, India. Appl. Geochem. 23, 863–892.

Mukherjee, A., Bhattacharya, P., Savage, K., Foster, A., Bundschuh, J., 2008a. Distribution of geogenic arsenic in hydrologic systems: controls and challenges. J. Contam. Hydrol. 99, 1–7.

Mukherjee, A., von Brömssen, M., Scanlon, B.R., Bhattacharya, P., Fryar, A.E., Hasan, M.A., et al., 2008b. Hydrogeochemical comparison and effects of overlapping redox zones on groundwater arsenic near the western(Bhagirathi sub-basin, India) and eastern (Meghna sub-basin, Bangladesh) of the Bengal basin. J. Contam. Hydrol. 99, 31–48.

Mukherjee, A., Fryar, A.E., O'Shea, B.M., 2009. Major occurrences of elevated arsenic in groundwater and other natural waters. In: Henke, K.R. (Ed.), Arsenic—Environmental Chemistry, Health Threats and Waste Treatment. John Wiley & Sons, Chichester, pp. 303–350.

Mukherjee, A., Scanlon, B.R., Fryar, A.E., Saha, D., Ghosh, A., Chaudhari, S., et al., 2012. Solute chemistry and fate of arsenic in the aquifers between the Himalayan foothills and Indian craton: influence of geology and geomorphology. Geochim. Cosmochim. Acta 90, 283–302.

Mukherjee, A., Verma, S., Gupta, S., Henke, K.R., Bhattacharya, P., 2014. Influence of tectonics, sedimentation and aqueous flow cycles on the origin of global groundwater arsenic: paradigms from three continents. J. Hydrol. 518, 284–299. <https://doi.org/10.1016/j.jhydrol.2013.10.044>.

Mukherjee, A., Fryar, A.E., Eastridge, E., Nally, R.S., Chakraborty, M., Scanlon, B.R., 2018. Controls on high and low groundwater arsenic on the opposite banks of the lower reaches of River Ganges, Bengal basin, India. Sci. Total Environ. vol. 645, 1371–1387.

Mukherjee, A. (Ed.), 2018. Groundwater of South Asia. first ed. Springer, Singapore.

Mukherjee, A., Gupta, S., Coomar, P., Fryar, A.E., Guillot, S., Verma, S., et al., 2019. Plate tectonics influence on geogenic arsenic cycle: from primary source to global groundwater enrichment. Sci. Total Environ. 683, 793–807. <https://doi.org/10.1016/j.scitotenv.2019.04.255>.

Mukherjee, A., Scanlon, B., Aureli, A., Langan, S., Guo, H., McKenzie, A., 2020. Global Groundwater: Source, Scarcity, Sustainability, Security and Solutions, first ed. Elsevier, ISBN: 9780128181720.

Murray, F.H., Brown, J.R., Fyfe, W.S., Kronberg, B.I., 1983. Immobilization of U-Th-Ra in mine wastes by phosphate mineralization. Can. Mineral. 21, 607–610.

Naftz, D.L., Rice, J.A., 1989. Geochemical processes controlling selenium in ground water after mining, Powder River Basin, Wyoming, USA. Appl. Geochem. 4, 565–575.

National Research Council (NRC), 1976. Selenium: Medical and Biological Effects of Environmental Pollutants. National Academy of Sciences Press, Washington, DC, p. 203.

National Research Council (NRC), 1983. Selenium in Nutrition. National Research Council, National Academy Press, Washington, DC.

National Research Council (NRC), 1989. Irrigation-Induced Water Quality Problems: What Can be Learned From the San Joaquin Valley Experience. National Academy Press, Washington, DC, p. 157.

Naseem, S., Rafique, T., Bashir, E., Bhanger, M.I., Laghari, A., Usmani, T.H., 2010. Lithological influences on occurrence of high-fluoride groundwater in Nagar Parkar area, Thar Desert, Pakistan. Chemosphere 78 (11), 1313–1321. Available from: https://doi.org/10.1016/j.chemosphere.2010.01.010.

Näslund, J., Snell, I., 2005. GIS-Mapping of Fluoride Contaminated Groundwater in Nakuru & Baringo District, Kenya. Department of Civil and Environmental Engineering, Lulea University of Technology, Lulea.

Nativ, R., 1988. Hydrogeology and Hydrochemistry of the Ogallala Aquifer, Southern High Plains, Texas Panhandle and Eastern New Mexico. Bureau of Economic Geology, Austin, TX.

Nazarenko, I.I., Ermakov, A.N., 1972. Analytical Chemistry of Selenium and Tellurium. John Wiley and Sons, New York.

Neal, R.H., 1995. Selenium. In: Alloway, B.J. (Ed.), Heavy Metals in Soils. Blackie Academic & Professional, London, pp. 260–283.

Nelson, K.H., Thompson, T.G., 1954. Deposition of salts from seawater by frigid concentration. J. Mar. Res. 13, 166–182.

Nicolli, H.B., Suriano, J.M., Gomez Peral, M.A., Ferpozzi, L.H., Baleani, O.A., 1989. Groundwater contamination with arsenic and other trace-elements in an area of the Pampa, Province of Cordoba, Argentina. Environ. Geol. Water Sci. 14 (1), 3–16.

Nicolli, H.B., Bundschuh, J., Garcia, J.W., Falcon, C.M., Jean, J.S., 2010. Sources and controls for the mobility of arsenic in oxidizing groundwaters from loess-type sediments in arid/semi-arid dry climates – Evidence from the Chaco-Pampean plain (Argentina). Water Res. 44, 5589–5604.

Nicolli, H.B., Bundschuh, J., Blanco, M.D.C., Tujchneider, O.C., Panarello, H.O., Dapena, C., et al., 2012. Arsenic and associated trace-elements in groundwater from the Chaco-Pampean plain, Argentina: results from 100 years of research. Sci. Total Environ. 429, 36–56. Available from: https://doi.org/10.1016/j.scitotenv.2012.04.048.

Nimick, D.A., 1998. Arsenic hydrogeochemistry in an irrigated river valley—a reevaluation. Ground Water 36 (5), 733–735.

Noll Jr., P.D., Newsom, H.E., Leeman, W.P., Ryan, J.G., 1996. The role of hydrothermal fluids in the production of subduction zone magmas: evidence from siderophile and chalcophile trace elements and boron. Geochim. Cosmochim. Acta 60, 587–611. <https://doi.org/10.1016/0016-7037%2895%2900405-X>.

Nordstrom, D.K., 2002. Worldwide occurrences of arsenic in ground water. Science 296, 2143−2145.
Nriagu, J.O., 1989. Occurrence and distribution of selenium. CRC Press, Boca Raton, FL.
Oldfield, J.E., 2002. Selenium world atlas. Selenium-Tellurium Development Association. Grimbergen, Belgium.
Orloff, K.G., Mistry, K., Charp, P., Metcalf, S., Marino, R., Shelly, T., et al., 2004. Human exposure to uranium in groundwater. Environ. Res. 94, 319−326.
Ortiz, D., Castro, L., Turrubiartes, F., Milan, J., Diaz-Barriga, F., 1998. Assessment of the exposure to fluoride from drinking water in Durango, Mexico, using a geographic information system. Fluoride 31 (4), 183−187.
Oruc, N., 2008. Occurrence and problems of high fluoride waters in Turkey: an overview. Environ. Geochem. Health: Med. Geol. Dev. Countries, 2 30 (4), 315−323.
Otero, N., Soler, A., Corp, R.M., Mas-Pla, J., Garcia-Solsona, E., Masque, P., 2011. Origin and evolution of groundwater collected by a desalination plant (Tordera, Spain): a multi-isotopic approach. J. Hydrol. 397, 37−46.
Ozsvath, D.L., 2006. Fluoride concentrations in a crystalline bedrock aquifer Marathon County, Wisconsin. Environ. Geol. 50 (1), 132−138. Available from: https://doi.org/10.1007/s00254-006-0192-6.
Paoloni, J.D., Fiorentino, C.E., Sequeira, M.E., 2003. Fluoride contamination of aquifers in the southeast subhumid pampa, Argentina. Environ. Toxicol. 18 (5), 317−320. Available from: https://doi.org/10.1002/tox.10131.
Park, S.C., Yun, S.T., Chae, G.T., Yoo, I.S., Shin, K.S., Heo, C.H., et al., 2005. Regional hydrochemical study on salinization of coastal aquifers, western coastal area of South Korea. J. Hydrol. 313, 182−194.
Parkhurst, B.R., Elder, R.G., Meyer, J.S., Sanchez, D.A., Pennak, R.W., Waller, W.T., 1984. An environmental-hazard evaluation of uranium in a rocky-mountain stream. Environ. Toxicol. Chem. 3, 113−124.
Paschke, S.S., Walton-Day, K., Beck, J.A., Webber, A., Dupree, J.A., 2014. Geologic sources and concentrations of selenium in the West-Central Denver Basin, including the Toll Gate Creek watershed, Aurora, Colorado, 2003−2007. In: United States Geological Survey Scientific Investigations Report 2013−5099, p. 30. <https://doi.org/10.3133/sir20135099>.
Peters, S.C., Blum, J.D., 2003. The source and transport of arsenic in a bedrock aquifer, New Hampshire, USA. Appl. Geochem. 18 (11), 1773−1787.
Pettenati, M., Perrin, J., Pauwels, H., Ahmed, S., 2013. Simulating fluoride evolution in groundwater using a reactive multicomponent transient transport model: application to a crystalline aquifer of Southern India. Appl. Geochem. 29, 102−116. Available from: https://doi.org/10.1016/j.apgeochem.2012.11.001.
Pickering, W.F., 1985. The mobility of soluble fluoride in soils. Environ. Pollut. (Ser. B) 9, 281−308.
Planer-Friedrich, B., 2000. Hydrogeological and Hydrochemical Investigations in the Rioverde Basin, Mexico. Institute of Geology, University of mining and technology, Freiberg.
Plank, T., Langmuir, C.H., 1993. Tracing trace-elements from sediment input to volcanic output at subduction zones. Nature 362 (6422), 739−743. <https://doi.org/10.1038/362739a0>.
Plant, J.A., Kinniburgh, D., Smedley, P.L., Fordyce, F.M., Klinck, B., 2004. Arsenic and selenium. In: Lollar, S.B. (Ed.), Environmental Geochemistry. In: Holland, H.D., Turekain, K.K. (Eds.), Treatise on Geochemistry Series, vol. 9. Elsevier, Amsterdam, pp. 17−66.
Podgorski, J.E., Eqani, S.A.M.A.S., Khanam, T., Ullah, R., Shen, H., Berg, M., 2017. Extensive arsenic contamination in high-pH unconfined aquifers in the Indus Valley. Sci. Adv. 23 (8), e1700935. Available from: https://doi.org/10.1126/sciadv.1700935.
Porcelli, D., Swarzenski, P.W., 2003. The behavior of U- and Th-series nuclides in groundwater. Uranium-Series Geochem., Rev. Mineral. Geochem. 52, 317−361.
Postma, D., Larsen, F., Nguyen, T.M.H., Mai, T.D., Pham, H.V., Pham, Q.N., et al., 2007. Arsenic in groundwater of the Red River floodplain, Vietnam: controlling geochemical processes and reactive transport modeling. Geochim. Cosmochim. Acta 71, 5054−5071.
Postma, D., Jessen, S., Nguyen, T.M.H., Mai, T.D., Koch, C.B., Pham, H.V., et al., 2010. Mobilization of arsenic and iron from Red River floodplain sediments, Vietnam. Geochim. Cosmochim. Acta 74, 3367−3381.
Postma, D., Pham, T.K.T., Søa, H.U., Vi, M.L., Jakobsen, R., 2017. Reactive transport modeling of arsenic mobilization in groundwater of the Red River floodplain, Vietnam. In: 15th Water-Rock Interaction International Symposium, WRI-15. Procedia Earth Planet. Sci. 17, 85−87. Available from: https://doi.org/10.1016/j.proeps.2016.12.003.
Postel, S.L., 1999. Pillar of Sand: Can the Irrigation Miracle Last? The Worldwatch Institute. W. W. Norton and Company, New York, p. 313, ISBN 0-393-31937-7.
Presser, T.S., Barnes, I., 1984. Selenium concentrations in waters tributary to and in the vicinity of the Kesterson National Wildlife Refuge, Fresno and Merced Counties, California. In: United States Geological Survey Water Resources Investigations Report 84-4122, 26 p.
Presser, T.S., Swain, W.C., 1990. Geochemical evidence for Se mobilization by the weathering of pyritic shale, San Joaquin Valley, California, U.S.A. Appl. Geochem. 5, 703−717.
Price, R.M., Herman, J.S., 1991. Geochemical investigation of salt-water intrusion into a coastal carbonate aquifer: Mallorca, Spain. GSA Bull. 103 (10), 1270−1279. <https://doi.org/10.1130/0016-7606(1991)103<1270:GIOSWI>2.3.CO;2>.
Prikryl, J.D., Jain, A., Turner, D.R., Pabalan, R.T., 2001. Uranium(VI) sorption behavior on silicate mineral mixtures. J. Contam. Hydrol. 47, 241−253.
Qafoku, N.P., Zachara, J.M., Liu, C., Gassman, P.L., Qafoku, O.S., Smith, S.C., 2005. Kinetic desorption and sorption of U (VI) during reactive transport in a contaminated Hanford sediment. Environ. Sci. Technol. 39 (9), 3157−3165.
Queste, A., Lacombe, M., Hellmeier, W., Hillerman, F., Bortulussi, B., Kaup, M., et al., 2001. High concentrations of fluoride and boron in drinking water wells in the Muenster region—results of a preliminary investigation. Int. J. Hyg. Environ. Health 203, 221−224.

Radespiel-Tröger, M., Meyer, M., 2013. Association between drinking water uranium content and cancer risk in Bavaria, Germany. Int. Arch. Occup. Environ. Health 86, 767−776.

Raanan, H., Vengosh, A., Paytan, A., Nishri, A., Kabala, Z., 2009. Quantifying saline groundwater flow into a freshwater lake using the Ra isotope quartet: A case study from the Sea of Galilee (Lake Kinneret), Israel. Limnol. Oceanogr. 54, 119−131.

Rafique, T., Naseem, S., Bhanger, M.I., Usmani, T.H., 2008. Fluoride ion contamination in the groundwater of Mithi sub-district, the Thar Desert, Pakistan. Environ. Geol. 56 (2), 317−326. Available from: https://doi.org/10.1007/s00254-007-1167-y.

Rafique, T., Naseem, S., Usmani, T.H., Bashir, E., Khan, F.A., Bhanger, M.I., 2009. Geochemical factors controlling the occurrence of high fluoride groundwater in the Nagar Parkar area, Sindh, Pakistan. J. Hazard. Mater. 171, 424−430.

Rango, T., Bianchini, G., Beccaluva, L., Ayenew, T., Colombani, N., 2009. Hydrogeochemical study in the Main Ethiopian Rift: new insights to the source and enrichment mechanism of fluoride. Environ. Geol. 58, 109−118.

Rango, T., Bianchini, G., Beccaluva, L., Tassinari, R., 2010a. Geochemistry and water quality assessment of central Main Ethiopian Rift natural waters with emphasis on source and occurrence of fluoride and arsenic. J. Afr. Earth Sci. 57, 479−491.

Rango, T., Colombani, N., Mastrocicco, M., Bianchini, G., Beccaluva, L., 2010b. Column elution experiments on volcanic ash: geochemical implications for the main Ethiopian rift waters. Water Air Soil Pollut. 208, 221−233.

Rango, T., Petrini, R., Stenni, B., Bianchini, G., Slejko, F., Beccaluva, L., et al., 2010c. The dynamics of central Main Ethiopian Rift waters: evidence from δD, $\delta^{18}O$ and $^{87}Sr/^{86}Sr$ ratios. Appl. Geochem. 25, 1860−1871.

Rao, N.S., 2002. Geochemistry of groundwater in parts of Guntur District, Andhra Pradesh, India. Environ. Geol. 41, 552−562.

Ravenscroft, P., Brammer, H., Richards, K.S., 2009. Arsenic Pollution: A Global Synthesis. Wiley-Blackwell Publication, p. 588.

Raychowdhury, N., Mukherjee, A., Bhattacharya, P., Johannesson, K., Bundschuh, J., Bejarano, G., et al., 2014. Provenance and fate of arsenic and other solutes in the Chaco-Pampean Plain of the Andean foreland, Argentina: from perspectives of hydrogeochemical modeling and regional tectonic setting. J. Hydrol. 518, 300−316.

Rayman, M.P., 2000. The importance of selenium to human health. Lancet 356, 233−241.

Razo, L.M.D., Corona, J.C., García-Vargas, G., Albores, A., Cebrián, M.E., 1993. Fluoride levels in well-water from a chronic arsenicism area of northern Mexico. Environ. Pollut. 80 (1), 91−94. Available from: https://doi.org/10.1016/0269-7491(93)90015-G.

Reddy, D.V., Nagabhushanam, P., Sukhija, B.S., Reddy, A.G.S., Smedley, P.L., 2010. Fluoride dynamics in the granitic aquifer of the Wailapally watershed, Nalgonda District, India. Chem. Geol. 269, 278−289.

Reimann, C.B.K., Tekle-Haimanot, R., Melaku, Z., Siewers, U., 2002. Drinking Water Quality, Rift Valley, Ethiopia. In: NGU-Report, Report No. 2002 033. NGU, Trondheim.

Reimann, C., Bjorvatn, K., Frengstad, B., Melaku, Z., Tekle-Haimanot, R., Siewers, U., 2003. Drinking water quality in the Ethiopian section of the East African Rift Valley I—data and health aspects. Sci. Total Environ. 311 (1−3), 65−80.

Reimann, C., Otteson, R.T., Cramer, J., 2005. Uranium in Drinking-Water? NGU, Geological Survey of Norway, Trondheim.

Rengasamy, P., 2006. World salinization with emphasis on Australia. J. Exp. Bot. 57, 1017−1023.

République du Sénégal, Ministère de l'Energie et de l'Hydraulique, Service de Gestion et de Planification des Ressources en Eau (SGPRE), 2001. Synthèse des données géochimiques. In: Interprétations en terme de modèle conceptuel des écoulements, 70.

Richardson, C., 1976. Phase relationships in sea ice as a function of temperature. J. Glaciol. 17, 507−519.

Robertson, F.N., 1985. Solubility controls of fluorine, barium and chromium in ground water in alluvial basins of Arizona. In: Practical Applications of Ground-Water Geochemistry Conference, Banff, June 1984, pp. 96−102.

Robertson, F.N., 1989. Arsenic in groundwater under oxidizing conditions, south-west United-States. Environ. Geochem. Health 11 (3−4), 171−185.

Roman, S., Peuraniemi, V., 1999. The effect of bedrock, glacial deposits and a waste disposal site on groundwater quality in the Haukipudas area, Northern Finland. In: Chilton, J. (Ed.), Groundwater in the Urban Environment. Balkema, Rotterdam, pp. 329−334.

Rosenthal, E., Jones, B.F., Weinberger, G., 1998. The chemical evolution of Kurnob Group paleowater in the Sinai-Negev province − a mass-balance approach. Appl. Geochem. 13, 553−569.

Rotiroti, M., Sacchi, E., Fumagalli, L., Bonomi, T., 2014. Origin of arsenic in groundwater from the multilayer aquifer in Cremona (Northern Italy). Environ. Sci. Technol. 48 (10), 5395−5403.

Rowland, H.A.L., Omoregie, E.O., Millot, R., Jimenez, C., Mertens, J., Baciu, C., et al., 2011. Geochemistry and arsenic behaviour in groundwater resources of the Pannonian Basin (Hungary and Romania). Appl. Geochem. 26, 1−17.

Sajidu, S.M.I., Masamba, W.R.L., Thole, B., Mwatseteza, J.F., 2008. Groundwater fluoride levels in villages of Southern Malawi and removal studies using bauxite. Int. J. Phys. Sci. 3 (1), 1−11.

Salameh, E., 1996. Water Quality Degradation in Jordan. The Higher Council of Science and Technology. Royal Society for the Conservation of Nature, Jordan.

Sanding, A., Bruno, J., 1992. The solubility of $(UO_2)_3(PO_4)_2 \cdot 4H_2O(s)$ and the formation of U(VI) phosphate complexes: their influence in uranium speciation in natural waters. Geochim. Cosmochim. Acta 56 (12), 4135−4145.

Sánchez-Martos, F., Pulido-Bosch, A., 1999. Boron and the origin of salinization in an aquifer in southeast Spain. C.R. Acad. Sci., Ser. IIA: Earth Planet. Sci. 328, 751−757.

Sánchez-Martos, F., Pulido-Bosch, A., Molina-Sánchez, L., Vallejos-Izquierdo, A., 2002. Identification of the origin of salinization in groundwater using minor ions (Lower Andarax, Southeast Spain). Sci. Total Environ. 297, 43−58.

Sanford, W.E., Konikow, L.F., 1989. Simulation of calcite dissolution and porosity changes in saltwater mixing zones in coastal aquifers. Water Resour. Res. 25, 655−667.

Saunders, J.A., Lee, M.K., Mohammad, S., 2005. Natural arsenic contamination of Holocene alluvial aquifers by linked tectonic, weathering, and microbial processes. Geochem. Geophys. Geosyst. 6 (4), Q04006. <https://doi.org/10.1029/2004GC000803>.

Saxena, V., Ahmed, S., 2001. Dissolution of fluoride in groundwater: a water-rock interaction study. Environ. Geol. 40 (9), 1084–1087. Available from: https://doi.org/10.1007/s002540100290.

Saxena, V., Ahmed, S., 2003. Inferring the chemical parameters for the dissolution of fluoride in groundwater. Environ. Geol. 43 (6), 731–736. Available from: https://doi.org/10.1007/s00254-002-0672-2.

Scanlon, B.R., Nicot, J.P., Reedy, R.C., Kurtzman, D., Mukherjee, A., Nordstrom, D.K., 2009. Elevated naturally occurring arsenic in a semiarid oxidizing system, Southern High Plains aquifer, Texas, USA. Appl. Geochem. 24, 2061–2071.

Schoups, G., Hopmans, J.W., Young, C.A., Vrugt, J.A., Wallender, W.W., Tanji, K.K., et al., 2005. Sustainability of irrigated agriculture in the San Joaquin Valley, California. Proc. Natl. Acad. Sci. U.S.A. 102, 15352–15356.

Schreiber, M.E., Sino, J.A., Freiberg, P.G., 2000. Stratigraphic and geochemical controls on naturally occurring arsenic in groundwater, eastern Wisconsin, USA. Hydrogeol. J. 8, 161–176.

Schreiber, M.E., Gotkowitz, M.B., Simo, J.A., Freiberg, P.G., 2003. Mechanisms of arsenic release to ground water from naturally occurring substances, Eastern Wisconsin. In: Welch, A.H., Stollenwerk, K.G. (Eds.), Arsenic in Groundwater: Geochemistry and Occurrence. Springer-Verlag, New York, pp. 259–280.

Schroeder, R.A., Orem, W.H., Kharaka, Y.K., 2002. Chemical evolution of the Salton Sea, California: nutrient and selenium dynamics. Hydrobiologia 473, 23–45.

Schwartz, M.O., Friedrich, G.H., 1973. Secondary dispersion patterns of fluoride in the Osor area, Province of Gerona, Spain. J. Geochem. Explor. 2, 103–114.

Scott, T.B., Allen, G.C., Heard, P.J., Randell, M.G., 2005. Reduction of U(VI) to U(IV) on the surface of magnetite. Geochim. Cosmochim. Acta 69, 5639–5646.

Seiler, R.L., 1995. Prediction of areas where irrigation drainage may induce selenium contamination of water. J. Environ. Qual. 24 (5), 973–979.

Seiler, R.L., 1997. Methods to Identify Areas Susceptible to Irrigation Induced Selenium Contamination in the Western United States. United States Geological Survey, Reston, VA, p. 4.

Seldén, A.I., Lundholm, C., Edlund, B., Högdahl, C., Ek, B.M., Bergström, B.E., et al., 2009. Nephrotoxicity of uranium in drinking water from private drilled wells. Environ. Res. 109, 486–494.

Shiklomanov, I.A., 1997. Comprehensive Assessment of the Freshwater Resources of the World. United Nation Commission for Sustainable Development, World Meteorological Organization and Stockholm Environment Institute, Stockholm.

Shrift, A., 1954. Sulfur-selenium antagonium: I. Antimetabolite action of selenate on the growth of *Chlorella vulgaris*. Am. J. Bot. 41 (3), 223–230.

Sidle, W.C., Wotten, B., Murphy, E., 2001. Provenance of geogenic arsenic in the Goose River basin, Maine, USA. Environ. Geol. 41 (1–2), 62–73.

Skeppström, K., Olofsson, B., 2007. Uranium and radon in groundwater—an overview of the problem. Eur. Water 17 (18), 51–62.

Smedley, P.L., Edmunds, W.M., Pelig-Ba, K.B., 1995. Groundwater vulnerability due to natural geochemical environment: 2: health problems related to groundwater in the Obuasi and Bolgatanga areas, Ghana. In: BGS Technical Report.

Smedley, P.L., 1996. Arsenic in rural groundwater in Ghana. J. Afr. Earth Sci. 22 (4), 459–470.

Smedley, P.L., Kinniburgh, D.G., 2002. A review of the source, behavior and distribution of arsenic in natural waters. Appl. Geochem. 17 (5), 517–568.

Smedley, P.L., Nkotagu, H., Pelig-Ba, K., McDonald, A.M., Tyler-Whittle, R., Whitehead, E.J., et al., 2002a. Fluoride in groundwater from high fluoride areas of Ghana and Tanzania. Br. Geol. Surv. 61.

Smedley, P.L., Nicolli, H.B., Macdonald, D.M.J., Barros, A.J., Tullio, J.O., 2002b. Hydrogeochemistry of arsenic and other inorganic constituents in groundwaters from La Pampa, Argentina. Appl. Geochem. 17 (3), 259–284. Available from: https://doi.org/10.1016/S0883-2927(01)00082-8.

Smedley, P.L., Zhang, M., Zhang, G., Luo, Z., 2003. Mobilisation of arsenic and other trace elements in fluviolacustrine aquifers of the Huhhot Basin, Inner Mongolia. Appl. Geochem. 18, 1453–1477.

Smedley, P.L., 2005. Arsenic occurrence in groundwater in South and East Asia: Scale, causes and mitigation. Towards a more effective operational response: Arsenic contamination of groundwater in South and East Asian countries II. Technical Report, World Bank Report No. 31303. World Bank, Washington, DC.

Smedley, P.L., Smith, B., Abesser, C., Lapworth, D., 2006. Uranium occurrence and behaviour in British groundwater. In: British Geological Survey Commissioned Report, CR/06/050N, 60 pp.

Smith, B., Powell, J., Gedeon, R., Amro, H., 1996. Groundwater pollution by natural radionuclides: an evaluation of natural and mining contamination associated with phosphorite (Jordan). In: Proceedings of the Conference: Second IMM Conference on Minerals, Metals and the Environment, Prague.

Srikanth, R., Viawanatham, K.S., Kahsai, F., Fisahatsion, A., Asmellash, M., 2002. Fluoride in groundwater in selected villages in Eritrea (North East Africa). Environ. Monit. Assess. 75, 169–177.

Stalder, E., Blanc, A., Haldimann, M., Dudler, V., 2012. Occurrence of uranium in Swiss drinking water. Chemosphere 86, 672–679.

Starinsky, A., 1974. Relationship Between Calcium-Chloride Brines and Sedimentary Rocks in Israel (Ph.D. thesis). The Hebrew University.

Stillings, L.L., Amacher, M.C., 2010. Kinetics of selenium release in mine waste from the Meade Peak Phosphatic Shale, Phosphoria formation, Wooley Valley, Idaho, USA. Chem. Geol. 269 (1–2), 113–123.

Stolz, J.F., Oremland, R.S., 1999. Bacterial respiration of arsenic and selenium. FEMS Microbiol. Rev. 23, 615–627.

Stolz, J.F., Basu, P., Oremland, R.S., 2002. Microbial transformation of elements: the case of arsenic and selenium. Int. Microbiol. 5 (4), 201–207.

Stone, J.J., Stetler, L.D., Schwalm, A., 2007. Final Report: North Cave Hills Abandoned Uranium Mines Impact Investigation. South Dakota School of Mines and Technology, Rapid City, SD.

Stotler, R.L., Frape, S.K., Ruskeeniemi, T., Ahonen, L., Onstott, T.C., Hobbs, M.Y., 2009. Hydrogeochemistry of groundwaters in and below the base of thick permafrost at Lupin, Nunavut, Canada. J. Hydrol. 373, 80–95.

Stüben, D., Berner, Z., Chandrasekharam, D., Karmakar, J., 2003. Arsenic enrichment in groundwater of West Bengal, India: geochemical evidence for mobilization of As under reducing conditions. Appl. Geochem. 18, 1417–1434.

Sturchio, N.C., Banner, J.L., Binz, C.M., Heraty, L.B., Musgrove, M., 2001. Radium geochemistry of ground waters in Paleozoic carbonate aquifers, midcontinent, USA. Appl. Geochem. 16, 109–122.

Su, C., Wang, Y., Xie, X., Zhu, Y., 2015. An isotope hydrochemical approach to understand fluoride release into groundwaters of the Datong Basin, Northern China. Environ. Sci. Process. Impacts 17 (4), 791–801. Available from: https://doi.org/10.1039/C4EM00584H.

Suma Latha, S., Anbika, S.R., Prasad, S.J., 1999. Fluoride contamination status of groundwater in Karnataka. Curr. Sci. 76, 730–734.

Susheela, A.K., Kumar, A., Bhatnagar, M., Bahudur, R., 1993. Prevalence of endemic fluorosis with gastro-intestinal manifestations in people living in some North-Indian villages. Fluoride 26 (2), 97–104.

Tandukar, N., Bhattacharya, P., Jacks, G., Valero, A.A., 2005. Naturally occurring arsenic in groundwater of Terai region in Nepal and mitigation options. In: Bundschuh, J., Bhattacharya, P., Chandrasekharam, D. (Eds.), Natural Arsenic in Groundwater. Proceedings of the Pre-Congress Workshop "Natural Arsenic in Groundwater", 32nd International Geological Congress, Florence, Italy, August 18–19, 2004. Third ed. Taylor and Francis, London.

Tanji, K., Valoppi, L., 1989. Groundwater contamination by trace elements. Agric. Ecosyst. Environ. 26, 229–274.

Tatsumi, Y., 1989. Migration of fluid phases and genesis of basalt magmas in subduction zones. J. Geophys. Res. 94 (B4), 4697–4707. <https://doi.org/10.1029/JB094iB04p04697>.

Taylor, S.R., McLennan, S.M., 1985. The Continental Crust: Its Composition and Evolution. Blackwell, Oxford.

Tekle-Haimanot, R., Melaku, Z., Kloos, H., Reimann, C., Fantaye, W., Zerihun, L., et al., 2006. The geographic distribution of fluoride in surface and groundwater in Ethiopia with an emphasis on the Rift Valley. Sci. Total Environ. 367, 182–190.

Thole, B., 2013. Ground water contamination with fluoride and potential fluoride removal technologies for East and Southern Africa. In:Ahmad Dar, I., Ahmad Dar, M. (Eds.), Perspectives in water pollution. IntechOpen. <http://www.intechopen.com>.

Tokunaga, T., Pickering, I., Brown, G., 1996. Selenium transformations in ponded sediments. Soil Sci. Soc. Am. J. 60, 781–790.

Tomita, J., Satake, H., Fukuyama, T., Sasaki, K., Sakaguchi, A., Yamamoto, M., 2010. Radium geochemistry in Na-Cl type groundwater in Niigata Prefecture, Japan. J. Environ. Radioact. 101, 201–210.

Travi, Y., 1993. Hydrogéologie et hydrochimie des aquifères du Sénégal. In: Sciences Géologiques, Memoire 95. Université de Paris-Sud, Paris.

Trelease, S.F., Beath, O.A., 1949. Selenium: Its Geological Occurrence and Its Biological Effects in Relation to Botany, Chemistry, Agriculture, Nutrition, and Medicine. Champlain Printers, Burlington, VT, 292 p.

Turner, B.D., Binning, P., Stipp, S.L.S., 2005. Fluoride removal by calcite: evidence for fluorite precipitation and surface adsorption. Environ. Sci. Technol. 39, 9561–9568.

Tyrovola, K., Nikolaidis, N.P., Veranis, N., Kallithrakas-Kontos, N., Koulouri-dakis, P.E., 2006. Arsenic removal from geothermal waters with zero-valent iron—effect of temperature, phosphate and nitrate. Water Res. 40, 2375–2386.

United States Environmental Protection Agency (US EPA), 1980. Ambient water quality criteria for selenium. In: NTIS, PB81-117814, 123 p.

United States Environmental Protection Agency (US EPA), 1987. Ambient water quality criteria for selenium—1987. In: Publication EPA-440/5-87-006. United States Environmental Protection Agency, Office of Water Regulations and Standards, Washington, DC, 121 p.

United States Environmental Protection Agency (US EPA), 1992. Rulemaking: water quality standards: establishment of numeric criteria for priority toxic pollutants. In: States' Compliance: Final Rule, 57 FR 60848, 166 p.

United States Environmental Protection Agency (US EPA), 1998a. Locating and estimating air emissions from sources of arsenic and arsenic compound. In: Publication EPA-454/R-98-013. United States Environmental Protection Agency, Office of Air Quality Planning and Standards, Research Triangle Park, NC, 279 p.

United States Environmental Protection Agency (US EPA), 1998b. Report on the Peer Consultation Workshop on Selenium Aquatic Toxicity and Bioaccumulation. United States Environmental Protection Agency, Columbus, OH, 59 p., 6 appendices.

United States Environmental Protection Agency (US EPA), 2012, Environmental Protection Agency Report. OH. <www.epa.state.oh.us/Portals/28/documents/gwqcp/fluoride_ts.pdf>.

United States Geological Survey (USGS), 1991–2003. NWQAD Warehouse. <infotrek.er.usgs.gov>.

Valenzuela-Vasquez, L., Ramirez-Henandez, J., Reyes-Lopez, J., Sol-Uribe, A., Lazaro-Mancilla, O., 2006. The origin of fluoride in groundwater supply to Hermosillo city, Sonora, Mexico. Environ. Geol. 51, 17–27.

van Geen, A., Win, K.H., Zawb, T., Naing, W., Mey, J.L., Mailloux, B., 2014. Confirmation of elevated arsenic levels in groundwater of Myanmar. Sci. Total Environ. 478, 21–24.

van der Hoek, W., Ekanayake, L., Rajasooriyar, L., Karunaratne, R., 2003. Source of drinking water and other risk factors for dental fluorosis in Sri Lanka. Int. J. Environ. Health Res. 13 (3), 285–293. Available from: https://doi.org/10.1080/0960312031000122433.

van Weert, F., van der Gun, J., Reckman, J., 2009. Global Overview of Saline Groundwater Occurrence and Genesis. In: IGRAC Report Nr. GP 2009-1. International Groundwater Resources Assessment Center, Utrecht.

Varsányi, I., Kovács, L.Ó., 2006. Arsenic, iron and organic matter in sediments and groundwater in the Pannonian Basin, Hungary. Appl. Geochem. 21 (6), 949–963.

Vasquez, L.V., Ramirez-Hernandez, J., Reyes-Lopez, J., Sol-Uribe, A., Lazaro-Mancilla, O., 2006. The origin of fluoride in groundwater supply to Hermosillo City, Sonora, Mexico. Environ. Geol. 51 (1), 17–27.

Vengosh, A., Heumann, K.G., Juraske, S., Kasher, R., 1994. Boron isotope application for tracing sources of contamination in groundwater. Environ. Sci. Technol. 28, 1968–1974.

Vengosh, A., Benzvi, A., 1994. Formation of salt plume in the coastal plain aquifer of Israel – the Beer Toviyya region. J. Hydrol. 160, 21–52.

Vengosh, A., Rosenthal, E., 1994. Saline groundwater in Israel: its bearing on the water crisis in the country. J. Hydrol. 156, 389–430.

Vengosh, A., Barth, S., Heumann, K.G., Eisenhut, S., 1999a. Boron isotopic composition of freshwater lakes from central Europe and possible contamination sources. Acta Hydrochim. Hydrobiol. 27, 416–421.

Vengosh, A., Spivack, A.J., Artzi, Y., Ayalon, A., 1999b. Geochemical and boron, strontium, and oxygen isotopic constraints on the origin of the salinity in groundwater from the Mediterranean coast of Israel. Water Resour. Res. 35, 1877–1894.

Vengosh, A., Gill, J., Davisson, M.L., Hudson, G.B., 2002. A multi-isotope (B, Sr, O, H, and C) and age dating (^{3}H–^{3}He and ^{14}C) study of groundwater from Salinas Valley, California: Hydrochemistry, dynamics, and contamination processes. Water Resour. Res. 38 (1), 9-1–9-17. <http://dx.doi/10.1029/2001WR000517>.

Vengosh, A., Kloppmann, W., Marei, A., Livshitz, Y., Gutierrez, A., Banna, M., et al., 2005. Sources of salinity and boron in the Gaza strip: natural contaminant flow in the southern Mediterranean coastal aquifer. Water Resour. Res. 41 (1), W01013. <http://dx.doi:10.1029/2004WR003344>.

Vengosh, A., Hening, S., Ganor, J., Mayer, B., Weyhenmeyer, C.E., Bullen, T.D., et al., 2007. New isotopic evidence for the origin of groundwater from the Nubian Sandstone Aquifer in the Negev, Israel. Appl. Geochem. 22, 1052–1073.

Vengosh, A., 2014. Salinization and saline environments. In: second ed. Holland, H.D., Turekian, K.K. (Eds.), Treatise on Geochemistry, vol. 4. Elsevier, Amsterdam, pp. 325–387.

Verma, S., Mukherjee, A., Mahanta, C., Choudhury, R., Mitra, K., 2016. Influence of geology on groundwater–sediment interactions in arsenic enriched tectonomorphic aquifers of the Himalayan Brahmaputra river basin. J. Hydrol. 540, 176–195.

Verma, S., Mukherjee, A., Mahanta, C., Choudhury, R., Badoni, R.P., Joshi, G., 2019. Arsenic fate in the Brahmaputra river basin aquifers: controls of geogenic processes, provenance and water-rock interactions. Appl. Geochem. 107, 171–186.

Viero, A.P., Roisenberg, C., Roisenberg, A., Vigo, A., 2009. The origin of fluoride in the granitic aquifer of Porto Alegre, Southern Brazil. Environ. Geol. 56, 1707–1719.

Villalobos, M., Trotz, M.A., Leckie, J.O., 2001. Surface complexation modeling of carbonate effects on the adsorption of Cr(VI), Pb(II) and U(VI) on goethite. Environ. Sci. Technol. 35, 3849–3856.

Vinson, D.S., 2011. Radium Isotope Geochemistry in Groundwater Systems: The Role of Environmental Factors (Ph.D. thesis). Duke University, Durham, NC.

Wagner, S.E., Burch, J.B., Bottai, M., Puett, R., Porter, D., Bolick-Aldrich, S., et al., 2011. Groundwater uranium and cancer incidence in South Carolina. Cancer Causes Control 22, 41–50.

Wang, L.F.M., Huang, J.Z., 1995. Outline of control practice of endemic fluorosis in China. Soc. Sci. Med. 41 (8), 1191–1195.

Wang, Y., Chen, X., Meng, G., Wang, S., Wang, Z., 2000. On changing trends of dD during seawater freezing and evaporation. Cold Reg. Sci. Technol. 31, 27–31.

Wang, Z., Gao, Y., 2001. Biogeochemical cycling of selenium in Chinese environments. Appl. Geochem. 16 (11–12), 1345–1351.

Wang, Z., Zachara, J.M., McKinley, J.P., Smith, S.C., 2005. Cryogenic laser induced U(VI) fluorescence studies of a U(VI) substituted natural calcite: implications to U(VI) speciation in contaminated Hanford sediments. Environ. Sci. Technol. 39 (8), 2651–2659.

Warner, N., Lgourna, Z., Boutaleb, S., Tagma, T., Vinson, D.S., Ettayfi, N., et al., 2010. A geochemical approach for the evaluation of water availability and salinity in closed basins: the Draa Basin, Morocco. In: American Geophysical Union, Fall Meeting, San Francisco, CA.

Warner, R., Meadows, J., Sojda, S., Price, V., Temples, T., Arai, Y., et al., 2011. Mineralogic investigation into occurrence of high uranium well waters in upstate South Carolina, USA. Appl. Geochem. 26 (2011), 777–788.

Warner, N.R., Jackson, R.B., Darrah, T.H., Osborn, S.G., Down, A., Zhao, K., et al., 2012. Geochemical evidence for possible natural migration of Marcellus Formation brine to shallow aquifers in Pennsylvania. Proc. Natl. Acad. Sci. U.S.A. 109, 11961–11966.

Waseem, A., Ullah, H., Rauf, M.K., Ahmad, I., 2015. Distribution of natural uranium in surface and groundwater resources: a review. Crit. Rev. Environ. Sci. Technol. 45 (22), 2391–2423. Available from: https://doi.org/10.1080/10643389.2015.1025642.

Weaver, T.R., Frape, S.K., Cherry, J.A., 1995. Recent cross-formational fluid-flow and mixing in the shallow Michigan Basin. Geol. Soc. Am. Bull. 107, 697–707.

Welch, A.H., Lico, M.S., 1998. Factors controlling As and U in shallow ground water, southern Carson Desert, Nevada. Appl. Geochem. 13, 521–539.

Welch, A.H., Westjohn, D.B., Helsel, D.R., Wanty, R.B., 2000. Arsenic in ground water of the United States: occurrence and geochemistry. Groundwater 38, 589–604.

Wenzel, W.W., Blum, W.E.H., 1992. Fluorine speciation and mobility in F-contaminated soils. Soil Sci. 153, 357–364.

Wiedemeier, T.H., Wilson, J.T., Kampbell, D.H., Miller, R.N., Hansen, J.E., 1995. Technical protocol for implementing intrinsic remediation with long term monitoring for natural attenuation of fuel contamination dissolved in groundwater. In: Defense Technical Information Center Document, vol. 2.

Winkel, L.H.E., Johnson, C.A., Lenz, M., Grundl, T., Leupin, O.X., Amini, M., et al., 2012. Environmental selenium research: from microscopic processes to global understanding. Environ. Sci. Tech. 46, 571–579. https://doi.org/10.1021/es203434d.

Wilber, C.G., 1983. Selenium, a potential environmental poison and a necessary food constituent. Charles C. Thomas. Publishers, Springfield, IL, p. 126.

Williams, K.T., Byers, H.G., 1934. Occurrence of selenium in pyrites. Ind. Eng. Chem. Anal. Ed. 6 (8), 296–297.

Williams, M., Fordyce, F., Paijitprapapon, A., Charoenchaisri, P., 1996. Arsenic contamination in surface drainage and groundwater in part of the southeast Asian tin belt, Nakhon Si Thammarat Province, southern Thailand. Environ. Geol. 27, 16–33.

Williams, M., 1997. Mining-related arsenic hazards: Thailand case-study. In: British Geological Survey Technical Report, WC/97/49.
Williams, W.D., 2001a. Anthropogenic salinization of inland waters. Hydrobiologia 466, 329–337.
Williams, W.D., 2001b. Salinization: unplumbed salt in a parched landscape. Water Sci. Technol. 43, 85–91.
Williams, J., Walker, G.R., Hatton, T.J., 2002. Dryland salinization: a challenge for land and water management in the Australian landscape. In: Haygarth, P.M., Jarvis, S.C. (Eds.), Agriculture, Hydrology, and Water Quality. CAB International, Wallingford.
World Health Organization (WHO), 1996. Trace Elements in Human Nutrition and Health. World Health Organization, Geneva.
World Health Organization (WHO), 2001. WHO Guidelines for Drinking-Water Quality: Arsenic in Drinking Water. Fact Sheet No. 210. World Health Organization, Geneva.
World Health Organization (WHO), 2004. Guidelines for Drinking Water Quality, third ed. World Health Organization, Geneva.
World Health Organization (WHO), 2011. Guidelines for Drinking Water Quality, fourth ed. World Health Organization, Geneva.
Water Research Commission (WRC), 2001. Distribution of fluoride-rich groundwater in Eastern and Mogwase region of Northern and North-west province. In: WRC Report No. 526/1/011.1-9.85. Water Research Commission, Pretoria.
Wright, W.G., 1999. Oxidation and mobilization of selenium by nitrate in irrigation drainage. J. Environ. Qual. 28, 1182–1187.
Wuyi, W., Ribang, L., Jiańan, T., Kunli, L., Lisheng, Y., Hairong, L., et al., 2002. Adsorption and leaching of fluoride in soils of China. Fluoride 35, 122–129.
Yang, G., Wang, S., Zhou, R., Sun, S., 1983. Endemic selenium intoxication of humans in china. Am. J. Clin. Nutr. 37, 872–881.
Yidana, S.M., Yakubo, B.B., Akabzaa, T.M., 2010. Analysis of groundwater quality using multivariate and spatial analyses in the Keta basin, Ghana. J. Afr. Earth Sci. 58, 220–234.
Young, S.M., Pitawala, A., Ishiga, H., 2010. Factors controlling fluoride contents of groundwater in north-central and northwestern Sri Lanka. Environ. Earth Sci. 63, 1333–1342. Available from: https://doi.org/10.1007/s12665-010-0804-z.
Zack, A.L., 1980. Unites States Geological Survey Water-Supply Paper, 2067, pp. 1–40.
Zamora, M.L., Tracy, B.L., Zielinski, J.M., Meyerhof, D.P., Moss, M.A., 1998. Chronic ingestion of uranium in drinking water: a study of kidney bioeffects in humans. Toxicol. Sci. 43, 68–77.
Zeng, Z., 1997. The control factors of the F formation in ground-water and its distributive regularity. Jinlin Geol. 16, 26–31.
Zhang, Y., Moore, J.N., 1997. Reduction potential of selenate in wetland sediment. J. Environ. Qual. 26 (3), 910–916.
Zielinski, R.A., Bloch, S., Walker, T.R., 1983. The mobility and distribution of heavy metals during the formation of first cycle red beds. Econ. Geol. 78, 1574–1589.
Zielinski, R.A., Asher-Bolinder, S., Meier, A.L., Johnson, C.A., Szabo, B.J., 1997. Natural or fertilizer-derived uranium in irrigation drainage: a case study in southeastern Colorado, USA. Appl. Geochem. 12, 9–21.
Zielinski, R.A., Simmons, K.R., Orem, W.H., 2000. Use of ^{234}U and ^{238}U isotopes to identify fertilizer-derived uranium in the Florida Everglades. Appl. Geochem. 15, 369–383.
Zingaro, R.A., Dufner, C.D., Murphy, A.P., Moody, C.D., 1997. Reduction of oxoselenium anions by iron (III) hydroxide. Environ. Int. 23 (3), 299–304.

Chapter 15

Out of sight, but not out of mind: Per- and polyfluoroalkyl substances in groundwater

Ruth Marfil-Vega[1], Brian C. Crone[2], Marc A. Mills[2] and Susan T. Glassmeyer[2]

[1]Shimadzu Scientific Instruments, Columbia, MD, United States, [2]United States Environmental Protection Agency, Office of Research and Development, Center of Environmental Solutions and Emergency Response, Cincinnati, OH, United States

15.1 Introduction

Globally, groundwater is an important source of water for irrigation and direct human consumption and is a resource currently under stress (Mukherjee et al., 2020). In the United States, 87% of the population gets their water from public supplies and 13% from private sources (Dieter et al., 2018). Regardless of the public or private source, nearly half of the water consumed comes from groundwater (23,849 MGD surface water vs 18,410 MGD groundwater; Fig. 15.1). Both surface and groundwater can become contaminated with chemicals and microorganisms. Though many of these contaminants are currently regulated, there are some chemical and biological contaminants that are not currently regulated but may be of public health concern. Collectively, these are known as contaminants of emerging concern (CECs) and include classes of chemicals such as pharmaceuticals, hormones and other endocrine-disrupting compounds, and per- and polyfluoroalkyl substances (PFAS). Sources of these CECs to drinking water supplies may include municipal and industrial wastewater, decentralized wastewater systems, agricultural and livestock runoff, industrial discharges, landfill leachate, and atmospheric deposition (Ahrens and Bundschuh, 2014; Guelfo et al., 2018; Hatton et al., 2018). Once these contaminants make it to groundwater, the subsurface nature of groundwater makes it more difficult to determine if it is impacted by a contamination source. In addition, with the minimal treatment, if any, that groundwater undergoes for some drinking water sources, there exists a greater probability of exposures to CECs through consumption than from surface water sourced drinking water.

PFAS as a class include at least 5000 chemicals (FDA, 2019), but less than 1% of these chemicals have been included in studies published in the peer-reviewed literature (Wang et al., 2017). The most commonly produced and studied PFAS are perfluorooctanoic acid (PFOA) and perfluorooctanesulfonic acid (PFOS). These compounds belong to two major classes of PFAS: PFOA is a perfluoroalkyl carboxylic acid (PFCA) and PFOS is a perfluoroalkane sulfonic acid (PFSA). The base structures of these two classes are $C_nF_{2n+1}COOH$ and $C_nF_{2n+1}SO_3H$, respectively. PFCAs and PFSAs can have linear and branched isomers. Linear isomers exist where all carbons are bonded to no more than two other carbons. Branched isomers are where at least one carbon is bonded to more than two other carbons. Generally, unless specified, PFAS measurements are considered to be of mixtures of linear and branched compounds.

PFCAs with $n \geq 7$ and PFSAs with $n \geq 6$ are called long-chain PFAS (IITRC, 2017a). Due to their environmental persistence, the long-chain PFAS, also referred to as legacy PFAS, are being phased out in most of the world, though their production may be continuing in China (Land et al., 2018). The longer chain PFAS have been measured in archived human blood plasma, although concentrations have decreased in samples collected between 2001 and 2015 (Olsen et al., 2017). Generally, replacement chemicals have been and are being developed to replace many of the legacy PFAS. New formulations and processes, such as the ammonium salt of hexafluoropropylene oxide (HFPO) dimer acid (commonly referred to as Gen X), are replacing the longer chain PFAS in commercial products, but there is limited knowledge of their fate and toxicity (Hopkins et al., 2018).

In the United States the current US Environmental Protection Agency (US EPA) drinking water health advisory level for PFOA and PFOS is 70 ng/L individual or as the sum of PFOS and PFOA (US EPA, 2016a,b). Per US EPA's website, "Health advisories provide information on contaminants that can cause human health effects and are known or

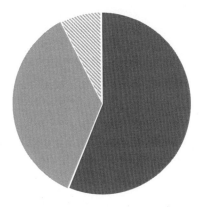

FIGURE 15.1 Relative volume of daily water withdrawals in the United States. *Adapted from Dieter, C.A., Maupin, M.A., Caldwell, R.R., Harris, M.A., Ivahnenko, T.I., Lovelace, J.K., et al., 2018, Estimated use of water in the United States in 2015: U.S. Geological Survey Circular 1441. In: Supersedes USGS Open-File Report 2017–1131. 65 p. https://doi.org/10.3133/cir1441. Volume data in millions of gallons per day (MGD).*

- Surface water public (56%)
- Surface water private (0.12%)
- Groundwater public (36%)
- Groundwater private (7.6%)

anticipated to occur in drinking water." Advisory levels provide technical information on health effects, measurement methods, and treatment technologies for drinking water. US EPA advisory levels are nonenforceable and nonregulatory. In addition, some states have set lower advisory levels (ITRC, 2017b).

Reporting on the presence of PFAS in water sources and treated drinking water has gone beyond the peer-reviewed science literature. PFAS contamination issues are frequently reported in the popular press and social media (SSEHRI, 2019). This heightened public awareness and concern sets PFAS apart from other CECs and increases the need for research on the occurrence and management of PFAS through all parts of the water cycle. This chapter is focused on summarizing the existing knowledge of PFAS in groundwater.

15.2 Analytical methods for monitoring per- and polyfluoroalkyl substances

Occurrence studies of PFAS in groundwater are limited by the availability of analytical methodologies, dependent on development of suitable technologies for achieving the needed sensitivity and method robustness, as well as harmonized lists of compounds of interest. Despite the recent more rapid advance in the development of analytical methods used in research studies, the lack of standardized methods increases the complexity for understanding the extent of groundwater contamination by PFAS. Liquid chromatography with mass spectrometry (LC/MS) detection (either triple quadrupole or high-resolution instruments) is the technology required for the extremely low detection limits for quantification of known PFAS and the identification of new ones in groundwater and other environmental samples. In the United States, LC/MS-based methods have had limited application in the environmental field, outside of research laboratories. However, due to the current public concern regarding PFAS, commercial and compliance laboratories are implementing this technology. Hence, increased information about the occurrence of PFAS is expected in the next years.

The US EPA published the first standardized method for drinking water [EPA 537: Determination of Selected Perfluorinated Alkyl Acids in Drinking Water by Solid Phase Extraction and Liquid Chromatography/Tandem Mass Spectrometry (LC/MS/MS); Shoemaker et al., 2009] for the analysis of 14 PFAS compounds (Table 15.1). This method was updated in 2018 [EPA 537.1: Determination of Selected Per- and Polyfluorinated Alkyl Substances in Drinking Water by Solid Phase Extraction and Liquid Chromatography/Tandem Mass Spectrometry (LC/MS/MS); Shoemaker and Tettenhorst, 2018] to include four additional compounds: HFPO-DA (GenX), ADONA, 11Cl-PF3OUdS, and 9Cl-PF3ONS. These methods were designed only for finished drinking water (from groundwater or surface water sources), so multiple stakeholders, including water utilities and commercial laboratories, adapted these methods for groundwater monitoring. Currently, the US EPA is working on promulgating two additional methods that will address more matrices and analytes. Draft method EPA 8327 [Per-and Polyfluoroalkyl Substances (PFAS) Using External Standard Calibration and Multiple Reaction Monitoring (MRM) Liquid Chromatography/Tandem Mass Spectrometry (LC/MS/MS)] was published in the summer of 2019 (US EPA, 2019); this method is similar to ASTM 7979 [Standard Test Method for Determination of Per- and Polyfluoroalkyl Substances in Water, Sludge, Influent, Effluent, and Wastewater by Liquid Chromatography Tandem Mass Spectrometry (LC/MS/MS)], originally published in 2016 (ASTM, 2019). Draft method EPA 8327 expands the list of target compounds as well as the type of samples that can be analyzed, including groundwater and other nonpotable waters. Method EPA 533 Determination of Per- and Polyfluoroalkyl Substances in Drinking Water by Isotope Dilution Anion Exchange Solid Phase Extraction and Liquid Chromatography/Tandem Mass Spectrometry (Rosenblum

TABLE 15.1 Comparison of analytes in US EPA methods 537, 537.1, 533 and 8327.

Chemical	Acronym	CAS number	EPA 537	EPA 537.1	EPA 533	EPA 8327
			Drinking Water	Drinking Water	Drinking Water	Reagent water, surface water, groundwater, and wastewater effluent
N-Ethyl perfluorooctanesulfonamidoacetic acid	NEtFOSAA		X	X		X
N-Methyl perfluorooctanesulfonamidoacetic acid	NMeFOSAA		X	X		X
Perfluorobutanesulfonic acid	PFBS	375-73-5	X	X	X	X
Perfluorodecanoic acid	PFDA	335-76-2	X	X	X	X
Perfluorododecanoic acid	PFDoA	307-55-1	X	X	X	X
Perfluoroheptanoic acid	PFHpA	375-85-9	X	X	X	X
Perfluorohexanesulfonic acid	PFHxS	355-46-4	X	X	X	X
Perfluorohexanoic acid	PFHxA	307-24-4	X	X	X	X
Perfluorononanoic acid	PFNA	375-95-1	X	X	X	X
Perfluorooctanesulfonic acid	PFOS	1763-23-1	X	X	X	X
Perfluorooctanoic acid	PFOA	335-67-1	X	X	X	X
Perfluorotetradecanoic acid	PFTeDA	376-06-7	X	X		X
Perfluorotridecanoic acid	PFTrDA	72629-94-8	X	X		X
Perfluoroundecanoic acid	PFUnA	2058-94-8	X	X	X	X
Hexafluoropropylene oxide dimer acid	HFPO-DA	13252-13-6		X	X	
11-Chloroeicosafluoro-3-oxaundecane-1-sulfonic acid	11Cl-PF3OUdS	763051-92-9		X	X	
9-Chlorohexadecafluoro-3-oxanone-1-sulfonic acid	9Cl-PF3ONS	756426-58-1		X	X	
4,8-Dioxa-3H-perfluorononanoic acid	ADONA	919005-14-4		X	X	
Perfluoropentanoic acid	PFPeA	2706-90-3			X	X
Perfluorobutanoic acid	PFBA	375-22-4			X	X
Perfluorodecyl sulfonate	PFDS	335-77-3				X
Perfluorononane sulfonate	PFNS	17202-41-4				X
Perfluoroheptyl sulfonate	PFHpS	375-92-8			X	X
Perfluoropentane sulfonate	PFPeS	68259-08-5			X	X
Perfluorooctanesulfonamide	FOSA	754-91-6				X
Fluorotelomer sulfonate 8:2	8:2 FTS	39108-34-4			X	X
Fluorotelomer sulfonate 6:2	6:2 FTS	27619-97-2			X	X
Fluorotelomer sulfonic acid 4:2	4:2 FTS	757124-72-4			X	X
Nonafluoro-3,6-dioxaheptanoic acid	NFDHA	151772-58-6			X	
Perfluoro(2-ethoxyethane)sulfonic acid	PFEESA	113507-82-7			X	
Perfluoro-3-methoxypropanoic acid	PFMPA	377-73-1			X	
Perfluoro-4-methoxybutanoic acid	PFMBA	863090-89-5			X	

and Wendelken, 2019) focuses on shorter chain PFAS; however, it will be only applicable for the analysis of finished drinking water, similarly to EPA 537 and EPA 537.1.

The Department of Defense (DOD) is also invested in the standardization of analytical methods for the assessment of PFAS in groundwater. While for the analysis of drinking water samples DOD requires laboratories to use method EPA 537 or EPA 537.1, DOD allows in-house developed methods for the analysis of other environmental samples provided the laboratories demonstrate adherence and compliance with the quality criteria established in the current version of DOD's Quality Systems Manual (currently version 5.3; DOD, 2019).

From research purposes, there is a large list of methods that differ in the target compounds, matrices, sample preparation, chromatography conditions, and/or detection instruments. As mentioned earlier in this chapter, the majority of the measurement methods rely on LC/MS for the detection and quantification, although there are also a lesser number using gas chromatography tandem MS (GC−MS) (Shiwaku et al., 2016). Nakayama et al. (2019) recently published an extensive review focused on the analysis of PFAS in different environmental matrices. In this review, 11 publications were reported for the analysis of PFAS in aqueous samples, with 3 of them being specific for groundwater samples (Ciofi et al., 2018; Janda et al., 2019; Wei et al., 2018). The methods developed by Janda et al. (2019) and Wei et al. (2018) included a solid phase extraction step for the sample preconcentration, while Ciofi et al. (2018) used a direct large-volume injection method (100 μL). These three methods employed LC/MS/MS and reported limits of detection from 0.013 ng/L (Ciofi et al., 2018) to 3.3 ng/L (Janda et al., 2019). The varied lists of target compounds were analyzed with electrospray source ionization in negative mode, typical conditions for legacy PFAS. Other recent methods include Dasu et al. (2017), which achieved method detection limits for 14 PFAS in drinking water of 0.59−3.4 ng/L using solid phase extraction with a 10 mL sample volume.

Applications of two complimentary approaches are quickly expanding the understanding of the presence of less common and novel PFAS as well as precursor compounds: (1) methods to analyze total oxidizable precursors TOP of the PFCAs and PFSAs (Houtz and Sedlak, 2012) and total organic fluorine (Trojanowicz et al., 2011) and (2) methods focused on the identification of unknown PFAS using nontargeted analysis (NTA) techniques (Xiao et al. 2017a; Ruan and Jiang, 2017). As mentioned previously, only a small fraction of the PFAS in production have been analyzed (Wang et al., 2017), so these types of studies are essential for a complete understanding of the impact of PFAS in the environment. The total PFAS have been less commonly applied to drinking water due to the methods' lack of sensitivity compared to targeted methods. In addition, precursor methods are generally focused more on contaminated sites where higher levels of precursor materials may be expected. The NTA approaches using high-resolution MS are useful to identify unknown compounds and transformation products. However, though rapidly improving, NTA approaches are generally considered lower throughput methods and require specialized equipment and highly trained personnel, which limits broad application.

15.3 Sources of per- and polyfluoroalkyl substances to the environment

The sources of PFAS in groundwater can be direct and indirect sources. Direct sources include landfill leachate (Lang, et al., 2017) and intrusion from discharges of PFAS in manufacturing, military/fire training, and other industrial uses/sites, such as printing and paper production, textile manufacturing, and metal electroplating (Clara et al., 2008). Indirect sources include infiltrated rainwater and atmospheric deposition (Eschauzier et al., 2013). Table 15.2, adapted from Guelfo et al. (2018), illustrates the pathways PFAS follows from various sources before reaching the groundwater. Given the persistence and mobility of these chemicals (Hale et al., 2017), their occurrence in groundwater may result in long-term exposures and risks and require expensive remediation costs (Cousins et al, 2016).

15.3.1 Aqueous film-forming foam

Aqueous film-forming foams (AFFFs) are effective in extinguishing hydrocarbon fires. The combined hydrophilic and hydrophobic properties of PFAS molecules make them an effective ingredient to isolate the fuel from the oxygen (Hale et al., 2017). A study completed by Weber et al. (2017) evaluated the geochemical and hydrologic processes controlling subsurface transport (e.g., sorption, desorption, and chemical transformation) of PFAS at a well-characterized site contaminated from infiltration of treated domestic wastewater and the application of AFFFs. Sixteen target PFAS, including PFCAs, PFSAs, perfluoroalkyl sulfonamides, and fluorotelomer sulfonates were analyzed in groundwater and sediment cores. Groundwater samples were also subjected to the total TOP assay (Houtz and Sedlak, 2012). It was determined that the unsaturated zone where AFFFs were applied and the unsaturated zone hydraulically downgradient from a former domestic wastewater treatment facility (with effluent infiltration beds) were both continuing sources to the groundwater despite almost two decades of inactivity. Different PFAS compositions were found in samples collected near the AFFF site and the former wastewater facility. In addition, perfluoroalkyl precursors in the plume that could be transformed to perfluorinated compounds were also confirmed. These findings reaffirmed the persistence of these chemicals in groundwater and were in agreement with the findings of Houtz et al. (2013). Houtz et al. (2013) confirmed the presence of residual PFAS precursors in groundwater where firefighting training was conducted between 1942 and 1990. They found that the fraction of perfluorinated carboxylates and sulfonates in the groundwater was larger than in archived AFFF formulations. These data suggest that previously released precursors are being biotransformed to persistent end products.

TABLE 15.2 Relevant groundwater pathways, and affected receptors resulting from groundwater per- and polyfluoroalkyl substance (PFAS) source types

Source type	Groundwater pathways	Receptors impacted
Fluoropolymers and PFAS manufacturing	VZ, Atm, SW	DW, GW, SW, B
AFFF—use DOD	VZ	DW, GW, SW, B
AFFF—use airport	VZ	DW, GW, SW, B
AFFF—use fire training	VZ	DW, GW, SW
AFFF—use petroleum	VZ	DW, GW
Fluoropolymers coating (e.g., plastics, textiles, and metals)	Not specified	DW, GW
Electronics	Not specified	DW, GW
Waste streams (landfills)	VZ, Atm	DW, GW
Waste streams (biosolids)	VZ	DW, GW, SW, B
Waste streams (septic systems)	VZ	DW, GW

AFFF, Aqueous film-forming foam; *Atm*, atmospheric deposition; *B*, biota; *DOD*, Department of Defense; *DW*, drinking water; *GW*, groundwater; *SW*, surface water; *VZ*, vadose zone.
Adapted from *Guelfo et al. (2018)*.

15.3.2 Landfill leachate

As mentioned earlier, landfill leachate may be a direct source of PFAS as many consumer and industrial products containing these chemicals are discarded in landfills at the end of their life cycle (Lang et al., 2017). Based on the results from 95 samples collected from 18 publicly owned landfills receiving mostly municipal solid waste (hence, less impacted by the presence of high-PFAS-containing industrial products), Lang et al. (2017) generated a conservative estimate of the amount of PFAS in collected leachate that was sent to wastewater treatment in 2013 in the United States of ~600 kg/year. The dominant PFAS was 5:3 fluorotelomer carboxylic acid that is a compound not normally included in "standardized" target lists. This release seems small when compared with the quantities of PFAS introduced in the market on a yearly basis. Nevertheless, this estimate needs to be put in perspective through consideration of the following factors: PFAS already in landfills will continuously be released into the environment for several decades and new PFAS-containing materials will be disposed. Though this study was limited in the number and types of landfills and included on 19 PFAS, this does indicate that the presence of PFAS in leachate warrants further study to better characterize the extent and characterize the relationship between leachates and the landfill contents and design. It is also noteworthy that there are still a large number of unlined landfills (<6000) in the United States from which PFAS-containing leachate may potentially infiltrate into groundwater.

15.3.3 Industrial sources

Recent studies conducted over several years in Asia (Shiwaku et al., 2016; Bao et al., 2019) highlight the impact PFAS manufacturing sites can have on the occurrence of these chemicals in groundwater. These studies monitored different target PFAS in groundwater samples and other matrices in the vicinity of fluoropolymer production sites. The common compounds in these studies were the PFCAs with carbon chain lengths from 4 to 10 atom. PFOA was among the most prominent compound in both studies with concentrations ranging between 45 to 7400 ng/L in Japan (Shiwaku et al., 2016) and 105 to 2510 ng/L in China (Bao et al., 2019). Perfluorohexanoic acid (PFHxA) was the second most prevalent compound in the study conducted by Shiwaku et al. (2016) with reported concentrations from 9.7 to 970 ng/L. The concentrations of PFOA and PFHxA presented different trends over the course of this study: while PFOA decreased over the years, the concentration of PFHxA increased. These may reflect the changes in the manufacturing processes similar to the trends in the United States (Houtz et al., 2016). In China, PFHxA was also detected at high concentrations (13–614 ng/L) in groundwater' however, perfluorobutanesulfonic acid (PFBS) presented higher concentrations than PFOA and PFHxA, with concentrations ranging from 64 to 21,200 ng/L. These results highlight the impact from industrial sources, as the study by Bao et al. (2019) was conducted in the vicinity of a factory producing PFBS and PTFE.

15.3.4 Other sources

Atmospheric depositions have also been confirmed as source of PFAS in groundwater (Wei et al., 2018). Emissions from manufacturing locations are deposited via precipitation across potentially larger areas. This poses a challenge for source control, as depositions and detections may occur far from the sources (Gewurtz et al., 2019). This may also present other major difficulties in the United States as PFOA and PFOS were phased out in 2015, but these chemicals could still reach groundwater from emissions in countries where they continue to be manufactured.

15.4 Occurrence studies

Occurrence studies of PFAS in environmental samples and, more specifically, groundwater, conducted to date, are limited in scope. Several factors explain this: (1) lack of unequivocal structural identification of specific chemicals within PFAS in the market, (2) very limited publicly available information regarding their historic use and production, and (3) limited availability of standardized analytical methods. While the literature review behind Fig. 15.2 is limited (only 21 papers), it illustrates the range of concentrations found in the environment (note log scale for concentration), as well as the comparatively high concentrations found in groundwater relative to other water matrices across several studies. A more comprehensive literature review that is beyond the scope of this chapter should be conducted to verify the trend of higher groundwater concentrations observed in this limited analysis.

In the United States, there are three major focus areas determining the occurrence of PFAS (including PFAS precursors) in groundwater: (1) studies centered around the presence in groundwater of well-studied long-chain and legacy classes (i.e., PFOA and PFOS); (2) evaluations of remediation projects aimed at mitigating PFAS contamination in highly impacted, or suspected highly impacted areas such as manufacturing sites, military sites, and locations near other PFAS industrial users; and (3) research efforts to advance the identification of new PFAS and sources present in the environment.

The Unregulated Contaminant Monitoring Rule 3 (UCMR3) required monitoring for six PFAS using EPA Method 537 (Shoemaker et al., 2009) in the treated drinking water from approximately 6000 utilities across the country between 2013 and 2015. The results have been widely publicized in peer-reviewed articles (Hu et al., 2016; Guelfo and Adamson, 2018; Eaton et al., 2018). The dissemination of the outcomes from UCMR3 in news, magazines, and social media resulted in increased public awareness of the potential for PFAS in drinking water (SSEHRI, 2019).

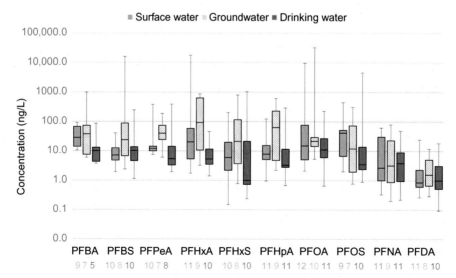

FIGURE 15.2 Summary of literature concentrations of PFAS in surface water, groundwater and drinking water. Belt is median, shoulders 25th and 75th percentile, whiskers are 5th and 95th percentile. Maximum concentrations reported in 21 peer-reviewed journal articles were used. Numbers under chemical name indicate the number of measurements for surface water (*green*, left bar in each group), groundwater (*gray stripe*, center bar), and drinking water (*blue*, right bar), respectively. *PFAS*, Per- and polyfluoroalkyl substances. *Adapted from Glassmeyer, S.T., Burns, E.E., Focazio, M.J., Furlong, E.T., Gribble, M.O., Jahne, M.A., et al., 2020. Water, water everywhere, but every drop unique: Emerging challenges in the science to understand the role of contaminants in management of drinking water supplies, GeoHealth. In preparation (Glassmeyer et al., 2020).*

There were a limited number of studies focused on the occurrence of PFAS in groundwater in the United States before UCMR3 was completed. Post el al. (2013) published on the occurrence of 10 PFAS (PFCAs and PFSAs and with carbon chains up to 10 carbons) in 18 groundwaters collected across New Jersey. The analytical method for this study was developed in-house by a commercial laboratory with method reporting limits of 5 ng/L. PFAS were detected in 55% (10 of 18) of the wells monitored. In four locations, several targets were determined at total concentrations ranging from 70 to 140 ng/L; in the remaining sites (six) only single and different PFAS were quantified, at concentrations between 5 and 80 ng/L. In another study, Schaider et al. (2014) characterized an aquifer heavily impacted by intrusion from septic tanks in Massachusetts. The main goal of this study was to evaluate trace organic contaminants, including PFOA and PFOS, with 10 and 1 ng/L, respectively, as method reporting limits. PFAS were detected in 4 of the 10 wells sampled in the study. In two of the wells the relatively low nitrate levels suggest that the source of the PFAS was not the local septic systems.

The state of Minnesota has included PFAS as an analyte in their Ambient Groundwater Monitoring Network. In the 2013 monitoring campaign, nearly 200 wells across the state and suspected to be subjected to anthropogenic influences were included in the study (Kroening, 2017). PFAS was detected in 69% of the wells. Although outcomes may be biased by the selection of sampling location, the study still demonstrated how pervasive PFAS from probable human sources can be in groundwater in both shallow and deep aquifer systems. The concentrations in the Minnesota study were typically low, but the maximum concentrations of PFHxS, PFOS and PFOA exceeded the health values established by the Minnesota Department of Health.

In addition to the increased interest in specific PFAS compounds and simultaneously to the UCMR3, Backe et al. (2013) and Barzen-Hanson and Field (2015) published on the identification and quantification of newly discovered PFAS from various AFFF formulations. In the study by Backe et al. (2013), 26 newly identified zwitterionic, cationic, and anionic PFAS and 21 legacy PFAS were analyzed in groundwater from five military bases where AFFFs had been used. PFOA and PFOS were quantified at concentrations up to 220,200 and 19,000 ng/L, respectively. In addition, newly identified compounds associated with the usage of different AFFFs were quantified in the groundwater samples. Barzen-Hanson and Field (2015) discovered in same samples collected in the study by Backe et al. (2013) the presence of two ultrashort PFSAs with two and three carbon chains at concentrations up to 7,500 ng/L [perfluoroethanesulfonate (PFEtS)] and 63,000 ng/L [perfluoropropanesulfonate (PFPrS)]. These high concentrations confirmed the persistence of the ultrashort-chain compounds since the use of AFFFs had been discontinued for decades. The results compiled in these two works illustrate the complexity of AFFF formulations and chemistry and how the PFAS compositions change over time as precursor compounds degrade into terminal products.

15.5 Removal of per- and polyfluoroalkyl substances from groundwater

Potential health and ecological risks from exposure to PFAS-contaminated water has led to increased research into treatment and remediation technologies to clean up impacted sites. Recent reviews of treatment technologies for PFAS-contaminated water provide a greater scope and depth for specific applications or media than what is covered in this chapter (Appleman et al., 2014; Du et al., 2014; Rahman et al., 2014; Cummings et al., 2015; Dickenson and Higgins, 2016; Kucharzyk et al., 2017; Ross et al., 2018; Crone et al., 2019). Most research related to the treatment of PFAS-contaminated groundwater has been conducted at the laboratory scale and is focused on PFOS and PFOA. It is difficult to provide specific guidance on selection of treatment methodologies because the composition of PFAS compounds present in groundwater is so broad, PFAS may occur with cocontaminants, and interactions with the matrices are complex and not well elucidated, (Buck et al., 2011; Kucharzyk et al., 2017; Hatton et al., 2018). With the strength of the C−F bond and the high energy requirements to break it, treatment typically focuses on partitioning PFAS from the water phase rather than on degradation. Although some destructive technologies, such as electron-beam, are being evaluated currently (Kucharzyk et al., 2017; Ross et al., 2018). Advanced oxidation processes (AOPs) can be used to treat PFAS-contaminated groundwater but face several important challenges, including: the generation of recalcitrant end products, short-chain PFAS as a by-product of degradation, and significant energy requirements (Ross et al., 2018; Trojanowicz et al., 2018). Sonochemical destruction has also been shown to degrade PFAS but faces many of the same challenges as AOPs (Kucharzyk et al., 2017). Conventional treatment technologies evaluated for removal of PFAS include sand filtration, UV degradation, coagulation/flocculation/sedimentation, oxidation, disinfection, aeration, and biodegradation. Typically, these technologies remove less than 5% of the measured PFAS (Appleman et al., 2014; Pan et al., 2016; Kucharzyk et al., 2017). Currently, the most effective treatment technologies for removing PFAS from water are activated carbon adsorption, ion-exchange resins (IEX), and filtration, via nanofiltration (NF), and reverse osmosis (RO) systems (Rahman et al., 2014; Kucharzyk et al., 2017; Ross et al., 2018).

PFAS-contaminated groundwater can be treated in situ or ex situ via pump and treat technologies (Kucharzyk et al., 2017). Currently, more information is available on ex situ technologies arising from the need to treat contaminated groundwater for drinking water production and several full-scale evaluations have been conducted (Appleman et al., 2014; Kucharzyk et al., 2017). In situ treatment has the potential to be faster, more cost-effective, and have reduced lifecycle impacts but development of these methodologies is still in the early stages and full-scale evaluations are lacking (Kucharzyk et al., 2017). Currently, the limited in situ treatments either use immobilization (sorbing PFAS in place) or chemical oxidation or reduction, or a combination of these technologies (Kucharzyk et al., 2017; Aly, 2019).

15.5.1 Granular activated carbon

Due to its large adsorption capacity, activated carbon and specifically granular activated carbon (GAC) has been widely used for the removal of a broad range of traditional and emerging contaminants, including PFAS from water (Xiao et al., 2017b). PFAS adsorption onto activated carbon is complex and many mechanisms, including acid−base interactions, electrostatic and hydrophobic interactions, micelle and hemimicelle formation, and multilayer adsorption, influence removal efficacy (Du et al., 2014; Zhi and Liu, 2015; Chen et al., 2017). Common GAC source materials include coconut shells, bituminous coal, wood, and bamboo (Qiu et al., 2007; Deng et al., 2010; Zhi and Liu, 2015; Xiao et al., 2017b). Of these, bituminous coal−based GAC is generally considered to be the most effective for adsorbing PFAS due to its smaller pore structure (Appleman et al., 2013; Inyang and Dickenson, 2017; McNamara et al., 2018). The Freundlich isotherm model has been found to better describe PFAS sorption than the Langmuir isotherm model. It is important to note, however, that capacity values (K) vary by over seven orders of magnitude for a single PFAS compound (Deng et al., 2015; Zhi and Liu, 2015) potentially indicating a need for an alternative isotherm model for this very broad and unique class of chemicals. These variations may arise from many factors, including differences in the water matrix such as pH, ionic strength, differences in the GAC used and its source material, and differences in the physical−chemical properties of the individual PFAS chemicals. In addition, the presence or absence of competing compounds such as natural organic matter (NOM) or cocontaminants may compete with PFAS for binding sites on the GAC (Dudley et al., 2015; Kothawala et al., 2017; Liu, 2017).

Though much uncertainty exists for the performance of GAC for these chemicals, in general, longer chain PFAS are more hydrophobic and have higher removal efficiencies with GAC than short-chain PFAS (Rahman et al., 2014; Xiao et al., 2017b). High removal efficiencies of a broad suite of both short and long-chain PFAS are possible but may require frequent bed replacement or regeneration of the GAC depending on the factors previously mentioned. Current partitioning models are ineffective at predicting adsorption capacity and batch isotherm or column studies (e.g., rapid small-scale column tests) using the intended sorbent and sorbate combination are needed to develop accurate carbon usage rates. Reports on the removal effectiveness in full-scale column studies vary widely, as well, and GAC replacement frequencies from 3 to 24 months have been reported in drinking water applications (Rumsby et al., 2009; Cummings et al., 2015). While little work has been conducted on the regeneration and reactivation of spent GAC, Watanabe et al. (2016) reported reactivation is possible for PFOS, PFOA, and PFHxA and they recommend that off-gas be heated to 1000°C to ensure mineralization of PFAS. However, there is limited information regarding the formation of by-products during this process. Overall, further evaluations of the performance, cost-effectiveness, and impacts of water quality, PFAS mixtures, and cocontaminants are needed.

15.5.2 Ion-exchange resins

IEX resins are typically polymeric spheres with a charged functional group capable of exchanging a preloaded counter ion, such as chloride, with PFAS (Dudley et al., 2015; Zaggia et al., 2016). The primary removal mechanism is ionic interaction but adsorption, agglomeration, hydrophobic interactions, PFAS functional group interactions, and resin functional group interactions play a role as well (Zaggia et al., 2016; Ross et al., 2018). Variations in IEX removal efficiencies of different PFAS species are likely driven by secondary removal mechanisms such as hydrophobicity as demonstrated by Zaggia et al. (2016). Results of pilot and full-scale evaluations of IEX PFAS treatment effectiveness reported vary significantly from less than 10% to greater than 90% (Appleman et al., 2014; McCleaf et al., 2017). This is likely due to the complexity of both the matrix and the broad suite of PFAS compounds being evaluated concurrently. Several authors made side-by-side comparisons of IEX and GAC and found that IEX generally outperforms GAC in time to breakthrough when treating long-chain PFAS compounds but that GAC performs better when treating short-chain PFAS compounds (Senevirathna et al., 2010; Chularueangaksorn et al., 2014; McCleaf et al., 2017; Woodard et al., 2017). A full life cycle and cost comparison is needed to inform decision makers. IEX resin treatment of PFAS-contaminated groundwater faces several complications, including competition with NOM and other naturally occurring ions, such as sulfate, nitrate, and bicarbonate, for binding sites (Tripp et al., 2003; Kothawala et al., 2017). Reports on the effectiveness of regeneration vary significantly but even when effective, they produce an additional highly concentrated waste stream that requires further treatment (Carter and Farrell, 2010; Deng et al., 2010; Liu, 2017). Further research is needed to fully understand efficacy across a broad range of PFAS, varying water quality conditions, existing and innovative IEX media, waste disposal, and cost-effectiveness.

15.5.3 Nanofiltration and reverse osmosis

Filtration technologies show mixed results depending primarily on the pore size of the membranes. Minimal studies have been conducted to examine efficacy of membrane processes for PFAS. However, size exclusion appears to be the dominate rejection mechanism, but membrane–PFAS interactions, such as charge rejection and hydrophobicity, play an important role as well (Appleman et al., 2014; Kucharzyk et al., 2017). Low-pressure microfiltration and ultrafiltration membranes show limited effectiveness since PFAS molecules are typically small enough to pass through the larger pores used in these lower cost technologies. Typical removal efficiencies of these systems range from 0% to 23% for PFOS and PFOA and are likely due to adsorption to particles, which are large enough to be retained by the membrane (McLaughlin et al., 2011; Takagi et al., 2011; Dickenson and Higgins, 2016).

High-pressure membranes such as NF and RO systems have smaller pores and are thereby able to retain a larger fraction of PFAS compounds but have higher energy requirements to operate. Typical PFAS removal efficiencies are greater than 90% and 99% for NF and RO, respectively (Appleman et al., 2013, 2014). One notable exception is for neutral PFAS such as perfluorooctanesulfonamide; Steinle-Darling and Reinhard (2008) reported removals as low as 42% when using a NF membrane. These results highlight the need for more research on mechanisms, membrane pore size and materials, impacts of water quality, and the impacts on PFAS properties, such as charge and electrostatic interactions, on removal. In addition to being costly because of the high pressure required for operation, membrane treatment also faces the challenges of performance degradation due to fouling and the generation of highly concentrated waste streams (up to 20% of the production flow depending on total dissolved solids concentrations) that will require subsequent treatment (Baruth, 2005).

Though groundwater treatment technologies have been long studied and generally well understood for many conventional contaminants, PFAS-contaminated groundwaters pose a unique challenge due to the limited physical–chemical characterization of the chemicals themselves and their unique behaviors as surfactants in complex mixtures. Table 15.3 summarizes the treatment technologies discussed in this chapter. Further research to develop and optimize (performance and cost) existing and innovative technologies, both in situ and ex situ, is needed to protect human health and the environment.

TABLE 15.3 Summary of per- and polyfluoroalkyl substance (PFAS) treatment technologies.

Treatment technology	Efficient for drinking water treatment for PFAS	Primary advantages	Primary disadvantages
Conventional treatment[a]	No	Well-established technologies for water treatment	Limited removal
Granular activated carbon	Yes	Easy to incorporate into conventional treatment system	Breakthrough of shorter chained PFAS, periodic regeneration needed
Ion-exchange resin	Yes	Better treatment of long-chain PFAS relative to GAC	Regeneration produces concentrated waste
Nanofiltration and reverse osmosis	Yes	Potential for highly efficient removal	Removal efficiency analyte dependent, high pressure required, concentrated wastes

GAC, Granular activated carbon.
[a]Sand filtration, UV, coagulation/flocculation/sedimentation, oxidation, disinfection, aeration, and biodegradation.

15.6 Conclusion

There are many data gaps in our understanding of fate of PFAS in the environment and their management, especially in groundwater. Analytical methods only exist for a fraction of the PFAS that have been produced, limiting our ability to accurately quantify the known chemicals. The environmental persistence of PFAS results in transport from areas of use to both surface waters and groundwater. Selected treatments such as GAC, ion-exchange, NF, and RO have been demonstrated to remove PFAS from water, but the efficacy of the treatment is compound dependent and may have undesirable consequences, such as high energy costs and/or generation of concentrated wastes. Focused research efforts are needed to better characterize fate and transport of PFAS in groundwater as well as to optimize remediation and treatment technologies.

References

Ahrens, L., Bundschuh, M., 2014. Fate and effects of poly- and perfluoroalkyl substances in the aquatic environment: a review. Environ. Toxicol. Chem. 33, 1921–1929.

Aly, Y., 2019. Enhanced Adsorption of Perfluoro Alkyl Substances in Groundwater; Development of a Novel In-Situ Groundwater Remediation Method. Available from: https://conservancy.umn.edu/handle/11299/202916.

Appleman, T.D., Dickenson, E.R., Bellona, C., Higgins, C.P., 2013. Nanofiltration and granular activated carbon treatment of perfluoroalkyl acids. J. Hazard. Mater. 260, 740–746.

Appleman, T.D., Higgins, C.P., Quiñones, O., Vanderford, B.J., Kolstad, C., Zeigler-Holady, J.C., et al., 2014. Treatment of poly- and perfluoroalkyl substances in U.S. full-scale water treatment systems. Water Res. 51, 246–255.

ASTM, 2019. D7979-19, Standard Test Method for Determination of Per- and Polyfluoroalkyl Substances in Water, Sludge, Influent, Effluent, and Wastewater by Liquid Chromatography Tandem Mass Spectrometry (LC/MS/MS). ASTM International, West Conshohocken, PA.

Backe, W.J., Day, T.C., Field, J.A., 2013. Zwitterionic, cationic, and anionic fluorinated chemicals in aqueous film forming foam formulations and groundwater from US military bases by nonaqueous large-volume injection HPLC-MS/MS. Environ. Sci. Technol. 47, 5226–5234.

Bao, J., Yu, W.J., Liu, Y., Wang, X., Jin, Y.H., Dong, G.H., 2019. Perfluoroalkyl substances in groundwater and home-produced vegetables and eggs around a fluorochemical industrial park in China. Ecotoxicol. Environ. Saf. 171, 199–205.

Baruth, E.E., 2005. Water Treatment Plant Design. American Water Works Association, and American Society of Civil Engineers.

Barzen-Hanson, K.A., Field, J.A., 2015. Discovery and implications of C2 and C3 perfluoroalkyl sulfonates in aqueous film-forming foams and groundwater. Environ. Sci. Technol. Lett. 95–99.

Buck, R.C., Franklin, J., Berger, U., Conder, J.M., Cousins, I.T., De Voogt, P., et al., 2011. Perfluoroalkyl and polyfluoroalkyl substances in the environment: terminology, classification, and origins. Integr. Environ. Assess. Manage. 74, 513–541.

Carter, K.E., Farrell, J., 2010. Removal of perfluorooctane and perfluorobutane sulfonate from water via carbon adsorption and ion exchange. Sep. Sci. Technol. 456, 762–767.

Chen, W., Zhang, X., Mamadiev, M., Wang, Z., 2017. Sorption of perfluorooctane sulfonate and perfluorooctanoate on polyacrylonitrile fiber-derived activated carbon fibers: in comparison with activated carbon. RSC Adv. 72, 927–938.

Chularueangaksorn, P., Tanaka, S., Fujii, S., Kunacheva, C., 2014. Batch and column adsorption of perfluorooctane sulfonate on anion exchange resins and granular activated carbon. J. Appl. Polym. Sci. 1313, 39782.

Clara, M., Scharf, S., Weiss, S., Gans, O., Scheffknecht, C., 2008. Emissions of perfluorinated alkylated substances (PFAS) from point sources—identification of relevant branches. Water Sci. Technol. 58, 59–66.

Ciofi, L., Renai, L., Rossini, D., Ancillotti, C., Falai, A., Fibbi, D., et al., 2018. Applicability of the direct injection liquid chromatographic tandem mass spectrometric analytical approach to the sub-ng L^{-1} determination of perfluoro-alkyl acids in waste, surface, ground and drinking water samples. Talanta 176, 412–421.

Cousins, I.T., Vestergren, R., Wang, Z., Scheringer, M., McLachlan, M.S., 2016. The precautionary principle and chemicals management: the example of perfluoroalkyl acids in groundwater. Environ. Int. 94, 331–340.

Crone, B.C., Speth, T.F., Wahman, D.G., Smith, S.J., Abulikemu, G., Kleiner, E.J., et al., 2019. Occurrence of per- and polyfluoroalkyl substances PFAS in source water and their treatment in drinking water. Crit. Rev. Environ. Sci. Technol. 4924, 2359–2396.

Cummings, L., Matarazzo, A., Nelson, N., Sickels, F., Storms, C., 2015. Recommendation on perfluorinated compound treatment options for drinking water. In: New Jersey Drinking Water Quality Institute Treatment Subcommittee Report 2015.

Dasu, K., Nakayama, S.F., Yoshikane, M., Mills, M.A., Wright, J.M., Ehrlich, S., 2017. An ultra-sensitive method for the analysis of perfluorinated alkyl acids in drinking water using a column switching high-performance liquid chromatography tandem mass spectrometry. J. Chromatogr. A 1494, 46–54.

Deng, S., Yu, Q., Huang, J., Yu, G., 2010. Removal of perfluorooctane sulfonate from wastewater by anion exchange resins: effects of resin properties and solution chemistry. Water Res. 4418, 5188–5195.

Deng, S., Nie, Y., Du, Z., Huang, Q., Meng, P., Wang, B., et al., 2015. Enhanced adsorption of perfluorooctane sulfonate and perfluorooctanoate by bamboo-derived granular activated carbon. J. Hazard. Mater. 282, 150–157.

Dickenson, E., Higgins, C., 2016. Treatment mitigation strategies for poly-and perfluoroalkyl substances. In: Project# 4322. Water Research Foundation, Denver, CO.

Dieter, C.A., Maupin, M.A., Caldwell, R.R., Harris, M.A., Ivahnenko, T.I., Lovelace, J.K., et al., 2018, Estimated use of water in the United States in 2015: U.S. Geological Survey Circular 1441. In: Supersedes USGS Open-File Report 2017–1131. 65 p. https://doi.org/10.3133/cir1441.

DOD, 2019. Department of Defense consolidated quality systems manual (QSM) for environmental laboratories, version 5.3. Available from: https://www.denix.osd.mil/edqw/documents/manuals/qsm-version-5-3-final/. (accessed 24.11.19.).

Du, Z., Deng, S., Bei, Y., Huang, Q., Wang, B., Huang, J., et al., 2014. Adsorption behavior and mechanism of perfluorinated compounds on various adsorbents—a review. J. Hazard. Mater. 274, 443–454.

Dudley, L.-A., Arevalo, E.C., Knappe, D.R., 2015. Removal of Perfluoroalkyl Substances by PAC adsorption and anion exchange. In: Project# 4344. The Water Research Foundation, Denver, CO.

Eaton, A., Bartrand, T., Rosen, S., 2018. Detailed analysis of the UCMR 3 DATABASE: implications for future groundwater monitoring. J. Am. Water Works Assoc. 110, 13–25.

Eschauzier, C., Klaasjan, J.R., Stuyfzand, P.J., De Voogt, P., 2013. Perfluorinated alkylated acids in groundwater and drinking water: identification, origin and mobility. Sci. Total Environ. 458, 477–485.

FDA, 2019. Per and polyfluoroalkyl substances (PFAS). Available from: https://www.fda.gov/food/chemicals/and-polyfluoroalkyl-substances-pfas. (accessed 17.11.19.).

Gewurtz, S.B., Bradley, L.E., Backus, S., Dove, A., McGoldrick, D., Hung, H., et al., 2019. Perfluoroalkyl acids in great lakes precipitation and surface water (2006–2018) indicate response to phase-outs, regulatory action, and variability in fate and transport processes. Environ. Sci. Technol. 53, 8543–8552.

Glassmeyer, S.T., Burns, E.E., Focazio, M.J., Furlong, E.T., Gribble, M.O., Jahne, M.A., et al., 2020. Water, water everywhere, but every drop unique: Emerging challenges in the science to understand the role of contaminants in management of drinking water supplies. GeoHealth. In preparation.

Guelfo, J.L., Adamson, D.T., 2018. Evaluation of a national data set for insights into sources, composition, and concentrations of per-and polyfluoroalkyl substances PFAS in US drinking water. Environ. Pollut. 236, 505–513.

Guelfo, J.L., Marlow, T., Klein, D.M., Savitz, D.A., Frickel, S., Crimi, M., Suuberg, E.M., 2018. Evaluation and management strategies for per- and polyfluoroalkyl substances (PFASs) in drinking water aquifers: perspectives from impacted US Northeast communities. Environ. Health Perspect. 126 (6), 505–513. 065001p.

Hale, S.E., Arp, H.P.H., Slinde, G.A., Wade, E.J., Bjørseth, K., Breedveld, G.D., et al., 2017. Sorbent amendment as a remediation strategy to reduce PFAS mobility and leaching in a contaminated sandy soil from a Norwegian firefighting training facility. Chemosphere 171, 9–18.

Hatton, J., Holton, C., DiGuiseppi, B., 2018. Occurrence and behavior of per- and polyfluoroalkyl substances from aqueous film-forming foam in groundwater systems. Rem. J. 282, 89–99.

Hopkins, Z.R., Sun, M., DeWitt, J.C., Knappe, D.R., 2018. Recently detected drinking water contaminants: GenX and other per- and polyfluoroalkyl ether acids. J. Am. Water Works Assoc. 110 (7), 13–28.

Houtz, E.F., Sedlak, D.L., 2012. Oxidative conversion as a means of detecting precursors to perfluoroalkyl acids in urban runoff. Environ. Sci. Technol. 46, 9342–9349.

Houtz, E.F., Higgins, C.P., Field, J.A., Sedlak, D.L., 2013. Persistence of perfluoroalkyl acid precursors in AFFF-impacted groundwater and soil. Environ. Sci. Technol. 47, 8187–8195.

Houtz, E.F., Sutton, R., Park, J.S., Sedlak, M., 2016. Poly-and perfluoroalkyl substances in wastewater: significance of unknown precursors, manufacturing shifts, and likely AFFF impacts. Water Res. 95, 142–149.

Hu, X.C., Andrews, D.Q., Lindstrom, A.B., Bruton, T.A., Schaider, L.A., Grandjean, P., et al., 2016. Detection of poly-and perfluoroalkyl substances (PFAS) in US drinking water linked to industrial sites, military fire training areas, and wastewater treatment plants. Environ. Sci. Technol. Lett. 3, 344–350.

Interstate Technology & Regulatory Council (ITRC), 2017a. Naming conventions and physical and chemical properties of per- and polyfluoroalkyl substances PFAS. Available from: https://pfas1.itrcweb.org/wpcontent/uploads/2017/10/pfas_fact_sheet_naming_conventions_11_13_17.pdf. (accessed 17.11.19.).

Interstate Technology & Regulatory Council (ITRC), 2017b. PFAS Fact Sheets PFAS-1. Available from: http://pfas-1.itrcweb.org/fact-sheets/. (accessed 17.11.19.).

Inyang, M., Dickenson, E.R., 2017. The use of carbon adsorbents for the removal of perfluoroalkyl acids from potable reuse systems. Chemosphere 184, 168–175.

Janda, J., Nödler, K., Brauch, H.J., Zwiener, C., Lange, F.T., 2019. Robust trace analysis of polar (C2-C8) perfluorinated carboxylic acids by liquid chromatography-tandem mass spectrometry: method development and application to surface water, groundwater and drinking water. Environ. Sci. Pollut. Res. 26, 7326–7336.

Kothawala, D.N., Köhler, S.J., Östlund, A., Wiberg, K., Ahrens, L., 2017. Influence of dissolved organic matter concentration and composition on the removal efficiency of perfluoroalkyl substances PFAS during drinking water treatment. Water Res. 121 (Suppl. C), 320–328.

Kroening, S., 2017 Perfluorinated chemicals in Minnesota's ambient groundwater, 2013 (wq-am4-02). Available from: https://www.pca.state.mn.us/water/groundwater-data or https://www.pca.state.mn.us/sites/default/files/wq-am4-02.pdf. (accessed 24.11.19.).

Kucharzyk, K.H., Darlington, R., Benotti, M., Deeb, R., Hawley, E., 2017. Novel treatment technologies for PFAS compounds: a critical review. J. Environ. Manage. 204, 757–764.

Land, M., de Wit, C.A., Bignert, A., Cousins, I.T., Herzke, D., Johansson, J.H., et al., 2018. What is the effect of phasing out long-chain per-and polyfluoroalkyl substances on the concentrations of perfluoroalkyl acids and their precursors in the environment? A systematic review. Environ. Evid. 7, 1–32.

Lang, J.R., Allred, B.M., Field, J.A., Levis, J.W., Barlaz, M.A., 2017. National estimate of per-and polyfluoroalkyl substance PFAS release to US municipal landfill leachate. Environ. Sci. Technol. 51, 2197–2205.

Liu, C., 2017. Removal of Perfluorinated Compounds in Drinking Water Treatment: A Study of Ion Exchange Resins and Magnetic Nanoparticles. Ontario, Canada.

McCleaf, P., Englund, S., Östlund, A., Lindegren, K., Wiberg, K., Ahrens, L., 2017. Removal efficiency of multiple poly-and perfluoroalkyl substances PFASs in drinking water using granular activated carbon GAC and anion exchange AE column tests. Water Res. 120, 77–87.

McLaughlin, C.L., Blake, S., Hall, T., Harman, M., Kanda, R., Foster, J., et al., 2011. Perfluorooctane sulphonate in raw and drinking water sources in the United Kingdom. Water Environ. J. 251, 13–21.

McNamara, J.D., Franco, R., Mimna, R., Zappa, L., 2018. Comparison of activated carbons for removal of perfluorinated compounds from drinking water. J. Am. Water Works Assoc. 1101, E2–E14.

Mukherjee, A., Scanlon, B., Aureli, A., Langan, S., Guo, H., McKenzie, A., 2020. Global Groundwater: Source, Scarcity, Sustainability, Security and Solutions, first ed. Elsevier. ISBN: 9780128181720.

Nakayama, S.F., Yoshikane, M., Onoda, Y., Nishihama, Y., Iwai-Shimada, M., Takagi, M., et al., 2019. Worldwide trends in tracing poly-and perfluoroalkyl substances (PFAS) in the environment. Trends Anal. Chem. 151, 115410.

Olsen, G.W., Mair, D.C., Lange, C.C., Harrington, L.M., Church, T.R., Goldberg, C.L., et al., 2017. Per-and polyfluoroalkyl substances PFAS in American Red Cross adult blood donors, 2000–2015. Environ. Res. 157, 87–95.

Pan, C.-G., Liu, Y.-S., Ying, G.-G., 2016. Perfluoroalkyl substances PFAS in wastewater treatment plants and drinking water treatment plants: removal efficiency and exposure risk. Water Res. 106, 562–570.

Post, Gloria, B., Louis, Judith, B., Lee Lippincott, R., Procopio, Nicholas, A., 2013. Occurrence of perfluorinated compounds in raw water from New Jersey public drinking water systems. Environ. Sci. Technol. 47, 13266–13275.

Qiu, Y., Fujii, S., Tanaka, S., 2007. Removal of perfluorochemicals from wastewater by granular activated carbon adsorption. Environ. Eng. Res. 44, 185–193.

Rahman, M.F., Peldszus, S., Anderson, W.B., 2014. Behaviour and fate of perfluoroalkyl and polyfluoroalkyl substances PFAS in drinking water treatment: a review. Water Res. 50, 318–340.

Rosenblum, L. Wendelken, S.C., 2019. Method 533: Determination of Per- and Polyfluoroalkyl Substances in Drinking Water by Isotope Dilution Anion Exchange Solid Phase Extraction and Liquid Chromatography/Tandem Mass Spectrometry. Available from: https://www.epa.gov/sites/production/files/2019-12/documents/method-533-815b19020.pdf. (accessed 22.02.20.).

Ross, I., McDonough, J., Miles, J., Storch, P., Thelakkat Kochunarayanan, P., Kalve, E., et al., 2018. A review of emerging technologies for remediation of PFAS. Rem. J. 282, 101–126.

Ruan, T., Jiang, G., 2017. Analytical methodology for identification of novel per-and polyfluoroalkyl substances in the environment. Trends Anal. Chem. 95, 122–131.

Rumsby, P.C., McLaughlin, C.L., Hall, T., 2009. Perfluorooctane sulphonate and perfluorooctanoic acid in drinking and environmental waters. Philos. Trans. R. Soc A: Math. Phys. Eng. Sci. 3671904, 4119–4136.

Schaider, L.A., Rudel, R.A., Ackerman, J.M., Dunagan, S.C., Brody, J.G., 2014. Pharmaceuticals, perfluorosurfactants, and other organic wastewater compounds in public drinking water wells in a shallow sand and gravel aquifer. Sci. Total Environ. 468, 384–393.

Senevirathna, S.T.M.L.D., Tanaka, S., Fujii, S., Kunacheva, C., Harada, H., Shivakoti, B.R., et al., 2010. A comparative study of adsorption of perfluorooctane sulfonate PFOS onto granular activated carbon, ion-exchange polymers and non-ion-exchange polymers. Chemosphere 806, 647–651.

Shiwaku, Y., Lee, P., Thepaksorn, P., Zheng, B., Koizumi, A., Harada, K.H., 2016. Spatial and temporal trends in perfluorooctanoic and perfluorohexanoic acid in well, surface, and tap water around a fluoropolymer plant in Osaka, Japan. Chemosphere 164, 603–610.

Shoemaker, J.A., Tettenhorst, D.R., 2018. Method 537.1: Determination of Selected Perfluorinated Alkyl Acids in Drinking Water by Soild Phase Extraction and Liquid Chromatography/Tandem Mass Spectrometry (LC/MS/MS). Available from: https://cfpub.epa.gov/si/si_public_record_report.cfm?dirEntryId = 343042&Lab = NERL&simpleSearch = 0&showCriteria = 2&searchAll = Determination + of + Selected + Per- + and + Polyfluorinated + Alkyl + Substances + &TIMSType = &dateBeginPublishedPresented = 11%2F02%2F2016. (accessed 24.11.19.).

Shoemaker, J.A., Grimmett, P.E., Boutin, B.K., 2009 Method 537: Determination of Selected Perfluorinated Alkyl Acids in Drinking Water by Solid Phase Extraction and Liquid Chromatography/Tandem Mass Spectrometry (LC/MS/MS). Available from: https://cfpub.epa.gov/si/si_public_record_report.cfm?Lab = NERL&dirEntryId = 198984&simpleSearch = 1&searchAll = EPA%2F600%2FR-08%2F092 + . (accessed 24.11.19.).

SSEHRI, 2019. Per- and polyfluoroalkyl substances: the social discovery of a class of emerging contaminants. Social Science Environmental Health Research Institute at Northwestern University. https://pfasproject.com/. (accessed 19.11.19.).

Steinle-Darling, E., Reinhard, M., 2008. Nanofiltration for trace organic contaminant removal: structure, solution, and membrane fouling effects on the rejection of perfluorochemicals. Environ. Sci. Technol. 4214, 5292–5297.

Takagi, S., Adachi, F., Miyano, K., Koizumi, Y., Tanaka, H., Watanabe, I., et al., 2011. Fate of perfluorooctanesulfonate and perfluorooctanoate in drinking water treatment processes. Water Res. 4513, 3925–3932.

Tripp, A., Clifford, D., Roberts, D., Cang, Y., Aldridge, L., Gillogly, T., et al., 2003. Treatment of perchlorate in groundwater by ion exchange technology. In: Project #2532. The Water Research Foundation, Denver, CO.

Trojanowicz, M., Musijowski, J., Koc, M., Donten, M.A., 2011. Determination of total organic fluorine (TOF) in environmental samples using flow-injection and chromatographic methods. Anal. Methods 3, 1039–1045.

Trojanowicz, M., Bojanowska-Czajka, A., Bartosiewicz, I., Kulisa, K., 2018. Advanced oxidation/reduction processes treatment for aqueous perfluorooctanoate PFOA and perfluorooctanesulfonate PFOS—a review of recent advances. Chem. Eng. J. 336, 170–199.

US EPA, 2016a. Drinking water health advisory for perfluorooctane sulfonate (PFOS). In: EPA 822-R-16-004 May 2016. U.S. Environmental Protection Agency, Office of Water. Available from: https://www.epa.gov/sites/production/files/2016-05/documents/pfos_health_advisory_final_508.pdf. (accessed on 24.11.19.).

US EPA, 2016b. Drinking water health advisory for perfluorooctanoic acid (PFOA). In: EPA 822-R-16-005 May 2016. U.S. Environmental Protection Agency, Office of Water. Available from: https://www.epa.gov/sites/production/files/2016-05/documents/pfoa_health_advisory_final_508.pdf. (accessed on 24.11.19.).

US EPA, 2019. Method 8327: Per- and polyfluoroalkyl substances (PFAS) using external standard calibration and multiple reaction monitoring (MRM) liquid chromatography/tandem mass spectrometry (LC/MS/MS). Available from: https://www.epa.gov/sites/production/files/2019-06/documents/proposed_method_8327_procedure.pdf. (accessed 24.11.19.).

Watanabe, N., Takemine, S., Yamamoto, K., Haga, Y., Takata, M., 2016. Residual organic fluorinated compounds from thermal treatment of PFOA, PFHxA and PFOS adsorbed onto granular activated carbon (GAC). J. Mater. Cycles Waste Manage. 18, 625–630.

Wang, Z., DeWitt, J.C., Higgins, C.P., Cousins, I.T., 2017. A never-ending story of per-and polyfluoroalkyl substances PFAS? Environ. Sci. Technol. 51, 2508–2518.

Weber, A.K., Barber, L.B., LeBlanc, D.R., Sunderland, E.M., Vecitis, C.D., 2017. Geochemical and hydrologic factors controlling subsurface transport of poly-and perfluoroalkyl substances, Cape Cod, Massachusetts. Environ. Sci. Technol. 51, 4269–4279.

Wei, C., Wang, Q., Song, X., Chen, X., Fan, R., Ding, D., et al., 2018. Distribution, source identification and health risk assessment of PFAS and two PFOS alternatives in groundwater from non-industrial areas. Ecotoxicol. Environ. Saf. 152, 141–150.

Woodard, S., Berry, J., Newman, B., 2017. Ion exchange resin for PFAS removal and pilot test comparison to GAC. Rem. J. 273, 19–27.

Xiao, F., Golovko, S.A., Golovko, M.Y., 2017a. Identification of novel non-ionic, cationic, zwitterionic, and anionic polyfluoroalkyl substances using UPLC–TOF–MSE high-resolution parent ion search. Anal. Chim. Acta. 988, 41–49.

Xiao, X., Ulrich, B.A., Chen, B., Higgins, C.P., 2017b. Sorption of poly-and perfluoroalkyl substances PFAS relevant to aqueous film-forming foam AFFF-impacted groundwater by biochars and activated carbon. Environ. Sci. Technol. 5111, 6342–6351.

Zaggia, A., Conte, L., Falletti, L., Fant, M., Chiorboli, A., 2016. Use of strong anion exchange resins for the removal of perfluoroalkylated substances from contaminated drinking water in batch and continuous pilot plants. Water Res. 91, 137–146.

Zhi, Y., Liu, J., 2015. Adsorption of perfluoroalkyl acids by carbonaceous adsorbents: Effect of carbon surface chemistry. Environ. Pollut. 202, 168–176.

Chapter 16

Geogenic-contaminated groundwater in China

Yongfeng Jia[1,2]

[1]State Key Laboratory of Environmental Criteria and Risk Assessment, Chinese Research Academy of Environmental Sciences, Beijing, P.R. China,
[2]State Environmental Protection Key Laboratory of Simulation and Control of Groundwater Pollution, Chinese Research Academy of Environmental Sciences, Beijing, P.R. China

16.1 Introduction

Global groundwater faces great challenges for sustainable use, especially in the densely populated China (Mukherjee et al., 2020; Jia et al., 2018). Groundwater accounts for one-third of China's total water resources. Among the 655 major cities in China, more than 400 cities use groundwater as drinking water. In northern China, 65% of drinking water, 50% of industrial water, and 33% of irrigation water are groundwater [MEP (Ministry of Environmental Protection), 2011]. Groundwater from private well is the main source of drinking water in rural areas. Groundwater quality has deteriorated nationwide due to pollution and geological background (Wen et al., 2013; Guo et al., 2014a). The geogenic-contaminated groundwater (GCG) refers to the groundwater rich or deficient in certain component caused by geological reason rather than anthropogenic activity, which cannot meet the requirements of drinking water or other guidelines, thus affecting human consumption (Jia et al., 2018). The GCG is widely distributed in China. Among the five categories of groundwater quality classification, 86.2% of the 10,168 monitoring wells in China are undrinkable categories IV and V [MEE (Ministry of Ecology and Environment), 2018]. The main indicators affecting groundwater quality are manganese, iron, total hardness, total dissolved solids (TDS), iodine, chloride, nitrogen, and sulfate, of which a considerable part is GCG. Similar results are revealed by a national groundwater quality survey conducted by China Geological Survey from 2005 to 2015, covering 4.4 million square kilometers and more than 30,000 monitoring wells. The results show that the groundwater with $Mn > 100$ μg/L, $Fe > 200$ μg/L, $TDS > 1000$ mg/L, $F > 1.0$ mg/L, $I > 80$ μg/L, and $As > 10$ μg/L accounts for 33.9%, 28.5%, 23.0%, 15.0%, 14.0%, and 7.83%, respectively [Chinese Academy of Sciences (CAS), 2018]. The GCG is more likely to have the regional distribution pattern than groundwater pollution that is controlled by the location of pollution source and therefore shows more adverse effects (Fendorf et al., 2010; Guo et al., 2014a).

The shortage of groundwater resources and endemic disease are two distinguished problems caused by GCG. High-salinity groundwater is widely distributed in northwestern China that reduces the available groundwater for drinking and irrigation purpose. The consumption of high-As, -F groundwater can lead to endemic arsenic poisoning and endemic fluorosis. Excessive iodine intake can cause iodine excess disorders, including thyroid autoimmunity, goiter, cretinism, and thyroid cancer (Li et al., 2016). Report shows that 0.2 million, 87 million, and 30 million people still use high-As, -F, and -I drinking water, most of which are groundwater [MOH (Ministry of Health), 2012]. Although much attention has been focused on the studies of high-As, -F, -Fe, -Mn, and -I groundwater, the general situation of China's GCG is rarely documented. The baseline quality of groundwater differs as the multiple climatic zones in China, with different landforms and sedimentary conditions. Some GCG problems may not be fully realized yet, because drinking-related diseases take years or even decades to appear. Therefore to identify the distribution and formation of GCG as well as its evolution affected by human activities is a prerequisite for ensuring drinking water safety.

16.2 The distribution and formation of geogenic-contaminated groundwater

16.2.1 High-salinity groundwater

High-salinity groundwater reduces the available water resources for consumption. It prevails in the inland basins of northwestern China and coastal regions in eastern China. Generally, in coastal regions due to high precipitation, surface water is abundant to provide drinking water; however, in the northwestern part with semiarid/arid climate and limited precipitation, groundwater is the dominant water resource. The highest TDS of >30, 27, and 145 g/L was documented in the North China Plain (NCP), Pearl River Delta, and Laizhou Bay, respectively, (Zhang et al., 2009; Currell et al., 2012; Wang and Jiao, 2012; Han et al., 2014). The Tarim Basin, Junggar Basin, and Qaidam Basin in the west are well known for their high-salinity or even brackish groundwater. Fortunately, fresh groundwater with TDS <1000 mg/L could also be found in these inland basins (Ma et al., 2010, 2013; Zhou et al., 2017). High-salinity groundwater is found in both shallow unconfined and deep confined aquifer, however, more common in shallow ones (Wen et al., 2005; Currell et al., 2011; Han et al., 2013; Zhang et al., 2013).

In the semiaird/arid inland basin the lack of recharge water as well as the salinity enrichment in the soil/sediment caused by strong evaporation favors the groundwater salinization. The groundwater recharge rate of 48 mm in Tengger Desert (near Hexi Corridor), 48−66 mm in Shanxi Provinces (Datong Basin), and 36−209 mm in the NCP was calculated by ^{3}H model (Lin and Wei, 2001; Kendy et al., 2003; Wang et al., 2004b; Scanlon et al., 2006). Evaporation could be an order of magnitude higher than precipitation in the inland basins (Jia et al., 2018). However, transpiration attributed by plants and/or dissolution of deposited evaporates rather than direct evaporation plays a major role in contributing high salinity in groundwater (Jia et al., 2017, 2018). This is also evidenced by plenty of soluble salts and minerals deposited in the lower Tarim Basin, Qaidam Basin, and Hetao Basin (Inner Mongolia Institute of Hydrogeology, 1982; Zhu, 1984; Liu et al., 2014; Jia et al., 2017). Seawater intrusion and/or geological deposited brine or evaporates in the sediment are the main reasons for the occurrence of high-salinity groundwater in coastal areas. The mixed seawater proportion could range from 0.07% to 94.4% (Zhao et al., 2017)

16.2.2 High-Fe and -Mn groundwater

The cooccurrence of high-Fe and -Mn groundwater is frequently the case due to their accompaniment in the minerals and similar chemical characteristic. Iron is thought to have no apparent health damage while it is an objectionable impurity in water supplies (Hem, 1985). High Mn in drinking water has been raising attention these years since it is believed to be harmful to children's intellectual performance and cause neurotoxic effects (Wasserman et al., 2006; McArthur et al., 2012). The densely distributed areas with high-Fe and -Mn are the Sanjiang Plain in the northern China and the middle and lower reaches of the Yangtze River in the south. The occurrence in other areas includes NCP, Taiyuan Basin, Hetao Basin, Yinchuan Basin, Songnen Plain, and lower Liaohe Plain in northern China and Huai River Plain, Sichuan Basin, Jianghan Plain, Dongting Lake Plain, and Pearl River Delta in southern China. These areas are distinguished with its organic matter−rich fluvial, lacustrine, or marine sedimentary environment. Groundwater Fe, with its concentration frequently lower than 10 mg/L and higher ones reach several decades, is generally several times to an order of magnitude higher than Mn, which is typically below 5.0 mg/L (Fig. 16.1).

The piedmont areas around the basin or plain host a primary Fe/Mn source, the weathering of rocks, and its migration to the lower lying areas make the sediment rich in Fe/Mn minerals. Generally, this Fe/Mn source is quite common in the sediment, so not being a limiting factor controlling Fe/Mn enrichment in groundwater. The solubility and chemical behavior of Fe and Mn in groundwater are controlled by redox potential and pH (Hem, 1985). Due to frequent neutral or near-neutral pH in groundwater, the redox condition plays a dominant role in regulating groundwater Fe and Mn mobilization (Appelo and Postma, 2005). As the sequence of redox reactions in groundwater shows the reduction of Fe/Mn oxides/hydroxides occurs between the previous O_2/NO_3^- reduction and subsequent SO_4^{2-} reduction. Groundwater Fe starts to rise during the iron-reducing stage and continue elevated at the SO_4^{2-}-reducing stage. The average concentrations of Fe(II) are 0.62, 1.25 mg/L in iron-reducing stage and SO_4^{2-}-reducing stage, respectively, in the Hetao Basin (Jia et al., 2014). However, the distribution of Mn shows no significant pattern in different reducing conditions. This may be controlled by the integrated effects of reduction and adsorption/coprecipitation (Buschmann and Berg, 2009; Jia et al., 2014). The groundwater Fe/Mn could be dropped down by the precipitation of Fe/Mn minerals. In strong reducing conditions, abundant HCO_3^- and HS^- are produced leading to the formation of siderite ($FeCO_3$) and mackinawite (FeS). Their oversaturation status and possible precipitation were documented in the Jianghan Plain, Datong Basin,

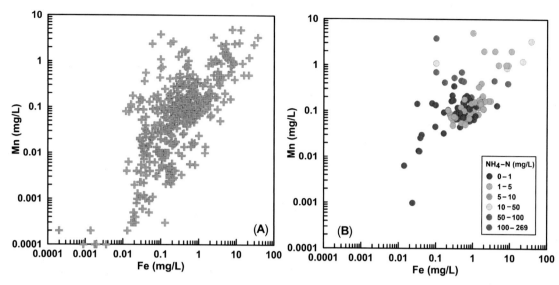

FIGURE 16.1 (A) The plot of groundwater Fe and Mn in the major basins/plains and (B) the cooccurrence of high NH$_4$−N with Fe and Mn (Jia et al., 2018).

Yinchuan Basin, and Hetao Basin (Jia et al., 2014; Guo et al., 2014c; Duan et al., 2015; Xie et al., 2015). This explains groundwater with very high NH$_4$−N does not show the highest Fe/Mn concentration (Fig. 16.1).

16.2.3 High-As groundwater

In terms of its toxicity and wide distribution, high-As groundwater could be the most harmful GCG in China. It has been documented in both arid/semiarid inland basins and river deltas under humid tropical climate (Guo et al., 2008; Jia and Guo, 2013). Around 19 provinces, 19.6 million populations might be at risk (Jin et al., 2003; Rodríguez-Lado et al., 2013). High-As groundwater (> 10 μg/L) has been found in major river basins such as the Yangtze River Basin, Yellow River Basin, and Pearl River Basin. The Xinjiang and Qinghai Provinces and Songnen Plain are also affected (Fig. 16.2). Geothermal groundwater in Yunnan and piedmont areas in part of Sichuan and Gansu Provinces is reported with high As as well.

The documented high-As groundwater (> 10 μg/L) is dominated by more toxic and less affinity As(III) in China (Rodríguez-Lado et al., 2013; Guo et al., 2014a, 2017). This indicates groundwater As enrichment is controlled by reduction processes. Arsenic mobility and enrichment are mainly controlled by dissimilatory reduction of As-bound Fe oxides/hydroxides (Fendorf et al., 2010; Guo et al., 2014a; Wang et al., 2019). The available As pool is determined by the reactivity of Fe oxides/hydroxides. The reactivity of As-bearing Fe oxides concluded by kinetic experiment is between lepidocrocite (γ-FeOOH) and poorly crystalline goethite (α-FeOOH) in the Hetao Basin (Shen et al., 2018). Ferrimagnetic minerals such as hematite (α-Fe$_2$O$_3$) and maghemite (γ-Fe$_2$O$_3$) were thought to be major carriers for As in the Datong Basin (Xie et al., 2009). Arsenic bound to amorphous Fe oxyhydroxide accounts for 30% in the Huhhot Basin (Smedley et al., 2003). The organic matter decomposition controls overall electron flow and the state of the redox reactions, which determine As levels as well (Fig. 16.3). In the Hetao Basin, groundwater As shows average of 89.5 and 265 μg/L in iron-reducing and sulfate reduction stages, respectively, (Jia et al., 2014). The combined effect of reductive dissolution, adsorption, and coprecipitation controls As and Fe cycling in aquifer, which leads to the poor correlation between aqueous Fe and As (Guo et al., 2008; Deng et al., 2009; Han et al., 2013; Wang and Jiao, 2012). Case study in the Hetao Basin shows that reductive dissolution of Fe(III) accounts for 70% As release, while desorption under elevated pH and competitive adsorption by HCO$_3^-$ and PO$_4^{3-}$ contributed less than 30% (Gao et al., 2020).

16.2.4 High-fluoride groundwater

It is reported that around 29 provinces, 26 million populations are affected by high-F groundwater (Wang and Huang, 1995; Fawell et al, 2006; Wen et al., 2013). Generally, groundwater F ranges from <1 mg/L to several mg/L,

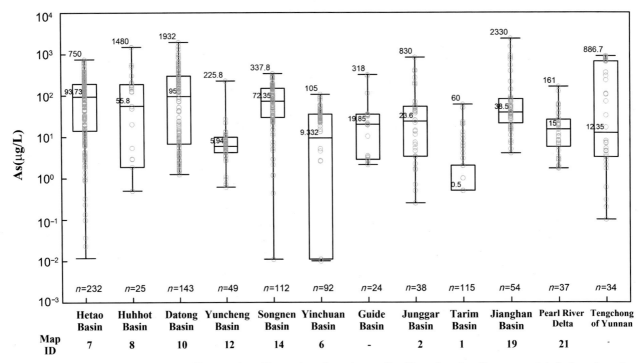

FIGURE 16.2 Arsenic concentrations in different regions. The number of samples was listed in each region. Data are overlain by box plots where the bars are the maximum and minimum values, the centerline is the median value, and box edges are the 25th and 75th percentiles (Jia et al., 2018).

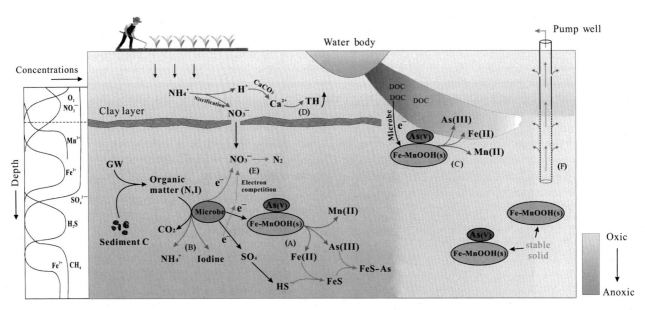

FIGURE 16.3 The conceptual model of the formation of GCG components related to reduction reactions as well as human-induced evolution of GCG (Jia et al., 2018). (A) The release of Fe, Mn, As contributed by reductive dissolution of Fe/Mn oxides; (B) the generation of NH_4^+ and iodine caused by OC decomposition; (C) the infiltration of reactive OC triggers the reductive dissolution of stable As-bearing Fe/Mn oxides; (D) nitrification of N fertilizers promotes groundwater total hardness; (E) agriculture-sourced NO_3^- competes electron originally flows to Fe/Mn oxides and SO_4^{2-} so as to mitigate As contamination; and (F) wells with long screen and imperfect seal may cause cross contamination among aquifers. *GCG*, Geogenic-contaminated groundwater.

in some cases such as thermal groundwater could be >10 mg/L (Fig. 16.4). High-F groundwater prevails in most basin/plain in northern China, known as the Tarim Basin, Junggar Basin, Hexi Corridor, Qaidam Basin, Guide Basin, Ejina Basin, Yinchuan Basin, Hetao Basin, Huhhot Basin, Loess Plateau, Guanzhong Basin, Datong−Taiyuan−Yuncheng Basin, Songnen Plain, NCP, and Huai River Plain (Jia et al., 2018). In southern China, high fluoride was documented in

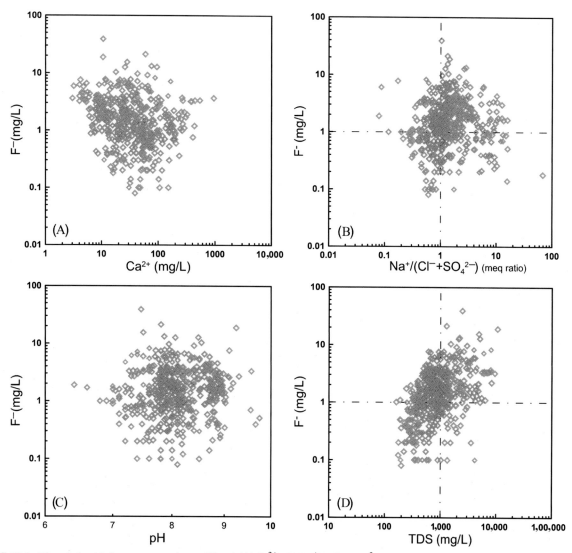

FIGURE 16.4 The relationship between groundwater F⁻ and (A) Ca^{2+}, (B) $Na^+/(Cl^- + SO_4^{2-})$, (C) pH, and (D) TDS (Jia et al., 2018). *TDS*, Total dissolved solids.

the geothermal groundwater such as Tengchong in Yunnan Province. High F in the northern China is frequently related to soda groundwater that is rich in sodium under the arid/semiarid climate (Liu et al., 1980; Wen et al., 2013). Besides, high-F groundwater tends to occur in shallow aquifer located in the central basin or flat downstream with long groundwater residence time.

Fluorite commonly distributed in igneous and sedimentary rock is the major fluoride mineral in the aquifer sediment. The average F content in sedimentary rock, igneous rock and metamorphic rock in the Hetao Basin is 580 ($n = 21$), 673 ($n = 30$), and 952 ($n = 43$) mg/kg, respectively (Zhao et al., 2007a). Aquifer sediments F with average of 608.9 mg/kg was reported in the Datong Basin (Zhao et al., 2007b; Su et al., 2013). Soil F content is 478, 463, 520, and 583 mg/kg in the Hetao, Datong, Taiyuan, and Yuncheng basins, respectively, all higher than average 440 mg/kg nationwide (Wei et al., 1984; Zhang, 2001; Zhao et al., 2007a). The dissolution of F-bearing minerals is the main process of releasing F (Wen et al., 2013). Desorption from sediment surface and ion exchange like the hydrolysis work as well. In semiarid/arid climate, groundwater frequently with high pH and alkaline condition favors the desorption as well as the hydrolysis of F-bearing silicates (OH^- in water exchanges for F^-) such as muscovites and biotites. Higher F (> 2.5 mg/L) in the Guanzhong Basin was believed to be combined with evaporation effect besides the dissolution of F-bearing minerals (Sun et al., 2013). Evapotranspiration dramatically enriches F not only due to its concentration process but also the formation of soda groundwater [$Na^+/(Cl^- + SO_4^{2-}) > 1$] that favors the further dissolution of CaF_2 minerals. Significantly, high-F^- groundwater (> 10 mg/L) is generally controlled by evaporation effect or thermal process (Zhang et al., 2010; Shi et al., 2010; Sun et al., 2015; Liang et al., 2015).

16.2.5 High-/low-iodine groundwater

Global geochemical surveys have found out the iodine excess or deficiency in drinking water both have adverse effects on health. Inadequate iodine intake can cause goiter, cretinism, and even infertility. On the other hand, excessive iodine intake can cause iodine excess disorders, including goiter, cretinism, thyroid autoimmune disease, and even thyroid cancer (Li et al., 2016). Multiple regions refer to 12 provinces were documented to have high-iodine groundwater, specifically the Huai River Plain (Anhui, Henan, and Jiangsu provinces), NCP (Beijing, Tianjin, Hebei, Henan, and Shandong provinces), Datong and Taiyuan Basin (Shanxi Provinces), Hetao Basin (Inner Mongolia), and Xinjiang and Fujian Provinces (Shen et al., 2007). Within the region the NCP and Huai River Plain are the most affected high-iodine areas. On the contrary, around 400 million populations are affected by low iodine (Dan and Li, 1994). Except coastal areas, most parts of the country are to some extent affected by low-iodine groundwater, especially in the mountain areas (Li et al., 2000).

The iodine mobilization is mainly controlled by biogeochemical processes (Hem, 1985; Amachi et al., 2005). Degradation of organic matter (OC) or organic iodine compounds as well as reductive dissolution of iron oxides/oxyhydroxides under reducing condition contributes to its enrichment (Fig. 16.3). Evaporation works as well in shallow groundwater. Microorganism-mediated OC decomposition plays a dominant role than evaporation on iodine enrichment in the Hetao Basin (Xu et al., 2013). The fluvial, lacustrine, or marine sedimentation with abundant organic matter offer possible iodine sources. Six marine transgressions occur in the NCP since Quaternary provides iodine-rich sea creatures in aquifers (Zhang et al., 2009). Low-iodine groundwater is more likely to distribute in mountain area. Average concentration of groundwater iodine is 2.1, 3.6, 76.7, and 105.9 μg/L in mountain area, hill, flat plain, and coastal region, respectively (Li et al., 2000). The lack of iodine in sediment as well as the high groundwater flow rate in mountain area is the main reason.

16.2.6 High-nitrogen groundwater

The natural enrichment of groundwater NO_3^- was found in semiarid/arid region. The elevated level of NH_4^+ was observed in OC-rich-reducing aquifers. High NO_3^- was observed in shallow groundwater (2.84–113 mg/L, average 37.6 mg/L, $n = 58$) as well as soil water (up to 1174 mg/L) in the Badain Jaran Desert (Gates et al., 2008; Pan, 2014; Jin et al., 2015). Similar cases were documented in other semiarid/arid regions worldwide (Barnes et al. 1992; Hartsough et al. 2001; Walvoord et al. 2003; Stone and Edmunds, 2014). High concentration of NH_4-N was reported in the Hetao Basin (<0.01–4.80 mg/L, $n = 103$), Datong Basin (<0.01–1.59 mg/L, $n = 82$), Yinchuan Basin (<0.01–3.50 mg/L, average 0.72 mg/L, $n = 92$), Songnen Plain (<0.01–8.49 mg/L, average 1.34 mg/L, $n = 87$), Jianghan Plain (<1–22 mg/L, average 3 mg/L, $n = 186$), and Sanjiang Plain (Jia et al., 2014; Li et al., 2014; Guo et al., 2014b,c; Gan et al., 2014). Abnormally high NH_4^+ (10–390 mg/L, $n = 40$) was documented in the aquifers with increase trend toward the coast in the Pearl River Delta (Jiao et al., 2010).

Elevated NO_3^- in the Badain Jaran Desert mainly is contributed by atmospheric N deposition and further microbial N fixation (Pan, 2014; Jin et al., 2015), the latter is contributed by leguminous plants or cyanobacteria (Edmunds, 2009; Huang et al., 2013). Its accumulation is favored by the limited denitrification due to the absence of OC in the desert (Edmunds and Gaye 1997; Gates et al. 2008). Naturally occurring NH_4^+ is mainly linked to reducing environment with abundant OC. Nitrogen compounds are frequently contained in natural OC, and therefore NH_4^+ becomes a good indicator for the intensity of OC decomposition as well as the extent of reducing conditions (McArthur et al., 2001; Postma, et al., 2007; Jia et al., 2014; Fig. 16.3). Good correlation was found between NH_4^+ and As in the Hetao Basin and Datong Basin (Jia et al., 2014; Zhang et al., 2017). Studies in the Jianghan Plain and Pearl River Delta show that high groundwater NH_4^+ was contributed by the decomposition of sediment sourced DOM (Jiao et al., 2010; Huang et al., 2016).

16.2.7 Other trace elements

Elevated toxic trace elements such as U and Cr were observed in oxic groundwater. High-U groundwater was documented in the Datong Basin (<0.02–288 μg/L, average 24 μg/L) and western Hetao Basin (<0.01–323 μg/L, average 23.8 μg/L) (Wu et al., 2014; Guo et al., 2015). High-U groundwater is confined in the alluvial fans in the Hetao Basin and shows its decrease trend toward flat plain in comparison with As distribution. Groundwater Cr ranging from under detection to 551.13 μg/L was documented in the Malian River Basin in Gansu, Loess Plateau of Shaanxi, Yuncheng Basin in Shanxi, and Pearl River Delta (Li, 2006; Zhu et al., 2014b; Wang et al., 2016).

Chemical weathering of U-bearing rocks such as carbonate veins, schist, or phyllite would mobilize U into groundwater (Banning et al., 2012; Alam and Cheng, 2014; Regenspurg et al., 2010). Redox conditions controlled groundwater U concentrations with high-U groundwater generally have relatively high Eh values in the alluvial fans. Under oxic environment, U(VI) as the soluble form was the dominant specie (Smedley et al., 2006). High-Cr groundwater is frequently dominated by Cr(VI). The transformation of Cr(III) to Cr(VI) at circumneutral pH in groundwater is its oxidation by Mn(III, IV) oxides (Fendorf and Zasoski, 1992; Mills et al., 2011). This is performed on the surface of Cr–Mn-rich minerals and favored by oxic condition (Kazakis et al., 2015).

16.3 Cooccurrence of different geogenic-contaminated groundwater components

The formation of GCG involves organic redox processes or inorganic water–rock interaction. In this case, component with similar release processes could coexist in the groundwater. Most common ones such as high salinity and fluoride, arsenic and fluoride, iron, manganese, and ammonia were documented in China.

16.3.1 High salinity and fluoride

From the large dataset, it is obvious to observe the cooccurrence of high-salinity and F^- groundwater (Fig. 16.4), which is widely distributed in semiarid/arid inland basins of northern China, known as the Hexi Corridor, Yinchuan Basin, Hetao Basin, Loess Plateau, Datong Basin, Taiyuan Basin, and Yuncheng Basin. The sufficient water–rock interaction due to long residual time of groundwater and concentration effect caused by evapotranspiration simultaneously promotes groundwater salinity and F^-. High-salinity groundwater characterized by high Na and low Ca favors the further dissolution of fluorite (Wang et al., 2009).

16.3.2 Arsenic and fluoride

Inland basins such as the Songnen Plain, Datong Basin, Hetao Basin, and Junggar Basin show significantly coexisting high As and F^- (Fig. 16.5). Geothermal-affected groundwater such as the Guide Basin and Tibet also shows the cooccurrence of As and F^-. Reported coexisting areas are distributed in arid/semiarid regions where groundwater shows high pH and high alkalinity. Desorption and evaporation effects may both work. Desorption triggers both As(V) and F^- releasing from Fe oxides (Smedley and Kinniburgh, 2002; Kim et al., 2012). In reducing environment the OC decomposition leading to the reductive dissolution of As could promote the dissolution of F-bearing minerals due to the production of high CO_2/HCO_3^- (Jia et al., 2018).

16.3.3 Iron, manganese, and ammonia

The occurrence of groundwater with high-Fe, -Mn, and -NH_4^+ was found in the Hetao Basin, Yinchuan Basin, Songnen Plain, Sanjiang Plain, Jianghan Plain, and Pearl River Delta (Wang et al., 2004a; Jia et al., 2014; Guo et al., 2014c). The reductive dissolution of Fe-/Mn-bearing oxides/hydroxides triggered by the OC decomposition controls the release of groundwater Fe and Mn. The OC is frequently rich in N component, and therefore its degradation generates NH_4^+. The variation of NH_4^+ concentrations among basins should be the combination of N content difference in OC and varied adsorption level on different minerals.

16.4 Geogenic-contaminated groundwater affected by anthropogenic activities

16.4.1 Further salinization of groundwater

The enhanced groundwater salinization has been reported in some major cities/basins (Zhang and Li, 2005; Zhang et al., 2009; Zhu et al., 2014a; Fig. 16.6). The further groundwater salinization could be triggered by following scenarios, including groundwater pollution, mixture among aquifers due to abstraction, overirrigation, and seawater intrusion deteriorated by excessive groundwater exploitation. Groundwater pollution is a straightforward way to input salinity to groundwater. Domestic and industrial wastewater is characterized with high levels of Na^+, Cl^-, and SO_4^{2-}, its common application for irrigation in water-stressed northern China in the past few decades has accelerated the groundwater

236 Theme | 3 Groundwater scarcity: quantity and quality

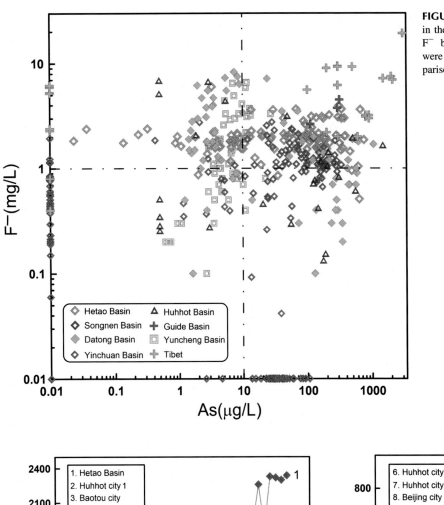

FIGURE 16.5 The cooccurrence of As and F⁻ in the major basins/regions. Samples with As and F⁻ below detection limit of 0.01 μg/L (mg/L) were plotted as 0.01 μg/L (mg/L) to make comparison (Jia et al., 2018).

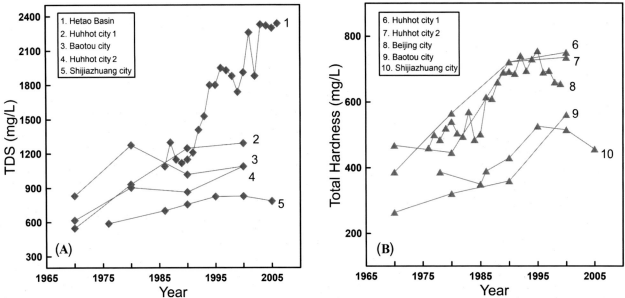

FIGURE 16.6 The evolution trend of (A) groundwater TDS and (B) total hardness of groundwater from major cities/basin along the year. 1—water sample from main drainage channel recharged by shallow groundwater, 2–10—groundwater (Jia et al., 2018). *TDS*, Total dissolved solids.

salinization (Tang et al., 2004; Heeb et al., 2012). Groundwater TDS shows rising trend in the large-scale groundwater depression cone of the NCP, Junggar Basin, and Northeastern China, which could be caused by groundwater mixture among aquifers and extended recharge path (Zhang et al., 2009; Chen et al., 2010). Overirrigation in the semiarid/arid areas accelerates groundwater salinization. The drainage water recharged by shallow groundwater in the heavily irrigated Hetao Basin, its TDS increases from 1200 mg/L in the 1980s to average 2000 mg/L during 2000−06 (Zhu et al., 2014a; Fig. 16.6). Extended seawater intrusion was observed in the densely populated eastern NCP where groundwater has been excessively exploited. From the 1970s to 2003 the extended areas with groundwater TDS of >5, 3−5, and 2−3 g/L are 739, 987, and 414 km^2, respectively, in Cangzhou (Zhang et al., 2009).

16.4.2 Elevated groundwater hardness

Elevated groundwater hardness with planar pattern has been reported in Beijing as well as other regions (Fig. 16.6). The production of acidity inputs H$^+$ to the vadose zone and aquifer could result in the elevated groundwater hardness (Pacheco et al., 2013; Fig. 16.3). The nitrification of intensive used N fertilizers generates H$^+$ ions and nitrite (NO$_2^-$). The import of H$^+$ triggers the dissolution of carbonate minerals, such as calcite or dolomite (Perrin et al., 2008; Pacheco et al., 2013). The intensive agricultural activity and heavily distributed groundwater NO$_3^-$ indicate that nitrification should play a major role in regulating groundwater hardness in China (Chen et al., 2005; Jia et al., 2018; Xin et al., 2019).

16.4.3 Cross contamination of aquifers

In order to abstract groundwater in its full capacity for irrigation, long-screened well is widely used in rural areas. This provides opportunity for salinity or other GCG contaminants to cross aquitard through the well (Han et al., 2016; Fig. 16.3). Even in standard wells, seals along well casings especially in older wells may be imperfect so to create conduit mixing groundwater among different aquifers (Santi et al., 2006; Han et al., 2016; Fig. 16.3). The cross contamination of aquifers could be partly responsible for the 10−20 m decline of boundary between upper saline and underlying fresh groundwater in the NCP since the 1970s (Fei et al., 2009). The distribution of groundwater iodine in the NCP was reported to be obviously altered by groundwater overexploitation (Zhang et al., 2013). The cross contamination of groundwater with different redox conditions results in the mobilization of As and U in the United States (Ayotte et al., 2011).

16.4.4 Trace element release/sequester due to redox change

Human activities such as pumping and discharge waste could affect groundwater flow or chemical component further to change redox state in aquifers. Temporal variation of arsenic concentrations was found along with the fluctuation of groundwater table caused by infiltration of surface water or irrigation or in the Jianghan Plain and Hetao Basin (Guo et al., 2013; Duan et al., 2015; Schaefer et al., 2016; Zhang et al., 2018). The infiltration of surface-originated reactive OC leads to the widely distributed groundwater As remains controversy in Bangladesh (Harvey et al., 2002; Fendorf et al., 2010; Fig. 16.4). However, the sure thing is reactive OC does play a crucial role in triggering microbial respiration coupled with the dissolution of As-bearing Fe/Mn oxides/hydroxides. Therefore As, Fe, Mn hazard could be aggravated if reactive OC was introduced to aquifer. In China, similar case is seldom reported; however, it is possible to occur since the widespread BOD pollution in river, lake (Pernet-Coudrier et al., 2012). On the other hand, As, Fe, Mn problem could be mitigated if oxidizing components being introduced to aquifer. The nitrification of N fertilizers brings oxidant NO$_3^-$ to groundwater might make this happen (Fig. 16.4). The As-affected areas such as Songnen Plain, Datong Basin, Hetao Basin, and Jianghan Plain are all developed with intensive agricultural activities, so the further impact of agriculture-sourced NO$_3^-$ on As, Fe, Mn behaviors merit studying.

16.5 Conclusion

The widespread GCG leads to the formation of endemic disease as well as reduces the available groundwater resources. Due to the variation of groundwater utilization, the GCG was mostly documented in northern China where groundwater accounts for >65% water supply, while insufficient use of groundwater may cause GCG not to be fully recognized in southern China.

The occurrence of GCG shows two types. One is inorganic enrichment process such as the dissolution and evaporation effect, which are responsible for groundwater with high TDS, hardness, and F$^-$. The other is known as organic redox reactions coupled with microbial activities that cause the mobilization of redox-sensitive component, such as As, Fe, Mn, U, Cr, and the by-product such as NH$_4^+$ and iodine. Favored by arid/semiarid climate, the GCG occurrence of first type prevails in northern China; however, it is not obvious in southern China with humid climate. However, the second type of GCG was reported in both southern and northern China. This is because groundwater redox state is mainly related to the biogeochemical characteristics of sediment. The newly deposited one with abundant reactive OC favors the formation of reducing conditions so as to release As, Fe, and Mn. The degradation of sediment OC could mobilize NH_4^+ and iodine. Therefore the second type of GCG is frequently distributed in aquifers with young Holocene sedimentation that prevails in both southern and northern China.

The widespread GCG threatens safe drinking water supply especially in rural areas where groundwater quality in private wells is mostly unknown. The concerning thing is drinking related disease could take years or decades to show. Although this chapter provides overall GCG situation in China, there is a great potential to improve. High-salinity groundwater could easily be identified, and As/F-affected areas have already attracted much attention. The emerging concern should be toxic trace elements such as Mn, U, Cr, and iodine, which are not easily confirmed. From a sustainable development perspective, the long-term impact of human activities on GCG evolution merits further study.

References

Alam, M.S., Cheng, T., 2014. Uranium release from sediment to groundwater, influence of water chemistry and insights into release mechanisms. J. Contam. Hydrol. 164, 72–87.

Amachi, S., Mishima, Y., Shinoyama, H., Muramatsu, Y., Fujii, T., 2005. Active transport and accumulation of iodide by newly isolated marine bacteria. Appl. Environ. Microbiol. 71, 741–745.

Appelo, C.A.J., Postma, D., 2005. Geochemistry, Groundwater and Pollution, second ed. Taylor & Francis Group PLC, Amsterdam, The Netherlands.

Ayotte, J.D., Szabo, Z., Focazio, M.J., Eberts, S.M., 2011. Effects of human-induced alteration of groundwater flow on concentrations of naturally-occurring trace elements at water-supply wells. Appl. Geochem. 26 (5), 747–762.

Banning, A., Cardona, A., Rüde, T.R., 2012. Uranium and arsenic dynamics in volcano-sedimentary basins—an exemplary study in north-central Mexico. Appl. Geochem. 27, 2160–2172.

Barnes, C.J., Jacobson, G., Smith, G.D., 1992. The origin of high-nitrate groundwaters in the Australian arid zone. J. Hydrol. 137, 181–197.

Buschmann, J., Berg, M., 2009. Impact of sulfate reduction on the scale of arsenic contamination in groundwater of the Mekong, Bengal and Red River deltas. Appl. Geochem. 24 (7), 1278–1286.

Chen, J.Y., Tang, C.Y., Sakura, Y., Yu, J., Fukushima, Y., 2005. Nitrate pollution from agriculture in different hydrogeological zones of the regional groundwater flow system in the North China Plain. Hydrogeol. J. 13, 481–492.

Chen, Z.Y., Wang, Y., Liu, J., Wei, W., 2010. Groundwater changes of selected groundwater systems in Northern China in recent fifty years. Quat. Sci. 30 (1), 115–126.

Chinese Academy of Sciences (CAS), 2018. Groundwater Sciences. Science Press, Beijing. (in Chinese).

Currell, M.J., Cartwright, I., Raveggi, M., Han, D.M., 2011. Controls on elevated fluoride and arsenic concentrations in groundwater from the Yuncheng Basin, China. Appl. Geochem. 26, 540–552.

Currell, M.J., Han, D., Chen, Z., Cartwright, I., 2012. Sustainability of groundwater usage in northern china, dependence on palaeowaters and effects on water quality, quantity and ecosystem health. Hydrol. Process. 26 (26), 4050–4066.

Dan, D., Li, P., 1994. Contents of iodine in environmental geochemistry and its relationship with iodine deficiency disorders in China. J. Mineral. Petrol. 14 (4), 69–75 (in Chinese with English Abstract).

Deng, Y., Wang, Y., Ma, T., 2009. Isotope and minor element geochemistry of high arsenic groundwater from Hangjinhouqi, the Hetao plain, Inner Mongolia. Appl. Geochem. 24, 587–599.

Duan, Y., Gan, Y., Wang, Y., Deng, Y., Guo, X., Dong, C., 2015. Temporal variation of groundwater level and arsenic concentration at Jianghan Plain, central China. J. Geochem. Explor. 149 (12), 106–119.

Edmunds, W.M., 2009. Geochemistry's vital contribution to solving water resource problems. Appl. Geochem. 24, 1058–1073.

Edmunds, W.M., Gaye, C.B., 1997. Naturally high nitrate concentrations in groundwaters from the Sahel. J. Environ. Qual. 26, 1231–1239.

Fawell, J., Bailey, K., Chilton, J., Dahi, E., Fewtrell, L., Magara, Y., 2006. Fluoride in drinking-water. In: Fewtrell, L., Bartram, J. (Eds.), Water Quality, Guidelines, Standards and Health. World Health Organization, Geneva.

Fei, Y.H., Zhang, Z.J., Song, H.B., Qian, Y., Chen, J.S., Meng, S.H., 2009. Discussion of vertical variations of saline groundwater and mechanism in North China Plain. Water Resour. Prot. 26, 21–23 (in Chinese with English abstract).

Fendorf, S., Michael, H.A., van Geen, A., 2010. Spatial and temporal variations of groundwater arsenic in south and southeast Asia. Science 328 (5982), 1123.

Fendorf, S., Zasoski, R.J., 1992. Chromium(III) oxidation by d-MnO$_2$. 1. Characterization. Environ. Sci. Technol. 26, 79–85.

Gan, Y., Wang, Y., Duan, Y., Deng, Y., Guo, X., Ding, X., 2014. Hydrogeochemistry and arsenic contamination of groundwater in the Jianghan Plain, central China. J. Geochem. Explor. 138 (3), 81–93.

Gao, Z.P., Jia, Y.F., Guo, H.M., Zhang, D., Zhao, B., 2020. Quantifying geochemical processes of arsenic mobility in groundwater from an inland basin using a reactive transport model. Water Resour. Res., 56, e2019WR025492. https://doi.org/10.1029/2019WR025492

Gates, J.B., Böhlke, J.K., Edmunds, W.M., 2008. Ecohydrological factors affecting nitrate concentrations in a phreatic desert aquifer in Northwestern China. Environ. Sci. Technol. 42 (10), 3531–3537.

Guo, Q., Guo, H., Yang, Y., et al., 2014c. Hydrogeochemical contrasts between low and high arsenic groundwater and its implications for arsenic mobilization in shallow aquifers of the northern Yinchuan Basin, P.R. China. J. Hydrol. 518, 464–476.

Guo, H., Jia, Y., Wanty, R.B., Jiang, Y., Zhao, W., Xiu, W., et al., 2015. Contrasting distributions of groundwater arsenic and uranium in the western Hetao basin, Inner Mongolia, implication for origins and fate controls. Sci. Total Environ. 541, 1172–1190.

Guo, H., Wen, D., Liu, Z., Jia, Y., Guo, Q., 2014a. A review of high arsenic groundwater in mainland and Taiwan, China, distribution, characteristics and geochemical processes. Appl. Geochem. 41 (1), 196–217.

Guo, H., Zhang, D., Ni, P., Cao, Y., Li, F., Jia, Y., et al., 2017. On the scalability of hydrogeochemical factors controlling arsenic mobility in three major inland basins of PR China. Appl. Geochem. 77, 15–23.

Guo, H., Zhang, D., Wen, D., Yang, W., Ni, P., Jiang, Y., et al., 2014b. Arsenic mobilization in aquifers of the southwest Songnen Basin, P.R. China, evidences from chemical and isotopic characteristics. Sci. Total Environ. 490, 590–602.

Guo, H., Zhang, Y., Jia, Y., Zhao, K., Li, Y., Tang, X., 2013. Dynamic behaviors of water levels and arsenic concentration in shallow groundwater from the Hetao Basin, Inner Mongolia. J. Geochem. Explor. 135 (6), 130–140.

Guo, H.M., Yang, S., Tang, X., Li, Y., Shen, Z., 2008. Groundwater geochemistry and its implications for arsenic mobilization in shallow aquifers of the Hetao Basin, Inner Mongolia. Sci. Total Environ. 393, 131–144.

Han, D., Currell, M.J., Cao, G., 2016. Deep challenges for China's war on water pollution. Environ. Pollut. 218, 1222–1233.

Han, D.M., Song, X.F., Currell, M.J., Yang, J.L., Xiao, G.Q., 2014. Chemical and isotopic constraints on evolution of groundwater salinization in the coastal plain aquifer of Laizhou Bay, China. J. Hydrol. 508 (2), 12–27.

Han, S.B., Zhang, F., Zhang, H., An, Y., Wang, Y., Wu, X., et al., 2013. Spatial and temporal patterns of groundwater arsenic in shallow and deep groundwater of Yinchuan Plain, China. J. Geochem. Explor. 135, 71–78.

Hartsough, P., Tyler, S.W., Sterling, J., Walvoord, M., 2001. A 14.6 kyr record of nitrogen flux from desert soil profiles as inferred from vadose zone pore waters. Geophys. Res. Lett. 28, 2955–2958.

Harvey, C.F., Swartz, C.H., Badruzzaman, A.B., et al., 2002. Arsenic mobility and groundwater extraction in Bangladesh. Science 298 (5598), 1602–1606.

Heeb, F., Singer, H., Pernet-coudrier, B., Qi, W., Liu, H., Longrée, P., et al., 2012. Organic micropollutants in rivers downstream of the megacity Beijing: sources and mass fluxes in a large-scale wastewater irrigation system. Environ. Sci. Technol. 46 (16), 8680–8688.

Hem, J.D., 1985. Study and interpretation of the chemical characteristics of natural water. In: U.S. Geological Survey Water-Supply Paper 2254[C] Geological Survey.

Huang, T.M., Pang, Z.H., Yuan, L., 2013. Nitrate in groundwater and the unsaturated zone in (semi)arid northern China, baseline and factors controlling its transport and fate. Environ. Earth Sci. 70 (1), 145–156.

Huang, S., Wang, Y., Ma, T., Wang, Y., Zhao, L., 2016. Fluorescence spectroscopy reveals accompanying occurrence of ammonium with fulvic acid-like organic matter in a fluvio-lacustrine aquifer of Jianhan Plain. Environ. Sci. Pollut. Res. Int. 23 (9), 1–10.

Inner Mongolia Institute of Hydrogeology, 1982. Hydrogeological Setting and Remediation, Approaches of Soil Salinity in the Hetao Basin, Inner Mongolia. Scientific Report (in Chinese).

Jia, Y.F., Guo, H., 2013. Hot topics and trends in the study of high arsenic groundwater. Adv. Earth. Science. 28 (1), 51–61 (in Chinese with English abstract).

Jia, Y.F., Guo, H., Jiang, Y., Wu, Y., Zhou, Y., 2014. Hydrogeochemical zonation and its implication for arsenic mobilization in deep groundwaters near alluvial fans in the Hetao Basin, Inner Mongolia. J. Hydrol. 518, 410–420.

Jia, Y.F., Guo, H., Xi, B., Jiang, Y., Zhang, Z., Yuan, R., et al., 2017. Sources of groundwater salinity and potential impact on arsenic mobility in the western Hetao basin, Inner Mongolia. Sci. Total Environ. 601-602, 691–702.

Jia, Y.F., Xi, B.D., Jiang, Y.H., Guo, H.M., Yang, Y., et al., 2018. Distribution, formation and human-induced evolution of geogenic contaminated groundwater in China: a review. Sci. Total. Environ. 643, 967–993.

Jiao, J.J., Wang, Y., Cherry, J.A., Wang, X., Zhi, B., Du, H., et al., 2010. Abnormally high ammonium of natural origin in a coastal aquifer-aquitard system in the Pearl River Delta, China. Environ. Sci. Technol. 44 (19), 7470–7475.

Jin, L., Edmunds, W.M., Lu, Z., Ma, J., 2015. Geochemistry of sediment moisture in the Badain Jaran Desert, implications of recent environmental changes and water-rock interaction. Appl. Geochem. 63, 235–247.

Jin, Y.L., Liang, C.H., He, G.L., Cao, J.X., Ma, F., Wang, H.Z., et al., 2003. Study on distribution of endemic arsenism in China. J. Hyg. Res. 23 (6), 519–540 (in Chinese with English abstract).

Kazakis, N., Kantiranis, N., Voudouris, K.S., Mitrakas, M., Kaprara, E., Pavlou, A., 2015. Geogenic Cr oxidation on the surface of mafic minerals and the hydrogeological conditions influencing hexavalent chromium concentrations in groundwater. Sci. Total Environ. 514, 224.

Kendy, E., Gerard-Marchant, P., Walter, M.T., Zhang, Y., Liu, C., Steenhuis, T.S., 2003. A soil-water-balance approach to quantify groundwater recharge from irrigated cropland in the North China Plain. Hydrol. Process. 17, 2011–2031.

Kim, S.H., Kim, K., Ko, K.S., Kim, Y., Lee, K.S., 2012. Co-contamination of arsenic and fluoride in the groundwater of unconsolidated aquifers under reducing environments. Chemosphere 87 (8), 851–856.

Li, J.X., Wu, G.J., Huang, H.Z., Ren, F.H., Gong, Z.T., Tan, J.A., et al., 2000. Regional geochemistry with agriculture and human health [M]. People's Medical Publishing House, Beijing. (in Chinese with English Abstract).

Li, J.X., Wang, Y., Guo, W., Xie, X., Zhang, L., Liu, Y., et al., 2014. Iodine mobilization in groundwater system at Datong Basin, China, evidence from hydrochemistry and fluorescence characteristics. Sci. Total Environ., 468–469 (2), 738–745.

Li, J.X., Wang, Y., Xie, X., Depaolo, D.J., 2016. Effects of water-sediment interaction and irrigation practices on iodine enrichment in shallow groundwater. J. Hydrol. 543, 293–304.

Li, W.D., Zou, Z., Zhao, L.S., Zhang, J.Q., Wang, X., Wang, Y.D., 2006. Investigation of endemic arsenism in areas Anhui province of China. Anhui J. Prev. Med. 12 (4), 193−196 (In Chinese with English abstract).

Li, Q., 2006. Analysis on distribution and formation causes of high-chrome groundwater in Yuncheng City. Groundwater. 28 (4), 31−33 (in Chinese with English abstract).

Liang, L., Zhu, M., Zhu, S., Zhang, L., Xie, X., 2015. Spatial distribution and enrichment of fluoride in geothermal water from eastern Guangxi, China. Saf. Environ. Eng. 22 (1), 1−6.

Lin, R., Wei, K., 2001. Environmental isotope profiles of the soil water in loess unsaturated zone in semi-arid areas of China. In: Yurtsever, Y. (Ed.), Isotope Based Assessment of Groundwater Renewal in Water Scarce Regions, IAEA Tecdoc-1246. IAEA, Vienna, pp. 101−118.

Liu, D., Chen, Q., Yu, Z., Yuan, Z., 1980. Geochemical environment problems concerning the endemic fluorine disease in China. Geochimica 1, 13−21 (in Chinese with English Abstract).

Liu, W., Liu, Z., An, Z., Sun, J., Chang, H., Wang, N., et al., 2014. Late Miocene episodic lakes in the arid Tarim basin, western China. Proc. Natl Acad. Sci. U.S.A. 111 (46), 16292.

Ma, J.Z., Pan, F., Chen, L.H., Edmunds, W.M., Ding, Z.Y., He, J.H., et al., 2010. Isotopic and geochemical evidence of recharge sources and water quality in the Quaternary aquifer beneath Jinchang city, NW China. Appl. Geochem. 25 (7), 996−1007.

Ma, J., He, J., Qi, S., Zhu, G., Zhao, W., Edmunds, W.M., et al., 2013. Groundwater recharge and evolution in the Dunhuang basin, northwestern China. Appl. Geochem. 28 (28), 19−31.

McArthur, J.M., Ravenscroft, P., Safiulla, S., Thirlwall, M.F., 2001. Arsenic in groundwater: Testing pollution mechanisms for sedimentary aquifers in Bangladesh. Water Resour. Res. 37, 109−117.

McArthur, J.M., Sikdar, P.K., Nath, B., Grassineau, N., Marshall, J.D., Banerjee, D.M., 2012. Sedimentological control on Mn, and other trace elements, in groundwater of the Bengal Delta. Environ. Sci. Technol. 46, 669−676.

MEE (Ministry of Ecology and Environment), 2018. China Ecological and Environmental Status Bulletin 2018. MEE (Ministry of Ecology and Environment), Beijing (in Chinese).

MEP (Ministry of Environmental Protection), 2011, National Plan for Groundwater Pollution Prevention and Control (2011−2020). MEP (Ministry of Environmental Protection), Beijing (in Chinese).

Mills, C.T., Morrison, J.M., Goldhaber, M.B., Ellefsen, K.J., 2011. Chromium(VI) generation in vadose zone soils and alluvial sediments of the southwestern Sacramento valley, California: a potential source of geogenic Cr(VI) to groundwater. Appl. Geochem. 26 (8), 1488−1501.

MOH (Ministry of Health), 2012. China's 12th Five-Year Plan for the Prevention and Control of Endemic Diseases. <http://www.gov.cn/zwgk/2012-01/29/content_2053487.htm> (in Chinese).

Mukherjee, A., Scanlon, B., Aureli, A., Langan, S., Guo, H., McKenzie, A., 2020. Global Groundwater: Source, Scarcity, Sustainability, Security and Solutions, first ed. Elsevier, ISBN: 9780128181720.

Pacheco, F.A.L., Landim, P.M.B., Szocs, T., 2013. Anthropogenic impacts on mineral weathering, a statistical perspective. Appl. Geochem. 36 (3), 34−48.

Pan, Y.H., 2014. Nitrate Circulation in Vadose Zone and Its Response to Paleo-Hydrology and Environment of Badain Jaran (Doctorial Dissertation). Lanzhou University (in Chinese with English Abstract).

Pernet-Coudrier, B., Qi, W.X., Liu, H.J., Müller, B., Berg, M., 2012. Sources and pathways of nutrients in the semi-arid region of Beijing−Tianjin, China. Environ. Sci. Technol. 46 (10), 5294−5301.

Perrin, A.S., Probst, A., Probst, J.L., 2008. Impact of nitrogenous fertilizers on carbonate dissolution in small agricultural catchments, implications for weathering CO_2 uptake at regional and global scales. Geochim. Cosmochim. Acta 72, 3105−3123.

Postma, D., Larsen, F., Hue, N.T.M., Mai, T.D., Viet, P.H., Nhan, P.Q., et al., 2007. Arsenic in groundwater of the red river floodplain, Vietnam, controlling geochemical processes and reactive transport modeling. Geochim. Cosmochim. Acta 71 (21), 5054−5071.

Regenspurg, S., Margot-Roquier, C., Harfouche, M., Froidevaux, P., Steinmann, P., Junier, P., et al., 2010. Speciation of naturally-accumulated uranium in an organic-rich soil of an alpine region (Switzerland). Geochim. Cosmochim. Acta 74, 2082−2098.

Rodríguez-Lado, L., Sun, G.F., Berg, M., Zhang, Q., Xue, H.B., Zheng, Q.M., et al., 2013. Groundwater arsenic contamination throughout China. Science 341, 866−868.

Santi, P.M., McCray, J.E., Martens, J.L., 2006. Investigating cross-contamination of aquifers. Hydrogeol. J. 14 (1−2), 51−68.

Scanlon, B.R., Keese, K.E., Flint, A.L., Flint, L.E., Gaye, C.B., Edmunds, W.M., et al., 2006. Global synthesis of groundwater recharge in semiarid and arid regions. Hydrol. Process. 20 (15), 3335−3370.

Schaefer, M.V., Ying, S.C., Benner, S.G., Duan, Y., Wang, Y., Fendorf, S., 2016. Aquifer arsenic cycling induced by seasonal hydrologic changes within the Yangtze river basin. Environ. Sci. Technol. 50 (7), 3521.

Shen, H.M., Zhang, S.B., Liu, S.J., Su, X.H., Shen, Y.F., Han, H.P., 2007. Study on geographic distribution of national high water iodine areas with contours of water iodine in high iodine areas. Chin. J. Endemiol 26 (6), 658−661 (in Chinese with English abstract).

Shen, M., Guo, H., Jia, Y., Cao, Y., Zhang, D., 2018. Partitioning and reactivity of iron oxide minerals in aquifer sediments hosting high arsenic groundwater from the Hetao basin, PR China. Appl. Geochem. 89, 190−201.

Shi, W.D., Guo, J.Q., Zhang, S.Q., Ye, C.M., Li, J., Ma, X.H., 2010. The distribution and geochemistry of geothermal groundwater bearing F and As in the Guide Basin. Hydrogeol. Eng. Geol. 37 (2), 36−41 (in Chinese with English abstract).

Smedley, P.L., Kinniburgh, D.G., 2002. A review of the source, behaviour and distribution of arsenic in natural waters. Appl. Geochem. 17 (5), 517−568.

Smedley, P.L., Zhang, M., Zhang, G., Luo, Z., 2003. Mobilisation of arsenic and other trace elements in fluviolacustrine aquifers of the Huhhot basin, Inner Mongolia. Appl. Geochem. 18 (9), 1453−1477.

Smedley, P., Smith, B., Abesser, C., Lapworth, D., 2006. Uranium Occurrence and Behavior in British Groundwater. British Geological Survey, Keyworth, Nottingham.

Stone, A.E.C., Edmunds, W.M., 2014. Naturally-high nitrate in unsaturated zone sand dunes above the Stampriet Basin, Namibia. J. Arid. Environ. 105 (6), 41−51.

Su, C.L., Wang, Y., Xie, X., Li, J., 2013. Aqueous geochemistry of high-fluoride groundwater in Datong Basin, northern China. J. Geochem. Explor. 135 (6), 79−92.

Sun, H., Ma, F., Liu, Z., Liu, Z., Wang, G., Nan, W., 2015. The distribution and enrichment characteristics of fluoride in geothermal active area in Tibet. China Environ. Sci. 35 (1), 251−259 (in Chinese with English Abstract).

Sun, Y.B., Wang, W.K., Zhang, C.C., Duan, L., Wang, Y.H., Li, H., 2013. Evolution mechanism of shallow high fluoride groundwater in the Guanzhong basin. Hydrogeol. Eng. Geol. 40 (6), 117−122 (in Chinese with English Abstract).

Tang, C., Chen, J., Shindo, S., Sakura, Y., Zhang, W., Shen, Y., 2004. Assessment of groundwater contamination by nitrates associated with wastewater irrigation: a case study in Shijiazhuang region, China. Hydrol. Process. 18 (12), 2303−2312.

Walvoord, M.A., Phillips, F.M., Stonestrom, D.A., Evans, R.D., Hartsough, P.C., Newman, B.D., et al., 2003. A reservoir of nitrate beneath desert soils. Science 302, 1021−1024.

Wang, L.F., Huang, J.Z., 1995. Outline of control practice of endemic fluorosis in China. Soc. Sci. Med. 41 (8), 1191−1195.

Wang, X.P., Berndtsson, R., Li, X.R., Kang, E.S., 2004b. Water balance change for a re-vegetated xerophyte shrub area. Hydrol. Sci. J.—J. Des. Sci. Hydrol. 49, 283−295.

Wang, Y., Bai, Y.C., Yin, X.L., Yang, W., Zuo, A.G., 2004a. Assessment and zoning of ecologic-geologic environment of Sanjiang Plain in Heilongjiang province. Hydrogeol. Eng. Geol. (in Chinese with English Abstract).

Wang, Y., Jiao, J.J., Zhang, K., Zhou, Y., 2016. Enrichment and mechanisms of heavy metal mobility in a coastal quaternary groundwater system of the Pearl River delta, China. Sci. Total Environ. 493, 545−546.

Wang, Y., Jiao, J.J., 2012. Origin of groundwater salinity and hydrogeochemical processes in the confined Quaternary aquifer of the Pearl River Delta, China. J. Hydrol. s438−439 (7), 112−124.

Wang, Y., Pi, K., Fendorf, S., Deng, Y., Xie, X., 2019. Sedimentogenesis and hydrobiogeochemistry of high arsenic Late Pleistocene-Holocene aquifer systems. Earth-Sci. Rev. 189, 79−98.

Wang, Y., Shvartsev, S.L., Su, C., 2009. Genesis of arsenic/fluoride-enriched soda water, a case study at Datong, Northern China. Appl. Geochem. 24 (4), 641−649.

Wasserman, G.A., Liu, X.H., Parvez, F., Ahsan, H., Levy, D., Factor-Litvak, P., et al., 2006. Water manganese exposure and children's intellectual function in Araihazar, Bangladesh. Environ. Health Perspect. 114, 124−129.

Wei, F.S., Chen, J.S., Wu, Y.Y., 1984. The soil background value in China. Geological Publishing House, Beijing (in Chinese).

Wen, D., Zhang, F., Zhang, E., Wang, C., Han, S., Zheng, Y., 2013. Arsenic, fluoride and iodine in groundwater of China. J. Geochem. Explor. 135 (6), 1−21.

Wen, X., Wu, Y., Su, J., Zhang, Y., Liu, F., 2005. Hydrochemical characteristics and salinity of groundwater in the Ejina Basin, northwestern China. Environ. Geol. 48 (6), 665−675.

Wu, Y., Wang, Y., Xie, X., 2014. Occurrence, behavior and distribution of high levels of uranium in shallow groundwater at Datong basin, northern China. Sci. Total Environ. 472, 809−817.

Xie, X., Wang, Y., Li, J., Yu, Q., Wu, Y., Su, C., et al., 2015. Effect of irrigation on Fe(III)-SO$_4^{2-}$, redox cycling and arsenic mobilization in shallow groundwater from the datong basin, china, evidence from hydrochemical monitoring and modeling. J. Hydrol. 523, 128−138.

Xie, X.J., Wang, Y.X., Duan, M.Y., Xie, Z.M., 2009. Geochemical and environmental magnetic characteristics of high arsenic aquifer sediments from Datong Basin, northern China. Environ. Geol. 58, 45−52.

Xin, J., Liu, Y., Chen, F., Duan, Y.J., Wei, G.L., Zheng, X.L., et al., 2019. The missing nitrogen pieces: a critical review on the distribution, transformation, and budget of nitrogen in the vadose zone-groundwater system. Water Res. 165, 114977.

Xu, F., Ma, T., Shi, L., Zhang, J.W., Wang, Y.Y., Dong, Y.H., 2013. The hydrogeochemical characteristics of high iodine and fluoride groundwater in the Hetao plain, Inner Mongolia. Procedia Earth Planet. Sci. 7, 908−911.

Zhang, F.C., Wen, D.G., Guo, J.Q., Zhang, E.R., Hao, A.B., An, Y.H., 2010. Research progress and prospect of geological environment in main endemic disease area. Geol. China 37 (3), 551−562 (in Chinese with English Abstract).

Zhang, Z.J., Fei, Y.H., Chen, Z.Y., et al., 2009. Investigation and assessment of sustainable utilization of groundwater resources in the North China Plain. Geological Publishing House, Beijing (in Chinese).

Zhang, Z.H., Li, L.R., 2005. Groundwater resources of China, Inner Mongolia Volume [M]. China Cartographic Publishing House, Beijing (in Chinese with English Abstract).

Zhang, E., Wang, Y., Qian, Y., Ma, T., Zhang, D., Zhan, H., et al., 2013. Iodine in groundwater of the North China Plain, spatial patterns and hydrogeochemical processes of enrichment. J. Geochem. Explor. 135 (6), 40−53.

Zhang, J., Ma, T., Feng, L., Yan, Y., Abass, O.K., Wang, Z., et al., 2017. Arsenic behavior in different biogeochemical zonations approximately along the groundwater flow path in Datong Basin, northern China. Sci. Total Environ. 584, 458−468.

Zhang, N., 2001. Distribution of fluorine and its affecting factors in soil in Shanxi. Acta Pedologica Sin. 38 (2), 284−287.

Zhang, Z., Guo, H., Zhao, W., Liu, S., Cao, Y., Jia, Y., 2018. Influences of groundwater extraction on flow dynamics and arsenic levels in the western Hetao Basin, Inner Mongolia, China. Hydrogeol. J. 1−14.

Zhao, Q., Su, X.S., Kang, B., et al., 2017. A hydrogeochemistry and multi-isotope (Sr, O, H, and C) study of groundwater salinity origin and hydrogeochemcial processes in the shallow confined aquifer of northern Yangtze River downstream coastal plain, China. Appl. Geochem. 86, 49–58.

Zhao, L.S., Wu, S., Zhou, J.H., Wang, J.H., Wang, J.W., 2007b. Eco-geochemical investigation on the endemic As and F poisoning in Datong Basin. Earth Sci. Front. 14 (2), 225–235 (in Chinese with English abstract).

Zhao, S.Z., Wang, X.K., Huang, Z.F., Li, S.B., Wang, Z., Su, M.X., et al., 2007a. Study on formation causes of high fluorine groundwater in Hetao area of Inner Mongolia. Rock. Miner. Anal. 26 (4), 320–324 (in Chinese with English Abstract).

Zhou, Y., Zeng, Y., Zhou, J., Guo, H., Li, Q., Jia, R., et al., 2017. Distribution of groundwater arsenic in XinJiang, P.R. China. Appl. Geochem. 77, 116–125.

Zhu, Z.D., 1984. Aeolian landforms in the Taklimakan Desert. In: El-Baz, F. (Eds.), Deserts and Arid Lands. Remote Sensing of Earth Resources and Environment, vol. 1. Springer, Dordrecht.

Zhu, D., Ryan, M.C., Sun, B., Li, C., 2014a. The influence of irrigation and Wuliangsuhai lake on groundwater quality in eastern Hetao basin, Inner Mongolia, China. Hydrogeol. J. 22 (5), 1101–1114.

Zhu, L., Sun, J.C., Zhang, Y.X., Liu, J.T., 2014b. Analysis of chemical characteristics and factors for anomalous chromium-rich groundwater in an area, Shanxi Province. Environ. Chem. 33 (11), 1864–1970 (in Chinese with English abstract).

Chapter 17

Screening of emerging organic pollutants in the typical hygrogeological units of China

Xiaopeng Qin[1], Tian Zhou[2], Shengzhang Zou[3] and Fei Liu[2]

[1]*Department of Technology Assessment, Technical Centre for Soil, Agricultural and Rural Ecology and Environment, Ministry of Ecology and Environment, Beijing, P.R. China,* [2]*MOE Key Laboratory of Groundwater Circulation and Environmental Evolution, Beijing Key Laboratory of Water Resources and Environmental Engineering, School of Water Resources and Environment, China University of Geosciences (Beijing), Beijing, P.R. China,* [3]*Institute of Karst Geology, CAGS, Karst Dynamics Laboratory, MLR & GZAR, Guilin, P.R. China*

17.1 Introduction

Antibiotics are naturally occurring or human-made compounds, which are effective against the negative and positive bacteria and are widely used in human, veterinary, and aquaculture treatments. Generally, antibiotics are divided into eight groups, includTheme 3ing quinolones, sulfonamides, tetracyclines (TCs), macrolides, β-lactams, aminoglycosides, cephalosporins, and carbapenems. Many antibiotics are still active, when they enter the environment after use. Their presence in waters and soils could affect the number or/and type of microbial communities in the natural environment (Girardi et al., 2011; Xiang et al., 2020) and lead to the occurrence of antibiotic resistance genes (Deng et al., 2018).

Groundwater is commonly treated and used for drinking, irrigation, and industry (Mukherjee et al., 2020). In recent years, some emerging organic pollutants (e.g., antibiotics) are observed in groundwater, which may have an influence on the quality of groundwater. As reported previously, many antibiotics were found in groundwater all over the world, especially in those areas with animal husbandry, aquaculture, or in Karst regions. Sulfonamides (e.g., sulfamethoxazole) or quinolones (e.g., ciprofloxacin) were the most abundant antibiotics in Catalonia of Spain (García-Galán et al., 2010a), Rhône-Alpes region of France (Vulliet, Cren-Olivé 2011), Yverdon-les-Bains of Switzerland (Morasch, 2013), Barcelona of Spain (López-Serna et al., 2013), Cluj-Napoca of Romania (Szekeres et al., 2018), and Gelderse Valley in Noord-Brabant Limburg of the Netherlands (Kivits et al., 2018). In addition, trimethoprim and carbamazepine were the most frequently detected antibiotics in Lower Saxony of Germany (Burke et al., 2016) and Alberta of Canada (Van Stempvoort et al., 2013), respectively.

In China, norfloxacin (NOR), ciprofloxacin, chlortetracycline, sulfamethazine, and other antibiotics were also found in groundwater of Guangzhou (Peng et al., 2014; Chen et al., 2016), Beijing (Chen et al., 2016), Xuzhou (Gu et al., 2019), Jianghan Plain (Tong et al., 2014; Yao et al., 2017), Guilin (Qin et al., 2020), and northern and southwestern regions of China (Chen et al., 2018; Li et al., 2018), which had the potential to harm human health. However, in "Standards for drinking water quality (GB 5749-2006)" and "Standard for groundwater quality (GB 14848-2017)" of China, there are no standard values for antibiotic until now, which is difficult to assess the quality of groundwater (or drinking water) that contains antibiotics. In this work, 107 groundwater samples were collected from the largest Karst areas in southwest China, 35 types of antibiotics were analyzed. The aims of our study are (1) to determine the concentrations of antibiotics in groundwater from the Karst regions of China; (2) to assess their risks to aquatic organisms; and (3) to screen typical antibiotics in the regions.

17.2 Materials and methods

17.2.1 Study area and sample collection

Three typical Karst regions in southwest China were investigated in this study. A total of 107 sampling sites were selected in Guizhou province, Yunnan province, and Chongqing. After collection the groundwater samples were covered with the ice bag and transferred to our laboratory immediately. Then the samples were stored at 4°C in the dark until they were analyzed.

17.2.2 Chemicals

The companies of antibiotics standards used in our study were shown in Table 17.1. The surrogate standards ofloxacin (OFL)-D$_3$ and sulfadimethoxine-D$_6$ were obtained from Witega (Berlin, Germany). Other analytical reagent grade or higher chemicals were purchased as reported by Chen et al. (2018) and Huang et al. (2019). Ultrapure water (MilliQ, Germany) was used in the following experiments.

TABLE 17.1 The physicochemical properties and analytical results of antibiotics in groundwater samples ($n = 107$) in this study.

No.	Antibiotics	Abb.	CAS no.[a]	K_{ow}	MDLs (ng/L)[b]	DF (%)[c]	Standards production company
I Sulfonamides							
1	Sulfacetamide	SA	144-80-9	0.11	1.06	6.5	Dr. Ehrenstorfer
2	Sulfapyridine	SPD	144-83-2	2.24	0.29	35.5	
3	Sulfisoxazole	SIZ	127-69-5	10.23	1.91	1.9	
4	Sulfachloropyridazine	SCPD	80-32-0	2.04	0.38	17.8	
5	Sulfadoxine	SDO	2447-57-6	5.01	0.42	7.5	
6	Sulfadimethoxine	SDM	122-11-2	42.66	1.39	2.8	
7	Sulfameter	SMT	651-06-9	2.57	0.54	12.2	
8	Sulfadimidine	SM2	57-68-1	1.55	1.24	2.8	
9	Sulfamethizole	SMTZ	144-82-1	3.47	0.86	2.8	
II Chloramphenicols							
10	Chloramphenicol	CAP	56-75-7	13.80	1.42	24.3	Dr. Ehrenstorfer
III Tetracyclines							
11	Tetracycline	TC	60-54-8	0.05	3.08	19.6	Dr. Ehrenstorfer
12	Chlortetracycline	CTC	57-62-5	0.24	3.72	5.6	
13	Oxytetracycline	OTC	79-57-2	0.13	3.17	19.6	
14	Doxycycline	DOX	564-25-0	0.95	3.28	2.8	
IV Lincosamides							
15	Lincomycin	LIN	859-18-7	1.95	1.06	46.7	Dr. Ehrenstorfer
V Macrolides							
16	Erythromycin	ERY	114-07-8	1148.2	0.68	43.0	Dr. Ehrenstorfer
17	Spiramycin	SPI	8025-81-8	74.13	2.99	3.7	
18	Roxithromycin	ROX	80214-83-1	562.3	0.84	38.3	
19	Josamycin	JOS	16846-24-5	1445.4	1.13	1.9	Sigma-Aldrich
VI Quinolones							
20	Lomefloxacin	LOM	98079-51-7	0.50	2.63	1.9	Sigma-Aldrich
21	Difloxacin	DIF	98106-17-3	7.76	2.37	1.9	
22	Moxifloxacin	MOX	151096-09-2	8.91	2.84	8.4	
23	Nalidixic acid	NDA	389-08-2	38.90	1.72	24.3	
24	Oxolinic acid	OXA	14698-29-4	8.71	3.19	2.8	
25	Ciprofloxacin	CIP	85721-33-1	1.91	2.66	29.9	
26	Fleroxacin	FLE	79660-72-3	1.74	3.00	0[d]	
27	Sparfloxacin	SPA	110871-86-8	0.95	2.59	4.7	

(Continued)

TABLE 17.1 (Continued)

No.	Antibiotics	Abb.	CAS no.[a]	K_{ow}	MDLs (ng/L)[b]	DF (%)[c]	Standards production company
28	Ofloxacin	OFL	82419-36-1	0.41	4.03	54.2	
29	Norfloxacin	NOR	70458-96-7	0.09	4.01	34.6	
30	Enrofloxacin	ENR	93106-60-6	5.01	2.71	7.5	
31	Danofloxacin	DAN	112398-08-0	–[e]	3.98	5.6	
32	Cinoxacin	CIN	28657-80-9	38.90	2.70	9.4	
33	Enoxacin	ENO	74011-58-8	0.63	3.97	25.2	
34	Flumequine	FLU	42835-25-6	39.81	0.82	30.8	
35	Pipemidic acid	PPA	51940-44-4	0.0071	2.57	9.4	

[a]CAS means Chemical Abstracts Service.
[b]MDLs means method detect limits.
[c]DF means the detection frequency.
[d]The concentrations of fleroxacin in all the 107 samples are below its MDLs.
[e]The K_{ow} values of spiramycin and moxifloxacin are obtained from the website (https://pubchem.ncbi.nlm.nih.gov/), and that of danofloxacin is unavailable. The K_{ow} values of other antibiotics were obtained from the ECOSAR V2.0 software

17.2.3 Analytical method

The target 35 antibiotics were analyzed using the combination of solid phase extraction method and ultraperformance liquid chromatography method, as reported in our previous studies (Chen et al., 2018; Huang et al., 2019). Quality assurance (Q_A) and quality control (Q_C) measures during the field sampling and laboratory analysis processes were also conducted as reported by Huang et al. (2019).

17.2.4 Risk characterization

The risk quotient (*RQ*) values of the antibiotic to different aquatic organisms are calculated with the following equations:

$$RQ = \frac{MEC}{PNEC} \tag{17.1}$$

$$PNEC = \frac{LC_{50}}{AF} \tag{17.2}$$

where *MEC* is the maximum concentration of the antibiotic detected in groundwater samples (mg/L), *PNEC* is the predicted no effect concentration on nontarget organisms (mg/L), LC_{50} (or EC_{50}) is the lowest median effective concentration value of the aquatic organism obtained from the Ecological Structure Activity Relationships (ECOSAR) V2.0 software of US Environmental Protection Agency (ECOSAR Predictive Model, https://www.epa.gov/tsca-screening-tools/ecological-structure-activity-relationships-ecosar-predictive-model), and *AF* (10–1000) is an appropriate assessment factor. In this study the *AF* (1000) is chosen and designed to be conservative and protective, due to the available limited data [HHS (U.S. Department of Health and Human Services), 2006].

The *RQ* values of different antibiotics were classified into the following four risk levels: negligible ($RQ \leq 0.01$), low ($0.01 < RQ \leq 0.1$), medium ($0.1 < RQ \leq 1$), and high ($RQ > 1$) (Sánchez-Bayo et al., 2002).

17.3 Results and discussion

17.3.1 Presence of antibiotics in groundwater

At least one antibiotic has been found in 86.9% of the groundwater sites sampled in this study (Fig. 17.1). The mixtures of antibiotics are very common. More than one antibiotic is detected at 79 of 107 wells, and 10 or more antibiotics are detected at 22 wells. As reported previously, different types of antibiotics were simultaneously observed in the sludge (Jia et al., 2012; Cheng et al., 2014), manure (Hu et al., 2010; Van den Meersche et al., 2019), and surface water (Zhang et al., 2015; Ngigi et al., 2020), which would enter the soil and groundwater during the application and infiltration processes.

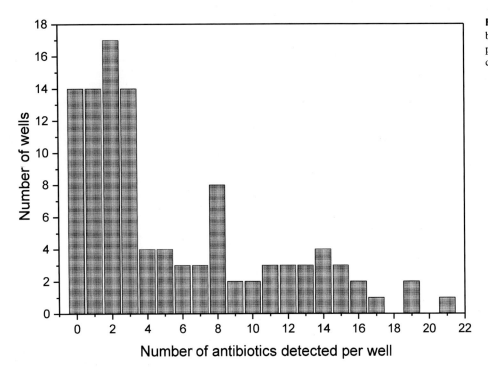

FIGURE 17.1 The statistics of numbers of different antibiotics observed per well and numbers of wells with certain number of antibiotics observed.

In this study, 15 or more antibiotics were observed in eight wells located in Kaiyang of Guizhou province. Sulfapyridine (SPD), oxytetracycline, erythromycin (ERY), lincomycin (LIN), roxithromycin (ROX), OFL, and NOR were all detected in these eight wells. As reported in our previous studies, these antibiotics were also frequently detected in river samples collected from the same Karst region in Kaiyang, and their maximum concentrations were in the ranges of 11.7–1200 (Zou et al., 2018) and 9.2–308.4 ng/L (Huang et al., 2019). According to results of chemical tracer tests (i.e., hexaammonium molybdate, rhodamine B, and fluorescein sodium), the groundwater in this Karst region always receives the seepage recharge from the river water, and thus the same contaminant has been detected both in the river water and groundwater.

17.3.2 Statistical analysis

The results of 35 kinds of antibiotics in the groundwater samples are shown in Table 17.1. The detection frequency (*DF*) values of 34 antibiotics [except fleroxacin (FLE)] are in the range of 1.9%–54.2%. OFL, LIN, ERY, ROX, and SPD are the top five most frequently detected antibiotics. FLE is not detected (below its method detect limit (MDL)) in all the 107 groundwater samples.

The *DF* of OFL in groundwater is 54.2% (Table 17.1), which is similar to 57% in Beijing (Liu et al., 2019) and 54.1% in northern and southwestern regions of China (Chen et al., 2018), but smaller than that (68.0%–100%) in other areas (López-Serna et al., 2013; Tong et al., 2014; Ma et al., 2015; Chen et al., 2016). The *DF* of LIN in groundwater is 46.7%, and is lower than that (63.5%) in north and southwest China (Chen et al., 2018). The *DF* (43.0%) of ERY is similar to 39.2% in northern and southwestern regions of China (Chen et al., 2018), and is smaller than that (100%) in Besòs River Delta of Spain (López-Serna et al., 2013). The *DF* (38.3%) of ROX is smaller than that in Barcelona of Spain (López-Serna et al., 2013) and some areas in China (Tong et al., 2014; Yao et al., 2017; Chen et al., 2018). The *DF* (35.5%) of SPD is also smaller than that (56.4%–72%) in Catalonia of Spain (García-Galán et al., 2010a,b).

In this study the maximum content of OFL in groundwater is 611 ng/L, which is smaller than that (1199.7 ng/L) in northern and southwestern regions of China (Chen et al., 2018), and is much higher than the reported values (7.6–178 ng/L) in Barcelona of Spain (Candela et al., 2016) and some areas in China (Peng et al., 2014; Tong et al., 2014; Ma et al., 2015; Chen et al., 2016; Liu et al., 2019). The maximum concentration of cinoxacin (CIN) is 509 ng/L, which is 33–74 times higher than that in north and southwest China (Chen et al., 2018) and Xuzhou (Gu et al., 2019). The maximum concentration of LIN is 243 ng/L, which is much higher than 8.3 ng/L in north China (Hu et al., 2010), and lower than 860.7 ng/L in northern and southwestern regions of China (Chen et al., 2018).

Generally, for the nonvolatile organic compounds, the octanol–water partition coefficient (K_{ow}) is a very important parameter, which could affect and predicate their environmental fate in the soil, sediment, and groundwater. The K_{ow} values of 34 antibiotics in this study range from 0.0071 to 1445.4 (Table 17.1), indicating that the properties of various antibiotics are quite different. There is no significant linear relationship ($R^2 < 0.01$, $n = 34$) between K_{ow} and DF values of the antibiotics (except danofloxacin). The appearance and concentration of the antibiotics in groundwater are mainly influenced by their physicochemical properties (e.g., K_{ow}) and usage in different areas.

17.3.3 Environmental risk assessment

The effects of different antibiotics on three aquatic organisms (i.e., fish, daphnid, green algae) are assessed using the RQ values of individual antibiotic as shown in Fig. 17.2. For fish the RQ values of all the antibiotics are below 0.01, and thus the risk is negligible. For daphnid the RQ values of ERY and ROX are in the range of 0.01–0.1, indicating that the risk is low. We should notice that risk of CIN to daphnid is high ($RQ > 1$). For green algae the risk of ERY and ROX is low ($0.01 < RQ \leq 0.1$), and that of CIN is medium ($0.1 < RQ \leq 1$). The RQ is an important parameter for the risk assessment of antibiotics, so it is also selected as an index in Section 17.3.4.

17.3.4 Screening of antibiotics in groundwater

As mentioned previously, 35 antibiotics in 107 groundwater samples have been analyzed and assessed. In China the relevant standards of antibiotics in waters are deficient, so it is difficult to distinguish that which antibiotic should be paid attention first. In our study a new method has been introduced to assess the risks of different antibiotics in the groundwater, and to screen the typical antibiotics among them. The screening score (SS) of the antibiotic is calculated using the following equation:

$$SS = \log\left(\frac{RQ \times DF \times 10^9}{K_{ow}}\right) \tag{17.3}$$

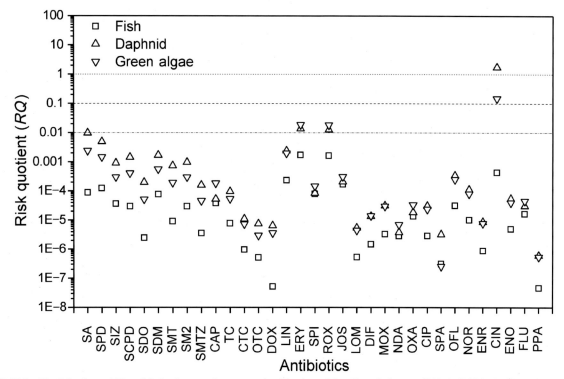

FIGURE 17.2 The RQ values of 33 antibiotics in groundwater samples. The data of danofloxacin is unavailable in ECOSAR V2.0 software, fleroxacin is not detected in groundwater, and they are excluded from the figure. *ECOSAR*, Ecological Structure Activity Relationships; *RQ*, risk quotient.

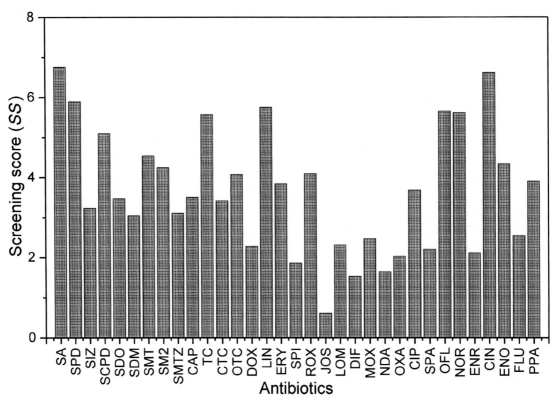

FIGURE 17.3 The *SS* values of 33 antibiotics in groundwater samples. *SS*, Screening score.

where *RQ* is the maximum value among that of fish, daphnid, or green algae and *DF* is the detection frequency of the antibiotic.

As shown in Fig. 17.3, the *SS* values of the antibiotics range from 0.61 to 6.76, indicating that their risk, *DFs*, and physicochemical properties are quite different. The *SS* values of sulfacetamide (SA), CIN, SPD, LIN, OFL, NOR, and TC are high among the 33 antibiotics, and thus they are considered the typical antibiotics in this region. These antibiotics in groundwater (or other aquatic systems) might contribute to hazardous effects on aquatic organism and subsequently on the health of human. To avoid the possible adverse effects, further detailed studies about their environmental toxicities are needed.

17.4 Conclusion and further research

At least one antibiotic has been found in 86.9% of the groundwater sites ($n = 107$) sampled in this study. OFL, LIN, ERY, ROX, and SPD were the top five most frequently detected ($DF > 35.0\%$) antibiotics. The *RQ* of CIN to daphnid was high ($RQ > 1$), and that to green algae was medium ($0.1 < RQ < 1$). We have introduced a new method considering *RQ*, *DF* and K_{ow} of the antibiotics, to assess their risks in groundwater. Among the 33 antibiotics, SA, CIN, SPD, LIN, OFL, NOR, and TC were considered as the typical antibiotics in these Karst regions.

Results shown in our study would provide some useful information for the protection of groundwater in typical Karst regions in the future. However, in this study, only 35 antibiotics were considered, and their toxicity data to aquatic organisms was obtained from the ECOSAR V2.0 software, which might be simplified. In the future, some new antibiotics and the environmental toxicity data should be concerned. The systematical investigation of antibiotics in groundwater all over China is needed. Moreover, the policies and standards of different antibiotics in the sludge, manure, sewage, surface water, soil, and groundwater are also needed in China.

Acknowledgments

This work was supported by the National Natural Science Foundation of China (nos. 41731282 and 41602259) and the program from China Geological Survey (DD20190323).

References

Burke, V., Richter, D., Greskowiak, J., Mehrtens, A., Schulz, L., Massmann, G., 2016. Occurrence of antibiotics in surface and groundwater of a drinking water catchment area in Germany. Water Environ. Res. 88 (7), 652–659.

Candela, L., Tamoh, K., Vadillo, I., Valdes-Abella, J., 2016. Monitoring of selected pharmaceuticals over 3 years in a detrital aquifer during artificial groundwater recharge. Environ. Earth Sci. 75, 244.

Chen, L., Lang, H., Liu, F., Jin, S., Yan, T., 2018. Presence of antibiotics in shallow groundwater in the northern and southwestern regions of China. Groundwater 56 (3), 451–457.

Chen, L., Wang, Y., Tong, L., Deng, Y., Li, Y., Gan, Y., et al., 2016. Risk assessment of three Fluoroquinolone antibiotics in the groundwater recharge system. Ecotox. Environ. Safe. 133, 18–24.

Cheng, M., Wu, L., Huang, Y., Luo, Y., Christie, P., 2014. Total concentrations of heavy metals and occurrence of antibiotics in sewage sludges from cities throughout China. J. Soil Sediment 14, 1123–1135.

Deng, W., Li, N., Ying, G., 2018. Antibiotic distribution, risk assessment, and microbial diversity in river water and sediment in Hong Kong. Environ. Geochem. Health 40 (5), 2191–2203.

García-Galán, M.J., Díaz-Cruz, M.S., Barceló, D., 2010b. Determination of 19 sulfonamides in environmental water samples by automated on-line solid-phase extraction-liquid chromatography–tandem mass spectrometry (SPE-LC–MS/MS). Talanta 81 (1–2), 355–366.

García-Galán, M.J., Garrido, T., Fraile, J., Ginebreda, A., Díaz-Cruz, M.S., Barceló, D., 2010a. Simultaneous occurrence of nitrates and sulfonamide antibiotics in two ground water bodies of Catalonia (Spain). J. Hydrol. 383 (1), 93–101.

Girardi, C., Greve, J., Lamshöft, M., Fetzer, I., Miltner, A., Schäffer, A., et al., 2011. Biodegradation of ciprofloxacin in water and soil and its effects on the microbial communities. J. Hazard. Mater. 198, 22–30.

Gu, D., Feng, Q., Guo, C., Hou, S., Lv, J., Zhang, Y., et al., 2019. Occurrence and risk assessment of antibiotics in manure, soil, wastewater, groundwater from livestock and poultry farms in Xuzhou, China. Bull. Environ. Contam. Toxicol. 103 (3), 590–596.

HHS (U.S. Department of Health and Human Services), 2006. Environmental Impact Assessments (EIA's) for Veterinary Medicinal Products (VMP's) – Phase II. 9–10.

Hu, X., Zhou, Q., Luo, Y., 2010. Occurrence and source analysis of typical veterinary antibiotics in manure, soil, vegetables and groundwater from organic vegetable bases, northern China. Environ. Pollut. 158, 2992–2998.

Huang, F., Zou, S., Deng, D., Lang, H., Liu, F., 2019. Antibiotics in a typical karst river system in China: spatiotemporal variation and environmental risks. Sci. Total Environ. 650, 1348–1355.

Jia, A., Wan, Y., Xiao, Y., Hu, J., 2012. Occurrence and fate of quinolone and fluoroquinolone antibiotics in a municipal sewage treatment plant. Water Res. 46 (2), 387–394.

Kivits, T., Broers, H.P., Beeltje, H., van Vliet, M., Griffioen, J., 2018. Presence and fate of veterinary antibiotics in age-dated groundwater in areas with intensive livestock farming. Environ. Pollut. 241, 988–998.

Li, X., Liu, C., Chen, Y., Huang, H., Ren, T., 2018. Antibiotic residues in liquid manure from swine feedlot and their effects on nearby groundwater in regions of North China. Environ. Sci. Pollut. Res. 25 (4), 11565–11575.

Liu, X., Zhang, G., Liu, Y., Lu, S., Qin, P., Guo, X., et al., 2019. Occurrence and fate of antibiotics and antibiotic resistance genes in typical urban water of Beijing, China. Environ. Pollut. 246, 163–173.

López-Serna, R., Jurado, A., Vázquez-Suñé, E., Carrera, J., Petrovic, Barceló, D., 2013. Occurrence of 95 pharmaceuticals and transformation products in urban groundwaters underlying the metropolis of Barcelona, Spain. Environ. Pollut. 174, 305–315.

Ma, Y., Li, M., Wu, M., Li, Z., Liu, X., 2015. Occurrences and regional distributions of 20 antibiotics in water bodies during groundwater recharge. Sci. Total Environ. 518–519, 498–506.

Morasch, B., 2013. Occurrence and dynamics of micropollutants in a karst aquifer. Environ. Pollut. 173, 133–137.

Mukherjee, A., Scanlon, B., Aureli, A., Langan, S., Guo, H., McKenzie, A., 2020. Global Groundwater: Source, Scarcity, Sustainability, Security and Solutions, first ed. Elsevier, ISBN: 9780128181720.

Ngigi, A.N., Magu, M.M., Muendo, B.M., 2020. Occurrence of antibiotics residues in hospital wastewater, wastewater treatment plant, and in surface water in Nairobi County, Kenya. Environ. Monit. Assess. 192, 18.

Peng, X., Qu, W., Wang, C., Wang, Z., Huang, Q., Jin, J., et al., 2014. Occurrence and ecological potential of pharmaceuticals and personal care products in groundwater and reservoirs in the vicinity of municipal landfills in China. Sci. Total Environ. 490, 889–898.

Qin, L., Pang, X., Zeng, H., Liang, Y., Mo, L., Wang, D., et al., 2020. Ecological and human health risk of sulfonamides in surface water and groundwater of Huixian karst wetland in Guilin, China. Sci. Total Environ. 708, 134552.

Sánchez-Bayo, F., Baskaran, S., Kennedy, I.R., 2002. Ecological relative risk (EcoRR): another approach for risk assessment of pesticides in agriculture. Agr. Ecosyst. Environ. 91 (1–3), 37–57.

Szekeres, E., Chiriac, C.M., Baricz, A., Szöke-Nagy, T., Lung, I., Soran, M.L., et al., 2018. Investigating antibiotics, antibiotic resistance genes, and microbial contaminants in groundwater in relation to the proximity of urban areas. Environ. Pollut. 236, 734–744.

Tong, L., Huang, S., Wang, Y., Liu, H., Li, M., 2014. Occurrence of antibiotics in the aquatic environment of Jianghan Plain, central China. Sci. Total Environ. 497–498, 180–187.

Van den Meersche, T., Rasschaert, G., Haesebrouck, F., Van Coillie, E., Herman, L., Van Weyenberg, S., et al., 2019. Presence and fate of antibiotic residues, antibiotic resistance genes and zoonotic bacteria during biological swine manure treatment. Ecotox. Environ. Safe. 175, 29–38.

Van Stempvoort, D.R., Roy, J.W., Grabuski, J., Brown, S.J., Bickerton, G., Sverko, E., 2013. An artificial sweetener and pharmaceutical compounds as co-tracers of urban wastewater in groundwater. Sci. Total Environ. 461–462, 348–359.

Vulliet, E., Cren-Olivé, C., 2011. Screening of pharmaceuticals and hormones at the regional scale, in surface and groundwaters intended to human consumption. Environ. Pollut. 159, 2929–2934.

Xiang, S., Wang, X., Ma, W., Liu, X., Zhang, B., Huang, F., et al., 2020. Response of microbial communities of karst river water to antibiotics and microbial source tracking for antibiotics. Sci. Total Environ. 706, 135730.

Yao, L., Wang, Y., Tong, L., Deng, Y., Li, Y., Gan, Y., et al., 2017. Occurrence and risk assessment of antibiotics in surface water and groundwater from different depths of aquifers: a case study at Jianghan Plain, central China. Ecotox. Environ. Safe. 135, 236–242.

Zhang, Q., Ying, G., Pan, C., Liu, Y., Zhao, J., 2015. Comprehensive evaluation of antibiotics emission and fate in the river basins of China: source analysis, multimedia modeling, and linkage to bacterial resistance. Environ. Sci Technol. 49 (11), 6772–6782.

Zou, S., Huang, F., Chen, L., Liu, F., 2018. The occurrence and distribution of antibiotics in the Karst river system in Kaiyang, Southwest China. Water Sci. Technol.: Water Supply 18 (6), 2044–2052.

Chapter 18

Groundwater pollution of Pearl River Delta

Guanxing Huang[1], Lingxia Liu[1], Chunyan Liu[1], Wenzhong Wang[1] and Dongya Han[1,2]

[1]*Institute of Hydrogeology and Environmental Geology, Chinese Academy of Geological Sciences, Shijiazhuang, P.R. China,*
[2]*Hebei GEO University, Shijiazhuang, P.R. China*

18.1 Introduction

Global groundwater is undergoing immense stress in recent times (Mukherjee et al., 2020). For example, the Pearl River Delta (PRD) is a rapidly UA in China. It has undergone three decades of urbanization expansion. Specifically, the UA in the PRD was more than double from 1988 to 2006 (Huang et al., 2018a). Unfortunately, the wholesale transformation from natural and agricultural areas to UAs in the PRD is accompanied by the anthropogenic pollution, such as huge illegal discharge of domestic sewage and industrial wastewater (Huang et al., 2020a). Therefore the water pollution in the PRD occurs. For instance, Cheung et al. (2003) reported that the elevated P, nickel, ammonium−nitrogen, and zinc concentrations in surface water in the PRD were found because of the effects of industrialization and urbanization.

Groundwater has become an important water resource for domestic and agricultural uses after the occurrence of surface water pollution in the PRD (Lu et al., 2009; Huang et al., 2013). However, the groundwater contamination in the PRD was also reported recently. For example, Zhang et al. (2020) reported that the industrialization accompanied by the wastewater leakage of township−village enterprises was mainly responsible for the elevated nitrate concentration in groundwater in the PRD. Similarly, elevated phosphate concentration in groundwater in UAs of the PRD was also mainly attributed to anthropogenic pollution, such as animal wastes and industrial wastewater (Huang et al., 2020a). In addition, Hou et al. (2020) reported that high concentration of groundwater Mn in fissured aquifer in the PRD ascribed mainly to the infiltration of industrial wastewater and domestic sewage during the industrialization and urbanization. These studies focused on one or several contaminants in groundwater in the PRD, a comprehensive understanding of the groundwater contamination in this area remains unclear.

The objectives of this chapter are to (1) delineate the distribution of groundwater chemistry and quality in the PRD, (2) extract the main impact chemicals for poor-quality groundwater, and (3) discuss the sources and driving forces of these chemicals in groundwater. It will be helpful for groundwater resource management in the PRD and as a reference for the investigation of groundwater contamination in UAs in other countries.

18.2 Study area

18.2.1 Hydrogeological and geological conditions

The PRD was formed as a result of the Tibetan Plateau uplift during the Tertiary and Quaternary Periods. Quaternary deposits cover the central and southern parts of the PRD and compose the PRD plain. Deposits in the PRD plain mainly consist of four stratigraphic units, including two marine units and two terrestrial units (Fig. 18.1). The younger marine unit has an elevation above −20 m and was formed during the Holocene, while the other three units were formed during the Pleistocene. The younger terrestrial unit can be sandy fluvial deposits or clayey silt and becomes a local aquifer, while the older terrestrial unit is dominated by sand and gravel and becomes the basal aquifer. Groundwater in porous aquifer is mainly recharged by vertical infiltration of precipitation and agricultural irrigation, and the lateral flow of rivers. In addition, porous aquifers in coastal areas are often intruded by seawater (Huang et al., 2018b). The PRD plain is

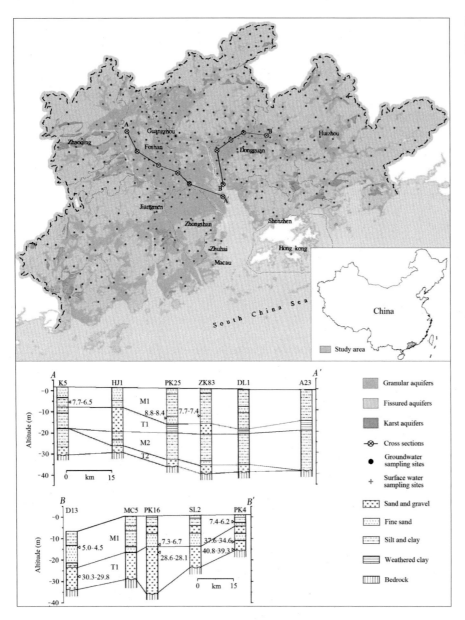

FIGURE 18.1 Hydrogeological setting and sampling sites in the PRD. *PRD*, Pearl River Delta.

surrounded by hills on east, west, and north, and the maximum elevation of hilly areas is 1210 m. Fissured aquifer is distributed in hilly areas where bedrocks (e.g., limestone, mudstone, sandstone, shale, dolomite, granite, and gneiss) ranging in age from Cambrian to Tertiary crop out. Karst aquifer is less than 10% of the total area. The regional groundwater flow in the PRD is from northeast and northwest to the coast (Zhang et al., 2019).

18.2.2 Characteristics of urbanization and industrialization in the Pearl River Delta

According to the different degrees of urbanization, land use in the PRD can be divided into three types comprising urbanized areas (UAs), peri-urban areas (PUAs), and non-UAs (NUAs) (Fig. 18.2). The UAs are associated with large-scale construction land, and the illegal discharge of domestic sewage and industrial wastewater in these areas often occurs. The PUAs are the epitaxial areas (~2 km outside UAs) of the UAs, and the illegal discharge of domestic sewage and industrial wastewater in these areas also occurs. Areas in the PRD except UAs and PUAs are called NUAs, and these areas mainly refer to agricultural lands and natural lands.

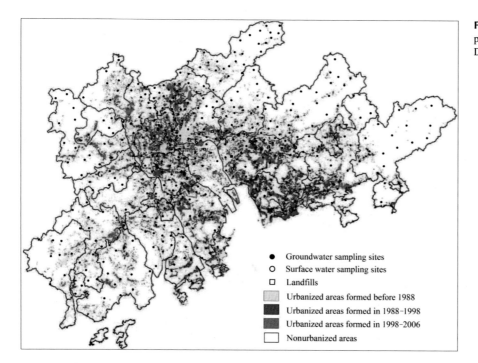

FIGURE 18.2 Urban expansion and sampling sites in the PRD. *PRD*, Pearl River Delta.

18.3 Materials and methods

In this study, 399 groundwater samples (granular aquifers—258, fissured aquifers—132, karst aquifers—9) and 9 surface water samples were collected (Fig. 18.1). Eighty-six chemicals comprising 31 inorganic chemicals and 55 organic chemicals were measured in waters, and the relevant analytical methods are shown in our previous studies (Huang et al., 2018a,b; Zhang et al., 2019). The relative errors for inorganic chemicals were $< \pm 6\%$. Quality control of organic chemicals followed the US EPA protocols.

18.4 Results and discussion

18.4.1 Groundwater chemistry

In the PRD, groundwater pH was in a range of 3.7–7.7 and shown mainly acidic (Table 18.1), and the groundwater acidification in this area was probably because of the acid deposition resulting from industrialization (Huang et al., 2018b). As shown in Table 18.1, HCO_3^- exhibited the highest median value of 71.6 mg/L in major anions, followed by Cl^-, NO_3^-, and SO_4^{2-}. By contrast, Ca^{2+} exhibited the highest median value of 28.1 mg/L in major cations, followed by Na^+, K^+, and Mg^{2+}. Among trace metal(loid)s in groundwater, Fe, Mn, Ba, and Zn exhibited median values between 0.01 and 0.1 mg/L, while others exhibited median values below 0.002 mg/L (Huang et al., 2018a). Elevated values of groundwater NH_4^+, I^-, and NO_2^- were found in the PRD (Zhang et al., 2020). Nineteen inorganic chemicals in groundwater in this area exceeded allowable limits in comparison with the standards of Chinese groundwater quality [General Administration of Quality Supervision Inspection and Quarantine of the People's Republic of China (GAQSIQPRC), 2017]. Table 18.1 shows the proportions of groundwaters with the values of chemicals exceeded allowable limits (PEL) of China. The PEL values of Mn, Fe, NH_4^+, and I^- are above 10%; NO_3^-, As, Pb, NO_2^-, Cl^-, Na^+, TH, and Be with the PEL values are in a range of 1%–10%; and others with the PEL values are below 1%.

Eighteen organic chemicals were found in groundwater in this area (Table 18.1). About 5% groundwater was detected one or more organic chemicals; in contrast, nearly 95% groundwaters were free of organic chemicals. A maximum number of 10 organic chemicals were found in one groundwater sampling site. Groundwater naphthalene in this area showed the highest detection rate of 2.51% in all detected organic chemicals. Another two organic chemicals, including chloroform and 2-methylnaphthalene, showed the detection rates in a range of 1%–2%, while others showed the detection rates below 1%. Eight organic chemicals, including 1,3,5-trimethylbenzene, trichloroethylene, *o*-xylene, 1,2,4-trimethylbenzene, chloroform, naphthalene, *n*-propylbenzene, and 2-methylnaphthalene, in groundwater in this area showed the highest values of >10 μg/L. Except for the detected organic chemicals without allowable limits, the

TABLE 18.1 Concentrations and PEL values of groundwater chemicals in the Pearl River Delta.

Item	Min	Med	Max	Detection rate (%)	Allowable limits	PEL (%)
Mn (mg/L)	—	0.05	3.41		0.1	35.6
Fe (mg/L)	—	0.06	41.10		0.3	19.3
NH_4^+ (mg/L)	—	0.02	60.00		0.64	15.0
I^- (µg/L)	—	—	1620		80	11.3
NO_3^- (mg/L)	—	19.7	333.5		88.6	7.3
As (µg/L)	—	0.5	303.0		10	6.8
Pb (µg/L)	—	1.4	50.0		10	4.8
NO_2^- (mg/L)	—	0.01	33.20		3.29	3.0
Cl^- (mg/L)	2.2	29.9	2043		250	2.8
Na^+ (mg/L)	0.5	19.1	1009		200	2.0
TH (mg/L)	4.9	93.5	1041		450	1.8
Be (µg/L)	—	—	38.0		2	1.3
SO_4^{2-} (mg/L)	—	18.1	324.3		250	0.5
F^- (mg/L)	—	0.08	1.29		1	0.5
Ni (µg/L)	—	—	26.5		20	0.5
Ba (mg/L)	—	0.05	0.86		0.7	0.3
Zn (mg/L)	—	0.02	2.55		1	0.3
Se (µg/L)	—	0.1	12.2		10	0.3
Hg (µg/L)	—	—	1.8		1	0.3
HCO_3^- (mg/L)	2.9	71.6	748.1			
Ca^{2+} (mg/L)	1.6	28.1	262.5			
Mg^{2+} (mg/L)	—	4.1	118.5			
K^+ (mg/L)	0.2	10.9	119.9			
pH	3.7	6.1	7.7			
Naphthalene (µg/L)	—	—	12.0	2.51	100	0
Chloroform (µg/L)	—	—	18.0	1.25	60	0
2-Methylnaphthalene (µg/L)	—	—	10.2	1.00		
Benzene (µg/L)	—	—	2.5	0.75	10	0
Ethylbenzene (µg/L)	—	—	2.4	0.75	300	0
o-Xylene (µg/L)	—	—	21.3	0.75	500	0
Bromodichloromethane (µg/L)	—	—	3.2	0.75	60	0
Toluene (µg/L)	—	—	8.4	0.50	700	0
m-,p-Xylene (µg/L)	—	—	2.7	0.50	500	0
Isopropylbenzene (µg/L)	—	—	3.2	0.50		
n-Propylbenzene (µg/L)	—	—	11.7	0.50		
1,3,5-Trimethylbenzene (µg/L)	—	—	59.0	0.50		
1,2,4-Trimethylbenzene (µg/L)	—	—	19.3	0.50		
1,1-Dichloroethylene (µg/L)	—	—	2.3	0.50	30	0
1,1,1-Trichloroethane (µg/L)	—	—	1.7	0.25	2000	0
Dibromochloromethane (µg/L)	—	—	1.0	0.25	100	0
Trichloroethylene (µg/L)	—	—	58.0	0.25	70	0
cis-1,2-Dichloroethylene (µg/L)	—	—	7.2	0.25	50	0

—, Below the detection limit; allowable limits: data from General Administration of Quality Supervision Inspection and Quarantine of the People's Republic of China (GAQSIQPRC) (2017); PEL, proportions of groundwaters with the values of chemicals exceeded allowable limits of China.

values of other detected organic chemicals in groundwater in this area were less than allowable limits in comparison with the Chinese groundwater quality standards [General Administration of Quality Supervision Inspection and Quarantine of the People's Republic of China (GAQSIQPRC), 2017].

18.4.2 Groundwater quality and main impact chemicals

The fuzzy synthetic evaluation approach was employed to assess the groundwater quality in this area in one of our previously studies (Zhang et al., 2019), and the groundwater quality was classified into five classes (classes I, II, and III are good quality, classes IV and V are poor quality). The results showed that groundwaters with classes I, II, III, IV, and V appeared in 54.4%, 7.3%, 21.3%, 3.5%, and 13.5%, respectively. In other words, more than 80% groundwater in this area was shown good quality. The poor-quality groundwater (classes IV and V) did not occur in karst aquifers but appeared in 9.1% of fissured aquifers and in 17.0% of granular aquifers, respectively. The proportions of poor-quality groundwater in fissured aquifers in UAs and PUAs were significantly higher in comparison with those in NUAs. Similarly, the proportion of poor-quality groundwater in granular aquifers in NUAs was also significantly lower in comparison with those in UAs and PUAs.

The main impact chemicals for poor-quality groundwater in various aquifers in this area were identified in one of our previous studies by using a principal component analysis (Zhang et al., 2019). In fissured aquifers, groundwaters with poor quality in UAs and NUAs were mainly due to the high concentrations of NO_3^- and I^-, respectively. By contrast, high levels of NO_3^- and Pb were mainly responsible for the groundwater with poor quality in fissured aquifers in PUAs. In granular aquifers, high levels of Mn and Fe were mainly responsible for the groundwater with poor quality in granular aquifers in UAs. Groundwater with poor quality in granular aquifers in PUAs was mainly because of the high concentrations of I^-, Ni, and Mn. By contrast, groundwater with poor quality in granular aquifers in NUAs was mainly due to the high values of Mn, As, and I^-. In addition, Zhang et al. (2019) also investigated the main impact chemicals for poor-quality groundwater in UAs developed at different periods in the PRD. Groundwater with poor quality in UAs developed before 1988 was mainly because of the high concentration of Mn. High levels of NH_4^+, NO_2^-, and Mn were mainly responsible for the groundwater with poor quality in UAs developed in 1988−98. By contrast, high concentrations of Fe, Mn, and Hg were mainly responsible for the groundwater with poor quality in UAs developed in 1998−2006.

18.4.3 Groundwater contamination

As mentioned previously, six heavy metal(loid)s (e.g., Mn, Fe, As, Pb, Ni, and Hg) and four other inorganic chemicals (e.g., NO^-_3, NH^+_4, NO^-_2, and I^-) were the main impact chemicals for groundwater with poor quality in various areas. Thus as far as groundwater contamination in the PRD is concerned, this chapter would focus on the abovementioned chemicals. Besides, some detectable organic contaminants in groundwater were also discussed.

18.4.3.1 Arsenic, manganese, and iron contamination in groundwater

As shown in Table 18.2, high-As (>0.01 mg/L) groundwater did not occur in karst aquifers but appeared in 9.3% of granular aquifers and in 2.3% of fissured aquifers. Similarly, the proportion of high-Mn (>0.1 mg/L) groundwater occurred in granular aquifers was approximately double and four times those in fissured and karst aquifers, respectively; the proportion of high-Fe (>0.3 mg/L) groundwater occurred in fissured aquifers was about 0.5 times that in granular aquifers and did not appear in karst aquifers. These indicate that hydrogeological conditions are probably important factors affecting the distributions of As, Mn, and Fe in groundwater in this area. Wang et al. (2012) reported that fine-grained Quaternary sediments related to granular aquifers in this area are enriched with As and Fe/Mn (oxyhydr)oxides, and the solid As is mainly bound to Fe/Mn (oxyhydr)oxides. Meanwhile, the younger marine stratum overlaying granular aquifers is enriched with organic matter (Jiao et al., 2010). Thus a reducing condition in marine strata in the PRD is formed resulting from the decomposition of organic matter, and the reduction of Fe/Mn (oxyhydr)oxides and the release of As adsorbed onto Fe/Mn (oxyhydr)oxides in aquifers occur. In this case, more proportion of dissolved Fe/Mn and released As from Quaternary sediments will enter into groundwater (Huang et al., 2018b). This is supported by the results reported by Hou et al. (2018) that high concentrations of Fe, Mn, and NH_4^+ (mineralization of organic nitrogen) are in favor of the enrichment of groundwater As in granular aquifers in this area. Therefore we believe that the reductive dissolution of Fe/Mn (oxyhydr)oxides and the release of As adsorbed onto Fe/Mn (oxyhydr)oxides are probably responsible for the much more proportion of high-As, -Fe, and -Mn groundwaters in granular aquifers in comparison with those in other two types of aquifers.

TABLE 18.2 PEL of heavy metal(loid)s, inorganic nitrogen compounds, and iodide in various aquifers and in areas with different degrees of urbanization.

PEL (%)	GA	FA	KA[a]	GA-U	GA-P	GA-N	FA-U	FA-P	FA-N
As	9.3	2.3	0	10.3	10.7	6	7.5	0	0
Mn	43.4	22	11.1	46.7	44	37.3	37.5	21.1	13.7
Fe	24	11.4	0	23.4	25	23.9	12.5	10.5	11
Pb	3.5	6.8	11.1	4.7	3.6	1.5	7.5	15.8	4.1
Ni	0.8	0	0	0	2.4	0	0	0	0
Hg	0.4	0	0	0.9	0	0	0	0	0
NO_3^-	7.0	8.3	0	11.3	6.0	1.5	12.5	10.5	5.5
NO_2^-	4.3	0.8	0	5.7	3.6	3.0	2.5	0	0
NH_4^+	20.5	5.3	0	24.3	21.4	13.4	12.5	5.3	1.4
I^-	13.9	6.8	0	14.9	13.1	13.4	15.0	0.0	4.1

FA-N, fissured aquifers in NUAs; *FA-P*, fissured aquifers in PUAs; *FA-U*, fissured aquifers in UAs; *GA, FA,* and *KA,* granular aquifers, fissured aquifers, and karst aquifers, respectively; *GA-N,* granular aquifers in NUAs; *GA-P,* granular aquifers in PUAs; *GA-U,* granular aquifers in UAs; *PEL,* proportions of groundwaters with the values of chemicals exceeded allowable limits of China.
[a]Karst aquifers with high-Mn and -Pb groundwaters were located at PUAs.

According to our field investigation, numerous factories were distributed in UAs in the PRD, and the leakage of industrial wastewater (enrich As and Mn) often occurred in UAs (Sun et al., 2009). For example, some river waters were enriched with high concentrations of As and Mn in the PRD because of the illegal discharge of industrial wastewater (Table 18.3). This indicates that industrialization is one of the important driving forces affecting the occurrence of high-As and -Mn groundwater in the PRD, because the proportions of high-As and -Mn groundwaters occurred in both of granular and fissured aquifers in UAs were much higher in comparison with those in NUAs (Hou et al., 2018, 2020). In addition, note that the sewage irrigation resulted from polluted rivers and the leakage of landfill leachate were also important sources for the occurrence of high-Mn groundwater in NUAs (Hou et al., 2020), because many landfills lacking antiseepage measures have been built in NUAs, and agricultural lands near rivers are often irrigated by sewage (Sun et al., 2009). Unlike high-As and -Mn groundwaters, the occurrence of high-Fe groundwater in this area is almost probably unaffected by the urbanization and industrialization, because the proportions of high-Fe groundwater in various areas with different degrees of urbanization were roughly equivalent (Table 18.2).

18.4.3.2 Lead, nickel, and mercury contamination in groundwater

As shown in Table 18.2, high-Pb (>0.01 mg/L) groundwater appeared in 11.1% of karst aquifers and were 3.2 and 1.6 times those in granular aquifers and fissured aquifers, respectively. This indicates that karst aquifers have the highest groundwater vulnerability for anthropogenic Pb contamination, followed by fissured aquifers, while granular aquifers have the lowest groundwater vulnerability. The proportions of high-Pb groundwater in both of fissured aquifers and granular aquifers in UAs and PUAs were nearly or more than two times those in NUAs. Similarly, all high-Pb groundwaters in karst aquifers were also found in PUAs. These indicate that the high concentration of Pb in groundwater in this area is mainly attributed to the infiltration of industrial wastewater accompanied by industrialization, because many township–village enterprises producing wastewater with high concentration of Pb were located at UAs and PUAs (Sun et al., 2009), and wastewater from these township–village enterprises was often directly discharged into the nearby ground without treatment in these areas (Huang et al., 2013). Note that high-Pb groundwater in NUAs appeared in 1.5% and 4.1% of granular aquifers and fissured aquifers, respectively, and were nearby rivers or landfills (Zhang et al., 2019). This indicates that high-Pb groundwater in NUAs in this area ascribed mainly to the irrigation with Pb-rich river waters and the leakage of landfill leachate (Huang et al., 2011).

Meanwhile, high-Ni and -Hg groundwaters appeared in less than 1% of granular aquifers and did not appear in fissured aquifers and karst aquifers (Table 18.2). These indicate that broader pollution sources in the PRD were for groundwater Pb in comparison with groundwater Ni and Hg, because the occurrence of high-Pb, -Ni, and -Hg

TABLE 18.3 Concentrations of surface water chemicals.

Item	Minimum	Median	Maximum	Detection rate (%)
Mn (mg/L)	0.06	0.21	0.43	
Fe (mg/L)	0.34	0.79	2.68	
NH_4^+ (mg/L)	0.4	3.6	42	
I^- (mg/L)	–	0.01	0.18	
NO_3^- (mg/L)	–	4.4	31.8	
As (μg/L)	1.9	4	16.7	
Pb (μg/L)	2	6.2	13	
NO_2^- (mg/L)	–	0.02	1.2	
Ni (μg/L)	5.7	15.5	770	
Hg (μg/L)	–	0.1	0.2	
Naphthalene (μg/L)	–	–	1.2	44.4
Chloroform (μg/L)	–	0.6	3.8	55.6
2-Methylnaphthalene (μg/L)	–	–	1.4	22.2
Benzene (μg/L)	–	–	1.8	33.3
Ethylbenzene (μg/L)	–	0.7	8	55.6
o-Xylene (μg/L)	–	–	3.6	44.4
Bromodichloromethane (μg/L)	–	–	1.5	11.1
Toluene (μg/L)	–	2.9	43.9	66.7
m-,p-Xylene (μg/L)	–	–	10	44.4
Isopropylbenzene (μg/L)	–	–	–	0
n-Propylbenzene (μg/L)	–	–	0.9	11.1
1,3,5-Trimethylbenzene (μg/L)	–	–	1.5	22.2
1,2,4-Trimethylbenzene (μg/L)	–	–	3.1	22.2
1,1-Dichloroethylene (μg/L)	–	–	–	0
1,1,1-Trichloroethane (μg/L)	–	–	–	0
Dibromochloromethane (μg/L)	–	–	–	0
Trichloroethylene (μg/L)	–	–	2.6	44.4
cis-1,2-Dichloroethylene (μg/L)	–	–	1.6	44.4

–, Below the detection limit.

groundwaters in the PRD ascribed to the human activities, not the geologic sources (Huang et al., 2018b; Zhang et al., 2019). High-Ni and -Hg groundwaters were located at PUAs and UAs, indicating that the occurrence of high-Ni and -Hg groundwaters in the PRD is likely due to the infiltration of industrial wastewater accompanied by industrialization, because many factories were located at UAs and PUAs, and the infiltration of wastewater from these factories often occurs (Huang et al., 2018b).

18.4.3.3 Nitrate, nitrite, ammonium, and iodide in groundwater

As shown in Table 18.2, high-NO_3^- (>88.6 mg/L) groundwater did not appear in karst aquifers but appeared in 8.3% of fissured aquifers and in 7% of granular aquifers. The proportions of high-NO_3^- groundwater in fissured aquifers in UAs and PUAs were 2.3 and 1.9 times those in NUAs. Similarly, the proportions of high-NO_3^- groundwater in granular aquifers in UAs and PUAs were 7.5 and 4.0 times those in NUAs. In addition, Zhang et al. (2020) reported that the leakage of sewage containing high concentrations of nitrogen often occurs in UAs and PUAs in this area. These indicate that the occurrence of high-NO_3^- groundwater in the PRD is probably attributed to the infiltration of sewage during the industrialization and urbanization (Zhang et al., 2015). Furthermore, Zhang et al. (2020) investigated the relationship between groundwater NO_3^- concentration and various socioeconomic parameters in the PRD by using the principal component analysis. The results showed that the infiltration of sewage resulting from township–village enterprises is mainly responsible for the high-NO_3^- groundwater in the PRD on a regional scale. In addition, high-NO_3^- groundwater in the PRD appeared in about 6% of NUAs, including agricultural lands, indicating that the use of fertilizers and the irrigation of sewage from polluted rivers were also important sources affecting the occurrence of high-NO_3^- groundwater in the PRD, because a large amount of nitrogen-rich fertilizers was used in agricultural lands of the PRD, and nitrogen-rich sewage irrigation often occurred in agricultural lands where polluted rivers nearby (Sun et al., 2009). Note that more high-NO_3^- groundwater was in UAs developed after 1998 in comparison with those in UAs developed before 1998, indicating that housing construction is also an important driving force for NO_3^- pollution in groundwater in newly UAs in this area (Zhang et al., 2020).

Unlike NO_3^-, the proportions of high-NO_2^- (>3.29 mg/L) and high-NH_4^+ (>0.64 mg/L) groundwaters in fissured aquifers were much lower in comparison with those in granular aquifers (Table 18.2), indicating that granular aquifers in favor of denitrification while fissured aquifers in favor of nitrification, because granular aquifers showed much lower Eh and DO values in comparison with those in fissured aquifers (Hou et al., 2020). In granular aquifers the proportion of high-NO_2^- groundwater in UAs was 1.6 times and 1.9 times those in PUAs and NUAs, respectively. Similarly, the proportions of high-NH_4^+ groundwater in both granular aquifers and fissured aquifers were increased with the increase of urbanization levels (Table 18.2). These indicate that urbanization should be a main driving force affecting the occurrence of high-NO_2^- and -NH_4^+ groundwaters in the PRD. Correspondingly, the leakage of domestic sewage was more often with the increase of urbanization levels in this area, and domestic sewage is commonly enriched with NH_4^+ and NO_2^- (Sun et al., 2009; Zhang et al., 2019). Therefore the leakage of domestic sewage accompanied by urbanization is mainly responsible for the much higher proportions of high-NH_4^+ and -NO_2^- groundwaters in UAs in comparison with those in NUAs in the PRD. In addition, the mineralization of organic nitrogen in marine sediments is another main driving force for the occurrence of high-NH_4^+ groundwaters in granular aquifers, because marine sediments in the PRD are characterized by anoxic environments and contain abundant organic nitrogen, which convert it to NH_4^+ and HCO_3^- (Jiao et al., 2010). Correspondingly, the proportion of high-NH_4^+ groundwater in granular aquifers in NUAs where the leakage of domestic sewage lacks was still more than 10% (Table 18.2). Similarly, Zhang et al (2020) reported that the mineralization of organic nitrogen in carbon-rich strata mainly responsible for the high levels of NH_4^+ in fissured aquifers was enhanced by a relatively anoxic environment due to the urbanization.

High-I^- (>0.08 mg/L) groundwater in granular aquifers was more than two times that in fissured aquifers but did not appear in karst aquifers (Table 18.2), indicating that hydrogeological conditions are probably main factors affecting the distribution of high-I^- groundwater in the PRD. Jiao et al. (2010) reported that marine strata overlaying granular aquifers are characterized by abundant organic matter such as iodine-rich organic matter and reducing environment in this area. Wang et al. (2012) reported that fine-grained Quaternary sediments related to granular aquifers in the PRD are enriched with Fe/Mn (oxyhydr)oxides, and Fe/Mn (oxy)hydroxides are commonly considered as the primary reservoirs for coastal sediment iodine because they have high adsorption capacities for iodine (Shetaya et al., 2012). In this case the decomposition of organic matter and the reductive dissolution of Fe/Mn (oxy)hydroxide in Quaternary sediments of the PRD would occur. Thus it can be concluded that the reductive dissolution of iodine-loaded Fe/Mn (oxy) hydroxides and the decomposition of iodine-rich organic matter in sediments are probably responsible for the occurrence of high-I^- groundwater in granular aquifers (Huang et al., 2020b). In addition, the leakage of wastewater accompanied by urbanization may also be important driving forces for high-I^- groundwater in UAs of the PRD, because the leakage of wastewater in UAs of the PRD is common (Huang et al., 2018b), and some surface water samples with high levels of I^- were collected from polluted rivers indicating that the wastewater in the PRD often enriches iodide (Table 18.3). However, the leakage of wastewater accompanied by urbanization is not the main origin for high-I^- groundwater in the PRD, because the proportions of high-I^- groundwater in granular aquifers in UAs and PUAs were close to those in NUAs (Table 18.2), where the wastewater leakage is scarce.

18.4.3.4 Organic contaminants in groundwater

Huang et al. (2018b) reported that the detection rates of organic contaminants in groundwater in UAs and PUAs were much higher than those in NUAs. Nearly, 86% groundwater samples with detected organic contaminants were located at UAs and PUAs. Thus the urbanization and industrialization are probably responsible for the occurrence of organic contaminants in groundwater in the PRD, because a large number of factories were built on the UAs and PUAs and discharged wastewater into nearby ground illegally (Sun et al., 2009). Moreover, the detection rates of groundwater organic contaminants in UAs developed at different periods were further investigated by Huang et al. (2018b). Results showed that the detection rates of groundwater organic contaminants in UAs developed during 1988−98 and 1998−2006 were markedly higher than those developed before 1988. In addition, Sun et al. (2009) reported that many more factories in the PRD were built on the newly formed UAs developed after 1988 in comparison with that on the UAs developed before 1988. Thus it can be speculated that groundwater organic contaminants in the PRD mainly originated from the industrialization, instead of from other human activities.

18.5 Conclusion

In the PRD, 19 inorganic components in some groundwater samples exceeded allowable limits of China and the PELs of 6 inorganic chemicals, including Mn, Fe, NH_4^+, I^-, NO_3^-, and As, were more than 5%. Eighteen organic chemicals were found in groundwater in this area. Groundwater naphthalene in this area showed the highest detection rate of 2.51% in all detected organic chemicals. About 5% groundwater was detected organic contaminants. Poor-quality groundwater appeared in 17.0% of granular aquifers and in 9.1% of fissured aquifers, respectively. In granular aquifers, poor-quality groundwaters in UAs, PUAs, and NUAs were mainly due to the high concentrations of Mn + Fe, I^- + Ni + Mn, and Mn + As + I^-, respectively. In contrast, poor-quality groundwaters in fissured aquifers in UAs, PUAs, and NUAs were mainly due to the high concentrations of NO_3^-, NO_3^- + Pb, and I^-, respectively.

The reductive dissolution of Fe/Mn (oxyhydr)oxides and the release of As adsorbed onto Fe/Mn (oxyhydr)oxides are mainly responsible for the occurrence of high-As, -Fe, and -Mn groundwaters in the PRD. In addition, the leakage of industrial wastewater accompanied by industrialization is also an important factor controlling the occurrence of high-As and -Mn groundwaters in UAs and PUAs. The infiltration of industrial wastewater accompanied by industrialization is mainly responsible for the occurrence of high-Pb, -Ni, and -Hg groundwaters in the PRD. In addition, the irrigation with Pb-rich river waters and the leakage of landfill leachate are also important factors for high-Pb groundwater in NUAs. The leakage of domestic and industrial wastewater is the main driving force for groundwater NO_3^- pollution in the PRD, and the use of fertilizers and the irrigation of sewage from polluted rivers are also important factors controlling the occurrence of high-NO_3^- groundwater in NUAs. Similarly, urbanization accompanied with the leakage of domestic sewage is also an important factor controlling the occurrence of high-NH_4^+ and -NO_2^- groundwaters in UAs, while the main one for the occurrence of high-NH_4^+ groundwater is the mineralization of organic nitrogen in strata. The reductive dissolution of iodine-loaded Fe/Mn (oxy)hydroxides and the decomposition of iodine-rich organic matter in sediments are probably responsible for the occurrence of high-I^- groundwater in this area. Groundwater organic contaminants in the PRD originated mainly from the industrialization.

Acknowledgments

This research was supported by the China Geological Survey Grant (DD20160309, DD20160308) and the Fundamental Research Funds for Central Public Welfare Research Institutes, CAGS (SK202005, SK201611, SK201410).

References

Cheung, K.C., Poon, B.H.T., Lan, C.Y., Wong, M.H., 2003. Assessment of metal and nutrient concentrations in river water and sediment collected from the cities in the Pearl River Delta, South China. Chemosphere 52, 1431−1440.

General Administration of Quality Supervision Inspection and Quarantine of the People's Republic of China (GAQSIQPRC), 2017. Standard for Groundwater Quality. Standards Press of China, Beijing.

Hou, Q., Sun, J., Jing, J., Liu, C., Zhang, Y., Liu, J., et al., 2018. A regional scale investigation on groundwater arsenic in different types of aquifers in the Pearl River Delta, China. Geofluids 2018, Article ID 3471295.

Hou, Q., Zhang, Q., Huang, G., Liu, C., Zhang, Y., 2020. Elevated manganese concentrations in shallow groundwater of various aquifers in a rapidly urbanized delta, South China. Sci. Total Environ. 701. Available from: https://doi.org/10.1016/j.scitotenv.2019.134777.

Huang, G., Liu, C., Li, L., Zhang, F., Chen, Z., 2020b. Spatial distribution and origin of shallow groundwater iodide in a rapidly urbanized delta: a case study of the Pearl River Delta. J. Hydrol. 585, 124860.

Huang, G., Liu, C., Sun, J., Zhang, M., Jing, J., Li, L., 2018a. A regional scale investigation on factors controlling the groundwater chemistry of various aquifers in a rapidly urbanized area: a case study of the Pearl River Delta. Sci. Total Environ. 625, 510−518.

Huang, G., Liu, C., Zhang, Y., Chen, Z., 2020a. Groundwater is important for the geochemical cycling of phosphorus in rapidly urbanized areas: a case study in the Pearl River Delta. Environ. Pollut. 260. Available from: https://doi.org/10.1016/j.envpol.2020.114079.

Huang, G., Sun, J., Zhang, Y., Chen, Z., Liu, F., 2013. Impact of anthropogenic and natural processes on the evolution of groundwater chemistry in a rapidly urbanized coastal area, South China. Sci. Total Environ. 463–464, 209–221.

Huang, G., Sun, J., Zhang, Y., Jing, J., Zhang, Y., Liu, J., 2011. Distribution of arsenic in sewage irrigation area of Pearl River Delta, China. J. Earth Sci. 22 (3), 396–410.

Huang, G., Zhang, M., Liu, C., Li, L., Chen, Z., 2018b. Heavy metal(loid)s and organic contaminants in groundwater in the Pearl River Delta that has undergone three decades of urbanization and industrialization: distributions, sources, and driving forces. Sci. Total Environ. 635, 913–925.

Jiao, J., Wang, Y., Cherry, J.A., Wang, X., Zhi, B., Du, H., et al., 2010. Abnormally high ammonium of natural origin in a coastal aquifer-aquitard system in the Pearl River Delta, China. Environ. Sci. Technol. 44 (19), 7470–7475.

Lu, F., Ni, H., Liu, F., Zeng, E., 2009. Occurrence of nutrients in riverine runoff of the Pearl River Delta, South China. J. Hydrol. 376, 107–115.

Mukherjee, A., Scanlon, B., Aureli, A., Langan, S., Guo, H., McKenzie, A., 2020. Global Groundwater: Source, Scarcity, Sustainability, Security and Solutions, first ed. Elsevier, ISBN: 9780128181720.

Shetaya, W.H., Young, S.D., Watts, M.J., Ander, E.L., Bailey, E.H., 2012. Iodine dynamics in soils. Geochim. Cosmochim. Acta 77, 457–473.

Sun, J., Jing, J., Huang, G., Liu, J., Chen, X., Zhang, Y., 2009. Report on the Investigation and Assessment of Groundwater Contamination in the Pearl River Delta Area. The Institute of Hydrogeology and Environmental Geology, Chinese Academy of Geological Sciences, Shijiazhuang, China.

Wang, Y., Jiao, J.J., Cherry, J.A., 2012. Occurrence and geochemical behavior of arsenic in a coastal aquifer–aquitard system of the Pearl River Delta, China. Sci. Total Environ. 427–428, 286–297.

Zhang, F., Huang, G., Hou, Q., Liu, C., Zhang, Y., Zhang, Q., 2019. Groundwater quality in the Pearl River Delta after the rapid expansion of industrialization and urbanization: distributions, main impact indicators, and driving forces. J. Hydrol. 577. Available from: https://doi.org/10.1016/j.jhydrol.2019.124004.

Zhang, M., Huang, G., Liu, C., Zhang, Y., Chen, Z., Wang, J., 2020. Distributions and origins of nitrate, nitrite, and ammonium in various aquifers in an urbanized coastal area, South China. J. Hydrol. 582. Available from: https://doi.org/10.1016/j.jhydrol.2019.124528.

Zhang, Q., Sun, J., Liu, J., Huang, G., Lu, C., Zhang, Y., 2015. Driving mechanism and sources of groundwater nitrate contamination in the rapidly urbanized region of South China. J. Contam. Hydrol. 182, 221–230.

Chapter 19

Hydrochemical characteristics and quality assessment of water from different sources in Northern Morocco

Lahcen Benaabidate[1], Ahmed Zian[2] and Othman Sadki[3]

[1]*Laboratory of Functional Ecology and Environment Engineering, University of Sidi Mohammed Ben Abdellah, Fez, Morocco,* [2]*National School of Applied Sciences of Al Hoceima, University Abdelmalek Essaadi, Tétouan, Morocco,* [3]*Department of Geochemistry, National Office of Hydrocarbons and Mines, Rabat, Morocco*

19.1 Introduction

The availability of adequate freshwater has become a limiting factor of the quality of life. More than availability, the problem is often the rational use of water than ensure its continuality. In the semiarid and arid regions, water scarcity was always a dominant problem (Ramadan, 2015). In poor arid and semiarid states, water failure encourages migration to cities, increasing stress on basic utilities (Robins and Fergusson, 2014). Furthermore, groundwater sustains rural communities wherever surface water is ephemeral, so keeping people on the land and retaining their livelihoods (Chopra and Gulati, 2001). According to Pereira et al. (2009), water scarcity is an increasing and irreversible problem that derives from a diverse set of causes. The more obvious of these are land use change and desertification, climate cycles and variability, demography and occupation of marginal lands, interception and diversion of surface water, groundwater mining and pollution.

The Northern Morocco, due to its geographical location, is subject to a semiarid climate. This climatic characteristic makes water a resource in high demand and poorly distributed both in time and in space. Indeed, the problem of water shortage in Northern Morocco is becoming more and more obvious despite the fact that Morocco adopting advanced water policy. The rapid increase in demand for water, particularly for agricultural and domestic needs, has led to an intense exploitation of the water resources available in aquifers of this region.

The global groundwater is undergoing immense stress in recent times (Mukherjee et al., 2020). Indeed, in the North of Morocco, although the region's water resources are relatively limited, they are found to be overexploited. This will certainly have negative repercussions on the chemical quality of the water. This work aims to assess the source waters quality in terms of hydrochemical and geochemical characteristics with some water-quality indexes and methods.

Groundwater, by its perpetual circulation through the voids of the rock and by its physicochemical properties, plays a vehicle role of dissolved elements and temperature. Geochemical investigation revealed that the most of those sources are hosted in carbonate and evaporite allochthonous rocks (Benaabidate, 2000).

The Northern Morocco and according to its complex geology is an area where rise different sources. The study area is divided into four geological domains (Cirak, 1987) (Fig. 19.1): (1) the Rif basin, consisting of an internal zone on the Mediterranean side and another external overlooking the Middle Atlas and called Pre-rif. This area contains different geological formations, such as Jurassic and Cretaceous limestones, Triassic evaporites, and Miocene marls and (2) the South Rif corridor bypasses the Rif area from the Atlantic to the Touahar pass. It encompasses three very distinct morphological units; the Rharb plain to the West, the Fez-Meknes plain (Saïss) and the Fes-Taza plain to the East. This corridor is a large depression filled with Miocene marls with a thickness up to 1000 m and overlay the Liassic hydrogeological aquifer. (3) The Rides basin made essentially of reliefs dominating the Neogenic deposits of the Rharb and Saïss plains and the Pre-rif domain. This basin rocks are mainly made of Jurassic limestones and sandstones and finally, (4) the Middle Atlas domain, which is covered by Jurassic carbonate formations, constituted of limestones and dolomites.

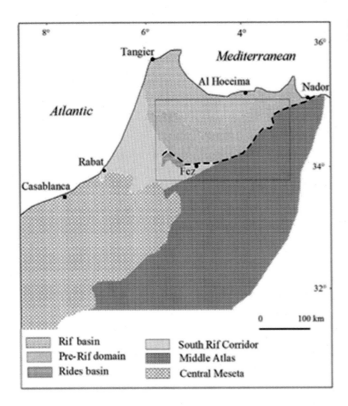

FIGURE 19.1 Geological setting of the Northern Morocco (Cirak, 1987).

19.2 Material and methods

Physical parameters, pH, electrical conductivity, and total dissolved solids (TDS) were measured in the field using a conductivity meter and a pH meter. Major elements and silica have been analyzed in the Laboratory of Geochemistry of the Ministry of Water, Energy, and Mines in Rabat, Morocco (Table 19.1).

The analyzes of dissolved gases in the source waters of some warm sources of the North of Morocco were made within the framework of the research project of native sulfur in Morocco launched earlier by the Ministry of Energy and Mines. The campaign of sampling and analysis of these gases interested only a few interesting emergences previously chosen. These analyses were done only on dissolved gases. This study was carried on 42 sources located in the North of Morocco (Fig. 19.2).

19.3 Hydrochemistry

19.3.1 Source water chemical facies

The plotting of the source chemical analyses of the three geographic domains on Piper diagram (Fig. 19.3) shows chemical facies that differ as much from one domain to another, as within the same domain:

1 Sodium chloride facies with three subfacies:
   a. hyperchlorinated sodium facies (Cl^-, Na^+); sources 7, 11, and 14;
   b. carbonate facies, which contains the sources 1, 6, 10, 12, 18, 29, and 38; and
   c. sulfate facies; sources 15, 26, 28, 30, 33, 35 40, 41, and 42.
2 Calcium bicarbonate (Ca^{2+}, HCO_3^-); sources 2, 3, 4, 9, 16, 17, 19 20, 21, 22, 23, 24, 32, and 34;
3 Bicarbonate sodium facies (Na^+, HCO_3^-); sources 27 and 39;
4 Calcium chloride facies (Cl^-, Ca^{2+}); sources 5, 8, 25, 36, and 37; and
5 Calcium sulfated facies (SO_4^{2-}, Ca^{2+}); sources nos. 13 and 31.

Fig. 19.3, although it allows a global visualization of the samples, did not clearly evoke the geographical distribution of the water samples used in this study. The obtained facies are very heterogeneous with a predominance of sodium chlorinated waters. These waters would have acquired this facies as a result of possible leaching of the Triassic salt

TABLE 19.1 The source chemical analyses.

Sources	No.	X	Y	T (°C)	pH	Cond (µs/cm)	TDS	NA$^+$	K$^+$	Mg^{2+}	Ca^{2+}	HCO$_3^-$	SO$_4^{2-}$	Cl$^+$	SiO$_2$	IB (%)
Matmata	1	580.2	388	31	7.2	311	1283.9	315.1	3.9	36.5	64.1	347.8	24	482.2	10	0.5
Bouzemlane	2	578	379.85	25	7	1443	650.1	59.8	1.2	38.9	62.1	372.2	9.6	99.3	7	1.0
Fellaj	3	575	383.5	27	6.8	674	533.5	36	1.5	6	105	304.76	12.04	60.88	7.3	2.6
Skhira	4	575.2	382.4	25	6.9	1258	516.2	36	3	7	104	281.33	13.2	64.62	7	5.0
Sidi Harazem	5	547.69	381.25	33.5	7.2	1260	828.9	130	3.8	36.6	82.4	317	15.3	231	9.4	3.4
Skhinat	6	549.65	382	25.5	7.1	2400	1619.3	346.5	9.8	63.4	113.8	386	43.6	642.5	11.1	1.7
My. Yacoub	7	5201.1	387.7	54	6.7	50200	29254.8	9354.1	242.4	289.4	1076.1	195.3	39.4	17747.7	30.8	1.4
Ain Allah	8	524.35	382.35	45	7.2	571	766.6	51.75	8.211	48.032	100.6	313	79.68	153.187	11.3	1.4
Skhounate	9	503.75	377.6	27	7.1	1324	681.6	50.6	3.9	23.1	102.2	360	33.6	95.7	12.5	0.0
Zerga	10	508.8	377.6	34	7.2	934	1028.1	207	3.9	35.3	66.1	341.7	38.4	322.7	13	0.6
Zalagh	11	542.2	392	34	7.2	8350	4428	1260	100	45	225	298.9	146.45	2324.35	17.1	0.8
Trhat	12	533.4	386.25	27	7.2	1470	1088.9	232.3	3.9	30.4	52.1	329.5	168.1	248.2	21	1.9
My. Idriss	13	488.8	483.5	31	6.7	38300	3018.1	282.9	7.8	107	464.9	280.7	1378.5	475.2	12.8	2.4
Tiouka	14	475.5	404.5	26	7.3	44900	20678.7	6782.7	93.8	317.4	585.2	335.6	235.3	12141.5	16.3	0.0
Outita	15	439.6	392.7	41	6.9	11110	6434.1	1704.3	39.1	124	350.7	292.9	1138.3	2734	28	1.3
Es Skhoun	16	488.7	395.5	25	7.5	750	573.1	50.6	2	25.5	74.1	280.7	33.6	88.6	18	1.6
Anseur	17	497.4	389.9	24	6.9	800	657.7	46	11.7	21.8	98.2	366.1	24	74.5	13	2.2
Maaser	18	496.1	390.25	21.5	6.9	857	725.2	125	2	18	73	256.2	62.5	177.5	11	0.5
Beida	19	488.8	396.15	22	6.9	1485	718.1	60	14	9	117	374.54	31.97	99.26	12.3	0.3
Robbani	20	488.7	397.7	23	7	1296	738	32	6	32	118	455.79	26.96	60.03	7.2	1.7
Sidi Boutmine	21	488.2	402	22	6.9	841	663.2	64	2	4	92	458.64	12.14	22.79	7.6	4.1
Ksob	22	540.6	391.08	21	7.2	657	511.8	27.2	5	14	83	382.84	13.6	18.87	7.3	3.6
Sarij	23	539.3	390.5	20	7	622	590.2	31	2	12	92	411.14	19.18	16.11	6.8	4.2
Fendès	24	538.2	390.3	20	6.9	740	567.6	15	11	9	115	370.64	20.83	19.43	6.7	2.5
Bou Draa	25	472.6	400.6	22.5	7.9	714	691.7	75.9	3.9	30.4	88.2	268.5	91.2	120.6	13	3.0
Khanza	26	456.9	460.6	19.6	6.4	11012	6931.8	2291.3	31	65	360.9	467.9	508	3170	18	7.0
Ghouzzal	27	503.8	469.75	19.8	6.8	1145	852.7	110.3	20	8	84.3	402.6	190	21.3	16.4	4.6

(Continued)

TABLE 19.1 (Continued)

Sources	No.	X	Y	T (°C)	pH	Cond (µs/cm)	TDS	NA$^+$	K$^+$	Mg^{2+}	Ca^{2+}	HCO$_3^-$	SO$_4^{2-}$	Cl$^+$	SiO$_2$	IB (%)
Kebbata	28	579.8	440.6	21.2	7.2	16100	4671	1450	19	79	275	257	645	1925	12.5	7.6
Merset	29	543.4	428.5	19.7	7.1	4720	1521.2	310.5	15.6	77.9	105.7	432.6	125.3	441.2	11.5	7.2
Mekkouch	30	512.9	421.25	21	6.7	3770	2474.3	800	13.9	57	21.9	315.3	459	791	15.2	5.0
Mourra	31	539	420.03	22	6.7	3700	3739.1	354.2	3.9	71.7	671.3	219.7	1878	521.3	19	2.2
Rmel	32	522.6	448.1	26	6.9	800	662.4	41.4	1.2	7.3	132.3	378.3	14.4	74.5	13	2.4
Tarmast	33	630.8	430.7	18.6	6.6	31500	19268.2	6249.3	11.5	107	998	514	2925	8423.1	15	3.7
M'hamed	34	590.9	447.8	19.4	7	940	783.2	75	9	18.5	115	426	78.1	42	18	4.7
Mouilha	35	511.8	423.9	18.7	6.8	20900	12618.3	3550	27	91	988	338	2252	5325	25	2.2
Kerma	36	534	423.5	21.5	6.8	1930	1627.9	117.3	89.9	45	240.5	488.2	360.2	251.8	35	1.1
Basra	37	457.2	467.2	20	6.3	1620	1081.9	34.6	67	33	132.1	294.3	58	197.4	12.5	3.8
Zeroual	38	519.6	454.1	22.3	6.6	12754	7859.9	2250	270	43	259	402.6	100	4522.7	12.6	5.8
Tafrant	39	518.2	454.6	22.5	6.8	1750	1052.5	251	6	9	64.2	431	194	84.3	13	5.4
Bouabelli	40	543.3	426.05	22	6.7	770	1773.2	530	20	15	29	195.2	339	630	15	3.3
Harra	41	543.1	426.1	20	6.9	23232	1435.2	241.6	19	61	94	241.2	567.2	201.3	11.9	1.8
Kebrita	42	514.9	420.9	21	6.8	9590	7971.2	2450	250	30	176	414.8	607	3976	13.7	2.9

FIGURE 19.2 Situation of studied sources.

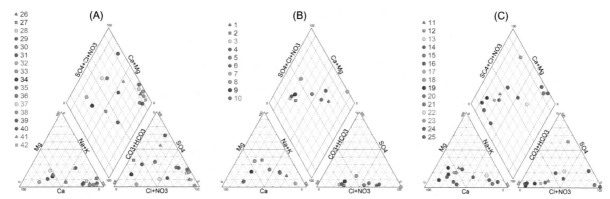

FIGURE 19.3 Chemical facies of source waters. (A) Rif Basin, (B) South Rifan Corridor, (C) Rides Basin.

deposits which mainly mark out the Rif basin (Fig. 19.3A). The calcium bicarbonate facies comes second in the order of importance. We note, for this facies, that apart from the two sources nos. 32 and 34, the other sources belong geographically to the South Rifan corridor (Fig. 19.3B) and to the area of the Rides basin (Fig. 19.3C).

19.3.2 Quality of source waters for irrigation

The water mineral salts have effects on the soil and plants. These salts can disturb the physical development of plants by absorbing water and act on the osmotic or chemical process through metabolic reactions. The effects of salts cause changes in its structure, permeability, and its aeration, indirectly affecting the development of plants (Bouchaou, 1988).

The irrigation water salinity quality of is generally expressed by relative convenience classes. Most classifications take into account the conductivity (expressing all of the dissolved salts) and especially the sodium content. The concentration of sodium is, therefore, important in the classification of irrigation water because it reacts with the soil and reduces its permeability (Djabri, 1987).

To quality of these waters for irrigation was assesses by using the Richards (1954) classification. This classification distinguishes different types of water, taking into account the electrical conductivity (μs/cm) at 25°C and the sodium adsorption ratio (S.A.R) calculated as follows:

$$S.A.R = \frac{Na^+}{\sqrt{(Ca^+ + Mg^{2+})/2}}$$

The concentrations of the ions are in meq/L.

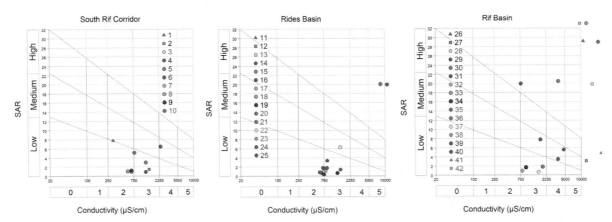

FIGURE 19.4 Application Richards diagram.

The plot of all water samples on the Richards diagram (Fig. 19.4) highlights that about 79% of all samples seem to be not suitable for irrigation. The C2-S1 class indicating poor quality water and highly mineralized, which can only be suitable for species well tolerant to salts and on well-drained and leached soils. This category contains four sources in the South Rif corridor and five sources in the Rides basin. The remaining sources and the Rif domain are in the category of waters of poor quality (C3, C4, and C5) because of their high mineralization. So, these waters are not recommended for use in the agricultural field.

19.4 Control of chemical element concentrations

In order to identify the factors and phenomena that control the chemical composition of the source waters in Northern Morocco, the study was focused on the relationships between the contents of major elements as well as the reactions that involve these elements and the minerals which can be present in the corresponding hydrothermal systems. The use of concentrations between major elements allows highlighting the origin of these elements and specify the factors that control their concentrations.

19.4.1 Binary ion correlations

19.4.1.1 Na^+-Cl^- correlation

In the study area, chlorine and sodium are the dominant anion and cation for 15 samples. The logmithmic diagram $logm(Na^+)-logm(Cl^-)$ (Fig. 19.5) shows that the representative points of the studied source waters are on a line with slope $+1$ ($r = 0.94$). This suggests that Na^+ and Cl^- share practically the same origin. These would be introduced into solution by dissolving the halite.

19.4.1.2 K^+-Cl^- correlation

The K^+-Cl^- diagram (Fig. 19.6) does not show a very good correlation ($r = 0.67$). Since Cl^- would be exclusively linked to the dissolution of halite, K^+ and Cl^- would not come from the dissolution of sylvite (KCl) which is, moreover, very rare in the sedimentary series of the Northern Morocco region (Robins and Fergusson, 2014). This would indicate that the K^+ concentration would be controlled by an equilibrium reaction, such as that described by Shikazono (1978) and Ellis (1969):

$$Feldspath - K + Na^+ \leftrightarrow Albite + K^+$$

19.4.1.3 $Ca^{2+}-HCO_3^-$ correlation

The Fig. 19.7 illustrates the correlation between the content of calcium and that of Bicarbonates in mmol/L. The graphic highlights that these two ions are not correlable in this study ($r = -0.22$).

This no correlation could be related to the fact that these two elements are not issued from the same origin. The sources that have a Ca^{2+}/HCO_3^- molar ratio close to/or slightly less than 0.5 are mainly due to the dissolution of

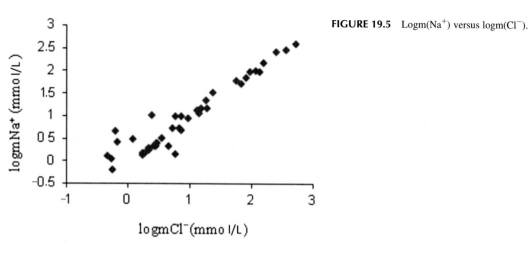

FIGURE 19.5 Logm(Na^+) versus logm(Cl^-).

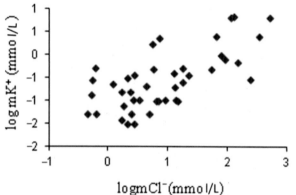

FIGURE 19.6 Logm(K^+) versus logm(Cl^-).

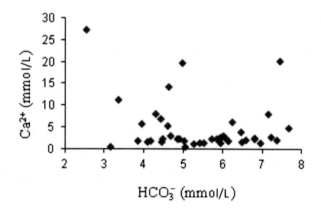

FIGURE 19.7 Logm(Ca^{2+}) versus logm(HCO_3^-).

calcite. Adversely, for those with a ratio greater than 0.5, calcium would be attributed to the dissolution of calcite as well as other calcium minerals such as gypsum.

19.4.1.4 Ca^{2+}–SO_4^{2-} correlation

Fig. 19.8 ($r = 0.6$) shows that the majority of samples are projected around the line Ca^{2+}/SO_4^{2-} which slope is equal to 1. For this group of sources the Ca^{2+} and SO_4^{2-} ions come, therefore, from the dissolution of gypsum. For the remaining sources that Ca^{2+}/SO_4^{2-} ratio is greater than 1, the excess of calcium would probably result from the dissolution of calcite, while those with a Ca^{2+}/SO_4^{2-} ratio <1, the excess of ion SO_4^{2-} could come from another sulfated mineral, probably anhydrite.

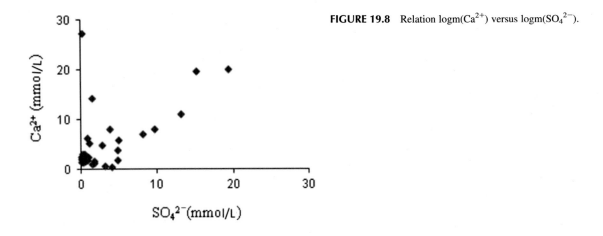

FIGURE 19.8 Relation logm(Ca^{2+}) versus logm(SO$_4^{2-}$).

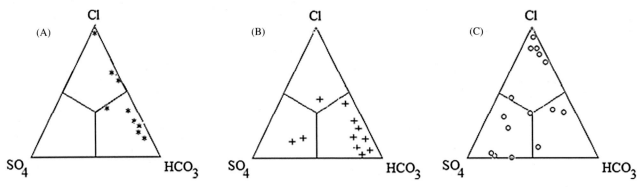

FIGURE 19.9 Cl–SO$_4$–HCO$_3$ diagram.

19.4.2 Cl–SO$_4$–HCO$_3$ diagram

The Cl–SO$_4$–HCO$_3$ diagram is a graphic representation that allows the classification of waters according to the Cl$^-$, SO$_4^{2-}$, and HCO$_3^-$ anions into three families (chlorinate, sulfate, and bicarbonate waters) which can provide information on their origin. Giggenbach (1991) showed that the majority of geochemical techniques give good results for chlorinated, bicarbonate waters placed on the Cl$^-$–HCO$_3$ axis of the diagram, and that the composition of sulfated water is generally affected by water–rock interactions in surface areas.

The application of this diagram to the sources of the Northern Morocco (Fig. 19.9) highlights that most of the South Rif corridor waters (Fig. 19.9A) is of the bicarbonate type. This is linked to the carbonate lithological nature (limestone and dolomite) of the deep Liassic reservoir that constitutes the origin of all these sources (Benaabidate, 2000).

However, the source of My Yacoub is projected into the chlorine pole according to its excessive content of this element. This strong mineralization would be due to a prolonged leaching of the salt deposits of the Pre-rif zone by the rising water from the deep reservoir. The deposits of the Pre-rif saliferous complex of and those of evaporate Triassic induce the sources emerging in the Rifan domain (Fig. 19.9B) to be generally of the chlorinated or sulfate type. Finally, the Rides basin sources present the three types of facies with the dominance of bicarbonate waters (Fig. 19.9C).

19.4.3 Index of base exchange

The index of base exchange (IBE) index was proposed by Schoeller (1965) for describing the geochemical reactions taking place in groundwater. There are substances that absorb and exchange their cations with anions present in groundwater. These substances are called permutolites, for example, clay minerals like kaolinite, illite, chlorite, halloysite, glauconite, zeolites, and organic substances (Thilagavathi et al., 2012). This index allows the determination of exchanges occurring between the aqueous and the solid phases (mainly the clay minerals). Indeed, the clay minerals exchange Na$^+$ and K$^+$ ions for Ca^{2+} and Mg^{2+} ions present in solution.

$$\text{IBE} = m\text{Cl}^- - \frac{m\text{Na}^+ + m\text{K}^+}{m\text{Cl}^-}$$

In this study the calculation of the IBE index gives values all negative. The cation exchange, that would, therefore, be highlighted, is likely given the significant thickness of the clay-marly layer that covers almost all of places where the sources studied emerge.

19.4.4 Water standards and potability

The waters emerging in the North Morocco region are in high demand for domestic consumption. The concentrations of the various chemical elements, involved in groundwater, directly affect the quality and the potability of this water. The World Health Organization established standards for drinking water according to the content of dissolved elements (Bremond and Vuichard, 1973).

These standards are applied to source waters of the study area in order to discuss the concentration of the different elements and their consequences on the potability of the source waters.

19.4.5 Sodium and potassium

According to WHO drinking water should not contain more than 150 mg/L of sodium. Analysis of our water samples has shown that the content of this element is very fluctuating. It can reach 9354.1 mg/L at the source of My Yacoub. Overall, 50% of samples have contents that exceed the standard set by WHO and can thus generate some health hazards including hypertension (UNSCO, 1987 in Younsi, 1994).

Potassium has a laxative effect but without important physiological disturbances except a slight taste that appears from 34 mg/L (Younsi, 1994). The standard for drinking water is 12 mg/L. In the case of the studied waters, only the charged sources in dissolved salts have higher potassium concentrations. These are sources nos. 7, 11, 14, 15, 19, 26, 27, 28, 29, 30, 35, 36, 38, 40, 41, and 42.

19.4.6 Calcium and magnesium

The WHO has set the acceptable limit for calcium at 75 mg/L and at 200 mg/L the maximum admissible concentration. As for magnesium, its maximum admissible content is of the order of 150 mg/L, while its minimum concentration is fixed at 30 mg/L if the water contains more sulfates. For the sources studied, 28% of these sources exceed the limit of 200 mg/L of the element calcium and only two sources (nos. 7 and 14) exceed that fixed for magnesium.

19.4.7 Chlorides

The standards set by the O.M.S. for the chloride ion content in waters intended for human consumption are 250 mg/L (750 mg/L for Morocco). Although high chloride content does not pose a high health risk (except for people with a poor diet) (Younsi, 1994), it gives an unpleasant taste to water and can cause a corrosion of pipes and makes this water unsuitable for irrigation. For the sources of Northern Morocco, 45% exceed the standard set by the WHO, that is, 250 mg/L, and 31% exceed the limit of admissible concentrations, that is, 600 mg/L.

19.4.8 Sulfates and bicarbonates

The drinking standards laid down by the WHO are 250 mg/L of sulfates and the maximum admissible concentration can reach 400 mg/L. As for bicarbonates, the presence of HCO_3^- ions in water gives it a pleasant odor and does not present any risk to human health.

Over the 42 sources listed, only 12 sources have sulfate concentrations greater than 250 mg/L. These sources are nos. 13, 15, 26, 28, 30, 31, 33, 35, 36, 40, 41, and 42.

The following Table 19.2 gives the classification of waters according to their potability according to their contents in major dissolved elements. The potability interval has been extended to the maximum acceptability limit for calcium.

This table illustrates that only 16 sources, about 38% of total sources, meet the drinking water standards set by the WHO and would, therefore, be suitable for water supply.

TABLE 19.2 Source water potability classification according to their major ions contents.

Ions	Sodium	Potassium	Calcium	Magnesium	Choloures	Sulfates
Suitable for irrigation	2, 3, 4, 5, 8, 9, 16–25, 27, 32, 34, 36, 37	1–6, 8, 9, 10, 12, 13, 16–25, 31, 32, 33, 34, 39	1–6, 8, 9, 10, 16–25, 27, 29, 30, 32 34, 37, 39–42	1–6 and 8–13, 15–42	1–5, 9, 10, 12, 13, 17–25, 27, 29, 31, 32, 33, 34, 36, 37, 39, 42	1–12, 14, 16–25, 27, 29, 32, 34, 37–40
Not suitable for irrigation	1, 6, 7, 10–15, 26, 28, 29, 30, 31, 33, 35, 38, 39, 40, 41, 42	7, 11, 14, 15, 26, 27, 28, 29, 30, 35, 36, 37, 38, 40, 41, 42	7, 11, 12, 13, 14, 15, 26, 28, 31, 33, 35, 36, 38	7, 14	6, 7, 8, 11, 14, 15, 16, 26, 28, 30, 35, 38, 40, 41	13, 15, 26, 28, 30, 31, 33, 35, 36, 41, 42

19.5 Principal component analysis

The principal component analysis (PCA) is used in order to interpret and grouping the water quality parameter (Gajbhiye et al., 2015). The physicochemical data measured in the field (pH, TDS, temperature) and the results of the major elements analyses (Na^+, K^+, Ca^{2+}, Mg^{2+}, Cl^-, HCO_3^-, and SO_4^{2-}) and the silica content were treated using the STAT-ITCF computer software.

The purpose of this PCA is to draw conclusions about the association of chemical elements and to characterize the distribution of individuals. To avoid false correlations and interpretations, some samples were eliminated. These eliminated waters generally have a total dissolved charge greater than 10 g/L. This PCA, therefore, brings together 38 individuals or statistical units and 11 variables. For the interpretation of the PCA the basic statistics have been determined. Table 19.3 gives the means and standard deviations of each variable. The correlation coefficients were first calculated for all of the sampled waters (Table 19.4).

This table reveals that the waters have the following characteristics:

- A strong link between TDS and Na^+, K^+, and Cl^-.
- The chlorine contents are correlated with those of the Na^+ and K^+ ions.
- A good correlation between the sulfate and magnesium contents and the calcium.
- No significant correlation between temperature, pH, silica content and other parameters.
- No link between the bicarbonate contents and the other parameters.

19.5.1 Variable space

Fig. 19.10A represents the distribution of the variables on the main plane "correlation circle." From this figure, we deduce that the axis F1 (44.8%) is determined, with the exception of HCO_3^-, by the major elements Na^+, Cl^-, SO_4^{2-}, Mg^{2+}, Ca^{2+}, K^+, and TDS. This allows considering the F1 axis as the mineralization axis. Around this axis the highly mineralized species Na^+, K^+, and Cl^- are opposed to the least mineralized species. Furthermore, the F1 axis is defined by SiO_2 but in less importance. There is opposition, along the same axis F1, between the mineralization and the pH. In fact, the pH and the rate of mineralization are inversely proportional; the lower pHs (acids) favorite rock dissolution and then release into the waters dissolved species that participate in the elevation of water mineralization.

The F2 axis (18.7%) is characterized by temperature and HCO_3^- but in opposite directions. The association between these two elements has no interpretation, because there is a weak correlation between the two ($r = -0.229$).

19.5.2 Individual space

The spatial study of individuals was carried out on the factorial plane F1 and F2. These individuals are, therefore, distributed according to their affinities in domains defined by the variable space (Fig. 19.10B). This configuration allowed envisaging four groups in the F1–F2 axis.

The first group is characterized by the sources of Outita (no. 15), Khanza (no. 26), Zeroual (no. 38), and Kebrita (no. 42). The respective water samples in this group are characterized by strong mineralization (greater than 6 g/L) and by a high content of chlorine and sodium. These sources are shifted downwards because they are pulled by the HCO_3^- pole. The strong mineralization of this group is responsible for the stretch of the F1 axis:

- The second group is represented by the sources of My Driss (no. 13) and Mourra (no. 31). The distribution of this group is made following their excessive contents of calcium and sulfate.

TABLE 19.3 Variable means and standard deviation.

Variables	Means	Standard deviation
TDS	1940.8	2114.6
T	25.2	5.8
pH	7	0.3
SiO$_2$	13.4	5.7
Na	435.3	677.9
K	28.6	59
Mg	36.8	28.1
Ca	147.5	128.4
HCO$_3$	345.5	73.4
SO$_4$	252.9	408.9
Cl	970.6	1131.7

TABLE 19.4 Variable correlation matrix.

	TDS	T	pH	SiO$_2$	Na	K	Mg	Ca	HCO$_3$	SO$_4$	Cl
TDS	1.000										
T	0.105	1.0000									
pH	−0.42	0.195	1.000								
SiO$_2$	0.354	0.123	−0.22	1.000							
Na	0.972	0.071	−0.36	0.284	1.000						
K	0.728	−0.01	−0.34	0.211	0.724	1.000					
Mg	0.549	0.340	−0.21	0.403	0.443	0.095	1.000				
Ca	0.595	0.090	−0.36	0.389	0.416	0.211	0.629	1.000			
HCO$_3$	0.052	−0.23	−0.19	0.066	0.098	0.217	−0.19	−0.13	1.000		
SO$_4$	0.533	0.084	−0.37	0.431	0.367	0.09	0.705	0.858	−0.33	1.000	
Cl	0.969	0.120	−0.34	0.262	0.981	0.81	0.416	0.419	0.106	0.315	1.000

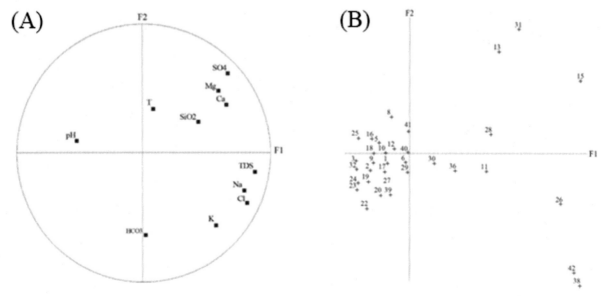

FIGURE 19.10 Variable distribution (A) and Biplot scores (B).

- The third group is intermediate between the first two groups. It includes the sources of Zalagh (no. 11) and Kebbata (no. 28). These sources have TDS values between those of the first group and the second group. Their mineralization is relatively high in Na^+, Cl^-, Ca^{2+}, and SO_4^{2-}.
- Group 4 contains 30 samples. It is characterized by sources that do not have well-individualized structures. Each water point "pulls" toward a variable without really showing an evolutionary trend.

19.6 Water minerals equilibrium

19.6.1 Carbonates equilibrium

The equilibrium diagram of carbonate minerals was carried out by Carpenter, 1962 in Daoud (1995) and repeated by Stumm and Morgan (1981). The projection of the two pCO_2 parameters and the logarithmic ratio of calcium and magnesium activities on the diagram (Fig. 19.11) shows that the sources are in equilibrium with dolomite in the domain of stable minerals and with aragonite in the domain of unstable minerals. Only one source is in balance with huntite in the area of unstable. This equilibrium means that dolomite is the main carbonate mineral which constitutes the original reservoir, or else the waters, during their ascent to the surface, have sufficiently leached out a dolomitic host. However, this balance diagram only considers pure mineral species. Consequently, a strongly magnesian calcite can be assimilated to a dolomite.

The characterization of water involves determining the ionic distribution of elements in solution, the activity coefficients and the activities of aqueous species (Fritz, 1991). Several calculation models have been developed in order to determine the possible relationship between the chemical composition of source waters and the thermometric expressions relating to ions. In this perspective the WateqF model (Plummer et al., 1976) makes it possible to calculate the ionic distribution of water by taking into account major elements and some trace elements, at the temperature of the water at emergence.

The WateqF also makes it possible to calculate the water saturation indices with regard to a large number of minerals in order to identify the mineral phases that control the concentrations of major elements. The data provided to this program are the physicochemical parameters (pH and temperature) and the concentrations of major elements and of some trace elements.

19.6.2 Silica equilibrium

Fig. 19.12 gives the plot of the studied source in the fields of mineral stability in the systems $CaO-Al_2O_3-SiO_2-H_2O$ (Fig. 19.12A) and $MgO-Al_2O_3-SiO_2-H_2O$ (Fig. 19.12B). The data used for making these diagrams are those compiled by Fritz (1991). The WateqF program (Plummer et al., 1976) calculates the Ca^{2+} and H^+ activities, as well as the molar contents of H_4SiO_4. The studied water samples show a greater or lesser degree of chemical intimacy with the mineralogical matrix (Fig. 19.12). Indeed, the water chemical composition always corresponds for these systems to the stability domain of kaolinite, which is the domain of most of the natural surface and underground waters (Garrels and Christ, 1965 in Grunberger, 1989).

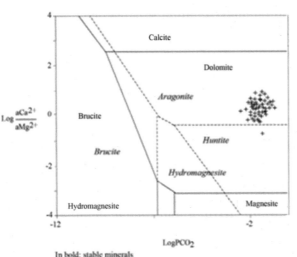

FIGURE 19.11 Source water equilibrium with carbonate minerals.

In bold: stable minerals
In italic: metastable minerals
In dash line: metastable domain

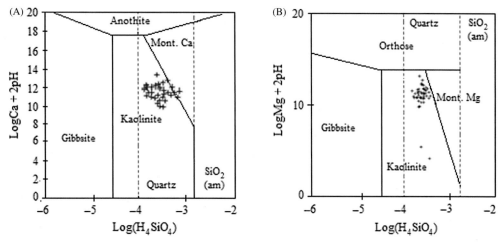

FIGURE 19.12 Source water equilibrium with silica minerals. (A) CaO–Al$_2$O$_3$–SiO$_2$–H$_2$O System, (B) MgO–Al$_2$O$_3$–SiO$_2$–H$_2$O System.

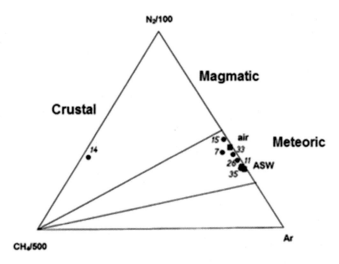

FIGURE 19.13 N$_2$–Ar–CH$_4$ diagram.

19.6.3 N$_2$–Ar–CH$_4$ gases diagram

Since isotopic data are not available for all the studied sources, the N$_2$–Ar–CH$_4$ diagram (Norman et al., 1998) was used to approach the deep origin of water emanating at the surface as sources. The plot of the six north Moroccan source gaz analyzes on this diagram (Fig. 19.13) highlights that the representative points of these sources are between the two lines indicating the minimum and the maximum of the meteoric waters N$_2$/Ar ratio, near the ASW (water Saturated Air). This would indicate a probable meteoric origin to these elements. However, the source no. 14 is plotted into the mixing zone of gaseous species originating primarily from the magmatic and crustal origins (Norman et al., 1998).

19.7 Conclusion

The interpretation of the physicochemical parameters of some northern Morocco sources allows concluding that these sources have pH values oscillating around neutrality and according to their TDS and Electrical conductivity, most of the sources are of the soft type.

These sources have variable facies due to the geological diversity of the studied area. The facies were determined both on the Piper and binary diagrams. These diagrams allowed classifying these waters as predominantly sodium–calcium chloride or sodium–calcium bicarbonate. The sulfated facies was encountered only for two sources with is in relation with the Triassic deposits. The obtained facies are very heterogeneous and do not allow a global visualization of samples in the study area.

Static processing of the chemical data by the PCA has shown that the first axis F1 is defined by mineralization because most of the elements are crowded around this axis. The application of this PCA did not allow grouping of individuals (sources) according to their origins linked to the geological diversity in the study area.

The source water quality highlighted that 71% of source waters are not suitable for agriculture uses and only 38% could be used as drinking waters. Nevertheless, the bad quality of source waters and rainfall scarcity in the region is becoming an important concern for the region that the decision-makers have to overcome by different solotions such as interbasin water transferts and the construction of small dams since the area is mostly mountainous.

References

Benaabidate, L., 2000. Caractérisation du bassin versant de Sebou: Hydrogéologie, qualité des eaux et géochimie des sources thermales. Thèse Doc. Es.Sc., USMBA, Fès., 250p (in French).

Bouchaou, L., 1988. Hydrogéologie du basin des sources karstiques du complexe calcaire haut-atlasien du Dir de Béni Mellal (Maroc). Thèse Univ. Franche — Comté. Sc. Terre, Besançon, France (in French).

Bremond, R., Vuichard, R., 1973. Parameters of Water Quality. Ministry of Nature Protection and Environment, Paris, p. 179.

Chopra, K., Gulati, S.C., 2001. Migration, Common Property Resources and Environmental Degradation. Sage, New Delhi.

Cirak, P., 1987. Le bassin sud rifain occidental au Néogène supérieur. Evolution de la dynamique sédimentaire de la paléogéographie au cours d'une phase de comblement. Mém. Inst. Géol. Bassin d'Aquitaine. No. 21, Bordeaux (in French).

Daoud, D., 1995. Caractérisation géochimique et isotopique des eaux souterraines et estimation du taux d'évaporation dans le bassin de Chott Chergui (zone semi — aride) Algérie. Thèse Univ. Pars Sud. Centre d'Orsay (in French).

Djabri, L., 1987. Contribution à l'étude hydrogéologique de la nappe alluviale de la plaine d'effondrement de Tébessa. Essai de modélisation. Thèse. Univ. Franche Comté, UFR. Sc. Tech., Besançon, 171 p (in French).

Ellis, A.J., 1969. Present day hydrothermal systems and mineral deposition. In: Proc. Ninth Common — Weather Mining and Met Congress (Mining and Petroleum Section) London. Inst. Mining and Metallurgy, pp. 1–30.

Fritz, B., 1991. Etude thermodynamique et modélisation des réactions hydrothermales et diagénétiques. Mém. Sc. Géol., Strasbourg, 65 174p (in French).

Gajbhiye, S., Sharma, S.K., Awasthi, M.K., 2015. Application of principal components analysis for interpretation and grouping of water quality parameters. Int. J. Hybrid. Inf. Technol. 8 (4), 89–96. <https://doi.org/10.14257/ijhit.2015.8.4.11>.

Giggenbach, W.F., 1991. Chemical techniques in geothermal exploration. In: D'amore, F. (Ed.), UNITAR-UNDP Guide book: An Application of Chemistry in Geothermal Reservoir Development. pp. 253–273.

Grunberger, O., 1989. Etude géochimique et isotopique de l'infiltration sous climat tropical contrasté — Massif du Piton des neiges. Ile de la Réunion. Thèse doct. Sc. Univ. Pars Sud. Centre d'Orsay, 269p (in French).

Mukherjee, A., Scanlon, B., Aureli, A., Langan, S., Guo, H., McKenzie, A., 2020. Global Groundwater: Source, Scarcity, Sustainability, Security and Solutions, first ed. Elsevier, ISBN: 9780128181720.

Norman, D.I., Chomiac, B.A., Moore, J.N., 1998. Approaching equilibrium from the hot and cold sides in the $FeS_2-FeS-Fe_3O_4-H_2S-CO_2$ system in the light of fluid inclusion gas analysis. In: Arehart, Hudson (Eds), *Water Rock Interaction*. A. A. Balkema, Rotterdam, pp. 565–568.

Pereira, L.S., Cordery, I., Iocavades, I., 2009. Coping with Water Scarcity: Addressing the Challengers. Springer, Heidelberg.

Plummer, L.N., Fontes, B.F., Truesdell, A.H., 1976. WateqF, a Fortran IV version of Wateq, a computer program for calculating chemical equilibrium of waters. In: USGS. Water Resources Investigations 13, 61 p.

Ramadan, E., 2015. Sustainable water resources management in arid environment: the case of Arabian Gulf. Int. J. Waste Resour. 5, 179. Available from: https://doi.org/10.4172/2252-5211.1000179.

Richards, L.A., 1954. Diagnosis and Improvement of Saline and Alkali Soils. US Salinity Laboratory Staff, US Department of Agriculture, Washington, DC.

Robins, N.S., Fergusson, J., 2014. Groundwater scarcity and conflict — managing hotspots. Earth Perspect. 1, 6. Available from: https://doi.org/10.1186/2194-6434-1-6.

Schoeller, H., 1965. Qualitative evaluation of groundwater resources. Methods and Techniques of Ground-Water Investigations and Development. UNESCO, pp. 54–83.

Shikazono, N., 1978. Possible cation in chloride rich geothermal waters. Chem. Geol. 23, 239–254.

Stumm, W., Morgan, J.J., 1981. Aquatic chemistry, An Introduction Emphasizing Chemical Equilibria in Natural Waters, second ed. Wiley & Sons, New York, p. 780.

Thilagavathi, R., Chidambaram, S., Prasanna, M.V., Thivya, C., Singaraja, C., 2012. A study on groundwater geochemistry and water quality in layered aquifers system of Pondicherry region, southest India. Appl. Water sci. 2, 253–269. Available from: https://doi.org/10.1007/s13201-012-0045-2.

Younsi, A., 1994. Etude des mécanismes des intrusions marines vers le système aquifère compris entre Oum-er-Rbia et Bir Jdid, provine d'El Jadida Maroc. Hydrogéologie, hydrochimie et modélisaion mathématique. Thèse 3ème Cycle. Univ. Couaïb Doukkali. Fac. Sc. El Jadida (in French).

Chapter 20

Arsenic in groundwater in the United States: research highlights since 2000, current concerns and next steps

Madeline E. Schreiber
Department of Geosciences, Virginia Tech, Blacksburg, VA, United States

20.1 Introduction

Groundwater supplies are under increasing stress due to both diminishing quantity and quality (Mukherjee et al., 2020). One contaminant that has impacted groundwater supplies on a global scale is arsenic. Arsenic (As) released from naturally occurring sources has caused extensive contamination of groundwater worldwide, impacting more than 140 million people (Michael, 2013), with the majority of exposures occurring in countries in Asia (e.g., Bangladesh, India, Vietnam, Cambodia, Nepal, and China). Arsenic poisoning linked to groundwater in Asia was identified in West Bengal (India) in 1984, although it took several decades for the extent to be recognized (Ravenscroft et al., 2009) and ongoing studies are still identifying new regions of contamination. The As crisis in Bangladesh has been described as "the largest poisoning of a population in history" (Smith et al., 1992).

Although nowhere near as widespread as in Asia, groundwater contamination by naturally occurring sources of As is an environmental problem that affects many parts of the US. Regional and national assessments conducted by the US Geological Survey (USGS) have suggested that trace elements, including As, are more widespread in groundwater than other contaminant groups (Ayotte et al., 2011; DeSimone et al., 2014). Arsenic, of particular importance due to its carcinogenicity, is present in elevated concentrations in groundwater in many regions; more than 5% of 6000 samples in one nationwide assessment exceeded As human health benchmarks (DeSimone et al., 2014). Recent work (Ayotte et al., 2017) suggests that over 2 million people in the United States are exposed to elevated As concentrations through drinking water from domestic wells.

Because As is a naturally occurring element that is found at detectable concentrations in many rocks and sediments, water–rock interactions impact its release to, and transport in, groundwater systems. These interactions are complex, involving not only the specific chemistries of water and minerals within the rock or sediment but also microbial activity, groundwater flow, and recharge/precipitation patterns as well as the human activity

Since the lowering of the US EPA drinking water standard (the maximum contaminant level, MCL) from 50 to 10 parts-per-billion (ppb) in 2001, which became enforceable in 2006 (the revision is known as the Final Arsenic Rule), hundreds, if not thousands, of studies have been conducted and published on various aspects of As in groundwater in the United States. In the past two decades, scientists, engineers, and health professionals have contributed substantial knowledge about As behavior in natural waters and effects on human health. These knowledge gains have led to many practical outcomes, including advances in water treatment and remediation, development of new methods to analyze As in different forms and different media, and improvements in delineating areas of risk.

The goal of this chapter is to review what has been learned about As in groundwater in the United States over the past 20 years, how that knowledge has improved fundamental understanding of As behavior and has led to practical applications for As mitigation, and to outline current concerns and research needs. Because so much research has been done and many summaries have been published, the reader is referred to additional literature for information on specific topics, including books (Ravenscroft et al., 2009; Welch and Stollenwerk, 2003; Ahuja, 2008; O'Day et al., 2005; Bundschuh et al., 1997; Jean et al., 2010, 2014; Deschamps and Matschullat, 2011; Kabay et al., 2010; Santini and Ward, 2018), review papers (e.g., Smedley and Kinniburgh, 2002; Nordstrom, 2002; Welch et al., 2000; Bowell et al., 2014; Mandal and Suzuki, 2002), and USGS publications (e.g., Ayotte et al., 1999, 2011; DeSimone et al., 2014; Focazio et al., 2000, 2006; DeSimone et al., 2009; Warner et al., 2015).

20.2 Research on arsenic in groundwater: 2000–20

20.2.1 Sources of Arsenic in groundwater

One of the focal topics over the past two decades has been to identify As sources to groundwater. Because As has both naturally occurring and anthropogenic sources, identifying the specific As source to groundwater is often challenging. Industrial sources of As [glass, munitions, insecticides/herbicides, wood preservatives, animal feed additives (see Smedley and Kinniburgh, 2002; Welch et al., 2000; Bowell et al., 2014)] generally cause contamination on a site-specific scale. Mining related activities, most importantly those related to sulfide and gold mining, can cause local to regional scale contamination of soils (Reimann et al., 2009) and surface waters (Nordstrom and Alpers, 1999). Although anthropogenic sources of contaminants such as As exist in the United States and can cause contamination of waters, geogenic sources are the "most fundamental risk factor for the potential mobilization of trace elements" (Ayotte et al., 2011).

Geogenic sources of As have been discussed in many journal articles and books (see Ravenscroft et al., 2009; Welch and Stollenwerk, 2003; Smedley and Kinniburgh, 2002; Bowell et al., 2014). Arsenic is present in over 500 minerals, including sulfides, oxides, arsenates, and others. Arsenic can also occur in solid solution with minerals or amorphous phases from trace (<1000 ppm) to atom percent concentrations (Foster, 2003) and can also be coprecipitated with minerals such as carbonate and phosphates. Importantly, As can adsorb strongly onto mineral surfaces (discussed in more detail later). Concentrations of As in minerals, rocks, sediments, and soils are reported in several papers (see Ravenscroft et al., 2009; Smedley and Kinniburgh, 2002; Bowell et al., 2014; O'Day, 2006). One key observation for As release to groundwater is that the solid-phase concentrations do not need to be elevated over "average" concentrations to cause As exceedances in groundwater (Cozzarelli et al., 2016; Hering and Kneebone, 2002). In addition, increasing sediment As does not necessarily result in increasing groundwater As (Erickson and Barnes, 2005; Thomas, 2003). The As concentration in the source is only one piece of the puzzle. Other important considerations include the biogeochemical processes that influence release, mobility, and ultimate fate of As in natural waters.

20.2.2 Key biogeochemical processes that influence As cycling

Biogeochemical processes that can affect As cycling in aquifers include oxidation/reduction, adsorption/desorption, precipitation/dissolution, among others. Organisms mediate many of these processes and are thus an integral piece of the As biogeochemical cycle.

20.2.2.1 Arsenic species in water

The most common As species in natural waters are As(III) and As(V), which form oxyanions in water. At pH 6–8, the dominant oxyanions of As(V) are $H_2AsO_4^-$ and $HAsO_4^{2-}$. In the same pH range the dominant oxyanion of As(III) is H_3AsO_3. As(III) is considered the more toxic form because it binds to sulfhydryl groups, impacting the function of many proteins (NRC, 1999). Arsenic also forms organic species, such as monomethylarsonate, and dimethylarsinate (Cullen and Reimer, 1989), but those tend to be of lower concentration than inorganic species in natural waters.

20.2.2.2 Adsorption reactions

Significant advances have been made in evaluating adsorption mechanisms of As to mineral surfaces, including hydrous Fe oxides (HFOs), Al oxides, Mn oxides, and clays. Adsorption studies involving As began in the 1980s (e.g., Dzombak and Morel, 1987; Goldberg, 1986), expanded in the 1990s (e.g., DeVitre et al., 1991; Fuller et al., 1993; Grossl et al., 1997; Wilkie and Hering, 1996; Manning et al., 1998; Manning and Goldberg, 1995, 1996a, 1996b, 1997; Fendorf et al., 1997; Jain et al., 1999; Raven et al., 1998; Waychunas et al., 1993), and flourished in the 2000s (e.g., Dixit and Hering, 2003; Dixit and Hering, 2006; Goldberg, 2002; Goldberg and Johnston, 2001; Grafe et al., 2002; Lin and Puls, 2000; Beaulieu and Savage, 2005; Manning and Suarez, 2000; Ona-Nguema et al., 2005; Smith and Naidu, 2009). Review papers and book chapters have been written on this topic (Ravenscroft et al., 2009; Smedley and Kinniburgh, 2002; Stollenwerk, 2003). In brief, results of these studies suggest the following:

- Arsenic adsorbs strongly to metal oxides; the mechanism is through inner-sphere surface complexes. Arsenic has a higher sorption affinity to HFOs than to Al hydroxides and clays.

- Arsenic adsorption is strongly dependent on the As species, the mineral and pH. For example, As(V) has a higher affinity than As(III) onto HFOs at lower pH (<5), whereas As(III) has a higher affinity at higher pH.
- Arsenic adsorption is also influenced by competitive oxyanions (e.g., phosphate) and natural organic matter.

Because there are multiple influences on As adsorption, building connections between observations made in laboratory experiments with those made by spectroscopic measurements and those inferred by surface complexation modeling can be challenging (Hering and Dixit, 2005).

20.2.2.3 Redox processes

Oxidation−reduction reactions influence the oxidation state, bioavailability, and mobilization of As(III) and As(V). These reactions can occur abiotically, such as the oxidation of As(III) by Mn oxide (Oscarson et al., 1981; Brannon and Patrick, 1987; Chiu and Hering, 2000; Manning et al., 2002) and green rust (Jönsson and Sherman, 2008; Su and Puls, 2004). However, many of the important redox reactions that impact As are mediated by microorganisms (Santini and Ward, 2018; Cullen and Reimer, 1989; Ahmann et al., 1994; Inskeep et al., 2002; Newman et al., 1998; Oremland and Stolz, 2003, 2005; Stolz and Oremland, 1999; Santini et al., 2002). Bacteria can reduce As(V), oxidize As(III), and methylate As for detoxification and generation of energy. These organisms are phylogenetically diverse and can be found in a wide range of habitats (Santini and Ward, 2018; Oremland and Stolz, 2003). Because As(III) and As(V) exhibit different adsorption characteristics (Dixit and Hering, 2003), reduction or oxidation of As can impact the ability of As species to either remain attached to or mobilize from mineral surfaces (Zobrist et al., 2000; Langner and Inskeep, 2000).

Organisms that reduce Fe and S can also affect As mobility. Fe-reducing bacteria reductively dissolve HFOs to which As is adsorbed. The importance of Fe-reducing bacteria in the process of As mobilization in Bangladesh and West Bengal was recognized in the early 2000s (Islam et al., 2004; McArthur et al., 2001; Nickson et al., 1998, 2000; van Geen et al., 2004, and many others). Some bacteria can reduce both Fe(III) and As(V), both as individual strains and as communities (Campbell et al., 2006). Reductive dissolution of HFOs can also change the surface area and properties of the mineral, resulting in changes in sorption/desorption mechanisms (e.g., Tadanier et al., 2005). In addition, rates of reductive dissolution are affected by Fe mineralogy and the accumulation of reaction products, including Fe(II) and As species, on the surface (Ford, 2002; Hansel et al., 2004; Roden and Urrutia, 2002). Bacterial sulfate reduction can also impact As mobility, as this process generates sulfide, which can bind with both Fe and As to remove As from water (Kirk et al., 2004; Sun et al., 2016; Moore et al., 1988; Rittle et al., 1995; Keimowitz et al., 2007).

20.2.3 Tools for studying arsenic

20.2.3.1 Analytical tools

20.2.3.1.1 Measuring arsenic speciation

Because As(III) and As(V) have different toxicities and transport properties, methods have been developed to identify and measure these species separately. Many papers describing As analytical methods, published in the 1990s and 2000s (e.g., Bednar et al., 2004; Melamed, 2005; Ali and Jain, 2004), provide thorough reviews of methods for analyzing As in water. Quantification of As to concentrations below 10 ppb can be accomplished using methods such as graphite furnace-atomic absorption spectroscopy and inductively coupled plasma-mass spectrometry. The 2000s brought advances in methodology, including coupling separation techniques (e.g., HPLC) with detection methods, to analyze both inorganic and organic As species in one sample (Garbarino et al., 2002). Studies also focused on optimal preservation of As species in samples (see Bednar et al., 2002, 2004). Field-based speciation methods using solid-phase extraction columns were also developed during this time (Le et al., 2000).

20.2.3.1.2 Sequential extraction

Another focus in the 2000s was to identify the solid phases that host As in complex soils and aquifer sediments to yield information on As ecotoxicological and mobility characteristics. Methods for sequential extraction of metals and trace elements from different fractions, including exchangeable, bound to carbonates, bound to Fe−Mn oxides, bound to organic matter, and residual phases were developed in the 1970s (e.g., Tessier et al., 1979). These methods were amended in the 2000s to specifically extract As (e.g., Keon et al., 2001), including As(III) and As(V) (Georgiadis et al., 2006).

Although sequential extraction methods continue to be widely used, these methods have limitations, including (1) the lack of consensus on the best methods to use; (2) the lack of selectivity of chemical solutions to perfectly extract As from a particular mineral; (3) the problem of mass transfer of As between solid phases between phases (Gruebel et al., 1988); (4) the wide differences in mineral solubility and dissolution kinetics, which can affect the efficacy of the method (Foster, 2003); and (5) in situ conditions such as redox, which must be preserved for accurate extractions. Despite the limitations, sequential extraction methods are useful for examining general patterns of As partitioning in complex solids and are particularly helpful for characterizing fine-grained sediments or organic phases that are difficult to characterize using other methods (Smedley and Kinniburgh, 2002).

20.2.3.1.3 Spectroscopic methods

Spectroscopic tools were fine-tuned in the 2000s to examine As speciation and bonding mechanisms to solids. These techniques, particularly infrared, X-ray photoelectron, and X-ray absorption spectroscopy (XAS), can provide information about As speciation and surface reactions in both synthetic and natural materials. Reviews of the methods provide additional information (Foster, 2003; Wang and Mulligan, 2008). Infrared (IR) and Raman spectroscopy can examine molecules on mineral surfaces such as surface hydroxyl groups and As oxyanions on mineral surfaces, and can also examine As in crystalline and X-ray amorphous solids (Foster, 2003; Goldberg and Johnston, 2001). X-ray photoelectron spectroscopy can analyze As oxidation state and to characterize mineral surfaces (Nesbitt and Muir, 1998). XAS, which includes extended X-ray absorption fine structure and X-ray absorption near edge structure, can examine As oxidation state and structural information about As atoms and nearest neighbors. Many studies have utilized these methods to investigate As(III) and As(V) at mineral surfaces, including HFOs, Al oxides, and clays [e.g., (Manning et al., 1998; Fendorf et al., 1997; Waychunas et al., 1993; Foster et al., 2003) and others]. The results have highlighted specific mechanisms of As bonding with mineral surfaces, providing quantitative information on surface complexation/adsorption reactions. Although some of these methods are limited by detection limits that are too high to measure As in natural materials, a recent study successfully used spectroscopic methods in combination with other analytical tools to examine As at low concentrations (<10 mg/kg) in glacial aquifer sediments, showing the promise of a multimethod approach (Nicholas et al., 2017).

20.2.3.2 Spatial maps

Spatial mapping of As concentrations in groundwater in the United States has been conducted at both regional and national scales. The USGS has been instrumental in collecting data on As and publishing datasets, maps, and statistical models, which have provided critical information for knowing where As occurs and identifying variables that influence As occurrence. The USGS provided the first nationwide picture of As in groundwater in 2000, using a dataset with 18,850 samples from public supply wells collected from 1973 to 1997 (Welch et al., 2000; Focazio et al., 2000). The map continues to be updated as new data are available (Ryker, 2001; Gronberg, 2011).

20.2.3.3 Modeling

20.2.3.3.1 Reactive transport models

Reactive transport models that simulate As transport started with 1D simulations with limited geochemical reactions and have developed into fully coupled flow and reactive transport models (Jung et al., 2009; Wallis et al., 2011). One model that has been used to evaluate As reactive transport is PHT3D (Prommer et al., 2003), a code based on PHREEQC-2 and MT3DMS, which simulates a suite of geochemical reactions, including equilibrium-controlled aqueous complexation and speciation, kinetic reactions, mineral precipitation/dissolution, ion exchange, and surface complexation reactions. PHT3D has been used to simulate As release and transport due to geochemical reactions induced by managed aquifer recharge (Wallis et al., 2011; Prommer et al., 2018). The code has also been successfully applied to simulate the transport of dissolved Fe(II) and As in an anaerobic plume and the subsequent oxidation of Fe(II) and As(III) during mixing as the plume discharges (Jung et al., 2009). In one recent study, PHT3D examined reactive transport of As from field injection of organic carbon in an aquifer in the Bengal Basin (Rawson et al., 2017).

Constructing reactive transport models to simulate As release from aquifer sediments and subsequent transport is challenging due to the complexity of biogeochemical reactions that impact As. Decisions must be made about how much of this complexity is needed in the model to reflect As behavior in the field, and as complexity (and the reactions/equations needed to describe the complexity) increases, so do the number of parameters to describe those reactions. Further work on "bridging the gap" between the detailed information learned in the laboratory to developing reactive transport models to simulate As at the field scale is needed.

20.2.3.3.2 Statistical models

Analysis of datasets using statistical models can identify areas susceptible to As contamination and factors that are associated with elevated As in groundwater. Logistic regression models have been particularly useful as they can predict the probability of binary outcomes (e.g., As concentration is above a threshold such as the MCL) and can evaluate the importance of independent variables, such as geologic formation, soil series, groundwater chemistry, land use, and hydrologic/climatic characteristics, for predicting As concentrations. Other statistical models, including machine learning methods, have also been used to identify areas of risk. Statistical models have been applied to nationwide data (Ayotte et al., 2017) and within specific regions of the United States, including New England (Ayotte et al., 2006), the Central Valley of California (Ayotte et al., 2016), Pennsylvania (Gross and Low, 2013), the Southwest (Thiros et al., 2015), Louisiana (Yang et al., 2014), Minnesota (Erickson et al., 2018), and Virginia (VanDerwerker et al., 2018).

20.2.4 Mechanisms of arsenic release to groundwater

Even when As is contained in geologic materials, it may be stable in solid phase and not be released to water. Arsenic concentrations are highly variable in minerals, sediments, and rocks and are generally a poor predictor of groundwater contamination (Ravenscroft et al., 2009). Another critical variable is the "geochemical trigger(s)" that can cause As to be released to groundwater. The four main mechanisms for As release, discussed in detail by others (e.g., Ravenscroft et al., 2009; Smedley and Kinniburgh, 2002; Nordstrom, 2002; Welch et al., 2000), include:

1. reductive dissolution of As-bearing Fe oxides;
2. oxidation of As-bearing sulfides;
3. pH-dependent desorption of As from oxides and clays; and
4. leaching of As from host rocks by geothermal waters.

Reductive dissolution resulting in As release has been a source of much study. An analysis of 230 cases worldwide of As contamination of groundwater (Ravenscroft et al., 2009) suggested that reductive dissolution is the dominant mechanism for As mobilization to groundwater. Because a key component of reductive dissolution of Fe is the source of organic matter that drives reduction, many studies have focused on identifying the sources of organic carbon, especially in the Bengal Basin and other regions of Asia that have extensive As contamination of groundwater (McArthur et al., 2001, 2004; Harvey et al., 2002, 2006; Neumann et al., 2010). Oxidation of sulfides requires an oxidant such as dissolved oxygen, nitrate (Senn and Hemond, 2002; Höhn et al., 2006) or Fe(III). The process of pH-dependent desorption results from release of As from mineral surfaces as pH increases; as discussed above, adsorption of As on HFOs, particularly As(V), decreases with increasing pH. Regions where pH-dependent desorption has been invoked as an As mobilization mechanism have baseline alkaline pH, although desorption can also be triggered by biogeochemical process and by mixing of waters. Leaching of As from host rocks by geothermal waters occurs in specific regions of active or former volcanic settings, including colliding plate boundaries, intraplate hot spots, and rift zones (Ravenscroft et al., 2009; Langner et al., 2001; Webster and Nordstrom, 2003).

20.3 Hydrogeochemical settings for arsenic in groundwater in the United States

Other hydrogeological factors, including age of groundwater and aquifer, historical and current hydraulic gradient, degree of flushing, and climate, contribute to create conditions where As is released to groundwater (Smedley and Kinniburgh, 2002). One key hydrogeological factor that has been shown to result in high-As groundwater on a regional scale is a lack of flushing (Smedley and Kinniburgh, 2002), which could be due to long residence time or to chemical evolution of groundwater along a flowpath (see Peters, 2008; Anning et al., 2012; Johannesson et al., 2019; Scanlon et al., 2009). Examples of As contamination of groundwater from geogenic sources have been documented in almost every state in the United States (Table 20.1) within the following aquifers.

20.3.1 Sand and gravel aquifers

20.3.1.1 Alluvial aquifers

Example: Mississippi River Valley Aquifer (Yang et al., 2014; Johannesson et al., 2019; Sharif et al., 2008; Borrok et al., 2018).

TABLE 20.1 Studies of geogenic arsenic contamination of groundwater by US state.

State[a]	Where	Aquifer materials	Source
AL	West–Central	SE Coastal Plain Eutaw Formation	Lee et al. (2007)
AK	Fairbanks area	Glacial-fluvial aquifers	Munk et al. (2011)
–	Fairbanks area	Crystalline (schist, gold veins)	Verplanck et al. (2008)
AR	Southeastern	Mississippi River Valley alluvial aquifer	Sharif et al. (2008)
AZ	South–Central–West	Basin-fill aquifers	Thiros et al. (2015), Anning et al. (2012)
CA	Throughout	Basin-fill aquifers	Ayotte et al. (2016), Thiros et al. (2015), Anning et al. (2012)
–	Eastern Sierra Nevada	Geothermal input	Wilkie and Hering (1998)
CO	South–Central	Basin-fill aquifers (San Luis Valley)	Anning et al. (2012), James et al. (2014)
CT	Several regions	Crystalline	Flanagan and Brown (2017)
FL	Central	Floridan Aquifer	Pichler et al. (2011), Price and Pichler (2006), Jones and Pichler (2007)
IA	North–Central	Glacial–Northwest provenance Wisconsin lobe	Erickson and Barnes (2005)
ID	Western	Columbia Plateau basin fill and basaltic rock aquifers	Frans et al. (2012)
–	South–Central, Southwestern	Snake River Plain basin fill, basalt rocks	Frans et al. (2012), Busbee et al. (2009), Bartolino and Hopkins (2016)
IL	Central	Lower Illinois River Basin, glacial deposits	Thomas (2003), Warner (2001), Kelly et al. (2005), Holm et al. (2004)
–	Northeastern	Upper Illinois River Basin, glacial deposits	Thomas (2003)
IN	Northwestern	Upper Illinois River Basin, glacial deposits	Thomas (2003)
–	Southeastern	Great and Little Miami River Basins, glacial deposits, valley fill	Thomas (2003)
KS	Although As is detected in High Plains aquifer system in other states, data on As in groundwater from geogenic sources specifically in Kansas could not be located for this chapter		
KY	Eastern	Sandstone, shale, coal	Fisher (2002), Shiber (2005)
LA	Throughout	Mississippi River Valley alluvial aquifer	Yang et al. (2014), Johannesson et al. (2019), Borrok et al. (2018)
MA	Coastal	Glacial aquifers (and confining units)	Jung et al. (2009), Bone et al. (2006), Kent and Fox (2004)
ME	Throughout	Crystalline	Peters (2008), Ayotte et al. (2003), Lipfert et al. (2006), Yang et al. (2009), Robinson and Ayotte (2006)
–	South–Central	Glacial deposits	Keimowitz et al. (2005)
MD	Coast	Coastal Plain	Drummond and Bolton (2010), Haque et al. (2008)
MI	Southeastern	Glacial deposits, outwash plain; till	Erickson and Barnes (2005), Kim et al. (2002), Szramek et al. (2004)
–	Southeastern	Sandstone, glacial ills	Kolker et al. (2003)
MN	Northwest, central	Glacial, alluvial sediments	Erickson and Barnes (2005, 2006), Erickson et al. (2018)
MO		Alluvial aquifers	Korte (1991)
MT		Alluvial aquifers recharged from Madison and Missouri Rivers (geothermal source)	Nimick et al. (1998), Nimick (1998), Sonderegger and Ohguchi (1988)
NC	South–central	Carolina Slate belt	Kim et al. (2011), Sanders et al. (2012), Reid et al. (2010)
ND	Eastern, Central, Northwestern	Glacial–Northwest provenance Wisconsin lobe	Erickson and Barnes (2005)
NE	Northern High Plains	Sand Hills eolian deposits; Ogallala/Arikaree/White River groups	Gosselin et al. (2006), McMahon et al. (2007), Gurdak et al. (2009), Stanton and Qi (2006)

(Continued)

TABLE 20.1 (Continued)

State[a]	Where	Aquifer materials	Source
NJ	Inner Coastal Plain	Coastal Plain aquifer	Barringer et al. (2010, 2011)
–		Mesozoic Rift Basins	Serfes et al. (2010), Senior and Sloto (2006), Serfes et al. (2005)
NH	Throughout	Crystalline, Metasedimentary	Peters (2008), Ayotte et al. (2003), Robinson and Ayotte (2006), Peters et al. (1999), Peters and Blum (2003)
–	Coastal	Glacial marine deposits	deLemos et al. (2006)
NM	Central	Basin-fill, Rio Grande aquifer	Anning et al. (2012), Bexfield and Plummer (2003), Camacho et al. (2011)
NV	Throughout	Basin-fill deposits	Anning et al. (2012), Welch and Lico (1998), Steinmaus et al. (2005)
NY	*Although As is detected in crystalline rocks in surrounding areas, data on As in groundwater from geogenic sources specifically in New York could not be located for this chapter*		
OH	Northeastern	Glacial, alluvial sediments	Thomas (2003), Matisoff et al. (1982)
–	Southeastern	Great and Little Miami River Basins, glacial buried valley sand and gravel	Thomas (2003)
OK	Central	Sandstone with mudstone (Central Oklahoma Aquifer)	Schlottmann and Breit (1992), Smith (2005)
OR	Willamette Basin	Volcanic ash; tuff deposits	Hinkle and Polette (1999)
–	Northern	Columbia Plateau basin fill and basaltic rock aquifers	Frans et al. (2012)
PA	Southeast	Mesozoic Rift Basins	Peters and Burkert (2008), Blake and Peters (2015)
–	North/Northeast	Glacial aquifers, sedimentary rocks	Gross and Low (2013), Low and Galeone (2007)
RI	*Although As is detected in crystalline rocks in other areas of New England, data on As in groundwater from geogenic sources specifically in Rhode Island could not be located for this chapter*		
SD	South-West	Arikaree aquifer	Bird et al. (2020)
–	South-Central	Arikaree aquifer	Carter et al. (1998), Powers et al. (2019)
–	East-central	Glacial—Northwest provenance Wisconsin lobe	Erickson and Barnes (2005)
TX	Northern/Northwest	Southern High Plains Aquifer; Ogallala Formation	Scanlon et al. (2009), Gurdak et al. (2009), Hudak (2006), Hudak (2000)
–	Southeast	Southern Gulf Coast Aquifer	Gates et al. (2011), Hudak (2003), Glenn and Lester (2010)
–	South	Carrizo-Wilcox Aquifer	Haque and Johannesson (2006), Willis et al. (2011)
UT	Several areas	Basin Fill (Cache/Salt Lake Valley)	Anning et al. (2012), Meng et al. (2017)
VA	North-Central	Mesozoic Rift Basins	VanDerwerker et al. (2018)
VT	Northern	Ultramafic rocks; serpentinite, magnesite	Peters (2008), Ryan et al. (2011)
–	Southwest	Metasedimentary rocks, slate, phyllite	Mango and Ryan (2015), Ryan et al. (2013), Ryan et al. (2015)
WA	South–Central–East	Columbia Plateau basin fill and basaltic rock aquifers	Frans et al. (2012)
WI	Southeastern	Glacial sediments, sand and gravel, till	Root et al. (2005, 2010)
–	Fox River Valley	St Peter Sandstone	Schreiber et al. (2000, 2003), Gotkowitz et al. (2004)
WV	West–Central	Permian Sandstone	Law et al. (2017)
WY	Northwestern	Aquifers recharged from Madison River; Spring waters from geothermal sources	Webster and Nordstrom, (2003), Stauffer and Thompson (1984), Nimick et al. (1998), Ball et al. (2010)

[a]*Groundwater As contamination from obvious mining sources are not included. No information was found for DE, GA, HI, MS, SC, or TN.*

20.3.1.2 Basin-fill aquifers

Examples: Central Valley, CA (Ayotte et al., 2016; Thiros et al., 2015; Anning et al., 2012); Carson Valley NV (Anning et al., 2012; Welch and Lico, 1998); Rio Grande Valley, NM (Thiros et al., 2015; Anning et al., 2012; Bexfield and Plummer, 2003); San Luis Valley, CO (Anning et al., 2012; James et al., 2014); Cache Valley, UT (Meng et al., 2017); Willamette Valley, OR (Hinkle and Polette, 1999); Basin-fill aquifers of the Columbia Plateau and Snake River Plain (Frans et al., 2012; Busbee et al., 2009).

20.3.1.3 High Plains aquifer

Examples: Northern High Plains (Gosselin et al., 2006; Gurdak et al., 2009; McMahon et al., 2007; Bird et al., 2020); Southern High Plains (Scanlon et al., 2009; Hudak, 2006).

20.3.2 Basaltic rock aquifers

Examples: Basaltic rock aquifers of the Columbia Plateau and Snake River Plain (Frans et al., 2012).

20.3.3 Glacial aquifers

Examples: *Northwest Provenance Glacial Drift* (MN, ND, SD, IA) (Erickson and Barnes, 2005); Lower Illinois and Upper Illinois River Basins (Thomas, 2003; Warner, 2001); Glacial alluvial aquifers in OH (Thomas, 2003); Great and Little Miami River Basins (Thomas, 2003); Glacial aquifers in MI (Kim et al., 2002; Szramek et al., 2004); Alaska, Cook Inlet (Munk et al., 2011); Northwest PA (Gross and Low, 2013).

20.3.4 Sedimentary rock aquifers

Examples: Sandstones [St Peter Sandstone, WI (Schreiber et al., 2000, 2003); Central Oklahoma Aquifer (Schlottmann and Breit, 1992), Permian Sandstone WV (Law et al., 2017)], Carbonates [Floridan Aquifer (Lazareva et al., 2015; Pichler et al., 2011; Price and Pichler, 2006)].

20.3.4.1 Special case: Mesozoic Rift Basins

Examples: Newark and Gettysburg Basins of NJ, PA (Peters and Burkert, 2008; Serfes et al., 2010; Senior and Sloto, 2006; Blake and Peters, 2015); Culpeper Basin of VA (VanDerwerker et al., 2018; Chapman et al., 2013).

20.3.5 Crystalline and meta-sedimentary rock aquifers

Examples: New Hampshire (Peters, 2008; Ayotte et al., 2003; Zheng and Ayotte, 2015; Lipfert et al., 2006); Maine (Yang et al., 2009, 2012); Vermont (Mango and Ryan, 2015; Ryan et al., 2011; Ryan et al., 2013); Connecticut (Flanagan and Brown, 2017); North Carolina (Kim et al., 2011; Sanders et al., 2012).

20.3.6 Coastal plain (semiconsolidated) aquifers

20.3.6.1 Atlantic coastal plain

Examples: New Jersey (Barringer et al., 2010, 2011; Mumford et al., 2012); Maryland (Drummond and Bolton, 2010; Haque et al., 2008).

20.3.6.2 Southeastern/Gulf Coastal Plain

Examples: Southeastern/Gulf Coastal Plain including Coastal Lowlands and Highlands. Examples: Carrizo-Wilcox Aquifer (Haque and Johannesson, 2006); Eutaw Aquifer (Gates et al., 2011).

20.3.7 Geothermal areas (western United States)

Examples: Yellowstone (Langner et al., 2001; Webster and Nordstrom, 2003; Stauffer and Thompson, 1984); Montana (Nimick et al., 1998); California (Wilkie and Hering, 1998).

20.4 Research highlights from 2000 to 2020

The extensive study of As over the past 20 years has shed much light on spatial and temporal patterns of As contamination of groundwater in the United States, including development of statistical models that can predict drivers of As contamination. Research on the impacts of human activities, such as groundwater pumping, managed aquifer recharge and introduction of bioavailable organic carbon sources, on release of naturally occurring As into groundwater are other themes that have been explored in the literature. Examples of these highlights are presented in the following sections.

20.4.1 Nationwide datasets show statistical and spatial patterns of groundwater As

Nationwide datasets have been critical for evaluating the magnitude of As contamination of groundwater in the US. Ryker (2003) compiled a summary table of datasets of As concentrations in public water supply and domestic wells in the United States prior to 2001. Analysis of these datasets shows that between 2% and 12% of the sources/systems tested had As concentrations >10 ppb. More recent studies have shown similar percentages, but within a more narrow range. Using NAWQA data from monitoring and drinking water wells from 1992 and 2003, Ayotte et al. (2011) reported that 7.0% of wells had As > 10 ppb. Using NAQWA data from domestic and public supply wells collected from 1991 to 2010, DeSimone et al. (2014) reported that 6.7% of the samples contain concentrations of As >10 ppb. Although these datasets reflect different wells, sampling methods, analytical methods, and detection limits, at the broad scale, they reflect similar pattern of As exceedances of groundwater samples.

Large datasets have allowed for extensive statistical analysis of As data. In the Ayotte et al. (2011) study, NAWQA data were analyzed to evaluate relationships of As (and other trace elements) in relation to climate, geology of aquifer materials, aquifer geochemistry, and redox potential and pH of the groundwater. Comparing aquifer types, As exceedances were highest in unconsolidated sand and gravel aquifers (12%), glacial unconsolidated sand and gravel aquifers (6.8%), basaltic/volcanic rock aquifers (6.2%), and crystalline rock (5.4%) and lowest in semiconsolidated sand (0.8%), sandstone (1.4%), sandstone and carbonate rock (1.1%), and carbonate rock (1.3%). The analysis also showed increasing As occurrence under anoxic conditions. Interestingly, the occurrence of As in samples generally increased from oxic to anoxic conditions when pH <7 but was consistently >50% when pH >7. In the unconsolidated sand and gravel aquifers, As occurrence increased as pH increased, but only for oxic and mixed conditions. Samples with high pH showed no difference with respect to redox. These results suggest that redox may be a more important driver of As mobility at lower pH than at higher pH.

DeSimone et al. (2014) examined specific aquifers within these broader groupings based on lithology, showing variations within the groups. For example, although As exceedances were generally low in carbonate rock (1.3%), specific carbonate aquifers, such as the Basin and Range carbonate aquifer, had close to 50% As exceedances. In addition, although sandstone aquifers overall have low exceedances (1.4%), sandstones associated with the Mesozoic Rift Basins have exceedances between 5% and 10%. Furthermore, within the unconsolidated sand and gravel aquifers, some aquifers, such as the Basin and Range basin-fill and the Rio Grande, have exceedances between 25% and 30%, while others, such as the Alluvial System in CO, have close to 0 exceedances. Thus, although generalizations can be made at broader scales, analysis of risk and exposure at regional scales must consider the drivers that influence As at those scales.

Embedded in the spatial variations of As in groundwater is the fact that wells may be open to different aquifers, or may be open to multiple aquifers. Patterns of depth dependence have observed in the fluviodeltaic aquifers of SE Asia; many studies have shown that the shallow Holocene aquifers have higher As than the deeper Pleistocene aquifers, but that is not always the case (Mukherjee and Fryar, 2008). In the United States, due to the diversity of geologic and aquifer units, depth dependence of As has been observed in many aquifers, but the detailed relationships are often unclear because they are influenced not only by geology/aquifer but also details of well construction. One limitation in assessing depth dependence is that information on well construction can be hard to find and may not be associated with samples in older datasets. Domestic wells are particularly challenging as homeowners often do not have information or knowledge about well construction.

20.4.1.1 The public has more access to arsenic data

In recent years the scientific community has put great effort and resources for making data publicly available. Federal agencies, such as the USGS, have clearinghouses for data, including the National Water Information System (https://waterdata.usgs.gov/nwis/). Scientists who receive grant funding from the National Science Foundation and other federal agencies are required to develop a data management plan, which includes plans on when/where data will be published. Organizations such as the Coalition for Publishing Data in the Earth and Space Sciences provide a directory of repositories for publishers and recommended best practices for publishing datasets. Data portals, including those supported by CUAHSI (https://www.cuahsi.org/data-models/publication/) and EDI (https://portal.edirepository.org/nis/home.jsp), provide infrastructure for publishing environmental data.

Open access codes, such as R and Python, have user communities that have developed specific packages, such as Shiny, to create data visualizations and interactive websites (https://shiny.rstudio.com/gallery/#user-showcase). For example, the Water Quality Portal Data Discovery tool (https://www.epa.gov/waterdata/water-quality-portal-data-discovery-tool), provides an easy to use interface allowing users to search, process, and display data from the water quality portal. The tool uses R to visualize the data selected from the portal and to assist users in data analysis.

Globally, researchers have developed online mapping tools, such as the Groundwater Assessment Platform (https://www.gapmaps.org/Home/Public) to allow for visualization of global databases of As and F concentrations and hazard maps; users can also overlay maps of climate, geology/lithology, population density, and soil characteristics. This easy-to-use mapping tool, which is also associated with a wiki platform to exchange information about As contamination of groundwater, allows anyone interested to examine datasets from across the globe and is an outstanding tool for both education and research.

20.4.2 Statistical models yield can predict drivers of arsenic release to groundwater

Statistical modeling has been applied at different scales to identify variables that influence As. At a nationwide scale, statistical modeling conducted by Ayotte et al. (2017) found that geologic sources, geochemical, hydrologic, and physical characteristics were significant predictors of elevated As. Drivers of As concentrations vary in regional analyses. For example, glacial sediment and aquifer characteristics, surficial and soil characteristics, and well construction variables, including screen length and distance of screens to overlying confining units (Erickson et al., 2018), were found to be drivers of elevated As in complex glacial aquifers in MN. In New England, specific rock types, high As concentrations in stream sediments, geochemical factors related to areas of Pleistocene marine inundation and proximity to intrusive granitic plutons, and hydrologic and landscape variables relating to groundwater residence time, were identified drivers of As occurrence in groundwater (Ayotte et al., 2006). In the Central Valley of CA, statistical modeling revealed that hydrologic position (reflecting distance along a flowpath), air temperature, soil geochemical characteristics, water table depth, and vertical fluxes, among others, were found to be important for predicting elevated As (Ayotte et al., 2016).

Combined, these studies have demonstrated the utility of statistical models for evaluating the relative importance of geologic, hydrologic, climatic, soils, and land use factors influencing As in groundwater at different scales. However, with any model, there are limitations of statistical models, including lack of datasets for potential influential variables. Small datasets, especially those with few observations of elevated As, are also challenging to utilize as they will likely yield poor model fits. Thus these models should not be used to predict As concentrations in a particular well, due to local variables that are often unknown, such as well construction and details of aquifer and geochemical conditions around the well, among others.

20.4.3 Statistical models can produce probability maps of arsenic risk

Statistical modeling can also be used to evaluate probability of elevated As. For example, using nationwide data from over 20,000 domestic wells, Ayotte et al. (2017) constructed a logistic regression model to quantify the probability of elevated As (>10 ppb) in groundwater across the United States. These results are presented as a map, showing different levels of probabilities of having As >10 ppb. These maps are especially useful for identifying areas of high risk. In addition, Ayotte et al. (2017) applied the mean probability of As exceedance to census data to estimate the potential population exposed to elevated As. The results suggested that there are potentially 2.1 million people in the United States using water from domestic wells that have As >10 ppb. States with a high percentage of domestic wells with As >10 ppb include Maine, New Hampshire, and Nevada. However, states with the highest population using domestic well water with As >10 ppb include Michigan, Ohio, and Indiana (Ayotte et al., 2017). This application greatly improves our ability to identify areas of risk in the United States where well testing and human health studies should be prioritized.

20.4.4 Arsenic concentrations may (but do not always) change over time

Temporal variability of As concentrations in well water has been the subject of many studies, most notably in regions of SE Asia. In the United States, studies on short-term variations have shown that As concentrations are influenced by sampling practices (Barcelona et al., 2005) and well construction (Erickson and Barnes, 2006).

Few generalizations can be gleaned from longer term studies. An evaluation of temporal patterns in 759 wells in Nevada over 20 years found a strong correlation between As concentrations from the same well, suggesting that As concentrations were stable over time (Steinmaus et al., 2005). A related study (Thundiyil et al., 2007) found little variation of As in groundwater between wet and dry seasons. In contrast, an analysis of 1245 wells by Ayotte et al. (2015) showed that As concentrations mostly showed small variations but in many wells, As concentrations varied dramatically, with As concentrations in public supply wells showing more temporal variation than domestic wells. The authors suggested that temporal variability is an important issue to consider and "has been overshadowed by datasets with large numbers of nondetect data."

Temporal trends in wells installed in glacial and bedrock aquifers in MN were analyzed to assess the impact of well construction on disrupting the aquifer geochemical equilibrium (Erickson et al., 2019). The results showed that well construction can alter the redox conditions of glacial aquifer for at least 1 year; over the study, As concentrations increased by 16% or more in 25% of wells in the glacial aquifers. In contrast, there was no significant change in As concentration in the bedrock wells. These results suggest that initial samples collected right after well construction should not be used to reflect long-term As concentration.

Seasonal patterns have also been observed in some datasets. For example, a recent study examined temporal variability of As in glacial and bedrock wells in NH (Levitt et al., 2019). In groundwater from the bedrock wells, As concentrations increased with recharge, suggesting a seasonal influence. Temporal patterns of As concentrations in wells have also been observed as relating to pumping and other human activities. For example, a study of As concentrations relating seasonal operation of public supply wells in Albuquerque NM (Bexfield and Jurgens, 2014) showed that wells are more likely to produce higher As concentrations in winter than in summer.

Making generalizations about temporal changes of As in well water is difficult to do when studies differ with respect to sampling design. Because the controls on temporal variability are not well-understood at this time and thus cannot be predicted, temporal variability should neither be assumed nor discounted for any individual system unless site-specific data are collected.

20.4.5 Human activities can promote arsenic release to groundwater

Human activity can influence As release to groundwater. Even the installation of a well can disturb the equilibrium redox conditions, which can affect As. Examples of human activity that can promote As release to groundwater, including well pumping, managed aquifer recharge, and introduction of human sources of organic carbon, are presented in the following subsections.

20.4.5.1 Impact of well pumping

Well pumping can affect As mobilization by changing flow paths, creating short-circuit pathways, and altering hydraulic gradients (Ayotte et al., 2011). These processes can physically introduce As to aquifers or change the prevailing biogeochemical conditions in the aquifer which promote As release.

In the Upper Floridan Aquifer, pumping of public supply wells induces a vertical gradient, causing oxic water of surficial aquifer to enter the more reducing aquifer (Ayotte et al., 2011; Price and Pichler, 2006; Mirecki et al., 2013). Because there are units with the aquifer, including the Suwanee Limestone, that contain As-rich pyrite (Price and Pichler, 2006), the introduction of oxygen through pumping has been shown to oxidize the pyrite, releasing As (Ayotte et al., 2011; Jones and Pichler, 2007; Mirecki et al., 2013). Pumping of domestic wells within the aquifer can also impact As release (Pichler et al., 2017), which is attributed to changing flow and geochemical conditions, introduction of oxygen, or mixing of fluids. In addition, just the presence of the wells, especially ones that cross multiple aquifers, can induce mobilization of As, even in the absence of significant pumping (Pichler et al., 2017).

Another example is from the St. Peter Aquifer in eastern Wisconsin, where the primary source of As to groundwater is As-rich sulfides. Although recharge can introduce oxygen to the sulfides in the recharge area where the St Peter subcrops, causing As release to groundwater, well installation can create the geochemical conditions needed to cause oxidation (Schreiber et al., 2000, 2003). Sulfide oxidation results in formation of HFOs, which can then sequester As through adsorption. Thus there is a secondary source of As that can form within the borehole environment that is then vulnerable to releasing As under reducing conditions. Field experiments conducted by Gotkowitz et al. (2004) examined the impact of domestic well pumping on redox changes and As mobilization. Under pumping rates and schedules for typical domestic wells, redox conditions changed rapidly during pumping, introducing oxygen, and increasing ORP. The authors hypothesized that the increasing oxygen would increase As concentrations, but that was not observed. Instead, As concentrations increased significantly during nonpumping periods, when reducing conditions prevailed.

These results suggest that both oxidation of sulfides and reduction of As-bearing Fe hydroxides can mobilize As to groundwater, depending on redox conditions of the vicinity of the well (Gotkowitz et al., 2004).

A third example is from the basin-fill aquifers of the San Joaquin Valley, CA. A 2018 study showed that well pumping increases As concentrations in groundwater in the Tulare Basin (Smith et al., 2018). Through modeling, the authors attribute the increased As to aquifer compaction, which results in draining of As-bearing pore fluids within clays.

20.4.5.2 Managed aquifer recharge

Managed aquifer recharge uses water management methods to recharge an aquifer for use in periods of need and is being increasingly used in many municipalities across the United States for storage of water resources (NGWA, 2020). However, mobilization of As and other trace elements into groundwater in these systems in a variety of settings has caused concern (Neil et al., 2012).

In Florida, managed aquifer recharge projects have injected treated drinking and reclaimed water into the Floridan Aquifer during the wet season for storage and later use in the dry season (Mirecki et al., 2013). Mobilization of As to groundwater has been observed in the aquifer as a result of managed aquifer recharge and has been attributed to introduction of oxygen in the recharged water, promoting oxidation of As-bearing sulfides (Pichler et al., 2011; Jones and Pichler, 2007; Arthur et al., 2002), as described above. However, although As is released to groundwater, it can undergo reactions in the aquifer, including sorption to HFOs (Vanderzalm et al., 2011) or coprecipitation within Fe sulfides (Mirecki et al., 2013), which can attenuate As in the aquifer in the short term.

In the San Joaquin Valley of California, managed aquifer recharge has been conducted near Stockton, where surplus water from reservoirs has been infiltrated via surface impoundments into an alluvial aquifer (McNab et al., 2009). A study combining sampling of recharge and groundwater and analysis of stable isotopes and tracers with geochemical modeling, suggests that release of As and other trace elements is likely due to desorption induced by infiltration of higher pH recharge (McNab et al., 2009).

A third example of As mobilized through aquifer recharge projects is from Orange Co., California (Fakhreddine et al., 2015). In this setting, highly purified water is infiltrated via surface recharge basins into an unconsolidated aquifer. Although recharge and the shallow aquifer are oxic, As is mobilized in groundwater, suggesting that redox processes are not responsible for As release. The recharge does not contain competitive ions and has a pH similar to that of groundwater. The results from lab experiments suggest that As, which resides in clay fraction of the sediments, is desorbed from clays due to changes in ionic composition. Laboratory column experiments using cation pretreatments were successful in limiting As desorption, underscoring that research on fundamental biogeochemical processes can improve treatment technology (Fakhreddine et al., 2015).

20.4.5.3 Introduction of anthropogenic organic carbon can drive reductive dissolution

Organic carbon drives reductive dissolution of HFOs, which can promote As release into groundwater. Identifying the primary organic carbon source driving reductive dissolution in SE Asia was the subject of a heated discussion in the 2000s (see McArthur et al., 2001, 2004; Harvey et al., 2002, 2006; Neumann et al., 2010). In the United States, sources of organic carbon that drive Fe reduction and As release were also studied in the 2000s within glacial sediments in the northern United States (Erickson and Barnes, 2005; Root et al., 2005, 2010).

Anthropogenic sources of organic carbon can also drive As release. Research on elevated As in landfill leachate suggested that biodegradation of organic waste triggered release of naturally occurring As from reductive dissolution of HFOs (deLemos et al., 2006; Hering et al., 2009). Similar observations of As release have been made at sites with sewage plumes (Pinel-Raffaitin et al., 2007; Stollenwerk and Colman, 2004), petroleum spills (Cozzarelli et al., 2016; Burgess and Pinto, 2005; Ghosh et al., 2003; Brown et al., 2010; Hering et al., 2009; Ziegler et al., 2017), and enhanced reductive bioremediation, which involves addition of organic carbon to facilitate biodegradation of contaminants (Hering et al., 2009; Suthersan and Horst, 2008; McLean et al., 2006; Ziegler et al., 2015). These studies show that introduction of bioavailable organic carbon to aquifers with natural-occurring As can cause unanticipated As release and should be monitored closely and over the long term (Cozzarelli et al., 2016; Ziegler et al., 2017; Ziegler et al., 2017).

20.4.6 Research leads to improved technology for arsenic detection and treatment

20.4.6.1 Improved remediation methods

Remediation of As utilizes a variety of processes, including adsorption, lime softening, anion exchange, oxidation, precipitation/coagulation, and membrane filtration. Treatment technologies have benefited from the plethora of information from scientific studies, including quantification of adsorption properties and oxidation kinetics, and testing and development of nanoparticles and novel materials, including membranes.

Design of permeable reactive barriers for in situ remediation of As-contaminated groundwater has also greatly benefitted from knowledge of As interaction with solids, including zero-valent iron (ZVI) (Kanel et al., 2005, 2006; Lackovic et al., 2000; Farrell et al., 2001; Su and Puls, 2001). During reaction, as ZVI oxidizes and corrodes to form hydrous FeII/FeIII oxides, As adsorbs and coprecipitates with the oxides (Su and Puls, 2004; Kanel et al., 2006; Su and Puls, 2003; Manning et al., 2002; Leupin and Hug, 2005; Klas and Kirk, 2013). Nanotechnology has also been an area of intense research over the past decade, providing applications to water treatment, including use of nanoparticles and nanocomposites (Crane and Scott, 2012; Tesh and Scott, 2014). Nanoparticles, including nano-sized ZVI (nZVI), can be injected into the ground as a dry powder or slurry to treat As-contaminated groundwater (Stefaniuk et al., 2016). The nZVI can be utilized within a permeable reactive barrier or mobilized within the plume, allowing the particles to treat As during transport (see examples in Stefaniuk et al., 2016; Cundy et al., 2008; Thiruvenkatachari et al., 2008).

Despite the promise of nanoparticles for in situ treatment of As in groundwater, further research on potential ecotoxicity of the nZVI is needed. One recent study (Crampon et al., 2019) found that nZVIs induced changes in the bacterial community composition, but changes in bacterial communities from nZVIs have not been observed in all studies (Kirschling et al., 2010). Another issue is the potential for uncontrolled release of nZVIs (e.g., Kanel et al., 2007). Use of nanocomposites has been proposed as an approach to stabilize nZVIs within a structure to avoid this problem (Tesh and Scott, 2014).

20.4.6.2 Development of biosensors to detect As

Advances in portable testing methods in the 2000s included development of biosensors for real-time detection of As (Stocker et al., 2003; Van Der Meer et al., 2004; Diesel et al., 2009; Chen and Rosen, 2014; Berberich et al., 2019). There are different kinds of biosensors, including cell-free (protein or DNA based) or whole cell-based (Chen and Rosen, 2014). Advantages of using biosensors to detect As include sensitivity, accuracy, low manufacturing cost, low detection limit, fast response time, ease of use, and portability (Chen and Rosen, 2014). However, there are disadvantages, such as the detection of only inorganic forms of As and other issues such as shelf life (Chen and Rosen, 2014). Research on incorporating sensors with microfluidics has resulted in miniaturized, integrated lab-on-a-chip devices (Yogarajah and Tsai, 2015) that utilize small volumes of fluids. These systems have the advantage of faster reaction times, less consumption of reagents, and reduced waste. In time, biosensors may become an effective screening tool for making "real-time" measurements of As levels in well water, which would be an improvement over test kits that produce hazardous arsine gas and generate hazardous waste.

20.5 Current concerns about arsenic in groundwater in the United States

Despite the many advances that have been made in addressing As in groundwater, there are still people in the United States who drink water with unsafe levels of As. Although public water supplies are regulated under the Safe Drinking Water Act, domestic wells are not, leaving homeowners with the responsibility of having their wells tested.

20.5.1 Most, but not all, public water supplies are meeting the drinking water standard

After an initial spike in 2008, violations of the As MCL by public water systems have decreased since the Final Arsenic Rule was enforced in 2006 (Foster et al., 2019). The percentage of public water system violations declined from 1.3% in 2008 to 0.55% in 2017; the population served by public water systems with MCL violations has decreased by more than 1 million people during this time period (Foster et al., 2019). However, violations still occur; California and Texas have the highest mean annual number of violations and the largest populations served by violating systems (Foster et al., 2019). Thus there are regions of the country where populations are still being exposed to elevated As through public water supplies, even close to 15 years after the enforcement of the Final Arsenic Rule.

20.5.2 Homeowners are responsible for testing of private well water

There are over 44 million people in the conterminous United States that rely on private wells; recent studies have estimated over 2 million of them may be exposed to elevated As (Ayotte et al., 2017). There are no federal regulations on well testing. Well sampling and analysis can be expensive; thus many homeowners do not regularly sample and test their wells, if at all (Flanagan et al., 2018; Flanagan et al., 2015; Flanagan et al., 2015). Even with testing, homeowners are left with making decisions about what action to take to treat As. For example, if a household treatment system is installed, the performance of the systems is difficult for homeowners to assess and the system can fail without homeowner knowledge (Walker et al., 2008; Zheng and Flanagan, 2017). Public health efforts to promote testing have had varied effectiveness and reveal disparities (Flanagan et al., 2016).

Few states mandate well testing. One exception is the New Jersey Private Well Testing Act (PWTA) of 2002, which requires testing of well water for As during real estate transactions in 12 counties of NJ. Flanagan et al. (2016) used survey data to examine how the PWTA has impacted household testing and mitigation behavior in these counties. The PWTA has increased testing rates, resulting in identification of more wells with As, and more treatment for wells with elevated As (Flanagan et al., 2016). However, results also indicate that residents often do not remember what type of treatment they are using and are thus not likely performing needed maintenance and monitoring, underscoring the need for additional support for homeowners even after action is taken. Another survey examined rates of testing and correlation of testing rates with education and socioeconomic status in NJ (Flanagan et al., 2016). Well owners with more education were 10 times more likely to participate in testing; when free testing was offered, about half participated and those who participated were of higher income and education (Flanagan et al., 2016). These results suggest that although efforts to promote testing are successful in increasing testing rates, socioeconomic status remains a barrier to testing.

Comparative survey experiments in NJ and ME were conducted to identify predictors of well water testing, treatment behavior, and routes of As exposure (Flanagan et al., 2016). The results show that socioeconomic status is not correlated with spatial location of high As. However, the study did show that there are socioeconomic disparities with respect to with differing rates of well testing, treatment, and choices of alternative water sources, suggesting that risk assessment should be based not only on spatial occurrence of As but also on social vulnerability of populations exposed to elevated As.

20.6 Next steps

As Zheng and Ayotte (2015) suggest, the United States is "at a crossroads" for reducing As exposure from private well water. Below are several recommendations, gleaned from the recent literature, on methods to reduce As exposure from drinking water in the United States.

20.6.1 Required testing would improve identification of wells with elevated As

Zheng and Flanagan (2017) argue that universal testing of As in private wells, along with community engagement, would greatly reduce the exposure of the United States population to As. They argue that a national screening requirement for private wells for As is the optimal way to reduce exposure and improve public health. Enacting legislation to require well testing, however, has been challenging. In New Jersey, legislation requiring well testing was passed in 2002. Other states have requirements for well testing for new construction and/or real estate transactions, but only five states include As in the testing (Zheng and Flanagan, 2017).

20.6.2 More support is needed for homeowners, especially in areas of high risk

Additional support for homeowners is needed to manage well water quality, including choices of water treatment, and continued monitoring and maintenance. This is especially needed in areas with socially vulnerable populations (Flanagan et al., 2016). For example, elevated As concentrations have been measured in drinking water supplies of the Navajo Nation in the Southwest and on Native American reservations in South Dakota (Bird et al., 2020; Carter et al., 1998; Powers et al., 2019). The results from the Navajo Nation study show that 15% of tested water sources had As >10 ppb (Hoover et al., 2017). Poor water quality has been a pervasive issue on tribal lands, as nationally, 12% of tribal public water systems have health-based violations, compared to 6% of nontribal water systems (Hoover et al., 2017). The Strong Heart Water Study was recently initiated as an intervention study to reduce As exposure in native communities in North and South Dakota (Powers et al., 2019). In this study, over 350 households are included in a water testing program. In addition, a pilot study examined the effectiveness of a point-of-use filter for As removal. Although the overall study is still underway, with expected completion in late 2021, results from the pilot study suggest that the filter was effective for reducing As burden for a period of 9 months (Powers et al., 2019). If filters are determined to be effective for protection, long-term financial and social support is needed to make sure that the filters have timely replacement and maintenance, and that testing is done routinely to ensure that the filters remain effective.

In areas with elevated As in groundwater, Zheng and Ayotte (2015) ask "does it still make sense to drill wells into rock formations where we know that there is a high likelihood of encountering unsafe levels of arsenic?" Locating alternative water sources, such as connecting to a public supply source where feasible; developing localized public community water supplies that would be under the jurisdiction of the Safe Drinking Water Act; or retrofitting wells to withdraw water from As-free aquifers, are all possibilities that should be explored.

20.6.3 More data are needed for prediction of spatial and temporal patterns

Statistical modeling has greatly improved spatial mapping of where As is (and may be) a concern and also can identify the important controls on elevated As. Despite these advances, and the parallel advances on fundamental processes controlling As release, we still do not have a clear picture of how these processes interact with each other at the field scale to cause the spatial variability observed in As concentrations in wells. Depth dependent controls on As are often unresolved in specific areas, as they involve complex interplay of geology, well hydraulics, fluid flow, and reactive transport. Related to this is our lack of knowledge on drivers of temporal variability that is observed in some datasets. This is of utmost concern for well testing, as if the concentrations change over days to weeks to months to seasons to years, providing guidance to homeowners on how frequently to test their well is critical, assuming that the homeowners are amenable to testing and have the resources to cover the cost. More field-scale studies, perhaps at sites where conditions are well-studied and variables are well-constrained, would help address these two challenges of spatial and temporal variability.

Currently, reactive transport models that simulate As release from geogenic sources into groundwater at the field scale are limited by the complexities of models and lack of data to support calibration. Developing screening methods to identify (at the field scale) which processes and variables are most important for each site could be a first step, followed by more complex modeling if it is needed. And, as has been mentioned in other papers (see O'Day et al., 2005), we need to improve the integration between laboratory results and field observations.

Statistical models have shown that predictive variables from national data sets can be used to estimate high As, even in regions where there are few data on As concentrations (Ayotte et al., 2017). For example, the Ayotte et al. (2017) statistical model predicts high As in parts of eastern New Mexico and northern Wisconsin where private well maps indicate no or sparse data, thus identifying areas of concern where domestic well water sampling efforts can be focused. The authors state, "Estimates of the population in the United States using domestic well water with high concentrations of arsenic may not accurately represent the population at risk if they do not account for unsampled areas," underscoring the importance of continued well data collection.

Last, research has focused on the current conditions of As in groundwater, but it is important to think about the future. Recent work (Ayotte et al., 2017) showed that climate variables (precipitation, recharge) are strong predictors of elevated As concentrations in groundwater across the United States. As we are experiencing climate change that impacts precipitation patterns in the United States, future research is needed to examine how these changes may affect As in groundwater.

20.6.4 Education and effective communication can improve awareness and action

20.6.4.1 Tools and training for analysis of big data sets

Over the past 20 years, massive amounts of data have been collected, and with much of that data being made publicly available, there are opportunities using meta-analysis methods to extract even more information about As in groundwater. However, the data are in different places and may be hard to find, making "datamining" a daunting task. Online tools such as the Groundwater Assessment Platform are wonderful ways to visualize big data sets to look at patterns and relationships. Students in K-12 and higher education programs would greatly benefit from learning how to access and analyze these and other large datasets using open source codes and data visualization applications in R and Python.

20.6.4.2 Communication with the public about the risks of As

Medical professionals are the first line of defense for protecting health, especially of children. The American Association of Pediatricians issued a policy statement in 2009 calling on pediatricians to encourage households on private wells to test their water (AAP, 2009). However, the extent to which this recommendation is given and then followed has not been determined.

Researchers who study As have gathered substantial knowledge, but we must also remember the importance of communicating our findings to the public, especially with homeowners, medical professionals, extension agents, and others at the front line of public health and local, regional, and state officials who make policy. As scientists, we can make the greatest difference if we can clearly explain how research on As in groundwater contributes to societal improvements, and that our results can be used to inform decision-making on health and environmental issues.

References

AAP, 2009. Drinking water from private wells and risks to children. Pediatrics 123 (6), 1599–1605.
Ahmann, D., Roberts, A.L., Krumholz, L.R., Morel, F.M., 1994. Microbe grows by reducing arsenic. Nature 371 (6500), 750.
Ahuja, S., 2008. Arsenic Contamination of Groundwater: Mechanism, Analysis, and Remediation. Wiley.
Ali, I., Jain, C.K., 2004. Advances in arsenic speciation techniques. Int. J. Environ. Anal. Chem. 84 (12), 947–964.
Anning, D.W., A.P. Paul, T.S. McKinney, J.M. Huntington, L.M. Bexfield, S.A. Thiros, 2012. Predicted nitrate and arsenic concentrations in basin-fill aquifers of the southwestern United States. Scientific Investigations Report 2012–5065. U.S. Geological Survey. 78 p.
Arthur, J.D., Dabous, A.A., Cowart, J.B., 2002. Mobilization of arsenic and other trace elements during aquifer storage and recovery, southwest Florida. US Geological Survey Artificial Recharge Workshop Proceedings. US Geological Survey, Denver, CO.
Ayotte, J., Gronberg, J., Apodaca, L., 2011. Trace elements and radon in groundwater across the United States, 1992-2003. Scientific Investigations Report 2011-5059. US Geological Survey, Reston, VA.
Ayotte, J., N. MH, G. Robinson, and R. Moore, 1999. Relation of arsenic, iron, and manganese in ground water to aquifer type, bedrock lithogeochemistry, and land use in the New England coastal basin, Water-Resources Investigations Report 99-4162. U.S. Geological Survey: Pembroke, NH. p. 61.
Ayotte, J.D., Nolan, B.T., Gronberg, J.A., 2016. Predicting arsenic in drinking water wells of the Central Valley, California. Environ. Sci. Technol. 50 (14), 7555–7563.
Ayotte, J.D., Nolan, B.T., Nuckols, J.R., Cantor, K.P., Robinson, G.R., Baris, D., et al., 2006. Modeling the probability of arsenic in groundwater in New England as a tool for exposure assessment. Environ. Sci. Technol. 40 (11), 3578–3585.
Ayotte, J.D., Montgomery, D.L., Flanagan, S.M., Robinson, K.W., 2003. Arsenic in groundwater in eastern New England: occurrence, controls, and human health implications. Environ. Sci. Technol. 37 (10), 2075–2083.
Ayotte, J.D., Medalie, L., Qi, S.L., Backer, L.C., Nolan, B.T., 2017. Estimating the high-arsenic domestic-well population in the conterminous United States. Environ. Sci. Technol. 51.
Ayotte, J.D., Belaval, M., Olson, S.A., Burow, K.R., Flanagan, S.M., Hinkle, S.R., et al., 2015. Factors affecting temporal variability of arsenic in groundwater used for drinking water supply in the United States. Sci. Total Environ. 505, 1370–1379.
Ayotte, J.D., Szabo, Z., Focazio, M.J., Eberts, S.M., 2011. Effects of human-induced alteration of groundwater flow on concentrations of naturally-occurring trace elements at water-supply wells. Appl. Geochem. 26 (5), 747–762.
Ball, J.W., R.B. McMleskey, and D.K. Nordstrom, 2010. Water-chemistry data for selected springs, geysers, and streams in Yellowstone National Park, Wyoming, 2006-2008. Open-File Report 2010-1192. U.S. Geological Survey, Reston, VA.
Barcelona, M., Varljen, M., Puls, R., Kaminski, D., 2005. Ground water purging and sampling methods: history vs. hysteria. Groundw. Monit. Rem. 25 (1), 52–62.
Barringer, J.L., Mumford, A., Young, L.Y., Reilly, P.A., Bonin, J.L., Rosman, R., 2010. Pathways for arsenic from sediments to groundwater to streams: biogeochemical processes in the Inner Coastal Plain, New Jersey, USA. Water Res. 44 (19), 5532–5544.
Barringer, J.L., Reilly, P.A., Eberl, D.D., Blum, A.E., Bonin, J.L., Rosman, R., et al., 2011. Arsenic in sediments, groundwater, and streamwater of a glauconitic Coastal Plain terrain, New Jersey, USA—chemical "fingerprints" for geogenic and anthropogenic sources. Appl. Geochem. 26 (5), 763–776.
Bartolino, J.R., Hopkins, C.B., 2016. Ambient water quality in aquifers used for drinking-water supplies, Gem County, southwestern Idaho, 2015. Scientific Investigations Report 2016-5170, U.S. Geological Survey.
Beaulieu, B.T., Savage, K.S., 2005. Arsenate adsorption structures on aluminum oxide and phyllosilicate mineral surfaces in smelter-impacted soils. Environ. Sci. Technol. 39 (10), 3571–3579.
Bednar, A.J., Garbarino, J.R., Ranville, J.F., Wildeman, T.R., 2002. Preserving the distribution of inorganic arsenic species in groundwater and acid mine drainage samples. Environ. Sci. Technol. 36 (10), 2213–2218.
Bednar, A.J., Garbarino, J.R., Burkhardt, M.R., Ranville, J.F., Wildeman, T.R., 2004. Field and laboratory arsenic speciation methods and their application to natural-water analysis. Water Res. 38 (2), 355–364.
Berberich, J., Li, T., Sahle-Demessie, E., 2019. Chapter 11—Biosensors for monitoring water pollutants: a case study with arsenic in groundwater. In: Ahuja, S. (Ed.), Separation Science and Technology. Academic Press, pp. 285–328.
Bexfield, L.M., Jurgens, B.C., 2014. Effects of seasonal operation on the quality of water produced by public-supply wells. Groundwater 52 (S1), 10–24.
Bexfield, L.M., Plummer, L.N., 2003. Occurrence of arsenic in ground water of the Middle Rio Grande Basin, central New Mexico. In: Welch, A.H., Stollenwerk, K.G. (Eds.), Arsenic in Ground Water: Geochemistry and Occurrence. Kluwer Academic Publishers, Boston, MA, pp. 295–327.
Bird, K.S., Navarre-Sitchler, A., Singha, K., 2020. Hydrogeological controls of arsenic and uranium dissolution into groundwater of the Pine Ridge Reservation, South Dakota. Appl. Geochem. 104522.

Blake, J.M., Peters, S.C., 2015. The occurrence and dominant controls on arsenic in the Newark and Gettysburg basins. Sci. Total Environ. 505, 1340–1349.

Bone, S.E., Gonneea, M.E., Charette, M.A., 2006. Geochemical cycling of arsenic in a coastal aquifer. Environ. Sci. Technol. 40 (10), 3273–3278.

Borrok, D.M., Lenz, R.M., Jennings, J.E., Gentry, M.L., Steensma, J., Vinson, D.S., 2018. The origins of high concentrations of iron, sodium, bicarbonate, and arsenic in the Lower Mississippi River Alluvial Aquifer. Appl. Geochem. 98, 383–392.

Bowell, R.J., Alpers, C.N., Jamieson, H.E., Nordstrom, D.K., Majzlan, J., 2014. The environmental geochemistry of arsenic—an overview. Rev. Mineral. Geochem. 79 (1), 1–16.

Brannon, J.M., Patrick, W.H., 1987. Fixation, transformation, and mobilization of arsenic in sediments. Environ. Sci. Technol. 21 (5), 450–459.

Brown, R.A., K.E. Patterson, M.D. Zimmerman, G.T. Ririe, 2010. Attenuation of naturally occurring arsenic at petroleum hydrocarbon–impacted sites. In: Fields, K.A., Wickramanayake, G.B. (Eds.), Seventh International Conference on Remediation of Chlorinated and Recalcitrant Compounds. Battelle Memorial Institute, Columbus, OH.

Bundschuh, J., Holländer, H.M., Ma, L.Q., 2014. In-situ remediation of arsenic-contaminated sites. In: Bundschuh, J., Bhattacharya, P. (Eds.), Arsenic in the Environment. CRC Press.

Bundschuh, J., Armienta, M.A., Birkle, P., Bhattacharya, P., Matschullat, J., Mukherjee, A.B., 1997. Natural arsenic in groundwaters of Latin America. In: Bundschuh, J., Bhattacharya, P. (Eds.), Arsenic in the Environment. CRC Press.

Burgess, W.G., Pinto, L., 2005. Preliminary observations on the release of arsenic to groundwater in the presence of hydrocarbon contaminants in UK aquifers. Mineral. Mag. 69 (5), 887–896.

Busbee, M.W., Kocar, B.D., Benner, S.G., 2009. Irrigation produces elevated arsenic in the underlying groundwater of a semi-arid basin in Southwestern Idaho. Appl. Geochem. 24 (5), 843–859.

Camacho, L.M., Gutierrez, W., Teresa Alarcon-Herrera, M., et al., 2011. Occurrence and treatment of arsenic in groundwater and soil in northern Mexico and southwestern USA. Chemosphere 83 (3), 211–225.

Campbell, K.M., Malasarn, D., Saltikov, C.W., Newman, D.K., Hering, J.G., 2006. Simultaneous microbial reduction of iron(III) and arsenic(V) in suspensions of hydrous ferric oxide. Environ. Sci. Technol. 40 (19), 5950–5955.

Carter, J., Sando, S., Hayes, T., Hammond, R., 1998. Source, occurrence, and extent of arsenic contamination in the Grass Mountain area of the Rosebud Indian Reservation, South Dakota. Water-Resources Investigations Report 97-4286, U.S. Geological Survey.

Chapman, M.J., Cravotta III, C.A., Szabo, Z., Lindsay, B.D., 2013. Naturally occurring contaminants in the Piedmont and Blue Ridge crystalline-rock aquifers and Piedmont Early Mesozoic basin siliciclastic-rock aquifers, eastern United States, 1994–2008. Scientific Investigations Report 2013-5072, U.S. Geological Survey.

Chen, J., Rosen, B.P., 2014. Biosensors for inorganic and organic arsenicals. Biosensors 4 (4), 494–512.

Chiu, V.Q., Hering, J.G., 2000. Arsenic adsorption and oxidation at manganite surfaces. 1. Method for simultaneous determination of adsorbed and dissolved arsenic species. Environ. Sci. Technol. 34 (10), 2029–2034.

Cozzarelli, I.M., Schreiber, M.E., Erickson, M.L., Ziegler, B.A., 2016. Arsenic cycling in hydrocarbon plumes: secondary effects of natural attenuation. Ground Water 54 (1), 35–45.

Crampon, M., Joulian, C., Ollivier, P., Charron, M., Hellal, J., 2019. Shift in natural groundwater bacterial community structure due to zero-valent iron nanoparticles (nZVI). Front. Microbiol. 10, 533.

Crane, R.A., Scott, T.B., 2012. Nanoscale zero-valent iron: future prospects for an emerging water treatment technology. J. Hazard. Mater. 211-212, 112–125.

Cullen, W., Reimer, K., 1989. Arsenic speciation in the environment. Chem. Rev. 89, 713–764.

Cundy, A.B., Hopkinson, L., Whitby, R.L.D., 2008. Use of iron-based technologies in contaminated land and groundwater remediation: a review. Sci. Total Environ. 400 (1), 42–51.

deLemos, J.L., Bostick, B.C., Renshaw, C.E., Sturup, S., Feng, X.H., 2006. Landfill-stimulated iron reduction and arsenic release at the Coakley Superfund Site (NH). Environ. Sci. Technol. 40 (1), 67–73.

Deschamps, E., Matschullat, J., 2011. Arsenic: natural and anthropogenic. In: Bundschuh, J., Bhattacharya, P. (Eds.), Arsenic in the Environment, 4. CRC Press, Boca Raton, FL.

DeSimone, L.A., Hamilton, P.A., 2009. Quality of Water From Domestic Wells in Principal Aquifers of the United States, 1991-2004. Scientific Investigations Report 2008–5227, U.S. Geological Survey.

DeSimone, L.A., McMahon, P.B., Rosen, M.R., 2014. The quality of our Nation's waters—Water quality in Principal Aquifers of the United States, 1991–2010, Circular, 1360. 151, U.S. Geological Survey.

DeVitre, R., Belize, N., Tessier, A., 1991. Speciation and adsorption of arsenic on diagenetic iron hydroxides. Limnol. Oceanogr. 36, 1480–1485.

Diesel, E., Schreiber, M., van der Meer, J.R., 2009. Development of bacteria-based bioassays for arsenic detection in natural waters. Anal. Bioanal. Chem. 394 (3), 687–693.

Dixit, S., Hering, J.G., 2003. Comparison of arsenic(V) and arsenic(III) sorption onto iron oxide minerals: implications for arsenic mobility. Environ. Sci. Technol. 37 (18), 4182–4189.

Dixit, S., Hering, J.G., 2006. Sorption of Fe(II) and As(III) on goethite in single- and dual-sorbate systems. Chem. Geol. 228 (1–3), 6–15.

Drummond, D.D., Bolton D.W., 2010. Arsenic in groundwater in the Coastal Plain aquifers of Maryland. Maryland Geological Survey Report of Investigations No. 78.

Dzombak, D.A., Morel, F.M.M., 1987. Adsorption of inorganic pollutants in aquatic systems. J. Hydraul. Eng.—ASCE 113 (4), 430–475.

Erickson, M.L., Barnes, R.J., 2006. Arsenic concentration variability in public water system wells in Minnesota, USA. Appl. Geochem. 21 (2), 305–317.

Erickson, M.L., Barnes, R.J., 2005. Glacial sediment causing regional-scale elevated arsenic in drinking water. Ground Water 43 (6), 796–805.

Erickson, M.L., Malenda, H.F., Berquist, E.C., Ayotte, J.D., 2019. Arsenic concentrations after drinking water well installation: time-varying effects on arsenic mobilization. Sci. Total Environ. 678, 681–691.

Erickson, M.L., Elliott, S.M., Christenson, C., Krall, A.L., 2018. Predicting geogenic arsenic in drinking water wells in glacial aquifers, north-central USA: accounting for depth-dependent features. Water Resour. Res. 54 (12), 10,172–10,187.

Fakhreddine, S., Dittmar, J., Phipps, D., Dadakis, J., Fendorf, S., 2015. Geochemical triggers of arsenic mobilization during managed aquifer recharge. Environ. Sci. Technol. 49 (13), 7802–7809.

Farrell, J., Wang, J., O'Day, P., Conklin, M., 2001. Electrochemical and spectroscopic study of arsenate removal from water using zero-valent iron media. Environ. Sci. Technol. 35 (10), 2026–2032.

Fendorf, S., Eick, M.J., Grossl, P., Sparks, D.L., 1997. Arsenate and chromate retention mechanisms on goethite.1. Surface structure. Environ. Sci. Technol. 31 (2), 315–320.

Fisher, R., 2002. Groundwater quality in Kentucky: arsenic. Kentucky Geological Survey Information Circular 5, Series XII.

Flanagan, S.M., Brown C.J., 2017. Arsenic and uranium in private wells in Connecticut, 2013-15. Open-File Report 2017-1046, U.S. Geological Survey.

Flanagan, S.V., Gleason, J.A., Spayd, S.E., Procopio, N.A., Rockafellow-Baldoni, M., Braman, S., et al., 2018. Health protective behavior following required arsenic testing under the New Jersey Private Well Testing Act. Int. J. Hyg. Environ. Health 221 (6), 929–940.

Flanagan, S.V., Marvinney, R.G., Zheng, Y., 2015. Influences on domestic well water testing behavior in a Central Maine area with frequent groundwater arsenic occurrence. Sci. Total Environ. 505, 1274–1281.

Flanagan, S.V., Marvinney, R.G., Johnston, R.A., Yang, Q., Zheng, Y., 2015. Dissemination of well water arsenic results to homeowners in Central Maine: influences on mitigation behavior and continued risks for exposure. Sci. Total Environ. 505, 1282–1290.

Flanagan, S.V., Spayd, S.E., Procopio, N.A., Marvinney, R.G., Smith, A.E., Chillrud, S.N., et al., 2016. Arsenic in private well water part 3 of 3: socioeconomic vulnerability to exposure in Maine and New Jersey. Sci. Total Environ. 562, 1019–1030.

Flanagan, S.V., Spayd, S.E., Procopio, N.A., Chillrud, S.N., Ross, J., Braman, S., et al., 2016. Arsenic in private well water part 2 of 3: who benefits the most from traditional testing promotion? Sci. Total Environ. 562, 1010–1018.

Flanagan, S.V., Spayd, S.E., Procopio, N.A., Chillrud, S.N., Braman, S., Zheng, Y., 2016. Arsenic in private well water part 1 of 3: impact of the New Jersey Private Well Testing Act on household testing and mitigation behavior. Sci. Total Environ. 562, 999–1009.

Focazio, M.J., Welch, A.H., Watkins, S.A., Helsel, D.R., Horn, M.A. 2000. A retrospective analysis on the occurrence of arsenic in ground-water resources of the United States and limitations in drinking-water-supply characterizations, Water Resources Investigations Report 99-4279. U.S. Geological Survey, Reston VA.

Focazio, M.J., Tipton, D., Dunkle Shapiro, S., Geiger, L.H., 2006. The chemical quality of self-supplied domestic well water in the United States. Groundw. Monit. Rem. 26 (3), 92–104.

Ford, R.G., 2002. Rates of hydrous ferric oxide crystallization and the influence on coprecipitated arsenate. Environ. Sci. Technol. 36 (11), 2459–2463.

Foster, A.L., Brown, J., Gordon, E., Parks, G.A., 2003. X-ray absorption fine structure study of As(V) and Se(IV) sorption complexes on hydrous Mn oxides. Geochim. Cosmochim. Acta 67 (11), 1937–1953.

Foster, A.L., 2003. Spectroscopic investigations of arsenic species in solid phases. In: Welch, A.H., Stollenwerk, K.G. (Eds.), Arsenic in Ground Water. Kluwer Academic Publishers, Boston, MA.

Foster, S.A., Pennino, M.J., Compton, J.E., Leibowitz, S.G., Kile, M.L., 2019. Arsenic drinking water violations decreased across the United States following revision of the maximum contaminant level. Environ. Sci. Technol. 53 (19), 11478–11485.

Frans, L.M., Rupert, M.G., Hunt Jr, C.D., Skinner, K.D., 2012. Groundwater quality in the Columbia Plateau, Snake River Plain, and Oahu basaltic-rock and basin-fill aquifers in the Northwestern United States and Hawaii, 1992-2010. Scientific Investigations Report 2012-5123. U.S. Geological Survey, Reston, VA.

Fuller, C.C., Davis, J.A., Waychunas, G.A., 1993. Surface-chemistry of ferrihydrite 2. Kinetics of arsenate adsorption and coprecipitation. Geochim. Cosmochim. Acta 57 (10), 2271–2282.

Garbarino, J.R., Bednar, A.J., Burkhardt, M.R., 2002. Methods of analysis by the U.S. Geological Survey National Water Quality Laboratory—arsenic speciation in natural-water samples using laboratory and field methods. Water Resources Investigations Report 02-4144. U.S. Geological Survey, Denver, CO.

Gates, J.B., Nicot, J.P., Scanlon, B.R., Reedy, R.C., 2011. Arsenic enrichment in unconfined sections of the southern Gulf Coast aquifer system, Texas. Appl. Geochem. 26 (4), 421–431.

Georgiadis, M., Cai, Y., Solo-Gabriele, H.M., 2006. Extraction of arsenate and arsenite species from soils and sediments. Environ. Pollut. 141 (1), 22–29.

Ghosh, R., W. Deutsch, S. Geiger, K. McCarthy, and D. Beckmann. 2003. Geochemistry, fate and transport of dissolved arsenic in petroleum hydrocarbon-impacted groundwater. Proceedings of the NGWA/API 20th Annual Conference and Exposition of Petroleum Hydrocarbons and Organic Chemicals in Groundwater. National Groundwater Association.

Glenn, S.M., Lester, L.J., 2010. An analysis of the relationship between land use and arsenic, vanadium, nitrate and boron contamination in the Gulf Coast aquifer of Texas. J. Hydrol. 389 (1–2), 214–226.

Goldberg, S., Johnston, C.T., 2001. Mechanisms of arsenic adsorption on amorphous oxides evaluated using macroscopic measurements, vibrational spectroscopy, and surface complexation modeling. J. Colloid Interface Sci. 234 (1), 204–216.

Goldberg, S., 1986. Chemical modeling of arsenate adsorption on aluminum and iron-oxide minerals. Soil Sci. Soc. Am. J. 50 (5), 1154–1157.

Goldberg, S., 2002. Competitive adsorption of arsenate and arsenite on oxides and clay minerals. Soil Sci. Soc. Am. J. 66 (2), 413–421.

Gosselin, D.C., Klawer, L.M., Joeckel, R.M., Harvey, F.E., Reade, A.R., McVey, K., 2006. Arsenic in groundwater and rural public water supplies in Nebraska, USA. Gt. Plains Res. 137–148.

Gotkowitz, M., Schreiber, M., Simo, J., 2004. Effects of water use on arsenic release to well water in a confined aquifer. Ground Water 42 (4), 568–575.

Grafe, M., Eick, M.J., Grossl, P.R., Saunders, A.M., 2002. Adsorption of arsenate and arsenite on ferrihydrite in the presence and absence of dissolved organic carbon. J. Environ. Qual. 31 (4), 1115–1123.

Gronberg, J., 2011. Map of arsenic in groundwater of the United States. <https://water.usgs.gov/GIS/metadata/usgswrd/XML/arsenic_map.xml>.

Gross, E.L., Low D.J., 2013. Arsenic concentrations, related environmental factors, and the predicted probability of elevated arsenic in groundwater in Pennsylvania. U.S. Geological Survey Scientific Investigations Report 2012-5257.

Grossl, P.R., Eick, M., Sparks, D.L., Goldberg, S., Ainsworth, C.C., 1997. Arsenate and chromate retention mechanisms on goethite. 2. Kinetic evaluation using a pressure-jump relaxation technique. Environ. Sci. Technol. 31 (2), 321–326.

Gruebel, K.A., Davis, J.A., Leckie, J.O., 1988. The feasibility of using sequential extraction techniques for arsenic and selenium in soils and sediments. Soil Sci. Soc. Am. J. 52 (2), 390–397.

Gurdak, J.J., McMahon, P.B., Dennehy, K., Qi, S.L., 2009. Water quality in the High Plains aquifer, Colorado, Kansas, Nebraska, New Mexico, Oklahoma, South Dakota, Texas, and Wyoming, 1999–2004. Circular 1337, U.S. Geological Survey 63.

Hansel, C.M., Benner, S.G., Nico, P., Fendorf, S., 2004. Structural constraints of ferric (hydr)oxides on dissimilatory iron reduction and the fate of Fe (II). Geochim. Cosmochim. Acta 68 (15), 3217–3229.

Haque, S., Johannesson, K.H., 2006. Arsenic concentrations and speciation along a groundwater flow path: the Carrizo Sand aquifer, Texas, USA. Chem. Geol. 228 (1–3), 57–71.

Haque, S., Ji, J., Johannesson, K.H., 2008. Evaluating mobilization and transport of arsenic in sediments and groundwaters of Aquia aquifer, Maryland, USA. J. Contam. Hydrol. 99 (1), 68–84.

Harvey, C.F., Swartz, C.H., Badruzzaman, A.B.M., Keon-Blute, N., Yu, W., Ali, M.A., et al., 2002. Arsenic mobility and groundwater extraction in Bangladesh. Science 298 (5598), 1602–1606.

Harvey, C.F., Ashfaque, K.N., Yu, W., Badruzzaman, A.B.M., Ali, M.A., Oates, P.M., et al., 2006. Groundwater dynamics and arsenic contamination in Bangladesh. Chem. Geol. 228 (1–3), 112–136.

Hering, J.G., Kneebone, P.E., 2002. Biogeochemical controls on arsenic occurrence and mobility in water supplies. In: Frankenberger, W.T. (Ed.), Environmental Chemistry of Arsenic. Marcel Dekker Inc, New York, pp. 175–202.

Hering, J.G., Dixit, S., 2005. Contrasting sorption behavior of arsenic (III) and arsenic(V) in suspensions of iron and aluminum oxyhydroxides. In: O'Day (Ed.) Advances in Arsenic Research. American Chemical Society, pp. 8–24.

Hering, J.G., O'Day, P.A., Ford, R.G., He, Y.T., Bilgin, A., Reisinger, H.J., et al., 2009. MNA as a remedy for arsenic mobilized by anthropogenic inputs of organic carbon. Groundw. Monit. Rem. 29 (3), 84–92.

Hinkle, S.R., Polette, D.J., 1999. Arsenic in ground water of the Willamette Basin, Oregon. Water-Resources Investigations Report, 98-4205, U.S. Geological Survey.

Höhn, R., Isenbeck-Schröter, M., Kent, D., Davis, J., Jakobsen, R., Jann, S., et al., 2006. Tracer test with As(V) under variable redox conditions controlling arsenic transport in the presence of elevated ferrous iron concentrations. J. Contam. Hydrol. 88 (1–2), 36–54.

Holm, T.R., Scott, J.W., Wilson, S.D., Kelly, W.R., Talbott, J.L., Roadcap, G.S., 2004. Arsenic Geochemistry and Distribution in the Mahomet Aquifer, Illinois. Illinois Waste Management and Research Center.

Hoover, J., Gonzales, M., Shuey, C., Barney, Y., Lewis, J., 2017. Elevated arsenic and uranium concentrations in unregulated water sources on the Navajo Nation, USA. Exposure Health 9 (2), 113–124.

Hudak, P., 2003. Arsenic, nitrate, chloride and bromide contamination in the Gulf Coast Aquifer, South-Central Texas, USA. Int. J. Environ. Stud. 60 (2), 123–133.

Hudak, P.F., 2000. Distribution and sources of arsenic in the southern High Plains Aquifer, Texas, USA. J. Environ. Sci. Health, A 35 (6), 899–913.

Hudak, P.F., 2006. Spatial and temporal patterns of arsenic concentration in the High Plains Aquifer of Texas, USA. Int. J. Environ. Stud. 63 (2), 201–209.

Inskeep, W.,P., McDermott, T.R., Fendorf, S., 2002. Arsenic (V)/(III) cycling in soils and natural waters: chemical and microbiological processes. In: Frankenberger, W.T. (Ed.), Environmental Chemistry of Arsenic. Marcel Dekker, Inc, New York, pp. 183–216.

Islam, F.S., Gault, A.G., Boothman, C., Polya, D.A., Charnock, J.M., Chatterjee, D., et al., 2004. Role of metal-reducing bacteria in arsenic release from Bengal delta sediments. Nature 430 (6995), 68–71.

Jain, A., Raven, K.P., Loeppert, R.H., 1999. Arsenite and arsenate adsorption on ferrihydrite: surface charge reduction and net OH-release stoichiometry. Environ. Sci. Technol. 33 (8), 1179–1184.

James, K.A., Meliker, J.R., Buttenfield, B.E., Byers, T., Zerbe, G.O., Hokanson, J.E., et al., 2014. Predicting arsenic concentrations in groundwater of San Luis Valley, Colorado: implications for individual-level lifetime exposure assessment. Environ. Geochem. Health 36 (4), 773–782.

Jean, J.S., Bundschuh, J., Chen, C.J., Guo, H.R., Liu, C.W., Lin, T.F., et al., 2010. The Taiwan crisis: a showcase of the global arsenic problem. In: Bundschuh, J., Bhattacharya, P. (Eds.), Arsenic in the Environment. CRC Press.

Johannesson, K.H., Yang, N., Trahan, A.S., Telfeyan, K., Mohajerin, T.J., Adebayo, S.B., et al., 2019. Biogeochemical and reactive transport modeling of arsenic in groundwaters from the Mississippi River delta plain: an analog for the As-affected aquifers of South and Southeast Asia. Geochim. Cosmochim. Acta 264, 245–272.

Jones, G.W., Pichler, T., 2007. Relationship between pyrite stability and arsenic mobility during aquifer storage and recovery in southwest central Florida. Environ. Sci. Technol. 41 (3), 723–730.

Jönsson, J., Sherman, D.M., 2008. Sorption of As(III) and As(V) to siderite, green rust (fougerite) and magnetite: implications for arsenic release in anoxic groundwaters. Chem. Geol. 255 (1), 173–181.

Jung, H.B., Charette, M.A., Zheng, Y., 2009. Field, laboratory, and modeling study of reactive transport of groundwater arsenic in a coastal aquifer. Environ. Sci. Technol. 43 (14), 5333–5338.

Kabay N., Bundschuh J., Hendry B., Bryjak M., Yoshizuka K., Bhattacharya P., et al., 2010. The global arsenic problem: challenges for safe water production. In: Bundschuh J, Bhattacharya P. (Eds). Arsenic in the Environment, CRC Press.

Kanel, S.R., Manning, B., Charlet, L., Choi, H., 2005. Removal of arsenic(III) from groundwater by nanoscale zero-valent iron. Environ. Sci. Technol. 39 (5), 1291–1298.

Kanel, S.R., Nepal, D., Manning, B., Choi, H., 2007. Transport of surface-modified iron nanoparticle in porous media and application to arsenic(III) remediation. J. Nanopart. Res. 9 (5), 725–735.

Kanel, S.R., Grenèche, J.-M., Choi, H., 2006. Arsenic(V) removal from groundwater using nano scale zero-valent iron as a colloidal reactive barrier material. Environ. Sci. Technol. 40 (6), 2045–2050.

Keimowitz, A.R., Mailloux, B.J., Cole, P., Stute, M., Simpson, H.J., Chillrud, S.N., 2007. Laboratory investigations of enhanced sulfate reduction as a groundwater arsenic remediation strategy. Environ. Sci. Technol. 41 (19), 6718–6724.

Keimowitz, A.R., Simpson, H.J., Stute, M., Datta, S., Chillrud, S.N., Ross, J., et al., 2005. Naturally occurring arsenic: mobilization at a landfill in Maine and implications for remediation. Appl. Geochem. 20 (11), 1985–2002.

Kelly, W.R., Holm, T.R., Wilson, S.D., Roadcap, G.S., 2005. Arsenic in glacial aquifers: sources and geochemical controls. Ground Water 43 (4), 500–510.

Kent, D.B., Fox, P.M., 2004. The influence of groundwater chemistry on arsenic concentrations and speciation in a quartz sand and gravel aquifer. Geochem. Trans. 5 (1), 1–12.

Keon, N.E., Swartz, C.H., Brabander, D.J., Harvey, C., Hemond, H.F., 2001. Validation of an arsenic sequential extraction method for evaluating mobility in sediments. Environ. Sci. Technol. 35 (13), 2778–2784.

Kim, D., Miranda, M.L., Tootoo, J., Bradley, P., Gelfand, A.E., 2011. Spatial modeling for groundwater arsenic levels in North Carolina. Environ. Sci. Technol. 45 (11), 4824–4831.

Kim, M.J., Nriagu, J., Haack, S., 2002. Arsenic species and chemistry in groundwater of southeast Michigan. Environ. Pollut. 120 (2), 379–390.

Kirk, M.F., Holm, T.R., Park, J., Jin, Q., Sanford, R.A., Fouke, B.W., et al., 2004. Bacterial sulfate reduction limits natural arsenic contamination in groundwater. Geology 32 (11), 953–956.

Kirschling, T.L., Gregory, K.B., Minkley, J.E.G., Lowry, G.V., Tilton, R.D., 2010. Impact of nanoscale zero valent iron on geochemistry and microbial populations in trichloroethylene contaminated aquifer materials. Environ. Sci. Technol. 44 (9), 3474–3480.

Klas, S., Kirk, D.W., 2013. Advantages of low pH and limited oxygenation in arsenite removal from water by zero-valent iron. J. Hazard. Mater. 252–253, 77–82.

Kolker, A., Haack, S.K., Cannon, W.F., Westjohn, D., Kim, M.-J., Nriagu, J., et al., 2003. Arsenic in southeastern Michigan. In: Welch, A.H., Stollenwerk, K. (Eds.), Arsenic in Ground Water. Kluwer Academic Publishers, pp. 281–294.

Korte, N., 1991. Naturally occurring arsenic in groundwaters of the Midwestern United States. Environ. Geol. Water Sci. 18 (2), 137–141.

Lackovic, J.A., Nikolaidis, N.P., Dobbs, G.M., 2000. Inorganic arsenic removal by zero-valent iron. Environ. Eng. Sci. 17 (1), 29–39.

Langner, H., Inskeep, W., 2000. Microbial reduction of arsenate in the presence of ferrihydrite. Environ. Sci. Technol. 34, 3131–3136.

Langner, H.W., Jackson, C.R., Mcdermott, T.R., Inskeep, W.P., 2001. Rapid oxidation of arsenite in a hot spring ecosystem, Yellowstone National Park. Environ. Sci. Technol. 35 (16), 3302–3309.

Law, R.K., Murphy, M.W., Choudhary, E., 2017. Private well groundwater quality in West Virginia, USA–2010. Sci. Total Environ. 586, 559–565.

Lazareva, O., Druschel, G., Pichler, T., 2015. Understanding arsenic behavior in carbonate aquifers: implications for aquifer storage and recovery (ASR). Appl. Geochem. 52, 57–66.

Le, X.C., Lu, X., Ma, M., Cullen, W.R., Aposhian, H.V., Zheng, B., 2000. Speciation of submicrogram per liter levels of arsenic in water: on-site species separation integrated with sample collection. Environ. Sci. Technol. 34 (11), 2342–2347.

Lee, M.K., Griffin, J., Saunders, J., Wang, Y., Jean, J.S., 2007. Reactive transport of trace elements and isotopes in the Eutaw coastal plain aquifer, Alabama. J. Geophys. Res.: Biogeosci. 112 (G2).

Leupin, O.X., Hug, S.J., 2005. Oxidation and removal of arsenic (III) from aerated groundwater by filtration through sand and zero-valent iron. Water Res. 39 (9), 1729–1740.

Levitt, J.P., Degnan, J.R., Flanagan, S.M., Jurgens, B.C., 2019. Arsenic variability and groundwater age in three water supply wells in southeast New Hampshire. Geosci. Front. 10 (5), 1669–1683.

Lin, Z., Puls, R.W., 2000. Adsorption, desorption and oxidation of arsenic affected by clay minerals and aging process. Environ. Geol. 39 (7), 753–759.

Lipfert, G., Reeve, A.S., Sidle, W.C., Marvinney, R., 2006. Geochemical patterns of arsenic-enriched ground water in fractured, crystalline bedrock, Northport, Maine, USA. Appl. Geochem. 21 (3), 528–545.

Low, D.J., Galeone, D.G., 2007. Reconnaissance of arsenic concentrations in ground water from bedrock and unconsolidated aquifers in eight northern-tier counties of Pennsylvania. Open-File Report. U.S. Geological Survey, Reston, VA, p. 39.

Mandal, B.K., Suzuki, K.T., 2002. Arsenic round the world: a review. Talanta 58 (1), 201−235.

Mango, H., Ryan, P., 2015. Source of arsenic-bearing pyrite in southwestern Vermont, USA: sulfur isotope evidence. Sci. Total Environ. 505, 1331−1339.

Manning, B.A., Suarez, D.L., 2000. Modeling arsenic(III) adsorption and heterogeneous oxidation kinetics in soils. Soil Sci. Soc. Am. J. 64 (1), 128−137.

Manning, B.A., Goldberg, S., 1997. Arsenic(III) and arsenic(V) absorption on three California soils. Soil Sci. 162 (12), 886−895.

Manning, B.A., Goldberg, S., 1996b. Modeling arsenate competitive adsorption on kaolinite, montmorillonite and illite. Clays Clay Miner. 44 (5), 609−623.

Manning, B.A., Goldberg, S., 1996a. Modeling competitive adsorption of arsenate with phosphate and molybdate on oxide minerals. Soil Sci. Soc. Am. J. 60 (1), 121−131.

Manning, B.A., Goldberg, S., 1995. Surface complexation modeling of arsenate adsorption on soil minerals. Abstr. Pap. Am. Chem. Soc. 209, 2−GEOC.

Manning, B.A., Hunt, M.L., Amrhein, C., Yarmoff, J.A., 2002. Arsenic(III) and arsenic(V) reactions with zerovalent iron corrosion products. Environ. Sci. Technol. 36 (24), 5455−5461.

Manning, B.A., Fendorf, S.E., Goldberg, S., 1998. Surface structures and stability of arsenic(III) on goethite: spectroscopic evidence for inner-sphere complexes. Environ. Sci. Technol. 32 (16), 2383−2388.

Manning, B.A., Fendorf, S.E., Bostick, B., Suarez, D.L., 2002. Arsenic(III) oxidation and arsenic(V) adsorption reactions on synthetic birnessite. Environ. Sci. Technol. 36 (5), 976−981.

Matisoff, G., Khourey, C.J., Hall, J.F., Varnes, A.W., Strain, W.H., 1982. The nature and source of arsenic in northeastern Ohio ground water. Ground Water 20 (4), 446−456.

McArthur, J., Ravenscroft, P., Safiullah, S., Thirlwall, M., 2001. Arsenic in groundwater: testing pollution mechanisms for sedimentary aquifers in Bangladesh. Water Resour. Res. 37, 109−117.

McArthur, J.M., Banerjee, D.M., Hudson-Edwards, K.A., Mishra, R., Purohit, R., Ravenscroft, P., et al., 2004. Natural organic matter in sedimentary basins and its relation to arsenic in anoxic ground water: the example of West Bengal and its worldwide implications. Appl. Geochem. 19 (8), 1255−1293.

McLean, J., Dupont, R., Sorensen, D., 2006. Iron and arsenic release from aquifer solids in response to biostimulation. J. Environ. Qual. 35 (4), 1193−1203.

McMahon, P.B., Dennehy, K.F., Bruce, B.W., Gurdak, J.J., Qi S.L., 2007. Water-quality assessment of the High Plains Aquifer, 1999−2004. Professional Paper 1749. U.S. Geological Survey.

McNab, W.W., Singleton, M.J., Moran, J.E., Esser, B.K., 2009. Ion exchange and trace element surface complexation reactions associated with applied recharge of low-TDS water in the San Joaquin Valley, California. Appl. Geochem. 24 (1), 129−137.

Melamed, D., 2005. Monitoring arsenic in the environment: a review of science and technologies with the potential for field measurements. Anal. Chim. Acta 532 (1), 1−13.

Meng, X., Dupont, R.R., Sorensen, D.L., Jacobson, A.R., McLean, J.E., 2017. Mineralogy and geochemistry affecting arsenic solubility in sediment profiles from the shallow basin-fill aquifer of Cache Valley Basin, Utah. Appl. Geochem. 77, 126−141.

Michael, H.A., 2013. An arsenic forecast for China. Science 341 (6148), 852−853.

Mirecki, J.E., Bennett, M.W., López-Baláez, M.C., 2013. Arsenic control during aquifer storage recovery cycle tests in the Floridan Aquifer. Groundwater 51 (4), 539−549.

Moore, J.N., Ficklin, W.H., Johns, C., 1988. Partitioning of arsenic and metals in reducing sulfidic environments. Environ. Sci. Technol. 22, 432−437.

Mukherjee, A., Fryar, A.E., 2008. Deeper groundwater chemistry and geochemical modeling of the arsenic affected western Bengal basin, West Bengal, India. Appl. Geochem. 23 (4), 863−894.

Mukherjee, A., Scanlon, B., Aureli, A., Langan, S., Guo, H., McKenzie, A., 2020. Global Groundwater: Source, Scarcity, Sustainability, Security and Solutions. Elsevier ISBN: 9780128181720.

Mumford, A.C., Barringer, J.L., Benzel, W.M., Reilly, P.A., Young, L.Y., 2012. Microbial transformations of arsenic: mobilization from glauconitic sediments to water. Water Res. 46 (9), 2859−2868.

Munk, L., Hagedorn, B., Sjostrom, D., 2011. Seasonal fluctuations and mobility of arsenic in groundwater resources, Anchorage, Alaska. Appl. Geochem. 26 (11), 1811−1817.

Neil, C.W., Yang, Y.J., Jun, Y.-S., 2012. Arsenic mobilization and attenuation by mineral−water interactions: implications for managed aquifer recharge. J. Environ. Monit. 14 (7), 1772−1788.

Nesbitt, H.W., Muir, I.J., 1998. Oxidation states and speciation of secondary products on pyrite and arsenopyrite reacted with mine waste waters and air. Mineral. Petrol. 62, 123−144.

Neumann, R.B., Ashfaque, K.N., Badruzzaman, A.B.M., Ashraf Ali, M., Shoemaker, J.K., Harvey, C.F., 2010. Anthropogenic influences on groundwater arsenic concentrations in Bangladesh. Nat. Geosci. 3 (1), 46−52.

Newman, D.K., Ahmann, D., Morel, F.M.M., 1998. A brief review of microbial arsenate respiration. Geomicrobiol. J. 15 (4), 255−268.

NGWA, 2020. Managed Aquifer Recharge. https://www.ngwa.org/what-is-groundwater/groundwater-issues/managed-aquifer-recharge

Nicholas, S.L., Erickson, M.L., Woodruff, L.G., Knaeble, A.R., Marcus, M.A., Lynch, J.K., et al., 2017. Solid-phase arsenic speciation in aquifer sediments: a micro-X-ray absorption spectroscopy approach for quantifying trace-level speciation. Geochim. Cosmochim. Acta 211, 228−255.

Nickson, R., McArthur, J., Ravenscroft, P., Burgess, W., Ahmed, K., 2000. Mechanism of arsenic release to groundwater, Bangladesh and West Bengal. Appl. Geochem. 15 (4), 403–413.

Nickson, R., McArthur, J., Burgess, W., Ahmed, K.M., Ravenscroft, P., Rahman, M., 1998. Arsenic poisoning of Bangladesh groundwater. Nature 395 (6700), 338.

Nimick, D.A., 1998. Arsenic hydrogeochemistry in an irrigated river valley—a reevaluation. Ground Water 36 (5), 743–753.

Nimick, D.A., Moore, J.N., Dalby, C.E., Savka, M.W., 1998. The fate of geothermal arsenic in the Madison and Missouri Rivers, Montana and Wyoming. Water Resour. Res. 34 (11), 3051–3067.

Nordstrom, D.K., Alpers, C.N., 1999. Geochemistry of acid mine waters. In: Plumlee, G., Logsdon, M. (Eds.), Environmental Geochemistry of Mineral Deposits: Part A: Processes, Techniques, and Health Issues. Society of Economic Geologists, pp. 133–155.

Nordstrom, D.K., 2002. Worldwide occurrences of arsenic in ground water. Science 296 (5576), 2143–2145.

NRC, 1999. Arsenic in Drinking Water. National Academy Press, Washington, DC.

O'Day, P.A., 2006. Chemistry and mineralogy of arsenic. Elements 2 (2), 77–83.

O'Day, P.A., Vlassopoulos, D., Meng, X., Benning, L., 2005. Introductory Remarks, In: O'Day, P.A., et al., (Eds.), Advances in Arsenic Research: Integration of Experimental and Observational Studies and Implications for Mitigation. ACS Publications.

O'Day, P.A., Vlassopoulos, D., Meng, X., Benning, L.G., 2005. Advances in Arsenic Research: Integration of Experimental and Observational Studies and Implications for Mitigation. ACS Publications.

Ona-Nguema, G., Morin, G., Juillot, F., Calas, G., Brown, G.E., 2005. EXAFS analysis of arsenite adsorption onto two-line ferrihydrite, hematite, goethite, and lepidocrocite. Environ. Sci. Technol. 39 (23), 9147–9155.

Oremland, R.S., Stolz, J.F., 2005. Arsenic, microbes and contaminated aquifers. Trends Microbiol. 13 (2), 45–49.

Oremland, R.S., Stolz, J.F., 2003. The ecology of arsenic. Science 300 (5621), 939–944.

Oscarson, D., Huang, P., Defosse, C., Herbillon, A., 1981. Oxidative power of Mn(IV) and Fe(III) oxides with respect to As(III) in terrestrial and aquatic environments. Nature 291 (5810), 50–51.

Peters, S., Blum, J., Klaue, B., Karagas, M., 1999. Arsenic occurrence in New Hampshire ground water. Environ. Sci. Technol. 33 (9), 1328–1333.

Peters, S.C., Blum, J.D., 2003. The source and transport of arsenic in a bedrock aquifer, New Hampshire, USA. Appl. Geochem. 18 (11), 1773–1787.

Peters, S.C., Burkert, L., 2008. The occurrence and geochemistry of arsenic in groundwaters of the Newark basin of Pennsylvania. Appl. Geochem. 23 (1), 85–98.

Peters, S.C., 2008. Arsenic in groundwaters in the Northern Appalachian Mountain belt: a review of patterns and processes. J. Contam. Hydrol. 99 (1), 8–21.

Pichler, T., Renshaw, C.E., Sültenfuß, J., 2017. Geogenic As and Mo groundwater contamination caused by an abundance of domestic supply wells. Appl. Geochem. 77, 68–79.

Pichler, T., Price, R., Lazareva, O., Dippold, A., 2011. Determination of arsenic concentration and distribution in the Floridan Aquifer System. J. Geochem. Explor. 111 (3), 84–96.

Pinel-Raffaitin, P., Le Hecho, I., Amouroux, D., Potin-Gautter, M., 2007. Distribution and fate of inorganic and organic arsenic species in landfill leachates and biogases. Environ. Sci. Technol. 41 (13), 4536–4541.

Powers, M., Yracheta, J., Harvey, D., O'Leary, M., Best, L.G., Bear, A.B., et al., 2019. Arsenic in groundwater in private wells in rural North Dakota and South Dakota: water quality assessment for an intervention trial. Environ. Res. 168, 41–47.

Price, R.E., Pichler, T., 2006. Abundance and mineralogical association of arsenic in the Suwannee Limestone (Florida): implications for arsenic release during water-rock interaction. Chem. Geol. 228 (1–3), 44–56.

Prommer, H., Barry, D.A., Zheng, C., 2003. MODFLOW/MT3DMS-based reactive multicomponent transport modeling. Ground Water 41 (2), 247–257.

Prommer, H., Sun, J., Helm, L., Rathi, B., Siade, A.J., Morris, R., 2018. Deoxygenation prevents arsenic mobilization during deepwell injection into sulfide-bearing aquifers. Environ. Sci. Technol. 52 (23), 13801–13810.

Raven, K.P., Jain, A., Loeppert, R.H., 1998. Arsenite and arsenate adsorption on ferrihydrite: kinetics, equilibrium, and adsorption envelopes. Environ. Sci. Technol. 32 (3), 344–349.

Ravenscroft, P., Brammer, H., Richards, K., 2009. Arsenic Pollution: A Global Synthesis, vol. 94. John Wiley & Sons.

Rawson, J., Siade, A., Sun, J., Neidhardt, H., Berg, M., Prommer, H., 2017. Quantifying reactive transport processes governing arsenic mobility after injection of reactive organic carbon into a Bengal Delta Aquifer. Environ. Sci. Technol. 51 (15), 8471–8480.

Reid, J.C., Haven, W.T., Eudy, D.D., Milosh, R.M., Stafford, E.G., 2010. Arsenic in groundwater in the North Carolina Eastern slate belt (Esb): Nash and Halifax counties, North Carolina. Southeast. Geol. 47 (3), 117–122.

Reimann, C., Finne, T.E., Nordgulen, O., Saether, O.M., Arnoldussen, A., Banks, D., 2009. The influence of geology and land-use on inorganic stream water quality in the Oslo region, Norway. Appl. Geochem. 24 (10), 1862–1874.

Rittle, K.A., Drever, J.I., Colberg, P.J.S., 1995. Precipitation of arsenic during bacterial sulfate reduction. Geomicrobiol. J. 13 (1), 1–11.

Robinson, G.R., Ayotte, J.D., 2006. The influence of geology and land use on arsenic in stream sediments and ground waters in New England, USA. Appl. Geochem. 21 (9), 1482–1497.

Roden, E.E., Urrutia, M.M., 2002. Influence of biogenic Fe(II) on bacterial crystalline Fe(III) oxide reduction. Geomicrobiol. J. 19 (2), 209–251.

Root, T.L., Bahr, J.M., Gotkowitz, M.B., 2005. Controls on arsenic concentrations in groundwater near Lake Geneva, Wisconsin. In: O'Day, P.A., et al., (Eds.), Advances in Arsenic Research: Integration of Experimental and Observational Studies and Implications for Mitigation, ACS Publications. pp. 161–174.

Root, T.L., Gotkowitz, M.B., Bahr, J.M., Attig, J.W., 2010. Arsenic geochemistry and hydrostratigraphy in Midwestern U.S. Glacial Deposits. Ground Water 48 (6), 903–912.

Ryan, P.C., West, D.P., Hattori, K., Studwell, S., Allen, D.N., Kim, J., 2015. The influence of metamorphic grade on arsenic in metasedimentary bedrock aquifers: a case study from Western New England, USA. Sci. Total Environ. 505, 1320–1330.

Ryan, P.C., Kim, J., Wall, A.J., Moen, J.C., Corenthal, L.G., Chow, D.R., et al., 2011. Ultramafic-derived arsenic in a fractured bedrock aquifer. Appl. Geochem. 26 (4), 444–457.

Ryan, P.C., Kim, J.J., Mango, H., Hattori, K., Thompson, A., 2013. Arsenic in a fractured slate aquifer system, New England, USA: influence of bedrock geochemistry, groundwater flow paths, redox and ion exchange. Appl. Geochem. 39, 181–192.

Ryker, S.J., 2003, Arsenic in ground water used for drinking water in the United States, In: Welch, A., Stollenwerk, K. (Eds). Arsenic in Ground Water, 2003, Kluwer Academic Publishers, 165–178.

Ryker, S.J., 2001. Mapping arsenic in groundwater. Geotimes 46 (11), 34–36.

Sanders, A.P., Messier, K.P., Shehee, M., Rudo, K., Serre, M.L., Fry, R.C., 2012. Arsenic in North Carolina: public health implications. Environ. Int. 38 (1), 10–16.

Santini, J.M., Ward, S.A., 2018. The metabolism of arsenite. In: Bundschuh, J., Bhattacharya, P. (Eds.), Arsenic in the Environment. CRC Press.

Santini, J.M., Stolz, J.F., Macy, J.M., 2002. Isolation of a new arsenate-respiring bacterium-physiological and phylogenetic studies. Geomicrobiol. J. 19 (1), 41–52.

Scanlon, B.R., Nicot, J.P., Reedy, R.C., Kurtzman, D., Mukherjee, A., Nordstrom, D.K., 2009. Elevated naturally occurring arsenic in a semiarid oxidizing system, Southern High Plains aquifer, Texas, USA. Appl. Geochem. 24 (11), 2061–2071.

Schlottmann, J., Breit, G., 1992. Mobilization of As and U in the central Oklahoma aquifer. In: Kharaka, Y., Maest, A. (Eds.), Water-Rock Interaction. Balkema, Rotterdam, pp. 835–838.

Schreiber, M., Simo, J., Freiberg, P., 2000. Stratigraphic and geochemical controls on naturally occurring arsenic in groundwater, eastern Wisconsin, USA. Hydrogeol. J. 8 (2), 161–176.

Schreiber, M., Gotkowitz, M., Simo, J., Freiberg, P., 2003. Mechanisms of arsenic release to ground water from naturally occurring sources, eastern Wisconsin. In: Welch, A.H., Stollenwerk, K.G. (Eds.), Arsenic in Ground Water. Kluwer Academic Publishers, pp. 259–280.

Schreiber, M.E., Simo, J.A., Freiberg, P.G., 2000. Stratigraphic and geochemical controls on naturally occurring arsenic in groundwater, eastern Wisconsin, USA. Hydrogeol. J. 8 (2), 161–176.

Senior, L.A., Sloto, R.A., 2006. Arsenic, boron, and fluoride concentrations in ground water in and near diabase intrusions, Newark Basin, Southeastern Pennsylvania. Scientific Investigations Report 2006-5261, U.S. Geological Survey.

Senn, D., Hemond, H.F., 2002. Nitrate controls on iron and arsenic in a suburban lake. Science 296, 2373–2376.

Serfes, M., Spayd, S., Herman, G., 2005. Arsenic Occurrence, Sources, Mobilization, and Transport in Groundwater in the Newark Basin of New Jersey. In: O'Day, P.A., et al., (Eds.), Advances in Arsenic Research: Integration of Experimental and Observational Studies and Implications for Mitigation. ACS Publications.

Serfes, M.E., Herman, G.C., Spayd, S.E., Reinfelder, J., 2010. Sources, mobilization and transport of arsenic in groundwater in the Passaic and Lockatong formations of the Newark Basin, New Jersey. N.J. Geol. Surv. Bull. 77, Chapter E.

Sharif, M., Davis, R., Steele, K., Kim, B., Hays, P., Kresse, T., et al., 2008. Distribution and variability of redox zones controlling spatial variability of arsenic in the Mississippi River Valley alluvial aquifer, southeastern Arkansas. J. Contam. Hydrol. 99 (1–4), 49–67.

Shiber, J., 2005. Arsenic in domestic well water and health in central Appalachia, USA. Water Air Soil Pollut. 160, 327–341.

Smedley, P.L., Kinniburgh, D.G., 2002. A review of the source, behaviour and distribution of arsenic in natural waters. Appl. Geochem. 17 (5), 517–568.

Smith, A.H., Hopenhayn-Rich, C., Bates, M.N., Goeden, H.M., Hertz-Picciotto, I., Duggan, H.M., et al., 1992. Cancer risks from arsenic in drinking water. Environ. Health Perspect. 97, 259–267.

Smith, E., Naidu, R., 2009. Chemistry of inorganic arsenic in soils: kinetics of arsenic adsorption–desorption. Environ. Geochem. Health 31, 49–59.

Smith, R., Knight, R., Fendorf, S., 2018. Overpumping leads to California groundwater arsenic threat. Nat. Commun. 9 (1), 2089.

Smith, S.J., 2005. Naturally occurring arsenic in ground water, Norman, Oklahoma, 2004, and remediation options for produced water. Fact Sheet 2005-3111. U.S. Geological Survey.

Sonderegger, J.L., Ohguchi, T., 1988. Irrigation related arsenic contamination of a thin, alluvial aquifer, Madison River Valley, Montana, U.S.A. Environ. Geol. Water Sci. 11 (2), 153–161.

Stanton, J.S., QiS.L., 2006. Ground-water quality of the northern High Plains aquifer, 1997, 2002-04. Scientific Investigations Report 2006–5138, U.S. Geological Survey.

Stauffer, R., Thompson, J., 1984. Arsenic and antimony in geothermal waters of Yellowstone National Park, Wyoming, USA. Geochem. Cosmochim. Acta 48, 2547–2561.

Stefaniuk, M., Oleszczuk, P., Ok, Y.S., 2016. Review on nano zerovalent iron (nZVI): from synthesis to environmental applications. Chem. Eng. J. 287, 618–632.

Steinmaus, C.M., Yuan, Y., Smith, A.H., 2005. The temporal stability of arsenic concentrations in well water in western Nevada. Environ. Res. 99 (2), 164–168.

Stocker, J., Balluch, D., Gsell, M., Harms, H., Feliciano, J., Daunert, S., et al., 2003. Development of a set of simple bacterial biosensors for quantitative and rapid measurements of arsenite and arsenate in potable water. Environ. Sci. Technol. 37 (20), 4743–4750.

Stollenwerk, K., 2003. Geochemical processes controlling transport of arsenic in groundwater: A review of adsorption. In: Welch, A., Stollenwerk, K. (Eds.), Arsenic in Groundwater. Kluwer Academic Publishers, Boston, MA, pp. 67–100.

Stollenwerk, K.G., Colman, J.A., 2004. Natural Remediation of Arsenic Contaminated Ground Water Associated With Landfill Leachate. Fact Sheet 2004-3057. U.S. Geological Survey.

Stolz, J.F., Oremland, R.S., 1999. Bacterial respiration of arsenic and selenium. FEMS Microbiol. Rev. 23 (5), 615−627.

Su, C., Puls, R.W., 2001. Arsenate and arsenite removal by zerovalent iron: kinetics, redox transformation, and implications for in situ groundwater remediation. Environ. Sci. Technol. 35 (7), 1487−1492.

Su, C., Puls, R.W., 2004. Significance of iron(II, III) hydroxycarbonate green rust in arsenic remediation using zerovalent iron in laboratory column tests. Environ. Sci. Technol. 38 (19), 5224−5231.

Su, C.M., Puls, R.W., 2003. In situ remediation of arsenic in simulated groundwater using zerovalent iron: laboratory column tests on combined effects of phosphate and silicate. Environ. Sci. Technol. 37 (11), 2582−2587.

Sun, J., Quicksall, A.N., Chillrud, S.N., Mailloux, B.J., Bostick, B.C., 2016. Arsenic mobilization from sediments in microcosms under sulfate reduction. Chemosphere 153, 254−261.

Suthersan, S., Horst, J., 2008. Aquifer minerals and in situ remediation: the importance of geochemistry. Groundw. Monit. Rem. 28 (3), 153−160.

Szramek, K., Walter, L.M., McCall, P., 2004. Arsenic mobility in groundwater/surface water systems in carbonate-rich Pleistocene glacial drift aquifers (Michigan). Appl. Geochem. 19 (7), 1137−1155.

Tadanier, C., Schreiber, M., Roller, J., 2005. Arsenic mobilization through microbially mediated deflocculation of ferrihydrite. Environ. Sci. Technol. 39 (9), 3061−3068.

Tesh, S.J., Scott, T.B., 2014. Nano-composites for water remediation: a review. Adv. Mater. 26 (35), 6056−6068.

Tessier, A., Campbell, P., Bisson, M., 1979. Sequential extraction procedure for the speciation of particulate trace metals. Anal. Chem. 51 (7), 844−851.

Thiros, S.A., Paul, A.P., Bexfield, L.M., Anning, D.W., 2015. The Quality of Our Nation's Waters: Water Quality in Basin-Fill Aquifers of the Southwestern United States: Arizona, California, Colorado, Nevada, New Mexico, and Utah, 1993-2009. Circular 126, U.S. Geological Survey Reston, VA, p. 126.

Thiruvenkatachari, R., Vigneswaran, S., Naidu, R., 2008. Permeable reactive barrier for groundwater. J. Ind. Eng. Chem. 14 (2), 145−156.

Thomas, M.A., 2003. Arsenic in Midwestern glacial deposits—occurrence and relation to selected hydrogeologic and geochemical factors. Water Resources Investigations Report 03-4228, U.S. Geological Survey.

Thundiyil, J.G., Yuan, Y., Smith, A.H., Steinmaus, C., 2007. Seasonal variation of arsenic concentration in wells in Nevada. Environ. Res. 104 (3), 367−373.

Van Der Meer, J.R., Tropel, D., Jaspers, M., 2004. Illuminating the detection chain of bacterial bioreporters. Environ. Microbiol. 6 (10), 1005−1020.

van Geen, A., Rose, J., Thoral, S., Garnier, J.M., Zheng, Y., Bottero, J.Y., 2004. Decoupling of As and Fe release to Bangladesh groundwater under reducing conditions. Part II: Evidence from sediment incubations. Geochim. Cosmochim. Acta 68 (17), 3475−3486.

VanDerwerker, T., Zhang, L., Ling, E., Benham, B., Schreiber, M., 2018. Evaluating Geologic Sources of Arsenic in Well Water in Virginia (USA). Int. J. Environ. Res. Public. Health 15 (4), 787.

Vanderzalm, J.L., Dillon, P.J., Barry, K.E., Miotlinski, K., Kirby, J.K., Le Gal La Salle, C., 2011. Arsenic mobility and impact on recovered water quality during aquifer storage and recovery using reclaimed water in a carbonate aquifer. Appl. Geochem. 26 (12), 1946−1955.

Verplanck, P.L., Mueller, S.H., Goldfarb, R.J., Nordstrom, D.K., Youcha, E.K., 2008. Geochemical controls of elevated arsenic concentrations in groundwater, Ester Dome, Fairbanks district, Alaska. Chem. Geol. 255 (1−2), 160−172.

Walker, M., Seiler, R.L., Meinert, M., 2008. Effectiveness of household reverse-osmosis systems in a Western U.S. region with high arsenic in groundwater. Sci. Total Environ. 389 (2−3), 245−252.

Wallis, I., Prommer, H., Pichler, T., Post, V., Norton, S.B., Annable, M.D., et al., 2011. Process-based reactive transport model to quantify arsenic mobility during aquifer storage and recovery of potable water. Environ. Sci. Technol. 45 (16), 6924−6931.

Wang, S.L., Mulligan, C.N., 2008. Speciation and surface structure of inorganic arsenic in solid phases: a review. Environ. Int. 34 (6), 867−879.

Warner, K.L., Ayotte, J.D., 2015. The Quality of our Nation's Waters: Water Quality in the Glacial Aquifer System, Northern United States, 1993-2009. Circular 1352, US Geological Survey.

Warner, K.L., 2001. Arsenic in glacial drift aquifers and the implication for drinking water—lower Illinois River Basin. Ground Water 39 (3), 433−442.

Waychunas, G., Rea, B., Fuller, C., Davis, J., 1993. Surface chemistry of ferrihydrite: Part 1. EXAFS studies of the geometry of coprecipitated and adsorbed arsenate. Geochim. Cosmochim. Acta 57, 2251−2269.

Webster, J.G., Nordstrom, D.K., 2003. Geothermal arsenic. In: Welch, A.H., Stollenwerk, K.G. (Eds.), Arsenic in Ground Water: Geochemistry and Occurrence. Kluwer Academic Publishers, Boston, MA, pp. 101−125.

Welch, A.H., Stollenwerk, K.G., 2003. Arsenic in Ground Water: Geochemistry and Occurrence. Kluwer Academic Publishers.

Welch, A.H., Lico, M.S., 1998. Factors controlling As and U in shallow ground water, southern Carson Desert, Nevada. Appl. Geochem. 13 (4), 521−539.

Welch, A.H., Westjohn, D., Helsel, D.R., Wanty, R.B., 2000. Arsenic in ground water of the United States: occurrence and geochemistry. Ground Water 38 (4), 589−604.

Wilkie, J.A., Hering, J.G., 1996. Adsorption of arsenic onto hydrous ferric oxide: effects of adsorbate/adsorbent ratios and co-occurring solutes. Colloids Surf., A: Physicochem. Eng. Aspects 107, 97−110.

Wilkie, J.A., Hering, J.G., 1998. Rapid oxidation of geothermal arsenic(III) in streamwaters of the eastern Sierra Nevada. Environ. Sci. Technol. 32 (5), 657−662.

Willis, S.S., Haque, S.E., Johannesson, K.H., 2011. Arsenic and antimony in groundwater flow systems: a comparative study. Aquat. Geochem. 17 (6), 775−807.

Yang, N., Winkel, L.H., Johannesson, K.H., 2014. Predicting geogenic arsenic contamination in shallow groundwater of South Louisiana, United States. Environ. Sci. Technol. 48 (10), 5660–5666.

Yang, Q., Jung, H.B., Culbertson, C.W., Marvinney, R.G., Loiselle, M.C., Locke, D.B., et al., 2009. Spatial pattern of groundwater arsenic occurrence and association with bedrock geology in Greater Augusta, Maine. Environ. Sci. Technol. 43 (8), 2714–2719.

Yang, Q., Jung, H.B., Marvinney, R.G., Culbertson, C.W., Zheng, Y., 2012. Can arsenic occurrence rates in bedrock aquifers be predicted? Environ. Sci. Technol. 46 (4), 2080–2087.

Yogarajah, N., Tsai, S.S.H., 2015. Detection of trace arsenic in drinking water: challenges and opportunities for microfluidics. Environ. Science: Water Res. Technol. 1 (4), 426–447.

Zheng, Y., Ayotte, J.D., 2015. At the crossroads: hazard assessment and reduction of health risks from arsenic in private well waters of the northeastern United States and Atlantic Canada. Sci. Total Environ. 505, 1237–1247.

Zheng, Y., Flanagan, S.V., 2017. The case for universal screening of private well water quality in the US and testing requirements to achieve it: evidence from arsenic. Environ. Health Perspect. 125 (8), 085002.

Ziegler, B.A., McGuire, J.T., Cozzarelli, I.M., 2015. Rates of As and trace-element mobilization caused by Fe reduction in mixed BTEX–ethanol experimental plumes. Environ. Sci. Technol. 49 (22), 13179–13189.

Ziegler, B.A., Schreiber, M.E., Cozzarelli, I.M., 2017. The role of alluvial aquifer sediments in attenuating a dissolved arsenic plume. J. Contam. Hydrol. (204), 90–101.

Ziegler, B.A., Schreiber, M.E., Cozzarelli, I.M., Crystal Ng, G.H., 2017. A mass balance approach to investigate arsenic cycling in a petroleum plume. Environ. Pollut. 231, 1351–1361.

Zobrist, J., Dowdle, P.R., Davis, J.A., Oremland, R.S., 2000. Mobilization of arsenite by dissimilatory reduction of adsorbed arsenate. Environ. Sci. Technol. 34 (22), 4747–4753.

Chapter 21

Hydrogeochemical characterization of groundwater quality in the states of Texas and Florida, United States

Shama E. Haque
North South University, Dhaka, Bangladesh

Groundwater provides 25% of the freshwater used in the United States [USGS (United States Geological Survey), 2010]. The National Water-Quality Assessment Program by US Geological Survey assessed trace element concentrations in groundwater samples collected between 1992 and 2003 from various aquifers within the United States. The United States Geological Survey (USGS) report indicates that roughly 19% (962 of 5097) wells across the United States have trace element concentrations that surpass the human-health benchmark for at least one parameter. Varying geochemical conditions within an aquifer can influence release of metals/metalloids above their allowable limits. The suitability of groundwater for public supply and other end uses depends on the concentrations of certain chemical constituents of water. Dissolved chemical constituents in groundwater are a growing worldwide concern due to possible human health hazards at high concentrations (Mukherjee et al., 2020).

Due to increasing growth in groundwater consumption, aquifers across the United States are being studied and assessed in great detail. Assessment of aquifer-based water quality provides valuable insight into the hydrogeochemical processes that are important in making informed decision for groundwater-management activities. This chapter highlights the research findings of previous investigations carried on groundwater quality in various aquifer of the coastal states of Texas and Florida. Both states have significant groundwater resources, which have been influenced by factors, such as population growth, water shortage and aquifer overdraft, water quality degradation, and salt water intrusion.

21.1 Groundwater quality in Texas

Numerous aquifers underlie most of the state of Texas. The Texas Water Development Board (TWDB) has identified 9 major (Pecos Valley, Seymour, Gulf Coast, Carrizo–Wilcox, Hueco–Mesilla Bolsons, Ogallala, Edwards–Trinity Plateau, and Edwards and Trinity aquifer) and 22 minor aquifers in the Texas according to the amount of water supplied by each groundwater resource. The aquifers produce roughly 62% of water used within the state. Approximately 80% of the extracted groundwater is used for agriculture, and the remaining 20% is for municipal and industrial usage (Kaiser and Skiller, 2001). As the state's population level rises, municipal, industrial, and agricultural demands for water continue to increase. The continuous long-term development and use of groundwater is limited by the fact that the withdrawal of groundwater is exceeding the recharge rate (TDWR, 1984). Frequent occurrence of droughts has also renewed the awareness of water availability as one of the most urgent economic issues facing Texas.

21.1.1 Edwards–Trinity plateau aquifer

The Edwards–Trinity plateau aquifer is the main aquifer stretching throughout most of southwestern part of the state of Texas (Mace et al., 2000). The aquifer has hydraulic connection with four main aquifers: (1) Pecos Valley, (2) Ogallala, (3) Trinity (Hill Country), and (4) Edwards (Balcones Fault Zone; Anaya, 2001). In addition, the Edwards–Trinity system has hydraulic connection to numerous minor aquifers: (1) Dockum, (2) Capitan, (3) Rustler,

(4) Hickory, (5) Ellenburger–San Saba, (6) Lipan, and (7) Marble Falls (Anaya, 2001). According to the TWDB, the water bearing unites comprise primarily limestone and dolomite of the Edwards Group, and sands of the Trinity Group. The groundwater contained in the aquifer system is primarily freshwater; however, a portion of the aquifer contains water that is slightly saline. The Edwards–Trinity aquifer meets majority of the growing water demands for 38 counties of which agriculture accounts for 70% of the overall withdrawal and 15% of the pumped water meets municipal needs (Ashworth and Hopkins, 1995; Barker and Ardis, 1992).

Bush et al. (1994) report that the main hydrochemical facies in the Edwards–Trinity system is calcium hydrogen carbonate along with the presence of at least seven different water types. Each water type has a distinct dissolved solids range and is distributed in a similar pattern to dissolved solids concentrations (Bush et al., 1994). Between 1996 and 1998 the USGS investigated the upper and middle zones of the Trinity aquifer in the southeast part of the Edwards Plateau and the unconfined and confined portions of the Edwards Aquifer in the San Antonio segment of the Balcones fault zone. The investigation concluded that the Trinity Aquifer groundwater contains higher levels of dissolved solids, sulfate, and chlorides compared to that of the Edwards Aquifer. The larger chemical variation in the Trinity Aquifer groundwater possibly reflects the local heterogeneity of the host rock (Fahlquist and Ardis, 2004). Hudak (2003) examined chloride levels and variations in bromide/chloride ratios from 198 groundwater wells tapping the Edwards–Trinity plateau aquifer. In particular, the study focused on eight west-central Texas counties, where land-use is dominated by agricultural and oil-production activities. In ~25% well water samples, chloride levels exceeded the 250 mg/L secondary drinking water standard, ~11% samples had values above 500 mg/L, and ~5% samples surpassed 1000 mg/L. The results indicate that dissolving evaporite deposits along with oil-field brine affect the surrounding area's water quality. In addition, the aquifer is susceptible to contamination in the recharge area and due to the presence of geological features, such as fractures, faults, and joints in the rocks (Standen and Opdyke, 2004). In the aquifer system, trace elements were detected at low concentrations; several pesticides and volatile organic compounds (VOCs) were found within the Edwards Aquifer, less commonly in the Trinity Aquifer, usually at much lower levels (<1 μg/L; Fahlquist and Ardis, 2004). Hudak (2018) analyzed uranium, nitrate, calcium, iron, and manganese levels in groundwater samples collected from 108 wells tapping the Edwards–Trinity aquifer of southwest Texas. The study reported that in ~16% of wells, uranium levels surpassed the drinking water standard (i.e., 30 μg/L) and reported that uranium and both nitrate and calcium levels were positively associated.

21.1.2 Ogallala aquifer

The Ogallala aquifer, extending from South Dakota to Texas, is the largest aquifer in North America. The aquifer, also known as the High Plains aquifer, lies much below of the High Plains region of the United States and has been the primary source of irrigation, domestic, and industrial water in the area. The Ogallala is a shallow water table aquifer that is composed of sand, gravel, and silt deposits of Tertiary age (Rajagopalan et al., 2006). The Ogallala typically comprises calcium bicarbonate to mixed-cation-bicarbonate type water in the north, with increased amounts of chloride and sulfate in the south (Schriver and Hopkins, 1998). According to the TWDB, the water quality within the aquifer varies with generally freshwater in the northern region (TDS <400 mg/L) whilst water quality deteriorates toward the southern region (TDS >1000 mg/L). Schriver and Hopkins (1998) report that in south Texas, dissolved nitrate and fluoride levels are in excess of their primary MCLs of 44.3 and 4.0 mg/L in 18% and 77% of wells completed in the aquifer, respectively. The Ogallala aquifer shows large percentage of MCL exceedance for drinking water arsenic levels (10 μg/L; Reedy et al., 2007). The higher concentrations of arsenic at shallower water table depth are likely associated with long-term phosphate fertilizers applications.

21.1.3 Seymour aquifer

The Seymour Aquifer, an unconfined aquifer, extends through 23-county area in north-central Texas and the Texas Panhandle (Jensen, 2007). Extensive areas of sandy soils at shallow depth to the groundwater table have led to water-quality degradation within the aquifer. In particular, regional groundwater quality has been impacted by elevated levels of nitrate through natural processes and anthropogenic activities. Chaudhuri and Ale (2014) reported that high levels of nitrate (>MCL) occurred in the calcium bicarbonate and calcium magnesium bicarbonate facies. The study found that in a number of wells nitrate nitrogen levels surpassed 35 mg/L. Agricultural chemicals are the possible sources of nitrate in the region. Note that high concentrations of nitrate can negatively impact uses of water for different purposes. In addition, between 2001 and 2004, Hudak (2008) investigated arsenic levels in 64 water samples collected from the

Seymour and found that only one water sample was above the primary MCL for arsenic in drinking water. The study did not observe any association between arsenic levels and well depth.

21.1.4 Pecos Valley Aquifer

The Pecos Valley Aquifer underlies an area of ~22,403 km^2 in the Trans-Pecos region of west Texas and eastern New Mexico. The aquifer consists of alluvial sediment formations and eolian deposits (Hutchison, 2016). The Pecos Valley Aquifer is a major aquifer in west Texas and provides majority of the freshwater supply (Meyer et al., 2012). According to the TWDB, over 80% of withdrawn water is used to meet the water demand for agriculture, and the remainder is used for public water supply along with industrial activities and power supply.

Chaudhuri and Ale (2014) evaluated the areal distribution and concentrations of total dissolved solids (TDS) in shallow (<50 m), intermediate (50–150 m), and deep (>150 m) municipal wells tapping the Pecos Valley Aquifer during the period of 1960s–70s and 1990s–2000s by analyzing data collected from the TWDB. During both intervals, over 60% of analyzed samples from shallow wells surpassed the secondary MCL for TDS. In parts of central and western Texas, brackish water appears to be more widespread with TDS concentrations ranging from less than 200 mg/L to above 10,000 mg/L (Meyer et al., 2012). Concentrations of dissolved chloride and sulfate exceeded their respective secondary MCLs, which possibly stemmed from past oil field activities (Hutchison, 2016). Hudak (2018) determined the concentrations of nitrate, arsenic, and selenium levels in 79 water wells that tap the aquifer and observed that only five arsenic observations, three selenium observations, and 18% of nitrate observations exceeded their respective drinking water standards. Agricultural and oilfield activity along with naturally sources possibly contribute contaminants, if any, into these groundwater.

21.1.5 Carrizo aquifer

The Carrizo aquifer, a major groundwater flow system in southern Texas, comprises sand locally interbedded with gravel, silt, clay, and lignite deposited in a fluviodeltaic environment. For the most part the extracted groundwater is used for agriculture and municipal water supply. Recharge to the Carrizo occurs primarily through precipitation in the aquifer outcrop in northern Atascosa County, and discharge from the aquifer occurs through cross-formational upward flow (Castro et al., 2000; Castro and Goblet, 2003).

Haque and Johannesson (2006a,b) studied dissolved arsenic levels in the Carrizo Aquifer along a well-studied flow path in southeastern Texas. These investigators found that the aquifer contains well-defined ferric iron and sulfate-reduction zones along the flow path. Furthermore, complex coupling of hydrobiogeochemical processes influence groundwater behavior in the zones of recharge. Dissolved arsenic concentrations are well below the MCL for arsenic in drinking water. The reductive dissolution of arsenic-bearing ferric oxyhydroxides/oxides in aquifer matrix is likely the most important geochemical trigger, which release arsenic into the Carrizo groundwater. Dissolved arsenic concentrations are typically lower where sulfate-reduction prevails. Basu et al. (2007) found that dissolved selenium are in nanomolal levels and vary in speciation along the flow path. Changing redox conditions in conjunction with varying groundwater pH along the flow path likely governs the distribution of selenium species in the aquifer. Dissolved antimony increases in the sulfidic groundwater where arsenic levels are low and pH-controlled release of antimony from mineral surfaces possibly play an important role in antimony mobilization (Willis et al., 2011). Speciation modeling of antimony using Eh values suggests that trivalent antimony will likely dominate in Carrizo groundwater. Biogeochemical reactive transport modeling indicates that mercury is released into the groundwater by dissimilatory reduction of goethite and hematite, which primarily occur as coating on sand grains (Johannesson and Neumann, 2013).

21.1.6 Barnett Shale aquifer

The Barnett Shale is a hydrocarbon-producing geological formation, which lies around the Dallas–Fort Worth area of Texas. In recent years, gas-production activity experienced significant growth due to successful developments in hydraulic fracturing treatments (Byrd, 2007). In 2007 the Barnett produced roughly 28 billion m^3 natural gas and was the second highest US gas producing field (Tian and Ayers, 2010). Hildenbrand et al. (2015) analyzed 550 groundwater samples collected from wells tapping aquifers overlying the Barnett Shale formation. These investigators identified several VOCs across the region, including several alcohols, BTEX (Benzene, Toluene, Ethylbenzene, and Xylenes) compounds and organochlorides. In addition, Fontenot et al. (2013) evaluated water quality in 100 private drinking water wells drawing from aquifers located above the Barnett Shale formation in North Texas. These investigators detected methanol and ethanol in 29% of water samples. In addition, dissolved arsenic, selenium, strontium, and TDS levels were over the USEPA's relevant MCL for drinking water in a number of groundwater samples collected from wells that are within ~3 km of a producing gas well. The high concentrations may result from mobilization and transport of chemical constituents from aquifer sediments into groundwater, hydro-geochemical changes produced by aquifer exploitation, or incidents resulting from defective gas well casing.

21.2 Aquifers in Florida

Communities across Florida rely heavily on freshwater from aquifers (Prinos, 2016). According to the Florida Department of Environmental Protection (FDPE), the state has unconfined, semiconfined or confined types of aquifers in various combinations throughout Florida. The aquifers supply over 8 billion gallons of groundwater per day; globally, the Florida's aquifers are among the most productive. In particular, the Floridan aquifer system (FAS), located in southeastern Florida, is one of the most prolific aquifers in the world. The FAS is the principal source of freshwater supply for much of Florida (Miller, 1986). According to the FDPE, deeper saline groundwater dominates in the far-west Panhandle and in south Florida. The shallower (or surficial) Sand-and-Gravel Aquifer (in the west) and the highly productive Biscayne Aquifer (in the south) are pumped for public water supplies. The Surficial Aquifer System and the Intermediate Aquifer System usually yield smaller quantities of groundwater and are commonly pumped for smaller public water supplies. The impermeable clays within the intermediate aquifer restrict the flow of groundwater and, possibly, retards contaminant migration from land surface.

21.2.1 Floridan aquifer system

The FAS underlies the entire state and comprises carbonate rocks that form a vertical continuous sequence (Miller, 1986). Based on permeability, the aquifer has been divided into the upper Floridan aquifer (UFA) and lower Floridan aquifer (LFA). The two types of aquifers are separated by a low-permeability carbonate-confining unit, known as the middle confining unit (Wicks and Herman, 1994). During the last glacial period, recharge of the entire FAS occurred through the infiltration of atmospheric precipitation (Morrissey et al., 2010). Furthermore, sea level rise led to increase in hydraulic head, which consequently lowered the rate of groundwater movement and confined freshwater within the upper aquifers, which subsequently allowed saline water to flow in the lower aquifers.

The UFA, the most productive part of FAS, is the principal water supply source in most cities in northern and north-central Florida. The aquifer is primarily composed of carbonate minerals (calcite and dolomite), and lesser amounts of gypsum, anhydrite, chert, quartz, apatite, metal oxides, sulfides, lignite, and clay minerals (Hanshaw et al., 1965; Rye et al., 1981; Plummer and Sprinkle, 2001). In the west-central region of the state, parts of the UFA range from confined to semiconfined to unconfined. The impermeable layers of the Miocene Hawthorn Formation acts as confining units (Wilson and Gerhart, 1979). Tibbals (1990) indicate that the presence of the confining units along with the proximity to recharge and discharge areas influences the water quality in the UFA. The study further found that low levels of dissolved solids (<250 mg/L) usually occur in the area of recharge. Moreover, along the St. Johns River and the coastal area of the Atlantic Ocean, dissolved-solid concentration of discharging groundwater generally exceeded 1000 mg/L, and in places surpassed 25,000 mg/L (Tibbals, 1990).

Numerous investigators (e.g., Tang and Johannesson, 2005; Haque and Johannesson, 2006a,b; Basu et al., 2007; Willis et al., 2011) have extensively characterized the chemical composition of groundwater of UFA along a well-studied flow path, located within west-central Florida. In the study area the total thickness of the UFA varies between 60 and 640 m (Miller 1986). Recharge to the UFA occurs near Polk City, Florida, where the aquifer system is mostly unconfined (Back and Hanshaw 1970; Miller 1986; Plummer and Sprinkle 2001). From the zone of recharge the resulting groundwater flows radially outward to the coast in the chiefly confined UFA and eventually discharges into the Gulf of Mexico (Wicks and Herman 1994; Plummer and Sprinkle 2001). Along the flow path the Hawthorn Group (sands, marl, clay, limestone, dolostone, and phosphatic deposits) overlies the UFA and provide confinement, whereas the confinement of the aquifer from below is due to the presence of the middle confining unit (dolomite with intergranular anhydrite and gypsum; Wilson and Gerhart 1979; Miller, 1986; Sacks et al., 1995). Tang and Johannesson (2005) report that near the area of recharge the water is calcium—magnesium—bicarbonate and become calcium—magnesium—sulfate-bicarbonate-type water at the middle reaches of the groundwater flow path and subsequently changes into calcium—magnesium—sulfate-type water in discharge areas. The primary sources of calcium ion in the groundwater are due to dissolution of calcite, dolomite, and gypsum; however, in some areas, the subsurface mixing of fresh groundwater with seawater also supplies higher concentrations of calcium ion to the groundwater (Katz, 1992).

The main source of magnesium ions in the water is due to dolomite dissolution along with mixing of seawater with fresh groundwater (Lawrence and Upchurch, 1982). Gypsum dissolution and mixing of freshwater with saline water contributes sulfate to the groundwater (Sacks et al., 1995). Katz (1992) identified that bicarbonate ions in the UFA groundwater results from (1) CO_2 and recharge water equilibration that infiltrates through the vadose zone; (2) carbonate minerals (calcite and dolomite) dissolution; and (3) organic matter oxidation by sulfate-reducing microbes. In addition, increasing TDS trends were associated with increasing depth and degree of aquifer confinement. Recharge to the unconfined parts of the aquifer occurs generally from precipitation and downward leakage from the overlying surficial aquifer.

Haque and Johannesson (2006a,b) found that along the flow path, more oxidizing groundwater govern in the vicinity of the recharge zone and further downflow the water becomes increasingly reducing in areas of ferric iron reduction and subsequently sulfate reduction. In addition, elevated levels of dissolved arsenic are observed in the oxidizing groundwater of the recharge area compared to further downflow. In particular, arsenate predominates in the recharge area, and arsenite dominates in the more reducing groundwater. The study suggests that dissolved arsenic mobilization is likely controlled by reductive dissolution of ferric iron, sulfate reduction, and possibly precipitation of pyrite. Basu et al. (2007) studied the concentrations and speciation of dissolved selenium along the same flow path in the UFA and found that selenate and selenite species predominates in the recharge zone. Beyond the area of recharge, selenate and organic selenide dominate in the reducing groundwater. Changing selenium speciation near the area of groundwater recharge appears to coincide with the initiation of ferric iron reducing conditions. Further down gradient, where sulfate reducing conditions govern, selenide occur locally. Willis et al. (2011) found that dissolved antimony levels are elevated in areas where reductive dissolution of ferric oxides/oxyhydroxides dominates and lower where sulfate reducing aquifer conditions exist.

Withdrawal of water from the LFA has increased over the recent years for municipal water supply. In northeast Florida the LFA holds fresh to brackish water, whereas in the south, the water is saline, and wastewater treatment effluent has been disposed of into the aquifer (Walsh, 2012). Elevated levels of dissolved chloride have been observed in wells that tap the UFA and in the upper region of the LFA (Spechler and Phelps, 1997). Data on trace metals, radionuclides, and man-made contaminants are limited.

21.2.2 Sand-and-gravel aquifer

Miller (1990) report that groundwater in the sand-and-gravel aquifer is potable nearly everywhere. The aquifer is primarily composed of quartz sediments that has very low solubility in water, and thus dissolved solids levels in the circulating groundwater are generally below 50 mg/L. In the sand-and-gravel aquifer, groundwater samples are characterized by chloride levels that are below 50 mg/L; however, significantly higher chloride levels (>1000 mg/L) occurred in some wells located in the freshwater-saltwater transition zone of the coastal region. Pereira et al. (1987) report that the sand-and-gravel aquifer is vulnerable to contamination, and this has already occurred in some portions of the upper region. In addition, in some areas, the groundwater has been contaminated by infiltration of liquid wastes that contain wood preservatives (such as pentachlorophenol and creosote) from a wood-preservation plant near Pensacola, Florida.

21.2.3 Biscayne aquifer

The Biscayne aquifer, located in south Florida, is a highly transmissive unconfined karst aquifer that is chiefly composed of highly permeable limestone and less-permeable sandstone (Renken et al., 2008). According to the FDEP, the residents of Dade and Broward counties and southern Palm Beach County depend solely on the Biscayne aquifer for potable water. In addition, a large quantity of water from the aquifer is transported to the Florida Keys through transmission pipeline. In south Florida, ~ 3 million people rely on the aquifer as their sole source of municipal and irrigation water (Marella, 2004). Precipitation received during the wet season along with downward leakage from drainage canals mainly controls the groundwater recharge for the Biscayne Aquifer (Bradner et al., 2005). In general, high calcium and bicarbonate content is typical of waters that are primarily controlled by carbonate reactions (Habtemichael and Fuentes, 2016). The Florida Ground-Water Quality Monitoring Network Program assessed the concentrations of major ions and selected trace metal in water from the Biscayne (Radell and Katz, 1991). The study found that levels of major ions along with calcium, sodium, bicarbonate, and dissolved solids increased with depth, possibly due to reduced circulation at depth. Dissolved potassium and nitrate levels are substantially lower at depth and median concentrations for barium, chromium, copper, lead, and manganese were less than the US Environmental Protection Agency relevant maximum contaminant levels.

The Biscayne Aquifer contains saltwater where it partially merges with floor of the Biscayne Bay and the Atlantic Ocean. Some of the saltwater has encroached inland owing to the lowering of inland groundwater table resulting from period of droughts and construction of drainage canals (Bradner et al., 2005). In addition, Habtemichael and Fuentes (2016) report that water quality near the saltwater front of the aquifer mostly attributed to seawater/freshwater mixing, reactions with carbonate minerals, and cation exchange.

The water quality in the Biscayne Aquifer is under threat of deterioration due to various anthropogenic activities. The highly permeable karstified limestone, in conjunction with the shallow water table augments the groundwater's susceptibility to contamination by land surface, leachate from landfills, saltwater intrusion, and from septic tanks (Marella, 2004). Mobilization of chemical contaminants is further enhanced by eogenetic karst characteristics of the aquifer, where limestone is near land surface enlarging conduits within these porous rocks (Renken et al, 2008).

Acknowledgments

The author is deeply grateful to Mahbub Haque, Lailun Nahar, and Sophie L. Haque for all their assistance and helpful comments. Special thanks are extended to Ahsan Saif and Faria Tabassum for assistance with editing the manuscript.

References

Anaya, R., 2001. An overview of the Edwards-Trinity aquifer system, central-west Texas. In: Mace, R.E., Mullican III, W.F., Angle, E.S. (Eds.), Aquifers of West Texas. Texas Water Development Board Report, 356. pp. 100–119.

Ashworth, J.B., Hopkins, J., 1995. Aquifers of Texas: Texas Water Development Board Report, vol. 345, p. 69.

Back, W., Hanshaw, B.B., 1970. Comparison of chemical hydrogeology of the carbonate peninsulas of Florida and Yucatan. J. Hydrol. 10, 330–368.

Barker, R.A., Ardis, A.F., 1992. Configuration of the Base of the Edwards–Trinity Aquifer System and Hydrogeology of the Underlying Pre-Cretaceous Rocks, West Central Texas: U.S. Geological Survey Water Resources Investigation Report 91-4071, 25.

Basu, R., Haque, S.E., Ji, J., Johannesson, K.H., 2007. Evolution of selenium concentrations and speciation in groundwater flow systems: upper Floridan (Florida) and Carrizo Sand (Texas) aquifers. Chem. Geol. 246, 147–169.

Bradner, A., McPherson, B.F., Miller, R.L., Kish, G., Bernard, B., 2005. Quality of groundw in the Biscayne Aquifer in Miami-Dade, Broward, and Palm Beach Counties, Florida, 1996-1998, with Emphasis on Contaminants. U.S. Geological Survey, Open-File Report 2004-1438.

Bush, P.W., Ulery, R.L., Rittmaster, R.L., 1994. Dissolved-solids concentrations and hydrochemical facies in water of the Edwards-Trinity aquifer system, west-central Texas. Geology. Available from: https://doi.org/10.3133/wri934126.

Byrd, C.L., 2007. Updated evaluation for the Central Texas–Trinity aquifer–priority groundwater management area. In: Priority Groundwater Management Area File Report December.

Castro, M.C., Goblet, P., 2003. Calibration of regional groundwater flow models: working toward a better understanding of site specific systems. Water Resour. Res. 39 (6), 1172.

Castro, M.C., Stute, M., Schlosser, P., 2000. Comparison of ^{4}He ages and ^{14}C ages in simple aquifer systems: implications for groundwater flow and chronologies. Appl. Geochem. 15, 1137–1167.

Chaudhuri, S., Ale, S., 2014. Temporal evolution of depth-stratified groundwater salinity in municipal wells in the major aquifers in Texas, USA. Sci. Total Environ. 472, 370–380.

Fahlquist, L., Ardis, A.F., 2004. Quality of water in the Trinity and Edwards aquifers, south-central Texas, 1996-98. In: USGS Series Scientific Investigations Report 2004–5201.

Fontenot, B.E., Hunt, L.R., Hildenbrand, Carlton Jr., D.D., Oka, H., Walton, J.L., et al., 2013. An evaluation of water quality in private drinking water wells near natural gas extraction sites in the Barnett Shale formation. Environ. Sci. Tech. 47, 10032–10040. Available from: https://doi.org/10.1021/es4011724. 2013.

Habtemichael, Y.T., Fuentes, H.R., 2016. Hydrogeochemical analysis of processes through modeling of seawater intrusion impacts in Biscayne Aquifer water quality, USA. Aquat. Geochem. 22, 197.

Hanshaw, B.B., Back, W., Rubin M., 1965. Carbonate equilibria and radiocarbon distribution related to groundwater flow in the Floridian limestone aquifer, U.S.A. In: Proceeding of the Dubrovnik Symposium, Hydrology of Fractured Rocks, vol. 1, pp. 601–614.

Haque, S., Johannesson, K.H., 2006a. Arsenic concentrations and speciation along a groundwater flow path: The Carrizo Sand aquifer, Texas, USA. Chem. Geol. 228, 57–71.

Haque, S.E., Johannesson, K.H., 2006b. Concentrations and speciation of arsenic along a groundwater flow-path in the Upper Floridan aquifer, Florida, USA. Environ. Geol. 50, 219–228. Available from: https://doi.org/10.1007/s00254-006-0202-8. 2006.

Hildenbrand, Z.L., Carlton Jr., D., Fontenot, B.E., Meik, J.M., Walton, J.L., Taylor, J.T., et al., 2015. A comprehensive analysis of groundwater quality in the Barnett Shale region. Environ. Sci. Technol. 49, 8254–8262.

Hudak, P.F., 2003. Oil production and groundwater quality in the Edwards-Trinity Plateau aquifer, Texas. Sci. World J. 3, 1147–1153.

Hudak, P.F., 2008. Distribution of arsenic concentrations in groundwater of the Seymour Aquifer, Texas, USA. Int. J. Environ. Health Res 18 (1), 79–82.

Hudak, P.F., 2018. Associations between dissolved uranium, nitrate, calcium, alkalinity, iron, and manganese concentrations in the Edwards-Trinity Plateau Aquifer, Texas, USA. Environ. Process 5, 441–450.

Hutchison, W.R., 2016. Edward-Trinity (Plateau), Pecos Valley and Trinity Aquifers: nine factor documentation and predictive simulations. In: GMA 7 Technical Memorandum, 15-06.

Jensen, R., 2007. Influences of Natural and Man-Made Sources of Contamination on Water Quality Trends in the Seymour Aquifer: A 2007 Status Report. Texas Water Resources Institute and University of Texas Bureau of Economic Geology, Texas.

Johannesson, K., Neumann, K., 2013. Geochemical cycling of mercury in a deep, confined aquifer: Insights from biogeochemical reactive transport modeling. Geochim. Cosmochim. Acta 106, 25–43.

Kaiser, R., Skiller, F.F., 2001. Deep trouble: options for managing the hidden threat of aquifer depletion in Texas. Texas Tech. Law Rev. 32, 249.

Katz, B.G, 1992, Hydrochemistry of the Upper Floridan Aquifer, Florida. In: Water-Resources Investigations Report 91-4196.

Lawrence, F.W., Upchurch, S.B., 1982. Identification of recharge areas using geochemical factor analysis. J. Ground Water 20 (6), 680–687.

Mace, R.E., Chowdhury, A.H., Anaya, R., Way, S.-C., 2000. Groundwater availability of the Trinity Aquifer, Hill Country area, Texas: numerical simulations through 2050. Texas Water Development Board Report 353.

Marella, R.L., 2004, Water withdrawals, use, discharge, and trends in Florida, 2000: U.S. Geological Survey Scientific Investigations Report 2004-5151, p. 136.

Meyer, J.E., Wise, M.R., Kalswad, S., 2012. Pecos Valley Aquifer, West Texas: structure and brackish groundwater. Texas Water Development Board Report 382.

Miller, J.A., 1986. Hydrogeologic framework of the Floridan aquifer system in Florida, Georgia, South Carolina and Alabama.

Miller, J.A., 1990. United States Geological Survey. Ground Water Atlas of the United States: Alabama, Florida, Georgia, South Carolina. Hydrologic Atlas 730-G.

Morrissey, S., Clark, J., Bennett, M., Richardson, E., Stute, M., 2010. Groundwater reorganization in the Floridan aquifer following Holocene sea-level rise. Nat. Geosci. 3, 683–687.

Mukherjee, A., Scanlon, B., Aureli, A., Langan, S., Guo, H., McKenzie, A., 2020. Global Groundwater: Source, Scarcity, Sustainability, Security and Solutions, first ed. Elsevier, ISBN. 9780128181720.

Pereira, W.E., Rostad, C., Updegraff, D.M., Bennett, J.L., 1987. Fate and Movement of Azaarenes and their anaerobic biotransformation products in an aquifer contaminated by wood-treated chemical. Environ. Toxicol. Chem. 6 (3), 163–176.

Plummer, L.N., Sprinkle, C.L., 2001. Radiocarbon dating of dissolved inorganic carbon in groundwater from confined parts of the Upper Floridan aquifer, Floridan, USA. Hydrogeol. J. 9, 127–150.

Prinos, S., 2016. Saltwater intrusion monitoring in Florida. Special Issue: Status Florida's Groundw. Resour 79, 269–278.

Radell, M.J., Katz, B.G., 1991. Major-ion and selected trace-metal chemistry of the Biscayne Aquifer, Southeast Florida. US Geological Survey. Water Resour. Invest. Rep. 91, 4009.

Rajagopalan, S., Anderson, T.A., Fahlquist, L., Rainwater, K.A., Ridley, M., Jackson, W., 2006. Widespread presence of naturally occurring perchlorate in High Plains of Texas and New Mexico. Environ. Sci. Technol. 40, 3156–3162.

Reedy, R.C., Scanlon, B.R., Nicot, J.-P., Tachovsky, J.A., 2007. Unsaturated zone arsenic distribution and implications for groundwater contamination. Environ. Sci. Technol. 41, 6914–6919.

Renken, R.A., Cunningham, K.J., Shapiro, A.M., Harvey, R.W., Zygnerski, M.R., Metge, D.W., et al., 2008. Pathogen and chemical transport in the karst limestone of the Biscayne aquifer: 1. Revised conceptualization of groundwater flow. Water Resour. Res 44.

Rye, R.O., Back, W., Hanshaw, B.B., Rightmire, C.T., Pearson Jr., F.J., 1981. The origin and isotopic composition of dissolved sulfide in ground water from carbonate aquifers in Florida and Texas. Geochim. Cosmochim. Acta 45 (10), 1941–1950.

Sacks, L.A., Herman, J.S., Kauffman, S.J., 1995. Controls on high sulfate concentrations in the Upper Floridan Aquifer in Southwest Florida. Water Resour. Res 31 (10), 2541–2552.

Schriver, S., Hopkins, J., 1998. Updated water-quality evaluation of the Ogallala Aquifer including selected metallic and non-metallic inorganic constituents. Texas Water Development Board.

Spechler, R.M., Phelps, G.G., 1997. Saltwater intrusion in the Floridan aquifer system, northeastern Florida. In: Georgia Water Resources Conference, Athens, GA.

Standen, A.R., Opdyke, D.R., 2004. Contamination migration, characteristics, and responses for the Edwards–Trinity (Plateau) aquifer. In: Mace, R. E., Angle, E.S., Mullican, W.F. III (Eds.), Aquifers of the Edwards Plateau. Texas Water Development Board Report 360, pp. 211–234.

Tang, J., Johannesson, K.H., 2005. Rare earth element concentrations, speciation, and fractionation along groundwater flow paths: The Carrizo Sand (Texas) and Upper Floridan Aquifers. In: Johannesson, K.H. (Ed.), Rare Earth Elements in Groundwater Flow Systems. Water Science and Technology Library, 51. Springer, Dordrecht.

TDWR (Texas Department of Water Resources), 1984. Water for Texas: a comprehensive plan for the future. In: Pub No. GP-4-1, p. 37.

Tian, Y., Ayers, W., 2010. Barnett Shale (Mississippian), Fort Worth Basin, Texas: regional variations in gas and oil production and reservoir properties. In: Society of Petroleum Engineers, SPE-137766-MS.

Tibbals, C.H., 1990. Hydrology of the Floridan Aquifer System in East-Central Florida, United States. United States Geol. Survey, Professional Paper; 1403-E.

USGS (United States Geological Survey), 2010. Estimated Use of Water in the United States.

Walsh, V.M., 2012. Geochemical Determination of the Fate and Transport of Injected Fresh Wastewater to a Deep Saline Aquifer (FIU Electronic Theses and Dissertations). p. 692.

Wicks, C.M., Herman, J.S., 1994. The effect of a confining unit on the geochemical evolution of ground water in the upper Floridan aquifer system. J. Hydrol. 153, 139–155.

Willis, S.S., Haque, S., Johannesson, K., 2011. Arsenic and antimony in groundwater flow systems: a comparative study. Aqua. Geochem 17, 775–807.

Wilson, W.E., Gerhart, J.M., 1979. Simulated changes in potentiometric levels resulting from groundwater development for phosphate mines, West-Central Florida. Dev. Water Sci 12, 491–515.

Chapter 22

Groundwater pollution in Pakistan

Noshin Masood, Shehla Batool and Abida Farooqi
Department of Environmental Sciences, Faculty of Biological Sciences, Quaid-i-Azam University, Islamabad, Pakistan

22.1 Introduction

Global groundwater is undergoing immense stress in recent times (Mukherjee et al., 2020). Pakistan is also facing adverse groundwater pollution issues along with the groundwater shortage. The major water resources in Pakistan can be divided into three basic categories: precipitation, surface water resources, and groundwater.

Groundwater is a major freshwater source in Pakistan after surface water especially since the last decade when Pakistan has experienced some drastic changes in temperature and precipitation due to climate change. The groundwater of the country is governed by Indus Basin. Indus Basin represents an extensive groundwater aquifer covering an area of 16.2 million hectares (mha). However, it is important to understand that groundwater resource has gone through three major changes in reference to both quality and quantity. The first was the poststorage phase (since 1947 to 1970s), the groundwater contribution to agriculture was around 12 billion m^3, which is 11% of the total water available for agriculture. The prestorage phase was marked with waterlogging and salinity due to the introduction of canal water irrigation system that was completed and became operational in the 1980s. Most of the waterlogging and salinity data was obtained by basin wide survey conducted by the Water and Power Development Authority in 1979 to determine early post-Tarbela conditions. The survey indicated 42% of the area under Indus Basin with less than 3 m water table depth (Ali et al., 2017), hence, classified as waterlogged.

Ahmad and Rashida (2001) also reported this water imbalance. However, they reported it in the form of river system losses and canal conveyance losses. It was reported that 12 billion m^3 (50%) of water loss was from the river systems, whereas the canal conveyance losses were about 31 billion m^3. The total water loss of Indus Basin Irrigation System (IBIS) was estimated to be 61 billion m^3. Two important assumptions were made in this study. It was stated that the field application of water only contributed toward high water recharge (waterlogging and salinity) where over irrigation was practiced. The fields that were normally under irrigated do not significantly contribute toward water recharge. Previously, Zuberiand Sufi (1992) also estimated a recharge of 56 billion m^3. The additional conveyance losses o IBIS due to Tarbela contributed 10% to the total groundwater recharge (Ahmed et al., 2001a). To combat this situation government introduced Salinity Control and Reclamation Program (SCARP) under which publicly owned tube wells were constructed all over Pakistan. The aim of SCARP was to increase agricultural production by reducing soil salinity and providing irrigation water. By the end of 1990s various reports claimed groundwater potential for exploitation of around 8 billion m^3.

The third phase that markedly changes the groundwater resource situation in Pakistan was the prolonged drought from 1999 to 2002. It was estimated that around 60,000 additional tube wells were installed in Sindh and Punjab provinces alone to meet the shortfall in the canal water supply. The drastic changes in the precipitation patterns, Pakistan's vulnerability to climate change, overexploitation of water resource, and exponential population growth has resulted in severe water imbalances. The water imbalance that is quite significant for Pakistan can be used to explain a very large gap between the minimum river inflow of 98.6 million acre feet (MAF) (realized in 2001–02) and the maximum inflow of 186.8 MAF (in 1959–60) (Ali et al., 2017). It has been reported that water table is receding 3 ft./year in Punjab while in Quetta water table depleted about 95 ft. during 2006–13 (Raza et al., 2017). Similar trends have been observed in other areas of the country.

Another aspect of the groundwater resources in Pakistan is their quality. Generally, there is a difference in quality from one hydro-geologic basin to another. Within the Indus Basin the water quality generally deteriorates from north to south. Among the various reports published so far show that areas close to recharging sources (rivers and aquifers recharged through rainfall) have good water quality.

It is estimated that around 40% of all reported diseases and deaths in Pakistan are attributed to poor water quality in the country (Ali et al., 2017). Moreover, the leading cause of deaths in infants and children up to 10 years of age is that of contaminated water. The mortality rate of 136 per 1000 live births due to diarrhea is reported, while every fifth citizen suffers from illness caused due to unsafe water. In Karachi, more than 10,000 people die annually of renal infection due to polluted drinking water.

The following sections describe the groundwater pollution in Pakistan.

22.2 Groundwater quality

Groundwater quality parameters include biological and physiochemical parameters. The biological parameters include microbial and bacterial contamination. Physiochemical parameters are the physical and chemical parameters (organic and inorganic ions). The physical parameters comprise pH, electrical conductivity, total dissolved solids, total suspended solids, salinity, and dissolved oxygen. The chemical parameters are divided into organic and inorganic ions, which can be further subdivided into cations and anions. However, an important aspect while considering water quality is its end use. Generally, the criteria for determining drinking water quality are more stringent. In Pakistan, drinking water quality parameters are defined by Pakistan Standard Quality Control Authority (PCSQA). Table 22.1 shows drinking water quality standards specified by PSCQA of the most common drinking water quality parameters.

22.2.1 Biological contamination of groundwater

The biological contamination is also referred as microbial contamination. It poses a serious human health concern (Jain et al., 2005). The most common waterborne diseases caused by microbes are nausea, diarrhea, typhoid, dysentery, gastrointestinal infections, and pneumonia (Shar et al., 1970). In Pakistan waterborne diseases are the one of major reasons of child mortality, about 230,000 infants (having age under 5 years) die every year due to water-related illnesses. In 2006 the number of reported diarrhea cases was about 4.5 million in the country. An estimate reveals that water-related diseases in Pakistan cause national income losses of pakistani rupees 25−58 billion (US$0.25−0.58 billion) annually, which is approximately 0.6%−1.44% of the country's GDP (Tahir and Rasheed, 2013).

The water contamination has been prevalent since last two decades in Pakistan's' water supply. However, the research focus was on inorganic water pollutants. The shift toward biological contamination happened in last decade. The major causes of microbial contamination in water in Pakistan are *Escherichia coli* and total coliform. Fig. 22.1 represents province-wise contamination of total coliform, fecal coliform, and *E. coli* reported in literature since 1990s to 2018 in Pakistan.

22.2.1.1 Punjab

Acinetobacter sp. has been reported in the groundwaters of Sargodha, Sahiwal, and Lahore. It is responsible for pneumonia, urinary tract bloodstream, and nosocomial infections (Rasheed et al., 2009; Tassadaq et al., 2013). *Aeromonas hydrophila* was found in Lahore and Sargodha (Ahad et al., 2000; Nabeela et al., 2014). *Enterobacter* sp. and *Edwardsiella tarda* have been found in Lahore, Sargodha, and Sahiwal (Rasheed et al., 2009). These microbes are responsible for gastroenteritis, diarrhea, and unitary tract infections (Nabeela et al., 2014). Other researchers have also reported species such as *Proteus* sp., *Salmonella* sp., *Shigella* sp., *Staphylococcus* sp., *Stenotrophomonas* sp., and *Streptococcus* sp. in Punjab (Nabeela et al., 2014).

22.2.1.2 Sindh

Karachi, Garhi Habibullah, Sahiwal, and Sukhur have been extensively studied in Sindh. *Serratia* sp. along with *Acinetobacter* sp. has been found in Sukhur and Karachi, respectively. Both species are responsible for nosocomial infections (Tassadaq et al., 2013). *Chryseobacterium meningosepticum*, *Citrobacter* sp., and *Pseudomonas aeruginosa* have been reported in Sukhur. *Chryseobacterium* sp. causes infant meningitis and immunocompromised infections in adults (Shar, 2012), whereas *Citrobacter* sp. and *P. aeruginosa* are responsible for urinary tract, respiratory tract, and blood sepsis infections; infant meningitis; and progressive pulmonary infections, respectively. The microbial species found in Karachi include *Micrococcus* sp., causing arthritis, endocarditis, meningitis, and pneumonia (Tassadaq et al., 2013); *Streptococcus* sp. that leads to streptococcal pharyngitis, bacterial pneumonia, endocarditis, erysipelas, and

TABLE 22.1 Drinking water quality standards by Pakistan Standard Quality Control Authority.

Parameters	Unit	MAC[a]	MAC[b]	WHO
Physical characteristics				
pH	–	7.0–8.5	6.5–9.2	6.5–7.5
Turbidity	NTU[c]	5	25	–
Color	TCU[d]	5	50	–
Odor and taste	–			
Chemical characteristics				
Calcium (Ca^{2+})	mg/L	75	200	200
Magnesium (Mg^{2+})	mg/L	50	150	150
Total hardness	mg/L	20	150	60–200
Sulfate (SO_4^{2-})	mg/L	200	400	250
Nitrate (NO_3^-)	mg/L	–	10	50
Chloride (Cl^-)	mg/L	200	600	250
Fluoride (F^-)	mg/L	–	0.7	1.5 and 0.7[c]
Iron (Fe)	mg/L	0.3	1.0	0.3
Zinc (Zn)	mg/L	5	15	5
Copper (Co)	mg/L	–	1.0	2.0
Manganese (Mn)	mg/L	–	0.5	0.5
Arsenic (As)	mg/L	–	0.01	0.01
Cadmium (Cd)	mg/L	–	0.003	0.003
Chromium (Cr)	mg/L	–	0.05	0.05
Lead (Pb)	mg/L	–	0.01	0.01
Bacteriological contamination				
Escherichia coli	Ml	–	0/250	0/250
Total coliform	Ml	–	0/250	0/250
Enterococci	Ml	–	0/250	0/250
Pseudomonas	Ml	–	0/250	0/250

[a]Maximum acceptable concentration.
[b]Maximum allowable concentration.
[c]Revised limit foe Asian region is 0.7 mg/L.
[d]MAC, maximum acceptable concentration; NTU, nephelometric turbidity units; TCU, true color unit.

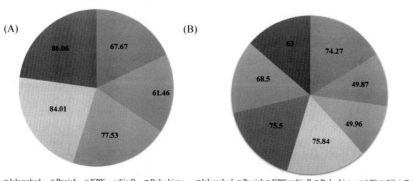

FIGURE 22.1 Bacterial contamination status of Pakistan: (A) total coliform (%) in groundwater sample and (B) fecal coliform/*Escherichia coli* in groundwater samples.

necrotizing fasciitis (Sarwar et al., 2004); *Klebsiella* sp. that is the leading cause of pneumonia; and *Stenotrophomonas* sp. that causes pulmonary infections (Shar, 2012; Rasheed et al., 2009).

22.2.1.3 Khyber Pakhtunkhwa

High levels of *E. coli* have been reported in the groundwater of Peshawar (Nabeela et al., 2014). Abbottabad, Mansehra, and Balakot have also been studied extensively for microbial contamination. The microbes identified in these cities include *Citrobacter* sp., *Enterobacter* sp., *Klebsiella* sp., *Shigella* sp., and *Vibrio* sp., *Staphylococcus* sp., and *Shigella* sp. have been found in the flood-affected areas of Khyber Pakhtunkhwa (KPK); they cause skin sepsis, postoperative wound infections, enteric infections, septicemia, and bacillary dysentery, respectively (Nabeela et al., 2014).

22.2.1.4 Azad Kashmir and Gilgit Baltistan

Muzaffarabad has been studied in Azad Kashmir region. *Vibrio* sp., *Staphylococcus* sp., *P. aeruginosa*, *Proteus* sp., *Klebsiella* sp., *Acinetobacter* sp., and *A. hydrophila* have been identified in these cities (Shar, 2012; Tassadaq et al., 2013).

22.3 Chemical contamination

22.3.1 Organic pollution of groundwater

Organic pollution in water is caused by application of herbicides, insecticides, and fungicides that have been used for pest control and agriculture. Major concern regarding organic pollution in groundwater is "emerging organic contaminants"(EOCs). These are the compounds that are newly developed or discovered due to analytical developments and have been recently classified as contaminants. However, EOCs have been characterized in the surface water and wastewater better than the groundwater environments.

The leading cause of organic pollution in Pakistan is pesticides. Though the pesticides agrochemical have been used since 1957, there has been no monitoring program for pesticide use and regulation carried out by Pakistani authorities, except for Punjab Private Groundwater Study in 2006. According to this study, more than 30 types of fungicides, 5 types of acaricides, 39 types of weedicides, 6 different types of rodenticides, and 108 types of insecticides are being used in Pakistan (IARC, 2012; Mehmood et al., 2017). Pesticides are also a leading cause of cancer, according to a report by, about 10–14 million people have suffered from water poisoning due to pesticide application worldwide (McGuire, 2016). Due to its unabated use in the developing countries, it has been reported that the 37,000 annual registered cases of cancer in the developing countries prove its direct or indirect link with pesticides applications.

The traces of pesticides have been found in groundwater along the agricultural land of Punjab (Faisalabad, Lahore, Multan, DG Khan, Bahawalnagar, and Muzaffargarh) and KPK (Swabi, Mardan, Thaktbai, Peshawar, and Charsadah). Most of these areas are used for cultivating cotton and tobacco. Among the pesticides found in groundwater also included aldrin, chlorpyriphos, dichlorvos, dieldrin, endosulfan, heptachlor, mevinphos, monocroptophos, phosphamidon, and propetamphos. These pesticides have been included in the "Pesticide Red List." The list classifies pesticides according to their potential hazard that have been banned by WHO or by other international agreements. The presence of these pesticides in the shallow aquifers indicates their use in the country. However, it cannot be ascribed as pesticide pollution due to lack in proper data. The soil characteristics, irrigation practices, and intensive spraying can be the reason of the leaching of these pesticides in shallow wells (Tariq et al., 2004). Regarding the use of these hazardous agrochemicals, the Agricultural Pesticide Technical Advisory Committee has finally prohibited the import chemicals of listed in WHO list in the first phase. However, the ban on their use will be effective from August, 2022.

Another major organic pollutant in groundwater is persistent organic pollutants and their metabolites. Their main sources are insecticides that have been administered on various crops especially cotton. It must be noted that some of these insecticides have been banned due to their hazardous nature. However, due to their persistent nature and secondary metabolites, they are still present in the environment. The earliest record of these was found in the cattle drinking water of Cattle Colony, Karachi, its metabolite in 1984. Insecticides detected were benzene hexachloride , dichlorodiphenyl-trichloroethane (DDT), and dichlorodiphenyldichloroethylene (Parveen and Masud, 1988). It was also detected in the cotton-growing areas of Samundri and Faisalabad, which reflected its large-scale use (Jabbar et al., 1993). In more recent studies, DDT and its metabolites were also found in Lahore and Hyderabad in trace amounts (Asi et al., 2008). However, higher concentrations were found in the cotton-growing belts of Punjab and Sindh, namely,

Bahawalpur, Khanewal, Sheikhupura, Baidan, Chakdenawala, and Mian Channu. Very high concentrations of DDT (70−400 μg/L) were found in a demolished DDT factory in Nowshera (Jan et al., 2009).

22.4 Inorganic pollution of groundwater

The inorganic pollution status of Pakistan as reported in various studies is summed up in Table 22.2. Among the inorganic water contaminants, trace metals, including heavy metals and anions, are of major concern. Following are the main heavy metals found in the groundwaters of Pakistan.

22.4.1 Trace and heavy metals

Trace metals are metals such as iron (Fe), cobalt (Co), and zinc (Zn), which are required for healthy human body development and function in very small amounts. Excess of these can cause harmful effects, whereas heavy metal generally refers to a metallic element with relatively high density or is toxic at low concentrations. Common examples of heavy metals include lead (Pb), mercury (Hg), and arsenic (As). To date countless research articles have been published emphasizing on the harmful effects of these metals present in the drinking water especially groundwater. However, the challenge to protect human population from the adverse effects of metal contamination is far greater in the developing world compared to the developed nations because of lack in both economic and technological resources. The following sections give a brief description of metals (heavy and trace) found in the groundwater of Pakistan.

22.4.1.1 Arsenic

Arsenic has been reported in the groundwater of the United States, Hungry, China, India, Chile, Bangladesh, Argentina, and Vietnam (Jadhav et al., 2015). It has been reported that out of 150 million people globally, 90 million are affected by arsenic in South East Asia (Pakistan, India, and Bangladesh) (Jadhav et al., 2015). Pakistan ranks fourth globally with regards to arsenic groundwater contamination (Jadhav et al., 2015). Sindh and Punjab provinces are the worst arsenic-affected areas of Pakistan. Districts Tharparkar, Sukhur, and Hyderabad been reported with arsenic groundwater pollution in Sindh (Khan et al., 2016). A cross-boundary study between Pakistan and India shows the extent of arsenic contamination in Punjab province of the both the countries. This study shows that districts Kasur, Lahore, Multan, Sheikhupura, GI Khan, Mianwali, and Vehari have high arsenic concentrations in groundwater of Punjab, Pakistan (van Geen et al., 2019). Under this blanket study almost 30,000 wells have been tested. 79% of the wells have been found with As < WHO limits (0.01 mg/L), 11% were between 0.01 and 0.05 mg/L, and 10% of the samples contain more than 0.05 mg/L As concentration. Results of this blanket testing along with other reported As concentrations are shown in Fig. 22.2. As groundwater contamination is prevalent along River Ravi in Pakistan. Villages such as Kullanlawalla, Kudpur, KotMagia, Devamal, and Chafatehwala have been identified as severely polluted. The inhabitants have visible signs of arsenicosis and various cancers are also prevalent. However, small projects such as water filtration plants, supply from safe wells, and public awareness campaigns under public−private partnerships have been carried in these areas in order to combat the situation.

22.4.1.2 Cadmium

Cadmium (Cd) is also Group 1 human carcinogen (IARC, 2012). Cd toxicity can lead to lung, kidney, and skeletal damage. It is also known for itai-itai disease that is one of the four major pollution diseases of Japan (Nordberg et al., 2002). Cd is mostly released in the groundwater environment through anthropogenic influences mainly mining activities. Major sources also include effluent discharges from marble, steel, and metal plating industries. Since Pakistan is a developing country with agriculture economy, the industrial sector is not well-developed comparatively. Therefore Cd contamination has been found in samples near industrial centers of the provincial capitals. High Cd concentrations (0.21 mg/L) have been reported in Hayatabad Industrial Estate, KPK. The average concentration in this area is 0.02 mg/L (Manzoor et al., 2006). The results of various groundwater studies on Cd pollution are shown in Fig. 22.2. Cd pollution in surface water shows a different trend, studies have reported undetected Cd concentrations in March to April and showed seasonal variations (0.01−0.05 mg/L) in other months. This indicates effect of environmental factors on Cd dissolution. However, to determine a certain trend, further investigations are needed. High Cd concentrations of 5.53 mg/L and 184 mg/kg have been observed in industrial wastewater and sediments near mining sites (Waseem et al., 2014), also, indicating anthropogenic influence in Cd pollution.

TABLE 22.2 Sources of inorganic contaminants in drinking water sources of Pakistan.

Region	Area	N	Inorganic contamination	Source of contamination	Remarks	References
Pakistan	Punjab, Baluchistan	747	High F^-	Calcium-poor aquifers, fluoride-bearing minerals, and cation exchange processes can be the source of F^- contamination	16% of Punjab and 22% of Baluchistan water samples had high F^- content	Tahir and Rasheed (2013)
	Indus plains of Pakistan (Punjab)		Elevated As	Geogenic source	Drinking untreated well water can pose a significant health risk, particular in the case of As along the Ravi River	van Geen et al. (2019)
	Pakistan		Elevated F^-	Geogenic source	Naturally occurring F^- had serious health concerns	Rasool et al. (2017)
Baluchistan	Quetta, Pishin, and Mastung	30	High As, Hg, Ni, Cd, Cr, Fe, Pb, F^-, and SO_4^{2-}	Geological composition of rocks (mafic and ultramafic) and anthropogenic activities such as agricultural and coal combustion of coal in brick kilns	Drinking water is of poor quality	Khanoranga and Khalid (2019)
	Loralai		High Pb		Due to weathering and erosion of mafic–ultramafic rocks (ophiolites) and clastic sedimentary rocks	Ullah et al. (2019)
	Zhoab		High Fe^{2+}, Pb^{2+}, Ni^{2+}, Co^{2+}, Cr^{3+}, and Cd^{2+}	Due to weathering and erosion of mafic–ultramafic rocks (ophiolites) and clastic sedimentary rocks		
	Northern Suleiman fold belt		High Fe, Ni, Pb, Co, Cr, and Cd	Major source of contamination is geogenic with little input from anthropogenic		
	29 districts of Baluchistan		High F^- and NO_3^-	Geogenic sources		Ghoraba and Khan (2013)
	Quetta valley		High NO_3^-, SO_4^{2-}, Cr, and Ni	Rock alteration and mining activity in the area are responsible for contamination		Khan et al. (2010)

Punjab	District Vehari	41	Elevated NO_3^-	Saline and brackish nature of underground water might be the cause of higher Na^+ levels	Unfit for drinking purposes	Khalid et al. (2018)
	Muzaffargarh	49	Elevated As, B, Fe, SO_4^{2-}, PO_4^{3-}	Canal irrigation has led to widespread waterlogging	Groundwater is highly contaminated to As	Nickson et al. (2005)
	Tehsil Mailsi	44	High As, Cd, Fe, Pb, NO_3^-, and SO_4^{2-}	Agrochemicals increase the As and trace elements concentration	Groundwater of the area is not safe for drinking purpose	Rasool et al. (2016)
	Low-lying Indus plain region	146	Elevated F^- and NO_3^-	Halite dissolution process is the major contributor for F^- enrichment in groundwater		Ali et al. (2019)
	Chichawatni, Vehari, Rahim Yar Khan, Bahawalpur, and Multan	123	Elevated As, SO_4^{2-}	Well depth, pH, salinity, CO_3 possibly controlled arsenic hydro-geochemical behavior	75% of wells exceeded WHO limit of 10-ppb As concentration	Shakoor et al. (2018)
	Ghazi Ghatt to Umar Kot, Southern Punjab	112	High NO_3^- and SO_4^{2-}	70% of the tested samples had high NO_3^-	Shallow wells had higher nitrate concentrations than deep wells indicated that anthropogenic count, that is, fertilizer applications	Khan et al. (2019)
	Lahore and Kasur		Elevated As, SO_4^{2-}, and NO_3^-, PO_4^{3-}	Desorption of As from metal surfaces under alkaline environment might be the source of As enrichment		Mushtaq et al. (2018)
	D G Khan	32	Slightly high As, SO_4^{2-}, NO_3^-,	Mild enrichment of As in few sites/wells is most probably due to microbial contamination, metal leaching from the mine wastes and reduction of HFO		Malana and Khosa (2011)
	Gujrat	70	Elevated As and Ni		Both geogenic and anthropogenic inputs are accountable for As and Ni contamination	Masood et al. (2019)
	Rahim Yar Khan	51	Elevated As, F^-, and NO_3^-	Higher levels of As were categorized by higher HCO_3, NO_3, SO_4^{2-}, and PO_4^{3-} at high pH, which was an indication of oxidizing environment	Exposed population is at high risk to As and F^- contamination	Farooqi et al. (2017)

(Continued)

TABLE 22.2 (Continued)

Region	Area	N	Inorganic contamination	Source of contamination	Remarks	References
	Kalalanwala		Elevated As, SO_4^{2-}, and F		91% of wells exceeded the WHO recommended limit of As in groundwater	Farooqi et al. (2007)
	Multan		High Pb, Cr, Ni, and Mn	Irrigation practices added heavy metals in water	Groundwater is highly contaminated to heavy metals	Nickson et al. (2005)
Sindh	Jaamshoro	153	Elevated As, Fe, and SO_4^{2-}	Due to widespread waterlogging from Indus river irrigation system	Anthropogenic count (coal combustion in brick kiln, power plant) and waterlogging are the main reasons of water contamination	Baig et al. (2009)
	Nagarparkar	32	High F^- and SO_4^{2-}	Groundwater quality reflects the influences of silicate mineral weathering and evaporation		Naseem et al. (2010) and Rafique et al. (2009)
	Umarkot	152	High F^- and SO_4^{2-}			
	Tharparkar	380	High As	High SO_4^{2-}	Groundwater is unfit for drinking and consumption purpose	Brahman et al. (2016)
	Tando Allahyar	79	High As and Mn	Seepage of irrigation canals contaminate drinking water sources		Naseem and McArthur (2018)
	Larkana		Elevated As	Seepage of irrigation canals contaminate drinking water sources		Ali et al. (2019)
KPK	Charsadda district		Elevated Pb, Cd, Ni, Fe, SO_4^{2-}, and NO_3^-		Drinking water contamination causes gastroenteritis, dysentery and diarrhea, hepatitis-A, hepatitis-B, and hepatitis-C	Khan et al. (2013)

Location	Samples	Contaminants	Sources	Findings	Reference
District Bannu	197	High Fe and SO_4^{2-}	Presence of the limestone, dolomite, gypsum, and the seams of sulfides, salts and coals within the quaternary sediments	Greatly influence the health conditions of the residents of this area	Arain et al. (2014)
Dargai	75	High F^-	Ion exchange, weathering, and mining activities enhance F^- contamination	51% groundwater samples exceeded the WHO guideline of F^- 1.5 mg/L.	Rashid et al. (2019) and Rashid et al. (2020)
Chitral	39	High NO_3^-, SO_4^{2-}, and PO_4^{3-}	Water–rock interaction and minerals dissolution is responsible for contamination		
Sawat	48	High F^-, SO_4^{2-} NO_3^-, and PO_4^{3-}	62.2% groundwater samples had high F^-	Fluorite mining and other geogenic sources are responsible for high F^-	Rashid et al. (2018)
Kohistan	45	Elevated Cd, Ni, Pb, and Zn	Pb–Zn sulfide deposits and the related mining activities are responsible for higher Pb, Zn, and Cd contents	Geogenic sources affect the drinking water quality	Muhammad et al. (2011)
Haripur	98	Elevated Pb		Discharge of industrial effluents can be the main source of heavy metals accumulation in shallow groundwater	Sial et al. (2006)
Bajaur agency	44	High SO_4^{2-} and NO_3^-	Dissolution of minerals, such as feldspar, biotite, muscovite, calcite, and dolomite, agrochemicals, and waste seepage affect groundwater quality	Geogenic and anthropogenic sources that affect the drinking water quality	Jehan et al. (2019)
Lower Dir	63	High Cd, Cr, Pb, Co, and Fe		Natural and anthropogenic activities are responsible for drinking water contamination in the area	Rashid et al. (2019)

HFO, Hydrous ferric oxide; *KPK*, Khyber Pakhtunkhwa.

22.4.1.3 Lead

Lead (Pb) is also another human carcinogen especially in its inorganic form. It is found in the Earth's crust as galena (lead sulfide) in low concentrations. However, its occurrence in environment is mainly due to anthropogenic activities. The major source of Pb in drinking water is due to Pb-based plumbing. such as paints, solder, stained glass, toys, lead-based batteries, ceramic glazes, and traditional medicines. Due to its widespread use, Pb is present in all environments. EPA and Center for Disease Control, United States, agree on the fact that there is no safe level for Pb consumption. The permissible limit for Pb in drinking water is 10 μg/L by WHO. Young children, infants, and fetuses are particularly vulnerable to Pb poising. It is a leading cause of central and peripheral nervous system damage in children. Although Pb poisoning is a major hazard, however, not enough research has been done on the subject in Pakistan. The main reason for this is the use of Pb-based pipes. Among the areas affected by Pb groundwater contamination include Azad Jammu and Kashmir, Hattar Industrial Estate, KPK, and Sialkot, Punjab. A detailed description of Pb contamination in Pakistan is shown on Fig. 22.2.

22.4.1.4 Nickel

Nickel (Ni) is widespread in environment and found in animals, plants, as well as soil. Its main sources are both natural and anthropogenic activities such as fossil fuel combustion and industrial and mining activities. It is widely used in alloys (especially in coatings, batteries, electroplating, mobiles, and packaging industry) (IARC, 2010). In Pakistan its major source of entry in human food chain is through its use in edible oil industry.

Ni is also among the Group 1 human carcinogens. Its common routes for human exposure include ingestion through food and water and adsorption through skin. Ni causes cancers of respiratory tract (lung and nasal cavity). Ni is found in the groundwater samples of all the major cities of Pakistan (Lahore and Karachi being most effected) ranging from 0.001 to 3.66 mg/L (Khan et al., 2010). Its permissible limit by PCSQA is 0.02 mg/L while the limit prescribed by WHO is 0.07 mg/L. Fig. 22.2 shows Ni pollution in Pakistan as reported in literature.

22.4.1.5 Iron

Iron (Fe) is a micronutrient. It acts as a catalyst in many enzymatic reactions. It is also an important component of number of proteins as well as hemoglobin (Trumbo et al., 2001). However, its daily intake must not exceed the Recommended Daily Allowance (RDA) values (8 mg/day and tolerable upper limit for adults is 45 mg/day). If ingested in higher amounts, it might lead to gastrointestinal distress. High Fe values have mainly been reported in rural Punjab, Kasur being highest (11.8 mg/L). Fig. 22.2 shows extent of Fe pollution in groundwaters of Pakistan. Various studies have also reported high Fe concentrations (up to 3 mg/L) with As-affected areas of rural Punjab (van Geen et al., 2019). However, high Fe concentrations can also be attributed to Fe-based pipes installed in boreholes. Therefore the contribution of geogenic sources in dissolution of Fe needs to be investigated further. Another major reason for Fe in groundwater can be its high concentrations in soil. Fe concentration (ranging from <1 to 196 mg/kg) has been reported in various parts of the country. However, there has been no observed adverse effect on plant growth. Though data is available regarding Fe concentrations in soil, its effect/extent of Fe leaching on groundwater needs further investigation to draw a conclusion.

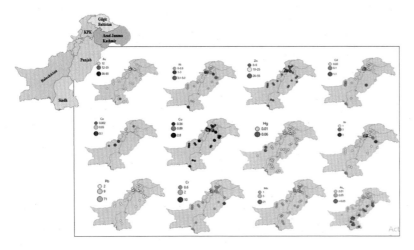

FIGURE 22.2 Metal concentrations as reported in literature in various parts of Pakistan.

22.4.1.6 Zinc

Zinc (Zn) is also a micronutrient. It plays an important role in enzyme activities. Zn is also a part of protein structure and performs vital roles during gene expression. Its RDA values are 8−11 mg/(male/female) and upper tolerable limits are 40 mg/day (Trumbo et al., 2001). Chronic Zn exposure leads to disruption in lipoprotein and cholesterol levels. It also causes problems in gastrointestinal tract and impairment of immune functions in human body.

Zn can be present in groundwater naturally due to rock water interactions. However, its ores are only slightly soluble in water so it's dissolved in only small amounts. Higher concentrations of Zn in groundwater can be attributed its concurrence with PB and Cd ores. Its anthropogenic sources may include fertilizers, steel production, coal combustion, and galvanized pipes. In Pakistan reported Zn concentrations are well below PSCQA limits (5−15 mg/L). Fig. 22.2 shows reported Zn concentrations in Pakistan.

22.4.2 Major anions

Major anions, including NO_3^-, SO_4^{2-}, PO_4^{3-}, and F^-, are the major contaminants in groundwater and have the potential to causes serious health concerns.

22.4.2.1 Nitrates

Among the nitrogenous anions, nitrates are the most stable and important part of nitrogen cycle. With WHO and national environmental quality standards (PAK-EPA) permissible limit of 10 mg/L, it is considered as potential contaminant in the country. Elevated levels of NO_3^- in the drinking water sources can cause serious health concerns such as blue baby syndrome (methemoglobinemia), brain disorder, thyroid disorder, genotoxicity, and colorectal cancer. Groundwater of Pakistan being an agricultural country is highly contaminated to nitrates (Table 22.2). Different studies of Pakistan reveal that apart from the natural sources, intensive fertilizers' application in agricultural sector, untreated sewage ponds, manure, and mining activities also manipulate its groundwater contamination (Table 22.2). Overall, highest groundwater nitrate concentrations are found in Baluchistan and Punjab. Nitrate contamination is traced in both shallow and deep aquifers. Its concentration may vary around the year and mainly dependent on precipitation, scope of fertilizers' application, drainage pattern, and soil types. Shallow aquifers with depth less than 120 ft. near the intensive agriculture areas are more susceptible to nitrate pollution (Tahir and Rasheed, 2016).

22.4.2.2 Phosphates

Phosphorus is an essential element for growth and development of biota. In water, it occurs as phosphate ion. Both natural (phosphorus-bearing minerals) and anthropogenic sources added and concentrated phosphorus into the various ecosystems. Intensive applications of fertilizers are the main source of phosphorus, potash, and nitrogen in agricultural area of Central Punjab and increased phosphate content in groundwater (Nickson et al., 2005). Canal irrigation and waterlogging prevailing in these agricultural hubs of Punjab and Sindh are the primary causes of fertilizer-based phosphate leaching (Baig et al., 2009; Nickson et al., 2005). This excessive leachable phosphate upon discharge into water bodies can cause algal bloom and eutrophication of aquatic environment. To prevent such damages US-EPA has recommended that concentration of phosphates in discharge water should be limited to <0.05 mg/L (USEPA, 1986).

22.4.2.3 Sulfates

Naturally, occurring sulfates in groundwater can have the potential to become health risk to both human and animals such as dehydration and diarrhea. WHO recommended level of sulfate in drinking water is 200 mg/L. Groundwater having sulfate content >250 mg/L develops bitter and medicinal taste in it. Higher concentration of SO_4^{2-} in groundwater is reported in arid and semiarid areas of Punjab and Sindh. Elevated sulfates in Layyah, Dera Ghazi Khan, and Faisalabad exhibited with dehydration and diarrhea in humans and animals (Daud et al., 2017; Mubarak et al., 2015). Application of fertilizers in these areas is one of the major reasons behind higher SO_4^{2-} concentrations, whereas in the case of Baluchistan and KPK mining of coal, fluorite, and other natural resources can also have the potential to increase sulfate contamination in their groundwater (Arain et al., 2014; Baig et al., 2009) (Table 22.2).

22.4.2.4 Fluoride

Fluoride with its dual effect is a critical anion. Naturally occurring F^- is one of the major water contaminants in Pakistan. Various studies have reported a wider range of groundwater F^- concentrations in different areas (Table 22.2), where most of the areas had geological source of groundwater F^- contamination. A detailed review of groundwater F^- data in various cities of Pakistan revealed that maximum concentration is traced in Lahore (23.60), Tehsil Mailsi (24.48), and Quetta (5.5 mg/L) (Rasool et al., 2017). WHO allowable limit of F^- in drinking waters is 1.5 mg/L. It is an essential element for teeth and bone strength. Its deficiency as well as excess can cause serious health concerns. Chronic exposure of fluoride may cause cardiovascular disorder fluorosis, bone deformations, and loss of intelligence quotient. Its concentration in the range of 1.5–3 mg/L may cause dental fluorosis, whereas concentration >3 mg/L is responsible for skeletal fluorosis. Dental fluorosis prevailing in Kasur, Raiwind, and Pattoki (Punjab) is such example (Abida Farooqi et al., 2007). Main sources of fluoride in Pakistan are natural (leaching of fluoride based parent minerals) and anthropogenic includes coal combustion, fertilizers, industrial wastes, and mining (Farooqi et al., 2007). An intensive study conducted in Punjab and Baluchistan revealed that calcium-poor aquifers under cation exchange retain more F^- (Tahir and Rasheed, 2013), whereas in the case of lower Indus Basin halite dissolution under arid condition is the major mechanism for F^- mobilization (Rasool et al., 2016). With some exceptions mining of coal, fluorite and other rock sources are also accountable for high F^- concentrations in drinking water sources of Sindh, Baluchistan, and KPK (Baig et al., 2009; Rashid et al., 2019).

References

Ahad, K., Anwar, T., Ahmad, I., Mohammad, A., Tahir, S., Aziz, S., et al., 2000. Determination of insecticide residues in groundwater of Mardan Division, NWFP, Pakistan: a case study. Water SA 26.

Ahmad, S., Rashida, M., 2001. Indus basin irrigation system water budget and associated problems. J. Eng. Appl. Sci. 20 (1), 9–75.

Ahmad, S., Yasin, M., Ahmad, M., Roohi, R., 2001a. GIS application for spatial and temporal analysis of groundwater in Mona SCARP area. Proceedings of the 2nd National Seminar on "Drainage in Pakistan". National Drainage Programme, WAPDA and University of Agriculture, Faisalabad. April 18–19, p. 343–363.

Ali, W., Aslam, M.W., Feng, C., Junaid, M., Ali, K., Li, S., et al., 2019. Unraveling prevalence and public health risks of arsenic, uranium and co-occurring trace metals in groundwater along riverine ecosystem in Sindh and Punjab, Pakistan. Environ. Geochem. Health . Available from: https://doi.org/10.1007/s10653-019-00278-7.

Ali, S.S., Baloch, K.A., Masood, S., 2017. Water sustainability in Pakistan: key issues and challenges. In: State Bank of Pakistan's Annual Report 2016-17. pp. 93–103. Retrieved from: <http://www.sbp.org.pk/reports/annual/arFY17/Chapter-07.pdf>.

Arain, M.B., Ullah, I., Niaz, A., Shah, N., Shah, A., Hussain, Z., et al., 2014. Evaluation of water quality parameters in drinking water of district Bannu, Pakistan: multivariate study. Sustain. Water Qual. Ecol. Available from: https://doi.org/10.1016/j.swaqe.2014.12.005.

Asi, M.R., Hussain, A., Muhmood, S.T., 2008. Solid phase extraction of pesticide residues in water samples: DDT and its metabolites. Int. J. Environ. Res. 145.

Baig, J.A., Kazi, T.G., Arain, M.B., Afridi, H.I., Kandhro, G.A., Sarfraz, R.A., et al., 2009. Evaluation of arsenic and other physico-chemical parameters of surface and ground water of Jamshoro, Pakistan. J. Hazard. Mater. Available from: https://doi.org/10.1016/j.jhazmat.2008.11.069.

Brahman, K.D., Kazi, T.G., Afridi, H.I., Baig, J.A., Arain, S.S., Talpur, F.N., et al., 2016. Exposure of children to arsenic in drinking water in the Tharparkar region of Sindh, Pakistan. Sci. Total Environ . Available from: https://doi.org/10.1016/j.scitotenv.2015.11.152.

Daud, M.K., Nafees, M., Ali, S., Rizwan, M., Bajwa, R.A., Shakoor, M.B., et al., 2017. Drinking water quality status and contamination in Pakistan. Biomed. Res. Int. Available from: https://doi.org/10.1155/2017/7908183.

Farooqi, A., Masuda, H., Firdous, N., 2007. Toxic fluoride and arsenic contaminated groundwater in the Lahore and Kasur districts, Punjab, Pakistan and possible contaminant sources. Environ. Pollut. 145 (3), 839–849. Available from: https://doi.org/10.1016/j.envpol.2006.05.007.

Farooqi, A., Sultana, J., Masood, N., 2017. Arsenic and fluoride co-contamination in shallow aquifers from agricultural suburbs and an industrial area of Punjab, Pakistan: Spatial trends, sources and human health implications. Toxicol. Ind. Health 33 (8). Available from: https://doi.org/10.1177/0748233717706802.

Ghoraba, S.M., Khan, A.D., 2013. Hydrochemistry and groundwater quality assessment in Balochistan province, Pakistan. Int. J. Res. Rev. Appl. Sci. 17.

IARC, 2010. IARC monographs on the evaluation of carcinogenic risks to humans. Carbon Black, Titanium Dioxide, and Talc. IARC Monogr. Available from: https://doi.org/10.1136/jcp.48.7.691-a.

IARC, 2012. Monograph Vol 100F Chemical agents and related occupations. In: IARC Monographs on the Evaluation of Carcinogenic Risks to Humans/World Health Organization, International Agency for Research on Cancer.

Jabbar, A., Masud, S.Z., Parveen, Z., Ali, M., 1993. Pesticide residues in cropland soils and shallow groundwater in Punjab Pakistan. Bull. Environ. Contam. Toxicol. Available from: https://doi.org/10.1007/BF00198891.

Jadhav, S.V., Bringas, E., Yadav, G.D., Rathod, V.K., Ortiz, I., Marathe, K.V., 2015. Arsenic and fluoride contaminated groundwaters: A review of current technologies for contaminants removal. J. Environ. Manage. Available from: https://doi.org/10.1016/j.jenvman.2015.07.020.

Jain, P., Sharma, J.D., Sohu, D., Sharma, P., 2005. Chemical analysis of drinking water of villages of Sanganer Tehsil, Jaipur District. Int. J. Environ. Sci. Technol. 2 (4), 373–379.

Jan, M.R., Shah, J., Khawaja, M.A., Gul, K., 2009. DDT residue in soil and water in and around abandoned DDT manufacturing factory. Environ. Monit. Assess. Available from: https://doi.org/10.1007/s10661-008-0415-2.

Jehan, S., Khan, S., Khattak, S.A., Muhammad, S., Rashid, A., Muhammad, N., 2019. Hydrochemical properties of drinking water and their sources apportionment of pollution in Bajaur agency, Pakistan. Measurement Available from: https://doi.org/10.1016/j.measurement.2019.02.090.

Khalid, S., Murtaza, B., Shaheen, I., Ahmad, I., Ullah, M.I., Abbas, T., et al., 2018. Assessment and public perception of drinking water quality and safety in district Vehari, Punjab, Pakistan. J. Clean. Prod. Available from: https://doi.org/10.1016/j.jclepro.2018.01.178.

Khan A.D., Iqbal, N., Ashraf, M., Sheikh, A.A., 2016. Groundwater investigations and mapping in the upper indus plain. Pakistan Council Research and Water Resources (PCRWR), Islamabad, pp.72.

Khanoranga, Khalid, S., 2019. An assessment of groundwater quality for irrigation and drinking purposes around brick kilns in three districts of Balochistan province, Pakistan, through water quality index and multivariate statistical approaches. J. Geochem. Explor. Available from: https://doi.org/10.1016/j.gexplo.2018.11.007.

Khan, H.M., Chaudhry, Z.S., Ismail, M., Khan, K., 2010. Assessment of radionuclides, trace metals and radionuclide transfer from soil to food of Jhangar Valley (Pakistan) using gamma-ray spectrometry. Water Air Soil Pollut. Available from: https://doi.org/10.1007/s11270-010-0390-4.

Khan, S.D., Mahmood, K., Sultan, M.I., Khan, A.S., Xiong, Y., Sagintayev, Z., 2010. Trace element geochemistry of groundwater from Quetta Valley, western Pakistan. Environ. Earth Sci. Available from: https://doi.org/10.1007/s12665-009-0197-z.

Khan, S., Shahnaz, M., Jehan, N., Rehman, S., Shah, M.T., Din, I., 2013. Drinking water quality and human health risk in Charsadda district, Pakistan. J. Clean. Prod. Available from: https://doi.org/10.1016/j.jclepro.2012.02.016.

Khan, S.N., Yasmeen, T., Riaz, M., Arif, M.S., Rizwan, M., Ali, S., et al., 2019. Spatio-temporal variations of shallow and deep well groundwater nitrate concentrations along the Indus River floodplain aquifer in Pakistan. Environ. Pollut. Available from: https://doi.org/10.1016/j.envpol.2019.07.019.

Malana, M.A., Khosa, M.A., 2011. Groundwater pollution with special focus on arsenic, Dera Ghazi Khan-Pakistan. J. Saudi Chem. Soc. Available from: https://doi.org/10.1016/j.jscs.2010.09.009.

Manzoor, S., Shah, M.H., Shaheen, N., Khalique, A., Jaffar, M., 2006. Multivariate analysis of trace metals in textile effluents in relation to soil and groundwater. J. Hazard. Mater. Available from: https://doi.org/10.1016/j.jhazmat.2006.01.077.

Masood, N., Farooqi, A., Zafar, M.I., 2019. Health risk assessment of arsenic and other potentially toxic elements in drinking water from an industrial zone of Gujrat, Pakistan: a case study. Environ. Monit. Assess. Available from: https://doi.org/10.1007/s10661-019-7223-8.

McGuire, S., 2016. World Cancer Report 2014. Geneva, Switzerland: World Health Organization, International Agency for Research on Cancer, WHO Press, 2015. Adv. Nutr. Available from: https://doi.org/10.3945/an.116.012211.

Mehmood, A., Mahmood, A., Eqani, S.A.M.A.S., Ishtiaq, M., Ashraf, A., Bibi, N., et al., 2017. A review on emerging persistent organic pollutants: current scenario in Pakistan. Hum. Ecol. Risk Assess. Available from: https://doi.org/10.1080/10807039.2015.1133241.

Mubarak, N., Hussain, I., Faisal, M., Hussain, T., Shad, M.Y., AbdEl-Salam, N.M., et al., 2015. Spatial distribution of sulfate concentration in groundwater of South-Punjab, Pakistan. Water Qual. Exposure Health. Available from: https://doi.org/10.1007/s12403-015-0165-7.

Muhammad, S., Shah, M.T., Khan, S., 2011. Health risk assessment of heavy metals and their source apportionment in drinking water of Kohistan region, northern Pakistan. Microchem. J. Available from: https://doi.org/10.1016/j.microc.2011.03.003.

Mukherjee, A., Scanlon, B., Aureli, A., Langan, S., Guo, H., McKenzie, A., 2020. Global Groundwater: Source, Scarcity, Sustainability, Security and Solutions, first ed. Elsevier, ISBN: 9780128181720.

Mushtaq, N., Younas, A., Mashiatullah, A., Javed, T., Ahmad, A., Farooqi, A., 2018. Hydrogeochemical and isotopic evaluation of groundwater with elevated arsenic in alkaline aquifers in Eastern Punjab, Pakistan. Chemosphere. Available from: https://doi.org/10.1016/j.chemosphere.2018.02.154.

Nabeela, F., Azizullah, A., Bibi, R., Uzma, S., Murad, W., Shakir, S.K., et al., 2014. Microbial contamination of drinking water in Pakistan—a review. Environ. Sci. Pollut. Res. Available from: https://doi.org/10.1007/s11356-014-3348-z.

Naseem, S., McArthur, J.M., 2018. Arsenic and other water-quality issues affecting groundwater, Indus alluvial plain, Pakistan. Hydrol. Processes 32 (9), 1235–1253. Available from: https://doi.org/10.1002/hyp.11489.

Naseem, S., Rafique, T., Bashir, E., Bhanger, M.I., Laghari, A., Usmani, T.H., 2010. Lithological influences on occurrence of high-fluoride groundwater in Nagar Parkar area, Thar Desert, Pakistan. Chemosphere 78 (11), 1313–1321. Available from: https://doi.org/10.1016/j.chemosphere.2010.01.010.

Nickson, R.T., McArthur, J.M., Shrestha, B., Kyaw-Myint, T.O., Lowry, D., 2005. Arsenic and other drinking water quality issues, Muzaffargarh District, Pakistan. Appl. Geochem. 20 (1), 55–68. Available from: https://doi.org/10.1016/j.apgeochem.2004.06.004.

Nordberg, G., Jin, T., Bernard, A., Fierens, S., Buchet, J.P., Ye, T., et al., 2002. Low bone density and renal dysfunction following environmental cadmium exposure in China. Ambio. Available from: https://doi.org/10.1579/0044-7447-31.6.478.

Parveen, Z., Masud, S.Z., 1988. Organochlorine pesticide residues in cattle drinking water. Pak. J. Sci. Ind. Res. 31 (1), 53–56.

Rafique, T., Naseem, S., Usmani, T.H., Bashir, E., Khan, F.A., Bhanger, M.I., 2009. Geochemical factors controlling the occurrence of high fluoride groundwater in the Nagar Parkar area, Sindh, Pakistan. J. Hazard. Mater. 171 (1–3), 424–430. Available from: https://doi.org/10.1016/j.jhazmat.2009.06.018.

Rasheed, F., Khan, A., Kazmi, S.U., 2009. Bacteriological analysis, antimicrobial susceptibility and detection of 16S rRNA gene of Helicobacter pylori by PCR in drinking water samples of earthquake affected areas and other parts of Pakistan. Malays. J. Microbiol. Available from: https://doi.org/10.21161/mjm.18609.

Rashid, A., Guan, D.X., Farooqi, A., Khan, S., Zahir, S., Jehan, S., et al., 2018. Fluoride prevalence in groundwater around a fluorite mining area in the flood plain of the River Swat, Pakistan. Sci. Total Environ. Available from: https://doi.org/10.1016/j.scitotenv.2018.04.064.

Rashid, A., Khan, S., Ayub, M., Sardar, T., Jehan, S., Zahir, S., et al., 2019. Mapping human health risk from exposure to potential toxic metal contamination in groundwater of Lower Dir, Pakistan: application of multivariate and geographical information system. Chemosphere. Available from: https://doi.org/10.1016/j.chemosphere.2019.03.066.

Rasool, A., Farooqi, A., Xiao, T., Ali, W., Noor, S., Abiola, O., et al., 2017. A review of global outlook on fluoride contamination in groundwater with prominence on the Pakistan current situation. Environ. Geochem. Health. Available from: https://doi.org/10.1007/s10653-017-0054-z.

Rasool, A., Farooqi, A., Xiao, T., Masood, S., Kamran, M.A., Bibi, S., 2016. Elevated levels of arsenic and trace metals in drinking water of Tehsil Mailsi, Punjab, Pakistan. J. Geochem. Explor. Available from: https://doi.org/10.1016/j.gexplo.2016.07.013.

Raza, M., Hussain, F., Lee, J.Y., Shakoor, M.B., Kwon, K.D., 2017. Groundwater status in Pakistan: a review of contamination, health risks, and potential needs. Crit. Rev. Environ. Sci. Technol. 47 (18), 1713–1762. Available from: https://doi.org/10.1080/10643389.2017.1400852.

Sarwar, G., Khan, J., Iqbal, R., Afridi, A.K., Khan, A., Sarwar, R., 2004. Bacteriological analysis of drinking water from urban and peri-urban areas of Peshawar. J. Postgraduate Medical Institute (Peshawar-Pakistan), 18(1).

Shakoor, M.B., Bibi, I., Niazi, N.K., Shahid, M., Nawaz, M.F., Farooqi, A., et al., 2018. The evaluation of arsenic contamination potential, speciation and hydrogeochemical behaviour in aquifers of Punjab, Pakistan. Chemosphere. Available from: https://doi.org/10.1016/j.chemosphere.2018.02.002.

Shar, A.H., 2012. Bacterial community patterns of municipal water of Sukkur city in different seasons. Afr. J. Biotechnol. Available from: https://doi.org/10.5897/ajb10.2056.

Shar, A.H., Kazi, Y.F., Soomro, I.H., 1970. Impact of seasonal variation on bacteriological quality of drinking water. Bangladesh J. Microbiol. Available from: https://doi.org/10.3329/bjm.v25i1.4862.

Sial, R.A., Chaudhary, M.F., Abbas, S.T., Latif, M.I., Khan, A.G., 2006. Quality of effluents from Hattar Industrial Estate. J. Zhejiang Univ. Sci. B. Available from: https://doi.org/10.1631/jzus.2006.B0974.

Tahir, M.A., Rasheed, H., 2013. Fluoride in the drinking water of Pakistan and the possible risk of crippling fluorosis. Drinking Water Eng. Sci. 6 (1), 17–23. Available from: https://doi.org/10.5194/dwes-6-17-2013.

Tahir, M.A., Rasheed, H., 2016. Distribution of nitrate in the water resources of Pakistan. Afr. J. Environ. Sci. Technol. Available from: https://doi.org/10.4314/ajest.v2i11.

Tariq, M.I., Afzal, S., Hussain, I., 2004. Pesticides in shallow groundwater of Bahawalnagar, Muzafargarh, D.G. Khan and Rajan Pur districts of Punjab, Pakistan. Environ. Int. 30 (4), 471–479. Available from: https://doi.org/10.1016/j.envint.2003.09.008.

Tassadaq, H., Aneela, R., Shehzad, M., Iftikhar, A., Jafar, K., Veronique, E.H., et al., 2013. Biochemical characterization and identification of bacterial strains isolated from drinking water sources of Kohat, Pakistan. Afr. J. Microbiol. Res. Available from: https://doi.org/10.5897/ajmr12.2204.

Trumbo, P., Yates, A.A., Schlicker, S., Poos, M., 2001. Dietary reference intakes: vitamin A, vitamin K, arsenic, boron, chromium, copper, iodine, iron, manganese, molybdenum, nickel, silicon, vanadium, and zinc. J. Am. Diet. Assoc. Available from: https://doi.org/10.1016/S0002-8223(01)00078-5.

Ullah, R., Muhammad, S., Jadoon, I.A.K., 2019. Potentially harmful elements contamination in water and sediment: evaluation for risk assessment and provenance in the northern Sulaiman fold belt, Baluchistan, Pakistan. Microchem. J. sb:host>Available from: https://doi.org/10.1016/j.microc.2019.04.053.

U.S. Environmental Protection Agency, 1986, Quality criteria for water 1986: Washington, D.C., U.S. Environmental Protection Agency Report 440/5-86-001, Office of Water, variously paged.

van Geen, A., Farooqi, A., Kumar, A., Khattak, J.A., Mushtaq, N., Hussain, I., et al., 2019. Field testing of over 30,000 wells for arsenic across 400 villages of the Punjab plains of Pakistan and India: implications for prioritizing mitigation. Sci. Total Environ. 654. Available from: https://doi.org/10.1016/j.scitotenv.2018.11.201.

Waseem, A., Arshad, J., Iqbal, F., Sajjad, A., Mehmood, Z., Murtaza, G., 2014. Pollution status of Pakistan: a retrospective review on heavy metal contamination of water, soil, and vegetables. Biomed. Res. Int. 2014, 813206.

Zuberi, F.A., Sufi, A.B., 1992. State of art of groundwater exploration, exploitation, management and legislation. IWASRI, WAPDA, Lahore, 26 p.

Chapter 23

Groundwater of Afghanistan (potential capacity, scarcity, security issues, and solutions)

Abdul Qayeum Karim and Sayed Hashmat Sadat
Department of Civil Engineering, Faculty of Engineering, Kabul University, Kabul, Afghanistan

23.1 Introduction

Afghanistan, a landlocked mountainous country with plains in the north and southwest, is described as being located within South and Central Asia (USGS, 2013; UNdata, 2011; Encyclopædia Britannica, 2010). The latitudes and longitudes of this country are 29° and 39°N, 60 and 75°E, respectfully. At 625,230 km² (241,402 sq. mi) (CIA, 1991), it is the world's 41st largest country (Gladstone, 2001). Its borders are with Pakistan in the south and east, Iran in the west, Turkmenistan, Uzbekistan, and Tajikistan in the north, and in the Far East with China (Qureshi, 2002).

This country has a continental climate with hot summers and cold winters, and in spite of having numerous surface-water resources, large parts of the country have often been plagued by drought. The large part of southwestern Afghanistan is one of the driest regions on the earth (NASA, 2006). During winter, in addition to the usual rainfalls, snowfalls in the Hindu Kush and Pamir Mountains form most water of the rivers, lakes, and streams after melting in spring (Mukherjee et al., 2020; The Christian Science Monitor, 2010). However, about 66% of the country's surface water flows to the neighboring countries (Crone, 2007).

Nevertheless, it is still wealthy in water resources essentially due to the high mountain ranges such as Hindu Kush and Baba, which are covered by snow. Over 80% of the country's water resources originate from the mountain ranges at altitudes over 2000 m.

It is not absolutely understood that what quantity of water resources, which are definitely underused, would be used while not straining the livelihoods and ecosystem in a changing climate. An example of Afghanistan is that we do not absolutely know what quantity of the groundwater may be extracted while not resulting in an excessive drawdown in groundwater levels (Akhtar, 2017).

In this country, groundwater is mostly utilized for irrigation purposes through the use of Qanats (Karezes), springs, and hand-dug shallow open wells. Recently, deep digging, driving, or drilling wells in the ground have become more common means of extraction for agricultural usage. There is a concern that the problems may arise and will pose a major challenge for the whole country, especially in the capital city of Kabul.

23.2 Topography and hydrogeology of Afghanistan

Afghanistan is described by broad desert fields, high mountain runs, and dispersed fertile valleys along the significant streams as it could be found in Fig. 23.1.

Asan annual average, snowmelt forms about 150,000 Mm³ precipitation and rainfall about 30,000 Mm³. Total precipitation is in the range of 180,000 Mm³/year. There were 174 stations for river gauging and/or rainfall measurements back in 1960–80. Based on Ministry of Energy and Water of Afghanistan, now there are a total of 127 stations that are functioning throughout the country at present and other 47 stations are in the process to be established in near future.

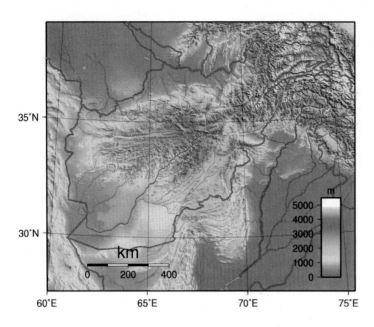

FIGURE 23.1 Topographic map of Afghanistan. *Courtesy: From https://commons.wikimedia.org/wiki/File:Afghanistan_Topography.png.*

The Registan and Dasht-i-Margo deserts in the south (comprise about 5% of the country) get very limited precipitation that mostly evaporates. The only perennial rivers are the Kabul River and its tributaries the Konar, Laghman, Logar, and Panjshir Rivers; the Helmand and the Arghandab; the Hari Rud; the Kunduz and the Kokcha; and the Amu Darya, which are mainly originating from the Hindu Kush.

In most places of the country, the aquifers—especially the quaternary aquifer, which is composed of alluvial medium and coarse sediments such as gravel, pebbles, cobbles, and boulders with various thickness and hydraulic properties—mainly have fresh groundwater, while in other parts of the country the aquifers mineralize the groundwater.

The geological formations are as follows:

1. Quaternary (Q) sediments
2. Neogene (N) sedimentary rocks
3. Paleogene (Pg) sedimentary rocks
4. Cretaceous–Paleogene (Cr–Pg) sedimentary rocks
5. Cretaceous (Cr) sedimentary rocks
6. Jurassic (Jr) sedimentary rocks
7. Triassic (Tr) sedimentary rocks
8. Permian–Triassic (Pr–Tr) sedimentary rocks
9. Palaeozoic (Pz) sedimentary and metamorphic rocks
10. Proterozoic (Pr) metamorphic rocks.
11. Volcanic Rocks
12. Acidic igneous rocks
13. Basic igneous rocks

23.3 Scarcity of groundwater quality and quantity

23.3.1 Quality challenges of groundwater in Afghanistan

Principle factors of groundwater contamination in Afghanistan are:

- groundwater overexploitation—bringing down groundwater level;
- permeability of overlaying layers (overlaying cover of aquifers);
- legal loopholes and lack of principles to protect groundwater resources and in addition, lack of control and mastery in its management due to political, economic, and social inefficiencies in the country; and
- lack of awareness about appropriate groundwater resources and up to date technology and knowledge about fitting of groundwater assets.

TABLE 23.1 Properties of Kabul groundwater.

Types of parameters	Parameter	USGS findings	DACAAR findings	MoPH findings	AWQS	WHO standards
Physical parameters	Color				Unobjectionable	
	Odor				Unobjectionable	
	Electrical conductivity (μS/cm)	992.8	7102	947	1250	
Chemical parameters	Arsenic (mg/L)	...	...	...	0.05	0.01
	Boron (mg/L)	...	...	...	1.0	0.3
	Chloride	...	...	625	250	250
	Fluoride (mg/L)	...	...	0.5	1.5	1.5
	Nitrate (mg/L)	5.7		528.15	50	50
	pH	5.8–8.4	5.8–8.4	7.35 (6.9–7.8)	6–9	No guideline
	SO$_4$ (ppm)	...		927	500	500
Bacteriological parameters	Total coliform (most probable number) in count/100 mL	780		460	5	
	Fecal coliform (5 days, 20C) al coliform (most probable number)	...		240	0	
	Escherichia coli	53		242.5	...	
	BOD	...		...	5	

Different studies of groundwater quality in Afghanistan have been carried out by different organizations such as USGS, DACAAR, MoPH, and Interim Water Quality Standards (IWQS) of Afghanistan, during the past few decades. These studies in the Kabul River Basin show that water quality is deteriorating due to a large number of erosions, which imposes a threat to the people. Table 23.1 shows a comparison of findings of different studies.

The parameters found to range out of the permissible limits set by WHO as well as the Afghanistan IWQS include mainly chemical and bacteriological parameters. The rest of the parameters were found within the permissible limits set by WHO.

Main qualitative concerns of Kabul Basin groundwater are:

1. Progressive increase in microbiological contamination, for example, coliform bacteria with time.
2. Progressive increase in nitrate concentrations with time.
3. Presence of arsenic and fluoride concentrations, for example, in three areas of the Kabul River Basin groundwater sources are found having arsenic, which is beyond the permissible limits. These regions are Logar, Panjsher, and Ghazni.

The presence of high rate of fecal coliform bacteria and high concentration level of nitrate indicates that Kabul Basin's drinking water systems are contaminated by fecal coliform (microbial pathogens) and nitrate (anthropogenic) contamination, which pose to the well-being of nearby occupants (Saffi, 2011). The presence of arsenic indicates strong toxicity, and a long period of time consumption of arsenic-bearing water results in necrosis and cancer for the inhabitants of Kabul Basin (Aini, 2007).

23.3.2 Quantity challenges of groundwater in Afghanistan

Numerous natural and anthropogenic factors such as precipitation, evapotranspiration, surface runoff, land use, and overexploitation affect the fluctuation of groundwater level. The fluctuation or change in groundwater level can be considered as (1) long term, (2) seasonal, and (3) short term. In overgrown cities with inadequate resource protection measures, where the groundwater extraction exceeds the recharge, the groundwater table may continue to drop down for several years almost irreversibly that sometimes took tens of thousands of years to recover the lost water. This trend is

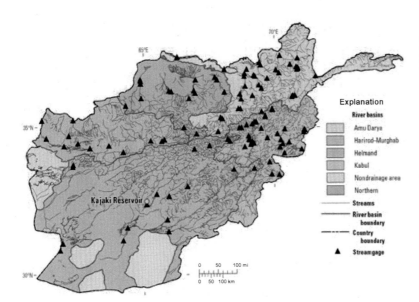

FIGURE 23.2 Afghanistan's major drainage basins, reinstalled streamgage stations and the Kajaki Reservoir, a hydropower and irrigation in the Helmand River Basin. *Courtesy: Thomas J. Mack, USGS.*

defined as long-term groundwater dropping. The seasonal fluctuation usually results from the influence of precipitation and water use, for example, irrigation and pumping for irrigation, which are all defined as seasonal cycle or seasonal fluctuations of groundwater. Short-term fluctuation that is called monthly fluctuation of groundwater level is estimated in alluvial aquifers for any extraordinary reason such as district water supply and pumping for other water systems. Numerous factors affect groundwater recharge, including but not limited to, evapotranspiration, precipitation, irrigation, and water supply, geologic formation of soils, surface flow, and urbanization (FAO, 1996a).

23.4 Afghanistan groundwater sustainability

Afghanistan in spite of having the blessings of five major river basins is still facing a severe shortage of water supplies.

Based on some reports, natural water storage capacity in the form of snow at elevations higher than 2000 m provides around 80% of Afghanistan's water resources (excluding the paleowater) (FAO, 2016; Saffi et al., 2014). In a study, the amount of precipitation water (327 mm/year) in the country is estimated to be around 213.5 km^3 in a year (Uhl and Tahiri, 2003). Current estimates show that Afghanistan's potential water resources are 65.3 km^3 produced annually from which 55.7 km^3 are surface water and 10.65 km^3 are groundwater (Fig. 23.2).

Kabul River Basin contributes about 11.5 km^3, Helmand River Basin 9.3 km^3, Hari Rod–Murghab River Basins 3.1 km^3, Northern River Basin 1.9 km^3, and Amu Darya (Panj) River Basin 11.7 km^3 to the surface-water resources produced in the country. And, the Kabul River Basin contributes around 1.92 km^3, Helmand and Western River Basins 2.98 km^3, Northern and Murghab River Basins 2.14 km^3, Hari Rod River Basin 0.64 km^3, and Amu Darya (Panj) River Basin is 2.97 km^3 to the groundwater resources produced in the country. Roughly, 1 km^3 overlap is between surface and groundwater (Uhl and Tahiri, 2003).

The country's perennial rivers flow imperceptibly throughout the year and most of the small streams only flow for 3–4 months during spring season and after that dry out through the rest of year. As the rainfall distribution, with regard to time and space, is greatly uneven and unbalanced, this affects the availability and access to water resources. Eventually, groundwater has become the only and the major source undergoes overstressed condition each year, with its crucial role in nourishment, household water supply, fisheries, power, jobs, combating drought, and eventually socioeconomic development and natural resources security in Afghanistan (Fig. 23.3). Nevertheless, groundwater has not a significant potential for development due to its limited nature that is also acutely impotent and sensitive to overexploitation or pollution of sources (FAO, 1996b).

Afghanistan is a poor country in the world, with an economy largely dependent on global aid and livelihoods through agriculture. Working in such difficult conditions and with not so significant returns and sustenance, reliable year-round sources of surface water and groundwater are crucial for farmers to irrigate their crops and watering their animals. Ephemeral streams and rivers (with seasonal flows) fed by rain and melting snow high up in Afghanistan's mountains ranges recharge alluvial aquifers found in populated valleys and give citizens with drinking water. As estimated, the total amount of groundwater recharge in the country is in the range of 10.65 km^3 (Broshears et al., 2005) to 16.5 km^3 (Rout, 2008).

FIGURE 23.3 Villagers line up for water at a community well near Kabul. *Courtesy: Robert E. Broshears, USGS.*

In developing countries, such as Afghanistan, essential information of water resources that are substantial for assuring a reliable urban water supply, potential for growing the economy of agriculture and industry, and for improving sanitation is basically very limited or even lacking at all. After decades of conflict, Afghanistan lacks the technical capacity, a framework for communication, infrastructure, and the modern equipment necessary for effectual hydrogeological analysis of its water resources.

As Afghanistan rises from a long time of struggle, as institutional capacities revive and develop, and as the desire for making balanced management decisions continues, sufficient data and a brimful understanding of the groundwater resource and its nature in the Kabul Basin will be indispensable (FAO, 2015). The authors believe that other cities in the country are also more or less at the same category.

As the key indicators of groundwater resources with regard to the sustainable development goals are its utilization, management, and sustainability, in Afghanistan, still access to reliable data of water resources is a challenge, and lack of reliable and continuous research and studies sometimes lead the policy development process to an unknown direction.

23.5 Afghanistan groundwater security

The total yearly water withdrawal in Afghanistan, as of 1987, is assessed to be around 26.11 km^3 from which 25.8 km^3 (99%) were designated as it were for agricultural purposes. But the foremost updated figure for the water withdrawal is that of 1998 whereby the overall yearly withdrawal volume of water for irrigation reason was assessed to be around 20 km^3 (Safi and Buerkert, 2011).

It is obvious that there is a high uncertainty in accessible information on water withdrawals over the country without giving a clear clarification for the contrasts having been appeared in a relatively short period of time, around 10 years. It subsequently underlines the desires for a significance of investigation on water availability and demand to be conducted for making possible the integral water management across the country. Out of the total water extracted, 3 km^3 (15%) are the groundwater withdrawal (Safi and Buerkert, 2011) whereas the remaining 17 km^3 (85%) are contributed by the surface water (Safi and Buerkert, 2011). Around 98% of the entire water withdrawn is utilized for agriculture and 1% each for domestic and industrial purposes over the country.

23.6 Solutions

With regards to the abovementioned challenges and barriers for sustainable development of Afghanistan, considering the lack of worldwide useful, availability of up-to-date, and sustainable development goals-relevant groundwater data, which makes it troublesome to create all inclusive, and even locally, pertinent suggestions and recommendations for groundwater utilization, management, and sustainability within the sustainable development goals framework, the authors recommend the followings for the development of groundwater in Afghanistan as a crucial mean for achieving the goals related to poverty, health, economy, education, and equal rights:

- Study the water-related hazards in the different regions in the country to understand how climate variability and climate change impact on people and identifying potential sites for groundwater recharge.
- Develop realistic scenarios of the potential impact of climate change on populations of different cultural and living conditions context to help decision makers manage trade-offs between immediate and long-term concerns in the country.
- Strengthen the multisectoral collaboration across academic disciplines, government departments, and service providers for improved safe and clean water supply services.
- Promote water use through decentralized decision-making, compelling management of urban contaminations, improved infrastructure planning, and strengthen regional cooperation to counter the horrific and immediate threats to water security postured by mankind drivers, including climate change, fair, and impartial sharing of resources, productivity, and sustainability.
- Data sharing among researchers and services or organization workforce; conflict management through institutional platforms; and public investment and private-sector funds for generating and knowledge sharing, improving public awareness services, and acting proactively.
- Manage carefully trade-offs among environment, economy, and political wells and other sectors in order to improve water security, achieve the Sustainable Development Goals, and ensure safe and adequate water availability.

References

Encyclopædia Britannica. Afghanistan. Encyclopædia Britannica. Archived from the original on 25 February 2010 (Retrieved 17.03.10.).

The Christian Science Monitor. Afghanistan's woeful water management delights neighbors. Csmonitor.com. 15 June 2010. Archived from the original on 14 November 2010 (retrieved 14.11.10.).

Aini, A., 2007. Water Conservation in Afghanistan. J. Dev. Sustain. Agric. 2 (1), 51–58.

Akhtar, F., 2017. Water Availability and Demand Analysis in the Kabul River Basin. Zentrum für Entwicklungsforschung, Afghanistan.

Broshears, R.E., Akbari, M.A., Chornack, M.P., Mueller, D.K., Ruddy, B.C., 2005. Inventory of ground-water resources in the Kabul Basin, Afghanistan. In: U.S. Geological Survey Scientific Investigations Report 2005-5090, 34 p.

UNdata. Composition of macro geographical (continental) regions, geographical sub-regions, and selected economic and other groupings. UNdata. 26 April 2011. Archived from the original on 13 July 2011 (retrieved 13.07.11.).

CIA. 26 November 1991. Archived from the original on 31 January 2014 (retrieved 04.02.12.).

Crone, A.J., April 2007. Earthquakes Pose a Serious Hazard in Afghanistan (PDF) (Technical report). US Geological Survey. Fact Sheet FS 2007–3027. Archived from the original on 27 July 2013 (retrieved 14.10.11.).

FAO, 1996a. Afghanistan, promotion of agricultural rehabilitation and development programmes. In: Water Resources and Irrigation (FAO Project TCP/AFG/4552. AREV).

FAO (also Klemm, W., 1997), 1996b. Promotion of Agricultural Rehabilitation and Development Programs in Afghanistan: A Report Part of the Afghanistan Agricultural Strategy, Food and Agriculture Organization of the United Nations (FAO Project TCP/AFG 4552).

FAO, 2015. Afghanistan: Geography, Climate and Population AQUASTAT website, Food and Agriculture Organization of the United Nations (FAO) from <http://www.fao.org/nr/water/aquastat/countries_regions/afg/index.stm> (retrieved 29.04.15.).

FAO, 2016. Computation of Long-Term Annual Renewable Water Resources (RWR) by Country (in km^3/year, Average). From <http://www.fao.org/nr/water/aquastat/data/wrs/readPdf.html?f = AFG-WRS_eng.pdf> (retrieved 21.12.16.).

Gladstone, C., 2001. Afghanistan Revisited. Nova Publishers, p. 121.

Mukherjee, A., Scanlon, B., Aureli, A., Langan, S., Guo, H., McKenzie, A., 2020. Global Groundwater: Source, Scarcity, Sustainability, Security and Solutions, first ed. Elsevier. ISBN: 9780128181720.

Qureshi, A.S., 2002. Water Resources Management in Afghanistan: The Issues and Options. International Water Management Institute.

Rout, B., 2008. How the Water Flows: A Typology of Irrigation Systems in Afghanistan, Afghanistan Research and Evaluation Unit Kabul.

Saffi, M.H., 2011. Groundwater Natural Resources and Quality Concern in Kabul Basin. DACAAR Kabul, Afghanistan, Kabul.

Saffi, M.H., Hydro-geologist, A.J. and Bhandari, B., 2014. Project Summary Report: 2003-January 2014 National Groundwater Monitoring Wells Network Finding Challenges and Recommended Solutions in Afghanistan.

Safi, Z., Buerkert, A., 2011. Heavy metal and microbial loads in sewage irrigated vegetables of Kabul, Afghanistan. J. Agric. Rural. Dev. Trop. Subtropics (JARTS) 112 (1), 29–36.

NASA. Snow in Afghanistan: Natural Hazards. NASA. 3 February 2006. Archived from the original on 30 December 2013 (retrieved 06.05.12.).

Uhl, V.W., Tahiri, Q.M., 2003. Afghanistan: An Overview of Groundwater Resources and Challenges. Uhl, Baron, Rana Associates, Inc. Washington Crossing, PA. From <http://www.vuawater.com/Case-Study-Files/Afghanistan/Afghanistan_Overview_of_GW_Resources_Study-2003.pdf> (retrieved 12.12.16.).

USGS. U.S. Maps. Pubs.usgs.gov. Archived from the original on 25 December 2013 (retrieved 19.05.12.).

Theme 4

Groundwater sustainability and security

Chapter 24

Groundwater resources sustainability

Jac van der Gun
Van der Gun Hydro-Consulting, Schalkhaar, The Netherlands

24.1 Sustainability and sustainable development

The Earth's natural resources are essential for human life and development. The biosphere and soils provide food and fibers; subsurface mining resources offer raw materials for construction and all kinds of industrial activities; various natural resources (fossil, renewable, and perpetual) supply energy; while water and fresh air meet physiological and other vital needs of living organisms. In principle, all human activities produce modifications in the natural environment and the state of natural resources. Single individuals usually contribute only very little to such modifications and most people have little or no knowledge of the underlying cause-and-effect relationships. Therefore they tend to remain unaware of such side effects of their activities and in most cases they just take for granted what nature offers. However, the individual contributions to unintended side effects add up and may eventually lead to larger scale depletion of natural resources or other forms of environmental degradation.

Throughout history, there have been eminent scholars who perceived and understood interactions between humans and nature, and who warned the general public on potentially harmful side effects of their activities. For instance, Plato in the 5th century BCE, Strabo and Columella in the 1st century BCE, and Pliny the Elder in the CE 1st century discussed different types of environmental degradation resulting from human activities such as farming, logging, and mining; they also recommended "sustainable practices" that would maintain the "everlasting youth" of the Earth (Du Pisani, 2006). Much later, by the end of the 18th century, Thomas Malthus called attention for the impact of population growth on the consumption of natural resources, stating that uncontrolled increase in population threatened to outstrip food production (Du Pisani, 2006). Around 1800 it was Alexander von Humboldt who revolutionized the way westerners see the natural world, by focusing attention on the interactions between the elements of this world, in his view connected in "a net-like intricate fabric." Not only did he describe the disastrous environmental effects of colonial plantations (characterized by cash crops, monocultures, irrigation, and deforestation) observed during his travels in Venezuela, he was also the first scientist to talk about harmful human-induced climate change and to explain the environmental importance of forests in terms of enriching the atmosphere with moisture, their cooling effect, and their role in water retention and protection against soil erosion (Wulf, 2015). His ideas inspired Charles Darwin, who in his famous book *On the Origin of Species* (Darwin, 1859) argued that in the struggle for life between different species or individual living organisms those best adapted to local environmental conditions have the best chances to survive. Twentieth century milestones that have greatly influenced thinking about the environment and natural resources are Carson's (1953) book *Silent Spring*, Hardin's (1968) *The Tragedy of the Commons*, Meadow's (1972) *Limits to Growth* (report of the Club of Rome), and the report *Our Common Future* [also known as *The Brundtland Report*, produced by the United Nations World Commission on Environment and Development (1987)]. The mentioned authors and many more pioneers that remain unmentioned here have in common that they recognize the importance of the sustainability of the services offered by natural resources and the natural environment; and that they claim that this sustainability may be threatened by human activities.

The term "sustainability" refers to "a state or condition that can be maintained over an indefinite period of time" (Du Pisani, 2006), or to "a capacity to maintain some entity, outcome or process over time" (Jenkins, 2009, cited by Klarin, 2018). The roots of this concept can be traced back to ancient times, but population growth, increase in consumption after the Industrial Revolution, and the risk of depleting crucial natural resources have in recent times deepened the conviction that these resources need to be used in a sustainable way (Du Pisani, 2006). Some thinkers on the subject go wider and even include the suggestion to abandon the prevailing anthropocentric perspective on sustainability and to redefine sustainability in terms of harmony between all life on Earth (Horton and Horton, 2019).

The term "development" is closely related to the old believe in "human progress, the idea that civilization has moved, is moving, and will move in a desirable direction" (Bury, 1932; Du Pisani, 2006). According to Remenyi (2004, cited by Klarin, 2018), "development is a process whose output aims to improve the quality of life and to increase the self-sufficient capacity of economies that are technically more complex and depend on global integration."

Since development tends to be inseparable from the exploitation of natural resources, while sustainability puts emphasis on their conservation, it may seem that development and sustainability are to a certain extent conflicting. This conflict was very prominently highlighted in the 1970s by the report *Limits to Growth* of the Club of Rome:

If the present growth trends in world population, industrialization, pollution, food production and resources depletion continue unchanged, the limits to growth on this planet will be reached sometime within the next one hundred years. The most probable result will be a rather sudden and uncontrollable decline in both population and industrial capacity.

Meadows (1972).

As an outcome of numerous debates on this alarming message, the concept of "sustainable development" has emerged—a compromise between development (or growth) and conservation. It was presented by the report *Our Common Future* (United Nations World Commission on Environment and Development, 1987; *"Brundtland Report"*), where it was defined as follows:

Sustainable development is development that meets the needs of the present without compromising the ability of future generations to meet their own needs.

United Nations World Commission on Environment and Development (1987).

Since 1987 this paradigm has been amply discussed, questioned, and criticized, but also widely embraced, and its interpretation and scope are still subject to evolution. It has been at the heart of many international conferences, congresses, summits, declarations, agreements, and programs of action, both national and international. Among the latter, the United Nations global program on Sustainable Development Goals, adopted in 2015, is prominent (United Nations, 2015). Although before the 1970s development used to be associated mainly with economic growth, sustainable development has become much wider in scope and nowadays deals with the natural environment (natural resources, ecosystems, pollution, and climate); the economy (economic growth, jobs, and living standards); and society (poverty eradication, food security, promotion of equity, gender equality, human rights, peace and justice, etc.). These three main facets of sustainable development (ecology, economy, and society) have become known as the "triple bottom line" (Du Pisani, 2006). Overall sustainable development is pursued by addressing these three "pillars" and striking a balance between them (Klarin, 2018).

24.2 Sustainability of groundwater services

24.2.1 Groundwater services

Groundwater is one of our main planet's natural resources and under conditions of appropriate use, governance, and management it contributes to sustainable development. People involved in using and managing groundwater need to be aware that groundwater is more than only a source of water to satisfy human needs: in many areas it has multiple functions and provides simultaneously different services. Fig. 24.1 shows common groundwater services, grouped under the four distinct categories of ecosystem services defined by the Millennium Ecosystem Assessment (MEA) (2005): (1) provisioning services, (2) regulatory services, (3) supporting services, and (4) cultural services. Note that some of the services require groundwater to be abstracted, while other ones are in situ services.

The services offered vary from one groundwater system to another, depending on their particular properties and setting, such as geology, topography, climatic conditions, hydrogeological characteristics, and hydraulic connection with surface water bodies. Among others, sizeable provisioning services require the presence of an aquifer capable of storing and transmitting substantial quantities of water; while these services can only be sustainable if the aquifer is receiving significant recharge. Regulatory services are controlled by reservoir capacity and properties of the aquifer rock and its overburden. Supporting services such as supply of groundwater to streams, springs, or wetlands are linked to particular combinations of topographical, geological, and hydrological conditions.

It is obvious that within the category of provisioning services competition is likely to occur if the demands for water are high and the quantities of available groundwater comparatively limited. In addition, provisioning services are even

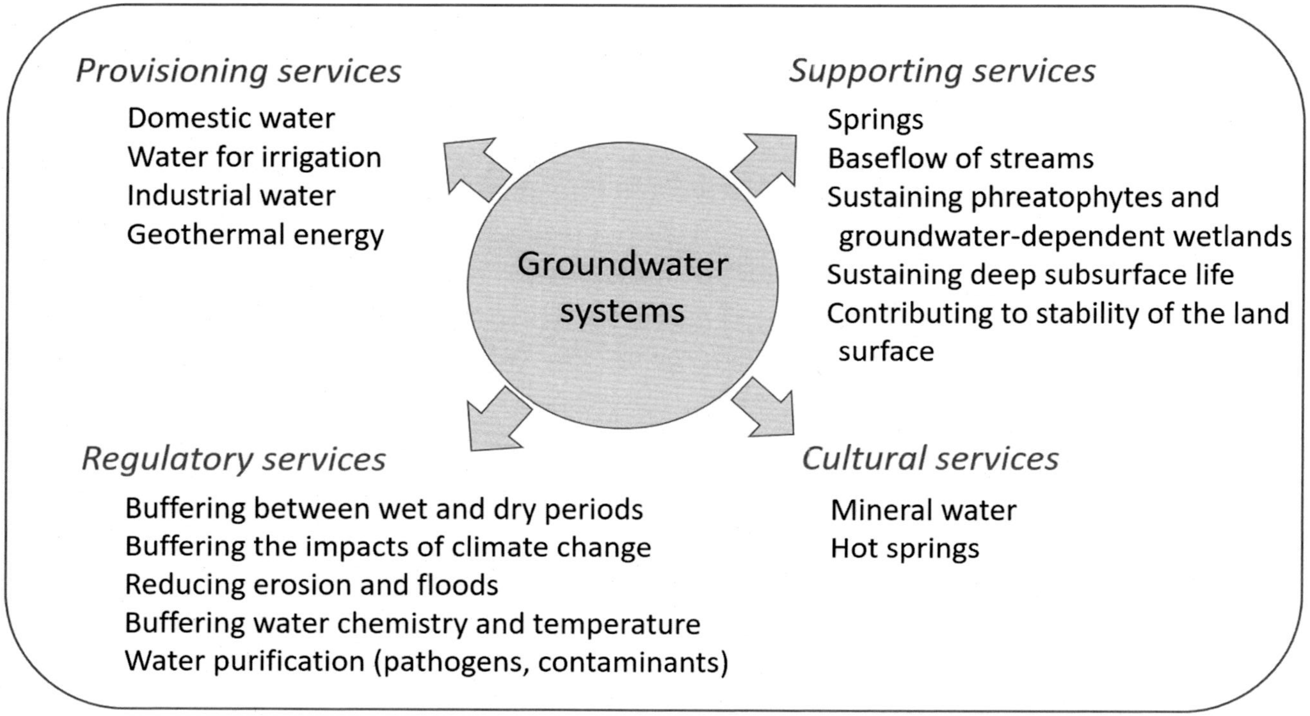

Provisioning services

Domestic water
Water for irrigation
Industrial water
Geothermal energy

Supporting services

Springs
Baseflow of streams
Sustaining phreatophytes and groundwater-dependent wetlands
Sustaining deep subsurface life
Contributing to stability of the land surface

Cultural services

Mineral water
Hot springs

Regulatory services

Buffering between wet and dry periods
Buffering the impacts of climate change
Reducing erosion and floods
Buffering water chemistry and temperature
Water purification (pathogens, contaminants)

Groundwater systems

FIGURE 24.1 Groundwater services and functions.

conflicting with most of the supporting services. In general, each single human interaction with a groundwater system—in the form of abstraction, drainage, enhancing or reducing recharge, pollution, or otherwise—modifies the state of the corresponding system and tends to affect the services it offers. As long as the capacity to provide essential services is preserved, the groundwater resources are considered to be sustainable. What is included under "essential services" leaves some room for subjectivity in the interpretation of groundwater resources sustainability. In practice, groundwater resources sustainability is predominantly associated with the sustainability of the provisioning services, whether or not taking into account the constraints imposed by the demand to maintain locally relevant supporting and other in situ services.

24.2.2 Potential threats to groundwater services

In recent times, global groundwater has become exposed to immense stress (Mukherjee et al., 2020), which means that in many areas around the world the sustainability of groundwater services is threatened. A brief outline of the main threats is provided in the following sections.

24.2.2.1 Intensive groundwater abstraction

Groundwater abstraction directly affects the hydrological regime of the aquifer concerned. The abstracted flow is initially balanced by depletion of stored groundwater, but on a medium term this storage depletion is gradually being replaced by *capture*, which is the sum of intercepted natural groundwater discharge and induced increases in groundwater recharge. Depending on the rate of abstraction from the aquifer compared to the potential maximum capture, on the longer term either a new hydrological equilibrium is reached or groundwater level declines continue until abstraction will stop due to physical, technical, or economic exhaustion of the aquifer. Intensive groundwater abstraction may produce a diversity of negative side effects: declining groundwater levels; increasing cost of groundwater abstraction; diminishing or even disappearing spring discharges and baseflows; degradation of wetlands; land subsidence; and intrusion of saline, brackish, or other low-quality water. It may have also positive side effects: by contributing to drainage of water-logged zones, enhancing groundwater recharge and creating more storage capacity for buffering between wet and dry periods.

24.2.2.2 Artificial drainage

Drainage activities with the purpose to evacuate groundwater, either for shallow water-table control or for facilitating activities at greater depth (in particular mining), affect groundwater services in a similar way like groundwater abstraction for any intended water use. Artificial drainage and removal of surface water, either by formal technical drainage provisions or by expanding the areas of impermeable land surface, tend to reduce groundwater recharge.

24.2.2.3 Salinization and pollution

When water in a groundwater system gets salinized or polluted, it may become unsuitable for certain groundwater services. This is evidently a major concern for domestic water supply, but also other services (provisioning as well as in situ services) may become affected.

Several processes may cause water in an originally fresh-groundwater aquifer to become brackish or saline. Some of them have natural causes (e.g., flooding by seawater), while other ones are induced by human action (intrusion of seawater or neighboring saline/brackish groundwater caused by groundwater abstraction; downward seepage of mineralized irrigation return flows; and injection of saline or brackish water).

Causes and mechanisms of groundwater pollution are numerous, and there is a large variation in pollutants. Households, industries, mining, and agriculture produce enormous quantities of waste and wastewater. Fetter (1993) distinguishes six categories of sources of groundwater contamination: (1) sources designed to discharge substances (septic tanks, injection wells, land application of wastewater); (2) sources designed to store, treat, and/or dispose of substances (landfills, open dumps, residential disposal, surface impoundments, mining waste and stock piles, graveyards, storage tanks, incineration and detonation sites, and radioactive waste disposal sites); (3) sources to retain substances during transport (pipelines, trucks, and trains); (4) sources discharging substances as a consequence of other planned activities (irrigation, use of pesticides and fertilizers, farm animal wastes, road salting, percolation of atmospheric pollutants, mine drainage, etc.); (5) sources providing a conduit for contaminated water to enter aquifers (wells, construction excavations); and (6) naturally occurring sources, the discharge of which is created or acerbated by human activity (interaction with polluted surface water, natural leaching enhanced by acid rain and saltwater intrusion).

Natural groundwater pollution, for example by dissolution of arsenic or fluoride, takes place at a very slow pace, thus is unlikely to affect groundwater services in the short to medium term.

24.2.2.4 Climate change and sea-level rise

Significant climate change and associated sea-level rise, predicted to occur during the present century and beyond, will undoubtedly affect the services offered by groundwater systems in the future.

Changes in climate variables are likely to vary prominently between geographical regions, but future patterns are difficult to predict with confidence (IPCC, 2014). Nevertheless, studies show that average temperatures will increase in the majority of the world's regions and that rainfall is likely to become concentrated more often than at present in showers or storms of high intensity. Higher temperatures usually tend to reduce groundwater recharge and—indirectly—lead to increased groundwater abstraction; both are contributing to higher groundwater development stress. Nevertheless, higher temperatures may enhance recharge in some specific areas, for example, where permafrost conditions are likely to disappear by thaw. Whether more intensive rainstorms will enhance or reduce groundwater recharge depends also on local conditions. Therefore predicting or assessing the impacts of climate change on the groundwater regimes in any particular region should be based on area-specific information and knowledge.

Sea-level rise—by IPCC (2019) estimated to become 0.43−0.84 m over the present century—will affect groundwater systems in coastal regions. Low-lying, unprotected coastal land underlain by fresh groundwater may be flooded by seawater, causing groundwater to become brackish or saline; sea-level rise will lead to a higher risk of flooding. Furthermore, a higher seawater level will cause salt water intrusion to move further inland and endanger fresh aquifer zones that are hydraulically connected to the sea or to coastal estuaries and rivers. Note that climate change—producing thermal expansion of the oceans and melt of glaciers and ice caps—is not the only cause of sea-level rise: groundwater abstraction does also contribute to it (Wada et al., 2010; Konikow, 2011), while storing surface water in reservoirs has the opposite effect (Pakhrel et al., 2012).

24.3 Approaches to pursuing, restoring, or enhancing groundwater resources sustainability

24.3.1 The umbrella: groundwater governance and management

Groundwater systems are exposed both to natural phenomena and to the individual activities and influences of numerous people. Typical characteristics of groundwater systems are their often large size, significant buffer capacity, open access to people living within their horizontal boundaries, and poorly defined groundwater ownership. Human actors may have competing or even conflicting interests regarding groundwater, and several of their activities—even those unrelated to groundwater—may unintentionally affect groundwater systems by negative side effects (externalities). Controlling the groundwater resources optimally and preventing them from being adversely affected by potential threats (as described in Section 24.2.2) is therefore evidently beyond reach of single individuals. Concerted efforts of entire communities, in the form of groundwater governance and management, are needed to pursue groundwater resources sustainability. This idea has spread and gradually gained wider acceptance, in particular during the later decades of the 20th century, after human interactions with groundwater systems had intensified in an unprecedented way, which has fundamentally changed the views on groundwater (Van der Gun, 2019). Groundwater management focuses on the interventions to be implemented in the field, groundwater governance on the enabling framework (laws, regulations, institutions, information and knowledge, policy, and planning) and on guiding principles (Villholth et al., 2018). It needs to be emphasized that assessing and monitoring the local groundwater systems and their context are indispensable for proper selection and design of the management interventions to be implemented, as well as for assessing their effects. Furthermore, awareness raising among stakeholders is essential to gain their acceptance of intended interventions and to enhance their willingness to adopt desired changes in their behavior in relation to groundwater. Governance and management approaches related to a number of key issues on groundwater resources sustainability are briefly described next.

24.3.2 Hydrogeological approaches to defining sustainability limits of abstraction

One of the questions often to be answered by a hydrogeological assessment of a region or aquifer system with renewable groundwater is how much groundwater can be abstracted from it in a sustainable way. Obviously, the hydrological regime of the aquifer system considered provides important guidance toward the answer, but different approaches are in use.

The most simple approach observed to addressing this question is the misconception that the rate of recoverable groundwater would be equal to the *mean rate of groundwater recharge*. This approach ignores the internal dynamics of the groundwater system and assumes—without proving it—that abstracting groundwater at such a rate (or lower) will be followed sooner or later by a new dynamic hydrological equilibrium. In addition, it does not pay attention to storage depletion and its impacts, nor to any of the nonprovisioning groundwater services. In spite of its inadequacies, this approach is still very popular among hydrogeologists, who use a variety of methods to estimate mean groundwater recharge rates.

Hydrodynamically more correct approaches make use of the *groundwater capture* concept. As outlined by Theis (1940), water artificially withdrawn from an aquifer is derived from a decrease in storage in the aquifer, a reduction of the previous discharge from the aquifer, an increase in the recharge, or a combination of these changes. The decrease in discharge plus the increase in recharge is termed capture (Lohman et al., 1972). As explained also by Bredehoeft et al. (1982) and by Konikow and Leake (2014), among others, the hydraulic head in the aquifer will continue to decline until the new withdrawal is balanced by capture. The maximum hydrologically sustainable rate of pumping from an aquifer equals the maximum capture that can be produced. The latter depends not only on the mean recharge rate, but also on the specific properties and setting of the aquifer, as well as on the pumping configuration (Van der Gun and Lipponen, 2010).

Safe yield came into use more than a century ago as a term indicating the maximum quantity of groundwater that can be abstracted from an aquifer in a sustainable way. The interpretation of this concept has evolved over time, as described by Alley and Leake (2004). Originally, the concept took the groundwater budget as the only constraint to groundwater abstraction; gradually, however, it was expanded by adding other constraints, such as economic feasibility of pumping, water quality degradation, and water rights. Consequently, Todd (1959) defined the safe yield of an aquifer system broadly as "the amount of water which can be withdrawn from it annually without producing an undesirable result."

In more recent decades, during which integrated water resources management and more comprehensive approaches to sustainable development have emerged, the external linkages of groundwater systems have come more to the forefront. As a result, many aquifer development and management plans nowadays take also into account the potential impacts of groundwater abstraction on surface water systems, ecosystems, and the environment. Ideally, the full range of identified groundwater services (see Fig. 24.1) is incorporated, as well as any potential negative impacts beyond these services. If the aquifer systems considered have been sufficiently assessed, then numerical modeling tools and capability are very helpful for identifying optimal strategies. They enable to simulate aquifer behavior under different groundwater abstraction patterns and intensities, and to compare the results in terms of performance gains and losses of the identified groundwater services and any other impacts. Selecting in this way a favorite strategy is a multicriteria decision problem, guided by political preferences, thus inherently subjective.

Nonrenewable groundwater resources form a special case. Absence of significant recharge implies that groundwater abstraction from nonrenewable groundwater necessarily produces progressive depletion of the stored water volume, which finally leads to exhaustion. In other words, groundwater abstraction from nonrenewable groundwater is hydrologically not sustainable. Nevertheless, carefully planned exploitation of nonrenewable groundwater may pursue an optimal balance between the needs of the present and the needs of future generations, and in this way it may contribute to overall sustainable development. Al-Eryani et al. (2006) use the expression "socially sustainable use of nonrenewable resources" for cases where the use of nonrenewable groundwater resources is reconciled with the "sustainability of human life." If the quantities of abstracted nonrenewable groundwater are used to generate an equivalent amount of "manufactured capital,"—which means that the sum of natural and manufactured capital remains constant—, then the designation *weak sustainability* may apply, a term defined in neoclassical economics (Klarin, 2018; Du Pisani, 2006).

24.3.3 Enhancing groundwater recharge

Groundwater resources sustainability is positively affected by a variety of human interventions and other activities that result in increased amounts of groundwater recharge. Enhanced recharge occurs in some cases unintentionally, for instance, as a side effect of surface water irrigation or changes in land use, by leakage from water pipes and sewers, or by drainage provisions that create storage capacity for receiving recharge that previously would have been rejected. On the other hand, also many activities, approaches, and techniques can be observed that intentionally enhance groundwater recharge, with the purpose to contribute to the capacity and performance of groundwater services and functions. These activities and techniques are referred to under the name "managed aquifer recharge" (MAR). Dillon et al. (2018) report on a worldwide survey on the steadily increasing application of MAR over the last 60 years. The different techniques used include streambed modifications, bank filtration, water spreading, and recharge wells. Recharge provisions range from very small local systems (e.g., using water from roof catchments) to very large schemes that require huge investments and advanced engineering technology (e.g., large recharge dam schemes).

24.3.4 Water demand management

One of the main strategies to prevent or combat groundwater overdraft—and thus contributing to groundwater resources sustainability—is water demand management (Tate, 1999; Savenije and van der Zaag, 2002). It aims at water conservation by controlling or reducing the abstracted quantities of water, in particular through more efficient allocation, conveyance, and use. Various opportunities to economize on water consumption in the water chain are almost everywhere present. For instance, free access to groundwater can be replaced by controlled access (restricted by allocation based on permits) and taxes based on metered abstraction may motivate people to reduce the quantities of groundwater pumped. Losses of water in irrigation conveyance systems or leaking from public water distribution networks—often considerable quantities— may be reduced by technical improvements. More efficient water use can be achieved in the agricultural sector by adopting water-efficient irrigation practices and methods (sprinkler and drip irrigation) and in industry by proper design of all water-using processes. If water scarcity is pressing, production may move into the direction of crops and industrial products that are less water-intensive, while low-value products are abandoned. Households can save water by lower pressure in their water distribution systems, use of low-volume flush toilets, waterless sanitation and other water-saving devices, and, in general, by being keen on avoiding wasting water. For purposes where water quality does not matter, low-quality water may be used instead of water of high quality. Finally, industrial on-site used water recycling, and recycling treated urban wastewater for irrigation purposes may reduce significantly the need for tapping new quantities of fresh groundwater. Awareness programs, legal provisions, taxes, water pricing, energy supply and pricing, subsidies, penalties, and fines are instrumental in pursuing a desired behavior of all stakeholders, including the general public. On a macro-level, local water demands may be reduced by importing water-intensive products (virtual water importation) and by a transition to diets with a relatively low water footprint (Mekonnen and Hoekstra, 2011; Hoekstra and Mekonnen, 2012).

24.3.5 Groundwater quality management

Groundwater quality management covers two main purposes: (1) protection against pollution and (2) remediation of polluted sites or aquifer zones.

Protection has the objective to prevent pollutants entering an aquifer or part of it (Foster et al., 2004). Some of the protective measures are generic in nature, for instance, a ban on selling and using certain harmful chemicals; prohibiting the application of liquid manure to agricultural fields during recharge seasons; implementing adequate sewerage and wastewater treatment provisions; and promoting appropriate well protection practices. Other ones are location-specific and focus on protecting existing groundwater abstraction sites (single wells or wellfields). In between these two extremes in terms of spatial extent are the prevention of seawater intrusion and the migration of other poor-quality waters, and interventions such as spatial planning of land use, which protects certain aquifers or parts of them by a local ban on certain uses and locating the uses with comparatively highest pollution risks in zones where they would do least harm.

Remediation of polluted sites or aquifer zones (Fetter, 1993) deals in the first place with source control: removal of contaminated soil (if feasible), unproperly dumped waste, leaking subsurface storage tanks, or buried drums with hazardous substances. If removal is technically or economically not feasible, then the pollution may be physically contained or hydraulically isolated. Pump-and-treat systems represent a second type of remediation: polluted groundwater is pumped, treated, and reinjected into the subsurface or diverted to a surface water system. Other remediation activities include recovery of light nonaqueous phase liquids (LNAPLs) and in situ bioremediation.

Saline water intrusion and brackish water upconing in coastal areas are usually caused by groundwater abstraction. However, groundwater quality degradation is in the majority of the cases a negative externality of activities unrelated to groundwater abstraction and use, which implies that in such cases there is no direct feedback from polluted groundwater to its polluter.

24.3.6 Adaptation to climate change and sea-level rise

Positive responses to climate change can be subdivided into two categories: mitigation and adaptation. *Mitigation* addresses the root causes and attempts to reduce climate change, in particular by measures that reduce CO_2 emissions, for which global support is markedly on the rise. *Adaptation* seeks to reduce vulnerabilities to climate change risks and the corresponding potential damage, as well as to benefit from new opportunities climate change may offer. Adaptation can take place at different scale levels, even at the level of single individuals, but it is likely to become more effective if dedicated policies and strategies at national and other aggregated spatial levels are developed and implemented (Lim and Spanger-Siegfried, 2004; EU, 2013). Adaptation in the water sector includes a wide range of aspects, facets, and approaches (Ludwig et al., 2009; Jones, 2011; Treidel et al., 2012). Among others, the uncertainty of climate change predictions needs to be taken into account explicitly as an additional risk in water resources planning and management. Next, the design of water infrastructure can no longer be based on historical hydrometeorological time series of the past, since the assumption of statistical stationarity is not valid any more. Furthermore, the water yielding capacity of many vulnerable water sources around the world is expected to decrease, which will lead to responses such as changes in water demands, substitution by supplies from other sources, conjunctive water management, MAR, and use of non-conventional water sources (desalinated water, recycled used waters). As far as replacement of supplies is concerned, it is likely that many surface-water based sources will be substituted by more resilient groundwater sources. Climate change will in some areas offer new opportunities rather than restrictions, for instance, for expanding agriculture to areas where climatic conditions are changing from unsuitable (too cold) to suitable (warmer) for crop growth.

Adaptation to sea-level rise will include construction or heightening of coastal defense infrastructure. Even if the defense works function satisfactorily and prevent low-lying land from being flooded by seawater, measures are needed to counteract increased risks of seawater intrusion into coastal aquifers.

24.3.7 Environmental management

Most groundwater management interventions are designed primarily with groundwater as an extractable resource in mind, which bears the risk of overlooking and neglecting supporting services and other in situ groundwater functions (see Fig. 24.1). This risk can be controlled by adopting an integrated water resources management approach and linking up with environmental management. Such an approach urges policymakers and planners to consider the trade-offs between the value of abstracted water and the area's environmental values; or, in cases where the protection of specific environmental values or objects has been declared priority, to deal with environmental criteria as a hard constraint in planning.

24.4 Geographic variation of groundwater resources sustainability

24.4.1 General comments

Geographic variation of groundwater resources sustainability depends on natural and anthropogenic factors. The main natural factors are geology, climate, and geographic setting. These factors, separately or in combination, define variations in features such as size of groundwater systems; their structure in terms of stacked aquifers and aquitards; permeability and storage capacity of individual aquifers; their exposure to sources of recharge; their hydraulic connection with rivers and the sea; their degree of confinement; their vulnerability to pollution; their susceptibility to seawater intrusion and other forms of salinization; groundwater recharge intensity and patterns; mean groundwater residence times; and patterns and magnitude of the evapotranspiration deficit or surplus.

Important anthropogenic factors are groundwater exploitation, deliberate modification of groundwater storage (by artificial drainage or by artificial recharge), land use type and land use practices (influencing groundwater quantity and quality), and the production and management of waste and wastewater. The variations in these anthropogenic factors, combined with those in the natural factors, generate an almost endless diversity of local conditions, with widely divergent rankings on the groundwater resources sustainability scale. The latter ranges from strongly sustainable to indisputably unsustainable, with intermediate ratings such as "sustainability at risk" and "weak sustainability."

Groundwater resources sustainability is a complex concept incorporating multiple facets and subject to differences in interpretation, depending on the interpreter and the specific local conditions. There is no simple metric to assess it unambiguously. Furthermore, for the majority of the world's groundwater systems, there is not enough information and knowledge available to classify them in terms of groundwater sustainability. Nevertheless, field observations, case studies, and mapped patterns of relevant variables allow to make some provisional steps toward sketching the geographic distribution of aquifers or zones where current groundwater abstraction regimes are indisputably or most probably not sustainable, or where other factors threaten the sustainability of groundwater services. The results are summarized next.

24.4.2 Groundwater resources sustainability endangered or disrupted by progressive storage depletion

If an aquifer system, exposed to a certain rate of abstraction, is not resilient enough to develop sooner or later a new dynamic hydrological equilibrium, then progressive groundwater storage depletion will take place, until economic, technical, or physical exhaustion follows. The corresponding state of "hydrological overexploitation" or even "mining" (in absence of significant recharge) is the most clearly defined type of unsustainability.

Aquifers with *nonrenewable groundwater resources* form the most obvious category. There is a lack of consensus among groundwater professionals on the definition of this category. Here, we consider it to comprise aquifers receiving no or only negligible quantities of recharge under current climatic conditions (Margat and Van der Gun, 2013), which implies that their possible exploitation derives water from storage rather than from capture. Fig. 24.2 shows the world's largest aquifer systems of this type. They include seven large aquifer systems in Northern Africa (Nubian Aquifer System, North Western Sahara Aquifer System, Senegalo–Mauritanian Basin, Taoudeni–Tanezrouft Basin, Iullemeden–Irhazer Aquifer System, Murzuk–Djado Basin, and Lake Chad Basin); two in Southern Africa (Stampriet Basin and Karoo Basin); one in the Middle East (Arabian Aquifer System); and two in Australia (Great Artesian Basin and Canning Basin). All these aquifer systems are located in zones with a dry climate, where negligible quantities of water are available to recharge them. In a large zone in Northern Asia, however, aquifers fail to be recharged because of permafrost soil conditions preventing surplus water to percolate downwards to the underlying aquifers (parts of the West Siberian Basin, Tunguss Basin, and Yakut Basin). Numerous other—usually smaller—aquifers with nonrenewable resources are scattered over the globe, often located below actively recharged aquifers but hydraulically unconnected to these. Although abstracting groundwater from such aquifers (mining) is in a strict sense not sustainable, it may under certain conditions be considered as "socially sustainable use of nonrenewable resources" (Al-Eryani et al., 2006) or qualify as "weakly sustainable" (Klarin, 2018; Du Pisani, 2006). In spite of their lack of hydrological sustainability fed by groundwater renewal, some of these aquifers contain such a large volume of nonrenewable groundwater resources that under current abstraction rates it would require up to tens of thousands of years to exhaust them physically (Richey et al., 2015). The resilience of such aquifers to abstraction thus depends on stored volume rather than on recharge. Policies and management activities aiming to protect the stored volumes in this category of aquifers are still rare, except for the Great Artesian Basin, where rehabilitation projects during the last few decades produced remarkable improvements of the aquifer's state (Habermehl, 2018). The groundwater resources of that aquifer could be classified as "weakly renewable.".

Long-term trends of groundwater storage depletion are not only observed in aquifers with nonrenewable groundwater resources, but also in a much larger number of aquifers containing renewable groundwater. Well-known examples are the Californian Central Valley, the Western alluvial aquifers, the Gulf Coastal Plain, and the High Plains in the United States; the Atacama groundwater system in Chile; the Hermosillo coastal aquifer and Guanajuato upland aquifers in Mexico; the upper Guadiana and other aquifers in South East Spain; the Sana'a Basin in Yemen; the aquifers of Iran's arid plains; the Indo-Gangetic Plain aquifers in Pakistan and India; and the North China Plain aquifer (EU, 2007; Motagh et al., 2008; Custodio, 2012; Konikow, 2011; Margat and Van der Gun, 2013; Scheihing, 2018; Bierkens and Wada, 2019). In addition to these examples, progressive storage depletion is observed in numerous other aquifers with renewable resources, varying from small to large in size. Like most of the aquifers mentioned previously, they are located particularly in parts of the world where dry climates prevail (see Fig. 24.2).

Long-term trends of declining groundwater levels do not necessarily imply that the current abstraction regime is unsustainable. In principle, the aquifer can still be in transition to a new hydrodynamic equilibrium, although the longer it takes, the less probable this will be. The diagnosis "unsustainable abstraction regime" or "hydrological overexploitation" can only be confirmed by local investigations showing that under the current abstraction regime the aquifer's resilience is insufficient for establishing a new hydrological equilibrium (by reducing natural outflow and/or inducing additional recharge).

Attempts to reduce, stop, or even reverse depletion of renewable groundwater resources can be observed in several countries. The main approaches are reducing groundwater abstraction and augmenting groundwater resources by artificial recharge. Reducing groundwater abstraction is usually very difficult since it requires people to change their behavior and to forgo their currently enjoyed benefits from groundwater use partly or entirely, which is generally conflicting with their personal short-term interests and sometimes with their formal or perceived rights. Successful control of groundwater abstraction is observed only in a limited number of countries, most of them in Europe. The second approach (artificial recharge) is less controversial and usually more successful, but it can be applied only if there is a local source of water available to feed the aquifer system concerned. Total worldwide reported artificial recharge

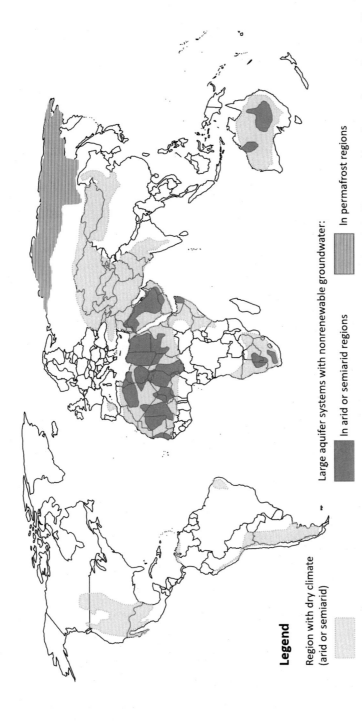

FIGURE 24.2 Global sketch map showing natural characteristics that preclude or strongly limit the sustainable abstraction of significant quantities of groundwater. Based on Margat, J., 2008. *Les eaux souterraines dans le monde*, UNESCO & BRGM, 187 p. (Margat, 2008); Nijsten, G.-J., Christelis, G., Villholth, K., Braune, E., Bécaye Gaye, C., 2018. Transboundary aquifers of Africa: review of the current state of knowledge and progress towards sustainable development and management. *J. Hydrol: Reg Stud.* https://doi.org/10.1016/j.ejrh.2018.03.004 (Nijsten et al., 2018); various climate maps.

reached 9.9 km^3/year in 2015, equivalent to 1% of the global groundwater abstraction. Countries with the largest volumes artificially recharged are India, United States, Germany, Italy, Australia, Spain, The Netherlands, Slovakia, Israel, and China (Dillon et al., 2018). There are also numerous areas and countries where considerable depletion of renewable groundwater resources does occur, but where significant efforts to control it are missing so far.

24.4.3 Groundwater resources sustainability endangered or disrupted by water quality degradation

Fresh groundwater resources endangered by quality degradation are scattered around the globe. If quality degradation is actually going to take place, then after some period of time these resources will be no longer suitable for certain provisioning services, hence will such services not be sustainable. Highest groundwater quality degradation risks are present in those zones where high vulnerability to pollution (or salinization) concurs with high concentrations of potentially infiltrating pollutants (or with the presence of saline/brackish water).

Pollution-related sustainability risks are prominent in shallow unconfined aquifers, at least in parts of such aquifers underneath urban areas or overlain by intensively cultivated agricultural lands. Local sanitation, sewage leaks, and many other forms of wastewater tend to produce below the surface of urban areas zones of polluted groundwater that slowly migrate downward. Groundwater inside these zones is not suitable anymore for uses that require a high water quality (such as drinking water), while neighboring parts of the shallow aquifer (immediately outside the urban area) and fresh groundwater resources in hydraulically connected deeper strata may be at risk as well, depending on the extent to which the migration of pollutants is controlled by natural factors and by human action. Diffuse shallow groundwater pollution under intensively cultivated agricultural land is characterized by other dominant pollutants (fertilizers and pesticides), but its migration follows a similar pattern. The majority of the sources of pollution mentioned in Section 24.2.2.3 are particularly a threat to shallow groundwater resources, but some sources form a risk for deep groundwater resources. Examples of such sources are disposal sites for radioactive waste and other substances stored permanently at great depth, as well as mining activities, and the recovery of oil and gas.

Fresh groundwater resources in low-lying coastal areas have special sustainability restrictions because of their exposure to the risk of saline water encroachment in response to groundwater abstraction, climate change, and sea-level rise. Fig. 24.3 shows global hotspots in this category. River deltas are of particular interest (Coleman, 1981): they usually enjoy considerable fresh-water recharge, but in view of the salinization hazard only a fraction of it can be abstracted sustainably.

A model study for the Rhine−Meuse−Scheldt delta in The Netherlands by Oude Essink et al. (2010) predicted the impact of sea-level rise on groundwater heads during the present century to remain limited to areas within 10 km from the coastline and the main rivers, but to produce progressive salinization of shallow groundwater, putting the quantities of fresh groundwater available for provisioning and ecosystem services under pressure. Although other river deltas have their own specific properties, they have in common that the fresh groundwater resources they contain is threatened by salinization, already under current conditions, but even more in the future, due to sea-level rise.

Many of the small islands in the world's oceans and seas are endowed with only thin fresh groundwater lenses (IGRAC, 2019); their exploitation requires utmost care, in order to avoid groundwater salinization. These islands, in particular the topographically very flat ones such as Tarawa atoll (Kiribati), Malé (Maldives), and Tongatapu (Tonga), are extremely vulnerable to sea-level rise, among others due to the related limited sustainability of groundwater as a source of domestic water.

24.4.4 Groundwater resources sustainability constrained by environmental considerations

Among the various competing or even conflicting groundwater services (Fig. 24.1), highest priority is usually given to provisioning services, often at the expense of environmental and other in situ services. Obviously, very low priority for nonprovisioning services prevails in water-scarce areas of the arid and semiarid zones, where it is already a major effort to provide and secure water in the quantities as demanded for domestic, agricultural, and industrial purposes. This leaves little or no room—physically nor in the mind of people—for allocating water to any of the nonprovisioning groundwater services. Consequently, several in situ groundwater services that used to be available before the onset of intensive groundwater development have nowadays disappeared in most of the severely water-stressed areas of the arid and semiarid zones: groundwater levels have declined, springs and baseflows have dwindled, and wetlands have dried up.

The situation is different in areas more favorably endowed with water resources. People there can to a certain extent afford to make a choice between different types of groundwater services and opt for those they prefer. Nevertheless, a basic understanding of how groundwater systems behave and good coordination between the many stakeholders are required, in order to prevent that the state of the groundwater systems changes in an uncontrolled way, with the risk of undermining the

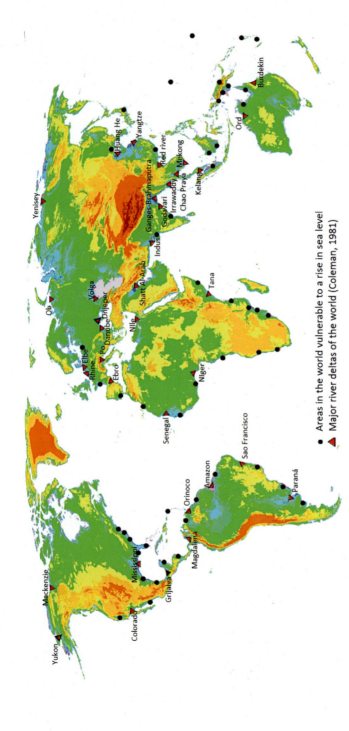

FIGURE 24.3 Global hotspots of groundwater vulnerability to sea-water intrusion and sea-level rise. *Modified from Oude Essink, G.H.P., van Baren, E.S., de Louw, P.G.B., 2010. Effects of climate change on coastal groundwater systems: a modelling study in The Netherlands. Water Resour. Res. 46, W00F04. doi:10.1029/2009WR008719.*

TABLE 24.1 Summary of Ramsar site statistics, by October 2019.

Continent	Number of sites	Aggregated area of sites (ha)	Share in total number of sites (%)	Share in total aggregated area (%)
North America	218	23,605,427	9	9
Central America and the Caribbean	75	3,706,034	3	1
South America	129	57,788,677	5	23
Europe (including the Russian Federation)	1092	27,172,150	46	11
Africa	413	110,042,647	17	43
Asia	363	22,118,653	15	9
Oceania	82	9,171,913	4	4
Total	2372	253,605,501	100	100

Based on Ramsar, 2019a. The List of Wetlands of International Importance. https://www.ramsar.org/document/the-list-of-wetlands-of-international-importance-the-ramsar-list.

sustainability of the groundwater services opted for. Where these and other elements of good groundwater governance are available, it is a political decision how to balance the different groundwater services in practice. A growing awareness of the value and vulnerability of the groundwater dependent environment, and a steadily increasing appreciation for nature and the environment has in several countries produced a shift in this balance in favor of in situ groundwater services. Environmental targets there have been introduced in groundwater management plans as constraints to groundwater abstraction. Typical examples of such targets are the prevention of land subsidence, the preservation of baseflows and springs, and the conservation of recognized valuable nature areas. Assigning the status of protected area to sites or areas with valuable nature, for instance, in national environmental policies or under international conventions such as Ramsar, contributes to ensuring that environmental protection and nature conservation there gets highest priority.

By October 2019 Ramsar's (2019a) *List of Wetlands of International Importance* included 2372 wetlands in 170 countries (see Table 24.1). Together, they cover almost 2% of the Earth's land surface. Europe leads the list in terms of number of sites, Africa in terms of aggregated area covered. The size of single Ramsar sites varies from 1 ha to more than 60,000 km^2. The contracting parties are expected to manage their Ramsar sites so as to maintain their ecological character and retain their essential functions and values for future generations (Ramsar, 2019b).

24.5 Conclusion

Groundwater resources sustainability is a complex concept. It is not an intrinsic property of a physical groundwater system, but rather it combines properties of such a system (including its setting) with current stresses acting upon it. These stresses may be anthropogenic (such as groundwater abstraction or pollution) and natural (e.g., climatic variation). If a groundwater system is resilient enough to continue providing the currently enjoyed groundwater services for indefinite time, then its groundwater resources are rated as sustainable. Often such groundwater services are associated only with groundwater abstraction (provisioning services), but in principle, it is also possible to link groundwater resources sustainability to in situ groundwater services.

As mentioned before, the exploitation of nonrenewable groundwater resources is inherently unsustainable, while the sustainability of renewable groundwater resources is endangered by various potential threats, widely spread over the globe. Most of these threats are related to human activities, which implies that it often lies in the hands of human communities to prevent groundwater services from becoming unsustainable, or to restore groundwater systems of which the services have become unsustainable. Reducing groundwater abstraction, augmenting the groundwater resources, and preventing pollutants from entering aquifers are the most common strategic options. Implementing measures and changing human behavior according to these options is usually very difficult in practice, due to conflicts of interests, poor understanding of the longer term consequences of adverse trends, and the lack of political will and/or institutional capabilities to ensure collective compliance with adopted policies and plans. Furthermore, groundwater governance and management are likely to develop and flourish only under conditions of peace, social stability, and adequate financing.

The enormous and vital importance of groundwater for the human society is beyond dispute. Groundwater plays a major role in supplying drinking water, ensuring food security by irrigation, adapting to climate variability, supporting biodiversity, sustaining surface water bodies, and pursuing poverty alleviation. Stresses on the groundwater systems tend to increase over time. Ensuring groundwater resources sustainability is therefore crucial. Nevertheless, groundwater ranks usually low on the political agendas and relatively few people are aware of groundwater sustainability hazards. As a result, many aquifers or aquifer zones around the world have been lost already by irreversible pollution or show severe groundwater storage depletion, while the sustainability of current groundwater services provided by numerous other aquifers is at risk and thus requires immediate protective action. In this connection a group of scientists and practitioners active in the groundwater sector met recently (October, 2019) at a Chapman Conference in Valencia and formulated a statement on global groundwater sustainability, entitled *Global Groundwater Sustainability: A Call to Action* (IAH, 2019). The statement urges international and national governmental and nongovernmental agencies, development organizations, corporations, decision-makers, and scientists to support global groundwater sustainability, and it calls for three items of action: (1) put the spotlight on global groundwater sustainability; (2) manage and govern groundwater sustainably from local to global scales; and (3) invest in groundwater governance and management.

References

Al-Eryani, M., Appelgren, B., Foster, S., 2006. Social and economic dimensions of non-renewable resources. In: Foster, S.S.D., Loucks, P. (Eds.), Non-Renewable Groundwater Resources. A Guidebook On Socially-Sustainable Management for Water Policy-Makers. UNESCO-IHP, Paris, pp. 25–34.

Alley, W.M., Leake, S.A., 2004. The journey from safe yield to sustainability. Ground Water 42 (1), 12–16.

Bierkens, M., Wada, Y., 2019. Non-renewable groundwater use and groundwater depletion: a review. Environ. Res. Lett. Available from: https://doi.org/10.1088/1748-9326/ab1a5f (Submitted and accepted for publication).

Bredehoeft, J.D., Papadopoulos, S.S., Cooper, H.H., 1982. Groundwater: the water-budget myth. Scientific Basis of Water-Resource Management, Studies in Geophysics. National Academic Press, Washington, DC, pp. 51–57.

Bury, J.B., 1932. The Idea of Progress. Dover, New York.

Carson, R., 1953. Silent spring. Crest Reprint. Fawcett Publications, Inc., Greenwich, CT, p. 155.

Coleman, J.M. (1981). Deltas: Processes of Deposition and Models for Exploration. International Human Resources Development Corporation, Boston, MA, 124 p.

Custodio, E., 2012. Intensive groundwater development: a water cycle transformation, a social revolution, a management challenge. In: Martinez, L., Garrido, A., López-Gunn, E. (Eds.), Rethinking Water and Food Security. CRC Press, Boca Raton, FL, pp. 259–298.

Darwin, C., 1859. On the Origin of Species. 448 p. (Reprint by The Pennsylvania State University (2001). A Penn State Electronic Classics Series Publication.

Dillon, P., Stuyfzand, P., Grischek, T., et al., 2018. Sixty years of global progress in managed aquifer recharge (31 authors)Hydrogeol. J. Available from: https://doi.org/10.1007/s10040-018-1841-z.

Du Pisani, J.A., 2006. Sustainable development – historical roots of the concept. Environ. Sci. 3 (2), 83–96.

EU, 2007. Mediterranean Groundwater Report. Technical Report on Groundwater Management in the Mediterranean and the Water Framework Directive, Produced by the MED-EUWI WG on Groundwater.

EU, 2013. The EU strategy on adaptation to climate change. https://ec.europa.eu/clima/sites/clima/files/docs/eu_strategy_en.pdf.

Fetter, C.W., 1993. Contaminant Hydrogeology. Prentice-Hall., New Jersey.

Foster, S., Garduño, H., Kemper, K., Tuinhof, A., Nanni, M., Dumars, C., 2004. Groundwater quality protection: defining strategy and setting priorities. In: GW-MATE Briefing Note Series, Note 8. The World Bank, Washington, DC.

Habermehl, M.A., 2018. Groundwater governance in the Great Artesian Basin, Australia. In: Villholth, K.G., López-Gunn, E., Conti, K.I., Garrido, A., Van der Gun, J. (Eds.), Advances in Groundwater Governance. CRC Press/Balkema, Boca Raton–London–New York–Leiden, pp. 411–442.

Hardin, G., 1968. The tragedy of the commons. Science 162, 1243–1248.

Hoekstra, A.Y., Mekonnen, M.M., 2012. The water footprint of humanity. Proc. Natl. Acad. Sci. U.S.A. 109 (9). Available from: www.pnas.org/cgi/doi/10.1073/pnas.1109936109.

Horton, P., Horton, B.P., 2019. Re-defining sustainability: living in harmony with life on Earth, One Earth, 1. pp. 86–94.

IAH, 15 Nov 2019. Global groundwater sustainability: a call to action. IAH News. https://iah.org/news/global-groundwater-sustainability-a-call-to-action and https://www.groundwaterstatement.org/.

IGRAC, 2019. GGIS-TWAP, information on small island developing states. Available from: https://www.un-igrac.org/global-groundwater-information-system-ggis.

IPCC, 2014. Climate change 2014: synthesis report, summary for policymakers. In: Contribution to the Fifth Assessment Report of the Intergovernmental Panel on Climate Change. IPCC, Geneva, Switzerland. https://www.ipcc.ch/site/assets/uploads/2018/02/AR5_SYR_FINAL_SPM.pdf.

IPCC, 2019. Summary for policymakers. In: IPCC Special Report on the Ocean and Cryosphere in a Changing Climate. IPCC, Geneva, Switzerland. https://www.ipcc.ch/site/assets/uploads/sites/3/2019/11/03_SROCC_SPM_FINAL.pdf.

Jenkins, W., 2009. first ed. Berkshire Encyclopaedia of Sustainability: The Spirit of Sustainability, vol. 1. Berkshire Publishing Group, Berkshire.

Jones, J.J.A. (Ed.), 2011. Sustaining Groundwater Resources. Springer, Books on Environmental Sciences, Initiative of the International Year of Planet Earth.
Klarin, T., 2018. The concept of sustainable development: from its beginning to the contemporary issues. Zagreb Int. Rev. Econ. Bus. 21 (1), 67–94.
Konikow, L., 2011. Contribution of global groundwater depletion since 1900 to sea-level rise. Geophys. Res. Lett. 38, L17401. Available from: https://doi.org/10.1029/2011GL048604.
Konikow, L.F., Leake, S.A., 2014. Depletion and capture: revisiting "the source of water derived from wells". Ground Water 52, 100–111.
Lim, B., Spanger-Siegfried, E. (Eds.), 2004. Adaptation Policy Frameworks for Climate Change: Developing Strategies, Policies and Measures. Cambridge University Press, Cambridge.
Lohman, S.W., Bennett, R.R., Brown, R.H., Cooper Jr., H.H., Drescher, W.J., Ferris, J.G., et al., 1972. Definitions of Selected Ground-Water Terms – Revisions and Conceptual Refinements. Water Supply Paper 1988. US Geological Survey (USGS), Reston, VA.
Ludwig, F., Kabat, P., van Schaik, H., van der Valk, M., 2009. Climate Change Adaptation in the Water Sector. Earthscan, London, Cooperative Programme on Water and Climate (CPWC).
Margat, J., 2008. Les eaux souterraines dans le monde. UNESCO & BRGM, p. 187.
Margat, J., Van der Gun, J., 2013. Groundwater Around the World: A Geographic Synopsis. CRC Press/Balkema, Leiden, The Netherlands, p. 348.
Millennium Ecosystem Assessment (MEA), 2005. Ecosystems and Human Well-Being: Synthesis. Island Press, Washington, DC, Millennium Ecosystem Assessment.
Meadows, D., 1972. The Limits to Growth. A Report for the Club of Rome Project on the Predicament of Mankind. Universe Books, New York.
Mekonnen, M.M., Hoekstra, A.Y., 2011. National Water Footprint Accounts: The Green, Blue and Grey Water Footprint of Production and Consumption. UNESCO-IHE Research Report Series No. 50, vol. 2. UNESCO-IHE.
Motagh, N., Walter, T.R., Sharifi, M.A., Fielding, E., Schenk, A., Anderssohn, J., et al., 2008. Land Subsidence in Iran caused by widespread water reservoir overexploitation. Geophys. Res. Lett. 35, L16403.
Mukherjee, A., Scanlon, B., Aureli, A., Langan, S., Guo, H., McKenzie, A., 2020. Global Groundwater: Source, Scarcity, Sustainability, Security and Solutions, first. ed. Elsevier, ISBN: 9780128181720.
Nijsten, G.-J., Christelis, G., Villholth, K., Braune, E., Bécaye Gaye, C., 2018. Transboundary aquifers of Africa: review of the current state of knowledge and progress towards sustainable development and management. J. Hydrol.: Reg. Stud. Available from: https://doi.org/10.1016/j.ejrh.2018.03.004.
Oude Essink, G.H.P., van Baren, E.S., de Louw, P.G.B., 2010. Effects of climate change on coastal groundwater systems: a modelling study in The Netherlands. Water Resour. Res. 46, W00F04. Available from: https://doi.org/10.1029/2009WR008719.
Pakhrel, Y.N., Hanasaki, N., Yeh, P.J.-F., Yamada, T.J., Kanae, S., Oki, T., 2012. Model estimates of sea-level change due to anthropogenic impacts on terrestrial water storage. Nat. Geosci. Available from: https://doi.org/10.1038/NGEO1476, Published online 20 May 2012.
Ramsar, 2019a. The list of wetlands of international importance. https://www.ramsar.org/document/the-list-of-wetlands-of-international-importance-the-ramsar-list.
Ramsar, 2019b. Ramsar portal. https://www.ramsar.org/.
Remenyi, J., 2004. What is development? In: Kingsbury, D., Remenyi, J., McKay, J., Hunt, J. (Eds.), Key Issues in Development. Palgrave Macmillan, Hampshire, New York, pp. 22–44.
Richey, A.S., Thomas, B.F., Lo, M.-H., Famiglietti, J.S., Swenson, S., Rodell, M., 2015. Uncertainty in global groundwater storage estimates in a total groundwater stress framework. Water Resour. Res. 5198–5216.
Savenije, H., van der Zaag, P., 2002. Water as an economic good and demand management: paradigms with pitfalls. Water Int. 27 (1), 98–104.
Scheihing, K.W., 2018. Water Resources Management in the Atacama Desert: Pivotal Insights into Arid Andean Groundwater Systems of Northern Chile (Ph.D. thesis). Technische Universität Berlin.
Tate, D., 1999. An Overview of Water Demand Management and Conservation. Vision 21: Water for People. Water Supply and Sanitation Collaborative Council, Geneva.
Theis, C.V., 1940. The sources of water derived from wells: essential factors controlling the response of an aquifer to development. Civ. Eng. 1940 (10), 277–280.
Todd, D.K., 1959. Ground Water Hydrology. J. Wiley & Sons, New York.
Treidel, H., Martin-Bordes, J.L., Gurdak, J.J. (Eds.), 2012. Climate Change Effects on Groundwater Resources. CRC Press/Balkema, Leiden, The Netherlands, IAH International Contributions to Hydrogeology, vol. 27.
United Nations, 2015. Transforming our world: the 2030 agenda for sustainable development. Available from: https://sustainabledevelopment.un.org/content/documents/21252030%20Agenda%20for%20Sustainable%20Development%20web.pdf.
United Nations World Commission on Environment and Development, 1987. Our Common Future. Report of the United Nations World Commission on Environment and Development (Brundtland Report). United Nations World Commission on Environment and Development. 300 p.
Van der Gun, J., 2019. The global groundwater revolution. Oxford Research Encyclopedia, Environmental Science. Oxford University Press. Available from: https://doi.org/10.1093/acrefore/9780199389414.013.632.
Van der Gun, J., Lipponen, A., 2010. Reconciling groundwater storage depletion due to pumping with sustainability. Sustainability. Available from: https://doi.org/10.3390/su2113418. http://www.mdpi.com/2071-1050/2/11/3418/ (Special issue Sustainability of Groundwater, November 2010).
Villholth, K.G., López-Gunn, E., Conti, K.I., Garrido, A., Van der Gun, J. (Eds.), 2018. Advances in Groundwater Governance. CRC Press/Balkema, Boca Raton, London, New York, Leiden, xxv + 594 pages.
Wada, Y., van Beek, L., van Kempen, C., Reckman, J., Vasak, S., Bierkens, M., 2010. Global depletion of groundwater resources. Geophys. Res. Lett. 37, L20402. Available from: https://doi.org/10.1029/2010GL044571.
Wulf, A., 2015. The Invention of Nature: Alexander von Humboldt's New World. London: John Murray, p. 473.

Chapter 25

Sustainability of groundwater used in agricultural production and trade worldwide

Carole Dalin
University College London, London, United Kingdom

25.1 Introduction

Water resources are renewable but limited, with reservoirs (oceans, ice caps, groundwater, lakes, rivers, and the atmosphere) that get replenished more or less quickly. Among these reserves, groundwater stocks almost all of the *liquid freshwater available* for human use (99%), and it is also the most slowly renewed reservoir. On a global average the same drop of water will remain approximately 1400 years in the same aquifer (Shiklomanov, 1997), but this time varies widely among and within aquifers (Befus et al., 2017; Wada et al., 2010). This slow renewal rate makes groundwater an asset relative to surface water resources, because it is less affected by short-term events, such as droughts. In the face of climate change, under which droughts are becoming more frequent globally, aquifers will be vital to cope with unreliable surface water supply, especially for farming but also for drinking and power generation. However, because groundwater is generally much more slowly renewed than surface water, adequate groundwater management is the key, and ensuring sustainable use of groundwater resources is a major challenge.

Exploitation of groundwater resources, including the development of extensive irrigation systems, has often led to excessive abstraction, causing drastic lowering of groundwater levels and associated storage. This overuse of groundwater resources can be considered *unsustainable*, because the resource will not be physically (or economically) available after a certain period of time if groundwater levels continue to decline. Groundwater overuse may induce long-term *groundwater depletion (GWD)* and also lead to reduced river flow and degradation of water quality, including increased salinity.

The effects of GWD are primarily local: GWD affects the local water resources and environment, and the water cycle within the concerned water basins. Thus analysis and possible solutions to this issue require regional hydrogeological studies. However, water-intensive commodities such as grains have been increasingly exchanged worldwide since the 1980s (Dalin et al., 2012). This global virtual water trade (VWT) has led to an increased hidden globalization of water resources. Thus understanding GWD requires international information, going beyond biophysical processes and the natural water cycle. In addition, groundwater sustainability is mediated by another global phenomenon—climate change. Projections of future climatic conditions are uncertain but important to include when estimating global and regional sustainability of groundwater use.

Considering these aspects and notwithstanding their challenges, the sustainability of water resources—including groundwater—has been assessed globally, with a perspective of population and climate changes (Vörösmarty et al., 2000), water sources (Rost et al., 2008; Siebert and Döll, 2010), agricultural sustainability (Dalin et al., 2017; Scanlon et al., 2012; Wada et al., 2012), and overall GWD (Gleeson et al., 2012; Konikow and Kendy, 2005). The storage and distribution of groundwater has been examined at the global scale with concerning findings (Gleeson et al., 2016). At the regional level, studies have noted the need for historical understanding (Alley and Alley, 2017; Rodell et al., 2018) and consideration of future alterations of the water cycle due to climate change (Green et al., 2011; Taylor et al., 2013).

Agriculture is by far the most water-intensive activity: on a global average, 90% of the freshwater (surface water and groundwater) we consume is for irrigation. Water is considered consumed not only when it is abstracted from its natural reservoir (with a well, pump, canal, dam, etc.) but also when its use leads to the water being immediately unavailable locally. In agriculture, consumptive use is thus the water evaporating from the soil and through crops. Population and economic growth are continuing to increase global food demand, and it is expected that, by 2050, we will require 60% more food production than in 2010 (OECD-FAO, 2012). Consequently, irrigation, including from slowly renewed aquifers, is expected to continue rising. This chapter outlines state-of-the-art knowledge of water—and particularly groundwater—resources used for food production and food trade globally. It reviews recent work on the key issue of groundwater sustainability in agriculture and highlights how and where groundwater overuse affects both food-producing and food-importing regions.

25.1.1 Water use for global food production and virtual water flows via international food trade

About 40% of agricultural water consumption is now from groundwater, and this share has been rapidly increasing in the last decades, particularly in the United States and India (Konikow, 2011; Siebert et al., 2010). Studies of GWD (Rodell et al., 2018; Wada et al., 2010) have shown that fast depletion mostly occurs in great food-producing regions of the world. Groundwater irrigation is relatively less developed in Africa, where the area equipped from groundwater irrigation is only 19% of the total irrigated land area, while this share goes up to 44% in Asia (Siebert et al., 2010). Since the late 1990s, rapid economic globalization has led to increasingly interdependent supply chains across most of the world, notably food supply chains. In parallel, economic and population growth has put more pressure on natural resources, such as water, to produce goods and services consumed globally.

The scientific literature has increasingly studied the link between water and food; notably the amount of water embedded in the most water-intensive commodities: agri-food products. Allan (1997) coined the term *virtual water* to refer to this volume of water consumed for producing food, for example, water evaporated by a crop during its growing period. His work particularly focused on the VWT coming into dry countries in the Middle East, via their imports of agri-food commodities. Other terms have been used to refer to this water consumption associated with food commodities, such as "water footprint" or "embedded water," which are equivalent to the *virtual water content* (VWC; in $m^3_{water}/tonne_{crop}$) of a product. The VWC of an agricultural crop is estimated as the ratio of evapotranspiration (ET; m^3_{water}/m^2_{land}) over the crop yield (Y; $tonne_{crop}/m^2_{land}$) during the crop's growing season (Hanasaki et al., 2010). It can be split into "blue" and "green" components, if one can determine how much ET is satisfied by rainwater (green ET) and how much additional water is applied via irrigation (driving blue ET). For most groundwater irrigation systems the applied volume is estimated as withdrawal (i.e., pumped/abstracted volume) minus "irrigation return flow": the portion of applied water infiltrating back into the aquifer systems (e.g., in Wada and Bierkens, 2014).

VWC of different agricultural products have been compared (e.g., Konar et al., 2011; Mekonnen and Hoekstra, 2011) and studies found variability of up to one order of magnitude across food commodities. On a global average, for example, soybean is shown to have a higher total VWC than wheat (Fig. 25.1, right panel). Similarly, meat products, in

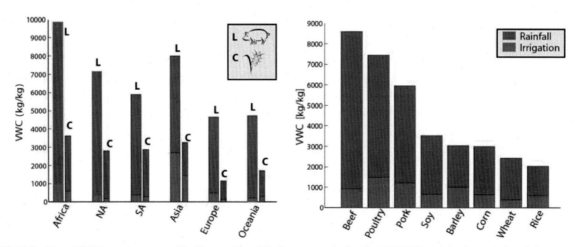

FIGURE 25.1 Mean VWC by water source. The blue portion of the bar represents the blue VWC (from irrigation water) and the green portion shows the green VWC (from soil moisture). Left panel: Mean VWC for each of the six regions: Africa, North America, South America, Asia, Europe, and Oceania. Thick bars represent livestock (L) products: beef, pork, and poultry. Thin bars show crops (C): barley, maize (corn), rice, soy, and wheat. Right panel: Global average VWC for each of the unprocessed livestock and crop products. The units are kilograms of water per kilogram of crop, equivalent to liters of water per kilogram of crop or m^3/Mg (1 Mg = 1 t). *VWC*, Virtual water content. *Data from Konar, M., Dalin, C., Suweis, S., Hanasaki, N., Rinaldo, A., Rodriguez-Iturbe, I., 2011. Water for food: the global virtual water trade network. Water Resour. Res., 47, W05520.*

particular beef, have a much higher VWC (by mass, but also per kilocalorie) than grains. Indeed, the VWC of animal products, like meat, accounts for the water consumed to produce all of the feed crops required for the animal throughout its lifetime, and animals convert calories from plants with a low efficiency (e.g., 10–40 cal of plant are required to obtain just 1 cal of beef meat; Eshel et al., 2014).

Importantly, and often less known by general consumers, the VWC of a given agricultural product may also significantly vary across regions, mainly because of the different climates and agricultural practices: high aridity and low yields lead to high crop VWC (e.g., in Africa, about 3500 m^3/t; Fig. 25.1), while a more humid climate and high yields enable to harvest crops with a low VWC (e.g., in Europe, about 1200 m^3/t; Fig. 25.1).

While most analyses of crop water use have considered "green" (rainfed only) and "blue" (irrigation) water sources, because irrigated and rainfed production have different socioeconomic and biophysical characteristics, groundwater is most often lumped with other irrigation sources. Blue and green VWC estimates reflect that, overall, irrigation is most important for food produced in Asia (Fig. 25.1, left panel) and represents, for example, a larger fraction of water use for maize than for wheat (Fig. 25.1 right panel).

VWT refers to the transfers of virtual water between different regions via trade of goods and services, such as agricultural products. This concept was introduced to highlight the hidden water resources consumed by countries that import food produced abroad, and thus virtually import some of their trade partners' water resources. VWT flows are generally estimated as follows: $\text{VWT}_{i,j,c} = \text{VWC}_{i,c} \times T_{i,j,c}$, where $\text{VWT}_{i,j,c}$ is the VWT (i.e., water embedded in trade, m^3/year) of crop c from region i to j, $\text{VWC}_{i,c}$ is the VWC (i.e., amount of water embedded in each unit of crop c, m^3/t) in region i, and $T_{i,j,c}$ is the trade of crop c from region i to j (t/year).

Studies of global VWT have confirmed that exchanges of water resources via international food trade have been rapidly increasing: both the number of country pairs trading food with each other and the global volume of international VWT embedded in food trade approximately doubled from 1986 to 2007 (Dalin et al., 2012, Fig. 25.2). Because of this, water resource management, particularly groundwater management, is shifting from local or regional issues to increasingly international issues (Aeschbach-Hertig and Gleeson, 2012). Some regions have played a larger role in this growth of global VWT volume (Fig. 25.2). Exports from South America to Asia contributed the most to the VWT volume increase between 1986 and 2007 (30%), followed by internal trade in North America (11%). South America has become a major virtual water exporter, mainly because of soybean exports from Brazil and Argentina, to all other main regions except North America and with negligible imports. Asia more than doubled its imports, importing mostly from South America (39%) and North America (25%), with an important internal VWT (29%). In contrast, virtual water exports from North America to Europe have decreased at the expense of increasing exports from North America to Asia (by 60%) and VWT growth within North America—mostly between the United States, Canada, and Mexico (by 310%, Dalin et al., 2012) (Fig. 25.2).

The patterns of VWT are dependent on the type of water source (blue or green): Konar et al. (2012) found that the United States, India, and Pakistan are the largest exporters of blue virtual water via food, while Argentina and Brazil fall from positions 2 and 3 for green water VWT to positions 6 and 9 for blue VWT, respectively. These differences are due to both climate and irrigation infrastructure in the countries of export. Some countries with extensive irrigation use

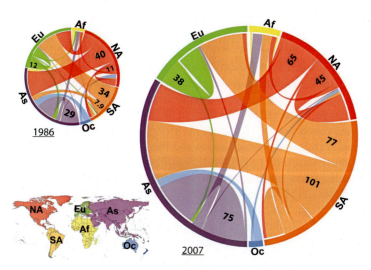

FIGURE 25.2 Virtual water trade via international food trade (in billions of cubic meters) in the years 1986 and 2007, grouped by region (Af = Africa, NA = North America, SA = South America; Oc = Oceania; As = Asia; and Eu = Europe). The map in the lower left provides a key to the regional color scheme. The color of each ribbon indicates the region of export. *Taken and modified from Dalin, C., Konar, M., Hanasaki, N., Rinaldo, A., Rodriguez-Iturbe, I., 2012. Evolution of the global virtual water trade network. Proc. Natl. Acad. Sci. U.S.A., 109, 5989–5994.*

and arid climates, such as Iraq and Morocco, are among the largest exporters of blue virtual water. While they do export crops, it is really their exceptionally high blue VWC (i.e., high use of irrigation water per unit crop) that drives them to be amongst the top exporters of blue virtual water.

Although the distinction between "green water" and "blue water"—that is, the water naturally stored in the soil versus the water supplied as irrigation from surface water bodies and groundwater—was made early on in VWT studies (Mekonnen and Hoekstra, 2011), groundwater has rarely been distinguished from other sources of irrigation (e.g., considering only groundwater use; Marston et al., 2015). Only a few recent VWT studies have started to include, beyond volumes of water consumed, the concept of environmental impact or sustainability of water use (e.g., Mekonnen and Hoekstra, 2016; Yano et al., 2015; Dalin et al., 2017; see Section 25.1.3).

In the next section, we cover why groundwater sustainability is a key concern and how this sustainability has been evaluated, both for all groundwater uses and for agricultural production in particular.

25.1.2 Sustainability of groundwater use overall and in particular for global food production

Water is the key for sustainable development, as it is essential for human life, ecosystems, and socioeconomic development. With the need to significantly increase food production by 2050, and recognition of water overexploitation in several regions, sustainable water management for agriculture has become a major concern. Agricultural products, in particular, water-intensive food commodities, such as pig meat and sugarcane, are increasingly produced to satisfy the demands of the growing and wealthier global population. Irrigation is the major contributor to physical water scarcity globally, as it drives 70% of the global gross water demand (Döll, 2009).

Water stress indicators have been developed with various approaches (e.g., Hanasaki et al., 2008; Vanham et al., 2018; Wada et al., 2010; Wada and Bierkens, 2014) and their applications at different temporal and spatial scales—for example, at the country (Seckler et al., 1999) or grid cell scale (Mekonnen and Hoekstra, 2016)—have identified severe pressure on water resources in several regions. Wada and Bierkens (2014) provided water scarcity estimates at the annual scale, while others analyzed it on a monthly basis (Hoekstra et al., 2012; Scherer and Pfister, 2016), or integrated both annual and monthly scales (Brauman et al., 2016). Wada and Bierkens (2014) also developed a range of future scenarios and modeled their consequences on water availability and sustainability. Generally, water scarcity metrics are categorical, identifying thresholds, such as a water use-to-availability ratio greater than 0.4 indicating "high water stress," and greater than 0.8 indicating "very high water stress" (Alcamo et al., 2000). Others, such as the Falkenmark indicator (Falkenmark, 1989), are also categorical but based on the local population: this indicator measures the freshwater flow available per person in a given region, considering values below the 1700 m^3/cap/year threshold to reflect water scarcity. More recent methods, such as life cycle assessment, focus on the possible environmental impacts of water use (Kounina et al., 2013), using freshwater use inventory (Boulay et al., 2011) weighted by local characterization factors that transform inventory flows into environmental impacts (Pfister et al., 2011). Focusing on the sustainability of water used for food production at the global scale, Yano et al. (2016) proposed a local "water unavailability" factor to weight the water footprint of agriculture. This study was the first agricultural water sustainability analysis to distinguish green, surface water and groundwater resources at the global scale, but their study lacks crop-specific analyses, as well as subnational and subbasin analyses. Similarly, Brauman et al. (2016) developed a water depletion metric, which highlighted regions vulnerable to water shortage; however, they did not consider green water, even though green and blue water sources are highly connected (Falkenmark, 2013; Vanham et al., 2018).

Depending on the local context, surface water and groundwater systems can be strongly connected, with groundwater discharging into surface water bodies and vice-versa. These connections make groundwater sustainability a subset of overall irrigation water sustainability (also see Fig. 25.3). The US National Groundwater Association defines *groundwater sustainability* as "the development and use of groundwater resources to meet current and future beneficial uses without causing unacceptable environmental or socioeconomic consequences" (Alley et al., 1999), together with a definition of resilience, which encompasses the "ability of a system to resist long-term damage and the time taken to recover from a perturbation" (URL). A range of metrics of groundwater sustainability exist but it is difficult to find a consensus, due to a combination of hydro-geophysical and socioeconomic factors influencing the possibility of accurately quantifying the chosen variables (Maimone, 2004). More recently, it has also been suggested that a "coherent, overarching framework of groundwater sustainability is more important for groundwater governance and management than the concepts of safe yield, renewability, depletion, or stress" (Gleeson et al., 2020a). These authors suggest the following definition: "Groundwater sustainability is maintaining long-term, dynamically stable storage of high-quality groundwater using inclusive, equitable, and long-term governance and management" (Gleeson et al., 2020a).

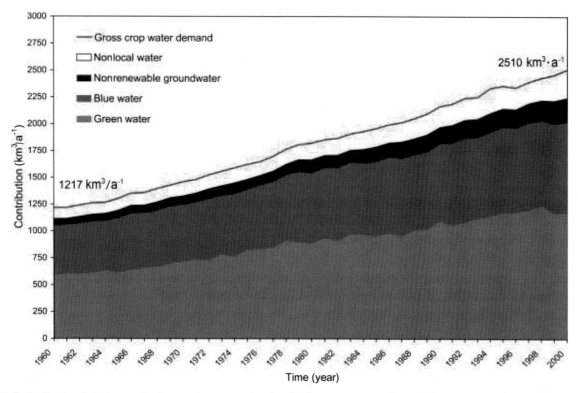

FIGURE 25.3 Past trends in the contribution per water resource to the global gross crop water demand. Green, green water; Blue, blue water; dark blue, nonrenewable groundwater. The white area between gross crop water demand and the three available water resources denotes the estimated contribution of nonlocal water resources (all in km^3/year). *Reproduced with permission from Wada, Y., van Beek, L., Bierkens, M., 2012. Nonsustainable groundwater sustaining irrigation: a global assessment. Water Resour. Res., 48, W00L06.*

Notwithstanding these practical and conceptual challenges, significant improvements in tools and models to track groundwater's storage and fluxes have recently been made and applied to sustainability studies. The sustainability of groundwater use has been estimated at different scales and with various methods, although uncertainties in these estimates remain significant. A range of terrestrial water storage changes estimates (including GWD), at regional to global scales, has been compared (Wada et al., 2017). Methods of estimation included regional statistics of on-site borehole water measurements, satellite-derived storage changes (from the NASA GRACE mission), numerical groundwater flow models (e.g., MODFLOW), and combination of these three types of evaluation methods. The study found that these estimates of GWD rates globally show a quite wide variability, as they ranged from 27 to 455 km^3/year, depending on the method, the temporal and regional coverage of the studies (Aeschbach-Hertig and Gleeson, 2012; de Graaf et al., 2017; Mukherjee et al., 2020). Taylor et al. (2013) reviewed existing estimates of GWD and associated sea-level rise at the global scale, cautioning, "a dearth of ground-based observations (...) not only limits our understanding of localized groundwater storage changes but also our ability to constrain evidence from GRACE satellite observations at larger scales ($\geq$150,000 km^2)." Despite this, there is good overall agreement across estimates for the spatial distribution of GWD: the regions with fastest depletion are areas of high groundwater pumping for irrigated agriculture, such as the southern Ogallala (or High Plains) aquifer in Texas, United States, northern India, and the North China Plain (Table 25.1).

A global, model-based study found that the world's total gross irrigation water demand in 2000 was satisfied at 20% from nonrenewable groundwater abstraction (Wada et al., 2012). This nonrenewable groundwater abstraction for irrigation was found to be most important in India (68 km^3/year), followed by Pakistan (35 km^3/year), the United States (30 km^3/year), Iran (20 km^3/year), China (20 km^3/year), Mexico (10 km^3/year), and Saudi Arabia (10 km^3/year). The study found that globally, this contribution more than tripled over the period 1960–2000 (from 75 to 234 km^3/year, Fig. 25.3). Another global study estimated that, on a global average, we use groundwater 3.5 times faster than the sustainable rate (Gleeson et al., 2012). This work introduced the concept of "groundwater footprint," which, unlike previous water footprint concepts, refers to the sustainability of water (here, groundwater) use: it is defined as the area-averaged ratio of abstraction to recharge (where the contribution to environmental streamflow is first deduced from the recharge). Similarly to the study by Wada et al. (2010), Gleeson et al. (2012) used national groundwater use statistics as

TABLE 25.1 Continental-level metrics from various global studies: groundwater use in irrigation, groundwater depletion, and groundwater depletion embedded in traded crops.

	Groundwater use for irrigation (in km³/year)	Groundwater depletion (GWD, in km³/year)	Virtual GWD exports via food export (in km³/year)
	Siebert et al. (2010) (Table 2)	Taylor et al. (2013) (Table 1)	Dalin et al. (2017)
Reference year	Average around 2000	2001–08	2010
Africa	17.86	5.5 ± 1.5	0.32
America	107.36	26.9 ± 7	7.34
Asia	398.63	111 ± 30	11.81
Europe	18.21	1.3 ± 10.7	0.69
Oceania	3.3	0.4 ± 0.2	0.11
World	545.36	145 ± 39	20.26

Groundwater use for irrigation (Siebert et al., 2010); groundwater depletion (overall) (Taylor et al., 2013); and virtual GWD exports in year 2010 via crop trade (Dalin et al., 2017).

input data sources and the PCR-GLOBW global hydrological model. Tuninetti et al. (2019) developed a new global indicator, the "water debt" indicator, which compares, for individual agricultural crops, their crop water use to water availability, for three types of water sources separately and together: green, surface water, and groundwater resources. This study analyzed sustainability of crop water use from the grid cell to the watershed level, and at the national scale.

Importantly, the sustainability of groundwater use in agriculture requires to evaluate long-term GWD, but it is also important to consider and estimate the ecosystem effects of a reduced groundwater storage on discharge to wetlands, lakes, and rivers, which may be highly dependent on groundwater supply.

25.1.3 Quantification of groundwater depletion for food trade

These indicators of crop water use sustainability, not always groundwater-specific, have been linked to global agri-food trade, using the VWT concept, in a handful of recent studies (e.g., Mekonnen and Hoekstra, 2016; Yano et al., 2015). This approach leads to new understanding of the environmental consequences of water use and food trade, as it considers local conditions of water sources and renewability, which affect the environmental sustainability and impacts of agricultural water use. This is important because consuming a given volume of water from a fossil aquifer in an arid region does not have the same environmental consequences as consuming this same volume from an abundant, little exploited river elsewhere. Hoekstra and Mekonnen (2016) estimated the sustainability of UK imports with regards to surface water irrigation used in producing countries, by comparing this irrigation volume to runoff net of other upstream uses; however, the study did not consider groundwater use or green water. The first study linking international food trade and GWD was Dalin et al. (2017), although it did not include the estimates of sustainability for other water sources than groundwater (i.e., surface and green water).

Importantly, for water sustainability indicators to be linked to trade, they need to be crop-specific: as for VWT analysis in general (see the VWT equation above), to know how much water is embedded in maize exported from the United States to China, one needs to estimate how much water is used to produce this crop in the United States, and not just the amount of water used for all crops in the United States. To obtain the volumes of GWD embedded in global food trade, Dalin et al. (2017) provided the first crop-specific estimates of GWD for irrigation globally. Previously available metrics of GWD were rarely linked to the specific use for the abstracted groundwater, and never to individual agricultural products. In a study of groundwater embedded in food transfers within the United States, Marston et al. (2015) also used the term "volume of groundwater per commodity unit"; however, although they focused on stressed aquifers, the link between traded products and depletion was not made explicit.

In their study of GWD and global food trade, Dalin et al. (2017) used a vertical water balance to estimate GWD, as the volume of groundwater abstraction for irrigation minus natural groundwater recharge and irrigation return flow (i.e., the additional recharge due to irrigation use). Confirming studies that had analyzed GWD in general (e.g., Gleeson et al., 2012; Wada et al., 2010), Dalin et al. (2017) found that GWD for irrigation is growing globally (by 22% from

Sustainability of groundwater used in agricultural production and trade worldwide **Chapter | 25** 353

2000 to 2010, with a doubling in China), and that this is largely concentrated in a few hotspot regions, notably North America, the Middle East and North Africa (MENA) region, Pakistan, India, and China. Overall, GWD for irrigation is mostly due to wheat and rice production.

It was estimated that 11% of the global volume of GWD for irrigation of crops was embedded in international trade (Dalin et al., 2017). For overall water use by crops (including rainfed crops), this number is around 18% (e.g., Hoekstra et al., 2012, Konar et al., 2011). A lower proportion of GWD being traded relative to all crop water use can be because countries with larger national GWD (e.g., the MENA region, India) export their crops relatively less than other producers (e.g., France or Brazil) do. Importantly, this volume of virtual GWD trade is growing rapidly over time: Dalin et al. (2017) found that it increased by 44% from 2000 to 2010 globally, with even faster increases in some regions, such as the exports from India (doubling in 10 years) and the imports by China (tripling in 10 years). Because of the concentration of GWD in a few areas, the virtual trade of GWD is much more heterogeneous than the total VWT (see, e.g., Konar et al., 2011). Indeed, GWD exchanges via food are dominated by exports from only three countries: Pakistan (33% of total GWD trade flows), the United States (25%), and India (10%) (Fig. 25.4). On the receiving end

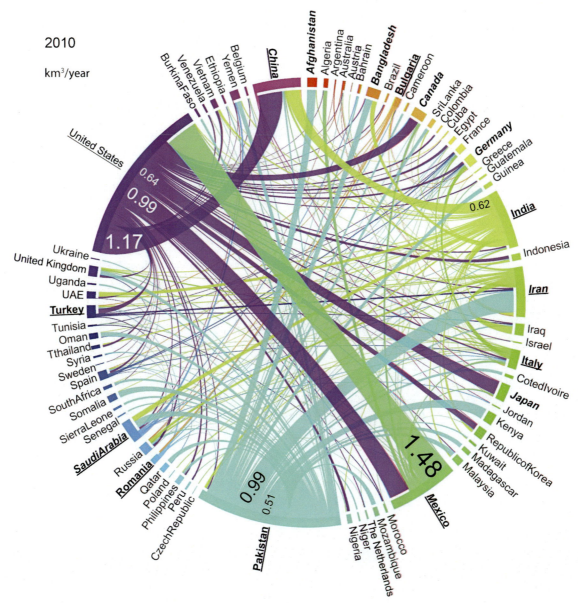

FIGURE 25.4 Embedded groundwater depletion in the international trade of crop commodities in 2010. Flows are in units of cubic kilometers per year. The top 10 importers are shown in bold font and the top 10 exporters are underlined. Ribbon colors indicate the country of export. *Reproduced with permission from Dalin, C., Wada, Y., Kastner, T., Puma, M., 2017. Groundwater depletion embedded in international food trade. Nature, 543, 700–704.*

of these GWD trade flows, China is the largest net importer (10% of the total flows), and other key importers include Iran, Mexico, and the United States (Dalin et al., 2017).

Several countries—such as the United States, Mexico, China, Iran, and Saudi Arabia—were found to be both top exporters (i.e., producers and exporters of crops irrigated from overexploited aquifers) and top importers (i.e., dependent on foreign crops also produced unsustainably; Dalin et al., 2017). This makes these countries exposed to risks to their food supply both domestically and via imports. In addition, it was found that most countries in the world receive large shares of their food imports from countries that irrigate those crops with water from overexploited aquifers. For example, approximately 90% of sunflower trade and 77% of maize trade originate from countries using overexploited aquifers to grow these crops (Dalin et al., 2017).

A limitation of this study, like all global VWT studies to date, is that the GWD content for each commodity needs to be averaged at the national level (even if it was computed on, e.g., a 10 by 10 km grid) before being combined with the bilateral international trade data, which is only available at the national scale worldwide. This limitation means that the food exports from one country may come from one producing region and not another (growing the same crop with a distinct GWD content), but this information is not available for global-scale studies.

In a follow-up study to Dalin et al. (2017), Gumidyala et al. (2020) used this approach and groundwater data applied to the United States for years 2002 and 2012. They combined estimates of GWD of crops produced in the United States with a dataset on state-level domestic and foreign transfers to evaluate the amount of nonrenewable groundwater embedded in the agricultural commodities transferred domestically and exported internationally. Although this study covers only one producing country (albeit a major global food producer and exporter), it provides finer scale estimates of the transfers of agricultural commodities produced from overexploited aquifers in the United States. Gumidyala et al. (2020) found that 26.3 km^3 of nonrenewable groundwater was transferred domestically in 2002 and 2.7 km^3 was sent abroad (Fig. 25.5A,C,E). In 2012 34.8 km^3 was transferred domestically (a 32% increase in 10 years) and 3.7 km^3 was exported (a 38% increase in 10 year, Fig. 25.5B,D,F). Over this period the agricultural commodity group with the highest increase in total domestic GWD transfers was cereal grains, with a 58.5% increase. Animal products group shows the highest increase in total GWD foreign exports (by 144.4%).

Among the country's states, California has the largest agricultural production, leading it to also have the largest total volume of GWD, despite not having the largest GWD per unit commodity (i.e., other states produce more

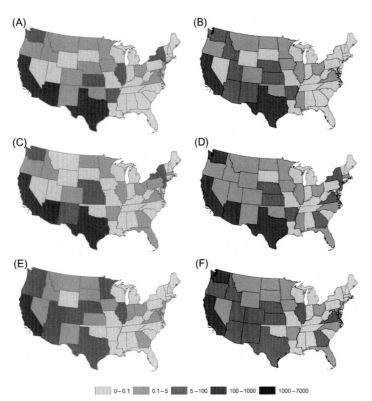

FIGURE 25.5 Maps of groundwater depletion exports from the United States. Exports of groundwater depletion (10^{6}m^3/year) are provided at the state spatial scale and for each agricultural commodity class considered in this study. Panels (A, B) show groundwater depletion exports of grains, panels (C, D) the groundwater depletion exports of fresh produce, and panels (E and F) the groundwater depletion exports of animal feed (SCTG 4). Panels (A), (C), and (E) show groundwater depletion exports in 2002. Panels (B), (D), and (F) show groundwater depletion exports in 2012. *Reproduced with permission from Gumidyala S., Ruess P., Konar M., Marston L., Dalin C. and Wada Y., Groundwater depletion embedded in domestic transfers & international exports of the United States. Water Resources Research 56. 2020.*

GWD-intensive products). However, California actually uses most of its own GWD. In fact, all of the major GWD transfer states retain the majority of their GWD, meaning the commodities remain inside the state where they are produced. It is important to note that only raw crop products and animal feed are included in this study. These products are often sourced locally as input into higher value products (i.e., meat, textiles, and processed foods), which may then be shipped elsewhere for final consumption. The importance of GWD to the California economy is consistent with other studies (Marston and Konar, 2017; Marston et al., 2015). After California [13.1 and 1.8 km^3 GWD embedded in domestic outflows and in foreign exports in 2012, respectively (Fig. 25.5)], the largest GWD exporting state is Texas (in 2012 4.7 and 0.9 km^3 of GWD embedded in domestic and foreign outflows, respectively, Fig. 25.5). The regions receiving most GWD via imports of the food commodities considered in 2012 from the United States are East Asia (1.62 km^3), Canada (0.57 km^3), and Mexico (0.44 km^3). These broad findings are consistent with those from Dalin et al. (2017) for year 2010, as shown in Fig. 25.4: the largest foreign exports of GWD via food from the United States go to, in order of importance, China, Mexico, Japan, Canada, and South Korea.

These studies of VWT focused on GWD (Dalin et al., 2017; Gumidyala et al., 2020) confirm the findings from the general VWT literature that water resources have become globalized. They show how the critical issue of GWD for irrigation can impact food systems worldwide: it affects not only producing countries but also, via trade linkages, food-importing regions.

These trade connections add a layer of complexity to the task of establishing future projections of groundwater-depletion globally. Such projections require to understand where future food will be grown, how much groundwater it will consume (sustainability or not), and they also require to project future trade patterns, and how potential shocks in production or prices due to groundwater exhaustion will propagate throughout the global food system (Puma et al., 2015).

25.2 Conclusion

This chapter has analyzed the place of groundwater and groundwater sustainability in the recent scientific literature, both in the area of VWT via food trade, which highlights a key global characteristics of otherwise largely local hydrological systems, and in the area of agricultural water use sustainability more generally.

From this discussion, one can draw the following key conclusions:

The first key conclusion concerns spatial scale at which groundwater is considered. This chapter shows that groundwater needs to be considered both at the local scale and at the regional and global scale. At the local scale, it is important to understand the local context affecting groundwater use and management. At the regional and global scales, wider drivers of groundwater use, such as climate change—which is affecting recharge rates, crop water requirements, and surface water availability—and, even less frequently considered, the global demand for food products.

The second key conclusion is that future research is needed at both of these scales to *improve our understanding of groundwater processes*. VWT studies and water sustainability metrics have only started to consider groundwater. Some future research needs to be done in terms of data collection and modeling approaches to better understand groundwater use and its socioeconomic and environmental sustainability and its ecological consequences. Further advances are required, in particular, to evaluate the links between aquifers and the local water cycle and ecosystems, and to quantify the effects of groundwater abstraction on the provision of ecosystem services, including via natural streamflow.

Last but not least, along with the necessary scientific efforts, it has become increasingly apparent that, while groundwater mismanagement is a growing concern in the scientific community, it is crucial to further raise awareness of these issues in other communities—including food consumers, food growers and retailers, and policy-makers. Integrating groundwater into *governance frameworks* is essential to achieve several of the United Nations' *Sustainable Development Goals*, including the goals of ending hunger, providing water security, ending poverty, and protecting biodiversity, which all rely on adequately managed water resources. While some recent crises have changed local-level agriculture or urban planning policies, the concept of groundwater management is absent from most policies and laws globally. This is due, among other issues, to challenges in defining groundwater rights and in accurately monitoring groundwater flows and stocks.

The knowledge of groundwater and the development of policies need to advance in parallel, with the aim to eventually enable the internalization of environmental externalities of the activities consuming groundwater. Unlike for many resources, there is no substitute for freshwater. Ways around this include desalination or improvements in water use efficiency. However, desalination is currently only effective in coastal regions because water volumes are too large to be cost-effectively transported over long distances, and the desalination process requires very large energy consumption and creates toxic waste. One suggested perspective to *better integrate groundwater sustainability in management from the local to the global context* is a revisited planetary boundary, which "could complement existing tools for water-resource management by offering a unique approach for assessing water-cycle modifications as part of the wider human impact on the Earth System" (Gleeson et al., 2020B).

Financial support

C. Dalin acknowledges the funding support of the UK Natural Environment Research Council (NERC Independent Research Fellowship NE/N01524X/1).

References

Aeschbach-Hertig, W., Gleeson, T., 2012. Regional strategies for the accelerating global problem of groundwater depletion. Nat. Geosci. 5, 853.

Alcamo, J., Henrichs, T., Rosch, T., 2000. World Water in 2025. World Water Series Report 2. Center for Environmental Systems Research University of Kassel. Global Sustainability 9.

Allan, J.A., 1997. 'Virtual Water': A Long Term Solution for Water Short Middle Eastern Economies? Occasional Paper 3. School of Oriental and African Studies (SOAS), University of London.

Alley, W., Alley, R., 2017. High and Dry: Meeting the Challenges of the World's Growing Dependence on Groundwater. Yale University Press.

Alley, W.M., Reilly, T.E., Franke, O.L., 1999. Sustainability of GroundWater Resources. US Geological Survey Circular 1186. US Geological Survey.

Befus, K.M., Jasechko, S., Luijendijk, E., Gleeson, T., Bayani Cardenas, M., 2017. The rapid yet uneven turnover of Earth's groundwater. Geophys. Res. Lett. 44, 5511–5520.

Boulay, A.-M., Bouchard, C., Bulle, C., Deschenes, L., Margni, M., 2011. Categorizing water for LCA inventory. Int. J. Life Cycle Assess. 16, 639–651.

Brauman, K.A., Richter, B.D., Postel, S., Malsy, M., Florke, M., 2016. Water depletion: an improved metric for incorporating seasonal and dry-year water scarcity into water risk assessments. Elementa: Sci. Anthropocene 4, 000083.

Dalin, C., Konar, M., Hanasaki, N., Rinaldo, A., Rodriguez-Iturbe, I., 2012. Evolution of the global virtual water trade network. Proc. Natl. Acad. Sci. U.S.A. 109, 5989–5994.

Dalin, C., Wada, Y., Kastner, T., Puma, M., 2017. Groundwater depletion embedded in international food trade. Nature 543, 700–704.

de Graaf, I.E.M., van Beek, R.L.P.H., Gleeson, T., Moosdorf, N., Schmitz, O., Sutanudjaja, E.H., et al., 2017. A global-scale two-layer transient groundwater model: development and application to groundwater depletion. Adv. Water Resour. 102, 53–67.

Döll, P., 2009. Vulnerability to the impact of climate change on renewable groundwater resources: a global-scale assessment. Environ. Res. Lett. 4, 035006.

Eshel, G., Shepon, A., Makov, T., Milo, R., 2014. Land, irrigation water, greenhouse gas, and reactive-nitrogen burdens of meat, eggs, and dairy production in the United States. Proc. Natl. Acad. Sci. U.S.A. 111, 11996–12001.

Falkenmark, M., 1989. The massive water scarcity now threatening Africa: why isn't it being addressed? Ambio 18, 112–118.

Falkenmark, M., 2013. Growing water scarcity in agriculture: future challenge to global water security. Philos. Trans. R. Soc. Lond. A: Math., Phys. Eng. Sci. 371, 20120410.

Gleeson, T., Cuthbert, M., Ferguson, G., Perrone, D., 2020a. Global groundwater sustainability, resources, and systems in the anthropocene. Annu. Rev. Earth Planet. Sci. 48, 1.

Gleeson, T., et al., 2020b. The water planetary boundary: interrogation and revision. One Earth 2 (3), 223–234.

Gleeson, T., Befus, K.M., Jasechko, S., Luijendijk, E., Cardenas, M.B., 2016. The global volume and distribution of modern groundwater. Nat. Geosci. 9, 161–167.

Gleeson, T., Wada, Y., Bierkens, M.F.P., van Beek, L.P.H., 2012. Water balance of global aquifers revealed by groundwater footprint. Nature 488, 197–200.

Green, T.R., Taniguchi, M., Kooi, H., Gurdak, J.J., Allen, D.M., Hiscock, K.M., et al., 2011. Beneath the surface of global change: impacts of climate change on groundwater. J. Hydrol. 405, 532–560.

Gumidyala S., Ruess P., Konar M., Marston L., Dalin C. and Wada Y., 2020. Groundwater depletion embedded in domestic transfers & international exports of the United States. Water Resources Research 56.

Hanasaki, N., Inuzuka, T., Kanae, S., Oki, T., 2010. An estimation of global virtual water flow and sources of water withdrawal for major crops and livestock products using a global hydrological model. J. Hydrol. 384, 232–244.

Hanasaki, N., Kanae, S., Oki, T., Masuda, K., Motoya, K., Shirakawa, N., et al., 2008. An integrated model for the assessment of global water resources – Part 1: model description and input meteorological forcing. Hydrol. Earth Syst. Sci. 12, 1007–1025.

Hoekstra, A.Y., Mekonnen, M.M., 2016. Imported water risk: the case of the UK. Environ. Res. Lett. 11, 055002.

Hoekstra, A.Y., Mekonnen, M.M., Chapagain, A.K., Mathews, R.E., Richter, B.D., 2012. Global monthly water scarcity: blue water footprints versus blue water availability. PLoS One 7, e32688.

Konar, M., Dalin, C., Suweis, S., Hanasaki, N., Rinaldo, A., Rodriguez-Iturbe, I., 2011. Water for food: the global virtual water trade network. Water Resour. Res. 47, W05520.

Konar, M., Dalin, C., Hanasaki, N., Rinaldo, A., Rodriguez-Iturbe, I., 2012. Temporal dynamics of blue and green virtual water trade networks. Water Resour. Res. 48, W07509.

Konikow, L.F., 2011. Contribution of global groundwater depletion since 1900 to sea-level rise. Geophys. Res. Lett. 38, L17401.

Konikow, L.F., Kendy, E., 2005. Groundwater depletion: a global problem. Hydrogeol. J. 13, 317−320.

Kounina, A., Margni, M., Bayart, J.-B., Boulay, A.-M., Berger, M., Bulle, C., et al., 2013. Review of methods addressing freshwater use in life cycle inventory and impact assessment. Int. J. Life Cycle Assess. 18, 707−721.

Maimone, M., 2004. Defining and managing sustainable yield. Groundwater 42, 809−814.

Marston, L., Konar, M., Cai, X., Troy, T.J., 2015. Virtual groundwater transfers from overexploited aquifers in the United States. Proc. Natl. Acad. Sci. U.S.A. 112, 8561−8566.

Marston L. and Konar M., 2017. Drought impacts to water footprints and virtual water transfers of the Central Valley of California. Water Resources Research.

Mekonnen, M.M., Hoekstra, A.Y., 2011. The green, blue and grey water footprint of crops and derived crop products. Hydrol. Earth Syst. Sci. 15, 1577−1600.

Mekonnen, M., Hoekstra, A., 2016. Four billion people facing severe water scarcity. Sci. Adv. 2, e1500323.

Mukherjee A., Scanlon B., Aureli A., Langan S., Guo H., and McKenzie A., 2020. Global Groundwater: Source, Scarcity, Sustainability, Security and Solutions. Elsevier, 1st. Ed., ISBN. 9780128181720.

OECD-FAO, 2012. OECD FAO Agricultural Outlook 2013−2022. OECD, Paris. Available from: www.oecd.org/site/oecd-faoagriculturaloutlook/highlights-2013-EN.pdf.

Pfister, S., Bayer, P., Koehler, A., Hellweg, S., 2011. Environmental impacts of water use in global crop production: hotspots and trade-offs with land use. Environ. Sci. Technol. 45, 5761−5768.

Puma, M.J., Bose, S., Chon, S.Y., Cook, B.I., 2015. Assessing the evolving fragility of the global food system. Environ. Res. Lett. 10, 024007.

Rodell, M., Famiglietti, J.S., Wiese, D.N., Reager, J.T., Beaudoing, H.K., Landerer, F.W., et al., 2018. Emerging trends in global freshwater availability. Nature 557, 651−659.

Rost, S., Gerten, D., Bondeau, A., Lucht, W., Rohwer, J., Schaphoff, S., 2008. Agricultural green and blue water consumption and its influence on the global water system. Water Resour. Res. 44, W09405.

Scanlon, B.R., Faunt, C.C., Longuevergne, L., Reedy, R.C., Alley, W.M., McGuire, V.L., et al., 2012. Groundwater depletion and irrigation sustainability in the US High Plains and Central Valley. Proc. Natl. Acad. Sci. U.S.A. 109, 9320−9325.

Scherer, L., Pfister, S., 2016. Dealing with uncertainty in water scarcity footprints. Environ. Res. Lett. 11, 054008.

Seckler, D., Barker, R., Amarasinghe, U., 1999. Water scarcity in the twenty-first century. Int. J. Water Resour. Dev. 15, 29−42.

Shiklomanov, I.A., 1997. Assessment of Water Resources and Water Availability in the World. SEI and WMO.

Siebert, S., Döll, P., 2010. Quantifying blue and green virtual water contents in global crop production as well as potential production losses without irrigation. J. Hydrol. 384, 198−217.

Siebert, S., Burke, J., Faures, J.M., Frenken, K., Hoogeveen, J., Doll, P., et al., 2010. Groundwater use for irrigation − a global inventory. Hydrol. Earth Syst. Sci. 14, 1863−1880.

Taylor, R.G., Scanlon, B., Döll, P., Rondell, M., van Beek, R., Wada, Y., et al., 2013. Groundwater and climate change. Nat. Clim. Change 3, 322−329.

Tuninetti, M., Tamea, S., Dalin, C., 2019. Water debt indicator reveals where agricultural water use exceeds sustainable levels. Water Resour. Res. 55, 2464−2477.

Vanham, D., Hoekstra, A.Y., Wada, Y., Bouraoui, F., de Roo, A., Mekonnen, M., et al., 2018. Physical water scarcity metrics for monitoring progress towards SDG Target 6.4: an evaluation of indicator 6.4.2 'Level of water stress'. Sci. Total Environ. 613, 218−232.

Vörösmarty, C.J., Green, P., Salisbury, J., Lammers, R.B., 2000. Global water resources: vulnerability from climate change and population growth. Science 289, 284−288.

Wada, Y., Bierkens, M.F.P., 2014. Sustainability of global water use: past reconstruction and future projections. Environ. Res. Lett. 9, 104003.

Wada, Y., Reager, J.T., Chao, B.F., Wang, J., Lo, M.-H., Song, C., et al., 2017. Recent changes in land water storage and its contribution to sea level variations. Surv. Geophys. 38, 131−152.

Wada, Y., van Beek, L.P., van Kempen, C.M., Reckman, J.W., Vasak, S., Bierkens, M.F., 2010. Global depletion of groundwater resources. Geophys. Res. Lett. 37, L20402.

Wada, Y., van Beek, L., Bierkens, M., 2012. Nonsustainable groundwater sustaining irrigation: a global assessment. Water Resour. Res. 48, W00L06.

Yano, S., Hanasaki, N., Itsubo, N., Oki, T., 2015. Water scarcity footprints by considering the differences in water sources. Sustainability 7, 9753−9772.

Yano, S., Hanasaki, N., Itsubo, N., Oki, T., 2016. Potential impacts of food production on freshwater availability considering water sources. Water 8, 163.

Chapter 26

Groundwater and society: enmeshed issues, interdisciplinary approaches

Flore Lafaye de Micheaux[1,2,3] and Mukherjee Jenia[4]

[1]*University of Lausanne, Lausanne, Switzerland,* [2]*International Union for Conservation of Nature, Gland, Switzerland,* [3]*French Institute of Pondicherry, Puducherry, India,* [4]*Indian Institute of Technology, Kharagpur, India*

26.1 Introduction

The groundwater crisis is far from being homogeneous across the planet. As for millennia, the spatial distribution of aquifers is uneven due to physical features such as climate and precipitation, geologic formations, and rocks and soils. However, human-induced groundwater overuses and various anthropogenic impacts on soils, surface waters, or landscapes lead to the extension of groundwater shortage and issues of pollution. In addition, they exacerbate the variability of water tables, already affected by climate change (Mukherjee et al., 2020). Reversely, these trends affect human uses of groundwater, agricultural practices, as well as livelihoods, health and well-being, economic patterns, governance systems and more generally societies, at local but also country and international levels.

Several social studies of water highlight the reciprocal making of water and society, in particular within the hydrosocial literature. The concept of hydrosocial cycle, developed in political ecology of water, calls attention to "socio-natural process by which water and society make and remake each other, over space and time" (Linton and Budds, 2014, 175). In contrast to the hydrological cycle, it insists on the internal link between water and society, with the former being not merely "physical water," but also "discursive water" or in an approximation, how people and institutions speak, think and know about water. Other approaches also aim to address the complexity of the water−society interactions, here groundwater, as further presented in this chapter. They apply cross-disciplinary perspectives to unpack these processes acknowledging the significance of interdisciplinarity for conducting research in this domain (Evers et al., 2017; Rusca and Di Baldassarre, 2019; Wesselink et al., 2017). As emphasized at the occasion of the recent merge of the International Council for Science and the International Social Science Council to form the International Science Council, "the importance of scientific understanding to society has never been greater, as humanity grapples with the problems of living sustainably and equitably on planet Earth."[1]

This chapter aims to introduce the reader to key bodies of literature that address the links between groundwater and society. The objective is to contribute to a better acknowledgment of social realities that build and intersect with legislation, governance, resource management, and implementation of sustainable initiatives, which reversely transform physical dimensions of aquifers. Sharing insights from socio-hydrology and hydrosocial case studies, the chapter sheds light on unexpected outcomes of implementing approaches that may overlook knowledge−power relations, uncertain causal chains or politics of scales. Finally, the authors argue that these "social" approaches call attention to meanings, values, and paradigms that significantly shape the hydrosocial cycle. A comprehensive understanding comprising these multi-layered and complex approaches is imperative to inform policy formulation geared toward a just, democratic, and resilient future.

The chapter has been structured in four sections. Section 26.2 presents socio-hydrology and socio-geohydrology approaches, with insights from applied cases around groundwater. These disciplines mainly rely on physical sciences and quantitative modeling, accounting for the mutual interactions between water and society. Section 26.3 introduces the hydrosocial cycle concept, emphasizing the specificity of the epistemological lens that hydrosocial analyses use,

---

1. International Science Council website: accessed on 30 March 2020, <https://council.science/publications/high-level-strategy/>.

such as power equations involved in water management and governance. Section 26.4 develops several hydrosocial case studies, highlighting what can be concretely learned from a relational and power-laden approach to water and society. Finally, Section 26.5 discusses how to think and conduct interdisciplinary research in the context of groundwater.

26.2 Socio-hydrology and socio-geohydrology: modeling of the groundwater−society interactions improved with stakeholders' perspectives

26.2.1 Introduction to socio-hydrology

Socio-hydrology emerged in the 2010s as a "new science of people and water" (Sivapalan et al., 2012, 1270), within the broader literature related to socio-ecological systems (Troy et al., 2015). In 2009 Elinor Ostrom characterized socio-ecological systems as a combination of resource, user, and governance subsystems, among which complex interactions prevail, particularly nonlinear ones (Ostrom, 2009). Socio-hydrology seeks to model such complexity in water context to study trends and changes and do predictions. This literature, in particular, aims to "understand the co-evolution of human and water systems and thus posits that a two-way coupling exists between these systems" (Troy et al., 2015, 3668).

As pointed by Sivapalan et al. (2014), conventional approaches to water sustainability tend to overlook the water−society two-way interactions, feedback loops, and threshold behaviors. This undermines their explanatory and predictive power, for instance in the three following paradoxes: the virtual water trade that persists to entail "irrational water resource outcomes"; efficiency-driven initiatives that lead to increasing water demand, up to further groundwater depletion in a Mexican case; and the decoupling of economic growth and increasing water demand such as in the Murrumbidgee basin (Australia) due to changing social norms (Sivapalan et al., 2014, 227−228). Therefore in order to improve predictions about water, socio-hydrology invites hydrogeologists to incorporate human activities within the water cycle. The human factor is not anymore seen as a mere external forcing (Di Baldassarre et al., 2013; Sivapalan et al., 2012; Troy et al., 2015). Elshafei et al. (2014), for example, develop a framework that identifies six key functional variables of human−water feedbacks, such as catchment hydrology, population, economics, environment, socioeconomic sensitivity, and collective response (p. 2141).

With the hydrological science as a background the methods of socio-hydrology often rely on quantitative or mathematical modeling approaches to represent the water−human relationship. Case studies elaborate coupled human−water models, using approaches varying from exploitation of empirical data such as human population dynamics, hydrological data, remote-sensing, ethnographic surveys to historical analysis, statistical analysis, and systems of differential equations (Di Baldassarre et al., 2013; Troy et al., 2015). A review of 69 socio-hydrological papers in 2017 highlights that many cases develop a mathematical model based on coupled, nonlinear differential equations (Wesselink et al., 2017). However, in some cases, only one-way influence can be modeled such as the effect of land use change on hydrologic flows, because of the complexity of water−society dynamics (see Troy et al., 2015 for a review).

26.2.2 Socio-hydrology and groundwater

The difficulty for socio-hydrological research to address the entanglement of social and water systems is particularly high in the case of groundwater resources. Human uses of groundwater often intervene within broader patterns of water consumption, while often being regulated and managed independently from surface waters.

In some socio-hydrological cases, researchers choose to ease the incorporation of groundwater processes in using a simple water balance approach. Zhou et al. (2015), for instance, design a "socio-hydrological water balance" that they apply to a catchment of the Murray−Darling River basin, Australia over a 100-year period. The proposed calculation of water balance includes changes in groundwater storage, as well as soil and surface water storage. Interestingly, the authors differentiate evapotranspiration according to ecological and social systems, with ecological evapotranspiration reflecting the one from precipitation, surface runoff and groundwater in native vegetation areas, while social evapotranspiration reflects evapotranspiration from croplands, grasslands, and water used by households and industries. This study of the socio-hydrological water balance help to characterize historical changes in water storage and water allocation toward either societal uses or ecosystem needs. Zhang et al. (2014) also use a socio-hydrological approach of water balance analysis to demonstrate the significant alteration of water exchanges and groundwater trends induced by water-saving practices of irrigation in an oasis of the Tarim River basin, China.

Some socio-hydrological models encompass a greater range of parameters of the human−water coupling. The researchers' purpose is then to identify the key ones on which managers should focus their actions, in order to achieve the desired outcomes. In this regard, Elshafei et al. (2015) develop a socio-hydrological model in the context of land use change and soil salinization entailed by rising water tables, in a semiarid area of Australia. The model include parameters linked to population, agricultural economic gain, life-style ecosystem services—with the proxy of the proportion of degraded land, community sensitivity state to water availability, and behavioral response (Elshafei et al., 2015). The application of this framework in the Lake Toolibin subcatchments is successful in isolating the behavior of the feedbacks of land use management (human system feedback) and land degradation (natural system feedback), and understanding some threshold behavior characteristics (Elshafei et al., 2015, 6464).

26.2.3 Incorporating stakeholders' perspectives: a "public" turn for socio-hydrology

Some socio-hydrological approaches emphasize the need for groundwater-related research and decisions to take into account stakeholders' perspectives, in contrast to the previous approaches that ultimately seek to refine the mathematical modeling, for "more science." Hund et al. (2018), for instance, propose a groundwater recharge indicator that they build through several exchanges with local communities and water managers in a field in Costa-Rica. This indicator aims to help decision-makers in their efforts to improve the resilience of communities to droughts, particularly seasonal ones. This approach illustrates the need to "think prediction differently" that Srinivasan et al. (2016) call for, while reflecting on the socio-hydrology discipline and its methodological and epistemological challenges. In this paper the authors emphasize that models and scenarios should not prioritize perfect accuracy of quantitative water accounts. They instead should focus on presenting "alternative, plausible and coevolving trajectories of the socio-hydrological system," to help social actors to access portfolios of options and adaptive responses and to better identify "safe operating space" for their actions (Srinivasan et al., 2016, 340–341). It may provoke and enable socio-hydrologists to gather "unconventional data" sources such as citizen science, Big data, or participatory monitoring to bridge the gap in data "that matters," instead of the ones that are usually easy to collect and to model.

In a similar vein an approach of *socio-hydrogeology* has emerged from socio-hydrology to address the specific challenges of integrated and sustainable groundwater management (Re, 2015). Socio-hydrogeologists are expected to advocate for stakeholder engagement in water-management decisions, while bringing their skills and knowledge of trained and experimented hydrogeologists (Hynds et al., 2018; Re, 2015). The rationale is that groundwater, in contrast to surface waters, is often managed and used at an individual level (private well owners, small farmers using groundwater-based irrigation, etc.), and not in an institutional and structured manner. Therefore finely tuned participatory approaches are required to implement the integrated water resources management principles. For example, Hynds et al. (2018) propose to precede socio-hydrogeological investigations with stakeholder network analysis. These efforts will also allow sustaining bottom-up management on longer term (Re, 2015).

In addition, and in contrast to the initial definition of socio-hydrology (Sivapalan et al., 2012), socio-hydrogeology encompasses the following three objectives that go beyond reflecting the mutual relationships between (ground)water and society (Re, 2015):

1. profiling the research questions in relation to the needs and potentially conflictual interests of stakeholders;
2. improving communication and cooperation between scientists and local communities through "demystifying science and scientists" for better knowledge sharing and appropriation;
3. creating knowledge and advice that are not only relevant and credible but also legitimate in regard to the values of the communities, even while they diverge.

The following sections aim to present another body of literature in social sciences and human geography that addresses these dimensions while bringing new angles in the analyses, that is, political ecology of water and the hydrosocial literature.

26.3 Political ecology and the hydrosocial cycle: paying attention to power relations and discourses embedded in water circulation

26.3.1 Political ecology of water

Political ecology is a field that critically interrogates the nature—society relations, particularly looking at the power relations that intersect and affect access to natural resources, in order to reveal disparities and injustices in the distribution of costs and benefits (Robbins, 2012). It emerged in the 1980s to address land degradation and development/

environmental issues, with a proposed combination of political economy and ecology (Blaikie and Brookfield, 1987). The discipline mainly scrutinizes specific fields and communities, while replacing local environmental change in broader sets of economic or political patterns, such as the ones led by capitalist or neoliberal decisions. Several studies also develop to deconstruct unquestioned perspectives related to environmental problems, including knowledge, in order to demystify self-interested strategies of dominant social groups. Political ecology borrows its methods and analytical tools from various disciplines such as geography, anthropology, political economy, political sciences, social sciences, history, or environmental sciences.

Hydrosocial analyses aim to reveal how power infuses the water—society connections and how water shapes and is shaped by society. The hydrosocial cycle concept emerged in the 2000s in the field of political ecology of water. This strand of political ecology emphasizes the political dimension of water—society relations. For example, a body of literature highlights the gaining strength over the 20th century of a state-led "hydraulic mission" that imposed a hegemonic paradigm of "river control" relying on dams and levees, which, in turn, fueled the administration's social and political power (Baghel, 2014; Molle, 2009; Swyngedouw, 2007). Other works effectively investigate the social distribution of benefits and costs in the context of interwoven perimeters of river-management policies (e.g., Matthews, 2012; Molle, 2005; Vogel, 2012) or in the distribution and access to urban potable water (e.g., Kaika, 2003; Swyngedouw et al., 2002).

26.3.2 The hydrosocial cycle: a critical rethinking of "water"

In addition, political ecology of water adopts a critique of approaches to water that overlook water's social construction. For instance, a river today cannot be merely interpreted as a natural object as it all together encompasses the physical transformations brought by humans, as well as their discourses on "what a river should be." Similarly, the "hydrosocial cycle," in contrast to the "hydrological cycle," aims to reflect the material as well as discursive coproduction of water and society (Bakker, 2000; Swyngedouw et al., 2002; Linton and Budds, 2014). Therefore in such approach, the hegemonic "meanings of water" or the dominant discourses, ideas or representations of water are the object of scholarly attention (Linton and Budds, 2014, 176). The ultimate goal of such analysis, that is common to poststructural political ecology, is to deconstruct how and by whom a "water problem" has been defined, as posing this problem as a given may entail further social and environmental injustices (Lafaye de Micheaux , 2019).

Finally, the hydrosocial cycle concept proposes a renewed understanding of "what water is." It draws on the "socionature" conceptualization, a neologism coined by Geographer Erik Swyngedouw that offers a relational approach of nature—society interactions (Castree et al., 2013; see also Swyngedouw, 1999, 2003). In the "socionature" perspective, society and nature are hybrids, internally connected and coproduced. This approach considers nature as coconstructed by humans, but it also points to the agency "i.e., the capacity to act, even without intentionality" of nonhuman entities, drawing from Bruno Latour's work on the concept of "actant" in the actor—network theory (Lafaye, 2019, 19). In this regard, some authors argue that the meanings of water are not somehow "attached" by humans to a physical reality, but that they emerge from the water—human interactions in which the agency and the materiality of water play a role (Strang, 2004; Krause and Strang, 2016). In the same vein the hydrosocial approach considers the mutual construction of water and society as iteration cycles through an *internal* and *dialectical* relationship, that is, as one unique system led by processes belonging to different spatial and temporal scales, mixing systems and subsystems, and that produce changing sets of configurations that are continuously reshaped/disrupted (Linton and Budds, 2014).

As shown in the next section, the hydrosocial lens may help to capture social phenomena and stakeholders' perspectives that may otherwise remain invisible, as groundwater often is, though they contribute to shape water uses, policies, and responses to policies. The hydrosocial lens is also instrumental in capturing the heterogeneity of water along its material, discursive, and symbolic aspects.

26.4 Mobilizing hydrosocial analyses to capture ground (water) realities

When political ecological and hydrosocial analyses are extended in the domain of groundwater, Bear and Bull's (2011, 2262) argument "politics played out not around water, but through water and often driven by water" assumes greater significance, especially within the context of the Global South. This is largely validated in empirical investigations conducted by Birkenholtz (2008, 2016) and Sultana (2013) in India and Bangladesh, as well as by other authors in South America (Da Silva and Hussein, 2019), up to the developed world (Fernandez, 2014).

26.4.1 Dispossession of irrigating farmers through institutions and infrastructures

Birkenholtz (2016) advances water political ecology on "accumulation by dispossession" and "water grabbing" by capturing tensions between statist accumulation and dispossession of irrigating farmers. Focusing on Jaipur, the capital of Rajasthan and the rural dam-reservoir complex of the Tonk District, the work studies unequal water transfer dynamics between urban infrastructure development project (meant for extra-economic uses: domestic, commercial and industrial) and rural irrigation reservoir (manifesting surface water–groundwater interactions). It brings to the fore complex processes of alienation of water from peasants, finally resulting into peasant resistance. The study examines the multifaceted processes of dispossession through the reallocation of water from irrigation to new centers of capital, which produce higher marginal returns and cost recovery in the state. Based on empirical findings evolved out of scoping analysis of policy documents and field ethnography, Birkenholtz (2016) concludes that "accumulation by dispossession" should be understood as a process that combines institutions and infrastructure to effectively dispossess farmers.

26.4.2 State and "scientific" versus local knowledge of water

Political ecology and hydrosocial approaches are also crucial to unravel power relations in scientific knowledge and discourse formation. Birkenholtz (2008) deploys the historical political ecological analysis of water to trace coevolution of knowledge and technologies analyzed across long-term temporal scales (precolonial, colonial and postindependence periods) in Rajasthan, an Indian State where the situation of overexploitation of groundwater is grim. In Rajasthan in 2008, there were over 1.4 million tube wells for agricultural purpose (out of total 20 million tube wells in India) and groundwater extraction currently surpassed recharge by 410 million cubic meters per year. Birkenholtz demonstrates how the binary state actions versus local responses have shaped the present situation. "Colonial hydrology" (D'Souza, 2006) was the major moment of departure when the state intervention was formalized through the implementation of groundwater lifting technologies and irrigation bureaucracies and its continued legacy during the contemporary times. However, while the state attempted to "assimilate, reorganize, plagiarize, or disparage local knowledge" in-between historical moments, farmers sought out other nonstatist channels such as tube well drilling firms. In addition, Birkenholtz (2008) reveals the heterogeneities of local communities, as he explores constellations of power among array of stakeholders also including the local Hindu water diviners (*Sunghas*). He particularly shows how animosities among differently situated actors relating to the source and reliable availability of groundwater and its practitioners continue, negatively impacting strategies and scenarios of groundwater governance in the near future.

Applying the hydrosocial analysis in a far different context of a French river basin, Fernandez (2014) also reveals the political processes behind the scientifically produced "Minimum Flow Requirements" and the related decisions surrounding water control using the Garonne River as the case study. Minimum flow requirements are presented as the outcomes of objective hydrological models, including groundwater circulation and water use data. However, through historical investigation, Fernandez (2014) shows how they "have been black-boxed to become the 'natural and desired' state of a river," while they are the product of coalitions formed by situated actors in order to control water resources (p. 269).

26.4.3 Groundwater and politics of scale

Other hydrosocial works point to the "politics of scale" in the context of groundwater. Da Silva and Hussein's (2019) work on the La Plata River Basin and Guarani Aquifer System, South America provides detailed insights on the production of scales and interactions among them. The authors establish the need to investigate scale in hydropolitics, that is, who are involved in scale construction, how do they perform this complex exercise, and why. The authors also study the implications of scales for hydropolitical dynamics across regional levels. Here, the spatial and political scales get extended from single case of conflict or collaboration over water resources to multiple historical and political interactions involving heterogeneous water and social actors. Particularly, some research findings drawn in the 1990s had a lasting impact on the production of a regional scale (from the previously operational local scale). These findings highlighted that the Guarani groundwater resources were part of a unified and integrated aquifer system. They made the interconnection between overexploitation and pollution in one country and its spiral implications on other three countries, clear. This new knowledge triggered the development of new modes of governance across four South American countries: Argentina, Brazil, Paraguay, and Uruguay, such as issuing regional policies, through treaties, and agreements, of groundwater monitoring. In particular, such monitoring targeted water extraction in Brazil for industries, Paraguay for irrigation, and Argentina and Uruguay for thermal tourism. However, the authors depict how Brazilian authorities made use of specious forms of power to establish their control over the Guarani Aquifer System. They bargained to add clauses or delayed ratification process according to their national interest; they took advantage of geography (a major part of the Guarani Aquifer System lies within the Brazilian territory) and of Brazil's economic, geopolitical, and military prominence to impose Brazilian control over the regional resources.

The Guarani Aquifer System example is an interesting yet complex manifestation of positional shifts between local and regional scales at crucial historical and political moments shaping environmental governance scales. In addition, the interactions between surface and groundwater, that is, between the river basin and the aquifer system, await the emergence and functioning of new scales that will shape and in turn be shaped by newer political equations and social realities. This validates the significance to explore multiscalar structures, including aquifer ones, in understanding hydropolitical trajectories.

26.4.4 Trajectories from "safe and good" groundwater to "bad" citizens

While political ecology of water literature sheds light on how water and power relations interact to determine differential access among divergent communities, producing uneven waterscapes, Sultana (2013) complicates the story of access where beneficiaries and nonbeneficiaries evolve along complex assemblages of technology, nature, and developmentalist visions. Drawing inspiration from works on technonatures (White and Wilbert, 2009) and materialities as well as agencies of things (Appadurai, 1986; Bennett, 2010; Ingold, 2007; Whatmore, 2002) as in the hydrosocial cycle approach, Sultana (2013) analyses the ways through which water, tube well technologies, arsenic and differentiated social power "disrupt, reconfigure, and reposition processes and notions of development in the Bengal Delta" (p. 344). It explores mutual enrollments of technologies (tube wells), ecologies (aquifers, arsenic), discourses (progress), and subjects (social actors) that shape the "technonatural" trajectories of development. The work traces the transforming trajectories through which perception on tube wells shifted from icon of progress to icon of poison. The popularization of tube well technology has to be contextualized within the larger development alternatives of providing "safe" and "good" groundwater to people against "unsafe" and "bad" surface water. But the implementation of the technology at scales led to the counter narrative of safety and development when arsenic-contaminated groundwater from deeper water tables transformed the status of tube wells from icons of progress to carrier of poison across the Bengal delta.

The study problematizes the continued emphasis on technical solutions to deliver water to the marginalized with no provisions for social and cultural contextual analyses and interpretations. The hydrosocial changes through the tube wells technologies, arsenic challenges, and groundwater have crafted development narratives with contrasting understandings of "good-and-bad citizens" against "good-and-bad water" (p. 345). While users of green-painted tube wells signifying safe water provision remain "developed subjects," communities relying on water from red-painted tube wells are perceived as "unruly development subjects" not paying heed to official restrictions of not drinking arsenic-laced water. But the financial, social, and political compulsions behind this unruliness are an area awaiting research explorations, Sultana (2013) argues.

26.5 Discussion: what interdisciplinarity for enmeshed issues?

The previous sections highlighted the analytical power of two interdisciplinary approaches to account for and ultimately serve groundwater—society relationships. These approaches are the outcomes of several conceptual and empirical investigations. However, interdisciplinary research is "both advocated and antagonized" (Rusca and Di Baldassarre, 2019, 1). For instance, though there is a clarion call for "a new social contract for global change research" (Castree, 2016, 12), Wesselink et al. (2017) lay out key challenges in bringing together socio-hydrologists and hydrosocial researchers to come up with shared methodologies and course of actions. "The two communities do not just have incompatible paradigmatic positions, they also have different societal and scientific prestige" (p. 9). Nevertheless, they suggest "narrative" as the binding thread for both these domains. Thick descriptions by hydrosocial researchers can offer the context and premise for socio-hydrologists to design and develop conceptual and mathematical models, which, in turn, can make theoretical posturing in hydrosocial more meaningful, valid, and grounded. Socio-hydrology also has to learn from hydrosocial that no single model fits all cases as each case is shaped by its cultural, geographical, and historical trajectories.

Rusca and Di Baldassarre (2019) make the case for "the multiple ways in which critical geographies of water and socio-hydrology are convergent, compatible, and complementary" (p. 1). Through detailed examples of large-scale projects and research across historical, spatial, and cultural scales, they formulate "critical water resource geography" or "interdisciplinary resource geography" as a framework that integrates "disciplines which are marked by epistemological and methodological differences," but these differences "do not necessarily entail disciplinary incompatibility" (p. 9). Rusca and Di Baldassarre (2019) explain why and how critical geographers can systematically mobilize numerical data and findings to remain ideologically oriented toward environmental and social justice. A critical geography "of numbers" can both test and challenge quantitative datasets by examining multilayered social realities, finally geared to unveil complex forms of distributive injustice (Langford and Winkler, 2014). Showing practical pathways to integrate case studies, Geographic Information System, predictive modeling designs, and social justice frameworks, Rusca and Di Baldassarre (2019) propounded that "interdisciplinary resource geography" has the potential to address scientific questions and problems that combine water political economy and changes in the hydrological system.

One illustration of such approach may be the ATCHA research project based upon long-term (15 years) and large-scale collaboration among multiple groundwater users and stakeholders. This program targets meaningful integration through assimilation of numbers, models, and in-depth narratives to design pathways to justice and sustainability within the diverse and complex scene of peasantry and access to water in South India (see Box 26.1).

Box 26.1 Sharing groundwater? A "common"-based approach to better understand conditions for effective participatory governance

How to assess the sustainability of agricultural systems under climate change, in a context of proliferated borewells and worsening groundwater depletion? The Indo-French Water Cell in Bangalore, India and its partners developed an interdisciplinary research project to address this broad question in Southern India. The ATCHA project[2] integrates technical design including a biophysical model with a multistakeholder (participatory) approach that can be useful in terms of adapting farming systems to climate change implemented across a network of experimental watersheds in Karnataka, India.

More specifically, a French Institute of Pondicherry's team explored the reasons for the current absence of concertation and collective action around groundwater management in the studied areas. Using an historical and political economy approach based on Ostrom's work related to "commons" (Ostrom, 1990), the research team highlights the difficulty to think the groundwater as a common in the study context. Since colonial times in India, the groundwater belongs to the owners of the land, unlike surface water that is controlled by state governments with a few exceptions. Short-term individual and competitive strategies for access to groundwater thus prevail. However, aquifers physically—not institutionally—function as a commons: the borewells are doomed to dry up if the water table strongly declines, unless deeper ones or ones in farther areas are created, thus worsening the situation for all. The researchers also point to the general context of increased "individualization of risks" in agrarian practices, also linked to small peasants' high-level indebtedness. With the Ostrom's approach, they show that several hindrances in the studied context, such as social hierarchy, economic dependencies, or lack of understanding of the horizontal fluxes of groundwater, led farmers neither to identify the need for changes of operational rules for aquifers, nor wish them. The fact is that after 5 or 6 years of groundwater decline, an abundant rainy season can recharge the aquifers. Yet not every farmer is able to wait for such a long period. Unequal access to water increases class differentiation, and reversely. While referring to the Groundwater (sustainable management) Model Act that was enacted in India in 2016 to trigger conducive legal environments at the state level with the recognition of groundwater as commons (Cullet and Koonan, 2017), the authors warned of implementation difficulties that remain, particularly, around issues of institutions versus users' representation and administrative versus natural resource's contrasted perimeters. Another critical question is the consequence of surface water depletion due to the excessive pumping. This particularly affects the livelihoods of marginalized people, especially poor women, who do not own land nor borewell, but a few cattle.

In their conclusion the researchers called for more attention to the conditions for participatory action, as well as to gender issues: "What is the point of creating a groundwater committee if its members cannot collaborate for lack of democracy, sociability, or shared rules?". Laws may establish a "substantial" common that is a common pool resource, but collective practices have to shape and confirm a related "normative" common, or a more abstract aim for collective action, in order to sustain the effective comanagement of the "substantial" common (Brédif and Christin, 2009). For instance, decision-making processes should engage borewell owners as well as landless households.

Sources: Guétat-Bernard, H., J., Deschamps-Rébéré, M., Oger-Marengo, L., Ruiz, 2019. Groundwater depletion and environmental injustice in South Indian Deccan: the death of the river, care lacking and ecosystem destruction. In:International Symposium Environmental Justice 2019: Transformative Connections, EJ 2019, 2−4 July 2019, University of East Anglia, Norwich, UK (Guétat-Bernard et al. 2019); Landy et al. (personal communication, forthcoming paper).

Advancing a similar idea, Evers et al. (2017) talk about "reciprocal learning" across the various ongoing water research domains and formulate "pluralistic water research" concept as "an integrative and interdisciplinary approach which aims to coherently and comprehensively integrate human-water dimensions" (p. 1). Engaging the perspective of

---

2. ATCHA project website: accessed on 30 March 2020, <https://www6.inrae.fr/atcha/Presentation>.

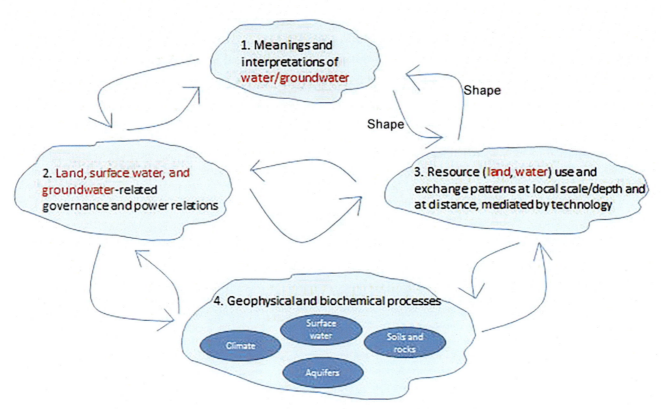

FIGURE 26.1 Groundwater-attuned hydrosocial cycle (inspired by Lafaye de Micheaux, F., Mukherjee, J., Kull, C.A., 2018. When hydrosociality encounters sediments: transformed lives and livelihoods in the lower basin of the Ganges River. Environ. Plann. E: Nat. Space 1, 641–663., Fig. 26.1).

both natural and social sciences and with "reciprocal learning" as the key component, Evers et al. (2017) depict how "pluralistic water research" "is defined by reciprocal physical and human boundary conditions which enable the analysis of multiple meanings and alternative framings" (p. 6).

This approach is particularly significant in the context of groundwater. As touched upon in the case studies in Sections 26.2 and 26.4, the feedback loops are particularly complex in systems where surface and groundwater intersect, with their respective users, regulations, and scales. Spatial and temporal processes do not match either with longer response from aquifers in comparison to surface water flows or with aquifer extensions that are disconnected from surface watersheds. In addition, groundwater extraction is often performed by individual well owners, with dispersed expertise, in contrast to structured river basin management based on centralized knowledge. Finally, groundwater extraction is also necessarily mediated by technology, such as tube wells. The framework of socionature, and particularly technonature, mobilized by Sultana (2013), invites us to propose a reframed "hydrosocial cycle" in groundwater context, drawing from Linton and Budds's conceptualization (p. 2014) (see Fig. 26.1). In this visualization, we choose to insist on the physical dimension that particularly interlaces groundwater and surface water over space and time. As in (Rusca and Di Baldassarre, 2019), this insistence on the material dimension of water draws from the approach of critical physical geography that aims toward the integration of human and physical geographies (Lave et al., 2014, see also Lafaye de Micheaux et al., 2018 in the context of rivers).

This representation may help hydrologists, when confronted to a "groundwater problem," to conceptualize the exact domain of their interventions within the groundwater–society system and their limitations, prior to their modeling work. It also could help them seek for other disciplines to come on board of their research projects, if the aim is to develop "sustainability science." Finally, it would also draw their attention to the roles of "meanings of water" (Linton and Budds, 2014, 176) that indirectly affect the core of their studies and that should be the object of fine-tuned investigations. For instance, Richard-Ferroudji (2019), in the context of India, highlights the "plurivalence" (plural ambivalence) of groundwater into the surface delineating compromises between preservation and necessity of this water resource.

26.6 Conclusion

The chapter has presented state-of-the-art on groundwater and society along two major emerging domains in contemporary water research: socio-hydrology and hydrosocial. This detailed review establishes how groundwater is as much social as material and technical. The empirical case studies surrounding socio-hydrology and hydrosocial demonstrate that while nuanced unraveling of water–society dynamics is the key agenda for both these paradigms, yet they are distinct in terms of methodological design and implementation, and hence epistemological foundations.

Drawing its roots from the socio-ecological systems, socio-hydrologists took the bold step to incorporate human activities within the water cycle. However, with hydrological science as the academic context, it relies on quantitative or mathematical modeling approaches to represent the water–human relationship. This leads researchers confront with several challenges including methodological difficulties to isolate factors within complex web of causal chains, and to design accurate quantitative models of human and social behaviors, even with a limited number of factors. On the epistemological side the researchers confront to ethical issues that are not familiar to hydrologists—in contrast to social scientists—such as dealing with social norms and values that may directly affect the socio-hydrological research. This becomes evident with empirical investigations around groundwater that the chapter has brought to the fore. More recently, the need to incorporate perspectives of different social stakeholders has been realized with socio-hydrogeology emanating as a subdomain within socio-hydrology, radically advocating for the cause.

The nonhyphenated hydrosocial understanding within political ecology of water is the provocation to perceive socionature as nonhyphenated or rather address the intersections between nature and society where they operate as hybrids, internally connected to each other and coproduced across space and time. The complex story of infrastructures and institutions, politics of scale, etc., get unveiled when hydrosociality is applied on groundwater. The divergent cases from the Global South expose these complexities, clearly crafting the pathway to perceive groundwater using the socionature and technonature lenses.

Interdisciplinarity through "reciprocal learning" (Evers et al., 2017) is a way forward. Yet, there are challenges and potentials to cross-feed and cross-fertilize socio-hydrology and hydrosocial. Rusca and Di Baldassarre (2019) have proposed the "spectrum theory" which postulates that "rather than opposing fields, natural and social scientists are on a spectrum, where academics holding different positionalities align in some epistemological or methodological approaches and also differ in others" (p. 7). Here, we wish to advance this theory beyond academia through mutual learning among academia and "nonacademic" (policy circles, civil society, user groups, and others) sectors toward just, democratic and resilient transformation along challenging yet possible lines of transition from epistemology (epistemic pluralities) to (multiple) axiologies. Moreover, as researchers committed to water studies from a holistic perspective, we advocate to not only disseminate scientific knowledge to every social stakeholder but also to involve these actors across every stage of research (analysis, design, development, implementation, and evaluation) and from research to action. Transdisciplinarity to evolve comprehensive water–society frameworks through spontaneous and organic integration of emerging water research approaches remains the crucial step to forge cross-sectoral exchanges and actions.

References

Appadurai, A., 1986. The Social Life of Things: Commodities in Cultural Perspective. Cambridge University Press, New York.

Baghel, R., 2014. River Control in India, Advances in Asian Human-Environmental Research. Springer International Publishing, Cham.

Bakker, K., 2000. Privatizing water, producing scarcity: The yorkshire drought of 1995. Economic Geo. 76 (1), 4–27. https://doi.org/10.1111/j.1944-8287.2000.tb00131.x.

Di Baldassarre, G., Kooy, M., Kemerink, J.S., Brandimarte, L., 2013. Towards understanding the dynamic behaviour of floodplains as human-water systems. Hydrol. Earth Syst. Sci. 17, 3235–3244.

Bear, C., Bull, J., 2011. Water matters: agency, flows and frictions. Environ. Plann. A 43, 2261–2266.

Bennett, J., 2010. Vibrant Matter: A Political Ecology of Things. Duke University Press, Durham, NC.

Birkenholtz, T., 2016. Dispossessing irrigators: water grabbing, supply-side growth and farmer resistance in India. Geoforum 69, 94–105.

Birkenholtz, T., 2008. Contesting expertise: the politics of environmental knowledge in northern Indian groundwater practices. Geoforum 39, 466–482.

Blaikie, P.M., Brookfield, H.C., 1987. Land Degradation and Society. Methuen, London; New York.

Brédif, H., Christin, D., 2009. La construction du commun dans la prise en charge des problèmes environnementaux: menace ou opportunité pour la démocratie? VertigO 9, 1. online.

Castree, N., 2016. Geography and the new social contract for global change research. Trans. Inst. Br. Geogr. 41, 328–347.

Castree, N., Kitchen, R., Rogers, A., 2013. A Dictionary of Human Geography. Oxford University Press, Oxford.

Cullet, P., Koonan, S., 2017. Water Law in India: An Introduction to Legal Instruments, second ed. Oxford University Press, New Delhi.

D'Souza, R., 2006. Water in British India: the making of a 'colonial hydrology'. Hist. Compass 4 (4), 621–628.
Elshafei, Y., Coletti, J.Z., Sivapalan, M., Hipsey, M.R., 2015. A model of the socio-hydrologic dynamics in a semiarid catchment: isolating feedbacks in the coupled human-hydrology system: socio-hydrologic dynamics in a semi-arid catchment. Water Resour. Res. 51, 6442–6471.
Elshafei, Y., Sivapalan, M., Tonts, M., Hipsey, M.R., 2014. A prototype framework for models of socio-hydrology: identification of key feedback loops and parameterisation approach. Hydrol. Earth Syst. Sci. 18, 2141–2166.
Evers, M., Höllermann, B., Almoradie, A., Garcia Santos, G., Taft, L., 2017. The pluralistic water research concept: a new human-water system research approach. Water 9, 933.
Fernandez, S., 2014. Much ado about minimum flows... unpacking indicators to reveal water politics. Geoforum 57, 258–271.
Guétat-Bernard, H., J., Deschamps-Rébéré, M., Oger-Marengo, L., Ruiz, 2019. Groundwater depletion and environmental injustice in South Indian Deccan: the death of the river, care lacking and ecosystem destruction. In:International Symposium Environmental Justice 2019: Transformative Connections, EJ 2019, 2–4 July 2019, University of East Anglia, Norwich.
Hund, S.V., Allen, D.M., Morillas, L., Johnson, M.S., 2018. Groundwater recharge indicator as tool for decision makers to increase socio-hydrological resilience to seasonal drought. J. Hydrol. 563, 1119–1134.
Hynds, P., Regan, S., Andrade, L., Mooney, S., O'Malley, K., DiPelino, S., et al., 2018. Muddy waters: refining the way forward for the "sustainability science" of socio-hydrogeology. Water 10, 1111.
Ingold, T., 2007. Materials against materiality. Archaeol. Dialogues 14, 1–16.
Kaika, M., 2003. Constructing scarcity and sensationalising water politics: 170 days that shook Athens. Antipode 35, 919–954.
Krause, F., Strang, V., Thinking relationships through water. Society Nat. Res. 29 (6), 2016, 633–38. https://doi.org/10.1080/08941920.2016.1151714.
Lafaye de Micheaux, F.L., 2019. Political Ecology of a Sacred River: Hydrosocial Cycle and Governance of the Ganges, India (Ph.D. thesis). University of Lausanne.
Lafaye de Micheaux, F., Mukherjee, J., Kull, C.A., 2018. When hydrosociality encounters sediments: transformed lives and livelihoods in the lower basin of the Ganges River. Environ. Plann. E: Nat. Space 1, 641–663.
Langford, M., Winkler, I., 2014. Muddying the water? Assessing target-based approaches in development cooperation for water and sanitation. J. Hum. Dev. Capabilities 15, 247–260.
Lave, R., Wilson, M.W., Barron, E.S., Biermann, C., Carey, M.A., Duvall, C.S., et al., 2014. Intervention: critical physical geography: critical physical geography. Can. Geogr./Le. Géographe canadien 58, 1–10.
Linton, J., Budds, J., 2014. The hydrosocial cycle: defining and mobilizing a relational-dialectical approach to water. Geoforum 57, 170–180.
Matthews, N., 2012. Water grabbing in the Mekong basin – an analysis of the winners and losers of Thailand's hydropower development in Lao PDR. Water Altern. 5, 392–411.
Molle, F., 2005. Elements for a Political Ecology of River Basins Development: The Case of the Chao Phraya River Basin. Thailand.
Molle, F., 2009. River-basin planning and management: the social life of a concept. Geoforum 40, 484–494.
Mukherjee, A., Scanlon, B., Aureli, A., Langan, S., Guo, H., McKenzie, A., 2020. Global Groundwater: Source, Scarcity, Sustainability, Security and Solutions, first ed. Elsevier, ISBN. 9780128181720.
Ostrom, E., 2009. A general framework for analysing sustainability of social-ecological systems. Science 325, 419–422.
Ostrom, E., 1990. Governing the commons: the evolution of institutions for collective action. The Political Economy of Institutions and Decisions. Cambridge University Press, Cambridge; New York.
Re, V., 2015. Incorporating the social dimension into hydrogeochemical investigations for rural development: the Bir Al-Nas approach for socio-hydrogeology. Hydrogeology J. 23, 1293–1304.
Richard-Ferroudji, A., 2019. Ambivalence about groundwater: promoting conservation while justifying over-exploitation in an Indian newspaper. RILE, Rev. Interdisciplinar de. Literatura e Ecocrítica: Diálogos Ecocríticos 1 (2), 64–92. https://hal.archives-ouvertes.fr/hal-02295491v1 (accessed 30.03.2020.).
Robbins, P., 2012. Political ecology: a critical introduction, Critical Introductions to Geography, second ed. J. Wiley & Sons, Chichester, West Sussex; Malden, MA.
Rusca, M., Di Baldassarre, G., 2019. Interdisciplinary critical geographies of water: capturing the mutual shaping of society and hydrological flows. Water 11, 1973.
Da Silva, L.P.B., Hussein, H., 2019. Production of scale in regional hydropolitics: an analysis of La Plata River Basin and the Guarani Aquifer System in South America. Geoforum 99, 42–53.
Sivapalan, M., Konar, M., Srinivasan, V., Chhatre, A., Wutich, A., Scott, C.A., et al., 2014. Socio-hydrology: use-inspired water sustainability science for the Anthropocene: commentary. Earth's Future 2, 225–230.
Sivapalan, M., Savenije, H.H.G., Blöschl, G., 2012. Socio-hydrology: a new science of people and water. Hydrol. Process. 26, 1270–1276.
Srinivasan, V., Sanderson, M., Garcia, M., Konar, M., Blöschl, G., Sivapalan, M., 2016. Prediction in a socio-hydrological world. Hydrol. Sci. J. 1–8.
Strang, V., 2004. The Meaning of Water. Bloomsbury Academic.
Sultana, F., 2013. Water, technology, and development: transformations of development technonatures in changing waterscapes. Environ. Plan. D: Soc. Space 31, 337–353.
Swyngedouw, E., 2007. Technonatural revolutions: the scalar politics of Franco's hydro-social dream for Spain, 1939–1975. Trans. Inst. Br. Geogr., N. Ser. 32, 9–28.
Swyngedouw, E., 2003. Modernity and the production of the Spanish waterscape, 1890–1930. In: Bassett, T., Zimmerer, K. (Eds.), Geographical Political Ecology. pp. 94–112.

Swyngedouw, E., 1999. Modernity and hybridity: nature, regeneracionismo, and the production of the Spanish waterscape, 1890–1930. Ann. Assoc. Am. Geogr. 89, 443–465.

Swyngedouw, E., Kaika, M., Castro, J.E., 2002. Urban water: perspective from political ecology. Built Environ. 28 (2), 124–137.

Troy, T.J., Konar, M., Srinivasan, V., Thompson, S., 2015. Moving sociohydrology forward: a synthesis across studies. Hydrol. Earth Syst. Sci. 19, 3667–3679.

Vogel, E., 2012. Parcelling out the watershed: the recurring consequences of organising Columbia river management within a basin-based territory. Water Altern. 5, 161–190.

Wesselink, A., Kooy, M., Warner, J., 2017. Socio-hydrology and hydrosocial analysis: toward dialogues across disciplines. Wiley Interdiscip. Rev.: Water 4, e1196.

Whatmore, S., 2002. Hybrid Geographies. Sage, London.

White, D., Wilbert, C., 2009. Technonatures: Environments, Technologies, Space Place Twenty-First Century. Wilfrid Laurier University Press, Waterloo, ON.

Zhang, Z., Hu, H., Tian, F., Yao, X., Sivapalan, M., 2014. Groundwater dynamics under water-saving irrigation and implications for sustainable water management in an oasis: Tarim River basin of western China. Hydrol. Earth Syst. Sci. 18, 3951–3967.

Zhou, S., Huang, Y., Wei, Y., Wang, G., 2015. Socio-hydrological water balance for water allocation between human and environmental purposes in catchments. Hydrol. Earth Syst. Sci. 19, 3715–3726.

Chapter 27

Groundwater sustainability in cold and arid regions

Rui Ma and Yanxin Wang

School of Environmental Studies & State Key Laboratory of Biogeology and Environmental Geology, China University of Geosciences, Wuhan, Hubei 430074, China

27.1 Importance of groundwater in hydrological systems

27.1.1 Cold regions

Characterized by the existence of ice and snow in part of the year, cold regions in the earth were separated into polar and sub-polar regions (Fig. 27.1), and historically it is believed that a large part of these cold regions is rarely populated and interfered with humanactivities (Shen, 2018). Due to the effect of economic development and population growth, this situation has been changed (Mukherjee et al., 2020). Although the human population may exert an impact on hydrogeology in cold regions, its extent is still much less than that of climate change. The high latitude and lowland cold regions cover a very broad domain such as glaciers, permafrost and snow cover, which contains abundant water in various forms. They are often headwater regions for many large rivers such as the Yukon River of Alaska, USA, the Lena and Kolyma Rivers in Siberia, Russian, and the Yangtze, Yellow and Heihe Rivers in China. The Tibetan Plateau, the world's largest and highest plateau, is regarded as the "world's third pole" because it contains abundant ice and snow resources. Thus, the groundwater in these regions is crucial water sources for the downstream areas, and the carbon cycle and response of ecosystem and hydrological systems to climate change have been hot topics for the studies in the past decades (Woo, 2012; Chang et al., 2018; Walvoord et al., 2019).

Groundwater recharge was largely affected by thawing of snow and ice, while groundwater flow processes were controlled by the distribution and freeze-thaw process of permafrost in these cold regions. A special issue on Cold Regions Hydrology in a Changing Climate was initiated by the IAHS International Commission on Snow and Ice Hydrology and the Predictions in Ungauged Basins with the aim to address hydrological issues associated with snow and ice hydrology, especially changes in the characteristics of wetlands, lakes and rivers in cold regions, and their interactions with ecosystems and human (Yang et al., 2011). To better understand the groundwater flow under the impact of the snow, ice and permafrost in the cold regions, Hinzman et al. (2013) organized a special issue "Hydrogeology in cold regions" in Hydrogeology Journal to address the groundwater-related issues in the different parts of cold regions. In this thematic issue, particular emphasis was put on the groundwater interactions with permafrost and ground ice. In the mountainous headwater regions of cold regions, permafrost and seasonal frost act as an "aquitard" when frozen but as a more recharged "aquifer" when thawing (Evans, 2017). Climate warming changes the hydraulic connection of the seasonal frost and permafrost, which alters the hydrological cycle and impacts the water security for their downstream regions. Permafrost exerts a significant effect on groundwater flow, and therefore hydrological cycles of cold regions, which is especially applied to the headwater regions of large rivers in the mountainous area (Walvoord et al., 2012). In these regions, the interplay between groundwater and permafrost has a large impact on water resource management, biogeochemical processes, and engineering construction (Cheng and Jin, 2013). The hydrogeological studies incorporated the permafrost distribution and freeze-thaw processes have been stimulated by the demand for water supplies, groundwater issues associated with mining, and construction of railways, buildings, and highways, etc. (Ma et al., 2017).

FIGURE 27.1 The distribution of cold region and arid and semiarid regions in the world, superimposed with groundwater depletion distribution. The climate data are from Trabucco and Zomer (2019) (https://cgiarcsi.community/data/global-aridity-and-pet-database/), the glacier and permafrost data in cold regions are from Brown et al. (2002) (https://nsidc.org/data/GGD318/versions/2), and the groundwater depletion data from Dalin et al. (2017).

27.1.2 Arid and semi-arid regions

The cold regions normally provide waters for downstream semi-arid and arid area via both surface runoff and subsurface flows, which is especially true in inland basins in northwestern China (Ma et al., 2017; Fig. 27.2). Arid and semi-arid regions occupy approximately one-third of the land area of the world, providing the inhabitant for about 20% of the total world population (White and Nackoney, 2003). Drylands of the world were commonly defined by the ratio of mean annual precipitation to mean annual potential evapotranspiration by the United Nations Convention to Combat Desertification. Among all the countries, Australia has the largest distribution of dryland with an area about 6.6 million km^2, and other countries, which have greater than 2 million km^2 dryland area, include United States, Russia, China and Kazakhstan. Hydrological processes in arid and semi-arid regions can be largely distinguished from those in humid regions, and the water flow and precipitation show extremely tempo-spatial variabilities (Wheater et al., 2010). The spatial distributed, long-term data in the area are often limited, and the data quality is also restricted. For the last decades, the demand for water sources has increased dramatically due to the increase in water use for domestic, agricultural and industrial activities. Surface waters such as river and freshwater lakes are normally ephemeral, and thus groundwater in arid-semiarid regions provides important reliable water resources for drinking, industry and agriculture, especially during the drought seasons. The accessibility of vegetation to groundwater was controlled by groundwater depth, which is the controlling factor for sustaining groundwater-dependent terrestrial ecosystems (Xu and Su, 2019).

The exploitation has been largely intensified during the past several decades with the application of advanced drilling and pumping technologies, causing the decline of water levels. Over-pumping and ineffective regulation resulted in the depletion of groundwater resources and the worsening of water quality in many arid areas. Many adverse effects associated with over-exploitation occurred, including deterioration of ecosystem, decrease in surface water flows and disappear of wetlands (Lin et al., 2018; Xiao et al., 2015; Heintzman et al., 2017), groundwater and soil salinity (Sun et al., 2016), land subsidence and desertification (Wang et al., 2018a). In addition to above water quality issues caused by human pumping, the water resources in arid and semiarid regions are also threatened by climate change, which in combination with water quantity issues added the challenge for groundwater sustainability. More pressure has been from population growth, associated with more domestic and agricultural water use. As a result, groundwater sustainability has become a major concern across the different countries, no matter which are developed and wealthy economies or undeveloped and poorer countries (Wheater et al., 2010).

Water transfer projects were implemented in some countries to divert water from water-rich areas to regions of water scarcity. However, many studies also reported the large impact of these projects on the groundwater flow field and groundwater quality. Lin et al. (2018) found that the water transfer project in the Dunhuang basin of northwestern China has increased the risk of soil salinization. The water conservation projects in inland river basins have strengthened the exploitation of the groundwater in the middle reaches of rivers; meanwhile, river leakage to groundwater was reduced in the lower reaches (Schilling et al., 2014; Wang et al., 2015). The increase in water deficit and occurrence of many complex conflicts among the water resources use and environment, ecology and development were caused by low precipitation and high evapotranspiration as well as fast socio-economic development in past several decades. Human being has been facing the globally unprecedented challenges for water resources scarcity, which are specially pointed out in the arid and semi-arid regions of the world. Thus, we must realize the urgent need for sustainable groundwater use, coordinate between the groundwater exploitation and water availability for the long term, and learn how to best manage the water resource from traditional and new measures (Green et al., 2011).

27.2 The characteristics of the hydrological cycle

27.2.1 The effect of permafrost distribution, snow and /or ice on groundwater systems in cold regions

The groundwater discharge quantity and duration may be altered by the changes in the timing and length of seasonal thawing in the underground and the recharge of snow and/or ice melt. Chang et al. (2018) identified the important role of glaciers and porous aquifers in the hydrological cycle in a small alpine mountain catchment of northeastern Qinghai-Tibet Plateau. Model analyses of Evans et al. (2018) suggested that the groundwater discharge to hillslope faces was controlled by snowmelt timing, and discharge to streams was affected by the duration of seasonal freezing. In watersheds which were dominated by permafrost or seasonal frost, the distribution of seasonal frost and permafrost and the freeze-thaw process greatly affect the groundwater recharge, flow path and its dynamic and discharge, in such area as the Qinghai-Tibet Plateau (White et al., 2007; Ge et al., 2011; Ma et al., 2017). The hydrogeological

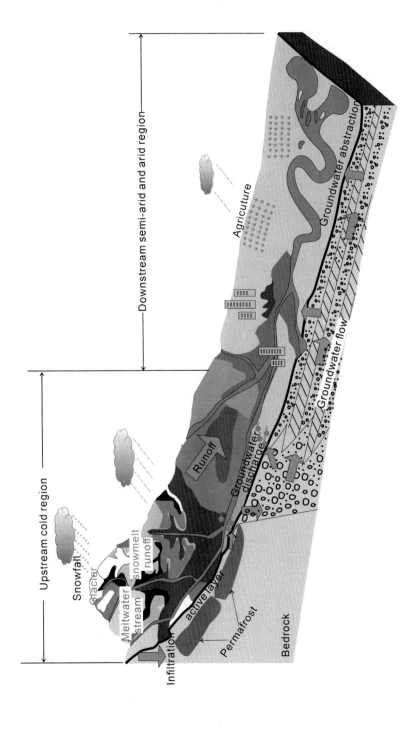

FIGURE 27.2 The schematic map for the hydrogeological cycle in a typical inland basin in northwestern China, where the cold region located in upstream provides important water resources for the downstream semi-arid and arid areas.

cycle can be affected by permafrost through changing the storage of ground ice, ice segregation and moisture migration. The occupation of ground ice in the interstitial voids decreases the permeability of the sediments matrix when permafrost freezes, leading the permafrost to act as an aquitard (Woo, 2012). When thawing, the active layer and seasonal frost act as an aquifer and could lead to the hydraulic connections between different water bodies, further altering the groundwater flow path and groundwater and surface water interaction (e.g., Zhang et al., 2013; Liao and Zhuang, 2017).

Some studies have reported an increase in arctic river baseflow and change in distributions of wetland and lake (Evans and Ge, 2017; Jepsen et al., 2016), while some other researches exhibited that lake surface area has been declining in both continuous and discontinuous permafrost regions (Smith et al., 2005; Carroll et al., 2011; Hu et al., 2019b). The hydrogeological processes that caused these changes were investigated with coupled heat and groundwater flow models with considering the dynamic freeze- thaw processes, and the results showed that the groundwater discharge shifted spatially from upslope to downslope and temporally increased during the winter season as a lateral suprapermafrost talik formed underlying the active layer (Lamontagne-Halle et al., 2018). O'Connor et al. (2019) explored how the combination of microtopography, macro topography, stratigraphy and thawing controlled suprapermafrost groundwater flow. Groundwater processes affected by permafrost were also shown to have a crucial role in affecting carbon release from the subsurface to streams in permafrost-dominated regions (Neilson et al., 2018; Walvoord et al., 2019).

Groundwater behavior is changing in permafrost terrain as permafrost continues to degrade due to climate change, which is becoming an important issue (Walvoord and Striegl, 2007). Although the recent advances, our knowledge of the hydrogeological cycle over these cold regions is still incomplete due to a paucity of hydrogeological information such as detailed hydrostratigraphy and hydraulic head data caused by the harsh weather and remote distance, and thus a lack of complete understanding of the controlling processes. Although numerical simulations have been used to explore the groundwater flow paths and discharges under different scenarios, the mechanism driving groundwater circulation and discharge in the impact of permafrost have not been unraveled in field studies. The quantitative substantiation is challenging because of the scarce hydrogeological and permafrost information in remote areas and complex hydrogeological processes in regional-scale permafrost regions.

27.2.2 Hydrological processes and its effect on groundwater quality in arid and semi-arid regions

The hydrological processes in arid and semi-arid regions greatly differ from those in humid regions due to the generally small precipitation and its highly spatiotemporal variability, as well as large potential evapotranspiration. Rainfall and flood events are thus hard to quantify, leading to difficulty in the calculation of groundwater recharge, which is essential to estimate the water balance and sustainable yield of the groundwater resource. The uncertainly in calculation of groundwater recharge in arid and semi-arid regions remains high due to the other uncertainties of the dominant inputs and outputs in the water-balance calculation since the groundwater recharge is normally low. The groundwater recharge in these arid and semi-arid regions is largely determined by the subsurface characteristics, the land surface properties, local climate (especially the duration and intensity of the precipitation) (Wheater et al., 2010).

The surface water and groundwater interaction also frequently occurred in the arid area although the surface water quantity is not rich. Wang et al. (2018b) summarised the several modes of surface water and groundwater interaction in the arid and semi-arid northwestern China. Lin et al. (2018) explored the change in groundwater and surface water exchange in an arid inland basin in northwestern China caused by the water transfer project, which is a common way to alleviate the water shortage in the arid basins. The studies on groundwater recharge from ephemeral channels and away from the active channel system indicated that the subsurface properties play an important role in determining the surface infiltration and groundwater recharge (Wheater et al., 2010). The increased demand for groundwater resource due to population growth, agricultural and economic expansion, as well as climate-induced pumping, resulted in the wide groundwater depletion in semi-arid and arid regions (Fig. 27.1 and Table 27.1).

Groundwater quality issue has become a major concern in semi-arid and arid regions, which exacerbated the water shortage situation. Under arid and semi-arid environments the salts are easy to accumulate in top soil and aquifer settings since the residual salts were concentrated after the freshwater in the soil was removed by evaporation and transpiration. Under this background, variable-density flow could occur, especially where groundwater with less density underlies the saline brines. Groundwater abstraction could result in water quality deterioration as the saline water which is common in shallow aquifers near-surface environments is drawn down to the target aquifer. The work of Hu et al. (2019a) indicated that groundwater salinity changed from mountainous areas to the alluvial fan, and further to the oasis and desert areas in the arid inland basin of China. The total dissolved solids in groundwater from the lower reaches of

TABLE 27.1 Groundwater storage depletion rate in different countries of semi-arid and arid regions.

Region or country	Location	Area	Period	Method	Groundwater storage depletion rate	Source
China	North China	370,000 km²	2003–2010	GRACE	22 ± 3 mm/yr	Feng et al., 2013
	NCP	140,000 km²	1971–2015	GRACE	17.8 ± 0.1 mm/yr	Gong et al., 2018
	Xinjiang	1,600,000 km²	2003–2016	GRACE	3.61 ± 0.85 ~ 3.10 ± 0.91 mm/yr	Hu et al., 2019b
	Loess Plateau	680,000 km²	2005–2014	GRACE	6.5 ± 0.7 mm/yr	Xie et al., 2018
India	Ganges River basin	32,753 km²	2002–2010	GRACE	48.85 mm/yr	Chinnasamy, 2017
	North India	-	2002-2013	GRACE	20 mm/yr	Asoka et al., 2017
	Southern India	-			10–20 mm/yr	
Australia	Canning Basin	-	2003–2013	GRACE	3.6 km³ yr⁻¹	Bierkensand Wada, 2019
USA	California's Central Valley	52,000 km²	2003–2010	GRACE	20.4 ± 3.9 mm yr⁻¹	Famiglietti et al., 2011
			1962–2003	Groundwater model	2.0 km³ yr⁻¹ (38.5 mm/yr)	Scanlon et al., 2012a
			2006–2010	GRACE	58.1 ± 5.6 mm/yr	Scanlon et al., 2012b
	High Plains Aquifer	450,000 km²	1950–2007	groundwater model	5.7 km³ yr⁻¹ (12.9 mm/yr)	Scanlon et al., 2012a
			1987–2007		7.0 km³ yr⁻¹	
Mexico	Central Valleys of Oaxaca	3744.64 km²	1986–2018	GRACE	12.9 mm/yr	Olivares et al., 2019
Iran	Whole country	-	2002-2012	GRACE and hydrological model	-8.9 mm/yr	Khaki et al., 2018
Northeast Africa	Dakhlasubbasin	660,000 km²	2003–2012	GRACE	6.73 ± 0.64 mm/yr	Mohamed et al., 2017
Saudi Arabia, Jordan, and Iraq	Saq aquifer system	560,000 km²	2002 - 2015	GRACE	-6.52 ± 0.29 mm/year	Fallatah et al., 2017

the watershed are usually high, especially in shallow aquifers. Sun et al. (2016) also addressed the mechanism of the formation of the high salinity groundwater in arid Dunhuang Basin, northwestern China, especially at the groundwater discharge area and irrigation district with a shallower water table. Salinization can also be caused by excessive irrigation for farming lands. Geogenic high As and F groundwater was widely detected in arid and semi-arid areas as well (Wang et al., 2018).

27.3 Groundwater modeling and challenges

27.3.1 Model development in the cold regions

Models are important for exploring the interplay among the snow, ice and permafrost and groundwater systems, and predict the effect of permafrost evolution on groundwater flow and its discharge into stream under the impact of climate changing scenarios in cold regions (Ge et al., 2011; Evans et al., 2018; Lamontagne-Halle et al., 2018). To account for the effect of freeze-thaw processes of permafrost on groundwater systems, a wide range of coupled flow and heat transport model tools have been developed, such as SUTRA, PermaFOAM, PFLOTRAN-ICE, and FEFLOW. Grenier et al. (2018) compared different numerical simulators for systems experiencing freeze-thaw processes with 2D test cases.

Painter et al. (2013) discussed challenges of modeling prediction for the hydrologic response to degrading permafrost, and suggested that models should consider the strategies for coupling different subsurface and surface processes and for integrating the change of topography and drainage network due to the evolution of the landscape. Coupled flow and heat transport numerical simulations were also applied to evaluate the safety of nuclear waste repositories under the impact of climate cooling at sites in Forsmark, Sweden (Bosson et al., 2013) and in the Fennoscandian Shield, Sweden, beneath the moving margin of an ice sheet (Vidstrand et al., 2013).

Since groundwater systems in the permafrost area are very sensitive to climate change, numerical models were used to explore how climate change affects the groundwater flow with incorporating the effect of snow, ice and permafrost. Ge et al. (2011) and Evans et al. (2018) used numerical model results to illustrate the deepening of active layer and increase in discharge of groundwater to stream under climate warming scenarios. The thawing of permafrost in cold regions could change the groundwater flow, which discharges to the streamflow, in turn, alters the flow magnitude and its seasonality, and chemical characteristics of stream water. The model results from Neilson et al. (2018) suggested that carbon export to rivers will be influenced by the groundwater contributions to surface water in arctic Alaska watersheds due to the increase of thaw depth in a warmer climate.

Despite the great efforts, our ability still needs to be improved to model the coupled effect of the changes in ice, snow and permafrost on the fluxes of water, energy and chemical components with considering the different hydrological processes. A particular concern for modeling is the inadequate model calibration resulted from the scattered field data and an incomplete understanding of permafrost-groundwater systems within cold regions.

27.3.2 Model application and challenges in the arid and semi-arid regions

Like the models applied in the other areas, physical-mechanism based models are normally required to capture the spatiotemporal variability of surface flow processes, to quantify groundwater recharge and characterize the groundwater flow and discharge, as well as to assess the water resources management in arid and semi-arid area (Wheater et al., 2010). As noted above, groundwater plays a key role in providing water to agriculture and vegetation ecosystems in arid and semiarid areas, and thus the interaction between groundwater and surface water is a key link for the water-ecosystem-agriculture system (Sun et al., 2018). Understanding and modeling the groundwater and surface water interaction and their connection to the agriculture-ecosystem in these regions is critical to groundwater sustainability. Integrated ecohydrological models, would provide strong technical support to the understanding of the complex system and agricultural water management (Sun et al., 2018). However, such comprehensive models are still underdeveloped. The agricultural activities such as surface water transfer, groundwater abstraction and irrigation could significantly affect the flow regimes of both surface water and groundwater, thus complicating the groundwater and surface water interaction (Tian et al., 2015). The available data and information need to be assimilated by the integrated modeling of surface and groundwater systems. Tian et al. (2015) coupled the Storm Water Management Model with Ground-Water and Surface-Water Flow Model to account for highly engineered flow systems and assessed the effect of agricultural water utilization on the hydrological cycle in a large inland basin, northwest China. A basin-scale integrated ecohydrological model was proposed and applied to the Heihe River Basin in China, and the model results could help balance the use of water resources between agricultural development and ecosystem conservation in the large arid and semiarid basins of the world (Sun et al., 2018).

As described above, variable-density flow is expected to be important in modeling groundwater flow in the arid area. Thus, attention should be also paid to the issues of modeling density-dependent flows which have been seldom tackled. The instruction on the modeling tools in arid regions is limited and needs to be strengthened for groundwater management. Models integrating physical mechanisms and distributed parameters are desirable for investigating the complicated interplay among water, agricultural system and ecosystems, but remain a big challenge (Sun et al., 2018).

27.4 The effect of climate change

Both human activity-induced and natural climate change can affect the spatiotemporal quantity and quality distribution of different components in the hydrologic cycle (Green et al., 2011). Understanding the effects of climate change on groundwater in arid and semi-arid as well as cold regions is crucial since the groundwater systems or groundwater-dependent ecosystems in these areas either fragile or sensitive to climate change.

27.4.1 Cold regions

In cold regions, hydrogeological changes can be always ascribed to the climate—related processes, such as freeze and thaw of soils, melting of snow and ice, transformation of snow and glacier cover. Climate warming could cause the thawing of permafrost, change the snowpack cover, and lead to the glacier retreat, which likely increase groundwater recharge (Dragoni and Sukhija, 2008) and also significantly alter the groundwater storage (Bhanja et al., 2018).

Permafrost warming and thawing are observed in many cold regions of the world (Anderson et al., 2013), and efforts have been made to explore the hydrological change associated with permafrost degradation (O'Donnell et al., 2012). Permafrost areal extent in China has decreased from 2.15×10^6 km^2 in the 1970s to 1.75×10^6 km^2 in 2006, and further to 1.59×10^6 km^2 in 2012, indicating a permafrost degradation of about 18.6% during 30 years (Cheng and Jin, 2013). Previous studies suggest that the hydrologic cycle could be significantly altered by the small increase in near-surface air temperatures in snowmelt-dominated regions through the seasonal change in streamflow (Tague and Grant, 2009). In a continuous permafrost region in the Qinghai-Tibet Plateau, China, Ge et al. (2011) projected that 2°C of warming could result in the increase of groundwater discharge to streams three-fold and the deepening down of active layers. Similarly, a general increasing trend in groundwater discharge to streamflow was indicated by more than 30 years streamflow data of the Yukon River in Alaska due to climate warming and permafrost thawing (Walvoord and Striegl, 2007). Another study in the Qinghai-Tibet Plateau indicated that permafrost degradation led to a decline in the regional groundwater table, further causing declines of lake levels and shrink of wetlands (Cheng and Wu, 2007).

The changes in groundwater discharge to stream in the downstream of catchment could be closely related to the variation of recharge in the upstream. To analyze how changes in recharge alter groundwater discharge, a site-specific study was conducted in a seasonally frozen ground catchment in the Rocky Mountains of Colorado, USA (Evans, 2017). The established model predicted that annual groundwater discharge increased at an average annual rate of 1% with an increase of 7% in the spring and decrease of 9% in the summer after 50 years when assuming climate is warming with 4.8°C/100 years.

27.4.2 Arid and semi-arid regions

The change in precipitation and evapotranspiration caused by climate warming directly influences groundwater recharge and indirectly influences groundwater demand and abstraction. The large changes in groundwater recharge may be induced by even small variations in precipitation in some arid and semi-arid areas (Woldeamlak et al., 2007). The study in the Central Valley, California USA showed that significant declines of groundwater table might be induced by light to severe drought and won't recover within the next 30-year prediction period (Miller et al., 2009). Climate change also has a substantial impact on groundwater and surface water interactions (Gosling et al., 2011). The less available surface-water will be caused by climate change, thus likely increasing the demand for groundwater abstraction (Russo and Lall, 2017). Groundwater-fed springs, lakes, streams and wetlands in some arid and semi-arid areas shank or disappeared due to the decrease of groundwater discharge and decline of groundwater table induced by intense human activities and climate change (Lin et al., 2018; Xiao et al., 2015; Heintzman et al., 2017).

Over the last several decades, groundwater depletion induced by both human activities and climate change has extended from local scales to much larger regions throughout the world (Russo and Lall, 2017; Asoka et al., 2017). More than 40 years groundwater levels records from over 15000 wells in US and their frequency analysis indicated that the response of water levels from both shallow (less than 30 m deep) and intermediate (30–150 m deep) wells is sensitive to interannual and decadal climate variability, and a strong connection existed between groundwater in deep aquifers and climate (Russo and Lall, 2017). Both satellite and local well data were used to identify the regional patterns in groundwater storage change and the relative contribution of groundwater pumping and monsoon precipitation in India (Asoka et al., 2017). The data suggested that more groundwater pumping was induced by precipitation variability that is related to the warming of the Indian Ocean surface temperatures. The above studies revealed that the groundwater storage could be either affected by climate-induced changes in recharge and evapotranspiration, or by the climate-enhanced groundwater pumping. However, the indirect effects of climate-induced water demand and abstraction on groundwater haven't drawn enough attention and should be considered in estimates of change of long-term groundwater-storage, especially in irrigated agricultural areas (Gurdak, 2017).

Global climate change may also affect groundwater quality in various ways. Solute transport may be affected by changes in recharge rates, flow path and discharge caused by atmospheric change. Changes in spatiotemporal groundwater recharge pattern that has been linked to interannual and multidecadal climate variability may mobilize pore-water reservoirs with high chloride and nitrate concentration in shallow aquifers in arid and semiarid regions (Green et al., 2011).

27.5 Integrated water management for groundwater sustainability

The hydrological cycle in cold regions can be affected by both human activities and climate change, which may also affect the surface water and groundwater resources in the downstream arid and semi-arid areas. Rapid population growth and economic development as well as increasing intensive agricultural activities, wastewater discharge without efficient water reuse have been putting immense pressure on the groundwater resources. There are optimal groundwater table depths for the growth of vegetation in arid and semi-arid areas, and thus more attention should be paid to the relationship between the groundwater and plant water uptake for the healthy vegetation ecosystem during water sources management. Given the relationship of the groundwater and its interaction with surface water to the ecosystem and agriculture system in the arid regions, the integrated assessment of groundwater and surface water resources is critical for reasonable water management and exploitation of groundwater from aquifer systems and protecting water from pollution. It is usually difficult for decision-making and optimization in water management since it encompasses multidisciplinary knowledge including engineering, social, economic and ecological constraints and objectives (Li et al., 2018; Wheater et al., 2010). The coordination of different aspects of the systems requires the use of numerical models as a support tool to integrate the multiple elements for groundwater decision-making and management. The groundwater resources should also be managed from the entire watershed-scale instead of local scale, considering both the upstream cold regions which mainly provides water to downstream arid and semi-arid areas which are the main water-consuming area.

Aquifer depletion found widely in arid and semi-arid regions is mostly attributed to socioeconomic induced groundwater over-pumping, and most of them are in irrigated agricultural areas of the world (Wang et al., 2018a; Gong et al., 2018; Scanlon et al., 2012a, 2012b), resulting in a global groundwater crisis. Climate variability is strongly correlated to vulnerable areas in terms of groundwater sustainability, especially for semi-arid and arid regions. However, many regions are lack long-term data of groundwater abstraction and storage. The evidence from the US, China and India (Russo and Lall, 2017; Wang et al., 2018a; Gong et al., 2018) unanimously reflects that the groundwater levels dropped widely across the large countries, including some areas where the water sources are not directly stressed by climate.

Existing researches have mostly focused on the direct impacts of climate change on groundwater via affecting hydrological processes. Great efforts are still required to fill the knowledge gaps of the indirect impact of climate variability and change on groundwater. Distinguishing the direct influence of climate change on groundwater from indirect impact is more complicated than on surface water, due to the large variation of the groundwater residence time from days to tens of thousands of years or more, and thus the response of groundwater to climate may be hysteretic, especially in deep aquifers (Gurdak, 2017). It is difficult to identify climatic stresses from human demand on groundwater since climate change can influence both the groundwater recharge and the water demand. Cautious groundwater management which considers global change must take into account the future projection of frequency and intensity of warming trends and dry period, as well as the socio-economic effects with constraints to avoid ecological disasters.

Acknowledgements

This work was financially supported by National Key Research and Development Program of China (2017YFC0406105) and National Natural Science Foundation of China (41772270).

References

Anderson, L., Birks, J., Rover, J., Guldager, N., 2013. Controls on recent Alaskan lake changes identified from water isotopes and remote sensing. Geophys. Res. Lett. 40, 3413–3418.

Asoka, A., Gleeson, T., Wada, Y., Mishra, V., 2017. Relative contribution of monsoon precipitation and pumping to changes in groundwater storage in India. Nat. Geosci. 10, 109–119.

Bhanja, S., Zhang, X., Wang, J., 2018. Estimating long-term groundwater storage and its controlling factors in Alberta Canada. Hydrol. Earth Syst. Sci. 22 (12), 6241–6255.

Bierkens, M., Wada, Y., 2019. Non-renewable groundwater use and groundwater depletion: a review. Environ. Res. Lett. 14, 063002.

Bosson, E., Selroos, J.O., Stigsson, M., Gustafsson, L., Destouni, G., 2013. Exchange and pathways of deep and shallow groundwater in different climate and permafrost conditions using the Forsmark site, Sweden, as an example catchment. Hydrogeol. J. 21, 225–237.

Brown, J., Ferrians, O., Heginbottom, J.A., Melnikov, E., 2002. Circum-Arctic Map of Permafrost and Ground-Ice Conditions, Version 2. [Indicate subset used]. NSIDC: National Snow and Ice Data Center, Boulder, Colorado USA.

Carroll, M.L., Townshend, J.R.G., DiMiceli, C.M., Loboda, T., Sohlberg, R.A., 2011. Shrinking lakes of the Arctic: spatial relationships and trajectory of change. Geophys. Res. Lett. 38, L20406.

Chang, Q., Ma, R., Sun, Z., Zhou, A., Hu, Y., Liu, Y., 2018. Using isotopic and geochemical tracers to determine the contribution of glacier-snow meltwater to streamflow in a partly glacierized alpine-gorge catchment in northeastern Qinghai-Tibet Plateau. J. Geophys. Res. Atmos. 123 (18), 10037–10056.

Cheng, G., Jin, H., 2013. Permafrost and groundwater on the Qinghai-Tibet Plateau and in northeast China. Hydrogeol. J. 21, 5–23.

Cheng, G., Wu, T., 2007. Responses of permafrost to climate change and their environmental significance, Qinghai-Tibet Plateau. J. Geophys. Res. 112, F02S03. Available from: https://doi.org/10.1029/2006JF000631.

Chinnasamy, P., 2017. Depleting groundwater - An opportunity for flood storage? A case study from part of the Ganges River basin, India. Hydrol. Res. 48 (2), 431–441. Available from: https://doi.org/10.2166/nh.2016.261.

Dalin, C., Wada, Y., Kastner, T., Puma, M.J., 2017. Groundwater depletion embedded in international food trade. Nature 543, 700–704.

Dragoni, W., Sukhija, B.S., 2008. Climate Change and Groundwater: A Short Review. Geol. Soc. Spec. Publ. 1–12.

Evans, S., 2017. The Hydrogeology of Cold Regions in a Warming World. Geol. Sci. Graduate Theses & Diss. 118.

Evans, S.G., Ge, S., 2017. Contrasting hydrogeologic responses to warming in permafrost and seasonally frozen ground hillslopes. Geophys. Res. Lett. 44 (4), 1803–1813. Available from: https://doi.org/10.1002/2016GL072009.

Evans, S.G., Ge, S., Voss, C.I., Molotch, N.P., 2018. The role of frozen soil in groundwater discharge predictions for warming alpine watersheds. Water Resour. Res. 54, 1599–1615. Available from: https://doi.org/10.1002/2017WR022098.

Fallatah, O.A., Ahmed, M., Save, H., Akanda, A., 2017. Quantifying temporal variations in water resources of a vulnerable middle eastern transboundary aquifer system. Hydrological Process. 31 (23), 4081–4091. Available from: https://doi.org/10.1002/hyp.11285.

Famiglietti, J.S., Ho, M., Bethune, J., Anderson, K.J., Syed, T.H., Swenson, S.C., et al., 2011. Satellites measure recent rates of groundwater depletion in California's Central Valley. Geophys. Res. Lett. 38 (3), 2–5. Available from: https://doi.org/10.1029/2010GL046442.

Feng, W., Zhong, M., Lemoine, J.M., Biancale, R., Hsu, H., Xia, J., 2013. Evaluation of groundwater depletion in North China using the Gravity Recovery and Climate Experiment (GRACE) data and ground-based measurements. Water Resour. Res. 49 (4), 2110–2118.

Ge, S.M., McKenzie, J., Voss, C., Wu, Q.B., 2011. Exchange of groundwater and surface-water mediated by permafrost response to seasonal and long term air temperature variation. Geophys. Res. Lett. 38, L14402. Available from: https://doi.org/10.1029/2011GL04791.

Gong, H., Pan, Y., Zheng, L., Li, X., Zhu, L., Zhang, C., et al., 2018. Long-term groundwater storage changes and land subsidence development in the North China Plain (1971–2015). Hydrogeology J. 26 (5), 1417–1427.

Gosling, S., Taylor, R.G., Arnell, N., Todd, M.C., 2011. A comparative analysis of projected impacts of climate change on river runoff from global and catchment-scale hydrological models. Hydrol. Earth Syst. Sci. 15 (1), 279–294.

Green, T.R., Taniguchi, M., Kooi, H., Gurdak, J.J., Allen, D.M., Hiscock, K.M., et al., 2011. Beneath the surface of global change: Impacts of climate change on groundwater. J. Hydrol. 405, 532–560.

Grenier, C., Anbergen, H., Bense, V., Chanzy, Q., Coon, E., Collier, N., et al., 2018. Groundwater flow and heat transport for systems undergoing freeze-thaw: Intercomparison of numerical simulators for 2D test cases. Adv. Water Resour. 114, 196–218.

Gurdak, J.J., 2017. Climate-induced pumping. Nat. Geosci. 10, 71.

Heintzman, L.J., Starr, S.M., Mulligan, K.R., Barbato, L.S., McIntyre, N.E., 2017. Using Satellite Imagery to Examine the Relationship between Surface-Water Dynamics of the Salt Lakes of Western Texas and Ogallala Aquifer Depletion. Wetlands 37 (6), 1055–1065.

Hinzman, L.D., Destouni, G., Woo, M.K., 2013. Preface: hydrogeology of cold regions. Hydrogeology J. 21, 1–4.

Hu, Y., Ma, R., Wang, Y., Chang, Q., Wang, Q., Ge, M., et al., 2019a. Using hydrogeochemical data to trace groundwater flow paths in a cold alpine catchment 33 (14), 1942–1960.

Hu, Z., Zhou, Q., Chen, X., Chen, D., Li, J., Guo, M., et al., 2019b. Groundwater depletion estimated from GRACE: A challenge of sustainable development in an arid region of Central Asia. Remote. Sens. 11 (16), 1908. Available from: https://doi.org/10.3390/rs11161908.

Jepsen, S.D., Walvoord, M.A., Voss, C.I., Rover, J., 2016. Effect of permafrost thaw on the dynamics of lakes recharged by ice-jam floods: case study of Yukon Flats, Alaska, Hydrol. Process. 30, 1782–1795.

Khaki, M., Forootan, E., Kuhn, M., Awange, J., van Dijk, A.I.J.M., Schumacher, M., et al., 2018. Determining water storage depletion within Iran by assimilating GRACE data into the W3RA hydrological model. Adv. Water Resour. 114, 1–18. Available from: https://doi.org/10.1016/j.advwatres.2018.02.008.

Lamontagne-Halle, P., McKenzie, J.M., Kurylyk, B.L., Zipper, S.C., 2018. Changing groundwater discharge dynamics in permafrost regions. Environ. Res. Lett. 13, 084017.

Li, X., Cheng, G., Lin, H., Cai, X., Fang, M., Ge, Y., et al., 2018. Watershed system model: The essentials to model complex human-nature system at the river basin scale. J. Geophys. Research: Atmospheres 123, 3019–3034.

Liao, C., Zhuang, Q., 2017. Quantifying the role of permafrost distribution in groundwater and surface water interactions using a three-dimensional hydrological model. Arctic, Antarctic, Alp. Res. 49 (1), 81–100.

Lin, J., Ma, R., Hu, Y., Sun, Z., Wang, Y., McCarter, C., 2018. Groundwater sustainability and groundwater/surface-water interaction in arid Dunhuang Basin, northwest China. Hydrogeol. J. 26, 1559–1572.

Ma, R., Sun, Z., Hu, Y., Chang, Q., Wang, S., Xin, W., et al., 2017. Hydrological connectivity from glaciers to rivers in the Qinghai-Tibet Plateau: Roles of suprapermafrost and subpermafrost groundwater. Hydrol. Earth Syst. Sci. 21 (9), 4803–4823.

Miller, N.L., Dale, L.L., Brush, C.F., Vicuna, S.D., Kadir, T.N., Dogrul, E.C., et al., 2009. Drought resilience of the California Central Valley surface-ground-waterconveyance system. J. Am. Water Resour. Assoc. 45 (4), 857–866.

Mohamed, A., Sultan, M., Ahmed, M., Yan, E., Ahmed, E., 2017. Aquifer recharge, depletion, and connectivity: Inferences from GRACE, land surface models, and geochemical and geophysical data. GSA Bull. 129 (5–6), 534–546. Available from: https://doi.org/10.1130/B31460.1.

Mukherjee, A., Scanlon, B., Aureli, A., Langan, S., Guo, H., and McKenzie, A., 2020. Global Groundwater: Source, Scarcity, Sustainability, Security and Solutions. Elsevier, 1st. Ed., ISBN. 9780128181720

Neilson, B.T., Cardenas, M.B., O'Connor, M.T., Rasmussen, M.T., King, T.V., Kling, G.W., 2018. Groundwater flow and exchange across the land surface explain carbon export patterns in continuous permafrost watersheds. Geophys. Res. Lett. 45, 7596–7605. Available from: https://doi.org/10.1029/2018GL078140.

O'Connor, M.T., Cardenas, M.B., Neilson, B.T., Nicholaides, K.D., Kling, G.W., 2019. Active layer groundwater flow: The interrelated effects of stratigraphy, thaw, and topography. Water Resour. Res. 55, 6555–6576.

O'Donnell, J.A., Jorgenson, M.T., Harden, J.W., McGuire, A.D., Kanevskiy, M.Z., Wickland, K.P., 2012. The Effects of Permafrost Thaw on Soil Hydrologic, Thermal, and Carbon Dynamics in an Alaskan Peatland. Ecosystems 15, 213–229.

Olivares, E.A.O., Torres, S.S.T., Jimenez, S.I.B., Enriquez, J.O.C., Zignol, F., Reygadas, Y., et al., 2019. Climate change, land use/land cover change, and population growth as drivers of groundwater depletion in the Central Valleys, Oaxaca, Mexico. Remote. Sens. 11 (11), 1290. Available from: https://doi.org/10.3390/rs11111290.

Painter, S.L., Moulton, J.D., Wilson, C.J., 2013. Modeling challenges for predicting hydrologic response to degrading permafrost. Hydrogeol. J. 21, 221–224.

Russo, T.A., Lall, U., 2017. Depletion and response of deep groundwater to climate-induced pumping variability. Nat. Geosci. 10, 105–108.

Scanlon, B.R., Faunt, C.C., Longuevergnec, L., Reedya, R.C., Alley, W.M., McGuired, V.L., et al., 2012a. Groundwater depletion and sustainability of irrigation in the US High Plains and Central Valley. PNAS. 109 (24), 9320–9325.

Scanlon, B.R., Longuevergne, L., Long, D., 2012b. Ground referencing GRACE satellite estimates of groundwater storage changes in the California Central Valley, USA. Water Resour. Res. 48 (4), W04520. Available from: https://doi.org/10.1029/2011WR011312.

Schilling, O.S., Doherty, J., Kinzelbach, W., Wang, H., Yang, P.N., Brunner, P., 2014. Using tree ring data as a proxy for transpiration to reduce predictive uncertainty of a model simulating groundwater–surface water–vegetation interactions. J. Hydrol. 519, 2258–2271.

Shen, H.H. (Ed.), 2018. Cold regions science and marine technology. Encyclopedia of Life Support Systems (EOLSS), volumes 3.

Smith, L.C., Sheng, Y., MacDonald, G.M., Hinzman, L.D., 2005. Disappearing Arctic Lakes. Science. 308, 1429.

Sun, Z., Ma, R., Wang, Y., Hu, Y., Sun, L., 2016. Hydrogeological and hydrogeochemical control of groundwater salinity in an arid inland basin: Dunhuang Basin, northwestern China. Hydrol. Process. 30, 1884–1902.

Sun, Z., Zheng, Y., Li, X., Tian, Y., Han, F., Zhong, Y., et al., 2018. The nexus of water, ecosystems, and agriculture in endorheic river basins: A system analysis based on integrated ecohydrological modeling. Water Resour. Res. 54, 7534–7556.

Tague, C., Grant, G.E., 2009. Groundwater dynamics mediate low-flow response to global warming in snow-dominated alpine regions. Water Resour. Res. 45 (7), W07421.

Tian, Y., Zheng, Y., Wu, B., Wu, X., Liu, J., Zheng, C., 2015. Modeling surface water-groundwater interaction in arid and semi-arid regions with intensive agriculture. Environ. Model. Softw. 63, 170–184.

Trabucco, A., Zomer, R., 2019. Global Aridity Index and Potential Evapotranspiration (ET0) Climate Database v2. figshare. Fileset. https://doi.org/10.6084/m9.figshare.7504448.v3.

Vidstrand, P., Follin, S., Selroos, J., Näslund, J., Rhén, I., 2013. Modeling of groundwater flow at depth in crystalline rock beneath a moving ice-sheet margin, exemplified by the Fennoscandian Shield, Sweden. Hydrogeol. J. 21, 239–255.

Walvoord, M.A., Striegl, R.G., 2007. Increased groundwater to stream discharge from permafrost thawing in the Yukon River basin: Potential impacts on lateral export of carbon and nitrogen. Geophys. Res. Lett. 34, L12402. Available from: https://doi.org/10.1029/2007GL030216.

Walvoord, M.A., Voss, C.I., Wellman, T.P., 2012. Influence of permafrost distribution on groundwater flow in the context of climate-driven permafrost thaw: Example from Yukon Flats Basin, Alaska, United States. Water Resour. Res. 48, W07524.

Walvoord, M.A., Voss, C.I., Ebel, B.A., Minsley, B.J., 2019. Development of perennial thaw zones in boreal hillslopes enhances potential mobilization of permafrost carbon. Environ. Res. Lett. 14, 015003.

Wang, J., Gao, Y., Wang, S., 2015. Land use/cover change impacts on water table change over 25 years in a desert-oasis transition zone of the Heihe River basin, China. Water 8, 11.

Wang, W., Wang, Z., Hou, R., Guan, L., Dang, Y., Zhang, Z., et al., 2018a. Modes, hydrodynamic processes and ecological impacts exerted by river–groundwater transformation in Junggar Basin, China. Hydrogeol. J. 26, 1547–1557.

Wang, Y., Zheng, C., Ma, R., 2018b. Review: Safe and sustainable groundwater supply in China. Hydrogeol. J. 26, 1301–1324.

Wheater, H.S., Mathias, S.A., Li, X. (Eds.), 2010. Groundwater Modelling in Arid and Semi-Arid Areas. Cambridge University Press.

White, D., Hinzman, L., Alessa, L., Cassano, J., Chambers, M., Falkner, K., et al., 2007. The arctic freshwater system: Changes and impacts. J. Geophys. Research: Biogeosciences 112, G04S54. Available from: https://doi.org/10.1029/2006JG000353.

White, R.P., Nackoney, J., 2003. Drylands, people, and ecosystem goods and services: A Web-Based Geospatial Analysis. World Resources Institute, Washington, DC, p. 58.

Woldeamlak, S.T., Batelaan, O., De Smedt, F., 2007. Effects of climate change on the groundwater system in the Grote-Nete catchment, Belgium. Hydrogeol. J. 15 (5), 891–901.

Woo, M.-K. (Ed.), 2012. Permafrost Hydrology. Springer, Berlin.

Xiao, S., Xiao, H., Peng, X., Song, X., 2015. Hydroclimate-driven changes in the landscape structure of the terminal lakes and wetlands of the China's Heihe River Basin. Env. Monit. Assess. 187, 4091.

Xie, X., Xu, C., Wen, Y., Li, W., 2018. Monitoring groundwater storage changes in the Loess Plateau using GRACE satellite gravity data, hydrological models and coal mining data. Remote. Sens. 10 (4), 605. Available from: https://doi.org/10.3390/rs10040605.

Xu, W., Su, X., 2019. Challenges and impacts of climate change and human activities on groundwater-dependent ecosystems in arid areas — A case study of the Nalenggele alluvial fan in NW China. J. Hydrol. 573, 376—385.

Yang, D., Marsh, P., Gelfan, A. (Eds.), 2011. Cold regions hydrology in a changing climate, 346. IAHs, p. 208. ISBN Number: 978-1-907161-21-6.

Zhang, X., He, J., Zhang, J., Polyakov, I., Gerdes, R., Inoue, J., et al., 2013. Enhanced poleward moisture transport and amplified northern high-latitude wetting trend, Nature. Clim. Change 3, 47—51.

Chapter 28

Groundwater in Australia—understanding the challenges of its sustainable use

Basant Maheshwari
Western Sydney University, Hawkesbury Campus, Penrith, NSW, Australia

28.1 Introduction

Groundwater is an important component of the hydrologic cycle, and its sustainability is influenced by its interaction with surface water, ecosystems, and people. It is a critical resource globally for food production, ecosystem services, and domestic and industrial water supplies, but the depletion of groundwater has the potential to result in significant interruption of societal and ecological functions. Groundwater being an invisible and common pool resource creates additional management and sustainability challenges and can cause conflict among different users and stakeholders. Groundwater is also emerging as a critical issue for cities and towns in Australia and other parts of the world. The global groundwater challenge we now face is not about developing the resource but its sustainable management (Shah et al., 2000; Fienen and Arshad, 2016; Mukherjee et al., 2020).

Australia is the driest inhabited continent on Earth occupying the vast land area with limited surface water resources and highly variable streamflows. Due to be being geologically an old continent, the groundwater in many parts the country is thousand years old. The surface water resources are limited and variable in most parts of the country. However, it is important to mention that the country has vast reserves of groundwater, mostly, it is very old groundwater. With a sound understanding of the characteristics of each aquifer and careful, long-term planning and management, the groundwater reserves can be sustained while deriving benefits for economic activities.

Groundwater is the main source of water for many urban, rural, and remote communities across Australia with some regions, more reliant on groundwater than others. Groundwater accounts for over 30% of Australia's total water used for drinking, irrigation, stock supply, bottling, mining, and many other uses (NWC, 2008; Johns, 2016). As the development of industrial and agricultural in Australia increases, the demand for water also steadily grows. In some parts of the country the current rate of groundwater extraction is depleting the resource faster than it is being recharged.

Over the last four decades, there has been a rapid development of groundwater resources for the various uses, and the groundwater use is expected to continue to increase with the growth of Australia's population and ongoing economic development. It is vital for water security and sustainability of urban areas, rural communities, industry, and our environment that we manage groundwater innovatively and effectively.

28.2 Aquifers in Australia

Aquifers in Australia can be grouped into two geologic formation types, sedimentary and fractured rocks (Fig. 28.1). In terms of spatial distribution of aquifers, they are grouped into 10 groundwater provinces or regions based on key hydrogeological features for understanding groundwater management options and strategies (Fig. 28.2). The major groundwater resources in Australia are located in six basins:

1. The Great Artesian Basin
2. The Murray—Darling Basin
3. The Perth Basin
4. The Canning Basin
5. The Daly Basin

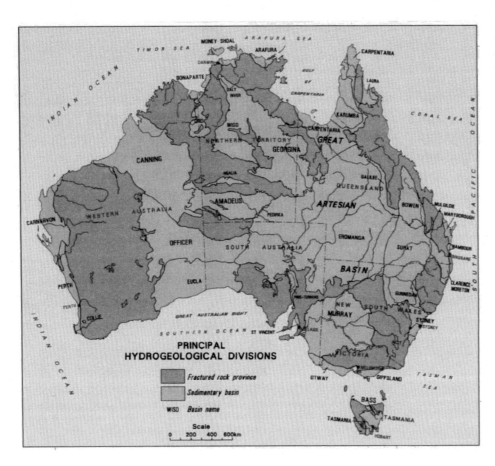

FIGURE 28.1 Hydrogeology map of Australia. *Adapted from Jacobson, G., Lau, J.E. 1987. Hydrogeology Map of Australia — 1:5m. Geoscience Australia, Canberra. <http://pid.geoscience.gov.au/dataset/ga/15629>.*

6. The Otway Basin

Apart from the abovementioned basins, there are numerous other local aquifer systems throughout Australia that have an important role in supplying water for domestic, farming, mining, and other industrial uses. During the early development, groundwater development occurred in sedimentary aquifers, and they are still the main source of groundwater in Australia (Harrington and Cook, 2014). Sedimentary aquifers are on in flatter area, mostly floodplains of rivers. Most parts of coastal areas of Queensland, New South Wales, and Western Australia have shallow sedimentary aquifers. The Great Artesian Basin, the Canning Basin, and the Perth Basin mostly have deep sedimentary formation and contain significant quantity of groundwater storage.

28.3 The Great Artesian Basin

The Great Artesian Basin in central Australia is one of the largest underground water reservoirs in the world (Johns, 2016). It covers one-fifth (22%) of the continent of Australia—occupying more than 1.7 million square kilometers area in the parts of Queensland, New South Wales, South Australia, and the Northern Territory (Department of Agriculture, Water and the Environment, 2020). The groundwater in the majority of the basin is under hydraulic pressure (artesian conditions), and its old groundwater is probably protected from climate change effects except some recharge areas. The groundwater of the basin is mostly old and nonrenewable resource, and for this reason, it needs to be used and managed carefully to minimize negative effects on spring flows and access to water for drinking and agricultural water supplies (Smerdon et al, 2012).

The groundwater in parts of the basin emerges naturally from the basin through cracks in the rock as springs, mostly on the edges of the basin where water table is close to the surface. Irrigation bores drilled in part of the basin have groundwater rising to the surface and flowing out without a pump. The groundwater from the basin supports pastoral industry, town water supplies, mining activities, and other industries located in the area. The basin contains large quantities of groundwater, an estimated 8,700,000 million m^3, with natural hydraulic pressures from the deeper aquifers

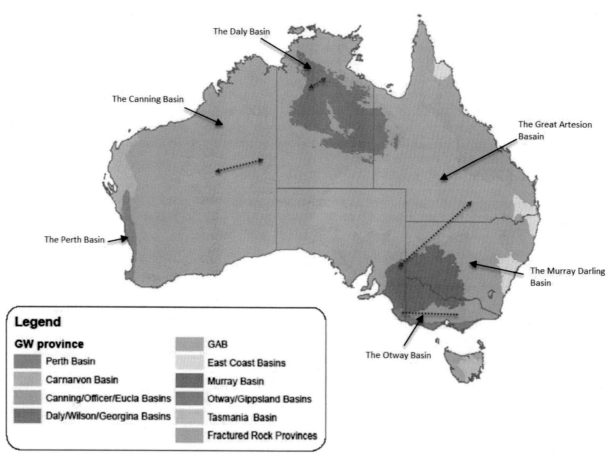

FIGURE 28.2 Groundwater provinces across Australia with the location of key basins. *Adapted from Bureau of Meteorology, 2020d. Australian Aquifer Boundary Grouping and Alignment With National Aquifer Framework. <http://www.bom.gov.au/water/groundwater/insight/documents/AquiferBoundariesMethod.pdf>.*

being aboveground surface in many places. The groundwater quality is generally good (830–1600 mg/L), but in some parts, it can be quite high in aquifers that contain very old groundwater (millions of years) (Ordens et al., 2020).

28.4 The Murray–Darling Basin

The Murray–Darling Basin is Australia's one of the most productive farming regions, and the region is home to 70% of Australia's irrigated land area. It covers an area of 1.06 million km^2 in southeastern Australia, 14% of Australia's total land area. Along with surface water, groundwater is an important water source for irrigated agriculture, urban water supplies, and environmental needs. Groundwater from sedimentary and alluvial aquifer systems in the basin accounts for most of the groundwater extractions for agricultural and other purposes in the basin (Bureau of Meteorology, 2020a–d). Basin consists of thousands of interconnected creeks and rivers flowing above a complex aquifers system. The southern part of the Great Artesian Basin covers the northern part of the Murray–Darling Basin. The groundwater used in the basin is mostly pumped from shallow aquifers. The basin aquifers include sedimentary formation within the flatter landscapes of the basin and the fractured rock and alluvial formation in the highlands bordering the basin (Ife and Skelt, 2004).

28.5 The Perth Basin

The Perth Basin covers an area of about 100,000 km^2 and is readily accessible, consisting of farming and natural lands in the central region. More than half of the basin area has been cleared for dryland agriculture, less than 5% is urban, and rest is natural land. Groundwater from the basin is the main source of water for domestic supplies for the Perth metropolitan area, irrigation for horticultural crops, mining, and other commercial needs in the area (Water for the Healthy

Country Flagship, 2009). Groundwater also provides baseflow for some rivers in the basin and lowers river salinity and keeps some flow during the dry periods.

The basin consists of flat coastal plains and is dominated by well-drained sandy soils that help easy percolation of rainwater for groundwater recharge in the basin. Any excess groundwater in the basin aquifer drains laterally toward the Indian Ocean, Swan–Canning River estuary system, and other regional boundaries. Most of the annual rainfall occurs during May–October period, and temperatures are at their lowest during this period. This means evapotranspiration is also low and the rainfall during this period is more effective in generating runoff and groundwater recharge.

The Perth metropolitan area in the basin is under high pressures of urbanization and groundwater pumping. Surface water resources in the basin are under stress due to the climate change, and so in the future, groundwater resources will be under additional stress to fill the gap to meet new water demands.

28.6 The Canning Basin

The Canning Basin lies in northern Western Australia and occupies about 430,000 km^2, and it is mostly covered by red sand plain and dunes from the Great Sandy Desert. The basin is about 500 km wide (NE–SW) and 800 km long (NW–SE) and has semiarid climate dominated by the monsoon. The basin is thinly populated, and groundwater extractions are small (less than 0.1 km^3/year), essentially for pastoral purposes. It does not have many watercourses, and most are short and all are ephemeral. They flow, occasionally, after a heavy rainfall, for a short period during the year. Three physiographic features of the basin include the Fitzroy Valley, that is, a moderately flat plain with recent alluvium, the Canning plain covered with sand dunes, and the coastal plain covered by coastal dunes and samphire marsh. The main rivers of the basin include the Fitzroy, Margaret, and Sturt in the north and the Oakover and De Grey in the south, and these rivers can become completely dry during droughts. The rainfall in the basin monsoonal with distinct wet and dry seasons in the north along the coast but are less defined inland. The mean annual rainfall varies from about 600 mm in north of Derby to less than 200 mm in the desert areas in the southeast (Ghassemi et al., 1992).

28.7 The Daly Basin

The Daly Basin is situated 200 km south of Darwin in northern Australia. It occupies 2,092,200 ha area and has average annual rainfall 1021 mm with tropical monsoonal climate. The main river of the basin is the Daly River while the Katherine River is the main tributary to the Daly River. The Daly River discharges into the Timor Sea in the northwest of the basin. Groundwater plays an important role in the region, and particularly the groundwater in the basin feeds the river systems and makes them perennial (Bureau of Meteorology, 2020a–d; Geoscience Australia, 2020a,b).

The main aquifers include Oolloo, Jinduckin, and Tindall aquifers that are in limestone and dolostone formations. The aquifers store significant volumes of groundwater (350,000,000 ML), provide baseflow in the Daly River and its tributaries, and maintain flow during dry season in the river system. During the dry period, most of the river flow source is groundwater from the aquifers. These rivers provide important benefits from environmental, cultural, and recreational point of view. The main uses for the water include agricultural irrigation, water for livestock, forestry, and industrial needs, and the region is currently going through an increased agricultural development. These aquifers also provide flow in springs that are culturally significant to aboriginal communities. The quality of the basin's groundwater falls within the drinking water guidelines in Australia (Tickell, 2009).

28.8 The Otway Basin

The Otway Basin is located in southeast of South Australia and southwest of Victoria. It is about 500 km long and has an area of 100,000 km^2 and has sedimentary rock formation. It includes major population centers such as Geelong and Warrnambool in Victoria and Mount Gambier in South Australia. Over the years the groundwater resources have been developed for agricultural irrigation and urban supply. Most farmers in the area have bores for extracting groundwater from the shallow aquifers for irrigation and stock. Water for town in the basin is often pumped from the deeper confined aquifers. Groundwater is an important resource in the western half of the Otway Basin (especially in South Australia) due to the lack of well-developed surface drainage systems over the karst-dominated limestone landscape. Compared to some other basins, this basin has low-to-moderate levels of groundwater development, and the risk of overpumping is limited. Most of the basin receives 600–800 mm/year. The basin has a diverse geological formation and source of valuable groundwater resource for the expanding agricultural and plantation forestry industry.

28.9 Groundwater uses

The total groundwater volume at the global scale is estimated to be about 13 times the volume of all surface water (lakes, rivers, and wetlands). Due to the arid nature of much of Australia, the ratio of the total volume of groundwater to the surface water is expected to be even higher. Groundwater plays an important role in Australia's water supply for domestic, agricultural, and industrial uses. In many towns and remote areas, groundwater is often the only reliable source of water supply. Groundwater accounts for up to 30% of the total water used in Australia and supports some groundwater-dependent ecosystems across the country (Bureau of Meteorology, 2019). In particular, agriculture is the major user of water, and the total water used from the various sources in agriculture is estimated to be 10.5 million ML (Fig. 28.3) and groundwater accounts for 2.2 million ML (20.9%) (ABS, 2019).

The sustainable use of groundwater is being affected by the pressures of climate change, groundwater pumping, and rising population. Groundwater sustainability depends on how we balance groundwater pumping with the annual recharge while meeting the demands of groundwater-dependent ecosystems. Under the arid and semiarid conditions prevailing in many parts of Australia, groundwater recharge volumes are often quite small compared to the volumes stored in the aquifer. Due to the low volumes of recharge, the underlying aquifer material, and the age of groundwater, the salinity of groundwater is often high, and for this reason, its use is restricted.

28.10 Groundwater entitlements and extractions

Groundwater planning and management in Australia are entrusted with states and territories, and the federal government is involved in some coordination role through broad national policy guidelines. For any groundwater use, groundwater plans need to be developed based on local water use, the environmental water needs, and sustainable groundwater yield.

Fig. 28.4 shows the distribution of bores (bore density) drilled across Australia to investigate geology, monitor groundwater, or pump out groundwater for agricultural or other uses. Groundwater levels and salinity from different sources are held by the Australian Bureau of Meteorology for over 228,000 bores across the country. A large number of bores are located in southeast and northeast of Australia, covering the Murray−Darling Basin and the Great Artesian Basin.

In Australia the access to groundwater depends on intended use of groundwater. There are two ways that one may be allowed to extract groundwater from an aquifer: licensed entitlement and nonlicensed entitlement. Licensed groundwater extraction is given for activities such as agriculture, mining, industrial, and other similar uses. The licensed entitlements volumes are typically less than 100 ML/year and often for agricultural activities. The entitlement in excess of 100 ML/year is more likely for town water supply and mining (Bureau of Meteorology, 2020a−d).

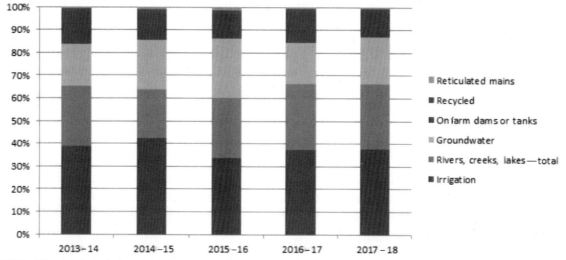

FIGURE 28.3 Water use in agriculture from the various sources, including groundwater during 2017−18. Reproduced with permission from *Bureau of Meteorology, 2020b. National water account 2016. <http://www.bom.gov.au/water/nwa/2016/daly/regiondescription/geographicinformation.shtml>); Bureau of Meteorology, 2020c. National water account 2018. <http://www.bom.gov.au/water/nwa/2018/index.shtml>).*

388 Theme | 4 Groundwater sustainability and security

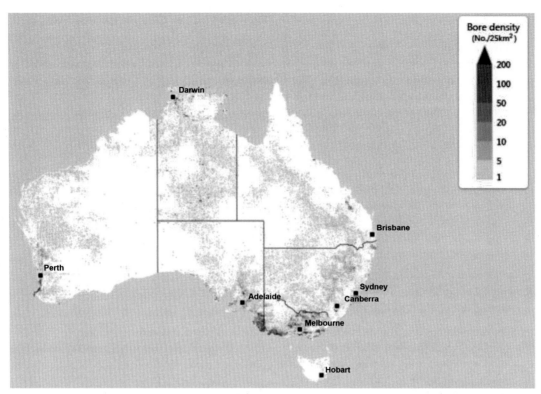

FIGURE 28.4 The distribution of bores drilled across Australia for groundwater monitoring and extraction. *Reproduced with permission from Bureau of Meteorology, 2020a. Australian groundwater insight. <http://www.bom.gov.au/water/groundwater/insight/#/overview/introduction>.*

Nonlicensed entitlement is groundwater is considered as basic right for domestic or stock water uses. When compared with licensed entitlement, the number of nonlicenced entitlements are far more but the volumes extracted per bore are <10 ML/year.

Fig. 28.5 shows the trends in groundwater extraction and the amount reflects the size of the population centers and the level of economic activities, including agriculture. There is a significant groundwater use in Western Australia and upper northeastern Australia. Fig. 28.6 shows longer term trends of groundwater level of upper, middle, and lower aquifer systems. The groundwater levels are affected by extraction patterns, but they are also influenced by local climate and land-use changes.

In Australia, mining plays an important role in the national economy and also it is a significant user of groundwater, and the industry can also impact groundwater quality. The amount of groundwater use depends on the type of mineral being extracted, local geology and geography of the mine. The volume of water used for the mining industry is about 4% of the total water use in Australia. It is also noted that metal ore mining activity uses more than half of the water used in mining industry (Australian Bureau of Statistics, 2017). Mine also extract water for dewatering the mine area if the water table is above the underground area to be mined.

28.11 Groundwater salinity

Depending upon the hydrogeology of the area, all groundwater contains some level of salt, but the salt concentration can change from fresh to saltier water. Old groundwater tends to become saltier as it passes from recharge area on the surface through the aquifer system picking up salts along the way. Groundwater use by plant and soil evaporation can increase salinity. During prolonged drought period the salinity of groundwater can increase as there are less recharge and more pumping, but groundwater salinity can get reduced when there is flooding, resulting in increased recharge and dilution of salty groundwater.

The salinity of groundwater in Australia is quite variable, varying from fresh, drinkable groundwater to highly saline water, and as such this can affect the types of uses of groundwater. In parts of Western Australian tablelands, groundwater is salty due to the combination of high evaporation rates and limited opportunities for refreshing groundwater. The groundwater in the Murray Basin is less salty, particularly where rainfall recharge is higher. On the other hand,

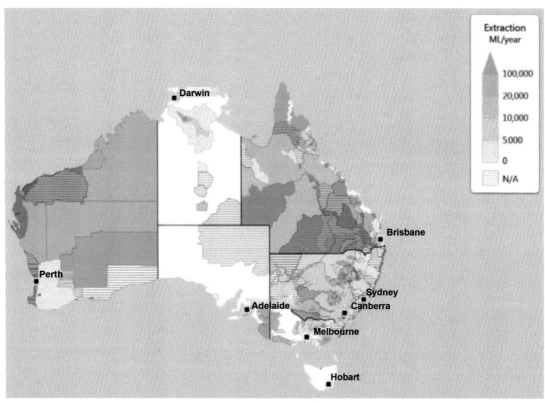

FIGURE 28.5 Groundwater extraction patterns in different parts of Australia. *Reproduced with permission from Bureau of Meteorology, 2020a. Australian groundwater insight. <http://www.bom.gov.au/water/groundwater/insight/#/overview/introduction>.*

groundwater in northern Australia is mostly fresh due to higher annual rainfall amounts when compared to other areas in Australia. Saline groundwater cannot be used for drinking and may not be suitable for crops but may still be a valuable resource in the absence of no other water sources in remote areas in Australia.

Fig. 28.7 shows average groundwater the salinity of monitored bores across Australia. The categories align with the beneficial use of groundwater. Fresh to marginal groundwater is suitable for all uses, including human consumption, but the use becomes limited if the water is saltier. Groundwater tends to saline to hypersaline in many parts of Victoria, South Australia, and Western Australia, whereas it tends to be fresh in Northern Australia and parts of the Murray Darling Basin.

28.12 Australian ecosystems and groundwater

Australia landscape has unique and fragile ecosystems due to its isolation from other countries. Water is a critical element in ecosystem and can affect its proper functioning. Australian ecosystems rely partly or mainly on the access to groundwater, and the dependency of Australian ecosystems on groundwater is well documented (Blewett, 2012). In many cases, groundwater determines the distribution of ecosystem types and their health in the landscape.

Six types of groundwater-dependent ecosystems are identified in Australia: (1) terrestrial vegetation that relies on shallow groundwater, (2) wetlands that are fed springs and groundwater flows, (3) waterways ecosystem where groundwater discharge provides a significant baseflow, (4) ecosystems with life that exists inside of aquifer and caves, (5) terrestrial fauna ecosystem that rely on groundwater for their drinking water needs, and (6) estuarine and nearshore marine systems that rely on the submarine discharge of groundwater (Geoscience Australia, 2020a,b).

From the ecosystems perspective, aquifers contribute to river flows, wetlands, and estuaries, and this contribution can be quite important, particularly during the drought years to sustain flora and fauna. In many parts of Australia, trees and some vegetative plants depend on shallow groundwater, especially during protracted droughts to meet their water needs. Roots of some Australian trees (e.g., jarrah trees in Western Australia) can go as deep as 40 m in search of groundwater. Many native animals and birds in the drier part of Australia depend on groundwater through local springs. Groundwater scarcity can also affect breeding cycle of fauna such as insects, worms, turtles, and habitats for native

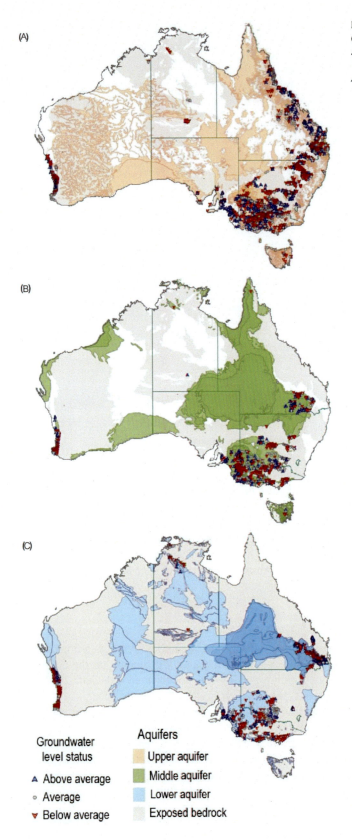

FIGURE 28.6 Groundwater level fluctuations across Australia in the (A) upper, (B) middle, and (C) lower aquifers. *Reproduced with permission from Bureau of Meteorology, 2019. Water in Australia 2017–18.* <http://www.bom.gov.au/water/waterinaustralia/files/Water-in-Australia-2017-18.pdf>.

Groundwater in Australia—understanding the challenges of its sustainable use **Chapter | 28** 391

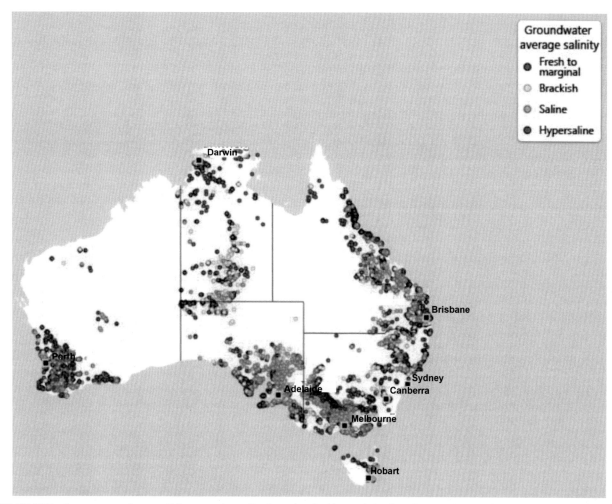

FIGURE 28.7 Average groundwater salinity across Australia. *Reproduced with permission from Bureau of Meteorology, 2020a. Australian groundwater insight. <http://www.bom.gov.au/water/groundwater/insight/#/overview/introduction>.*

animals. In general, groundwater availability in an area can affect the overall biodiversity, and changes in the groundwater availability can lead to disappearance of some species.

28.13 Concluding remarks

Groundwater use in Australia is increasing, and it is a significant or sometimes the only source of water for drinking, crop production, mining, and other activities in many parts of Australia. Australia's future water security depends on the sustainable use of groundwater along with surface water. The issue we face is the groundwater use in access to annual natural recharge, but such overuse may not be noticed for several decades especially if proper monitoring is not in place.

Groundwater is an invisible resource and often estimating its storage and movement is complex and uncertain, and therefore groundwater systems are difficult to conceptualize. There is a systematic monitoring of groundwater level and quality monitoring being facilitated by the Australian Bureau of Meteorology and data are made available publicly. Such data help to visualize this invisible resource, and as such, this can assist in dispelling many myths and misconceptions and allow evidence-based policy development.

Whether it is surface water or groundwater, we cannot manage them in isolation as both are the integral part of the water cycle. Groundwater in the area interacts with surface water supplies and so they affect each other. It is important to note that Australia being the driest inhabited continent and frequent and regular occurrence of droughts, many of its ecosystems (including river flow), plants, and animals depend upon groundwater availability for their existence. Due to the variable climate, the sustainable extraction limit of groundwater for an aquifer in Australia is generally much less than the rate of annual recharge or renewal.

References

ABS, 2019. Water Use on Australian Farms, 2017–18. Australian Bureau of Statistics. <https://www.abs.gov.au/ausstats/abs@.nsf/Latestproducts/4618.0Main%20Features22017-18?opendocument&tabname = Summary&prodno = 4618.0&issue = 2017-18&num = &view = > ANU E-Press, Canberra, p. 571. <https://press-files.anu.edu.au/downloads/press/p194981/pdf/book.pdf>.

Australian Bureau of Statistics, 2017. Water account, Australia, 2015–16. <http://www.abs.gov.au/AUSSTATS/abs@.nsf/DetailsPage/4610.02015-16?OpenDocument> (accessed 23.04.20.).

Blewett, R.S. (Ed.), 2012. Shaping a Nation: A Geology of Australia. Geoscience Australia.

Bureau of Meteorology, 2019. Water in Australia 2017–18. <http://www.bom.gov.au/water/waterinaustralia/files/Water-in-Australia-2017-18.pdf>.

Bureau of Meteorology, 2020a. Australian groundwater insight. <http://www.bom.gov.au/water/groundwater/insight/#/overview/introduction>.

Bureau of Meteorology, 2020b. National water account 2016. <http://www.bom.gov.au/water/nwa/2016/daly/regiondescription/geographicinformation.shtml>.

Bureau of Meteorology, 2020c. National water account 2018. <http://www.bom.gov.au/water/nwa/2018/index.shtml>.

Bureau of Meteorology, 2020d. Australian Aquifer Boundary grouping and alignment with national aquifer framework. <http://www.bom.gov.au/water/groundwater/insight/documents/AquiferBoundariesMethod.pdf>.

Department of Agriculture, Water and the Environment, 2020. The Great Artesian Basin. <https://www.agriculture.gov.au/water/national/great-artesian-basin> (accessed 10.04.20.).

Fienen, M.N., Arshad, M., 2016. The International Scale of the Groundwater Issue. In: Jakeman, A.J., Barreteau, O., Hunt, R.J., Rinaudo, J.D., Ross, A. (Eds.), Integrated Groundwater Management. Springer, Cham, <https://link.springer.com/chapter/10.1007/978-3-319-23576-9_2#citeas>.

Geoscience Australia, 2020a. Groundwater dependent ecosystems. <http://www.ga.gov.au/scientific-topics/water/groundwater/understanding-groundwater-resources/groundwater-dependant-ecosystems>.

Geoscience Australia, 2020b. Northern Stuart Corridor. <http://www.ga.gov.au/eftf/groundwater/northern-stuart-corridor>.

Ghassemi, F., Ethminan, H., Ferguson, J., 1992. A reconnaissance investigation of the major Palaeozoic aquifers in the Canning Basin, Western Australia, in relation to Zn-Pb mineralisation. BMR J. Aust. Geol. Geophys. 13, 37–54.

Harrington, N., Cook, P., 2014, Groundwater in Australia, National Centre for Groundwater Research and Training, Australia, p. 40.

Ife, D., Skelt, K., 2004. Murray-Darling Basin Groundwater Status 1990–2000: Summary Report, Murray-Darling Basin Commission, Canberra, p. 177.

Jacobson, G., Lau, J.E., 1987. Hydrogeology Map of Australia – 1:5m. Geoscience Australia, Canberra. <http://pid.geoscience.gov.au/dataset/ga/15629>.

Johns, C., 2016. The Forgotten Resource: Groundwater in Australia. <http://www.futuredirections.org.au/wp-content/uploads/2016/02/SAP-Groundwater-in-Australia.pdf>.

Mukherjee, A., Scanlon, B., Aureli, A., Langan, S., Guo, H., McKenzie, A., 2020. Global Groundwater: Source, Scarcity, Sustainability, Security and Solutions, first ed. Elsevier, ISBN: 9780128181720.

NWC, 2008. Groundwater Position Paper. National Water Commission. <http://www.connectedwaters.unsw.edu.au/articles/2008/04/national-water-commission-groundwater-position-statement>.

Ordens, C.M., McIntyre, N., Underschultz, J.R., Ransley, T., Moore, C., Mallants, D., 2020. Preface: advances in hydrogeologic understanding of Australia's Great Artesian Basin. Hydrogeol. J. 28 (1), 1–11. Available from: https://doi.org/10.1007/s10040-019-02107-8.

Shah, T., Molden, D., Sakthivadivel, R., Seckler, D., 2000. The Global Groundwater Situation: Overview of Opportunities and Challenges. International Water Management Institute, Colombo, <https://core.ac.uk/download/pdf/6472688.pdf>.

Smerdon, B.D., Ransley, T.R., Radke, B.M., Kellett, J.R., 2012. Water Resource Assessment for the Great Artesian Basin. A Report to the Australian Government from the CSIRO Great Artesian Basin Water Resource Assessment. CSIRO Water for a Healthy Country Flagship, Australia.

Tickell, S., 2009. Groundwater in the Daly Basin. Northern Territory Government Department of Natural Resources the Environment The Arts and Sport, Technical Report No. 27/2008D. <https://frackinginquiry.nt.gov.au/submission-library?a = 432959>.

Water for the Healthy Country Flagship, 2009. Groundwater Yields in South-West Western Australia, Factsheet 2. CSIRO, p. 3. <http://www.clw.csiro.au/publications/waterforahealthycountry/swsy/pdf/SWSY-Factsheet-Groundwater.pdf>.

Further reading

Clarke, J.D.A., Lewis, S.J., Fontaine, K., Kilgour, P.L., Stewart, G., 2015. Regional hydrogeological characterisation of the Otway Basin, Victoria and South Australia. In: Technical Report for the National Collaboration Framework Regional Hydrogeology Project, Geoscience Australia Record 2015/12, p. 129.

Laws, A.T., 1991. Outline of the groundwater resource potential of the Canning Basin, Western Australia. In: Proceedings Int. Conf. 'Groundwater in Large Sedimentary Basins', Perth, 1990. AWRC, Conference Series No 20, pp. 47–58.

The Senate Environment and Communications References Committee, 2018. An Inquiry Into the Adequacy of the Regulatory Framework Governing Water Use by the Extractive Industry, Commonwealth of Australia, p. 89. <https://www.aph.gov.au/Parliamentary_Business/Committees/Senate/Environment_and_Communications/WaterUseGovernance/Report>.

Chapter 29

Groundwater recharge and sustainability in Brazil

Paulo Tarso S. Oliveira[1], Murilo Cesar Lucas[2], Raquel de Faria Godoi[1] and Edson Wendland[3]

[1]*Faculty of Engineering, Architecture and Urbanism and Geography, Federal University of Mato Grosso do Sul, Campo Grande, Brazil,*
[2]*Department of Civil Engineering, Federal University of Technology-Paraná, Pato Branco, Brazil,* [3]*Department of Hydraulics and Sanitary Engineering, University of São Paulo, São Carlos, Brazil*

29.1 Insights from groundwater availability in Brazil

Brazil is known by its rich biodiversity and the huge surface and groundwater water availability. The country is divided in 12 hydrographic regions (Table 29.1 and Fig. 29.1), including the largest rivers (e.g., Amazon, Paraná, and São Francisco) and aquifers (Alter do Chão, Guarani, and Urucuia) in the world. However, water availability across the country is poorly distributed, leading to regions with scarcity and other with relative abundance. For instance, 83% and 67% of surface and groundwater availability, respectively, are located in the Amazon basin, where less than 5% of the population lives. On the other hand, the Paraná River Basin, the most populated region, has 6% and 10% of surface and groundwater availability, respectively (Table 29.1).

Groundwater plays an important role in water security, providing locally available water supplies in several regions of Brazil, and, often, with high-water quality. Further, the aquifers have a large storage capacity and resilience for long periods of drought, being a paramount alternative to face water scarcity in the country. The 37 outcropping areas in Brazil fall into three domains: granular, fissured, and karst, representing 54%, 45%, and 1% of the Brazilian territory, respectively (Fig. 29.1).

In the granular group, water is contained in the pore space between the grains (primary porosity). It is represented by sedimentary rocks and includes some of the largest worldwide aquifers, such as Guarani Aquifer System (GAS) (Botucatu and Pirambóia Formations) and Amazon Aquifer System (Alter do Chão, Solimões, and Içá Formations). In fissured aquifers, water is storage and flows preferentially through the rock discontinuities, which provides a secondary porosity associated with faults, fractures, and diaclases. It is represented by igneous and metamorphic rocks, and constitutes the so-called crystalline terrains. In karst aquifers, water is stored in large discontinuities in the rock formation, such as conduits and caves, associated with dissolution features. This group corresponds to the region of occurrence of sedimentary or metasedimentary rocks associated with limestone layers. The latter are related to the features of dissolution (ANA—Agência Nacional de Águas, 2007).

Despite to the importance and the large groundwater availability in Brazil, the aquifers are still poor monitored. However, some states have developed groundwater monitoring networks, such as São Paulo State, which began its activities in 1990. This network is composed by 180 wells in which 40 physical, chemical, and biological parameters are monitored every 6 months. In the Rio Grande do Norte State, groundwater is being monitored since 2010, and the number of wells has varied from 100 (at the beginning of the monitoring) to 81 wells, depending on field investigations, prioritizing issues related to the aquifer, access, site structure, and well type. In the Federal District, monitoring started in 2013, with semiannual measurements of the static groundwater level and 11 water quality parameters. They have monitored 42 wells, always one shallow (up to 30 m) close to a deep one (up to 150 m) (ANA—Agência Nacional de Águas, 2007). More recently, the Geological Survey of Brazil has implemented the Integrated Groundwater Monitoring Network (RIMAS), available at http://rimasweb.cprm.gov.br. This program counts on over 400 monitoring wells which may in future be upgraded to a national network. The program is fundamentally quantitative, that is, it focuses on recording groundwater levels. However, a quality control and alert system was designed with quarterly measurements

TABLE 29.1 Water availability in Brazilian hydrographic regions.

Brazilian hydrographic region	Mean discharge (m³/s)	Surface water availability (m³/s⁻¹)[a]	Groundwater availability (m³/s)[b]
Amazon	208,457	65,617	9809
East Atlantic	1556	271	137
Occidental Atlantic Northeast	3112	397	223
Oriental Atlantic Northeast	791	218	79
Southeast Atlantic	4843	1325	148
South Atlantic	2869	513	272
Paraguai	2836	1023	450
Paraná	12,398	4390	1479
Parnaíba	774	325	218
São Francisco	2914	875	334
Tocantins	14,895	3098	1064
Uruguai	4906	550	433

[a]*Surface water availability computed as the mean discharge of 95% probability of exceedance Q_{95}.*
[b]*Portion of the groundwater that can be exploited sustainably.*
Source: ANA—Agência Nacional de Águas, 2017. Conjuntura dos recursos hídricos no Brasil 2017 : relatório pleno [WWW Document]. <http://www.snirh.gov.br/portal/snirh/centrais-de-conteudos/conjuntura-dos-recursos-hidricos> (ANA—Agência Nacional de Águas, 2017).

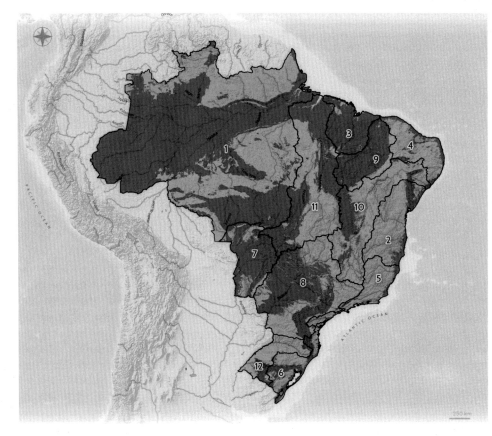

FIGURE 29.1 The 12 Brazilian Hydrographic Regions are composed of a single basin, groups of basins or contiguous watersheds, with homogeneous or similar natural, social, and economic characteristics. The hydrogeological domains (granular, fissured, and karst) are also presented. *ANA—Agência Nacional de Águas, 2013. Sistemas Aquíferos [Digital Map] [WWW Document].* Data from: *https://metadados.ana.gov.br/geonetwork/srv/pt/metadata.show?id = 150 (ANA—Agência Nacional de Águas, 2013).*

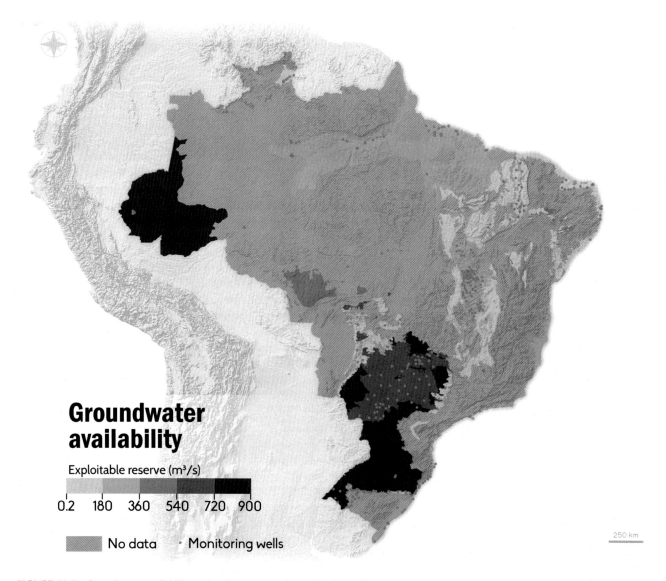

FIGURE 29.2 Groundwater availability and recharge areas of the 27 main aquifers. Total of 408 monitoring wells; however, they are poorly (spatially) distributed over IBGE—Instituto Brasileiro de Geografia e Estatística (2010).

of electrical conductivity and temperature, with chemical analyses of other 43 parameters (ANA—Agência Nacional de Águas, 2007). Therefore it is clear that groundwater monitoring in Brazil is still fairly incipient and may explain the few studies on groundwater recharge reported next in the present chapter (Fig. 29.2).

Most of the water resources studies in Brazil are concentrated on surface water; the groundwater recharge, potential use, and sustainability are still poorly addressed. In general, groundwater has become a solution for scarcity of surface water (in quantity and quality) in several Brazilian cities and also to be used as a primary source of water for crop irrigation across the country. For instance, during the 2013 and 2015 period when the worst drought reached the São Paulo State, the State and federal governments intensified studies to evaluate the possibility of increasing groundwater use, mainly in the GAS, the largest ($\sim$1.1 million km^2) transnational groundwater reservoir in South América to improve water availability in southeastern Brazil. However, an excessive groundwater pumping may lead to aquifer depletion and contamination and decrease rivers baseflow such as have been found in different regions across the world (Rodell et al., 2009; Scanlon et al., 2012; Oliveira et al., 2019; Mukherjee et al., 2020). Therefore it is important to better understand the groundwater recharge in different aquifers, land covers, and hydrometeorological conditions.

29.2 Overview of global groundwater recharge dynamics

Recharge is defined as the movement of water into the soil, flowing downward through the unsaturated zone, below the root zone of plants, until it arrives at the saturated zone, adding to groundwater storage. This definition meets to those proposed by Fitts (2012), Freeze and Cherry (1979) and is in accordance with the classical definition stated by Lerner et al. (1990) in the past. Water in the saturated zone flowing from an aquifer to another (inter aquifer flow) is not considered as recharge (Healy, 2010).

Recharge varies spatially and temporally driven by factors such as climatic conditions (Crosbie et al., 2012; Rossman et al., 2018; Xie et al., 2019), land cover and land use (Scanlon et al., 2007; Adelana et al., 2015; Mattos et al., 2019; Silva et al., 2019), thickening of the unsaturated zone (Cao et al., 2016) vegetation physiology and water-table depth (Oliveira et al., 2017), the age of vegetation (Natkhin et al., 2012), precipitation characteristics (frequency, duration, and intensity) (Tashie et al., 2016; Melo and Wendland, 2017), and soil hydraulic conductivity (Kim and Jackson, 2012). Thus because of the spatiotemporal variability in recharge rates, its estimating in a dynamic environmental relies on frequent updates.

The World-Wide Hydrogeological Mapping and Assessment Programme has been studying and providing global groundwater information for international discussion on water resources. According to Groundwater Resources Map of the World (Fig. 29.3), around 52% of the Brazilian territory comprehend large and quite uniform groundwater basins, that may offer good conditions for groundwater exploitation, while 47% is represented by local and shallow aquifers. Moreover, 27% of Brazilian aquifers and aquifer systems offer more than 300 mm/year, and 87% provide more than 100 mm/year. Groundwater recharge rates are estimated for the standard hydrologic period 1961–90 and derived from simulations with the global hydrological model WaterGAP (Döll and Fiedler, 2008).

Considering the intrinsic aquifers variability, to date, there is no method (or equipment) to measure recharge directly. Therefore all approaches used to estimate recharge lack in truth ground-based measures. The choice of recharge method has been based on measured data even though the method may not fit the particular hydrogeological conceptual model (Walker et al., 2019). Several methods are available for estimating recharge, being generally organized into six main groups (Healy, 2010): water-budget, models, Darcy methods, unsaturated/saturated zone methods, surface water based, and tracers. For further details on groundwater recharge methods, readers may refer to Scanlon et al. (2002) and Healy (2010).

The accurate recharge estimating is also a difficult task because no method has been considered as standard in groundwater science yet. Several researchers recommended applying multiple (three at least) methods to reduce uncertainty in recharge estimates (Coes et al., 2007; Delin et al., 2007). However, the use of multiple methods may not ensure accurate recharge rates (Healy, 2010). Contrary, it may lead to maintaining the uncertainty because of the different estimates produced by methods of distinct assumptions, data sources, and spatiotemporal scales (Walker et al., 2019). Therefore caution is necessary in using multiple recharge methods in order to avoid inappropriate comparison between different sources of recharge; for example, actual recharge (Scanlon et al., 2002) and potential recharge (Healy, 2010).

As the majority of recharge methods depends on measured time series data, spatiotemporal recharge estimation has been a challenge for many developing countries. In developing countries, such as Brazil, there is a lack of point ground-based hydrological data. To overcome this limitation the use of remote sensing (e.g., satellites) products coupled with Geographical Information Systems (GIS) have been increasing and becoming a common tool in hydrological science (Lettenmaier et al., 2015; Brocca et al., 2017) over the past two decades. One of the main advantages of remote sensing products is the possibility of dealing with spatially distributed hydrologic information over time (Műnch et al., 2013) to understand trends for water management and planning purposes. The use of remote sensing should grow in hydrological science as its systematic error decreases and finer spatial scales are achieved.

Many studies have used satellite products to estimate recharge using the water-budget method (Szilagyi et al., 2011; Khalaf and Donoghue, 2012; Műnch et al., 2013; Lucas et al., 2015; Coelho et al., 2017). Most of these studies applied the water-budget method in the unsaturated zone to estimate potential recharge. Moreover, the terrestrial water storage (TWS) product measured by the Gravity Recovery and Climate Experiment (GRACE) satellite, since mid-2002 through 2017 and now continued since May 2018 by the GRACE Follow-On, has risen the possibility to assess spatiotemporal groundwater levels across the globe. GRACE and GRACE-FO data provide vertically integrated estimates of changes in TWS, which include soil moisture, surface water, groundwater, and snow. The TWS data have been combined with models from the Global Land Data Assimilation System (Rodell et al., 2004), in situ measurements, and other remote sensing product, to evaluate groundwater storage changes (Scanlon et al., 2012; Voss et al., 2013). In addition, TWS has been used to estimate recharge worldwide (Henry et al., 2011; Gonçalvès et al., 2013; Wu et al., 2019).

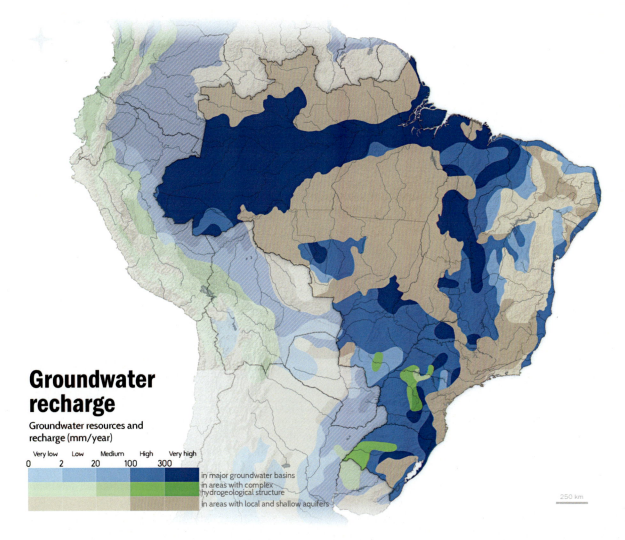

FIGURE 29.3 Groundwater Resources Map of the World, mapped by the UNESCO. The blue color is used for large and rather uniform groundwater basins (usually indicates large sedimentary basins which may offer good conditions for groundwater exploitation), green color indicates complex hydrogeological structure (occurrence of highly productive aquifers in heterogeneous folded or faulted regions close to nonaquifers) and brown color represents regions with limited groundwater resources in local and shallow aquifers. *BGR—Bundesanstalt für Geowissenschaften und Rohstoffe & UNESCO, 2008. Groundwater Resources of the World 1 : 25,000,000 [WWW Document]. https://www.whymap.org/ (BGR—Bundesanstalt für Geowissenschaften und Rohstoffe & UNESCO, 2008).*

29.3 Studies on recharge in Brazil

To search for information about estimating recharge in Brazilian aquifers, we used the Scopus scientific database. We also searched the main methods that have been used to estimate recharge in Brazil. Based on the results from search engines (title, abstract, and keywords), we found a total of 207 published papers between 1982 and 2019. The two keywords used for search were "groundwater recharge" and "Brazil." The studies that focuses in aquifer recharge estimating were selected, resulting in 26 among the 207 published papers. In addition, nine published papers from a Brazilian Groundwater Journal (Revista Águas Subterrâneas, RAS, available at https://aguassubterraneas.abas.org/asubterraneas) were also selected. Despite that the RAS is still not indexed in SCOPUS, we searched papers in this journal because it is the main Brazilian groundwater journal and has valuable information about recharge of Brazilian aquifers. The keyword *recharge estimate* (*estimativa de recarga* in Portuguese) was used for searching in RAS. Therefore we used in our evaluation 35 published papers in total.

There are few studies on groundwater recharge estimating in Brazil. We found that only 35 published papers have focused on understanding and/or to quantifying the recharge mechanism in Brazilian aquifers. The first studies were

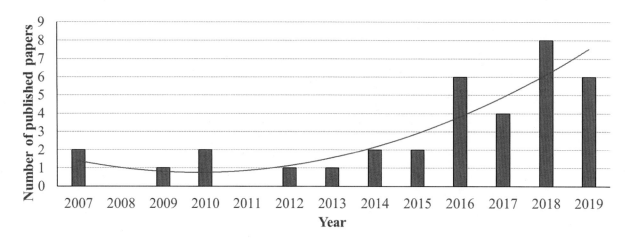

FIGURE 29.4 Number of published papers in the Scopus database about groundwater recharge in Brazilian aquifers. The solid blue trend line shows the polynomial fit to data.

published in 2007 and recently in the Scopus scientific database, since 2012, the number of publications has increased significantly (Fig. 29.4). Approximately 71% of the studies were published between 2016 and 2019. It is important to make clear that discussion on recharge values already exist in Brazil since 1960s, such as reports from the Departamento Nacional de Obras Contra as Secas (DNOCS) and the Departamento de Águas e Energia Elétrica do Estado de São Paulo (DAEE) together with consulting engineers of TAHAL-Water Planning for Israel Ltd, PhD dissertation (Rebouças, 1976) and the initial studies about the GAS (Rocha, 1997). However, most of these studies showed recharge values without explaining the scientific method that was used and the validations of it. Therefore our findings highlight the lack of information to deal with groundwater over Brazilian territory and enhance the need of developing more studies on recharge urgently.

One explanation for the scarcity of published papers is that hydrogeological data, mainly groundwater levels, of Brazilian aquifers are scarce or nonexistent. Some of the existing data are restricted to specific research groups at the universities and thus these data come up after some paper publication or academic thesis and dissertations. The RIMAS has been implemented since 2009 by Geologic Survey of Brazil and accounts for 408 monitoring wells drilled in 25 different aquifers. However, most of these wells have been monitored from 2012 to 2014. Thus RIMAS has no continuous long-term groundwater levels record yet. Furthermore, a decrease of field experimental investigations in Brazil has been observed due to the difficulty to provide continuous funding to keep long-term experimental monitoring sites. Unfortunately, this happened not only in Brazil but is also occurring around the world (Burt and McDonnell, 2015; Vidon, 2015; Tetzlaff et al., 2017).

We found that most studies of recharge estimating in Brazil were conducted in the GAS (eight papers) and in the Serra Geral Aquifer System (four). Some studies were conducted in the Rio Claro aquifer (three), Sete Lagoas Aquifer (two), Barreiras Aquifer (two), Urucuia Aquifer System (one), and Bauru Aquifer System (two), and 13 studies did not mention the aquifer's name. These studies were carried in nine of the 25 Brazilian states (Amazônia, Minas Gerais, Paraná, Pernambuco, Rio Grande do Norte, Rio Grande do Sul, Santa Catarina, and São Paulo) (Fig. 29.5). The geological and hydrogeological description of these aquifers can be found in Feitosa et al. (2016). Because most of the recharge studies were developed in the GAS a specific discussion about it is presented in the present chapter.

The GAS has gained much attention of the scientific community since 2002 because of its largest volume of freshwater and high-water quality for urban water supply. The GAS volume ranges from $29,550 \pm 4000 \text{ km}^3$ to $32,830 \pm 4400 \text{ km}^3$ and has an area of $1,088,000 \text{ km}^2$ (OAS—Organization of American States, 2009) across four countries in South America: Brazil, Argentina, Paraguay, and Uruguay. The GAS is formed by the eolian sandstones of the Jurassic (Botucatu Formation) and fluvio-eolian Triassic (Pirambóia Formation) periods (Sracek and Hirata, 2002). Recharge is supposed to occur in the outcrop areas of Pirambóia and/or Botucatu Formation, which corresponds to only ~11% of the GAS entire area (OAS—Organization of American States, 2009). In the outcrop areas the GAS is an unconfined aquifer, while the most area of the GAS is confined by the basaltic rocks of the SASG. The financial investment of the Guarani Project contributed to increase the hydrogeological knowledge of the GAS (PSAG, 2009),

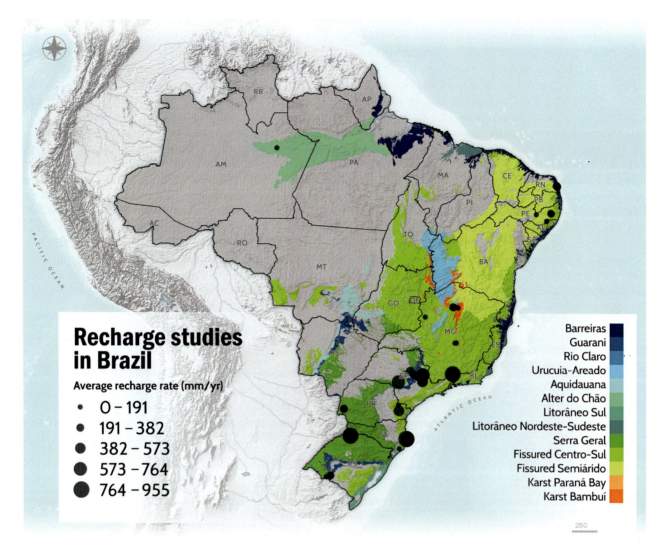

FIGURE 29.5 Density map of recharge studies and average recharge rates in annual basis according to Brazilian states and hydrographic regions. The average recharge rates were obtained from the results published in the 35 papers found in our search engine.

including recharge estimates. The Guarani Project was conducted by the Organization of American States (OAS) and supported by the Global Environmental Facility. The experience obtained in the GAS should be viewed as a benchmark for future studies in other Brazilian aquifers.

Despite the advances, the withdrawal rate that can sustainable be exploited in the GAS is still an unsolved question. In addition, the long-term recharge estimates in the GAS have been restricted to a few unconfined areas, such as, Onça Creek Basin (22°10′ to 22°15′ S and 47°55′ to 48°00′ W) (Lucas and Wendland, 2016; Mattos et al., 2019), and to both unconfined and confined areas, such as, Ibicuí Watershed (29°00′ to 31°0′ S and 56°00′ to 54°00′ W) (Borges et al., 2017). Consequently, researchers who handled with numerical groundwater models (Rodríguez et al., 2013) and also Brazilian water resources, decision-makers have assumed a single average recharge rate for the entire GAS. However, although the GAS may show similar geologic features in its outcrop area, differences in precipitation characteristics (intensity, duration, and frequency), unsaturated thickness, and land cover and land use changes (LCLUC) can lead to distinguish recharge rates in different locations. Thus future studies on recharge estimating are needed in different locations of the GAS outcrop areas.

It is also important to point out that although the recognized importance of the GAS outcrop areas to recharge, these areas have been covered by agricultural (annual and semiperennial crops) and livestock activities. These activities may affect the quantity and quality of groundwater. A few studies have reported the consequences of LCLUC on recharge in the GAS outcrop areas (see Lucas and Wendland, 2016; Mattos et al., 2019). For instance, Mattos et al. (2019) concluded that the increase in Eucalyptus plantation area affected seasonal and annual hydrology of the Onça Creek Basin by increasing evapotranspiration rates and, ultimately, leading to a decrease in recharge and groundwater levels. Oliveira et al. (2017) assessed groundwater recharge in different physiognomies of the Brazilian savanna (known as the Cerrado) located in an outcrop area of the GAS. They demonstrated that the replacement of undisturbed dense Cerrado with croplands will likely alter recharge dynamics. Lucas and Wendland (2016) presented average recharge rates for various land uses in the GAS outcrop area: 135 mm/year under Eucalyptus plantation, 248 mm/year under citrus, 296 mm/year under sugarcane, and 401 mm/year under grassland.

In the context of water use, although the exploitation of the GAS exceeds 1.0 km^3/year, overall it does not represent an overexploitation in comparison with its full capacity (Sindico et al., 2018). However, since 1970, groundwater drawdown between 30 and 40 m in the GAS were locally observed in the city of Ribeirão Preto, São Paulo State, Brazil (Foster et al., 2009). Ribeirão Preto (with a total area of 652 km^2) has a current (estimated) population of 703,000 habitants and its public water system is entirely supplied by groundwater from the GAS. The simulations of numerical groundwater flow model performed by researchers of the University of São Paulo (USP) at São Carlos School of Engineering showed a drawdown cone of about 70 m over the last 80 years in the central area of Ribeirão Preto. Moreover, the consequences of the GAS groundwater levels decline in Ribeirão Preto for the rivers have not been addressed yet. Therefore this intense overexploitation of the GAS in Ribeirão Preto should be considered a warning to the urgently need of studies on groundwater security in Brazil.

29.3.1 Recharge methods used in Brazilian studies

The methods used to estimate recharge in Brazilian aquifers were classified following Healy (2010) and the two most used of them were reviewed. The water-table fluctuation (WTF) method has been the most used method to estimate recharge in Brazilian unconfined aquifers (Fig. 29.6). However, because of the scarcity of groundwater levels record in Brazil, only five (among the 12) studies used the WTF method with a time series longer than 4 years (i.e., Andrade et al., 2014; Albuquerque et al., 2015; Lucas et al., 2015; Rama et al., 2018; Teramoto and Chang, 2018; Mattos et al., 2019).

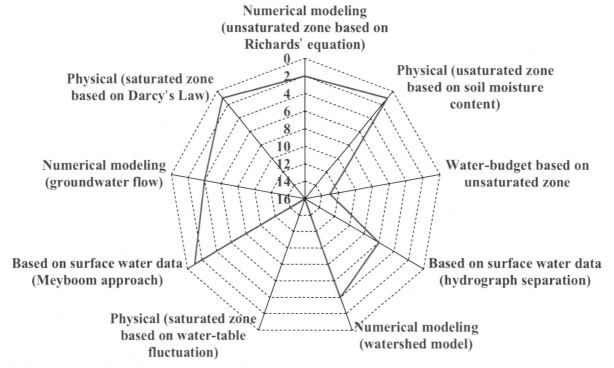

FIGURE 29.6 Number of methods used to estimate groundwater recharge in Brazilian aquifers.

The WTF method combines rises in water-table and the specific yield of the WTF depth to estimate recharge:

$$R = S_y \cdot \frac{\Delta h}{\Delta t} \quad (29.1)$$

where R is recharge (L), S_y is specific yield, Δh is the rise in the water-table (L), and t is time (T). The Δh is the difference between the peak of the water-table rise and lowest point of the extrapolated antecedent recession curve at the time of the peak. The WTF method is very attractive and popular among recharge methods because of its simplicity in mathematical application. Despite this mathematical simplicity, WTF method requires care to deal with the extrapolation of the water-table recession curve. The extrapolation of antecedent recession curve leads to inherent uncertainty of the method. A power-law (see Wendland et al., 2007) and linear functions have been often used to extrapolate recession curve in the recharge studies of the Brazilian aquifers.

S_y is the major source of uncertainty in the WTF method (Crosbie et al., 2019). Few efforts have been made to quantify recharge uncertainty based on the S_y uncertainty (i.e., Delottier et al., 2018). Besides the difficulty of acquiring reliable S_y estimates, it is intrinsic time-dependence, which implies a nonsteady-state condition (Childs, 1960; Nachabe, 2002). Because a reliable estimation of S_y in fractured aquifers (like the SASG) is still a challenge, physical methods based on WTF (such as WTF method) has not been used in Brazilian fractured aquifers. In this context, little attention has been given to measure S_y in bench-scale or field-scale of the Brazilian aquifers. As a consequence, a great number of studies have adopted some outdated value obtained from literature, average single values for extensive study areas, or have adopted some value without provide reasonable justification. In fact, the absence of measured S_y values is not an exclusive issue of Brazilian aquifers but has been seemed in other aquifers worldwide.

Braga et al. (2016) used $S_y = 0.02$ from Rebouças (1976); however, the citation was not provided in the reference list of the paper. Despite this, Rebouças (1976) proposed a range of S_y values based on the classification of hydrogeological reservoir (aquifer, aquiclude, and aquitard). Therefore because S_y can vary with depth to water-table and geometric arrangement of soil particles, the value used by Braga et al. (2016) can be considered oversimplified for the entire study area. Andrade et al. (2014) used an average (outdated) $S_y = 0.10$ based on CISAGRO (1991). Simon et al. (2017) used an average S_y value of 0.16 according to Rocha (1997), OAS—Organization of American States (2009) and Martelli (2012). However, Rocha (1997) did not detailed what method was used to measure S_y. Moreover, in the Results section, Simon et al. (2017) estimated recharge using a range of S_y values (between 0.14 and 0.18) that was not mentioned (or justified) in the Methods section.

Lucas and Wendland (2016) used measured S_y values ranging from 0.085 to 0.159 according to Gomes (2008) in the GAS outcrop area. It is important to observe the large difference among S_y values obtained by Gomes (2008) in this small study area (~65 km^2) that is predominantly composed by sandy soils (65.5% of fine sand, 13.9% of silt and clay, 20.5% of medium sand, and 0.1% of course sand). Therefore the use of S_y values from different study areas may lead to unrealistic recharge estimates. Future studies should be aware to use WTF method with reliable updated measurements of S_y in the WTF zone.

The water-budget based on unsaturated zone has been the second most used method to estimate recharge in Brazilian aquifers (Fig. 29.6). This can be justified by the more availability of ground-based hydrometeorological data (e.g., precipitation, solar radiation, wind speed, air humidity, and air temperature) in comparison with groundwater level records and aquifer's parameters (e.g., S_y and hydraulic conductivity) in Brazil. In addition, the ongoing evolution in temporal and spatial resolution of remote sensing products in hydrology and the advances on GIS techniques has increased the opportunity to use water-budget method (Lettenmaier et al., 2015). Following this trend, in Brazil, a growing number of studies have used remote sensing products to estimate recharge based on water-budget based on unsaturated zone (i.e., Lucas et al., 2015; Coelho et al., 2017; Melati et al., 2019).

Overall, water-budget method based on unsaturated zone has been handled by delimiting the control volume from the land surface to the aquifer water table in a one-dimensional soil column representative of the study area. In this case, recharge is viewed as a potential water (or potential recharge) that can eventually reach the water table and add to groundwater storage. Potential recharge is the subsurface water percolating beneath the plants root-zone through the unsaturated zone. Potential recharge is usually higher than actual recharge because a certain amount of water can be stored in the unsaturated zone space pore. Recharge is estimated as a residual of all hydrological components of the water-budget equation:

$$R = P - ET - Roff - \frac{\Delta S}{\Delta t} \quad (29.2)$$

where P is precipitation (L), ET is real evapotranspiration (L), $Roff$ is surface runoff (L), ΔS is the change in water storage (L), and t is time.

As a limitation, several studies have neglected the ΔS component (Eq. 29.2) to estimate recharge in monthly (Vestena and Kobiyama, 2007; Soares and Velásquez, 2013; Lucas et al., 2015; Braga et al., 2016; Coelho et al., 2017), in annual (Melati et al., 2019), and in daily basis (Silva et al., 2019). However, assuming a steady-state condition for ΔS can bring up large uncertainty in monthly recharge estimates (Eq. 29.2). The large uncertainty arises because water can take different residence times in different compartments inside the control volume. For example, water can take more than 1 month to pass through the unsaturated zone until reaching water-table as recharge. Storage subsurface water in the unsaturated zone and lag time between percolation and recharge increases in unsaturated zone thickening (Cao et al., 2016).

Moreover, some studies (Lucas et al., 2015) further simplified water-budget equation by neglecting the $Roff$ (Eq. 29.2) because of the occurrence of sandy soils and/or flat topography in the study area. However, consecutive precipitation events can increase soil moisture and, consequently, increasing significant amounts of $Roff$ component and thus decrease recharge rates. On the other hand, some studies (Coelho et al., 2017; Silva et al., 2019) incorporate spatial distribution of the $Roff$ over the study area and others (Albuquerque et al., 2015; Pontes et al., 2016; Galvão et al., 2018; Souza et al., 2019) have assumed both ΔS and $Roff$ components in the water-budget method. Furthermore, few studies have reported uncertainties of recharge estimates in the water-budget method. Errors in the individual hydrologic components (Eq. 29.2) contribute to the overall uncertainty in recharge. Thus future studies should move up to demonstrate the uncertainty of recharge estimates.

29.4 Challenges and future directions toward a groundwater sustainability in Brazil

The population growth has been intensified the need for agricultural and energy production. To produce more food and energy will require the use of more water, which are also linked to ecosystem services. This connection is known as water–food–energy–ecosystems (WFEE) nexus (Oliveira et al., 2019). Understanding the interconnected risks and vulnerability of these sectors is crucial for the development of sustainable resources management plans and for mitigating competition among them. Water is the main key of the WFEE nexus; therefore, one of the first steps to better understand these connections is to investigate the hydrological processes and water demands. However, in Brazil the investigation on water resources have been focused on surface water, despite the intensification of groundwater use across the country. Groundwater has been used in Brazil to water supply in several cities and mainly to sustain irrigation, and hence paramount for food security. About 39% of the Brazilian cities (total of 5565) are solely supplied by groundwater and 14% are supplied by both surface and groundwater (ANA—Agência Nacional de Águas, 2010).

It is important to highlight that the agribusiness is the main sector of Brazils' economy, representing one-third of Brazil's gross domestic product—in 2018 it was about US$ 379 billion—and 50% of its exports (China, European Union, and the United States being the major importers) (CEPEA—Centro de Estudos Avançados em Economia Aplicada, 2019). In addition, it is expected that over the coming decades, global food demands will rise further and food production will compete for space and water resources with crops used as biofuels, increasing the dependency on groundwater globally and highlighting the central place of groundwater resources within the WFEE nexus (Lawford et al., 2013; de Graaf et al., 2019). Therefore it is expected that the groundwater use will keep rising in the next years in Brazil. However, groundwater overexploitation may lead to groundwater depletion, also compromising the surface water (environmental flow requirements) and the ecosystems services.

Alarming declines in groundwater levels have been reported on different spatial scales, for example, in the California Central Valley (Scanlon et al., 2012), India (Rodell et al., 2009) and across the global (Gleeson et al., 2012; de Graaf et al., 2019). In Brazil, recent studies using the GRACE have found a significant decreasing trend in TWS in the São Francisco river basin and related it to an excessive groundwater pumping in the region, mainly for food production (Oliveira et al., 2014; Moreira et al., 2019; Oliveira et al., 2019). de Graaf et al. (2019) estimate that by 2050, environmental flow limits will be reached for ~42%–79% of the watersheds in which there is groundwater pumping worldwide.

In addition, changing climate also has the potential to alter the groundwater fluxes (both in recharge and discharge). Cuthbert et al. (2019) show that areas where water-table are more sensitive to changes in recharge are also those that have the longest groundwater response times. In particular, groundwater fluxes in arid regions are shown to be less responsive to climate variability than in humid regions. In the Brazilian aquifers, Hirata and Conicelli (2012) estimated that the impact of long-term climate changes by 2050 will lead to a severe reduction in 70% of recharge in the Northeast region aquifers (comparing to 2010 values), varying from 30% to 70% in the North region. Montenegro and Ragab (2010) used a numerical modeling (watershed model) and general circulation models to evaluate the impact of climate change on recharge in a Brazilian semiarid watershed. They showed an average decrease (between high and

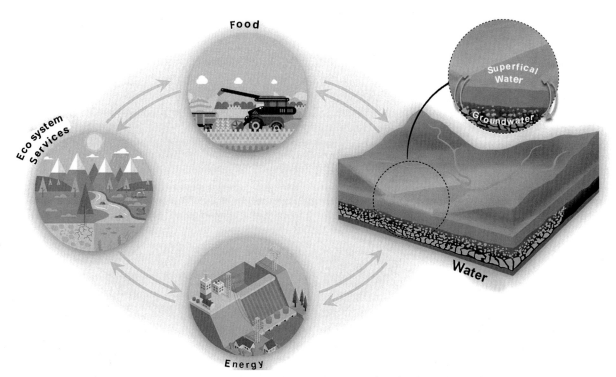

FIGURE 29.7 Water–food–energy–ecosystems nexus emphasizing the interconnection of surface and groundwater.

low emissions) by 35%, 68%, and 77% in recharge for three dry future scenarios: 2010–39, 2040–69, and 2070–99, respectively. The dry future scenarios correspond to a period of decrease in the precipitation amount.

In the context of groundwater sustainability, recharge is the most relevant process of groundwater replenishment. Estimates of recharge play a major role in quantifying levels of sustainable groundwater use (Richey et al., 2015; Bierkens and Wada, 2019; Cuthbert et al., 2019). Long-term (30 years) average recharge can be adopted as the maximum amount of groundwater abstraction (Döll and Fiedler, 2008). Further, recharge rates have been adopted in groundwater sustainability indicators (Gleeson et al., 2012; Richey et al., 2015; Thomas, 2019). Therefore to assure groundwater sustainability in Brazil is needed to better understand groundwater recharge and environmental limits for groundwater pumping.

Groundwater sustainability have often been related to the physical availability of groundwater for actual and future human needs and ecosystems health. In contrast, unsustainable groundwater can be achieved when groundwater abstraction surpasses recharge rates (Foster and MacDonald, 2014) and thus depleting groundwater levels and reduction volumes over time. Thus groundwater sustainability should be handled with nonstationarity assumption because both groundwater storage changes and groundwater use varies with time (Thomas, 2019).

The increase in water demand and the decrease in the surface water availability (in quantity and quality) will boost the use of groundwater as the main source of water for public or private uses in many regions across Brazil (Hirata and Conicelli, 2012). This expected increase in groundwater dependence and the negative effects related to overexploitation shows urgent need to identify the limits to groundwater pumping and determine where and when these limits will be reached (de Graaf et al., 2019).

In addition, an integrated use of surface and groundwater must be considered in the water-management planning (Fig. 29.7). Therefore to provide a Brazilian sustainable development, such as proposed by the 2030 agenda and adopted by all the United Nations, we need first to intensify the studies on groundwater recharge, environmental limits for groundwater pumping, interconnection of surface and groundwater, and their integration in the WFEE nexus.

Acknowledgments

This study was supported by grants from the Ministry of Science, Technology, Innovation and Communication (MCTIC) and National Council for Scientific and Technological Development (CNPq) (grants numbers 441289/2017-7 and 306830/2017-5). This study was also financed in part by the Coordenação de Aperfeiçoamento de Pessoal de Nível Superior—Brasil (CAPES)—Finance Code 001 and CAPES PrInt.

References

Adelana, S.M., Dresel, P.E., Hekmeijer, P., Zydor, H., Webb, J.A., Reynolds, M., et al., 2015. A comparison of streamflow, salt and water balances in adjacent farmland and forest catchments in south-western Victoria, Australia. Hydrol. Process. 29, 1630–1643. Available from: https://doi.org/10.1002/hyp.10281.

Albuquerque, C., Montenegro, S.M.G.L., Montenegro, A.A., de, A., Fontes Jr., R., 2015. Recarga de aquífero aluvial sob uso agrícola. Águas Subterrâneas 29. Available from: https://doi.org/10.14295/ras.v29i1.27931.

ANA—Agência Nacional de Águas, 2007. Disponibilidade e demandas de recursos hídricos no Brasil [WWW Document]. <https://www.ana.gov.br/textos-das-paginas-do-portal/publicacoes>.

ANA—Agência Nacional de Águas, 2010. Atlas de abastecimento urbano de água [WWW Document]. <http://atlas.ana.gov.br/Atlas/forms/Home.aspx>.

ANA—Agência Nacional de Águas, 2013. Sistemas Aquíferos [Digital Map] [WWW Document]. <https://metadados.ana.gov.br/geonetwork/srv/pt/metadata.show?id = 150>.

ANA—Agência Nacional de Águas, 2017. Conjuntura dos recursos hídricos no Brasil 2017: relatório pleno [WWW Document]. <http://www.snirh.gov.br/portal/snirh/centrais-de-conteudos/conjuntura-dos-recursos-hidricos>.

Andrade, T.S., Montenegro, S.M.G.L., Montenegro, A.A.D.A., Rodrigues, D.F.B., 2014. Estimation alluvial recharge semiarid. Eng. Agric. 34, 211–221. Available from: https://doi.org/10.1590/S0100-69162014000200003.

BGR—Bundesanstalt für Geowissenschaften und Rohstoffe & UNESCO, 2008. Groundwater Resources of the World 1: 25,000,000 [WWW Document]. <https://www.whymap.org/>.

Bierkens, M.F.P., Wada, Y., 2019. Non-renewable groundwater use and groundwater depletion: a review. Environ. Res. Lett. 14 (6). Available from: https://doi.org/10.1088/1748-9326/ab1a5f.

Borges, V.M., Fan, F.M., Reginato, P.A.R., Athayde, G.B., 2017. Groundwater recharge estimating in the Serra Geral aquifer system outcrop area—Paraná State, Brazil. Águas Subterrâneas 31, 338. Available from: https://doi.org/10.14295/ras.v31i4.28872.

Braga, L.T.P., Velásquez, L.N.M., Fleming, P.M., Rodrigues, P.C.H., 2016. Recarga do tipo localizada em região semiárida: estudo de caso em Dolinas da Bacia do Rio Verde Grande, Minas Gerais. Águas Subterrâneas 30, 153. Available from: https://doi.org/10.14295/ras.v30i2.27924.

Brocca, L., Crow, W.T., Ciabatta, L., Massari, C., De Rosnay, P., Enenkel, M., et al., 2017. A review of the applications of ASCAT soil moisture products. IEEE J. Sel. Top. Appl. Earth Obs. Remote. Sens. 10 (5). Available from: https://doi.org/10.1109/JSTARS.2017.2651140.

Burt, T.P., McDonnell, J.J., 2015. Whither field hydrology? The need for discovery science and outrageous hydrological hypotheses. Water Resour. Res. 51, 5919–5928. Available from: https://doi.org/10.1002/2014WR016839.

Cao, G., Scanlon, B.R., Han, D., Zheng, C., 2016. Impacts of thickening unsaturated zone on groundwater recharge in the North China Plain. J. Hydrol. 537, 260–270. Available from: https://doi.org/10.1016/j.jhydrol.2016.03.049.

CEPEA—Centro de Estudos Avançados em Economia Aplicada, 2019. PIB do agronegócio brasileiro [WWW Document]. <https://www.cepea.esalq.usp.br/br/pib-do-agronegocio-brasileiro.aspx>.

Childs, E.C., 1960. The nonsteady state of the water table in drained land. J. Geophys. Res. 65, 780–782. Available from: https://doi.org/10.1029/jz065i002p00780.

CISAGRO, 1991. Companhia Integrada de Servicos Agropecuários. Projeto de irrigação da Fazenda Nossa Senhora do Rosário, Pesqueira. Recife.

Coelho, V.H.R., Montenegro, S., Almeida, C.N., Silva, B.B., Oliveira, L.M., Gusmão, A.C.V., et al., 2017. Alluvial groundwater recharge estimation in semi-arid environment using remotely sensed data. J. Hydrol. 548, 1–15. Available from: https://doi.org/10.1016/j.jhydrol.2017.02.054.

Coes, A.L., Spruill, T.B., Thomasson, M.J., 2007. Multiple-method estimation of recharge rates at diverse locations in the North Carolina Coastal Plain, USA. Hydrogeol. J. 15, 773–788. Available from: https://doi.org/10.1007/s10040-006-0123-3.

Crosbie, R.S., Doble, R.C., Turnadge, C., Taylor, A.R., 2019. Constraining the magnitude and uncertainty of specific yield for use in the water table fluctuation method of estimating recharge. Water Resour. Res. 55, 7343–7361. Available from: https://doi.org/10.1029/2019WR025285.

Crosbie, R.S., McCallum, J.L., Walker, G.R., Chiew, F.H.S., 2012. Episodic recharge and climate change in the Murray-Darling Basin, Australia. Hydrogeol. J. 20, 245–261. Available from: https://doi.org/10.1007/s10040-011-0804-4.

Cuthbert, M.O., Gleeson, T., Moosdorf, N., Befus, K.M., Schneider, A., Hartmann, J., et al., 2019. Global patterns and dynamics of climate–groundwater interactions. Nat. Clim. Change 9 (2). Available from: https://doi.org/10.1038/s41558-018-0386-4.

de Graaf, I.E.M., Gleeson, T., (Rens) van Beek, L.P.H., Sutanudjaja, E.H., Bierkens, M.F.P., 2019. Environmental flow limits to global groundwater pumping. Nature. Available from: https://doi.org/10.1038/s41586-019-1594-4.

Delin, G.N., Healy, R.W., Lorenz, D.L., Nimmo, J.R., 2007. Comparison of local- to regional-scale estimates of ground-water recharge in Minnesota, USA. J. Hydrol. 334, 231–249. Available from: https://doi.org/10.1016/j.jhydrol.2006.10.010.

Delottier, H., Pryet, A., Lemieux, J.M., Dupuy, A., 2018. Estimating groundwater recharge uncertainty from joint application of an aquifer test and the water-table fluctuation method. Hydrogeol. J. 26, 2495–2505. Available from: https://doi.org/10.1007/s10040-018-1790-6.

Döll, P., Fiedler, K., 2008. Global-scale modeling of groundwater recharge. Hydrol. Earth Syst. Sci. 12, 863–885. Available from: https://doi.org/10.5194/hess-12-863-2008.

Feitosa, F.A.C., Diniz, J.A.O., Kirchheim, R.E., et al., 2016. Assessment of Groundwater Resources in Brazil: Current Status of Knowledge. Groundwater Assessment, Modeling, and Management. CRC Press, pp. 33–57, https://doi.org/10.1201/9781315369044-3.

Fitts, C.R., 2012. Groundwater Science. Academic Press, p. 492.

Foster, A., Hirata, R., Vidal, A., Schmidt, G., Garduño, H., 2009. The Guarani initiative—towards realistic groundwater management in a transboundary context. GW-MATE (Sustainable Groundwater Management: Lessons from Practice) Case Profile Collection, Number 9. The World Bank, Geneva, p. 28. Available from. Available from: http://siteresources.worldbank.org/INTWAT/Resources/GWMATE_English_CP_09.pdf.

Foster, S., MacDonald, A., 2014. The 'water security' dialogue: Why it needs to be better informed about groundwater. Hydrogeol. J. 22, 1489–1492. Available from: https://doi.org/10.1007/s10040-014-1157-6.

Freeze, R.A., Cherry, J.A., 1979. Groundwater. Prentice-Hall Inc, Englewood Cliffs, NJ, p. 624.

Galvão, P., Hirata, R., Conicelli, B., 2018. Estimating groundwater recharge using GIS-based distributed water balance model in an environmental protection area in the city of Sete Lagoas (MG), Brazil. Environ. Earth Sci. 77. Available from: https://doi.org/10.1007/s12665-018-7579-z.

Gleeson, T., Wada, Y., Bierkens, M.F.P., Van Beek, L.P.H., 2012. Water balance of global aquifers revealed by groundwater footprint. Nature 488, 197–200. Available from: https://doi.org/10.1038/nature11295.

Gomes, L.H., 2008. Determinação da recarga profunda na bacia-piloto do ribeirão da Onça em zona de afloramento do Sistema Aquífero Guarani a partir de balanço hídrico em zona saturada (Master's thesis). https://doi.org/10.11606/D.18.2008.tde-28042009-132417.

Gonçalvès, J., Petersen, J., Deschamps, P., Hamelin, B., Baba-Sy, O., 2013. Quantifying the modern recharge of the "fossil" Sahara aquifers. Geophys. Res. Lett. 40, 2673–2678. Available from: https://doi.org/10.1002/grl.50478.

Healy, 2010. Estimating groundwater recharge, 2010. Cambridge University Press, Cambridge, p. 245.

Henry, C.M., Allen, D.M., Huang, J., 2011. Groundwater storage variability and annual recharge using well-hydrograph and GRACE satellite data. Hydrogeol. J. 19, 741–755. Available from: https://doi.org/10.1007/s10040-011-0724-3.

Hirata, R., Conicelli, B.P., 2012. Groundwater resources in Brazil: a review of possible impacts caused by climate change. An. Acad. Bras. Cienc. 84, 297–312. Available from: https://doi.org/10.1590/S0001-37652012005000037.

IBGE—Instituto Brasileiro de Geografia e Estatística, 2010. Área de recarga dos principais sistemas aquíferos [WWW Document]. <http://geoftp.ibge.gov.br/atlas/nacional/>.

Khalaf, A., Donoghue, D., 2012. Estimating recharge distribution using remote sensing: a case study from the West Bank. J. Hydrol. 414–415, 354–363. Available from: https://doi.org/10.1016/j.jhydrol.2011.11.006.

Kim, J.H., and Jackson, R.B., 2012. A Global Analysis of Groundwater Recharge for Vegetation, Climate, and Soils. Vadose Zone Journal, 11. https://doi.org/10.2136/vzj2011.0021RA.

Lawford, R., Bogardi, J., Marx, S., Jain, S., Wostl, C.P., Knüppe, K., et al., 2013. Basin perspectives on the Water-Energy-Food Security Nexus. Curr. Opin. Environ. Sustain. Available from: https://doi.org/10.1016/j.cosust.2013.11.005.

Lerner, D., Issar, A., Simmers, I., 1990. Groundwater Recharge: A Guide to Understanding and Estimating Natural Recharge. International Contributions to Hydrogeology, Heise.

Lettenmaier, D.P., Alsdorf, D., Dozier, J., Huffman, G.J., Pan, M., Wood, E.F., 2015. Inroads of remote sensing into hydrologic science during the WRR era. Water Resour. Res. 51, 7309–7342. Available from: https://doi.org/10.1002/2015WR017616.

Lucas, M., Oliveira, P.T.S., Melo, D.C.D., Wendland, E., 2015. Evaluation of remotely sensed data for estimating recharge to an outcrop zone of the Guarani Aquifer System (South America). Hydrogeol. J. 23, 961–969. Available from: https://doi.org/10.1007/s10040-015-1246-1.

Lucas, M., Wendland, E., 2016. Recharge estimates for various land uses in the Guarani Aquifer System outcrop area. Hydrol. Sci. J. 1–10. Available from: https://doi.org/10.1080/02626667.2015.1031760.

Martelli, G.V., 2012. Monitoramento de flutuação de níveis de água em aqüíferos freáticos para avaliação do potencial de recarga em área de afloramento do sistema Aqüífero Guarani em Cacequi.

Mattos, T.S., Oliveira, P.T.S., de, Lucas, M.C., Wendland, E., 2019. Groundwater recharge decrease replacing pasture by eucalyptus plantation. Water 11, 1213. Available from: https://doi.org/10.3390/w11061213.

Melati, M.D., Fan, F.M., Athayde, G.B., 2019. Groundwater recharge study based on hydrological data and hydrological modelling in a South American volcanic aquifer. Comptes Rendus - Geosci. 351, 441–450. Available from: https://doi.org/10.1016/j.crte.2019.06.001.

Melo, D., de, C.D., Wendland, E., 2017. Shallow aquifer response to climate change scenarios in a small catchment in the Guarani Aquifer outcrop zone. An. Acad. Bras. Cienc. 89, 391–406. Available from: https://doi.org/10.1590/0001-3765201720160264.

Montenegro, A., Ragab, R., 2010. Hydrological response of a Brazilian semi-arid catchment to different land use and climate change scenarios: a modelling study. Hydrol. Process. 24, 2705–2723. Available from: https://doi.org/10.1002/hyp.7825.

Moreira, A.A., Ruhoff, A.L., Roberti, D.R., Souza, V., de, A., da Rocha, H.R., et al., 2019. Assessment of terrestrial water balance using remote sensing data in South America. J. Hydrol. 575, 131–147. Available from: https://doi.org/10.1016/j.jhydrol.2019.05.021.

Mukherjee, A., Scanlon, B., Aureli, A., Langan, S., Guo, H., McKenzie, A., 2020. Global Groundwater: Source, Scarcity, Sustainability, Security and Solutions, first ed. Elsevier, ISBN. 9780128181720.

Münch, Z., Conrad, J.E., Gibson, L.A., Palmer, A.R., Hughes, D., 2013. Satellite earth observation as a tool to conceptualize hydrogeological fluxes in the Sandveld, South Africa. Hydrogeol. J. 21, 1053–1070. Available from: https://doi.org/10.1007/s10040-013-1004-1.

Nachabe, M.H., 2002. Analytical expressions for transient specific yield and shallow water table drainage. Water Resour. Res. 38, 11-1-11–7. Available from: https://doi.org/10.1029/2001wr001071.

Natkhin, M., Steidl, J., Dietrich, O., Dannowski, R., Lischeid, G., 2012. Differentiating between climate effects and forest growth dynamics effects on decreasing groundwater recharge in a lowland region in Northeast Germany. J. Hydrol. 448–449, 245–254. Available from: https://doi.org/10.1016/j.jhydrol.2012.05.005.

OAS—Organization of American States, 2009. Aquífero Guarani - Programa estratégico de ação [WWW Document]. <http://www.oas.org/DSD/WaterResources/projects/Guarani/SAP-Guarani.pdf> (accessed 30.01.20.).

Oliveira, P.T.S., Almagro, A., Colman, C., Kobayashi, A.N.A., Rodrigues, D.B.B., Meira Neto, A.A., et al., 2019. Nexus of water-food-energy-ecosystem services in the Brazilian Cerrado. In: Rui, C., da Silva, V., Tucci, C.E.M., Scott, C.A. (Eds.), Org. Water and Climate Modeling in Large Basins 5, fifth ed. Porto Alegre, ABRHidro, pp. 7–30.

Oliveira, P.T.S., Leite, M.B., Mattos, T., Nearing, M.A., Scott, R.L., Oliveira Xavier, R., et al., 2017. Groundwater recharge decrease with increased vegetation density in the Brazilian cerrado. Ecohydrology 10, e1759. Available from: https://doi.org/10.1002/eco.1759.

Oliveira, P.T.S., Nearing, M.A., Moran, M.S., Goodrich, D.C., Wendland, E., Gupta, H.V., 2014. Trends in water balance components across the Brazilian Cerrado. Water Resour. Res. 50, 7100–7114. Available from: https://doi.org/10.1002/2013WR015202.

Pontes, L.M., Coelho, G., de Mello, C.R., Da Silva, A.M., Oliveira, G.C., Viola, M.R., 2016. Avaliação de modelo de balanço hídrico com base na estimativa da recarga potencial. Rev. Ambient. e Agua 11, 915–928. Available from: https://doi.org/10.4136/ambi-agua.1856.

PSAG, 2009. Report on the Project for the Environmental Protection and Sustainable Development of the Guarani Aquifer System: Advances on Guarani Aquifer System Knowledge 5 General Secretariat of the Project. Montevideo.

Rama, F., Miotlinski, K., Franco, D., Corseuil, H.X., 2018. Recharge estimation from discrete water-table datasets in a coastal shallow aquifer in a humid subtropical climate. Hydrogeol. J. 26, 1887–1902. Available from: https://doi.org/10.1007/s10040-018-1742-1.

Rebouças, A.C., 1976. Recursos hídricos subterrâneos da Bacia do Paraná: análise de pré-viabilidade. Universidade de São Paulo, São Paulo. https://doi.org/10.11606/T.44.2014.tde-02062014-141431.

Richey, A.S., Thomas, B.F., Lo, M., Reager, J.T., Famiglietti, J.S., Voss, K., et al., 2015. Quantifying renewable groundwater stress with GRACE. Water Resour. Res. 51, 5217–5238. Available from: https://doi.org/10.1002/2015WR017349.

Rocha, G.A., 1997. O grande manancial do Cone Sul. Estud. Avançados 11, 191–212. Available from: https://doi.org/10.1590/s0103-40141997000200013.

Rodell, M., Houser, P.R., Jambor, U., Gottschalck, J., Mitchell, K., Meng, C.J., et al., 2004. The global land data assimilation system. Bull. Am. Meteorol. Soc. 85, 381–394. Available from: https://doi.org/10.1175/BAMS-85-3-381.

Rodell, M., Velicogna, I., Famiglietti, J.S., 2009. Satellite-based estimates of groundwater depletion in India. Nature 460, 999–1002. Available from: https://doi.org/10.1038/nature08238.

Rodríguez, L., Vives, L., Gomez, A., 2013. Conceptual and numerical modeling approach of the Guarani Aquifer System. Hydrol. Earth Syst. Sci. 17, 295–314. Available from: https://doi.org/10.5194/hess-17-295-2013.

Rossman, N.R., Zlotnik, V.A., Rowe, C.M., 2018. Using cumulative potential recharge for selection of GCM projections to force regional groundwater models: a Nebraska Sand Hills example. J. Hydrol. 561, 1105–1114. Available from: https://doi.org/10.1016/j.jhydrol.2017.09.019.

Scanlon, B.R., Healy, R.W., Cook, P.G., 2002. Choosing appropriate techniques for quantifying groundwater recharge. Hydrogeol. J. 10, 18–39. Available from: https://doi.org/10.1007/s10040-001-0176-2.

Scanlon, B.R., Jolly, I., Sophocleous, M., Zhang, L., 2007. Global impacts of conversions from natural to agricultural ecosystems on water resources: quantity versus quality. Water Resour. Res. 43. Available from: https://doi.org/10.1029/2006WR005486.

Scanlon, B.R., Longuevergne, L., Long, D., 2012. Ground referencing GRACE satellite estimates of groundwater storage changes in the California Central Valley, USA. Water Resour. Res. 48. Available from: https://doi.org/10.1029/2011WR011312.

Silva, C., de, O.F., Manzione, R.L., Albuquerque Filho, J.L., 2019. Combining remotely sensed actual evapotranspiration and GIS analysis for groundwater level modeling. Environ. Earth Sci. 78. Available from: https://doi.org/10.1007/s12665-019-8467-x.

Simon, F.W., Reginato, P.A.R., Kirchheim, R.E., Troian, G.C., 2017. Estimativa de recarga do sistema aquífero guarani por meio da aplicação do método da variação da superfície livre na bacia do Rio Ibicuí-RS. Águas Subterrâneas 31, 12. Available from: https://doi.org/10.14295/ras.v31i2.28631.

Sindico, F., Hirata, R., Manganelli, A., 2018. The Guarani Aquifer System: from a Beacon of hope to a question mark in the governance of transboundary aquifers. J. Hydrol. Reg. Stud. 20, 49–59. Available from: https://doi.org/10.1016/j.ejrh.2018.04.008.

Soares, L.C., Velásquez, L.N.M., 2013. Estimativa da Recarga Aquífera na Bacia do Rio Riachão, Norte de Minas Gerais. Águas Subterrâneas 27. Available from: https://doi.org/10.14295/ras.v27i2.27359.

Souza, E., Pontes, L.M., Fernandes Filho, E.I., Schaefer, C.E.G.R., Dos Santos, E.E., 2019. Spatial and temporal potential groundwater recharge: The case of the Doce River Basin. Brazil. Rev. Bras. Cienc. do Solo 43. Available from: https://doi.org/10.1590/18069657rbcs20180010.

Sracek, O., Hirata, R., 2002. Geochemical and stable isotopic evolution of the Guarani Aquifer System in the state of São Paulo, Brazil. Hydrogeol. J. 10, 643–655. Available from: https://doi.org/10.1007/s10040-002-0222-8.

Szilagyi, J., Zlotnik, V.A., Gates, J.B., Jozsa, J., 2011. Mapping mean annual groundwater recharge in the Nebraska Sand Hills, USA. Hydrogeol. J. 19, 1503–1513. Available from: https://doi.org/10.1007/s10040-011-0769-3.

Tashie, A.M., Mirus, B.B., Pavelsky, T.M., 2016. Identifying long-term empirical relationships between storm characteristics and episodic groundwater recharge. Water Resour. Res. 52, 21–35. Available from: https://doi.org/10.1002/2015WR017876.

Teramoto, E.H., Chang, H.K., 2018. Métodos WTF e simulação numérica de fluxo para estimativa de recarga – exemplo Aquífero Rio Claro em Paulínia/SP. Águas Subterrâneas 32, 173–180. Available from: https://doi.org/10.14295/ras.v32i2.28943.

Tetzlaff, D., Carey, S.K., McNamara, J.P., Laudon, H., Soulsby, C., 2017. The essential value of long-term experimental data for hydrology and water management. Water Resour. Res. 53, 2598–2604. Available from: https://doi.org/10.1002/2017WR020838.

Thomas, B.F., 2019. Sustainability indices to evaluate groundwater adaptive management: a case study in California (USA) for the Sustainable Groundwater Management Act. Hydrogeol. J. 27, 239–248. Available from: https://doi.org/10.1007/s10040-018-1863-6.

Vestena, L.R., Kobiyama, M., 2007. Water balance in Karst: case study of the Ribeirão da Onça Catchment in Colombo City, Parana. State-Brazil. Braz. Arch. Biol. Technol. 50, 905–912. Available from: https://doi.org/10.1590/S1516-89132007000500020.

Vidon, P.G., 2015. Field hydrologists needed: a call for young hydrologists to (re)-focus on field studies. Hydrol. Process. 29, 5478–5480. Available from: https://doi.org/10.1002/hyp.10614.

Voss, K.A., Famiglietti, J.S., Lo, M., de Linage, C., Rodell, M., Swenson, S.C., 2013. Groundwater depletion in the Middle East from GRACE with implications for transboundary water management in the Tigris-Euphrates-Western Iran region. Water Resour. Res. 49, 904–914. Available from: https://doi.org/10.1002/wrcr.20078.

Walker, D., Parkin, G., Schmitter, P., Gowing, J., Tilahun, S.A., Haile, A.T., et al., 2019. Insights from a multi-method recharge estimation comparison study. Groundwater 57, 245–258. Available from: https://doi.org/10.1111/gwat.12801.

Wendland, E., Barreto, C., Gomes, L.H., 2007. Water balance in the Guarani Aquifer outcrop zone based on hydrogeologic monitoring. J. Hydrol. 342, 261–269. Available from: https://doi.org/10.1016/j.jhydrol.2007.05.033.

Wu, Q., Si, B., He, H., Wu, P., 2019. Determining regional-scale groundwater recharge with GRACE and GLDAS. Remote Sens. 11, 154. Available from: https://doi.org/10.3390/rs11020154.

Xie, Y., Crosbie, R., Simmons, C.T., Cook, P.G., Zhang, L., 2019. Uncertainty assessment of spatial-scale groundwater recharge estimated from unsaturated flow modelling. Hydrogeol. J. 27, 379–393. Available from: https://doi.org/10.1007/s10040-018-1840-0.

Chapter 30

Groundwater management in Brazil: current status and challenges for sustainable utilization

Prafulla Kumar Sahoo[1,2], Paulo Rógenes Monteiro Pontes[2], Gabriel Negreiros Salomão[2,3], Mike A Powell[4], Sunil Mittal[1], Pedro Walfir Martins e Souza Filho[2] and José Tasso Felix Guimarães[2]

[1]*Department of Environmental Science and Technology, Central University of Punjab, Bathinda, India,* [2]*Vale Institute of Technology (ITV), Belém, Brazil,* [3]*Geology and Geochemistry Graduate Program (PPGG), Geosciences Institute (IG), Federal University of Pará (UFPA), Belém, Brazil,* [4]*Department of Renewable Resources, Faculty of Agriculture, Life and Environmental Sciences (ALES), University of Alberta, Edmonton, AB, Canada*

30.1 Introduction

Groundwater, the largest usable freshwater, constitutes nearly 98% of all liquid (not frozen) freshwater available on Earth (Famiglietti, 2014; Margat and Gun, 2013; Eckstein, 2017). Because of its universal availability, low capital cost, and general quality, it has become one of the major water resources to meet domestic and irrigation demand for rural populations and small communities in developing countries (Foster et al., 2000; Stone et al., 2020). Concomitantly, in many nations, it supports a wide range of basic needs from public health to poverty alleviation and economic and environmental services [Foster et al., 2000; UNESCO-WWAP (World Water Assessment Programme), 2012]. This increasing demand has caused serious concern over sustainable use and has resulted in falling water tables, aquifer deterioration through saline intrusion, contamination, and failing public infrastructure (subsidence) [Foster et al., 2000; UNESCO-WWAP (World Water Assessment Programme), 2012; Stone et al., 2020]. In many places, groundwater management has been "out of sight and out of mind" (McCaffrey, 2001; Patole, 2015). Thus the global groundwater is undergoing immense stress in recent times (Mukherjee et al., 2020).

Brazil (which has 2.8% of the world's population) is a water-rich country that contributes around 12% of the total freshwater available in the world (based on 2011 World Bank data) (Hirata et al., 2017), resulting in the highest annual per capita water availability (43,027 m^3/year) in the world in 2017 (UN, 2017). However, the mismatch between its freshwater resources and locations of high population density, along with the growing risk of surface water pollution and critical climatic events result in water shortage in many parts of the country (Gesicki and Sindico, 2014). In this scenario, groundwater helps reduce the vulnerabilities of surface water to population forcing (Gesicki and Sindico, 2014). Concomitantly, in the semiarid Northeast regions of Brazil, it plays a crucial role in meeting various demands in urban, semiurban, and rural areas, including household, agriculture, irrigation, and industrial process [Agência Nacional de Águas (ANA), 2012, 2016; Hirata and Conicelli, 2012; Vasconcelos et al., 2013; Hirata et al., 2007; Hirata and Suhogusoff, 2019]. Also, in Brazil, groundwater responsible for maintaining 90% of the perennial rivers annually. In some regions (e.g., Northeast), several intermittent rivers occur due to crystalline aquifers, which do not provide consistent water [Agência Nacional de Águas (ANA), 2017]. Overall, agriculture is the main use of groundwater in several regions (68%), followed by domestic use (18%) and industry (14%) [Agência Nacional de Águas (ANA), 2009]. It is estimated that around 16% of the Brazil's population depends solely on groundwater resources, while 77.8% of cities and communities depend on both surface and groundwater [IBGE (Instituto Brasileiro de Geografia e Estatística), 2002]. This high demand leads to lowering water levels and consequently vulnerability to seawater intrusion (Cabral et al., 2004, 2008; Silva and Pizani, 2003) as well as deteriorating natural quality (Cabral et al., 2008; Menezes et al., 2014; Meng et al., 2015; Furtado et al., 2015). Also, climate change is expected to modify groundwater recharge and cause additional water scarcity (Hirata and Conicelli, 2012; Souza et al., 2018). For example, urbanized areas such as

São Paulo and Rio de Janerio are experiencing problems such as water pollution (Stollberg et al., 2014; Otto et al., 2015; Awange et al., 2016; Hu et al., 2017). Other areas such as Northeastern Brazil are suffering water shortages due to frequent drought-related events (e.g., Lemos et al., 2002; Rowland et al., 2015; Fleischmann et al., 2019). The Brazilian context requires now strategies and guidelines to facilitate sustainable policy in order to manage this resource and promote its sustainable growth (Cabral et al., 2008).

Recently, the Brazilian government has taken several steps to advance the management of its water resources (Braga, 2000; Porto and Porto, 2002; Campos and Studart, 2000; Cabral et al., 2008; Stollberg et al., 2014; Benjamin et al., 2005). The most effective groundwater management strategies include legislation, licensing, and monitoring and control of groundwater resources, along with other alternatives (Cabral et al., 2008; Patole, 2015, references therein). In view of this the present work consists of three sections. The first section addresses groundwater status, availability, and aquifer characterization. The second outlines the Brazilian legal system as related to groundwater governances toward sustainable use and the flaws in its management. Finally, conclusions are presented along with the current challenges needed toward sustainable models for the future. Data for this work is from scientific literature, governmental reports, legislation, and international documents found in *Scopus, Web of Science, PupMed, Medline, Sciencedirect, Springer, Wiley Online Library, and Google Scholar*.

30.2 Groundwater resources of Brazil

30.2.1 Physical and climatic characteristics

Brazil is the largest country in the South America that covers approximately half of the continent's landmass (Fig. 30.1A). A realistic estimation of groundwater availability of this country relative to use is crucial for groundwater resource management (Hirata et al., 2006). In Brazil, total groundwater reserves are approximately 112,000 km^3 (Rebouças, 1999), while the renewable reserve is estimated at 42,289 m^3/s (Hirata and Conicelli, 2012), wherein the major sedimentary aquifers occupy close to 20.43 m^3/s [Agência Nacional de Águas (ANA), 2005a,b]. The total volume of the available exploitable reserve of groundwater reserves in aquifers can provide approximately 14,649 m^3/s (Table 30.1), which is much lower than the availability of surface water reserves [91,300 m^3/s; Agência Nacional de Águas (ANA), 2015]. Agência Nacional de Águas (ANA) (2009) reported that annual internal renewable and exploitable groundwater resources are calculated at 645.6 and 129.1 km^3/year, respectively. Currently, the extraction of groundwater is over 17.5 km^3/year from more than 2.5 million tube wells (Hirata and Suhogusoff, 2019), with approximately 10,800 new wells being drilled annually [Agência Nacional de Águas (ANA), 2005a]. The current extraction only represents approximately 2% of the total effective groundwater recharge, thus, appears to be appropriate to meet future demands (Hirata and Suhogusoff, 2019).

Watershed management requires sound knowledge of both groundwater and surface water hydrogeology (Feitosa et al., 2016). According to the Federal Law 9.433/97, the "watershed" is defined as the territorial unit for the National Water Resources Policy and Water Laws implementation. Based on this, the Brazilian territory (8512 million km^2; Ferrarini et al., 2016) was divided into 12 hydrographic regions (Fig. 30.1B; Table 30.1) (CPRM—Serviço Geológico do Brasil, 2014; Feitosa et al., 2016). The Amazônia basin along with the Tocantins—Araguaia basins cover the largest portion (nearly 55%) of the country's total drainage area. The hydrographic regions of Amazônia include about 80% of the surficial water in Brazil, even though population density and demand for water is low.

Regional climate and rainfall patterns are the major factors that control Brazil's water resources (Marengo et al., 2017). Variations on the sea surface temperature of the Pacific, for instance, are responsible for climatic phenomena that affect the hydrological cycle in Brazil (e.g., El-Niño). Although the climate of the country varies considerably from tropical in the north to temperate in the south, a tropical and warm climate generally predominates, with an average air temperature higher than 20°C (Marengo et al., 2009). Precipitation is the main source of groundwater and aquifer recharge (Hu et al., 2017). Fig. 30.1C shows the annual rainfall patterns of the states. Table 30.1 also presents the annual mean precipitation in each hydrographic region for the period 1961–2007, showing large spatial variability, ranging from 1003 mm/year in São Francisco to 2205 mm/year in Amazônia [Agência Nacional de Águas (ANA), 2013], which is around 81% of the availability of Brazilian water resources distributed in 45% of the land area (Ferrarini et al., 2016). Northeastern Brazil experiences a semiarid climate with high potential evaporation (2000 mm/year) and receives low precipitation (average of 500 mm/year). The highest rainfall in the Amazon basin is consistent with the highest surface and groundwater availability (Table 30.1). In addition, the temporal distribution of precipitation (seasonality), with practically all rain occurs in only 2 or 3 months of each year, may aggravate the problem of water scarcity in some regions, as a significant part of precipitation ends up as runoff, rather than recharging aquifers (Silva

Groundwater management in Brazil: current status and challenges for sustainable utilization **Chapter | 30** 411

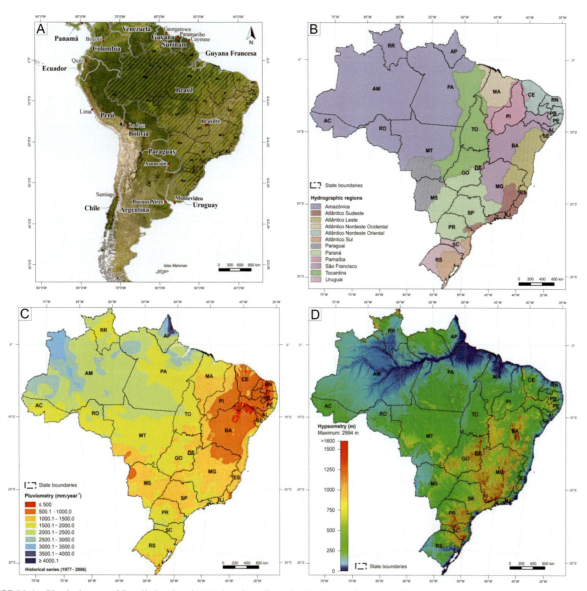

FIGURE 30.1 Physical maps of Brazil showing the (A) location, (B) major hydrographic regions, (C) annual precipitation, and (D) terrain elevation of the country. *Modified from CPRM—Serviço Geológico do Brasil, 2014. Mapa Hidrogeológico do Brasil ao Milionésimo. Organizadores: Diniz, J. A.O., Monteiro, A.B., De Paula, T.L.F., Silva, R.C. Recife. <http://www.cprm.gov.br/publique/Hidrologia/Mapas-e-Publicacoes/Mapa-Hidrogeologico-do-Brasil-ao-Milionesimo-756.html> (accessed January 2020).*

et al., 2020). Döll and Flörke (2005) demonstrated that the seasonal rainfall is the major factor influencing groundwater recharge. Elevation is also another major factor influences groundwater table level variation, that is, higher level of groundwater table at higher elevation areas (Hu et al., 2017). This is related with groundwater flow directions that follow the principle of hydraulic gradient—water flow from high gradients to low gradients (Freeze and Witherspoon, 1967). The elevation of the terrain all over Brazil is shown in Fig. 30.1D. Based on this, Hu et al. (2017) demonstrate the groundwater/surface water flow directions in the major river basins. In the Amazon basin the groundwater flow direction of both northern and southern areas points toward the Amazon River, while in the Paraná Basin the flow direction is from northeast to the southwest.

30.2.2 Hydrogeological features of aquifers

Aquifer properties such as lithology, porosity, and permeability (primary and secondary), and jointing/faulting can play a crucial role in groundwater distribution, flow, and storage (Farlin et al., 2013; Feitosa et al., 2016; Hu et al., 2017), so

TABLE 30.1 Average annual rainfall (data from 1961 to 2007) and estimation of surface and groundwater availability in the different hydrographic regions of Brazil.

Hydrographic regions	Area (km^2)	Total annual rainfall (mm)	Surface water availability (m^3/s)	Groundwater availability (m^3/s)
Amazônica	3,844,917.76	2205	73,748	9809
Paraguai	362,263.92	1359	782	450
Tocantins–Araguaia	918,273.16	1774	5447	1064
Paraná	877,513.54	1543	5792	1479
Uruguai	174,127.78	1623	567	433
Atlântico Sudeste	213,316.01	1401	1109	148
Paranaíba	331,808.82	1064	379	218
São Francisco	636,137.07	1003	1896	334
Atlântico Leste	386,068.13	1018	305	137
Atlântico Sul	186,079.86	1644	565	272
Nordeste Ocidental	268,906.01	1700	320	223
Nordeste Oriental	285,281.21	1052	91	79
Total			91,300	14,646

From Agência Nacional de Águas (ANA), 2013. Relatório de Conjuntura dos Recursos Hídricos no Brasil. (accessed 20.12.19.); Agência Nacional de Águas (ANA), 2017. Brazilian water resources report 2017. Accessed from: <http://www.snirh.gov.br/portal/snirh/centrais-de-conteudos/conjuntura-dos-recursos-hidricos/conj2017_rel_ingles-1.pdf>.

knowing the hydrogeological properties of all aquifers within a watershed is important for managing groundwater resources (Stollberg et al., 2014; Zagonari, 2010; Hu et al., 2017). The hydrogeology of Brazil is highly diversified (Hu et al., 2017). Mapping presented in the 2013 Brazilian Water Resources Report was simplified and updated in 2016 for 37 aquifers and outcrop aquifer systems, compared to the 181 addressed previously [Agência Nacional de Águas (ANA), 2017]. The groundwater potential of the major aquifers is categorized into two major hydrogeological domains, formed dominantly by sedimentary or igneous/metamorphic rocks and is divided into nine regions (Hu et al., 2017; CPRM—Serviço Geológico do Brasil, 2014; Fig. 30.2A; Table 30.2). Sedimentary rocks cover around 48% of the total area of Brazil and are mainly comprised of sandstones, shales, conglomerates, and carbonatic rocks (limestones and dolomites), while igneous/metamorphic rocks include mostly basalts, felsic and intermediate volcanic rocks, granitoid rocks, and their metamorphic equivalents (Ricardo and Bruno, 2011; CPRM—Serviço Geológico do Brasil, 2014). In Brazil the majority of the aquifers as well as the best aquifers in terms of productivity are associated with sedimentary rock formations (Fig. 30.2B). The widespread occurrence of sedimentary formations and prevailing climatic conditions give rise to favorable groundwater conditions in many parts of Brazil. However, some areas, where igneous and metamorphic rocks are largely dominant (Northeast region), exhibit lower groundwater potential [Agência Nacional de Águas (ANA), 2017]. The main Brazilian aquifers are located in the regions 2, 4, and 6 (Fig. 30.2), which correspond, respectively, to the Amazon, Paraná, and Parnaíba sedimentary basins. The zones 1, 3, 5, 7, 8, and 9 (Fig. 30.2A) are characterized by the total dominance of Precambrian igneous and metamorphic rocks and are not favorable for the occurrence of large aquifers.

The Amazon and Guarani aquifers located in the Amazon and Paraná basins, respectively, are the two largest groundwater reservoirs in Brazil (Alisson, 2014). High reservoir storage capacity within these basins is linked to the abundance of rainfall. Less reservoir capacity and reduced rainfall (annual average is rainfall <10 cm) is typical of the Northeast and Southeast coastal regions (Hu et al., 2017). The Amazon aquifer system (zone 2), formed by combining Solimões, Içá, and Alter do Chão aquifers, is located in Northern Brazil. The Guarani Aquifer System (GAS) includes the Bauru, Serra Geral, and Botucatu and Piramboia aquifers (zones 4 and 5). The major lithologies and water storage capacity of these aquifers are given in Table 30.3 (Hu et al., 2017). The Aquidauana, Bauru-Caiuá, and Pantanal aquifers (zone 4, Southeast to Central-West regions) are associated with sedimentary rock formations, while the Bambui and Serra Geral aquifers (zones 4, 5, and 7) occur, respectively, in carbonatic and basaltic rocks. There are no porous aquifers in the Northeastern and Southeastern coastal region of Brazil (zones 8 and 9, respectively).

Groundwater management in Brazil: current status and challenges for sustainable utilization Chapter | 30 413

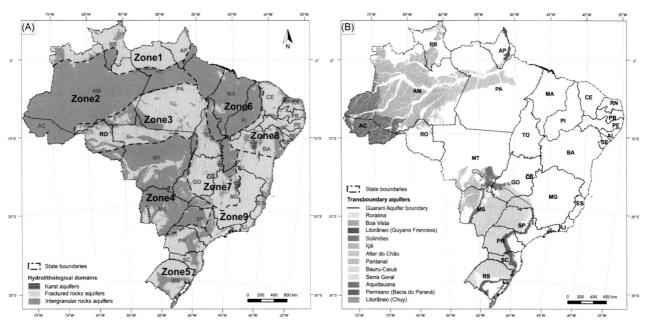

FIGURE 30.2 (A) Major hydrogeological domains and (B) the major aquifer systems in Brazil. The fractured rocks aquifers are related to igneous/metamorphic terranes, and intergranular rock aquifers are related to sedimentary/sandstones. *Modified from CPRM—Serviço Geológico do Brasil, 2014. Mapa Hidrogeológico do Brasil ao Milionésimo. Organizadores: Diniz, J.A.O., Monteiro, A.B., De Paula, T.L.F., Silva, R.C. Recife. <http://www.cprm.gov.br/publique/Hidrologia/Mapas-e-Publicacoes/Mapa-Hidrogeologico-do-Brasil-ao-Milionesimo-756.html> (accessed January 2020); Hu, K., Awange, J.L., Forootan, E., Goncalves, R.M., Fleming, K., 2017. Hydrogeological characterisation of groundwater over Brazil using remotely sensed and model products. Sci. Total Environ. 599–600, 372–386.*

TABLE 30.2 Dominant lithologies (which represent only the first rock layer under the surface) of the major aquifers present in the different regions of Brazil. Igneous/metamorphic terranes are related to fractured rocks aquifers and sedimentary/sandstones are related to intergranular rock aquifers.

Region	Rock type	Aquifer
1	Igneous/metamorphic	None
2	Sedimentary/sandstones	Alter do Chão, Icá and Solimões
3	Igneous/metamorphic	None
4	Sedimentary/sandstones	Bauru-Caiuá, Serra Geral, Botucatu and Piramboia, Pantanal
5	Basalts with associated sandstones	Serra Geral and Botucatu and Piramboia
6	Sedimentary/sandstones	Itapecuru, Piauí
7	Carbonatic rocks	Urucuia and Bambuí
8	Igneous/metamorphic	None
9	Igneous/metamorphic	None

Source: Modified from CPRM—Serviço Geológico do Brasil, 2014. Mapa Hidrogeológico do Brasil ao Milionésimo. Organizadores: Diniz, J.A.O., Monteiro, A.B., De Paula, T.L.F., Silva, R.C. Recife. <http://www.cprm.gov.br/publique/Hidrologia/Mapas-e-Publicacoes/Mapa-Hidrogeologico-do-Brasil-ao-Milionesimo-756.html> (accessed January 2020); Ricardo, H., Bruno, P.C., 2011. Groundwater resources in Brazil: a review of possible impacts caused by climate change. Ann. Braz. Acad. Sci. 84 (2), 297–312. http://dx.doi.org/10.1590/S0001-37652012005000037; Hu, K., Awange, J.L., Forootan, E., Goncalves, R.M., Fleming, K., 2017. Hydrogeological characterisation of groundwater over Brazil using remotely sensed and model products. Sci. Total Environ. 599–600, 372–386.

In terms of water resource management, aquifers may not be constrained to a particular watershed even though each aquifer presents the distribution of aquifer recharge areas in the 12 hydrographic regions of the country. For example, the Serra Geral aquifer, which occupies a significant area of Uruguai's hydrographic region (80%), also occurs in the hydrographic regions of Paraguai and Paraná. In Brazil the Guarani aquifer has a small outcrop (recharge area) located

TABLE 30.3 Descriptions of the rock types and their hydraulic features in the Amazon and Guarani aquifers.

Stratigraphic formation	Aquifer type	Rock type	Rock components	Water storage identification
Amazon aquifer system				
Içá	Unconfined	Sedimentary	Fine to medium sandstones and siltstones	Small
Solimões	Semiunconfined	Sedimentary	Greenish argillaceous sandstones	Small
Alter do Chão	Semiunconfined	Sedimentary	Coarse and friable sandstones	Large
Guarani Aquifer System				
Bauru	Unconfined	Sedimentary	Sandstone with quartz dominant and carbonatic	Small
Serra Geral	Semiunconfined	Basalts	Sandstone with quartz dominant and carbonatic	Small
Botucatu and Pirambóia	Semiunconfined	Sedimentary	Aeolian sandstone with quartz plus feldspars	Large

Source: From Ondra, R., 2002. Geochemical and stable isotopic evolution of the Guarani Aquifer System in the state of São Paulo, Brazil. Hydrogeol. J. 10, 643–655; Eliene, L.S., Paulo, H.F.G., Cleane, S.S.P., Marcus, P.M.B., José, G.A.D., Wilker, R.R.B., 2013. Sintese da hidrogeologia nas bacias sedimentares do amazonas e do solimões: Sistemas aquíferos icá-solimões e alter do chão. Geosci. USPUSP 13 (1), 107–117. https://doi.org/10.5327/Z1519-874X2013000100007; CPRM—Serviço Geológico do Brasil, 2014. Mapa Hidrogeológico do Brasil ao Milionésimo. Organizadores: Diniz, J.A.O., Monteiro, A.B., De Paula, T.L.F., Silva, R.C. Recife. <http://www.cprm.gov.br/publique/Hidrologia/Mapas-e-Publicacoes/Mapa-Hidrogeologico-do-Brasil-ao-Milionesimo-756.html> (accessed January 2020); Hu, K., Awange, J.L., Forootan, E., Goncalves, R.M., Fleming, K., 2017. Hydrogeological characterisation of groundwater over Brazil using remotely sensed and model products. Sci. Total Environ. 599–600, 372–386 (Ondra, 2002; Eliene et al., 2013).

in the Atlântico Sul hydrographic region (1.5%), but its main water resources are found in the subsurface of the Serra Geral basalts [Agência Nacional de Águas (ANA), 2015]. Therefore watershed/resource management and protection should consider the entire extent of any aquifer across adjoining regions even though the function of the aquifer is different (recharge vs extraction) [Agência Nacional de Águas (ANA), 2015]. The sedimentary rocks that mainly constitute the Amazon and Guarani aquifers have larger storage capability and higher permeability ($>10^{-4}$ m/s) than the igneous/metamorphic rocks found in most of the country (permeability $<10^{-7}$ m/s) (Hu et al., 2017).

30.3 Groundwater resource management in Brazil

30.3.1 Background of water resource management

Brazil is governed by civil law and has a federal system that contains 26 states. The Water Code established in July 1934 was the first legislation for water resource management in Brazil; thereafter several laws have been drafted and different institutions established by the federal government and states for its management (Patole, 2015). The federal constitution established in 1988 is the basis for current Brazilian water resource management (Cassuto and Sampaio, 2013 and reference therein) and designates surface waters as either federal or state property depending on whether or not the river crosses state boundaries; groundwater is part of the state domain. The 1988 constitution laid the foundation for establishment of the National Water Law of 1997, also known as the "Water Law" (Law 9.433/1997) (Brazil National Water Policy, 1997) which adopted integrated water resource management (IWRM) strategies based on the Dublin principles (1992).

The National Water Resources Council (CNRH) was created by Federal Law No. 9.433, with the responsibility of developing a National Water Resource Policy (Patole, 2015, references therein; Luiz, 2017). This body consists of representatives from relevant ministries and secretariats of the presidency of Brazil, representatives of the State Water Resources Councils (CERH), and representatives from water user associations and civil organizations. The National Water Resources Policy is responsible for (1) water permits, (2) water quality objectives into usage classes, (3) water charges, (4) national water resource information system, and (5) water resources plans. Their main focus with respect to groundwater resources includes coordination of the National Plan of Hydric Resources (PNRH) at the national, regional, and state levels; analyses of amendments to legislation and establishing supplementary guidelines for the PNRH; and establishing general criteria for awarding water rights and related charges [Patole, 2015, reference therein; Agência Nacional de Águas (ANA), 2011].

The National Council on the Environment (CONAMA) was established in 1981 by the National Environmental Policy Act (Federal Law 6938). The government established the Brazilian Water Agency through Federal Law no. 9.983 (2000), which is responsible for monitoring, control, and evaluation of activities related to water resources and the execution of the PNRH. In addition, an integrated management system of surface and groundwater was constituted in 2001.

30.3.2 National laws/legislation

Historically, little attention was paid to groundwater. According to Article 526 of the 1916 Civil Code, groundwater belonged to the landowner, who has the right to extract water free of constraints unless it caused damage to public waters or other users (Article 96, Federal Decree no. 24.643/1934). However, growing pressure over exploitation has forced policymakers to provide more formal legal structure (Victor et al., 2015). The Water Code was the first legislation to address groundwater issues but since then various legislation has been drafted for groundwater management (Patole, 2015). In the 1888 federal constitution, groundwater was declared as an asset of common use for the public (Article 225) under the jurisdiction of the state, not the union (Cassuto and Sampaio, 2013 and reference therein). Water rights were redefined under Article 1 of the Federal Water Law 9.433/1997 (Camargo and Ribeiro, 2009). This new law directed each state to take responsibility for conservation and protection within state boundaries. This was a move to decentralize water management, with the expectation of joint participation in the decision-making process of the governmental and nongovernmental sectors (water users and civil society organization) (Cabral et al., 2008). Over the past 20 years, Brazil has reformed their national water laws to include the Dublin principles and integrated water resources management (IWRM) and have properly addressed the specific components such as ownership, permits, monitoring, abstraction, and pollution control (Patole, 2015).

The states have control over some relevant issues, such as considering if aquifer use is sustainable given availability, recharge rates, and extraction rates [Agência Nacional de Águas (ANA), 2014, 2017]; usage permits are time-bound and vary state wise [Agência Nacional de Águas (ANA), 2011]. Federal Law no. 9.433/1997 provides the rules for drilling wells and includes licensing and means of monitoring groundwater level and discharge rates; owners of older wells are given an extension for obtaining the licensing (Cabral et al., 2004). Wells exclusively dedicated to the service of small rural settlements (Article 12, § 1°) are exempted from obtaining a license, but they have to follow some administrative procedure such as state registration. Drilling a well without authorization is a violation of Federal Law no. 9.433/1997. Some states have specific management plans; in Pernambuco state water taking is regulated by the Pernambuco State Environmental Organization and their Water Resources Agency (CPRH), which has zones with a maximum withdrawal limit in the central region of Recife. In some of these zones, they also impose bans on drilling or operating any new wells (Cabral et al., 2004).

30.3.3 Integrated management of surface water and groundwater

To integrate both groundwater and surface water management systems in Brazil, some regulations have been formulated. The CNRH in 2001 advocated a series of resolutions include Resolutions 9/2000, 15/2001, 22/2002, 48/2005, 76/2007, 91/2008, 107/2010, and 126/2011, to integrate ground and surface water management (Cassuto and Sampaio, 2013). Resolution 76/2007 developed a framework that includes the extraction of subsurface mineral water resources. Prior to Resolution 91/2008, surface water was only considered in some regulations while groundwater was considered a specific case, such as the National Groundwater Plan, which was later integrated into the National Water Plan by Resolution 99/2009. Planning and implementation of the Groundwater Integrated Qualitative and Quantitative Monitoring National Network was achieved by Resolution 107/2010. Resolution 126/2011 developed a national registry scheme for surface and groundwater users that can provide reliable information to all agencies improve management policy. CONAMA has also issued surface and groundwater water quality standards under Resolution 357/2005 and Resolution 396/2008, respectively (Cassuto and Sampaio, 2013).

30.3.4 Management of transboundary groundwater

Transboundary aquifers are increasingly important under resource management (Puri, 2003; Burchi and Mechlem, 2008; Cassuto and Sampaio, 2011). The GAS is one of the world's largest transboundary aquifers underlying Argentina, Brazil, Paraguay, and Uruguay. The GAS holds 30 trillion m^3 of water, or 1.2 million km^3 (Cassuto and

Sampaio, 2013). Although the majority of GAS's waters are utilized by public sectors, significant amounts are also used by agricultural and industry. The four overlying countries have launched the Protection and Sustainable Development of the GAS (PSAG, 2003−09) along with various international organizations to insure proper resource management (Cassuto and Sampaio, 2013; Sugg et al., 2015). In the Strategic Action Program of the GAS, four management contexts were prioritized: (1) global, (2) national, (3) state or province, and (4) municipal/local. The objective is a more accurate and integrated information system to help facilitate the management process, along with involving local stakeholders, to insure real solutions to water resource management and aquifer protection. According to the knowledge gained by the Guarani Aquifer Project, the most important concern is to promote groundwater governance at the local level because it is this level where most of the contamination and overexploitation occurs; this information is then taken to the national level. The four countries that overlie the GAS signed a management agreement (Acordo) in 2010, but problems that surround aquifer management still exist (Cassuto and Sampaio, 2011; Cassuto and Sampaio, 2013, and reference therein); any solution to multinational aquifer management will need to consider local, domestic laws, which is a very difficult situation.

Brazil overlies approximately 80% of the GAS and should be considered the leader in GAS management (Cassuto and Sampaio, 2013), but it has no clear provision to deal with the issue of transboundary groundwater, either at the state or federal levels. A constitutional amendment has been proposed to grand federal jurisdiction over transboundary water resources (Cassuto and Sampaio, 2013; Sugg et al., 2015), but this effort does not properly define the federal versus state administrative roles or how the "Acordo" will be enforced. This confusion has led São Paulo, which is the main consumer of water, to form a State Water Resources Council (CERH-SP) to manage its water resources. This body has identified, and regulates, contamination zones, but this is difficult given the transboundary nature of the aquifer.

30.3.5 Management of mineral water resources

Groundwater in Brazil is legally divided into three categories: mineral waters, potable water, and normal groundwater (Gesicki and Sindico, 2014). The Code of Mineral Waters (CMW) (1945) considers both mineral water and potable table water as mineral resources. Under Brazilian law, mineral water is defined as having distinct physical−−chemical characteristics compared to other categories of groundwater; it is considered a mineral substance and is managed as part of the federal government's assets; governed by the mining legal framework (Gesicki and Sindico, 2014). The National Department of Mineral Production regulates mineral water following specific mining exploitation laws, while normal groundwater comes under state domain and is regulated by the Brazilian System of Water Resource Management, which promotes a decentralized system for water resource management. These two systems of water management differ from both legal and management perspectives, causing conflict between federal and state agencies. These challenges have resulted in suggestions (Gesicki and Sindico, 2014) that legislation regarding mineral water be revised within the CMW to better align with regulations governing use of regular groundwater. This would require that the National Policy for Water Resources consider integrated regulation of mineral and other groundwaters at federal and state levels.

30.3.6 Groundwater monitoring and assessment

Effectively monitoring quantity and quality of groundwater is essential when assessing aquifer vulnerability, degradation tendency, identifying groundwater protection zones and extraction potential (Stollberg et al., 2014) and subsequent policy related to management and protection (Stollberg et al., 2014; Hirata et al., 2007; Melati et al., 2019). Monitoring is mandated in Article 2 of the National Environmental Policy, which aims to preserve, enhance, and restore the quality of environmental resources. CONAMA Resolution 357/396 considers monitoring an essential tool to evaluate the quality and quantity of groundwater as well as a mechanism to ensure compliance with standards set by the competent authorities. CONAMA Resolution 396, in accordance with Laws 6938 (August 31, 1981) and 9433 (January 8, 1997), classifies groundwater into six subcategories based on use and quality. It builds on the standards set by CONAMA Resolution 357 regarding the discharge of effluent and modifies them based on specific use. In addition, areas of groundwater recharge and important surface zones are identified relative to the maintenance of groundwater quality and procedures are outlined to manage these areas (Borges et al., 2017; Costa et al., 2019). On a national scale, groundwater availability is quite limited and outdated (Zoby and Matos, 2002). The Brazilian National Department of Mineral Production (DNPM) was the first organization to do hydrogeological mapping of the country, and later aquifer information was incorporated by Rebouças (1988) (Hirata et al., 2017). The National Water Agency [Agência Nacional de Águas (ANA), 2005a,b] also integrated regional data for the main aquifer systems, including water quality, reserves, and productivity. Currently, the Geological Survey of Brazil (CPRM) has an online database, which provides maps of the hydrogeological domains and subdomains in a geographic information system on a scale 1:2,500,000 (Hirata et al., 2017). CPRM also created the Integrated Groundwater Monitoring Network (RIMAS) in 2009; a nationwide network for monitoring the main Brazilian aquifers to increase hydrogeological information and concurrently the Resources Management Company of Ceará (COGERH) began a network for monitoring groundwater levels [IGRAC (International Groundwater Resources Assessment Centre) Publication, 2014]. Until 2014 CPRM managed a network of nearly 293 monitoring stations from 26 aquifers within 16 states [IGRAC (International Groundwater Resources Assessment Centre) Publication, 2014]. All these wells are equipped with automatic systems, including data recording and validation, and the information is entered in Groundwater Information System (SIAGAS) developed by Brazilian Geological Survey (CPRM) (1997). SIAGAS data outputs are being used to monitor and manage groundwater [Nascimento et al., 2008; IGRAC (International Groundwater Resources Assessment Centre) Publication, 2014]. In addition, some states have specific monitoring plans. For example, Sao Paulo state developed a groundwater quality monitoring network (CETESB) in 1990 for mainly public water supply well monitoring and for its groundwater quality and quantity monitoring integrated network (CETESB/DAEE) starting in 2009.

Groundwater contamination vulnerability mapping has played a key role in preserving aquifers as it identifies the role of natural aquifer properties and anthropogenic activities on groundwater quality (Foster et al., 2013; Jang et al., 2016). Important aquifer properties consist of several geological/hydrogeological characteristics such as lithology, texture, structure, depth to water table, recharge rate, hydraulic conductivity, and geogenic contaminants. Anthropogenic activities related to land use/land use change, cover, population centers, and industry are also important. Vulnerability mapping can be very crucial for understanding coastal aquifers. For example, Silva and Pizani (2003) studied vulnerability assessment on coastal aquifers in the eastern portion of Rio de Janeiro state, where the highest contamination potential is attributed to the deposition of colluvium/alluvium, high hydraulic conductivity, and low relief. These areas require specific studies in order to assist land occupation planning and to enhance aquifer protection, particularly in recharge areas where policy will impact development policies. Furthermore, hydroelectric development in Brazil is matter of serious concern for groundwater and public water supply protection (Gauthier et al., 2019).

30.4 Alternatives for groundwater management and water sourcing

The semiarid Northeastern regions of Brazil are frequently associated with water shortages and require special attention regarding water availability [Agência Nacional de Águas (ANA), 2015] especially when droughts diminish groundwater recharge for unconfined aquifers. It is predicted that by 2050 long-term climate change will lead to a 70% reduction in aquifer recharge in the Northeastern region of Brazil as compared to 2010 (Hirata and Conicelli, 2012). To combat this problem and promote the sustainable development of this region, some alternative water sources such as rainwater harvesting, wastewater reuse, and desalination have been proposed and implemented.

30.4.1 Adopting rainwater harvesting

Rainwater harvesting is one of the effective and sustainable water conservation methods encouraged by Brazilian legislation, particularly for semiarid regions (Gnadlinger, 2006; Teston et al., 2017, 2018; Ghisi et al., 2017). The federal government launched the "One Million Cisterns" program (P1MC) in 2001 to provide safe drinking water to 1 million rural people in semiarid regions of Brazil and has installed approximately 372,000 cisterns (Gomes et al., 2012) resulting in greater availability to water and time saved for women who were tasked with collecting water (Silva et al., 2006). Another government program, "One Land Two Waters" (P1 + 2), has redefined semiarid regions and complements the P1MC (Gnadlinger, 2005, 2014). In this program, farmers trained to catch and store rainwater for agriculture and other potable and nonpotable uses. The National Water Resources Plan also encourages the use of rainwater in urban areas. The NBR 15527 provides guidelines for designing rainwater harvesting systems for rooftops (Teston et al., 2017, 2018); however, it is not efficient enough for the collection of potable water. In July 1999 the Brazilian Association for Catchment, Management, and Utilization of Rainwater was formed with the intend to efficient use of rainwater in Brazil and at the national level, a bill was proposed in May 2015 regarding mandatory implantation of rainwater harvesting for all buildings >200 m^2 (Federal Law Project no. 1750 of 2015) (Teston et al., 2018). This legislation still needs the approval of political leaders, which is yet to materialize. Unfortunately, some proposed municipal legislation is so restrictive that it is negatively impacting the effective implementation and practice of rainwater harvesting in Brazil (Gomes et al., 2012).

30.4.2 Artificial groundwater recharge and reuse of wastewater

Artificial groundwater recharge is possible during seasonal dry periods, or in the case of overexploitation (Stollberg et al., 2014), especially in the case of using pretreated wastewater. The BRAMAR (Brazil Managed Aquifer Recharge) research team proposed strategies and technologies for Water Reuse and Managed Aquifer Recharge as part of IWRM (BRAMAR, 2014; Meon et al., 2016). This German–Brazilian research project focuses on water scarcity in semiarid regions of Brazil. BRAMAR is also looking for alternative water sources, such as rainwater harvesting and wastewater reuse to enhance local-level reservoir management (Meon et al., 2016). In the Mossoró, CSA, low-cost wastewater reclamation schemes were implemented and tested for reusing water for irrigating agricultural land and to control water pollution and water scarcity. Wastewater treatment plants have reached their limits, especially during heavy storm events, and groundwater recharge might be an effective method to relieve pressure on treatment plants (Stollberg et al., 2014) once the efficacy of the water is tested post filtration (via monitoring wells). The controlled infiltration of pretreated municipal wastewater in tropical soils may support the regional groundwater balance as a sustainable drinking water resource (Stollberg et al., 2014; Stepping, 2016).

30.4.3 Desalination

Desalination is an alternative for producing freshwater and expanding existing supplies, where appropriate (Bremere et al., 2001; Júnior et al., 2019). In Brazil, small desalination plants have been installed in the semiarid regions since the 1990s. However, they were installed without providing proper operational and maintenance guidelines and training. Due to this failure, the Brazilian federal government, with the coordination of the Environmental Ministry, implemented desalinization systems as part of the Fresh Water Program [Programa Água Doce (PAD), 2012]. This program promotes sustainable groundwater use and provides safe and clean potable water for human consumption in the semiarid and arid region of Brazil. This program was carried out through a network of 200 institutions from 10 Brazilian states (Alagoas, Bahia, Ceará, Minas Gerais, Maranhão, Piauí, Paraíba, Pernambuco, Rio Grande do Norte, and Sergipe) as well as other federal partners and has improved the health and quality of life of people in semiarid regions. Ferreira et al. (2017) reported that more than 460 desalination plants were installed up to June 2017, with estimates of generating a total of 400,000 gal of potable water/day, benefiting nearly 500 communities in semiarid regions. Also, this program trained over 1000 people, including state technicians and desalination plant operators. With the PAD, Brazil also contributes to its Sustainable Development Goals by eliminating poverty and putting the country on a sustainable path [Programa Água Doce (PAD), 2012; Júnior et al., 2019].

30.5 The hydroschizophrenia of groundwater management

Policies and legislation related to groundwater resources have reasonably advanced to accomplish safe exploitation or to protect aquifers from overexploitation, saltwater intrusion, and contamination. However, Jarvis et al. (2005) and Villar (2016) point out the "hydroschizophrenia" syndrome associated with the conundrum of surface versus groundwater management in transboundary hydrogeological systems, either between states within a country or between countries. In Brazil, overexploitation is mainly occurring in aquifers such as Inajá, Exu, Missão Velha, and Beberibe, which have low exploitable reserves, which cannot meet the demand in the long term [Agência Nacional de Águas (ANA), 2007]. On regional scales, aquifers like the Guarani aquifer are very much subject to international concern and so it is important to define groundwater restriction zones for drilling new wells (Villar and Ribeiro, 2009). Because groundwater in Brazil is exclusively under state domain, this obligation is the duty of states to implement integrated management of groundwater resources (Camargo and Ribeiro, 2009). However, as discussed earlier, at the state level, water management bodies are still facing serious challenges to facilitate groundwater management between states and with the federal government. This situation is exacerbated by a lack of effective database access to water quality and quantity monitoring networks, geophysical and geological databases, and water use statistics; the River Basin Management Plans have not achieved their mandates (Goetten, 2015). Furthermore, although there are some laws in place obtain a water permit for exploitation of groundwater and register the new wells with the competent authority (Silva et al., 2008), official figures of the number of wells and groundwater users are still unavailable (ANA, 2013; Villar, 2016). According to the Agência Nacional de Águas (ANA) (2017) there were more than 278,000 registered wells in 2016, but the estimated real number is approximately 1.2 million, which implies that the majority of wells are illegal since they are not licensed. This illegal activity is certainly impacting actual data on water use/extraction, the effect on ecosystem services and transboundary systems—adding to potential "hydroschizophrenia."

Separate information systems have been created at the federal level for surface waters (National Information System on Water Resources) and groundwater (Information System on Groundwater—SIAGAS), but they are not properly integrated and well coordinated with other environmental systems/programs. At the federal level, several legal acts were adopted to cover groundwater in their water management system along with other specific programs such as the National Groundwater Agenda and the National Program for Groundwater. However, the federal government needs to play a major role for empowering the states to support effective governance and management of their groundwater resources (Camargo and Ribeiro, 2009). Furthermore, because of hidden nature of groundwater, their management faces other crucial challenges such as limitations of available data on water quality, quantity, lithology, and vulnerability, besides difficulties to apply the water management tools and instruments established by Law 9.433/1997 (Bohn et al., 2014; Villar, 2016).

According to Federal Law no. 9.433/1997, drilling a well or exploiting groundwater without permission is an administrative violation. However, offenders are seldom penalized due to the lack of state enforcement. Knowing this, the number of illegal wells is increasing relative to licensed wells, with concomitant reduction in revenues, which adds to the cost of maintenance and replacement for municipal/state water systems (Custodio, 2002; Bertolo et al., 2015). This leads to significant overexploitation, for example, wells that do not produce water or will go dry quickly.

30.6 Final considerations and current challenges

Groundwater plays a key role in public water supply and it can be one solution to water scarcity, particularly when facing climate variability and supporting regions suffering from scarcity or pollution of surface water. Geological, hydrogeological, and climatic conditions could be main factors that affect the groundwater storage capacity and supplies of Brazil. The largest groundwater storage capacity of the country is the Amazon aquifer that has the rock layers of highest permeability.

Brazilian management of groundwater resources includes several efforts related to institutional framework and national water laws that address major challenges such as ownership rights, permitting systems, monitoring, abstraction, and pollution control; however, management systems are still not in accordance with Brazil's future strategic planning. The following would support better groundwater management and development opportunities and new technologies to help insure Brazil's future water security:

1. Since water is unevenly distributed, an integrated strategy for the use of both surface and groundwater should be incorporated into any groundwater management strategy (Hirata and Conicelli, 2012); regulatory gaps and inconsistencies between the implementation of surface and groundwater management must be taken seriously by policymakers. That is to say, state versus federal bias may be removed from regulatory management/protection paradigms.
2. Coordination and implementation between federal ministries, national councils, and legislative bodies are important to improve the effectiveness of the national laws to ensure abstraction and pollution control of groundwater.
3. Maintaining and upgrading groundwater monitoring networks is essential to the linkage between policymaking and regulatory implementation for safe exploitation of groundwater resources. In this regard, CPRM should optimize its groundwater monitoring networks and integrate with hydrometeorological data networks for building an effective IWRM system [IGRAC (International Groundwater Resources Assessment Centre) Publication, 2014].
4. Identifying groundwater recharge potential is critical for protection of groundwater recharge areas, which is missing in many regions of Brazil.
5. Land use change and surface water pollution controls must be incorporated to mitigate impact on groundwater resources (Patole, 2015).
6. Anthropogenic contamination from all sources must be monitored carefully, regardless of their geographic proximity within a watershed, and considering the transboundary nature of water resources.
7. Alternative sources such as rainwater harvesting and aquifer recharge enhancement using wastewater and desalinization techniques should be encouraged via legislation in semiarid regions.
8. Mineral water definitions and legislative regulations regarding its management need to be revised considering the ecosystem versus government definitions and subsequent management (Gesicki and Sindico, 2014).
9. Remote sensing with advance mathematical modeling could enhance knowledge regarding groundwater storage potential and management.
10. Environmental education and public awareness via policies and academic programs is critical to current/future groundwater protection and use.

References

Agência Nacional de Águas (ANA), 2005a. Panorama da qualidade das águas subterrâneas no Brasil. Brasília. <http://www.ana.gov.br/sprtew/recursoshidricos.asp> (accessed 20.12.19.).

Agência Nacional de Águas (ANA), 2005b. Disponibilidade e demandas de recursos hídricos no Brasil. Brasília. <http://www.ana.gov.br/sprtew/recursoshidricos.asp> (accessed 20.12.19.).

Agência Nacional de Águas (ANA), 2007. GEO Brazil Tematic Series, Water Resources Component of a Series of Reports on the Status and Prospects for the Environment in Brazil. ANA/PNUMA, Brasília. <http://arquivos.ana.gov.br/institucional/sge/CEDOC/Catalogo/2010/GEOBrasilResumoExecutivo_Ingles.pdf>.

Agência Nacional de Águas (ANA), 2009. Cojuntura dos Recursos Hídricos no Brasil.

Agência Nacional de Águas (ANA), 2011. Permit System for Water Resources Use. ANA, Brasília. (Cadernos de Capacitação em Recursos Hídricos, vol. 6). <http://arquivos.ana.gov.br/institucional/sge/CEDOC/Catalogo/2012/OutorgaDeDireitoDeUsoDeRecursosHidricos.pdf> (In Portuguese).

Agência Nacional de Águas (ANA), 2012. Cojuntura dos Recursos Hídricos no Brasil. Informe 2012. Ediçao especial. Agência Nacional de Águas. ANA.

Agência Nacional de Águas (ANA), 2013. Relatório de Conjuntura dos Recursos Hídricos no Brasil. (accessed 20.12.19.) https://www.ana.gov.br/noticias-antigas/relata3rio-de-conjuntura-dos-recursos-hadricos.2019-03-15.8720485917.

Agência Nacional de Águas (ANA). 2014. Agência destaca importância das águas subterrâneas para gestão de recursos hídricos. In: XVIII Congresso Brasileiro de Águas Subterrâneas, que acontece de 14 e 17 de outubro em Belo Horizonte. <https://www.ana.gov.br/noticias-antigas/agaancia-destaca-importac-ncia-das-a-guas.2019-03-15.6037406004>.

Agência Nacional de Águas (ANA), 2015. Superintendência de Planejamento de Recursos.

Agência Nacional de Águas (ANA), 2017. Brazilian water resources report 2017. Available from: <http://www.snirh.gov.br/portal/snirh/centrais-de-conteudos/conjuntura-dos-recursos-hidricos/conj2017_rel_ingles-1.pdf>.

Agência Nacional de Águas (ANA), 2016. Conjuntura Dos Recursos Hídricos No Brasil. <https://www.ana.gov.br/noticias-antigas/ana-publica-informe-2016-do-relata3rio-conjuntura.2019-03-15.2821446185> (accessed 15.12.2019).

Alisson, E., 2014. Amazonia has an "underground ocean". Available from: <http://agencia.fapesp.br/amazonia_has_an_underground_ocean/19679/> (accessed January 2020).

Awange, J.L., Mpelasoka, F., Goncalves, R.M., 2016. When every drop counts: ability equal to or over analysis of droughts in Brazil for the 1901–2013 period. Sci. Total Environ. 1472 (88), 566–567. Available from: http://doi.org/10.1016/j.scitotenv.2016.06.031.

Benjamin, A.H., Marques, C.L., Tinker, C., 2005. The water giant awakes: an overview of water law in Brazil. Tex. Law Rev. 83, 2185–2244.

Bertolo, R., et al., maio-agosto, 2015. Água subterrânea para abastecimento público na Região Metropolitana de São Paulo: é possível utilizá-la em larga escala?. Revista DAE 63, 6–18.

Bohn, N., Goetten, W.J., Primo, A.P., 2014. Governança da água subterrânea no Estado do Rio Grande do Sul. REGA 11 (1), 33–43.

Borges, V.M., Fan, F.M., Reginato, P.A.R., Athayde, G.B., 2017. Groundwater recharge estimating in the Serra Geral aquifer system outcrop area – Paraná State, Brazil. Águas Subterrâneas 31, 338–346. Available from: http://doi.org/10.14295/ras.v31i4.28872.

Braga, B.P.F., 2000. The management of urban water conflicts in the metropolitan region of São Paulo. Water Int. 25 (2), 208–213.

BRAMAR, 2014. Strategies and technologies for water scarcity mitigation in northeast of Brazil: water reuse, managed aquifer recharge and integrated water resources management. Descrição projeto de pesquisa de cooperação bilateral Brasil-Alemanha. UFCG, Campina Grande; RWTH Aachen University, Aachen. (Relatório Parcial).

Brazilian Geological Survey (CPRM), 1997. Brazil National Groundwater System – SIAGAS. Available from: <siagasweb.cprm.gov.br> (accessed 20.12.16.).

Brazil National Water Policy, 1997 Law 9433 of 1997. Available from: <www.planalto.gov.br/ccivil_03/leis/L9433.htm> (accessed 12.10.19).

Bremere, I., Kennedy, M., Stikker, A., Schippers, J., 2001. How water scarcity will effect the growth in the desalination market in the coming 25 years. Desalination 138, 7–15.

Burchi, S., Mechlem, K., 2008. Groundwater in International Law: Compilation of Treaties and Other Legal Instruments. SSRN, Rome, 2005. Available from: <https://ssrn.com/abstract = 1259042> (accessed 10.01.20.).

Cabral, J.J.S.P., Farias, V.P., Sobral, M.C., Paiva, A.L.R., Santos, R.B., 2008. Groundwater management in Recife. Water Int. 33, 86–99. Available from: https://doi.org/10.1080/02508060801927648.

Cabral, J.J.S.P., Montenegro, S.M.G.L., Paiva, A.L.R., Farias, V.P., 2004. Groundwater salinization in the central region of Recife (Brazil) due to brackish water in Capibaribe river at high tide. In: Araguás, Custodio, Manzano (Eds.), 18 SWIM. Cartagena 2004, Spain. IGME. <http://www.swim-site.nl/pdf/swim18/swim18_065.pdf> (accessed 20.12.19.).

Camargo, E., Ribeiro, E., 2009. A proteção jurídica das águas subterrâneas no Brasil. In: Ribeiro, W.C. (Ed.), Governança da água no Brasil: uma visão interdisciplinar. Annablume, FAPESP, CNPq, São Paulo.

Campos, J.N.B., Studart, T.M.C., 2000. An historical perspective on the administration of water in Brazil. Water Int. 25 (1), 148–156.

Cassuto, D.N., Sampaio, S.R., 2011. Keeping it legal: Transboundary management challenges facing Brazil and the Guarani. Water Int. 36 (5), 661–670. Available from: https://doi.org/10.1080/02508060.2011.599779.

Cassuto, D.N., Sampaio, S.R., 2013. Hard, soft & uncertain: the Guarani Aquifer and the challenges of transboundary groundwater. Colo. J. Int. Environ. Law Policy 24, 1. forthcoming, SSRN. Available from: <https://ssrn.com/abstract = 2212168>.

Costa, A.M., Salis, H.H.C., Viana, J.H.M., Pacheco, F.A.L., 2019. Groundwater recharge potential for sustainable water use in urban areas of the Jequitiba River Basin, Brazil. Sustainability 11, 2955.

CPRM—Serviço Geológico do Brasil, 2014. Mapa Hidrogeológico do Brasil ao Milionésimo. Organizadores: Diniz, J.A.O., Monteiro, A.B., De Paula, T.L.F., and Silva, R.C. Recife. <http://www.cprm.gov.br/publique/Hidrologia/Mapas-e-Publicacoes/Mapa-Hidrogeologico-do-Brasil-ao-Milionesimo-756.html> (accessed January 2020).

Custodio, E., 2002. Aquifer overexploitation: what does it mean. Hydrogeol. J. 10 (2), 257−277.

Döll, P., Flörke, M., 2005. Global-scale estimation of diffuse groundwater recharge. In: Frankfurt Hydrology Paper 03. Institute of Physical Geography, Frankfurt University, Frankfurt am Main, Germany, 21 p.

Eckstein, G., 2017. The International Law of Transboundary Groundwater Resources. Routledge, Abingdon, Oxon, New York.

Eliene, L.S., Paulo, H.F.G., Cleane, S.S.P., Marcus, P.M.B., José, G.A.D., Wilker, R.R.B., 2013. Sintese da hidrogeologia nas bacias sedimentares do amazonas e do solimões: Sistemas aquíferos icá-solimões e alter do chão. Geosci. USPUSP 13 (1), 107−117. Available from: http://doi.org/10.5327/Z1519-874X2013000100007.

Famiglietti, J.S., 2014. The global groundwater crisis. Nat. Clim. Change 4, 945−948.

Farlin, J., Drouet, L., Galle, T., Kies, A., 2013. Delineating spring recharge areas in a fractured sandstone aquifer (Luxembourg) based on pesticide mass balance. Hydrogeol. J. 21 (4), 799−812.

Feitosa, F.A.C., Diniz, J.A., Kirchheim, R.E., Kiang, C.H., Feitosa, E.C., 2016. Assessment of groundwater resources in Brazil: current status of knowledge. In: Groundwater Assessment, Modeling, and Management. CRC Press.

Ferrarini, A.F.S.F., Ferreira Filho, J.B.S., Horridge, M. 2016. Water demand prospects in Brazil: a sectoral evaluation using an inter-regional CGE model. In: 19th Annual Conference on Global Economic Analysis. Washington, DC.

Ferreira, R.S., Veiga, H.P., Santos, R.G.B., Saia, A., Rodrigues, S.C., Bezerra, A.F.M., et al., 2017. Empowering brazilian northeast rural communities to desalinated drinking water access: programa água doce. In: The International Desalination Association World Congress, 2017, São Paulo. Water Reuse & Desalination Ensure Your Water Future: [Proceedings...]. São Paulo: International Desalination Association. <http://ainfo.cnptia.embrapa.br/digital/bitstream/item/171288/1/2017AA53.pdf> (accessed 15.12.19.).

Fleischmann, A.S., Fan, F.M., Paiva, R.C.D., Athayde, G.B., 2019. Estimates of groundwater depletion under extreme drought in the Brazilian semi-arid region using GRACE satellite data: application for a small-scale aquifer. Hydrogeol. J. 27, 2789−2802.

Foster, S., Chilton, J., Moench, M., Cardy, F., Schiffler, M., 2000. Groundwater in rural development. Facing the challenges of supply and resource sustainability. In: World Bank Teohnical Paper No. 463.

Foster, S., Hirata, R., Andreo, B., 2013. The aquifer pollution vulnerability concept: aid or impediment in promoting groundwater protection? Hydrogeol. J. 21, 1389−1392.

Freeze, R.A., Witherspoon, P.A., 1967. Theoretical analysis of regional groundwater flow: 2. Effect of water-table configuration and subsurface permeability variation. Water Resour. Res. 3 (2), 623−624. Available from: http://doi.org/10.1029/WR003i002p00623.

Furtado, Z.N.C., Marteli, A.N., Lollo, J.A., Oliveira, J.N., 2015. Groundwater vulnerability assessment in a Brazilian urban área. In: 2014 International Conference on Civil, Urban and Environmental Engineering (CUEE 2014), Beijing, China.

Goetten W.J., Avaliação da Governança da Água Subterrânea nos Estados de São Paulo, Paraná, Santa Catarina e Rio Grande do Sul. Blumenau. 2015. 317f. Dissertação (Mestrado em Engenharia Ambiental) - Engenharia Ambiental, Fundação Universidade Regional de Blumenau.

Gauthier, C., Lin, Z., Peter, B.G., Moran, E.F., 2019. Hydroelectric infrastructure and potential groundwater contamination in the Brazilian Amazon: Altamira and the Belo Monte Dam. Prof. Geogr. 71, 292−300. Available from: https://doi.org/10.1080/00330124.2018.1518721.

Gesicki, A.L.D., Sindico, F., 2014. The environmental dimension of groundwater in Brazil: conflicts between mineral water and water resource management. J. Water Resour. Prot. 6, 1533−1545.

Ghisi, E., Colasio, B.M., Geraldi, M., Teston, A., 2017. Rainwater harvesting in buildings in Brazil: a literature review. Proceedings 2017, 1. Available from: https://doi.org/10.3390/ecws-2-04955. x.

Gnadlinger, J., 2005. O Programa Uma Terra Duas Águas (P1 + 2) e a Captação e o Manejo de Água da chuva [The One Land Two Waters Sources Programme (P1 + 2) and rainwater harvesting and management]. In: Proceedings of Simpósio Brasileiro de Captação e Manejo de Água de Chuva [Brazilian Symposium on Rainwater Harvesting and Management], Teresina, Brazil, 11−14 July 2005. (In Portuguese).

Gnadlinger, J., 2014. How can rainwater harvesting contribute to living with droughts and climate change in semi-arid Brazil? Waterlines 33 (2), 146−153. Available from: https://doi.org/10.3362/1756-3488.2014.015.

Gnadlinger J., Community water action in semiarid Brazil, an outline of the factors for success. In: Official Delegate Publication of the Fourth World Water Forum. Mexico City/Mexico, March 16−22, 2006, pp. 150−158.

Gomes, U.A.F., Heller, L., Pena, J.L.A., 2012. National program for large scale rainwater harvesting: an individual or public responsibility? Water Resour. Manage. 26, 2703−2714.

Hirata, B., Conicelli, B.P., 2012. Groundwater resources in Brazil: a review of possible impacts caused by climate change. An. Acad. Bras. Ciênc. 84 (2), 297−312.

Hirata, R., Suhogusoff, A., Fernandes, A., 2007. Groundwater resources in the State of São Paulo (Brazil): the application of indicators. An. Acad. Bras. Ciênc. 79 (1), 141−152.

Hirata, R., Suhogusoff, A.V., 2019. How much do we know about the groundwater quality and its impacto n Brazilian society today? Acta Limnol. Bras. 31, e109.

Hirata, R., Zobbi, J.L., Fernandes, A., Bertolo, R.A., 2006. Hidrogeología del Brasil: Una breve crónica de las potencialidades, problemática y perspectivas. Bol. Geol. Minero, Madrid 217 (1), 25−36.

Hirata R., Zoby J.L.G., Oliveira F.R., Ground Water: Strategic or Emergency Reserve. In: Bicudo C.E.M., Tundisi J.G., Scheuenstuhl M.C.B. (Eds.), Waters of Brazil: Strategic Analysis. 2017, pp 187, Springer.

Hu, K., Awange, J.L., Forootan, E., Goncalves, R.M., Fleming, K., 2017. Hydrogeological characterisation of groundwater over Brazil using remotely sensed and model products. Sci. Total Environ. 599–600, 372–386.

IBGE (Instituto Brasileiro de Geografia e Estatística), 2002. Pesquisa Nacional de Saneamento Básico – 2000. IBGE, CDROM, Rio de Janeiro.

IGRAC (International Groundwater Resources Assessment Centre) Publication, 2014. Groundwater Monitoring in Latin America. Summary report of information shared during the regional workshop on groundwater monitoring. <https://www.un-igrac.org/resource/groundwater-monitoring-latin-america> (accessed 20.12.19.).

Jang, C., Lin, C., Lianc, C., Chen, J., 2016. Developing a reliable model for aquifer vulnerability. Stoch. Environ. Res. Risk Assess. 30, 175–1872016. Available from: http://doi.org/10.1007/s00477-015-1063-z.

Jarvis, T.W., et al., 2005. International borders, groundwater flow and hydroschizophrenia. Groundwater 43 (5), 764–770.

Júnior, R.G.C., Freitas, M.A., Silva, N.F., Filho, F.R.A., 2019. Sustainable groundwater exploitation aiming at the reduction of water vulnerability in the Brazilian Semi-Arid Region. Energies 12, 904. Available from: https://doi.org/10.3390/en12050904.

Luiz A., Perspective on groundwater aquifer governance. In: Villholth et al. (Eds.), Advances in Groundwater Governance, 2017, CRC Press.

Lemos, M.C., Finan, T.J., Fox, R.W., Nelson, D.R., Tucker, J., 2002. The use of seasonal climate forecasting in policymaking: lessons from northeast Brazil. Clim. Change 55 (4), 479–597. Available from: http://doi.org/10.1023/A:1020785826029.

Marengo, J.A., Jones, R., Alves, L.M., Valverde, M.C., 2009. Future change of temperature and precipitation extremes in South America as derived from the PRECIS regional climate modeling system. Int. J. Climatol. 29, 2241–2255.

Marengo, J.A., Tomasella, J., Nobre, C., 2017. Climate change and water resources. In: de Mattos Bicudo, C.E., et al., (Eds.), Water of Brazil. Springer.

Margat, L., Gun, J., 2013. Groundwater around the World, A Geographic Synopsis. CRC Press. Taylor & Francis Group, p. 238.

McCaffrey, S.C., 2001. The Law of International Watercourses: Non-navigational Uses. Oxford University Press, Oxford, pp. 414–415.

Melati, M.D., Fleischmann, A.S., Fan, F.M., Paiva, R.C.D., Athayde, G.B., 2019. Estimates of groundwater depletion under extreme drought in the Brazilian semi-arid region using GRACE satellite data: application for a small-scale aquifer. Hydrogeol. J. Available from: https://doi.org/10.1007/s10040-019-02065-1.

Menezes, J.P.C.D., Bertossi, A.P.A., Santos, A.R., Neves, M.A., 2014. Correlação entre uso da terra e qualidade da água subterrânea. Eng. Sanit. Ambiental 19, 173–186. Available from: http://doi.org/10.1590/S1413-41522014000200008.

Meng, X., Deng, B., Shao, J., Yin, M., Liu, D., Hu, Q., 2015. Confined aquifer vulnerability induced by a pumping well in a leakage area. Proc. Int. Assoc. Hydrol. Sci. 368, 442–447.

Meon, G., Rêgo, J., Schoniger, M., Schimmelpfennig, S., Walter, F., 2016. Strategies and technologies for water scarcity mitigation in Northeast of Brazil: water reuse, managed aquifer recharge and integrated water resources management (BRAMAR). <http://www.watersciencealliance.org/wp-content/uploads/2016/11/GERBRA_workshop_book_of_abstracts.pdf> (accessed 15.12.19.).

Mukherjee, A., Scanlon, B., Aureli, A., Langan, S., Guo, H., McKenzie, A., 2020. Global Groundwater: Source, Scarcity, Sustainability, Security and Solutions, first ed. Elsevier, ISBN: 9780128181720.

Nascimento, F.M.F., Carvalho, J.E., Peixinho, F.C., 2008. Sistema de informações de água subterrânea – siagas histórico, desafios e perspectivas. In: XV Congresso Brasileiro de Águas Subterrâneas.

Ondra, R., 2002. Geochemical and stable isotopic evolution of the Guarani Aquifer System in the state of São Paulo, Brazil. Hydrogeol. J. 10, 643–655.

Otto, F.E.L., Coelho, C.A.S., King, A., Perez, E.C., Wada, Y., Oldenborgh, G.J., et al., 2015. Factors other than climate change, main drivers of 2014/15 water shortage in southeast Brazil [in "Explaining extreme events of 2014 from a climate perspective"]. Bull. Am. Meteorol. Soc. 96 (12), S1–S172.

Patole, M., 2015. Brazilian Groundwater Law – Abstraction and Pollution Controls. https://doi.org/10.13140/RG.2.1.2358.5365.

Porto, R.L.L., Porto, M.F.A., 2002. Planning as a tool to deal with extreme events. The new Brazilian water resources management system. Water Int. 27 (1), 14–19.

Programa Água Doce (PAD), 2012. Ministerio do Meio Ambiente. (s.d.) Áqua Doce. <http://www.gov.br/aqua-doce> (accessed 09.12.19.).

Puri, S., 2003. Transboundary aquifers: international water law & hydrogeological uncertainty. Water Int. 28 (2), 276–279.

Rebouças, A., 1988. Águas Subterrâneas. In: Rebouças, A., Braga, B., Tundisi, J. (Eds.), Águas Doces no Brasil: capital ecológico, uso e conservação. Escrituras Editora, São Paulo, pp. 117–150.

Ricardo, H., Bruno, P.C., 2011. Groundwater resources in Brazil: a review of possible impacts caused by climate change. Ann. Braz. Acad. Sci. 84 (2), 297–312. Available from: http://doi.org/10.1590/S0001-37652012005000037.

Rowland, L., Da Costa, A.C.L., Galbraith, D.R., Meir, P., 2015. Death from drought in tropical forests is triggered by hydraulics not carbon starvation. Nature 528 (7580), 119–122. Available from: http://doi.org/10.1038/nature15539.

Silva, M.M.A., Holz, J.F.D., Freire, C.C., A outorga de direito do uso da água subterrânea nos estados brasileiros. In: Congresso Brasileiro de Águas Subterrâneas XV, 2008. Natal. Anais... São Paulo: ABAS. 2008, CD.

Silva, D.D., Pereira, S.B., Vieira, E.O., 2020. Integrated water resources management in Brazil. In: Vieira, E., Sandoval-Solis, S., Pedrosa, V., Ortiz-Partida, J. (Eds.), Integrated Water Resource Management. Springer, Cham.

Silva, et al., 2006. Environmental Evaluation of the Performance of the Cisterns Program of MDS and ASA: Executive Summary. EMBRAPA, Brasilia.

Silva Jr, G.C., Pizani, T.C., 2003. Vulnerability assessment in coastal aquifers between Niterói and Rio das Ostras, Rio de Janeiro State, Brazil. Rev. Lat. Am. Hidrogeol. 3 (1), 93–99.

Souza, E., Pontes, L.M., Fernandes Filho, E.I., Schaefer, C.E.G.R., Santos, E.E., 2018. Spatial and temporal potential groundwater recharge: the case of the Doce River Basin, Brazil. Rev. Bras. Ciênc. Solo 43. Available from: http://doi.org/10.1590/18069657rbcs20180010.

Stepping, K. 2016. Urban Sewage in Brazil: Drivers of and Obstacles to Wastewater Treatment and Reuse. <https://www.die-gdi.de/uploads/media/DP_26.2016.pdf> (accessed 15.12.19.).

Stollberg, R., Campos, J.E.G., Gorges, W.R., Borges, W.R., Gonçalves, T.D., Gaffron, A., et al., 2014. In: Lorz, C., Makeschin, F., Weiss, H. (Eds.), Integrated Water Resource Management in Brazil. IWA Publishing, 2014.

Stone, A., Lanzoni, M., Smedley, P., 2020. Groundwater resources: past, present and future. In: Dadson, S.J., Garrick, D.E., Penning-Rowsell, E.C., Hall, J.W., Hope, R., Hughes, J. (Eds.), Water Science, Policy, and Management: A Global Challenge, first ed. John Wiley & Sons Ltd. Publisher, Wiley.

Sugg, Z.P., Varady, R.G., Gerlak, A.K., Grenade, R., 2015. Transboundary groundwater governance in the Guarani Aquifer System: reflections from a survey of global and regional experts. Water Int. 40 (3), 377–400. Available from: https://doi.org/10.1080/02508060.2015.1052939.

Teston, A., Colasio, B.M., Ghisi, E., 2017. State of the art on water savings in buildings in Brazil: a literature review. In: Ghisi, E. (Ed.), Frontiers in Civil Engineering, 2. Bentham Science Publishers, Sharjah, pp. 3–64.

Teston, A., Geraldi, M.S., Colasio, B.M., Ghisi, E., 2018. Rainwater harvesting in buildings in Brazil: a literature review. Water 10, 471. Available from: https://doi.org/10.3390/w10040471.

UN, 2017. O Direito Humano à Água e Saneamento. Comunicado. 8.

UNESCO-WWAP (World Water Assessment Programme), 2012. Groundwater and Global Change: Trends, Opportunities and Challenges. UNESCO, Paris/London, Side Publication Series: 01.

Vasconcelos, S.M.S., Teixeira, Z.A., Neto, J.A.C., Luna, R.M., 2013. Estimativa da Reserva Renovável dos Sistemas Aquíferos da Porção Oriental da Bacia Sedimentar do Araripe [Estimate of the renewable reserve of Araripe sedimentary basin eastern aquifer systems]. Rev. Bras. Recur. Hídricos 18, 99–109.

Victor, D.G., Almeida, P., Wong, L., 2015. Water management policy in Brazil. In: ILAR Working Paper.

Villar, P.C., 2016. Groundwater and the right to water in a context of crisis. Ambiente Soc. São Paulo XIX (1), 85–102. <http://www.scielo.br/pdf/asoc/v19n1/1809-4422-asoc-19-01-00085.pdf> (accessed 20.12.19.).

Villar, P.C., Ribeiro, W.C., 2009. Sociedade e gestão do risco: o aquífero Guarani em Ribeirão Preto-SP, Brasil. Rev. Geogr. Norte Gd. 51–64.

Zagonari, F., 2010. Sustainable, just, equal, and optimal groundwater management strategies to cope with climate change: insights from Brazil. Water Resour. Manage. 24, 3731–3756. Available from: https://doi.org/10.1007/s11269-010-9630-z.

Zoby, J.L.G., Matos, B., 2002. Águas subterrâneas no Brasil e sua inserção na Política Nacional de Recursos Hídricos. In: Congresso Brasileiro De Águas Subterrâneas, 12. Florianópolis, 2002. Florianópolis: ABAS, 2002. CD-ROM.

Chapter 31

Challenges of sustainable groundwater development and management in Bangladesh: vision 2050

K.M. Ahmed
Department of Geology, Faculty of Earth and Environmental Sciences, University of Dhaka, Dhaka, Bangladesh

31.1 Introduction

Water scarcity is rapidly growing around the world due to the overexploitation of freshwater resources in many parts of the world due to consumption of the maximum global potential for freshwater availability due to population increase, changes in water consumption patterns, and possible impacts of climate change (Kummu et al., 2016). Groundwater is the most extracted raw material with the current withdrawal rate of 982 km^3/year to meet the demands of various sectors (NGWA, 2016). On a global scale, groundwater exploitation covers approximately 50% of drinking water needs, 20% of the demand for irrigation water, and 40% of the needs of self-supplied industry (UNESCO, 2004). Due to increasing pressure on groundwater, major concerns have been raised during the last decade about global depletion of groundwater due to overexploitations (Wada et al., 2010; Gleeson et al., 2012). The situation has worsened further making sustainable supply of groundwater for various uses as one of the major challenges for the current world (Mukherjee et al., 2020).

Bangladesh is characterized by very high amount of renewable water resources mostly originating as runoff outside the country (Price et al., 2014). However, most of this water passes through the country during the monsoon when water demand is minimum. Consequently, dry season demand is met mainly from groundwater extracted from the extensive Bengal Aquifer System for meeting demands of freshwater for drinking, agricultural, and industrial sectors. Annual groundwater resources availability for the country is estimated at 65 billion m^3 where net groundwater recharge is estimated around 28–32 billion m^3/year (CSIRO, 2014b). Bangladesh has been making good economic progress in the recent years by increasing contributions of industrial and service sectors (PWC, 2019). But the sustainability of development activities, everywhere in the world, is intricately related to availability of adequate water with desired quality. This chapter aims to assess the extent of current groundwater uses, impacts of abstractions on resource, future demands projections, potential risk to groundwater resources, and options for management to ensure sustainable supplies until 2050 and beyond.

31.2 Groundwater occurrences in Bangladesh

The groundwater systems in Bangladesh are part of the Indo-Gangetic basin alluvial transboundary aquifer, one of the world's most important freshwater reservoirs (MacDonald et al., 2016). This extensive aquifer system is formed by sediments eroded from the Himalayas and transported by the rivers Indus, Ganges, and Brahmaputra and distributed across the Indo-Gangetic plain stretching in Pakistan, India, Nepal, and Bangladesh (Fig. 31.1). The aquifer system is composed of unconsolidated alluvium occurring mostly at low relief where groundwater is encountered at shallow depths (Bonsor et al., 2017; Mukherjee et al., 2015).

In Bangladesh, groundwater is abstracted over most of the country from the Holocene unconsolidated alluvial aquifer and Plio–Pleistocene Dupi Tila aquifer occurring under unconfined and confined conditions, respectively (Ahmed, 2011). The Tipam Sandstone Formation forms the third important aquifer, mostly in the folded hilly areas. Shallow alluvial aquifers are recharged through rainfall and monsoonal flooding and replenished depending on the landform and

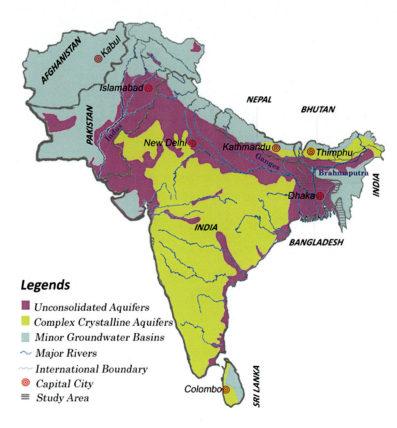

FIGURE 31.1 Map of the Indo-Gangetic aquifer system. Modified after Mukherjee, A., Saha, D., Harvey, C.F., Taylor, R.G., Ahmed, K.M., Bhanja, S.N., 2015. Groundwater systems of the Indian sub-continent. J. Hydrol.: Regional Stud., 4, 1–14.

geomorphology (Zahid and Ahmed, 2006; Shamsudduha et al., 2009). The deeper systems are recharged in the hilly areas to the north and east of the country (Michael and Voss, 2009).

There are a number of aquifer classifications proposed by various studies (Davies, 1994; BGS and DPHE, 2001; Zahid and Ahmed, 2006; Ahmed, 2011). Two approaches have been followed in these studies: one classification based on the depth of occurrences from the ground surface, irrespective of geological age and the other based on geological age irrespective of depth. Combining all proposed classifications, the following two classification schemes can be made, detailed descriptions of the aquifer systems are beyond the scope of this chapter.

Depth-based classification schemes are:

1. the upper or shallow aquifer occurring up to a depth of 60 m,
2. the intermediate aquifer occurring between 60 and 150 m,
3. the deep aquifer between 150 and about 350 m depth, and
4. the very deep aquifer occurring below 350 m.

Geological formation−based classification schemes are:

1. Holocene unconsolidated alluvial aquifer,
2. Pliocene Dupi Tila Sandstone aquifer, and
3. Mio−Pliocene Tipam Sandstone Aquifer.

Spatially, the country can be divided into six hydrogeological provinces (Fig. 31.2A), considering geomorphology and nature of Quaternary sediments (Ahmed, 2011). Groundwater quantity and quality vary significantly over these six regions. Fig. 31.2B presents a north−south section showing the depth of occurrences of various aquifers identified in BGS and DPHE, 2001.

31.3 Groundwater quality and concerns

Groundwater was introduced as the preferred source of drinking water in place of surface water in the 1960s due to its low total dissolved solid contents and superior microbiological quality. However, detailed assessment of chemical

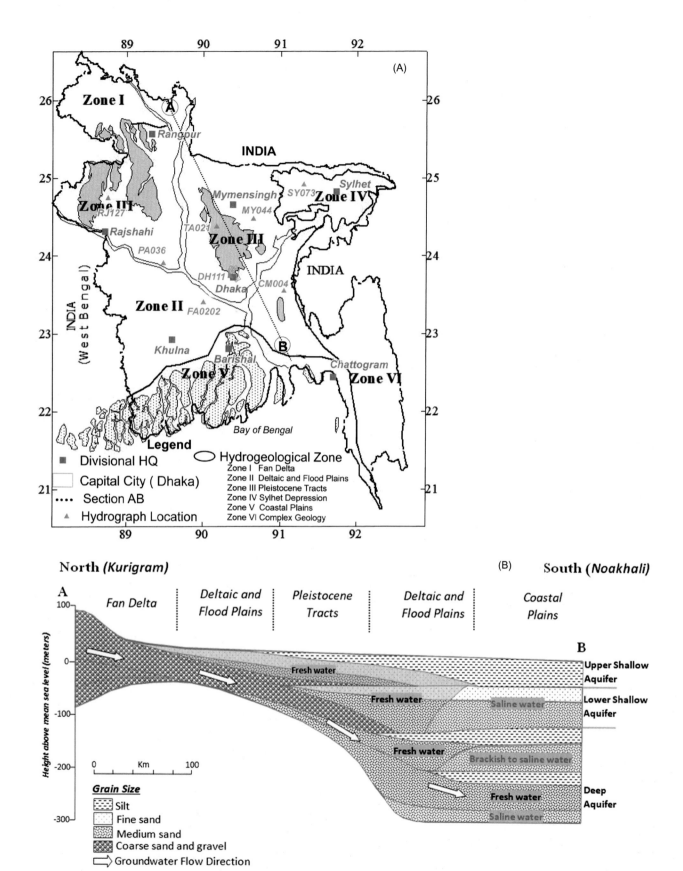

FIGURE 31.2 (A) Major hydrogeological provinces of Bangladesh and (B) generalized aquifer system shown in an N–S hydrostratigraphic section; alignment of the section is marked on part (A) as A–B. *(A) Modified after Ahmed, K.M., 2011. Groundwater contamination in Bangladesh. In: Grafton, R.Q., Hussey, K. (Eds.), Water Resources Planning and Management. Cambridge University Press, pp. 529–559. (B) BGS and DPHE, 2001. Arsenic contamination of groundwater in Bangladesh. In: Kinniburgh D.G., Smedley P.L. (Eds). British Geological Survey Technical Report WC/00/19. British Geological Survey: Keyworth.*

quality of groundwater was not done until the early 1990s when widespread arsenic occurrences in the shallow aquifer were detected (Nickson et al., 2000; Ahmed et al., 2004; Ravenscroft et al., 2005). High salinity was found to be a limiting factor in the coastal region for use of shallow groundwater, and people had to rely on surface water for drinking until deep fresh groundwater was put in use in the 1980s. Apart from natural quality issues, anthropogenic activities resulting from domestic, agricultural, and industrial practices also modify groundwater quality significantly. Microbiological contaminations of very shallow groundwater due to poor sanitation conditions have been a matter of concern as well (Macdonald et al., 1999; Saha et al., 2019). Major changes in recharge conditions and groundwater quality have been documented in different parts of Bangladesh, including Dhaka City (Burgess et al., 2011). BGS and DPHE, 2001 conducted the National Hydrochemical Survey (NHS) during 1998—99 for evaluating groundwater quality across the country, except for the Chittagong Hill Tracts. Apart from the NHS, many local and regional studies were prompted by arsenic and other natural water quality and anthropogenic pollution events.

31.3.1 Occurrences and distribution of arsenic

Since its first detection in 1993, many national, regional, and local studies were conducted by testing millions of water supply wells. There are strong geological controls on the spatial and vertical distributions of dissolved arsenic above the allowable drinking water limits (BGS and DPHE, 2001; Ahmed et al., 2004). High dissolved arsenic was present mainly in the Holocene unconsolidated sedimentary aquifers in the deltaic and floodplain areas within 100 m depths whereas groundwater abstracted from deeper than 150 m layers, with some exceptions, of Pleistocene Dupi Tila aquifers was safe from arsenic (BGS and DPHE, 2001). However, recent studies reported occurrences of arsenic at deeper levels under special geological settings (Mahmud et al., 2017; Khan et al., 2019). Arsenic occurrences in shallow aquifers prompted exploitation of the deep aquifers below 150 m for providing safe water, and thousands of deep tube well (DTW) have been installed in the arsenic-affected regions (Ravenscroft et al., 2013). Despite various efforts, many people are still exposed drinking water arsenic above 50 ppb, although it has reduced from initial 33 million in 1998—99 to about 18 million in 2019 (MICS, 2019).

31.3.2 Occurrences and distribution of salinity

High salinity in shallow groundwater occurs in the coastal areas of Bangladesh. This prompted the installation of thousands of DTW in the region after independence of Bangladesh. However, there are some localities where deep fresh aquifer is absent, and people are exposed to high salinity surface water and shallow groundwater (Islam et al., 2019). Apart from the coastal region, there occur some inland pockets of high salinity (Woobaidullah et al., 1998). Although groundwater was introduced mainly due to high microbiological contamination in surface water sources for drinking, it has appeared as a major issue of concerns for public health as major national surveys found large number of drinking water well containing *Escherichia coli* (MICS, 2019). Apart from the abovementioned three quality issues, iron and manganese occur extensively all over Bangladesh (BGS and DPHE, 2001). Iron is considered as a nuisance for water supply as tenements are needed when concentrations are high. Manganese also occurs at relatively high concentrations in Bangladesh and is a matter of concern for drinking water-quality management (BGS and DPHE, 2001).

31.4 Groundwater uses and impacts of abstractions

Over the last six decades, groundwater became the main source for domestic, irrigation, and industrial uses in Bangladesh because of its easy availability round the year and general good quality. Groundwater abstraction has increased manifold due mainly to dry season rice cultivation using minor irrigation (Zahid and Ahmed, 2006; Shamsudduha et al., 2009, 2011; CSIRO, 2014b; Mojid et al., 2019). Municipal abstraction has also increased all over the country, specifically in Dhaka City (Ahmed et al., 1999; Hoque et al., 2007; Burgess et al., 2011; Khan et al., 2016). Industrial abstractions have also increased in the major industrial belts around Dhaka and Chittagong (ARUP, 2016). Bangladesh ranks sixth in the global league of top groundwater user countries, whereas its south Asian neighbors India and Pakistan rank first and fourth, respectively (NGWA, 2016). Groundwater development in Bangladesh dates to the British time, but extensive abstractions started after its emergence as an independent country in 1971. Groundwater-level monitoring started in Bangladesh in the 1960s through a national network of observation wells established by then East Pakistan Water and Power Development Authority. Currently, the national groundwater monitoring is done by Bangladesh Water Development Board (BWDB). Also, Bangladesh Agricultural Development Corporation (BADC) has a less extensive monitoring system, whereas Department of Public Health Engineering (DPHE) monitors water level

Challenges of sustainable groundwater development and management in Bangladesh: vision 2050 Chapter | 31   429

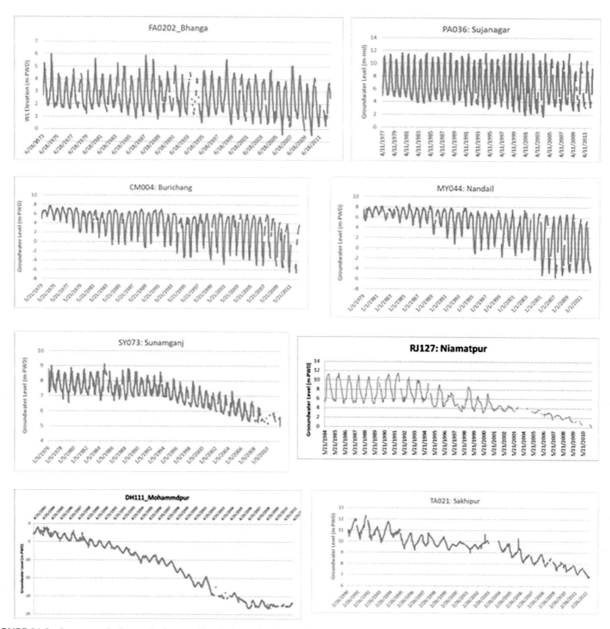

FIGURE 31.3   Long-term hydrographs from various hydrogeological regions of Bangladesh showing different trends and fluctuation patterns: the top two hydrographs show dynamic equilibrium of groundwater; next two hydrographs show accentuated fluctuations but almost full recovery as evidence of pumping creating options for additional recharge to the aquifer; and the last four hydrographs show depletion of storage due to agricultural, municipal, and industrial pumping. *Data from BWDB; locations of the observation wells are marked on Fig. 31.2A.*

once a year, during the driest time of the year. Trends of groundwater level have been analyzed by various authors for assessing impacts of pumping and sustainability of abstractions in relation to agricultural (Shamsudduha et al., 2009, 2011; CSIRO, 2014a; Kirby et al., 2015; Mojid et al., 2019) and municipal and industrial pumping (Ahmed et al., 1998; Hoque et al., 2007; Parvin, 2019; GWP and BWP, 2019). These studies reported declining trends in heavy abstraction areas such as the Dhaka City (municipal pumping), Barind Tract (agricultural pumping), and Gazipur (industrial pumping). However, in many other areas the dynamic equilibrium is maintained with or without accentuated fluctuation due to higher abstractions from the 1980s, mainly for irrigation. Fig. 31.3 presents several long-term hydrographs constructed with BWDB weekly groundwater-level monitoring data at shallow depths (locations are shown in Fig. 31.2A); the hydrographs depict different types of responses to irrigation, municipal, and industrial pumping. However, the hydrographs do not cover the most recent times as data was not available.

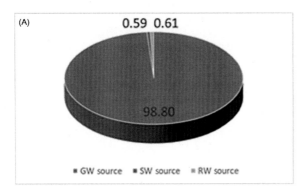
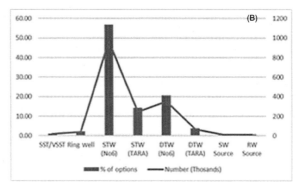

FIGURE 31.4 Status of public rural water supply options in Bangladesh: (A) Proportions of groundwater and surface water as source of drinking water as in 2018; (B) Proportions of different types of public water supply technologies as in 2018. *DTWs*, Deep tube wells. *Data from DWASA, 2019. Annual Report 2018−2019. Dhaka Water Supply and Sewerage Authority. Accessed from: <http://dwasa.org.bd/wp-content/uploads/2020/03/Annual-Report-Corrected-2018-19S.pdf>.*

31.4.1 Domestic uses in rural and urban areas

In the rural areas, 98% of people rely on groundwater for drinking purposes. There are approximately 10 million rural domestic water wells in Bangladesh, and about 90% of these are installed privately. DPHE installs rural public water supply options for ensuring optimum coverage, and as of the end of 2018, a number of such options are more than 1.7 million (DPHE, personal communications). Staggering 98.8% of these options use groundwater sources; 0.59% and 0.61% options use surface water and rainwater, respectively (Fig. 31.4A and B). Relative proportions of common types of rural safe water technologies are shown in Fig. 31.3B.

Installation of deep wells >150 m was started in the coastal area to avoid salinity after independence of Bangladesh. DTW installations got new momentum and outside the coastal area after the detection of arsenic in shallow wells. BGS and DPHE (2001) study reported that only 1% of the tested deep wells exceeded arsenic concentrations of 0.05 mg/L compared to around 27% of the shallow wells. Since 2002 deep wells have been installed at a faster rate in areas of arsenic contaminations and have become the most effective means of arsenic mitigation.

The urban population of Bangladesh increased from 7.6% in 1970 to 37.4% in 2019 growing at an average annual rate of 3.33%, and 39.4% of the population is urban in 2020. Capital Dhaka hosts almost half of countries urban population, whereas other major cities are Chattogram, Khulna, Rajshahi, Sylhet, Barisal, Rangpur, Gazipur, and Mymensingh (please refer to Fig. 31.2A for location). Groundwater-based urban water supply started in the 1940s by the DPHE, and the first Water Supply and Sewerage Authority (WASA) was formed in 1963 for dealing with Dhaka city. Major improvements in urban water supplies were made after the independence of Bangladesh. Currently, there are four WASAs for Dhaka (including Narayanganj), Chattogram, Khulna, and Rajshahi cities; water supply in other urban areas is managed by six city corporations and more than 300 Pauroshavas (municipalities) with the support from the DPHE. Dhaka grew very rapidly as the national capital where more than 95% of water supply in the 1980s was dependent on groundwater. Currently, about 79% of water for about 20 million people is provided from groundwater sources by the Dhaka WASA (DWASA, 2019). Fig. 31.5 presents the historical development of population and number of municipal deep wells in Dhaka city. Apart from Dhaka, most other major cities and all towns rely almost 100% on groundwater for municipal water supplies except Chattogram where the majority of the supply comes from surface water sources. Water supply in about 95% of smaller towns relies on groundwater, and only 5% uses surface water or both surface and groundwater (DPHE, pers.com).

Sustainability of groundwater water supply in Dhaka city has been a matter of concern for many years (Ahmed et al., 1998, 1999; Morris et al., 2003; Hoque et al., 2007). Recent studies also draw major concerns about Dhaka's groundwater in terms of quality and quantity (Burgess et al., 2011; Khan et al., 2016). Groundwater level underneath of Dhaka is declining alarmingly due to combined impact of groundwater overexploitation and land-use changes. There are reports of depletion of groundwater level in other major cities of Bangladesh. Mirdad and Palit (2014) reported that groundwater table was declining in Chattogram city continuously and recommended for use of alternative sources. Feasibility study to assess the sustainability of Khulna city water supply recommended to harness surface water or bringing in water by developing well field outside the city instead of full dependency on groundwater source abstracted from the aquifer underlying the city (JICA, 2011). Vulnerability assessments of the water sources of Khulna predicted potential saline intrusion of groundwater in the near future if abstraction rates are not curbed. Further, unsustainable groundwater exploitation resulting from a poorly regulated well drilling and overabstraction was becoming a threat for long-term water supply to the city (Mensah, 2018). Haque et al. (2012) reported decline of water level in Rajshahi city and suggested for conjunctive use of surface water and groundwater. Zafor et al. (2017) reported decline of water level in Sylhet region, including the city.

31.4.2 Irrigation uses

Surface water has been the main source of minor irrigation until the 1980, before the liberalization of irrigation through shallow tube wells as the numbers multiplied very quickly. By 1982 groundwater sources overtook the surface water sources in terms of percentages of areas irrigated. BADC conducts an annual survey of minor irrigation equipment during each Boro season and provides statistics of irrigation equipment and areas covered by each type. 2017−18 survey data (BADC, 2019) showed that 66 Mha land was irrigated, of which 73% were covered with groundwater and 27% with surface water (Fig. 31.5). Fig. 31.6A and B shows percentages of irrigated area covered by different types of minor irrigation equipment. It is evident that the shallow tube wells covered more than half of the total irrigated areas followed by surface water-based low-lift pumps. DTWs rank third followed by gravity flow and canal irrigations.

31.4.3 Industrial uses

Agriculture has been the main economic activity in the then East Pakistan and Bangladesh until the1908s. However, the establishment of jute industries started in the 1950s and has been the major industrial sector for more than three decades. The scenario started to change with the start of pharmaceutical, leather, and garments manufacturing in the 1980s. Rapid industrialization started in Gazipur, located northern edge of the greater Dhaka region. In the 1990s the area transformed from predominantly agricultural to urban and industrial types. More than 2000 industries are located in the greater Dhaka region, and most industries extract a huge amount of groundwater from the underlying Dupi Tila aquifer, the same aquifer being used for Dhaka city's water supply as well. It is estimated that the total groundwater abstraction in the Greater Dhaka area shall increase from 5.9 Mm^3/d in 2019 to 10 Mm^3/d in 2030, of which 40% is used by the

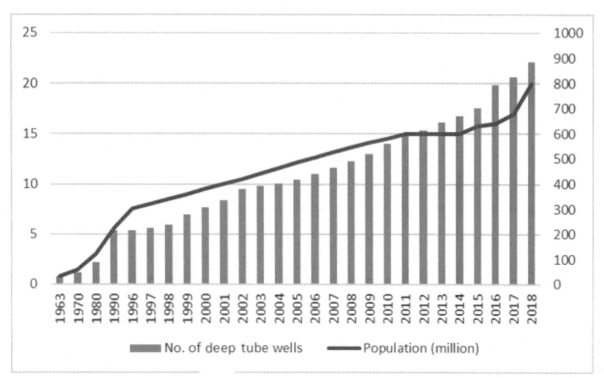

FIGURE 31.5 Status of minor irrigation in Bangladesh during 2017−18: (A) distribution of irrigable, irrigated areas with proportions of areas covered by surface and groundwater and (B) proportions of area covered by different modes of minor irrigation equipment using surface and groundwater sources. *Data from BADC, 2019. Minor Irrigation Survey Report 2017−18 (Rabi Season). Survey Conducted by Bangladesh Agricultural Development Corporation (BADC), Department of Agriculture Extension (DAE), Barind Multipurpose Development Authority (BMDA). Report prepared by Bangladesh Agricultural Development Corporation.*

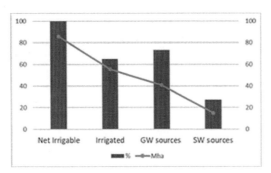
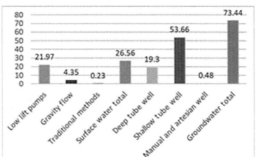

FIGURE 31.6 Status of minor irrigation in Bangladesh during 2017-18: (A) Distribution of irrigable, irrigated areas with proportions of areas covered by surface and groundwater; (B) proportions of area covered by different modes of minor irrigation equipment using surface and groundwater sources. *Data from BADC, 2019. Minor Irrigation Survey Report 2017–18 (Rabi Season). Survey Conducted by Bangladesh Agricultural Development Corporation (BADC), Department of Agriculture Extension (DAE), Barind Multipurpose Development Authority (BMDA). Report prepared by Bangladesh Agricultural Development Corporation.*

DWASA for municipal supplies in the megacity, and most of remaining 60% is extracted by various industries GWP (2019). Parvin (2019) analyzed groundwater levels in Gazipur and reported that depletion was maximum during the period 1998–2005 and rapid depletion pattern after the year 2000.

31.5 Major challenges

31.5.1 Meeting increased demands in 2050

Bangladesh's population will reach 190 million in 2020 and 218 million in 2050; about half of the people would live in urban areas (BDP, 2019). Increased amount of water would be needed for meeting domestic demands in rural and urban areas. Additional food production would be needed to feed the millions. Bangladesh's economy has witnessed steady growth in the past decade, with the country's annual gross domestic product growth rate at 7.86% in 2018. It is poised to increase by 7% on an average till 2033 (PWC, 2019). Bangladesh's economy ranked 31st in the world in 2016 and shall rise to 29th and 23rd in 2030 and 2050, respectively (PWC, 2017). Industrial sector is growing fast in Bangladesh that will withdraw more water in the near future. Dhaka is the fastest growing Megacity that will be joined by other megacities in 2050 resulting in very high demand for municipal water supply. It is projected that industrial water demand can increase by 109%, domestic water demand by 75%, and agricultural water demand by 43% in 2030 for taking the water demand to 52.96 billion m^3, from 35.87 billion m^3 in 2014 (Amin et al., 2018). Bangladesh Delta Plan (BDP, 2019) projected demands for water for various sectors as shown in Fig. 31.7.

31.5.2 Impacts of climate change

Climate change may have significant impacts on the hydrologic cycle and hydro-disasters may become more frequent resulting into quality- and quantity-related issues for the groundwater resources. A World Bank study (World Bank, 2010) concluded that average temperature in monsoon and winter will rise between 0.7°C and 1.3°C, respectively, by 2030 and 1.1°C and 1.8°C by 2050. The same study also concluded that total amount of rainfall shall rise by 300 and 500 mm in 2030 and 2050, respectively, but the monsoon rainfall will increase by 11% and 28% in 2030 and 2050, respectively, and winter rainfall for the same time will decrease by 3% and 37% for the same time. The rise in temperature and variations in rainfall would have detrimental impacts on groundwater recharge, and there will be greater demand for storing rain and floodwater. Higher temperature would also result in higher rate of evaporation and larger demand for irrigation water. However, according to CSIRO (2014b), the natural variability of rainfall and river flow is expected to dominate over climate change at least in the near future. MacDonald et al. (2015) reported that groundwater, particularly deep groundwater, in the Indo-Gangetic aquifer system would provide a cushion for climate change mitigation. But excessive pumping poses a greater threat to this deep groundwater than climate change. Hirji et al. (2017) stated that meeting challenges of increasing demands of water for food, people, energy, and industry in the south Asian region shall become more complicated due to the impacts of changing climate. More recently, Shamsudduha (2018) reported that the impacts of human development on groundwater resources in Bangladesh were evident, but it was not clear how changing climate would impact the quality and quantity of groundwater.

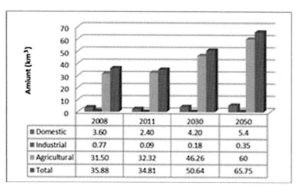 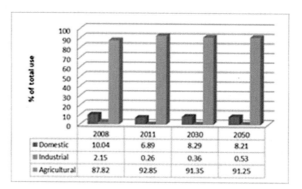

FIGURE 31.7 Estimated demands for water for various sectors of Bangladesh. *Data from BDP, 2019. Bangladesh Delta Plan 2100. Formulated by General Economics Division (GED), Bangladesh Planning Commission, Government of the People's Republic of Bangladesh.*

31.5.3 Arsenic and other contamination issues

Even after more than 25 years of its first detection, a recent national drinking water-quality survey reported about 10 million people were found to drink water concentrations above the national standard of 50 μg/L (MICS, 2019). Flanagan et al. (2012) estimated that arsenic exposures to concentrations >50 μg/L account for an annual 24,000 adult deaths in the country. The same study also reported that an arsenic-related mortality rate of 1 in every 18 adult deaths could represent an economic burden of 13 billion USD in lost productivity alone over the next 20 years. Ahmad et al. (2018) reported occurrences of dermatological manifestations, noncommunicable diseases, including cancer, adverse pregnancy outcomes, and decreased intelligence quotient among the children. Shamsudduha et al. (2019a,b) for the first-time developed groundwater risk maps at the national scale for Bangladesh combining data on arsenic, salinity, and water storage, using geospatial techniques linking hydrological indicators for water quality and quantity. The multihazard groundwater risk maps showed that a considerable proportion of land area (5%–24% under extremely high to high risks) in Bangladesh was under combined risk of arsenic and salinity contamination along with depletion of groundwater storage. The mapping estimated that 6.5–24.4 million people were exposed to a combined risk of high arsenic, salinity, and groundwater storage depletion. Apart from natural arsenic contaminations, the risk of anthropogenic contaminations from agricultural, municipal, and industrials sources is also high. Ahmed (2011) gave an overview of other natural and anthropogenic contaminants, along with arsenic, which should be carefully monitored for the protection of groundwater from degradation.

31.5.4 Transboundary issues

Bangladesh has the highest dependency on transboundary flow for its total renewable water resources compared to the other south Asian countries as shown in Fig. 31.8 (Price et al., 2014). The country also shares the extensive Indo-Gangetic transboundary aquifer (MacDonald et al., 2016; Khan et al., 2014). As there is an increase in demand in upper riparian countries, higher use of both surface and groundwater may have adverse impacts on the lower riparian countries. This can become a major challenge in sustaining freshwater supply in Bangladesh if regional cooperation is not fostered for water resources management.

31.6 Sustainable groundwater management: vision 2050

31.6.1 Surface water harnessing

Although about half of the country goes underwater during the monsoon, the availability of surface water is very low during the dry season when water demand is the highest. Quality of river water is also deteriorating due to uncontrolled discharge of industrial wastes and poor sewerage conditions. It is very important to harness surface water reservoirs for using in dry season irrigation. Treated surface water can also be used for drinking water supplies. Increased use of surface water can reduce pressure on groundwater.

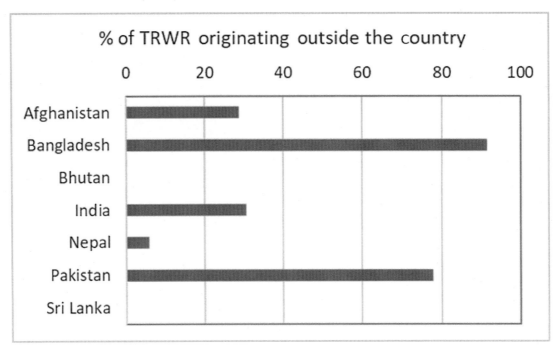

FIGURE 31.8 Percentage of total renewable water resources originating from outside the country in different south Asian countries. *Data from Price G., Alam R., Hasan S., Humayun F., Kabir M.H., Karki C.S., et al., Attitudes to water in South Asia, 2014, Royal Institute of International Affairs, London, ISBN: 978784130121.*

31.6.2 Better irrigation water management

Current irrigation practices result in huge wastage of water, and a significant amount of water can be saved by improving irrigation efficiency. Improved methods such as sprinkler and subsurface irrigation may be used to save water. Low water–consuming crops can also be introduced particularly in areas where groundwater levels are declining fast.

31.6.3 Groundwater monitoring, abstraction controls, and licensing

Groundwater abstraction is increasing every day without a proper management and monitoring plan. Conventional water pumps are becoming inoperative in many areas due to declining water levels and need to be replaced by alternative expensive pumping technologies. Groundwater governance is almost nonexistent in the country due mainly to a lack of proper institutional arrangements. There are rules, regulations, and policies to ensure proper management, but enforcement is lacking. The current National Monitoring Network focuses mainly on the shallow aquifer except in the coastal region. Multilevel monitoring should also include water quality alongside water level.

31.6.4 Pollution abatement and control

Shallow groundwater is being polluted by various domestic and industrial sources. High population density, poor sanitation conditions, and uncontrolled waste disposal make shallow groundwater particularly vulnerable to pollution. Efforts combining awareness and legal aspects are needed for pollution control and abatement to protect groundwater.

31.6.5 Applications of managed aquifer recharge

Urbanization and other infrastructural developments reduce natural recharge significantly resulting in groundwater over-exploitations or even mining conditions. Managed aquifer recharge (MAR) can be adopted to augment aquifers during monsoon when surface water is abundant. However, proper national policy and guidelines are necessary to protect groundwater being polluted by low-quality source water. In urban areas, MAR can augment declining groundwater resources as well as reduce waterlogging. MAR can also be applied to improve shallow groundwater quality such as reducing salinity and diluting natural or anthropogenic pollutants (Sultana et al., 2015). MAR is also a known tool for averting saltwater intrusion in coastal aquifers.

31.6.6 Wastewater reuse

Bangladesh will have to adapt various water purifying technologies, including reverse osmosis for salinity removal. Paradigm shift in people's perception about groundwater, which is now undervalued by the users, is necessary for proper evaluation of the resource out of sight out of mind. Recycle and reuse of wastewater has to be promoted by combining technological options and awareness of stakeholders. Regional cooperation on water has to be increased for basin-wide management by conjunctive management.

31.6.7 Awareness building

Full commitment toward adaptation to integrated water resources management needs to be ensured at all levels. Decentralized water management at planning area or basin scales should be introduced. MAR should be adopted at all levels to augment the declining water levels. Water-sensitive urban designs must be adopted and implemented for major urban areas and urban conglomerates for reducing impacts of urbanization on groundwater.

31.6.8 Judicial use of deep groundwater

The use of deep groundwater started as a source of freshwater in the coastal region but abstractions increased significantly over the last two decades as a source of arsenic-safe water. There are risks of vertical and lateral ingression of arsenic-contaminated water if abstractions are not properly managed (Knappett et al., 2016). Burgess et al. (2010) reported that vulnerability of arsenic contamination of deep wells by arsenic was controlled by the geometry of induced groundwater flow paths and the geochemical conditions. Recent studies highlighted the need for judicial use, monitoring, tracer studies for understanding recharge mechanisms, and undertaking management plans aided by groundwater modeling studies for ensuring sustainable use of deep groundwater (Lapworth et al., 2018; Shamsudduha et al., 2019a, b). However, due to its importance as a source of fresh and arsenic-safe water, Ravenscroft et al. (2013) suggested to continue using use it even if it is unsustainable.

31.6.9 Groundwater governance

Despite heavy reliance on groundwater resources, the status of management is in very poor conditions. There are many agencies involved in groundwater development and management in the country. Also, there exists good set of policy, strategy, law, and regulations to ensure management of water resources. But most of the organizations as well as legal instruments put more emphasis on surface water. There is no law specifically to protect groundwater from overexploitation and contaminations. Bangladesh requires an effective groundwater governance mechanism to be spearheaded by an institution with legal authority and technical capabilities Better Governance should be the main agenda for meeting the challenges groundwater demand by 2050.

31.6.10 Research and development activities

The main stresses on groundwater in Bangladesh are heavy pumping and risks of contaminations. Courses on groundwater sciences need to be introduced at tertiary levels to produce more groundwater professional. The research institutes should be equipped with state-of-the-art laboratory facilities. Predictive groundwater modeling can aid better in decision-making. The national monitoring system needs to be updated along with open access to data for all users and interested parties.

31.7 Groundwater: resource out of sight but not to be out of mind

Like today, groundwater shall remain one of the main sources of water for various uses in 2050 as well. It is evident from the estimation of future demands that groundwater alone shall not be able to meet the increasing demand. Ensuring sustainability of groundwater development under the trends for increasing population, urbanization, industrialization, and agricultural intensifications is a mammoth challenge. Possible impacts of climate change and/or climate variability would make the challenge even more complicated. The use of groundwater in Bangladesh is managed by various agencies, the and current status of management is an ideal example of "Tragedy of Commons." Despite the existence of national policies, strategies, and laws, there is no integrated strategy to ensure the beneficial use, protection, and sustainability of this vital natural resource. Current developments are mostly done on an ad hoc basis driven by demand. However, the National Water Management Plan adopted the principle of integrated water resources management instead of sectoral approach. Also, the Bangladesh Delta Plan is a long-term water-centric plan for ensuring water for all.

One major problem is general lack of appropriate capacities within the agencies dealing with groundwater. Country's water sector is overwhelmingly dominated by surface water professionals and general awareness about groundwater among the planners and policy makers is generally low. Proper institutional arrangements with adequate legal authority and technical capability are vital for ensuring sustainability of the strategic resource for meeting demands of future generations. A central groundwater institution should spearhead the functions such as monitoring, exploration and assessments, quantity and quality protection and regulation, integrated development, and research and knowledge generation.

Bangladesh is making remarkable progress toward attaining the status of a developed country by 2050. Ensuring sustainable supplies of water with adequate quantity and quality is key to reaching that goal. Groundwater can play a major role in achieving the goal if the strategic resource is governed properly for maintaining sustainability.

Acknowledgments

I would like to thank Dr. Anwar Zahid of Bangladesh Water Development Board (BWDB) for providing groundwater monitoring data and Mr. Mohammad Saifur Rahman of the Department of Public Health Engineering (DPHE) for providing data on public water supply options in rural and urban areas. I am grateful to my colleague Dr. Mahfuz Khan for going through the manuscript; his suggestions have been very useful in improving the quality of the chapter. Thanks are also due to my MPhil Student Mr. Md. Jahid Alam and Research Fellow Ms. Taspiya Hamid for their helps with diagrams and formatting texts. Finally, I would like to thank Dr. Abhijit Mukherjee, first, for inviting me to contribute to the book "Global Groundwater: Source, Scarcity, Sustainability, Security and Solutions" and, second, for his editorial review in finalizing the chapter.

References

Ahmed, K.M., 2011. Groundwater contamination in Bangladesh. In: Grafton, R.Q., Hussey, K. (Eds.), Water Resources Planning and Management. Cambridge University Press, pp. 529–559.

Ahmed, K.M., Bhattacharya, P., Hasan, M.A., Akhter, S.H., Alam, S.M., Bhuyian, M.H., et al., 2004. Arsenic enrichment in groundwater of the alluvial aquifers in Bangladesh: an overview. Appl. Geochem. 19 (2), 181–200.

Ahmed, K.M., Hasan, M.A., Burgess, W.G., Dottridge, J., Ravenscroft, P., van Wanderen, J., 1999. Dupi Tila aquifer of Dhaka: hydraulic and hydrochemical response to extensive exploitation. Groundwater in the Urban Environment: Selected City Profiles. AA. Balkema, pp. 19–30.

Ahmed, K.M., Hasan, M.A., Chowdhury, S.Q., Sharif, S.U., Haque, M.E., Hossain, A.K.M.S., 1998. Impact of urbanisation on the geoenvironment of Dhaka City. J. Asiatic Soc. Bangladesh, Sci. 24 (2), 339–352.

Ahmad, S.A., Khan, M.H., Haque, M., 2018. Arsenic contamination in groundwater in Bangladesh: implications and challenges for healthcare policy. Risk Manage. Healthc. Policy 2018 (11), 251–261.

Amin, Z., Chowdhury, I., Islam, B., 2018. Bangladesh Water Sector Network Study, Final Report. Light Castle Partners and Partners for Water, Dhaka, Bangladesh, p. 92. Light Castle, 2018. Prepared for Partners for Water.

ARUP, 2016. An analysis of industrial water use in Bangladesh with a focus on the leather and textile industries. In: Report Commissioned by IFC/WB With Support From 2030 Water Resources Group. Bangladesh Water Pact. <http://www.textilepact.net/publications.html>.

BADC, 2019. Minor Irrigation Survey Report 2017–18 (Rabi Season). Survey Conducted by Bangladesh Agricultural Development Corporation (BADC), Department of Agriculture Extension (DAE), Barind Multipurpose Development Authority (BMDA). Report prepared by Bangladesh Agricultural Development Corporation.

BDP, 2019. Bangladesh Delta Plan 2100. Formulated by General Economics Division (GED), Bangladesh Planning Commission, Government of the People's Republic of Bangladesh.

BGS and DPHE, 2001. Arsenic contamination of groundwater in Bangladesh. In: Kinniburgh D.G., Smedley P.L. (Eds). British Geological Survey Technical Report WC/00/19. British Geological Survey: Keyworth.

Bonsor, H.C., MacDonald, A.M., Ahmed, K.M., Burgess, W.G., Basharat, M., Calow, R.C., et al., 2017. Hydrogeological typologies of the Indo-Gangetic basin alluvial aquifer, South Asia. Hydrogeol. J. 25 (5), 1377–1406.

Burgess, W.G., Hasan, M.K., Rihani, E., Ahmed, K.M., Hoque, M.A., Darling, W.G., 2011. Groundwater quality trends in the Dupi Tila aquifer of Dhaka, Bangladesh: sources of contamination evaluated using modelling and environmental isotopes. Int. J. Urban Sustain. Dev. 3 (1), 56–76.

Burgess, W.G., Hoque, M.A., Michael, H.A., Voss, C.I., Breit, G.N., Ahmed, K.M., 2010. Vulnerability of deep groundwater in the Bengal Aquifer System to contamination by arsenic. Nat. Geosci. 3, 83–87. Available from: https://doi.org/10.1038/ngeo750. Progress Article.

Burgess W.G., Hasan M.K., Rihani E., Ahmed K.M., Hoque M.A., Darling W.G., 2011. Groundwater quality trends in the Dupi Tila aquifer of Dhaka, Bangladesh: sources of contamination evaluated using modelling and environmental isotopes. *International Journal of Urban Sustainable Development*, 3 (1), 56–76.

CSIRO, 2014a. Sustaining water resources for food security in Bangladesh. In: Mainuddin, M., Kirby, M., Walker, G., Connor, J. (Eds.), CSIRO Sustainable Development Investment Portfolio Project. CSIRO Land and Water Flagship, Australia, p. 108.

CSIRO, 2014b. Bangladesh Integrated Water Resources Assessment: Final Report. CSIRO Australia in Association With WARPO, BWDB, IWM, BIDS, and CEGIS.

Davies, J., 1994. The Hydro-geochemistry of Alluvial Aquifers in Central Bangladesh. Groundwater Quality. Published by Chapman and Hall, pp. 9–18, ISBN 0412586207.

DWASA, 2019. Annual Report 2018–2019. Dhaka Water Supply and Sewerage Authority. Accessed from: <http://dwasa.org.bd/wp-content/uploads/2020/03/Annual-Report-Corrected-2018-19S.pdf>.

Flanagan, S.V., Johnston, R.B., Zheng, Y., 2012. Arsenic in tube well water in Bangladesh: health and economic impacts and implications for arsenic mitigation. Bull. World Health Organ. 90, 839–846. Available from: https://doi.org/10.2471/BLT.11.101253. 2012.

Gleeson, T., Wada, Y., Bierkens, M.F.P., van Beek, L.P.H., 2012. Water balance of global aquifers revealed by groundwater footprint. Nature 488, 197. Available from: https://doi.org/10.1038/nature11295.

GWP, BWP, 2019. Rapid assessment of greater Dhaka groundwater sustainability. In: Report Prepared by Global Water Partnership and Bangladesh Water Partnership for 2030WRG, IFC/WB, 38 p.

GWP, 2019. Rapid Assessment of Greater Dhaka Groundwater Sustainability. Report commissioned by 2030WRG, IFC/WB.

Haque, M.A.M., Jahan, C.S., Mazumder, Q.H., Nawaz, S.M.S., Mirdha, G.C., Mamud, P., et al., 2012. Hydrogeological condition and assessment of groundwater resource using visual modflow modeling, Rajshahi city aquifer, Bangladesh. J. Geol. Soc. India 79 (1), 77–84.

Hirji, R., Nicol, A., Davis, R., 2017. South Asia Climate Change Risks in Water Management: Climate Risks and Solutions-Adaptation Frameworks for Water Resources Planning, Development, and Management in South Asia. World Bank.

Hoque, M.A., Hoque, M.M., Ahmed, K.M., 2007. Declining groundwater level and aquifer dewatering in Dhaka metropolitan area, Bangladesh: causes and quantification. Hydrogeol. J. 15 (8), 1523–1534.

Islam, M.A., Hoque, M.A., Ahmed, K.M., Butler, A.P., 2019. Impact of climate change and land use on groundwater salinization in Southern Bangladesh—implications for other Asian Deltas. Environ. Manage. 64, 640–649. Available from: https://doi.org/10.1007/s00267-019-01220-4.

JICA, 2011. Feasibility Study for Khulna Water Supply Improvement Project in the People's Republic of Bangladesh: Final Report. Khulna Water Supply Project. Japan International Cooperation Agency and NJS Consultants Co Ltd. Accessed from: <http://open_jicareport.jica.go.jp/pdf/12020541.pdf>.

Khan, M.R., Koneshloo, M., Knappett, P.S., Ahmed, K.M., Bostick, B.C., Mailloux, B.J., et al., 2016. Megacity pumping and preferential flow threaten groundwater quality. Nat. Commun. 7 (1), 1–8.

Khan, M.R., Michael, H.A., Nath, B., Huhmann, B.L., Harvey, C.F., Mukherjee, A., et al., 2019. High-arsenic groundwater in the southwestern Bengal basin caused by a lithologically controlled deep flow system. Geophys. Res. Lett. 46 (22), 13062–13071.

Khan, M.R., Voss, C.I., Yu, W., Michael, H.A., 2014. Water resources management in the Ganges Basin: a comparison of three strategies for conjunctive use of groundwater and surface water. Water Resour. Manage. 28 (5), 1235–1250.

Kirby, J.M., Ahmad, M.D., Mainuddin, M., Palash, W., Quadir, M.E., Shah-Newaz, S.M., et al., 2015. The impact of irrigation development on regional groundwater resources in Bangladesh. Agric. Water Manage. 159, 264–276.

Knappett, P.S.K., Mailloux, B.J., Choudhury, I., Khan, M.R., Michael, H.A., Barua, S., et al., 2016. Vulnerability of low-arsenic aquifers to municipal pumping in Bangladesh. J. Hydrol. 539, 674–686. Available from: https://doi.org/10.1016/j.jhydrol.2016.05.035.

Kummu, M., Guillaume, J.H.A., de Moel, H., Eisner, S., Flörke, M., Porkka, M., et al., 2016. The world's road to water scarcity: shortage and stress in the 20th century and pathways towards sustainability. Sci. Rep. 6 (38495), (2016). Available from: https://doi.org/10.1038/srep38495.

Lapworth, D.J., Zahid, A., Taylor, R.G., Burgess, W.G., Shamsudduha, M., Ahmed, K.M., et al., 2018. Security of deep, groundwater in the coastal Bengal Basin revealed by tracers. Geophys. Res. Lett. 45, 8241–8252. Available from: https://doi.org/10.1029/2018GL078640.

MacDonald, A.M., Bonsor, H.C., Ahmed, K.M., Burgess, W.G., Basharat, M., Calow, R.C., et al., 2016. Groundwater quality and depletion in the Indo-Gangetic Basin mapped from in situ observations. Nat. Geosci. 9 (10), 762–766.

MacDonald, A.M., Bonsor, H.C., Taylor, R., Shamsudduha, M., Burgess, W.G., Ahmed, K.M., et al., 2015. Groundwater resources in the Indo-Gangetic Basin: resilience to climate change and abstraction. In: British Geological Survey Open Report, p. 63.

Macdonald, D., Ahmed, K.M., Islam, M.S., Lawrence, A., Khandker, Z.Z., 1999. Pit latrines – a source of contamination in peri-urban Dhaka? Waterlines 17 (4), 6–8.

Mahmud, M.I., Sultana, S., Hasan, M.A., Ahmed, K.M., 2017. Variations in hydrostratigraphy and groundwater quality between major geomorphic units of the Western Ganges Delta plain, SW Bangladesh. Appl. Water Sci. 7, 2919–2932. Available from: https://doi.org/10.1007/s13201-017-0581-x.

Mensah, H.S.K., 2018. Khulna Water Supply Project, Bangladesh. Report Prepared by Asian Institute of Technology With Funding From Japan Water Research Centre Under the Network on Water Technology in Asia Pacific Project (New Tap Project), p. 11.

Michael, H., Voss, C., 2009. Controls on groundwater flow in the Bengal Basin of India and Bangladesh: regional modeling analysis. Hydrogeol. J. 17, 1561–1577. Available from: https://doi.org/10.1007/s10040-008-0429-4.

MICS, 2019. Progotir Pathey, Bangladesh Multiple Indicator Cluster Survey 2019, Survey Findings Report. Bangladesh Bureau of Statistics (BBS) and UNICEF Bangladesh, Dhaka, p. 564, ISBN: 978-984-8969-34-2.

Mirdad, M.A.H., Palit, S.K., 2014. Investigation of ground water table in the South-East (Chittagong) part of Bangladesh. Am. J. Civ. Eng. 2 (2), 53−59. Available from: https://doi.org/10.11648/j.ajce.20140202.17.
Mojid, M.A., Parvez, M.F., Mainuddin, M., Hodgson, G., 2019. Water table trend—a sustainability status of groundwater development in North-West Bangladesh. Water 2019 (11), 1182. Available from: https://doi.org/10.3390/w11061182.
Morris, B.L., Seddique, A.A., Ahmed, K.M., 2003. Response of the Dupi Tila aquifer to intensive pumping in Dhaka, Bangladesh. Hydrogeol. J. 11 (4), 496−503.
Mukherjee, A., Saha, D., Harvey, C.F., Taylor, R.G., Ahmed, K.M., Bhanja, S.N., 2015. Groundwater systems of the Indian sub-continent. J. Hydrol.: Regional Stud. 4, 1−14.
Mukherjee, A., Scanlon, B., Aureli, A., Langan, S., Guo, H., McKenzie, A., 2020. Global Groundwater: Source, Scarcity, Sustainability, Security and Solutions, first ed. Elsevier, ISBN: 9780128181720.
NGWA, 2016. Facts About Global Groundwater Usage, National Groundwater Association. <http://www.ngwa.org/Fundamentals/Documents/global-groundwater-use-fact-sheet.pdf>.
Nickson, R.T., McArthur, J.M., Ravenscroft, P., Burgess, W.G., Ahmed, K.M., 2000. Mechanism of arsenic release to groundwater, Bangladesh and West Bengal. Appl. Geochem. 15 (4), 403−413.
Parvin, M., 2019. The rate of decline and trend line analysis of groundwater underneath Dhaka and Gazipur City. J. Water Resour. Prot. 11 (3), 348−356.
Price, G., Alam, R., Hasan, S., Humayun, F., Kabir, M.H., Karki, C.S., et al., 2014. Attitudes to water in South Asia. Royal Institute of International Affairs, London, ISBN: 978784130121.
PWC, 2017. The long view how will the global economic order change by 2050? In: The World in 2050 − Summary Report. PricewaterhouseCoopers LLP, p. 72. <www.pwc.com>.
PWC, 2019. Destination Bangladesh. In: Report Prepared by Mamun Rashid, Salman Afsar Alam and Nahiyan Nasir from PwC Bangladesh Private Limited, Dhaka. p. 36.
Ravenscroft, P., Burgess, W.G., Ahmed, K.M., Burren, M., Perrin, J., 2005. Arsenic contamination of groundwater in the Bengal Basin, Bangladesh: distribution, field relations, origins and mitigation. Hydrogeol. J. 13, 727−751.
Ravenscroft, P., McArthur, J.M., Hoque, M.A., 2013. Stable groundwater quality in deep aquifers of Southern Bangladesh: the case against sustainable abstraction. Sci. Total Environ. 454−455, 627−638. Available from: https://doi.org/10.1016/j.scitotenv.2013.02.071.
Saha, R., Dey, N.C., Rahman, M., Bhattacharya, P., Rabbani, G.H., 2019. Geogenic arsenic and microbial contamination in drinking water sources: exposure risks to the coastal population in Bangladesh. Front. Environ. Sci. Available from: https://doi.org/10.3389/fenvs.2019.00057Published online on 08 May, 2109. Available from: https://doi.org/10.3389/fenvs.2019.00057.
Shamsudduha, M., 2018. Impacts of human development and climate change on groundwater resources in Bangladesh. Groundwater of South Asia. Springer, Singapore, pp. 523−544.
Shamsudduha, M., Chandler, R.E., Taylor, R.G., Ahmed, K.M., 2009. Recent trends in groundwater levels in a highly seasonal hydrological system: The Ganges-Brahmaputra-Meghna Delta. Hydrol. Earth Syst. Sc. 13 (12), 2373−2385.
Shamsudduha, M., Joseph, G., Haque, S.S., Khan, M.R., Zahid, A., Ahmed, K.M.U., 2019b. Multi-Hazard Groundwater Risks to the Drinking Water Supply in Bangladesh: Challenges to Achieving the Sustainable Development Goals. Policy Research Working Paper 8922, The World Bank.
Shamsudduha, M., Taylor, R.G., Ahmed, K.M., Zahid, A., 2011. The impact of intensive groundwater abstraction on recharge to a shallow regional aquifer system: evidence from Bangladesh. Hydrogeol. J. 19, 901−916. Available from: https://doi.org/10.1007/s10040-011-0723-4.
Shamsudduha, M., Zahid, A., Burges, W.G., 2019a. Security of deep groundwater against arsenic contamination in the Bengal Aquifer System: a numerical modeling study in southeast Bangladesh. Sustain. Water Resour. Manage. 5, 1073−1087. Available from: https://doi.org/10.1007/s40899-018-0275-z.
Sultana, S., Ahmed, K.M., Mahtab-Ul-Alam, S.M., Hasan, M., Tuinhof, A., Ghosh, S.K., et al., 2015. Low-cost aquifer storage and recovery: implications for improving drinking water access for rural communities in coastal Bangladesh. J. Hydrol. Eng. 20 (3), B5014007.
UNESCO, 2004. Groundwater resources of the world and their use. In: Zektser, I.S., Everett, L.G. (Eds.), IHP-VI, Series on Groundwater. UNESCO, Paris, ISBN: 92-9220-007-0p. 346. No. 6, ISBN.
Wada, Y., van Beek, L.P.H., van Kempen, C.M., Reckman, J.W.T.M., Vasak, S., Marc, F.P., et al., 2010. Global depletion of groundwater resources. Geophys. Res. Lett. 37, L20402. Available from: http://dx.doi.org/10.1029/2010GL044571.
Woobaidullah, A.S.M., Ahmed, K.M., Hasan, M.A., Hasan, M.K., 1998. Saline groundwater management in Manda thana of Naogaon District, NW Bangladesh. J. Geol. Soc. India 51, 49−56.
World Bank, 2010. Bangladesh Climate Change and Sustainable Development. Report No. 21104-BD. The World Bank, Rural Development Unit, South Asia Region.
Zafor, M., Alam, M., Bin, J., Rahman, M., Amin, M.N., Zafor, M.A., et al., 2017. The analysis of groundwater table variations in Sylhet region, Bangladesh. Environ. Eng. Res. 22 (4), 369−376.
Zahid, A., Ahmed, S.R.U., 2006. Groundwater resources development in Bangladesh: contribution to irrigation for food security and constraints to sustainability, In: No H039306, IWMI Books, Reports, International Water Management Institute, New Delhi.

Chapter 32

Integrating groundwater for water security in Cape Town, South Africa

G. Thomas LaVanchy[1], James K. Adamson[2] and Michael W. Kerwin[3]

[1]*Department of Geography, Oklahoma State University, Stillwater, OK, United States,* [2]*Northwater International, Chapel Hill, NC, United States,* [3]*Department of Geography & the Environment, University of Denver, Denver, CO, United States*

32.1 Introduction

Groundwater represents the largest reservoir of unfrozen freshwater, provides potable water to approximately 2 billion people globally, and enables 44% of irrigated food production worldwide (CGIAR, 2017). Yet, global groundwater usage represents a conundrum. Simultaneously, it is under immense stress (Mukherjee et al., 2020), while also heralded as a stabilizing force for water security amidst climate change (Healy, 2019). In the United States, where irrigation accounts for 70% of total fresh groundwater withdrawal, cycles of drought and unsustainable practices have contributed to aquifer overexploitation in many regionally significant aquifers across the country, including the High Plains and California Central Valley (Konikow, 2013; Scanlon et al., 2012; Wagner, 2017). Consequently, portions of these regions now face serious challenges to agricultural production, regional economies, and ecosystem services (CAST, 2019; Lauer et al., 2018). Similar scenarios are unfolding in the North China Plain and the central portion of the Indo-Gangetic Plain (Famiglietti, 2014).

Groundwater is increasingly valued as strategic to food security in its ability to help farmers buffer against the effects of shifting rainfall patterns (Foster and MacDonald, 2014). Likewise, it is viewed as a key element of water security for urban centers, where population growth and suburban expansion stress the capacity of current supplies and infrastructure. Flörke et al. (2018) analyzed 482 of the largest cities worldwide with respect to water security. Their findings suggested 27% of these cities will face surface water deficits by 2050. Given the heterogeneity of need and opportunity for groundwater usage, management of this precious resource will require scientifically informed, socially equitable, economically balanced, and environmentally sensitive efforts.

This chapter draws on the example of a recent water crisis in Cape Town (South Africa) to illustrate the opportunity for groundwater to support water security, as well as illuminate inherent challenges in governing and managing this critical dimension of the perpetually restless hydrologic cycle. The term "Day Zero" was used by Cape Town officials to signal a day on the calendar in 2018 when the taps would be turned off city-wide due to insufficient supply. This crisis was the result of 3 consecutive years of record drought and a water management plan designed upon stationarity expectations of Western Cape rainfall patterns (LaVanchy et al., 2019). The Western Cape Water Supply System (WCWSS) has historically relied upon surface water to meet its domestic, industrial, and agricultural water needs. Because the Western Cape experiences cycles of drought, numerous recommendations have repeatedly been made to encourage water managers in Cape Town to diversify their supply portfolio. Given the mounting evidence that suggests stationarity is dead (Milly et al., 2008), water managers in Cape Town must align water provisioning with an evolving landscape of both water supply and demand. This chapter primarily speaks to the drivers of Day Zero and the opportunity for groundwater to aid in diversification of supply to meet a growing demand. As such, we outline the major aquifers, current state of exploration, and limitations to the use of groundwater in supporting water security for Cape Town. The lessons, opportunities, and challenges discussed are relevant to other urban centers in the world, especially those fortunate to have the opportunity for diversified water sources in the shadow of surface water uncertainties.

32.2 Situating Cape Town

Cape Town (∼34°S, 18.5°E) is a growing, coastal city of 3.9 million people that lies at the southern tip of the African continent (Fig. 32.1). Year-round comfortable weather combined with an abundance of natural beauty, fertile agriculture, and unmatched biodiversity has helped Cape Town become one of the most popular tourist destinations on Earth (Rogerson and Visser, 2020). The City and surrounding Western Cape region is classified as a pure Mediterranean climate, characterized by moderate year-round temperatures, historically persistent winter rainfall, and warm, dry summers. As with other Mediterranean regions, Cape Town's climate is heavily influenced by the ocean. Adjacent to the city, the Atlantic Ocean Benguela Current moves cold water northward along the western coast of South Africa. Strong upwelling zones result in cool surface ocean temperatures that rarely exceed 18°C, even in the heat of summer. Presently, much of the region's precipitation falls during the winter months (June, July, and August), when expansions of the circumpolar vortex bring the westerly storm track and its associated frontal systems into closer contact with the Cape Peninsula (Fig. 32.2). Traversing inland, this system breaks down abruptly to the north and east of Cape Town, where the climate is classified as subtropical and dominated by summer rainfall in the Karoo and Savanna ecosystems.

As with many coastal cities, Cape Town has been impacted by human-induced climate change. Sea-level rise, although mostly imperceptible, has caused beach and dune erosion and made coastal aquifers more vulnerable to saltwater intrusion. Measurable alterations to terrestrial ecosystems, including the extraordinarily high biodiversity Cape Floral Kingdom, have been correlated to increasing temperatures and higher aridity linked to higher evapotranspiration (Lawal et al., 2019). Model simulations reinforce that changes to Cape Town's predictable Mediterranean rainfall system are likely just beginning and suggest that the region should brace for statistically significant reductions in precipitation (up to 30%) by 2070 and additional warming (up to 3°C) that will combine to increase severity of future droughts (Engelbrecht et al., 2015; Jury, 2013; Kusangaya et al., 2014). Of even greater concern to water managers in Cape Town is the rapid increase in surface ocean temperatures globally (Cheng et al., 2019) that may not be accounted for in some predictive climate models. Rapidly warming south Atlantic Ocean temperatures could be especially disruptive to

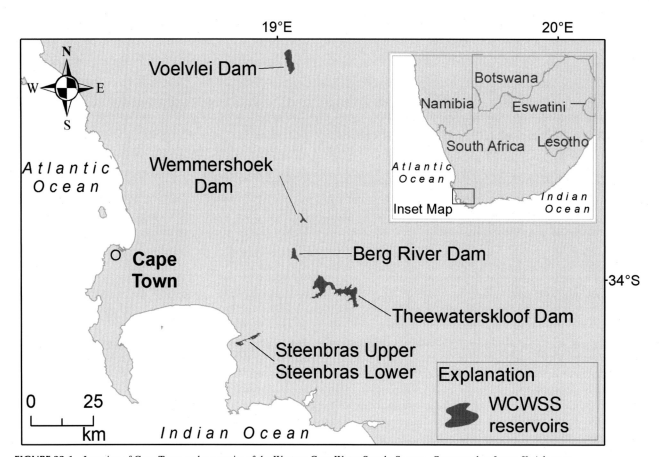

FIGURE 32.1 Location of Cape Town and reservoirs of the Western Cape Water Supply System. *Cartography: James K. Adamson.*

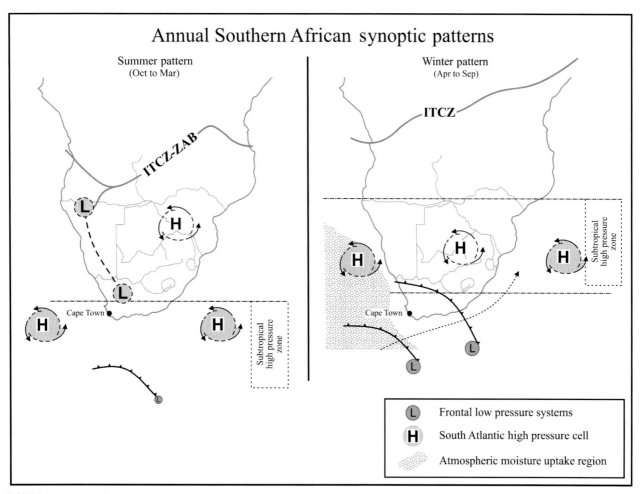

FIGURE 32.2 Annual synoptic patterns over the Western Cape in relation to migration pathways of airborne moisture. *Cartography: Michael P. Larson.*

Cape Town because the timing and position of wintertime storm fronts are partially driven by seasonal changes in sea surface temperatures.

32.2.1 The Day Zero drought

Cape Town garnered international attention in 2018 after the City introduced the phrase "Day Zero" to warn that its rapidly dwindling water supply would soon be incapable of supplying water to residents, tourists, industry, and agriculture. Reservoir levels dropped to ~21% of capacity in May 2018. Yet, Day Zero never arrived because of unmatched conservation and the relief of rainfall during the 2018 winter.

Although population growth and state negligence were identified as drivers of the Day Zero crisis (LaVanchy et al., 2019), it was 3 consecutive years of rainfall deficits that pushed the system to near failure. In fact, the 2015–17 drought preceding Day Zero was the most severe in over 100 years of instrumental records (Burls et al., 2019). Meteorologically, this drought was caused by the development of unusually strong South Atlantic anticyclones positioned further south than normal. In effect, this pressure anomaly weakened incoming winter storms that were otherwise capable of delivering much needed rainfall and steered other storms away from Cape Town entirely (See Fig. 32.2). Archival records from before the advent of weather recording devices (including archives of the Dutch East Indies Company) and high-resolution paleoclimate data are sparse in southern Africa (Cordova et al., 2019). Thus gaining a prehistoric or geologic perspective on the severity of the Day Zero drought is difficult, although the few available records suggest that the regional Mediterranean climate system was active throughout the late-Holocene, although occasionally disrupted by multiyear droughts of unknown severity (Chase et al., 2017).

For now the immediate threat of running out of water has passed, but Cape Town's water supply system remains fragile, especially as climate change and sea surface temperature warming threaten to further disrupt the timing and position of wintertime storm fronts needed for reservoir replenishment. In addition, it is certain that multiyear droughts, similar to or worse than the Day Zero drought, will impact Cape Town again and must be accounted for in any future water security and water management initiatives.

32.2.2 Water provision and security

The City of Cape Town (CCT) receives its water from a series of six reservoirs that capture and store rainfall and trace amounts of melted snowfall (Fig. 32.1). Nearby, several microwater supply systems support smaller urban contexts, though this chapter will only focus on the CCT. The WCWSS links the six reservoirs and supplies bulk water to the CCT (324 Mm^3/year), the agriculture sector (144 Mm^3/year), and smaller nearby municipalities (23 Mm^3/year) (Fig. 32.3). The system is managed by the National Department of Water and Sanitation in cooperation with the City. This shared space of water governance between national and local entities is politically contentious, given that the Western Cape is the only province in South Africa run by the opposition party.

As with other infrastructure projects in Cape Town, the WCWSS was designed for a modern, postapartheid, global city but remains hamstrung by its complete dependence on reservoir-filing rainstorms each and every winter. It has been documented for some time the City needs to augment its supply with groundwater, desalination, and reuse (LaVanchy et al., 2019). Beyond these physical dimensions of water, the system also faces water justice challenges due to the legacy of social and economic segregation during apartheid (Enqvist and Ziervogel, 2019). During the Day Zero event of 2018, all vulnerable aspects of Cape Town's water management system became painfully exposed and numerous efforts were initiated to secure the future of water for Cape Town (Taing et al., 2019).

South Africa's constitution expressly guarantees everyone the right to sufficient water, legally defined as 50 L/person/day (Government of South Africa, 1996). Further, the constitution requires the government to "take reasonable legislative and other measures within its available resources to achieve the progressive realization of each of these rights." In addition, UN General Assembly Resolution 64/292 (July 28, 2010) defines the right to water as "sufficient, safe, accessible, and affordable water for personal and domestic use"; in other words, a continuous supply of clean water to each person (without discrimination) that can be used for drinking, personal sanitation and hygiene, clothes washing, and food preparation (Matchaya and Nhemachena, 2018). Had household taps been turned off during the Day Zero crisis, the CCT would have tested enforcement of this constitutional right by defending how its emergency plan to provide only 25 L/person/day at central water stations fulfilled this human right.

As it became apparent the City was running out of water in 2018, allocation to the agriculture sector became more contentious. On the one hand, residents questioned how farms could still receive supply from the WCWSS, while their household taps were deemed expendable. (Agriculture constitutes 29% of the demand on the WCWSS.) Simultaneously, agriculture is a key economic driver in South Africa. In 2015 the wine industry contributed 36 billion ZAR (6.3%) of the gross domestic product and was responsible for ~300,000 direct and indirect jobs (Ferreira and Hunter, 2017). With future climate change expected to increase the water demand for existing vineyards globally (Hannah et al., 2013), augmenting the water supply of the WCWSS with groundwater may also be required for sustainable productivity of viticulture in South Africa.

32.3 Groundwater opportunities

The aquifer systems of regional significance to Cape Town are the Table Mountain Group (TMG) aquifers and Sandveld Group aquifers, the most notable of the latter include the Cape Flats and Atlantis aquifers (Fig. 32.4). These aquifer systems are diverse from hydrogeological, geographic, and socioeconomic perspectives, which makes their development and integration into Cape Town's water portfolio complex.

Significant investments have been made over the last 40 years by the academic community, researchers, and public sector into the study, characterization, and management of these aquifer systems. The aquifers represent significant resources for towns, rural areas, agriculture, and the environment throughout the Western Cape and play an integral role in water, economic, and food security across the province. The development, combined with integrated and adaptive management of these aquifers to supplement surface water shortages, is fundamental to achieving long-term water security, not only for Cape Town but also the Western Cape.

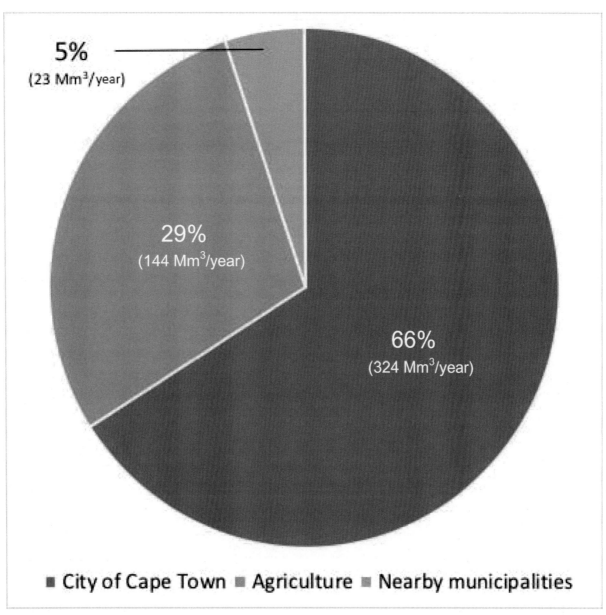

FIGURE 32.3 Water allocations for the Western Cape Water Supply System.

32.3.1 Table Mountain Group aquifers

The TMG geological formations are predominately sandstone and quartzite formations of Ordovician to Carboniferous age that support a vast system of discontinuous, fractured bedrock aquifers within the intermontane and the coastal plain. The formations outcrop across an area greater than 40,000 km^2 (Fig. 32.4) and are concealed beneath younger geological formations across a much greater spatial extent. Aquifer recharge has significant variability both spatially and temporally and may be as high as 50% of annual rainfall in some localities (Duah, 2010). The large areal extent, great thickness (up to 4000 m), extensive network of fractures, and benefit from high orographic precipitation (Rosewarne, 2002) make the aquifer a regionally significant resource. These characteristics support estimates that aquifer recharge is greater than 1 billion m^3/year and storage is perhaps on the order of tens of billions of cubic meters of water. Despite the big numbers, the aquifer is discontinuous and the zones that can store and yield water of regional significance are certainly not available everywhere. The resulting uncertainty and variability regarding water occurrence and sustainability combined with the remoteness and synonymous connection with sensitive ecosystems are factors that

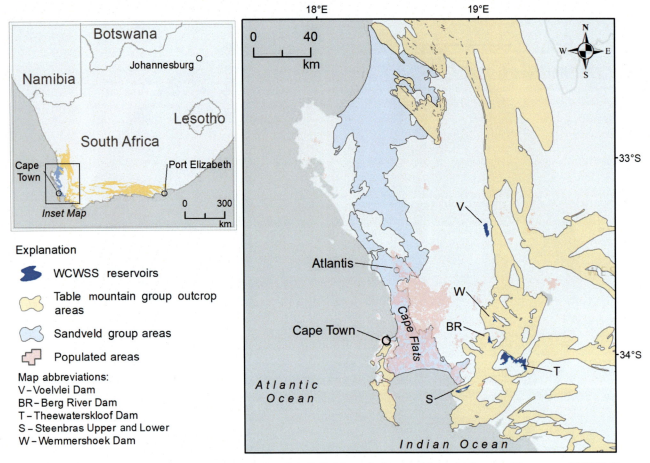

FIGURE 32.4 Map of South Africa with WCWSS reservoirs and the Table Mountain Group and Sandveld Group aquifer systems. *WCWSS*, Western Cape Water Supply System. *Cartography: James K. Adamson.*

have inhibited research and development investments to date. The TMG aquifers currently support the agricultural industry and towns throughout the Western Cape, more recently including Cape Town. Some towns extract as much as 4500 m^3/day from the aquifers (Rosewarne, 2002); however, from a bulk perspective, the current level of aquifer development is low considering the renewable resources potentially available.

32.3.2 Sandveld Group aquifers

The Sandveld Group is eolian fine-to-medium grained quartz sands of Cenozoic age that blanket over 6000 km^2 along a 150 km N–S trending coastal zone of low-lying areas bounded by the inland hills and Cape Peninsula (Fig. 32.4). The sands originate from post–Paleozoic erosion of the TMG rocks (Harris et al., 1999) and overlie Malmesbury Group graywackes, phyllite shales, and Cape Granites. The sands are up to 60 m thick in some areas such as the Atlantis Aquifer, and saturated thickness in the Sandveld Group aquifers rarely exceed 35 m (Bugan et al., 2016; Tredoux et al., 2009). Unlike the TMG, the Sandveld Group aquifers are unconsolidated, continuous, and more predictable in terms of groundwater behavior and water occurrence. Excluding artificial recharge, estimates of renewable groundwater resources are variable in the literature, but typically fall in the range of 60 Mm^3/year, with recharge ranging between 10% and 50% of mean annual precipitation (Adelana et al., 2010; Meyer, 2001). Aquifer storage in the Sandveld Group aquifers is estimated at over 8000 Mm^3 (Meyer, 2001), which includes the Cape Flats, Atlantis, Berg River, and Grootwater units. Delicate water balance, limited aquifer thickness, short residence time, poor water quality, and vulnerability to pollution are challenges these aquifers present with respect to their role in strengthening water security in the region.

The Cape Flats aquifer (CFA) is considered one of the largest aquifers of the Sandveld Group, encompassing approximately 400 km^2 in the low area between False Bay, the Cape Peninsula, and inland hills. The water table is typically less than 5 m below the surface and boreholes are not often deeper than 30 m. The potential renewable groundwater yield has been estimated at 15 Mm^3/year (Meyer, 2001) and recharge ranging from 16% to 47% of mean precipitation (Adelana et al., 2010). The water availability estimates in the literature are variable and recent data growth in the national archive and research suggests complex recharge dynamics that include anthropogenic sources such as water system leakage. Current water demand is limited to agriculture and private usage, and the research suggests significant opportunity for augmentation of the CCT supply when coupled with artificial recharge, from a water quantity standpoint.

The Atlantis aquifer, the second largest of the Sandveld Group, is approximately 130 km^2 in extent and extends from the Atlantic Ocean to the bedrock outcrops east of the town of Atlantis. It is a steep gradient, low slope system that is up to 60 m thick, and saturated thickness rarely exceeds 35 m (Bugan et al., 2016; Tredoux et al., 2009). Inclusive of artificial recharge, production capacity is documented to be in the range of 4.4–10 Mm^3/year [CCT (City of Cape Town), 2001; CMC (Cape Metropolitan Council) et al., 1995; Meyer, 2001] and storage in the range of 400 Mm^3 (Meyer, 2001). The Atlantis aquifer represents a premier case study for managed aquifer recharge, as the town of Atlantis has been implementing what should be considered a successful artificial recharge program since the mid-1970s with strict and disciplined management of the resource. Indirect recycling of stormwater runoff and treated wastewater into the aquifer have resulted in a system where it is estimated that 30% of water pumped from the aquifer is of artificial recharge origin [DWA (Department of Water Affairs, South Africa), 2010]. Recent studies have found that the water levels are increasing to levels that may impact infrastructure and ecological resources as water demands have decreased without subsequent modifications to the recharge scheme (Bugan et al., 2016).

There are several other named aquifers within the Sandveld Group, some of which include the Langebaan, Grootwater, Berg River, Elandsfontein, and Adamboerskraal systems. Groundwater potential is considered less regionally significant compared to their counterparts; however, they present opportunities to strengthen water security in the Western Cape, especially when abstraction is coupled with managed aquifer recharge.

32.4 Groundwater management challenges

During and following Day Zero, incorporating groundwater into Cape Town's system was increasingly viewed as a vital buffer against precipitation deficits. However, significant obstacles associated with groundwater pollution, borehole contamination and management, injection schemes, historical perceptions of groundwater, and differing recharge rates associated with climate change have hampered implementation efforts. Despite urgency and an apparent abundance, effective management of regional groundwater resources is not straight-forward. Coordinating the various physical dimensions of integrating groundwater into a centralized rainfall collection system requires sufficient human resources and structures which are simultaneously organized and like-minded. Numerous issues, both physical and human, present significant challenges to integrating groundwater into the current supply and in managing yield against various environmental and economic concerns. With respect to Cape Town, we consider human dimensions to be of greater potential challenge than physical dimensions. First, we outline the latter.

32.4.1 Physical dimensions

The issues and challenges to sustainable management fall along a couple of themes, namely, the complexity and sensitivity of the respective aquifers, and the conventional path dependencies that limit upscaling and integration of groundwater into a surface water designed system.

A significant amount of water appears available in both the TMG and Sandveld Group aquifers. However, the TMG is both geographically vast and heterogenous, effectively requiring a greater breadth of research to generate localized hydrogeological knowledge to inform sustainable yield and development and abstraction strategies. Too, the TMG is overlain with sensitive and protected biotic communities (locally known as *fynbos*), thus limiting legal access and inhibiting support for advancing exploration and development. Research has advanced, especially in the intermontane areas where winter precipitation is most reliable to recharge aquifers and more easily studied with cost-effective remote sensing methods. As one of several examples, Hartnady and Hay (2002) identified large potential volumes of groundwater in the Olifants and Doring catchments based on analysis of geological structure. Others have identified anomalous groundwater potential of the bedrock associated with alluvial basins, geological contact areas, and the well-mixed and fractured Peninsula aquifer (Diamond and Harris, 2019).

The greatest challenges for integrating TMG aquifers into Cape Town's water portfolio may not be technical in nature. Rather, utilizing the high-recharge units of the TMG will require socioeconomic, political, and environmental cooperation. For example, the TMG does not benefit from existing research and the feedback loops that are coupled with a history of management and development investments, as is the case with the Sandveld aquifers. Catchment-based water allocations and water rights do not enable adaptive and dynamic management warranted by advancements in resource knowledge uncovered by the experts. Further challenges lie in the establishment of water abstraction and allocation management scenarios across a range of stakeholders. Despite these challenges, the water security value of having access to a small portion of the huge groundwater reserves within the TMG aquifers makes it perhaps the greatest opportunity for Cape Town to achieve long-term water security. The smaller Western Cape community of Hermanus has sustainably accessed the TMG for nearly a decade and provides a promising example for managers of the WCWSS. Although the surrounding *fynbos* is highly sensitive to disturbance, it is possible to systematically protect its fragile ecology, while strategically developing much needed groundwater resources from the TMG. This undoubtedly would require continued research and coordination between scientific experts from various disciplines and City water managers, and likely more technical (and costly) approaches to infrastructure that satisfy the needs of multiple objectives. To do otherwise would be irresponsible to both societal and ecological needs.

Hydrologically, the Sandveld Group aquifers have been examined in great detail, particularly the Cape Flats and Atlantis aquifers. These intergranular aquifers have well-defined limits, are accessible, and have been developed, monitored, and managed for decades. Integrating the CFA into the water portfolio for Cape Town, however, comes with many challenges that are substantially different than the TMG. Water pollution associated with socioeconomic activities, population growth, and diverse land uses is inherent to urban settings. Treatment costs could be significant to achieve potable standards, and emerging contaminants may impact future suitability and feasibility of the resource. In addition, these aquifers are quite sensitive to over abstraction, resulting in seawater intrusion and altered chemistry. Use of the CFA would also mandate intensive management interventions such as artificial recharge and continuous research and monitoring to balance the integrated usage of the resource.

The Atlantis aquifer, which has been successfully utilized for over four decades, provides a promising template for aquifer management within the Sandveld Group (Bugan et al., 2016). In the suburban town of Atlantis, indirect recycling of urban stormwater runoff and treated domestic wastewater through Managed Aquifer Recharge (MAR) has been used to balance groundwater withdrawals. Challenges within the Atlantis system include sustaining strict resource management, occurrences of emerging contaminants, and decreasing capacity of production wells due to iron oxidation (Bugan et al., 2016). Also, decreases of water demand, not balanced with the appropriate artificial recharge scheme, have sometimes led to higher water tables, which potentially exasperate water quality issues, threaten sensitive ecosystems, and increase flooding risk. Despite these obstacles, the current level of research and knowledge of the aquifers, combined with established management strategies [DWAF (Department of Water Affairs and Forestry, South Africa), 2007] make the Cape Flats or other Sandveld Group aquifers an opportunity to strengthen the water security of Cape Town.

32.4.2 Human dimensions

Brown and Farrelly (2009) outlined 12 human factors that serve as barriers to effective water management, and ultimately undermine water security. Of these, the following are relevant to the context of Cape Town:

- Lack of information, knowledge in applying integrated forms of management
- Insufficient monitoring and evaluation
- Technocratic path dependencies
- Uncoordinated institutional framework
- Unclear, fragmented roles, and responsibilities
- Lack of political and/or public will

The first two barriers will need to be addressed if groundwater is to sustainably augment Cape Town's water supply. The existing knowledge of aquifer properties and safe yield will need to be coordinated and integrated with the expansion of research to enable the community of water agencies, institutions, and passionate experts to work collaboratively together towards aligned objectives. Advanced research and policy development that synthesizes academic disciplines with an objective to address the various technical and policy challenges is critical. The spirit of the initiative led by the Water Research Commission in 2002 (Pieterson and Parsons, 2002) should be resurrected and updated with a stronger emphasis on bridging science and policy to define solutions and guide research priorities.

TABLE 32.1 Development considerations for Table Mountain and Sandveld Group aquifers.

Development Considerations	Table Mountain Group aquifers		Sandveld Group aquifers		
	Intermontane	Coastal and Peninsula	Cape Flats	Atlantis	Others
Relative ratio of groundwater development at present	L	L	L–M	H	M
Gaps in synthesized research	M	H	L	L	M–H
Complexity in groundwater management	L	M	H	H	M
Access and water infrastructure challenges	H	M	L	M	M
Ecosystem and natural area concerns	H	M	H	L	M
Potable water treatment costs	L	L–M	M–H	M	L
Saltwater intrusion risk	L	M	H	H	M
Contamination/pollution risk (biological, emerging contaminants, agriculture)	M	M	H	M	M
Complexity of sociopolitical factors	H	H	H	H	M
Intensity of impacts from climate and land use changes	L	M	H	H	M

H, High; *L*, low; *M*, moderate.

Continued investments and commitment to building and maintaining the national groundwater archive will further support synthesis of knowledge and decrease the temporal and spatial data gaps needed to inform water security investments. As monitoring increases, the archive will further inform to guide safe yield and management progress. This is particularly important amidst the proliferation of private boreholes during and after Day Zero. (Presently, the South African Water Allocation Reform Strategy does regulate or require monitoring of boreholes installed for household use.)

The final three barriers noted above may prove the most challenging since they require improvement in collaborative management amongst hierarchical and vertical institutions and development of effective communication of goals and pathways. An example of this challenge was evident during the Day Zero crisis. As noted previously, bulk water supply within South Africa rests with the national government, whereas municipalities are responsible for local delivery. This shared water governance broke down during Day Zero as reservoir levels dropped alarmingly low and repeated requests by the opposition party for temporary agricultural transfers were ignored. Ultimately, the National Department of Water and Sanitation did free up water supply and Day Zero was averted; however, collaboration between rival political parties outside of a water crisis seems dubious.

Collectively, these barriers pose formidable, though not impossible, challenges to navigate. As noted earlier in the chapter, water security will only be achieved through scientifically informed, socially equitable, economically balanced, and environmentally sensitive efforts. Table 32.1 is provided for relative comparison among the aquifers with respect to various physical and human dimensions.

32.5 Conclusion

Future changes in oceanic and general atmospheric circulation will alter the stationarity of natural drought variability in the Western Cape and thus likely reduce available rainfall for reservoir storage in Cape Town. Despite this, it is evident that groundwater resources are available to support the dynamic water needs of the City. In a physical sense, augmenting surface water reservoirs with groundwater at some capacity with a strategic management scheme should reasonably achieve water security. However, it is not clear if the human dimensions of the water systems are actually capable of adapting to change. This skepticism is partially rooted in the dialog generated three decades prior to the Day Zero crisis when the scientific community (both physical and social) forecasted the need to build a more resilient water supply system. These warnings were largely unheeded by City water managers but resurfaced in 2017. Whereas the world viewed the Day Zero crisis voyeuristically as an intriguing experiment with drought, water supply, and conservation, the CCT was forced into an unwanted public audit of their failing water system and management decisions. Despite repeated warnings, adequate resources, and an advanced technical knowledge, the CCT was not equipped to ride out a 3 year drought. As post–Day Zero reports continue to coalesce around a future water security plan that must account for a growing population, agricultural demand, climate change, and mismanagement, it is unclear how Cape Town's water system will weather the next drought. As this chapter has argued, Cape Town's resilience to drought must include diversification with groundwater, operationalized with research, financing, and integrated approaches. To the point, diversification is clearly the answer, delay the enemy, and poor planning anathema to Cape Town. It is unclear if such a highly advanced water system and regional management scheme will actually materialize in contemporary South Africa when other, much needed infrastructure initiatives have stalled due to corruption and mismanagement.

References

Adelana, S., Xu, Y., Vrbka, P., 2010. A conceptual model for the development and management of the Cape Flats aquifer, South Africa. Water SA 36 (4), 461–474 (retrieved 06.02.19.) Available from <http://www.wrc.org.za>.

Brown, R.R., Farrelly, M., 2009. Delivering sustainable urban water management: a review of the hurdles we face. Water Sci. Technol. 59 (5), 839–846. Available from: https://doi.org/10.2166/wst.2009.028.

Bugan, R., Javanovic, N., Isreal, S., Tredoux, G., Genthe, B., Steyn, M., et al., 2016. Four decades of water recycling in Atlantis (Western Cape, South Africa): past, present and future. Water SA 42 (4).

Burls, N.J., Blamey, R.C., Cash, B.A., Swenson, E.T., al Fahad, A., Bopape, M.J.M., et al., 2019. The Cape Town "Day Zero" drought and Hadley cell expansion. NPJ Clim. Atmos. Sci. 2 (27), 1–8. Available from: https://doi.org/10.1038/s41612-019-0084-6.

CAST (Council for Agricultural Science and Technology), 2019. Aquifer Depletion and Potential Impacts on Long-term Irrigated Agricultural Productivity. Issue Paper 63. CAST, Ames, IA.

CCT (City of Cape Town), 2001. Bulk water supply infrastructure; background information for WSDP. <http://www.capetown.gov.za/water/wsdp/documents/chapter4.pdf> (accessed 02.02.20.).

CGIAR, 2017. Building resilience through sustainable groundwater use. In: CGIAR Research Program on Water, Land, and Ecosystems (WLE). In: WLE Towards Sustainable Intensification: Insights and Solutions Brief 1. International Water Management Institute (IWMI), Colombo. 12 p. doi:10.5337/2017.208.

Chase, B.M., Chevalier, M., Boom, A., Carr, A.S., 2017. The dynamic relationship between temperate and tropical circulation systems across South Africa since the last glacial maximum. Quat. Sci. Rev. 174, 54–62.

Cheng, L., Abraham, J., Hausfather, Z., Trenberth, K.E., 2019. How fast are the oceans warming? Science 363 (6423), 128–129. Available from: https://doi.org/10.1126/science.aav7619.

CMC (Cape Metropolitan Council), Liebenberg & Stander, CSIR, 1995. The Atlantis Water Resource Management Scheme – An Overview. Cape Metropolitan Council, Cape Town.

Cordova, C.E., Kirsten, K.L., Scott, L., Meadows, M., Lücke, A., 2019. Multi-proxy evidence of late-Holocene paleoenvironmental change at Princessvlei, South Africa: the effects of fire, herbivores, and humans. Quat. Sci. Rev. 221, 105896. Available from: https://doi.org/10.1016/j.quascirev.2019.105896.

Diamond, R., Harris, C., 2019. Annual shifts in O- and H-isotope composition as measures of recharge: the case of the Table Mountain springs, Cape Town, South Africa. Hydrogeol. J. 27, 2993–3008. Available from: https://doi.org/10.1007/s10040-019-02045-5.

Duah, A., 2010. Sustainable Utilization of the Table Mountain Group Aquifers (Ph.D. thesis). Department of Earth Sciences, University of the Western Cape, 182 p.

DWA (Department of Water Affairs, South Africa), 2010. Strategy and guideline development for national groundwater planning requirements. In: The Atlantis Water Resource Management Scheme: 30 Years of Artificial Groundwater Recharge. Report No. PRSA 000/00/11609/10 – Activity 17 (AR5. 1). Department of Water Affairs, Pretoria.

DWAF (Department of Water Affairs and Forestry, South Africa), 2007. Artificial Recharge Strategy: Version 1.3. Department of Water Affairs and Forestry, Pretoria.

Engelbrecht, F., Adegoke, J., Bopape, M.J., Naidoo, M., Garland, R., Thatcher, M., et al., 2015. Projections of rapidly rising surface temperatures over Africa under low mitigation. Environ. Res. Lett. 10 (8), 085004.

Enqvist, J.P., Ziervogel, G., 2019. Water governance and justice in Cape Town: an overview. WIREs Water 6 (4), e1354. Available from: https://doi.org/10.1002/wat2.1354.

Famiglietti, J., 2014. The global groundwater crisis. Nat. Clim. Change 4, 945–948.

Ferreira, S.L., Hunter, C.A., 2017. Wine tourism development in South Africa: a geographical analysis. Tour. Geogr. 19 (5), 676–698. Available from: https://doi.org/10.1080/14616688.2017.1298152.

Flörke, M., Schneider, C., McDonald, R.I., 2018. Water competition between cities and agriculture driven by climate change and urban growth. Nat. Sustainability 1, 51–58.

Foster, S., MacDonald, A., 2014. The 'water security' dialogue: why it needs to be better informed about groundwater. Hydrogeol. J. 22, 1489–1492. Available from: https://doi.org/10.1007/s10040-014-1157-6.

Government of South Africa, 1996. Constitution of the Republic of South Africa, Act 108.

Hannah, L., Roehrdanz, P.R., Ikegami, M., Shepard, A.V., Shaw, M.R., Tabor, G., et al., 2013. Climate change, wine, and conservation. PNAS 110 (17), 6907–6912. Available from: https://doi.org/10.1073/pnas.1210127110.

Harris, C., Oom, B.M., Diamond, R.E., 1999. A preliminary investigation of the oxygen and hydrogen isotope hydrology of the greater Cape Town area and an assessment of the potential of using stable isotopes as tracers. Water SA 25 (1), 15–24.

Hartnady, C.J.H., Hay, E.R., 2002. The use of structural geology and remote sensing in hydrogeological exploration of the Olifants and Doring River catchments. In: Pieterson, K., Parsons, R. (Eds.), A Synthesis of the Hydrogeology of the Table Mountain Group—Formation of a Research Strategy. Water Research Commission, Pretoria, pp. 19–32.

Healy, R., 2019. Groundwater resilience in sub-Saharan Africa. Nature 572, 185–186.

Jury, M.R., 2013. Climate trends in southern Africa. S. Afr. J. Sci. 109 (1–2), 1–11. Available from: https://doi.org/10.1590/sajs.2013/980.

Konikow, L.F., 2013. Groundwater depletion in the United States (1900–2008). In: U.S. Geological Survey Investigations Report 2013–5079.

Kusangaya, S., Warburton, M.L., Van Garderen, E.A., Jewitt, G.P., 2014. Impacts of climate change on water resources in southern Africa: a review. Phys. Chem. Earth 67, 47–54. Parts A/B/C.

Lauer, S., Sanderson, M.R., Manning, D.T., Suter, J.F., Hrozencik, R.A., Guerrero, B., et al., 2018. Values and groundwater management in the Ogallala Aquifer Region. J. Soil. Water Conserv. 73 (5), 593–600. Available from: https://doi.org/10.2489/jswc.73.5.593.

LaVanchy, G.T., Kerwin, M.W., Adamson, J.K., 2019. Beyond 'Day Zero': insights and lessons from Cape Town (South Africa). Hydrogeol. J. 27 (5), 1537–1540. Available from: https://doi.org/10.1007/s10040-019-01979-0.

Lawal, S., Lennard, C., Hewitson, B., 2019. Response of southern African vegetation to climate change at 1.5 and 2.0° global warming above the pre-industrial level. Clim. Serv. 16, 100134. Available from: https://doi.org/10.1016/j.cliser.2019.100134.

Matchaya, G., Nhemachena, C., 2018. Justiciability of the right to water in the SADC Region: a critical appraisal. Laws 7 (2), 18. Available from: https://doi.org/10.3390/laws7020018.

Meyer, P.S., 2001. An Explanation of the 1:500 000 Hydrogeological Map of Cape Town 3317. Department of Water Affairs and Forestry, Pretoria, p. 59.

Milly, P.C.D., Betancourt, J., Falkenmark, M., Hirsch, R.M., Kundzewicz, Z.W., Lettenmaier, D.P., et al., 2008. Stationarity is dead: whither water management? Science 319 (5863), 573–574. Available from: https://doi.org/10.1126/science.1151915.

Mukherjee, A., Scanlon, B., Aureli, A., Langan, S., Guo, H., McKenzie, A., 2020. Global Groundwater: Source, Scarcity, Sustainability, Security and Solutions, first ed. Elsevier, ISBN: 9780128181720.

Pieterson, K., Parsons, R., 2002. The need for appropriate research on the Table Mountain Group aquifer systems. In: Pieterson, K., Parsons, R. (Eds.), A Synthesis of the Hydrogeology of the Table Mountain Group—Formation of a Research Strategy. Water Research Commission, Pretoria, pp. 4–7.

Rogerson, J.M., Visser, G., 2020. Recent trends in South African tourism geographies. In: Rogerson, J.M., Visser, G. (Eds.), New Directions in South African Tourism Geographies. Springer, Switzerland, pp. 1–14.

Rosewarne, P., 2002. Hydrogeological characteristics of the Table Mountain Group aquifers. In: Pieterson, K., Parsons, R. (Eds.), A Synthesis of the Hydrogeology of the Table Mountain Group—Formation of a Research Strategy. Water Research Commission, Pretoria, pp. 33–46.

Scanlon, B.R., Faunt, C.C., Longuevergne, L., Reedy, R.C., Alley, W.M., McGuire, V.L., et al., 2012. Groundwater depletion and sustainability of irrigation in the US High Plains and Central Valley. PNAS 109 (24), 9320–9325. Available from: https://doi.org/10.1073/pnas.1200311109.

Taing, L., Chang, C.C., Pan, S., Armitage, N.P., 2019. Towards a water secure future: reflections on Cape Town's Day Zero crisis. Urban Water J. 16 (7), 530–536.

Tredoux, G., Genthe, B., Steyn, M., Engelbrecht, J.F.P., Wilsenach, J., Jovanovic, N., 2009. An assessment of the Atlantis artificial recharge water supply scheme (Western Cape, South Africa). In: Brebbia, C.A., Jovanovic, N., Tiezzi, E. (Eds.), Management of Natural Resources, Sustainable Development and Ecological Hazards II. Wit Press, Southampton, pp. 403–413.

Wagner, K.L., 2017. Assessing irrigation aquifer depletion: introduction. J. Contemp. Water Res. Educ. 162, 1–3.

Chapter 33

Drivers for progress in groundwater management in Lao People's Democratic Republic

Cécile A. Coulon[1,2], Paul Pavelic[1] and Evan Christen[3]

[1]*International Water Management Institute, Vientiane, Lao PDR,* [2]*Department of Geology and Geological Engineering, Université Laval, Québec, Québec, Canada,* [3]*Penevy Services Pty Ltd, Huskisson, NSW, Australia*

33.1 Introduction

Lao People's Democratic Republic (PDR) is a landlocked country of around 7 million people situated in the heart of the Mekong region. Listed amongst the 47 UN-designated Least Developed Countries, its level of socioeconomic development is comparable to neighboring Myanmar and Cambodia but significantly lower than that of China, Thailand, and Vietnam. Subsistence-based farming is the primary means of food security and income for almost 80% of households.

Poorly developed, water-rich countries such as Lao PDR have historically paid most attention to surface water resources, with limited consideration to groundwater. Development has traditionally been oriented toward the country's abundant surface water resources. However it is estimated that Lao PDR's groundwater resources are greatly underutilized and have a clear potential for expansion (Pavelic et al., 2014; Lacombe et al., 2017a; Viossanges et al., 2018).

Lao PDR has high rates of groundwater replenishment (5572 m^3/capita/year), significantly higher than most of its neighboring countries [Siebert et al., 2010; FAO (Food and Agriculture Organization of the United Nations), 2016; World Bank, 2016]. Its annual groundwater recharge generated internally is estimated at 37,900 Mm^3/year (2014 estimation; FAO Food and Agriculture Organization of the United Nations, 2016), whereas total groundwater use amounts to only 79.9 Mm^3/year, or about 0.2% of recharge. Borehole pumping is roughly estimated at 57.2 Mm^3/year[1] (World Bank, 2016), groundwater and spring water use for small town supply at 18.7 Mm^3/year (Chantavong, 2011) and groundwater use for irrigation is approximately 4.0 Mm^3/year[2] (Siebert et al., 2010). These estimations neglect industrial water use, which is thought to be small overall (GHD, 2015). A more detailed examination of groundwater use is given in Fig. 33.2.

Although the scale of current groundwater use is limited, this is changing for a variety of reasons (Pavelic et al., 2014). Groundwater is available year round, which is particularly advantageous over the long dry season (November–April), but also during prolonged dry spells within the wet season. This year-round availability, combined with the high level of control that farmers with wells have over their own water use, make groundwater valuable to support dry season cash cropping. Whilst the use of groundwater for irrigated agriculture has been discussed for decades (e.g., Johnson, 1983), the knowledge and expertise to develop appropriate technologies have been unavailable until recently and is now becoming more attractive to farmers [Vote et al., 2015; ACIAR (Australian Centre for International Agricultural Research), 2017; Clément et al., 2018]. Groundwater is also generally available in areas where there is no surface water. Provision of safe and reliable domestic water supplies in rural areas increasingly comes from groundwater accessed via boreholes or protected dug wells fitted with hand pumps. In some upland villages, wells are not needed as groundwater discharge into streams and rivers enables it to be collected throughout the year directly from the watercourse (Billving and Ågren, 2007; DAI Development Alternatives Inc., 2015).

The increasing accessibility of groundwater also likely accounts for rising groundwater use (Mukherjee et al., 2020). Rapid electrification of rural villages is enabling the use of cheap electric pumps with lower operating costs than using diesel. Studies

---

1. Assumes 50 L/capita/day and 75% of rural population using groundwater supplied through boreholes.
2. Assumes 2000 mm/year of annual irrigation demand as an upper limit.

in Lao PDR have consistently shown that the number of wells equipped with electric pumps for domestic and home gardening purposes has increased rapidly in recent years in the Vientiane Plain and other lowland areas (Serre, 2013; Vinckevleugel, 2015; Vote et al., 2015). With electrification and improved access to markets, it is likely that groundwater will also gradually be used beyond domestic household purposes (Vote et al., 2015). The number of private well drilling companies operating in rural areas has increased as a result of the improved market opportunities (DAI Development Alternatives Inc., 2015).

Another driver for the increase in groundwater use is that surface water sources, including major waterways such as the Mekong River (MRC Mekong River Commission, 2011), are becoming less reliable with climate change. Precipitation and temperature extremes are projected to increase with climate change (IPCC Intergovernmental Panel on Climate Change, 2012) and these will lead to an increased occurrence of flood and drought events. With the majority of the population dependent on natural resources for food security and livelihoods, Lao PDR faces significant threats from climate change. For example, an estimated 46% of Lao PDR's rural population was at risk of food insecurity due to drought in 2007 (World Bank & GFDRR 2011).

Although groundwater has been underrepresented in the water resources and related sectors in Lao PDR, this is steadily changing. This chapter serves to consolidate and document the latest knowledge, challenges, gaps, and outlook for groundwater resource development and governance.

33.2 Groundwater resources in Lao People's Democratic Republic

33.2.1 Groundwater systems

Groundwater is contained within shallow and deep aquifers that are composed of fractured hard rock, consolidated marine or terrestrial sediments, and alluvial/fluvial deposits along major watercourses (Fig. 33.1). In the northern and eastern uplands of the country the Annamian geological region includes granitic and metamorphic rocks, whilst the Indosinian region (mainly in the lowlands adjacent to the Mekong River) consists of sandstone, siltstone, shale, mudstone, limestone, conglomerate, and basalt (Johnson, 1983).

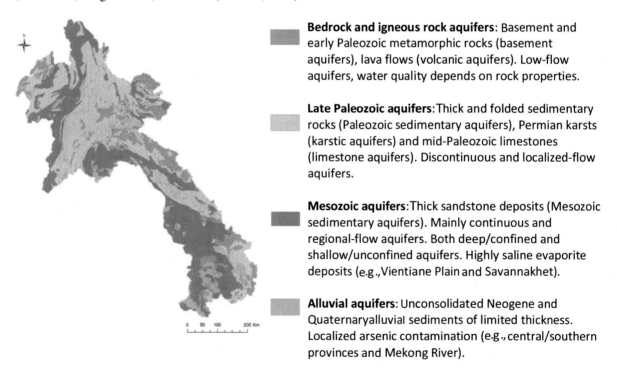

Bedrock and igneous rock aquifers: Basement and early Paleozoic metamorphic rocks (basement aquifers), lava flows (volcanic aquifers). Low-flow aquifers, water quality depends on rock properties.

Late Paleozoic aquifers: Thick and folded sedimentary rocks (Paleozoic sedimentary aquifers), Permian karsts (karstic aquifers) and mid-Paleozoic limestones (limestone aquifers). Discontinuous and localized-flow aquifers.

Mesozoic aquifers: Thick sandstone deposits (Mesozoic sedimentary aquifers). Mainly continuous and regional-flow aquifers. Both deep/confined and shallow/unconfined aquifers. Highly saline evaporite deposits (e.g., Vientiane Plain and Savannakhet).

Alluvial aquifers: Unconsolidated Neogene and Quaternary alluvial sediments of limited thickness. Localized arsenic contamination (e.g., central/southern provinces and Mekong River).

Sources: Johnson 1983; Landon 2011; Viossanges et al. 2018

FIGURE 33.1 Main aquifer types in Lao PDR. *PDR*, People's Democratic Republic. *Modified from Johnson, J.H., 1983. Preliminary appraisal of the hydrogeology of the Lower Mekong Basin. In: Interim Committee for Coordination of Investigations of the Lower Mekong Basin; Landon, M.K., 2011. Preliminary compilation and review of current information on groundwater monitoring and resources in the lower Mekong river basin. In: U.S. Geological Survey Report to the Mekong River Commission, Version Dated July 2011, pp. 1–34; Viossanges, M., Pavelic, P., Rebelo, L.-M., Lacombe, G., Sotoukee, T., 2018. Regional mapping of groundwater resources in data-scarce regions: the case of Laos. Hydrology 5 (1), 2, doi:10.3390/hydrology5010002.*

Groundwater is thought to be available in many areas, however aquifers are often very localized and variably productive (Landon, 2011; Viossanges et al., 2018). Well yields and water quality are highly dependent on the local geological conditions and are highly variable due to complex depositional and postdepositional process. For example, in the uplands of Khamounan Province, pumping rates for individual wells can vary from 1 to 2 L/s for sandstones and siltstones to 10−20 L/s for limestones. In the fresh alluvial aquifers of the Vientiane Plains that overly rock salt layers, pumping rates are modest (1−3 L/s). Even wells spaced less than 100 m apart can vary significantly in terms of their productivity and usability.

33.2.2 Groundwater use

Most of Lao PDR's rural population relies on groundwater for domestic purposes and to support livelihood generating activities. The National Agricultural Census showed that in 2010/11, 45% of villages used groundwater as their main domestic supply [MAF (Ministry of Agriculture and Forestry), 2012], and these are located mainly in lowland areas where groundwater availability and population density are highest (Fig. 33.2). Urban populations in towns and cities also use groundwater: at least 25 of the town water supply systems constructed by the provincial Nam Papa's (water

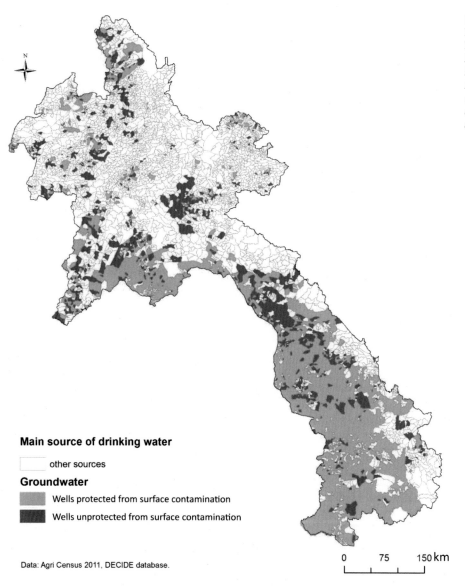

FIGURE 33.2 Groundwater uses in Lao PDR. *PDR*, People's Democratic Republic. *Modified from Viossanges, M., Pavelic, P., Rebelo, L.-M., Lacombe, G., Sotoukee, T., 2018. Regional mapping of groundwater resources in data-scarce regions: the case of Laos. Hydrology 5 (1), 2, doi:10.3390/hydrology5010002.*

Source: Viossanges et al. 2018

supply utilities) use spring water and groundwater, amounting to an extraction of nearly 30,000 m³/day (Chantavong, 2011). Groundwater is also used for industrial purposes by beverage companies (e.g., the Champasak plant of the Lao Beverage Company; Ha et al., 2015) and salt production (Pavelic et al., 2014).

Across the Mekong floodplains and some parts of the highlands, groundwater is accessed through lined dug wells and is commonly used for home gardening, year-round vegetable cropping, poultry and livestock [Suhardiman et al., 2015; ACIAR (Australian Centre for International Agricultural Research), 2016; Clément et al., 2018]. Beyond subsistence farming, groundwater from lined and unlined dug wells is starting to be used in more commercial farming systems to grow cash crops; for instance to meet dry-season water demand of coffee in the basaltic Bolaven Plateau in Champasak Province (Johnston et al., 2010), or dry-season cash cropping in the lowland provinces (Vote et al., 2015; Clément et al., 2018).

Communities in both urban and rural areas benefit indirectly from groundwater through a variety of ecosystem services. Groundwater flows maintain dry-season stream flows and wetlands, thus preserving freshwater habitats and fisheries which help local communities achieve food security and contribute to their livelihoods (Pavelic et al., 2014; Lacombe et al., 2017b).

33.3 Major groundwater challenges

Questions related to the development of groundwater resources have long been raised; however, these have been put aside as higher priorities were addressed and insufficient technical and financial resources inhibited progress. Surface water policies and associated institutional and legal frameworks are considerably more advanced than for groundwater. Groundwater can thus be considered to be slowly catching up with surface water. The main challenges and barriers to developing effective groundwater management in Lao PDR are referred to next.

33.3.1 Quantity and quality-related issues

Poor groundwater planning and management can result in falling water levels and deteriorating water quality that can, in turn, lead to reduced availability, increased pumping costs, and livelihood impacts (and conflicts for access in extreme cases). Lao PDR is a country with high groundwater recharge potential, relatively sparse population, and low levels of groundwater use; therefore risks associated with rampant groundwater overexploitation are likely to be lower than in many other countries. Protection of groundwater quality would arguably be as or more important than quantity-related issues.

Natural and anthropic groundwater quality issues have been found in Lao PDR. They include contamination by arsenic and other trace elements in central and southern Lao PDR (Jakariya and Deeble, 2008; Chanpiwat et al., 2011; Brindha et al., 2016), salinity originating from evaporite deposits in the Vientiane Plain and Champasak Province (Viossanges et al., 2018), and contamination of shallow groundwater from sewage systems and diffuse agricultural chemicals (Chantavong, 2011; GHK International, Halcrow Group, Lao Consulting Group, 2012; Noda et al., 2017; Brindha et al., 2019). Groundwater quality is closely linked to land use and land management practices. Increased inputs of fertilizers and pesticides associated with a push toward agricultural intensification (Brindha et al., 2019) are putting aquifers at greater risk of pollution. Rapid rates of urbanization (Okamoto et al., 2014), the emergence of numerous special economic zones of concentrated development (MPI Ministry of Planning and Investment, 2015) and waste management associated with mining leases, landfills, and industries are potential pollution sources at more localized scales (Ongkeo and Dam, 2015).

33.3.2 State of groundwater knowledge and information systems

There is still limited knowledge of the spatial distribution, availability, and quality of Lao PDR's groundwater resources. Local groundwater data is scarce and fragmented, data resolution for hydrogeological characteristics (including aquifer geometry, hydraulic properties, and water quality) is generally too low to determine locations and depths of potentially high-yielding wells (Viossanges et al., 2018). This makes groundwater development risky and curbs investment (Knudsen et al., 2004).

Information sharing mechanisms between stakeholders are inefficient. Existing information is extremely difficult to access as it is scattered across agencies (e.g., groundwater drilling activities led by the National Centre for Environmental Health and Water Supply—otherwise known as "Nam Saat"—are mainly held in provinces) and stakeholders' data [e.g., private operators, nongovernmental organizations (NGOs): in the case of industrial wells, well construction details, and information on groundwater extraction] are generally not easily accessible to government authorities (Chantavong, 2011).

No formal mechanism exists for collecting and compiling groundwater data. Furthermore, well-reporting standards are poor [GHD and ADB (Asian Development Bank), 2012] and current Environmental Impact Assessment (EIA) guidelines do not include groundwater impacts (Ongkeo and Dam, 2015). As a result, information systems are incomplete (Ongkeo and Dam, 2015; GHD, 2015). There is no long-term, national, or regional monitoring network to survey groundwater levels and groundwater quality [MoNRE (Ministry of Natural Resources and Environment), 2014]. Monitoring is not systematic and is mostly dispersed across local projects and is limited to the projects' lifetimes (Landon, 2011). Groundwater modeling supporting groundwater planning and decisions is limited by sparse datasets and lack of institutional and individual capacities.

33.3.3 Other barriers to groundwater management

In the past, policies and legislation have not addressed groundwater management specifically or adequately, due to the primary focus on surface waters and to the absence of any government MPI institution tasked specifically with groundwater management. Laws and regulations are inadequate and disconnected from field realities (e.g., no well permitting system or registration of well drillers—GHD, 2010); groundwater regulations and their enforcement are clearly lacking [GHD and ADB (Asian Development Bank), 2012; Vote et al., 2015].

There is little known about the physical, chemical, and microbiological aspects of Lao PDR's groundwater systems, and this equally applies to the economic, social, environmental, and legal aspects. At the national level, there is a lack of capacity of water resource managers in groundwater, and coordination between all concerned stakeholders (government institutions, private sector, communities, etc.) is inadequate. At the field level well-drilling standards remain low, indicating a clear capacity gap in well design, construction, and maintenance. It is estimated, for instance, that roughly 30%–40% of all hand-pumped wells are nonfunctional [GHD and ADB (Asian Development Bank), 2012]. An indicative list of the major groundwater stakeholders in Lao PDR is given in Table 33.1 as many of these will be referred to throughout this chapter.

33.4 Recent efforts to strengthen groundwater governance

33.4.1 Overview of policy, institutional, and legal changes

33.4.1.1 Changes in government policy

Lao PDR's government policies and strategies are increasingly recognizing the need for more systematic groundwater management. The 2004 National Strategy for the Rural Water Supply and Environmental Health Sector considered groundwater as a water source reserved for drinking purposes and broached the topic of groundwater management (with the attribution of mandates to different ministries). The current 5-year National Socio-Economic Development Plan (NSEDP) for 2016–20 refers to groundwater for the first time [MPI (Ministry of Planning and Investment), 2015], acknowledging at the highest level the importance of sustainable groundwater management for the country's socioeconomic development. This is an important symbolic step as the NSEDP specifies the directives for the emergent ministerial strategies and for provincial-, district-, and village-level action plans. In another symbolic step the draft National Water Resources Strategy until 2025 and Action Plan 2016–20 [MoNRE (Ministry of Natural Resources and Environment), 2014], which reflects the government's direction and vision in water resource management, includes groundwater management as an essential part of water resource management. Building upon certain elements of the Strategy and Action Plan (Part 4), the National Groundwater Action Plan (NGWAP) was finalized in June 2015, outlining the steps needed to improve groundwater management in the country (Box 33.1). This last point represents a significant milestone in systematizing progress in groundwater management nationally.

33.4.1.2 Changes in institutional arrangements

The Ministry of Natural Resources and Environment (MoNRE) was created in 2011 to deliver more effective and sustainable management and use of natural resources [MoNRE (Ministry of Natural Resources and Environment), 2014]. It was created by merging the former Water Resource and Environment Administration with departments and divisions related to natural resources and the environment such as land management, geology, and forest conservation (Sisouvanh et al., 2013). The Division of Groundwater Management (DGM) was then established within MoNRE's Department of Water Resources (DWR) in 2012. Its directives include developing and implementing groundwater management in Lao

TABLE 33.1 Indicative overview of groundwater stakeholders in Lao People's Democratic Republic.

	Users/practitioners	Management/planning	Research/investigation
Government	DHUP, Nam Saat *Establishing groundwater infrastructure for urban/rural groundwater use*	DWR, NREI, DIH, PCD, WaSRO Province and District offices (e.g., PoNRE), municipalities *Drafting/implementing policies, laws, and regulations* *Awareness raising, information sharing*	DoI, DWR, NAFRI NREI, DGM, DoGM *Data collection, monitoring, modeling*
International donors	Provide funds for projects in all sections		
NGOs	NGOs (e.g., ADRA, Plan International, SNV, World Vision) *Drilling for WASH projects*	—	—
Academia	—	FWR-NUoL *Groundwater management courses*	Academic research, Lao and international research partners *Groundwater assessments (availability/quality)*
Private	Beverage suppliers, salt manufacturers *Commercial use* Farmers *Irrigation* Rural householders *Domestic use* Drilling contractors *Drilling for other stakeholders*	Consultants (general, law, educational, etc.) *Capacity building and training, technical assistance for policy, legal and regulatory drafting* Rural householders *Community management*	Private consultants, engineers *Groundwater assessments, mapping (as part of various projects)*

Notes: Other stakeholders in addition to those listed here may be directly or indirectly associated with groundwater activities. Examples include the areas of land use management and energy provision. Private stakeholders, international donors and other entities are not specifically mentioned. The text in italics indicates the main function(s) of the stakeholders. *DGM*, Division of Groundwater Management; *DHUP*, Department of Housing and Urban Planning; *DIH*, Department of Industry and Handicraft; *DoGM*, Department of Geology and Minerals; *DoI*, Department of Irrigation; *DWR*, Department of Water Resources; *FWR*, Faculty of Water Resources; *NGO*, nongovernmental organization; *NREI*, Natural Resource Environment Institute; *NUoL*, National University of Laos; *PCD*, Pollution Control Department; *PoNRE*, Provincial Office of Natural Resources and Environment; *WaSRO*, Water Supply Regulatory Office.
Sources: Data from Ongkeo, O., Dam, R.A.C., 2015. National Integrated Water Resources Management Support Program (ADB TA 7780), Component 3: Groundwater Assessment, Updated National Groundwater Action Plan; Pavelic, P., Xayviliya, O., Ongkeo, O., 2014. Pathways for effective groundwater governance in the least-developed-country context of the Lao PDR. Water Int. 39, 469–485. Available from: <http://www.tandfonline.com/doi/abs/10.1080/02508060.2014.923971>.

PDR at all levels (national, provincial, and district), developing legislation and regulations for groundwater access and use, establishing a database and sharing information and technical support with other sectors and at all levels [MoNRE (Ministry of Natural Resources and Environment), 2012; Pavelic et al., 2014]. The DGM is the first government entity solely devoted to groundwater management and its mandate and activities are in line with the NGWAP (Box 33.1).

In July 2015 the Ministry of Education officially approved and endorsed the new Faculty of Water Resources (FWR) at the National University of Laos (NUoL), a teaching and research facility replacing the former Water Resource Engineering Department within the Faculty of Engineering. This institution is dedicated to developing technical capacities in water resources, with programs focusing on Integrated Water Resources Management (IWRM) and irrigation (AECOM Asia Company Ltd; Earth Systems Lao; Maxwell Stamp Ltd., 2015). As groundwater is relevant to these programs, FWR could contribute significantly toward enhanced academic and technical capacities in groundwater. The upgrading to faculty status, enabled, in part, by ADB Project 43114-012, also allows enhanced opportunities for groundwater-related research (GHD, 2015).

33.4.1.3 Revised legal arrangements

With support from IFC and the World Bank [DWR (Department of Water Resources) and IFC (International Finance Corporation), 2014], the DWR has revised the National Law on Water and Water Resources and the 2017 version, approved by the National Assembly, includes seven new articles dealing with groundwater. These articles address the

> **BOX 33.1 Overview of the National Groundwater Action Plan.**
>
> The National Groundwater Action Plan (NGWAP) was prepared by the National Research Environment Institute (NREI), the Department of Water Resources (DWR) and other agencies as a component of the larger National Integrated Water Resources Management Support Project led by the ADB. The NGWAP's principal aims are to enhance local capacities, develop a regulatory framework, raise awareness, and gather information on the country's groundwater resources. The expertise and information gained by implementing the Action Plan would then be used for developing more effective groundwater management in Lao PDR. It has two central objectives: (1) formulating and implementing groundwater regulations and (2) strengthening groundwater management planning and capacity. These are addressed through five components:
> - *Enhancing institutional capacity*: Defining and strengthening the capacity of the main implementing agencies of the plan, establishing a working group on groundwater, identifying issues amongst different stakeholders.
> - *Enhancing legal capacity*: Developing a National Groundwater Use and Management Policy, developing groundwater-related legislative and regulatory frameworks for industry, water quality, and overall operational use.
> - *Improving the knowledge base*: Encouraging research investigations and assessing the resource in priority areas, developing groundwater monitoring networks and a National Groundwater Management Information System, preparing information products for stakeholders and end-users.
> - *Increasing stakeholder, community, and end-user involvement*: Developing national level information exchange through the Working Group on Groundwater, exchanging data, awareness raising, establishing water user committees and a water fund for operation, and maintenance at the village level, providing technical training for well construction.
> - *Capacity building and training*: Developing research on the Groundwater Resources Assessment program, and a BSc–MSc groundwater management curriculum, implementing a development plan for the formation of a National Groundwater Research Center, training technicians at the provincial and community levels, providing technical, and policy training for government officials.
>
> The NGWAP is a long-term plan that requires financial investment and time. The Plan states that it could be implemented over a 3- to 5-year period with a little over a US$4.5 million budget (including inputs from local and international experts). Capacity building and training, being a relatively long-term process, may require longer time frames. In the absence of forthcoming donor support, the NGWAP has yet to be formally implemented, although informally different activities have been taking place that are in line with it. The DWR is currently drafting the National Groundwater Management Regulation, developing a so-called National Groundwater Profile (a national assessment of the country's groundwater resources), creating information sheets and posters intended for awareness raising; NREI is continuing its work on groundwater model development and monitoring of the Vientiane Plain initiated through the ACIAR Project LWR/2010/081.
>
> *Sources: Based on Ongkeo, O., Dam, R.A.C., 2015. National Integrated Water Resources Management Support Program (ADB TA 7780), Component 3: Groundwater Assessment, Updated National Groundwater Action Plan; ACIAR (Australian Centre for International Agricultural Research), 2016. Enhancing the resilience and productivity of rainfed dominated systems in Lao PDR through sustainable groundwater use. In: Final Report for ACIAR Project LWR/2010/081. Available from: <http://aciar.gov.au/publication/fr2016-35>.*

principles of groundwater use, groundwater quality protection, user rights, well drilling and permitting, groundwater monitoring and management [DWR (Department of Water Resources) and IFC (International Finance Corporation), 2014]. Another noteworthy development includes the drafting of a National Regulation for Groundwater Management by MoNRE.

33.4.1.4 Changes enhanced by projects and investments

Various projects, led by national and international experts and backed by international donors and technical assistance partners, are exploring groundwater-related topics in Lao PDR [e.g., agricultural groundwater use in Lao PDR—ACIAR (Australian Centre for International Agricultural Research), 2017; Suhardiman et al., 2015; Vote et al., 2015; urban groundwater use—Noda et al., 2017; the revision of the National Water Laws—DWR (Department of Water Resources) and IFC (International Finance Corporation), 2014; capacity building and climate change adaptation—Somphone and Xayviliya, 2017]. These projects can generate useful strands of knowledge, insight, expertise, and capacity. The outputs of several projects are referred to in more detail in Section 33.5 which provide examples of their contribution to groundwater management in Lao PDR.

33.4.2 Enhancing groundwater knowledge and data management

The advancements on resource knowledge, stakeholder involvement, and capacities are detailed in the following sections and summarized in Table 33.2. The future outlook and recommendations are provided in the last section.

TABLE 33.2 Groundwater governance indicators for Lao People's Democratic Republic.

Groundwater governance indicator		Implementation level			
		National scale		**Local scale**	
Technical	Groundwater body/aquifer delineation	Original maps (early 1990s)/groundwater potential maps (2018), low resolution	w	Local mapping underway, fragmented [ACIAR (Australian Centre for International Agricultural Research), 2016]	w
	Groundwater piezometric/quality-monitoring network	No national, long-lasting network; only at policy/regulatory level	n	In some areas as part of projects, Vientiane Plain monitoring under DWR/NREI mandate	y
	Groundwater pollution hazard assessment	No national wide-scale assessment	n	Some local assessments, for different pollution types	y
	Availability of aquifer numerical "management model"	–	na	Vientiane Plain model not at the stage to support management yet	w
Institutional	Groundwater agency as "Groundwater resource guardian"	Creation of MoNRE (2011), DGM (2012)	y	–	na
	Permanent stakeholder-engagement mechanism	Encouraged in recent policies but only policy-level; in Law on Water and Water Resources (2017), not enforced yet	w	DGM is piloting this in the upper Vientiane Plain	y
	Coordination with agricultural development	Attempt of MoNRE/MAF coordination	w	Coordination at provincial/district levels	n
	Coordination with urban/industrial development	Attempts of MoNRE/MPWT/private sector coordination	w	Coordination at provincial/district levels	n
Legal and fiscal	Water well drilling permits and groundwater use rights	Groundwater Management Regulation drafting and national Water Laws revision but unpublished; revised Law on Water Resources: regulating groundwater user groups and drilling/digging practices	w	-	na
	Instruments to constrain water well construction/use	–	n	–	na
	Sanctions for illegal water well operation	–	n	–	na
	Groundwater abstraction and use charging	Planning is underway at national scale (to regulate individuals, private industries etc.); towns using groundwater are paying for supplies	w	–	na
	Land-use control to reduce diffuse source pollution	–	n	–	na
	Constraints on ground discharge of waste (water)s	Instruction on the management of hazardous waste (MONRE 2015)/Env. Protection Law (2013)/Regulation on Control of Pesticides (MAF 2010); but in theory only	w	-	na

(Continued)

TABLE 33.2 (Continued)

Groundwater governance indicator		Implementation level			
		National scale		Local scale	
	Groundwater users registered and regulated	–	n	Few isolated cases (e.g., Beerlao)	w
Policy and planning	Public investment in groundwater management	Encouraged in recent policies/revised law, but only policy-level	w	Community aquifer-management organizations exists in several areas	y
	Financial policies encouraging groundwater sustainability	–	n	–	na
	Incentives for groundwater ecosystem services	–	n	–	na
	Existence of Groundwater Management Action Plan	Exists but is not being implemented yet	w	–	na

Notes: y: yes; w- weak; n: no; na: not applicable. Coordination with energy development is not included because the indicators used are from Groundwater Governance Project, 2016.
Source: Modified from Groundwater Governance Project, 2016. Global framework for action to achieve the vision on groundwater governance. In: Groundwater Governance – A Global Framework For Action, GEF (Global Environment Facility), UNESCO-IHP (International Hydrological Program), FAO (Food and Agriculture Organization of the United Nations), World Bank and IAH (International Association of Hydrogeologists). <http://www.groundwatergovernance.org>, Annex 1: Qualitative Indicators for Groundwater Governance Framework.

33.4.2.1 National- and local-scale assessments in priority areas increase local knowledge

There is ongoing research on groundwater quantity issues. Recent research, merging previously scattered global and local data, is reevaluating the potential of Lao PDR's aquifers in terms of storage, yield, and spatial coverage. As part of this research, maps have been developed at the national scale to provide a first-level assessment of high potential areas for groundwater development (Viossanges et al., 2018). Relatively high groundwater potential has been identified in central and southern Lao PDR; alluvial and Mesozoic sandstone aquifers seem to be promising, whereas karstic and volcanic aquifers equally show potential but are still poorly understood (Fig. 33.1). Maps have been published and discussed and shared with the government (Viossanges et al., 2018). A groundwater recharge map has been developed at the national scale (Lacombe et al., 2017a): this broad-level assessment of the aquifers' replenishment capacities provides a first indication of the amount of groundwater that may be sustainably extracted. While they do not represent local conditions accurately, these maps can still be used to support management by helping to define the upper limits for sustainable groundwater extraction and to identify priority areas for more in-depth research. Local research and development projects also exist, slowly reducing the knowledge gaps in certain areas. For instance, much research has been done in the Vientiane Plain, including groundwater assessments (e.g., Perttu et al., 2011) and water-quality studies (e.g., Chanpiwat et al., 2011; Brindha et al., 2019).

33.4.2.2 Recent policies and draft data systems aim at formalizing and systematizing data collection

The NGWAP and NWRSAP both have objectives of centralizing data in a national information system under the supervision of a dedicated government institution, of unifying database formulation and of developing piezometric networks for water quantity and quality monitoring (Ongkeo and Dam, 2015; GHD, 2010—Box 2). Standards and specifications for well construction were discussed in consulting reports (GHD, 2015) and under the NIWRMSP (National IWRM Support Project) progress has been made in producing first drafts of National Groundwater Well Log Standard Forms, a Water Well ID scheme for well identification and a well database (GHD, 2015). However these are not yet put into use by the concerned stakeholders. They have yet to be finalized, requiring external support (AECOM Asia

Company Ltd; Earth Systems Lao; Maxwell Stamp Ltd., 2015), and regulations would help in making their use more widespread. During the course of ACIAR Project LWR/2010/081, Natural Resource Environment Institute (NREI) established a GIS groundwater data system for the Vientiane Plain and a national-scale well database was created [ACIAR (Australian Centre for International Agricultural Research), 2016]. DWR has been undertaking routine groundwater-level monitoring since 2014 in the Vientiane Plain [ACIAR (Australian Centre for International Agricultural Research), 2016], which is directly in line with the NGWAP. A national regulatory framework is in the advanced stages of preparation by MoNRE [DGM (Division of Groundwater Management), 2017]. The current draft document specifies a host of principles, rules, and measures on well drilling and groundwater pumping. It spells out the minimum qualification requirements for the certification of drillers and the obligations expected according to the size of the company (e.g., fees, reporting, and water quality testing). Those wishing to use groundwater for commercial purposes of any kind (agriculture, animal husbandry, industry, etc.)—not including small-scale use to meet basic entitlements—are expected to apply for a permit and water use charges apply according to the volumetric level of withdrawal.

33.4.2.3 Top-down versus bottom-up approaches

While legal and regulatory instruments are being developed to monitor and control groundwater users through well licensing and water pricing, the influence of these on the majority of users, who are small scale, through direct intervention (possibly including punitive measures, e.g., penalties for rule-breakers) is likely to be limited and economically prohibitive for the Government.

Informal groundwater management systems at the field level, largely disconnected from policy-driven processes, should also be acknowledged. Several studies have revealed community groundwater management in Lao PDR. The degree of success reported from the few documented cases is mixed. In local communities practicing self-regulation under seasonally water-scarce conditions, water supplies were maintained during drought. Also, conflict across different ethnic groups was avoided (Vinckevleugel, 2015). In other cases, limited knowledge and capacity of local communities has exacerbated conditions of water scarcity [DAI (Development Alternatives Inc.), 2015]. Informal management systems should be fostered in situations where they are clearly warranted. In all likelihood a mix of so-called top-down and bottom-up approaches may best serve the interests of the country.

33.4.2.4 Modeling efforts aim at supporting groundwater planning

The first known groundwater flow model in Lao PDR was a catchment-level model developed in 2005 to better understand the effect of irrigation projects on land salinization in Savannakhet Province (Wiszniewski et al., 2005). A groundwater model for the Vientiane Plain was developed by NREI under ACIAR Project LWR/2010/081, with training from experts from the Groundwater Research Center of Khon Kaen University, Thailand [NREI (Natural Resources and Environment Institute), 2017] and support from government officials. This represents the first regional groundwater model for Lao PDR developed by government officials (Lao nationals) for groundwater planning and management purposes [ACIAR (Australian Centre for International Agricultural Research), 2016].

33.4.3 Mechanisms of stakeholder coordination and involvement

33.4.3.1 Existing issues in government coordination and overall stakeholder communication

In Lao PDR, decision-making in natural resource management has traditionally been undertaken with virtually no coordination between departments and sectors, each setting its own target and strategies with little consultation of others [MoNRE (Ministry of Natural Resources and Environment), 2014; Pavelic et al., 2014]. The key ministries within the water, agriculture, energy, and public health sectors have also operated quite independently on matters relating to groundwater. Consequently, ministries often have overlapping responsibilities relating to water resources (Sisouvanh et al., 2013).

Coordination between groundwater stakeholders is inadequate at all levels [ACIAR (Australian Centre for International Agricultural Research), 2016, 2017]. Consequently, at the community level, groundwater resources are not well understood and there is limited access to information and institutional support. Data collection is scant or nonexistent for many of the drilling projects carried out by the private sector and national and international NGOs. These gaps have led to misconceptions and costly or failed projects (Knudsen et al., 2004; Chithtalath, 2007; Vote et al., 2015).

33.4.3.2 Institutional mapping and consultation workshops as a preliminary form of cross-sector policy coordination

Institutional mapping for water governance and groundwater stakeholder mapping have been initiated as part of several projects [Pavelic et al., 2014; ACIAR (Australian Centre for International Agricultural Research), 2016] and groundwater institutional planning has advanced through the NGWAP process (GHD, 2015). This defines the responsibilities of different government entities and stakeholders in groundwater management (see Table 33.1). During the drafting of the NGWAP and the NWRSAP, concerted efforts were made to develop a dialog between government entities though cross-sectoral workshops including the Department of Groundwater Management (DGM), Department of Irrigation (DoI), Department of Mines (DoM), DWR, and NREI (Ongkeo and Dam, 2015). The NGWAP also sets out to establish a Working Group on Groundwater (see Box 33.1), which was informally initiated by an ACIAR funded project [ACIAR (Australian Centre for International Agricultural Research), 2016]. Attempts at developing cooperation with groundwater agencies in the region (e.g., cooperation between NREI and the GWRC-KKU to establish the Vientiane Plain model) and at regional cooperation (through the efforts of regional and international organizations such as UNESCO and CCOP—Khongsab, 2015) should also be acknowledged.

33.4.3.3 Consultation of stakeholders in recent groundwater policy and legislation drafting

Efforts were made to consult government agencies at different levels in the NWRSAP drafting and line agencies were integrated in the process through consultation workshops held at the provincial, ministerial, and national levels (Ongkeo and Dam, 2015). Beyond coordination between government agencies, a wider range of stakeholders (local authorities and communities, private sector operators, representatives from different sectors) was consulted in the process of revising groundwater regulations supportive of the revised Law on Water Resources, to take into account their views and needs [ACIAR (Australian Centre for International Agricultural Research), 2016]. Participatory policy making is a feature of the revised Law on Water Resources, as it also emphasizes every stakeholder's role in achieving effective groundwater management [DWR (Department of Water Resources) and IFC (International Finance Corporation), 2014].

33.4.3.4 Communities newly integrated in recent policy and legislation

Communities have been integrated into recent legislation and policy. The revised Law on Water Resources recognizes communities' ownership rights of groundwater and formulates the mechanisms for community participation in groundwater management. It defines community groundwater management as the establishment of groundwater user groups responsible for local management, maintenance, and monitoring of the resource and regulated by the MoNRE [DWR (Department of Water Resources) and IFC (International Finance Corporation), 2014]. Another example is that of the Groundwater Management Plan for the upper part of the Vientiane Plain, which includes awareness raising and dissemination of research activities to local communities [DGM (Division of Groundwater Management), 2017] and conforms to the NGWAP's aim of end-user involvement (see Box 33.1). A growing number of studies also focus on understanding local groundwater uses, perceptions, and needs [Agricultural Census, 2010/11; Suhardiman et al., 2015; Vote et al., 2015; DGM (Division of Groundwater Management), 2017]. Although their number remains limited, this is an important first step in assimilating communities' specific needs. Overall, it is likely too early to know if the stakeholder-consultation processes as indicated previously have stimulated genuine participation in decision-making process or otherwise.

33.4.4 Development of human resources and groundwater-management capacity

33.4.4.1 Existing issues with institutional and technical capacities

In Lao PDR government agencies and NGOs generally lack expertise and fundamental knowledge of the groundwater resource (GHD, 2015). For instance NREI, with expertise mainly in engineering and the environment, lacks knowledge in hydrogeology and groundwater resources and DoI's experience on groundwater irrigation is limited (Pavelic et al., 2014). Another example of lack of capacity is that of a well drilling project in the Bokeo province, which was unsuccessful because local conditions had not been sufficiently analyzed in advance by the concerned NGO (Chithtalath, 2007).

There is a strong need to improve teaching and research capacity (GHD, 2015). There are few Masters-level and no known PhD-level groundwater specialists teaching or practicing in Lao PDR (Pavelic et al., 2014). Groundwater practitioners lack technical capacity in both operation and maintenance activities [GHD and ADB (Asian Development Bank), 2012; Ongkeo and Dam, 2015]. Poor-quality drilling or pumping practices can lead to aquifer contamination (AECOM Asia Company Ltd; Earth Systems Lao; Maxwell Stamp Ltd., 2015) and high running and maintenance costs can lead to short life span of projects. Through the use of international experts the private sector has made detailed groundwater assessments associated with high-value investments such as mining (e.g., Gabora et al., 2014); however, this has achieved little to develop the local capacity.

33.4.4.2 Training programs generate institutional capacity for groundwater management

The knowledge and technical capacity of government staff (particularly from DWR, DGM, NREI, NUoL, and DoI) have been enhanced under different projects. For example, as part of the NIWRMSP a 2-week groundwater management training program took place in 2015. This provided training in hydrogeology, technical, and groundwater-management issues to staff from government institutions and from the national university (GHD, 2015). Under ACIAR Project LWR/2010/081, two short courses on groundwater took place in 2013. The first course covered basic issues such as well drilling, monitoring, management, and the second was on modeling (GWRC-KKU, 2013a,b). The knowledge and techniques acquired during these courses are currently being used in groundwater planning. After follow-up workshops and training held over an 18-month period, NREI staff have developed and are now able to apply a groundwater model for the Vientiane Plain at a basic level. This can serve to begin to help inform decision-making. However NREI has yet to reach the stage of independent modeling and still requires external support [ACIAR (Australian Centre for International Agricultural Research), 2016].

33.4.4.3 Teaching and research capacity building to develop a community of groundwater experts

The pace of research on groundwater has picked up. In 2014 three groundwater-related papers were published in peer-reviewed journals, six conference papers were published and two specialists were practicing full time in the county (Pavelic et al., 2014). As part of ACIAR Project LWR/2010/081, groundwater modeling workshops were held between 2013 and 2016 to develop a local community of practice and to encourage national researchers in the field of groundwater. Several projects have focused on improving teaching capacity within the NUoL and the FWR. A groundwater-management training program took place under the NIWRMSP (2015) for NUoL professors and undergraduate students and NIWRMSP/ACIAR project material are designed to serve as teaching material [ADB (Asian Development Bank) and MoNRE (Ministry of Natural Resources and Environment), 2015a; AECOM Asia Company Ltd; Earth Systems Lao; Maxwell Stamp Ltd., 2015; ACIAR (Australian Centre for International Agricultural Research), 2016]. FWR started providing hands-on training for students to learn about groundwater-based irrigation (Keokhamphui et al., 2016) under the ACIAR Project LWR/2010/081 and DoI is using the groundwater-based irrigation system at Ekxang village (implemented as part of ACIAR Project LWR/2010/081) as a field facility for training of students from the Thangone Irrigation College. FWR was provided with academic equipment and the quality of its curricula was improved through exchange with international consultants under the NIWRMSP project (GHD, 2015). One of FWR's future aims is to develop Groundwater Management and Water Governance programs [ADB (Asian Development Bank) and MoNRE (Ministry of Natural Resources and Environment), 2015b]. As a result of ACIAR Project LWR/2010/081, seven Bachelor's/Master's theses were produced by Lao nationals [and a further six Bachelor's/Master's/PhD studies were underway—ACIAR (Australian Centre for International Agricultural Research), 2016], developing a local community specialized in the field of groundwater research and providing a basis for future capacity building.

33.4.4.4 Technical capacity disseminated to groundwater practitioners

The NGWAP places increasing importance on disseminating and transferring technical knowledge to end-users (e.g., to provide technical training for well construction, see Box 33.1). Some projects provide expertise to groundwater practitioners, such as during ACIAR Project LWR/2010/081 where research implementation generated new capacity in groundwater-based irrigation and the findings were disseminated to policy-makers, farmers, and extension officers [ACIAR (Australian Centre for International Agricultural Research), 2017].

33.5 Outlook: pathways forward for Lao People's Democratic Republic

33.5.1 Effective policy making and implementation

As presented in Section 33.4, groundwater has recently begun to be more firmly included within policy, institutions, and legislation. However, there is still a wide gap between the finely crafted messages within the various guiding documents and realities on the ground. It is in bridging this gap that further intervention in policy and action can have most significant impacts. With this in mind, key priorities are discussed next.

33.5.1.1 Promoting cross-sectoral coordination

Establishing more effective linkages, with clear roles and responsibilities defined for each government institution, would ensure horizontal as well as vertical integration from the national through to the local level and be helpful in the planning and implementing of groundwater-related activities. The MoNRE, in partnership with the Ministry of Agriculture and Forestry (MAF), is well placed to provide effective coordination through their specialized departments and divisions. This most notably applies to the recently formed Groundwater Management Division (within MoNRE) and the DoI (within MAF). Specific agencies such as Nam Saat (Ministry of Health), the Department of Housing and Urban Planning (Ministry of Public Works and Transportation) and the DoM (Ministry of Energy and Mines) carry out groundwater activities and, therefore, also have specific interests.

Under the guidance of MoNRE and/or MAF, a cross-sector working group (WG) should coordinate the efforts of ministries and departments with groundwater management responsibilities across major functional areas such as policy formulation, project planning, resource assessment, monitoring, land use planning, regulation/enforcement and research. The WG could contribute through activities as per the following:

- Develop and strengthen standards for development agencies, private industry, and government in groundwater project implementation.
- Disseminate, monitor, and enforce adoption of the abovementioned standards (for more effective service delivery, knowledge generation, and capacity building).
- Develop regulations and enforcement controlling large-scale commercial development, to ensure that risks posed by polluting or overpumping groundwater are minimized.
- Introduce reasonable pricing on groundwater withdrawals for large-scale investments, and consider the benefits of subsidies on energy (solar power as well as conventional fuel) for groundwater pumped by small-scale farmers.

33.5.1.2 Promoting strategic planning of groundwater resources

The WG could also support more strategic planning of groundwater resources through the following activities:

- Delineate and carry out technical assessments of aquifers considered highest priority by the WG, recognizing aquifer suitability as well as community needs for groundwater (as detailed later).
- Incrementally develop and implement groundwater management plans by DGM in priority aquifers and/or areas (plans inclusive of groundwater protection mechanisms, through strategies that protect most vulnerable aquifers from pollution and overuse and in line with land- and river basin–management plans).
- Establish regional/national groundwater monitoring networks covering water quantity and quality, as the NGWAP sets out to do (build upon the monitoring infrastructure in place on the Vientiane Plain, earlier managed by DGM and NREI).
- Develop an effective groundwater information–management system hosting all data relating to groundwater and providing information products specific to stakeholders, regions, and sectors.
- Strengthen standards for project contractors and facilitate the provision of documentation by implementing agencies (government agencies, donors, NGOs, consultants). This would include the following:
  - Licensing and registration of drilling contractors for projects exceeding a minimum scale.
  - Mandatory completion of simple, standardized-format drilling reports, and well-log sheets and submission to local authorities (to be included in a centralized information system).
  - Inclusion of groundwater in EIAs associated with new industrial, mining, and hydropower companies. Any proposed revisions to EIAs would need to be negotiated through MoNRE as the administering authority (according to Government of Lao PDR Decree of 2010, Decree No. 112/PM).

33.5.2 Strengthening institutional and human resource capacity

The capacity of individuals and organizations from various government (and other) institutions involved with groundwater still remains low. Setting up capable institutions and individuals who can contribute effectively to resource governance according to their mandates and specific areas of expertise is an important prerequisite. Some of the short and longer term approaches for achieving this are:

- Improving academic curricula and developing formal degree training in groundwater at Bachelor and Masters levels within the key faculties and departments of NUoL (in the fields of water resources, engineering, agriculture, environmental science, and others). This might also be applicable to some of the smaller, regional universities in the country.
- Developing cooperation between NUoL and ASEAN (or other) universities or organizations with more established capabilities. Cooperative programs for postgraduate study, vocational training, and staff exchange are already in place in some departments through university networks and/or donor capacity building programs: although there is little in relation to groundwater. These can provide foundations to build upon (e.g., the NUoL Faculty of Engineering previously had in place a Master's Program on Environmental Engineering and Management through EU and German support since 2008, in cooperation with universities from Germany, Vietnam and Thailand).
- Conducting targeted training of groundwater professionals including (1) training and accreditation for well-drilling professionals and groundwater technicians from government and larger private organizations, for improved groundwater exploration and well completion, and (2) training for government officers within MoNRE (at the national to district levels) in data collection and analysis and data management systems.

33.5.3 Continuing efforts in applied research

Research investments are needed to expand the broad, multidisciplinary information, and knowledge base that should underpin the planning, development, and management of groundwater resources. A broad indication of prospective research topics are offered below, demonstrating the more obvious gaps based on findings and discussions of the ACIAR and ADB supported projects:

- Conduct regional and local groundwater mapping, particularly in areas where existing information suggests a likely potential for development (e.g., lowland alluvial plains and Bolaven Plateau).
- Delineate groundwater dynamics at local to regional scales, covering aquifer structure, properties and boundaries, water balances, and flow systems (including interactions with surface water).
- Assess groundwater quality and public health risks, particularly in areas where groundwater reliance is high (Fig. 33.2) and vulnerable to anthropogenic and/or natural pollution.
- Map areas of groundwater dependence at the national scale (e.g., starting with existing Ramsar-listed wetlands; Lacombe et al., 2017b) and establish the hydro-ecological functioning and future threats within these priority areas.
- Evaluate the cost-effectiveness and related issues (e.g., supply chains) for alternative drilling and pumping technologies in different settings (particularly alluvium, sandstone, limestone/karst, and basalt terrain).
- Mainstream local and gendered participation in rural groundwater planning, for remote upland communities.
- Develop institutional and financing arrangements for long-term maintenance of groundwater infrastructure for rural areas.
- Evaluate the socioecological trade-offs and impacts associated with possible large-scale groundwater developments.
- Evaluate the impacts of future climate on groundwater demand, seasonal water availability, and dry-season surface water flows.

The findings of the research topics described previously can bring about direct and indirect benefits for policy formulation and implementation. A few specific examples are listed next:

- Better decision-making on drilling technologies, well locations, and target depths through resource assessments.
- Reduced costs for choosing, setting up, or running groundwater pumping infrastructure.
- Better informed land-use policies and planning through the improved knowledge of water quality. This can assist resource managers in tailoring groundwater plans to local contexts.

33.5.4 Participation of stakeholders

33.5.4.1 Awareness raising

Awareness raising of key stakeholder groups from within government agencies, donors, private industry, and NGOs is needed to create knowledge and understanding of groundwater issues, and their importance in establishing effective groundwater management. Entry points to actively raise awareness on groundwater include:

- Relevant sector and subsector working groups of the Government bringing groundwater onto the agenda to establish how groundwater can best serve their mandates;
- The formal media (newspaper, TV, and radio) and informal media (social networks and other online platforms such as Facebook, Tholakhong, and Laofab) disseminating news and other information at the national to local level;
- Projects and programs focusing on developing issues such as health and sanitation, climate change adaption, green agriculture and others; and
- World Water Day, which is widely celebrated in Lao PDR, soliciting more widespread engagement and participation from civil society groups, schools, and businesses.

Enablement for these and similar activities could come from government and academia with established expertise, notably DWR on policies and regulations, NREI on environmental and climate-resilience related issues, Nam Saat on community water supplies, Nam Papa on town water supply, DoI on irrigation, NUoL-FWR on education, and the Ministry of Energy and Mines on the geological environment.

33.5.4.2 Sustainable financing arrangements

Adequate and reliable funding is a necessary prerequisite for sustaining the work described previously. Research and capacity building–related activities, in particular, are strongly dependent on funding sources external to government. Funding may be forthcoming from the donors that have traditionally supported research projects and scholarship programs in Lao PDR (e.g., ACIAR, CIRAD and others). Donors that have a focus on rural development (e.g., ADB, WB, JICA, EU, USAID, GIZ, and DFAT) may have a role to play. This also applies to large hydropower, mining, or commercial agriculture companies who often rely on international experts for impact assessments or to develop new water supplies.

Donors and large private sector players, along with the local and international NGOs and consultants that drive development activities, represent as yet an untapped resource. Adjusting approaches to include modest—yet important—budget allocations would yield good returns on investment (generation of stronger development outcomes through local-based professionals who could, in turn, work to serve the interests of these stakeholder groups better). Smaller ongoing investments are likely to be more fruitful than large one-offs (e.g., the nurturing of young Lao professionals interested in a career in groundwater through scholarships, to provide formal training and gain new knowledge and techniques). The creation of expertise within government, academia, and industry brings about benefits that go beyond research alone. This would include improvements in the various groundwater management functional areas such as resource assessments, monitoring and strategic planning for new supplies, operation and maintenance of existing supplies, resource protection, and policy formulation. Encouraging potential funders to shift from the *business as usual* mode to one that is more supportive of developing higher capacity in groundwater remains the major challenge.

Acknowledgments

This work is derived from the findings of the Australian Centre for International Agricultural Research (ACIAR) supported project LWR/2010/081 (*Enhancing the resilience and productivity of rainfed-dominated systems in Lao PDR through sustainable groundwater use*— http://aciar.gov.au/project/lwr/2010/081) as well as relevant components of the ADB projects 40193-012 (*Lao People's Democratic Republic: Updating the National Water Policy and Strategy*— https://www.adb.org/projects/40193-012/main) and 43114-012 (Lao People's Democratic Republic: National Integrated Water Resources Management Support Project— https://www.adb.org/projects/43114-012/main). The authors would like to thank Dr. Alvar Closas and Dr. Karen Villholth from the International Water Management Institute for their comments on the manuscript. This research was carried out as part of the CGIAR Research Program on Water, Land, and Ecosystems (WLE) and supported by CGIAR Fund Donors.

Acronyms

ACIAR	Australian Centre for International Agricultural Research
ADB	Asian Development Bank
ASEAN	Association of Southeast Asian Nations

CCOP	Coordinating Committee for Geoscience Programs in East and Southeast Asia
DGM	Division of Groundwater Management (DWR-MoNRE)
DHUP	Department of Housing and Urban Planning (Ministry of Public Works and Transportation)
DIH	Department of Industry and Handicraft (Ministry of Industry and Commerce)
DoGM	Department of Geology and Minerals (Ministry of Energy and Mines)
DoI	Department of Irrigation (Ministry of Agriculture and Forestry)
DWR	Department of Water Resources (MoNRE)
EIA	Environmental Impact Assessment
EU	European Union
FWR	Faculty of Water Resources (NUoL)
GWRC-KKU	Groundwater Research Center, Khon Kaen University
IFC	International Finance Corporation
IWMI	International Water Management Institute
IWRM	Integrated Water Resources Management
Lao PDR	Lao People's Democratic Republic
MAF	Ministry of Agriculture and Forestry
MoNRE	Ministry of Natural Resources and the Environment
MPWT	Ministry of Public Works and Transport
Nam Saat	National Centre for Environmental Health and Water Supply (Ministry of Health)
NGO	nongovernmental organization
NGWAP	National Groundwater Action Plan
NIWRMSP	National Integrated Water Resources Management Support Project
NREI	Natural Resource Environment Institute (MoNRE)
NSEDP	National Socioeconomic Development Plan
NUoL	National University of Laos
NWRSAP	National Water Resources Strategy until 2025 and Action Plan 2016–2020
PCD	Pollution Control Department (MoNRE)
PoNRE	Provincial Office of Natural Resources and Environment (MoNRE)
UN	United Nations
UNESCO	United Nations Educational, Scientific and Cultural Organization
WASH	Water, sanitation and hygiene
WaSRO	Water Supply Regulatory Office
WG	Cross-Sector Working Group
WREA	Water Resource and Environment Administration

References

ACIAR (Australian Centre for International Agricultural Research), 2016. Enhancing the resilience and productivity of rainfed dominated systems in Lao PDR through sustainable groundwater use. In: Final Report for ACIAR Project LWR/2010/081. Available from: <http://aciar.gov.au/publication/fr2016-35>.

ACIAR (Australian Centre for International Agricultural Research), 2017. Groundwater for irrigation in Lao PDR: promoting sustainable farmer-managed groundwater irrigation technologies for food security, livelihood enhancement and climate resilient agriculture. ACIAR Policy Brief. Available from: <http://aciar.gov.au/publication/fs2017>.

ADB (Asian Development Bank) and MoNRE (Ministry of Natural Resources and Environment), 2015a. Faculty of Water Resources – NUoL Teaching Manuals: Report and Reference guide. In: National Integrated Water Resources Management Support Program (ADB TA 7780), Component 4: Technical support to National University of Laos, Faculty of Water Resources in Developing IWRM Education.

ADB (Asian Development Bank) and MoNRE (Ministry of Natural Resources and Environment), 2015b. Research Collaboration Challenges & Opportunities for the Faculty of Water Resources, National University of Laos. Workshop Report. National Integrated Water Resources Management Support Program (ADB TA 7780). Component 4: Technical Support to National University of Laos, Faculty of Water Resources in Developing IWRM Education.

AECOM Asia Company Ltd; Earth Systems Lao; Maxwell Stamp Ltd., 2015. Technical Assistance Consultant's Report, December 2015: National Integrated Water Resources Management Support Project (ADB TA 7780), Package 1, Capacity Building: Final Report. Project No. 43114.

Billving, A., Ågren, A., 2007. Water Resources in the Cultivated Uplands of Lao PDR: A Case Study of Huay Maha Village (Master thesis). Ref. LWR-EX-07-28, KTH Royal Institute of Technology, Department of Land and Water Resources Engineering, Stockholm, Sweden, 65p.

Brindha, K., Pavelic, P., Sotoukee, T., 2019. Environmental assessment of water and soil quality in the Vientiane Plain, Lao PDR. Groundw. Sustain. Dev. 8, 24–30.

Brindha, K., Pavelic, P., Sotoukee, T., Douangsavanh, S., Elango, L., 2016. Geochemical characteristics and groundwater quality in the Vientiane plain, Laos. Exposure Health 1–16. Available from: https://link.springer.com/article/10.1007/s12403-016-0224-8?shared-article-renderer.

Chanpiwat, P., Sthiannopkao, S., Cho, K., Kim, K., San, V., Suvanthong, B., et al., 2011. Contamination by arsenic and other trace elements of tube-well water along the Mekong River in Lao PDR. Environ. Pollut. 159 (2), 567–576. Available from: <https://doi.org/10.1016/j.envpol.2010.10.007>.

Chantavong, P., 2011. Groundwater for Water Supply in Laos. On behalf of the DHUP (MPWT). In: Presented at the launch meeting, Regional Water Knowledge Hub for Groundwater Management, 2–3 June 2011, Bangkok, Thailand. Available from: <http://www.iges.or.jp>.

Chithtalath, S., 2007. Community development projects and the status of ethnic minority women in the Moksuk-Tafa area, Bokeo province, Laos. Commun. Dev. J. 42 (3), 299–316. Available from: <https://doi.org/10.1093/cdj/bsl018>.

Clément, C., Vinckevleugel, J., Pavelic, P., Xiong, K., Valee, L., Sotoukee, T., et al., 2018. Community-managed groundwater irrigation on the Vientiane Plain of Lao PDR: Planning, implementation and findings from a pilot trial. In: IWMI Working Paper 183. International Water Management Institute (IWMI), Colombo, Sri Lanka, 52p.

DAI (Development Alternatives Inc.), 2015. Water supply feasibility assessment in Khammouan, Lao PDR. In: Report Prepared for USAID Mekong Adaptation and Resilience to Climate Change (USAID Mekong ARCC) Program, February 2015.

DGM (Division of Groundwater Management), 2017. Groundwater Profile of Keo Oudom, Viengkham, Thoulakhom and Phonhong districts, Vientiane Plain, Lao PDR. Prepared by the Division of Groundwater Management (in Lao language).

DWR (Department of Water Resources) and IFC (International Finance Corporation), 2014. Revision of the Law on Water and Water Resources – Draft Concept Paper for the Revised Law. Available from: <http://www.ifc.org/wps/wcm/connect/region__ext_content/regions/east+asia+and+the+pacific/countries/eap-hydro-waterlawconceptnote>.

FAO (Food and Agriculture Organization of the United Nations), 2016. AQUASTAT Website. Retrieved from: <http://www.fao.org/nr/water/aquastat/main/index.stm>.

Gabora, M., Martin, N., Clements, N., 2014. Application of the Null Space Monte Carlo Method in a groundwater flow model of mine pit dewatering. In: Sui, Sun, Wang (Eds.), An Interdisciplinary Response to Mine Water Challenges. China University of Mining and Technology Press, Xuzhou, pp. 14–18.

GHD, 2010. Updating the National Water Resources Policy and Strategy (ADB TA 7013-LAO): Mid-Term Report. March 2010.

GHD, 2015. National Integrated Water Resources Management Support Program (ADB TA 7780): Outcome Study Report. November 2015.

GHD and ADB (Asian Development Bank), 2012. National Integrated Water Resources Management Support Program (ADB TA 7780), Package 3: Groundwater Assessment, Working paper 2: Resource Assessment Report. December 2012.

GHK International, Halcrow Group, Lao Consulting Group, 2012. Updated Technical Report of: Solid Waste Management Improvements for Pakse Urban Environmental Improvement Project (RRP LAO 43316) Based on Final Report for ADB TA 7567-LAO, July 2011. March 2012.

Groundwater Governance Project, 2016. Global framework for action to achieve the vision on groundwater governance. In: Groundwater Governance – A global Framework for Action, GEF (Global Environment Facility), UNESCO-IHP (International Hydrological Program), FAO (Food and Agriculture Organization of the United Nations), World Bank and IAH (International Association of Hydrogeologists). <http://www.groundwater-governance.org>.

GWRC-KKU, 2013a. Short course on hydrogeology I: Fundamental of groundwater resources. In: Proceedings of the First Short Course held at Groundwater Research Center, Faculty of Technology, Khon Kaen University, Thailand, 22–25 April, 2013, 219 p.

GWRC-KKU, 2013b. Short course on hydrogeology II: Groundwater modeling and application. In: Proceedings of the Second Short Course held at Groundwater Research Center, Faculty of Technology, Khon Kaen University, Thailand, 4–8 November, 2013.

Ha, K., Ngoc, N., Lee, E., Jayakumar, R., 2015. Current status and issues of groundwater in the Mekong river basin. In: Published by KIGAM (Korea Institute of Geoscience and Mineral Resources), CCOP Technical Secretariat, UNESCO Bangkok Office.

IPCC (Intergovernmental Panel on Climate Change), 2012. Managing the risks of extreme events and disasters to advance climate change adaptation. In: A Special Report of Working Groups I and II of the IPCC. Cambridge University Press, Cambridge, UK, and New York, 582p. Available from: <http://ipcc-wg2.gov/SREX/>.

Jakariya, M., Deeble, S., 2008. Evaluation of arsenic mitigation in four countries of the greater Mekong region. In: Final Report – December 2008. Supported by UNICEF (United Nations Children's Fund) – AusAID (Australian Agency for International Development).

Johnson, J.H., 1983. Preliminary appraisal of the hydrogeology of the Lower Mekong Basin. In: Interim Committee for Coordination of Investigations of the Lower Mekong Basin.

Johnston, R., Lacombe, G., Hoanh, C., Noble, A., Pavelic, P., Smakhtin, V., et al., 2010. Climate change, water and agriculture in the greater Mekong subregion. IWMI Res. Rep. 136, 60. Available from: https://doi.org/10.5337/2010.212.

Keokhamphui, K., et al., 2016. Enhancing cash crop production through groundwater and agricultural waste applications. Sci. J. Natl Univ. Laos 11, 2–12 (written Lao Lang.).

Khongsab, S., 2015. Ground Water Management and National Water Resources in Lao PDR. In: Project Report of the CCOP-GSJ (Geological Survey of Japan)/AIST (National Institute of Advanced Industrial Science and Technology, Japan) – DGR (Department of Groundwater Resources, Thailand). Groundwater Phase III Project Kick-Off Meeting, 10–12 February 2015, Bangkok, Thailand.

Knudsen, J.B.S., Ruden, F., Smith, B.T., 2004. The online support and training project for the groundwater sector of Lao PDR. In: Proceedings of 30th WEDC International Conference 'People-Centred Approaches to Water and Environmental Sanitation', Vientiane, Lao PDR, 2004, pp. 434–437.

Lacombe, G., Pavelic, P., McCartney, M., Phommavong, K., Viossanges, M., 2017b. Hydrological assessment of the Xe Champone and Beung Kiat Ngong wetlands. In: Report to FAO for the Project: Climate Change Adaptation in Wetlands Areas (CAWA).

Lacombe, G., Douangsavanh, S., Vongphachanh, S., Pavelic, P., 2017a. Regional assessment of groundwater recharge in the lower Mekong basin. Hydrology 4 (4), 60. Available from: https://doi.org/10.3390/hydrology4040060.

Landon, M.K. 2011. Preliminary compilation and review of current information on groundwater monitoring and resources in the lower Mekong river basin. In: U.S. Geological Survey Report to the Mekong River Commission, Version Dated July 2011, pp. 1-34.

MAF (Ministry of Agriculture and Forestry), 2012. National Agricultural Census 2010/11 Highlights. In: Report Prepared by the Steering Committee for the Agricultural Census, Agricultural Census Office.

MoNRE (Ministry of Natural Resources and Environment), 2012. Decision on the Establishment and Activities of Department of Water Resources. Unofficial Translation Dated 17 May 2012.

MoNRE (Ministry of Natural Resources and Environment), 2014. MoNRE Vision Toward 2030 − Natural Resources and Environment Strategy, 10 years 2016-2025. Draft Version 3 October 2014.

MPI (Ministry of Planning and Investment), 2015. Lao People's Democratic Republic: Five Year National Socio-Economic Development Plan VIII (2016−2020).

MRC (Mekong River Commission), 2011. Impacts of climate change and development on Mekong flow regimes: first assessment − 2009. In: MRC Management Information Booklet Series No. 4.

Mukherjee, A., Scanlon, B., Aureli, A., Langan, S., Guo, H., McKenzie, A., 2020. Global Groundwater: Source, Scarcity, Sustainability, Security and Solutions, first ed. Elsevier, ISBN: 9780128181720.

Noda, K., Makino, T., Kimura, M., Douangsavanh, S., Keokhamphui, K., Hamada, H., et al., 2017. Domestic water availability in Vientiane, Lao PDR − the water quality variation in the rainy season. J. Agric. Meteorol. 73 (1). Available from: https://doi.org/10.2480/agrmet.D-16-00001.

NREI (Natural Resources and Environment Institute) 2017. Groundwater model development in Vientiane Plain. In: Draft Report. Ministry of National Resources and Environment, Lao PDR.

Okamoto, K., Sharifi, A., Chiba, Y., 2014. The impact of urbanization on land use and the changing role of forests in Vientiane. In: Integrated Studies of Social and Natural Environmental Transition in Laos. Springer, Japan, pp. 29−38.

Ongkeo, O., Dam, R.A.C., 2015. National Integrated Water Resources Management Support Program (ADB TA 7780), Component 3: Groundwater Assessment, Updated National Groundwater Action Plan.

Pavelic, P., Xayviliya, O., Ongkeo, O., 2014. Pathways for effective groundwater governance in the least-developed-country context of the Lao PDR. Water Int. 39, 469−485. Available from: <http://www.tandfonline.com/doi/abs/10.1080/02508060.2014.923971>.

Perttu, N., Wattanasen, K., Phommasone, K., Elming, S.-Å., 2011. Characterization of aquifers in the Vientiane Basin, Laos, using Magnetic Resonance Sounding and Vertical Electrical Sounding. J. Appl. Geophys. 73, 207−220. Available from: https://doi.org/10.1016/j.jappgeo.2011.01.003.

Serre, L.-A., 2013. Impact of Informal Irrigation on the Groundwater Resources at the Village Scale on the Vientiane Plain, Lao PDR (ISTOM Diploma Thesis) (in French).

Siebert, S., Burke, J., Faures, J.M., Frenken, K., Hoogeveen, J., Döll, P., et al., 2010. Groundwater use for irrigation − a global inventory. Hydrol. Earth Syst. Sci. 14, 1863−1880. Available from: <www.hydrol-earth-syst-sci.net/14/1863/2010/>.

Sisouvanh, A., Bouapao, L., Sayalath, C., Phosalath, S., Phounvixay, V., Ngonvorarath, V., et al. 2013. Institutional arrangements: Policies and administrative mechanisms for water governance in the Lao People's Democratic Republic. In: Mekong Project 4 on Water Governance, Challenge Program for Water and Food Mekong.

Somphone, K., Xayviliya, O., 2017. Climate change and groundwater resources in Lao PDR. J. Groundw. Sci. Eng. 5 (1), 53−58.

Suhardiman, D., Giordano, M., Leebouapao, L., Keovilignavong, O., 2015. Farmers' strategies as building block for rethinking sustainable intensification. Agric. Hum. Values 33 (3), 563−574.

Vinckevleugel, J. 2015. Institutional Arrangements in Local Groundwater Governance: A Case Study of the Groundwater Resource in Phousan Village, in Phonhong District in Vientiane Province, Laos (Master thesis). Department of Earth and Water and Environment Sciences, University of Montpellier II.

Viossanges, M., Pavelic, P., Rebelo, L.-M., Lacombe, G., Sotoukee, T., 2018. Regional mapping of groundwater resources in data-scarce regions: the case of Laos. Hydrology 5 (1), 2. Available from: https://doi.org/10.3390/hydrology5010002.

Vote, C., Newby, J., Phouyyavong, K., Inthavong, T., Eberbach, P., 2015. Trends and perceptions of rural household groundwater use and the implications for smallholder agriculture in rain-fed Southern Laos. Int. J. Water Resour. Dev. 31 (4), 558−574. Available from: <http://www.scopus.com/inward/record.url?eid = 2-s2.0-84945478114&partnerID = tZOtx3y1>.

Wiszniewski, I., Lertsirivorakul, R., Merrick, N., Milne-Home, W., Last, R., 2005. Groundwater flow section modelling of salinisation processes in the Champhone catchment, Savannakhet province, Lao PDR. Proceedings of the 2005 International Conference on Simulation and Modelling, Eds. V. Kachitvichyanukul, U. Purintrapiban & P. Utayopas.

World Bank, 2016. World Development Indicators. Retrieved from: <http://data.worldbank.org/country/lao-pdr>.

World Bank and GFDRR (Global Facility for Disaster Reduction and Recovery), 2011. Vulnerability, Risk Reduction, and Adaptation to Climate Change, Lao PDR. Country Profiles: World Bank Group.

Chapter 34

Groundwater sustainability and security in South Asia

Soumendra Nath Bhanja[1] and Abhijit Mukherjee[2,3]

[1]*Interdisciplinery Centre for Water Research, Indian Institute of Science, Bangalore, India,* [2]*Department of Geology and Geophysics, Indian Institute of Technology (IIT) Kharagpur, Kharagpur, India,* [3]*Applied Policy Advisory for Hydrosciences (APHA) group, School of Environmental Science and Engineering, Indian Institute of Technology (IIT) Kharagpur, Kharagpur, India*

34.1 Introduction

Groundwater is the largest source of freshwater across the globe and contributes significantly toward the food security (Aeschbach-Hertig and Gleeson, 2012). The per capita distribution of groundwater is extremely heterogeneous with large dependence on population density. Several past studies report groundwater depletion across the globe over the years (Rodell et al., 2009, 2018; Reager and Famiglietti, 2013; Voss et al., 2013; Richey et al., 2015; Bhanja et al., 2017b; Bhanja and Mukherjee, 2019; Bhanja et al., 2018). Groundwater processes are not so straightforward like the surface water. For example, flow is controlled by groundwater potential, which is a factor of land gradient, subsurface material's property, gravity, and viscosity. Groundwater recharge in the surficial soil layer can happen directly after a rainfall event though the unsaturated soil zone. Due to the recharge process complexity, deeper groundwater can be termed a nonrenewable resource in human timescale. However, groundwater resources are not properly managed in parts of the South Asia (Mukherjee et al., 2015). As a result, large parts of South Asia show diminishing groundwater trends in recent years linked to uncontrolled/unsustainable use (Mukherjee, 2018; Rodell et al., 2018; Bhanja et al., 2020). Groundwater depletion factors include incompetent water use practices, poor maintenance of irrigation systems, absence of water withdrawal policy, and very low pricing of water (Bhanja et al., 2017b; Mukherjee and Bhanja, 2019).

South Asia constitutes eight countries Afghanistan, Bangladesh, Bhutan, India, Myanmar, Nepal, Pakistan, and Sri Lanka (Fig. 34.1). Apart from hosting the densest global population, the region is characterized by highly heterogeneous topography with highest places on the globe to large planes extending to thousands of kilometers (Fig. 34.1). Groundwater availability and recharge in South Asia is controlled by various factors, including population-linked withdrawal, availability of rechargeable water, and subsurface property (Bhanja et al., 2019a). Population density lies within <100 to as high as 10,000 persons/km^2 in significant number of cities in South Asia (Fig. 34.2). Keeping in mind rapid population growth, quantification of groundwater availability and proper management strategies are required to address groundwater security in near future (Bhanja and Mukherjee, 2019). Further, groundwater provides the baseflow for sustenance of most rivers, for example, Ganges river (Mukherjee et al., 2018b) and sustainable drinking water sources (Mukherjee et al., 2019). Therefore groundwater security issue in South Asia is a crucial component of global groundwater issues that may aggravate in near future (Mukherjee et al., 2020).

34.2 Data

34.2.1 Study region

Several major rivers flow through the study region and that are supporting the water demand over the years (Fig. 34.3). The Indus−Ganges−Brahmaputra basin regions support the densest global population comparing any other areas of similar size (Figs. 34.2 and 34.3). Extended regions with groundwater contamination with either arsenic, fluoride, or salinity are observed in parts of South Asia (Fig. 34.3). In addition, fecal coliforms are also reported in large tracts of India (Mukherjee et al., 2019). The areas also coincide with the densely populated regions. It is really a challenging

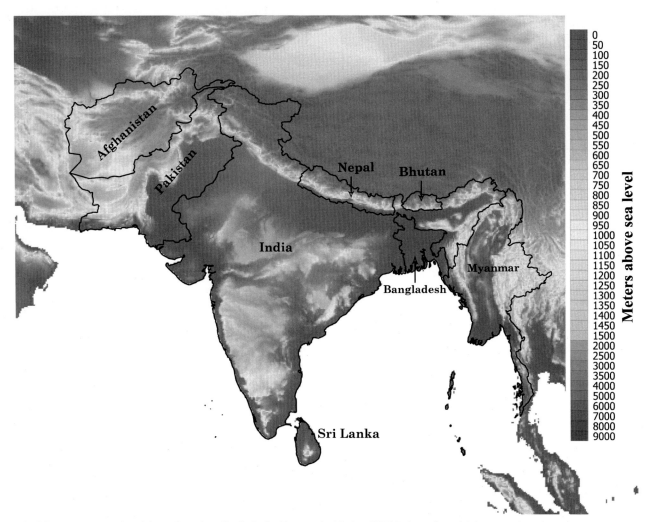

FIGURE 34.1 Topography of the study region. *Shuttle Radar Topography Mission (SRTM): https://www2.jpl.nasa.gov/srtm/.*

task to allocate adequate amount of contamination-free groundwater to every citizens in near future. Annual mean (1979–2012) precipitation rates show highly heterogeneous rates with values from as low as <200 to >5000 mm/year in parts of the study area (Fig. 34.4). Heterogeneity in precipitation rates gives rise to heterogeneity in meteoric inflow of rechargeable water.

34.2.2 WaterGAP3 model

In order to investigate the evapotranspiration rates and groundwater recharge, we used the simulation output of WaterGAP3 (Water—Global Assessment and Prognosis-3) global hydrological model. It is an integrated, grid-based, global-scale freshwater resource simulation model with multiple hydrological capabilities. WaterGAP3 comprises a rainfall–runoff model, a water quality model, and five water use models (Döll et al., 2009; Flörke et al., 2013). The model solves storage equations to simulate water storages in the following components: canopy, snowpack, soil, renewable groundwater, and surface water bodies.

34.3 Results and discussions

34.3.1 Evapotranspiration and groundwater recharge

Evapotranspiration controls the amount of infiltrating water. Higher rates of evapotranspiration reduce groundwater recharge. Annual evapotranspiration rates are obtained through WaterGAP3 model simulation. Linear regression

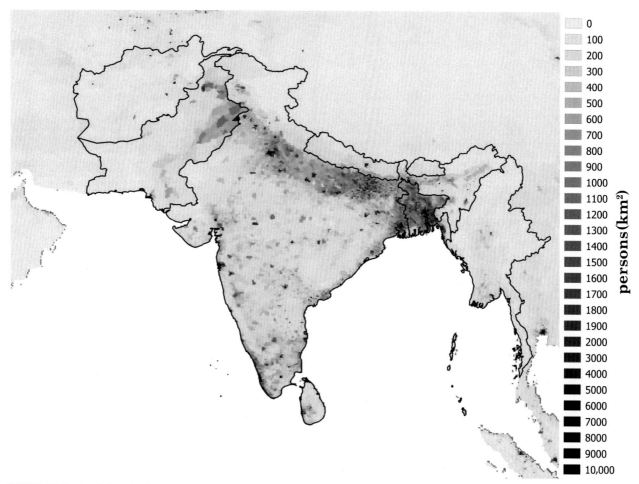

FIGURE 34.2 Population density map of the region during 2000. *Socioeconomic Data and Applications Center (SEDAC): https://sedac.ciesin. columbia.edu/data/set/gpw-v4-population-density-rev11.*

coefficient of evapotranspiration between 1979 and 2012 is shown in Fig. 34.5. The densely populated plains of Indus−Ganges−Brahmaputra rivers show increasing trend of evapotranspiration in 1979−2012. This infers further reduction of rechargeable water availability.

Modeled groundwater recharge data show heterogeneous rates with higher values in eastern parts (Fig. 34.6A). Modeled recharge rates more or less follow precipitation patterns (Fig. 34.4) due to lack of subsurface characterization in the model. Linear regression coefficient of recharge rates between 1979 and 2012 is shown in Fig. 34.6B. Again, the data show a net reduction in recharge between 1979 and 2012 in Indus−Ganges−Brahmaputra river basins. The estimates, if real, are the genuine concerns for the groundwater managers as some sustainable solutions are required in quick time to protect groundwater in those basins.

34.3.2 Contamination issues

The groundwater contamination issues are rising over the years. Large parts of the study area are already subjected to groundwater contamination (Fig. 34.3). Apart from the previously known contaminants (i.e., arsenic, fluoride, salinity, and coliforms), emerging contaminants linked to human activities are reported in recent studies (Saha and Alam, 2014). Synthetic organic substances are occurring now in natural water within the study region (Duttagupta et al., 2020). Recent estimates on Indus−Ganges−Brahmaputra basin show that approximately 60% of the groundwater in contaminated by arsenic and/or salinity: the water is unusable for day-to-day purpose (MacDonald et al., 2016).

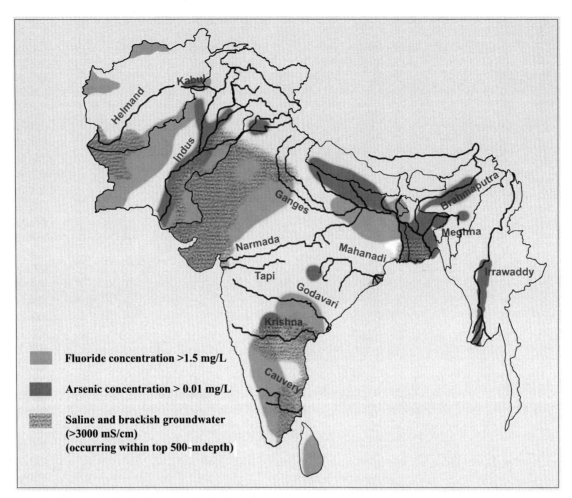

FIGURE 34.3 Groundwater contamination map and major rivers.

34.3.3 Population

Population rise and water demand related to it is the major concern to tackle groundwater issues in near future. With limited surface water availability, people are increasingly dependent upon groundwater (Mukherjee et al., 2018). Warming linked increasing evapotranspiration further reduces the water availability in the region (Fig. 34.5). Although some of the high-yielding aquifers are present in the region, per capita water availability is diminishing day by day.

34.4 Summary and way forward

The abovementioned facts of groundwater crises might exacerbate upon including future surface water scarcity linked with rapid melting of glaciers. Comparatively higher surface runoff is predicted using IPCC scenarios during 2090–99 on comparing data from last century (Bates et al., 2008), exposing the regions to higher flood occurrence. The estimates are sometimes highly uncertain on Himalayan region due to uncertain meteorological data (Yoon et al., 2019). Extreme precipitation and flash floods are increasing in the mid-altitude and foothills of Himalayas. East–west gradient of groundwater recharge is also required to be kept in mind on developing groundwater management policies. Moreover, it is very challenging to provide contaminant-free water to citizens. We envisage future researches should be directed toward the per capita groundwater availability and its prediction. Proper management strategies should be framed based on these analyses.

As a first step, groundwater monitoring should be conducted over the region with good spatial and temporal resolution. The Central Groundwater Board of India maintains an excellent groundwater monitoring network in the country (Bhanja et al., 2017a). Similar network is required to be built in other countries too. Gravity Recovery and Climate Experiment (GRACE) sensors were designed to monitor groundwater storage from space; however, the resolution is

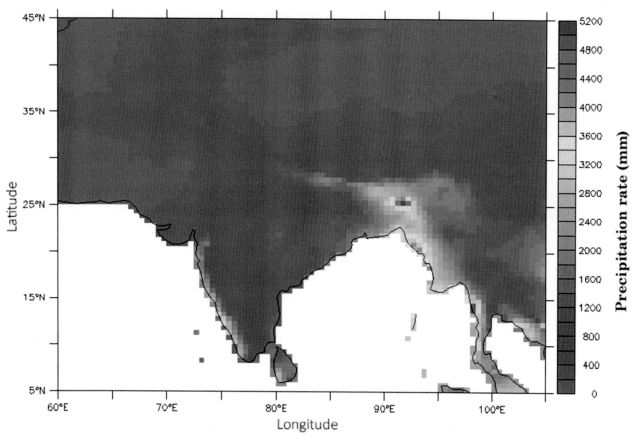

FIGURE 34.4 Annual mean precipitation rates (mm, 1979–2012) using the ECMWF ERA-Interim reanalysis data (Dee et al., 2011).

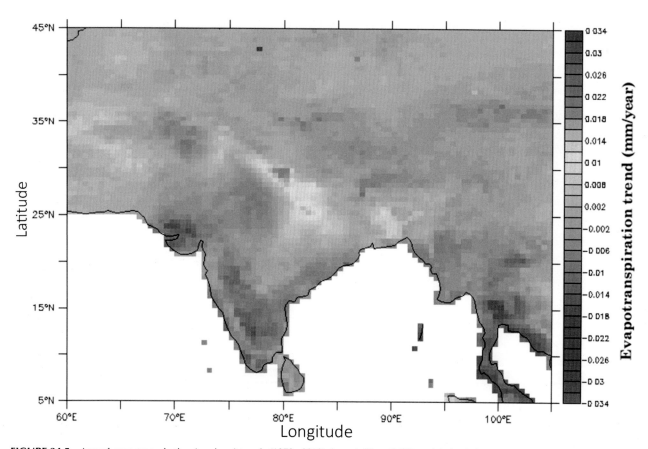

FIGURE 34.5 Annual evapotranspiration (mm/year) trends (1979–2012) through WaterGAP3 model simulation.

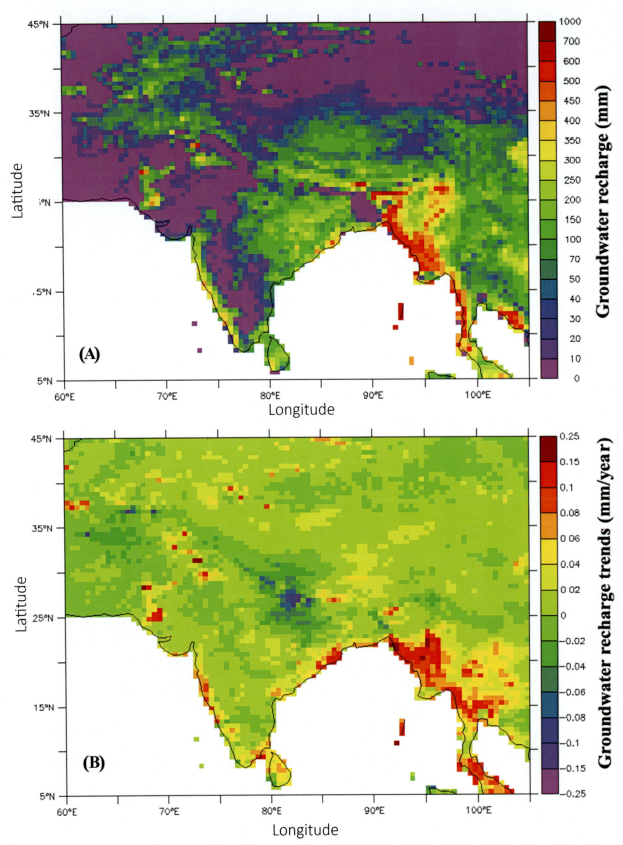

FIGURE 34.6 Maps of (A) annual mean groundwater recharge rates (mm, 1979–2012) through WaterGAP3 model simulation; (B) annual groundwater recharge (mm/year) trends (1979–2012).

very coarse (Landerer and Swenson, 2012). GRACE data provide promising estimates of groundwater storage variations in large river basin regions (Bhanja et al., 2016). Alternatively, other remote sensing techniques such as vegetation indices could be used for qualitative estimation of groundwater storage in high spatial resolution (Bhanja et al., 2019b). Data assimilation techniques using GRACE data on physical models (Girotto et al, 2017; Li et al., 2019) and artificial intelligence techniques (Sun et al., 2019) are proved to be promising to increase the spatial resolution of groundwater storage monitoring.

Acknowledgments

S.N.B. acknowledges support from the Indian Institute of Science in the form of CV Raman Postdoctoral Fellowship for partly carrying out the study. WaterGAP model simulation outputs are obtained from http://www.earth2observe.eu/.

References

Aeschbach-Hertig, W., Gleeson, T., 2012. Regional strategies for the accelerating global problem of groundwater depletion. Nat. Geosci. 5, 853–861.

Bates, B., Kundzewicz, Z., Wu, S., 2008. Climate change and water. In: Technical Paper of the Intergovernmental Panel on Climate Change. Intergovernmental Panel on Climate Change Secretariat, Geneva, Switzerland.

Bhanja, S.N., Mukherjee, A., Saha, D., Velicogna, I., Famiglietti, J.S., 2016. Validation of GRACE based groundwater storage anomaly using in-situ groundwater level measurements in India. J. Hydrol. 543, 729–738.

Bhanja, S.N., Rodell, M., Li, B., Saha, D., Mukherjee, A., 2017a. Spatio-temporal variability of groundwater storage in India. J. Hydrol. Available from: https://doi.org/10.1016/j.jhydrol.2016.11.052.

Bhanja, S.N., Mukherjee, A., Wada, Y., Chattopadhyay, S., Velicogna, I., Pangaluru, K., et al., 2017b. Groundwater rejuvenation in parts of India influenced by water-policy change implementation. Sci. Rep. 7.

Bhanja, S.N., Mukherjee, A., Rodell, M., 2018. Groundwater storage variations in India. Groundwater of South Asia. Springer, Singapore, pp. 49–59.

Bhanja, S.N., Mukherjee, A., 2019. In situ and satellite-based estimates of usable groundwater storage across India: implications for drinking water supply and food security. Adv. Water Resour. 126, 15–23.

Bhanja, S.N., Mukherjee, A., Rangarajan, R., Scanlon, B.R., Malakar, P., Verma, S., 2019a. Long-term groundwater recharge rates across India by in situ measurements. Hydrol. Earth Syst. Sci. 23 (2), 711–722.

Bhanja, S.N., Malakar, P., Mukherjee, A., Rodell, M., Mitra, P., Sarkar, S., 2019b. Using satellite-based vegetation cover as indicator of groundwater storage in natural vegetation areas. Geophys. Res. Lett. 46 (14), 8082–8092.

Bhanja, S.N., Mukherjee, A., Rodell, M., 2020. Groundwater storage change detection from in situ and GRACE-based estimates in major river basins across India. Hydrol. Sci. J. 65 (4), 650–659.

Dee, D.P., Uppala, S.M., Simmons, A.J., Berrisford, P., Poli, P., Kobayashi, S., et al., 2011. The ERA-Interim reanalysis: configuration and performance of the data assimilation system. Q. J. Roy. Meteor. Soc. 137, 553–597. Available from: https://doi.org/10.1002/qj.828.

Döll, P., Fiedler, K., Zhang, J., 2009. Global-scale analysis of river flow alterations due to water withdrawals and reservoirs. Hydrol. Earth Syst. Sci. 13, 2413–2432. Available from: https://doi.org/10.5194/hess-13-2413-2009.

Duttagupta, S., Mukherjee, A., Bhattacharya, A., Bhattacharya, J., 2020. Wide exposure of persistent organic pollutants (PoPs) in natural waters and sediments of the densely populated Western Bengal basin, India. Sci. Total Environ. 717, 137187.

Flörke, M., Kynast, E., Bärlund, I., Eisner, S., Wimmer, F., Alcamo, J., 2013. Domestic and industrial water uses of the past 60 years as a mirror of socio-economic development: a global simulation study. Glob. Environ. Change 23, 144–156. Available from: https://doi.org/10.1016/j.gloenvcha.2012.10.018.

Girotto, M., De Lannoy, G.J., Reichle, R.H., Rodell, M., Draper, C., Bhanja, S.N., et al., 2017. Benefits and pitfalls of GRACE data assimilation: A case study of terrestrial water storage depletion in India. Geophys. Res. Lett. 44 (9), 4107–4115.

Landerer, F.W., Swenson, S.C., 2012. Accuracy of scaled GRACE terrestrial water storage estimates. Water Resour. Res. 48, W04531. Available from: https://doi.org/10.1029/2011WR011453.

Li, B., Rodell, M., Kumar, S., Beaudoing, H.K., Getirana, A., Zaitchik, B.F., et al., 2019. Global GRACE data assimilation for groundwater and drought monitoring: advances and challenges. Water Resour. Res. 55 (9), 7564–7586.

MacDonald, A.M., Bonsor, H.C., Ahmed, K.M., Burgess, W.G., Basharat, M., Calow, R.C., et al., 2016. Groundwater quality and depletion in the Indo-Gangetic Basin mapped from in situ observations. Nat. Geosci. 9 (10), 762–766.

Mukherjee, A., Saha, D., Harvey, C.F., Taylor, R.G., Ahmed, K.M., Bhanja, S.N., 2015. Groundwater systems of the Indian sub-continent. J. Hydrol. Reg. Stud. 4, 1–14.

Mukherjee, A., 2018. Groundwater of South Asia. Springer.

Mukherjee, A., Bhanja, S.N., Wada, Y., 2018. Groundwater depletion causing reduction of baseflow triggering Ganges river summer drying. Sci. Rep. 8 (1), 1–9.

Mukherjee, A., Bhanja, S.N., 2019. An untold story of groundwater replenishment in India: impact of long-term policy interventions. In: Singh, A., et al., (Eds.), Water Governance: Challenges and Prospects. Springer Nature Singapore Pte Ltd, pp. 205–218.

Mukherjee, A., Duttagupta, S., Chattopadhyay, S., Bhanja, S.N., Bhattacharya, A., Chakraborty, S., et al., 2019. Impact of sanitation and socio-economy on groundwater fecal pollution and human health towards achieving sustainable development goals across India from ground-observations and satellite-derived nightlight. Sci. Rep. 9 (1), 1–11.

Mukherjee, A., Scanlon, B., Aureli, A., Langan, S., Guo, H., McKenzie, A., 2020. Global Groundwater: Source, Scarcity, Sustainability, Security and Solutions, first ed. Elsevier. ISBN: 9780128181720.

Reager, J.T., Famiglietti, J.S., 2013. Characteristic mega-basin water storage behavior using GRACE. Water Resour. Res. 49 (6), 3314–3329.

Richey, A.S., Thomas, B.F., Lo, M.-H., Famiglietti, J.S., Reager, J.T., Voss, K., et al., 2015. Quantifying renewable groundwater stress with GRACE. Water Resour. Res. 51, 5217–5238. Available from: https://doi.org/10.1002/2015WR017349.

Rodell, M., Velicogna, I., Famiglietti, J.S., 2009. Satellite-based estimates of groundwater depletion in India. Nature 460, 999–1002.

Rodell, M., Famiglietti, J.S., Wiese, D.N., Reager, J.T., Beaudoing, H.K., Landerer, F.W., et al., 2018. Emerging trends in global freshwater vailability. Nature 557 (7707), 651–659.

Saha, D., Alam, F., 2014. Groundwater vulnerability assessment using DRASTIC and Pesticide DRASTIC models in intense agriculture area of the Gangetic plains, India. Environ. Monit. Assess. 186 (12), 8741–8763.

Sun, A.Y., Scanlon, B.R., Zhang, Z., Walling, D., Bhanja, S.N., Mukherjee, A., et al., 2019. Combining physically based modeling and deep learning for fusing GRACE satellite data: can we learn from mismatch? Water Resour. Res. 55 (2), 1179–1195.

Voss, K.A., Famiglietti, J.S., Lo, M., Linage, C., de, Rodell, M., Swenson, S.C., 2013. Ground-water depletion in the Middle East from GRACE with implications for transboundary water management in the Tigris-Euphrates-Western Iran region. Water Resour. Res. 49, 904–914. Available from: https://doi.org/10.1002/wrcr.20078.

Yoon, Y., Kumar, S.V., Forman, B.A., Zaitchik, B.F., Kwon, Y., Qian, Y., et al., 2019. Evaluating the uncertainty of terrestrial water budget components over High Mountain Asia. Front. Earth Sci. 7, 120.

Chapter 35

Role of measuring the aquifers for sustainably managing groundwater resource in India

Dipankar Saha[1], Sanjay Marwaha[2] and S.N. Dwivedi[2]
[1]*Formerly at the Central Ground Water Board, Government of India, Faridabad, India,* [2]*Central Ground Water Board, Faridabad, India*

35.1 Introduction

In India, groundwater is a crucial resource for socioeconomic development of the country. Globally, India stands much ahead of other countries in groundwater extraction, at about 250 billion m^3 (BCM) annually, followed by the United States and China. The Indian subcontinent shows one of the most diverse hydrogeological framework in the world, considering range of hydraulic characteristics and wide variation in resource availability (Saha et al., 2017; Saha and Roy, 2018). In an assessment on the future water-stressed hotspots, OCED has identified India as one of the three leading countries susceptible to face most severe water risks, along with the United States (Central Valley) and China (Northern China Province), which is going to create a significant negative impact on food and drinking water security with collateral socioeconomic damages (OECD, 2017). Mukherjee et al. (2020) have discussed the nature and extent of the global groundwater stress, and Chen et al. (2014) assessed the decadal (between January 2003 and December 2012) net annual loss of 20.4 ± 7.1 km^3 groundwater resource from the northwestern India, known as the food bowl of the country.

Agriculture is the main consumer of groundwater in India. The latest resource assessment results issued by the Govt. of India reveal that 89% of our annual groundwater extraction of 431.8 BCM is consumed by irrigation (CGWB, 2017). Over the last four decades the area brought under irrigation has increased at the rate of 1.86%. Interestingly, the area irrigated by groundwater source is growing at 2.95% annually, while canal irrigated area is expanding at a rate of 0.6% only. Presently, about 62% of the net irrigated area (68 million ha) is catered from groundwater resources, besides meeting up ~85% in rural water supply and more than 50% of demand from urban sector (Saha et al., 2018). The most interesting part of increasing dependence on groundwater for irrigation is zooming numbers of wells owned by farmers, which is often referred as *atomistic development* (Kulkarni et al., 2014; Vijay Shankar et al., 2011). The value of groundwater used for irrigation per year is 7.6–8.3 billion US$ per year, and the size of the groundwater irrigation economy of India would be some US$ 75–80 billion (Shah, 2009).

The dependence on groundwater is taking toll also, mainly in two domains: (1) large areas have come under overexploitation, where annual extraction exceeds annual recharge and (2) deteriorating chemical quality and increasing pollution of groundwater such as increasing salinity and incidence of high fluoride, and arsenic (NITI Aayog, 2018; Saha et al., 2018; Saha and Sahu, 2015; Banerjee et al., 2012). Though considering the country as whole, extraction of groundwater is not increasing that rapidly in last decade; however, enhanced dependence on groundwater would be experienced at many places because of factors such as urbanization and changing lifestyle, rapid industrial growth, and lastly imposed uncertainty in rainfall pattern because of climate change (Sharma, 2009). There is an urgent need to manage this precious natural resource sustainably. This calls for holistic understanding of the aquifer systems, groundwater quality, and resource availability in space and time domain.

35.2 Regional aquifer framework

World-wide Hydrogeological Mapping and Assessment Programme (WHYMAP, 2008) indicated that the aquifer system in Indo–Ganga–Brahmaputra Plains is one of the most potential aquifer systems in the world. Other than this unit the remaining

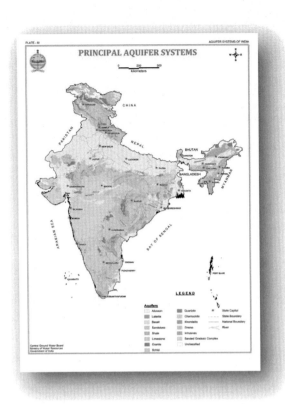

FIGURE 35.1 Principal aquifer systems in India (CGWB, 2012).

part of India is marked as "complex hydrogeological structure" (WHYMAP, 2008). However, the alluvial plains along large rivers and the deltaic regions on the east coast also hold potential aquifer systems (Saha et al., 2018). In tune with the wide variation of geology and terrain, the aquifers in India exhibit a rich diversity. Researchers proposed different classifications of the aquifer systems in India (Saha et al., 2019). CGWB has adopted a classification system after reviewing all existing classifications and based on the data generated so far through various pan India hydrogeological/geophysical/hydrogeochemical surveys and exploration activities (Fig. 35.1). This classification has proposed 14 *principal aquifer systems*, which has been further subdivided into 42 *major aquifer systems* based on lithologic and hydrogeologic characters (CGWB, 2012).

Alluvium is the most widespread aquifer system in the country, spanning over 31% geographical area (Table 35.1). This unit is made up of unconsolidated sediments laid down under fluvial, deltaic, lacustrine, or eolian activities. They are confined in the states of Assam, West Bengal, Bihar, Uttar Pradesh, Punjab, Haryana, and Rajasthan.

This unit forms the most prolific, multitiered aquifer system and is intensely extracted for groundwater resources. Geographically, the next most widely distributed aquifer system (17% of the country area) is *basalt* (Sl no. 3), which forms hard rock aquifers but uniquely characterized by layered aquifers made up of the upper vesicular zone of each flow (Saha and Agrawal, 2006). It is spread over mainly in the states of Maharashtra, Gujarat, Karnataka, and Madhya Pradesh and marked with low- to medium-yield potential.

Around 41% of the country area is covered by a wide array of hard rock aquifers, which includes *gneiss*, *Banded Gneissic Complex*, *schists*, *granites*, *quartzites*, *charnokites*, *khondalites*, and *intrusives* (CGWB, 2012). The aquifer systems from these groups are characterized by vertically interconnected, two-layered aquifers. The upper one is moderately potential weathered zone, followed by low to moderately potential fractured/jointed aquifer system, which continue even up to 500 m bgl or more (Shah, 2012). It covers mainly the states of Jharkhand, Chhattisgarh, Madhya Pradesh, Telangana, Andhra Pradesh, Tamil Nadu, Karnataka, and Kerala.

35.3 Spatiotemporal behavior of hydraulic heads and replenishable resources

Water levels in India are being monitored by the CGWB and the state departments dealing with groundwater. CGWB monitors water levels from permanently established network of about 20,000 observation wells across the length and breadth of the country. Considering that the southwestern monsoon is the main contributor for recharging groundwater resource, water-level measurements are planned four times a year: January, May (premonsoon), August (midmonsoon), and November (postmonsoon). The monitoring wells are both piezometers (borewells) and dug wells.

TABLE 35.1 Geographical coverage of different principal aquifers in the country.

Sl. no.	Principal aquifer system	Area (km²)	(% of total area)
1	Alluvium	9,45,754	29.82
2	Laterite	40,926	1.29
3	Basalt	5,12,302	16.15
4	Sandstone	2,60,416	8.21
5	Shale	2,25,397	7.11
6	Limestone	62,899	1.98
7	Granite	1,00,992	3.18
8	Schist	1,40,935	4.44
9	Quartzite	46,904	1.48
10	Charnockite	76,360	2.41
11	Khondalite	32,914	1.04
12	BGC	4,78,383	15.09
13	Gneiss	1,58,753	5.01
14	Intrusives	19,896	0.63

Source: After Saha, D., Marawaha, S., Dwivedi, S.N., 2018. National Aquifer Mapping and Management Plan—a step towards water security in India, 2018. In: Singh, A., Saha, D., Tyagi, A.C. (Eds.), Water Governance: Challenges and Opportunities, Springer. ISBN: 978-981-13-2699-8.

In an annual cycle, deepest water levels are recorded in May and the shallowest in August. During postmonsoon 2019 an analysis of CGWB hydrograph monitoring station data indicates that 27% of wells have water level <2 m bgl, whereas 39%, 21%, and 8% wells recorded water levels in the depth range of 2–5, 5–10, and 10–20 m bgl, respectively. A small percentage (5%) of wells show water level >20 m bgl (Fig. 35.2). While, during premonsoon 2019, 41% wells are showing water level in the depth range of 5–10 m bgl, while 23%, 5%, and 2% wells are showing water level in the range of 10–20, 20–40, and more than 40 m bgl (Fig. 35.3).

Large swath of the country is witnessing declining water levels because of increasing extraction of groundwater. The declining trend is particularly acute in overexploited areas, where annual recharge is less than extraction (Fig. 35.4).

However, the water levels measured and presented in the maps represent the shallow aquifer systems. In view of increasing extraction from deeper aquifers, there is an urgent need for dense and continual measurement of the deeper aquifers. This is particularly important as the recharge in deeper aquifer system follows long pathways and relatively complex process in comparison to the shallow aquifers. However, there are evidences that in many parts of India, both in soft and hard rock aquifer systems, the deeper aquifers are experiencing declining trend.

35.4 How much groundwater we are extracting

About 90% of our annual groundwater extraction is used for agriculture. Other societal uses include that consume remaining 10% are domestic consumption and also industrial demand that include manufacturing industries, packaged drinking water, infrastructures, and mining activities. CGWB estimates groundwater resource availability of entire India at regular interval (cgwb.gov.in). The latest resource estimation as on 2017 reveals that out of 6584 groundwater assessment units in the country, 1034 are overexploited, annual extraction is more than annual recharge (Saha et al., 2019). In another 253 units, extraction exceeds 90% of annual recharge. In total, 4520 units are such that extraction is less than 70% of the recharge, which is considered as "safe" as per the Govt. of India Guidelines (GEC, 1997).

An analysis of data of groundwater extraction for groundwater resources assessment since 2004, 2009, 2011, 2013, and 2017 reveals that groundwater extraction has increased from 231 BCM in 2004 to 249 BCM in 2017 (Table 35.2). The groundwater extraction for irrigation has, however, marginally reduced over the years. If we consider the groundwater draft for irrigation in percentage of total extraction, it has reduced from 92% in 2004 to 90% in 2013 and further to 89% in 2017 (Fig. 35.5).

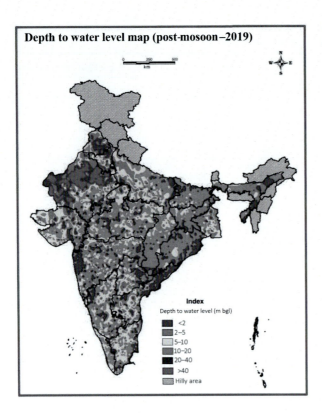

FIGURE 35.2 Postmonsoon 2019 water levels representing shallow aquifers in India.

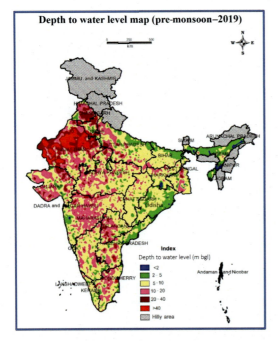

FIGURE 35.3 Premonsoon 2019 water levels representing shallow aquifers in India.

35.5 Expanding groundwater contamination

Both geogenic and anthropogenic contaminations are rampant in India. Fluoride and arsenic contaminations, two major geogenically sourced, are widely reported (CGWB, 2014). Further, salinity and iron that have excessive concentrations are also reported from various parts. High fluoride in groundwater is reported from the entire country, but more prevalent are in the states of Andhra Pradesh, Tamil Nadu, Uttar Pradesh, Gujarat, and Rajasthan, where 50%−100% of the

Role of measuring the aquifers for sustainably managing groundwater resource in India **Chapter | 35** 481

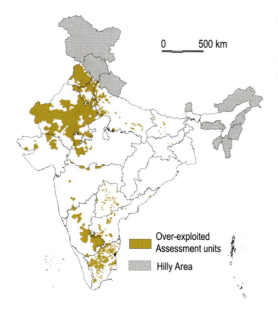

FIGURE 35.4 Areas showing long-term decline in water levels during either of the one or both premonsoon and postmonsoon periods. *Adapted from Saha, D., Roy, R., 2018. Groundwater resources of India: potential, challenges and management. In: Groundwater Development and Management, Issues and Challenges in South Asia. Groundwater Development and Management, New Delhi. doi:10.1007/978-3-319-75115-3_2.*

TABLE 35.2 Groundwater recharge and extraction in India.

	2004	2009	2011	2013	2017
Net ground water availability (BCM)	399	396	398	411	393
Annual GW draft for all uses (BCM)	231	243	245	253	249
Annual GW draft for irrigation (BCM)	213	221	222	228	222
	92	91	90%	90	89
Stage of GW extraction (%)	58	61	62	62	63

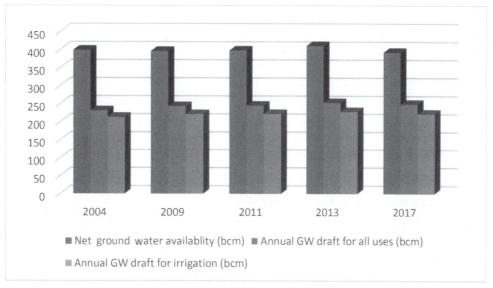

FIGURE 35.5 Groundwater extraction pattern between 2004 and 2017.

districts reported excess fluoride (CGWB, 2014). FRRDF (1999) estimated that about 66 million people in India are consuming water with fluoride level beyond the permissible limit (1.5 mg/L) prescribed by the Bureau of Indian Standards (IS 10500:2012). The limit of arsenic in drinking consumption was 50 parts per billion (ppb) (IS10500:2012), which vide an amendment in June 2015 that has been changed to 10 ppb (http://cgwb.gov.in/Documents/WQ-standards.pdf). A nationwide study by CGWB revealed arsenic concentration in excess of 10 ppb has been reported from 21 states in India (Saha et al., 2019). However, the contamination is mostly confined to the Indo–Ganga–Brahmaputra alluvial terrain covering parts of West Bengal, Uttar Pradesh, Bihar, Punjab, Haryana, and Assam (Saha, 2009; Saha and Sahu, 2015; Mukherjee et al., 2015; Banerjee et al., 2012). In these areas, arsenic concentration in deeper aquifers of the multitiered aquifer system generally remains within the limit (Saha and Sahu, 2015). Salinity, though not a serious health hazard, is present both sourced as inland salinity (caused by rock–water interaction) and coastal salinity (caused by seawater intrusion because of heavy freshwater pumping in coastal aquifers and spread of seawater during high tide or storm event) (Mukherjee et al., 2015). Inland salinity is reported mainly from western, northwestern, and southern parts of India, mainly in the states Rajasthan, Gujarat, and Haryana. Coastal salinity has been reported from the states of Tamil Nadu, West Bengal, Odisha, Gujarat, and Andhra Pradesh. Iron exceeding the limit has been reported from the entire India, covering from parts of 26 states (CGWB, 2014).

Anthropogenic contamination emanates from industrial effluents, untreated sewage and mining activities, overuse of fertilizers and pesticides, leachates from landfills, etc. The anthropogenic contaminations are localized in geographical extent. Nitrate contamination is most widely distributed throughout the world's aquifers and also in India (Saha and Alam, 2014). Nitrogenous fertilizers and domestic wastes are the most important sources of nitrate in groundwater. Instances of elevated concentrations of nitrate have been reported from parts of 21 states (Table 35.1). Besides nitrate, zinc (Zn), mercury (Hg), and manganese (Mn) have also been reported from some pockets (CGWB, 2014). Some researchers have reported high pesticides in groundwater in different parts of India (Ghose et al., 2009).

35.6 Measuring and understanding the aquifers

Traditionally, the hydrogeological surveys being carried out routinely by CGWB and different state government departments primarily focused on delineating prospective zones and construction of high-yielding exploratory wells (Saha et al., 2018). However, increasing overexploitation of aquifers and expanding groundwater contamination have compelled the policy makers to realize the need for mapping the aquifers in three-dimensional and understanding quality and resource availability in an aquifer-specific environment. There is also a growing realization on two broad aspects, (1) the huge effort and money spent by the state and the central governments on artificial recharge and rainwater harvesting would be effective if it entails a clear understanding of aquifers and (2) a policy shift is needed toward demand-side interventions, particularly for agriculture sector, which also warrants a space-time domain understanding of groundwater resources. Besides, it is also needed that millions of users are to be empowered with information on groundwater. In view of all these, Govt. of India has initiated the ambitious National Aquifer Mapping and Management Programme (NAQUIM), often referred to as *aquifer mapping*, in the year 2012 during the 12th Plan period of 2012–17. Out of 32 lakh km^2 area of the country, 24.9 lakh km^2 are taken up for mapping. Till date, ~12.5 lakh km^2 have been covered under this program. The focus was given to water-stressed areas in the states of Haryana, Punjab, Rajasthan, Gujarat, Tamil Nadu, Telangana, and Bundelkhand regions in Madhya Pradesh and Uttar Pradesh. The area covered till March 2017 is produced in Fig. 35.6.

There are four broad sets of activities under the NAQUIM programme:

1. *Data Compilation and Data Gap Analysis*: Synthesis of all existing data, placed in a GIS format, and analyses made on further data requirement as per the laid down guidelines. The quantum of data to be generated for various work items is determined.
2. *Data Generation and Integration*: Large array of data was generated by multiple activities through hydrogeological, geophysical, hydrochemical, and isotopic surveys. Besides, additional data generation through exploration through drilling, infiltration tests, pumping test, slug tests, and yield tests.
3. *Aquifers Map Preparation*: Based on data generation, lithologic models are prepared and clubbed with hydraulic parameters and hydraulic head data. Delineating aquifers, contaminated and volumetrically stressed aquifers, recharge mechanism, and capacity of the aquifers to accept recharge, etc.
4. *Aquifers Management Plan Formulation*: Sustainable management plan of groundwater resources by adopting demand and supply-side interventions. Quantification of different proposed interventions. Artificial recharge plans and alternate aquifer-based water supply in contaminated areas.

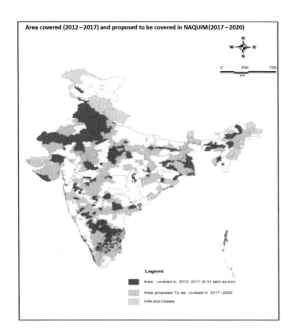

FIGURE 35.6 Areas covered under NAQUIM till March 2017 and proposed to be covered during 2017–20. *NAQUIM*, National Aquifer Mapping and Management Programme.

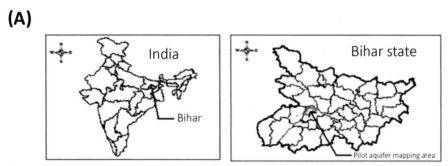

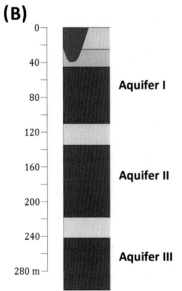

FIGURE 35.7 (A) Location of the study area, Maner block, Patna district, Bihar and (B) multilayered aquifers in the study area.

35.7 The sustainable management plan—an example

The study area covers about 600 km^2 area in the Indo-Gangetic Plain in the state of Bihar. The area is located in the confluence of the Ganga and its right-hand tributary Sone. The state capital Patna (population 3 million) is located in one corner of the area (Fig. 35.7A). The major groundwater issues in the area include:

1. existence of a potential aquifer system, with total dependence on groundwater, including urban demand of Patna;
2. high dependence on groundwater for agriculture in the semiurban and rural areas;
3. arsenic contamination of groundwater at shallow depth at places; and
4. reported decline of water levels in the urban area.

A three-tier aquifer system is delineated within 300 m bgl (Fig. 35.7B): (1) top 20–30 m thick sandy clay layer are marked with sand lenses (2–3 m thick) which sustain dug well and shallow hand pumps (water level—1.22–9.04 m bgl, electrical conductivity 380–1940 μS/cm); (2) *Aquifer I*, between 35 and 100 m bgl, holding groundwater under unconfined to semiconfined condition, exploited for irrigation and domestic purposes. Major source of recharge is rainfall and leakage from overlying formations. Water level ranges in 5.73–13.16 m bgl during premonsoon and 1.49–10.08 m bgl during postmonsoon. Transmissivity ranges between 3600 and 6000 m^2/day, and the aquifer can sustain a pumping of 200–300 m^3/h with a modest drawdown of 4–8 m.

The EC (310–1240 μS/cm) is lower than the shallow zone and is fit for irrigation and drinking uses except for arsenic beyond 50 ppb in the northeastern part bordering the Ganga River. (3) *Aquifer II*, at the depth range of 110–265 m bgl, groundwater occurs under confined condition with transmissivity ranging from 5890 to 15,480 m^2/day. The aquifer can sustain a yield of 350–500 m^3/h for a drawdown of 6–10 m. This aquifer is extensively tapped for Patna urban water supply. The hydraulic head generally remains within 10 m bgl during premonsoon, except in Patna urban area where it goes up to 16 m bgl. The water is significantly fresh (EC 315–556 μS/cm).

The management interventions have been formulated (Fig. 35.8) based on the output of numerical groundwater modeling and synthesis of the hydrogeologic information. The following strategies have been formulated:

1. Considering the increasing demand, extraction may be increased @ of 2%/annum up to year 2025 in urban zone; however, it is recommended to relocate the existing heavy-duty municipal water supply wells (present extraction 123 mcm/year) from zone B to zone C (Fig. 35.8) to avoid interference of cone of depression.
2. Reduction in groundwater draft in zone B by 50 mcm/year, by supplementing the supply from the Ganga River as envisaged by the Government of Bihar. This would have a significant positive impact on the hydraulic head for Aquifer II to the tune of 2–3 m/year in the next 5 years.
3. Increase in cropping intensity to 200% from the current 126% in zone A would lead to decline in hydraulic head in the peri-urban and rural agricultural belt by 6 m and even up to 10 m in certain pockets. A comprehensive artificial recharge plan is desirable in zone E (Fig. 35.8) where Aquifer I is connected to surface.
4. Tube wells tapping Aquifer I should be discouraged in zone D where shallow groundwater is found to be arsenic contaminated (Fig. 35.8). Increased draft would lead to lowering of hydraulic head and possible downward percolations of arsenic-rich water from the aquitard zone and spreading of contaminated areas into Aquifer I laterally and vertically.
5. Rooftop rainwater harvesting for Patna urban area (zone B) targeting recharge of Aquifer II using recharge wells. The normal injection rate in Aquifer II for a well of 6 in. diameter and 12 m long screen has been worked out as

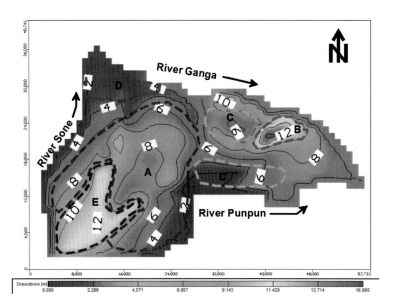

FIGURE 35.8 Management plan of the area.

~ 11 m^3/h. Considering this rate, one injection well in the urban area is sufficient to accommodate the peak recharge pulse arising out of rainfall intensity up to 40 mm/h from a roof area of ~ 400 m^2.

The management plans have been formulated so as to strike a balance between the present groundwater draft and the projected requirement with minimal impact on the hydrogeological regime in aquifer specific environment. Similarly, aquifer maps and management plans are being developed in 1:50,000 scale for the entire country for sustainable management of this precious natural resource.

35.8 Way forward

The solution toward sustainable management of groundwater resources primarily lies in addressing the issues of irrigation sector. The effective interventions are demand-side, such as enhancing irrigation efficiency and aligning the crops, cropping calendars as per the availability of water in watershed or subbasin level in space and time domain. NAQUIM outputs can help immensely in identifying the areas where demand-side interventions are essential to be implemented, as the supply-side interventions will have a limiting contribution. The Govt. of India has taken up an ambitious program to facilitate demand-side intervention through a scheme named Atal Jal Yojana. Similarly, the NAQUIM outputs also highlight the areas where surplus groundwater resource is available and aquifer hydraulics is conducive for additional supply for various societal uses.

The most sought after supply-side intervention is rainwater harvesting and artificial recharge, which have occupied the centerstage in combating the water scarcity of India. A number of government schemes promote these activities, particularly under MGNREGS, a lot of efforts are given and money put on water conservation and artificial recharge—related works. Any recharge scheme formulated should be dovetailed with the detailed understanding of the aquifer system brought out by NAQUIM to get optimum techno-economic benefit. Besides designing and locating the recharge structure, we should also take into consideration on source water for recharge, particularly in arid and semiarid regions. If required the interbasin-transferred water through the proposed river interlinking projects should be thought of as source of large-scale artificial recharge. In view of unpredictability of rainfall due to climate change, the recharge structures should also be designed to target high rainfall events. The quality of source water for artificial recharge should also be taken care of so that the aquifer is not contaminated. Remediating a contaminated aquifer is a time-consuming, costly, and often impossible job. The NAQUIM outputs also reveal evidences of high potential, large aquifer systems, enclosed in low-potential aquifers. If such aquifer system is found to be exhausted because of overexploitation, mega scale artificial recharge schemes can be taken up in a planned and coordinated manner. The urban areas are also increasingly dependent on groundwater thus creating localized overexploitation. A detailed aquifer mapping is needed, and optimally extractable groundwater resource should be worked out after all possible implementation of artificial recharge schemes for urban and peri-urban areas also.

Since groundwater is extracted by millions of farmers through privately owned wells, it is imperative that there should be a seamless flow of data from NAQUIM outputs to the farmers, in a web-based and user-friendly format. Adequate and reliable data can immensely facilitate community-based management, by helping the community in sharing the resources, planning artificial recharge and rainwater harvesting, and also imposing self-regulation if required. Central Ground Water Board and Ministry of Jal Shakti are looking this issue on priority, but it warrants further attention, and mobile app—based information dissemination system would be taken up on a large scale. In the next stage the aquifer mapping should be taken up in further detailed scale, in priority areas.

References

Banerjee, D.M., Mukherjee, A., Acharya, S.K., Chatterjee, D., Mahanta, C., Saha, D., et al., 2012. Contemporary groundwater pollution studies in India: a review. Proc. Indian Nat. Sci. Acad. 78 (3).

CGWB, 2012. Aquifer Systems of India, Ministry of Water Resources River Development and Ganga Rejuvenation. Govt. of India. Available from: <www.cgwb.gov.in/Aquifer-Atlas.html>.

CGWB, 2014. A Concept Note on Geogenic Contaminationof Ground Water in India With Special Reference to Nitrate. Ministry of Water Resources, Govt. of India, pp. 99.

CGWB. 2017. Dynamic Ground Water Resources of India as on 2013, Central Ground Water Board, Ministry of water resources River Development and Ganga Rejuvenation, Govt of India.

Chen, J., Jin, L., Zizhan, Z., Ni, S., 2014. Long-term groundwater variations in northwest India from satellite gravity measurements. Global Planet. Change 116, 130–138.

FRRDF, 1999. State of Art Report on the Extent of Fluoride in Drinking Water and the Resulting Endemicity in India. Fluorosis Research and Rural Development Foundation, New Delhi.

GEC, 1997. Ground Water Resource Estimation Methodology, Report of the Ground Water Resource Estimation Committee. Ministry of Water Resources, Govt. of India, New Delhi. https://doi.org/10.1007/978-3-319-75115-3_2.

Ghose, N.C., Saha, D., Gupta, A., 2009. Synthetic detergents (surfactants) and organoclorine pesticide signature in surface water and groundwater of Greater Kolkata, India. Journal of Water Resources and Protection 4, 290−298.

Kulkarni, H., Shah, M., Vijay Shankar, P.S., 2014. Shaping the contours of groundwater governance in India. J. Hydrol. Reg. Stud. 4. Available from: https://doi.org/10.1016/j.ejrh.2014.11.004.

Mukherjee, A., Saha, D., Harvey, C.F., Taylor, G., Ahmed, K.M., Bhanja, N., 2015. Groundwater systems of the Indian sub-continent. J. Hydrol.: Reg. Stud. 4. Available from: https://doi.org/10.1016/j.ejrh.2015.03.005.

Mukherjee, A., Scanlon, B., Aureli, A., Langan, S., Guo, H., McKenzie, A., 2020. Global Groundwater: Source, Scarcity, Sustainability, Security and Solutions, first ed. Elsevier. ISBN: 9780128181720.

NITI Aayog, 2018. Report of Working Group I, Inventory & Revival of Springs in the Himalayas for Water Security. NITI Aayog, New Delhi. Available from: <niti.gov.in>.

OECD, 2017. Water Risk Hotspots for Agriculture. Organisation for Economic Co-operation and Development (OECD).

Saha, D., Alam, F., 2014. Groundwater vulnerability assessment using DRASTIC and Pesticide DRASTIC models in intense agriculture area of the Gangetic plains, India. Environ Monit Assess. Doi:10.1007/s10661-014-4041-x.

Saha, D., Agrawal, A.K., 2006. Determination of specific yield using water balance approach—a case study of Torla Odha water shed in Deccan Trap Province, Maharashtra State, India. Hydrogeol. J. 14, 625−635.

Saha, D., 2009. Arsenic groundwater contamination in parts of middle Ganga plain, Bihar. Curr. Sci. 97 (6), 753−755.

Saha, D., Dwivedi, S.N., Senthilkumar, M., 2019. Groundwater: a critical and stressed resource in India and its sustainable management issues. In: Majumdar, P.P., Tiwari, V.M., (Eds.), Water Futures of India. IISc Press.

Saha, D., Marwaha, S., Mukherjee, A., 2017. Groundwater resources and sustainable management issues in India. In: Saha, D., Marwaha, S., Mukherjee, A. (Eds.), Clean and Sustainable Groundwater in India. SpringerISBN 978-981-10-4551-6. Available from: https://doi.org/10.1007/978-981-10-4552-3_1.

Saha, D., Marawaha, S., Dwivedi, S.N., 2018. National Aquifer Mapping and Management Plan—a step towards water security in India, 2018. In: Singh, A., Saha, D., Tyagi, A.C. (Eds.), Water Governance: Challenges and Opportunities. Springer.

Saha, D., Roy, R., 2018. Groundwater resources of India: potential, challenges and management. Groundwater Development and Management, Issues and Challenges in South Asia. Groundwater Development and Management, New Delhi. Available from: https://doi.org/10.1007/978-3-319-75115-3_2.

Saha, D., Sahu, S., 2015. A decade of investigations on groundwater arsenic contamination in Middle Ganga Plain, India. Environ. Geochem. Health 38. Available from: https://doi.org/10.1007/s10653-015-9730-z.

Shah, T., 2012. Community response to aquifer development: distinct patterns in India's alluvial and hard rock aquifer areas. Irrig. Drain. 61. Available from: https://doi.org/10.1002/ird.1656.

Shah, T., 2009. India's ground water irrigation economy: the challenge of balancing livelihoods and environment. Bhujal N. 24 (4).

Sharma, K.D., 2009. Groundwater management for food security. Curr. Sci. 96 (11).

Vijay Shankar, P.S., Kulkarni, H., Krishnan, S., 2011. India's groundwater challenge and the way forward. Econ. Pol. Wkly: EPW (2), January 8.

WHYMAP, 2008. World-Wide Hydrogeological Mapping and Assessment Programme. UNESCO. <https://en.unesco.org/themes/water-security/hydrology/programmes/whymap>.

Further reading

Saha, D., Ray, R.K., 2018. Groundwater resources of India. In: Sikdar, P.K. (Ed.), Groundwater Development and Management. Capital Publishing Company, New Delhi, India.

Saha, D., Sahu, S., Chandra, P.C., 2010. Arsenic-safe alternate aquifers and their hydraulic characteristics in contaminated areas of Middle Ganga Plain, Eastern India. Environ. Monit. Assess. 175. Available from: https://doi.org/10.1007/s10661-010-1535-z.

Saha, D., Zahid, A., Shrestha, S.R., Pavelic, P., 2016. Ground water resources in the Ganges River Basin. In: Bharati, L., Sharma, B.R., Smakthin, V. (Eds.), The Ganges River Basin: Status and Challenges in Water, Environment and Livelihoods. Routledge.

Chapter 36

Balancing livelihoods and environment: political economy of groundwater irrigation in India*

Tushaar Shah[1], Abhishek Rajan[2] and Gyan P Rai[2]
[1]*Institute of Rural Management Anand, Anand, India,* [2]*International Water Management Institute (IWMI)-Tata Water Policy Program, Anand, India*

36.1 Evolution of Indian irrigation

A stagnant agricultural economy and its role in weakening rural demand has emerged, during recent years, as a stumbling block in India's otherwise spectacular growth story. It is widely perceived that the lack of irrigation facilities and poor public investments in irrigation development are the major barriers behind slowdown in agriculture. In the last two decades, Government of India (GoI) has countered this by investing massively in Accelerated Irrigation Benefits Program (AIBP). Even in the flagship irrigation scheme of GoI—*Pradhan Mantri Krishi Sinchai Yojana* (Prime Minister's Agricultural Irrigation Scheme)—the central government has laid great stress on stepping up the pace of irrigation development by focusing on major and medium irrigation projects through AIBP. However, despite the massive investments in AIBP, the area irrigated by public irrigation systems in India has stagnated, even declined leaving irrigation planners bewildered (Shah et al., 2016).

Irrigation has always been central to life and society in the plains of South Asia (Shah, 2009). According to Alfred Deakin, a three-time Australian prime minister and an irrigation enthusiast of the early 20th century, British India was the world's irrigation champion with 12 million ha of irrigated land, which was higher than any other country of that time (The Age, 1891). This is not surprising. In a normal year, India receives 4000 km^3 of rainfall precipitation, large by any standards; but a large part of it falls in eastern India. P.R. Pisharoty, one of India's leading meteorologists, pointed out that almost the entire Indian rainfall is received in just around 100 hours of precipitation per year. This type of rainfall within short bursts of time makes storage and irrigation crucial for the survival of agrarian societies (Agarwal and Narain, 1997). Therefore water-managed agriculture has been the bedrock of civilization in this part of the world since the ancient days of Indus Valley Civilization. However, India's irrigation has undergone a profound transition over the millennia.

Since the Vedic period until the early 1800s, farming communities adapted their agrarian lives to the hydrology of river basins. There are records of numerous canal networks, tanks, and artificial lakes constructed by kings and managed by village communities and local overlords (Shah, 2009). However, according to the economist Angus Maddison, the irrigated land in the Mughal era was less than 5% of the cultivated and the public irrigation system contributed very little (Maddison, 1971). With the advent of British East India Company, restoration and construction of large irrigation projects were undertaken. Area under canal irrigation rapidly expanded in the last decades of the 19th century, converting large swathes of pastoral land into intensively cultivated ones, especially in Indus—Ganga Basin (Stone, 1984). Canal irrigation became the source of economic dynamism as irrigated lands were geared toward the production of commercial crops to serve British interests (McGinn, 2009). In the new irrigation regime of British India the state replaced village communities and local landlords and controlled the irrigation systems. The colonial era left India with around 8 million ha of canal irrigated area, well managed by a highly centralized, irrigation bureaucracy. However,

---

*This paper draws heavily on several years of research on groundwater governance by the IWMI-Tata Water Policy Program with which the authors are associated.

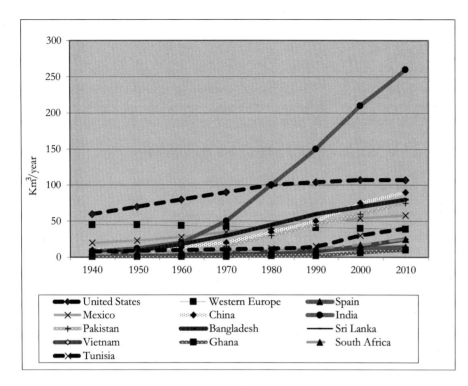

FIGURE 36.1 Growth in agricultural groundwater used in selected countries: 1940–2010. *Author's estimates based on various sources [Shah, T., 2014. Groundwater Governance and Irrigated Agriculture. Global Water Partnership, Stockholm, 2014. TEC Background Paper #19 (Shah, 2014)].*

post-independence, the collection of irrigation fee significantly shrunk and the performance of irrigation systems declined (Shah, 2011). Over time, large irrigation systems have been stuck in build-neglect-rebuild syndrome and have failed to meet the demands of Indian agriculture.

The colonial hydrology created pockets of agrarian prosperity in canal commands, which was less than 6% of total cultivated land. By the 1960s, the mounting population pressure and consecutive drought years in 1964 and 1965 heightened the need to intensify land use, and year-round irrigation access emerged as a top priority of smallholders as well as governments (Shah, 2009). Easy availability of small mechanical pumps and boring rigs provided the technological breakthrough to meet this priority. At the beginning of 1970, there was rapid expansion in groundwater irrigated areas. The number of energized irrigation wells increased from some 150,000 in 1950 to nearly 20.5 million by 2013. Around 1960, India was a relatively minor user of groundwater in agriculture compared to countries such as the United States and Spain; by 2000 the country had emerged as the global champion in groundwater irrigation—pumping around 220–230 km^3/year, as the chart in Fig. 36.1 shows. However, globally, unsustainable withdrawal and demographic pressure are threatening groundwater reserves (Mukherjee et al., 2020). Due to a meteoric rise in abstraction rates, India has emerged as the hotspot of global groundwater stress.

36.2 Changing organization of the irrigation economy

India's irrigation economy has undergone massive transformation since 1970. The state and village community have been dislodged as major irrigation players by millions of small farmers, each with this tiny captive irrigation system, as the driver of irrigation development. During 1830–1970, government was the provider of irrigation and the farmer a passive recipient. Under this model of irrigation development, irrigated areas were concentrated in canal command, and there was little irrigation outside the command area. However, come 1970s, and a new era of atomistic irrigation unfolded in which small-pump irrigation began to crowd out the large-scale gravity flow irrigation systems. Groundwater development through private tube wells made irrigation accessible within canal commands and outside.

Under the dominance of canal irrigation, crops had to wait for water to be released at the headworks and flow through a network of canals before getting irrigated; now, under private pump irrigation, water was scavenged on-demand and applied just-in-time when crops needed it most.

During 50 years since 1970, privately financed groundwater irrigation has created twice or more irrigated area in India than massive public investments in canal irrigation managed to create since 1830. According to the government

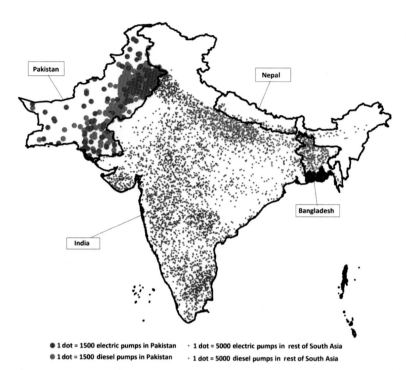

FIGURE 36.2 Energized irrigation wells in India, Pakistan, Bangladesh, and Nepal (c. 2013–14).

figures, over 60% of irrigated areas are today served by groundwater. Between 1990 and 2015, 90% of the newly irrigated area have been added by groundwater irrigation (MoA, 2015). Fig. 36.2 shows the pervasive spread of diesel and electric tube wells in the Indian subcontinent. Until 1960, Indian farmers owned just a few tens of thousands of mechanical pumps using diesel or electricity to pump water; today India has over 20 million modern water extraction structures [Government of India (GoI), 2017]. Every fourth cultivator household has a tube well, and two of the remaining three use purchased irrigation service supplied by tube well owners (Shah, 2009).

Rapid growth in population pressure on agricultural land, in turn generating pressure for intensification of land use, has been the primary demand-side driver of India's groundwater boom. But this boom has also been actively fed by supply-side factors and government policies. During early years after Independence, many states established large government run public tube well programs to provide subsidized irrigation to farmers. These generated awareness among the peasantry about benefits of groundwater irrigation—such as convenience, reliability, and timeliness. Subsidies and credit for tube wells and pumps and their easy availability too played a key role. However, arguably the most powerful supply-side factor has been the policy of many states to provide subsidized electricity to irrigation wells. It is around this nexus between subsidized farm power supply and irrigated agriculture that India's political economy of irrigation is precariously hinged upon.

36.3 Energy-irrigation nexus

The case of promoting rural electrification to accelerate agricultural production became stronger during the droughts in 1965 and 1966 when the country became embarrassingly dependent on foreign food aid to stave off mass-scale starvation. Policymakers were encouraged by productivity gains made from the Green Revolution technology by early tube well-irrigators of Punjab and Haryana. Taking a cue from this, from the late 1960s through 1970s, government policies aggressively encouraged the adoption of electric pumps and the maximum utilization of tube well water by reducing the electricity tariffs per kWh after a specified minimum use (Banerjee et al., 2014). All these measures kick-started a massive tube well irrigation economy that spread from western to eastern parts of the country. However, as groundwater irrigation accelerated, a host of second-generation problems arose which brought into bold relief the challenge of managing a highly decentralized, anarchic irrigation economy.

As the number of electric tube wells grew, State Electricity Boards (SEB) found the "transaction costs" of metering and billing the growing agricultural clientele, scattered across remote locations, burdensome (Dubash, 2007; Shah et al., 2004). Controlling malpractices by farmers, SEB's staff and the collusion of both in unethical behavior—such as under-

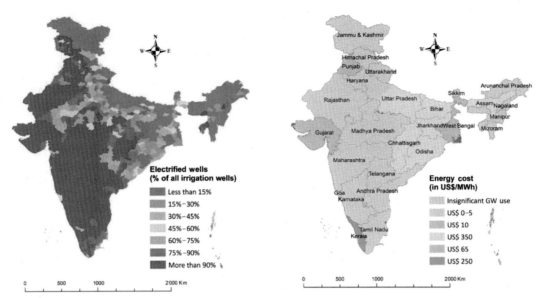

FIGURE 36.3 Energy divide in India (left) and energy cost of groundwater pumping in India (right). *Modified from Shah, T., Rajan, A., Rai, G.P., et al., 2018. Solar Pumps and South Asia's energy-groundwater nexus: exploring implications and reimagining future. Environ. Res. Lett. 13 (11), 1–13.*

reporting and under-billing of electricity consumption—further added to the staggering transaction costs. This led many SEBs to change the way power supply to tube wells was charged in between 1970s and 1980s, shifting from volumetric-pricing of electricity to flat tariffs based on the horse power (hp) rating of tube wells (Dubash, 2007; Shah et al., 2004).

This change from metered tariff to flat tariff for farm power supply set into motion a powerful chain reaction that has fundamentally reshaped Indian irrigation. The flat power tariff drastically reduced the marginal cost of pumping groundwater to near-zero levels, creating a "free power illusion" among farmers. Field studies during the 1980s showed that flat tariffs were pro-poor, making private tube well owners behave much like public utilities, offering quality irrigation service to the resource poor at an affordable price (Chambers et al., 1989; Shah, 1993; Dubash, 2007). Boom in groundwater markets enhanced equity in access to irrigation (Shah, 1993). But soon, politicians realized the potential of farm power subsidies for creating vote-banks. Flat tariffs became "upwardly sticky"; many states left flat tariffs unrevised for years, even decades (Shah et al., 2004).

All these in turn increased demand for electricity connections for tube wells. The number of electric tube wells increased from less than a million in 1980 to 8 million in the mid-1990s to 12 million around 2001 and to some 15–16 million by 2013 [Government of India (GoI), 2001, 2005, 2017]. Fig. 36.3 shows the association of electricity subsidies and adoption of electric pumps. Western and Southern Indian states—from Punjab down to Tamil Nadu—have massive density of electric pumps operating on free or nearly free electricity. This rapid adoption of grid-connected pumps has also led to massive subsidy burden on SEBs, pushing many of them to bankruptcy. In effect, it is the farm power subsidies—estimated in 2018 around US$14–15 billion—that made India the world's largest groundwater irrigation economy (Shally and Sharma, 2018).

36.4 Socioeconomic significance of the groundwater boom

Explosive growth in shallow tube wells and small pumps has democratized Indian irrigation and taken irrigation beyond command areas to the nook and corners of the country. Around 90 million rural households directly depend on groundwater for irrigation, including both pump owners and beneficiaries from vibrant groundwater markets [Government of India (GoI), 2014]. Several studies have tried to capture the scale and impact of the groundwater economy. As per one estimate (Daines and Pawar, 1987), groundwater contributed 70%–80% to the value of irrigated production. A World Bank study in the early 1990s estimated that groundwater irrigated agriculture contributed to 9% of India's GDP (World Bank, 1998). Debroy and Shah (2003) estimated the market value of groundwater used in agriculture per year to be US $ 10 billion in India and inferred the agriculture output supported by groundwater irrigation to be 4–5 times its market value in 2001–02. A recent study by Goswami et al. (2017) estimated the size of the groundwater irrigated economy to be US$55−60 billion in 2010—70% of the total irrigated agricultural output. While the bulk of public expenditure continues to flow in canal irrigation projects, groundwater irrigation has been the driving force behind agricultural transformation in the last four decades.

The pump irrigation revolution has purveyed myriad and widespread benefits to the country's farming communities. Among several things, groundwater irrigation (1) has brought spatial equality in irrigation by taking irrigated agriculture away from canal command areas; (2) offered irrigation benefits to millions of resource poor, pump-less farmers through irrigation markets; (3) has given greater water control, year-round access to the farmers leading to productivity gains; (4) mitigated the devastating impacts of drought on the region's agriculture; (5) improved farm wages and increased demand for farm labor year-round; and (6) facilitated the shift from subsistence wheat−rice cropping system to cultivation of high value crops such as fruits, vegetables, and milk, especially in dry land areas (Shah, 2009; Joshi et al., 2004). These benefits have been instrumental in alleviating poverty and stabilizing the rural livelihoods. The groundwater boom has ensured food sufficiency to the increasing population of more than one billion and made large-scale famines history.

The silent pump revolution has been the driver of the country's agricultural growth. Many parts of India, such as Telangana, which had little scope for canal irrigation experienced a booming agricultural economy based on groundwater (Vakulabharanam, 2004). West Bengal transformed from a rice-deficit to rice-surplus state due to groundwater supported *boro* (summer) paddy crop (Shah, 2009). In north-western India, intensive groundwater pumping has resolved the long-standing problems of water-logging and secondary salinization and saved massive investments in drainage and salinity management (Briscoe and Malik, 2006). In effect, the rise in groundwater irrigation has activated the subsurface storage—which remained underutilized in the past—but now captures and stores 250−270 km^3 of runoffs in a year, creating a massive physical stock of groundwater for agricultural utility.

36.5 The sustainability challenge

Against these myriad benefits, India's groundwater boom has produced many negative impacts and posed formidable policy and management challenges for (1) groundwater resource management, (2) electricity industry, and (3) public irrigation systems.

Rapid expansion in electric tube wells has resulted in inexorable increase in the stress on groundwater aquifers in western and southern India where irrigated agriculture has come to depend on unsustainable exploitation of groundwater (see Fig. 36.4). In vast areas of Punjab, Haryana, western UP, Rajasthan, and Gujarat with alluvial aquifers, groundwater levels have experienced a secular decline. NASA's GRACE data showed rapid groundwater depletion in the Indian states of Rajasthan, Punjab, and Haryana, where groundwater withdrawal exceeds recharge by 17.7 ± 4.5 km^3/year (Rodell et al., 2009). Other global studies have also identified north-western and southern India as the hotspots of groundwater depletion (Wada et al., 2010). Nationwide assessment by Central Groundwater Board (CGWB) (2017) found 30% of groundwater blocks in the country to be in the semicritical, critical, or overexploited category. A World Bank study estimated that 60% of groundwater resources will be in a critical state by 2030 in a business-as-usual scenario (World Bank, 2010).

Since water yields of diesel pumps dwindle rapidly at depth greater than 20 m, falling groundwater levels in the west and south has only deepened the farmers' dependence on electric submersible tube wells. Deepening tube wells and increasing the hp of the pump beyond the approved load became the dominant strategies of farmers. In hard rock peninsular India, where going deep was of little use in improving water availability in wells, adaptive strategies used by farmers were built around the "free power illusion" involving multiple lifts of water. This has further solidified tube well owners as a powerful interest group persistently demanding electricity subsidies (Shah et al., 2004). Farm power supply policies and groundwater depletion became locked in a vicious cycle in western and southern states where much irrigated agriculture came to depend on pumping depleted aquifers with free power (Dubash, 2007). This downward spiral made it challenging to control, much less reverse, groundwater depletion in western and southern India.

A related development has been a perverse division of India's groundwater economy in electricity-powered western and southern India and diesel-powered Eastern India—eastern UP, Bihar, Assam, West Bengal, coastal Orissa (Shah et al., 2018). Compared to west and south, eastern India has much more groundwater potential and deeper agrarian poverty but has little electricity to power the pumps. Smallholders of eastern India are obliged to use far more expensive diesel to pump groundwater. Eastern India needs and deserves energy subsidy for groundwater irrigation; yet, most of India's farm energy subsidy is captured by west and south where the future of groundwater irrigated agriculture itself is at peril.

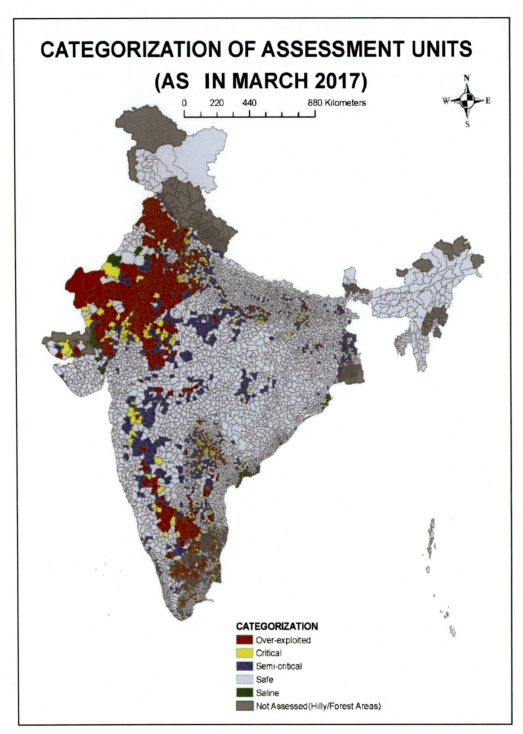

FIGURE 36.4 Pockets of groundwater over-exploitation in Western and Southern India. *From http://cgwb.gov.in/GW-Assessment/GWRA-2017-National-Compilation.pdf. Used with permission from Central Ground Water Board.*

Another major victim of the groundwater boom has been the electricity industry. A World Bank study reported that electricity tariff supplied to farmers amounted less than 10% of the cost of supply (Monari, 2002). Electricity subsidy burden to agriculture is reported to be US$14−15 billion in 2018 and has doubled since 2012. Agriculture accounts for 20%−45% of electricity supplied by many power utilities but less than 3%−7% of their revenues [Power Finance Corporation (PFC), 2013]. Finances of many power utilities are in the red for years. This has severely limited their ability to invest in new power plants and maintain transmission and distribution lines. Unmetered or free power supply to agriculture has over the decades disrupted operational efficiencies and internal accountability mechanisms within utilities, with a great deal of power theft passed off as agricultural consumption (Shah et al., 2004). Severe financial losses and lesser accountability of agriculture sector to power utilities have adverse impacts on rural power supply such as power outages, frequent voltage fluctuations, and phase imbalances (Gulati and Pahuja, 2015).

The groundwater boom has also hit the performance of public canal irrigation because it changed farmer attitudes toward surface irrigation from canals and tanks (reservoirs). Until the 1970s, tube well development in canal commands was viewed as "complementary" to surface irrigation. As vertical drains, the tube wells' role was lauded for promoting conjunctive management of surface and groundwater. By the 1990s, however, in many canal and tank commands, tube well irrigation was "substituting" surface irrigation. With a captive source of virtually free irrigation in electric tube wells, command area farmers were insulated from deteriorating management and O&M (operation and maintenance) of surface storages and canals. This resulted in weakening of the performance pressure on canal system managers from command area farmers (Shah, 2011). It also contributed to the slow progress in and indifferent results of a variety of interventions to promote Participatory Irrigation Management in canal and tank commands.

36.6 Sustainable groundwater governance

India's high dependence on groundwater has led to the depletion of groundwater resources at alarming rates in arid alluvial and hard-rock aquifers. The groundwater horror stories from the states of Punjab and Haryana, considered country's breadbaskets, are depressing. The looming groundwater crisis is a serious threat to food security and the agricultural economy. Seckler et al. (1999) wrote that a quarter of India's food harvest is at risk if she fails to manage her groundwater properly. Since the key agricultural regions are the worst hit by groundwater crisis, there are serious socioeconomic consequences to India's drying aquifers. The ongoing situation calls for immediate policy actions to manage groundwater resources, which have been either nonexistent or toothless so far.

Globally, policies and interventions to ensure sustainable groundwater use is a work in progress (Shah, 2009). The rapid groundwater development in the last two decades has outpaced the institutional responses to sustainably manage the groundwater. Different set of measures including (1) direct regulation through laws controlling groundwater abstraction in countries such as Oman, Jordan, and Israel; (2) economic instruments such as water-pricing in China and pump-taxes in western United States; (3) tradable groundwater rights in United States and Mexico; and (4) community-based resource management using groundwater user associations in Spain and Mexico and village communities in India have been tried worldwide for sustainable groundwater governance. India has piloted some of these measures; however, they have failed to scale up and combat the existing groundwater threat. Complex hydrogeological settings and the large-scale groundwater extraction by millions of borewell owners makes groundwater governance a unique challenge in India, requiring a range of protocols and strategies to effectively manage the groundwater. Across the years, different supply-side and demand-side instruments of groundwater management have been tested. Some of these have been briefly reviewed in this discussion.

36.6.1 Direct regulation through legal framework and administrative action

States use different legal and administrative instruments such as well licensing and registration, groundwater use permits, groundwater property, and usufructuary rights, in many countries, notably Oman, Iran, Saudi Arabia, Israel, and United States, to directly manage and protect groundwater (Shah, 2009). India does not have a formal legislation on regulating groundwater abstraction. The federal government drafted a "model" groundwater bill in 1970, which has been subsequently revised and circulated to the states for implementation [Kulkarni et al., 2015; Water Governance Facility (WGF), 2013]. Several states have adopted model bills to enact laws providing regulatory powers. However, the regulatory powers provided by these laws have remained ineffective in changing the status quo for a number of reasons such as lack of political will, public support, and resources for law enforcement [Water Governance Facility (WGF), 2013]. Tamil Nadu, a severely groundwater-stressed Indian state, passed "Tamil Nadu Groundwater Development and Management Act" to regulate groundwater extraction in 2003. However, the state government failed to implement the law and repealed it in 2013, fearing an outcry from the farming community (Ramakrishnan, 2013). The law had provisions which required farmers owning pump of over 1 hp to register with the Groundwater Authority, which was strongly opposed by the farmers.

The fundamental reason behind weak public regulation of groundwater is a colonial law from mid-19th century "India Easement Act 1882," which has not been altered since then (Kulkarni et al., 2015). This law grants private ownership to the groundwater resource, allowing landowners to freely extract groundwater from their land. The public perception of groundwater as a private good interferes with the compliance of government's regulations and hampers the transition from an unregulated to regulated regime (Mechlem, 2016). Also, the existing legal regime follows a centralized command and control approach and neglects the community knowledge and participation, who are the primary stakeholders, resulting in more failures than successes (Koonan, 2016; Cullet, 2014). Another major issue with enacting and enforcing any groundwater law in India is the prohibitive costs of registering and monitoring 20 million pump owners scattered across the countryside. A country like Spain, which has only 0.5 million groundwater wells, has been struggling without success, for the past two decades to build an inventory of its groundwater irrigators (Llamas et al., 2015). One could imagine the complexities of doing a similar exercise in India which has 40 times more pump owners.

In the wake of the failure of the past legal regimes in addressing the groundwater challenges, the Government of India has proposed a new regulatory framework "Groundwater (Sustainable Management) Bill, 2017" based on the current understandings of groundwater. The new bill proposes local-level regulation of groundwater through community-based approaches, regulating groundwater as a common pool resource by applying public trust doctrine and aquifer-level management to effectively control and regulate the groundwater use (Cullet, 2018). The adoption and uptake of this bill by the state governments would strengthen the role of groundwater legislations in achieving effective and sustainable groundwater governance.

36.6.2 Community-based groundwater management

The idea of community participation for groundwater management has been in discussion in India for several years. In the last two decades, several innovative examples have emerged, showing the effectiveness of community participation in groundwater supply and demand management.

World Bank reported two such case studies of demand-side management in Andhra Pradesh and Hivre Bazaar in Maharashtra. In Andhra Pradesh a group of NGOs, supported by the FAO worked with village communities in 700 drought-prone villages under the Andhra Pradesh Farmer Managed Groundwater System (APFAGMS) project. The project was an enabling initiative to build the capacity of farmers to sustainably manage groundwater, mostly by self-regulatory measures. The farmers were trained in collecting, measuring, and analyzing the groundwater level and rainfall data. Later, this collected information was used by farmer groups to plan their cropping pattern taking into account groundwater availability. The World Bank report argued that the intervention has reduced groundwater withdrawals in a significant number of villages (Garduno et al., 2009). This experiment demonstrated that improving hydrological understanding of farmers would lead to sustainable management of groundwater resource. World Bank study declared APFAGMS as an immediate success, calling it as the "first example globally of large-scale success in groundwater management by communities". However, postproject investigation by several researchers have highlighted the weakening of participatory norms and processes in APFAGMS villages, questioning the sustainability of such models in the absence of external support (Verma et al., 2012; Reddy et al., 2014). Another World Bank study reported the case of Hivre Bazaar, where the village community has adopted self-regulatory mechanisms to sustainably manage groundwater resource for a long time (Foster et al., 2009). The village community has used social control to enforce measures such as prohibition on the use of bore wells and ban on cultivation of water-intensive crops for the past three decades to arrest groundwater depletion, and combined these efforts with crop water budgeting, agricultural diversification, and produce marketing. The study reported that a strong local institution under the leadership of a charismatic village leader has been crucial in Hivre Bazaar's success in mobilizing the community and influencing their behavior change toward resource use. Nonetheless, the replication of such leaders is difficult but the model of Hivre Bazaar provides a template for interlinking groundwater-management practices with agricultural gains, which should be replicated across the country.

In Saurashtra-Kutch region of Gujarat, studies have reported about the "mass movements" to harvest excess rainwater for groundwater recharge through new check dams, percolation ponds, and open well recharge as a response to depleting groundwater conditions (Shah, 2000). These community movements were inspired by local spiritual leaders and supported by government, NGOs, and private industries. Studies have shown widespread increase in the availability of groundwater resource, especially after a good monsoon, in Saurashtra region due to mass-based groundwater recharge activities (Jain, 2012). Moreover, the socioeconomic impact of the entire recharge movement was significant as it stabilized the Kharif crop and offered winter and summer irrigation in the region (Shah, 2008). Alwar district in Rajasthan experienced a similar initiative of augmenting groundwater recharge with people's participation and contribution (Shah, 2008).

Each of these initiatives shows the merits of community participation in sustainable groundwater management. Learnings from these innovative approaches have inspired the current policy paradigm to move toward a decentralized control over groundwater, promoting local level groundwater management by village bodies. The Model Groundwater Bill 2017 recommends the management and regulation of groundwater at local level. In 2019, the Government of India launched a US$1 billion *Atal Bhujal Yojana* scheme, supported by the World Bank, aimed at improving groundwater management through community participation. The program will benefit 8350 Gram Panchayats of 78 groundwater-stressed districts in the states of Gujarat, Haryana, Karnataka, Madhya Pradesh, Maharashtra, Rajasthan, and Uttar Pradesh [Press Information Bureau (PIB), 2019].

36.6.3 Indirect instruments—energy pricing and rationing

In India the association between groundwater overdraft and electricity subsidies has been clear and direct. The decades-old energy subsidies regime seems unlikely to be replaced by volumetric-pricing of electricity as it can lead to farmers' backlash, which political leaders of governments dread. However, this energy-groundwater nexus offers indirect instruments for demand management of groundwater. Agricultural feeder separation and intelligent rationing of farmer power supply in Gujarat and Punjab have delivered some success in containing the deleterious impacts on groundwater-energy economy. Gujarat introduced *Jyotir Gram Yojana* (JGY) in 2003 which involved the separation of feeders serving irrigation tube wells and offered farmers 8 hours of daily power ration, down from 15 to 18 hours of poor quality power earlier (Shah and Verma, 2008). Punjab followed suit and now offers 7−8 hours of daily power ration in rice irrigation season and 4−5 hours of daily power during the remainder year.

Rationing of farm power in Gujarat and Punjab has demonstrated that the aggregate demand of groundwater and energy could be directly influenced by varying the energy supply. Between 2002 and 2014, the farm power consumption of Gujarat declined by 11 percent [Indian Express (IE), 2014]. However, the impact of JGY on reducing groundwater extraction has mixed evidence. Several studies claimed the JGY has reduced groundwater extraction as electricity consumption is directly proportional to groundwater abstraction. (Mukherji et al., 2010; Shah et al., 2004; Bhanja et al., 2017). On the contrary, recent study by Chindarkar and Grafton (2019) has found that groundwater storage has further depleted with JGY. Since more states are going for agricultural feeder separation and farm power rationing in the future, a detailed investigation of the impacts of JGY on groundwater extraction will help in ring fencing these policies to maintain environmental sustainability of groundwater irrigation.

Several other experiments are being piloted to overcome distortions created by farm power subsidies. The government of Punjab, with the support of IWMI and World Bank, is implementing two pilots. In the first pilot, Direct Benefit Transfer of Energy Subsidy, the project offers each tube well owner an annual pay-out equal to the cost of existing power consumption, meter the tube well to charge for actual power consumption, and then allow the farmer to encash power savings (Outlook, 2018). Another experiment will meter all tube wells, provide 24-hours power supply and give farmers a quota of free power based on a criterion. Farmers will be charged for any excess energy use at commercial tariffs, but they will be allowed to encash any saving from the free energy quota at the same tariffs (Gulati and Pahuja, 2015). Both the pilots attempt to manage the energy-groundwater demand by incentivizing the change in pumping behavior of farmers. Scaling up of these ideas will depend on the results of the impact assessment of these pilots, which is underway.

36.6.4 The advent of solar irrigation

Amidst the growing governance challenges of groundwater, India has been rapidly adopting solar irrigation pumps (SIPs). With increasing affordability due to falling prices of PV cells, heavy government subsidies on capital cost and a strong impetus on promoting renewable energy use, SIPs have gained a foothold in Indian irrigation in quick time. From around 1000 in 2012 the number of SIPs has soared to 150,000 in 2017−18, at a CAGR of 68% per year (Shah et al., 2018).

SIPs operate at nearly-zero marginal cost and provide 8−10 hours of uninterrupted daytime energy supply free to the farmers. There is real concern that with such a farm power regime, SIPs will intensify the cultivation of water-intensive crops, increase dry-season cropping and water selling, all of which can increase the groundwater extraction and further deepen the groundwater crisis in western India with high insolation but limited groundwater resource. India is experimenting with several innovative models of solar promotion, with some addressing this negative externality of groundwater overdraft.

In arguing that solar pump promotion strategy should be shaped by groundwater endowments of a region, IWMI-Tata Water Policy Program set up two solar pump promotion experiments, one in water-scarce Gujarat and another in water-abundant north Bihar.

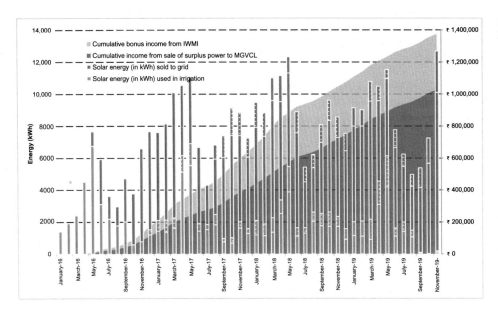

FIGURE 36.5 Operating report of Dhundi Experiment.

In Gujarat, IWMI's Dhundi solar pilot demonstrates the idea of "Solar Power as Remunerative Crop (SPaRC)." SPaRC shows the way of solarizing irrigation without affecting groundwater sustainability by giving farmers the option to "grow" solar energy that they can use either to draw groundwater or sell it to the electricity utility for cash. Nine farmers of Dhundi village were financially supported to adopt grid-connected solar pumps for irrigation. They were formed into a micro-grid managed by their cooperative. The electricity utility gave the cooperative a 25 years power purchase contract to purchase their pooled surplus solar energy metered at a single point. The experiment hypothesized that the additional income from selling surplus solar power will incentivize farmers to economize on energy and groundwater consumption. Fig. 36.5 presents the operating results of *Dhundi* experiment between January 2016 and November 2019. Dhundi solar farmers used just 35% of their solar power for irrigation; had they not got paid for selling their surplus energy at a remunerative price, chances are that they would have used more of their solar generation for pumping groundwater instead of letting it go waste. The *Dhundi* solar pilot presents a classic case of integrating livelihood solutions with the energy-groundwater nexus, suggesting that the right incentives can help shape a desirable behavior change. This experiment inspired the Government of India's new solar irrigation policy—*Kisan Urja Suraksha evam Utthaan Mahabhiyan* (KUSUM)—to solarize one million existing grid-connected SIPs and give the farmers an option to sell the surplus power to local power utilities [Business Today (BT), 2019]. The scheme has already kick-started in the state of Gujarat where some 2500 tube wells on 75 agricultural feeders have been solarized with a surplus power purchase contract.

IWMI's *Chakhaji* pilot in north Bihar aimed to demonstrate a model of promoting solar irrigation to create pro-poor water markets in groundwater rich eastern India. The experiment hypothesized that setting up entrepreneurial farmers as Solar Irrigation Service Providers (S-ISP) would crowd out the expensive diesel-pumping and create an affordable, pro-poor irrigation market for pump-less farmers. Eight farmers in *Chakhaji* were supported with 30% capital cost subsidy to own solar pumps and 1000 ft. of buried pipeline each to expand their command. Early evidence from Chakhaji showed that introduction of S-ISPs had crowded out 30–40 diesel water sellers in the village, reduced the price of irrigation water by 40%–60%, and rapidly expanded the irrigated areas in this groundwater rich landscape dominated by agrarian poverty (Shah et al., 2018).

The arrival of every new technology creates an opportunity to redesign economic institutions. In India's irrigation economy, arrival of solar pumps presents a sterling opportunity to transform a perverse energy-irrigation nexus into a virtuous one. IWMI's solar experiments show how this can happen.

36.7 Conclusion: from resource development to management mode

Worldwide the governance of agricultural groundwater use is presenting a wicked challenge. For a populous and poor agrarian economy like India's, this challenge is even more complex and formidable. In the business-as-usual scenario, problems of groundwater over-exploitation in India will only become more acute, widespread, serious, and visible in the years to come. The frontline challenge is not just supply-side innovations but to put in operation a range of corrective mechanisms before the problem becomes either insolvable or not worth solving. This involves a transition from resource "development" to resource "management" mode (Moench, 1994). Throughout Asia—where symptoms of over-exploitation are all too clear—groundwater administration still operates in the "development" mode, treating water availability to be unlimited, and directing their energies on enhancing groundwater production. A major barrier that prevents transition from the groundwater *development* to *management* mode is the lack of information. Many countries with severe groundwater depletion problems do not have any idea of how much groundwater occurs and who withdraws how much groundwater and where. Indeed, even in European countries where groundwater is important in all uses, there is no systematic monitoring of groundwater occurrence and draft (Hernandez-Mora et al., 1999). Moreover, compared to reservoirs and canal systems, the amount and quality of application of science and management to national groundwater sectors has been far less primarily because unlike the former, groundwater is in the private, "informal" sector, with public agencies playing only an indirect role. The challenge of creating order in an informal economy demands out-of-box responses. Some of the experiments outlined in this paper point to possible ways forward in this direction.

References

Agarwal, A., Narain, S., 1997. Dying Wisdom: Rise, Fall and Potential of India's Traditional Water Harvesting Systems. Centre for Science and Environment, New Delhi.
Banerjee, S.G., Barnes, D., Singh, B., et al., 2014. Power for All: Electricity Access Challenges in India. The World Bank, Washington, DC.
Bhanja, S.N., Mukherjee, A., Rodell, M., et al., 2017. Groundwater rejuvenation in parts of India influenced by water-policy change implementation. Sci. Rep. 7, 7453.
Briscoe, J., Malik, R.P.S., 2006. India's Water Economy: Bracing for a Turbulent Future. Oxford University Press, New Delhi.
Business Today (BT), 2019. PM Modi Approves KUSUM Scheme With Rs 34,000 Crore Central Aid. Business Today, New Delhi. Available from: <https://www.businesstoday.in/top-story/pm-modi-approves-kusum-scheme-with-rs-34000-crore-central-aid/story/320424.html>.
Central Groundwater Board (CGWB), 2017. Dynamic Groundwater Resources of India (As on 31 March 2013). Central Groundwater Board, Government of India, New Delhi.
Chambers, R., Saxena, N.C., Shah, T. 1989. To the Hands of the Poor: Water and Trees. Delhi, India: Oxford University Press and IBH, London: Intermediate Technology Publications.
Chindarkar, N., Grafton, R.Q., 2019. India's depleting groundwater: when science meets policy. Asia Pac. Policy Stud. 6, 108–124.
Cullet, P., 2014. Groundwater law in India: towards a framework ensuring equitable access and aquifer protection. J. Environ. Law 26 (1), 55–81.
Cullet, P., 2018. Model Groundwater (Sustainable Management) Bill, 2017. A new paradigm for groundwater regulation. Indian Law Rev. 2, 263–276.
Daines, S.R., Pawar, J.R., 1987. Economic Returns to Irrigation in India, SDR Research Groups Inc. & Development Group Inc. Report prepared for U.S. Agency for International Development Mission to India. New Delhi.
Debroy, A., Shah, T., 2003. Socio-Ecology of Groundwater Irrigation in India. In: Llamas, R., Custodio, E. (Eds.), Intensive Use of Groundwater: Challenges and Opportunities, Lise. Netherlands Swets and Zeitlinger Publishing Co, pp. 307–335.
Dubash, N.K., 2007. The Electricity-Groundwater Conundrum: case for a political solution to a political problem. Econ. Polit. Wkly. 42 (52), 45–55.
Foster, S., Limaye, S., Mandavkar, Y., et al., 2009. A Hydrogeologic and Socioeconomic Evaluation of Community-based Groundwater Resource Management – The Case of Hivre Bazaar in Maharashtra-India. Case Profile Collection Number 22, GWMATE Briefing Notes Series. The World Bank, Washington, DC.
Garduno, H., Foster, S., Raj, P., van Steenbergen, F., 2009. Addressing Groundwater Depletion Through Community-Based Management Actions in the Weathered Granitic Basement Aquifer of Drought-Prone Andhra Pradesh – India. Case Profile Collection Number 19, GWMATE Briefing Notes Series. The World Bank, Washington, DC.
Goswami, A., Rajan, A., Verma, S., Shah, T., 2017. Irrigation and India's Crop-Milk Economy: A Simple Recursive Model and Early Results.
Government of India (GoI), 2001. Report on Census of Minor Irrigation Schemes (1993–94). Ministry of Water Resources, Minor Irrigation Division, Government of India, New Delhi.
Government of India (GoI), 2005. Report on Third Census of Minor Irrigation Schemes (2000–01). Ministry of Water Resources, Minor Irrigation Division, Government of India, New Delhi.
Government of India (GoI), 2014. Key Indicators of Situation of Agricultural Households in India. National Sample Survey Office, Ministry of Statistics and Programme Implementation, New Delhi.
Government of India (GoI), 2017. Report on Fifth Census of Minor Irrigation Schemes (2000-01). Ministry of Water Resources, Minor Irrigation Division, Government of India, New Delhi.
Gulati, M., Pahuja, S., 2015. Direct delivery of power subsidy to manage energy–ground water–agriculture nexus. Aquat. Procedia 5, 22–30.
Hernandez-Mora, N., Llamas, R., Martinez-Cortina, L., 1999. Misconceptions in Aquifer Overexploitation Implications for Water Policy in Southern Europe, Paper for the 3rd Workshop SAGA, Milan.

Indian Express (IE), 2014. Farmers Pay 56 Paise per Unit of Electricity. Indian Express, New Delhi. Available from: <https://indianexpress.com/article/cities/ahmedabad/farmers-pay-56-paise-per-unit-of-electricity/>.

Jain, R.C., 2012. Role of Decentralized Rainwater Harvesting and Artificial Recharge in Reversal of Groundwater Depletion in the Arid and Semi-arid Regions of Gujarat, India. IMWI-Tata Water Policy Program. Water Policy Research Highlight #49.

Joshi, P.K., Gulati, A., Birthal, P.S., Tiwari, L., 2004. Agriculture Diversification in South Asia: Pattern, Determinants and Policy Implications. Econ. Pol. Wkly. 39, 2457–2467.

Koonan, S., 2016. Revamping the groundwater legal regime in India: towards ensuring equity and sustainability. Socio-Legal Rev. 12 (2), 45–73.

Kulkarni, H., Shah, M., Vijay Shankar, P.S., 2015. Shaping the contours of groundwater governance in India. J. Hydrol.: Regional Stud. 4 (Part A), 172–192.

Llamas, R., Custodio, E., d e La Hera, A., et al., 2015. Groundwater in Spain: increasing role, evolution, present and future. Environ. Earth Sci. 73, 2567–2578.

Maddison, A., 1971. Class Structure and Economic Growth: India and Pakistan Since the Moghuls. W. W. Norton, New York.

McGinn, P., 2009. Capital, Development, and Canal Irrigation in Colonial India. Bangalore, India. Institute for Social and Economic Change. Working Paper #209.

Mechlem, K., 2016. Groundwater governance: the role of legal frameworks at the local and national level-established practice and emerging trends. Water 8, 347.

MoA, 2015. "9th Agricultural Census 2010–11". Agriculture Census Division, Department of Agriculture, Cooperaon and Farmers Welfare, Ministry of Agriculture & Farmers Welfare, Government of India. Available Online: http://agcensus.dacnet.nic.in.

Moench, M., 1994. Approaches to groundwater management: to control or enable? Econ. Polit. Wkly. 29 (39), 135–146.

Monari, L., 2002. Power Subsidies: A Reality Check on Subsidizing Power for Irrigation in India. The World Bank, Washington, DC, Note Number #244.

Mukherjee, A., Scanlon, B., Aureli, A., Langan, S., Guo, H., McKenzie, A., 2020. Global Groundwater: Source, Scarcity, Sustainability, Security and Solutions, first ed. Elsevier, ISBN: 9780128181720.

Mukherji, A., Shah, T., Verma, S., 2010. Electricity reforms and their impact on ground water use in states of Gujarat, West Bengal and Uttarakhand, India. In: Lundqvist, J. (Ed.), On the Water Front: Selections From the 2009 World Water Week in Stockholm. Stockholm International Water Institute, Stockholm.

Outlook, 2018. Punjab CM Urges Big Farmers to Give Up Power Subsidy. Outlook, Chandigarh. Available from: <https://www.outlookindia.com/newsscroll/punjab-cm-urges-big-farmers-to-give-up-power-subsidy/1237954>.

Power Finance Corporation (PFC), 2013. PFC Report: Performance of State Power Utilities From the Years 2010-11 to 2012-13. Power Finance Corporation, New Delhi.

Press Information Bureau (PIB), 2019. Prime Minister Launches Atal Bhujal Yojana. Government of India, New Delhi.

Ramakrishnan, T., 2013. Adieu to Tamil Nadu Groundwater Law. The Hindu, Chennai. Available from: <https://www.thehindu.com/news/national/tamil-nadu/adieu-to-tamil-nadu-groundwater-law/article5147072.ece>.

Reddy, V.R., Mandala, S.R., Rout, S., 2014. Groundwater governance: a tale of three participatory models in Andhra Pradesh, India. Water Altern. 7 (2), 275–297.

Rodell, M., Velicogna, I., Famiglietti, J.S., 2009. Satellite based estimates of groundwater depletion in India. Nature 460, 999–1002.

Seckler, D., Barker, R., Amarasinghe, U., 1999. Water scarcity in the twenty-first century. Water Resour. Dev. 15 (1/2), 29–42.

Shah, T., 1993. Groundwater Markets and Irrigation Development: Political Economy and Practical Policy. Oxford University Press, Bombay.

Shah, T., 2008. India's master plan for groundwater recharge: an assessment and some suggestions for revision. Econ. Polit. Wkly. 43 (51), 41–49.

Shah, T., 2009. Taming the Anarchy? Groundwater Governance in South Asia. RFF Press, Washington, DC.

Shah, T., 2011. Past, present and the future of canal irrigation in India. In: India Infrastructure Report 2011. Oxford University Press, New Delhi, pp. 69–89.

Shah T., Verma S., Durga N., Rajan A., Goswami A. and Palrecha A., Har Khet Ko Pani (Water to every farm): Rethinking Pradhan Mantri Krishi Sinchayi Yojana (PMKSY), 2016, Policy Brief 1, IWMI-Tata Water Policy Program, Anand.

Shah, T., 2014. Groundwater Governance and Irrigated Agriculture. Global Water Partnership; 2014, Stockholm, TEC Background Paper #19.

Shah, T., 2000. Mobilizing social energy against environmental challenge: understanding the groundwater recharge movement in Western India. Nat. Resour. Forum 24 (3), 197–209.

Shah, T., Verma, S., 2008. Co-management of electricity and groundwater: an assessment of Gujarat's *Jyotirgram* scheme. Econ. Polit. Wkly. 43 (7), 59–66.

Shah, T., Rajan, A., Rai, G.P., et al., 2018. Solar Pumps and South Asia's energy-groundwater nexus: exploring implications and reimagining future. Environ. Res. Lett. 13 (11), 1–13.

Shah, T., Scott, C., Kishore, A., Sharma, A., 2004. Energy-Irrigation Nexus in South Asia: Improving Groundwater Conservation and Power Sector Viability. International Water Management Institute Research Report # 70, Colombo.

Shally, M., Sharma, Y.S., 2018. DBT route likely for agriculture power subsidy, farmers to pay by meter. The Economic Times, Mumbai. Available Online: <https://economictimes.indiatimes.com/news/economy/agriculture/dbt-route-likely-for-agriculture-power-subsidy-farmers-to-pay-by-meter/articleshow/66469724.cms?from = mdr>.

Stone, I., 1984. Canal Irrigation in British India. Perspectives on Technological Change in a Peasant Economy. Cambridge University Press, Cambridge.

The Age, 1891. Deakin on Irrigation. 7 February 1891. <http://150.theage.com.au/view_bestofarticle.asp?straction = update&inttype = 1& intid = 437> (accessed 01.11.06.).

Vakulabharanam, V., 2004. Agricultural growth and irrigation in Telangana: a review of evidence. Econ. Polit. Wkly. 39 (13), 1421–1426.

Verma, S., Krishnan, S., Reddy, A., Reddy, K.R., 2012. Andhra Pradesh Farmer Managed Groundwater Systems (APFAGMS): A Reality Check. IMWI-Tata Water Policy Program. Water Policy Research Highlight #37.

Wada, Y., van Beek, L.P.H., van Kempen, C.M., et al., 2010. Global depletion of groundwater resources. Geophys. Res. Lett. 37, L20402.

Water Governance Facility (WGF), 2013. Groundwater Governance in India: Stumbling Blocks for Law and Compliance. Stockholm International Water Institute, Stockholm, WGF Report No. 3.

World Bank, 1998. India: Water Resources Management Sector Review Groundwater Regulation and Management Report. The World Bank, Washington, DC.

World Bank, 2010. Deep Wells and Prudence: Towards Pragmatic Action for Addressing Groundwater Overexploitation in India. The World Bank, Washington, DC.

Theme 5

Future of groundwater and solutions

Chapter 37

The future of groundwater science and research

David K. Kreamer[1], David M. Ball[2], Viviana Re[3], Craig T. Simmons[4], Thomas Bothwell[5], Hanneke J.M. Verweij[6], Abhijit Mukherjee[7] and Magali F. Moreau[8]

[1]Department of Geosciences, University of Nevada, Las Vegas, NV, United States, [2]Independent Hydrogeological Consultant, Dublin, Ireland, [3]Department of Earth Sciences, University of Pisa, Pisa, Italy, [4]National Centre for Groundwater Research and Training, College of Science and Engineering, Flinders University, Adelaide, SA, Australia, [5]Rosetta Stone Consulting, Perth, WA, Australia, [6]Independent Expert Pressure and Fluid Flow Systems, Delft, The Netherlands, [7]Department of Geology and Geophysics, Indian Institute of Technology (IIT) Kharagpur, Kharagpur, India, [8]GNS Science, Wairakei Research Center, Taupo, New Zealand

37.1 Introduction

The future of hydrogeology holds many challenges and opportunities for scientists, engineers, social scientists, and decision makers. On one hand, tomorrow's pressures on groundwater may be akin but will not be the same as today's challenges. With extreme global population growth being exceeded by even greater acceleration in water consumption, the predicted worldwide water scarcity will place increasing community reliance on groundwater. This, combined with continuing overexploitation and pollution of water resources and climate variability, can negatively impact groundwater supply for agriculture and food security (Taylor et al., 2013). The role of climate change on groundwater resources is not well defined and potentially broad-ranging, including recharge mechanisms, areas, and volumes (Meixner et al., 2016); discharge to sea (Cooper et al., 2015); groundwater quality; and implication on drinking water security (Aladejana et al., 2020). Local and regional conflicts over clean water availability are expected to rise as shortages push individuals and communities toward the ragged edge of civility and toward associated instability, an anticipated increase in political tensions, and perhaps even state failure (NIC, 2012; Kreamer, 2012; Schleussner et al., 2016). Global Groundwater is undergoing immense stress in recent times (Mukherjee et al., 2020; Siebert et al. 2010). These challenges are likely to affect more traditional groundwater supply and quality projects for practitioners in the future.

On the other hand, there still remain new and largely unexplored areas of inquiry, including the hydrogeology of the oceanic lithosphere and extraterrestrial hydrogeology (Davis and Elderfield, 2004; Baker et al., 2005; Kastner et al., 2015). With more precise and varied means of groundwater assessment, new perspectives on groundwater science are constantly emerging (Streetly and Heathcote, 2018). Finally, how hydrogeology interacts with societal decisions, social equity, and the changing requirements of people and ecosystems has a huge impact on future water-related research (Villholth et al., 2019; Re, 2015; Hynds et al., 2018).

The future and frontiers of groundwater science and research have broad potential and provide a very positive outlook for continuing hydrogeological advancements. These innovations and discoveries can help humanity, preserve ecosystems, support sustainable resource management, and provide intellectually challenging avenues of investigation for generations to come. Perspectives on the nature and value of the subterranean world are changing, fueled by discoveries in multiple fields, spurred on by expanding needs of numerous economic markets, influenced by global changes in weather and climate, and built on greater understanding provided by past groundwater pioneers. Because hydrogeology is one of the most interdisciplinary of sciences, as the world moves forward in many fields, so does groundwater science and research.

37.2 How are fundamental groundwater perspectives changing?—"Darcy is dead"

This provocative title is not meant to disparage the pioneering contributions of Henry Darcy, rather to address prevalent overuse and misapplication of his work on flow through porous media. Scientific hydrogeology can be said to have started with Darcy. He identified two types of flow: "sometimes there are actual underground watercourses that flow with appreciable velocities in fissures, crevices, or natural cavities, and sometimes there are aquifers derived from infiltration through sand that flows slowly between two impermeable beds" (Darcy, 1856). He performed extensive lab experiments and derived a laminar flow law (now called Darcy's law) to describe groundwater flow and a nonlinear flow law (now called the Darcy−Weisbach equation) to describe turbulent flow in pipes; and the two distinct types of flow were widely recognized over the next 80 years (Worthington, 2013; Fetter, 2004). While Darcy clearly recognized that flow through consolidated, indurated, and hard rock media exists and is not directly related to flow in porous media, his mathematics for nonporous media flow was derived on the basis of uniform pipes, not tortuous fracture systems that change in aperture size. In the early 1900s, Meinzer (1928) recognized that mathematical description of flow in artesian aquifers was "relatively meager." However, in the following years, it was suggested that all rocks could be treated as "insoluble and chemically inert" and that all aquifers could be treated as porous media (Hubbert, 1940). This and other influential papers stimulated a great expansion of mathematical calculations and later of computer modeling. The contributions of Darcy, Hubbert, and others were giant leaps forward in quantifying groundwater flow, but this early work skewed quantitative groundwater science toward the homogeneous and isotropic.

However, now another paradigm shift is underway in the field of hydrogeology. While flow through porous media will always be important in alluvial sediments that are often near surface and the first to be exploited, there is growing realization that a huge fraction of groundwater flow and storage normally is in, and through, breaks or conduits in impermeable or low-permeability rock (Sharp, 2014; Worthington, 2013; Worthington et al., 2016). Fundamental analytical groundwater flow equations were first established with an assumption of homogeneity and isotropy, but their continued, widespread use is often for their simplicity and ease of application, not because of their appropriateness to actual field conditions. The lack of a practitioner's ability to measure and understand a complicated subsurface system does not substantiate an assumption that a complex, heterogeneous system can be represented as an equivalent porous medium.

There is a growing breakaway from the perspective that is biased toward porous media groundwater resources and groundwater flow, with recognition that trying to fit the common heterogeneous groundwater flow systems into a porous media conceptual model is holding back hydrogeology and, sometimes, expensively and severely divorcing its application from reality. Porous medium analysis worked well in unconsolidated sediments, where permeability is frequently proportional to grain size. However, this is not the case in bedrock aquifers, where permeability is proportional to the solubility and solution rates of the minerals that the aquifer is composed of (Worthington et al., 2016). In hard rock and karst groundwater systems, flow follows preferential pathways, following discrete conduits of fractures or solution cavities, and structural geology combined with weathering and dissolution is often of primary importance. The ubiquity and importance of discrete conduit subsurface systems underlines current weaknesses in applying traditional thinking, broad analytical models, and numerical models originally geared toward porous media (like the widely used MODFLOW, described by Langevin et al., 2017), without appreciation of the possibility of vast uncertainties (Sharp, 2014). Lesser-used approaches, such as in situ tracer and hydraulic tests, can be superior to standard aquifer testing approaches in hard rock geological sites. Often "standard methods" are not appropriate methods.

Adding to this break with blind application of traditional conceptual models and groundwater mathematics is the realization, typified by the preferential flow of nonaqueous-phase fluids, that fluid movement can follow discrete pathways and exhibit Rayleigh−Taylor instabilities even in "homogeneous" porous media (Kalisch et al., 2016; Glass and Nicholl, 1996). The effects of varying fluid density arising from differences in temperature and solute concentration can also influence fingering flow and macrodispersion (Simmons, 2005). The rebalancing of basic hydrogeological approaches away from homogeneous, isotropic porous media and toward preferential flow and appreciation of discontinuities in hard rock is more authentic and is becoming increasingly easier to quantify, particularly as groundwater monitoring and our ability to track and trace groundwater advances. For example, identification of regional groundwater flow is often supplemented with chemical analysis and dating the average residence times of subsurface flow (IAEA, 2013). Many new projects require this modification of conventional thinking (Mádl-Szőnyi and Tóth, 2015). For example, questions of nuclear waste disposal can necessitate projections of flow at different spatial and temporal scales (Tóth, 2016). As more basin-wide studies are done, the understanding of regional groundwater flow systems is becoming essential.

A consequence of this reappraisal will be that use of the term "aquifer," which implies that groundwater flow and storage are properties of the rock itself, will be found to be inadequate in some cases, particularly where flow is not associated with one geologic unit but alternates among several units according to their ensemble fracture, matrix, and structural properties. The term aquifer will become increasingly replaced by the more widely applicable term or concept, "groundwater flow system" (Engelen and Kloosterman, 1996; Verweij, 1999, 2003). The latter term embraces a system of storage and flow through pores in a rock matrix but equally embraces storage and flow in breaks or gaps in the rock that can bear little or no relation to lithology or stratigraphic boundaries. It also brings to the fore a greater emphasis on understanding the interaction between groundwater and structural geology heterogeneity as well as sedimentary, tectonic, geothermal, geochemical, weathering, and biochemical processes (Ingebritsen et al., 2006). Different processes alter rock; some enlarging the aperture size, frequency, and configuration of fractures in some rock, affecting fracture spacing and connectivity, and clogging gaps (healing fractures with chemical precipitates) in other rock.

37.3 Fossil fuel energy, geothermal energy, and mineral resources—the groundwater connection and the future

For many years a mainstay of geological employment and research has been in the field of petroleum, fossil fuels, and mineral resources. Hydrogeologists have played their part, with calculations of dewatering underground and pit mines, in calculating water supply for these and other industrial applications, in assessing the hydrodynamic influence for the identification of petroleum and mineral resources, for a variety of contaminant studies and a diversity of subsurface fluid injection activities. These areas of interest for groundwater practitioners and researchers will continue to grow with new directions and lines of inquiry.

In the emerging field of geothermal energy, groundwater scientists and engineers are projected to have greater influence on the identification of resources. The percentage of geothermal used in the world is projected to increase relative to other sources and has many uses for electrical generation, comfort heating, and recreational uses. According to the International Geothermal Association (IGA) (2020), resources are utilized nowadays in 83 countries. The United States is the top geothermal energy–producing country in the world, followed by Indonesia, the Philippines, Turkey, New Zealand, Mexico, and Italy, with a total of greater than 12,000 MW generated in over 80 countries (Butts, 2018; Nelskamp and Verweij, 2012). There are several categories of low and high energy geothermal production, including dry steam, flashing steam, binary cycles, heat pumps, and enhanced geothermal systems. Perhaps where future hydrogeologists can make the most impact is with "blind systems," that is, systems with no discernible ground surface indication of hot water. According to James Faulds, Director of the Nevada Bureau of Mines and Geology and Nevada State Geologist, the United States, geological methods for identifying conditions and fault structures commensurate with "blind" geothermal potential are being well established and outstripping associated hydrogeological methods for recognizing successful development (Faulds, 2019). A recent review shows that there is a large global geothermal resource base in sedimentary aquifers for direct heat use (Limberger et al., 2018). The fields of deeper "hot rock" geothermal that has both great risk and enormous potential and enhanced geothermal systems are also quickly evolving in research activity. With increasing interest in the use of geothermal resources in sedimentary basins from near surface to depths several kilometers, the need to evaluate the possible hazards of its use is also increasing. The assessment of hazards such as ground surface subsidence (Békési et al., 2019), seismicity (Buijze et al., 2020), and thermal pollution of groundwater resources all require understanding of the geothermal production on the groundwater system (Mádl-Szőnyi and Simon, 2016).

Fluid and waste injection and circulation associated with hydraulic fracturing, wastewater disposal, geothermal energy, and other industries has become increasing important in the last few decades. For example, Ferguson (2015) reports that in the decades of the 1980s, 1900s, 2000s, and early 2010s, over 23 km^3 of water have been injected into the Western Canada Sedimentary Basin alone, and even more subsurface injection has occurred between 1998 and 2013 in the State of Texas, the United States (Fetter et al., 2018). Enhanced oil recovery involves injection of fluids and different compounds to increase pressure on, and lower the viscosity of, underground petroleum reserves. Chemical additives in this process are considered regarding their effect on wettability, interfacial tension alteration, and rheology by employing both field experiments and laboratory investigations (e.g., core flooding experiments) (Goudarzi et al., 2015; Kamal et al., 2017). These activities of fluid injection directly confront basic questions of effective porosity and permeability, and the physics of a given geological framework. Combating climate change with underground sequestration of carbon dioxide is one more area that can require subsurface injection and will involve interdisciplinary collaboration with groundwater scientists and engineers (Gibson-Poole et al., 2006; Hortle et al., 2016). A significant research effort has been funded to understand and optimize these injection practices, and the associated responses in subsurface fluid movement and likely will continue in the future.

The relationship between subsurface fluid injection and earthquake generation has been another interesting area of research in the last few years (Castro et al., 2019; Skoumal et al., 2018; Keranen and Weingarten, 2018). The US Geological Survey reported a 5.6 earthquake was generated by fluid injection on November 6, 2011 in central Oklahoma, the United States and that similar quakes have been recorded in several US states and throughout the world (Fetter et al., 2018). Conventionally, hazard assessment has considered the impact of pore pressure increase on stress fields in rocks connected to injection activities, but the consequences are now being realized for additional poroelastic mechanisms in extending fully coupled stress fields without exhibiting a pore pressure increases (Goebel and Brodsky, 2018). The extraction of fossil fluids over the past decades has led to large pressure reduction of the gas and oil and the adjacent groundwater in the reservoirs. This pressure reduction and associated changes in stress may lead to compaction of the reservoir itself, subsidence of the ground surface above the reservoir and, in some cases, may also induce seismicity (NJG, 2017). Research challenges for hydrogeologists involve all these aspects of the impact fossil fluid extraction, for example, the establishment of the 3D extension of the pressure depletion in time, the compaction during and after production, the quantification of surface subsidence, and its influence on shallow groundwater flow systems.

Groundwater flow systems at different temporal and spatial scales influence our understanding of resource characterization, estimation of reserves, future production, and environmental and societal impact. This includes not only the search for water but also for minerals, coals, fossil fluids, geothermal, and for the potential for subsurface storage. Community understanding and social license to operate are often driven by perceptions of aquifer and seal integrity, protection of groundwater and surface water, and the impact of energy resource development and subsurface storage (e.g., CO_2 sequestration) on the environment. Worldwide growth in multiple uses of the subsurface, as well as environmental concerns related to these activities, has increased the need for better understanding of underground conditions, including its hydraulic properties, and pressure, and groundwater flow conditions at greater depths than those conventionally studied for water supply purposes (Verweij et al., 2019).

37.4 Groundwater can be a deep subject

As shallow groundwater resources are exploited, the attention of groundwater scientists and engineers have increasingly turned to the deep subsurface. Extraction and recharge of these zones are complex and often accompanied by expensive learning curves for those unfamiliar with groundwater at depth. For many conventional groundwater specialists who have dealt only with near surface flow systems, there are many new challenges. Deep groundwater systems may be part of a deep reaching meteoric cycle influenced by surface processes (gravity-induced groundwater flow systems) or systems that are driven entirely by geological (sedimentation, erosion) and tectonic processes. With increasing depth, the amount of data and their distribution decreases.

Data acquisition at profound depths, modeling the deep subsurface, drilling at depth, and pumping or recharge operations far below ground surface are unique challenges for many traditional hydrogeologists. Groundwater found at depth typically has long residence time underground, and the same tools used for petroleum, geothermal, and some mineral exploration (e.g., geophysics) are used for deep groundwater investigation (Boubaya, 2017; Sultan et al., 2019; Kumar et al., 2014; Chandra et al., 2019; Verweij et al., 2012).

Information and techniques from other industries that have considerable history in dealing with fluid flow at depth can be acquired to help, as well as implementation of innovative regulations and policies, utilization of hybrid technologies, and clear, multidisciplinary communication of ideas, concepts, and methodologies. Many commercial operations (e.g., oil and gas, geothermal, and mining) create a wealth of geological and hydrogeological information in their everyday exploration activities. Important information for hydrogeologists can be lost, as boreholes that are unproductive for oil, hot water, or other resources can be closed and sealed, as the companies move on to more "productive" sites. But in arid lands and especially in economically developing countries, the geological and hydrogeological information gleaned from this "unsuccessful" exploration, and even the abandoned boreholes themselves, could be of great value to the local communities. A future move toward requiring international industries to identify and develop groundwater resources for host nations, in the course of their exploration efforts, would benefit humankind. International adoption of government data open-file legislative standards such as those in place in New Zealand (data release after 5 years, https://wiki.seg.org/wiki/Open_data) would benefit many countries by not only promoting the development of resources but also allowing the data to be used for humanitarian efforts.

Several commercial operations can be detrimental to deep groundwater. Fluid and waste injection is often done at depth, and the interaction of injected with native, deep groundwater has become of increasing concern. Chemical additives for fracking fluids are being closely examined, and disposal of waste fluids from many industries may eventually corrupt aquifers that were previously thought to be too deep to draw upon. The depth, long residence time, and subsequent mineralization of deep groundwater can create brines too saline for commercial use (Ferguson et al., 2018). Coal bed methane (CBM) operations, also known as "coal seam gas" ventures, often extract large volumes of water from depth to release relatively clean methane or "sweet gas," but in some cases, this extraction can deplete or contaminate a future resource. Recent hydrogeological regulatory oversight in Australia has reduced the potential for damaging effects of the CBM Industry. Despite this additional security, operations in this deep zone are still often conducted with considerably less capital cost than counterparts in the more conventional oil and gas industry. Highly efficient optimized workflows utilized in factory mode drilling campaigns have enabled this significant cost reduction.

Deep groundwater is expensive to locate, has large associated drilling expenses, and has higher pumping cost, but has been increasingly relied upon for water supply. More and more, exploration and research is turning to this hidden resource. To deal with future challenges related to increasing subsurface activities and possible undesired effects at the surface, it is important to be able to use and build on the wealth of existing, but at present not always publicly available, data, knowledge, and methods for future research.

37.5 The subterranean biological world and groundwater-dependent ecosystems

Groundwater science and engineering is certainly a realm of interdisciplinary research, and far-reaching connections and relationship to the living world are hugely important. Groundwater-dependent ecosystems in springs, on the banks of rivers, in caves, and in groundwater-fed estuaries directly affect other aquatic and terrestrial species on earth (Kreamer et al., 2015). Hydrogeologists have studied the underground migration of contaminants and pathogens for years, with a classic example being the London cholera outbreak of 1854. There is renewed interest in the movement and viability of viruses in different media, with the Covid-19 pandemic reinforcing the notion that the security of drinking water and wellhead protection is not restricted to isolated community boreholes.

It is anticipated that there will be a greater awareness and level of research into the biological and biochemical ecosystems in the subsurface and in surface waters near the issuance of springs. Some have considered aquifers themselves to contain the "ultimate groundwater-dependent ecosystems" (Humphreys, 2006). Groundwater flow systems contain abundant, self-sustaining, and diverse populations of crustaceans, helminths, bacteria, protozoans, and fungi (and in extreme cases, fish). Stygobites, which are obligate groundwater fauna, can occur in karst dissolution cavities and the voids and interstitial spaces in fractured rock and alluvium. They derive imported energy from above ground (except in special cases where chemoautotrophic fixation of energy occurs in situ) and have convergent morphologies, including slow metabolism, impairment or loss of eyes, reduced pigment, enhanced nonoptic senses, ability to survive in low carbon and oxygen environments, and vermiform body form (Humphreys, 2006). Investigating and understanding the life cycles and metabolism of these populations and the biochemical processes in the groundwater system will reveal their obvious importance in the context of water quality, water chemistry, groundwater protection and vulnerability, and pollution remediation. They will also be investigated in order to understand the current and ancient development of solution conduits or dissolution of carbonates, evaporates, and sandstones by autochthonous biochemical processes.

Springs, wetlands, and other groundwater-dependent ecosystems are often critical sources of biodiversity within landscapes and coastlines, providing indispensable habitat for many endangered species and sustaining "untold hundreds of rare or unique plant, macroinvertebrate, fish, amphibian, and some reptile and mammal species" (Kreamer et al., 2015). The biota in isolated springs can also tell a story of long-term climate changes. High concentrations of endemic species present in these surface manifestations of groundwater are likely ancient systems, often with strong trophic structure (Kodrick-Brown and Brown, 2007; Blinn, 2008). Endemism evolves regularly in groundwater-dependent ecosystems in arid regions where springs may be far apart or isolated island environments, developing through relicturalization and adaptation cycles related to environmental change. Long-term stable conditions can produce paleoendemic species, whereas the onset of an altered environmental regime (e.g., drought, desertification, marine or freshwater transgression, periods of extreme climate, or other large-scale environmental changes) may reduce dispersal and accessibility of habitat, imposing isolation, allele fixation, and local adaptation (Kreamer et al., 2015).

From a water quantity standpoint, groundwater overexploitation and drought can lower water tables; dry up springs, wetlands, and rivers; and destroy biological habitat (Devitt and Bird, 2016; Mukherjee et al., 2018a; Kath et al., 2018). The misnomer "safe yield" has been used for many years to justify pumping as much water as is calculated that recharges an aquifer. This can lead to eradication of groundwater-dependent ecosystems. The antiquated idea of "safe yield" now is replaced with studies of "sustainable yield," that is, how much groundwater can be withdrawn without negatively impacting-dependent biota. The study of water stress on the subterranean biological world extends up into the vadose zone as well. There are both long historical and much expanding, cross-disciplinary research by agronomists, biologists, and hydrologists concerning water uptake by plants and the limits of water stress on flora and fauna.

The subterranean microbiological transformation of chemical contaminants in groundwater has long been known and studied (Domenico and Swartz, 1990; Sirisena et al., 2013; Fetter et al., 2018). These transformations can occasionally result in the daughter products being more harmful than the parent compounds, but on the contrary, the ability to transform harmful pollutants into innocuous by-products has opened up a continually advancing field of in situ contaminant bioremediation (Hazen, 2010). Contaminant groundwater studies can range from essential and basic monitoring of organics and inorganics in the subsurface (Stelma and Wymer, 2012; Moreau et al., 2019) to specific studies of sorption and transformation (Franzblau et al., 2016; Sirisena et al., 2018; Fisher, 2019).

37.6 Coast to coast

With rising sea level each decade the question of the extent of seawater intrusion into coastal aquifers has become an immense concern, and saline groundwater intrusion has been called the "invisible flood" and the "leading edge of sea-level rise" (Tully et al., 2019; Werner and Simmons, 2009; Jiao and Post, 2019). This invasion affects many living communities. According to the World Ocean Review, approximately 200 million people live within 5 m of sea level, with an expected increase to 400−500 million by the end of the century. Adjacent freshwater wetlands and estuaries, critical habitats for the growth and development of many species, can be affected (Luijendijk et al., 2020). Encroachment of saline water along coastlines has the potential to cause damage to communities and ecosystems, with cascading effects potentially causing decreased agricultural production and upland forest retreat (Kaushal et al., 2018). The protection of coastal aquifers has intensified in recent years with more seaside communities opting to inject municipal wastewater as a barrier between intruding ocean water and freshwater aquifers. Submarine discharge of contaminants can also be a problem (Lu et al., 2020).

Persisting challenges include the lack of monitoring to determine the location of, and changes in, the saltwater−freshwater interface (Werner et al., 2013). There are novel and upcoming methods for monitoring and analysis such as coastal airborne electromagnetic surveys (Goebel et al., 2019; King et al., 2018). These innovative tools herald new ways to evaluate saline water intrusion.

37.7 Under the ocean

Climate scientists have confidently advanced the notion, with ample evidence, that atmospheric changes are impacting global warming and extreme weather events. Yet compared to its vastness, little information on the quantity and constancy of heat exchange through the ocean floor has been gathered, nor is its quantitative impact on global climate known. Dedicated scientists have been investigating the oceanic lithosphere for many years, but its shear extent and remoteness make comprehensive monitoring difficult. The ocean covers the majority (71%) of the earth, but what is known about the hydrogeology of the ocean floor is minor compared to terrestrial groundwater. Offshore nonsaline groundwater could be a significant source of freshwater supply in the future (Cohen et al., 2009; Post et al., 2013). The hydrogeology of oceanic lithosphere has gathered steadily increasing interest because of its potential for freshwater, mineral and petroleum development, and potential environmental connections.

Hydrothermal flux through the oceanic lithosphere is highly variable and is normally ignored with observations of climate change. Oceanic scientists jokingly refer to "Bullard's Law," that is, "Never duplicate an oceanic heat flow measurement for fear it might differ from the first by two orders of magnitude" (Sclater, 2004). Generally, it is thought that heat flux is greatest near mid-oceanic ridges, and that fluid permeability is generally low but variable in pelagic sea sediments with greater chance of fluid movement where sediments thin (e.g., on the flanks of ridges and seamounts), and that some underlying basalts and fractures may be able to transmit fluids horizontally for kilometers (Davis and Elderfield, 2004; Daigle and Screaton, 2015). While a tremendous amount of seafloor information has been gathered, a tremendous amount still needs to be done. With more discoveries and insights every day, the field of oceanic hydrogeology seems boundless.

37.8 Extraterrestrial hydrology—the sky's not the limit

A burgeoning field of inquiry is the search for water in the universe. Because water is associated with the possibility of life, intense interest has focused on its existence in planets, moons, and other objects both inside and outside our solar system. According to Schneider (2020), as of March 1, 2020, 4241 planets beyond our solar system have been confirmed (exoplanets) in 3139 systems. Some exoplanets are the right distance from their star to have water, for example, two of the five identified planets around the star Kepler-62. In a study of more than 4000 exoplanets, approximately 35% seems to have over half their mass represented by water (Zeng et al., 2019).

The discovery of evidence that water exists in different forms on other planets below planetary surfaces means that hydrogeologists will be (or should be) working with astronomers and space scientists to assess these resources and bring forward practical ideas and methods for future exploration and exploitation of these resources, prior to manned exploration missions. Inside our solar system, there is ample evidence of water on moons and planets. In addition to earth, planets, and moons with the possibility of groundwater, ice or even oceans include Mars, Europa, Enceladus, Ganymede, Callisto, Ceres, and several of the moons and ice planets (Hall, 2016; Baker et al., 2005). There are indications that both Mars and Venus had large bodies of surface water in the past (Chan et al., 2018; ESA, 2019). With Mars in particular where no surface water is present, groundwater would be essential to any future colonization. Salese et al. (2019) recently suggested that there is evidence of planet-wide groundwater systems on Mars. One of the challenges is that absorption spectroscopy and geochemical methods, typical for confirming the presence of ice and atmospheric water vapor, are not as effective for liquid water, and certainly not for groundwater. The field of extraterrestrial groundwater science has just begun.

Chris Hadfield (the Canadian commander of the International Space Station—ISS) said on a recent trip to Dublin, Ireland,

I used to think that space was the great exploration frontier, but actually space is easy. With a bit of math and physics it is entirely predictable. It is empty. There's nothing there. I realized, looking down from the ISS cupola, as I flew round the earth 16 times a day that the real new exploration frontier was below the surface of Earth and the other planets. We know so little. The subsurface of our planets is far more complex than space. I had to go to space to understand our lack of knowledge about the planet.

Hadfield (2019).

Hydrogeologists and groundwater professionals will have a fundamental role in planetary exploration and future settlements.

37.9 Groundwater quality and emerging contaminants

Arguably, some of the most rapid and profound changes in groundwater science have occurred in the realm of subsurface contamination and groundwater quality. Novel techniques for contaminant site characterization and remediation are constantly moving forward (Fetter et al., 2018). New, previously unrecognized emerging pollutants are being acknowledged and documented, driven in part by constantly improving analytical chemistry techniques.

At the same time, both affluent countries and poorer countries are similarly struggling with the legacy of traditional pollutants. Agricultural by-products such as pesticides and high nitrates associated with fertilizers plague shallow European and Asian groundwater alike. Although the European Union issued the European Nitrate Directive three decades ago, nitrate remains a serious danger to human health and ecosystems throughout parts of Europe (Musacchio et al., 2020). Subsurface petroleum leaks cause problems in South America, Australia, the Middle East, and on other continents. Many of the same mining wastes, surfactants, and mineral refining by-products generated in North America can be problematic in African mining operations and elsewhere.

In addition, processes, products, and perspectives are evolving. Hundreds of new pesticides and other industrial chemicals are developed each year, and their use has steadily grown, increasing the risk to vulnerable aquifers and communities that depend on them. According to the Food and Agriculture Organization of the United Nations, annual use of pesticides worldwide is estimated at about 3.5 billion kg (7.7 billion lbs.) and a decade ago use was only about 72% of that amount (Alavanja, 2009). Industrial processes are not static either opening up a potential for new contaminants and new contamination routes. For example, some mining operations are moving toward in situ leaching methods where a liquid mobilizing agent (lixiviant) is pumped underground, mixes with an ore body, then the pregnant solution is pumped to the surface for mineral extraction. The potential for excursion of contamination from these mine sites must be carefully evaluated, particularly as many of these systems are in hard rock, have preferential pathways, and typical groundwater mathematical analogs used to assess porous media do not apply. Natural, nonanthropogenic pollutants (e.g., arsenic, fluoride) are also a worldwide concern. Groundwater can be critically polluted by natural background mineralization and have aqueous pollutant concentration and controls that depend on the geologic provenance of regional groundwater flow and fluctuating groundwater conditions (Mukherjee et al., 2018b; Moumouni and Fryar, 2017; Tobin et al., 2018). And, in vadose zone studies, attention is not only on contaminant liquids, but the movement and intrusion of contaminant vapors and issuance of gases might affect climate change (Kreamer, 2001; Etiope et al., 2019).

The opportunities for groundwater scientists and researchers in groundwater quality studies are numerous and on many scales. The physics of contaminant sorption, the chemistry and biology of contaminant transformation, and the mathematics of contaminant migration can involve investigations ranging from the microscopic to the macroscopic. Numerical models of both contaminant flow and plume remediation scenarios are constantly being improved. Case studies from around the world will continue to engage practitioners for many years to come.

Importantly, environmental professionals are constantly recognizing new groundwater contaminants that were previously not on the radar screen (Lapworth et al., 2019; Gaston et al., 2019). The list of emerging contaminants is long, including many categories of substances and transformation products (Stuart and Lapworth, 2014). Pharmaceuticals alone, used for human, livestock, and illegal purposes, comprise a large number of potentially mobile compounds such as nonsteroidal antiinflammatory/analgesics, antibiotics, and psychiatric drugs (Paiqa and Delerue-Matos, 2016). Hormones and sterols are being extensively studied (e.g., estrogen compounds, testosterone, androsterone, cholesterol, and 4-androstenedione) as well as food additives/preservatives such as butylated hydroxyanisole and butylated hydroxytoluene (BHA, BHT) (Bartelt-Hunt et al., 2011).

There are many other classes of emerging contaminants. Pesticide metabolites, water and wastewater treatment products (e.g., *N*-nitrosodimethylamine, trihalomethane compounds), and flame retardants (e.g., triazoles, alkyl phosphates, polybrominated diphenyl ethers) have had negative impact on the environment and continue to produce degradation of groundwater. Other groundwater studies are being conducted on the movement and impact of personal-care products (e.g., parabens, diethyltoluamide (DEET), triclosan, and sunscreen compounds) and "lifestyle" compounds (e.g., caffeine, artificial sweeteners, and nicotine). There are a plethora of industrial by-products and additives that have and are being researched, including methyl *tert*-butyl ether (MTBE), ethylene dibromide (EDB), *N*-butyl benzene sulfonamide (BBSA), phthalates, bisphenols, bis(2-chloroethyl)ether, and dioxanes. Surfactants of great interest are alkyl ethoxylates, and recently, perfluoralkyl and polyfluoralkyl substances (PFAS) have taken center stage as their harmfulness, ubiquity, and persistence in the environment have been recognized. This latter is a broad group of chemical compounds used for a multitude of purposes, including nonstick cookware, firefighting foam, waterproof clothing, and stain-resistant fabrics. Of the PFAS compounds, considerable attention has been focused on perfluorooctane sulfonate and perfluorooctanoic acid as they have been frequently found in groundwater and throughout the environment (Wei et al., 2018). Plastic microfibers are a new, important emerging groundwater contaminant. Synthetic microfiber pollution can come from many sources and can "adsorb persistent bioaccumulative and toxic chemicals, which include persistent organic pollutants and metals" (Re, 2019). The impact of emerging contaminants on groundwater science and research has been immense and will continue to be a focal point in the future.

Another mushrooming aspect of groundwater-quality investigations is in the field of environmental forensics. For decades, water-quality indicators have aided hydrogeologists in tracing and tracking groundwater movement and mixing, in calculation of average groundwater residence time (groundwater dating), and in litigation determining the culpability or innocence of potential polluters. Now, better analytical techniques have opened up entirely new ways of tracking regional groundwater and contaminant migration and "fingerprinting" pollutants (Morrison and Murphy, 2015).

37.10 The new tools

Advances in monitoring, field operational systems, remedial techniques, nanotechnology, and analytical methods have effectively augmented and revolutionized groundwater studies and research. Satellites and drones have become new and essential tools for evaluating the geological framework of an area. These same tools can define the land use and vegetative cover, helping to quantify evapotranspiration and overall water balance. Visible and invisible spectra can be recorded, aerial radiometric measurements can be made, and mass changes attributed to groundwater level changes can be made. Although the Gravity Recovery and Climate Experiment satellites have very poor resolution, they have played a major role in inferring major groundwater changes, particularly in large plains (Bhanja and Mukherjee, 2019). How these tools will be embedded and their possible impact on how we monitor groundwater systems for short and long term remain to be defined.

The next generation of groundwater flow and solute transport models (e.g., FEFLOW, HydroGeoSphere, and latest versions of MODFLOW) and new models yet to be developed poses challenges for modelers and mathematicians alike. The quest for efficient algorithms that more accurately simulate a complex underground reality is ongoing. Parallel issues include input data assimilation and reduction strategies, modeling uncertainty, and visualization and communication schemes for results. Models by their very nature are simplifications of reality, but Schwartz et al. (2017) point out the importance of a balance between simplicity and complexity in groundwater modeling and the dangers of "naïve simplicity."

New techniques for accessing groundwater condition and evolution include a new arsenal of tools that were not available or affordable earlier. For example, microbial ecology, DNA and lipid analysis, genome sequencing, and tryptophan-like fluorence techniques are all moving the microbiological groundwater field forward (David et al., 2014; Sorensen et al., 2018). Two relatively new, sensitive, high-resolution techniques for trace analysis in groundwater are inductively coupled plasma optical emission spectrometry and total reflection X-ray spectrometry. With chemical analysis of trace elements the use of disk-based solid-phase extraction and large-volume injection gas chromatography—mass spectrometry has become a more useful instrument. Enhancement of reversed-phase high-performance liquid chromatography has contributed to the detection of persistent, mobile microorganic compounds particularly with advanced separation techniques (Schmidt, 2018). Time-of-flight technology also opens possibilities of analyzing for large suites, allowing pilot sampling to inform the design or the review of targeted monitoring programs (Moreau et al., 2019). All these continuing monitoring and analytical advances facilitate the expansion of groundwater science.

37.11 Laws, regulation, guidance, and governance of groundwater

Many groundwater professionals have been called upon to serve as technical expert witnesses in groundwater court cases, as members of government or industrial advisory boards, as experts at public hearings, and/or giving testimony to government entities or agencies. These opportunities continue to grow and often require the state-of-the-art knowledge and experience of practitioners and researchers. The massive amount of groundwater legislation around the world is impossible to list or innumerate here, but its interface with science is usually site specific and requires adaptation of general concepts to unique field situations. Well-known examples of the groundwater professional involvement in litigation are groundwater pollution in Woburn, Massachusetts, the United States with the solvent trichloroethylene (made famous by the subsequent book and movie *A Civil Action*) and hexavalent chromium pollution of groundwater in Hinckley, California, the United States (recounted in the movie *Erin Brockovich*). Legal testimony requires sometimes facile and broad-brush, sometimes complex and detailed, articulation of groundwater concepts and procedures in formal and informal settings. Often scrutiny of opposing augments is required. Legal disputes can range from regional multinational transboundary groundwater issues to smaller scale litigation of groundwater quality and quantity.

The scale of professional guidance to communities can vary as well. Planning can range from the United Nation's worldwide effort of improvement embodied in the Strategic Development Goals that include many connections with groundwater, to a small community's goals for sustainable clean water supplies or the rising issues of managing transboundary aquifers (Eckstein and Eckstein, 2005). And even in places where unambiguous groundwater laws and regulations exist, a lack of proper enforcement and poor governance can hobble the clearest regulatory goals and strategic plans (Villholth et al., 2019).

37.12 Socio-hydrogeology in the future of groundwater science

Socio-hydrogeology is a recently developed discipline that fosters the structured incorporation of the social dimension into hydrogeological investigations (Re, 2015). This new approach aims at improving information sharing, maximizing the understanding of the socioeconomic implications of any groundwater investigations, and ensures the dissemination of hydrogeological findings/research outcomes to communities and decision makers. Socio-hydrogeological investigation can also address questions of social equity, scientific ethics, poverty reduction, risk minimization, groundwater management, and hydrophilanthropy (Wilson et al., 2013; Kreamer, 2016; Bhanja et al., 2017; Mukherjee et al., 2019). Originally applied to investigations in economically developing countries (Re, 2015; Limaye, 2017; Re et al., 2017; Frommen and Ambrus, 2019), this approach is currently being adopted also in more affluent regions (Musacchio et al., 2020) and subject to an ongoing discussion on how to refine the science behind it (Hynds et al., 2018; Re, 2020).

The approach is centered on the role of hydrogeologists and other groundwater scientists and engineers, as advocates for groundwater protection, and it fosters the understanding of reciprocity between groundwater, its final users, and comanagement. To do so, groundwater professionals are called upon to be directly acquainted with the social value of groundwater resources, thus assessing how scarce or polluted resources can affect (or are already affecting) human and ecological well-being.

A new paradigm that allows hydrogeologists to overcome the isolation resulting from ultraspecialization is therefore required and is opening up different narratives that allow hydrogeologists to effectively interact with other disciplines (e.g., social sciences and economics) and with the civil society. In addition, a crucial challenge is learning the most effective communication strategies for outreach and capacity building, two activities that too often have a marginal role in groundwater investigations. The burgeoning field of integrated groundwater management is an attempt to bring all the social, economic, technical, environmental, political, cultural matters together (Jakeman et al., 2016).

37.13 Education and outreach

One of the major contributions that groundwater professionals can make to the sustainability of our world, in the present and the future, is to make a determined effort to explain our understanding of the subsurface to people, ordinary people. Groundwater scientists and engineers can carry on talking to each other, and a few open-minded professionals and decision makers, with the hope that at some point society will wake up and recognize the worldwide value of hydrogeological skills and knowledge. Conversely, we can recognize that the world will not educate itself; individuals, and societies will remain content with an "out of sight out of mind" attitude. Therefore as responsible scientists working for a sustainable world, it is incumbent on us to demystify the subsurface and groundwater within it so that ordinary people everywhere can make rational decisions about the resource upon which they, and ecosystems, depend (Re and Misstear, 2017).

A challenge is making sure that our education is keeping pace with the field itself—there is extraordinary breadth and depth in groundwater activities, and they are increasing rapidly, but unfortunately, many of the tools developed in research are not used in practice. The growing research—practice gap is real and on the increase, and education has an important role to play. It is hard to argue for more research when so much of what has already been developed is not used in our daily hydrogeologist's toolbox. Stochastic hydrogeology, environmental tracers, uncertainty analysis, inverse methods, and optimization are among the tools groundwater researchers have developed, but practitioners do not use daily. A call for more research cannot be made without an acknowledgment of the growing research—practice gap, the reasons for that, and possible remedies (Simmons et al., 2012; Miller and Gray, 2008).

37.14 The unexpected challenges

Then there are always the unexpected events that groundwater scientists and engineers might be challenged with, exemplified by the largest marine petroleum disaster in history. When the Deepwater Horizon exploded and sank on April 20, 2010 in the Gulf of Mexico and leaked an estimated 780,000 m^3 of oil from the sea bottom over several months, by June 2010, capping operations were about to begin, to seal off an undersea borehole that had been belching oil. But there was legitimate concern that capping the leak would build up pressure and cause the entire ocean floor to blow out (Betancourt, 2020). A hydrogeologist, Paul Hsieh from the US Geological Survey, was called on to model the problem in Menlo Park, California, the United States. He was given less than 24 hours to estimate whether capping would cause an even larger disaster. Hsieh worked all night converting water viscosities and densities to those of crude, using groundwater models converted on the spot to models of oil, with only blurry cell phone photographs of unfolding data from the ocean floor via Houston, United States, over 2650 km away. My biggest problem Hsieh joked, "that was the coffee machine was in another building" (Hsieh, 2016). By the next morning, Hsieh reported to the assembled team in Houston that according to his modeling calculations the cap would hold without blowing out the seafloor. The capping was instituted and the seafloor held. There are few groundwater scientists and engineers who have not had short time frames thrust upon them, testing their knowledge and experience. The adaptability to carefully evaluate each situation, to "think outside the box," and to integrate data and input from a wealth of different disciplines will continue to characterize the actively engaged groundwater professional for years to come.

The future of groundwater science and research may take unexpected turns and twists in the coming years, but armed with an arsenal of new analytical and monitoring techniques, practitioners and researchers have the opportunity to continue to improve global conditions. The authors fully acknowledge that this chapter cannot represent all the forthcoming challenges in groundwater investigations. Entire fields of inquiry have not been addressed here, such as managing aquifer recharge, karst hydrogeology, impact of climate change on groundwater, soil water relations, urban water issues, and an increasing community emphasis on applying hydrogeology for the betterment of humankind and the planet. We hope that at least a flavor of the multitude of future possibilities and opportunities has been conveyed. The interdisciplinary nature of the science allows groundwater specialists to follow their intellectual curiosities in many directions of their choosing and advance our understanding of this precious underground resource. The prognosis for the future of groundwater science and research is bright, and its frontiers continue to expand.

Acknowledgments

This chapter was written with the support of the International Association of Hydrogeologists Working Group on the Future and Frontiers of Hydrogeology, as part of the IAH "Forward Look" initiative. The support of IAH is gratefully acknowledged. The helpful suggestions of Stephen H.R. Worthington, the reviewers and editors of this manuscript, are also appreciatively recognized.

References

Aladejana, J.A., Kalin, R.M., Sentenac, P., Hassan, I., 2020. Assessing the impact of climate change on groundwater quality of the shallow coastal aquifer of eastern Dahomey basin, southwestern Nigeria. Water 12, 224.

Alavanja, M.C., 2009. Introduction: pesticides use and exposure extensive worldwide. Rev. Environ. Health 24 (4), 303−309. Available from: https://doi.org/10.1515/reveh.2009.24.4.303.

Baker, V., Dohm, J.M., Fairén, A.G., Ferre, P.A., Ferris, J.C., Miyamoto, H., et al., 2005. Extraterrestrial hydrogeology. Hydrogeol. J. 13 (1), 51−68. Available from: https://doi.org/10.1007/s10040-004-0433-2.

Bartelt-Hunt, S., Snow, D.D., Damon-Powell, T., Miesbach, D., 2011. Occurrence of steroid hormones and antibiotics in shallow groundwater impacted by livestock waste control facilities. J. Contam. Hydrol. 123 (2011), 94−103. Available from: https://doi.org/10.1016/j.jconhyd.2010.12.010.

Békési, E., Fokker, P.A., Martins, J.E., Limberger, J., Damien Bonté, D., Van Wees, J.-D., 2019. Production-induced subsidence at the Los Humeros geothermal field inferred from PS-InSAR. Geofluids 2019, 12Article ID 2306092. Available from: https://doi.org/10.1155/2019/2306092.

Betancourt, M., 2020. Modeling under pressure. Eos 101, . Available from: https://doi.org/10.1029/2020EO140138. Published on 25 March 2020.

Bhanja, S.N., Mukherjee, A., 2019. In situ and satellite-based estimates of useable groundwater storage across India: implications for drinking water supply and food security. Adv. Water Resour. 126 (2019), 15−23. Available from: https://doi.org/10.1016/j.advwatres.2019.02.001.

Bhanja, S.N., Mukherjee, A., Rodell, M., Wada, Y., Chattopadhyay, C., Velicogna, I., et al., 2017. Groundwater rejuvenation in parts of India influenced by water-policy change implementation. Sci. Rep. 7, 7453. Available from: https://doi.org/10.1038/s41598-017-07058-2.

Blinn, D.W., 2008. The extreme environment, trophic structure, and ecosystem dynamics of a large, fishless desert spring: Montezuma Well, Arizona. In: Stevens, L.E., Meretsky, V.J. (Eds.), Aridland Springs in North America: Ecology and Conservation. University of Arizona Press, Tucson, AZ, pp. 98−126.

Boubaya, D., 2017. combining resistivity and aeromagnetic geophysical surveys for groundwater exploration in the Maghnia Plain of Algeria. J. Geol. Res. 2017, 14|Article ID 1309053. Available from: https://doi.org/10.1155/2017/1309053.

Buijze, L., Van Bijsterveldt, L., Cremer, H., Paap, B., Veldkamp, H., Wassing, B.B.T., et al., 2020. Review of induced seismicity in geothermal systems worldwide and implications for geothermal systems in the Netherlands. Neth. J. Geosci. 98, E13. Available from: https://doi.org/10.1017/njg.2019.6.

Butts, E., 2018. Geothermal system design. Water Well J. <https://waterwelljournal.com/geothermal-system-design/> (last accessed 02.04.20.).

Castro, A.P., Dougherty, S.L., Harrington, R.M., Cochran, E.S., 2019. Delayed dynamic triggering of disposal-induced earthquakes observed by a dense array in Northern Oklahoma. J. Geophys. Res. Solid Earth 124 (4), 3766−3781. Available from: https://doi.org/10.1029/2018JB017150.

Chan, N., Perron, J.T., Mitrovica, J.X., Gomez, N.A., 2018. new evidence of an ancient Martian ocean from the global distribution of valley networks. JGR Planets 123 (8), 2138−2150. Available from: https://doi.org/10.1029/2018JE005536.

Chandra, S., Auken, E., Maurya, P.K., Ahmed, S., Verma, S.K., 2019. Large scale mapping of fractures and groundwater pathways in crystalline hardrock by AEM. Sci. Rep. 9 (1), 398. Available from: https://doi.org/10.1038/s41598-018-36153-1.

Cohen, D., Person, M., Wang, P., Gable, C., Hutchinson, D., Marksamer, A., et al., 2009. Origin and extent of fresh paleowaters beneath the Atlantic continental shelf. Groundwater 48 (1), 143−158.

Cooper, H.M., Zhang, C., Selch, D., 2015. Incorporating uncertainty of groundwater modeling in sea-level rise assessment: a case study in South Florida. Clim. Change 129, 281−294. Available from: https://doi.org/10.1007/s10584-015-1334-1.

David, L.A., Maurice, C.F., Carmody, R.N., Gootenberg, D.B., Button, J.E., Wolfe, B.E., et al., 2014. Diet rapidly and reproducibly alters the human gut microbiome. Nature 505, 559−563. Available from: https://doi.org/10.1038/nature12820

Daigle, H., Screaton, E.J., 2015. Evolution of sediment permeability during burial and subduction. Geofluids 15 (1−2), 84−105. Available from: https://doi.org/10.1002/2015GL064542.

Darcy, H., 1856. Les fontaines publiques de la ville de Dijon. Victor Dalmont, Paris.

Davis, E.E., Elderfield, H. (Eds.), 2004. Hydrogeology of the Oceanic Lithosphere. Cambridge University Press, New York, ISBN 0521819296HB.

Devitt, D.A., Bird, B., 2016. Changes in groundwater oscillations, soil water content and evapotranspiration as the water table declined in an area with deep rooted phreatophytes. Ecohydrology 9 (6), 1082−1093. Available from: https://doi.org/10.1002/eco.1704.

Domenico, P.A., Schwartz, F.W., Physical and Chemical Hydrogeology, 824 pp. John Wiley & Sons. ISBN 0 471 50744 X; 0 471 52987 7.

Eckstein, Y., Eckstein, G.E., 2005. Transboundary aquifers: conceptual models for development of international law. Ground Water 43 (5), 679−690.

Engelen, G.B., Kloosterman, F.H., 1996. Hydrological systems analysis; methods and applications, Water Science and Technology Library, 20. Kluwer Academic Publishers, Dordrecht, The Netherlands, p. 152.

ESA, 2019. 7. Water loss. European Space Agency Venus Express. <https://sci.esa.int/web/venus-express/-/54068-7-water-loss> (last accessed 05.04.20.).

Etiope, G., Ciotoli, G., Schwietzke, S., Schoell, M., 2019. Gridded maps of geological methane emissions and their isotopic signature. Earth Syst. Sci. Data 11 (1), 1–22. Available from: https://doi.org/10.5194/essd-11-1-2019.

Faulds, J., 2019. Personal Communication, February 7, 2019.

Ferguson, G., McIntosh, J.C., Perrone, D., Jasechko, S., 2018. Competition for shrinking window of low salinity groundwater. Environ. Res. Lett. 13, 114013. Available from: https://doi.org/10.1088/1748-9326/aae6d8.

Ferguson, G.A.G., 2015. Deep injection of waste water in the western Canada sedimentary basin. Groundwater 53 (2), 187–194. Available from: https://doi.org/10.1111/gwat.12198.

Fetter Jr., C.W., 2004. Hydrogeology: a short history, Part 2. Ground Water 42 (6/7), 949–953.

Fetter, C.W., Boving, T., Kreamer, D.K., 2018. Contaminant Hydrogeology, third ed. Waveland Press, ISBN: 978-157766-583-0, p. 647ISBN. Available from: https://doi.org/10.1007/s12665-018-7921-5.

Fisher, N.K., 2019. The influence of Fe(III) and Mn(IV) on the biotransformation of hydrocarbons in groundwater. Geoecology. Eng. Geology, Hydrogeology, Geocryology 2019 (3), 21–31. Available from: https://doi.org/10.31857/S0869-78092019321-31.

Franzblau, R.E., Daughney, C.J., Swedlund, P.J., Weisener, C.G., Moreau, M., Johannessen, B., et al., 2016. Cu(II) removal by *Anoxybacillus flavithermus*-iron oxide composites during the addition of Fe(II)$_{aq}$. Geochim. Cosmochim. Acta 172, 139–158. Available from: https://doi.org/10.1016/j.gca.2015.09.031. Bibcode: 2016GeCoA.172.139F.

Frommen, T., Ambrus, K., 2019. Pani Check and Panni Doctors. <http://www.avbstiftung.de/projekte/artikel/news/pani-check-pani-doctors/> (last accessed 11.09.19.).

Gaston, L., Lapworth, D.J., Stuart, M., Arnscheidt, J., 2019. Prioritisation approaches for substances of emerging concern in groundwater: a critical review. Environ. Sci. Technol. 53 (11), 6107–6122.

Gibson-Poole, C.M., Svendsen, L., Underschultz, J., Watson, M.N., Ennis-King, J., Van Ruth, P.J., et al., 2006. Gippsland basin geosequestration: a potential solution for the Latrobe Valley brown coal CO_2 emissions. APPEA J. 46, 413–434. Available from: https://doi.org/10.1071/AJ05024.

Glass, R.J., Nicholl, M.J., 1996. Physics of gravity driven fingering of immiscible fluids within porous media: an overview of current understanding and selected complicating factors. Geoderma 70 (2–4), 133–163.

Goebel, M., Knight, R., Halkjær, M., 2019. Mapping saltwater intrusion with an airborne electromagnetic method in the offshore coastal environment, Monterey Bay, California. J. Hydrol. Reg. Stud. 23, 100602. Available from: https://doi.org/10.1016/j.ejrh.2019.100602.

Goebel, T.H.W., Brodsky, E.E., 2018. The spatial footprint of injection wells in a global compilation of induced earthquake sequences. Science 361, 899–904. Available from: https://doi.org/10.1126/science.aat5449. PMID 30166486.

Goudarzi, A., Delshad, M., Mohanty, K.K., Sepehrnoori, K., 2015. Surfactant oil recovery in fractured carbonates: experiments and modeling of different matrix dimensions. J. Pet. Sci. Eng. 125, 136–145.

Hadfield, 2019. Personal Communication.

Hall, S., 2016. Our solar system is overflowing with liquid water. Sci. Am. 314 (6), 14–15. <https://www.scientificamerican.com/article/our-solar-system-is-overflowing-with-liquid-water-graphic/> (accessed 05.04.20.).

Hazen, T.C., 2010. In situ: groundwater bioremediation. In: Timmis, K.N. (Ed.), Handbook of Hydrocarbon and Lipid Microbiology. Springer, Berlin, Heidelberg, ISBN: 978-3-540-77584-3. Available from: https://doi.org/10.1007/978-3-540-77587-4_191, ISBN.

Hortle, A., Xu, J., Dance, T., 2016. Integrating hydrodynamic analysis of flow systems and induced pressure decline at the Otway CO_2 storage site to improve reservoir history matching. Mar. Pet. Geol. 45, 159–170.

Hsieh, P., 2016. Personal Communication on May 4, 2016.

Hubbert, M.K., 1940. The theory of ground-water motion. J. Geol. 48, 785–944.

Humphreys, W.F., 2006. Aquifers: the ultimate groundwater-dependent ecosystems. Aust. J. Bot. 54, 115–132.

Hynds, P., Regan, S., Andrade, L., Mooney, S., O'Malley, K., DiPelino, S., et al., 2018. Muddy waters: refining the way forward for the "sustainability science" of socio-hydrogeology. Water 10 (9), 1111. Available from: https://doi.org/10.3390/w10091111.

IAEA, 2013. Isotope Methods for Dating Old Groundwater. International Atomic Energy Agency, Vienna, ISBN: 978-92-0-137210-9, p. 357, *Publication* 1587, ISBN.

International Geothermal Association (IGA), 2020. <https://www.geothermal-energy.org/> (last accessed 13.04.20.).

Ingebritsen, S., Sanford, W., Neuzil, C., 2006. Groundwater in Geologic Processes, second ed. Cambridge University Press, p. 536.

Jakeman, A.J., Barreteau, O., Hunt, R.J., Rinaudo, J.-D., Ross, A. (Eds.), 2016. Integrated Groundwater Management, Concepts, Approaches and Challenges. Springer NatureISBN 978-3-319-23576-9. Available from: http://doi.org/10.1007/978-3-319-23576-9.

Jiao, J., Post, V., 2019. Coastal Hydrogeology. Cambridge University Press, p. 418, EAN: 9781107030596.

Kalisch, H., Mitrovic, D., Nordbotten, J.M., 2016. Rayleigh–Taylor instability of immiscible fluids in porous media. Continuum Mech. Thermodyn. 28, 721–731. Available from: https://doi.org/10.1007/s00161-014-0408-z.

Kamal, M.S., Adewunmi, A.A., Sultan, S.S., Al-Hamad, M.F., Mehmood, U., 2017. Recent advances in nanoparticles enhanced oil recovery: rheology, interfacial tension, oil recovery, and wettability alteration. J. Nanomater. 2017, 15Article ID 2473175. Available from: https://doi.org/10.1155/2017/2473175.

Kastner, M., Becker, K., Davis, E.E., Fisher, A.T., Jannasch, H.W., Solomon, E.A., et al., 2015. New insights into the hydrogeology of the oceanic crust through long-term monitoring. Oceanography 19 (4), 46–57.

Kath, J., Boulton, A.J., Harrison, E.T., Dyer, F.J., 2018. A conceptual framework for ecological responses to groundwater regime alteration (FERGRA). Ecohydrology 11 (7). Available from: https://doi.org/10.1002/eco.2010.

Kaushal, S.S., Likens, G.E., Pace, M.L., Utz, R.M., Haq, S., Gorman, J., et al., 2018. Freshwater salinization syndrome on a continental scale. Proc. Natl. Acad. Sci. 115, E574–E583. Available from: https://doi.org/10.1073/pnas.1711234115.

Keranen, K.M., Weingarten, M., 2018. Induced seismicity. Annu. Rev. Earth Planet. Sci. 46, 149–174.

King, J., Oude Essink, G., Karaolis, M., Siemon, B., Bierkens, M.F.P., 2018. Quantifying geophysical inversion uncertainty using airborne frequency domain electromagnetic data: applied at the province of Zeeland, The Netherlands. Water Resour. Res. 54, 8420–8441. Available from: https://doi.org/10.1029/2018WR023165.

Kodrick-Brown, A., Brown, J.H., 2007. Native fishes, exotic mammals, and the conservation of desert springs. Front. Ecol. Environ. 5, 549–553.

Kreamer, D.K., 2001. Down the rabbit hole with Alice – sucking soil gas all the way. Ground Water Monit. Rem. 21 (4), 52–56.

Kreamer, D.K., 2012. The past present and future of water conflict and international security. J. Contemp. Water Res. Educ. 149, 87–94.

Kreamer, D.K., 2016. Hydrophilanthropy gone wrong—how well-meaning scientists, engineers, and the general public can make the worldwide water and sanitation situation worse. In: Wessel, G.R., Greenberg, J.K. (Eds.), Geoscience for the Public Good and Global Development: Toward a Sustainable Future. pp. 1–15. *Geological Society of America Special Paper 520*, doi: 10.1130/2016.2520(19).

Kreamer, D.K., Stevens, L.E., Ledbetter, J.D., 2015. Chapter 9: Groundwater dependent ecosystems – policy challenges and technical solutions. In: Adelana, S. (Ed.), Groundwater, Hydrochemistry, Environmental Impacts and Management Impacts. Nova Publishers, New York, pp. 205–230. ISBN 978-1-63321. Online access: <https://www.novapublishers.com/catalog/product_info.php?products_id = 52986&osCsid = be410bfe49edb2ea0ea3239891d33244>.

Kumar, D., Rao, V.A., Sarma, V.S., 2014. Hydrogeological and geophysical study for deeper groundwater resource in quartzitic hard rock ridge region from 2D resistivity data. J. Earth Syst. Sci. 123 (3), 531–543. <https://www.ias.ac.in/article/fulltext/jess/123/03/0531-0543> (accessed 05.04.20.).

Langevin, C.D., Hughes, J.D., Banta, E.R., Niswonger, R.G., Panday, S., Provost, A.M., 2017. Documentation for the MODFLOW 6 groundwater flow model. U.S. Geological Survey Techniques and Methods. , p. 197*book 6*, chap. A55. Available from: https://doi.org/10.3133/tm6A55.

Lapworth, D.J., Lopez, B., Laabs, V., Kozel, R., Wolter, R., Ward, R., et al., 2019. Developing a groundwater watch list for substances of emerging concern: a European perspective. Environ. Res. Lett. 14, 035004. <https://iopscience.iop.org/article/10.1088/1748-9326/aaf4d7/pdf>.

Limaye, S.D., 2017. Socio-hydrogeology and low-income countries: taking science to rural society. Hydrogeol. J. (25), 1927–1930. Available from: https://doi.org/10.1007/s10040-017-1656-3.

Limberger, J., Boxem, T., Pluymaekers, M., Bruhn, D., Manzella, A., Calcagno, P., et al., 2018. Geothermal energy in deep aquifers: a global assessment of the resource base for direct heat utilization. Renew. Sustain. Energy Rev. 82, 961–975.

Lu, J., Wu, J., Zhang, C., Zhang, Y., 2020. Possible effect of submarine groundwater discharge on the pollution of coastal water: occurrence, source, and risks of endocrine disrupting chemicals in coastal groundwater and adjacent seawater influenced by reclaimed water irrigation. Chemosphere 250, Article 126323. Available from: https://doi.org/10.1016/j.chemosphere.2020.126323.

Luijendijk, E., Gleeson, T., Moosdorf, N., 2020. Fresh groundwater discharge insignificant for the world's oceans but important for coastal ecosystems. Nat. Commun. 11, 1260. Available from: https://doi.org/10.1038/s41467-020-15064-8.

Mádl-Szőnyi, J., Simon, S., 2016. Involvement of preliminary regional fluid pressure evaluation into the reconnaissance geothermal exploration – example of an overpressured and gravity driven basin. Geothermics 60, 156–174.

Mádl-Szőnyi, J., Tóth, Á., 2015. Basin-scale conceptual groundwater flow model for an unconfined and confined thick carbonate region. Hydrogeol. J. 23, 1359–1380. Available from: https://doi.org/10.1007/s10040-015-1274-x.

Meinzer O.E., 1928. Compressibility and elasticity of artesian aquifers, Economic Geology, 23 (3):263. Available from: http://dx.doi.org/10.2113/gsecongeo.23.3.263.

Meixner, T., Manning, A.H., Stonestrom, D.A., Allen, D.M., Ajami, H., Blasch, K.W., et al., 2016. Implications of projected climate change for groundwater recharge in the western United States. J. Hydrol. 534, 124–138. Available from: https://doi.org/10.1016/j.jhydrol.2015.12.027.

Miller, C.T., Gray, W.G., 2008. Hydrogeological research, education, and practice: a path to future contributions. J. Hydrol. Eng. 13 (1). Available from: https://doi.org/10.1061/(ASCE)1084-0699(2008)13:1(7).

Moreau, M., Hadfield, J., Hughey, J., Sanders, F., Lapworth, D.J., White, D., et al., 2019. A baseline assessment of emerging organic contaminants in New Zealand groundwater. Sci. Total Environ. 686, 425–439. Available from: https://doi.org/10.1016/j.scitotenv.2019.05.210.

Moumouni, A., Fryar, A.E., 2017. Controls on groundwater quality and dug-well asphyxiation hazard in Dakoro area of Niger. Groundw. Sustain. Dev. 5, 235–243. Available from: https://doi.org/10.1016/j.gsd.2017.08.004.

Mukherjee, A., Bhanja, S.N., Wada, Y., 2018a. Groundwater depletion causing reduction of baseflow triggering Ganges river summer drying. Sci. Rep. 8, 12049. Available from: https://doi.org/10.1038/s41598-018-30246-7.

Mukherjee, A., Duttagupta, S., Chattopadhyay, S., 2019. Impact of sanitation and socio-economy on groundwater fecal pollution and human health towards achieving sustainable development goals across India from ground-observations and satellite-derived nightlight. Sci. Rep. 9, 15193. Available from: https://doi.org/10.1038/s41598-019-50875-w.

Mukherjee, A., Fryar, A.E., Eastridge, E.M., Nally, R.S., Chakraborty, M., Scanlon, B.R., 2018b. Controls on high and low groundwater arsenic on the opposite banks of the lower reaches of River Ganges, Bengal basin, India. Sci. Total Environ. 645, 1371–1387. Available from: https://doi.org/10.1016/j.scitotenv.2018.06.376.

Mukherjee, A., Scanlon, B., Aureli, A., Langan, S., Guo, H., McKenzie, A., 2020. Global Groundwater: Source, Scarcity, Sustainability, Security and Solutions, first ed. Elsevier, ISBN: 9780128181720.

Murphy, B., Morrison, R., 2015. Introduction to Environmental Forensics, third ed. Academic Press, p. 704, Hardcover ISBN: 9780124046962, eBook ISBN: 9780124047075.

Musacchio, A., Re, V., Mas-Pla, J., Sacchi, E., 2020. EU Nitrates Directive, from theory to practice: environ-mental effectiveness and influence of regional governance on its performance. AMBIO 49 (2), 504–516. Available from: https://doi.org/10.1007/s13280-019-01197-8.

Nelskamp, S., Verweij, J.M., 2012. Using Basin Modeling for Geothermal Energy Exploration in the Netherlands – An Example from the West Netherlands Basin and Roer Valley Graben. Ministry of Economic Affairs, 4DMod/Geothermie Project Number 034.24607/034.20566 TNO-060-UT-2012-00245, <https://www.nlog.nl/sites/default/files/west%20netherlands%20basin_nelskamp%20and%20verweij_tno-060-ut-2012-00245.pdf>.

NIC, 2012. National Intelligence Council, Special Report – Intelligence Community Assessment, Global Water Security, U.S. National Intelligence Council, Intelligence Community Assessment. Defense Intelligence Agency, pp. 1–16. <https://www.dni.gov/files/documents/Special%20Report_ICA%20Global%20Water%20Security.pdf>.

NJG, 2017. Induced seismicity in the Groningen gas field, the NetherlandsSpec. issue Neth. J. Geosci. 96, 284.

Paiqa, P., Delerue-Matos, C., 2016. Determination of pharmaceuticals in groundwater collected in five cemeteries' areas (Portugal). Sci. Total Environ. 569 (2016), 16–22. Available from: https://doi.org/10.1016/j.scitotenv.2016.06.090.

Post, V., Groen, J., Kooi, H., Person, M., Ge, S., Edmunds, W.M., 2013. Offshore fresh groundwater reserves as a global phenomenon. Nature 504, 71–78. Available from: https://doi.org/10.1038/nature12858.

Re V., Sacchi E., Kammoun S., Tringali C., Trabelsi R., Zouari K., Daniele, S., 2017. Integrated socio-hydrogeological approach to tackle nitrate contamination in groundwater resources. The case of Grombalia Basin (Tunisia). Science of the Total Environ. 593–594: 664–676. Accessed 03.11.19.

Re, V., 2020. Socio-hydrogeology and geoethics. State of the art and future challenges. In: proceedings of the first congress on Groundwater and Geoethics, Porto (Portugal), 18–22 May 2020. In press.

Re, V., 2015. Incorporating the social dimension into hydrogeochemical investigations for rural development: the Bir Al-Nas approach for socio-hydrogeology. Hydrogeol. J. 23, 1293–1304.

Re, V., 2019. Shedding light on the invisible: addressing the potential for groundwater contamination by plastic microfibers. Hydrogeol. J. 27, 2719–2727. Available from: https://doi.org/10.1007/s10040-019-01998-x.

Re, V., Misstear, B., 2017. Education and capacity development for groundwater resources management. In: Villholth, K.G., Lopez-Gunn, E., Conti, K., Garrido, A., Van Der Gun, J. (Eds.), Advances in Groundwater Governance. CRC Press, London, pp. 212–228.

Salese, F.F., Pondrelli, M., Neeseman, A., Schmidt, G., Ori, G.G., 2019. Geological evidence of planet-wide groundwater system on Mars. J. Geophys. Res. Planets 124 (2019), 374–395. Available from: https://doi.org/10.1029/2018je005802.

Schmidt, T.C., 2018. Recent trends in water analysis triggering future monitoring of organic micropollutants. Anal Bioanal Chem 410, 3933–3941. Available from: https://doi.org/10.1007/s00216-018-1015-9.

Schleussner, C.F., Donges, J.F., Donner, R.V., Schellnhuber, H.J., 2016. Enhanced conflict risks by natural disasters. Proc. Natl. Acad. Sci. 113 (33), 9216–9221. Available from: https://doi.org/10.1073/pnas.1601611113.

Schneider, J., 2020. Interactive extra-solar planets catalog. In: The Extrasolar Planets Encyclopedia. (last accessed 02.04.20.).

Schwartz, F.W., Liu, G., Aggarwal, P., Schwartz, C.M., 2017. Naïve simplicity: the overlooked piece of the complexity – simplicity paradigm. Groundwater 55 (5), 703–711. Available from: https://doi.org/10.1111/gwat.12570.

Sclater, J.G., 2004. Variability of heat flux through the seafloor: discovery of hydrothermal circulation in the oceanic crust. Hydrogeology of the Oceanic Lithosphere. Cambridge University Press, Cambridge, pp. 3–27.

Sharp, J.M., 2014. In: Sharp, J.M. (Ed.), Fractured Rock Hydrogeology. Taylor Francis books, CRC Press, ISBN: 9780429227530p. 408. eBook ISBN. Available from: https://doi.org/10.1201/b17016.

Siebert S., Burke J., Faures J.M., Frenken K., Hoogeveen J., Doll P., et al., 2010. Groundwater use for irrigation – a global inventory, Hydrol. Earth Syst. Sci. 14, 1863–1880.

Simmons, C.T., 2005. Variable density groundwater flow: from current challenges to future ossibilities. Hydrogeol. J. 13, 116–119. Available from: https://doi.org/10.1007/s10040-004-0408-3.

Simmons, C.T., Hunt, R.J., Cook, P.G., 2012. Using every tool in the toolbox. Ground Water 50 (3), 323.

Sirisena, K., Daughney, C.J., Moreau, M., Sim, D.A., Lee, C.K., Cary, S.C., et al., 2018. Bacterial bioclusters relate to hydrochemistry in New Zealand groundwater. FEMS Microbiol. Ecol. 94 (11). Available from: https://doi.org/10.1093/femsec/fiy170. article fiy170.

Sirisena, K.A., Daughney, C.J., Moreau-Fournier, M., Ken, G., Ryan, K.G., Chambers, G.K., 2013. National survey of molecular bacterial diversity of New Zealand groundwater: relationships between biodiversity, groundwater chemistry and aquifer characteristics. FEMS Microbiol. Ecol. 86 (3), 490–504. Available from: https://doi.org/10.1111/1574-6941.12176.

Skoumal, R.J., Ries, R., Brudzinski, M.R., Barbour, A.J., Currie, B.S., 2018. Earthquakes induced by hydraulic fracturing are pervasive in Oklahoma. J. Geophys. Res. Solid Earth 123 (12), 10,918–10,935. Available from: https://doi.org/10.1029/2018JB016790.

Sorensen, J.P.R., Baker, A., Cumberland, S.A., Lapworth, D.J., MacDonald, A.M., Pedley, S., et al., 2018. Real-time detection of faecally contaminated drinking water with tryptophan-like fluorescence: defining threshold values. Sci. Total Environ. 622-623, 1250–1257. Available from: https://doi.org/10.1016/j.scitotenv.2017.11.162.

Stelma (Jr), G.N., Wymer, L.J., 2012. Research considerations for more effective groundwater monitoring. J. Water Health 10 (4), 511–521. Available from: https://doi.org/10.2166/wh.2012.016.

Streetly, M., Heathcote, J., 2018. Advances in groundwater system measurement and monitoring documented in 50 years of QJEGH. Q. J. Eng. Geol. Hydrogeol. 51 (2), 139–155. Available from: https://doi.org/10.1144/qjegh2016-140.

Stuart, M.E., Lapworth, D.J., 2014. Chapter 2: Transformation products of emerging organic compounds as future groundwater and drinking water contaminants. In: Lambropoulou, Nollet (Eds.), Transformation *Products of Emerging Contaminants in the Environment: Analysis, Processes, Occurrence, Effects and Risks.* Wiley, ISBN: 978-1-118-33959-6964. Available from: https://doi.org/10.1002/9781118339558.ch02.

Sultan, A.S.A., Mohamadin, M.I., Sabet, H.S., Takey, M.S., 2019. Geophysical interpretation for groundwater exploration around Hurghada area, Egypt. NRIAG J. Astron. Geophys. 8 (1), 171−179. Available from: https://doi.org/10.1080/20909977.2019.1647389.

Taylor, R.G., et al., 2013. Groundwater and climate change. Nat. Clim. Change 3, 322−329.

Tobin, B., Springer, A., Kreamer, D.K., Schenk, E., 2018. Review: the distribution, flow, and quality of Grand Canyon springs, Arizona (USA). Hydrogeol. J. 26 (3), 721−732. Available from: https://doi.org/10.1007/s10040-017-1688-8. ISSN 1431-2174.

Tóth, J., 2016. The evolutionary concepts and practical utilization of the Tóthian theory of regional groundwater flow. Int. J. Earth Environ. Sci. 1, 111. Available from: https://doi.org/10.15344/2456-351X/2016/111.

Tully, K., Gedan, K., Epanchin-Niell, R., Strong, A., Bernhardt, E.S., BenDor, T., et al., 2019. The invisible flood: the chemistry, ecology, and social implications of coastal saltwater intrusion. BioScience 69 (5), 368−378. Available from: https://doi.org/10.1093/biosci/biz027.

Verweij, H., Underschultz, J., Mádl-Szőnyi, J., Wendebourg, J., 2019. Basin hydrodynamicsSpec. IssueMar. Pet. Geol. Available from: https://www.journals.elsevier.com/marine-and-petroleum-geology/special-issues.

Verweij, J.M., 1999. Application of fluid flow systems analysis to reconstruct the post Carboniferous hydrogeohistory of the onshore and offshore Netherlands. Mar. Pet. Geol. 16, 561−579.

Verweij, J.M., 2003. Fluid Flow Systems Analysis On Geological Timescales in Onshore and Offshore Netherlands; With Special Reference to the Broad Fourteens Basin. NSG Contribution, 278 p., 2003.09.05; ISBN 90-5986-035-7.

Verweij, J.M., Simmelink, H.J., Underschultz, Witmans, N., 2012. Pressure and fluid dynamic characterisation of the Dutch subsurface. Neth. J. Geosci. 91 (4), 465−490.

Villholth, K.G., Lopez-Gunn, E., Conti, K., Garrido, A., Van Der Gun, J. (Eds.), 2019. Advances in Groundwater Governance. CRC Press, Taylor and Francis, ISBN 9781138029804 − CAT# K30061.

Wei, C., Wang, Q., Song, X., Chen, X., Fan, R., Ding, D., et al., 2018. Distribution, source identification and health risk assessment of PFASs and two PFOS alternatives in groundwater from non-industrial areas. Ecotoxicol. Environ. Saf. 152, 141−150. Available from: https://doi.org/10.1016/j.ecoenv.2018.01.039.

Werner, A.D., Bakker, M., Post, V.E.A., Vandenbohede, A., Lu, C., Ataie-Ashtiani, B., et al., 2013. Seawater intrusion processes, investigation and management: recent advances and future challenges. Adv. Water Resour. 51, 3−26. Available from: http://doi.org/10.1016/j.advwatres.2012.03.004.

Werner, A.D., Simmons, C.T., 2009. Impact of sea-level rise on seawater intrusion in coastal aquifers. Ground Water 47, 197−204. Available from: https://doi.org/10.1111/j.1745-6584.2008.00535.x.

Wilson, S., Zhang, H., Burwell, K., Samantapudi, A., Dalemarre, L., Jiang, C., et al., 2013. Leaking underground storage tanks and environmental injustice: is there a hidden and unequal threat to public health in South Carolina? Environ. Justice 6 (5), 175−182. Available from: https://doi.org/10.1089/env.2013.0019.

Worthington, S.R.H., 2013. Development of ideas on channel flow in bedrock in the period 1850-1950. Groundwater 51 (5), 804−808.

Worthington, S.R.H., Davies, G.J., Alexander Jr., E.C., 2016. Enhancement of bedrock permeability by weathering. Earth Science Rev. 160, 188−202.

Zeng, L., Jacobsen, S.B., Sasselov, D.D., Petaev, M.I., Vanderburg, A., Lopez-Morales, M., et al., 2019. Growth model interpretation of planet size distribution. PNAS 116 (20), 9723−9728. <https://doi.org/10.1073/pnas.1812905116>.

Further reading

Siebert, S., Burke, J., Faures, J.M., Frenken, K., Hoogeveen, J., Doll, P., et al., 2010. Groundwater use for irrigation − a global inventory. Hydrol. Earth Syst. Sci. 14, 1863−1880.

Chapter 38

Technologies to enhance sustainable groundwater use

Roger Sathre[1,2]
[1]*Institute for Transformative Technologies, Berkeley, CA, United States,* [2]*Linnaeus University, Växjö, Sweden*

38.1 Technology levers to enhance groundwater security

Groundwater is increasingly important to human welfare, and dependence on groundwater resources is growing. However, renewable groundwater supply is ultimately limited, and overexploitation is leading to depletion and contamination of aquifers (Mukherjee et al., 2020). This chapter discusses a wide range of technology interventions aimed at improving groundwater security.

First, technologies for groundwater mapping are described, with the goal of understanding aquifer processes leading to improved groundwater management. Knowledge gained from groundwater mapping can be used to manage the recharging of aquifers, enabling higher rates of rainwater capture and more renewable groundwater. Groundwater mapping also informs the management of saline groundwater intrusion, to eliminate aquifer contamination that will be an increasing problem due to sea-level rise.

A range of technologies can reduce groundwater demand by improving water-use efficiency, so that less water is needed to achieve the same ends. In the agricultural sector, many irrigation efficiency improvements are available to reduce the quantities of water extracted and applied to fields. However, the net groundwater implications of these technologies are complex, as excess irrigation water often contributes to groundwater recharge. In the household and municipal sector, end-use water efficiency improvements can reduce final water demand, and distribution improvements can reduce water losses and extraction requirements. In the industrial sector, global best practice efficiency improvements can strongly reduce water demand for many industrial processes.

Many technologies are available to improve the quality of groundwater that is contaminated by impurities. Many regions possess abundant brackish groundwater that is presently unutilized, and emerging technologies can desalinate this water at much lower cost and energy use than seawater desalination. In regions where groundwater is contaminated by arsenic and fluoride, purification technologies are available to remove these natural toxins. Groundwater in some regions contains biological pathogens due to inadequate disposal of fecal waste, and a range of technologies can be used to make this water safe for consumption.

While many regions suffer from inadequate groundwater quantity or quality, other regions, including much of sub-Saharan Africa, possess abundant clean groundwater that is currently underutilized due to economic water scarcity. In these areas, improved technologies are needed to access and extract the groundwater. Low-cost methods for digging or drilling wells, and for pumping groundwater to the surface, can improve the well-being and economic development of people in these regions.

Many technology levers already exist that could enhance the sustainable use of groundwater resources, and others are being developed to meet growing needs. Widespread deployment of these technologies has the potential to significantly improve water security conditions for people around the globe. Concerted effort is needed to enable the economic, political, and logistical requirements of such deployment, so that renewable groundwater resources contribute maximally to long-term water security and resilience.

38.2 Groundwater mapping and management

The technology set for groundwater mapping seeks to understand the broad hydrogeological landscape, to inform better management of groundwater resources. Comprehensive groundwater mapping can be used to determine the location and quality of groundwater, as well as aquifer recharge mechanisms that affect the sustainable quantities of groundwater available for extraction and use. A primary focus of groundwater mapping is on the mechanisms of sustainable groundwater cycling, to facilitate long-term utilization and enhancement [through, e.g., managed aquifer recharge (MAR)] of sustainable groundwater resources. It may also be used to temporarily increase water supply by enabling the precise siting and one-time extraction of remaining fossil water aquifers. India [CGWB (Indian Central Ground Water Board), 2019] and Ethiopia [ATA (Ethiopian Agricultural Transformation Agency), 2019] have conducted pilot groundwater mapping studies in areas with different hydrogeological terrains, with intentions to map the entire countries and improve resource usage.

A range of techniques is used to provide primary data for groundwater mapping (Klee et al., 2015). Data are gathered across many scales, from in situ underground sampling, to surface-based sensing techniques, to remote sensing from aircraft or satellites. Multiple techniques are commonly used in parallel to combine information on different characteristics, to generate more robust maps. Numerous sensing technologies have been developed based on acoustic, electric, or magnetic principles, to provide data on geological structures, surface morphology, and their hydrologic characteristics:

1. Seismic surveys are made by propagating acoustic energy through the ground, and tracking seismic refraction of compression waves that show increasing velocity with density. Seismic surveys provide information on the internal structure of aquifers such as clay and silt layers that limit the overall vertical permeability.
2. Electrical resistivity techniques (such as electrical resistivity tomography) induce an electrical current in the ground, and the resulting electrical potential at different locations is used to measure the variation in ground conductivity, or its inverse, resistivity. Different materials, and the fluids within them, will show different abilities to conduct an electric current. Electrical resistivity methods are often used for well siting and for locating suitable sites for percolation fields in hard rock areas.
3. Ground penetrating radar emits high-frequency electromagnetic waves into the ground and receives and interprets the microwave energy reflected back to the surface from different underground materials, to provide detailed subsurface cross sections.
4. Airborne transient electromagnetic systems can be fitted to helicopters or fixed-wing aircraft or drones and are used to map the apparent conductivity of the ground.
5. Time domain electromagnetic methods can map the shallow subsurface but are susceptible to interference from pipelines and power lines. In this method, current pulses are sent through a large square wire loop on the ground. The decay of current at the end of each pulse generates a magnetic field that enters the ground. Eddy currents induced by this changing magnetic field generate secondary magnetic fields in the ground. The amplitude and rate of decay of these secondary fields are measured at the surface and analyzed to determine underground characteristics.
6. Frequency domain electromagnetics are typically used to measure variations in lateral conductivity along linear or gridded profiles.
7. Direct physical surveys can be made of existing wells to determine the quality of groundwater as well as the depth and fluctuation of the water table. This is typically done manually (if at all), providing only intermittent data points. The development and deployment of a distributed network of digital sensors to provide real-time monitoring of well water characteristics throughout the region, could be an important advance for sustainable groundwater management. Hydrogeological data from multiple sources could then be integrated to form the basis of an online map of real-time groundwater quality and quantity.

38.3 Managing aquifer recharge

As groundwater mapping provides greater understanding of subsurface conditions, this knowledge can be applied by MAR technologies that aim to increase the rate of groundwater recharge to allow greater groundwater extraction without risk of water table decline. While some deep aquifers contain fossil water that was stored long ago and does not circulate unless accessed and extracted by wells, most aquifers are dynamic and receive newer water via recharge mechanisms while losing older water via groundwater pumping and natural discharge to rivers and oceans. While groundwater extraction is increasing, the rate of natural aquifer recharge is diminishing in many regions, due to rapid urbanization and land use changes that reduce infiltration of rainwater into the soil.

MAR is achieved by reducing the fraction of rainwater that runs off the land surface, thus increasing the fraction that infiltrates through the land surface and enters the soil. This is typically implemented through engineered structures

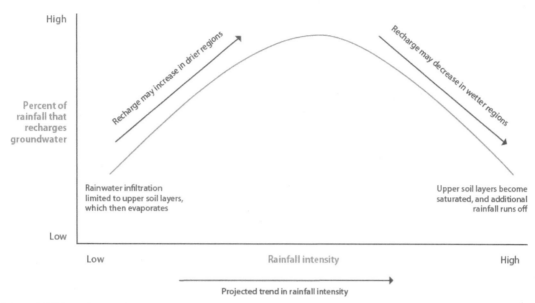

FIGURE 38.1 Rainfall intensity is projected to increase due to global climate change, which may have variable effects on groundwater recharge. *Credit: ITT (Institute for Transformative Technologies), 2018. Technology Breakthroughs for Global Water Security: A Deep Dive into South Asia. Institute for Transformative Technologies, Berkeley, CA [ITT (Institute for Transformative Technologies), 2018].*

that slow the downstream flow of surface runoff water, allowing more of it to infiltrate into the ground [CGWB (Indian Central Ground Water Board), 2007]. A wide range of structures is available at varying scales, including farm-level swales, check dams, percolation tanks and ponds, dams, and barrages (Gale, 2005).

MAR is suitable only in particular locations, because it requires three conditions: the availability of uncommitted surface water, underground storage space, and the demand for groundwater (Shah, 2008). In drier regions the amount and timing of rainfall limit the amount of runoff that may be harnessed to recharge groundwater. In regions with unsuitable geology, even if seasonal water is plentiful, there may be inadequate aquifer porosity to store significant water or there may be natural barriers between the surface runoff and underground aquifer.

In closed river basins, where all available surface water is allocated and used, MAR will not increase total water supply even if local runoff and geology are suitable, because an upstream user's gain will lead to a downstream user's loss. Furthermore, the effects of future climate change on aquifer recharging are uncertain. The amount, timing, and intensity of precipitation are projected to change, though its effect on the partition of rainwater into runoff and infiltration will vary by location (see Fig. 38.1).

Nevertheless, where conditions are suitable, MAR may contribute significantly to regional groundwater security by increasing allowable sustainable groundwater extraction rates. Ideally, groundwater should be considered a storage reservoir to smooth fluctuations and allow flexible access, not a stock to be depleted.

38.4 Managing saline groundwater intrusion

The technology set for managing groundwater salinity is an increasingly important application of groundwater mapping. Fresh groundwater aquifers are often surrounded by saltwater on one or more sides or underneath. Since freshwater is less dense than saline water, it tends to flow on top of the surrounding or underlying saline groundwater. Under natural conditions the boundary between freshwater and saltwater maintains a stable equilibrium. Under some circumstances the saltwater can move (or intrude) into the freshwater aquifer, making the water nonpotable. When freshwater is pumped from an aquifer that is near saline groundwater, the boundary between saltwater and freshwater moves in response to the pumping. If this continues, unusable saline water will be pumped up from the well.

Freshwater aquifers are naturally recharged by rainwater, and the recharge rate can be manipulated by MAR. Thus there is dynamic interplay between freshwater withdrawals, freshwater recharge, and surrounding saline aquifers. Rising sea levels due to climate change are slowly increasing the gradient of saline water, although coastal aquifers are more vulnerable to groundwater extraction than to predicted sea-level rise (Ferguson and Gleeson, 2012).

Techniques are under development to manage saline groundwater intrusion. Actions, such as controlling the rate and depth of groundwater extractions and augmenting freshwater recharge by MAR, can reduce or eliminate undesired saline intrusion. Skimming wells can also be used to sustainably exploit fresh groundwater lenses that overlie native saline groundwater (Saeed and Ashraf, 2005). The freshwater lenses are renewed through percolation of rain and irrigation water. Skimming wells are designed and operated to minimize the mixing between overlying freshwater and underlying saline water.

Successful management of saline groundwater intrusion requires a deep understanding of aquifer dynamics and how they may be manipulated. The tools of groundwater mapping provide this essential knowledge of the hydrogeological landscape.

38.5 Improving groundwater-use efficiency

By increasing the efficiency with which water resources are used, more utility can be obtained from each unit of available groundwater. Different technology sets are relevant for the agricultural, household, and industrial sectors.

38.5.1 Improving irrigation and agricultural efficiency

Irrigation of cropland is the greatest user of groundwater in many regions. Irrigation efficiency is typically measured in terms of the percentage of applied water that is taken up by the roots of growing crops. However, the net groundwater implications of improving irrigation efficiency are complex, because much of the applied water that is not used by crops infiltrates into deeper soil horizons and recharges groundwater stocks [Perry, 2007; FAO (Food and Agriculture Organization of the United Nations), 2017].

A range of existing technologies and methods can be used to increase water-use efficiency in irrigation, with varying costs. The most common practice globally is surface irrigation, where water is applied directly to the surface of a flat or gently sloped field. Two promising options for improving efficiency of surface irrigation are precision land leveling and tensiometer-based irrigation scheduling. Two important alternatives to surface irrigation are sprinkler irrigation and drip irrigation, which are increasingly used for higher value crops for increased control and production as well as for water savings.

With simple surface irrigation techniques, less than half of the applied water is typically transpired by the growing crops, and a small percentage is lost to nonproductive evaporation. The remaining water infiltrates into deeper soil layers and becomes groundwater that can be recycled by wells. Depending on rainfall and soil characteristics, farm field infiltration can comprise a significant source of groundwater recharge (see Fig. 38.2). From a basin-level perspective, therefore, the leaked water is not truly lost, because it can be abstracted as groundwater and used (Grafton et al., 2018).

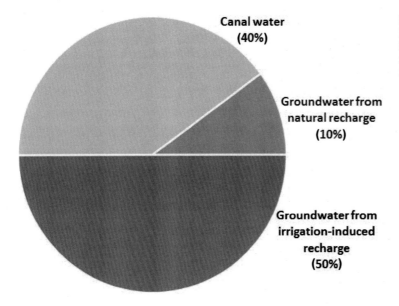

FIGURE 38.2 Half of irrigation water in Pakistan's Punjab province is from groundwater recharged through irrigation water infiltration. *Credit: data from World Bank, 2005. Pakistan's water economy running dry (World Bank, 2005).*

While irrigation water-use efficiency improvements focus on reducing water inputs to agriculture, there are numerous other agricultural improvements that increase crop production without directly affecting water use. This indirectly enhances groundwater security by increasing overall agricultural production, thus reducing the need for additional irrigation water to satisfy the growing demand for food. Farm-level management may be improved by stronger agricultural extension services to inform about and advocate higher yielding, water conserving agricultural crops and techniques. Foresighted policy instruments, including appropriate subsidies and taxes on inputs such as seeds, fertilizers, and water, can stimulate more rational consumption patterns within the agricultural sector.

There is great scope for increasing agricultural yields in low-achieving irrigated regions, via appropriate agricultural extension activities and other interventions. Many farmers could raise water productivity by adopting proven agronomic practices such as soil fertility maintenance and pest management (Molden and Oweis, 2007). The highest gains in water productivity are likely in areas where yields are still low, warranting a development focus on regions with lower agricultural performance.

38.5.2 Improving household water distribution and use efficiency

In regions that rely on groundwater for household water supply, improving the efficiency of delivering and using groundwater resources can be an important lever for sustainability. Much water is lost to leakage during municipal distributions. Effectively managing physical losses (leakage) in distribution systems requires active control measures, speed and quality of repairs, and effective pressure management. Globally, substantial experience has been accumulated in successfully distributing continuous water supply throughout large cities. Best practice recommendations have been developed for broad actions to reduce nonrevenue water and intermittent water supply, including complete metering of production and consumption, improved billing and collection, and identification and repair of visible and invisible leaks [ADB (Asian Development Bank), 2010].

Metering of water flows at multiple points throughout a municipal water utility system is important to managing municipal water flows and identifying leakage. Modern flow metering, pressure management, and data capture technologies can quickly identify burst pipes and estimate the gradual accumulation of smaller leaks (Simbeye, 2010). Flow meters can only detect the general area of leakage but cannot pinpoint the exact location of a leak. For this, sensors such as ultrasonic noise loggers, leak noise correlators, and ground microphones are used to detect the exact location of the leakage for repair.

Another important tool to reduce urban water loss is pressure management, as leakage rates are very sensitive to system pressure. The rate of leakage in water distribution networks is a function of the pressure applied by pumps or gravity. There is a direct physical relationship between the water pressure, and both slow leakage rate and the frequency of burst pipes. The most common and cost-effective measure is automatic pressure reducing valves that are installed at strategic points in the network to reduce or maintain network pressure at a set level. Other pressure management measures include air relief valves to release negative pressures or air bubbles in a pipeline, variable speed controllers, and break-pressure tanks.

Notwithstanding the accumulated global best practices for urban water distribution, many municipal utilities are currently challenged to provide their inhabitants with continuous supply of high-quality water. Intermittent water supply, in particular, is impeding the adoption of improved water utility management practices such as metering and automated control systems. These best practice approaches require continuous water supply and cannot be applied where pressure is supply-driven rather than demand-driven (Kumpel and Nelson, 2016).

Another technology set for reducing groundwater demand aims to use less household water. Improved appliances are commercially available that use less water than conventional appliances that perform the same function. Examples include low-flush toilets, low-flow sinks and showers, and water-efficient clothes washing machines. These water-efficient devices are gaining increasing attention in some industrialized regions facing water constraints. In the current emerging market context, water conserving flush toilets may play a role in reducing groundwater demand, through replacing older flush mechanisms in existing buildings and installing high-efficiency flush toilets in new construction. Though challenging, increasing the distribution and usage efficiency of household water is a significant lever for improving groundwater sustainability.

38.5.3 Improving industrial water-use efficiency

In addition to farm and household use, another important demand for groundwater is the industrial sector. Groundwater is commonly used by industries, due to its accessible cost and reliable supply (Fig. 38.3). Particularly water-intensive

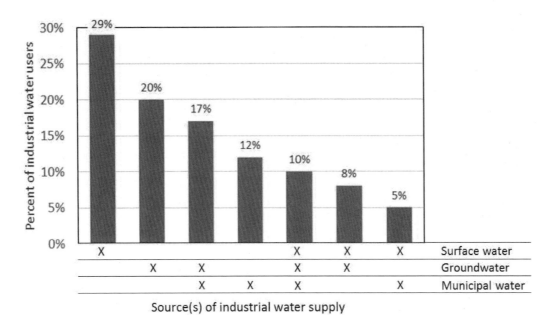

FIGURE 38.3 Sources of industrial water in India. Groundwater is used by 55% of industrial water users in India. *Credit: data from Perveen, S., et al., 2012. Water Risks for Indian Industries: A Preliminary Study of 27 Industrial Sectors. Federation of Indian Chambers of Commerce and Industry (FICCI) and Columbia University Water Center (CWC) (Perveen et al., 2012).*

industries include steel, textiles, pulp, and paper. In regions of groundwater stress, industrial water demand competes with household and irrigation water demand for limited supply.

Industrial production and its associated water use are increasing rapidly. Many industrial processes require water for various purposes such as cooling and washing. Industrial water-use efficiency in many groundwater-stressed regions is quite low, and the adoption of global best practices would significantly reduce industrial water use. In recent decades, as water constraints have been felt in various parts of the world, much effort has been expended globally to devise industrial processes that conserve water. Implementation of this global best practice could significantly reduce future demand for industrial groundwater, while reducing the quantity and improving the quality of industrial wastewaters.

38.6 Purifying contaminated groundwater

Groundwater quality problems can be caused by chemical, physical, microbiological, or aesthetic issues. Here we discuss removing salt from brackish groundwater, removing arsenic and fluoride from groundwater, and killing biological pathogens that are present in groundwater due to inadequate sanitation methods.

38.6.1 Removing salt from brackish groundwater

Desalination is a technology set that seeks to make freshwater from saline water sources such as seawater or brackish water. The salt content of water is typically measured in grams of total dissolved solids (TDS) per liter of water. While there are no definitive standards, water is generally considered potable when it contains TDS less than about 1 g/L. The salinity of ocean water averages 35 g/L globally. Many regions possess abundant groundwater that is brackish, with TDS up to about 5 g/L.

There are numerous desalination technologies, which can be divided into four major categories depending on the driving force of the process: thermal, pressure, electrical, and chemical (Miller, 2003; Youssef et al., 2014; Subramani and Jacangelo, 2015).

1. Thermally driven systems use evaporation and condensation at different temperatures and pressures as the main process to separate salts from water. In these systems, heat transfer is used to either boil or freeze the feedwater to convert it to vapor or ice, so that salts are separated from the water. The most common thermal processes include the multistage flash process and the multieffects distillation process (Shatat and Riffat, 2014). Other thermally activated systems include vapor compression distillation, humidification–dehumidification, solar distillation, and freezing.

2. Pressure-activated systems use a pressure gradient to force water through a semipermeable membrane, leaving salts behind. In recent decades, membrane technologies have matured and most new desalination installations use membranes. Of these, the reverse osmosis (RO) process is the most common; others include forward osmosis and nanofiltration.
3. Electrically activated systems take advantage of the charged nature of salt ions in solution, by using an electric field to remove ions from water. The most common configuration is electrodialysis (ED), which currently accounts for about 4% of global desalinated water production. An emerging technology is capacitive deionization (CDI).
4. Chemically activated desalination systems include ion-exchange desalination, liquid–liquid extraction, and gas hydrate or other precipitation schemes. There are numerous alternate desalination processes that are technically possible but are limited by economic or practical issues (Miller, 2003).

Cost of energy supply strongly affects the cost of desalination, and a major source of variability is the form of energy that drives the desalination process, such as heat, pressure, electrical, or chemical (Rao et al., 2016). In general, thermal desalination uses large amounts of heat, RO uses much smaller amounts of electricity, and ED uses even less electricity but is limited to low-salinity feedwater (see Fig. 38.4). The overall cost of the various processes also varies and is heavily dependent on scale. Larger facilities are far less expensive per cubic meter of freshwater.

Membrane-based seawater desalination technologies are approaching theoretical limits of energy efficiency and are already used at commercial scale for industrial and domestic use (Elimelech and Phillip, 2011). Although minor incremental efficiency improvements may still be gained, it is unlikely that major technology breakthroughs will fundamentally alter the seawater desalination landscape.

For brackish groundwater, there are major opportunities for significant reductions in desalination cost and energy use through innovative electrochemical or other emerging techniques. The minimum theoretical energy requirement for desalination varies with the salinity of the feedwater—less energy is fundamentally needed to desalinate brackish water, compared to seawater. Electrically driven techniques, such as ED and CDI, are limited to low-salinity feedwater, but potentially cost less and require less energy than pressure or thermal techniques (see Fig. 38.4). ED and CDI technologies use less energy because they transport the (relatively few) dissolved salt ions out of the feedwater, rather than

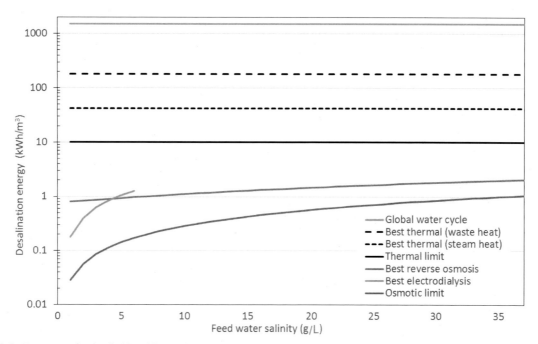

FIGURE 38.4 Energy use for desalination (kilowatt-hour per cubic meter of freshwater) as a function of feedwater salinity (grams TDS per liter of feedwater) for various desalination processes. Note vertical axis is logarithmic. *TDS*, Total dissolved solids. *Credit: data from Cerci, Y., et al., 2003. Improving the thermodynamic and economic efficiencies of desalination plants: minimum work required for desalination and case studies of four working plants. In: Program Final Report No. 78. Mechanical Engineering, University of Nevada, Reno (Cerci et al., 2003); Fritzmann, C., et al., 2007. State-of-the-art of reverse osmosis desalination. Desalination 216, 1–76 (Fritzmann et al., 2007); Elimelech, M., Phillip, W.A., 2011. The future of seawater desalination: energy, technology, and the environment. Science 333, 712–717; Shatat, M., Riffat, S.B., 2014. Water desalination technologies utilizing conventional and renewable energy sources. Int. J. Low Carbon Technol., 9, 1–19.*

transporting the (plentiful) water molecules away from the salt as in thermal and pressure technologies (Suss et al., 2015). The electrical current required for ED and CDI is proportional to the amount of salt removed (Knust et al., 2014). ED and CDI are highly efficient for desalinating feedwater on the dilute end of the brackish water range (0.6–4 g/L TDS).

38.6.2 Removing arsenic from groundwater

In parts of the world the underground geological formations comprise minerals containing arsenic, resulting in groundwater containing this natural toxin. Consumption of water, including both drinking water and cooking water, with elevated arsenic levels over a prolonged period can result in serious health conditions, including skin lesions, hyperkeratosis, melanosis, and cancer in different organs, which in some cases is fatal. The probability and severity of health effects increase with exposure level and duration.

The most common current arsenic removal technologies can be grouped into five categories: oxidation, ion-exchange, activated alumina, membrane, and coagulation/coprecipitation/adsorption (Rahman and Al-Muyeed, 2009). Some promising technologies, such as electrocoagulation, are emerging. Each of these technologies has trade-offs in terms of feed water characteristics (i.e., pH, concentrations of arsenic, iron, phosphate, silicate, and calcium); operation and maintenance complexity; and aesthetic water quality. The arsenic concentration of the feed water is a key factor influencing the removal efficiency and cost. Technologies that demonstrate high removal efficiencies when treating moderately arsenic-contaminated water may not be as efficient when treating highly contaminated water (Shan et al., 2018).

In terms of effectiveness, oxidation–filtration and ion-exchange technologies have shown poor efficacy, while zerovalent iron and other adsorption technologies work well. Users give coagulation–coprecipitation–filtration technologies mixed reviews (Amrose et al., 2015). A variety of arsenic removal technologies are available at the community and household level. The most widely used household arsenic removal systems use zerovalent iron, such as the SONO filter. However, many household users complain of low-flow rate and occasional clogging. Community-level treatment typically exists as column filters containing media such as activated alumina, granular ferric hydroxide, or hybrid anion exchange.

A major concern regarding arsenic removal technologies is that the collected arsenic must be disposed of after it has been removed from a water source. Unfortunately, this disposal practice is often unregulated and arsenic waste is sometimes dumped in ponds or open fields. The arsenic concentration of these wastes varies widely but can reach 7.5 g/kg (Amrose et al., 2015), roughly a million times more concentrated than the maximum allowable arsenic concentration in safe drinking water. An effective large-scale arsenic removal program will require a method for responsible disposal of the collected arsenic.

38.6.3 Removing fluoride from groundwater

Fluoride is another naturally occurring element that is present in some groundwater. Fluoride contamination is highly prevalent in hyperarid and humid areas of Asia and Africa. Various defluoridation techniques have been developed, including coagulation, adsorption, ion-exchange, electrochemical, and membrane-based methods (Mumtaz et al., 2015). The coagulation technique uses reagents, such as aluminum salts, lime, calcium and magnesium salts, polyaluminum chloride and alum to precipitate fluoride through a chemical reaction in which the precipitated fluoride coagulates and can then be removed. The Nalgonda technique is a coagulation–defluoridation technique that has seen limited acceptance because it is relatively difficult to maintain and operate. This is a common problem with defluoridation technologies.

Adsorption processes involve continuously cycling fluoride-contaminated water through columns containing an adsorbent, such as bone char, activated alumina, activated carbon, activated bauxite, ion-exchange resins, fly ash, super phosphate and tricalcium phosphate, clays and soils, synthetic zeolites, and other adsorbent minerals. The cyclic sorption tends to aggregate and concentrate the fluoride, which can then be disposed of safely.

Electrochemical processes include electrocoagulation and electrosorptive techniques. During electrocoagulation processes, Al^{3+} ions are released from aluminum electrodes through an anodic reaction, which generate aluminum hydroxides that adsorb fluoride ions near the electrodes, resulting in a fluoride complex that can then be easily removed. Electroadsorptive techniques only differ from adsorption techniques in that an electric field is applied to the adsorbent bed, which increases its adsorption capability.

The use of membrane technologies in defluoridation is relatively new and includes RO, nano- and ultrafiltration, ED, and Donnan dialysis. These processes typically have high operational costs compared to other defluoridation techniques (Ayoob et al., 2008).

38.6.4 Killing biological pathogens in groundwater

Microbial pathogens are often present in groundwater due to inadequate sanitation practices that allow fecal pathogens to enter groundwater. Pathogens that are present in groundwater must be killed, deactivated, or physically removed before the water can be safely consumed. There are numerous ways this can be achieved at various scales, including through chemical, thermal, radiation, and filtration methods (Gadgil, 1998; Amrose et al., 2015)

Several of these technologies are often combined together to achieve better drinking water quality. For example, ceramic filters can be lined with silver and copper nanoparticles to ensure all pathogens are killed and filtered out. Another example is chloramination, which is the combined use of chlorine and ammonia. Chlorine is the most widely used water disinfectant, while ozone is the second most widely used.

Chemicals used as disinfectants will sometimes react with naturally occurring chemicals in a water source to produce by-products that are harmful to human health. Some common by-products are bromate, chlorite, haloacetic acids, and trihalomethanes. Many of these by-products are toxic and/or carcinogenic (US EPA, 2018). Activated carbon filters may be used secondarily to adsorb and remove some of these by-products.

38.7 Improving groundwater access

While physical constraints to water quality or quantity limit some people's access to groundwater, economic constraints limit access by other households and farms. Of particular concern is the lack of access to groundwater for agricultural irrigation in sub-Saharan Africa. Most farmers in this region use low-yielding and highly variable subsistence rain-fed farming methods, despite the presence of abundant shallow groundwater. Access to irrigation is a critical constraint to increasing agricultural productivity for smallholder farmers in sub-Saharan Africa. With irrigation, farmers can increase crop yields, produce more consistent harvest, diversify their portfolio toward higher income crops, and increase the total number of harvests in a given year. While eliminating economic water scarcity in this context will require broader socioeconomic interventions, appropriate technologies for low-cost well drilling and water pumping may make important contributions.

38.7.1 Well digging and drilling

Well drilling to access groundwater played a major role in the Green Revolution and is an important part of water supply development in many regions. Millions of borewells have been drilled since the 1960s, and many regions have developed strong local expertise in siting and drilling wells, as well as in producing and maintaining drilling equipment. This expertise is currently lacking in most parts of sub-Saharan Africa, as well as in parts of South Asia with limited agricultural intensification. A variety of drilling technologies are available, depending on soil characteristics and required depth.

The difficulty and cost of digging wells vary as functions of water table depth and soil formation. When groundwater is available at depths less than 4 m, manual digging or drilling is adequate. Shallow hand-dug wells are inexpensive, with the primary cost being the digger's time. For deeper wells, numerous manual drilling techniques have been developed to produce shallow borewells in favorable geology, including sludging and augering. Manual drilling techniques, often involving community participation as labor, are typically slow and limited in the geological strata they can drill. Mechanized drilling of deeper wells is typically unaffordable for subsistence farmers. This type of drilling is typically conducted with portable diesel-powered rigs, such as percussion and rotary percussion methods. Powered mechanical rigs are expensive (>$100,000) and have limited mobility to reach remote areas. They are able to effectively drill through most geological features and are relatively quick to create a well but are expensive and have high costs for capital equipment, fuel, and labor.

Current well drilling technologies suffer from high cost, limited portability, slow drilling rate, and/or limited geologic suitability. To expand groundwater opportunities to rural populations facing economic water scarcity, a drilling technology is needed that combines the speed and capability of powered equipment with the portability and low-cost of manual techniques. Such a technology could enable more accessible borewells in regions that suffer from economic water scarcity, including sub-Saharan Africa and parts of South Asia.

38.7.2 Groundwater pumping

The technology set for lifting water is fundamental to groundwater access. An important groundwater pump distinction is based on the source of motive power: human muscles or mechanically powered. Manual-powered pumps are commonly used to lift groundwater from boreholes and shallow wells for household use and low-lift irrigation in rural areas. Technologies for shallow and deep hand pumps were significantly advanced during the International Drinking Water Decade from 1981 to 1990. Robust community-scale pumps, such as the India Mark III and the Afridev, were designed and widely deployed, with attention not only to technical efficiency but also user ergonomics and practical maintenance. The treadle pump, developed in Bangladesh during the 1970s and 1980s, is a low-cost shallow pump that is actuated by strong leg muscles and can lift sufficient water for irrigating smallholder farms. While marginal improvements may be made to manual pump technology, no radical innovations are expected. Rather, water providers can improve access by increasing coverage and ensuring timely maintenance and repairs.

Most groundwater pumps are powered by an external energy source, usually grid electricity or diesel fuel. Efficiency studies of electrical pumpsets in South Asia found average efficiencies of 30% or less (World Bank, 2001; Singh, 2009; Kaur et al., 2016). These efficiency levels are much lower than typical best practice pumpset efficiencies of greater than 50%, and well below the practical efficiency limit of about 85%. Efficiencies can be increased by matching the size of pumps and motors to their tasks, and replacing foot valves and suction and delivery piping to reduce frictional losses. Diesel-powered groundwater pumps are more often used where grid electricity is not available, such as parts of sub-Saharan Africa and South Asia. The cost of diesel necessary to run these pumps is variable and increases the overall operating cost for farmers.

Solar-powered electric pumpsets, which use photovoltaic (PV) arrays to convert sunlight to electricity that then power submersible or surface-mounted electrical pumps, are at an earlier stage of development and deployment. Direct solar pumping can be quite efficient, as all harvested power is used for pumping and there is no need for batteries and associated losses. Modern positive displacement pumps have efficiencies of up to 70% [GIZ (Gesellschaft für Internationale Zusammenarbeit), 2013]. There are significant barriers to the scale-up of PV-powered irrigation pumps, including the high upfront cost of PV systems, which are typically 10 times that of conventional pumps (KPMG, 2014). This cost difference is diminishing over time as PV system costs decline. As there is zero marginal cost for additional water pumping, there is a risk of unrestrained aquifer depletion if the technology is scaled up in the absence of rational water allocation systems.

38.8 Conclusion

The absolute demand for water is increasing due to demographic, industrial, and agricultural growth. Meanwhile, local water resources are constrained, based on precipitation, topology, and geology. The hydraulic boundaries of water basins seldom align with the political boundaries of social discourse, thus water conflicts arise. Groundwater storage is abundant and unutilized in some regions, such as much of sub-Saharan Africa, and overexploited and depleting in others, such as parts of South Asia and North America.

The deployment of select technologies holds promise to enhance the sustainable use of groundwater resources. Groundwater mapping is an essential first step, to understand the subsurface landscape. This knowledge can then be practically applied to manage and increase groundwater recharge, and to reduce saline groundwater intrusion into freshwater aquifers. Improving water-use efficiency in the agriculture, household, and industrial sectors can increase utility from each available unit of groundwater. However, the net groundwater implications of irrigation efficiency improvements are complex, as "wasted" irrigation water often contributes to groundwater recharge. Technologies can be used to improve the quality of groundwater contaminated by salt, arsenic, fluoride, and organic pathogens. Finally, in regions suffering from economic water scarcity, improved technologies for creating wells and pumping groundwater can increase access to needed groundwater.

References

ADB (Asian Development Bank), 2010. The issues and challenges of reducing non-revenue water.

Amrose, S., et al., 2015. Safe drinking water for low-income regions. Annu. Rev. Environ. Resour. 17 (51), 9.1–9.29.

ATA (Ethiopian Agricultural Transformation Agency), 2019. Shallow ground water mapping. Web-accessed from: <http://www.ata.gov.et/programs/highlighted-deliverables/input-voucher-sales-system-ivs/>.

Ayoob, S., et al., 2008. A conceptual overview on sustainable technologies for the defluoridation of drinking water. Crit. Rev. Environ. Sci. Technol. 38 (6), 401–470.

Cerci, Y., et al., 2003. Improving the thermodynamic and economic efficiencies of desalination plants: minimum work required for desalination and case studies of four working plants. In: Program Final Report No. 78. Mechanical Engineering, University of Nevada, Reno.

CGWB (Indian Central Ground Water Board), 2007. Manual on Artificial Recharge to Groundwater. Ministry of Water Resources, Govt. of India.

CGWB (Indian Central Ground Water Board), 2019. Aquifer mapping & management. Web-accessed from: <http://cgwb.gov.in/Aquifer-mapping.html>.

Elimelech, M., Phillip, W.A., 2011. The future of seawater desalination: energy, technology, and the environment. Science 333, 712–717.

FAO (Food and Agriculture Organization of the United Nations), 2017. Does improved irrigation technology save water? A review of the evidence.

Ferguson, G., Gleeson, T., 2012. Vulnerability of coastal aquifers to groundwater use and climate change. Nat. Clim. Change 2, 342–345.

Fritzmann, C., et al., 2007. State-of-the-art of reverse osmosis desalination. Desalination 216, 1–76.

Gadgil, A., 1998. Drinking water in developing countries. Annu. Rev. Energy Environ. 23, 253–286.

Gale, I., 2005. Strategies for Managed Aquifer Recharge (MAR) in Semi-arid Areas. UNESCO.

GIZ (Gesellschaft für Internationale Zusammenarbeit), 2013. Solar water pumping for irrigation: opportunities in Bihar, India.

Grafton, R.Q., et al., 2018. The paradox of irrigation efficiency. Science 361, 748–750.

ITT (Institute for Transformative Technologies), 2018. Technology Breakthroughs for Global Water Security: A Deep Dive into South Asia. Institute for Transformative Technologies, Berkeley, CA.

Kaur, S., et al., 2016. Assessment and mitigation of greenhouse gas emissions from groundwater irrigation. Irrig. Drain. 65, 762–770.

Klee, P., et al., 2015. Greater Water Security with Groundwater: Groundwater Mapping and Sustainable Groundwater Management. Rethink Water.

Knust, K.M., et al., 2014. Electrochemical desalination for a sustainable water future. ChemElectroChem 1, 850–857.

KPMG, 2014. Feasibility analysis for solar agricultural water pumps in India.

Kumpel, E., Nelson, K.L., 2016. Intermittent water supply: Prevalence, practice, and microbial water quality. Environ. Sci. Technol. 50, 542–553.

Miller, J.E., 2003. Review of water resources and desalination technologies. In: Report SAND 2003-0800. Sandia National Laboratories.

Molden, D., Oweis, T.Y., 2007. Pathways for increasing agricultural water productivity. In: Molden, D., et al., (Eds.), Water for Food, Water for Life: A Comprehensive Assessment of Water Management in Agriculture. Routledge, London, Chapter 7.

Mukherjee, A., et al., 2020. Global Groundwater: Source, Scarcity, Sustainability, Security and Solutions. Elsevier, ISBN: 9780128181720.

Mumtaz, N., et al., 2015. Global fluoride occurrence, available technologies for fluoride removal, and electrolytic defluoridation: a review. Crit. Rev. Environ. Sci. Technol. 45, 2357–2389.

Perry, C., 2007. Efficient irrigation, inefficient communication, flawed recommendations. Irrig. Drain. 56, 367–378.

Perveen, S., et al., 2012. Water Risks for Indian Industries: A Preliminary Study of 27 Industrial Sectors. Federation of Indian Chambers of Commerce and Industry (FICCI) and Columbia University Water Center (CWC).

Rahman, M.H., Al-Muyeed, A., 2009. Arsenic crisis of Bangladesh and mitigation measures. J. Water Supply Res. Technol. 58 (3), 228–245.

Rao, P., et al., 2016. Survey of available information in support of the energy-water bandwidth study of desalination systems. In: Report LBNL-1006424. Lawrence Berkeley National Laboratory.

Saeed, M.M., Ashraf, M., 2005. Feasible design and operational guidelines for skimming wells in the Indus basin, Pakistan. Agric. Water Manage. 74, 165–188.

Shah, T., 2008. India's master plan for groundwater recharge: an assessment and some suggestions for revision. Econ. Pol. Wkly. 43 (51), 41–49.

Shan, Y., et al., 2018. Cost and Efficiency of Arsenic Removal from Groundwater: A Review. United Nations University Institute for Water, Environment and Health.

Shatat, M., Riffat, S.B., 2014. Water desalination technologies utilizing conventional and renewable energy sources. Int. J. Low Carbon Technol. 9, 1–19.

Simbeye, I., 2010. Managing Non-Revenue Water: NRW Sourcebook for Trainers. Internationale Weiterbildung und Entwicklung.

Singh, A., 2009. A Policy for Improving Efficiency of Agriculture Pump sets in India: Drivers, Barriers and Indicators. Climate Strategies.

Subramani, A., Jacangelo, J.G., 2015. Emerging desalination technologies for water treatment: a critical review. Water Res. 75, 164–187.

Suss, M.E., et al., 2015. Water desalination via capacitive deionization: what is it and what can we expect from it? Energy Environ. Sci. 8, 2296–2319.

US EPA (United States Environmental Protection Agency), 2018. National primary drinking water regulations. Web-accessed from: <https://www.epa.gov/>.

World Bank, 2001. India: power supply to agriculture. Report No. 22171-IN.

World Bank, 2005. Pakistan's water economy running dry. Report No. 44375.

Youssef, P.G., et al., 2014. Comparative analysis of desalination technologies. Energy Procedia 61, 2604–2607.

Chapter 39

Applications of Gravity Recovery and Climate Experiment (GRACE) in global groundwater study

Jianli Chen[1] and Matt Rodell[2]

[1]Center for Space Research, University of Texas at Austin, Austin, TX, United States, [2]NASA Goddard Space Flight Center, Greenbelt, MD, United States

39.1 Introduction

The Earth's gravity field is governed by mass distribution in the Earth system, including the atmosphere, ocean, hydrosphere, cryosphere, and solid Earth. The static portion of the gravity field mainly reflects mass distribution in the solid Earth. Spatial variations of the static gravity field are mostly controlled by the topography of land, bathymetry of the ocean, and internal density structure of the solid Earth, and temporal variations of the gravity field are mainly driven by water and air mass redistributions within the atmosphere, ocean, hydrosphere, and cryosphere (Wahr et al., 1998; Cazenave and Chen, 2010; Tapley et al., 2019). Solid Earth geophysical processes, such as the glacial isostatic adjustment (GIA), also affect the time-variable gravity field at long-term timescales (Peltier, 2004). In addition, seismic deformations (earthquakes) lead to time-variable gravity changes from instantaneous (coseismic) to several years (postseismic) timescales (Han et al., 2006; Chen et al., 2007).

Observations of the time-variable gravity field can be used for estimating and monitoring mass redistribution in the Earth system. This is especially valuable for detecting and quantifying sea level change and terrestrial water storage (TWS) change, including floods, droughts, groundwater change, and ice mass loss in polar regions (Cazenave and Chen, 2010; Tapley et al., 2019). Observation of gravitational change is also an important means for studying solid Earth geophysical processes, for example, quantification of the GIA effects (Tamisiea et al., 2007) and co- and postseismic deformations of major earthquakes (Han et al., 2006; Chen et al., 2007). However, accurately mapping the time-variable gravity field is difficult. Due to the lack of in situ measurements, it is impossible to determine temporal variations of the gravity field (using in situ data) on global or even large regional scales.

Launched in 2002, the Gravity Recovery and Climate Experiment (GRACE) satellite gravity mission began a completely new era of measuring time-variable gravity field with unprecedented accuracy and offered a revolutionary means for studying large-scale mass variations and redistributions in the climate system (Tapley et al., 2019). GRACE was decommissioned in November 2017, and the GRACE Follow-On (GFO) mission was launched in May 2018 to continue the endeavor. The nearly two-decade long record of GRACE/GFO time-variable gravity observations has revolutionized the studies of mass variations in different components of the climate system, including TWS changes at regional to global scales (Wahr et al., 2004; Landerer and Swenson, 2012; Rodell et al., 2018), flood and drought monitoring (Papa et al., 2008; Reager and Famiglietti, 2009; Chen et al., 2010; Houborg et al., 2012), groundwater change and depletion (Rodell et al., 2007, 2009; Scanlon et al., 2012; Döll et al., 2014), water storage changes in snow and surface reservoirs (Niu et al., 2007; Chen et al., 2017), terrestrial water-balance assessment (Rodell et al., 2004a; Sheffield et al., 2009; Rodell et al., 2015), ice melting of polar ice sheets and mountain glaciers (Velicogna and Wahr, 2006; Luthcke et al., 2013), and global ocean mass change (Chambers et al., 2010; Chen et al., 2018).

These successful applications of GRACE time-variable gravity measurements have demonstrated their value for monitoring and studying various components of the global water cycle. However, there are three major limitations of the technique. One is the coarse spatial resolution (no better than about 150,000 km^2 at midlatitudes), which is controlled by a few factors, including the truncation of GRACE spherical harmonic (SH) solutions at certain degrees (e.g., at degree 60 or 96, which are the highest degrees and orders of SH coefficients provided in GRACE gravity solutions), and the need for spatial filtering to suppress the dominant spatial noise at high degrees and orders (Wahr et al., 1998). Therefore GRACE gravity measurements are mostly applied to study large regional or basin-scale water mass changes. The second limitation is the vertical integration of the observed mass changes. GRACE measures gravitation effects on satellite orbit (mainly intersatellite range and range-rate change), which reflects the total contribution from all mass components on the Earth surface (i.e., from deep groundwater to the top of the atmosphere). Over land regions, after atmospheric mass circulation effects are removed, GRACE-observed time-variable gravity change represents the integrated contribution of mass variations in all forms of TWS, including soil moisture, snow, surface water (in lakes and rivers), and groundwater. GRACE itself cannot separate the different contributions. Independent estimates from other sources are needed when the focus is on a particular hydrological variable (e.g., groundwater). The third limitation is that each gravity solution is a monthly average, and there is a data latency of several weeks to a few months, which makes it challenging to use the observations for operational applications.

This chapter provides a brief description of GRACE satellite gravimetry, a state-of-the-art review of GRACE applications in quantifying groundwater storage change, and an in-depth discussion of the major challenges in using GRACE time-variable gravity measurements and how to address these challenges.

39.2 GRACE and GFO missions and data products

The first GRACE was a twin satellite dedicated gravity mission, jointly sponsored by NASA and German Aerospace Center (DLR) (Tapley et al., 2004). The GRACE satellites were launched on March 17, 2002, with an initial altitude of ~500 km and a near-polar inclination of 89 degrees. The two satellites were separated by ~220 km along orbit track. GRACE observed variations in Earth's gravity field by tracking the intersatellite range and range-rate between two coplanar, low altitude satellites (see Fig. 39.1) via a K-band ranging (KBR) system (Tapley et al., 2004). Each satellite was equipped with a SuperSTAR Accelerometer, a GPS receiver, Star Cameras, and Laser Retro Reflectors. The GRACE Science Data System used the KBR range and range-rate data, along with other ancillary data, to estimate variations of the gravity field on monthly basis with respect to a background gravity model used in GRACE data processing. The GFO mission is basically a duplicate of GRACE, with two satellites flying in the same polar orbit and carrying the same science instruments with upgraded hardware. GFO also carries an experimental laser-ranging interferometer instrument that may become the standard for future GRACE-like missions.

Monthly GRACE/GFO (noted as GRACE hereafter, unless GFO is specifically used) global gravity solutions are routinely produced by three GRACE data processing centers, which include the University of Texas at Austin Center

FIGURE 39.1 The GRACE twin satellites orbited the Earth at an initial altitude of ~500 km in a near-polar orbit (89 degrees inclination). GRACE was launched on March 17, 2002 with a design lifetime of 5 years and was decommissioned in November 2017. *GRACE*, Gravity Recovery and Climate Experiment. *Courtesy of NASA and CSR.*

for Space Research (CSR), NASA Jet Propulsion Laboratory (JPL), and GeoForschungsZentrum (GFZ) in Potsdam. GRACE time-variable gravity solutions are distributed by NASA's Physical Oceanography Distributed Active Archive Center (http://podaac.jpl.nasa.gov/grace/), and GFZ's Information System and Data Center (http://isdc.gfz-potsdam.de/grace-isdc/). GRACE monthly gravity solutions are provided in the form of fully normalized SH (also called Stoke) coefficients. During GRACE data processing, to minimize aliasing errors from high-frequency signals, tidal effects and atmospheric and oceanic mass changes have been removed using predictions from atmospheric and ocean circulation models (Bettadpur, 2018). After an extended period of data calibration and validation (Cal/Val), the GFO project released its first year of gravity solutions in June 2019, with accuracy similar to that of the original GRACE solutions.

In addition, GRACE mass concentration (mascon) solutions are also generated by a number of institutes, including NASA Goddard Space Flight Center (GSFC), JPL, and CSR. These mascon solutions are derived from an alternative processing approach (Rowlands et al., 2005) that offer improved spatial resolution and do not require Gaussian smoothing or destriping filters to suppress the longitudinal stripes that are seen in SH solutions (Swenson and Wahr, 2006). The JPL mascon solutions are distributed by the NASA GRACE Tellus project (https://grace.jpl.nasa.gov/data/get-data/jpl_global_mascons/). The CSR and GSFC mascon solutions are also available online (http://www2.csr.utexas.edu/grace/RL06_mascons.html, https://earth.gsfc.nasa.gov/geo/data/grace-mascons).

39.3 Quantification of groundwater change using Gravity Recovery and Climate Experiment

Improved monitoring of various components of the global water cycle (e.g., precipitation, evapotranspiration, river runoff/discharge, soil water, and groundwater; see Fig. 39.2) is important for understanding the Earth's climate and ecosystems. Groundwater supports agricultural, industrial, and domestic activities worldwide, especially in heavily irrigated agricultural regions [e.g., northern India and the North China Plain (NCP)] and arid and semiarid regions that lack alternative water supplies (e.g., North Africa and Middle East). Groundwater storage change is controlled by the balance between recharge (i.e., inflows to the groundwater aquifers from above), natural discharge (i.e., outflows to surface water systems or the ocean), and groundwater abstraction. Extended droughts or excessive groundwater abstraction can lead to groundwater depletion and regional water resource scarcity, often impacting ecosystems, the local economy, and people's livelihood (Foster and Loucks, 2006; Gleeson et al., 2010; Famiglietti, 2014; Mukherjee et al., 2020).

A good understanding of groundwater storage changes, in particular its long-term variability, is critical for maintaining sustainable economic developments and healthy ecosystems. Accurate quantification of groundwater storage

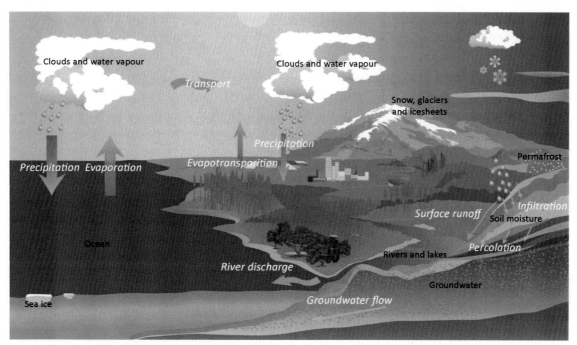

FIGURE 39.2 Illustration of the global water cycle. *Courtesy of NASA.*

changes also play an important role in understanding the global and regional hydrological cycles and connections with climate change. Despite its importance, groundwater resources are often poorly monitored, and accurate quantification of groundwater storage change is difficult due to the lack of adequate in situ observations (Famiglietti, 2014). Even though dense networks of monitoring wells exist in some developed countries (such as the United States and Australia), accurate quantification of regional groundwater storage changes using in situ well data is difficult due to temporal undersampling and uncertainties associated with converting water level changes to storage volumes (Rodell et al., 2007). The spatial and temporal sampling of monitoring networks is often inadequate for accurate quantification of groundwater storage change on large regional or basin scales. Furthermore, restrictive data sharing policies in many nations (e.g., China and Russia) also impede studies of groundwater storage change using in situ well data.

In a given region or river basin, TWS change is determined by the water budget, that is, the balance of the rates of precipitation (P), evapotranspiration (ET), surface, and subsurface runoff (R). TWS change (ΔTWS) over a unit time (Δt) can be computed as a function of water fluxes:

$$\frac{\Delta \text{TWS}}{\Delta t} = P - \text{ET} - R \tag{39.1}$$

TWS change can also be expressed as the sum of the component changes, that is, in nonglacierized regions, ΔTWS integrates changes in the water stored in soil moisture (ΔM_{SM}), snow (ΔM_{SN}), groundwater aquifers (ΔM_{GW}), and surface water reservoirs (ΔM_{SW}):

$$\Delta \text{TWS} = \Delta M_{SM} + \Delta M_{SN} + \Delta M_{GW} + \Delta M_{SW} \tag{39.2}$$

Changes in the water stored in vegetation are typically small enough to be ignored (Rodell et al., 2005). TWS change is an important measure of integrated water fluxes (i.e., P, ET, and R) or the sum of water storage changes in various components (i.e., soil water, snow water, groundwater, and surface reservoirs). Accurate quantification of TWS change is also difficult owing to the lack of in situ observations of all of the fluxes or all of the components at large regional or basin scales. Global land surface models (LSMs) offer an alternative means for studying TWS changes. An advantage of using LSMs is that they employ knowledge of the relevant water and energy cycle processes, as embodied in physical equations, to fill observational gaps and produce spatially and temporally continuous fields of the states and fluxes of interest. However, the results are only as good as the land surface parameters and meteorological forcing inputs, both of which contain errors (Kato et al., 2007; Güntner et al., 2007).

GRACE satellite gravity measurements provide indirect measurements of TWS changes on a global basis (Tapley et al., 2004). Early GRACE hydrological applications mainly focused on studying seasonal TWS changes (e.g., Wahr et al., 2004; Ramillien et al., 2005). The extended series of GRACE time-variable gravity solutions (over 17 years now) have enabled more and more studies at interannual and long-term timescales and on a wider range of topics, from interannual TWS anomalies driven by extreme floods (Reager and Famiglietti, 2009; Chen et al., 2010), TWS deficit due to major droughts (Leblanc et al., 2009; Thomas et al., 2014; Abelen et al., 2015), to long-term groundwater depletion (Rodell et al., 2009; Scanlon et al., 2012; Chen et al., 2016), and climate change (Rodell et al., 2018; Tapley et al., 2019).

The water-balance Eq. (39.2) can be rewritten as,

$$\Delta M_{GW} = \Delta \text{TWS} - (\Delta M_{SM} + \Delta M_{SN} + \Delta M_{SW}) \tag{39.3}$$

This equation provides a means to quantify regional groundwater storage change (ΔM_{GW}) by combining GRACE TWS estimates with estimates of water storage changes in the other components (ΔM_{SM}, ΔM_{SN}, ΔM_{SW}) from independent sources (e.g., model output and/or in situ observations). Soil and snow water storage changes (ΔM_{SM}, ΔM_{SN}) are typically estimated using LSMs, and surface water storage changes (ΔM_{SW}) can be estimated using either models or observations (e.g., in situ gauges or satellite altimetry; Rodell and Famiglietti, 2001; Getirana et al., 2017; Rodell et al., 2018). However, any errors in GRACE ΔTWS and the other water storage change estimates (ΔM_{SM}, ΔM_{SN}, and ΔM_{SW}) will accumulate as errors in the resulting groundwater storage estimates.

39.4 Gravity recovery and climate experiment applications in groundwater storage change

Since around 2009, a series of studies have successfully applied GRACE satellite gravity measurements to quantify regional groundwater storage changes. Two pioneering studies, Rodell et al. (2009) and Tiwari et al. (2009), used GRACE TWS estimates, combined with soil and snow water storage estimates from the Global Land Data Assimilation

System (GLDAS) model (Rodell et al., 2004b) to assess groundwater depletion in the Ganges-Brahmaputra river basins (Northwest and North India). The two studies reported that the GRACE gravity measurements captured a significant TWS decline in the region during the period August 2002 to October 2008. Considering the close to normal level precipitation in the region during the period, and the widespread irrigated agriculture in the region, the TWS decrease was attributed to groundwater depletion mainly due to pumping for irrigation. The groundwater depletion rate averaged over the Northwest Indian (NWI) states of Rajasthan, Punjab, and Haryana (438,000 km^2) was estimated to be 17.7 ± 4.5 km^3/year over the period August 2002 to October 2008 (Rodell et al., 2009). During the same period the estimated groundwater depletion rate for a broader region covering Northwest to Northeast India (2,700,000 km^2) is up to 54 ± 9 km^3/year (Tiwari et al., 2009). Using an extended record of GRACE solutions, and a different data-processing method, Chen et al. (2014) suggest that the groundwater depletion rate in the NWI region is 20.4 ± 7.1 km^3/year averaged over the 10-year period January 2003 to December 2012. During the first 5 years (2003–07) the estimated rate (29.4 ± 8.4 km^3/year) from Chen et al. (2014) is significantly larger than previous estimates (Rodell et al., 2009) for approximately the same period. The large difference is partly attributed to the improved treatment of leakage error using global forward modeling (Chen et al., 2014).

GRACE-based groundwater depletion in the NWI region has been corroborated by in situ well data (Bhanja et al., 2020). Northern India is one of the most populous regions in the world. Excessive groundwater pumping for agricultural irrigation and domestic consumption in response to the growing demand for water and food has exceeded the replenishable groundwater supply, causing a steady decline in the water table (Hoque et al., 2007; Rodell et al., 2009; Tiwari et al., 2009) which has continued into recent years (Rodell et al., 2018; Bhanja et al., 2020). Fig. 39.3 shows a preliminary analysis of TWS changes in the NWI region (the same area as studied in Chen et al., 2014) observed by the GRACE gravity solutions (covering the whole GRACE mission, April 2002 to June 2017), plus the first year of GFO solutions (June 2018 to August 2019). Long-term TWS decrease is evident over the entire 17-year period (with a brief slow down during 2011–13). Consistent with previous studies, GLDAS Noah model predicted soil and snow water changes do not show any notable decrease during period. The differences between GRACE TWS and GLDAS soil

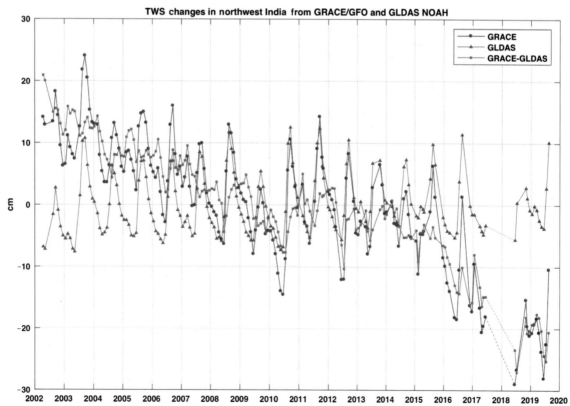

FIGURE 39.3 Comparisons of GRACE/GFO-observed TWS anomalies, GLDAS Noah model estimates of soil and snow water storage anomalies, and groundwater anomalies estimated as the difference (all in cm equivalent water height) in Northwest India for the period April 2002 to August 2020 (the same region as in Chen et al., 2014). The gaps between GRACE and GFO are connected by dashed lines. *GFO*, GRACE Follow-On; *GLDAS*, Global Land Data Assimilation System; *GRACE*, Gravity Recovery and Climate Experiment; *TWS*, terrestrial water storage.

moisture and snow (GRACE-GLDAS), reflecting mainly groundwater storage changes, indicate a substantial decreasing trend continuing into the GFO period.

The Central Valley in California is one of the most productive agricultural regions in the United States, accounting for ~8% of the food grown in the United States and ~17% of the country's irrigated land (Faunt, 2009). The underlying aquifer has been a major source of water for agricultural activities in the Central Valley region, particularly during droughts when surface water supplies are limited. Excessive groundwater pumping has led to steadily decreasing water levels in recent decades (Faunt, 2009). Famiglietti et al. (2011) combined GRACE gravity measurements, snow data from remote sensing, in situ surface water storage estimates, and soil moisture estimates from GLDAS to quantify groundwater change in the Central Valley. They concluded that TWS across the Sacramento and San Joaquin River Basins was decreasing at a rate of 4.8 ± 0.4 km^3/year during the period October 2003 to March 2010, and the majority of the water losses were due to groundwater depletion in the Central Valley. Scanlon et al. (2012) carried out a similar study to estimate long-term groundwater storage change in the Central Valley by comparing GRACE estimates with in situ well observations and reported that over the drought period from April 2006 through September 2009, the groundwater depletion rate in the Central Valley region was 7.7 ± 0.7 km^3/year, based on GRACE, in remarkable agreement with their in situ well based estimate of 7.7 ± 0.1 km^3/year. The difference between the estimates from these two studies is mostly due to their different study periods.

GRACE has also detected significant groundwater depletion in a number of other regions, including the High Plains Aquifer (HPA) in the central United States, the NCP, and the Middle East (see Fig. 39.4). The HPA, with an area of ~450,000 km^2, is ranked first in terms of groundwater withdrawals among aquifers in the United States (Maupin and Barber, 2005). A few previous studies (Strassberg et al., 2009; Longuevergne et al., 2010) determined that GRACE-estimated groundwater storage changes were highly correlated with those estimated from in situ well observations, indicating that GRACE is capable of detecting long-term groundwater storage changes in the HPA region. However, Brookfield et al. (2018) cautioned that comparisons between GRACE and in situ observations in the HPA are complicated by the complex hydrogeology of the region. Combining GRACE gravity measurements and GLDAS simulated

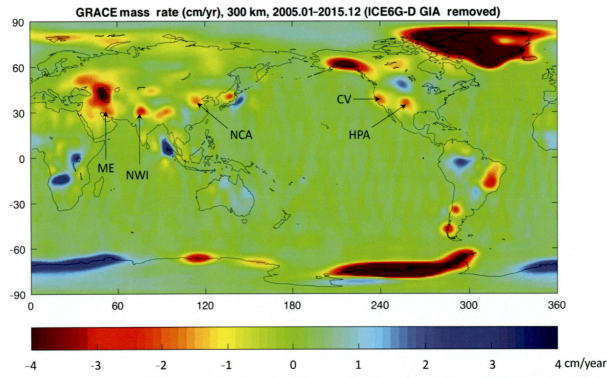

FIGURE 39.4 GRACE-observed global mass rates (in cm/year of equivalent water height) over the period January 2005 to December 2015, derived from CSR RL06 GSM solutions with 300 km Gaussian smoothing. C20 coefficients are replayed by SLR data (TN11), and GIA effects have been removed using ICE6G-D. Several regions with apparent groundwater depletion, including NWI, Central Valley, HPA, NCP, and Middle East (ME) are marked by black arrows. *CSR*, Center for Space Research; *GIA*, glacial isostatic adjustment; *GRACE*, Gravity Recovery and Climate Experiment; *HPA*, High Plains Aquifer; *NCP*, North China Plain; *NWI*, Northwest Indian; *SLR*, Satellite Laser Ranging.

soil moisture and snow water change, Breña-Naranjo et al. (2014) showed that the average groundwater depletion rate of the HPA was 12.5 ± 0.4 km^3/year during the period 2003–13.

The NCP is one of the most populous areas of the world, and it has also supported intensive groundwater-based crop irrigation for the past several decades. Excessive groundwater abstractions have caused the water table to drop significantly throughout the NCP (Yang et al., 2015). Even though local governments have enforced some conservation measures, significant groundwater is still expected to decline due to the lack of adequate surface water resources in the region. Following a similar GRACE—model output approach, Feng et al. (2013) estimated that the average NCP groundwater depletion rate in the NCP was 8.3 ± 1.1 km^3/year over the period 2003–10. This depletion appears to have continued at least through 2015 years (see Fig. 39.4).

Huge mass losses in the Caspian Sea and surrounding regions mainly reflect water level decline in the Caspian Sea (Chen et al., 2017), and groundwater depletion in the adjacent land. The Caspian Sea is the largest enclosed inland body of water on Earth, with a surface area of ∼371,000 km^2. Satellite altimeter measurements show that the average water level of the Caspian Sea is dropping at an alarming rate of ∼9 cm/year since 2006, which is ∼30 times larger than the global mean sea level rate (Chen et al., 2017). However, some of the mass losses in the surrounding regions, especially in regions south, southwest and northwest to the Caspian Sea are clearly not related to Caspian Sea level drop and more likely reflect groundwater depletion (see Fig. 39.4). Based on GRACE measurements and LSM model estimates, Joodaki et al. (2014) concluded that groundwater storage in the Middle East region decreased significantly during the period 2003–12, with an estimated rate of up to 25 ± 3 km^3/year, while Voss et al. (2013) reported a TWS depletion rate of 20.5 km^3/year during 2003–09. Quantification of groundwater storage change in the Middle East using GRACE is challenging, as in addition to uncertainty of model predicted soil and snow water storage changes, appropriate removal of contributions from the Caspian Sea and other surface water bodies is essential for calculating groundwater storage change in the region (Longuevergne et al., 2013; Mulder et al., 2015).

39.5 Major error sources of Gravity Recovery and Climate Experiment—estimated groundwater change

There are two major error sources of GRACE-estimated long-term groundwater storage change. The first is the uncertainty of GRACE-derived TWS changes, and the other is the uncertainty estimates of water storage changes in soil moisture, snow, and surface reservoirs used to isolate groundwater from total TWS. Groundwater depletion often occurs on scales too small to be resolved by GRACE. Due to the limitations of the satellite gravimetry technique, quantification of TWS changes over small basins or regions smaller than about 150,000 km^2 is highly uncertain. The spatial resolution of GRACE gravity solutions is mainly controlled by two factors: (1) the altitude of the GRACE satellites' (∼500 km at launch) and the intersatellite distance (∼220 km), and (2) the need for spatial filtering and smoothing to suppress noise in the GRACE data (i.e., the longitudinal stripes and other noise) and thus to extract meaningful mass change signals.

In theory, it is possible to use GRACE to calculate TWS change in a region of any size. However, the derived TWS change may not represent the true mass change signal in the region of interest, due to signal leakage from adjacent regions and signal attenuation caused by truncation and spatial smoothing associated with GRACE data processing (Chen et al., 2014). Minimizing these effects and restoring the "true" magnitude of the TWS changes is crucial for proper interpretation of GRACE observations. A number of methods have been proposed help to reduce the signal leakage and attenuation. Fitting and applying scale factors is one approach. For a given grid point or region a scale-factor can be calculated as the ratio between model simulated mass changes before and after applying the same truncation and spatial smoothing that are used in GRACE data processing (Velicogna and Wahr, 2006; Landerer and Swenson, 2012). These scale factors are then multiplied by the GRACE TWS estimates to correct for the leakage/attenuation. However, the scale factor method may only be valid when model predicted TWS changes resemble the true TWS variability reasonably well (or at least its spatial patterns) (Chen et al., 2016).

Forward modeling is another method to help reduce GRACE leakage. This method was originally developed for studying long-term polar ice sheet mass losses (Chen et al., 2006; Wouters et al., 2008) and mountain glaciers (Chen et al., 2007), when the geographical locations of the ice mass losses are fairly well known. The forward modeling method has been extended to studies of global ocean mass change (Chen et al., 2013) and regional groundwater depletions (Chen et al., 2014), in which the locations of the mass changes are less well known. In forward modeling, iterative numerical simulations are carried out to find a "true" mass change field that best matches the GRACE-observed mass changes, after applying the same data-processing procedures as applied to GRACE data to the modeled field, which include the truncation of SHs, treatments of low-degree terms, and spatial filtering and smoothing. One of the major advantages of forward modeling is that it is solely dependent on GRACE data, and thus it is not reliant on models as in the scale-factor approach. The details of the forward modeling algorithm and procedures with and without constraint are discussed in Chen et al. (2015).

GRACE mascon solutions offer improved spatial resolution and are not affected by longitudinal striping (as in the GSM solutions). Using high-resolution GRACE mascon solutions can improve the accuracy of GRACE TWS estimates. However, due to fundamental limitations of the GRACE observation method (mainly controlled by the satellites altitude and intersatellite distance), leakage is still an issue in GRACE mascon solutions (Watkins et al., 2015; Save et al., 2016; Loomis et al., 2019). The JPL RL06 mascons are solved on roughly $3° \times 3°$ equal-area elements, and the product is then distributed on a $0.5° \times 0.5°$ grid for convenience. To help reduce leakage effects, separately determined scale factors derived using the CLM4.0 LSM are provided with the mascon solutions (Wiese et al., 2016). The CSR RL06 mascons are solved on $1° \times 1°$ equal-area elements, and the products are provided on a $0.25° \times 0.25°$ grid. When examining regional mass change, appropriate leakage corrections are again recommended (Save et al., 2016; Chen et al., 2017). GSFC mascons are also solved on $1° \times 1°$ equal-area elements, but they are defined using different geophysical categories (e.g., land, ocean, glaciers, and ice sheets) (Loomis et al., 2019). Leakage across geophysical boundaries (e.g., land/ocean and ice sheet/ocean) is expected to be relatively small. However, similar to the CSR mascons, leakage is still expected at region or basin scales.

As previously described, LSM output is useful for removing soil moisture and snow (and surface water when available) water storage changes from GRACE TWS time series. However, any uncertainty or bias in the model output will direct to errors in the resulting groundwater estimates. The accuracy of LSM output depends on the quality of the input meteorological data and the ability of the models to simulate hydrological processes effectively. Differences are common among output time series from different LSMs and among output time series from a single LSM forced by different meteorological inputs (Rodell et al., 2004a,b). Differences between estimates from multiple models provide are a reasonable proxy for uncertainty in the model output (Kato et al., 2007). Fig. 39.5 compares soil moisture and snow

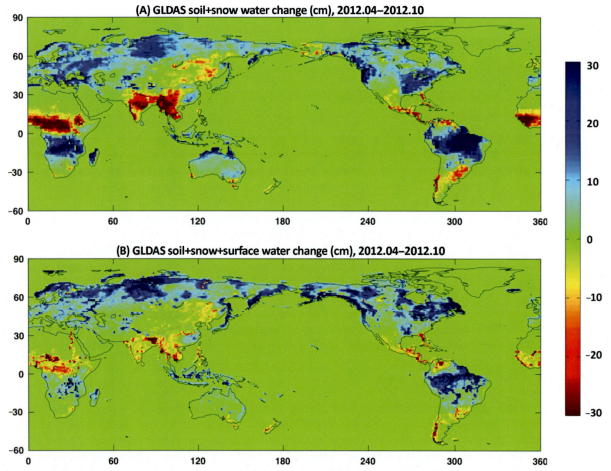

FIGURE 39.5 (A) GLDAS soil and snow water changes (in cm of equivalent water height; surface water, a typically minor component outside of humid tropical regions, is not included in GLDAS) between April 2012 and October 2012. (B) WGHM soil and snow (plus surface) water changes between the same 2 months. *GLDAS*, Global Land Data Assimilation System; *WGHM*, WaterGAP Global Hydrology Model.

water change between April 2012 and October 2012 from the GLDAS Noah model (Rodell et al., 2004b) with similar water storage estimates from the WaterGAP Global Hydrology Model (WGHM) (Döll et al., 2003; Güntner et al., 2007) over the same time span. Surface water storage change, a relatively minor component in most of the land areas (except for the floodplains in humid tropical regions and areas where lakes dominate the landscape, such as parts of central Canada; Getirana et al., 2017) is included in the WGHM estimates.

The differences in the output from these two commonly used LSMs as seen in Fig. 39.4 illustrates the challenge in using model results to separate the GRACE-observed TWS components. The discrepancies are mostly attributable to differences in the model inputs (meteorological forcing and parameters) and their treatments of complex hydrological processes. In some cases, these differences accumulate over long timescales, particularly where and when TWS trends are present (Chen et al., 2016). Nevertheless, many studies have applied model simulated soil moisture and snow for isolating groundwater storage changes from GRACE TWS changes, particularly when a multimodel mean is used (e.g., Rodell et al., 2009). Groundwater depletion is most commonly driven by excessive abstraction for agricultural and domestic usage (e.g., in NWI, NCP, HPA, and Central Valley), as opposed to surface climate conditions (e.g., extended droughts). Further, long-term soil moisture storage trends tend to be much smaller in magnitude than groundwater depletion trends. In these cases the approach can be applied with GRACE-derived TWS data to estimate groundwater depletion with negligible contamination from colocated hydrological signals (soil moisture, snow, and surface waters). On seasonal timescales, surface hydrometeorological conditions are likely to be a bigger driver of groundwater storage changes and soil moisture and/or snow water storage changes can dominate the TWS variations, making isolation of groundwater variations more difficult.

39.6 Gravity Recovery and Climate Experiment data assimilation

An alternative to subtracting model simulated soil moisture and snow (and surface water) time series from GRACE TWS in order to isolate groundwater storage is to integrate the GRACE data into a model simulation using data assimilation. Zaitchik et al. (2008) first introduced this capability using an ensemble Kalman smoother and demonstrated that the resulting groundwater, soil moisture, and snow water equivalent time series were more accurate (compared with in situ data) than the open loop (no data assimilation) results. Zaitchik et al. (2008) assimilated basin-averaged TWS time series from GRACE, but recent investigations have taken advantage of the additional spatial information available in current gridded GRACE products by assimilating those gridded products directly (Girotto et al., 2016; Kumar et al., 2016). Data assimilation within a LSM ensures that all TWS components and fluxes are consistent with each other and with the GRACE observations. Other advantages include the following: (1) data assimilation harnesses the higher spatial resolutions of the other model inputs and produces output TWS fields with significantly higher resolution than GRACE alone, (2) temporal resolution is likewise increased from monthly to daily or even hourly, (3) data latency can be reduced from multiple months to within a day or two of real time. As a result of these improvements, data assimilation enables GRACE data to be used for practical applications that require high-resolution, near-real time data. For example, Houborg et al. (2012) developed soil moisture and groundwater drought/wetness monitoring products for the contiguous United States that have been generated and disseminated routinely since 2011, as well as being incorporated into the United States Drought Monitor (see https://nasagrace.unl.edu/). Building on this capability, a global GRACE/GFO data assimilation based drought/wetness product has recently been deployed (Li et al., 2019) as well as a seasonal soil moisture and groundwater drought/wetness forecast product for the contiguous United States (Getirana et al., 2020).

39.7 Summary

Over the past 18 years, GRACE and GFO time-variable gravity measurements have been widely used to study large-scale mass variations in different components of the Earth system, including land hydrology, ocean, cryosphere, and solid Earth, thus greatly improving our understanding of the Earth system. GRACE offered the first opportunity for accurately quantifying large-scale TWS changes, associated with the global water cycle, and provided a tool for monitor major floods and droughts on global basis. GRACE is unique in its ability to observed mass changes in the entire column from the solid Earth surface to the top of the atmosphere. When combined with other independent data (e.g., other available remote sensing and in situ measurements or model simulations), GRACE can be used to study various components of TWS and the terrestrial water cycle, including groundwater, soil moisture, snow, evapotranspiration, and river discharge (Cazenave and Chen, 2010; Tapley et al., 2019).

GRACE identified significant long-term TWS trends in several regions of the world, many of which are attributed to groundwater depletion due to excessive pumping of groundwater for agricultural use. Using GRACE gravity measurements and LSM model predictions of soil and snow water (plus other available data), many studies have successfully quantified long-term groundwater storage changes in important regions, including NWI (Rodell et al., 2009; Tiwari et al., 2009; Chen et al., 2014), California's Central Valley (Famiglietti et al., 2011; Scanlon et al., 2012), the HPA (Strassberg et al., 2009; Longuevergne et al., 2010; Breña-Naranjo et al., 2014), the NCP (Feng et al., 2013), and the Middle East (Longuevergne et al., 2013; Joodaki et al., 2014; Mulder et al., 2015).

Despite the great potential of GRACE satellite gravimetry, accurate quantification of groundwater storage change using GRACE data is challenging. GRACE TWS and independent estimates of soil moisture, snow, and surface water storage changes all are subject to considerable uncertainty. The uncertainty of GRACE TWS estimates depends on the size of the region, how well the data processing addresses striping, signal leakage, and residual noise in GRACE SH solutions, and errors in the background atmosphere and ocean circulation models used in GRACE data processing. The uncertainty in GRACE low-degree SH coefficients and geocenter motion series is believed to play a minor role in GRACE regional TWS estimates (Chen et al., 2005). The uncertainty of model simulated soil and snow water change is mainly attributed to errors in the input meteorological data and the simplifications required to simulate complex physical processes economically. When groundwater depletion is multiannual and primarily driven by anthropogenic factors (as in most of the case studies discussed previously), uncertainty in the LSM output becomes less of a concern. Further, GRACE data assimilation enables spatial and temporal downscaling and extrapolation to near-real time, improving the fidelity of the results and making them useful for operational applications such as drought monitoring and forecasting.

Together, the GRACE and GFO missions are expected to provide more than two decades of satellite gravity measurements. Continuous improvements of GRACE and GFO background geophysical models and data processing methods are also expected, which will improve the spatial resolution and accuracy of future generations of GRACE products and enable a wider range of applications. Groundwater depletion shows no signs of abating in certain regions (e.g., in NWI and NCP), making continued monitoring by GFO into the future crucial.

References

Abelen, S., Seitz, F., Abarca-del-Rio, R., Güntner, A., 2015. Droughts and floods in the La Plata basin in soil moisture data and GRACE. Remote Sens. 7, 7324–7349. Available from: https://doi.org/10.3390/rs70607324.

Bettadpur, S., 2018. CSR Level-2 Processing Standards Document for Product Release 06, GRACE 327-742, The GRACE Project, Center for Space Research, University of Texas at Austin.

Bhanja, S.N., Mukherjee, A., Rodell, M., 2020. Groundwater storage change detection from in situ and GRACE-based estimates in major river basins across India. Hydrol. Sci. J. 65 (4), 650–659.

Breña-Naranjo, J.A., Kendall, A.D., Hyndman, D.W., 2014. Improved methods for satellite-based groundwater storage estimates: a decade of monitoring the high plains aquifer from space and ground observations. Geophys. Res. Lett. 41, 6167–6173. Available from: https://doi.org/10.1002/2014GL061213.

Brookfield, A.E., Hill, M.C., Rodell, M., Loomis, B.D., Stotler, R.L., Porter, M.E., et al., 2018. In situ and GRACE-based groundwater observations: similarities, discrepancies, and evaluation in the High Plains aquifer in Kansas. Water Resour. Res. 54 (10), 8034–8044.

Cazenave, A., Chen, J., 2010. Time-variable gravity from space and present-day mass redistribution in the Earth system. Earth Planet. Sci. Lett. 298, 263–274. Available from: https://doi.org/10.1016/j.epsl.2010.07.035.

Chambers, D.P., Wahr, J., Tamisiea, M.E., Nerem, R.S., 2010. Ocean mass from GRACE and glacial isostatic adjustment. J. Geophys. Res. (Solid Earth) 115 (B14), B11415. Available from: https://doi.org/10.1029/2010JB007530.

Chen, J., Famiglietti, J.S., Scanlon, B.R., Rodell, M., 2016. Groundwater storage changes: present status from GRACE observations. Surv. Geophys. 37, 397–417. Available from: https://doi.org/10.1007/s10712-015-9332-4.

Chen, J.L., Rodell, M., Wilson, C.R., Famiglietti, J.S., 2005. Low degree spherical harmonic influences on Gravity Recovery and Climate Experiment (GRACE) water storage estimates. Geophys. Res. Lett. 32, L14405. Available from: https://doi.org/10.1029/2005GL022964.

Chen, J.L., Li, J., Zhang, Z.Z., Ni, S.N., 2014. Long-term groundwater variations in Northwest India from satellite gravity measurements. Glob. Planet. Change 116, 130–138. Available from: https://doi.org/10.1016/j.gloplacha.2014.02.007.

Chen, J.L., Tapley, B., Save, H., Tamisiea, M.E., Bettadpur, S., Ries, J., 2018. Quantification of ocean mass change using gravity recovery and climate experiment, satellite altimeter, and Argo floats observations. J. Geophys. Res. (Solid Earth) 123 (B12), 10. Available from: https://doi.org/10.1029/2018JB016095.

Chen, J.L., Wilson, C.R., Li, J., Zhang, Z., 2015. Reducing leakage error in GRACE-observed long-term ice mass change: a case study in West Antarctica. J. Geodesy 89, 925–940. Available from: https://doi.org/10.1007/s00190-015-0824-2.

Chen, J.L., Wilson, C.R., Tapley, B.D., 2006. Satellite gravity measurements confirm accelerated melting of Greenland Ice Sheet. Science 313, 1958–1960. Available from: https://doi.org/10.1126/science.1129007.

Chen, J.L., Wilson, C.R., Tapley, B.D., 2010. The 2009 exceptional Amazon flood and interannual terrestrial water storage change observed by GRACE. Water Resour. Res. 46, W12526. Available from: https://doi.org/10.1029/2010WR009383.

Chen, J.L., Wilson, C.R., Tapley, B.D., 2013. Contribution of ice sheet and mountain glacier melt to recent sea level rise. Nat. Geosci. 6, 549–552. Available from: https://doi.org/10.1038/NGEO1829.

Chen, J.L., Wilson, C.R., Tapley, B.D., Blankenship, D.D., Ivins, E.R., 2007. Patagonia Icefield melting observed by Gravity Recovery and Climate Experiment (GRACE). Geophys. Res. Lett. 34, L22501. Available from: https://doi.org/10.1029/2007GL031871.

Chen, J.L., Wilson, C.R., Tapley, B.D., Save, H., Cretaux, J.-F., 2017. Long-term and seasonal Caspian Sea level change from satellite gravity and altimeter measurements. J. Geophys. Res. (Solid. Earth) 122, 2274–2290. Available from: https://doi.org/10.1002/2016JB013595.

Döll, P., Kaspar, F., Lehner, B., 2003. A global hydrological model for deriving water availability indicators: model tuning and validation. J. Hydrol. 270, 105–134.

Döll, P., Müller Schmied, H., Schuh, C., Portmann, F.T., Eicker, A., 2014. Global-scale assessment of groundwater depletion and related groundwater abstractions: Combining hydrological modeling with information from well observations and GRACE satellites. Water Resour. Res. 50, 5698–5720. Available from: https://doi.org/10.1002/2014WR015595.

Famiglietti, J.S., 2014. The global groundwater crisis. Nat. Clim. Change 4 (11), 945–948.

Famiglietti, J.S., Lo, M., Ho, S.L., Bethune, J., Anderson, K.J., Syed, T.H., et al., 2011. Satellites measure recent rates of groundwater depletion in California's Central Valley. Geophys. Res. Lett. 38, L03403. Available from: https://doi.org/10.1029/2010GL046442.

Faunt, C.C. (Ed.), 2009. Groundwater Availability of the Central Valley Aquifer. US Geological Survey, CA, Prof. Paper 1766.

Feng, W., Zhong, M., Lemoine, J.-M., Biancale, R., Hsu, H.-T., Xia, J., 2013. Evaluation of groundwater depletion in North China using the Gravity Recovery and Climate Experiment (GRACE) data and ground-based measurements. Water Resour. Res. 49, 2110–2118. Available from: https://doi.org/10.1002/wrcr.20192.

Foster, S.S.D., Loucks, D.P. (Eds.), 2006. Non-Renewable Groundwater Resources: A Guidebook on Socially-Sustainable Management for Water-Policy Makers. UNESCO.

Getirana, A., Kumar, S., Girotto, M., Rodell, M., 2017. Rivers and floodplains as key components of global terrestrial water storage variability. Geophys. Res. Lett. 44 (20), 10–359.

Getirana, A., Rodell, M., Kumar, S., Beaudoing, H.K., Arsenault, K., Zaitchik, B., et al., 2020. GRACE improves seasonal groundwater forecast initialization over the U.S. J. Hydrometeor. 21 (1), 59–71. Available from: https://doi.org/10.1175/JHM-D-19-0096.1.

Girotto, M., De Lannoy, G.J., Reichle, R.H., Rodell, M., 2016. Assimilation of gridded terrestrial water storage observations from GRACE into a land surface model. Water Resour. Res. 52 (5), 4164–4183.

Gleeson, T., et al., 2010. Groundwater sustainability strategies. Nat. Geosci. 3, 378–379. Available from: https://doi.org/10.1038/ngeo881.

Güntner, A., Stuck, J., Werth, S., Döll, P., Verzano, K., Merz, B., 2007. A global analysis of temporal and spatial variations in continental water storage. Water Resour. Res. 43, W05416. Available from: https://doi.org/10.1029/2006WR005247.

Han, S.-C., Shum, C.K., Bevis, M., Ji, C., Kuo, C.-Y., 2006. Crustal Dilatation Observed by GRACE After the 2004 Sumatra-Andaman Earthquake. Science 313, 658–662. Available from: https://doi.org/10.1126/science.1128661.

Hoque, M.A., Hoque, M.M., Ahmed, K.M., 2007. Declining groundwater level and aquifer dewatering in Dhaka metropolitan area, Bangladesh: causes and quantification. Hydrogeol. J. 15, 1523–1534. Available from: https://doi.org/10.1007/s10040-007-0226-5.

Houborg, R., Rodell, M., Li, B., Reichle, R., Zaitchik, B., 2012. Drought indicators based on model assimilated GRACE terrestrial water storage observations. Water Resour. Res. 48, W07525. Available from: https://doi.org/10.1029/2011WR011291.

Joodaki, G., Wahr, J., Swenson, S., 2014. Estimating the human contribution to groundwater depletion in the Middle East, from GRACE data, land surface models, and well observations. Water Resour. Res. 50, 2679–2692. Available from: https://doi.org/10.1002/2013WR014633.

Kato, H., Rodell, M., Beyrich, F., Cleugh, H., van Gorsel, E., Liu, H., et al., 2007. Sensitivity of land surface simulations to model physics, land characteristics, and forcings, at four CEOP sites. J. Meteorol. Soc. Jpn. 85, 187–204. Ser. II.

Kumar, S.V., Zaitchik, B.F., Peters-Lidard, C.D., Rodell, M., Reichle, R., Li, B., et al., 2016. Assimilation of gridded GRACE terrestrial water storage estimates in the North American Land Data Assimilation System. J. Hydrometeorol. 17 (7), 1951–1972.

Landerer, F.W., Swenson, S.C., 2012. Accuracy of scaled GRACE terrestrial water storage estimates. Water Resour. Res. 48, W04531. Available from: https://doi.org/10.1029/2011WR011453.

Leblanc, M.J., Tregoning, P., Ramillien, G., Tweed, S.O., Fakes, A., 2009. Basin-scale, integrated observations of the early 21st century multiyear drought in southeast Australia. Water Resour. Res. 45, . Available from: https://doi.org/10.1029/2008WR0073334408–440.

Li, B., Rodell, M., Kumar, S.V., Beaudoing, H.K., Getirana, A., Zaitchik, B.F., et al., 2019. Global GRACE data assimilation for groundwater and drought monitoring: advances and challenges. Water Resour. Res. 55, 7564–7586. Available from: https://doi.org/10.1029/2018WR024618.

Longuevergne, L., Scanlon, B.R., Wilson, C.R., 2010. GRACE Hydrological estimates for small basins: Evaluating processing approaches on the High Plains Aquifer, USA. Water Resour. Res. 46, 11517. Available from: https://doi.org/10.1029/2009WR008564.

Longuevergne, L., Wilson, C.R., Scanlon, B.R., Crétaux, J.F., 2013. GRACE water storage estimates for the Middle East and other regions with significant reservoir and lake storage. Hydrol. Earth Syst. Sci. 17, 4817–4830. Available from: https://doi.org/10.5194/hess-17-4817-2013.

Loomis, B.D., Luthcke, S.B., Sabaka, T.J., 2019. Regularization and error characterization of GRACE mascons. J. Geodesy 93 (9), 1381–1398. Available from: https://doi.org/10.1007/s00190-019-01252-y.

Luthcke, S.B., Sabaka, T.J., Loomis, B.D., Arendt, A.A., McCarthy, J.J., Camp, J., 2013. Antarctica, Greenland and Gulf of Alaska land-ice evolution from an iterated GRACE global mascon solution. J. Glaciology 59, 613–631. Available from: https://doi.org/10.3189/2013JoG12J147.

Maupin, M.A., Barber, N.L., 2005. Estimated withdrawals from principal aquifers in the United States, 2000. : U.S. Geol. Surv. Circular 1279, 46.

Mukherjee, A., Scanlon, B., Aureli, A., Langan, S., Guo, H., McKenzie, A., 2020. Global Groundwater: Source, Scarcity, Sustainability, Security and Solutions, first ed. Elsevier, ISBN: 9780128181720, ISBN.

Mulder, G., Olsthoorn, T.N., Al-Manmi, D.A.M.A., Schrama, E.J.O., Smidt, E.H., 2015. Identifying water mass depletion in northern Iraq observed by GRACE. Hydrol. Earth Syst. Sci. 19, 1487–1500. Available from: https://doi.org/10.5194/hess-19-1487-2015.

Niu, G.-Y., Seo, K.-W., Yang, Z.-L., Wilson, C., Su, H., Chen, J., et al., 2007. Retrieving snow mass from GRACE terrestrial water storage change with a land surface model. Geophys. Res. Lett. 34, L15704. Available from: https://doi.org/10.1029/2007GL030413.

Papa, F., Prigent, C., Rossow, W.B., 2008. Monitoring flood and discharge variations in the large Siberian rivers from a multi-satellite technique. Surv. Geophys. 29, 297–317. Available from: https://doi.org/10.1007/s10712-008-9036-0.

Peltier, W.R., 2004. Global glacial isostasy and the surface of the ice-age earth: the ice-5g (Vm2) model and GRACE. Annu. Rev. Earth Planet. Sci. 32, 111–149. Available from: https://doi.org/10.1146/annurev.earth.32.082503.144359.

Ramillien, G., Frappart, F., Cazenave, A., Güntner, A., 2005. Time variations of land water storage from an inversion of 2 years of GRACE geoids [rapid communication]. Earth Planet. Sci. Lett. 235, 283–301. Available from: https://doi.org/10.1016/j.epsl.2005.04.005.

Reager, J.T., Famiglietti, J.S., 2009. Global terrestrial water storage capacity and flood potential using GRACE. Geophys. Res. Lett. 36, . Available from: https://doi.org/10.1029/2009GL04082623402–2340.

Rodell, M., Beaudoing, H.K., L'Ecuyer, T.S., Olson, W.S., Famiglietti, J.S., Houser, P.R., et al., 2015. The observed state of the water cycle in the early twenty-first century. J. Clim. 28 (21), 8289–8318.

Rodell, M., Chao, B.F., Au, A.Y., Kimball, J.S., McDonald, K.C., 2005. Global biomass variation and its geodynamic effects: 1982–98. Earth Interact. 9 (2), 1–19.

Rodell, M., Chen, J., Kato, H., Famiglietti, J.S., Nigro, J., Wilson, C.R., 2007. Estimating groundwater storage changes in the Mississippi River basin (USA) using GRACE. Hydrogeol. J. 15, 159–166. Available from: https://doi.org/10.1007/s10040-006-0103-7.

Rodell, M., Famiglietti, J.S., 2001. An analysis of terrestrial water storage variations in Illinois with implications for the Gravity Recovery and Climate Experiment (GRACE). Water Resour. Res. 37 (5), 1327–1339.

Rodell, M., Famiglietti, J.S., Chen, J., Seneviratne, S.I., Viterbo, P., Holl, S., et al., 2004a. Basin scale estimates of evapotranspiration using GRACE and other observations. Geophys. Res. Lett. 31 (20).

Rodell, M., Famiglietti, J.S., Wiese, D.N., Reager, J.T., Beaudoing, H.K., Landerer, F.W., et al., 2018. Emerging trends in global freshwater availability. Nature 557 (7707), 651–659.

Rodell, M., Houser, P.R., Jambor, U., Gottschalck, J., Mitchell, K., Meng, C.-J., et al., 2004b. The global land data assimilation system. Bull. Am. Meteorol. Soc. 85 (3), 381–394.

Rodell, M., Velicogna, I., Famiglietti, J.S., 2009. Satellite-based estimates of groundwater depletion in India. Nature 460, 999–1002. Available from: https://doi.org/10.1038/nature08238.

Rowlands, D.D., Luthcke, S.B., Klosko, S.M., Lemoine, F.G., Chinn, D.S., McCarthy, J.J., et al., 2005. Resolving mass flux at high spatial and temporal resolution using GRACE intersatellite measurements. Geophys. Res. Lett. 32 (4).

Save, H., Bettadpur, S., Tapley, B.D., 2016. High-resolution CSR GRACE RL05 mascons. J. Geophys. Res. (Solid Earth) 121, 7547–7569. Available from: https://doi.org/10.1002/2016JB013007.

Scanlon, B.R., Longuevergne, L., Long, D., 2012. Ground referencing GRACE satellite estimates of groundwater storage changes in the California Central Valley, USA. Water Resour. Res. 48, 4520. Available from: https://doi.org/10.1029/2011WR011312.

Sheffield, J., Ferguson, C.R., Troy, T.J., Wood, E.F., McCabe, M.F., 2009. Closing the terrestrial water budget from satellite remote sensing. Geophys. Res. Lett. 36 (7).

Strassberg, G., Scanlon, B.R., Chambers, D., 2009. Evaluation of groundwater storage monitoring with the GRACE satellite: case study of the High Plains aquifer, central United States. Water Resour. Res. 45, W5410. Available from: https://doi.org/10.1029/2008WR006892.

Swenson, S., Wahr, J., 2006. Post-processing removal of correlated errors in GRACE data. Geophys. Res. Lett. 33, L08402. Available from: https://doi.org/10.1029/2005GL025285.

Tamisiea, M.E., Mitrovica, J.X., Davis, J.L., 2007. GRACE gravity data constrain ancient ice geometries and continental dynamics over Laurentia. Science 316, 881–883. Available from: https://doi.org/10.1126/science.1137157.

Tapley, B.D., Bettadpur, S., Watkins, M.M., Reigber, C., 2004. The Gravity Recovery and Climate Experiment; mission overview and early results. Geophys. Res. Lett. 31 (9), L09607. Available from: https://doi.org/10.1029/2004GL019920.

Tapley, B.D., Watkins, M.M., Flechtner, F., Reigber, C., Bettadpur, et al., 2019. Contributions of GRACE to understanding climate change. Nat. Clim. Change 9, 358–369. Available from: https://doi.org/10.1038/s41558-019-0456-2.

Thomas, A.C., Reager, J.T., Famiglietti, J.S., Rodell, M., 2014. A GRACE-based water storage deficit approach for hydrological drought characterization. Geophys. Res. Lett. 41 (5), 1537–1545.

Tiwari, V.M., Wahr, J., Swenson, S., 2009. Dwindling groundwater resources in northern India, from satellite gravity observations. Geophys. Res. Lett. 36, . Available from: https://doi.org/10.1029/2009GL03940118401–1840.

Velicogna, I., Wahr, J., 2006. Acceleration of Greenland ice mass loss in spring 2004. Nature 443, 329–331. Available from: https://doi.org/10.1038/nature.05168.

Voss, K.A., Famiglietti, J.S., Lo, M., Linage, C., Rodell, M., Swenson, S.C., 2013. Groundwater depletion in the Middle East from GRACE with implications for transboundary water management in the Tigris-Euphrates-Western Iran region. Water Resour. Res. 49, 904–914. Available from: https://doi.org/10.1002/wrcr.20078.

Wahr, J., Molenaar, M., Bryan, F., 1998. Time variability of the Earth's gravity field: hydrological and oceanic effects and their possible detection using GRACE. J. Geophys. Res. 103, 30205–30230. Available from: https://doi.org/10.1029/98JB02844.

Wahr, J., Swenson, S., Zlotnicki, V., Velicogna, I., 2004. Time-variable gravity from GRACE: first results. Geophys. Res. Lett. 31, L11501. Available from: https://doi.org/10.1029/2004GL019779.

Watkins, M.M., Wiese, D.N., Yuan, D.-N., Boening, C., Landerer, F.W., 2015. Improved methods for observing Earth's time variable mass distribution with GRACE using spherical cap mascons. J. Geophys. Res. (Solid. Earth) 120, 2648–2671. Available from: https://doi.org/10.1002/2014JB011547.

Wiese, D.N., Landerer, F.W., Watkins, M.M., 2016. Quantifying and reducing leakage errors in the JPL RL05M GRACE mascon solution. Water Resour. Res. 52, 7490–7502. Available from: https://doi.org/10.1002/2016WR019344.

Wouters, B., Chambers, D., Schrama, E.J.O., 2008. GRACE observes small-scale mass loss in Greenland. Geophys. Res. Lett. 35, L20501. Available from: https://doi.org/10.1029/2008GL034816.

Yang, X., Chen, Y., Pacenka, S., Gao, W., Zhang, M., Sui, P., et al., 2015. Recharge and groundwater use in the North China Plain for six irrigated crops for an eleven year period. PLoS One 10 (1), e0115269. Available from: https://doi.org/10.1371/journal.pone.0115269.

Zaitchik, B.F., Rodell, M., Reichle, R.H., 2008. Assimilation of GRACE terrestrial water storage data into a land surface model: Results for the Mississippi River basin. J. Hydrometeorol. 9, 535–548. Available from: https://doi.org/10.1175/2007JHM951.1.

Chapter 40

Use of machine learning and deep learning methods in groundwater

Pragnaditya Malakar[1], Soumyajit Sarkar[2], Abhijit Mukherjee[1,2], Soumendra Bhanja[3] and Alexander Y. Sun[4]

[1]*Department of Geology and Geophysics, Indian Institute of Technology (IIT) Kharagpur, Kharagpur, India,* [2]*Applied Policy Advisory for Hydrosciences (APHA) group, School of Environmental Science and Engineering, Indian Institute of Technology (IIT) Kharagpur, Kharagpur, India,* [3]*Interdisciplinary Centre for Water Research, Indian Institute of Science, Bangalore, India,* [4]*Bureau of Economic Geology, The University of Texas at Austin, Austin, TX, United States*

extendedKeywords: Groundwater prediction; groundwater quantity; groundwater quality; artificial intelligence (AI); machine learning (ML); deep learning (DL)

40.1 Introduction

Groundwater is the largest storage of freshwater resources, which serves as the major inventory for most of the human consumption through agriculture, industrial, and domestic water supply (Taylor et al., 2013; Famiglietti, 2014), sustaining the global water and food security. Hence, it is a very critical task to regulate the balance between supply and demand for the large population. The main two aspects of groundwater management are to secure usable groundwater storage and to assure safe quality of the available groundwater. In the past few decades, rapid groundwater depletion is reported globally in most of the aquifers due to the increasing stress from the growing population (Siebert et al., 2015; Bhanja et al., 2016, 2017; Rodell et al., 2018; Mukherjee, 2018; Mukherjee et al., 2020). Furthermore, groundwater is prone to contamination from various nonpoint and point sources (National Research Council, 1993). Thus identifying the zones of contamination and their future distribution is becoming a major global concern.

In situ measurements of groundwater level (GWL) from the observation wells provide a high-resolution direct quantification of subsurface groundwater storage (Bhanja and Mukherjee, 2019). However, recent advancements of satellite and land surface model (LSM) based water storage measures such as gravity recovery and climate experiment (GRACE) based terrestrial water storage (TWS) data, global land data assimilation system (Rodell et al., 2004) and other measures, such as normalized difference vegetation index (NDVI), have been reported to be a suitable proxy of groundwater storage (Bhanja et al., 2019). Thus these available networks of in situ monitoring wells and satellite-based products could be effectively used to model and/or predict groundwater quantity. Moreover, the pollution caused by natural and anthropogenic activities in groundwater may be persistent for prolonged periods (Alcalá and Custodio, 2014). Therefore it is necessary to take effective environmental measures for monitoring and management of groundwater quality, and restraining the pollution sources.

40.1.1 Importance of advanced data-driven methods in groundwater resources

Numerical, physically based, statistical, and conceptual modeling techniques are some of the most popular tools used for groundwater modeling over the years. Some of the limitations associated with these conventional methods include data dependency on various parameters, with some of the parameters being more difficult to obtain. Thus data-driven black-box models have the inherent benefits of accurately predicting the concerning variables without the particulars on the underlying mechanisms. It should be noted that despite having considerable advantages over traditional methods, these advanced methods have few pitfalls, such as low generalization ability, overtraining, using irrelevant data as input, using improper modeling methods and tuning parameters, and others (Rajaee et al., 2019). Regardless of these weaknesses, these data-driven methods have been chosen more often by researchers in different parts of the world for groundwater resource modeling and quality estimation due to their simplistic approach and satisfactory results.

40.2 Global literature review

The heterogeneous complex hydrologic systems are governed by several natural- and human-induced triggers. Thus data-driven methods such as artificial intelligence (AI), including machine learning (ML) and deep learning (DL), can serve as suitable modeling techniques without requiring a full understanding of the underlying mechanism. Researchers around the globe have applied these methods to predict (simulate) GWLs and TWS, and to assess the distribution of major contaminants. In the next section, we provide a brief overview of the applications of AI methods for groundwater modeling.

40.2.1 Groundwater quantity

Coulibaly et al. (2001) used GWL, temperature, and precipitation as the inputs to develop three different types of artificial neural network (ANN) models to simulate GWL in the Gondo aquifer of Burkina Faso. They reported that the recurrent neural network (RNN) is the most appropriate modeling method than the radial basis function—ANN and ANN (Coulibaly et al., 2001). In another study, Yoon et al. (2011) built ANN and support vector machine (SVM) models to predict GWL fluctuations using tide level, precipitation, and past GWLs as inputs at a South Korean coastal aquifer. They further reported that GWL, tide level, and precipitation in descending order are the most effective input variable to model GWL, and SVM performs better than ANN (Yoon et al., 2011). Mohanty et al. (2010) develop three different ANN models with Levenberg—Marquardt (LM), Bayesian regularization (BR), and adaptive learning rate back-propagation (GDX) (GDX) as training algorithms for weekly GWL prediction in eastern India. Pumping rates of previous weeks were used as input in addition to precipitation, evaporation, river stage, drainage water level, and GWL. The BR algorithm was proven to be better than the other algorithms (Mohanty et al., 2010). Sun (2013) used GRACE TWS data, precipitation, maximum and minimum temperature to predict GWL variation and concluded that incorporation of meteorological variable with GRACE TWS data improves the ANN model performances (Sun, 2013). Shiri et al. (2013) examined the capabilities of genetic programing (GP), adaptive neuro-fuzzy inference system (ANFIS), ANN, SVM, and autoregressive integrated moving average methods, using precipitation, evapotranspiration, and GWL for GWL forecasting at daily scale in Korea. The results indicated that GP models were the best models compared to others (Shiri et al., 2013).

To assess the performances of 1—4 months ahead GWL forecasting, four different model structures, that is, ANN, wavelet-ANN, ANFIS, and wavelet-ANFIS models, were used by Moosavi et al. (2013) in Mashhad plain of Iran. It was confirmed that modified ANN and ANFIS structures incorporating wavelet transformation increase the ability of forecasting, and among all the models, wavelet-ANFIS models are found to be superior (Moosavi et al., 2013). Barzegar et al. (2017a, 2017b) attempted to forecast the GWL (monthly) in Azarbaijan, Iran using wavelet with ANN and group method of data handling models. The wavelets based decomposed GWL time series used as an input with appropriate lag and stepwise selection process. They reported that the best performances are obtained from the boosting multiwavelet-ANN models (Barzegar et al., 2017a). Mukherjee and Ramachandran (2018) predicted GWLs for a number ($n = 5$) of in situ observation wells in India using the SVM, ANN, and linear regression model (LRM) models with GRACE derived TWS anomaly data and meteorological variables. Their findings suggest that SVM performs better, among others, and the performance of the models increases significantly when climatic data are included with GRACE data (Mukherjee and Ramachandran, 2018). Chen et al. (2011) used ANN and GRACE data to reconstruct the past TWS and validate the model's performances with in situ GWL data in Songhua River Basin, Northern China.

A brief list of previous studies is presented in Table 40.1.

40.2.2 Groundwater quality

Amini et al. (2008) predicted global groundwater arsenic contamination using observed arsenic concentrations around the world, using soil parameters, geology, climate, and elevation as the predictor variables. ANFIS and logistic regression models were used to develop the arsenic probability maps for two arsenic mobilization zones. The models were able to predict 77% arsenic concentrations in reducing zones and 68% in oxidizing zones correctly (Amini et al., 2008). Moreover, Erickson et al. (2018) used boosted regression trees (BRTs) to predict elevated arsenic concentrations

TABLE 40.1 List of previous studies, in which machine learning and deep learning methods were used to model groundwater.

Reference	Input variables	Used AI models	Region of study
Coulibaly et al. (2001)	GWL, P, T	ANN	Gondo aquifer, Republic of Burkina Faso
Lallahem et al. (2005)	GWL, P, mean T, effective P, potential ET	ANN	Chalky aquifer Of northern France
Daliakopoulos et al. (2005)	GWL, T, P, Q	ANN	Island of Crete, Greece
Nayak et al. (2006)	GWL, neighboring wells GWL, P, canal releases	ANN	Central Godavari Delta System, India
Krishna et al. (2008)	GWL, P, ET	ANN	Andhra Pradesh state, India
Feng et al. (2008)	GWL, P, E, Q, population, irrigation ratio, irrigation area	ANN	Shiyang river basin, northwest China
Nourani et al. (2008)	GWL, P, mean T, Q	ANN	Tabriz aquifer, Iran
Mohanty et al. (2010)	GWL, P, E, river stage, SWL, pumping rate	ANN	Orissa, India
Chen et al. (2011)	GWL, neighboring wells GWL	ANN	Southern Taiwan
Adamowski and Chan (2011)	GWL, P, T	Wavelet-ANN, ANN	Quebec, Canada
Yoon et al. (2011)	GWL, P, tide level	SVM, ANN	Beach of the Donghae City, Korea
Sreekanth et al. (2011)	GWL, P, E, T, H	ANN, ANFIS	Maheshwaram watershed, India
Taormina et al. (2012)	GWL, P, ET	ANN	Lagoon of Venice, Italy
Shirmohammadi et al. (2013)	P	ANFIS	Mashhad plain, Iran
Sahoo and Jha (2013)	GWL, P, T, river stage, seasonal dummy variables	ANN	Konan basin, Kochi, Japan
Sun (2013)	GRACE satellite data, P, min and max T	ANN	United States
Shiri et al. (2013)	GWL, P, ET	GP, ANN, ANFIS, SVM	Hoengchon, south Korea
Moosavi et al. (2013)	GWL, P, E, average Q	Wavelet-ANFIS, wavelet-ANN, ANFIS, ANN	Mashhad plain, Iran
Emamgholizadeh et al. (2014)	P, recharge, irrigation returned flow, pumping rates	ANFIS, ANN	Bastam plain, Iran
Suryanarayana et al. (2014)	GWL, P, max T, mean T	Wavelet-SVR, SVR, ANN	Visakhapatnam, India
Tapoglou et al. (2014)	GWL, SWL, T, P	ANN-neurofuzzy-GS	Bavaria, Germany
He et al. (2014)	GWL	Wavelet-ANN, ANN	Ganzhou region, China
Jha and Sahoo (2015)	GWL, P, T, river stage, seasonal dummy variables	ANN-GA	Konan basin, Kochi, Japan
Khalil et al. (2015)	Recharge, P, T	Wavelet-ensemble-ANN, wavelet-ANN, ANN	Manitou mine site, Quebec, Canada
Juan et al. (2015)	GWL, T, P	ANN	Qinghai-Tibet Plateau, China
Gholami et al. (2015)	P, tree-rings	ANN	Caspian southern coasts, Iran
Khaki et al. (2015)	P, H, E, min and max T	ANFIS, ANN	Langat basin, Malaysia

(Continued)

TABLE 40.1 (Continued)

Reference	Input variables	Used AI models	Region of study
Nourani et al. (2015)	GWL, P, runoff	Wavelet-ANN, ANN	Ardabil plain, Iran
Gong et al. (2016)	GWL, SWL, P, T	ANFIS, SVM, ANN	Shore of Lake Okeechobee, Florida, United States
Sun et al. (2016)	SWL, P	ANN	Nee Soon swamp forest, Singapore
Chang et al. (2016)	GWL, Q, P	ANN (SOM-NARX)	Zhuoshui River basin, Taiwan
Han et al. (2016)	GWL, climate conditions, well extractions, Q, reservoir operations	ANN (SOM-statistical model)	western Hexi Corridor, northwest China
Nourani and Mousavi (2016)	GWL, P, Q	Wavelet-ANFIS, wavelet-ANN	Miandoab plain, Iran
Yoon et al. (2016)	GWL, P	SVM, ANN	South Korea
Ebrahimi and Rajaee (2017)	GWL	Wavelet-ANN, wavelet-SVR, ANN, SVR	Qom plain, Iran
Barzegar et al. (2017a, 2017b)	GWL	Wavelet-ANN	Azarbaijan, Iran
Yu et al. (2018)	GWL, ET, Q	Wavelet-ANN, wavelet-SVR	Northwest China
Wunsch et al. (2018)	P, T	ANN	Southwest Germany
Mukherjee and Ramachandran (2018)	GRACE satellite data, P, min and max T, H, wind	LRM, SVR, ANN	India
Lee et al. (2019)	SWL, pumping rates	ANN	South Korea
Kouziokas et al. (2018)	P, T, H	ANN	Pennsylvania, United States
Sun et al. (2019)	GRACE TWSA, NOAH LSM, GWL	CNN	India
Chen et al. (2011)	GRACE, LSM, GWL	ANN	Songhua River Basin, China

AI, Artificial intelligence; *ANFIS*, adaptive neuro-fuzzy inference system; *ANN*, artificial neural network; *CNN*, convoluted neural network; *E*, evaporation; *ET*, evapotranspiration; *GA*, genetic algorithm; *GS*, geostatistics; *GP*, genetic programing; *GRACE*, gravity recovery and climate experiment; *GWL*, groundwater level; *H*, humidity; *LRM*, linear regression model; *LSM*, land surface model; *NARX*, nonlinear autoregressive with exo-genous inputs neural network; *P*, precipitation; *Q*, river flow/discharge/runoff; *R*, rainfall; *SOM*, self-organizing map; *SVM*, support vector machine; *SVR*, support vector regression; *SWL*, surface water level; *T*, temperature; *TWSA*, terrestrial water storage anomalies.
Source: Modified from Rajaee et al. (2019).

(≥ 10 μg/L) in groundwater for the complex central and northwestern Minnesota, United States, using arsenic measurement data and total set of 74 predictor variables, that is, aquifer characteristics, soil parameters, well construction data. They reported that aquifer characteristics, soil parameters, and certain well construction such as smaller screen lengths were significant to describe high arsenic concentration (Erickson et al., 2018). In a different study, a comparative approach of different ML methods was used by Barzegar et al. (2017a, 2017b) to predict groundwater fluoride in the Maku region of northwest Iran. The performance of extreme learning machines (ELMs) was compared with two other ML techniques, for example, multilayer perception (MLP) and SVMs using fluoride and major ions such as Na^+, K^+, Ca^{2+}, and HCO_3^- as predictor variables. The study found that the ELM model performed well than the other two models (Barzegar et al., 2017b). Nolan et al. (2015) predicted groundwater nitrate concentration for the central valley region of California, United States, in the shallow well depths using three ML models such as BRT, ANN, and Bayesian network models. A set of 41 independent variables were used to develop the models, which includes soil parameters, proxies for groundwater age, aquifer characteristics, vertical water movement, and land use. They reported that the BRT model outperformed the other two models (Nolan et al., 2015).

Random forests (RFs) were successfully applied by researchers to predict the distribution pattern of contaminants such as arsenic, fluoride, and nitrate from regional to country scales around the globe. Wheeler et al. (2015) used RF to predict nitrate concentration in groundwater using nitrate measurements from 34,084 private wells in the state of Iowa, United States. They found that RF was predicting better than the conventional LRM. Depth of wells, slope length in 1 km radius of well, and distance to nearest livestock firms were found to be some of the most important variables in the final model. Some recent notable studies are Podgorski et al. (2018), Bindal and Singh (2019) and Chakraborty et al. (2020) who, all used RF to perform predictions. Podgorski et al. (2018) used RF to develop a prediction map of groundwater fluoride for the entire country. They had used 12,600 groundwater fluoride measurements data collected from the Central Groundwater Board (CGWB) to develop the prediction model along with 25 surficial variables related to geology, soil characteristics, and climate. RF model showed greater prediction accuracy when compared with a logistic regression model. The study concluded that the states/union territories northwest India and southern India were at a high risk of fluoride contamination in groundwater. The study also delineated that high fluoride conditions are prevalent in arid regions. The other study by Bindal and Singh (2019) has predicted the spatial distribution pattern of arsenic in groundwater for the state of Uttar Pradesh, India, using arsenic data and 20 other predictor variables from 1473 measurements. The RF model achieved an overall prediction accuracy of 84.67%, performing better than univariate, logistic, and various fuzzy inference models. Chakraborty et al. (2020) modeled groundwater arsenic hazard on a regional scale across the transboundary Ganges River delta of India and Bangladesh, where they, for the first time, integrated a RF model with a physical, hydrogeological model of the study area. Unlike the other studies, which mostly emphasis on surficial parameters, Chakraborty et al. (2020) were able to provide more realistic, influence of subsurface geology on water quality.

Reference	Study region	Models used	Pollutant predicted	Input variables
Menezes et al. (2020)	Brazil	GLM, ridge regression, MARS, bagged Tree, XGBoost, cubist, and RF	Arsenic	Geology, soil parameters, land cover, topography, minerals, temperature, precipitation
Winkel et al. (2008)	Bangladesh, Cambodia, Thailand, Vietnam	MLR	Arsenic	Topography, soil, geology, climate, hydrology
Cho et al. (2011)	Cambodia, Laos, and Thailand	Multiple linear regression, PCR, ANN, and PC-ANN	Arsenic	pH, redox potential, EC, TDS, and temperature
Rodríguez Lado et al. (2013)	Inner Mongolia, Gansu, Shanxi, Ningxia, Henan, and Heilongjiang of China	MLR	Arsenic	Topography, EVI, the density of rivers, distance from rivers, Earth's gravitational force, subsoil texture, salinity, geology, and population density
Chakraborty et al. (2020)	Ganges river delta, India and Bangladesh	BRT, RF	Arsenic	Geology, hydrogeology, aquifer connectivity, topography, soil parameters, geology, land use, temperature, precipitation, runoff, irrigation
Bindal and Singh (2019)	Uttar Pradesh, India	Univariate logistic regression, fuzzy, AFR, and ANFIS, RF	Arsenic	Soil parameters, land cover, slope, distance from rivers, drainage, TWI, GRACE data, GWL
Erickson et al. (2018)	Minnesota, United States	BRT	Arsenic	Location, aquifer material properties, landscape, soil geochemistry, and well construction attributes.
Jangle et al. (2016)	Vaishali district, Bihar, India	MLR	Arsenic	Elevation, slope, distance from rivers, NDVI, land use, hydraulic conductivity, net GW recharge and depth of wells

Nadiri et al. (2013)	Maku, Iran	Sugeno fuzzy logic, Mamdani fuzzy logic, ANN, neuro-fuzzy coupled to SCMAI	Fluoride	Major ions, pH, and silicon (SiO_2)
Podgorski et al. (2018)	India, Nepal, Bangladesh, Bhutan	RF and MLR	Fluoride	Precipitation, temperature, evapotranspiration, AET, PET, aridity, cropland, irrigation amounts, irrigated area, geology, and soil parameters
Barzegar et al. (2016)	Maku, Iran	ELM, MLP, and SVM	Fluoride	Na^+, Ca^{2+}, K^+, and HCO_3^-
Mohammadi et al. (2016)	Khaf, Iran	ANN	Fluoride	pH, temperature, EC, TDS, TH, alkalinity, Cl^-, SO_4^{2-}, HCO_3^-, Na^+, Ca^{2+}, Mg^{2+}
Amini et al. (2008)	Global	MLR and ANFIS	Arsenic	Geology, evapotranspiration, soil pH, Na^+, and HCO_3^- concentration
Yesilnacar et al. (2008)	Harran plain, Turkey	ANN	Nitrate	pH, temperature, EC, and GWL
Wheeler et al. (2015)	Iowa, United States	RF, regression tree, linear regression, Kriging models, and GAM	Nitrate	Agricultural land use, nitrogen input to the land surface, soil parameters, geology, climate, irrigation and aquifer characteristics
Nolan et al. (2015)	California, United States	BRT, ANN, and BN	Nitrate	Soil parameters, land use, groundwater age, and aquifer texture, and vertical water fluxes
Arabgol et al. (2016)	Arak plain, Iran	SVM	Nitrate	Water temperature, EC, groundwater depth, TDS, DO, pH, land use, and season of the year
Wang et al. (2006)	Huantai County, China	ANN	Nitrate	Crop yields and nitrogen inputs from fertilizer and manure, soil sand content, organic content, land use

AET, Actual evapotranspiration; *AFR*, adaptive fuzzy regression; *ANFIS*, adaptive neuro-fuzzy inference system; *ANN*, artificial neural network; *BN*, Bayesian networks; *BRT*, boosted regression tree; *DO*, dissolved oxygen; *EC*, electrical conductivity; *ELM*, extreme learning machine; *EVI*, enhanced vegetation index; *GAM*, generalized additive model; *GLM*, generalized linear model; *GRACE*, gravity recovery and climate experiment; *GWL*, groundwater level; *MARS*, multivariate adaptive regression splines; *MLP*, multilayer perceptron; *MLR*, multivariate logistic regression; *NDVI*, normalized difference vegetation index; *PC-ANN*, principal component combined with ANN; *PCR*, principal component regression; *PET*, potential evapotranspiration; *RF*, random forest; *RF*, random forests; *SCMAI*, supervised committee machine with artificial intelligence; *SVM*, support vector machine; *TDS*, total dissolved solids; *TH*, total hardness; *TWI*, topographic wetness index; *XGBoost*, extreme gradient boosted tree.

40.3 Application of some of the widely used artificial intelligence methods in India

In recent times a host of AI methods such as ANN, SVM, RF, GP, convoluted neural network (CNN), ANFIS, their hybrid methods and variants, have been widely used in groundwater hydrology for both quantity and qualitative predictions or simulations. It should be noted that all the methods that follow a technology based on simulating human behavior is AI, whereas ML is a subset of AI which trains the system to learn from past data without explicit programing. DL is the networks that are capable of learning and adaptation from the data itself. In this section, we provide a brief overview of three popular methods, that is, ANN, SVM, and RF.

40.3.1 Methods description

40.3.1.1 Machine learning–based methods

40.3.1.1.1 Artificial neural networks

ANNs are models that replicate the human biological neural networks. Based on the data on past variables, ANN estimate unknown functions and predict future time series of the variables. ANNs consist of elements that define the network function (Rajaee et al., 2019). A typical ANN structure is composed of processing elements called neurons, and similar neurons reside in one single layer. A standard ANN structure has three types of layers: input, hidden, and output layers. The data are introduced in the system with the input layer, then the data processing is executed in the hidden layer(s), and finally, the output layer is where the results are produced through a preferred transfer function. The network is being trained with data by adjusting and optimizing the network parameters (i.e., weights and biases). The BP algorithm, the LM algorithm, BR algorithm are some of the common training algorithm used.

The ANN is categorized based on the flow direction of information and processing: (1) feed forward neural network (FNN); (2) RNN.

The FNNs propagate the input signal in the forward direction, whereas the RNN uses the outputs back as inputs of the network.

40.3.1.1.2 Random forests

RF is an ensemble learning machine that consists of a large number of decision tree algorithms governed by randomly sampled vectors to generate repeated predictions of a given problem. A decision tree denotes a hierarchical arrangement of a set of conditions, implemented successively from the root node to the terminal node (Quinlan, 2014). Randomness in the trees is introduced by the selection of a random set of data with replacement, used to build individual tree (bagging), and select a random subset of predictor variables at the individual node of splitting (split selection). The dependent variable is then split by each of the trees, which is controlled by the distribution of the dependent variable over the predictors and, therefore, achieving the highest variance of the dependent variable. The data splitting is repeated from the root node to each internal node by the rule of the tree until the previously specified stop condition is achieved. The output of each tree is a binary class, depending on their predicted probabilities by the probability cut-off and successively selecting the final class based on the most votes (Breiman, 2017) described that the first step of the induction of a decision tree is the selection of optimal splitting starts by the splitting of the dependent variable (parent node) into binary classes, where any child node is "purer" than their parent nodes.

40.3.1.2 Deep learning–based methods

40.3.1.2.1 Convolutional neural network (CNN)

CNN is a deep-learning model architecture that broadly follows the human visual cortex architecture and principle. CNN is capable of extracting embedded subtle features from the inputs that are provided into the system. In CNN, discrete convolution operations are applied based on somewhat similar to a nonlinear transformation of the system input to project the input image(s) onto a hierarchy of feature maps. A typical CNN architecture comprises of the input layer followed by a series of hidden layers (convolutional layers that extract spatial features from each layer's input) and, output layer. The advantage of CNN over the traditional multilayer perception neural networks (have limitations on scaling images) is its ability to learn multiscale spatial patterns from multisource gridded data. CNN involves a convolution operation, as a filter or kernel to scan along each dimension of the input image. A widely used activation function for hidden CNN layers is the Rectified Linear Unit (ReLU) function, which is less costly to compute than other nonlinear functions (e.g., sigmoid) and is previously reported to improve the training speed significantly in CNN (Sun et al., 2019).

40.3.2 Case studies from India

40.3.2.1 Application of machine learning

40.3.2.1.1 Prediction of GWL in India based on GWL and NDVI as input in ANN modeling

In this section, we will discuss some of our recent studies in which we develop an ANN model to predict GWL using past GWL and NDVI. Here, we have divided the total quarterly temporal data set (2005–13) into the training set (2005–10) and test set (2011–13). In other words, we trained the model using data for the first 24 seasons, and the last

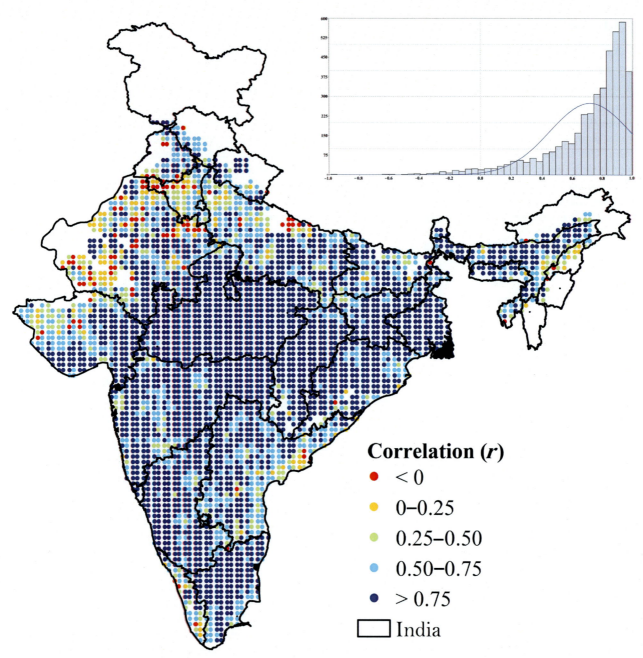

FIGURE 40.1 The correlation coefficient between observed and ANN predicted GWL and percentile distribution of correlation coefficient (top left). *ANN*, Artificial neural network; *GWL*, groundwater level.

12 seasons were used to test the models. Please note that all the analysis is executed at a spatial resolution of $0.25° \times 0.25°$. A FNN with logistic sigmoidal activation function to build the model, whereas LM optimization is employed as the training function (Bhanja et al., 2019). The spatial distribution of the correlation coefficient between the observed and predicted GWL between 2011 and 2013 is demonstrated in Fig. 40.1. The figure suggests that overall good correlation is found between the observed and predicted GWL. However, the relationship is weakened mostly in North-West India and a few isolated places, where the human-induced groundwater abstraction is dominated (MacDonald et al., 2016; Malakar et al., 2018; Rodell et al., 2009). The results indicate the poor performances of the models in irrigated-agriculture regions.

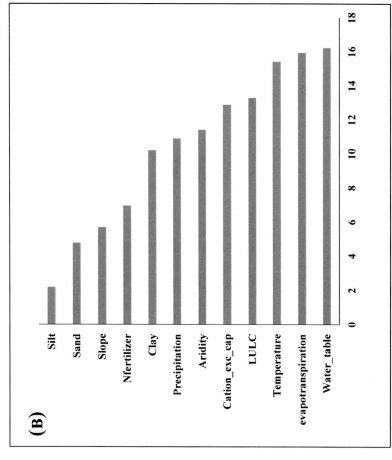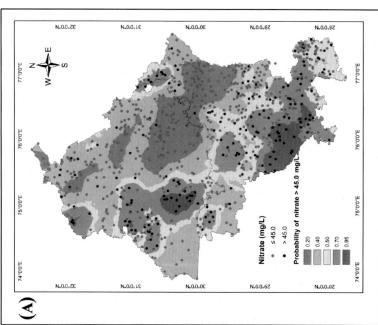

FIGURE 40.2 (A) Predicted probability map of nitrate concentration exceeding 45 mg/L with measured nitrate concentration points, (B) Mean decrease in accuracy with predictor variables.

40.3.2.1.2 Application of random forest in groundwater contamination prediction in India

In this section, we show our present work of predicting groundwater nitrate for the two northern states of India. RF classification (RFC) model was used to predict the probability of groundwater nitrate for the states Haryana and Punjab, India. Nitrate concentrations of 850 locations were collected from the CGWB, and a set of 16 predictor variables from hydrogeologic, climate, and anthropogenic categories was used to develop the RFC model for total 815 location points. Our findings reveal that 196 points (24%) exceeded the threshold value of 45 mg/L, and rest 619 points (76%) were equal or less than 45 mg/L. Then the data was converted to binary classes, 1 for the points with nitrate >45 mg/L and 0 for nitrate $\leq$ 45 mg/L. The data was split randomly into training (80%) and testing (20%) sets, maintaining the ratio of 24% of high nitrate points in both. We grew 1000 trees to produce the RFC using the R programing environment. The final RF model achieved an accuracy of 91% for the entire data set, and the sensitivity and specificity values were 0.91 and 0.89, respectively, with a very high prediction ability (AUC = 95). Our model suggests that groundwater table, evapotranspiration, temperature, and land use, and land cover are important variables for predicting nitrate in groundwater. The predicted probability map indicated that the southern part of Haryana and the western part of Punjab have a greater probability of nitrate concentrations above the permissible 45 mg/L (Fig. 40.2).

40.3.2.2 Application of deep learning

40.3.2.2.1 Using a combination of physically based modeling and CNN to learn the spatiotemporal pattern from satellite and land surface model based terrestrial water storage

In a recent study, Sun et al. (2019) provide a hybrid method by combining physically-based modeling and convolutional neural network (CNN), a deep-learning method, to predict the spatial-temporal variability of TWS anomalies (TWSA) in India. The CNN-based deep-learning models, such as the Unet, VGG16, and SegnetLite are trained to learn the spatial-temporal mismatch between the GRACE based and simulated (by LSM as inputs) TWSA. Moreover, the NOAH-simulated TWSA is corrected using the learned mismatch. In general, the three deep-learning models used here, are found to improve the NOAH TWSA significantly at both the country and grid levels (40° × 40° square regions). The findings are particularly promising considering the significantly small training sample size used in the study. Finally, they compared the in situ GWS anomaly with the mismatch between the observed and simulated model values (Sun et al., 2019). Correlation analysis reveals a good correlation between the two, suggesting the mismatch patterns efficiently compensate for the missing groundwater storage in the LSM based on NOAH for many parts of India. The data break between GRACE and its follow-on mission (GRACE-FO) can be successfully filled with the trained CNN models using and correcting the NOAH-simulated TWSA.

References

Adamowski, J., Chan, H.F., 2011. A wavelet neural network conjunction model for groundwater level forecasting. J. Hydrol. 407, 28–40. <https://doi.org/10.1016/j.jhydrol.2011.06.013>.

Alcalá, F.J., Custodio, E., 2014. Spatial average aquifer recharge through atmospheric chloride mass balance and its uncertainty in continental Spain. Hydrol. Process. 28, 218–236.

Amini, M., Abbaspour, K.C., Berg, M., Winkel, L., Hug, S.J., Hoehn, E., et al., 2008. Statistical modeling of global geogenic arsenic contamination in groundwater. Environ. Sci. Technol. 42, 3669–3675.

Arabgol, R., Sartaj, M., Asghari, K., 2016. Predicting Nitrate Concentration and Its Spatial Distribution in Groundwater Resources Using Support Vector Machines (SVMs) Model. Environ. Model. Assess 21, 71–82. Available from: https,//doi.org/10.1007/s10666-015-9468-0.

Barzegar, R., Fijani, E., Asghari Moghaddam, A., Tziritis, E., 2017a. Forecasting of groundwater level fluctuations using ensemble hybrid multi-wavelet neural network-based models. Sci. Total Environ. 599–600, 20–31. <https://doi.org/10.1016/j.scitotenv.2017.04.189>.

Barzegar, R., Moghaddam, A.A., Adamowski, J., Fijani, E., 2017b. Comparison of machine learning models for predicting fluoride contamination in groundwater. Stochastics Environ. Res. Risk Assess. 31, 2705–2718.

Bhanja, S.N., Malakar, P., Mukherjee, A., Rodell, M., Mitra, P., Sarkar, S., 2019. Using satellite-based vegetation cover as indicator of groundwater storage in natural vegetation areas. Geophys. Res. Lett. 46, 8082–8092. <https://doi.org/10.1029/2019gl083015>.

Bhanja, S.N., Mukherjee, A., 2019. In situ and satellite-based estimates of usable groundwater storage across India: implications for drinking water supply and food security. Adv. Water Resour. 126, 15–23.

Bhanja, S.N., Mukherjee, A., Rodell, M., Wada, Y., Chattopadhyay, S., Velicogna, I., et al., 2017. Groundwater rejuvenation in parts of India influenced by water-policy change implementation. Sci. Rep. 7 (1), 7453.

Bhanja, S.N., Mukherjee, A., Saha, D., Velicogna, I., Famiglietti, J.S., 2016. Validation of GRACE based groundwater storage anomaly using in-situ groundwater level measurements in India. J. Hydrol. 544, 428–437.

Breiman, L., 2017. Classification and Regression Trees. Routledge.

Bindal, S., Singh, C.K., 2019. Predicting groundwater arsenic contamination, Regions at risk in highest populated state of India. Water Res 159, 65–76. Available from: https,//doi.org/10.1016/j.watres.2019.04.054.

Barzegar, R., Asghari Moghaddam, A., Adamowski, J., Fijani, E., 2017. Comparison of machine learning models for predicting fluoride contamination in groundwater. Stoch. Environ. Res. Risk Assess 31, 2705–2718. Available from: https,//doi.org/10.1007/s00477-016-1338-z.

Chakraborty, M., Sarkar, S., Mukherjee, A., Shamsudduha, M., Ahmed, K.M., Bhattacharya, A., et al., 2020. Modeling regional-scale groundwater arsenic hazard in the transboundary Ganges River Delta, India and Bangladesh: infusing physically-based model with machine learning. Sci. Total Environ . Available from: https://doi.org/10.1016/j.scitotenv.2020.141107.

Chen, L.H., Chen, C.T., Lin, D.W., 2011. Application of Integrated Back-Propagation Network and Self-Organizing Map for Groundwater Level Forecasting. J. Water Resour. Plan. Manag. 137, 352–365. Available from: https://doi.org/10.1061/(ASCE)WR.1943-5452.0000121.

Cho, K.H., Sthiannopkao, S., Pachepsky, Y.A., Kim, K.W., Kim, J.H., 2011. Prediction of contamination potential of groundwater arsenic in Cambodia, Laos, and Thailand using artificial neural network. Water Res 45, 5535–5544. Available from: https,//doi.org/10.1016/j.watres.2011.08.010.

Chang, F.J., Chang, L.C., Huang, C.W., Kao, I.F., 2016. Prediction of monthly regional groundwater levels through hybrid soft-computing techniques. J. Hydrol. 541, 965–976. <https://doi.org/10.1016/j.jhydrol.2016.08.006>.

Chen, H., Zhang, W., Nie, N., Guo, Y., 2019. Long-term groundwater storage variations estimated in the Songhua River Basin by using GRACE products, land surface models, and in-situ observations. Sci. Total Environ. 649, 372–387. <https://doi.org/10.1016/j.scitotenv.2018.08.352>.

Coulibaly, P., Anctil, F., Aravena, R., Bobée, B., 2001. Artificial neural network modeling of water table depth fluctuations. Water Resour. Res. 37, 885–896. <https://doi.org/10.1029/2000WR900368>.

Daliakopoulos, I.N., Coulibaly, P., Tsanis, I.K., 2005. Groundwater level forecasting using artificial neural networks. J. Hydrol. 309, 229–240. <https://doi.org/10.1016/j.jhydrol.2004.12.001>.

Ebrahimi, H., Rajaee, T., 2017. Simulation of groundwater level variations using wavelet combined with neural network, linear regression and support vector machine. Global Planet. Change 148, 181–191. <https://doi.org/10.1016/j.gloplacha.2016.11.014>.

Emamgholizadeh, S., Moslemi, K., Karami, G., 2014. Prediction the groundwater level of Bastam plain (Iran) by artificial neural network (ANN) and adaptive neuro-fuzzy inference system (ANFIS). Water Resour. Manage. 28 (15), 5433–5446.

Erickson, M.L., Elliott, S.M., Christenson, C., Krall, A.L., 2018. Predicting geogenic arsenic in drinking water wells in glacial aquifers, North-Central USA: accounting for depth-dependent features, Water Resour. Res., 54. pp. 10–172.

Famiglietti, J.S., 2014. The global groundwater crisis. Nat. Clim. Change 4, 945–948. <https://doi.org/10.1038/nclimate2425>. Ground Water Vulnerability Assessment: Predicting Relative Contamination Potential Under Conditions of Uncertainty, 1993. National Academies Press.

Feng, S., Kang, S., Huo, Z., Chen, S., Mao, X., 2008. Neural networks to simulate regional ground water levels affected by human activities. Ground Water 46, 80–90. <https://doi.org/10.1111/j.1745-6584.2007.00366.x>.

Gholami, V., Chau, K.W., Fadaee, F., Torkaman, J., Ghaffari, A., 2015. Modeling of groundwater level fluctuations using dendrochronology in alluvial aquifers. J. Hydrol. 529, 1060–1069. <https://doi.org/10.1016/j.jhydrol.2015.09.028>.

Gong, Y., Zhang, Y., Lan, S., Wang, H., 2016. A comparative study of artificial neural networks, support vector machines and adaptive neuro fuzzy inference system for forecasting groundwater levels near Lake Okeechobee, Florida. Water Resour. Manage. 30, 375–391. <https://doi.org/10.1007/s11269-015-1167-8>.

Han, J.C., Huang, Y., Li, Z., Zhao, C., Cheng, G., 2016. Groundwater level prediction using a SOM-aided stepwise cluster inference model. J. Environ. Manage. 182, 308–321.

He, Z., Zhang, Y., Guo, Q., Zhao, X., 2014. Comparative study of artificial neural networks and wavelet artificial neural networks for groundwater depth data forecasting with various curve fractal dimensions. Water Resour. Manage. 28, 5297–5317.

Jangle, N., Sharma, V., Dror, D.M., 2016. Statistical geospatial modelling of arsenic concentration in Vaishali District of Bihar, India. Sustain. Water Resour. Manag. 2, 285–295. Available from: https,//doi.org/10.1007/s40899-016-0049-4.

Jha, M.K., Sahoo, S., 2015. Efficacy of neural network and genetic algorithm techniques in simulating spatio-temporal fluctuations of groundwater. Hydrol. Process. 29 (5), 671–691.

Juan, C., Genxu, W., Tianxu, M., 2015. Simulation and prediction of suprapermafrost groundwater level variation in response to climate change using a neural network model. J. Hydrol. 529, 1211–1220.

Khaki, M., Yusoff, I., Islami, N., 2015. Simulation of groundwater level through artificial intelligence system. Environ. Earth Sci. 73 (12), 8357–8367.

Khalil, B., Broda, S., Adamowski, J., Ozga-Zielinski, B., Donohoe, A., 2015. Short-term forecasting of groundwater levels under conditions of mine-tailings recharge using wavelet ensemble neural network models. Hydrogeol. J. 23, 121–141.

Kouziokas, G.N., Chatzigeorgiou, A., Perakis, K., 2018. Multilayer feed forward models in groundwater level forecasting using meteorological data in public management. Water Resour. Manage. 32, 5041–5052. <https://doi.org/10.1007/s11269-018-2126-y>.

Krishna, B., Satyaji Rao, Y.R., Vijaya, T., 2008. Modelling groundwater levels in an urban coastal aquifer using artificial neural networks. Hydrol. Process. 22, 1180–1188. <https://doi.org/10.1002/hyp.6686>.

Lallahem, S., Mania, J., Hani, A., Najjar, Y., 2005. On the use of neural networks to evaluate groundwater levels in fractured media. J. Hydrol. 307, 92–111. <https://doi.org/10.1016/j.jhydrol.2004.10.005>.

Lee, S., Lee, K.K., Yoon, H., 2019. Using artificial neural network models for groundwater level forecasting and assessment of the relative impacts of influencing factors. Hydrogeol. J. 27, 567–579. Available from: https://doi.org/10.1007/s10040-018-1866-3.

MacDonald, A.M., Bonsor, H.C., Ahmed, K.M., Burgess, W.G., Basharat, M., Calow, R.C., et al., 2016. Groundwater quality and depletion in the Indo-Gangetic Basin mapped from in situ observations. Nat. Geosci. 9, 762–766. <https://doi.org/10.1038/ngeo2791>.

Mohanty, S., Jha, M.K., Kumar, A., Sudheer, K.P., 2010. Artificial neural network modeling for groundwater level forecasting in a river Island of eastern India. Water Resour. Manage. 24, 1845–1865.

Moosavi, V., Vafakhah, M., Shirmohammadi, B., Behnia, N., 2013. A wavelet-ANFIS hybrid model for groundwater level forecasting for different prediction periods. Water Resour. Manage. 27, 1301–1321.

Mukherjee, A. (Ed.), 2018. Groundwater of South Asia. Springer, ISBN 978-981-10-3888-4, 799 p.

Mukherjee, A., Ramachandran, P., 2018. Prediction of GWL with the help of GRACE TWS for unevenly spaced time series data in India: analysis of comparative performances of SVR, ANN and LRM. J. Hydrol. 558, 647–658.

Mukherjee, A., Scanlon, B., Aureli, A., Langan, S., Guo, H., McKenzie, A., 2020. Global Groundwater: Source, Scarcity, Sustainability, Security and Solutions, first ed. Elsevier. ISBN: 9780128181720.

Malakar, P., Mukherjee, A., Sarkar, S., 2018. Potential application of advanced computational techniques in prediction of groundwater resource of india. In: Mukherjee, A. (Ed.), Groundwater of South Asia. Springer, Singapore, pp. 643–655. Available from: https://doi.org/10.1007/978-981-10-3889-1_37.

Mohammadi, A.A., Ghaderpoori, M., Yousefi, M., Rahmatipoor, M., Javan, S., 2016. Prediction and modeling of fluoride concentrations in groundwater resources using an artificial neural network, a case study in Khaf. Environ. Heal. Eng. Manag 3, 217–224. Available from: https,//doi.org/10.15171/ehem.2016.23.

Menezes, M.D., Bispo, F.H.A., Faria, W.M., Gonçalves, M.G.M., Curi, N., Guilherme, L.R.G., 2020. Modeling arsenic content in Brazilian soils, What is relevant? Sci. Total Environ 712. Available from: https,//doi.org/10.1016/j.scitotenv.2020.136511.

National Research Council, 1993. Ground water vulnerability assessment, Predicting relative contamination potential under conditions of uncertainty. National Academies Press.

Nadiri, A.A., Fijani, E., Tsai, F.T.C., Moghaddam, A.A., 2013. Supervised committee machine with artificial intelligence for prediction of fluoride concentration. J. Hydroinformatics 15, 1474–1490. Available from: https,//doi.org/10.2166/hydro.2013.008.

Nayak, P.C., Satyaji Rao, Y.R., Sudheer, K.P., 2006. Groundwater level forecasting in a shallow aquifer using artificial neural network approach. Water Resour. Manage. 20, 77–90. <https://doi.org/10.1007/s11269-006-4007-z>.

Nolan, B.T., Fienen, M.N., Lorenz, D.L., 2015. A statistical learning framework for groundwater nitrate models of the Central Valley, California, USA. J. Hydrol. 531, 902–911.

Nourani, V., Alami, M.T., Vousoughi, F.D., 2015. Wavelet-entropy data pre-processing approach for ANN-based groundwater level modeling. J. Hydrol. 524, 255–269. <https://doi.org/10.1016/j.jhydrol.2015.02.048>.

Nourani, V., Asghari Mogaddam, A., Nadiri, A.O., 2008. An ANN-based model for spatiotemporal groundwater level forecasting. Hydrol. Process. 22, 5054–5066.

Nourani, V., Mousavi, S., 2016. Spatiotemporal groundwater level modeling using hybrid artificial intelligence-meshless method. J. Hydrol. 536, 10–25. <https://doi.org/10.1016/j.jhydrol.2016.02.030>.

Podgorski, J.E., Labhasetwar, P., Saha, D., Berg, M., 2018. Prediction Modeling and Mapping of Groundwater Fluoride Contamination throughout India. Environ. Sci. Technol. 52, 9889–9898. Available from: https,//doi.org/10.1021/acs.est.8b01679.

Quinlan, J.R., 2014. C4. 5: Programs for Machine Learning. Elsevier.

Rajaee, T., Ebrahimi, H., Nourani, V., 2019. A review of the artificial intelligence methods in groundwater level modeling. J. Hydrol. 572, 336–351. Available from: https,//doi.org/10.1016/j.jhydrol.2018.12.037.

Rodríguez-Lado, L., Sun, G., Berg, M., Zhang, Q., Xue, H., Zheng, Q., Johnson, C.A., 2013. Groundwater arsenic contamination throughout China. Science 341, 866–868. Available from: https,//doi.org/10.1126/science.1237484.

Rodell, M., Famiglietti, J.S., Wiese, D.N., Reager, J.T., Beaudoing, H.K., Landerer, F.W., et al., 2018. Emerging trends in global freshwater availability. Nature 557 (7707), 651–659.

Rodell, M., Houser, P.R., Jambor, U.E.A., Gottschalck, J., Mitchell, K., Meng, C.J., et al., 2004. The global land data assimilation system. Bull. Am. Meteorol. Soc. 85 (3), 381–394.

Rodell, M., Velicogna, I., Famiglietti, J.S., 2009. Satellite-based estimates of groundwater depletion in India. Nature 460, 999–1002. <https://doi.org/10.1038/nature08238>.

Sahoo, S., Jha, M.K., 2013. Groundwater-level prediction using multiple linear regression and artificial neural network techniques: a comparative assessment. Hydrogeol. J. 21, 1865–1887.

Shiri, J., Kisi, O., Yoon, H., Lee, K.K., Nazemi, A.H., 2013. Predicting groundwater level fluctuations with meteorological effect implications – a comparative study among soft computing techniques. Comput. Geosci. 56, 32–44.

Shirmohammadi, B., Vafakhah, M., Moosavi, V., Moghaddamnia, A., 2013. Application of several data-driven techniques for predicting groundwater level. Water Resour. Manage. 27 (2), 419–432.

Siebert, S., Kummu, M., Porkka, M., Döll, P., Ramankutty, N., Scanlon, B.R., 2015. Historical Irrigation Dataset (HID). Hydrol. Earth Syst. Sci. 19, 1521–1545. <https://doi.org/10.13019/M20599>.

Sreekanth, P.D., Sreedevi, P.D., Ahmed, S., Geethanjali, N., 2011. Comparison of FFNN and ANFIS models for estimating groundwater level. Environ. Earth Sci. 62, 1301–1310.

Sun, A.Y., 2013. Predicting groundwater level changes using GRACE data. Water Resour. Res. 49, 5900–5912. <https://doi.org/10.1002/wrcr.20421>.

Sun, A.Y., Scanlon, B.R., Zhang, Z., Walling, D., Bhanja, S.N., Mukherjee, A., et al., 2019. Combining physically based modeling and deep learning for fusing GRACE satellite data: can we learn from mismatch? Water Resour. Res. 55, 1179–1195. <https://doi.org/10.1029/2018WR023333>.

Sun, Y., Wendi, D., Kim, D.E., Liong, S.Y., 2016. Technical note: application of artificial neural networks in groundwater table forecasting-a case study in a Singapore swamp forest. Hydrol. Earth Syst. Sci. 20, 1405–1412. <https://doi.org/10.5194/hess-20-1405-2016>.

Suryanarayana, C., Sudheer, C., Mahammood, V., Panigrahi, B.K., 2014. An integrated wavelet-support vector machine for groundwater level prediction in Visakhapatnam, India. Neurocomputing 145, 324–335.

Taormina, R., Chau, K., Sethi, R., 2012. Artificial neural network simulation of hourly groundwater levels in a coastal aquifer system of the Venice lagoon. Eng. Appl. Artif. Intell. 25, 1670–1676.

Tapoglou, E., Karatzas, G.P., Trichakis, I.C., Varouchakis, E.A., 2014. A spatio-temporal hybrid neural network-Kriging model for groundwater level simulation. J. Hydrol. 519, 3193–3203.

Taylor, R.G., Scanlon, B., Döll, P., Rodell, M., Van Beek, R., Wada, Y., et al., 2013. Ground water and climate change. Nat. Clim. Change 3, 322–329. <https://doi.org/10.1038/nclimate1744>.

Wunsch, A., Liesch, T., Broda, S., 2018. Forecasting groundwater levels using nonlinear autoregressive networks with exogenous input (NARX). J. Hydrol. 567, 743–758. <https://doi.org/10.1016/j.jhydrol.2018.01.045>.

Yesilnacar, M.I., Sahinkaya, E., Naz, M., Ozkaya, B., 2008. Neural network prediction of nitrate in groundwater of Harran Plain, Turkey. Environ. Geol 56, 19–25. Available from: https,//doi.org/10.1007/s00254-007-1136-5.

Yoon, H., Hyun, Y., Ha, K., Lee, K.K., Kim, G.B., 2016. A method to improve the stability and accuracy of ANN- and SVM-based time series models for long-term groundwater level predictions. Comput. Geosci. 90, 144–155. <https://doi.org/10.1016/j.cageo.2016.03.002>.

Yoon, H., Jun, S.C., Hyun, Y., Bae, G.O., Lee, K.K., 2011. A comparative study of artificial neural networks and support vector machines for predicting groundwater levels in a coastal aquifer. J. Hydrol. 396, 128–138. <https://doi.org/10.1016/j.jhydrol.2010.11.002>.

Yu, H., Wen, X., Feng, Q., Deo, R.C., Si, J., Wu, M., 2018. Comparative study of hybrid wavelet artificial intelligence models for monthly groundwater depth forecasting in extreme arid regions, northwest China. Water Resour. Manage. 32 (1), 301–323.

Wheeler, D.C., Nolan, B.T., Flory, A.R., DellaValle, C.T., Ward, M.H., 2015. Modeling groundwater nitrate concentrations in private wells in Iowa. Science of the Total Environment 536, 481–488. Available from: https,//doi.org/10.1016/j.scitotenv.2015.07.080.

Winkel, L., Berg, M., Amini, M., Hug, S.J., Johnson, A.A., 2008. Predicting groundwater arsenic contamination in Southeast Asia from surface parameters. Nat. Geosci 1, 536–542. Available from: https,//doi.org/10.1038/ngeo254.

Wang, M.X., Liu, G.D., Wu, W.L., Bao, Y.H., Liu, W.N., 2006. Prediction of agriculture derived groundwater nitrate distribution in North China Plain with GIS-based BPNN. Environ. Geol 50, 637–644. Available from: https,//doi.org/10.1007/s00254-006-0237-x.

Chapter 41

Desalination of brackish groundwater to improve water quality and water supply

Yvana D. Ahdab and John H. Lienhard

Rohsenow Kendall Heat Transfer Laboratory, Massachusetts Institute of Technology, Cambridge, MA, United States

41.1 Introduction

Water scarcity around the world is leading to a greater dependence on groundwater to meet freshwater demand (Mukherjee et al., 2020). Despite the fact that most groundwater resources are brackish [500 mg/L ≤ total dissolved solids (TDS) ≤ 5000 mg/L] (Ahdab et al., 2018), brackish groundwater (BGW) remains a largely untapped resource in many parts of the world. In the United States, for example, the volume of BGW (1000 mg/L ≤ TDS ≤ 10,000 mg/L) was measured to be over 35 times the amount of fresh groundwater used annually (Stanton et al., 2017). Increased exploitation of brackish groundwater may relieve some of the mounting pressure on freshwater supplies, particularly in drier landlocked regions. Desalination can be employed to reduce the salt concentration in brackish groundwater to meet the needs of freshwater applications.

41.1.1 Brackish groundwater composition

All naturally occurring waters contain some level of TDS, a measure of the concentration of all inorganic and organic dissolved substances, including salts, minerals, and metals. TDS determines whether surface water or groundwater resource is fresh or brackish. Brackish groundwater contains a TDS greater than freshwater but less than seawater. A variety of classification schemes are used to categorize waters with different TDS. Brackish groundwater falls within the 500–20,000 mg/L TDS range, with some classifications placing the upper TDS limit at 10,000 mg/L (Stanton et al., 2017). In comparison, seawater typically contains a TDS greater than 25,000 mg/L.

Unlike seawater, both the TDS and major ion constituents of brackish water vary greatly with depth of the well below the land surface and with geographic location, as a result of local geologic, hydrologic, and climactic conditions. Groundwaters containing higher TDS are more often drawn from greater depth below the land surface (Stanton et al., 2017). These variations are critical in determining the feasibility, required treatment, and associated cost of brackish groundwater usage. Because a specific location may correspond to a particular BGW composition, location is crucial in water resource planning and treatment system selection and design.

Brackish groundwater must be treated before use in applications that require high water quality, such as drinking water and irrigation. Water used for public supply, which fulfills the majority of the population's daily water needs, must not include high dissolved solids concentration or significant concentrations of specific constituents: the US Environmental Protection Agency recommends that drinking water contain less than 500 mg/L of TDS to ensure public health (Office of Water of the U.S. Environmental Protection Agency, 2018). Water used for irrigation of agricultural crops, the largest consumer of our water supplies globally, is limited by dissolved solids concentration, the relative amount of solutes, and specific constituents that can be damaging to crops.

41.1.2 Desalination

Desalination is a water treatment that removes dissolved solids and other minerals from a water resource that would otherwise be unsuitable for use in freshwater applications. Desalination has been on the rise since the 1960s, and this trend is expected to continue into the 2020s (Fig. 41.1). Today, the global desalination operating capacity is approximately 70 million m^3/day (Global Water Intelligence, 2020). It is primarily used in treating seawater (61% of desalinated water) and brackish water (21% of desalinated water), the majority of which is groundwater (Global Water Intelligence, 2020; Jones et al., 2019). Its various end uses include municipal (e.g., drinking water), industry, power, and irrigation. Brackish water desalination requires less energy and can recover more product water for a given amount of feedwater than seawater desalination.

Commercial desalination technologies can be divided into two main categories: thermal and membrane. Thermal technologies, most commonly multistage flash (MSF) distillation and multiple-effect distillation (MED), mimic the hydrological cycle of evaporation and condensation by heating salty water to form water vapor that is then condensed into fresh water. Membrane processes, such as reverse osmosis (RO) and electrodialysis (ED, EDR), use a semipermeable membrane that prevents or allows the passage of certain salt ions. The driving force for transport can be a pressure, electrical potential, temperature, or concentration gradient. Due to their increased energy efficiency and cost-effectiveness, membrane processes have surpassed the once dominant thermal processes in terms of desalination capacity (Fig. 41.1).

To treat brackish groundwater, membrane desalination technologies, primarily RO and ED, are used without exception (Veerapaneni et al., 2011; Al-Karaghouli and Kazmerski, 2013; He and Bond, 2019). Thermal processes are not utilized, primarily because their energy requirements are independent of salinity, unlike membrane processes in which energy requirements decrease with salinity. RO is overwhelmingly the dominant desalination technology, including for brackish water treatment. In 2019 RO and ED produced 76% and 2.4%, respectively, of desalinated water globally (Global Water Intelligence, 2020). Around a quarter of RO-generated water and well over half of ED generated water originated from brackish water (Global Water Intelligence, 2020; Jones et al., 2019).

The remainder of this chapter is divided into two sections. The first section focuses on brackish groundwater desalination technologies, including associated energy consumption, cost, and environmental impact. The second section examines regional and national variations in desalination capacity, technology, feedwater type, and expenditure. In particular, trends in the United States, Saudi Arabia, Australia, China, Spain, and India are investigated.

41.2 Desalination process

A brackish groundwater desalination plant is typically composed of six key stages (Fig. 41.2): (1) groundwater is pumped from wells; (2) the raw water undergoes pretreatment, depending on its composition, to reduce membrane

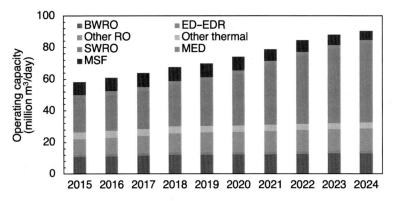

FIGURE 41.1 Operating desalination capacity from 2015 to 2019 and expected operating desalination capacity from 2020 to 2024. A breakdown of operating capacity by plant (RO, ED or EDR, MSF, MED, Other) and feedwater is shown. Data is from DesalData online database (Global Water Intelligence, 2020). *BW*, Brackish water; *ED*, electrodialysis; *EDR*, electrodialysis reversal; *MED*, multiple-effect distillation; *MSF*, multistage flash; *RO*, reverse osmosis; *SW*, seawater.

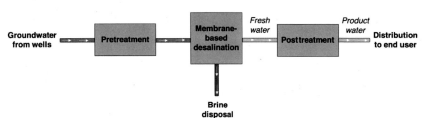

FIGURE 41.2 A process diagram for a typical brackish groundwater plant comprises groundwater pumping, pretreatment, electricity-driven (e.g., pump or power supply) desalination, brine disposal, posttreatment, and distribution stages.

fouling; (3) the pretreated water is fed into the desalination stage (RO, ED), which yields desalinated water (low in salinity) and brine (concentrated in salinity); (4) the brine is disposed of or further concentrated and dried to achieve zero liquid discharge (ZLD); (5) the desalinated water is posttreated; and (6) final product water is distributed to the end consumer or a storage tank using service pumps.

This section provides a detailed overview of these desalination plant stages. Membrane fouling, RO and ED systems, including corresponding pretreatment and posttreatment, plant energy consumption, and cost data, bring management, emerging desalination technologies, and renewable coupled desalination are discussed.

41.2.1 Membrane fouling and pretreatment

Depending on feedwater composition and membrane type, several constituents can result in membrane contamination, which is referred to as fouling. Fouling reduces membrane efficiency, resulting in a shorter membrane lifetime, more frequent cleaning, and a decrease in recovery rate. Recovery rate is defined as the fraction of freshwater produced from a given amount of feedwater. The primary types of fouling in membrane desalination systems are scaling and biofouling. Scaling occurs due to the precipitation of inorganic salts, such as carbonate, sulfates, and silica, from the feedwater onto the membrane surface. Brackish water RO (BWRO) is especially prone to membrane scaling (Veerapaneni et al., 2011; He and Bond, 2019). Biofouling arises from the growth of bacteria on the membranes, which depends on temperature, pH, dissolved oxygen, and composition of feedwater. In some cases, high concentrations of suspended and colloidal matter in the feedwater block the feed flow channels in the membrane elements. Pretreatment of desalination source water before it enters a membrane is required to minimize the fouling potential, extend membrane life, and maximize recovery rate. Pretreatment can involve chemical processes, physical processes, or a combination of the two. The extent of pretreatment required in brackish desalination facilities is less than that of seawater, due to the lower fouling potential of many groundwater sources.

41.2.2 Reverse osmosis

RO uses a semipermeable membrane that enables the passage of water, while rejecting salts, under an applied pressure. It represents the state-of-the-art desalination technology for brackish water applications, because it can reject a variety of contaminants in a single process with lower energy consumption. Aside from raw water intake and product water conveyance, an RO facility comprises pretreatment; desalination modules with RO membranes; a high-pressure pump to drive desalination; posttreatment; and, in some cases, an energy recovery device (ERD) that depressurizes the brine leaving the system.

41.2.2.1 Pretreatment

RO membranes are sensitive to pH, oxidizers, a wide range of organics, algae, bacteria, particulates, and other foulants. The most common pretreatment method is the chemical addition of antiscalants and acid to prevent the formation of pH-dependent membrane scaling, followed by cartridge filters to remove particulates that will plug or foul membranes. In some cases, more pretreatment may be necessary to control iron and manganese, using oxidation/filtration pretreatment, or to reduce sand loading from wells, using sand separators or strainers.

41.2.2.2 Desalination mechanism

A schematic drawing of the RO desalination mechanism is shown in Fig. 41.3. The natural osmotic pressure of a saline solution will drive water from the low-to-high solute (salt) concentration side of a semipermeable membrane. RO uses

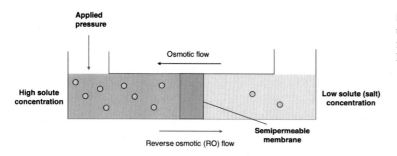

FIGURE 41.3 A schematic drawing of the RO desalination mechanism. An applied pressure forces water to flow from the high to the low solute concentration side. *RO*, Reverse osmosis.

a high-pressure pump to apply a hydraulic pressure greater than the osmotic pressure to the saltier side. The applied pressure required for brackish water typically ranges from 17 to 27 bars (seawater: 55 to 82 bar) (Al-Karaghouli and Kazmerski, 2013); water higher in salinity will require a higher hydraulic pressure and will consume more energy to overcome the osmotic pressure. Under the applied pressure, water is forced through the membrane to the low solute concentration side. Salt ions almost entirely remain on the high concentration side, although some salt leakage from high to low solute concentration will occur due to diffusion that results from the salinity gradient across the membrane. RO yields a freshwater stream (permeate) and a concentrated solution (brine or concentrate) on the high-pressure side of the membrane.

41.2.2.3 Membranes

RO membranes can be broadly categorized as low-pressure elements (brackish water) and high-pressure elements (seawater). Several types of membranes are available on the market. The two most commonly used membrane configurations are hollow fiber[1] and spiral wound[2] in a cross-flow filtration. In cross-flow the pressurized water flows parallel to, rather than perpendicular to, the membrane surface in order to assist in the removal of concentrated salts from the surface; this configuration reduces the rate of fouling and salt leakage into the permeate from diffusion. Key RO membrane parameters include permeability (i.e., the rate of salt diffusion across the membrane) and rejection (i.e., the quantity of salt rejected from the feedwater). These vary significantly with membrane type. Current brackish RO membranes remove between 98% and 99.2% of TDS from the feedwater and are designed to produce a permeate of approximately 500 mg/L (He and Bond, 2019). They have a life expectancy of 2–5 years (Veerapaneni et al., 2011; He and Bond, 2019). Improvements in these membranes continue to simultaneously reduce desalination energy requirements and the rate of fouling.

41.2.2.4 System design

Typically six to eight membrane modules are placed in series within a fiberglass pressure vessel. RO plants are often composed of two to three stages in order to maximize recovery rate. In these multistage configurations the brine from two first-stage pressure vessels will serve as feedwater to a single second stage and so on. The recovery rate of BWRO systems ranges from 75% to 85% (Faust and Aly, 2018), resulting in a concentrated brine stream that must be disposed of (see Section 41.2.6 for details). Almost all systems are single pass (i.e., the feedwater is sent through the RO unit once), with the exception of facilities treating highly brackish water.

41.2.2.5 Energy recovery devices

Applying an external pressure in excess of the osmotic pressure requires a significant amount of energy, some of which remains in the pressurized brine stream leaving the last RO stage. ERDs can be used to recover energy from this pressurized brine. While ERDs are used in almost all seawater facilities, their implementation in brackish facilities is not commonplace. BWRO has a lower pressure requirement and a higher recovery rate than SWRO, which results in a smaller amount of recoverable energy in the brine stream. However, recent developments in ERDs for low-pressure applications suggest that even a small amount of energy recovery would result in positive returns for BWRO plants (Veerapaneni et al., 2011). Consequently, ERDs are increasingly incorporated into brackish water facilities. The devices are either positive displacement, for example, pressure exchangers, or centrifugal, for example, the directly coupled turbocharger, which is the most widely adopted ERD in brackish water facilities.

41.2.2.6 Posttreatment

Following the desalination process, the product water is often low in alkalinity, hardness, and pH. Posttreatment may be required to remove dissolved gases (e.g., hydrogen sulfide), stabilize the product water, and/or further disinfect the product water. pH control minimizes corrosion of piping, tanks, and pumps in distribution networks. Air-stripping towers, also known as degasifiers or decarbonators, increase pH through the removal of dissolved carbon dioxide and remaining sulfides. The addition of lime or calcium chloride or blending with raw water provides stable hardness in the product water. Chlorine gas is used for primary disinfection and sodium hypochlorite for secondary disinfection.

---

1. Hollow fiber modules comprise bundles of fibers with diameters of 200–2,500 μm (Ho, 2007). The ends of these bundles are potted in an epoxy or polyurethane resin and cut open to expose the lumens of the fibers.
2. Spiral wound modules consist of multiple flat sheet membranes separated by a porous mesh sheet and wrapped together into a sandwich configuration.

41.2.3 Electrodialysis

ED reversal (referred to as ED or EDR throughout this chapter) is an electrochemical separation process that removes salt ions from a given feedwater, unlike RO which strives to keep salt ions in the feedwater. It relies on semipermeable, ion-exchange membranes that enable the passage of ions with a particular charge. Aside from source water intake and product water distribution, an EDR plant consists of pretreatment, a membrane stack for desalination, a direct-current power supply to drive desalination, a low-pressure circulation pump to flow water through the desalination system, and posttreatment. While RO is used across the BGW salinity range, EDR is typically limited to brackish waters containing a TDS that is less than 5000 mg/L for cost reasons (Veerapaneni et al., 2011; Al-Karaghouli and Kazmerski, 2013). In recent years, the market share of EDR brackish water desalination has diminished due to improvements in RO performance and decreases in RO membrane cost.

41.2.3.1 Pretreatment

Because EDR systems allow for salt transport and the reversal of the direction of such transport, they are generally more robust to fouling than RO. The polarity of the applied voltage potential, which determines the direction of ion transport, is periodically reversed (3–4 times per hour) to flush scalants from the membrane surface on the concentrating side. This reversal lessens the need for continuous chemical feeds and cleanses alternating electrodes (during anodic operation) of acid formation. EDR can also tolerate high concentrations of silica, which are present in many BGWs, without a significant effect on recovery, unlike RO. However, the addition of antiscalants to control the formation of inorganic scale and cartridge filters to remove suspended solids that can foul the membranes is still required. Depending on source water quality, there may be a need for additional pretreatment, such as conventional coagulation, sedimentation, and filtration.

41.2.3.2 Desalination mechanism

Fig. 41.4 shows an EDR membrane stack comprised two membrane pairs (number of membrane pairs in an actual EDR system is usually much larger than two). Each pair consists of two types of ion-exchange membranes in order of alternating charge between two electrodes. Cation-exchange membranes (CEMs) enable the passage of cations, or positively charged ions such as calcium and sodium. Conversely, anion-exchange membranes (AEMs) enable the passage of anions or negatively charged ions such as sulfate and chloride. Spacers are placed between the membranes, as well as the membranes and electrodes. A voltage potential difference, rather than a pressure as in RO, drives the desalination process. Ion

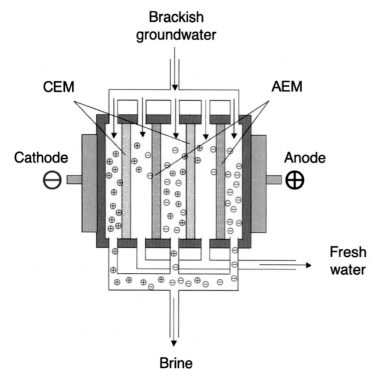

FIGURE 41.4 A simplified EDR stack comprising two electrodes, two CEMs, and two AEMs with brackish groundwater as the feedwater. An applied voltage across the electrodes yields a brine stream and a freshwater stream. *AEMs*, Anion-exchange membranes; *CEMs*, cation-exchange membranes; *EDR*, electrodialysis reversal.

transport through the membrane is induced, with cations and anions migrating toward the cathode and anode, respectively. The alternating membranes trap cations and anions in the brine channel. Caution must be taken in choosing an operating voltage and current for an EDR process below the operating limit that will cause water dissociation.[3]

41.2.3.3 Membranes

EDR systems use flat sheet membranes, reinforced with synthetic fiber, that are stacked in a module between electrodes. The number of membranes varies depending on the target membrane area for a given application. The key EDR membrane properties are charge-based ion selectivity (i.e., selection of specific ions for removal) and electrical conductivity. The membranes have a particularly high removal efficiency for multivalent ions, such as calcium and magnesium, although they also remove monovalent ions, such as sodium and chloride under a sufficient applied voltage. Ion selectivity is best at lower salinities (greater than 90% removal of TDS) and decreases at higher salinities, which hampers EDR performance for more saline feedwaters. Membranes with low electrical resistance are desired in order to consume less energy during the desalination process. The life expectancy of EDR membranes far exceeds that of RO, with an average of 10 years for AEMs and 15 years for CEMs (Veerapaneni et al., 2011).

41.2.3.4 System design

Similar to RO, staging provides the opportunity to achieve the desired level of desalination and to increase recovery rate. ED typically operates with three stages, in which brine from the first stage serves as feed to the subsequent stage. The first stack achieves approximately 60% salt removal, the second 85%, and the third up to 94% (Veerapaneni et al., 2011). ED systems can operate at high water recoveries of 85% to 94% (Faust and Aly, 2018).

41.2.3.5 Posttreatment

Posttreatment requirements for EDR depend on whether the product water will be used for industrial or potable purposes. In industrial applications, mixed-bed ion-exchange units serve as polishers by removing any remaining ions in the product water. In municipal or potable water applications, disinfection through a small chlorinator or corrosion control is often implemented.

41.2.4 Energy consumption using conventional energy sources

Electricity is the only form of energy required in RO and EDR. Table 41.1 compiles specific energy consumption (SEC) data in kWh/m^3 of produced water for RO and EDR brackish water facilities. SEC includes the energy required for groundwater pumping, pretreatment, desalination, posttreatment, and conveyance. SEC depends on feedwater salinity and temperature, membrane properties, age of the facility, conveyance of the raw and treated water, and pretreatment requirements. For example, higher salinities require a greater energy consumption to achieve desalination. For groundwaters containing a TDS of less than 10,000 mg/L and conventional energy sources, the SEC range for RO in

TABLE 41.1 Specific energy consumption in kWh/m^3 of produced water for reverse osmosis (RO) and electrodialysis (ED) desalination of brackish groundwater containing a total dissolved solids of 1000 to 10,000 mg/L.

	SEC (kWh/m^3)	Feedwater (mg/L)
RO	0.26–3	1,000–10,000
ED	0.5–5.5	1000–5000
	3–7	1,000–10,000

The data is obtained from review papers and reports that compile these values from numerous sources (Veerapaneni et al., 2011; Al-Karaghouli and Kazmerski, 2013; Rao et al., 2016; Semiat, 2008; Mezher et al., 2011; Avlonitis et al., 2003; Ziolkowska, 2015; United Nations, 2009; Quteishat, 2009; Lopez et al., 2017; Al-Karaghouli and Kazmerski, 2012; Fatima et al., 2014; Singh, 2011; MacHarg, 2011; Qiu and Davies, 2012). *SEC*, Specific energy consumption.

---

3. The dissociation of water occurs when a higher driving voltage must be applied to maintain a higher current density (applied current per membrane area). This phenomenon arises when ion depletion on the feedwater side of the membrane increases electrical resistance.

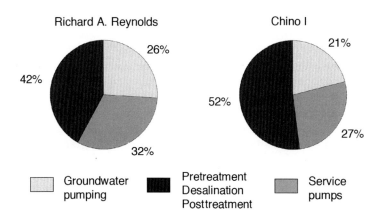

FIGURE 41.5 SEC breakdown of the Richard A. Reynolds and Chino I brackish groundwater RO plants in California. *RO*, Reverse osmosis; *SEC*, specific energy consumption. Data from *Veerapaneni, S., Klayman, B., Wang, S., Bond, R., 2011. Desalination Facility Design and Operation for Maximum Efficiency, Water Research Foundation Denver.* <http://www.sciencedirect.com/science/article/pii/S1364032113000208>.

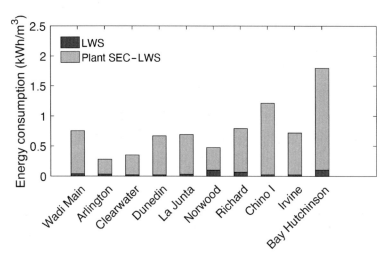

FIGURE 41.6 LWS compared to SEC of 10 brackish groundwater RO plants with complete feedwater composition data (Veerapaneni et al., 2011). *LWS*, Least work of separation; *RO*, reverse osmosis; *SEC*, specific energy consumption.

the surveyed literature is 0.26–3 kWh/m^3 and for ED is 0.5–7 kWh/m^3. The SEC of ED for lower salinity groundwaters (1000–5000 mg/L) ranges from 0.5 to 5.5 kWh/m^3, with recent sources (Veerapaneni et al., 2011; Rao et al., 2016) reporting a SEC of 0.5–1.8 kWh/m^3. SEC greatly increases with TDS in EDR systems. Up to a TDS of 1500 mg/L, EDR SEC is comparable to that of RO. At higher TDS concentrations, EDR energy consumption is significantly greater than that of RO. As a result, EDR is not typically used for brackish waters containing a TDS greater than 5,000 mg/L. Conversely, RO can be used across salinities spanning the brackish and seawater ranges.

The breakdown of plant SEC varies on a case-by-case basis depending on plant parameters, such as system design and size and fouling propensity of feedwater. Fig. 41.5 reflects the differences in the SEC breakdown of two brackish groundwater RO plants in California. Pretreatment, RO and posttreatment, which are lumped together in available datasets, dominate SEC relative to pumping and conveyance. It should be noted that groundwater pumping energy requirements may increase as fresher and shallower groundwater sources continue to be overextracted, demanding that wells be deeper. Moreover, conveyance SEC will vary depending on the distance the water must be pumped to the end user.

The typical SEC of brackish water plants far exceeds the theoretical minimum energy required for desalination. Depending on the desalination process used, SEC is usually 5–26 times greater than the theoretical minimum (Gude, 2018a,b). Fig. 41.6 shows a comparison between the least work of separation (LWS), which is equivalent to the theoretical minimum energy required for desalination based on a given input water, and plant SEC for various BWRO plants in the United States. Because LWS accounts for the minimum energy required by only the desalination phase, the differences in plant SEC and LWS are likely overestimated. Nonetheless, the disparity shows that much room still remains for improvement in terms of desalination energy efficiency.

41.2.5 Economics of desalination

The total desalination cost ($/m^3 of produced water) is a function of the capital cost (CAPEX) and operating cost (OPEX) needed to produce one unit (1 m^3) of freshwater. CAPEX comprises construction (direct capital) and nonconstruction (indirect capital) project costs. OPEX includes costs for operation and maintenance, energy, labor, chemicals, brine disposal, and plant management. As mentioned previously, RO production costs have decreased in recent years due to membrane advancements, and ED is generally not believed to be cost-effective in treating feedwater with TDS greater than 5,000 mg/L unless maximizing recovery rate is the priority. ED has some economic potential for partially desalting high salinity feeds, if a pure product is not required (McGovern et al., 2014a,b) or in hybrid RO−ED arrangements (McGovern et al., 2014a,b).

Table 41.2 includes cost data for BWRO and ED from the literature. Cost largely depends on feedwater salinity and desalination production capacity. Total RO plant expenses range from 0.20 to 1.33 $/m^3 in the surveyed literature. Large RO systems (capacity of 40,000 m^3/day) cost 0.26−0.54 $/m^3. Small RO systems (capacity of 20−1,200 m^3/day) cost 0.78−1.33 $/m^3. This cost increases drastically for RO systems operating at a capacity less than 20 m^3/day. Total ED plant expenses range from 0.6 to 1.05 $/m^3, where larger capacity plants also correspond to lower cost.

Fixed costs (e.g., capital amortization and insurance) dominate the total cost to produce water from brackish groundwater, whereas SEC dominates the cost to produce water from seawater. Greenlee et al. (2009) outline a typical cost distribution of BWRO plants: capital recovery (54%), SEC (11%, compared to seawater 44%); maintenance (9%), membrane replacement (7%), labor (9%), and chemicals (10%). Veerapaneni et al. (2011) report the following cost breakdown: capital recovery (27%), SEC (17%), maintenance (17%), membrane replacement (11%), labor (17%), and chemicals (10%). A key cost driver for inland brackish desalination plants is brine disposal (see Section 41.2.6 for details). A comparison of these cost breakdowns is shown in Fig. 41.7.

TABLE 41.2 Total cost of brackish groundwater (1,000 ≤ TDS ≤ 10,000 mg/L) desalination in $/m^3 of produced water, using reverse osmosis (RO) and electrodialysis (ED) and conventional energy sources for a comprehensive range of desalination capacities unless otherwise specified in the capacity column (Veerapaneni et al., 2011; Al-Karaghouli and Kazmerski, 2013; Rao et al., 2016; Semiat, 2008; Mezher et al., 2011; Avlonitis et al., 2003; Ziolkowska, 2015; United Nations, 2009; Quteishat, 2009; Lopez et al., 2017; Al-Karaghouli and Kazmerski, 2012; Fatima et al., 2014; Singh, 2011; Zotalis et al., 2014; Karagiannis and Soldatos, 2008; Greenlee et al., 2009; Miller et al., 2003; Arroyo and Shirazi, 2012).

	Total cost ($/m^3)	Capacity (m^3/day)
RO	0.10−1.33	Typical range
	0.26−0.54	40,000
	0.78−1.33	20−1,200
	0.56−12.99	Few (<20)
ED	0.60−1.05	Typical range

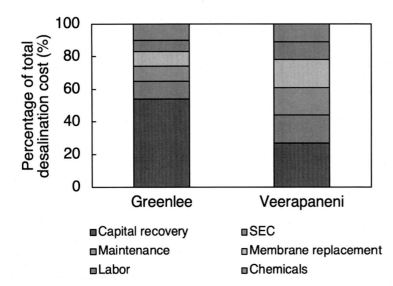

FIGURE 41.7 Typical BWRO plant cost breakdown from studies conducted by Veerapaneni et al. (2011) and Greenlee et al. (2009). *BWRO*, Brackish water reverse osmosis.

41.2.6 Brine management

Brine is the high salinity byproduct of the desalination process. Its characteristics and volume depend on source water and desalination technology used. For example, RO brackish groundwater desalination generates a brine stream that is 4–10 times as concentrated in salinity as the feedwater. Current brine disposal methods negatively impact the environment and are limited by high capital costs. The cost of brine disposal is 5%–33% the total cost of desalination, with inland brackish desalination plants lying in the upper echelon of this range (Jones et al., 2019; Ahmed et al., 2001). Consequently, cost-effective and efficient brine management is critical to address environmental pollution. A desirable alternative to liquid brine disposal is fully dewatering the brine to a solid product, so as to achieve ZLD. Table 41.3 includes information on these disposal and treatment methods, including the treatment principle and the cost. This section elaborates on the results in this table.

41.2.7 Brine disposal

Current methods for disposing of desalination brine are surface water discharge, sewer discharge, deep-well injection, evaporation ponds, and land application. A method is selected depending on a variety of factors, including brine composition and quantity, geographic location, availability of receiving site (e.g., surface body), and capital and operating costs. Over 90% of seawater desalination plants use surface water discharge back into the ocean, while sewer discharge, deep-injection wells, and land application are almost exclusively used by brackish water desalination plants.

The most common practice for inland brackish groundwater facilities is to dispose to surface water bodies (47%), sewer discharge (42%), and deep-well injection (9%) (Veerapaneni et al., 2011). The remaining 1% includes other methods, such as evaporation ponds and thermal treatment. Surface water discharge is proving to have very detrimental environmental effects. For example, annual economic damage due to increased salinity from brine disposal in the Central Arizona Valley and Southern California coastal plain basin has been estimated to be $15–$30 million and $95 million, respectively, for each 100 mg/L increase in product water TDS (Veerapaneni et al., 2011). Even if the brine is diluted using large wastewater effluent flows prior to discharge, the increase in surface water salinity over time results in salinization of surrounding land, which has economic and environmental repercussions. Deep-well injection, land application, and evaporation ponds may be suitable alternatives to surface water discharge, largely depending on the local climate and brine volume. Deep-well injection is cost-effective but risks groundwater pollution. Evaporation ponds are very pricey and can only be used in dry climates with high evaporation rates and land availability. For example, this approach has been used in the UAE and Oman (Ahmed et al., 2001). Land application may be useful for the disposal of small brine volumes to irrigate plants and grasses with high salinity tolerance (Panagopoulos et al., 2019).

41.2.8 Brine treatment

ZLD combines desalination technologies to produce freshwater and achieve zero liquid waste from a desalination plant. This approach consists of a concentration stage (membrane technologies), as well as evaporation and crystallization stages (thermal technologies). Together, they yield a pure water stream that can be used for drinking water, irrigation, etc., and a compressed solid waste for environmental friendly disposal or further processing into a useful material. However, as shown in Table 41.3, the ZLD approach is by far the most costly method and may have indirect

TABLE 41.3 Brine disposal and treatment principles and cost ($/m^3 of rejected brine) (Greenlee et al., 2009; Miller et al., 2003; Panagopoulos et al., 2019).

Method	Principle	Cost ($/m^3)
Surface water discharge	Discharged into surface water	0.03–0.30
Sewer discharge	Discharged into existing sewage collection system	0.30–0.66
Deep-well injection	Injected into porous subsurface rock formations	0.33–2.65
Evaporation ponds	Evaporated, resulting in salt accumulation at pond bottom	1.18–10.04
Land application	Irrigates salt-tolerant crops and grasses	0.74–1.95
ZLD	Concentrated and evaporated to yield freshwater and solid	0.66–26.41

ZLD, Zero liquid discharge.

environmental impact as a result of its large energy requirements. Further research is being conducted on reducing energy consumption of and incorporating renewable energy sources and low-grade waste heat in ZLD.

41.2.9 Desalination using renewable energy sources

Desalination processes typically rely on fossil fuel power plants, which emit greenhouse gases, to meet their energy intensive needs. Rahuy et al. reports that energy consumption accounts for 89%–99% of desalination's total environmental load (Veerapaneni et al., 2011). Renewable energy sources (wind, solar thermal, geothermal) provide alternatives to mitigate the environmental impact of desalination.[4]

Despite its promise, renewable energy powered desalination accounted for 1% of the total global installed desalination capacity as of 2016 (Gude, 2018b), with photovoltaics (PV) leading at 43%, followed by solar thermal 27%, wind turbine 20%, and hybrid 10% sources (Cavalcante Junior et al., 2019). The biggest barrier to adoption has historically been the high capital cost of renewable energy systems in comparison to conventional energy systems. A 2011 study (Bilton et al., 2011) reports water production costs of 2.17–2.41 $/m^3 for select brackish PV–RO systems (10 m^3/day) in Australia, Tunisia, Jordan, and the United States. A 2013 review paper (Al-Karaghouli and Kazmerski, 2013) finds that PV–RO (<100 m^3/day), PV–ED (<100 m^3/day), and wind-RO (50–2,000 m^3/day) systems require a SEC of 1.5–4 kWh/m^3 and water production costs of 6.50–9.10 $/m^3, 10.40–11.70 $/m^3, and 1.92–5.20 $/m^3, respectively. In comparison, RO (20–1200 m^3/day) powered by fossil fuels required a SEC of 1.5–2.5 kWh/m^3 and water production cost of 0.78–1.33 $/m^3. ED (small capacity) required a SEC of 2.64–5.5 kWh/m^3 and cost of 0.60 $/m^3 (Al-Karaghouli and Kazmerski, 2013). According to a 2014 study (Wright and Winter, 2014), PV–ED can cost significantly less than PV–RO for small-scale systems (6–15 m^3/day) using lower salinity feedwater (e.g., 50% cost reduction for 2,000 mg/L feedwater).

However, the variability of energy prices over the past decade is quickly changing the desalination landscape, as the electricity generated by new solar and wind power projects is becoming cheaper than the electricity generated by new coal and gas power plants around the world. According to a 2019 report from the business intelligence company Bloomberg NEF (Maisch, 2019), recent onshore wind and solar power plants have achieved parity with average wholesale prices in parts of Europe, California, and China, some of the world's largest markets. The expected levelized cost of electricity (LCOE) of recently financed solar projects ranges from 0.027 to 0.036 $/kWh in India, Chile, and Australia (Maisch, 2019) and is less than 0.020 $/kWh in California, the UAE, and Portugal (Dudley, 2019). In the United States the average LCOE of wind power dropped from 0.070 $/kWh in 2009 to less than 0.020 $/kWh in 2017 (Wiser and Bolinger, 2017). The cost-competitiveness of wind and solar power has motivated increased development of large-scale desalination plants powered by renewable energy, in addition to the already implemented small-scale systems[5] (Klaimi et al., 2019). Brazil's Agua Doce Program consists of brackish water desalination systems powered by PV that aim to provide high quality water to 500,000 people in the semiarid region of Brazil (Cavalcante Junior et al., 2019). The Arabian Gulf is increasingly shifting its entire desalination infrastructure to PV, with such RO plants operating at up to 100,000 m^3/day in Saudi Arabia (Gude, 2018a,b). Australia contains wind-powered RO plants with even larger desalination capacities [e.g., Kurnell-Sydney seawater RO plant with a capacity of 250,000 m^3/day (Gude, 2018a,b)].

Renewable energy coupled desalination at small and large scales is expected to only become more economically attractive as the price of fossil fuels continues to increase and that of renewable technologies continues to decline. The International Desalination Association has set a 2020–2025 target of using renewable energy in 20% of new desalination plants (Isaka, 2013).

41.2.10 Emerging desalination technologies

In addition to the development of new generation membrane materials for desalination, several desalination technologies that improve water recovery and/or energy consumption are emerging. These technologies are typically variations of RO, including nanofiltration (NF) and semibatch RO.

---

4. A 30 kW wind turbine coupled with an RO unit, with an SEC of 4.38 kWh/m^3 and water recovery of 30%, can reduce CO$_2$ emissions by 80,028 tons annually. Similarly, an analysis by the US National Renewable Energy Laboratory of a small PV–RO system (5 m^3/day driven by 5 kW PV system) in a remote region of Iraq has shown a 8170 kg reduction in CO$_2$ and other hazardous gases (Klaimi et al., 2019).

5. Locally available renewable energy sources for desalination can provide a cost-effective alternative in remote areas that have low population density and weak water and electricity infrastructure.

41.2.11 Nanofiltration

NF membranes and RO membranes are similar in many ways. Both are pressure-driven membrane desalination technologies that foul easily. Their differences stem from the size and charge of contaminants that each technology is capable of removing. RO membranes effectively remove most ions from product water, with the exception of dissolved gases and some weakly charged molecules that are low in molecular weight. NF is able to reject larger, strongly charged ions (e.g., 90% calcium removal), but it enables more passage of monovalent and smaller molecular weight ions (e.g., 70% sodium removal). The salt rejection of NF membranes is often inadequate in treating brackish groundwater. However, NF requires less energy than RO and consequently has been widely adopted in some parts of the United States for brackish groundwater desalination. NF is also often used for softening, that is, to remove calcium and magnesium (hardness) from a given solution.

41.2.12 Semibatch reverse osmosis

The RO systems discussed thus far operate in a continuous mode. In other words the membranes are treating the same feedwater, so the applied pressure to overcome the osmotic pressure is fixed. In semi-batch RO, also known as closed-circuit RO, the brine is recirculated and mixed with the pressurized feedwater in order to reduce the osmotic pressure of the feed over time and the overall energy consumption required for desalination. A variableoperating pressure is applied as the feed pressure changes. Brine recirculation allows for more than 90% recovery rate for brackish water desalination systems. A SEC of 0.64–0.76 kWh/m^3 has been reported in the literature (Subramani and Jacangelo, 2015). Desalitech, LLC, which has commercialized this technology, claims a 20% energy consumption reduction in semibatch RO compared to continuous RO (Subramani and Jacangelo, 2015).

41.3 Global and national trends in desalination

Location is of the utmost importance in desalination system design and selection for a variety of reasons, including the geographic variation in BGW composition and differences in regional water needs and in local costs of energy or electricity. For example, in much of the Middle East (e.g., Saudi Arabia and Israel), desalination is the primary, if not only, option to provide the required water supply. In other countries (e.g., China, Australia, and United States), desalination provides a water supply that is more reliable, albeit more expensive, than traditional river and aquifer systems. For island users of desalination, such as in the Caribbean, energy is often very expensive compared to the energy costs in large oil-producing countries. This section explores global and national differences in operating and contracted desalination capacities by plant type, feedwater type, and target end use and in capital and operating desalination expenditures annually. Results are based on the most up-to-date data from the Global Water Intelligence desalination database (Global Water Intelligence, 2020).

41.3.1 Global trends

Fig. 41.8 shows annual operating capacity and contracted capacity by plant type from 2015 to 2024. The desalination market as a whole is expected to continue on an upward trajectory. BWRO and ED–EDR together comprised 20% of the total operating desalination capacity in 2019. BWRO operating capacity has been and is expected to continue growing from year-to-year, although 2019 experienced a decrease in the annual growth rate from 4.3% (2018 vs. 2017) to 0.9% (2019 vs. 2018). Similarly, BWRO contracted capacity in 2019 reached the lowest level since 2015 (0.37 million m^3/day). This drop may be caused by market changes in China and Saudi Arabia, which are both major players in the desalination space; China is pivoting toward wastewater reuse, while Saudi Arabia is pivoting toward enhanced transmission and storage infrastructure. ED–EDR annual growth rate in terms of operating capacity was positive in 2015–2018 and negative in 2018–2019. Its growth rate is expected to remain negative, as RO becomes more energy efficient and cost-effective.

41.3.1.1 Annual desalination expenditures

Annual desalination capital and operating expenditures by plant type from 2015 to 2024 can be found in Fig. 41.9. In 2019, BWRO and ED CAPEX was $276 million and $54.8 million, respectively; BWRO and ED OPEX was $1,455 million and $216 million, respectively. The OPEX of both technologies remains relatively constant, while the CAPEX varies, across the surveyed time period.

570 Theme | 5 Future of groundwater and solutions

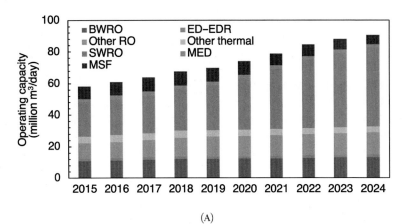
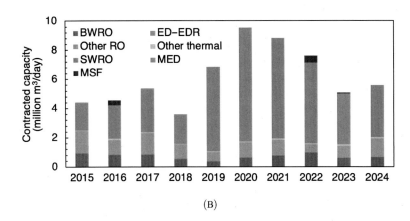

FIGURE 41.8 Annual desalination (A) operating capacity and (B) contracted capacity by plant type from 2015 to 2024.

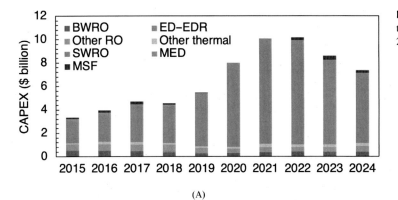
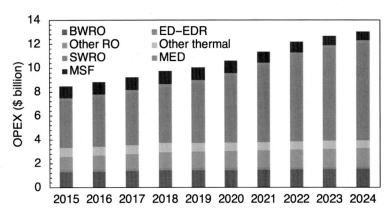

FIGURE 41.9 Annual desalination (A) capital expenditures and (B) operating expenditures by plant type from 2015 to 2024.

41.3.1.2 Geographic region

Large numbers of desalination facilities are located in the Middle East and North Africa (MENA), Arab States of the Gulf, Asia/Pacific, Europe, and North America, while relatively few are based in Latin America, Caribbean, and sub-Saharan Africa (Fig. 41.10). Desalination plants are concentrated on or near the coastline for seawater desalination. Inland desalination plants tend to be smaller in capacity than coastal desalination plants. Today, the MENA and Arab States of the Gulf contain almost half of global desalination capacity, with Saudi Arabia, UAE, and Kuwait serving as major regional and global leaders. The Asia/Pacific region has the next largest desalination regional capacity as a result of China's market share, followed by North America, almost entirely due to US capacity, and Europe, where Spain is the leader.

41.3.1.3 Target end use

Desalination provides water for the following sectors in decreasing order: municipal (e.g., drinking water), industry, power, irrigation, and military. Municipal desalination plants are located worldwide, particularly in MENA. Compared to MENA, North America, Western Europe, and East Asia, and Pacific regions contain a larger proportion of nonmunicipal desalination plants, because industrial and power sectors constitute large market shares. The few desalination plants in South America and Africa are primarily for industrial use. In Eastern Europe and Central Asia, sub-Saharan Africa, and Southern Asia, desalination plants are predominantly designed to produce water for industrial and private applications. Fig. 41.11 demonstrates the sectoral use of desalinated water annually from 2015 to 2024 in terms of industry and utility/other, which includes municipal, power, irrigation, and military.

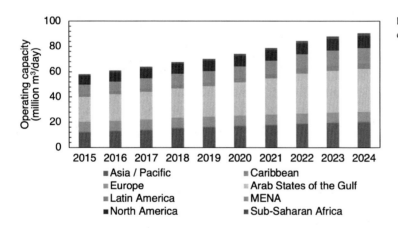

FIGURE 41.10 The breakdown of operating desalination capacity by geographic region from 2015 to 2024.

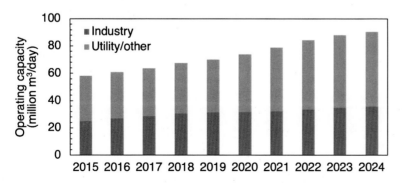

FIGURE 41.11 Sectoral use of desalinated water annually from 2015 to 2024 in terms of industry and utility/other, which includes municipal, power, irrigation, and military.

41.3.2 National trends

The United States, China, Saudi Arabia, Australia, and India are the top markets by contracted brackish water capacity (Fig. 41.12A), while the United States, Saudi Arabia, China, Spain, and Australia are the top markets by brackish water desalination expenditure (Fig. 41.12B). Consequently, these six countries are considered leaders in brackish water desalination. The contracted brackish water capacity is expected to decrease in India, Saudi Arabia, and China and increase in Australia and the United States, with the largest projected capacity in the United States. The annual OPEX from 2020 to 2024 is projected to range from 67% to 97% of the total annual desalination expenditure for brackish water desalination in Australia, Spain, China, Saudi Arabia, and the United States.

Fig. 41.13 shows the national operating desalination capacities of Australia, Spain, China, Saudi Arabia, United States, and India as a function of (1) feedwater type, (2) plant type, and (3) target end use in 2019. Together, the desalination capacity, independent of feedwater, of these nations constitutes over 45% of the global capacity: Saudi Arabia (15.8%), United States (11.4%), China (98.8%), Spain (6.1%), India (2.9%), and Australia (1.2%). Feedwater type is divided into three categories: seawater (20,000–50,000 mg/L), brackish water (3,000–20,000 mg/L), and other (<3,000 mg/L). The "other" category includes freshwater (<500 mg/L), wastewater, and low salinity brackish water (500–3,000 mg/L). The desalination feedwater breakdown varies from country to country. For instance, desalinated water in the US predominantly originates from brackish water (500–20,000 mg/L), the majority of which is groundwater, while desalinated water in Saudi Arabia primarily originates from seawater. Across the considered countries, RO is overwhelmingly the dominant technology for both brackish water and seawater, with the exception of Saudi Arabia in which thermal systems play a substantial role. India, China, and Australia primarily use their desalinated water for industrial purposes, while Spain, Saudi Arabia, and United States primarily use it for other purposes (e.g., municipal and agriculture).

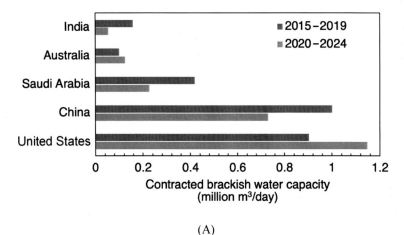

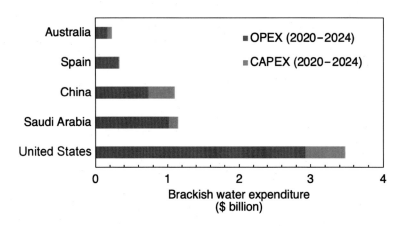

FIGURE 41.12 Top national markets by (A) contracted brackish water desalination capacity and (B) brackish water desalination CAPEX and OPEX. *CAPEX*, Capital cost; *OPEX*, operating cost.

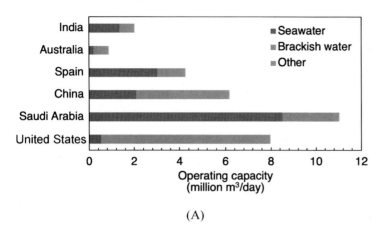
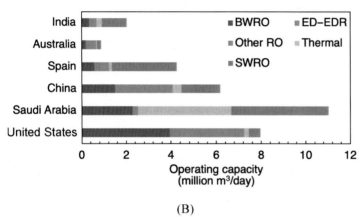
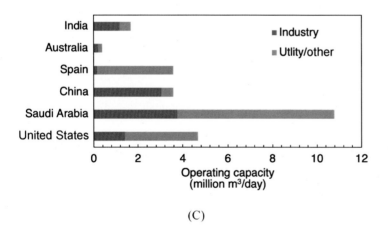

FIGURE 41.13 The national operating desalination capacities of the United States, China, Saudi Arabia, Spain, India, and Australia as a function of (A) feedwater type, (B) plant type, and (C) target end use in 2019.

Acknowledgments

The authors would like to thank the Bureau of Reclamation under Agreement Number R17AC00135 for funding much of this work and Tom Pankratz from Global Water Intelligence for helpful discussions.

References

Ahdab, Y.D., Thiel, G.P., Böhlke, J., Stanton, J., Lienhard, J.H., 2018. Minimum energy requirements for desalination of brackish groundwater in the united states with comparison to international datasets. Water Res. 141, 387–404. Available from: https://doi.org/10.1016/j.watres.2018.04.015. Available from: http://www.sciencedirect.com/science/article/pii/S0043135418302999.

Ahmed, M., Shayya, W.H., Hoey, D., Al-Handaly, J., 2001. Brine disposal from reverse osmosis desalination plants in Oman and the United Arab Emirates. Desalination 133 (2), 135–147. Available from: https://doi.org/10.1016/S0011-9164(01)80004-7. Available from: http://www.sciencedirect.com/science/article/pii/S0011916401800047.

Al-Karaghouli, A., Kazmerski, L., 2012. Economic and technical analysis of a reverse-osmosis water desalination plant using deep-3.2 software. J. Environ. Sci. Eng., A 1 (3), 318–328.

Al-Karaghouli, A., Kazmerski, L.L., 2013. Energy consumption and water production cost of conventional and renewable-energy-powered desalination processes. Renew. Sustain. Energy Rev. 24, 343–356. Available from: https://doi.org/10.1016/j.rser.2012.12.064. Available from: http://www.sciencedirect.com/science/article/pii/B9780444521149500086.

Arroyo, J., Shirazi, S., 2012. Cost of brackish groundwater desalination in Texas. Tech. Rep.

Avlonitis, S., Kouroumbas, K., Vlachakis, N., 2003. Energy consumption and membrane replacement cost for seawater RO desalination plants. Desalination 157 (1), 151–158 desalination and the Environment: Fresh Water for all. Available from: https://doi.org/10.1016/S0011-9164(03)00395-3. Available from: http://www.sciencedirect.com/science/article/pii/S0011916403003953.

Bilton, A.M., Wiesman, R., Arif, A., Zubair, S.M., Dubowsky, S., 2011. On the feasibility of community-scale photovoltaic-powered reverse osmosis desalination systems for remote locations. Renew. Energy 36 (12), 3246–3256. Available from: https://doi.org/10.1016/j.renene.2011.03.040. Available from: http://www.sciencedirect.com/science/article/pii/S0960148111001674.

Cavalcante Junior, R.G., Vasconcelos Freitas, M.A., da Silva, N.F., de Azevedo Filho, F.R., 2019. Sustainable groundwater exploitation aiming at the reduction of water vulnerability in the Brazilian semi-arid region. Energies 12 (5), 904. Available from: https://doi.org/10.3390/en12050904.

Dudley, D., 2019. Race Heats Up For Title Of Cheapest Solar Energy In The World. <https://www.forbes.com/sites/dominicdudley/2019/10/17/cheapest-solar-energy-in-the-world/#1a1418c34772>.

Fatima, E., Elazhar, M., Hafsi, M., Elmidaoui, A., 2014. Performances of electrodialysis process in desalination of brackish waters at various salinities and voltage. Int. J. Adv. Chem. 2 (2), 49–52. Available from: https://doi.org/10.14419/ijac.v2i2.1741. Available from: https://www.sciencepubco.com/index.php/IJAC/article/view/1741.

Faust, S.D., Aly, O.M., 2018. Chemistry of Water Treatment. CRC Press.

Global Water Intelligence, 2020. Desaldata online database. <https://www.desaldata.com/>.

Greenlee, L.F., Lawler, D.F., Freeman, B.D., Marrot, B., Moulin, P., 2009. Reverse osmosis desalination: water sources, technology, and today's challenges. Water Res. 43 (9), 2317–2348. Available from: https://doi.org/10.1016/j.watres.2009.03.010. Available from: http://www.sciencedirect.com/science/article/pii/S0043135409001547.

Gude, G., 2018b. Renewable Energy Powered Desalination Handbook: Application and Thermodynamics. Butterworth-Heinemann.

Gude, G., 2018a. Sustainable Desalination Handbook: Plant Selection, Design and Implementation. Butterworth-Heinemann.

He, C.Q., Bond, R.G., 2019. Inland Desalination and Concentrate Management. American Water Works Association.

Ho, C.C., 2007. Chapter 7 – Membranes for bioseparations. In: Yang, S.-T. (Ed.), Bioprocessing for Value-Added Products from Renewable Resources. Elsevier, Amsterdam, pp. 163–183. Available from: https://doi.org/10.1016/B978-044452114-9/50008-6.

Isaka, M., 2013. Water desalination using renewable energy. Tech. Rep. International Renewable Energy Agency.

Jones, E., Qadir, M., van Vliet, M.T., Smakhtin, V., mu Kang, S., 2019. The state of desalination and brine production: a global outlook. Sci. Total Environ. 657, 1343–1356. Available from: https://doi.org/10.1016/j.scitotenv.2018.12.076. Available from: http://www.sciencedirect.com/science/article/pii/S0048969718349167.

Karagiannis, I.C., Soldatos, P.G., 2008. Water desalination cost literature: review and assessment. Desalination 223 (1–3), 448–456.

Klaimi, R., Alnouri, S.Y., Al-Hindi, M., Azizi, F., 2019. Optimization techniques for coupling renewable/hybrid energy options with desalination systems for carbon footprint reduction. Chem. Eng. Res. Des. 151, 270–290. Available from: https://doi.org/10.1016/j.cherd.2019.09.010. Available from: http://www.sciencedirect.com/science/article/pii/S0263876219304253.

Lopez, A.M., Williams, M., Paiva, M., Demydov, D., Do, T.D., Fairey, J.L., et al., 2017. Potential of electrodialytic techniques in brackish desalination and recovery of industrial process water for reuse. Desalination 409, 108–114. Available from: https://doi.org/10.1016/j.desal.2017.01.010. Available from: http://www.sciencedirect.com/science/article/pii/S0011916416310463.

MacHarg, J.P., 2011. Energy Optimization of Brackish Groundwater Reverse Osmosis Desalination. Texas Water Development Board.

Maisch, M., 2019. Solar Electricity Can Retail for $0.027-0.036/kWh as Renewables Close in on Global Grid Parity. <https://www.pv-magazine.com/2019/11/01/solar-electricity-can-retail-for-0-027-0-036-kwh-as-renewables-close-in-on-global-grid>.

McGovern, R.K., Zubair, S.M., Lienhard, J.H., 2014b. The benefits of hybridising electrodialysis with reverse osmosis. J. Membr. Sci. 469, 326–335. Available from: https://doi.org/10.1016/j.memsci.2014.06. Available from: http://www.sciencedirect.com/science/article/pii/S0376738814004918.

McGovern, R.K., Zubair, S.M., Lienhard, J.H., 2014a. The cost effectiveness of electrodialysis for diverse salinity applications. Desalination 348, 57–65. Available from: https://doi.org/10.1016/j.desal.2014.06.010. Available from: http://www.sciencedirect.com/science/article/pii/S0011916414003312.

Mezher, T., Fath, H., Abbas, Z., Khaled, A., 2011. Techno-economic assessment and environmental impacts of desalination technologies. Desalination 266 (1–3), 263–273. Available from: https://doi.org/10.1016/j.desal.2010.08.035.

Miller, J.E., et al., 2003. Review of Water Resources and Desalination Technologies, 49. Sandia National Laboratories, Albuquerque, NM. Available from: http://dx.doi.org/10.2172/809106.

Mukherjee, A., Scanlon, B., Aureli, A., Langan, S., Guo, H., McKenzie, A., 2020. Global Groundwater: Source, Scarcity, Sustainability, Security and Solutions, first ed. Elsevier, ISBN: 9780128181720.

Office of Water of the U.S. Environmental Protection Agency, 2018. 2018 Edition of the drinking water standards and health advisories. Tech. Rep., EPA 822-F-18-001. Office of Water of the U.S. Environmental Protection Agency, Washington, DC.

Panagopoulos, A., Haralambous, K.-J., Loizidou, M., 2019. Desalination brine disposal methods and treatment technologies — a review. Sci. Total Environ. 693, 133545. Available from: https://doi.org/10.1016/j.scitotenv.2019.07.351. Available from: http://www.sciencedirect.com/science/article/pii/S0048969719334655.

Qiu, T., Davies, P.A., 2012. Comparison of configurations for high-recovery inland desalination systems. Water 4 (3), 690–706. Available from: https://doi.org/10.3390/w4030690.

Quteishat, K., 2009. Desalination and water affordability. In: SITeau International Conference, Casablanca, Morocco.

Rao, P., Aghajanzadeh, A., Sheaffer, P., Morrow, W.R., Brueske, S., Dollinger, C., et al., 2016. Volume 1: survey of available information in support of the energy-water bandwidth study of desalination systems. Tech. Rep., Lawrence Berkeley National Laboratory.

Semiat, R., 2008. Energy issues in desalination processes. Environ. Sci. Technol. 42 (22), 8193–8201. Available from: https://doi.org/10.1021/es801330u, PMID:. Available from: 19068794. Available from: https://doi.org/10.1021/es801330u.

Singh, R., 2011. Analysis of energy usage at membrane water treatment plants. Desalin. Water Treat. 29 (1–3), 63–72. Available from: https://doi.org/10.5004/dwt.2011.2988. Available from: https://doi.org/10.5004/dwt.2011.2988.

Stanton, J.S., Anning, D.W., Brown, C.J., Moore, R.B., McGuire, V.L., Qi, S.L., et al., 2017. Brackish groundwater in the United States. Tech. Rep., US Geological Survey.

Subramani, A., Jacangelo, J.G., 2015. Emerging desalination technologies for water treatment: a critical review. Water Res. 75, 164–187. Available from: https://doi.org/10.1016/j.watres.2015.02.032. Available from: http://www.sciencedirect.com/science/article/pii/S0043135415001050.

United Nations, 2009. ESCWA water development report 3 role of desalination in addressing water scarcity.

Veerapaneni, S., Klayman, B., Wang, S., Bond, R., 2011. Desalination Facility Design and Operation for Maximum Efficiency. Water Research Foundation Denver. Available from: http://www.sciencedirect.com/science/article/pii/S1364032113000208.

Wiser, R., Bolinger, M., 2017 Wind technologies market report. Tech. Rep. U.S. Department of Energy's Office of Energy Efficiency and Renewable Energy.

Wright, N.C., Winter, A.G., 2014. Justification for community-scale photovoltaic-powered electrodialysis desalination systems for inland rural villages in India. Desalination 352, 82–91. Available from: https://doi.org/10.1016/j.desal.2014.07.035.

Ziolkowska, J.R., 2015. Desalination leaders in the global market - current trends and future perspectives. Water Supply 16 (3), 563–578. Available from: https://doi.org/10.2166/ws.2015.184. arXiv: <https://iwaponline.com/ws/article-pdf/16/3/563/412113/ws016030563.pdf>.

Zotalis, K., Dialynas, E.G., Mamassis, N., Angelakis, A.N., 2014. Desalination technologies: Hellenic experience. Water 6 (5), 1134–1150. Available from: https://doi.org/10.3390/w6051134.

Chapter 42

Desalination of deep groundwater for freshwater supplies

Veera Gnaneswar Gude and Anand Maganti
Department of Civil and Environmental Engineering, Mississippi State University, Mississippi State, MS, United States

42.1 Introduction

Around one-third of the world's population accesses groundwater for freshwater supplies and the trend is increasing constantly (Margat, 2008; Margat and Van der Gun, 2013). This increased level of abstraction coupled with groundwater pollution has caused many issues for groundwater-dependent ecosystems (Mukherjee et al., 2020). Climate change and its variability, on the other hand, affect the groundwater recharge rates (Scanlon et al., 2012; Taylor et al., 2013). This situation has triggered many communities to access deep groundwater sources. These sources are often saline in nature and may require additional treatment considerations. Therefore it is crucial to develop and implement effective management practices to protect and properly utilize this rather finite source. Management practices may include both technical and nontechnical approaches. Understanding the importance of groundwater governance and its proper implementation of efficient management practices could enhance the accessibility of groundwater resources for long-term sustainability.

The purpose of this chapter is twofold. First, a discussion of the global groundwater depletion problems and groundwater desalination is presented. Next, groundwater desalination challenges, technology assessment, and/or management practices related to sustainable water supplies and their implementation in developed and developing countries are presented.

42.2 Groundwater desalination—influencing factors

Groundwater depletion and therefore the need for its protection is a local, regional, and global problem. Increasing withdrawal of groundwater for domestic water supplies and agricultural uses has become a major challenge for many local water management agencies. Despite the critical importance of the groundwater source for sustainable development, insufficient management or governance is given to these sources in comparison with surface water sources (Famiglietti, 2014). Groundwater sources are often poorly monitored and managed, and sometimes they do not receive any attention in some developing countries. Groundwater quality can be degraded by use. For example, pumping may cause saltwater or brackish water to encroach into fresh groundwater sources. Along the coastlines, seawater intrusion can occur in some areas due to groundwater overdraft. Water use can introduce contaminants into the groundwater. For example, irrigation water return flow may contain pesticides or nutrients that were either surface applied or added directly to the irrigation water. Irrigation return flow also generally has a higher dissolved solid content due to concentration by evapotranspiration. Conjunctive use of ground and surface water, including artificial recharge using surface water, can also affect the groundwater quality.

42.2.1 Motivation for groundwater desalination

Most of the arid and inland communities are left with a single option of accessing their local groundwater sources as the lift and conveyance option can be cost intensive and sometimes practically not feasible. The demand for groundwater extraction is inevitable in these regions as these communities are usually dependent on agriculture for their economy. This situation leads to overexploitation of the resource that is often termed as "mining" or "borrowing from the future." For example, in the United States, groundwater withdrawal rates are more than doubled in the period from 1965 to 1995 to supply domestic consumption, with substantially larger groundwater abstraction occurring to support industry such as mining and irrigated crops (Glennon, 2002). Recently, 43% of total global irrigation water was extracted from groundwater sources, with America and Asia extracting 48% and 45%, respectively (Siebert et al., 2010). Furthermore, approximately 25% of the world's population depends on groundwater pumping for drinking water, and many of these are in semiarid and arid zones (Glennon, 2002). Recent trends of groundwater aquifer depletion rates show that some level of groundwater depletion is persistent across the world in major groundwater reserves, which has caused water stress issues in these regions (Wada et al., 2010, 2016; Gude, 2018).

In coastal regions where the groundwater is confined by or is influenced by coastal aquifers, it presents an attractive option for desalination when compared with the seawater desalination that is energy and cost intensive. In addition, groundwater desalination is considered as a more cost-effective and environment-friendly option in comparison with seawater desalination. A study comparing brackish water groundwater desalination with seawater desalination through a life cycle impact analysis has reported that the brackish groundwater desalination has 50% less environmental impacts (Muñoz and Fernández-Alba, 2008). Potential impacts due to brine discharge are also lower for brackish groundwater desalination. Another study, considering a zero-liquid discharge process with greater than 95% salinity reduction, reported that the energy consumption is significantly lower than coastal seawater desalination option, regardless of the conveyance distance (Sobhani et al., 2012). In addition to lower specific energy consumption, higher water recovery and lower residual discharge volumes also found to be beneficial factors for adopting brackish groundwater desalination (Gude, 2016, 2018).

42.2.2 Considerations for groundwater desalination

Often groundwater sources include high alkalinity and high hardness due to the presence of polyvalent cations. Arsenic, chromium, fluoride, radium, and uranium can be present in concerning concentrations that require additional treatment schemes. Nonconventional processes such as ion exchange and membrane processes are often considered for treating these waters. Groundwater and other water sources that have significant total dissolved solid concentrations (especially salt) are called brackish water sources (Gude, 2011). The definition of brackish water is very flexible with saline water concentrations ranging from 3000 to 30,000 mg/L. Brackish waters lack the suspended sediment content that seawater has and will be less costly to treat. Thus brackish groundwater may serve as a source of water supply as water demand increases (particularly in arid and semiarid regions) and treatment costs for desalination continue to fall.

Several desalination techniques are applicable for achieving salt separation from the brackish water sources (Gude and Fthenakis, 2020). Electrodialysis and low-pressure (or low-energy) membrane processes are suitable for this type of water. Ion exchange can also be used to treat water that contains additional contaminants of concern. Nanofiltration and reverse osmosis (RO) technologies are suitable for saline waters with greater than 30,000 mg/L TDS. Blending is considered in most of the water treatment schemes as it eliminates the need for treating the entire required treatment volumes as long as the final water quality can be achieved by mixing the treated water with a fraction of untreated source water. This saves chemical- and energy-associated costs, including capital costs. Groundwater desalination is different from seawater desalination. Some of the differences between the groundwater desalination and seawater desalination options are shown in Table 42.1.

TABLE 42.1 Differences between groundwater and seawater desalination options.

Groundwater desalination	Seawater desalination
No need for conveyance	Potential need for lift and conveyance
Available on-site	Available off-site or consumer has to be near the source
Pretreatment is mild	Pretreatment is intensive
Low-energy desalination technologies are suitable	High energy-demanding desalination technologies required
Lower environmental impact due to lower energy consumption	Higher environmental impact due to high energy consumption
Concentrate reuse is possible	Concentrate management is both cost and energy intensive
Smaller footprint	Larger footprint

In comparison with seawater, brackish groundwater usually has low suspended solids, low bacterial count, and low content of organic matter, but high concentrations of sparingly soluble inorganic salts such as calcium and barium salts and silica can be found in these waters, which are the main reasons for scaling problems in the desalination process (Elsaid et al., 2012). Dissolved solids may originate from chemical weathering or dissolution of geological formations, that is, minerals, which can be attributed to the direct contact of groundwater with the calcium carbonate and calcium sulfate rocks forming the aquifer. Sulfates may also result from biological oxidation of reduced sulfur species causing scaling by calcium sulfate and calcium carbonate. Silica is another constituent from amorphous or crystalline SiO_2 and major clay minerals (Faust and Aly, 1998). Crystalline silica has a very low solubility in water; however, amorphous silica can have solubility up to 120 mg/L at pH 7, and the solubility increases with pH increase reaching around 889 mg/L at pH 10 (Hamrouni and Dhahbi, 2001).

42.2.3 Environmental impacts of groundwater desalination

Groundwater desalination option is not entirely environment-friendly despite the motivating factors discussed in Section 42.2.1. The act of introducing a remote underground resource into the surface environment is considered an environment-unfriendly act as this disrupts the natural recharge cycles. The deep underground water sources are often saline in nature, meaning that more salts are introduced into the surface environment with additional considerations for management.

42.3 Desalination technology assessment

The three commonly used technologies worldwide for seawater desalination are multistage flash (MSF), multieffect distillation (MED), and RO. MSF was developed in the 1950s and has advanced significantly since the 1980s (Gude, 2016). It was the predominant and robust thermal desalination technology. This technology is now employed for 26% of total seawater desalination capacity worldwide. MED is also another thermal desalination technology used for desalinating seawater, and this technology has been applied in larger desalination plants (Gude, 2017). RO is a relatively new membrane desalination technology that is applied in seawater and groundwater desalination and other various water treatment options. With the advancement of high-performance membranes and installation of energy recovery devices, the cost and energy consumption for this technology has decreased significantly during the past decades (Gude, 2011). The dominance of thermal technologies has faded away in the past two decades due to the advent and advancement of membrane technologies. Other promising desalination methods and details regarding the specific energy requirements for the desalination processes are discussed elsewhere (Gude and Fthenakis, 2020). Readers are referred to recent contributions on desalination and water reuse topics for detailed list of technological, energy, and environmental management advances and alternatives (Gude, 2015, 2016, 2017, 2018).

There are three important considerations for desalination plant design and proper operations: (1) plant design, (2) pretreatment scheme, and (3) proper operation and maintenance schedules; with proper pretreatment as the foundation for successful operation (Gude, 2011; Elsaid et al., 2012). The primary goal of pretreatment is to lower the fouling propensity during the desalination process, and the required pretreatment options depend mainly on the characteristics of the water source (Gude, 2011). Scale inhibitors or antiscalants are added to the source water during pretreatment to prevent scale formation and usually work synergistically with dispersant polymers (Gude, 2011, 2016).

Although salinity of the deep groundwater sources is lower than the seawater, the energy requirements to treat these waters can pose a challenge to local electric grids (Gude, 2015, 2016, 2017). Variable salinity feed is also possible with continued withdrawals in this region. The challenge of providing reliable potable water systems includes inventions in the design of flexible desalination systems to take better advantage of the capital equipment investments in these coastal regions. In view of the aforementioned potential vulnerabilities and energy−water nexus issues, blending both the seawater and groundwater sources for desalination application appears to mitigate the subsidence of land surface and provide recharge times for the groundwater naturally from precipitation and other discharges. As deep groundwater becomes the new potable water supply source, treated wastewaters as well as produced waters from oil and gas operations become part of the integrated water management scheme for these drought-stricken areas. Novel desalination technologies that can utilize high salinity and high hardness containing produced waters as feed are yet to be developed to address water scarcity most economically (Meng et al., 2016; Gude, 2018).

More aggravated water reuse options could be considered in these regions prior to considering desalination option. Ten percent of the total available freshwater, or 3.5 million km^3, is consumed around the world. Of this amount, about 330 km^3 are generated globally as municipal wastewater (NRC, 2012). For example, the 32 billion gal—or 121 million

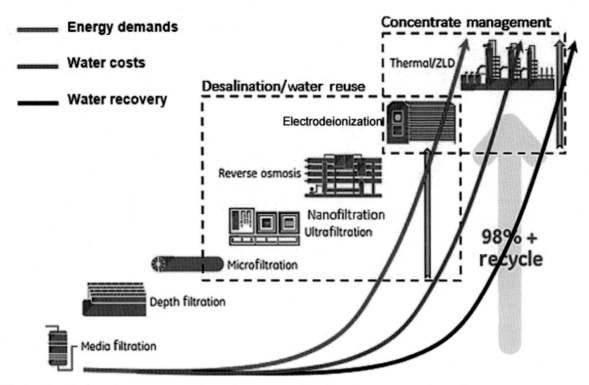

FIGURE 42.1 Desalination and water treatment options and water reuse and concentrate management options and cost recovery/reuse relationships.

m^3—of municipal wastewater are discharged nationwide in the United States each day, approximately 45.5 million m^3 are discharged to an ocean or estuary—an amount equivalent to 6% of total water use in the United States. Reusing this water would directly augment the nation's total water supply. However, water reuse may not be an affordable alternative in all scenarios. Desalination processes are mainly classified into membrane and thermal processes. Membrane processes are also commonly used for water reuse applications around the world. Desalination processes leave a concentrated stream called "brine" that is a major environmental concern. A comparison of possible recovery rates and representative costs for desalination, water reuse, and concentrate management technologies is shown in Fig. 42.1. It can be noted that water treatment costs increase with complexity of the process and the desired end water quality. For recycling and reuse options, zero-liquid discharge options based on electrodeionization and thermal evaporation and crystallization are being developed that are energy intensive and cost-prohibitive at present.

42.4 Groundwater desalination in the United States

The USGS (United States Geological Survey) has provided data on brackish water use in the United States. This report shows that saline groundwater use increased from 1985 to 2005, when roughly 3 billion gal of saline groundwater were being used every day in the United States. At the time (2005), this was roughly 3.7% of the nation's total groundwater use. An estimated 95% of that use occurred in eight states: Alaska, California, Hawaii, Louisiana, Oklahoma, Texas, Utah, and Wyoming. There are at least 650 active brackish water desalination plants in the United States in 2010, with a treatment capacity of 402 million gal a day, mostly used for municipal water use.

Groundwater depletion is highly pronounced in regions where there are increased withdrawals for irrigation as in the case of Mississippi. In some cases, it is compounded by the water needs due to excessive population growth, for example, the water crisis in the Central Valley of California. As the quantity declines in many groundwater reserves, the quality of the source becomes unsuitable for many purposes. In many parts of the world, groundwater is not suitable for direct consumption due to naturally occurring hazardous substances such as uranium, fluoride, and arsenic (Yarlagadda et al., 2011). In some cases the impairment is caused by illicit discharges by industrial sectors. Often, these groundwaters also contain high dissolved solids leading to their categorization as brackish waters. Traditional treatment techniques are not adequate to remove the dissolved solids.

For instance, two-thirds of the continental United States, including New Mexico, have large volumes of saline water sources. The total volume of groundwater in aquifers in New Mexico is estimated to be 20 billion ac-ft.; however, 75% of the groundwater is too saline (10,000–35,000 ppm) for most uses, and the remaining 25% of the groundwater contains dissolved concentrations of lower than 2000 mg/L. In Texas, there is approximately 2.7 billion ac-ft. of brackish groundwater in the aquifers of the state. More than 100 public water supply systems in Texas use brackish groundwater sources in their supply. Many of those are small units. The $67-million brackish water desalination plant in El Paso is the country's largest municipal inland desalination plant, producing 20 million gal/day. Excess withdrawal of easily accessible groundwater reserves creates number of challenges in the form of water quality impairment caused by induced leakage from the land surface, confining layers or adjacent aquifers containing saline or contaminated water (Konikow and Kendy, 2005). For coastal regions, this causes seawater intrusion into groundwater aquifers. Many large cities in the coastal areas currently experience this problem. This situation stems interest for exploring alternative water sources for freshwater supplies. In a recent study, Kang and Jackson (2016) presented deep groundwater sources as new untapped potential water sources for drought-stricken California. The deep groundwater source may be attractive for the local administrators to meet the increasing demands for water supplies in the region.

42.5 Groundwater desalination in developing countries

Many developing countries rely heavily on agriculture for food production and to support economic development. Exploitation or even overexploitation of groundwater reserves is a common and inevitable practice in these countries. India is home to 17.5% of the world's population yet has access to only 4% of the global freshwater resources, many of which are declining in both quantity and quality (Wright, 2014). The Ministry of Water Resources of India reported that approximately 75% of India's drinking water is sourced from aquifers (Soni and Pujari, 2010). However, 60% of these aquifers contain levels of total dissolved solids, which exceed the taste threshold of 500 mg/L, with aquifers in the coastal regions of Gujarat exceeding 1900 mg/L (Wright, 2014). In addition, the use of groundwater containing more than 2000 mg/L of dissolved salts results in reduced crop yield, and increased sodicity of the soil further reduces agricultural income (Nayar et al., 2017). Desalination of seawater sources is an attractive option for coastal communities through centralized plants. However, centralized desalination plants are not suitable for small communities in remote villages. A recent study explored the possibility of developing mobile desalination systems that can be transported to the place of need to meet the agricultural and domestic water needs by conveniently exploiting the local saline groundwater sources through advanced membrane desalination technologies (Li et al., 2018).

Similar to India, China is also dependent on groundwater sources for food supplies and to foster agriculture-based economy. This country also faces issues related to overexploitation of groundwater sources. In contrast to India, China has a centralized mechanism in which each village has a local government-employed (paid by local taxes) representative who is responsible for setting the water use fee and planning and implementation of irrigation practices (Shah, 2005). There were 136 desalination plants by 2017 with cumulative water production of 1.19×10^6 m^3/day. Due to the increasing groundwater issues, there is more focus on seawater desalination plants in this country (Gong et al., 2019).

42.6 Decision-making for municipal desalination plants

Brackish groundwater desalination plant should not be considered as the first option for continuing water supplies. A decision framework is required to determine when this option is appropriate. As shown in Fig. 42.2, the water supply problem should be assessed in terms of both quality and quantity to determine if it is a short- or long-term issue (Gude et al., 2010; Vedachalam and Riha, 2012). Short-term water shortage issues can be better managed with measures such as infrastructure upgrades to fix leaks and meter age issues. In addition, water conservation and reuse schemes can be implemented to address the water resource issues. If the water sources are abundant in neighboring regions, they can be transported to point of need. Water resource issues categorized as long-term need more careful evaluation of current regulations and quality standards prior to considering the desalination option. Long-term issues indicate that the economy of the region could be significantly impacted if new water sources are acquired. Concentrated brines will result from desalination that needs proper management or disposal indicating potential future environmental issues. Installation of a new desalination plant is cost intensive. This requires a detailed cost–benefit analysis to determine its feasibility. Finally, a planning and coordination scheme should be used to secure funds for building and operating desalination plants.

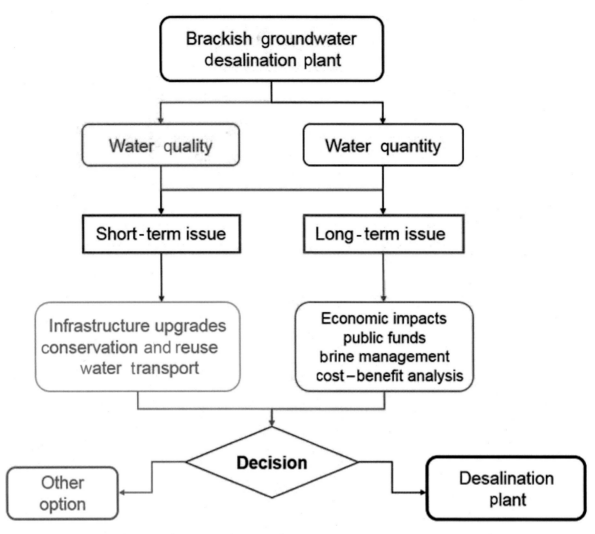

FIGURE 42.2 Decision-making framework for new municipal brackish groundwater desalination plants.

42.7 Conclusion

Deep groundwater sources require additional treatment considerations. It is crucial to develop and implement effective management practices to protect and properly utilize this rather finite source. Management practices may include both technical and nontechnical approaches. More research needs to be conducted to study the use of concentrate or their disposal options. Short- and long-term impacts of accessing deep groundwater sources should be evaluated prior to considering desalination plants. Proper characterization and resource distribution should be studied to develop suitable operating schemes for the required desalination capacity. Finally, monitoring and assessment of the short- and long-term effects of the removal of brackish groundwater, such as saltwater intrusion, land subsidence, or changes in water quality over time, should be considered seriously.

References

Elsaid, K., Bensalah, N., Abdel-Wahab, A., 2012. Inland desalination: potentials and challenges. Adv. Chem. Eng. 10, 2068.
Famiglietti, J.S., 2014. The global groundwater crisis. Nat. Clim. Change 4 (11), 945–948.
Faust, S.D., Aly, O.M., 1998. Chemistry of Water Treatment. CRC Press.
Glennon, R., 2002. Water Follies: Groundwater Pumping and the Fate of America's Fresh Waters. Island Press, Washington, DC.
Gong, S., Wang, H., Zhu, Z., Bai, Q., Wang, C., 2019. Comprehensive utilization of seawater in China: a description of the present situation, restrictive factors and potential countermeasures. Water 11 (2), 397.

Gude, V.G., 2011. Energy consumption and recovery in reverse osmosis. Desalin. Water Treat. 36 (1-3), 239–260.
Gude, V.G., 2015. Energy storage for desalination processes powered by renewable energy and waste heat sources. Appl. Energy 137, 877–898.
Gude, V.G., 2016. Desalination and sustainability—an appraisal and current perspective. Water Res. 89, 87–106.
Gude, V.G., 2017. Desalination and water reuse to address global water scarcity. Rev. Environ. Sci. Bio/Technol. 16 (4), 591–609.
Gude, V.G., 2018. Desalination of deep groundwater aquifers for freshwater supplies—challenges and strategies. Groundw. Sustain. Dev. 6, 87–92.
Gude, V.G., Fthenakis, V., 2020. Energy efficiency and renewable energy utilization in desalination systems. Prog. Energy 2, 022003.
Gude, V.G., Nirmalakhandan, N., Deng, S., 2010. Renewable and sustainable approaches for desalination. Renew. Sustain. Energy Rev. 14 (9), 2641–2654.
Hamrouni, B., Dhahbi, M., 2001. Thermodynamics description of saline waters-prediction of scaling limits in desalination processes. Desalination 137 (1-3), 275–284. ISSN: 0011-9164.
Kang, M., Jackson, R.B., 2016. Salinity of deep groundwater in California: water quantity, quality, and protection. Proc. Natl. Acad. Sci. U.S.A. 113 (28), 7768–7773.
Konikow, L.F., Kendy, E., 2005. Groundwater depletion: a global problem. Hydrogeol. J. 13 (1), 317–320.
Li, Q., Lian, B., Wang, Y., Taylor, R.A., Dong, M., Lloyd, T., et al., 2018. Development of a mobile groundwater desalination system for communities in rural India. Water Res. 144, 642–655.
Margat, J., 2008. Exploitations et utilisations des eaux souterraines dans le monde. Coédition: UNESCO et. BRGM 52.
Margat, J., Van der Gun, J., 2013. Groundwater Around the World: A Geographic Synopsis. CRC Press.
Meng, M., Chen, M., Sanders, K.T., 2016. Evaluating the feasibility of using produced water from oil and natural gas production to address water scarcity in California's Central Valley. Sustainability 8 (12), 1318.
Mukherjee, A., Scanlon, B., Aureli, A., Langan, S., Guo, H., McKenzie, A., 2020. Global Groundwater: Source, Scarcity, Sustainability, Security and Solutions, first ed. Elsevier, ISBN: 9780128181720.
Muñoz, I., Fernández-Alba, A.R., 2008. Reducing the environmental impacts of reverse osmosis desalination by using brackish groundwater resources. Water Res. 42 (3), 801–811.
Nayar, K.G., Sundararaman, P., O'Connor, C.L., Schacherl, J.D., Heath, M.L., Gabriel, M.O., et al., 2017. Feasibility study of an electrodialysis system for in-home water desalination in urban India. Dev. Eng. 2, 38–46.
NRC, National Research Council, 2012. Water Reuse: Potential for Expanding the Nation's Water Supply Through Reuse of Municipal Wastewater. National Academies Press.
Scanlon, B.R., Longuevergne, L., Long, D., 2012. Ground referencing GRACE satellite estimates of groundwater storage changes in the California Central Valley, USA. Water Resour. Res. 48 (4).
Shah, T., 2005. The new institutional economics of India's water policy. In: African Water Laws: Plural Legislative Frameworks for Rural Water Management in Africa, 26–28 January 2005.
Siebert, S., Burke, J., Faures, J.M., Frenken, K., Hoogeveen, J., Döll, P., et al., 2010. Groundwater use for irrigation – a global inventory? Hydrol. Earth Syst. Sci. 14 (10), 1863–1880.
Sobhani, R., Abahusayn, M., Gabelich, C.J., Rosso, D., 2012. Energy footprint analysis of brackish groundwater desalination with zero liquid discharge in inland areas of the Arabian Peninsula. Desalination 291, 106–116.
Soni, A.K., Pujari, P.R., 2010. Ground water vis-a-vis sea water intrusion analysis for a part of limestone tract of Gujarat Coast, India. J. Water Resour. Prot. 2 (5), 462.
Taylor, R.G., Scanlon, B., Döll, P., Rodell, M., Van Beek, R., Wada, Y., et al., 2013. Ground water and climate change. Nat. Clim. Change 3 (4), 322–329.
Vedachalam, S., Riha, S.J., 2012. Desalination in northeastern US: lessons from four case studies. Desalination 297, 104–110.
Wada, Y., Lo, M.H., Yeh, P.J.F., Reager, J.T., Famiglietti, J.S., Wu, R.J., et al., 2016. Fate of water pumped from underground and contributions to sea-level rise. Nat. Clim. Change 6, 777–780.
Wada, Y., van Beek, L.P., van Kempen, C.M., Reckman, J.W., Vasak, S., Bierkens, M.F., 2010. Global depletion of groundwater resources. Geophys. Res. Lett. 37 (20).
Wright, N.C., 2014. Justification for community-scale photovoltaic-powered electrodialysis desalination systems for inland rural villages in India. Desalination 352, 82–91.
Yarlagadda, S., Gude, V.G., Camacho, L.M., Pinappu, S., Deng, S., 2011. Potable water recovery from As, U, and F contaminated ground waters by direct contact membrane distillation process. J. Hazard. Mater. 192 (3), 1388–1394.

Chapter 43

Quantifying future water environment using numerical simulations: a scenario-based approach for sustainable groundwater management plan in Medan, Indonesia

Pankaj Kumar[1], Binaya Kumar Mishra[2], Ram Avtar[3] and Shamik Chakraborty[4]

[1]*Natural Resources and Ecosystem Services, Institute for Global Environmental Strategies, Hayama, Japan,* [2]*School of Engineering, Pokhara University, Lekhnath, Nepal,* [3]*Faculty of Environmental Earth Science, Hokkaido University, Sapporo, Japan,* [4]*Faculty of Sustainability Studies, Hosei University, Tokyo, Japan*

43.1 Introduction

Water is a basic necessity for sustainable and inclusive development of human society. Yet, access to clean and safe drinking water is a major challenge (Heinemann et al., 2002; FAO, 2016), which is well recognized under the global sustainable development goals (SDGs). To put things in perspective, more than 1.1 billion people do not have adequate access to clean drinking water and it is feared that nearly two-thirds of the nations will be under water stress by 2025 (Pink, 2016; Kumar, 2019). Researchers argue that water insecurity as a major hindrance for socioeconomic growth in the developing world. This is especially the case in Asia and Africa, where increasing demand for water and contamination of key water sources complicate the scenario further (Huizinga et al., 2017; Mukate et al., 2017). While it is estimated that about 3 billion people will reside in Asian cities by year 2050 (United Nations, Department of Economic and Social Affairs, Population Division (UN DESA), 2015), countries such as India, Pakistan, and Afghanistan have already reached the water stress level in 2015, and several other countries such as Nepal and Bangladesh are underway (Gareth et al., 2014). Thus ensuring sustainable water supply and meeting the SDGs in growing Asian cities remains a critical challenge. Although the concepts of water security vary within the academia (Hoekstra et al., 2018), most researchers now agree that the water crisis in the 21st century is much more a management issue rather than the actual scarcity of water (Tundisi, 2008).

Of the various issues that impound water scarcity, frequent extreme weather conditions, deteriorating water quality, and lack of water governance are main factors in most of the developing countries, which broadened the debates on water security over the last two decades (Mukherjee et al., 2020). Key functions such as water availability and runoff alter the flood regimes of rivers due to changes in climatic patterns. According to the Intergovernmental Panel on Climate Change (IPCC) (2014) report, a greater number of regions are likely to experience extreme weather events, including heavy precipitation and floods in the future. Consequently, the SDGs have put focus simultaneously on water quantity and quality, an approach that was largely missing in the Millennium Development Goals, which only provided quantity related targets (Angelo et al., 2018). Nonetheless, a multitude of other factors, including population growth, rapid urbanization, and climate change, are also expected to pose significant challenges in achieving water security in the future (Huizinga et al., 2017; Mukate et al., 2017).

Together with wastewater infrastructure improvement, sewerage network improvement, and building dams, nature-based solution, along with adaptive governance of water resources, has been identified as an important tool for achieving water security (UN, 2018; Kalantari et al., 2018; Wild et al., 2017). However, managing urban water environment still remains a significant challenge (UN Water, 2017; Ferguson et al., 2018); and the concept of urban water security is fundamentally different from the concept of water security (Hoekstra et al., 2018). A number of allied factors, such as high population density, concentrated demand of fresh water, distribution system, and wastewater recycling and taxing are particularly important for an urban context. To meet the challenges posed by such complexity, tapping reliable, all-weather sources remain an important first step for urban water planners.

Unfortunately, water resource planning in cities has been done largely in a piecemeal manner, without taking into account of larger socioeconomic factors, and this is the case especially in the developing countries (Downing, 2012). While planners in developing countries are often obsessed with meeting the basic water demands through adequate supply, a critical step to achieve future urban water security is to integrate both hydrological as well as socioeconomic factors. It is highly imperative to understand urban water security through a system perspective, including natural (i.e., source), social, economic, and infrastructural components.

Several holistic approaches have been conceptualized in the field of water resource management since the 1980s, of which the integrated water resource management (IWRM) model has received the highest attention (Hoekstra et al., 2018). Different components of water resource governance are targeted by the IWRM model (e.g., socioeconomic status, hydrometeorological factors, agriculture, industries, and wastewater), to aid science-led decision-making (Frija et al., 2015; Blanco-Gutiérrez et al., 2013). Several numerical IWRM models such as RIBASIM (River Basin Simulation Model), HEC-HMS, Flo-2D, MIKE, WEAP (Water Evaluation and Planning), and WBalMo (Water Balance Model) have been developed and widely used to address water security issues (Ingol-Blanco and McKinney, 2013; Slaughter et al., 2014; Kumar et al., 2018). Some of these models such as WEAP and HEC-HMS are widely used in water resource modeling for water resource planners in developing countries, because they are not data intensive and the software package comes free of cost.

Unplanned rapid urban expansion, in combination with high economic growth, results in unhealthy water environment around water bodies in most of the developing countries. Despite its importance, their current status and their management strategies for the near future is poorly documented. This study strives to apply integrated analysis to assess the current situation and the simulated future status of water quality and quantity (flooding) in the Deli River watershed crossing Medan city, Indonesia, with the ultimate goal to help to formulate plans for sustainable water resource management in the area. The research procedure applies systems analysis and includes studies of technical models and future scenarios that are affecting water infrastructure in a city. The study highlights practical recommendations that are based on the results and are important for successful implementation of sustainable water management strategies and improving decision-making process in water-related sector in Indonesia.

43.2 Study area

Medan is the capital city of North Sumatra province. It is located between $3°35'N$ latitude and $98°40'E$ longitude (Fig. 43.1). Medan city is divided into 21 districts. The total population of Medan is 2,191,140 (Statistics Indonesia, 2014), and its total area is 265 km². The climate is classified as equatorial, characterized by heavy and frequent rainfall. The annual rainfall is about 2263 mm and the temperature ranges from 24°C to 32°C. Medan city is characterized by three watersheds: Belawan watershed, Deli watershed, and Percut watershed.

43.3 Methodology

Systems approach can be a meaningful solution to solve issues in urban water management that requires understanding complexity associated with water resource management. Integrated systems approaches capture interdisciplinarity, associated with diverse natural, technical, and institutional dimensions that are associated with complex city infrastructure (Urban, 2015). Properly understanding and solution of these issues requires researchers and professionals from different professional disciplines, including academia, to exchange ideas, and learn from each other to explore new, robust solutions. In this work an integrated approach is used to assess the current situation as well as predict future situation of water environment and likely adaptation measures for urban water management as shown in Fig. 43.2.

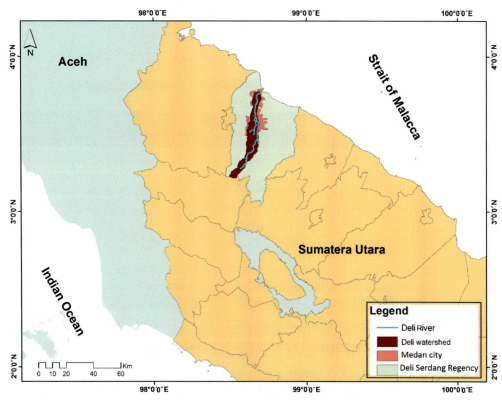

FIGURE 43.1 Deli watershed location.

43.3.1 Different drivers

43.3.1.1 Precipitation change

In this study, precipitation is a relevant indicator to determine the effect of climate change on flood and water quality. For analysis of food systems, study of precipitation change was carried out with emphasis on daily maximum rainfall and mean monthly rainfall. There are three rain-gauge stations: Tutungan in the upstream, Sampali in the center, and Belawan in the downstream of the watershed. Precipitation change assessment was started with screening of daily rainfall data of Sampali station. Selection of Sampali station was made considering its location and availability of long record (1980–2015) of rainfall data. Rainfall data of 1980–2004 were considered for the precipitation change assessment. Changes in daily maximum rainfall for different return periods and mean monthly rainfall were used for comparing flood inundation and river pollution conditions. Daily rainfall outputs of three GCMs: MRI-CGCM3, MIROC5, and HadGEM2-ES were extracted for assessing future climate conditions. Out of these three GCMs, MRI-CGCM3 precipitation outputs were found to be suitable considering rainfall characteristics of observation data. Future climate was characterized by MRI-CGCM3 rainfall data of 2020–44 period. Quantile mapping technique was applied for correcting biases in the GCM output. Finally, empirical frequency analysis was applied for estimating daily maximum rainfall (Figs. 43.3–43.5).

In order to evaluate the effects of climate change on water quality, the change in monthly average precipitation was evaluated. The GCM output is downscaled at the local level to aid reliable impact assessment (Sunyer et al., 2015). Statistical downscaling was followed by trend analysis to get climate variables at monthly scale. Trend analysis is a less computation demanding technique, which enables reduction of biases in the precipitation frequency and intensity (Elshamy et al., 2009). Regarding future simulation to assess the climate change impact, MRI-CGCM3 and MIROC5 precipitation output was used because of its wide use and high temporal resolution compared to other climate models. Our study is based on the RCP4.5 and RCP8.5 emission scenario, which assumes that global annual GHG emissions (measured in CO_2 equivalents) will peak around 2040 and then will decline (Intergovernmental Panel on Climate Change (IPCC), 2014). In this study the GCM data are from the 1985 to 2004 and 2020 to 2039 periods (each with a 20-year length) and represent the current and future (2030) climates, respectively. The result is shown in

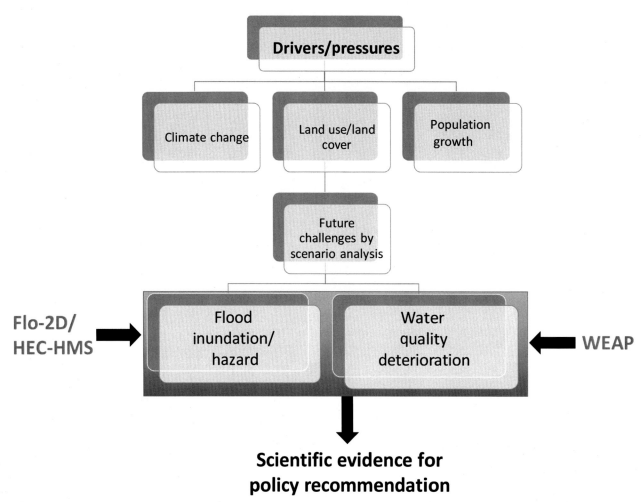

FIGURE 43.2 Research framework under this work.

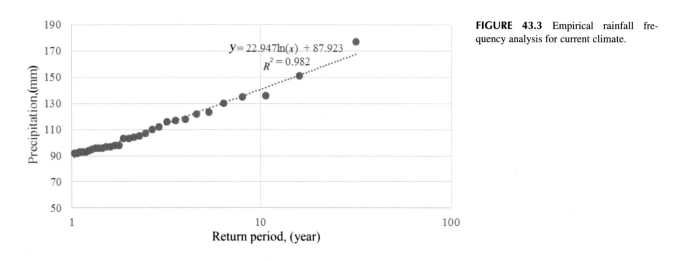

FIGURE 43.3 Empirical rainfall frequency analysis for current climate.

Fig. 43.6, which indicates that annual precipitation in the simulated GCM output is not much different than that of current observed one. The values for total annual precipitation in the case of observed_2015, Sim_2030_MRICGCM3_45, Sim_2030_MIROC5_45, Sim_2030_MRICGCM3_85, and Sim_2030_MIROC5_85 are 2061.6, 2139.1, 2187.1, 2114.7, and 2156.9 mm, respectively. We have fixed other parameters as constant like population growth for estimating the

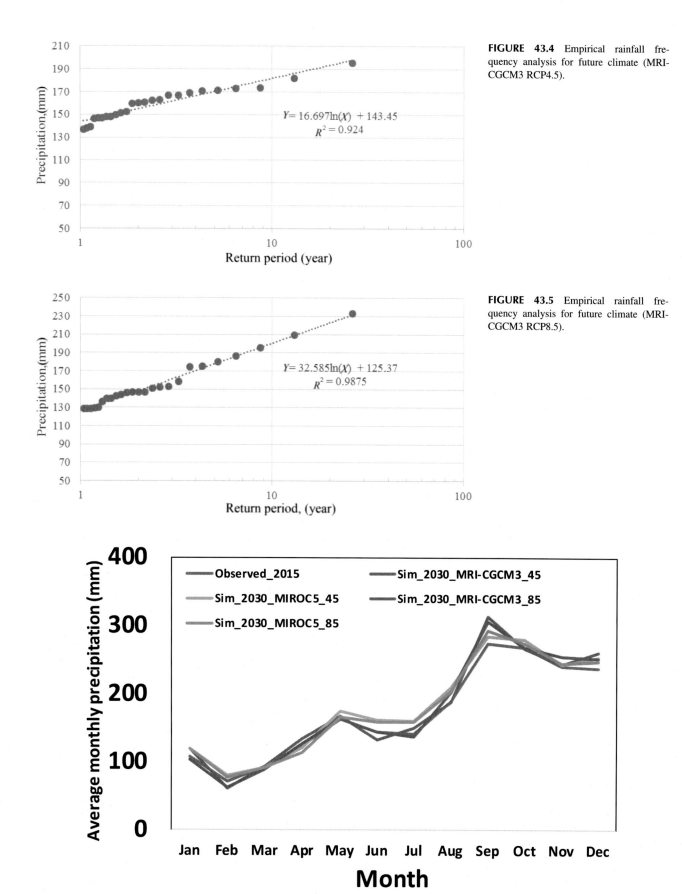

FIGURE 43.4 Empirical rainfall frequency analysis for future climate (MRI-CGCM3 RCP4.5).

FIGURE 43.5 Empirical rainfall frequency analysis for future climate (MRI-CGCM3 RCP8.5).

FIGURE 43.6 Graph showing a comparative study for current and future monthly rainfall.

effect of precipitation only on water quality. Once all other factors are kept constant, we have given simulated value of precipitation as an input on the water quality simulation for future, starting from year 2016, that is, right after year 2015. Finally, for water quality simulation, we have used MRI-CGCM3 with RCP_8.5.

43.3.1.2 Land use change

Two past land use maps are established and used for the analysis regarding prediction of future land use/land cover map of 2030. In order to prepare the land use maps for Deli watershed, four satellite images Landsat are applied (Table 43.1). It was observed that the watershed covers two different scenes located in various paths/rows. For this reason, appropriate processing is required before image analysis and classification. First two Landsat images were merged in a single raster image using Mosaic Dataset of ArcGIS. Then, band composites, clipping, and supervised classification were done to get the land use map of the watershed. Land Change Modeler for ArcGIS was used to predict the land use pattern based on the previous change trend. Through image classification, only four classes are identified.

43.3.1.3 Population growth

The whole study area has been divided into different demand sites for estimating the effect of population growth (one of our two major key drivers), on water quality status. These demand sites mainly represent the population of different cities on both sides of Deli River within the study area, and who have direct impact on river environment through discharging domestic sewerages into the river. Future population is estimated from ratio method using data for United Nations, Department of Economic and Social Affairs, Population Division (UN DESA) (2015) projection rate. The total population of 2,200,001 was considered at base year, that is, 2010 in our study area. For the future population projection, the annual growth rate was considered 0.96, 1.58, 2.25, and 2.04 during the period of 2011–15, 2016–20, 2021–25, and 2025–30, respectively. Henceforth, population considered for current year (2015) and target year (2030) was 2,307,648 and 3,085,883, respectively.

43.3.2 Urban flood

Of the three watersheds of Medan city, the Deli River has a big potential for causing flood inundation. Due to the small river flow capacity for the Deli, flood had occurred frequently in Deli River watershed, which flows through the central part of the city. The flood events were reported increasing due to urbanization of the city and its surrounding areas, enhanced by changing climate. This study aims to contribute to the reduction of flood damage, stabilization and enhancement of people's livelihoods, and promotion of local economy with improved city flood risk management plan. Medan floodway was constructed at Titi Kuning at Deli River to Percut River with current of design discharge 70 m^3/s, planned to be increased to 120 m^3/s and further.

Flood modeling consists of two parts: the hydrologic modeling estimates the peak discharge while hydraulic modeling flood inundation simulation. Total catchment area of Deli River basin is about 400 km^2. Out of this, upper catchment area (u/s of floodway) is about 160 km^2. HEC-HMS was run for peak discharge simulation in upper region with outlet at floodway diversion point. Flo-2D model was run in the lower watershed boundary. Topography data were extracted from SRTM with 30 m resolution (Fig. 43.7).

TABLE 43.1 Satellite images applied.

No.	Path/row	Data set	Acquisition data	Land cloud cover (%)
1	129/057	Landsat 7 ETM C1 Level 1	May 19, 2003	1
2	129/058	Landsat 7 ETM C1 Level 1	May 19, 2003	0
3	129/057	Landsat 8 OLI/TIRS C1 Level 1	February 21, 2015	12.87
4	129/058	Landsat 8 OLI/TIRS C1 Level 1	February 21, 2015	2.54

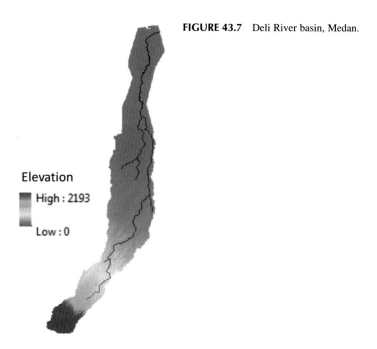

FIGURE 43.7 Deli River basin, Medan.

43.3.3 Water quality

43.3.3.1 Basic information regarding the model and data requirement

WEAP model was used to simulate future water quality variables in the year 2030 to assess alternative management policies in the Deli River basin. A wide range of input data such as point and nonpoint pollution sources, their locations and concentrations, past spatiotemporal water quality, wastewater treatment plant (WWTP) population, historical rainfall, evaporation, temperatures, drainage networks, river flow—stage—width relationships, river length, groundwater, surface water inflows, and land use/land cover is used for water quality modeling.

WEAP model was developed for the Deli River basin for four command areas with interbasin transfers. Hydrologic modeling requires the entire study area to be split into smaller catchments with consideration of the confluence points and physiographic and climatic characteristics (Fig. 43.8). The hydrology module within the WEAP tool enables modeling of the catchment runoff and pollutant transport processes into the river. Pollutant transport from a catchment accompanied by rainfall—runoff is enabled by ticking the water quality modeling option. Pollutants that accumulate on catchment surfaces during nonrainy days reach water bodies through surface runoff. The WEAP hydrology module computes catchment surface pollutants generated over time by multiplying the runoff volume and concentration or intensity for different types of land use. During simulation the land use information was broadly categorized into three categories: agricultural, forest, and built-up areas. The soil data parameters were identified using previous secondary data and the literature. Daily rainfall has been collected at Sampali meteorological station, for the period from 1990 to 2010. Daily average stream flow data from 2005 to 2014 were measured at five stations, Mangonsidi, Raden Saleh, Unibis, Simpang Kantor, and Medan Labuhan of the Deli River, and were utilized to calibrate and validate the WEAP hydrology module simulation. Data for the water quality indicators (biochemical oxygen demand (BOD) and *Escherichia coli*) were also collected at four of the abovementioned five stations and used for water quality modeling.

43.3.3.2 Model setup

The entire problem domain and its different components are divided into four catchments considering influent locations of major tributaries (Fig. 43.8). Other major considerations are the five demand sites and one WWTP to represent the problem domain. Here, demand sites are meant to identify domestic (population) defined with their attributes explaining water consumption and wastewater pollution loads per capita, water supply source, and wastewater return flow. Dynamic attributes are described as functions of time and include population and industries. WWTPs are pollution handling facilities with design specifications that include total capacity and removal rates of pollutants. The flow of wastewater into the Deli River and its tributaries mainly feeds through domestic, industrial, and stormwater runoff routes. In the absence of any precise information about the type of WWTP operational at current stage, UASB-SBR type of

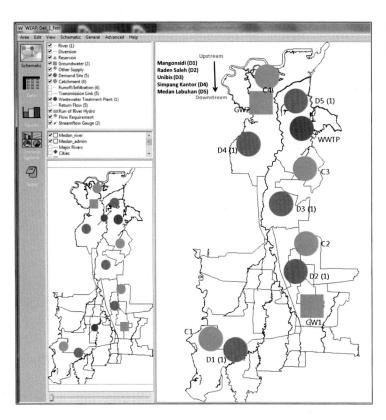

FIGURE 43.8 Schematic diagram showing the problem domain for water quality modeling in Medan using WEAP interface. *WEAP*, Water Evaluation and Planning.

TABLE 43.2 Comparison of daily maximum rainfall for 50- and 100-year return period.

Return period (year)	Current climate	MRI-CGCM3	
		RCP4.5	RCP8.5
50	177.7	208.8	252.9
100	193.6	220.4	275.4

WWTP is considered in the modeling and its treatment efficiency is assumed as 97% for BOD, 77% for total nitrogen, and 99.69% for fecal coliform (Khan et al., 2013). No precise data are available regarding the total volume of wastewater production from domestic sources. In the absence of detailed information, the daily volume of domestic wastewater generation is estimated 130 L of average daily consumption per capita based on literature review (WWAP, 2017). Future projection was made to observe the water quality of Deli River in 2030 considering the effect of climate change and population growth and current existing WWTPs of capacity of 18 MLD.

43.4 Results and discussion

43.4.1 Precipitation change

Using bias-corrected rainfall, daily maximum rainfall for different return periods was estimated. Table 43.2 provides comparison of daily rainfall for 50- and 100-year return period. However, in this study, 50-year daily maximum rainfall was employed for assessing flood inundation simulation. An increase of 17.5% and 43% was found for RCP4.5 and RCP8.5 emission scenarios, respectively, with respect to baseline period. This significant increase in daily maximum points out that climate change consideration should be incorporated into Medan future flood risk management plan for its sustainable urban water environment.

An assessment of the climate change impacts on precipitation over the Deli River Basin was conducted by using bias-corrected GCM data for the 1985–2004 and 2020–39 periods. The quantile-based bias corrections were used to identify the bias pattern in the GCM precipitation data. This bias pattern was identified by comparing the observations with the corresponding GCM data. A comparison of the daily precipitation values indicated that peaks in the GCM values were significantly smaller than the peaks in the observation values. The GCM precipitation data showed significantly larger wet days than those of the observed precipitation data. The performance of the quantile-based correction technique was evaluated by comparing the monthly average rainy days and daily precipitation amounts. Plots of the monthly average rainy days and daily rainfall magnitudes indicated a similar number of rainy days and extreme rainfall magnitudes, thereby demonstrating the effectiveness of the quantile-based bias correction technique.

Rainfall intensity duration frequency (IDF) curves can be used to estimate rainfall intensities for different durations and return periods. Frequency analysis using Gumbel extreme value method enabled the generation of rainfall IDF curves and extreme precipitation change assessment for the present and future climate scenarios over the Deli River watershed. In this study the 1-day maximum precipitation for the 50- and 100-year return period was determined for the current and future precipitation data sets. These values clearly indicate that extreme precipitation events for all return periods and all durations will be more frequent and intense in the future Table 43.3.

43.4.2 Land use change

The comparison between current and predicted land use maps show an increase of urban growth (Fig. 43.9). Our results indicate an expansion of built-up areas to about 20%, which may lead to a rise in runoff while reducing infiltration. Consequently, flood risk can be more significant in the future. The results indicate that a suitable land use plan is required to avoid increasing the vulnerability of people and buildings.

TABLE 43.3 Comparison of 1-day maximum rainfall for current and future climate conditions.

Return period (year)	One-day maximum rainfall (mm)		
	Current	Future average of three GCMs	Future extreme among three GCMs
50	207	227	330
100	228	249	365

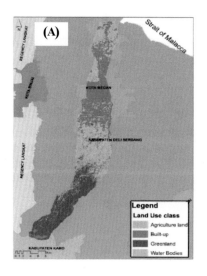
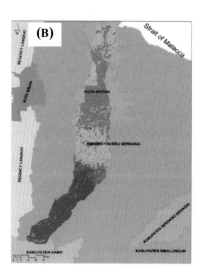

FIGURE 43.9 Land use change map: (A) land use 2015 and (B) land use 2030.

43.4.3 Urban flood

Although river improvement and drainage projects, including floodway connecting Deli River to Percut River, had been undertaken with utilization, flooding is still remain a considerable hazard in the Medan city. The HEC-HMS was calibrated for the flood event of November 2001 flood event. During this flood event a peak discharge of 290 m^3/s was reported (source: Mitsubishi UFJ Research and Consulting Co., Ltd. Report). On the other hand, simulated flood discharge was 314 m^3/s, which can be considered acceptably well. After diverting flood discharge of 70 m^3/s via floodway, inundation simulation task was proceeded for the current climate condition. Fig. 43.10 provides a spatial comparison of flood inundation simulation in lower region for current and future conditions. Table 43.4 indicates there will significant increase of flood inundation despite increased flood diversion. It is expected that the flood inundation will increase in 2030, especially marked by an estimated inundation of 3–4.5 m depth in the central and northern part of Medan, and in the south along the Deli River. Therefore flood risk master plan of Medan requires more attention to address the problem increased flooding.

43.4.4 Water quality

Future simulation of water quality was conducted using selected parameters (BOD and *E. coli*) under business as usual (BAU) scenario while including effect of population growth and climate change. The simulation considered the capacity of currently existing WWTP (capacity of 18 MLD and mere coverage of 16% of total population in the study area) and

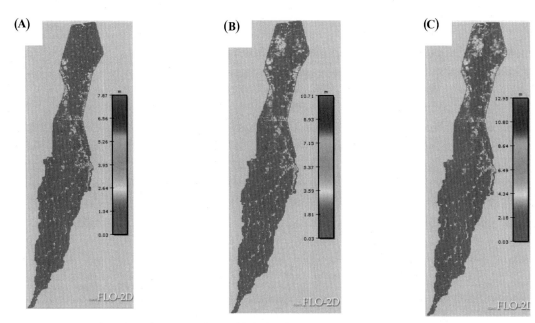

FIGURE 43.10 (A) Maximum flow depth for current climate (2015) for 50-year return period; (B) maximum flow depth (m) for 2030 for MRI RCP45 for 50-year return period; and (C) maximum flow depth (m) for 2030 for MRI RCP85 for 50-year return period.

TABLE 43.4 Comparison of flood inundation under current and future climate conditions.

Inundation depth (m)	Inundation area (km²)	
	Current	Future
0.2–0.5	17.4	23.8
0.5–1.5	21.3	32.6
>1.5	8.2	16.6

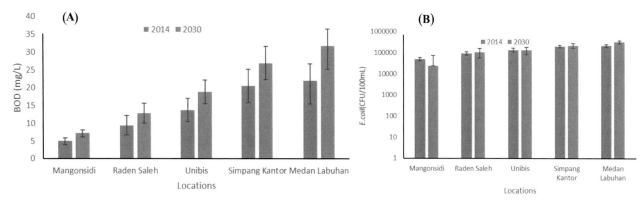

FIGURE 43.11 The simulation result of the annual average values of (A) BOD and (B) *Escherichia coli* for five different locations in 2014 and 2030 under different scenarios (current situation and business as usual).

that this capacity will continue by year 2030. The result for water quality is shown in Fig. 43.11. It can be observed that current status of water quality throughout the river is very poor as compared with local guideline for class 2 (BOD < 5 mg/L; *E. coli* < 1000 CFU/100 mL). And it is even worse in the case of downstream locations of the river because of the cumulative effect of waste disposal and injection of untreated wastewater coming from the upstream areas. Our results show that in the case of BAU scenario, effect of both climate change and population changes are prominent in water quality status. It is expected to deteriorate further in 2030 when compared to current situation. Due to climate change and population changes, the water quality parameters BOD and *E. coli* will be further deteriorating by 54.2% and 12.4%, respectively, on average in 2030 compared to the current situation. Furthermore, the individual contribution from population and climate change on BOD and *E. coli* was 86% and 14% and 92% and 8%, respectively. A high concentration of nitrate in water indicates the influence of untreated sewerage input. The value for BOD varies from 5.2 to 22.6 mg/L, which indicates that most of the water samples are moderately to extremely polluted, and not safe for a healthy aquatic system. The *E. coli* value, a commonly used biological indicator of water quality status, also shows that no significant improvement in future course of time, although this might be because of unavailability of data like rate of chlorination.

43.5 Conclusion and recommendation

Although river improvement and drainage projects, including floodway connecting Deli River to Percut River, had been undertaken, these projects have not been able to reduce flooding in the Medan city. In the present study the HEC-HMS model was used to provide spatial comparison of flood inundation simulation in lower reaches of the Medan river basin for current and future scenarios. Result shows that there will be significant increase of flood inundation despite increased flood diversion projects. Furthermore, the flood inundation is expected to increase in 2030. The extent of inundation can be 3–4.5 m in depth in the central and northern part of Medan, and in the south along the Deli River. Regarding the water quality component, it was observed that current status of water quality throughout the river is very poor as compared to local guideline for class 2 (BOD < 5 mg/L; *E. coli* < 1000 CFU/100 mL). In addition, the water quality status is even worse in the case of downstream locations of the river. From the result of BAU scenario, effects of both climate change and population change are prominent in water quality status. Water quality is expected to deteriorate further in 2030 when compared to current situation. Due to climate change and population changes, the two water quality parameters BOD and *E. coli* will be further deteriorating to 54.2% and 12.4%, respectively, on average by 2030 when compared to the current situation. The previous results suggest that current management policies and water resource management plans in near future are not enough to check the pollution level within the desirable limit and reduce the frequency of flooding in the study area and hence calls for transdisciplinary research in more holistic way for doing it sustainably.

References

Angelo, D.J., Rulli, M.C., D'Ordorico, P., 2018. The global water grabbing syndrome. Ecol. Econ. 143, 276–285.

Blanco-Gutiérrez, I., Varela-Ortega, C., Purkey, D.R., 2013. Integrated assessment of policy interventions for promoting sustainable irrigation in semi-arid environments: a hydro-economic modeling approach. J. Environ. Manage. 128, 144–160.

Downing, T.E., 2012. Views of the frontiers in climate change adaptation economics. WIREs Clim. Change 3, 161–170.

Elshamy, M.E., Seierstad, I.A., Sorteberg, A., 2009. Impacts of climate change on Blue Nile flows using bias-corrected GCM scenarios. Hydrol. Earth Syst. Sci. 13, 551–565.

FAO, 2016. AQUASTAT website. Food and Agriculture Organization of the United Nations (FAO). (accessed 15.05.016.).

Ferguson, L., Chan, S., Santelmann, M.V., Tilt, B., 2018. Transdisciplinary research in water sustainability: what's in it for an engaged researcher-stakeholder community? Water Altern. 11 (1), 1–18.

Frija, A., Dhehibi, B., Chebil, A., Villholth, K.G., 2015. Performance evaluation of groundwater management instruments: The case of irrigation sector in Tunisia. Groundw. Sustain. Dev. 1. Available from: https://doi.org/10.1016/j.gsd.2015.12.001.

Gareth, P. et al., 2014. Attitudes to water in South Asia. In: Chatham House Report. Chatham House, The Royal Institute of International Affairs, UK, Vinset Advertising, New Delhi, India, pp. 114.

Heinemann, A.B., Hoogenboom, G., Faria, R.T., 2002. Determination of spatial water requirements at county and regional levels using crop models and GIS: an example for the State of Parana, Brazil. Agric. Water Manage. 52, 177–196.

Hoekstra, A.Y., Buurman, J., van Ginkel, K.C., 2018. Urban water security: a review. Environ. Res. Lett. 13 (5), 053002.

Huizinga, J., Moel, H. de, Szewczyk, W., 2017. Global flood depth-damage functions. Methodol. Database Guidel. Available from: https://doi.org/10.2760/16510EUR 28552 EN.

Ingol-Blanco, E., McKinney, D., 2013. Development of a hydrological model for the Rio Conchos basin. J. Hydrol. Eng. 18 (3), 340–351.

Intergovernmental Panel on Climate Change (IPCC), 2014. Climate change 2014: synthesis report. In: Pachauri, R.K., Meyer, L.A. (Eds.), Contribution of Working Groups I; II and III to the Fifth Assessment Report of the Intergovernmental Panel on Climate Change; Core Writing Team. IPCC, Geneva, p. 151.

Kalantari, K., Ferreira, C.S.S., Keesstra, S., Destouni, G., et al., 2018. Nature based solutions for flood-drought risk mitigation in vulnerable urbanizing parts of East-Africa. Current Opinion Environ. Sci. Health 5, 73–78.

Khan, A.A., Gaur, R.Z., Diamantis, V., Lew, B., Mehrotra, I., Kazmi, A.A., et al., 2013. Continuous fill intermittent decant type sequencing batch reactor application to upgrade the UASB treated sewage. Bioprocess Biosyst. Eng 36, 627–634.

Kumar, P., Masago, Y., Mishra, B.K., Fukushi, K., 2018. Evaluating future stress due to combined effect of climate change and rapid urbanization for Pasig-Marikina River, Manila. Groundw. Sustain. Dev. 6, 227–234.

Kumar, P., 2019. Numerical quantification of current status quo and future prediction of water quality in eight Asian Mega cities: challenges and opportunities for sustainable water management. Environ. Monit. Assess. 191, 319. Available from: https://doi.org/10.1007/s10661-019-7497-x.

Mukate, S., Panaskar, D., Wagh, V., Muley, A., Jangam, C., Pawar, R., 2017. Impact of anthropogenic inputs on water quality in Chincholi Industrial area of Solapur, Maharashtra, India. Groundw. Sustain. Dev. Available from: https://doi.org/10.1016/j.gsd.2017.11.001.

Mukherjee, A., Scanlon, B., Aureli, A., Langan, S., Guo, H., McKenzie, A., 2020. Global Groundwater: Source, Scarcity, Sustainability, Security and Solutions, first ed. Elsevier. ISBN 9780128181720

Pink, R.M., 2016. Introduction. Water Rights in Southeast Asia and India. Palgrave Macmillan, New York, pp. 1–14.

Slaughter, A.R., Mantel, S.K., Hughes, D.A., 2014. Investigating possible climate change and development effects on water quality within an arid catchment in South Africa: a comparison of two models. In: Ames, D.P., Quinn, N.W.T., Rizzoli, A.E. (Eds.), Proceedings of the Seventh International Congress on Environmental Modelling and Software, June 15–19, 2014, San Diego, CA, ISBN: 978-88-9035-744-2.

Sunyer, M.A., Hundecha, Y., Lawrence, D., Madsen, H., Willems, P., Martinkova, M., et al., 2015. Intercomparison of statistical downscaling methods for projection of extreme precipitation in Europe. Hydrol. Earth Syst. Sci. 19, 1827–1847.

Statistics Indonesia, 2014. Statistical Yearbook of Indonesia 2014. Jakarta: BPS – Statistics Indonesia https://www.neliti.com/publications/49018/statistical-yearbook-of-indonesia-2014

Tundisi, J.G., 2008. Water resources in the future: problems and solutions. Estud. Avançados 22 (63), 7–16.

UN Water, 2017. Wastewater: the untapped resources. Facts and figure. In: The United Nations World Water Report 2017. Italy, p. 83.

United Nations, Department of Economic and Social Affairs, Population Division (UN DESA), 2015. World Urbanization Prospects: The 2014 Revision (ST/ESA/SER.A/366). p. 517.

United Nation, 2018. Special climate report: 1.5°C is possible but requires Unprecedented and urgent action. New York, USA, Accessed 20.05.19. https://www.un.org/sustainabledevelopment/blog/2018/10/special-climate-report-1-5oc-is-possible-but-requires-unprecedented-and-urgent-action/

Urban, M.C., 2015. Accelerating extinction risk from climate change. Science 348 (6234), 571–573.

Wild, M., Ohmura, A., Schär, C., Müller, G., Folini, D., Schwarz, M., et al., 2017. The Global Energy Balance Archive (GEBA) version 2017: a database for worldwide measured surface energy fluxes. Earth Syst. Sci. Data 9, 601–603.

World Water Assessment Programme (WWAP), 2017. The United Nations World Water Development Report 2017. In Wastewater: the untapped resource. Paris, France: UNESCO

Chapter 44

Managed aquifer recharge with various water sources for irrigation and domestic use: a perspective of the Israeli experience

Daniel Kurtzman[1] and Joseph Guttman[2]

[1]Institute of Soil, Water and Environmental Sciences, Agricultural Research Organization, The Volcani Center, Rishon LeZion, Israel,
[2]Mekorot, Israel National Water Company Ltd., Tel Aviv, Israel

44.1 Introduction

44.1.1 Why Israel has a significant managed aquifer recharge experience?

Recent estimations of global managed aquifer recharge (MAR) as percentage of worldwide groundwater withdraws are in the area of 1%, while in Israel, managed recharge accounts for more than 10% of the extractions from aquifers (Dillon et al., 2019). The percentage of managed recharge in Israel acceded 7% of the total groundwater production since the late 1960s. Nevertheless, the objectives of MAR, the water sources, and the major method of injection have changed significantly in the last 50 years (Sellinger and Aberbach 1973; Schwartz and Bear 2016). Whereas induced riverbank infiltration is a MAR method used since the late 19th century (Sprenger et al., 2017), which makes MAR significant in many surface water–rich countries (e.g., Hungary, Germany, the Netherlands, and Italy), the Israeli high-percentage MAR case is based on continuous development of new water resources that can be stored and/or treated mainly in the Israeli Coastal Aquifer.

Water is consumed in Israel according to the Water Law from 1959. This law's theme is that water (including wastewater) is a national resource, owned by the people and held in trust by the government for the benefit of the people (Laster and Livney, 2009). Together with this centralized attitude for managing water, another pillar of the Israeli water policy was to allocate large volumes of water to develop and sustain irrigated agriculture on its semiarid and arid lands (Nativ, 2004). To carry out the distribution of the water from the state to the people beside a water authority (water commission before 2007), the Israeli government has/had two strong water organizations: (1) the national water company—Mekorot to produce (treat) and conduct most of the water from water sources to consumers and (2) Tahal—Water Planning for Israel, planning, performing applicative research, pilot projects, etc., aimed at how to sustain and increase the water available for the people in the next decades (Tahal was privatized in 1995). In the 1960s–70s Tahal had employed about 1000 people, including water, construction, mechanical, electrical, and chemical engineers as well geologists, many good hydrologists (~70), and economists (Nevo and Shalev 2008). This mass of planners came out with many MAR plans, especially after efforts in constructing surface reservoirs failed due to high percolation rates, in the 1950s (Nevo and Shalev, 2008). Those MAR plans that were implemented make up most of the Israeli MAR experience (in volumes of recharged water), to date. As nation-scale investments in planning–building and operating large water systems are/were relatively high—the nature of most of the MAR systems in Israel is large operations [measured in million m^3 (MCM), many wells rather than few, large infiltration sites, conduction of water tens and hundreds kilometers, etc.]. Smaller scale MAR systems are developing recently mostly within the frame of water-sensitive urban planning (Page et al., 2018) and will not be discussed in this chapter.

44.1.2 The Israeli Coastal Aquifer

Coincidently, the Zionist movement of Jewish people settling in Israel (1880−present) started close to the time when the rotary drilling rigs were first used. The development of well drilling capabilities made it possible to populate this semiarid−arid land with scarce surface-water sources, rapidly in the 1st century of Zionism. The existence of a freshwater, unconfined aquifer in which one can drill almost anywhere to 10−60 m below surface and produce water, turned the Israeli Mediterranean coast the place where most of the agricultural and urban development occurred between 1880 and 1980. Cultivated land over this aquifer probably peeked in the 1980s−90s, also as a result of a major MAR project that moved allot of the agricultural production south to the Negev region (the Shafdan system, which will be described in Section 44.4). For the urban development, this is true until present (2020).

Nevertheless, the fast development resulted in overpumping, hydrological craters, and seawater intrusions peaked in the coastal aquifer in the 1950s (water table at 10 m below sea level at Tel Aviv in 1957), making MAR projects attractive.

Hydrologically, this coastal aquifer is a relatively homogenous and horizontally isotropic medium, with little preferential flow paths. The storage coefficient is high (mainly the specific yield of the aquifer), and the hydraulic conductivity is not extremely high in the range of 3−30 m/day, making it a good reservoir for seasonally and in some cases multiyear storage.

Another circumstance of historical and geomorphological that contributed too many MAR systems based on infiltration ponds in this aquifer is that 15% of the soil covering the coastal aquifers are sand dunes (Fig. 44.1). In the 21st century, irrigated agriculture on sands is conceivable due to the possibility to irrigate and fertigate at high frequency with low masses, so the retention capacity and nutrient contents in the soil are not a limitation. Nevertheless, the agricultural development on this land occurred mostly between 1880 and 1980, hence the sand dunes were not considered arable and large areas were left with no use. Tahal and Mekorot, the strong planning and operating water companies, were able to take control of significant areas in the sand dunes and use them for infiltration ponds for MAR systems (Fig. 44.1).

44.2 Managed aquifer recharge of ephemeral stream floods in the coastal aquifer through infiltration basins, increasing freshwater supply (1959−present)

The commitment of the state of Israel to sustain\enhance freshwater availability for the use of its people leads to the idea of diverting ephemeral streams' flood flow from their natural flow (and lose) to the Mediterranean, to keep them fresh by storage in aquifers. The existence of sand dunes a few kilometers from the coastline overlaying the coastal aquifer (Fig. 44.1) rose the possibility of diverting flood flows from basins outside the aquifer, toward the sand dunes for effective infiltration, recharge, storage in the aquifer, and recovery by wells surrounding the infiltration basins. Two MAR facilities were planned by Tahal and are operated by Mekorot in light of this idea: (1) The Shikma stream MAR facility that catches floods from a loess soil semiarid basin (average precipitation 350 mm/year), southeast of the aquifer (operating since 1959), and (2) the Menashe streams MAR facility catching flash flood of the Taninim ephemeral stream and its main tributaries located northeast of the aquifer (operating since 1967). Plans for catching flows of ephemeral streams flowing to the Mediterranean Sea near the central sand dunes, the Lachish and Sorek streams, were made by Tahal but were never implemented (Nevo and Shalev 2008).

Here we will elaborate on Menashe streams MAR facility in the northern part of the aquifer (Fig. 44.1), which is the more elegant and successful floodwater MAR operation in Israel.

What makes the Menashe streams MAR project successful for more than 50 years of operation is the strong and simple hydrogeological rational that stands behind it. Namely, collect water from floods in ephemeral streams that drain basins sitting on a relatively impermeable rock where the runoff coefficient is high (10%−30% of precipitation), and there is no exploitable aquifer beneath them, and conduct them to highly percolating sands (runoff coefficient <0.5%) overlaying a good aquifer (Fig. 44.2).

Looking on annual volumes of floodwater that entered the system (settling and infiltration basins) from 53 years of operation (1967−2019), we get on average 9.9 MCM/year with a standard deviation around 8 MCM/year and median of 7.8 MCM/year. On average about 15% of the precipitation over the drainage basin gets stored in the sandy aquifer at the southeast and recovered by wells mostly in the summer (Fig. 44.2). Maximum recharge to the system was 31.8 MCM/year while 0 managed recharge of floods occurred in 3 drought years out of the 53 years record. From the histogram of annual volumes of MAR we see that more than 1/3 of the years end up with volumes smaller than 4 MCM/year and 2/3 with volumes smaller than 13 MCM/year (Fig. 44.3).

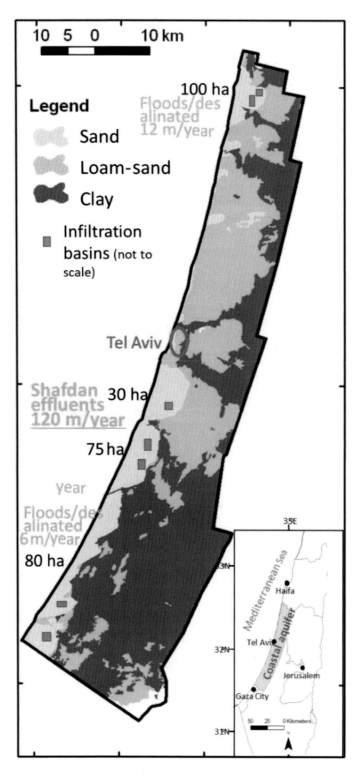

FIGURE 44.1 Soil type above the Israeli Mediterranean coastal aquifer and location of infiltration basins used for MAR facilities. Note the location of the infiltration ponds on the sands at distance of 3–7 km from the coastline enabling high filtration rates and large volumes of aquifer for storage. At the northern and southern facilities (Menashe and Shikma, respectively), MAR of ephemeral streams winter floods and surplus of desalinated seawater is performed only in the winter (settling and infiltration basins are included). At the Shafdan MAR facility in the center, effluents are injected for soil aquifer treatment and storage year round, and on average (2006–18), 120 m columns of secondary effluents percolate through the basins' surfaces each year. *MAR*, Managed aquifer recharge.

The diversion of the floodwater is done in an open channel of 16 km (gravitational flow with no energy invested). Because the diversion and infiltration occur only in the winter and basins are drained and dry up in summer, maintenance expenses are low [infiltration rates recover during the summer (Sellinger and Aberbach, 1973)]. Water quality of the diverted water is monitored, especially in concern for herbicides and pesticides due to significant agricultural areas in the drainage basin. Some traces of simazine (herbicide) are found in the diversion-channel water (below drinking

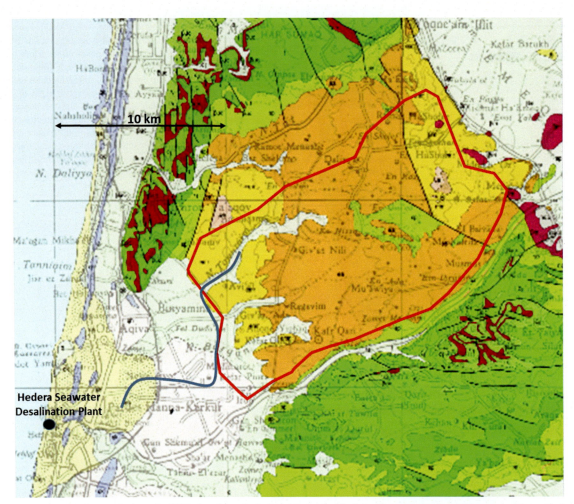

FIGURE 44.2 Geological map (Sneh et al., 1998) describing the hydrological rational of the Menashe streams MAR system. The red polygon contains (schematically) the basin from which the water drains to the ephemeral streams during heavy winter rains. The blue line represents the 16 km diversion channel that takes the flow toward the sandy area overlaying the northern part of the Israeli Coastal Aquifer (Fig. 44.1). The orange and beige colors within the polygon represent Eocene-age chalk with low-matrix hydraulic conductivity, and insignificant vertical-karst features that make deep percolation, small and surface runoff high. The yellow dotted area, where the southwestern end of the conversion channel is sand dunes of Holocene age with high infiltration rates. Note the Hadera desalination plant (operating since 2010) from which excess water that cannot be distributed directly is also injected into the aquifer through the same infiltration basins of the Menashe streams MAR system (described in Section 44.5). *MAR*, Managed aquifer recharge.

water standards) but none in the water abstracted from wells of the MAR system (e.g., Harel and Kostelitz, 2007). Potable water in Israel for household usage costs more than 2$ US/m^3 (January 2020), hence these, cheaply conducted and produced, 10 MCM/year of potable water (otherwise wasted to the sea) obtained through MAR of Menashe streams floods, is thought off as a successful MAR system.

44.3 Managed aquifer recharge of groundwater and especially lake water through wells for freshwater supply (1965–90 and reexamination 2012–20)

In the late 1950s the southern part of the national Israeli water system started to operate, mainly conveying water from the Turonian age carbonate aquifer (known also as the Western Mountain Aquifer) to the more arid areas in the Negev desert in the south of Israel (Fig. 44.4). Nevertheless, this system also enabled MAR from the carbonate aquifer to parts of the coastal aquifer that suffered from overpumping and seawater intrusion (Sellinger and Aberbach, 1973). However, the connection of the Sea of Galilee (Known also as Lake Kinneret) to this system to form the National Water Carrier (NWC) in 1964 boosted the operation of MAR through wells to be the major MAR activity in Israel in the period 1965–90 (Schwartz and Bear, 2016).

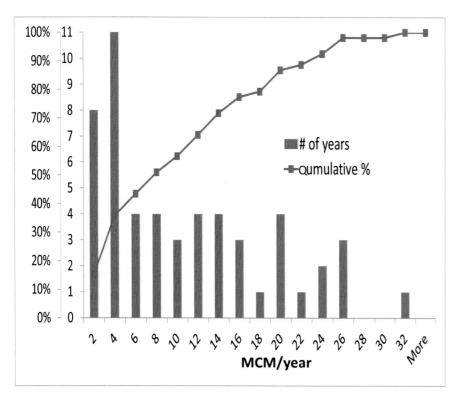

FIGURE 44.3 Histogram of annual volumes of floodwater that were diverted from the Menashe streams toward the basins in the sand dunes for MAR in 53 years 1967–2019 (bins of 2 MCM/year, i.e., 0–2 MCM—8 years, 2–4 MCM—11 years, etc.) Note that in more than half of the years the volume is between 4 and 26 MCM. More statistics are in the text. *MAR*, Managed aquifer recharge.

Since 1965 injection of Sea of Galilee water into the calcareous sandstone coastal aquifer and the fractured, carbonate mountain aquifer in the central part of Israel was preformed (Fig. 44.4). Due to clogging problems in the coastal aquifer, the injection of direct Sea of Galilee water had decreased, and the injection of Sea of Galilee water continued in the fractured carbonate mountain aquifer only. The clogging and water-quality issues (discussed herein), and an extreme drought period that reduced sharply the available water in the Sea of Galilee, brought to the end of this operation at the beginning of the 1990s.

44.3.1 Technical considerations concerning managed aquifer recharge through wells

The major advantage in using wells for MAR is the ability to locate them in places where the subsurface geology is not suitable for infiltration ponds (e.g., clay layers close to ground surface), in small and densely populated areas, recharge confined aquifers, and in areas close to the water supply networks. Yet, the main disadvantage is the relatively low injection rate per well, and the need for many recharge wells in the event of total injection volumes is high.

In general, the capacity of injection well to absorb the quantities of water injected into the well depends on several factors:

- The capability of the formation to absorb the injected water. This factor is composed of two subfactors:
  - the lithology (unconsolidated or consolidated rocks) and their hydrological properties (water level, permeability, effective porosity, storativity) and
  - the well structure (casing, screen, gravel pack) and the length of the open section against the aquifer layers.
- The conveying capacity of the surface water network connecting to the well.
- The resistance of the well equipment installed inside the well (pump in dual-propose wells).
- Water type and its quality: when the injection water is of higher quality low turbidity and organic matter, then the injection capacity is better.
- Clogging effect: the resistance to inflow resulting from clogging of the screen section. This includes the development of physical clogging, trapped air bubbles, and biofouling.

The wells that were used between the 1960s and 1980s were a mixture of dual-purpose wells and recharge wells. Dual-purpose well means a pumping well equipped with a pump that is used also for injection. In these wells, during injection, the flow direction reverses and the water flows downward through the well column and the pump situated at the lower end of the tool column. The resistance to the backflow of water through the impellers inside the pump restricts the injection rate. Increasing the injection rate can be done by withdrawal of the pump prior to the injection. In such recharge-dedicated wells (no pump), injection is done through a conductor pipe that its bottom should be below the static water level.

A pilot research project focused on the relationship between flow rate, depth of injection point, and creation of air bubbles and their movement toward the screen was performed by Mekorot (Guttman et al., 2017). It showed that the bottom of the injection pipe should be at least a few meters below the static water level and not too close to the top of the screen. The pilot also shows that during the injection, different sizes of air bubbles are created. The larger air bubbles tend to move mainly upward, but the smaller air bubbles move downward to the screen with the injected water as dissolved air and may decrease the injection rate.

44.3.2 Some history and experience from the managed aquifer recharge through well period 1965−90

In the second half of the 1960s the new water source of the Sea of Galilee was used as a source for rehabilitation of the overpumped coastal aquifer especially in the Tel Aviv area. An example from one of many Tahal's reports considering this operation summarizes: between October 1966 and May 1967, 18.8 MCM water from the NWC were injected to 29 wells in Tel Aviv and surrounding area; 16 of the 29 wells are on the eastern border of the saline seawater intrusion advancing from the west, where the purpose of the injection is blocking this advance; and 13 wells were in the hydrologic depression created during the 1950s−1960s at eastern Tel Aviv, aiming at refilling the depression (Rabovsky, 1967).

Monitoring carried out between the 1960s and the 1980s shows that the injection rate declines generally with the cumulative injected volume and with the duration of injection. A good indicator to follow the changes in the well/aquifer performance is the specific capacity measured during pumping (pumping−discharge/drawdown) or the specific absorption capacity measured during injection (injection−discharge/water-level rise). It was found that in the sandy coastal aquifer the specific absorption capacity decreased in wells having high specific capacity (and high absorbance capacity) by about 20% from the initial value. In wells having low-to-medium specific capacity (and low-to-medium absorbance capacity), the decline was large up to 60% from the initial specific absorbance. In the limestone−dolomite mountain aquifer, the reduction was small up to 15% from the initial value (Guttman, 1994).

In dual-purpose wells, backwash (repumping) after the end of the injection period was necessary for redevelopment and restoration of the specific capacity of the well. The advantage of using a dual-purpose well is that the repumping and redeveloping can start immediately after the injection season ended.

Between the 1960s and 1980s the source of injection water from the Sea of Galilee contained high concentrations of suspended solids and organic matter (since 2007 this water is filtered and quality in these parameters is much better). Therefore, at that time, after switching from recharge to pumping, the water pumped during the first few hours contained excessive bacteria growth. In most of the wells, pumping to the drainage of few hours brought the quality of the pumped water to a level allowing to supply them. On the other hand, there were few wells in both aquifers (sandy coastal aquifer and limestone−dolomite aquifer) that the amount of water that was pumped out during the redevelopment was greater than the amount that was injected into the wells, turning the MAR operation nonprofitable.

Throughout that period, the solution to reduce the decrese in the specific-absorbance of the injection wells in the sandy coastal aquifer, was to inject the water from the Sea of Galilee into injection wells in the carbonate mountain aquifer, whereas water from this limestone−dolomite aquifer, which are of much higher quality, were used for injection in the sandy coastal aquifer (Fig. 44.4).

44.3.3 New thoughts and experiments on managed aquifer recharge through wells due to availability of water of better quality today (2012−20)

Between 2005 and 2015, five large seawater desalination plants along the Mediterranean coast were established (Fig. 44.4, total desalinated seawater production reached 600 MCM/year in 2019). An understanding that the water system will have surplus high-quality desalinated water at different periods leads to the decision to renew the capability of water injection through wells in the coastal aquifer. In 2012 Mekorot started a program to reevaluate the feasibility and profitability of recharge wells as part of a general MAR policy to handle surplus desalinated water. The program included drilling of three full-scale pilot sites. In two sites the pilot involved production, recharge, and two observation wells in the same well yard, and the third site contains a dual-purpose well and one observation well.

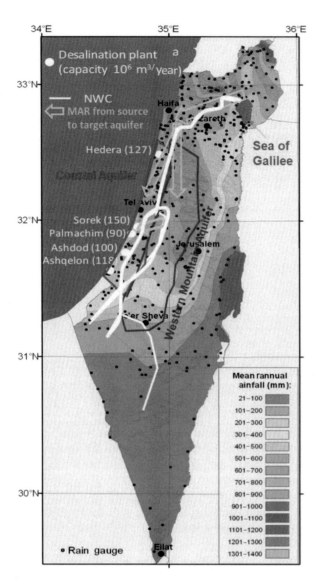

FIGURE 44.4 Managed aquifer recharge through wells. Water from the Sea of Galilee to the Western Mountain Aquifer (*purple polygon*) and the Coastal Aquifer (*red polygon*) and from the mountain to the coastal aquifer. The white line is schematically the backbone of the NWC that enabled this operation. Its northern part, between the Sea of Galilee and the split to east of Tel Aviv, was connected in 1964 enabling MAR through wells associated with the NWC. This winter managed recharge operation enabled moving water from the less depleted water resources to the more depleted aquifers close to the consumer's center. Seawater desalination plants started to operate in 2005 and became a major potable water resource flowing in the NWC. This brought to reconsidering MAR through wells in recent years. MAR of small fraction of the desalinated seawater through infiltration ponds will be discussed in Section 44.5. *MAR*, Managed aquifer recharge; National Water Carrier (NWC). *Modified from Stanhill, G., Kurtzman, D., Rosa, R., 2015. Estimating desalination requirements in semiarid climates: a Mediterranean case study. Desalination 355, 118–123 (Stanhill et al., 2015).*

Guttman et al. (2017) show the differences in the hydraulic parameters between the pumping test and the injection test in the recharge well despite both being conducted in the same well. They also show that the rise of the water level during the injection test is greater than the drawdown during the pumping test, and the specific absorbance during the injection test was only one-third of the specific capacity in the pumping test.

In 2018 a third pilot site for wells MAR was established about 15 km south to the previous sites on the same sandy aquifer. In this pilot, Mekorot constructed a new dual-purpose well using Muni-Pak screen. Muni-Pak screen is a dual stainless-steel screen and in-between glass beads as an inner gravel pack. This type of screen allowing more aggressive redevelopment as required after the injection had finished. Outside the screen, glass beads were installed as an outer gravel pack. The glass-bead gravel pack improves the well efficiency and gives better hydraulic conductivity by offering excellent roundness than conventional quartz (silica) gravel. Pumping and injection tests were conducted in this dual-purpose well. There were differences in the specific capacity rate between the pumping test and the specific absorbance in the injection test (Fig. 44.5). Nevertheless, in comparison to the pumping tests carried out in the first pilot (Guttman et al., 2017), where the specific absorption during the injection test was smaller by 66% from the specific capacity in the pumping test, the difference between the capacity during injection versus pumping in the new dual-purpose well was only 19% (Fig. 44.5). If we ignore possible differences in aquifer properties in the two pilot sites, then the better results in the last pilot (dual-purpose Muni-Pak screen) is a result of the new well design.

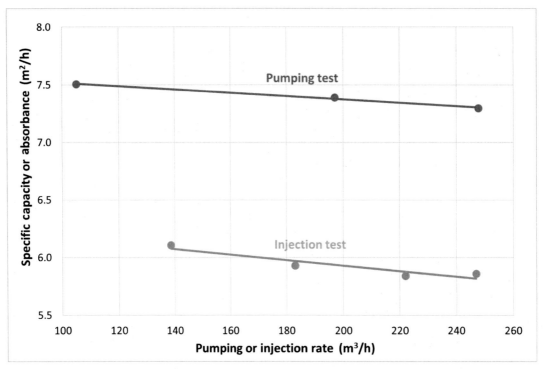

FIGURE 44.5 The specific capacity and specific absorbance during pumping and injection tests, respectively, in the Muni-Pak screen dual-purpose well.

There are still some debates about whether dual-purposes well or single injection well is the best solution to inject the surplus desalinated seawater. As said before, in dual-purpose well the injection rate is limited to the flow through the impellers inside the pump while in a single injection well, this constrain does not exist, and therefore the injection rate in a single injection well is greater. On the other hand, dual-purpose well is more effective than single-purpose injection wells because in dual-purpose wells, redevelopment and backwashing from the accumulated solid and organic contaminants are relatively easy since the pump already existed in the well. In a single-purpose well, it is necessary to install a pump after the injection season for the redevelopment, this is a very costly operation.

44.4 Managed aquifer recharge of secondary effluents in infiltration basins—the Shafdan water reclamation system for irrigation (1987–present)

The idea that wastewater is part of the state's water potential for supply to agricultural use was thought already in the 1950s, and hence wastewater is included as water in the Water Law from 1959. To date, more than 80% of the domestic wastewater in Israel is reused in agriculture, which is about 40% of all water used for irrigation, nationally. About one-third of the reclaimed wastewater ($\sim$125 MCM/year) is treated to the level of water that can be used for irrigation with no limitation (i.e., as if they are freshwater) through the Shafdan soil aquifer treatment (SAT) MAR system. This is, by far, the largest and most intensive MAR operation in Israel today, operating since 1987 (Fig. 44.1).

Concern regarding decrease in subsoil permeability under infiltration ponds due to soil structure deterioration drew the attention of Tahal planers because the relatively high salinity and sodium content in both water from the Sea of Galilee and wastewater effluents (Berend and Kary, 1966). Investigation of this issue in the 1960s revealed that the uppermost layer in the sandy infiltration ponds consists of sand rich in carbonates. Therefore the percolating water is significantly enriched with bivalent cations, thus improving the sodium adsorption ratio (SAR) and reducing the tendency to leach out cementing agents from the soil (Berend and Kary, 1966). Today, 33 years after the MAR operation of the Shafdan started, it can be said that, generally, if the flooding and drainage routine is kept, and topsoil is plowed from time to time, no substantial reduction in infiltration rates was recorded in the infiltration ponds, that a depth of more than 100 m of secondary effluents percolated through their surface each year for more than three decades.

A few biogeochemical aspects of the Shafdan SAT are discussed here. In 2018 the secondary effluents that infiltrated through the Shafdan basins contained roughly: 10 mg/L dissolved organic carbon (DOC), 0.5 mg/L phosphorus

(P), and 10 mg/L total nitrogen (N, including NH_4^+, NO_3^-, NO_2^-, and organic-N). Water pumped from Shafdan reclamation wells located close to the infiltration ponds (no dilution with natural groundwater) contain DOC <1 mg/L, P <0.1 mg/L, and total N ~5 mg/L practically all as NO_3-N (Aharoni et al., 2019). Calcium and magnesium concentrations increased by 50%–60% from secondary effluent to SAT product while alkalinity increases by about 40%, Na and Cl concentrations are generally unchanged in the SAT.

Oxidation of DOC and the nonoxidized N species to NO_3^- followed by denitrification of about half of the total N are the main processes. Mienis and Arye (2018) acknowledge also that keeping the total N input at 10 mg/L is essential for avoiding reduced forms of N in the aquifer. P concentrations are reduced mainly due to adsorption and high distribution coefficients ($20 < K_d < 55$ L/kg, Lin and Banin 2006). $CaCO_3$ dissolution of the calcareous sandstone–sand porous medium is responsible for the increase in Ca^{2+} and alkalinity (Goren et al., 2011). Increase in Mg^{2+} during SAT is probably a relatively new phenomenon (Goren et al., 2011 report a decrease until 2004) due to the large source of desalinated seawater that feeds the wastewater treatment plant especially after 2010. The desalinated seawater has very high Ca/Mg ratio due to remineralization with Ca only (Ronen-Eliraz et al., 2017). Hence, cation exchange in the Shafdan SAT system probably changed to Mg^{2+} desorption, causing the increase in Mg^{2+} we see in the reclaimed SAT water today. Kloppmann et al. (2018) show by using multiisotope tracing the increase of desalinated water in the mixture of the effluent of Shafdan SAT system as well as in the reclamation wells' water.

Shafdan's SAT operates with no disinfection prior to infiltration (or after pumping). Nevertheless, no positive tests for fecal coliforms were recorded in any of the SAT reclamation wells for 20 years, and practically all enterovirus tests were negative. The SAT provides efficient reagent-less physical pathogen removal, superior to chlorination, and without its adverse health effects (Elkayam et al., 2015). A very comprehensive survey of indicator bacteria, pathogen viruses, coliphages, microbial source tracking (MST) indicators, and antibiotic resistance genes (ARG) revealed complete elimination of all microorganisms and viruses. ARGs were found in levels similar to those in wells outside the Shafdan area (Elkayam et al., 2018).

44.5 Managed aquifer recharge of surplus desalinated seawater through infiltration basins (2014–present)

In recent years, there are short times when the drinking water national network cannot distribute part/all the desalinated seawater that is produced, and for operational/economic reasons the desalination plants are not partly/fully shut down (0%–3% of total production). In these instances, both the infiltration ponds at Shikma and Menashe MAR systems (Fig. 44.1) are used for infiltration of this surplus desalinated seawater (a few MCM per year).

Differently than most MAR water that is injected to the subsurface through infiltration basins (e.g., treated wastewater, floodwater), the desalinated seawater is of very high quality and surface clogging is not a significant issue (Ganot et al., 2017). Another feature regarding this MAR operation is the high uncertainty of the availability of the water, because unlike most MAR operations, it is a solution for some operational/economic disorder. Hence from the point of view of water availability, this MAR operation is relatively at high risk (Rodríguez-Escales et al., 2018).

Aquifer storage and recovery is the main purpose of this MAR operation; nevertheless, mineralization of the infiltrating water is gained through the process, as well (Ronen-Eliraz et al., 2017; Ganot et al., 2018a). The lack of Mg^{2+} in the reverse-osmosis desalinated water is a public health concern and a headache for some farmers that irrigate with this water (Rosen et al., 2018; Avni et al., 2013). The desalinated water is enriched with ~35 mg/L Ca^{2+} by fast limestone dissolution with acids, as a posttreatment done in the seawater desalination plants. Ganot et al. (2018a) show (by observations and simulations) that the infiltration process can enrich the water with Ca^{2+} to the same level without any need of acids and bases, and this process is sustainable for centuries due to the high $CaCO_3$ content of the unsaturated zone and aquifer sediments. These sediments enrich the desalinated water by about 3 mg/L Mg^{2+} in the fast infiltration process to the shallow groundwater under the infiltration pond. This process is less sustainable, as the exchangeable Mg^{2+} in the unsaturated zone is exhausting in a decadal timescale at typical desalinated-seawater injection rates (Ganot et al., 2018a). Further enrichment of this water with minerals in the long-term flow from below the infiltration ponds to the abstraction wells (years–decade) is much less certain and under investigation nowadays.

A very good tracer for following the spread and mixing of reverse-osmosis desalinated water in the aquifer is the stable isotopes of water (^2H and ^{18}O). The desalinated recharged water maintains the high isotopic signature of the eastern Mediterranean, hence has a high contrast comparing with naturally freshwater in the aquifer that went through an evaporation process. Therefore a calibrate flow and conservative transport model of these isotopes in the aquifer are a

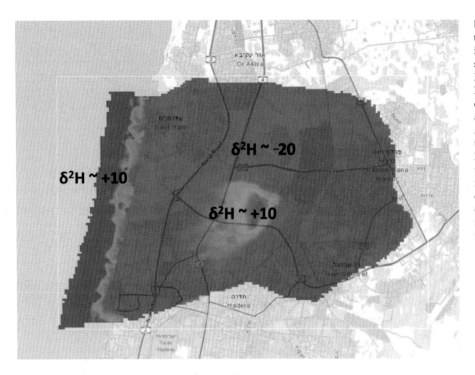

FIGURE 44.6 A snapshot from a transport model of δ^2H (water ^2H-isotope concentration in the well-known standardized ‰ units) in part of the Israeli Coastal Aquifer flowing into the Mediterranean. Note the reverse-osmosis desalinated seawater is injected to the aquifer about 4 km from the coastline. Although desalinated, these water have the same isotopic signature as eastern Mediterranean seawater ($\sim +10$‰), which are in contrast to the naturally freshwater in the aquifer (~ -20‰), enabling tracking and simulating development of the desalinated water plume in the aquifer (see Ganot et al., 2018b for details).

good tool for simulating the expected spread and mix of the desalinated seawater with the naturally freshwater (Fig. 44.6 here and Ganot et al., 2018b).

References

Aharoni, A., Negev, I., Cohen, E., Bar, O., Sherrer, D., Bar-Noy, N., et al., 2019. Monitoring the injection and reclamation of the Shafdan effluents: summary of 2018. In: Mekorot and Ecolog Eng. Report (in Hebrew).

Avni, N., Eben-Chaime, M., Oron, G., 2013. Optimizing desalinated seawater blending with other sources to meet magnesium requirements for potable and irrigation waters. Water Res. 47, 2164–2176.

Berend, J.E., Kary, S., 1966. Changes in soil structure affecting subsoil permeability of spreading grounds. In: Tahal Consulting Engineers, Tel Aviv, Israel. Project # 640 Report.

Dillon, P., Stuyfzand, P., Grischek, T., Lluria, M., Pyne, R.D.G., Jain, R.C., et al., 2019. Sixty years of global progress in managed aquifer recharge. Hydrogeol. J. 27, 1–30.

Elkayam, R., Aharoni, A., Vaizel-Ohayon, D., Sued, O., Katz, Y., Negev, I., et al., 2018. Viral and microbial pathogens, indicator microorganisms, microbial source tracking indicators, and antibiotic resistance genes in a confined managed effluent recharge system. J. Environ. Eng. 144 (3).

Elkayam, R., Michail, M., Mienis, O., Kraitzer, T., Tal, N., Lev, O., 2015. Soil aquifer treatment as disinfection unit. J. Environ. Eng. 141, 12.

Ganot, Y., Holtzman, R., Weisbrod, N., Bernstein, A., Siebner, H., Katz, Y., et al., 2018a. Geochemical processes during managed aquifer recharge with desalinated seawater. Water Resour. Res. 54, 978–994.

Ganot, Y., Holtzman, R., Weisbrod, N., Nitzan, I., Katz, Y., Kurtzman, D., 2017. Monitoring and modeling infiltration–recharge dynamics of managed aquifer recharge with desalinated seawater. Hydrol. Earth Syst. Sci. 21, 4479–4493.

Ganot, Y., Holtzman, R., Weisbrod, N., Bernstein, A., Siebner, H., Katz, Y., et al., 2018b. Managed aquifer recharge with reverse-osmosis desalinated seawater: modeling the spreading in groundwater using stable water isotopes. Hydrol. Earth Syst. Sci. 22, 6323–6333.

Goren, O., Gavrieli, I., Burg, A., Lazar, B., 2011. Cation exchange and $CaCO_3$ dissolution during artificial recharge of effluent to a calcareous sandstone aquifer. J. Hydrol. 400, 165–175.

Guttman, J., 1994. Artificial recharge to Israel's carbonate aquifer. In: Second International Symposium on Artificial Recharge of Groundwater. Orlando, USA, pp. 751–760.

Guttman, J., Negev, I., Rubin, G., 2017. Design and testing of recharge wells in a coastal aquifer: summary of field scale pilot tests. Water 9, 53.

Harel, Y., Kostelitz, S., 2007. Menashe Streams Recharge System: Water Quality Report 2006–2007, Reported to the Israel Water Authority (in Hebrew).

Kloppmann, W., Negev, I., Guttman, J., Goren, O., Gavrieli, I., Guerrot, C., et al., 2018. Massive arrival of desalinated seawater in a regional urban water cycle: a multi-isotope study (B, S, O, H). Sci. Total Environ. 619-620, 272–280.

Laster, R., Livney, D., 2009. Israel: the evolution of water law and policy. In: Dellapenna, J.W., Gupta, J. (Eds.), The Evolution of the Law and Politics of Water. Springer, Dordrecht.

Lin, C., Banin, A., 2006. Phosphorous retardation and breakthrough into well water in a soil-aquifer treatment (SAT) system used for large-scale wastewater reclamation. Water Res. 40, 1507–1518.

Mienis, O., Arye, G., 2018. Long-term nitrogen behavior under treated wastewater infiltration basins in a soil-aquifer treatment (SAT) system. Water Res. 134, 192–199.

Nativ, R., 2004. Can the desert bloom? Lessons learned from the Israeli case. Ground Water 42, 651–657.

Nevo, N., Shalev, Z., 2008. Tahal, the first 50 years. 378 p. Shenaar Communication, Tel Aviv (in Hebrew).

Page, D., Bekele, E., Vanderzalm, J., Sidhu, J., 2018. Managed aquifer recharge (MAR) in sustainable urban water management. Water 10, 239.

Rabovsky, Y., 1967. Recharge of water from the National Water Carrier (NWC) to the coastal aquifer in the Dan region (Tel Aviv metropolitan), a summary of the 1966/1967 winter. In: Tahal Consulting Engineers, Tel Aviv, Israel. Project # 566 Report (in Hebrew).

Rodríguez-Escales, P., Canelles, A., Sanchez-Vila, X., Folch, A., Kurtzman, D., Rossetto, R., et al., 2018. A risk assessment methodology to evaluate the risk failure of managed aquifer recharge in the Mediterranean Basin. Hydrol. Earth Syst. Sci. 22, 3213–3227.

Ronen-Eliraz, G., Russak, A., Nitzan, I., Guttman, J., Kurtzman, D., 2017. Investigating geochemical aspects of managed aquifer recharge by column experiments with alternating desalinated water and groundwater. Sci. Total Environ. 574, 1174–1181.

Rosen, V., Gal, V., Garber, O., Chen, Y., 2018. Magnesium deficiency in tap water in Israel: the desalination era. Desalination 426, 88–96.

Schwartz, J., Bear J.,. 2016. Artificial recharge of groundwater in Israel. Part of the international project: 60 years history of MAR lead by Dillon, P., and Stuyfzand, P., International Association of Hydrogeologists IAH. <https://recharge.iah.org/60-years-history-mar>.

Sellinger, A., Aberbach, S.H., 1973. Artificial recharge of coastal-plain aquifer in Israel. IAHS Publ. 111, 701–714.

Sneh, A., Bartov, Y., Weissbrod, T., Rosensaft, M., 1998. Geological Map of Israel, 1:200,000. Isr. Geol. Surv. <https://www.gov.il/he/departments/general/israel-map-1-200k> (December 2019).

Sprenger, C., Hartog, N., Hernández, M., Vilanova, M., Grützmacher, G., Scheibler, F., et al., 2017. Inventory of managed aquifer recharge sites in Europe: historical development, current situation and perspectives. Hydrogeol. J. 25, 1909–1922.

Stanhill, G., Kurtzman, D., Rosa, R., 2015. Estimating desalination requirements in semiarid climates: a Mediterranean case study. Desalination 355, 118–123.

Chapter 45

MAR model: a blessing adaptation for hard-to-reach livelihood in thirsty Barind Tract, Bangladesh

Chowdhury Sarwar Jahan[1], Md. Ferozur Rahaman[2], Quamrul Hasan Mazumder[1] and Md. Iquebal Hossain[3]

[1]*Department of Geology & Mining, University of Rajshahi, Rajshahi, Bangladesh,* [2]*Institute of Environmental Science, University of Rajshahi, Rajshahi, Bangladesh,* [3]*Barind Multi-Purpose Development Authority, Rajshahi, Bangladesh*

45.1 Introduction

Scarcity of freshwater is now a global concern where sustainable management of it is a major challenge. The groundwater studies present a compilation of compelling insights to groundwater scenarios affecting groundwater-stressed regions globally. It includes quantity, exploration, quality and pollution, economics, management and policies, groundwater and society, and sustainable sources and efficient solutions (Mukherjee et al., 2020). This most important natural resource on the Earth is under stress due to increasing population growth, intensification of agriculture, urbanization, and industrialization along with possible impacts of climate change. Bangladesh, a country of rivers, heavily depends on groundwater for domestic, irrigation, and commercial uses. High dependency on groundwater has happened as a result of limited surface water in rivers, canals (*Kharies*), swamps (*Beels*), etc. during dry season when demand is highest. The country ranks sixth of the global list in most groundwater extracting countries where about 97% people in urban and rural areas use for drinking and 75% for irrigation coverage (compared to 43% of global irrigation coverage). The public health and economic benefits of using groundwater are remarkable but issues of overexploitation and poor management make the vital resource increasingly vulnerable every year.

Historical data suggest that Bangladesh is a most vulnerable country to disaster globally, with negative consequences associated with various natural (e.g., flood, drought, tropical cyclone) and human-induced hazards [DMB (Disaster Management Bureau), 2010]. Environmental disasters are events of huge magnitude and negative impact on society and environment. The National Water Policy [NWPo (National Water Policy), 1999] declares drought as a major challenge to face water-scarcity issues in agro-based North-West Bangladesh known as "Barind Tract"—elevated landmass of Pleistocene age (Brammer, 1996). It has caused loss of food production leaving livelihood at risk and expected to be aggravated in future. Monitoring data identify the Tract as hotspot where continuous depletion of the storage due to agricultural intensifications is evident and warrants management actions to ensure sustainable drinking water supply to meet up the increasing demands.

Barind Tract covers Godagari, Tanore, Nachole, Gomastapur, Porsha, Sapahar, and Niamatpur *Upazilas* of Rajshahi, Chapai Nawabganj, and Naogaon districts (district and *Upazila* are the second and third lowest tiers of administrative units in Bangladesh) (Fig. 45.1) covering an area of 2586 km^2 and 1,417,928 inhabitants [BBS (Bangladesh Bureau of Statistics), 2017]. The Tract—a landmass of triangular wedges—has characteristics of dissected and undulatory nature, and edged by parallel rivers of seasonal and minor seasonal nature, such as the mighty Ganga (Padma) in the south; Mahananda and Purnabhaba in the west; and Atrai, Sib-Barnai, etc. in the east. The elevation of the tract is comparatively higher (47 m amsl) than the adjoining floodplains (11 m amsl). Topographically, the landform has undulating nature having slopes toward south-east and south-west, and runoff water after rainy season flows toward rivers and swamps (*Beels*) through *Kharies* resulting scarcity of surface water. As rainy season ends, rivers flow low, and streams and canals become almost dry resulting scarcity of surface water.

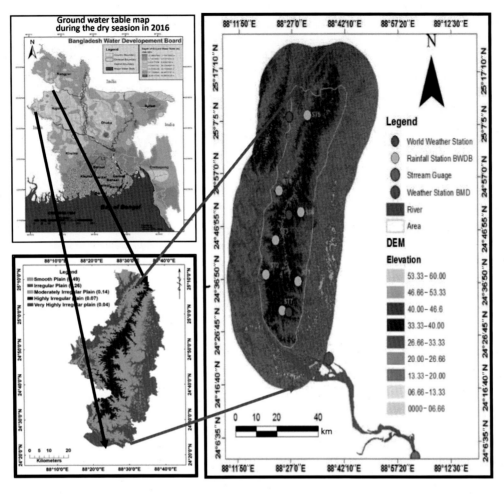

FIGURE 45.1 Groundwater hot spots (red and yellow color) of Bangladesh due to overexploitations of groundwater. *GWC of Bangladesh Water Development Board (2016) along with geomorphologic and elevation maps, and locations of meteorological and weathers stations in Barind Tract.*

Generally, the area considered as semiarid lies in the monsoon and summer dominant hemisphere. The climate is usually warm and humid. Based on rainfall, humidity, temperature, and wind pressure the weather condition is classified as premonsoon—dry summer; monsoon—rainy days; and winter—cold and dry. The rainfall is comparatively less with an average being about 1410 mm lower than the national average of 2550 mm that mainly (84%) occurs during the monsoon. This area has been designated as drought prone, with its annual average temperature ranges from 25° to 35°C in the hottest season and 9° to 15°C in the coolest season. So the tract experiences extremes that are clearly in contrast to the climatic condition of rest of the country (Banglapedia, 2014). The Pleistocene clay deposits (Barind Clay) comprising surface lithology is hydrogeologically semiimpervious, and is composed of clay-silt aquitard {low infiltration rate: 1—2 mm/day [UNDP (United Nations Development Programme), 1982] having thickness 3.0—47.5 m} of semiimpervious nature. The aquifer system is single to multiple layered (two-four) of Plio-Pleistocene age (thickness: 5.0—42.5 m) (Jahan et al., 2007) having suitability mostly for drinking and household purposes and scope for irrigation uses is limited (Jahan, 1997).

For achieving country's food security as declared by the Govt. of Bangladesh (GoB) with the policy of maintaining ecological balance in the Tract, the Barind Integrated Area Development Project (BIADP), later the Barind Multipurpose Development Authority (BMDA) under the Ministry of Agriculture (MoA), GoB came forward to play a central role in focusing water resources management to gear-up strong agricultural activity aiming at achieving country's food security and ensuring drinking water availability since the late 1980s. Groundwater based irrigation plays a vital role to cultivate high-yield variety of *Boro* rice during dry summer and supplementary irrigation for rain-fed *Aman* rice during rainy season. The multicropping agricultural practice has boosted cropping intensity from 117% in pre-BIADP time to more than 220% (national average: 185%). The Tract has potential for groundwater development through 8728 high capacity deep tube wells (DTWs) (depth > 80 m bgl) [BMDA (Barind Multipurpose Development Authority), 2001], but irrigation demand is so high that Barind farmers are withdrawing water through nearly 24,000 DTWs operated by BMDA and private owners. As a result, natural rainwater recharge in rainy season could not balance the overexploitation of groundwater resource for extended irrigated land areas, and as a consequence the groundwater table (GWT) has been depleting at drastic rate since the beginning of 2000's undermining resource sustainability of the area (Jahan et al., 2010; Rahman et al., 2016a, 2016b, 2017).

In Bangladesh and also in Barind Tract the management policy for water resource has an outlook of sectoral and top-down approach that results resource unsustainably paying more cost economically, socially, and ecologically. But the supply of safe and enough water for drinking and agriculture should not be subsectoral service, but must be an integrated management strategy. In this connection, GoB has adopted the Bangladesh Water Act [BWA (Bangladesh Water Act), 2013]—an umbrella act of legal provision of the NWPo (National Water Policy) (1999), which accordingly may declare any area of the country as water stress to protect water resource or aquifer for a specific period with a provision of priority allocation or abstraction for water use in water stress area. The GoB in 2018 adopted new ordinance named *"Krishi Kaje Bhugorvostho Pani Bebosthapona Ain, 2017"* (in *Bengali*) [Krishi Kaje Bhugorvostho Pani Bebosthapona Ain (Groundwater Management Law in Agriculture) (2017)] and framed the Bangladesh Water Rules [BWR (Bangladesh Water Rules), 2018] under BWA (Bangladesh Water Act) (2013) to apply at stakeholder level with the help of integrated water resource management (IWRM) strategy for their drinking, domestic and agricultural needs. But demand of water for drinking and irrigation purposes will be acute in coming days, so the enactment of this strategy is a demand of time.

Several countries in the world are working with IWRM strategy for balancing demand and supply side chain management. IWRM is an empirical concept that has built up from the on-the-ground experience of practitioners. Although many parts of the concept have been around for several decades—in fact in the World Summit on Sustainable Development in 1992 in Rio the concept has made the object of extensive discussions as to what it means in practice. The Global Water Partnership's (GWP) definition of IWRM is widely accepted. It states: IWRM is a process which promotes the coordinated development and management of water, land and related resources, in order to maximize the resultant economic and social welfare in an equitable manner without compromising the sustainability of vital ecosystems (UN-Water, GWP, 2007; UN-Water, 2008). In 2006 UN-Water formed a Task Force on IWRM which in May 2008 completed its mandate presenting the "Status Report on IWRM and Water Efficiency Plans" at the 16th session of the Commission on Sustainable Development.

For resolving adverse impact on water sector in Bangladesh, lot of studies and programs have been undertaken by the GoB during the past decades. But unfortunately country often adopts paradigm shifts in the management of their water resources primarily as a result of exogenous pressures (and to a limited extent endogenous factors), but lack of domestic ownership and leadership of the concept; limited resources; and institutional miss-matches often result in an implementation of the ideas limited to form rather than practice. Current global climate change and scarcity of surface water have made the water related problems more critical in country. Under these circumstances, water resources management needs more comprehensive and integrated approaches.

The issue of IWRM has been increasingly highlighted after the 1990s in the country. The NWPo (National Water Policy) (1999), National Water Management Plan [National Water Management Plan (NWMP), 2004], National Water Resources Database, Regional Technical Assistance (ADB-RETA, 2009), Bangladesh Water Act [BWA (Bangladesh Water Act), 2013] are the major initiatives of the water management sectors in Bangladesh. Although lot of constraints exist to develop IWRM plan in the country, yet the existing policy, plan, guideline, law, institution, and information system provide a basis for the IWRM implementation in considering the principles such as efficiency, equity and environmental sustainability. Nevertheless, advancing IWRM is a process of incremental steps with tools such as enabling environment, institutional framework, and management instrument as methodology.

Unfortunately, in Bangladesh no initiative has been taken to implement the IWRM strategy at field level especially in the water-stressed Barind Tract. The strategy for the Tract started implementing with "4R" principles (*Reuse*—harvesting of rainwater; *Recharge*—augmentation of groundwater; *Recycle*—recharging groundwater and use it again in need; and *Restoration*—reclamation of groundwater resource by protecting its depletion) as pilot study. The attempt has been initiated at national level by the Water Resources Planning Organization (WARPO), Ministry of Water Resources (MoWR), GoB with the support of the Swiss Agency for Development and Corporation (SDC) since 2015, and at subnational level by the Consortium of the Swiss Red Cross (SRC) and DASCOH Foundation, an NGO as breakthrough challenge to plan and implement the noble task for the people by providing safe drinking and irrigation water along with other facilities. Here basic "Pillars" adopted for IWRM strategy are enabling environment; institutional roles and framework; and management instruments including practices, technologies, and processes to apply on daily basis.

The present study only considers the scientific outcomes of IWRM strategy, and enabling environment and institutional roles and framework are kept outside purposefully. It has detailed the task on the management instruments of IWRM strategy in the Tract including identification of challenges of drought for sustainable water resource management; preparation of groundwater potential zonation map for effective identification of suitable area for extraction of water; identification of geomorphological, geological, hydrogeological, and engineering aspects to create scientific basis for applying rainwater harvesting (RWH) technique with managed aquifer recharge (MAR) model; and finally the impact assessment of MAR model application that included field visits, and focus group discussion comprising 155 beneficiaries with predesigned questionnaire. In this context, MAR model—an artificial groundwater recharge (AGR) process through pipes, dug well, shaft, etc. has beenadopted for augmentation of groundwater by rainwater and storm/runoff water from rooftop and *Kharies* for all beneficial uses in this water-scarce area which is relevant with the Sustainable Development Goal 6 (SDG 6) via at least five of the SDG 6 subgoals and their corresponding indicators.

45.2 Challenges of groundwater resource management plan

The average rainfall of the Tract is very low (44% less than the national average) with much seasonal nature. At the same time, high seasonal and annual rainfall variability (20–26%) has thrown a great challenge for framing its water management plan. The annual trend of rainfall pattern shows a declining character (2.76 mm/year) [using statistical methods of Mann–Kendall (*MK*) Trend Test and Sen's Slope Estimator—"MAKESENS"] (Ali et al., 2012) has necessitated the demand for groundwater. Consequently, the irrigation system in cropping seasons like winter (*Rabi*) and summer (*Boro* rice), that is, 7–8 months is completely dependent on groundwater. The precipitation concentration index (*PCI*) (Oliver, 1980) (average: 19.5) indicates rainfall distribution of irregular to strongly irregular nature with high seasonal variability. Moreover, the mean seasonality index ($\overline{SI}$) (Kanellopoulou, 2002) (0.94) also revealsmarkedly seasonal rainfall pattern with prolonged dry period. All these parameters indicate an alarming situation not only in summer and winter seasons but also for rain-fed rice agriculture (*Aman*) where groundwater based supplementary irrigation becomes inevitable due to dry spell for couples of rainless days in the rainy season.

Monthly water balance study reveals that utilization of soil moisture starts in October, that is, soil starts to dry and soil moisture deficit begins in December continuing till end of May. At the same time, estimated time series of potential evapo-transpiration (P_{ET}) shows increasing trend that is consistent with the rising trend of annual temperature. Analytical result for the period 1980–2017 shows that total number of dry episodes have higher occurrence in the decade of 2001–10, and number of drought events of mild-to-moderate nature along with SPI-3 (agricultural), SPI-6 (meteorological) and SPI-12 (hydrological) (Lloyd-Hughes and Saunders, 2002) have been increasing drastically in winter and premonsoon seasons with steady increasing trend in rainy season. Frequency and risk ranking of drought effect show the highest value belonging to moderate-to-high condition (class B), and must be addressed by adaptation measures (Ramamasy and Baas, 2007). During the period 2010–17, the depth of GWT runs below the suction level (7 m) both in summer and rainy seasons, and people do not get even drinking water. Fluctuation of GWT ranges from 2 to 17 m during the period of 1980–2017, but it is ofhigher rate in recent time. The annual depth of GWT has shown depletion at a rate of 0.23 m/year. Rainfall and GWT for the period 1980–2017 show general declining trends with rapid declining trend of later especially after 2004 (Fig. 45.2).

Moreover, the changing values of SPI-6 are compatible with the changing values of SPI-12 and ultimately SPI-3. However after 2004 due to huge groundwater withdrawal for irrigation, the depth of GWT has declined rapidly even with the positive SPI values during monsoon. Relationship between annual mean depth of GWT and that of SPI values shows that the depth of GWT is continuously increasing having positive correlation, and even during steady or reverse

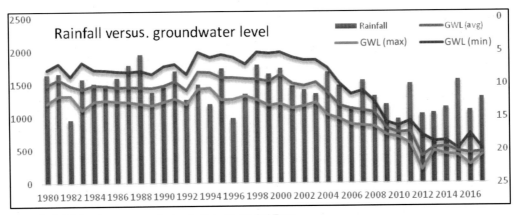

FIGURE 45.2 Plot of rainfall (mm) versus groundwater depth (m) in Barind Tract.

trend of SPI values its depth is continuously increasing (Fig. 45.3). Moreover, the area under groundwater irrigation by different pumping moods like low-capacity tube wells (depth <80 m bgl), that is, STWs; high-capacity tube wells (depth > 80 m bgl), that is, DTWs; and power pumps in Chapai Nawabganj and Rajshahi districts along with production of major crop such as rice (*Aman*, *Aus*, and *Boro*) and wheat for last couple of decades have steadily increased. So it is revealed that mild and moderate drought effects due to the characteristic rainfall pattern in the Tract along with extension of irrigated agriculture play an important role for huge withdrawal of groundwater in recent years aggravating hydrological drought situation in the Tract. All these factors together have put stress and threat for sustainability of water resource in the Tract that may be intensified in coming days.

45.3 Groundwater resource potentiality

Globally groundwater is considered as vulnerable resources, which is squeezing day by day. The delineation of groundwater potential zones is an urgent need for sustainable development of livelihood. In this context, the geographical information system (GIS) and remote sensing (RS) techniques have emerged as effective tools to demarcate potential zones of groundwater. The RS technique plays controlling role to gather information source to find out the relationship among geomorphology, geology, lineament, slope, land use, and land cover (LULC), etc. All these information can be entered as input in GIS environment and then overlaid with other information. For delineation of groundwater potentiality (GP) of an area, GIS and RS techniques have been used by researchers where thematic layers such as geomorphology, drainage, drainage density, lithology, lineament, geology, LULC, etc. are used (Muralitharan and Palanivel, 2015; Kumar et al., 2016; Rahaman et al., 2017). Here satellite technique gives inclusive information about ground surface monitoring that will contribute baseline information of groundwater resource potentiality (Tweed et al., 2007).

Geomorphologically an area has been categorized as smooth plain > smooth to irregular plain > moderately irregular > irregular and highly irregular plain. The major part of the Tract falls into highly irregular to irregular category, and naturally gets comparatively less time for rainfall infiltration. So it creates a hurdle to demarcate hydromorphological categorization, and hence the Tract is less potential for groundwater occurrence (Dey, 2014). Drainage map has been prepared with the help of topographic maps (scale: 1:50,000) and shuttle radar topography mission—digital elevation model (SRTM—DEM) where major part has drainage density values of 0.58–1.17 km/km^2 indicating less permeable area of infiltration of rainwater (Jaiswal et al., 2003). The drainage pattern is of subdendritic character with tree-branching nature of low homogeneity and uniformity, where major portion of runoff water after rainfall losses due to poor infiltration capacity of top surface soil (Barind Clay). It also indicates less potentiality for groundwater occurrence in the area. Based on availability of rainfall amount and rate (Rose and Krishnan, 2009) in long-run, map is prepared and categorized as 1401–1450, >1451–1500, >1501–1550, >1551–1600, and >1601–1650 mm. The less amount rainfall with poor infiltration capacity has resulted low potentiality of groundwater occurrence in the Tract, so for contemplating the GP, less weighted value of rainfall should be taken into consideration than that of the higher value (Fashae et al., 2014).

Based on the dominance of lithological unit, the top surface soil in the Tract is mostly covered by Barind clay, followed by alluvial sand, marshy clay, etc. including water bodies. On the basis of hydraulic property such as conductivity (from higher to lower values), lithological units have categorized as alluvial sand > water marsh clay > marshy clay > Barind clay. Accordingly, the calculated values of hydraulic properties such as hydraulic conductivity (K), transmissivity (T), and specific yield (S) [Jahan, 1997; IWM (Institute of Water Modelling), 2006] are 10–20 m/day, 229–3100 m^2/day, and 0.06–0.30, respectively. Lineament is the most important linear features showing subsurface elements or structural weakness such as faults and is usually extracted by visual analysis of enhanced Landsat ETM + panchromatic band (band 8) image data. It acts as passage way of rainfall for recharging the groundwater storage. The lineament density in the Tract varies up to 0.017 km/km^2 and has undeniable hydrogeological importance as favorable channel in aquifer recharge.

The transformation of rainfall as runoff water flow in the slope of the Tract is strictly linked with the process of infiltration potential for groundwater recharge. According to Berhanu et al. (2013), the prepared slope map from SRTM—DEM has categorized as 0%– <3%—flat, 3%– <8%—gentle, 8%– <15%—moderate, and 15%– <30%—steep. The Tract has poor potentiality for groundwater occurrence because its major part has flat-gentle slope (value < 8%) with poor infiltration capacity of rainwater through top soil cover (Barind Clay). For monitoring LULC pattern and its spatial change, RS approach has been used involving the usage of satellite images made in light of compatible spatial resolution (30 m) in Landsat—December 8, 2016 and has coregistered to subpixel accuracy in ERDAS IMAGINE (9.1) for removing geometric incongruity. The Enhanced Thematic Mapper Plus—ETM + of 7 (near

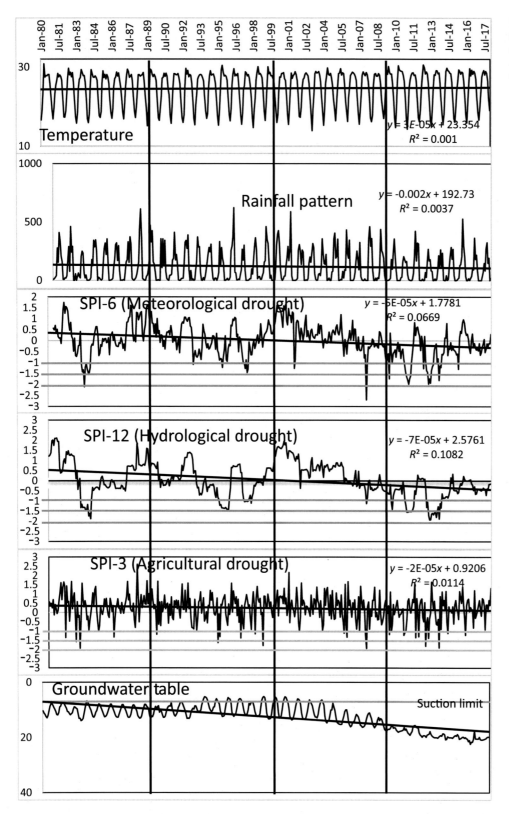

FIGURE 45.3 Comparison of annual average changes in depth of GWT and SPI-3, SPI-6, and SPI-12 in Barind Tract. *GWT*, Groundwater table.

infrared), 4 (red), and 2 (green) bands—has used for analyzing LULC classes as cultivated land (75%), water body (5%), settlement (19%) and sand bar (1%).

To demarcate the potential zones for groundwater, thematic layers such as geomorphology, drainage density, rainfall, lithology, lineament density, slope, and LULC have integrated by the weighted overlay method of ArcGIS software. For multicriteria evaluation by analytical hierarchy process (Saaty, 1980; Arivalagan et al., 2015), variables such as classes, weights, ranks having different impacts for the thematic layers have been used. The values of "consistency ratio (CR)" for the thematic layers vary from 0.01 to 0.09. Accordingly, the major influencing factors such as lithology, lineament density, geomorphology, and slope have shared 35, 21, 19, and 12%, respectively, along with minor factors such as drainage density, rainfall, and LULC of respective percentage as 6, 5, and 3. Based on thematic layers, GP zonation map has constructed and the Tract has divided into classes such as high (2%), moderate (4%), poor (77%), and very poor (17%) representing respective land area of 18, 102, 2016, and 450 km^2 (Fig. 45.4).

Sensitivity analysis for GP zonation has carried out by map-removal and single-parameter techniques. In map-removal technique, its index has varied widely due to possible removal of lineament density from the computation [mean variation index (MVI)—1.99%]. It may mainly contribute relatively to higher weight that has assigned to the lithology layer [empirical mean weight (EMW)—24%]. But it is moderately sensitive to rainfall (MVI—2.00) but less sensitive to LULC, geomorphology, drainage density, slope, and lithology with respective MVI of 2.16, 1.21, 1.28, 1.78, and 2.31. On the other hand, single-parameter technique has deviation in effective weights in comparison to the empirical weights. Here lithology is most effective parameter [mean effective weight (MEW)—27.47%)] and is followed by slope, lineament density, geomorphology, drainage density, rainfall, and LULC with MEW value of 24.49, 22.00, 12.87, 8.09, 3.37, and 1.44%, respectively, with respective EMW values of 11, 24, 15, 7, 4, and 2%. On the other hand, MEW and EMW values of drainage density, rainfall, and LULC are close to each other. Therefore it is revealed that lithology, lineament, and slope are the most influencing factors for GP in the Tract. Here GP has close relation with lineament density, but surface lithology (Barind Blay) is an important barrier for its potentiality.

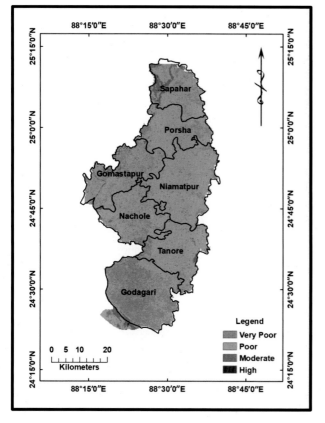

FIGURE 45.4 Groundwater potentiality zonation map of the Barind Tract.

45.4 Potential zones for groundwater recharge and selection of sites for artificial recharge of groundwater

Demarcation of potential zones for groundwater recharge (PZGR) and selecting sites of AGR have performed on the basis of small-scale topographic maps with an approach of RS and GIS. Here morphometric parameters have played an important role and have needed to be integrated with the information such as hydrology and hydrogeology. Thereafter, selection for appropriate sites for AGR in augmentation and conservation of groundwater has made for implementation of IWRM strategy in the Tract.

The map of the PZGR has been prepared using thematic layers such as lithology, geomorphology, drainage density, slope, and aquifer transmissivity. The geomorphological data have taken out from the United States Geological Survey (USGS) website and that of slope values have obtained from the SRTM—DEM. The drainage network map has prepared from the topographic and DEM, and its density has computed by spatial analyst tool. The inverse distance weight interpolation technique has used to analyze the temporal distribution of rainfall using point rainfall data as obtained from six rain- gauge stations (RGSs) of the Bangladesh Water Development Board (BWDB) in the Tract and has rectified and geo-referenced using the Universal Transverse Mercator (UTM)—a plane coordinate grid system for map projection, and the world geodetic system (WGS84) for global positioning system considering its reference coordinate system.

Aquifer transmissivity (T) values in the Tract calculated using the methods of Jacob (1950) and Hantush (1956) range from 229 to 3100 m^2/day (Jahan, 1997). Based on T values (Pitman, 1981) the area has divided into groundwater development potential zones such as Zone-I (<500 m^2/day): smaller portion in the central zone with poor GP having suitability only for drinking water supply only; Zone-II (500–1000 m^2/day): occupies mostly rest part of central part with well suitability for drinking purposes, but limited scope for irrigation uses; and Zone-III (>1000 m^2/day): occupies in smaller part in the rest part of the tract having good potentiality for groundwater development and water supply.

Based on hydrogeologic significance in AGR, weights of thematic layers, namely, geomorphology, lithology, drainage density, slope, and aquifer transmissivity, have calculated with respective values of 30%, 22%, 15%, 23%, and 10%, respectively. For these thematic layers, comparison of pair-wise matrix has done, and calculations of normalized weights by analytic hierarchy process (Saaty, 1980) and "CR" of assigned weights have performed accordingly. All normalized weights of individual distinctive polygons as integrated layer were used to calculate the groundwater recharge index (GWRI) as follows:

$$\text{GWRI} = (L_w L_{wi} + \text{GM}_w \text{GM}_{wi} + \text{DD}_w \text{DD}_{wi} + \text{SL}_w \text{SL}_{wi} + \text{AT}_w \text{AT}_{wi})$$

where L is the lithology, GM is the geomorphology, DD is the drainage density, SL is the slope, AT is the Aquifer transmissivity, subscript w is the normalized weight of a theme, and subscript wi is the normalized weight of the individual elements of a theme.

To demarcate AGR zones, individual thematic layer is integrated using Arc Map 10.5 software. Based on GWRI values, the integrated map is classified into zones as high (>0.5); moderate (0.25–0.5); and unsuitable (<0.25). The resultant AGR zonation map (Fig. 45.5A) has revealed that about 34% area belongs to category of unsuitable for percolation process (e.g., percolation tank), but has high potentiality for injected recharge process (e.g., MAR model); 65% area falls under moderate category for percolation process but has high potentiality for injected recharge process; and rest 1% area is suitable for percolation process (e.g., percolation tank) where injected process of recharge has little importance. Moreover, the AGR zonation has good accord with other information such as geography, geology, and landscape characteristics of the Tract. The suitability map for locating AGR sites was generated by superimposing AGR zonation, drainage (second- and third-order streams), and lineament with a 100-m buffer maps have helped to detect favorable sites (crossing of second- and third-order streams and lineament) (Fig. 45.5B). AGR structures such as percolation tank, and forced injection through pipes, dug well and shaft (e.g., MAR model) (Asano, 1985) operated for recharge processes are advocated in site suitability map. The precise location and techno-economic feasibility of MAR model have been detailed through field investigations.

45.5 Implementation of managed aquifer recharge model

45.5.1 Piloting of managed aquifer recharge model at household level—pioneer attempt during 2013–16

The MAR model through RWH technique can play an important role to reduce vulnerability to climate and hydrological variability; to control overabstraction; to recharge aquifers protecting declining yields; and to restore the

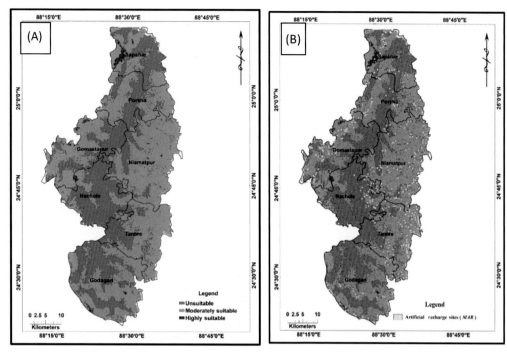

FIGURE 45.5 (A) Potential suitability map for artificial recharge zone and (B) suitability map for locating artificial groundwater recharge sites in Barind Tract.

groundwater balance. It may also be applied to sustain or improve the functioning of ecosystems and the quality of groundwater. The implementation of optimal AGR method requires knowledge of physical, chemical and biological phenomena involving water filtration in the aquifer, together with the engineering aspects related to plant design and maintenance operations. The MAR model has implemented in November 2013 at household level through pipe recharge technique in villages Mallickpur in Fatehpur *Union*, Khoribari in Nizampur *Union*, and Ganoir in Nachole *Union* of Nachole *Upazila* of Chapai Nawabganj district with the financial support from NGO Forum for Public Health, Bangladesh—an NGO as pioneer and promising adaptation measure to mitigate drinking water scarcity for livelihood in this drought-prone tract.

Geological and hydrogeological aspects: The vertical electrical sounding survey using resistivity meter (Model: SSR-MP-AT-S; and data analysis tool: software IGIS 2.0) has been carried out to delineate hydrogeological sections up to 166 m depth in Mallickpur, Ganoir, and Khoribari villages. The sections reveal that the Barind Clay (Zone-I) has thickness range from 16 to 22 m and underlain by the main water-bearing layer or aquifer (Zone-II) of fine and medium sand lithology (thickness: 8−27 m) having potentiality for groundwater development. A very thick (8−17 m) silty-clay aquitard (Zone-III) underlies Zone-II. Below Zone-III, a water-bearing layer or aquifer layer (Zone-IV) of fine to medium sand lithology (thickness: 13−41 m) exists up to 91−95 m depth.

Engineering aspect: RGSs have set up in each site of MAR model to measure daily rainfall amount. At household level, rainwater has collected from the roof of corrugated iron at the doorstep and poured directly via pipes into aquifer through recharge box (size: 1.5 m × 1.5 m) and filled up with 6, 10, and 20 mm size brick cheeps. Before injecting rainwater, it has made free from silt and debris present. The total catchment (roof of corrugated iron) is 200 m^2 with five recharge points in each village. The depth of the recharge box is 3 m from ground level lie in the top surface Barind clay layer. Depth of GWL is measured in the permanent hydrograph stations (PHSs) (well dia.: 15.25 cm) of 28 and 34 m length placed in the aquifer zone in Mallickpur and Ganoir villages, respectively. Two filters are fitted at depths 20 and 22.4 m in PHSs with respective lengths of 1 and 6 m. Schematic diagram of the MAR model at household level is shown in Fig. 45.6A.

Monitoring and impact assessment: The operation of MAR model and its impact to augment groundwater resource has assessed by monitoring the depth of GWT during the period of Jan 2014−Dec 2016, and reveals that:

- During the months of Feb-May in every year since the last two to three decades especially in summer season, the GWT had a declining trend but retrieved its original level in the following monsoon period up to 2004 due to sufficient amount of groundwater recharge in respect to withdrawal, but afterward it did not returned due to inadequate recharge amount in comparison to withdrawal. In the last week of April 2014 the amount of rainfall was 41 mm, so in the first and second week of May the declining trend of GWT has not observed.

- In 2014 (Jan–Oct) the total amount of rainfall was 687 mm in Ganoir village, and the depth of GWT was 33.82 m (Jan 2014), but the depths were 34.13 (end of April) and 26.0 m (September), that is, GWT has declined by 1.31 m and has reversed up to 7.82 m, respectively, in comparison to January 2014. On the other hand, the total amount of rainfall was 677 mm (during Jan–Oct) in Mallickpur village. In January 2014 the depth of GWT was 11.11 m, but the depths of GWT were 13.30 (end of April) and 6.10 m (September), that is, GWT has declined by 2.19 m and has reversed up 5.01 m, respectively, in comparison to January 2014. So after the implementation of MAR model, scenario of GWT has started to reverse due to injected or forced rainfall recharge to groundwater artificially. The GWT hydrographs in before and after MAR model application period during 2014–16 are shown in Fig. 45.6B.

Water quality in before and after MAR implementation: For suitability study of groundwater for drinking purposes due to MAR model application, collected water samples from nearby hand tube wells at before and after model implementation stages have analyzed for physical parameters such as pH, EC, and temperature; chemical elements such as Fe^{Total}, SO_4^{2-}, PO_4^-, NO_3^-, and As; and bacteriological content (*Fecal Coliform*) in the NGO Forum Laboratory, Dhaka. The result of analysis has disclosed that at both the stages groundwater is safe for drinking purposes according to ECR (Environment Conservation Rules) (1997), BDWS (Bangladesh Drinking Water Standard) (2004), and WHO (World Health Organization) (2008); and free from arsenic contamination and hazard from *Fecal Coliform* bacteria contamination.

45.5.2 Managed aquifer recharge model as integrated water resource management strategy in Barind Tract since 2015

RWH potentiality: For MAR model application in the Tract the preliminary task to find out the interrelation of rainfall trend, depth of GWL, and subsurface geology in Kakonhat village in Godagari *Upazila* as has played an important role for its potentiality of RHW (Fig. 45.7). Hydrogeological parameters such as thickness of Barind clay (aquitard) overlying the main aquifer, thickness, and depth of aquifer have delineated to identify potential zone for RWH. Here, the multilayered aquifer system is covered by thick Barind clay, and underlying by the main aquifer from where inhabitants are

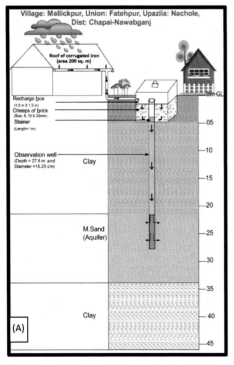
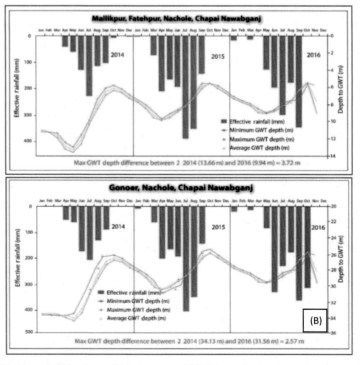

FIGURE 45.6 (A) Schematic diagram of piloted MAR model at household level and (B) hydrograph of GWT in pre- and postmodel operative stages during (2014–16) in villages Mallickpur and Ganoir of Nachole *Upazila* in Barind Tract. *GWT*, Groundwater table; *MAR*, managed aquifer recharge.

withdrawing water for drinking and irrigation use. In 1980 the average depth of GWT was 9.44 m, which had declined at depth of 21.19 m in 2017, that is, the area has potentiality of vertically vacuum space of 11.75 m that can be filled up by harvested amount of rainwater during rainy season using MAR model and its storage is enough for sustainable supply of water in subsequent days annually. Moreover, the rainfall pattern shows a decreasing trend and is consistent with the declining trend of GWT.

Implementation of the MAR model at mass scale started since 2016, where the number of household in different *Upazilas* varies from 19,770 to 72,186 with total daily/capita water consumption of 62.47 L per HH. The average consumption of water (daily/person) for drinking, cooking, bathing, domestic washing, and toileting in rural areas are 3.53, 6.71, 27.26, 12.18, and 12.75 L, respectively (Milton et al., 2006). The farmers experience, research field experience and water footprint study for total agriculture water demand for major crops vary from 264,906,890 to 589,883,365 m^3, 211,236,841 to 473,809,866 m^3, and 178,745,805 to 328,814,110 m^3, respectively (Islam, 2017). During rainy season, the amount of runoff water (calculated by SCS-CN method of USDA SCS) (Rallison, 1980) is 1215 mm. So, nearly 70% of drinking, domestic, and agriculture water demand in the Tract can be fulfilled with the harvested amount of rainwater (Rahaman et al., 2019).

MAR model at household and institution level: The MAR model has implemented through pipe recharge technique in the office building at Mondumala village in Tanore *Upazila* and that of dug well recharge technique in auditorium building at Kakonhat village and at Kadipur village in Godagari *Upazila* in household level in 2016. The design and management of MAR model involve the considerations of engineering aspects as well as geology, hydrology, monitoring of GWL, groundwater quality, and impact assessment. In the rural areas of Bangladesh, earth-made houses and one- or two-storied buildings of *Union /Upazila* office, schools, colleges, etc. have roof area of 40 and 350 m^2, respectively (Biswas and Mandal, 2014). The roof area of auditorium is 692 m^2 (RWH potentiality of 765 m^3). The rainwater is injected into aquifer through dug well via recharge box (size: 1.5 m × 1.5 m) which has filled up with 6, 10, and 20 mm size brick cheeps. The depth of the recharge box is 3 m from ground surface lie in the Barind clay layer. The schematic diagrams of the MAR model through pipes and dug well (diameter: 1 m) recharge technique are shown in Figs. 45.8 and 45.9 respectively.

Presently, 129 MAR models (by pipe recharge and dug well recharge technique) have implemented in Barind Tract by the SRC-DASCOH Foundation. Fig. 45.10 shows the locations of the constructed MAR models in HHs, industries, institutions, and office buildings until 2019.

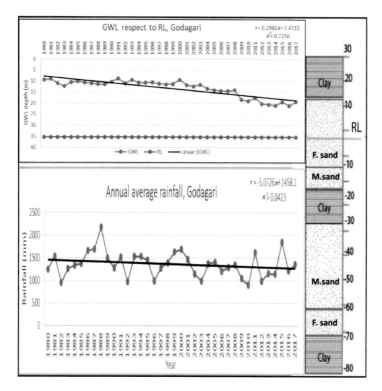

FIGURE 45.7 Relationship between rainfall, groundwater table and lithology from geoelectric resistivity survey in village Kakonhat in Godagari *Upazila* in Barind Tract.

MAR model for runoff water harvesting in canal (Khari): The Institute of Water Modelling, Bangladesh [IWM (Institute of Water Modelling), 2006] has pinpointed potentiality for MAR model application to recharge groundwater by runoff water flowing through the *Kharies*. The model has operated through well [recharge well (RW)] in April 2013; and shaft [recharge shaft (RS)] in March 2012 at Sormongla *Khari* in Godagari *Upazila* and Rasulpur *Khari* in Niamatpur *Upazila*, respectively (Hossain et al., 2019). The RS has been drilled up to depth of 36.5 m, which has filled up with sand. To prevent horizontal load of top Barind clay, cylindrical brick wall (dia.: 1.16 m and height: 0.15 m) was constructed at the *Khari* bed. The constructed filter pit (dia.: 3 m and depth: 3 m) was filled with coarse sand and gravel. In RW, uPVC blind pipe and strainer (slot width: 1.52 mm, and open area: 20%) were used. The circular space (between casing and wall of borehole) was filled by pea gravel (size: 4–10 mm). The high erosion mechanism of Barind Clay by runoff water decrease the infiltration rate of the recharge unit in *Khari* due to clogging effect in the model.

In dried-up *Khari*, quantity of runoff water in dry season has aided for volumetric calculation of recharge rate (RR) (L/min). In conventional MAR model application through RW the RR has declined from 28.64–48.0 L/min at starting time to 1.5–2.5 L/min after operation of few days and even months as a result of clogging effect due to weathering of Barind clay. At the same time due to this clogging effect, the GWT has started to rise for couple of weeks and even for months at the beginning during the period of 2013–15 and then has started to fall gradually and tends toward zero. During submergence of *Khari* by runoff water in rainy season, top of the sandy layer in recharge structures has clogged by thick clay deposits (thickness: 40–80 mm), and it is replaced from time to time by clean sandy materials. At the side of *Khari* near recharge units, PHSs are stationed for monitoring the depth of GWT. The GWT hydrograph for RS in Sormongla *Khari* during 2013–14 has similarly in declining situation of RR, increasing clogging effect, etc. as that for RW.

In this context the engineering design of the model (both for RW and RS) has modified to overcome operational constrains with better performance. Modified engineering design of MAR model through RW and RS has given in Fig. 45.11A and B, and the view of *MAR* model through RW at Sormongla *Khari* has shown in Fig. 45.12. In modified MAR model by RW and RS the clogged clay layer has easily removed even during available runoff water in *Khari* from time to time; and as a result, the GWT has rising trend when runoff water is flowing through the *Khari*. Analytical results of groundwater sample for the parameters such as pH, EC, TH, TDS, Ca^{2+}, Mg^{2+}, Fe^{Total}, Cl^-, As, SO_4^{2+}, NO_3, and PO_4^- during before and after MAR model implementation stages (both for conventional and modified engineering design of RW and RS) have shown the values varies as 7.0–7.6, 129–764 µS/cm, 169–244 mg/L, 133–313 mg/L, 50.51–76.89 mg/L, 8.31–18.93 mg/L, 0.01–0.98 mg/L, 43.50–56.95 mg/L, 0.00 mg/L, 3.50–18.0 mg/L, 1.26–6.8 mg/L, and <0.01–2.0, respectively. The analytical parameters are within the acceptable limit for drinking purpose according to standards of BDWS (Bangladesh Drinking Water Standard) (2004) and WHO (World Health Organization) (2008) (Hossain et al., 2019).

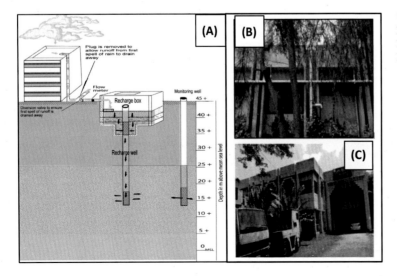

FIGURE 45.8 (A) Schematic diagram of well recharge in MAR model; and (B and C) office building in village Mondumala in Tanore *Upazila* in Barind Tract. *MAR*, Managed aquifer recharge.

MAR model: a blessing adaptation for hard-to-reach livelihood in thirsty Barind Tract, Bangladesh Chapter | 45 621

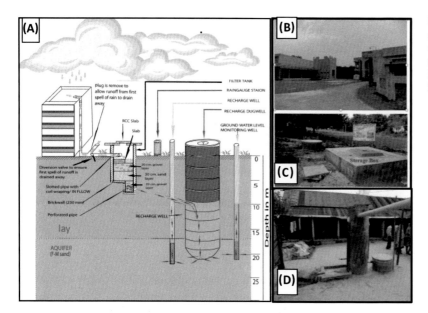

FIGURE 45.9 (A) Schematic diagram of dug well recharge in MAR model; (B) auditorium building in village Kakonhat; (C) dug well and recharge box; (D) dug well recharge at household level in village Kadipur in Godagari *Upazila* in Barind Tract. *MAR*, Managed aquifer recharge.

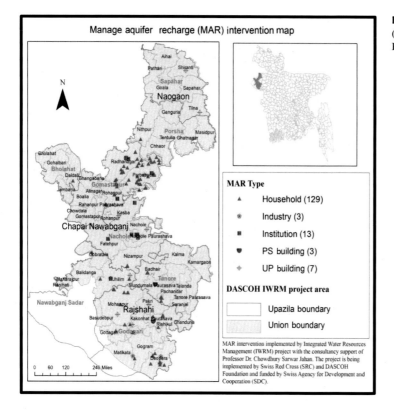

FIGURE 45.10 Locations of implemented MAR model (household, industry, institution, and office buildings) in Barind Tract. *MAR*, Managed aquifer recharge.

45.5.3 Impact assessment of managed aquifer recharge model as integrated water resource management strategy

The impact assessment of MAR model application has included citizens' pursuance, governance issues, and institutionalization of the strategy for better water supply system by the local government institutions (LGIs) such as the *Union* Council and *Upazila* Council (LGIs at grass root level in Bangladesh). The beneficiaries have opinion that after implementation of the MAR model the availability of groundwater scenario has changed, and drinking water has become

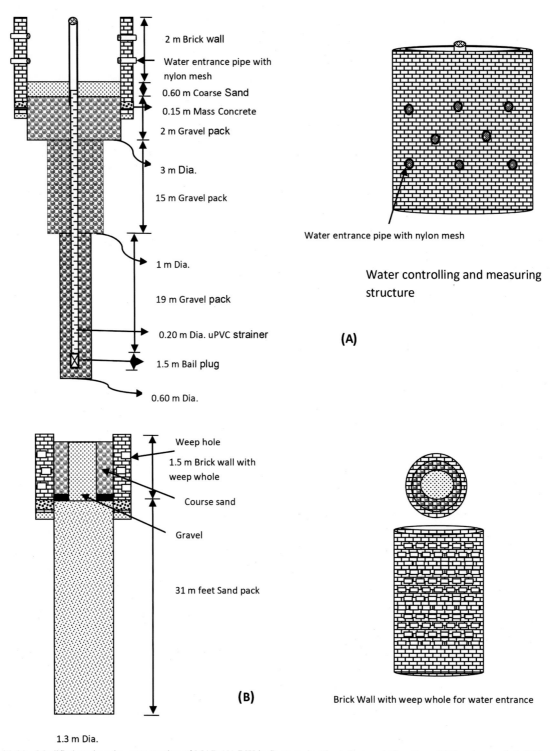

FIGURE 45.11 Modified engineering construction of MAR (A) RW in Sormongla *Khari*, Godagari *Upazila* and (B) recharge shaft (RS) at Rasulpur *Khari*, Niamatpur *Upazila* in Barind Tract. MAR, Managed aquifer recharge; RW, recharge well.

available to inhabitants round the year. They have asserted their collective responsibility in financial involvement for implementing MAR model and its maintenance, and mostly pleased about the good governance issues related to implementation of this model in hard-to-reach areas in rural Bangladesh. Finally, the MAR models as IWRM strategy are anthropologically, economically, and environmentally justifiable through inclusiveness of inhabitants.

FIGURE 45.12 View of modified *MAR model* at Sormongla *Khari* in Barind Tract (A) recharge well with water measuring/controlling structure; (B) water entrance pipe with double layer filter screen; (C) observation well at Khari side; and (D) recharge well in *Khari*.

45.6 Conclusion

The nexus between drought (moderate–high risk) and groundwater resources development issues has provided a sign for better management planning of later in the drought-prone Barind Tract. Here monthly, annual, and seasonal rainfall time series show declining trend with unusual seasonal nature. Long dry spell with few rainy days results drought events, and the Tract suffers from high soil moisture deficit for 7–8 months. So groundwater based irrigation demand has increased drastically over the years resulting groundwater withdrawal at alarming rate. The declining trend of rainfall amount, frequent droughts, extension of irrigated and cultivated area, increasing crop production, etc. are influencing factors for rapid declining trend of GWT. It adds vulnerability to the groundwater resource that creeps toward hardship in management plan—a burning issue for theTract.

Moreover, highly irregular landscape with the gentle slope facilitates flow of major portion of rainwater as runoff into rivers and *Kharies* due to less rainwater infiltration capacity of surface soil (Barind clay). The cultivated land and settlement areas cover 75% and 19%, respectively, and have thrust toward increasing demand of groundwater for drinking and irrigation uses, and ultimately its overexploitation comes out almost round the year. The combination effort of RS process and computer-based tool GIS reveals that moderate-to-high groundwater potential zone exists only in 6% area in the Tract, and rest 94% belongs to poor–very poor category. Accordingly, about 34% area belongs to unsuitable category for AGR through percolation process (e.g., percolation tank) but have high potentiality for injected recharge process (e.g., MAR model), that of 65% area under moderate category through percolation process but have high potentiality for injected recharge process. But bimodal distribution annual rainfall pattern in rainy season (>500 mm) shows high RWH potentiality during rainy season.

Considering all constrains and advantages, the IWRM strategy has been considered as the best practice for sustainable water resources management through "4R" principles. In this regard, RWH technique using MAR model was enacted as pioneer attempt for recharging groundwater artificially through pipes, dug well, and technologically modified RS and RW—a successful and accepted blessing adaptation measure to augment groundwater resource for drinking and agricultural use. The model is anthropologically, economically, and environmentally justifiable to inhabitants. Moreover, they are very much optimistic about sustainability of implemented MAR models in the Tract along with addressed issues of good governance. This positive attitude of beneficiaries is above other implemented water resource management projects in Bangladesh. Moreover, for sustainability of the implemented MAR model, effective maintenance work should be operated on a regular basis. Finally, MAR model of IWRM strategy with governance issues must be addressed for achieving the Vision 2021 of the GoB and visioning the SDGs and targets by 2030 as proclaimed by the General Assembly of the United Nations.

Acknowledgments

The authors wish to express their sincere acknowledgment and thanks for the generous support of the Consortium of SRC-DASCOH Foundation, Rajshahi, Bangladesh; NGO Forum for Public Health, Bangladesh; and the Barind Multipurpose Development Authority (BMDA), Rajshahi, Bangladesh for useful discussions and providing field data. The logistic and ultimately financial supports from the Consortium of SRC-DASCOH Foundation and NGO Forum for Public Health, Bangladesh are also gratefully acknowledged.

References

ADB-RETA, 2009. Regional Technical Assistance Supporting IWRM (Bangladesh). Water Resources Planning Organization, Ministry of Water Resources, GoB.

Ali, M.H., Abustan, I., Rahman, M.A., Haque, A.A.M., 2012. Sustainability of groundwater resources in the North-Eastern Region of Bangladesh. Water Resour. Manage. J. 26, 623–641. Available from: https://doi.org/10.1007/s11269-011-9936-5.

Arivalagan, J., Sleight, V.A., Thorne, M.A., Peck, L.S., Berland, S., Marie, A., et al., 2015. Characterization of the mantle transcriptome and biomineralisation genes in the blunt-gaper clam, Mya Truncata. Mar. Genom. 27, 47–55. Available from: https://doi.org/10.1016/j.margen.2016.01.003.

Asano, T., 1985. Artificial Recharge of Groundwater. Butterworth Publishers, Boston, MA, p. 767.

Banglapedia, 2014. Barind Tract, National Encyclopedia of Bangladesh, National Science. Asiatic Society of Bangladesh, Dhaka.

BBS (Bangladesh Bureau of Statistics), 2017. Bangladesh Statistics 2017. Statistics and Informatics Division (SID), Ministry of Planning, GoB.

BDWS (Bangladesh Drinking Water Standard), 2004. Bangladesh Drinking Water Standard, Dhaka, Bangladesh.

Berhanu, B., Melesse, A.M., Seleshi, Y., 2013. GIS-based hydrological zones and soil geo-database of Ethiopia. Catena 104, 21–31.

Biswas, B.K., Mandal, B.H., 2014. Construction and Evaluation of Rainwater Harvesting System for Domestic use in a Remote and Rural Area of Khulna, Bangladesh. Hindawi Publishing Corporation. Article ID: 751952. <http//dx.doi.org.10.1155/2014/751952>.

BMDA (Barind Multipurpose Development Authority), 2001. Project Proforma (Rebound) for the Barind Integrated Area Development Project, Phase-II, fourth revision Barind Multipurpose Development Authority, Rajshahi.

Brammer, H., 1996. The Geography of the Soils of Bangladesh, first ed. The University Press Limited, Dhaka, 287 p.

BWA (Bangladesh Water Act), 2013. Bangladesh Water Act 2013. Ministry of Law, Justice and Parliamentary Affairs, Legislatives and Parliamentary Affairs Division, Government of the People's Republic of Bangladesh.

BWR (Bangladesh Water Rules), 2018. Bangladesh Water Rules 2018. Ministry of Law, Justice and Parliamentary Affairs, Legislatives and Parliamentary Affairs Division, Government of the People's Republic of Bangladesh.

Dey, S., 2014. Delineation of ground water prospect zones using remote sensing, GIS techniques—a case study of Baghmundi development block of Puruliya district, West Bengal. Int. J. Geol. Ear. Environ. Sci. 4 (2), 62–72. ISSN: 2277-2081 (online). An open access, online international journal. Available from: <http://www.cibtech.org/jgee.htm>.

DMB (Disaster Management Bureau), 2010. Standing Order on Disaster. Ministry of Food and Disaster Management, Disaster Management and Relief Division, Disaster Management Bureau, Dhaka.

ECR (Environment Conservation Rules), 1997. The Environment Conservation Rules, 1997 [Bangla text of the rules was published in the Bangladesh Gazette, Extra-ordinary Issue of 28-8-1997 and amended by Notification SRO 29-Law/2002 of 16 February 2002]. In: Government of the People's Republic of Bangladesh Ministry of Environment and Forest Notification Date, 12 Bhadra 1404/27 August 1997.

Fashae, O.A., Tijani, M.N., Talabi, A.O., Adedeji, O.I., 2014. Delineation of groundwater potential zones in the crystalline basement terrain of SW-Nigeria: an integrated GIS and remote sensing approach. Appl. Water Sci. 4 (1), 19–38.

Hantush, M.S., 1956. Analysis of data from pumping tests in leaky aquifers. Trans. Am. Soc. Civ. Eng. 37, 702–714.

Hossain, M.I., Bari, N., Miah, S.U., Jahan, C.S., Rahaman, M.F., 2019. Performance of MAR model for stormwater management in Barind Tract, Bangladesh. Groundwater Sustain. Dev. 10. <https://doi.org/10.1016/j.gsd.2019.100285>.

Islam, M.R., 2017. Water Governance in Drought-prone Barind Area, North-west Bangladesh: Challenges of Sustainable Development (Unpublished Ph.D. thesis). Institute of Bangladesh Studies, University of Rajshahi.

IWM (Institute of Water Modelling), 2006. Groundwater Model Study for Deep Tubewell Installation Project in Barind Area. Final Report, vol. I (Main Report).

Jacob, C.E., 1950. Flow of ground water. In: Rouse, H. (Ed.), Engineering Hydraulics. John Wiley and Sons, New York, pp. 86–321.

Jahan, C.S., 1997. Origin of Groundwater and Prospects of Its Utilization in Barind Area, Bangladesh (Unpublished Ph.D. thesis). Moscow State Geological Prospecting Academy.

Jahan, C.S., Islam, M.A., Mazumder, Q.H., Asaduzzaman, M., Islam, M.M., Islam, M.O., et al., 2007. Evaluation of depositional environment and aquifer condition in the Barind Area, Bangladesh, using gamma ray well log data. J. Geol. Soc. India 70, 1070–1076.

Jahan, C.S., Mazumder, Q.H., Adham, M.I., Hossain, M.M.A., Haque, A.M., 2010. Study on groundwater recharge potentiality of Barind Tract, Rajshahi District, Bangladesh using GIS and remote sensing technique. J. Geol. Soc. India 75, 432–438.

Jaiswal, R.K., Mukherjee, S., Krishnamurthy, J., Saxena, R., 2003. Role of remote sensing and GIS techniques for generation of groundwater prospect zone towards rural development: an approach. Inter. J. Rem. Sen. 24 (5), 993–1008.

Kanellopoulou, E.A., 2002. Spatial distribution of rainfall seasonality in Greece. J. Weather. 57, 215–219.

Krishi Kaje Bhugorvostho Pani Bebosthapona Ain (Groundwater Management Law in Agriculture), 2017. Law of Groundwater Management in Agriculture. Ministry of Law, Justice and Parliamentary Affairs, Legislatives and Parliamentary Affairs Division, Government of the People's Republic of Bangladesh.

Kumar, P., Herath, S., Avtar, R., Takeuchi, K., 2016. Mapping of groundwater potential zones in Killinochi area, Sri Lanka, using GIS and remote sensing techniques. Sustain. Water Res. Manage. 2, 419–430. Available from: https://doi.org/10.1007/s40899-016-0072-5.

Lloyd-Hughes, B., Saunders, M.A., 2002. A drought climatology for Europe. Int. J. Climatol. 22, 1571–1592.

Milton, A.H., Rahman, H., Smith, W., Shrestha, R., Dear, K., 2006. Water consumption patterns in rural Bangladesh: are we underestimating total arsenic load? J. Water Health 4 (4), 431–436.

Mukherjee, A., Scanlon, B., Aureli, A., Langan, S., Guo, H., McKenzie, A., 2020. Global Groundwater: Source, Scarcity, Sustainability, Security and Solutions, first. ed. Elsevier, ISBN: 9780128181720.

Muralitharan, J., Palanivel, K., 2015. Groundwater targeting using remote sensing, geographical information system and analytical hierarchy process method in hard rock aquifer system, Karur district, Tamil Nadu, India. Earth Sci. Inform. 8, 827–842.

National Water Management Plan (NWMP), 2004. National Water Management Plan of Bangladesh. Prepared by WARPO Under the Supervision of the Ministry of Water Resources, GoB.

NWPo (National Water Policy), 1999. National Water Policy 1999. Water Resources Planning Organization, Ministry of Water Resource, People's Republic of Bangladesh.

Oliver, J.E., 1980. Monthly precipitation distribution: a comparative index. J. Professional Geographer 32, 300–309.

Pitman, G.T.K., 1981. Aquifer and Recharge Evaluation in Bangladesh, Technical Note No. 8 UNDP/UNDTCD Groundwater Survey, BGD/74/009, BWDB, GWC.

Rahaman, M.F., Jahan, C.S., Arefin, R., Mazumder, Q.H., 2017. Morphometric analysis of major watersheds in Barind Tract, Bangladesh: a remote sensing and GIS-based approach for water resource management. J. Hydrol. 5 (6), 86–95.

Rahaman, M.F., Jahan, C.S., Mazumder, Q.H., 2019. Rainwater Harvesting: Practiced Potential for Integrated Water Resource Management in drought-prone Barind Tract. Groundwater for Sustainable Development, Elsevier BV, <https://doi.org/10.1016/j.gsd.2019.100267>.

Rahman, A.T.M.S., Jahan, C.S., Mazumder, Q.H., Kamruzzaman, M., Hosono, T., 2017. Drought analysis and its implication in sustainable water resource management in Barind Area, Bangladesh. J. Geol. Soc. India 89 (1), 47–56. Available from: https://doi.org/10.1007/s12594-017-0557-3.

Rahman, A.T.M.S., Kamruzzaman, M., Jahan, C.S., Mazumder, Q.H., Hossain, A., 2016a. Evaluation of spatio-temporal dynamics of water table in NW Bangladesh—an integrated approach of GIS and statistics. Sustain. Water Resour. Manage 2 (3), 297–312. Available from: https://doi.org/10.1007/s40899-016-0057-4.

Rahman, A.T.M.S., Kamruzzaman, M., Jahan, C.S., Mazumder, Q.H., 2016b. Long-term trend analysis of water table using 'MAKESENS' model and sustainability of groundwater resources in drought prone Barind Area, NW Bangladesh. J. Geol. Soc. India 87, 179–193.

Rallison, R.E., 1980. Origin and evolution of SCS runoff equation. In: Symposium on Watershed Management, July 21–23, 1980, Boise, Idaho, pp. 912–924.

Ramamasy, S., Baas, S., 2007. Climate Variability and Change: Adaptation to Drought in Bangladesh. A Resource Book and Training Guide. Institution for Rural Development, FAO, Rome.

Rose, R.S., Krishnan, N., 2009. Spatial analysis of groundwater potential using remote sensing and GIS in the Kanyakumari and Nambiyar basins, India. J. Ind. Soc. Rem. Sen. 37 (4), 681–692.

Saaty, T.L., 1980. The Analytic Hierarchy Process. McGraw-Hill, New York, p. 278.

Tweed, S.O., Leblanc, M., Webb, J.A., Lubczynski, M.W., 2007. Remote sensing and GIS for mapping groundwater recharge and discharge areas in salinity prone catchments, south eastern Australia. Hydrogeol. J. 15, 75–96.

UNDP (United Nations Development Programme), 1982. Groundwater Survey: The Hydrogeological Conditions of Bangladesh. In: United Nations Development Programme (UNDP), Technical Report DP/UN/BGD-74-009/1. UNDP, New York, 113 p.

UN-Water, 2008. Status Report on Integrated Water Resources Management and Water Efficiency Plans.

UN-Water, GWP, 2007. Road Mapping for Advancing Integrated Water Resources Management (IWRM) Processes.

WHO (World Health Organization), 2008. third ed. Guidelines for Drinking-Water Quality, vol. 1. World Health Organization, Geneva.

Index

Note: Page numbers followed by "*f*" and "*t*" refer to figures and tables, respectively.

A

Accelerated Irrigation Benefits Program (AIBP), 487
ACIAR. *See* Australian Centre for International Agricultural Research (ACIAR)
Adaptive neuro-fuzzy inference system (ANFIS), 546
Adsorption processes, 526
Adsorption reactions, 276–277
Advanced data-driven methods in groundwater resources, 545–546
Advanced oxidation processes (AOPs), 221
AEMs. *See* Anion-exchange membranes (AEMs)
AFFFs. *See* Aqueous film-forming foams (AFFFs)
Afghanistan, 323
   groundwater security, 327
   groundwater sustainability, 326–327
   scarcity of groundwater quality and quantity, 324–326
      quality challenges of groundwater, 324–325
      quantity challenges of groundwater, 325–326
   solutions, 327–328
   topography and hydrogeology of, 323–324
Agência Nacional de Águas (ANA), 409–410, 413–415, 417–418
Agency of Toxic Substances and Disease Registry (ASTDR), 135
Agua Doce Program, 568
AI. *See* Artificial intelligence (AI)
AIBP. *See* Accelerated Irrigation Benefits Program (AIBP)
Alberta, Canada
   data and methods, 74–75
   groundwater level changes, 78
   rainfall and snowmelt water, 76–77
   surface water level changes, 77–78
Alluvial aquifers, 279–281
Alter do Chão, 393
Ammonia, 235
Ammonium in groundwater, 257–258
ANA. *See* Agência Nacional de Águas (ANA)
Andhra Pradesh Farmer Managed Groundwater System project (APFAGMS project), 494
ANFIS. *See* Adaptive neuro-fuzzy inference system (ANFIS)

Anion-exchange membranes (AEMs), 563–564
ANN. *See* Artificial neural network (ANN)
Anthropocene, 47
Anthropogenic activities, GCG by, 235–237
Anthropogenic organic carbon, 286
Antibiotics, 243–244
   in groundwater, 245–248
AOPs. *See* Advanced oxidation processes (AOPs)
APFAGMS project. *See* Andhra Pradesh Farmer Managed Groundwater System project (APFAGMS project)
AQUASTAT database, 4
Aqueous film-forming foams (AFFFs), 218
Aquifer(s), 113
   aquifer framework, GBM River delta, 131
   aquifer-based water quality, 301
   in Australia, 383–384
   in Florida, 304–306
   mapping, 482
   in Nile riparian countries, 88–91
ARF. *See* As removal filters (ARF)
Arid environments, 180–181
Arid regions
   climate change effect on groundwater, 378
   groundwater sustainability in, 373
   hydrological processes and effect on groundwater quality in, 375–376
   model application and challenges in, 377
Arid volcanic mountains, 180
Arsenic (As), 10, 129–130, 187–190, 235, 275, 313, 433
   concentrations, 284–285
   contamination, 134–135, 255–256
   detection and treatment, 286–287
   geogenic arsenic contamination, 280*t*
   in groundwater, 276–279, 287–288
      hydrogeochemical settings for, 279–282
      in United States, 288–290
   mitigation, 135–138
   occurrences and distribution, 428
   removing from groundwater, 526
   research from 2000 to 2020, 283–287
      human activities, 285–286
      nationwide datasets, 283–284
      statistical models, 284
Artesian spring, 29*f*
Artificial aquifer recharge, 166
Artificial drainage, 334
Artificial groundwater recharge, 418

Artificial intelligence (AI), 546
   methods in India, 550–554
Artificial neural network (ANN), 546, 551
As removal filters (ARF), 137–138
As-safe community, 137–138
ASTDR. *See* Agency of Toxic Substances and Disease Registry (ASTDR)
Atal Bhujal Yojana scheme, 494–495
ATCHA research project, 365
Atomistic development, 477
Australia, groundwater in
   aquifers in Australia, 383–384
   Australian ecosystems and, 389–391
   Canning Basin, 386
   Daly Basin, 386
   Great Artesian Basin, 384–385
   groundwater entitlements and extractions, 387–388
   groundwater salinity, 388–389
   groundwater uses, 387
   Murray–Darling Basin, 385
   Otway Basin, 386
   Perth Basin, 385–386
Australian Centre for International Agricultural Research (ACIAR), 451, 459–460
Awareness raising, 465
Azad Kashmir and Gilgit Baltistan, groundwater biological contamination in, 312

B

Back-propagation (BP), 546
BADC. *See* Bangladesh Agricultural Development Corporation (BADC)
BAMWSP. *See* Bangladesh Arsenic Mitigation Water Supply Project (BAMWSP)
Bangladesh, 609
   arsenic and other contamination issues, 433
   groundwater, 435–436
      occurrences in, 425–426
   groundwater quality and concerns, 426–428
      occurrences and distribution of arsenic, 428
      occurrences and distribution of salinity, 428
   groundwater uses and impacts of abstractions, 428–432
      domestic uses in rural and urban areas, 430–431

627

Bangladesh (*Continued*)
  industrial uses, 431–432
  irrigation uses, 431
  impacts of climate change, 432
  meeting increased demands in 2050, 432
  sustainable groundwater management, 433–435
  transboundary issues, 433
  water sector in, 611
Bangladesh Agricultural Development Corporation (BADC), 430
Bangladesh Arsenic Mitigation Water Supply Project (BAMWSP), 135–137
Bangladesh Water Act (BWA), 611
Bangladesh Water Development Board (BWDB), 430, 616
Barind Integrated Area Development Project (BIADP), 610–611
Barind Multipurpose Development Authority (BMDA), 610–611
"Barind Tract", 609
  MAR model as integrated water resource management strategy in, 618–620
Barnett Shale aquifer, 303–304
Basaltic rock aquifers, 282
Basin-fill aquifers, 282
Basin-scale integrated ecohydrological model, 377
BAU scenario. *See* Business as usual scenario (BAU scenario)
Bauru Aquifer System, 398
Bay of Bengal (BoB), 129
Bayesian regularization model (BR model), 546
BBSA. *See* N-butyl benzene sulphonamide (BBSA)
Bengal Aquifer System, 425
Benzene, Toluene, Ethylbenzene, and Xylenes compounds (BTEX compounds), 303–304
BIADP. *See* Barind Integrated Area Development Project (BIADP)
Bicarbonates, 269
Binary ion correlations, 266–267
Biogeochemical processes influencing As cycling, 276–277
Biological contamination, 310
  of groundwater, 310–312
    Azad Kashmir and Gilgit Baltistan, 312
    Khyber Pakhtunkhwa, 312
    Punjab, 310
    Sindh, 310–312
BIS. *See* Bureau of Indian Standards (BIS)
Biscayne aquifer, 305–306
Blue Nile river basin, 170–173
BMDA. *See* Barind Multipurpose Development Authority (BMDA)
BoB. *See* Bay of Bengal (BoB)
Bottom-up approaches, 460
BP. *See* Back-propagation (BP)
BR model. *See* Bayesian regularization model (BR model)
Brackish groundwater (BGW), 559
  composition, 559
  economics, 565–566
  emerging desalination technologies, 568
  process, 560–569
  using renewable energy sources, 568
  desalination, 560, 581
Brackish water RO (BWRO), 561–562, 566, 569
  plant cost breakdown, 566f
Brackish water upcoming, 337
BRAMAR. *See* Brazil Managed Aquifer Recharge (BRAMAR)
Brazil
  challenges and future directions toward groundwater sustainability in, 402–403
  global groundwater recharge dynamics, 396
  groundwater availability in, 393–395
  studies on recharge in, 397–402
    recharge methods used in Brazilian studies, 400–402
Brazil Managed Aquifer Recharge (BRAMAR), 418
Brazilian National Department of Mineral Production (DNPM), 416–417
Brine disposal and treatment principles, 567t
Brine management, 567
Brine treatment, 567–568
BTEX compounds. *See* Benzene, Toluene, Ethylbenzene, and Xylenes compounds (BTEX compounds)
"Bullard's Law", 508
Bureau of Indian Standards (BIS), 135–137
Business as usual scenario (BAU scenario), 594–595
BWA. *See* Bangladesh Water Act (BWA)
BWDB. *See* Bangladesh Water Development Board (BWDB)
BWRO. *See* Brackish water RO (BWRO)

C

Cadmium (Cd), 313–317
Calcium, 269, 564
Canning Basin, 386
Cape Flats aquifer (CFA), 445
Cape Metropolitan Council (CMC), 445
Capital cost (CAPEX), 565–566
Carbonate
  equilibrium, 272
  geochemistry, 23–25
  rocks, 23, 24f
Carbonate aquifers, groundwater of
  carbonate geochemistry, 23–25
  carbonate rocks, 23, 24f
  challenges in monitoring and modeling, 31–32
  environmental issues, 30–31
  hydrochemical evolution, 23–25
  permeability, 25–26
  porosity, 25–26
  recharge and flow, 26–30
  water supply, 30–31
Caribbean, 97
Carrizo aquifer, 303
Cation-exchange membranes (CEMs), 563–564
CBM. *See* Coal bed methane (CBM)
CCT. *See* City of Cape Town (CCT)
CECs. *See* Contaminants of emerging concern (CECs)
CEMs. *See* Cation-exchange membranes (CEMs)
Center for Space Research (CSR), 532–533
Central Dry Zone, 52
Central Groundwater Board (CGWB), 491, 548–550
CFA. *See* Cape Flats aquifer (CFA)
CGWB. *See* Central Groundwater Board (CGWB)
Chemical contamination, 312–313
  organic pollution of groundwater, 312–313
Chemically activated desalination systems, 525
China, desalination in, 569
Chlorides, 269
Chryseobacterium meningosepticum, 310–312
Citrobacter sp., 310–312
City of Cape Town (CCT), 442, 445
Climate change, 164, 335, 577
  adaptation to, 337–338
  effect on groundwater, 377–378
    arid and semi-arid regions, 378
    cold regions, 378
  on groundwater resources, 41, 503
  impact in NCP, 66
  impacts on groundwater in Bangladesh, 432
  to transboundary aquifers, 120
Climate experiment, quantification of groundwater change using, 533–534
Climate of Himalayan river basins, 49–52
Climate of KHB, 103
Climate warming, 378
  changes, 371
CLM4.0. *See* Community Land Model version 4.0 (CLM4.0)
Closed-circuit reverse osmosis. *See* Semibatch reverse osmosis
Cl–SO$_4$–HCO$_3$ diagram, 268
CMC. *See* Cape Metropolitan Council (CMC)
CMW. *See* Code of Mineral Waters (CMW)
CNN. *See* Convoluted neural network (CNN)
Coal bed methane (CBM), 506–507
Coal seam gas (CSG), 41
Cobalt (Co), 313
Code of Mineral Waters (CMW), 416
Cold region, 73, 76–77
  climate change effect on groundwater, 378
  groundwater sustainability in, 371
  model development in, 376–377
  effect of permafrost distribution, snow and/or ice on groundwater systems, 373–375
Colonial hydrology, 488
Community Land Model version 4.0 (CLM4.0), 68
Community-based groundwater management, 494–495
Consultation workshops, 461

Contaminants of emerging concern (CECs), 215
Contaminated groundwater purification, 524–527
  killing biological pathogens in groundwater, 527
  removing arsenic from groundwater, 526
  removing fluoride from groundwater, 526–527
  removing salt from brackish groundwater, 524–526
Convoluted neural network (CNN), 550–551
Crystalline and meta-sedimentary rock aquifers, 282
CSG. *See* Coal seam gas (CSG)
CSR. *See* Center for Space Research (CSR)
Customary water schemes, 183

D

DAEE. *See* Departamento de Águas e Energia Elétrica do Estado de São Paulo (DAEE)
Daly Basin, 386
Danish International Development Agency (DANIDA), 135–137
Darcy's law, 504
Darcy–Weisbach equation, 26–28, 504
"Day Zero" drought, 439, 441–442
Decision-making for municipal desalination plants, 581
Deep groundwater, judicial use of, 435
Deep learning (DL), 546. *See also* Machine learning (ML)
  DL-based methods, 551
    CNN, 551
Deep tube well (DTW), 428, 610–611
Deep-well injection, 567
DEET. *See* Diethyltoluamide (DEET)
Degradation of organic matter (OC), 234
DEM. *See* Digital elevation model (DEM)
Departamento de Águas e Energia Elétrica do Estado de São Paulo (DAEE), 397–398
Departamento Nacional de Obras Contra as Secas (DNOCS), 397–398
Department of Defense (DOD), 217
Department of Groundwater Management. *See* Division of Groundwater Management (DGM)
Department of Irrigation (DoI), 461
Department of Mines (DoM), 461
Department of Public Health Engineering. *See* Directorate of Public Health Engineering (DPHE)
Department of Water Affairs and Forestry (DWAF), 446
Department of Water Resources (DWR), 456–457
Depletion of groundwater, 430–431
Depth-based classification schemes, 426
Desalination, 15, 165, 418, 524, 560. *See also* Groundwater desalination
  capacity operation, 560f, 570f, 573f

economics, 565–566
emerging desalination technologies, 568
global trends, 569–571
  annual desalination expenditures, 569–570
  geographic region, 571
  target end use, 571
national trends, 572, 572f
process, 560–569
  brine disposal and treatment principles, 567t
  brine management, 567
  brine treatment, 567–568
  electrodialysis, 563–564
  energy consumption using conventional energy sources, 564–565
  membrane fouling and pretreatment, 561
  nanofiltration, 568
  reverse osmosis, 561–562
  semibatch reverse osmosis, 568
  using renewable energy sources, 568
sectoral use of desalinated water, 571f
Detection frequency (DF), 246
Developing countries, groundwater desalination in, 581
Development, 332
DF. *See* Detection frequency (DF)
DGM. *See* Division of Groundwater Management (DGM)
Dhundi solar pilot, 496
Diethyltoluamide (DEET), 510
Digital elevation model (DEM), 102–103
Directorate of Public Health Engineering (DPHE), 135–137, 430
Directorate of Public Health Engineering and United Nations Children's Fund (DPHE/UNICEF), 135–137
Dissolved organic carbon (DOC), 135, 604–605
Dissolved oxygen (DO), 190
Distributed-parameter models, 32
Division of Groundwater Management (DGM), 459–461
DL. *See* Deep learning (DL)
DMSe. *See* D-Methyl-selenide (DMSe)
DNOCS. *See* Departamento Nacional de Obras Contra as Secas (DNOCS)
DNPM. *See* Brazilian National Department of Mineral Production (DNPM)
DO. *See* Dissolved oxygen (DO)
DOC. *See* Dissolved organic carbon (DOC)
DOD. *See* Department of Defense (DOD)
DoI. *See* Department of Irrigation (DoI)
DoM. *See* Department of Mines (DoM)
Domestic sewage, 251–252
DPHE. *See* Directorate of Public Health Engineering (DPHE)
DPHE/UNICEF. *See* Directorate of Public Health Engineering and United Nations Children's Fund (DPHE/UNICEF)
Draft Articles, 126–127
Draft data systems, 459–460
Drilling success rate, 177–178, 180–181
Drought, 609. *See also* Groundwater drought

Dryland
  areas of Eastern Africa, 179
  salinity, 43
DTW. *See* Deep tube well (DTW)
Dublin principles, 415
DWAF. *See* Department of Water Affairs and Forestry (DWAF)
DWR. *See* Department of Water Resources (DWR)

E

Earth
  gravity field, 531
  natural resources, 331
  observation satellite monitoring, 53–55
  observation system, 48–49
Eastern Africa, groundwater status in
  characteristics of dryland areas, 179
  drinking water delivery practices, 181–182
  hydrogeology difficulties in, 179–181
  policy and practice implication, 184–185
  securing water in hydrogeological environments, 182–184
Ecological Structure Activity Relationships (ECOSAR), 245
Economics of desalination, 565–566
ED. *See* Electrodialysis (ED)
EDB. *See* Ethylene dibromide (EDB)
EDR. *See* Electrodialysis reversal (EDR)
Education, 512
Edwards–Trinity plateau aquifer, 301–302
Egypt, groundwater in, 88–90
EIA. *See* Environmental Impact Assessment (EIA)
El Niño–Southern Oscillation (ENSO), 148–151
Electrically activated systems, 525
Electrochemical processes, 526
Electrodialysis (ED), 560, 563–564, 569
  desalination mechanism, 563–564
  membranes, 564
  posttreatment, 564
  pretreatment, 563
  SEC, 564t
  system design, 564
Electrodialysis reversal (EDR), 560, 563–565
  simplified EDR stack, 563f
Elevated groundwater hardness, 237
Emerging organic contaminants (EOCs), 312
Empirical mean weight (EMW), 615
Energy
  consumption using conventional energy sources, 564–565
  energy-groundwater nexus, 495
  energy-irrigation nexus, 489–490
  pricing and rationing, 495
Energy recovery device (ERD), 561
  of RO, 562
ENSO. *See* El Niño–Southern Oscillation (ENSO)
Enterobacter sp., 312

Environmental controls on groundwater, 146–151
  large-scale climate phenomena, 148–151
  precipitation, 146–148
  subsurface hydrogeological conditions, 148
Environmental Impact Assessment (EIA), 455
Environmental management, 338
Environmental risk assessment, 247
EOCs. See Emerging organic contaminants (EOCs)
EPM. See Equivalent porous medium (EPM)
Equivalent porous medium (EPM), 31–32
ERD. See Energy recovery device (ERD)
Erythromycin (ERY), 246
Escherichia coli, 310, 428
ET. See Evapotranspiration (ET)
Ethiopia, groundwater in, 90–91
Ethylene dibromide (EDB), 510
Evaporation ponds, 567
Evapotranspiration (ET), 348, 470–471
Extraction limits, 38
Extraterrestrial hydrology, 508–509

F

Faculty of Water Resources (FWR), 456
FAO. See UN Food and Agricultural Organization (FAO)
FAS. See Floridan aquifer system (FAS)
FDPE. See Florida Department of Environmental Protection (FDPE)
Finite Element subsurface FLOW system (FEFLOW system), 69
Flood, 609
  inundation, 587
Florida, aquifers in, 304–306
  Biscayne aquifer, 305–306
  FAS, 304–305
  sand-and-gravel aquifer, 305
Florida Department of Environmental Protection (FDPE), 304
Florida Ground-Water Quality Monitoring Network Program, 305
Floridan aquifer system (FAS), 304–305
Fluoride (F), 10, 190–192, 235, 319–320
  removing from groundwater, 526–527
Fluorosis, 319–320
Food trade, groundwater depletion for, 352–355
Fossil fuel energy, 505–506
Fouling, 561
Fresh groundwater resources, 341
Freshwater, 73
FundiFix model, 184
FWR. See Faculty of Water Resources (FWR)

G

GAC. See Granular activated carbon (GAC)
Ganga−Brahmaputra−Meghna River delta (GBM River delta), 50–52, 129
  aquifer framework, 131
  arsenic mitigation, 135–138
  geologic and geomorphologic setting, 130–131
  groundwater arsenic contamination, 134–135
  groundwater flow system, 131–133
  hydrogeochemistry, 133–134
GAS. See Guarani Aquifer System (GAS)
GBM River delta.
  See Ganga−Brahmaputra−Meghna River delta (GBM River delta)
GCG. See Geogenic-contaminated groundwater (GCG)
GDEs. See Groundwater-dependent ecosystems (GDEs)
Genetic programing (GP), 546
GeoForschungsZentrum (GFZ), 532–533
Geogenic contaminants, 187
Geogenic groundwater pollutants, global distribution of, 188f
  arsenic, 187–190
  fluoride, 190–192
  salinity, 196–198
  selenium, 192–194
  uranium, 194–196
Geogenic-contaminated groundwater (GCG), 229
  by anthropogenic activities, 235–237
  cooccurrence of, 235
  distribution and formation of, 229–235
Geographical Information Systems (GIS), 396, 613
Geological formation−based classification schemes, 426
Geothermal energy, 505–506
GERD. See Grand Ethiopian Renaissance Dam (GERD)
GFO mission. See GRACE Follow-On mission (GFO mission)
GFZ. See GeoForschungsZentrum (GFZ)
GIA. See Glacial isostatic adjustment (GIA)
GIS. See Geographical Information Systems (GIS)
Glacial aquifers, 282
Glacial isostatic adjustment (GIA), 531
Global groundwater, 4, 7, 97, 177, 251, 309
  recharge dynamics, 396
Global Groundwater Information System, 114
Global Land Data Assimilation System (GLDAS) model, 534–535
GMAs. See Groundwater management areas (GMAs)
Goddard Space Flight Center (GSFC), 532–533
Govt. of Bangladesh (GoB), 610–611
GP. See Genetic programing (GP); Groundwater potentiality (GP)
GRACE Follow-On mission (GFO mission), 531
  and data products, 532–533
GRACE satellite. See Gravity Recovery and Climate Experiment satellite (GRACE satellite)
Grand Ethiopian Renaissance Dam (GERD), 81
Granular activated carbon (GAC), 222
Gravity Recovery and Climate Experiment satellite (GRACE satellite), 48–49, 53–55, 57, 87, 396, 531, 545
  data assimilation, 539
  and data products, 532–533
  error sources of GRACE−estimated groundwater change, 537–539
  GRACE-observed time-variable gravity change, 531–532
  for groundwater drought, 151–153
  in groundwater storage change, 534–537
  quantification of groundwater change using, 533–534
  terrestrial water storage, 55–56
Great Artesian Basin, 384–385
Green Revolution, 527
Groundwater, 3, 47, 75, 78, 243, 251, 301, 309, 383, 425, 435–436, 439, 451, 469, 545
  abstraction, 334
  advanced data-driven methods in groundwater resources, 545–546
  antibiotics in, 245–248
  application of deep learning, 554
  application of machine learning, 551–554
    GWL prediction in India, 551–553
    RF application in groundwater contamination prediction in India, 554
  arsenic in, 276–279
  artificial intelligence methods in India, 550–554
    deep learning−based methods, 551
    machine learning−based methods, 551
  in Australia, 36–38
  availability in Brazil, 393–395
  biological contamination, 310–312
  chemistry, 253–255
  contamination, 255–259
    expanding in India, 480–482
    issues in South Asia, 471
  crisis, 359
  for drinking and agricultural use, 53
  in Egypt, 88–90
  in Ethiopia, 90–91
  in Extended Lake Victoria basin, 91
  extraction on surface-water systems, 40
  flow system in GBM River delta, 131–133
  global literature review, 546–550
    groundwater quality, 546–550
    groundwater quantity, 546
  governance, 435
  in hydrological systems
    arid and semi-arid regions, 373
    effect of climate change, 377–378
    cold regions, 371
    hydrological cycle characteristics, 373–376
    integrated water management for groundwater sustainability, 379
  mapping, 519–520
  MAR of, 600–604
  model application and challenges in arid and semi-arid regions, 377
  model development in cold regions, 376–377

in NCP, 65–66
in Nile basin, 86–88
planning, 460
pollution in Pakistan
  biological contamination of groundwater, 310–312
  chemical contamination, 312–313
  groundwater quality, 310–312
  inorganic pollution of groundwater, 313–320
pumping, 528
recharge, 371, 469–471
  rates, 4f
rejuvenation, 15
removal of PFASs, 221–223
salinity, 388–389
salinization, 196
scarcity, 6–11
  quality, 9–11
  quantity, 6–8
  solutions, 14–15
  source and availability, 4–6
in Sudan and South Sudan, 90
sustainability. *See* Sustainable groundwater
Groundwater depletion (GWD), 347
  quantification for food trade, 352–355
Groundwater desalination, 577–579. *See also* Desalination
  considerations for, 578–579
  in developing countries, 581
  environmental impacts of, 579
  motivation for, 577–578
  in United States, 580–581
Groundwater drought, 145, 151–153
  characteristics of, 153–156
  environmental controls on groundwater, 146–151
  GRACE, 151–153
  indicators, 153
Groundwater irrigation in India
  changing organization of irrigation economy, 488–489
  energy-irrigation nexus, 489–490
  Indian irrigation evolution, 487–488
  socioeconomic significance of groundwater boom, 490–491
  sustainability challenge, 491–493
  sustainable groundwater governance, 493–496
Groundwater level (GWL), 545
  prediction in India, 551–553
Groundwater management, 301, 377, 379, 519–520
  alternatives for groundwater management and water sourcing, 417–418
  adopting rainwater harvesting, 417
  artificial groundwater recharge and reuse of wastewater, 418
  desalination, 418
  groundwater resource management in Brazil, 414–417
  groundwater resources of Brazil, 410–414
  hydrogeological features of aquifers, 411–414

hydroschizophrenia of groundwater management, 418–419
training programs generate institutional capacity for, 462
Groundwater management areas (GMAs), 40
Groundwater potentiality (GP), 613
Groundwater quality
  aquifers in Florida, 304–306
  challenges in Afghanistan, 324–325
  management, 337
  in Pakistan, 310–312
  in Texas, 301–304
    Barnett Shale aquifer, 303–304
    Carrizo aquifer, 303
    Edwards–Trinity plateau aquifer, 301–302
    Ogallala aquifer, 302
    Pecos Valley aquifer, 303
    Seymour aquifer, 302–303
Groundwater resource, 45
  approaches to pursuing, restoring, or enhancing
    adaptation to climate change and sea-level rise, 337–338
    enhancing groundwater recharge, 336–337
    environmental management, 338
    groundwater governance and management, 335
    groundwater quality management, 337
    hydrogeological approaches to defining sustainability limits of abstraction, 335–336
    water demand management, 337
  in Australia, 35–36
    evolution of groundwater management, 38
    groundwater management issues, 40–43
    groundwater usage, 38–40
    historical development of groundwater, 36–38
    hydro-illogical cycle, 43
    MAR, 44–45
  challenges of groundwater resource management plan, 612–613
  geographic variation
    endangered or disrupted by progressive storage depletion, 339–341
    endangered or disrupted by water quality degradation, 341
    general comments, 338
    sustainability constrained by environmental considerations, 341–343
  management in Brazil
    groundwater monitoring and assessment, 416–417
    integrated management of surface water and groundwater, 415
    mineral water resources management, 416
    national laws/legislation, 415
    transboundary groundwater management, 415–416
    water resource management, 414–415
  potentiality, 613–615
  sustainability

of groundwater services, 332–335
and sustainable development, 331–332
Groundwater science and research, 503
  coast to coast, 508
  education and outreach, 512
  extraterrestrial hydrology, 508–509
  fossil fuel energy, geothermal energy, and mineral resources, 505–506
  fundamental groundwater perspectives, 504–505
  groundwater quality and emerging contaminants, 509–510
  groundwater resources, 506–507
  laws, regulation, guidance, and governance of groundwater, 511
  new tools, 510–511
  under ocean, 508
  socio-hydrogeology in future of groundwater science, 511–512
  subterranean biological world and groundwater-dependent ecosystems, 507–508
  unexpected challenges, 512
Groundwater services, 332–334
  sustainability, 332–335
    potential threats to groundwater services, 334–335
Groundwater storage (GWS), 47, 57–59, 65
  dynamics
    gravity recovery and climate experiment, 53–56
    in Himalayan river basins, 53–59
    impacts of global change, 57–59
    mapping groundwater storage, 56–57
Groundwater table (GWT), 610–611
Groundwater-dependent ecosystems (GDEs), 40
Groundwater-dependent terrestrial ecosystems, 373
Groundwater-fed irrigation, 47–48
Groundwater-management capacity, 461–462
Groundwater–food–energy nexus, 12
Groundwater–society interactions modeling, 360–361
GSFC. *See* Goddard Space Flight Center (GSFC)
Guarani, 393
Guarani Aquifer Project, 415–416
Guarani Aquifer System (GAS), 364, 412, 415–416
GWD. *See* Groundwater depletion (GWD)
GWL. *See* Groundwater level (GWL)
GWS. *See* Groundwater storage (GWS)
GWT. *See* Groundwater table (GWT)

H

Hazardous substances, 580
HCA. *See* Hydrogeologically complex aquifers (HCA)
Heavy metals, 313–319
Hexafluoropropylene oxide (HFPO), 215
HFOs. *See* Hydrous Fe oxides (HFOs)
HFPO. *See* Hexafluoropropylene oxide (HFPO)

High Plains aquifer, 282
High-/low-iodine groundwater, 234
High-As groundwater, 231
High-Fe and-Mn groundwater, 230–231
High-fluoride groundwater, 231–234
High-salinity groundwater, 230
Himalayan region, 47
Himalayan river basins, 48f, 49
   groundwater for drinking and agricultural use, 53
   groundwater storage dynamics in, 53–59
   hydrology and climate of, 49–52
      GBM river basin, 50–52
      Indus river basin, 49–50
      Irrawaddy river basin, 52
Human activities on groundwater in NCP, 65–66
Human resources development, 461–462
Human-induced hazards, 609
Human–water feedbacks, 360
Hydro-economics, 13–14
Hydro-illogical cycle, 43
Hydrochemical evolution, 23–25
Hydrochemistry in Northern Morocco
   quality of source waters for irrigation, 265–266
   source water chemical facies, 262–265
Hydrogeochemical settings for arsenic, 279–282
Hydrogeochemistry, 133–134
Hydrogeological
   approaches, 335–336
   conditions, 148
   features of aquifers, 411–414
Hydrogeologically complex aquifers (HCA), 6
Hydrogeology, 45, 503
   of KHB, 102–103
Hydrological cycle characteristics
   hydrological processes and effect on groundwater quality, 375–376
   permafrost distribution, snow and /or effect ice on groundwater systems, 373–375
Hydrology of Himalayan river basins, 49–52
Hydroschizophrenia of groundwater management, 418–419
Hydrosocial cycle, 359, 362
Hydrous Fe oxides (HFOs), 276–277

I
IBE. *See* Index of base exchange (IBE)
IBIS. *See* Indus Basin Irrigation System (IBIS)
Ice effect on groundwater systems, 373–376
IED. *See* Iodine excess disorders (IED)
IEX. *See* Ion-exchange resins (IEX)
IFC. *See* International Finance Corporation (IFC)
IGB. *See* Indo-Gangetic basin (IGB)
IGIS 2.0 software, 617
Impact assessment of MAR model as IWRM, 621–622
Index of base exchange (IBE), 268–269
India
   amount of groundwater extraction, 479
   expanding groundwater contamination, 480–482
   groundwater in, 477
   measuring and understanding aquifers, 482
   regional aquifer framework, 477–478
   spatiotemporal behavior of hydraulic heads and replenishable resources, 478–479
   sustainable management plan, 483–485
India Easement Act (1882), 493–494
Indian irrigation evolution, 487–488
Individual space, 270–272
Indo-Gangetic basin (IGB), 47–48
Indus Basin Irrigation System (IBIS), 309
Indus river basin, 49–50
Industrial sources, 219
Industrial uses of groundwater, 431–432
Industrial wastewater, 251–252, 256
Industrial water-use efficiency improvement, 523–524
Industrialization in PRD, 252
Information sharing mechanisms, 454–455
Inorganic pollution of groundwater, 313–320
   anions, 319–320
      fluoride, 319–320
      nitrates, 319
      phosphates, 319
      sulphates, 319
   trace and heavy metals, 313–319
      arsenic, 313
      cadmium, 313–317
      iron, 318
      lead, 318
      nickel, 318
      zinc, 319
Institutional mapping, 461
Integrated ecohydrological models, 377
Integrated management of surface water and groundwater, 415
Integrated water management for groundwater sustainability, 379
Integrated water resource management (IWRM), 414–415, 456, 611
   MAR model as
      in Barind Tract, 618–620
      impact assessment of, 621–622
      model, 586
Intensive groundwater abstraction, 334
Interbasin water transfer, 184
Interdisciplinarity, 364–366
Intergovernmental Panel on Climate Change (IPCC), 585
International Finance Corporation (IFC), 456–457
Iodide in groundwater, 257–258
Iodine excess disorders (IED), 229, 234
Ion-exchange resins (IEX), 221–223
IPCC. *See* Intergovernmental Panel on Climate Change (IPCC)
Iron (Fe), 235, 313, 318
   contamination, 255–256
Irrawaddy river basin, 52
Irrigation, 15
   and agricultural efficiency, 522–523
   in Nile basin, 84–86
   salinity, 43
   source waters for, 265–266
   uses of groundwater, 431
   water-use efficiency, 523
Israel
   Israeli Coastal Aquifer, 598
   MAR
      of ephemeral stream floods, 598–600
      experience, 597
      of groundwater and especially lake water, 600–604
      of secondary effluents in infiltration basins, 604–605
      of surplus desalinated seawater through infiltration basins, 605–606
Itai-itai disease, 313
IWRM. *See* Integrated water resource management (IWRM)

J
JGY. *See* Jyotir Gram Yojana (JGY)
Joint Plan Of Action (JPOA), 135–137
Jordan River, 166–168, 167f
JPL. *See* NASA Jet Propulsion Laboratory (JPL)
JPOA. *See* Joint Plan Of Action (JPOA)
Jyotir Gram Yojana (JGY), 495

K
K-band ranging system (KBR system), 532
Karst terrains, recharge in, 26–28
Karstic rocks, 25
Karstification, 25–26
KBR system. *See* K-band ranging system (KBR system)
KHB. *See* Kingston Hydrologic Basin (KHB)
Khyber Pakhtunkhwa, groundwater biological contamination in, 312
Killing biological pathogens in groundwater, 527
Kingston Hydrologic Basin (KHB), 99–103
   climate of, 103
   hydrogeology of, 102–103
   methodology and analytical procedures, 103–110
   population and water supply, 99–102
   results and discussion, 110–111
Kisan Urja Suraksha evam Utthaan Mahabhiyan (KUSUM), 496
Klebsiella sp., 312

L
Lake Victoria basin, groundwater in, 91
Land
   application, 567
   and groundwater salinization, 42–43
   use change, 590, 593
   use in Nile basin, 84–86
Land Change Modeler for ArcGIS, 590
Land surface model (LSM), 545
Landfill leachate, 219

Index

Landlocked mountainous country, 323
Lao People's Democratic Republic (PDR), 451
   continuing efforts in applied research, 464
   effective policy making and implementation, 463
      promoting cross-sectoral coordination, 463
      promoting strategic planning of groundwater resources, 463
   groundwater challenges, 454–455
      other barriers to groundwater management, 455
      quantity and quality-related issues, 454
      state of groundwater knowledge and information systems, 454–455
   groundwater resources in
      groundwater systems, 452–453
      groundwater use, 453–454
   indicative overview of groundwater stakeholders in, 456t
   participation of stakeholders, 465
   recent efforts to strengthen groundwater governance
      changes enhanced by projects and investments, 456–457
      changes in government policy, 455
      changes in institutional arrangements, 455–456
      development of human resources and groundwater-management capacity, 461–462
      enhancing groundwater knowledge and data management, 457–460
      mechanisms of stakeholder coordination and involvement, 460–461
      revised legal arrangements, 456–457
      strengthening institutional and human resource capacity, 464
Large-scale climate phenomena, 148–151
LC/MS. *See* Liquid chromatography with mass spectrometry (LC/MS)
LC/MS/MS. *See* Liquid Chromatography/Tandem Mass Spectrometry (LC/MS/MS)
LCOE. *See* Levelized cost of electricity (LCOE)
Lead (Pb), 313, 318
   contamination, 256–257
Least work of separation (LWS), 565, 565f
Levelized cost of electricity (LCOE), 568
Levenberg–Marquardt model (LM model), 546
LFA. *See* Lower Floridan aquifer (LFA)
LGIs. *See* Local government institutions (LGIs)
Life cycle assessment, 350
Limestone, 23, 25–26, 29–31
LIN. *See* Lincomycin (LIN)
Lincomycin (LIN), 246
Linear regression model (LRM), 546
Liquid chromatography with mass spectrometry (LC/MS), 216
Liquid Chromatography/Tandem Mass Spectrometry (LC/MS/MS), 216–217
LM model. *See* Levenberg–Marquardt model (LM model)
Local and shallow aquifers (LSA), 5–6

Local government institutions (LGIs), 621–622
Local knowledge of water, 363
Local-scale assessments in priority areas increase local knowledge, 459
Long-chain PFAS, 215
Lower Floridan aquifer (LFA), 304
LRM. *See* Linear regression model (LRM)
LSA. *See* Local and shallow aquifers (LSA)
LSM. *See* Land surface model (LSM)
Lumped-parameter models, 32
LWS. *See* Least work of separation (LWS)

M

Machine learning (ML), 546. *See also* Deep learning (DL)
   approaches, 69
   ML-based methods, 551
      ANN, 551
      RFs, 551
MAF. *See* Ministry of Agriculture and Forestry (MAF)
Magnesium (Mg), 269, 564
Major basin aquifers (MBA), 5–6
MAKESENS. *See* Mann–Kendall Trend Test and Sen's Slope Estimator (MAKESENS)
Managed aquifer recharge (MAR), 15, 43, 278, 286, 434–435, 446, 597
   of ephemeral stream floods, 598–600
   of groundwater and especially lake water, 600–604
      history and experience from MAR through well period, 602
      new thoughts and experiments on MAR through wells, 602–604
      technical considerations concerning MAR through wells, 601–602
   Israel experience, 597
   model, 611–612
      challenges of groundwater resource management plan, 612–613
      groundwater resource potentiality, 613–615
      impact assessment as IWRM, 621–622
      implementation, 616–622
      as IWRM strategy in Barind Tract, 618–620
      piloting at household level, 616–618
      PZGR and selection of sites for artificial recharge of groundwater, 616
   of secondary effluents in infiltration basins, 604–605
   of surplus desalinated seawater through infiltration basins, 605–606
Management practices, 577
Managing aquifer recharge (MAR), 520–521
Manganese (Mn), 235
   contamination, 255–256
Mann–Kendall Trend Test and Sen's Slope Estimator (MAKESENS), 612
MAR. *See* Managed aquifer recharge (MAR); Managing aquifer recharge (MAR)

Maximum contaminant level (MCL), 275
MBA. *See* Major basin aquifers (MBA)
MCL. *See* Maximum contaminant level (MCL)
MEA. *See* Millennium Ecosystem Assessment (MEA)
Mean effective weight (MEW), 615
Mean variation index (MVI), 615
MED. *See* Multiple-effect distillation (MED)
Mekong River Commission (MRC), 452
Membrane
   contamination, 561
   of ED, 564
   fouling and pretreatment, 561
   membrane-based seawater desalination technologies, 525
   processes, 560
   of RO, 562
MENA. *See* Middle East and North Africa (MENA)
Mercury (Hg), 313
   contamination, 256–257
Methyl tert-butyl ether (MTBE), 510
D-Methyl-selenide (DMSe), 193
MEW. *See* Mean effective weight (MEW)
Microbial contamination. *See* Biological contamination
Middle East and North Africa (MENA), 163, 352–353, 571
   impacts of water scarcity, 164–165
   Jordan River, 166–168, 167f
   Nile River, 170–173
   Tigris–Euphrates river system, 168–169
   water resources, 163–164
   water resources management, 165–166
Middle East and North Africa region (MENA region)
Millennium Ecosystem Assessment (MEA), 332
Mineral resources, 505–506
Mineral water, 416
   resources management, 416
Mining, 577–578
   on groundwater resources, 41
Ministry of Agriculture (MoA), 610–611
Ministry of Agriculture and Forestry (MAF), 453–454, 463
Ministry of Natural Resources and Environment (MoNRE), 455–456
Ministry of Water Resources (MoWR), 611
ML. *See* Machine learning (ML)
MLP. *See* Multilayer perception (MLP)
MoA. *See* Ministry of Agriculture (MoA)
Model development in cold regions, 376–377
MoNRE. *See* Ministry of Natural Resources and Environment (MoNRE)
Mosaic Dataset of ArcGIS, 590
MoWR. *See* Ministry of Water Resources (MoWR)
MRC. *See* Mekong River Commission (MRC)
MRM. *See* Multiple Reaction Monitoring (MRM)
MSF. *See* Multistage flash (MSF)
MTBE. *See* Methyl tert-butyl ether (MTBE)
Multilayer perception (MLP), 546–548

Multiple Reaction Monitoring (MRM), 216–217
Multiple-effect distillation (MED), 560, 579
Multistage flash (MSF), 579
  distillation, 560
Municipal desalination plants, decision-making for, 581
Murray–Darling Basin, 385
MVI. See Mean variation index (MVI)

N

N-butyl benzene sulphonamide (BBSA), 510
N_2–Ar–CH_4 gases diagram, 273
Nano-sized ZVI (nZVI), 287
Nanofiltration (NF), 221, 223, 568
NASA Jet Propulsion Laboratory (JPL), 532–533
National Arsenic Policy and Mitigation Action Plan, 138
National Council on Environment, 415
National Dryland Salinity Program, 43
National Environmental Policy Act, 415
National Groundwater Action Plan (NGWAP), 457
National Groundwater Plan, 415
National Hydrochemical Survey (NHS), 426–428
National IWRM Support Project (NIWRMSP), 459–460
National Plan of Hydric Resources (PNRH), 414–415
National scale assessments in priority areas increase local knowledge, 459
National University of Laos (NUoL), 455
National Water Commission (NWC), 99–100
National Water Initiative (NWI), 38, 40
National Water Law, 414, 597
National Water Management Plan (NWMP), 611
National Water Policy (NWPo), 609
National Water Reform Framework Agreement, 38
National Water-Quality Assessment Program, 301
Natural hazards, 609
Natural Resources and Environment Institute (NREI), 460
Nazareth series ignimbrites, 180
NBI. See Nile Basin Initiative (NBI)
NCP. See North China Plain (NCP)
NDVI. See Normalized difference vegetation index (NDVI)
NF. See Nanofiltration (NF)
NGOs. See Nongovernmental organizations (NGOs)
NGWAP. See National Groundwater Action Plan (NGWAP)
NHS. See National Hydrochemical Survey (NHS)
Nickel (Ni), 318
  contamination, 256–257
Nile Basin Initiative (NBI), 81
Nile River, 170–173

Nile river basin, 81
  aquifers in, 88–91, 92*t*
  groundwater in, 86–88
  land use and irrigation in, 84–86
  surface water in, 81–83
Nitrate, 99–100, 106, 319
  in groundwater, 257–258
NIWRMSP. See National IWRM Support Project (NIWRMSP)
NIWRMSP project, 462
Non-UAs (NUAs), 252
Nongovernmental organizations (NGOs), 134, 454–455
Nonrenewable groundwater resources, 336
Nontargeted analysis (NTA), 218
Norfloxacin (NOR), 243
Normalized difference vegetation index (NDVI), 545
North China Plain (NCP), 65, 230, 533, 537
  China's South-to-North water diversion, 66–68
  climate change impact on groundwater in, 66
  groundwater storage assessment in, 68–70
  human activities on groundwater in, 65–66
Northern Morocco, 261
  control of chemical element concentrations, 266–269
  hydrochemistry, 262–266
  material and methods, 262
  PCA, 270–272
  source chemical analyses, 263*t*
  water minerals equilibrium, 272–273
NREI. See Natural Resources and Environment Institute (NREI)
NSAS. See Nubian Sandstone Aquifer System (NSAS)
NTA. See Nontargeted analysis (NTA)
NUAs. See Non-UAs (NUAs)
Nubian Sandstone Aquifer System (NSAS), 90
NUoL. See National University of Laos (NUoL)
NWC. See National Water Commission (NWC)
NWI. See National Water Initiative (NWI)
NWMP. See National Water Management Plan (NWMP)
NWPo. See National Water Policy (NWPo)
nZVI. See Nano-sized ZVI (nZVI)

O

OAS. See Organization of American States (OAS)
OC. See Degradation of organic matter (OC); Organic carbon (OC)
Occurrence studies of PFASs, 220–221
Ogallala aquifer, 302
"One Million Cisterns" program (P1MC), 417
Operating cost (OPEX), 565–566
Organic
  contaminants in groundwater, 258–259
  pollution of groundwater, 312–313
Organic carbon (OC), 278
Organization of American States (OAS), 398–399

Orographic effect, 49
Otway Basin, 386
Overallocation of groundwater, 40
Overexploitation, 477, 482, 485
Overuse of groundwater, 40
Oxidation–reduction reactions (Redox reactions), 277
  redox change, 237

P

Pacific Decadal Oscillation (PDO), 148–151
PAD. See Programa Água Doce (PAD)
Pakistan Standard Quality Control Authority (PCSQA), 310
Participation of stakeholders, 465
PCA. See Principal component analysis (PCA)
PCI. See Precipitation concentration index (PCI)
PCR-GLOBW global hydrological model, 351–352
PCR-GLOBWB. See PCRaster Global Water Balance (PCR-GLOBWB)
PCRaster Global Water Balance (PCR-GLOBWB), 69
PCSQA. See Pakistan Standard Quality Control Authority (PCSQA)
PDO. See Pacific Decadal Oscillation (PDO)
PDR. See Lao People's Democratic Republic (PDR)
Pearl River Delta (PRD), 251
  groundwater chemistry, 253–255
  groundwater contamination, 255–259
  groundwater quality and main impact chemicals, 255
  hydrogeological and geological conditions, 251–252
  industrialization in, 252
  materials and methods, 253
  urbanization in, 252
Pecos Valley aquifer, 303
Per-and polyfluoroalkyl substances (PFASs), 215
  analytical methods for, 216–218
  occurrence studies, 220–221
  removal from groundwater, 221–223
  sources of, 218–220
Perfluoroalkane sulfonic acid (PFSA), 215
Perfluoroalkyl carboxylic acid (PFCA), 215
Perfluorobutanesulfonic acid (PFBS), 219
Perfluorohexanoic acid (PFHxA), 219
Perfluorooctanesulfonic acid (PFOS), 215
Perfluorooctanoic acid (PFOA), 215
Peri-urban areas (PUAs), 252
Permafrost distribution effect on groundwater systems, 373–376
Permafrost in cold regions, 371
Permeability, carbonate aquifers, 25–26
Permutolites, 268–269
Perth Basin, 385–386
PFASs. See Per-and polyfluoroalkyl substances (PFASs)
PFBS. See Perfluorobutanesulfonic acid (PFBS)

PFCA. *See* Perfluoroalkyl carboxylic acid (PFCA)
PFHxA. *See* Perfluorohexanoic acid (PFHxA)
PFOA. *See* Perfluorooctanoic acid (PFOA)
PFOS. *See* Perfluorooctanesulfonic acid (PFOS)
PFSA. *See* Perfluoroalkane sulfonic acid (PFSA)
PHED. *See* Public Health Engineering Department (PHED)
Phosphates, 319
Photovoltaics (PV), 568
　arrays, 528
Piped water supply (PWS), 137–138
Pleistocene clay deposits, 609–610
PNEC. *See* Predicted no effect concentration (PNEC)
PNRH. *See* National Plan of Hydric Resources (PNRH)
Political ecology, 359
　of water, 361–362
Politics of scale, groundwater and, 363–364
Pollution, 334–335
　abatement and control, 434
Pond sand filter (PSF), 137–138
Population growth, 590
Porosity, carbonate aquifers, 25–26
Potassium (K), 269
Potential zones for groundwater recharge (PZGR), 616
　and selection of sites for artificial recharge of groundwater, 616
Pradhan Mantri Krishi Sinchai Yojana, 487
PRD. *See* Pearl River Delta (PRD)
Precipitation, 146–148
　change, 587–590, 592–593
Precipitation concentration index (PCI), 612
Predicted no effect concentration (PNEC), 245
Pressure-activated systems, 525
Principal component analysis (PCA), 270–272
Private Well Testing Act (PWTA), 288
Programa Água Doce (PAD), 418
Pseudomonas aeruginosa, 310–312
PSF. *See* Pond sand filter (PSF)
PUAs. *See* Peri-urban areas (PUAs)
Public Health Engineering Department (PHED), 135–137
Pump irrigation revolution, 491
Punjab, groundwater biological contamination in, 310
PV. *See* Photovoltaics (PV)
PWS. *See* Piped water supply (PWS)
PWTA. *See* Private Well Testing Act (PWTA)
Pyroclastics, 180
PZGR. *See* Potential zones for groundwater recharge (PZGR)

Q

Quantification
　of groundwater change, 533–534
　of groundwater depletion for, 352–355

R

Rain-gauge stations (RGSs), 616
Rainfall water, 75–77
Rainwater harvesting (RWH), 137–138, 166, 417, 611–612
　potentiality, 618–619
Random forests (RFs), 548–551
　application in groundwater contamination prediction in India, 554
RDA. *See* Recommended Daily Allowance (RDA)
Reactive transport models, 278
Recharge rate (RR), 620
Reciprocal learning, 365–366
Recommended Daily Allowance (RDA), 318
Recurrent neural network (RNN), 546
Red Queen effect, 116–117
Redox reactions. *See* Oxidation–reduction reactions (Redox reactions)
Reductive dissolution, 255, 258
Remediation methods for arsenic detection, 286–287
Remediation of polluted sites, 337
Remote sensing (RS), 73, 613
Renewable energy
　coupled desalination, 568
　desalination using renewable energy sources, 568
Reverse osmosis (RO), 221, 223, 525, 560–562
　desalination mechanism, 561–562, 561*f*
　energy recovery devices, 562
　membranes, 562
　posttreatment, 562
　pretreatment, 561
　SEC, 564*t*
　system design, 562
　technologies, 578
RFs. *See* Random forests (RFs)
RGSs. *See* Rain-gauge stations (RGSs)
RIBASIM. *See* River Basin Simulation Model (RIBASIM)
Rice, 612–613
Rift volcanics, 180
Risk quotient (RQ), 245
River Basin Simulation Model (RIBASIM), 586
RNN. *See* Recurrent neural network (RNN)
RO. *See* Reverse osmosis (RO)
Roxithromycin (ROX), 246
RQ. *See* Risk quotient (RQ)
RR. *See* Recharge rate (RR)
RS. *See* Remote sensing (RS)
RWH. *See* Rainwater harvesting (RWH)

S

S-ISP. *See* Solar Irrigation Service Providers (S-ISP)
S.A.R. *See* Sodium adsorption ratio (S.A.R)
Safe yield, 336
Saline groundwater intrusion management, 521–522
Saline water intrusion, 337
Salinity, 196–198, 235
Salinity Control and Reclamation Program (SCARP), 309
Salinization, 334–335
　of groundwater, 235–237
　land and groundwater, 42–43
Sand and gravel aquifers, 279–282, 305
Sandveld Group aquifers, 444–445
SAT. *See* Shafdan soil aquifer treatment (SAT)
Satellite gravimetry technique, 537
Saudi Arabia, desalination in, 569
Scarcity
　of freshwater, 609
　groundwater, 6–11
　　quality, 9–11, 324–326
　　quantity, 6–8, 324–326
SCARP. *See* Salinity Control and Reclamation Program (SCARP)
Scenario-based approach
　land use change, 593
　methodology, 586–592
　　land use change, 590
　　population growth, 590
　　precipitation change, 587–590
　　urban flood, 590
　　water quality, 591–592
　precipitation change, 592–593
　study area, 586
　urban flood, 594
　water quality, 594–595
"Scientific" knowledge of water, 363
Screening of emerging organic pollutants
　antibiotics, 243
　antibiotics in groundwater, 245–248
　environmental risk assessment, 247
　materials and methods, 243–245
　statistical analysis, 246–247
Screening score (SS), 247–248
SDC. *See* Swiss Agency for Development and Corporation (SDC)
SDGs. *See* Sustainable Development Goals (SDGs)
Sea-level rise, 335
　adaptation to, 337–338
Seawater intrusion, 43
SEB. *See* State Electricity Boards (SEB)
SEC. *See* Specific energy consumption (SEC)
Sedimentary rock aquifers, 282
Selenium (Se), 192–194
Selenomethionine (SeMet), 193
Semi-arid regions
　climate change effect on groundwater, 378
　groundwater sustainability in, 373
　hydrological processes and effect on groundwater quality in, 375–376
　model application and challenges in, 377
Semibatch reverse osmosis, 568
Sequential extraction, 277–278
Serra Geral Aquifer System, 398
Sewer discharge, 567
Seymour aquifer, 302–303
SGIs. *See* Standardized groundwater indices (SGIs)

SH. *See* Spherical harmonic (SH)
Shafdan soil aquifer treatment (SAT), 604
Shallow groundwater, 434
Shigella sp., 312
Shuttle radar topography mission–digital elevation model (SRTM–DEM), 613
SIAGAS system, 416–417
"Significant harm", 117
Silica equilibrium, 272
Sindh, groundwater biological contamination in, 310–312
Sinkhole, 23, 26–28
SIPs. *See* Solar irrigation pumps (SIPs)
Snow effect on groundwater systems, 373–376
Snowmelt water, 75–77
SNWD Project, 66
Society, groundwater and
  groundwater-attuned hydrosocial cycle, 366f
  hydrosocial cycle, 362
  incorporating stakeholders' perspectives, 361
  interdisciplinarity for enmeshed issues, 364–366
  mobilizing hydrosocial analyses to capture ground (water) realities, 362–364
    dispossession of irrigating farmers through institutions and infrastructures, 363
    groundwater and politics of scale, 363–364
    state and "scientific" *vs.* local knowledge of water, 363
    trajectories from "safe and good" groundwater to "bad" citizens, 364
  political ecology, 361–362
Socio-geohydrology, 360–361
Socio-hydrogeology, 361
  in future of groundwater science, 511–512
Socio-hydrology, 360
  and groundwater, 360–361
  "public" turn for, 361
Socio-natural process, 359
Socioeconomic significance of groundwater boom, 490–491
Sodium (Na), 269
Sodium adsorption ratio (S.A.R), 265
Soil freeze–thaw cycle, 73
Solar irrigation, 495–496
Solar irrigation pumps (SIPs), 495
Solar Irrigation Service Providers (S-ISP), 496
Solar Power as Remunerative Crop (SPaRC), 496
Solar-powered electric pumpsets, 528
Solid Earth geophysical processes, 531
Sound governance of transboundary aquifers, 119–120, 122–124
Source water
  chemical facies, 262–265
  for irrigation, 265–266
South Asia, 469
  contamination issues, 471
  evapotranspiration and groundwater recharge, 470–471
  population, 472
  study region, 469–470
  WaterGAP3 model, 470

South Sudan, groundwater in, 90
South-to-North water diversion, 66–68
SPaRC. *See* Solar Power as Remunerative Crop (SPaRC)
Spatial maps, 278
SPD. *See* Sulfapyridine (SPD)
Specific energy consumption (SEC), 564–565, 564t
  breakdown, 565f
Spectroscopic methods, 278
Spherical harmonic (SH), 531–532
SPI. *See* Standard Precipitation Index (SPI)
Spring, 23, 28–29
SRC. *See* Swiss Red Cross (SRC)
SRTM–DEM. *See* Shuttle radar topography mission–digital elevation model (SRTM–DEM)
SS. *See* Screening score (SS)
Stakeholder
  consultation in groundwater policy and legislation drafting, 461
  coordination and involvement, 460–461
  participation, 465
    awareness raising, 465
    sustainable financing arrangements, 465
Stampriet aquifer, 124–125
Standard Precipitation Index (SPI), 153
Standardized groundwater indices (SGIs), 153
Staphylococcus sp., 312
State Electricity Boards (SEB), 489–490
State knowledge of water, 363
Static water level (SWL), 106
Statistical Institute of Jamaica (STATIN), 97
Strategic planning of groundwater resources, 463
Subsurface hydrogeological conditions, 148
Subterranean biological world and groundwater-dependent ecosystems, 507–508
Sudan, groundwater in, 90
Sulfapyridine (SPD), 246
Sulfates, 269
Sulphates, 319
Support vector machine (SVM), 546
Surface water, 75, 77–78
  discharge, 567
  harnessing, 433
  interaction, 377–378
  in Nile basin, 81–83
  systems, 40
Surface water storage (SWS), 47
Sustainability, 331–332
  of groundwater services, 332–335
  science, 366
Sustainable development, 331–332
  of Afghanistan, 327–328
Sustainable Development Goals (SDGs), 3–4, 47, 119–120, 177, 332
Sustainable financing arrangements, 465
Sustainable groundwater, 350
  for global food production, 350–352
  governance, 493–496
    advent of solar irrigation, 495–496

    community-based groundwater management, 494–495
    direct regulation through legal framework and administrative action, 493–494
    indirect instruments, 495
  management, 433–435
    awareness building, 435
    better irrigation water management, 434
    groundwater governance, 435
    groundwater monitoring, abstraction controls, and licensing, 434
    judicial use of deep groundwater, 435
    MAR, 434–435
    pollution abatement and control, 434
    research and development activities, 435
    surface water harnessing, 433
    wastewater reuse, 435
  quantification of groundwater depletion for food trade, 352–355
  and security, 11–14
    groundwater–food–energy nexus, 12
    trade and hydro-economics, 13–14
    urbanization, 12–13
  water use for global food production and virtual water flows, 348–350
Sustainable management plan, 483–485
"Sustainable yield", 38
SVM. *See* Support vector machine (SVM)
Swiss Agency for Development and Corporation (SDC), 611
Swiss Red Cross (SRC), 611
SWL. *See* Static water level (SWL)
SWS. *See* Surface water storage (SWS)

T

Table mountain group aquifers (TMG aquifers), 443–444
Tamil Nadu Groundwater Development and Management Act, 493
TBA. *See* Transboundary aquifers (TBA)
TCs. *See* Tetracyclines (TCs)
TDS. *See* Total dissolved solids (TDS)
Technical capacity disseminated to groundwater practitioners, 462
Technology to enhance sustainable groundwater use
  groundwater mapping and management, 519–520
  improving groundwater access, 527–528
    groundwater pumping, 528
    well digging and drilling, 527
  improving groundwater-use efficiency, 522–524
    improving household water distribution and use efficiency, 523
    improving industrial water-use efficiency, 523–524
    improving irrigation and agricultural efficiency, 522–523
  managing saline groundwater intrusion, 521–522
  MAR, 520–521

purifying contaminated groundwater, 524–527
technology levers to enhance groundwater security, 519
Terrestrial water storage (TWS), 48–49, 151, 396, 531, 545
　gravity recovery and climate experiment, 55–57
Tetracyclines (TCs), 243
Texas, groundwater quality in, 301–304
Texas Water Development Board (TWDB), 301
Thermally driven systems, 524
Tigris–Euphrates river system, 168–169
TMG aquifers. *See* Table mountain group aquifers (TMG aquifers)
Top-down approaches, 460
Total desalination cost, 565–566
　brackish groundwater, 566t
Total dissolved solids (TDS), 15, 25, 303, 524, 559
Total water storage (TWS), 68
Trace elements, 234–235
　release/sequester, 237
Trace metals, 313–319
Transboundary aquifers (TBA), 6, 113
　climate change, 120
　collaboration, potential dispute resolution, 114
　Draft Articles, 126–127
　global inventory and classification of, 114–115
　under high developmental stress, 120–122
　"invisibility cape", 126
　in national priorities, 117–119
　Red Queen effect, 116–117
　SDGs, 119–120
　Stampriet aquifer, 124–125
　urgency of sound governance, 122–124
　water availability, 114
Transboundary groundwater management, 415–416
Transboundary river basins, 47–48, 53, 59
Treated wastewater reuse, 165–166
Tropical cyclone, 609
TWDB. *See* Texas Water Development Board (TWDB)
TWS. *See* Terrestrial water storage (TWS); Total water storage (TWS)
TWS anomalies (TWSA), 554

U

UAs. *See* Urbanized areas (UAs)
UCMR3. *See* Unregulated Contaminant Monitoring Rule 3 (UCMR3)
UFA. *See* Upper Floridan aquifer (UFA)
UN Food and Agricultural Organization (FAO), 4
United Nations Educational, Scientific and Cultural Organization (UNESCO), 4
United States (US)
　arsenic in groundwater, 276–279, 287–288
　groundwater desalination in, 580–581

United States Geological Survey (USGS), 145, 275, 301, 580
Universal Transverse Mercator (UTM), 616
Unregulated Contaminant Monitoring Rule 3 (UCMR3), 220
Upazila, 619
Upper Floridan aquifer (UFA), 304
Uranium (U), 194–196
Urban flood, 590, 594
Urban groundwater efficiency, 15
Urbanization, 12–13
　in PRD, 252
Urbanized areas (UAs), 252
Urucuia, 393
Urucuia Aquifer System, 398
US Environmental Protection Agency (US EPA), 216–217
USGS. *See* United States Geological Survey (USGS)
UTM. *See* Universal Transverse Mercator (UTM)

V

Variable space, 270
Vibrio sp., 312
Victoria Nile, 170
Virtual water content (VWC), 348
Virtual water flows, 348–350
Virtual water trade (VWT), 347, 349, 360

W

WAR. *See* Water Resources Authority (WAR)
WARPO. *See* Water Resources Planning Organization (WARPO)
WASA. *See* Water Supply and Sewerage Authority (WASA)
Wastewater
　reuse, 435
　reuse of, 418
Wastewater treatment plant (WWTP), 591
Water, 47, 585
　arsenic species in, 276
　conservation projects, 373
　demand, 360
　　management, 337
　entitlements, 39
　harvesting, 183
　minerals equilibrium, 272–273
　provision, 442
　for public supply, 559
　quality, 164–165, 594–595
　　analysis, 106–110
　　basic information regarding the model and data requirement, 591
　　management, 183–184
　　model setup, 591–592
　reform, 43
　resources, 163–166, 347
　　management, 413–415
　　mitigation to water scarcity, 165–166
　rights, 415
　sourcing, 417–418

　standards and potability, 269
　stress indicators, 350
　supply, 30–31
　transfer projects, 373
　use
　　efficiency, 519
　　for global food production, 348–350
Water Code, 415
"Water debt" indicator, 351–352
Water Law. *See* National Water Law
Water Resources Authority (WAR), 103
Water Resources Planning Organization (WARPO), 611
Water scarcity, 393, 425, 442, 559, 609
　groundwater management challenges, 445–447
　　human dimensions, 446–447
　　physical dimensions, 445–446
　groundwater opportunities, 442–445
　　Sandveld Group aquifers, 444–445
　　table mountain group aquifers, 443–444
　impacts of, 164–165
　situating Cape Town, 439–442
　　Day Zero drought, 441–442
　　water provision, 442
Water Supply and Sewerage Authority (WASA), 430
Water-budget equation, 401
Water-table fluctuation method (WTF method), 400
Waterborne diseases, 310
Water–food–energy–ecosystems (WFEE), 402
WaterGAP Global Hydrological Model (WGHM), 69, 538–539
Water—Global Assessment and Prognosis-3 model (WaterGAP3 model), 470
Watershed management, 410
WEAP model, 591
Well pumping, 285–286
WFEE. *See* Water–food–energy–ecosystems (WFEE)
WGHM. *See* WaterGAP Global Hydrological Model (WGHM)
White Nile, 170
World Health Organization (WHO), 134
World-wide Hydrological Mapping and Assessment Program (WHYMAP), 4, 477–478
WTF method. *See* Water-table fluctuation method (WTF method)
WWTP. *See* Wastewater treatment plant (WWTP)

X

X-ray absorption spectroscopy (XAS), 278

Z

Zero liquid discharge (ZLD), 560–561
Zero-valent iron (ZVI), 287
Zinc (Zn), 313, 319

Printed in the United States
By Bookmasters